Die Grundlehren der mathematischen Wissenschaften

in Einzeldarstellungen
mit besonderer Berücksichtigung
der Anwendungsgebiete

Band 198

Herausgegeben von J. L. Doob A. Grothendieck E. Heinz
F. Hirzebruch E. Hopf W. Maak
S. MacLane W. Magnus J. K. Moser
M. M. Postnikov F. K. Schmidt
D. S. Scott K. Stein

Geschäftsführende Herausgeber B. Eckmann und B. L. van der Waerden

Steven A. Gaal

Linear Analysis and Representation Theory

Springer-Verlag
New York Heidelberg Berlin 1973

AMS Subject Classifications (1970)

Primary: 22 D 05, 22 D 10, 22 D 15, 22 D 25, 22 D 30, 22 E 15, 22 E 60, 28 A 70, 43 A 35, 43 A 65, 43 A 85, 43 A 90

Secondary: 46 H 05, 46 J 05, 47 A 10, 47 B 05, 47 B 10, 58 A 05

ISBN 0-387-06195-9 Springer-Verlag New York Heidelberg Berlin

ISBN 3-540-06195-9 Springer-Verlag Berlin Heidelberg New York

---

 Library of Congress Catalog Card Number 72-95686. Printed in Germany. Monophoto typesetting and offset printing: Zechnersche Buchdruckerei, Speyer. Bookbinding: Konrad Triltsch, Würzburg.

Preface

In an age when more and more items are made to be quickly disposable or soon become obsolete due to either progress or other man caused reasons it seems almost anachronistic to write a book in the classical sense. A mathematics book becomes an indespensible companion, if it is worthy of such a relation, not by being rapidly read from cover to cover but by frequent browsing, consultation and other occasional use. While trying to create such a work I tried not to be encyclopedic but rather select only those parts of each chosen topic which I could present clearly and accurately in a formulation which is likely to last. The material I chose is all mathematics which is interesting and important both for the mathematician and to a large extent also for the mathematical physicist. I regret that at present I could not give a similar account on direct integrals and the representation theory of certain classes of Lie groups. I carefully kept the level of presentation throughout the whole book as uniform as possible. Certain introductory sections are kept shorter and are perhaps slightly more detailed in order to help the newcomer progress with it at the same rate as the more experienced person is going to proceed with his study of the details. On the one hand I made the chapters as independently readable as possible under the given conditions and on the other hand I gave as many cross references as seemed practical in order to increase the serviceability of the volume and reinforce its unity.

I offer my grateful thanks for the encouragement given to me by my mother, wife and daughters, Barbara and Dorothy during the years when this book was being written. I also thank the farmers, workers and industries of Minnesota whose taxes supported myself and my family during these years through the tenure of a professorship at their University.

My appreciation is due to Professor Robert Fakler of the University of Michigan who read almost the entire manuscript very carefully and pointed out a number of minor inaccuracies and many errors and omissions made by me while typing the manuscript and filling in the various symbols by hand. His voluntary efforts substantially improved the quality of the manuscript.

My sincere thanks to Springer-Verlag and the various people associated with it for the encouragement and help during the actual writing and for the efficient and accurate work during the production. I like to thank especially Professor B. Eckmann and Dr. K. Peters for their constant interest and gentle persuasion during the past years. I also thank Dr. E. Pannwitz of Zentralblatt für Mathematik who first recommended my works and brought me into contact with the publisher of the Grundlehren der mathematischen Wissenschaften series.

The employees of the Zechnersche Buchdruckerei in Speyer did an outstanding work during the typesetting and printing of this volume. It is a pleasure to express my appreciation for their cooperation and accuracy.

Finally I thank Dr. K.F. Springer for further strengthening my past association with Springer-Verlag and for promoting mathematics by publishing what follows.

Minneapolis, January 27, 1973 Steven Alexander Gaal

Table of Contents

Chapter I

Algebras and Banach Algebras

We now turn to a detailed discussion of a number of topics from pure algebra and the theory of Banach algebras. Most of the concepts and results treated here are essential for later use but from time to time we supplement these with other results which further illuminate these topics or make the discussion more complete.

The reader may already know most of this material or he may want to progress fast in spite of not being familiar with it. In either case he should start with one of the later chapters and use the present one as a reference. The same applies to Chapters II and III. Since some of the deeper results are used later only on a very limited number of occasions this approach might be very advantageous.

1. Algebras and Norms

An *algebra A over a field F* is a set with three binary operations $(a_1, a_2) \mapsto a_1 + a_2$, $(\lambda, a) \mapsto \lambda a$ and $(a_1, a_2) \mapsto a_1 a_2$ where $a_1, a_2, a \in A$ and $\lambda \in F$. Also it is required that A be a left vectorspace over F with respect to the first two operations and a ring with respect to the first and the third one. It is also supposed that $\lambda(a_1 a_2) = (\lambda a_1) a_2 = a_1(\lambda a_2)$ for all $\lambda \in F$ and $a_1, a_2 \in A$. In this definition the field F need not be commutative. For instance the set of all $n \times n$ matrices (α_{ik}) whose entries α_{ik} belong to the skewfield F form an algebra $M_n(F)$, called the *full matrix algebra over F*, with respect to the usual matrix operations. It contains various subalgebras such as the algebra of all $n \times n$ upper (lower) triangular matrices.

By an *ideal I* of an algebra A we mean a subalgebra of A which is a two sided ideal. The adjective two sided will be used only if we wish to stress this property of the ideal in question. Similarly a *left (right) ideal* means a subalgebra which is a left (right) ideal. Since an ideal or a one sided ideal I is a subalgebra we have $\lambda a \in I$ for every $a \in I$ and every scalar λ in F. Therefore an ideal in the ring theoretic sense need not be an ideal in the algebra sense.

In order to see an example for this one has to consider an algebra in which indecomposable elements exist. By this we mean elements a in the algebra A such that $a \neq xy$ for every x and y in A. The principal ideal generated by such an element a is not an ideal in the algebra theoretic sense. For instance let A be the convolution algebra

$$A = \{f\colon f \in L^1(\mathbb{R}) \text{ and } f(x) = f(-x) \text{ for all } x\}$$

where $\mathbb{R}$ is the field of reals and let a be a continuous function in A which is not bounded. Then a is indecomposable because $f*f$ is bounded for every f in A and so $f*g$ is bounded for every f and g in A. For a proof of the boundedness of $f*f$ see Proposition VII.2.14.

An algebra A is called *simple* if its only ideals are 0 and A itself.

Proposition 1. *The full matrix algebra $M_n(F)$ is simple.*

Proof. Denote by e_{ij} the matrix having entry 1 in the $i-j$ place and zero elsewhere. For λ in F let $\tilde{\lambda}$ denote the diagonal matrix having λ on its main diagonal and 0 everywhere else. Now if $a = (\alpha_{ij})$ is an arbitrary matrix from $M_n(F)$ then $e_{ki} a e_{jl}$ is the matrix containing α_{ij} in its $k-l$ place and zero everywhere else. Therefore $\tilde{\alpha}_{ij} = \sum e_{ki} a e_{jk}$. Now let I be an ideal in $M_n(F)$ and let a in I be different from 0, say $a_{ij} \neq 0$. Then $\tilde{1} = \tilde{\alpha}_{ij} \tilde{\alpha}_{ij}^{-1} \sum e_{ki} a e_{jk} \tilde{\alpha}_{ij}^{-1}$ where by $a \in I$ the right hand side belongs to I. Hence by $\tilde{1} \in I$ everything else is in I and $I = M_n(F)$.

Corollary. *If V is a finite dimensional vector space over F then $\mathscr{L}(V)$, the algebra of linear transformations of V into itself is a simple algebra.*

Let F be a field and let $p, q \in F$ be given. Then the *quaternion algebra* $F(p,q)$ is that subalgebra of $M_4(F)$ which consists of all matrices of the form

$$\begin{pmatrix} a_0 & a_1 p & a_2 q & -a_3 pq \\ a_1 & a_0 & a_3 q & -a_2 q \\ a_2 & -a_3 p & a_0 & a_1 p \\ a_3 & -a_2 & a_1 & a_0 \end{pmatrix}.$$

Clearly $F(p,q)$ can be identified with the set of all quadruples (a_0, a_1, a_2, a_3). Then the multiplication is most easily described by letting $1 = (1,0,0,0)$, $i = (0,1,0,0)$, $j = (0,0,1,0)$, $k = (0,0,0,1)$ and by defining the products of these basic elements as follows:

	1	i	j	k
1	1	i	j	k
i	i	p	k	pj
j	j	$-k$	q	$-qi$
k	k	$-pj$	qi	$-pq$

If $p=q=-1$ and $F=\mathbb{R}$ then $F(p,q)$ is the algebra of Hamilton quaternions. The quaternions $(a,0,0,0)\,(a\in F)$ form an isomorphic image of the field F in $F(p,q)$. Although $\mathbb{C}$ is also embedded in $\mathbb{R}(-1,q)$ one can easily check that $\mathbb{R}(-1,q)$ is not an algebra over $\mathbb{C}$.

An algebra A over a normed field F is called a *normed algebra* if a *norm* $\|\cdot\|$ is defined on A such that it becomes a normed vectorspace over F and if in addition the inequality $\|xy\|\leqslant\|x\|\cdot\|y\|$ holds for every $x,y\in A$. For example if A is an algebra of bounded complex valued functions defined on some set S then A endowed with $\|x\|=\operatorname{supr}|x(s)|$ is a normed algebra over $\mathbb{C}$. One also sees that if $p,q<0$ then the quaternion algebra $\mathbb{R}(p,q)$ is normed by

$$\|x\|=\sqrt{x_0^2-px_1^2-qx_2^2+pqx_3^2}\,.$$

Actually in any $\mathbb{R}(p,q)$ one has the identity $Nxy=NxNy$ where the algebraic norm Nx is defined by

$$Nx=x_0^2-px_1^2-qx_2^2+pqx_3^2.$$

An element e of the algebra A is called its *identity* if $ex=xe=x$ for every x in A. Clearly A has at most one identity element e. By $xe=x$ we have $\|e\|\geqslant 1$ in any normed algebra with identity e. If $e=0$ then $A=0$.

Proposition 2. *If A is a normed algebra with identity e and $\|e\|>1$ then A can be renormed such that the new norm defines the same topology on it and $\|e\|=1$.*

Proof. Define the new norm $|\cdot|$ by

$$|x|=\operatorname*{lub}_{y\neq 0}\frac{\|xy\|}{\|y\|}\,.$$

Then by $\|xy\|\leqslant\|x\|\cdot\|y\|$ we have $|x|\leqslant\|x\|$ and choosing $y=e$ we have $|x|\geqslant\|x\|/\|e\|$. Thus $\|x\|$ and $|x|$ induce the same topology on A. Moreover we have $\|x_1x_2y\|\leqslant|x_1|\cdot\|x_2y\|$ so $\|x_1x_2y\|\leqslant|x_1|\cdot\|x_2y\|\leqslant|x_1|\cdot|x_2|\cdot\|y\|$ and so y being arbitrary $|x_1x_2|\leqslant|x_1|\cdot|x_2|$.

Proposition 3. *Every algebra A can be extended to an algebra A_e with identity e such that A is a two sided maximal ideal in A_e.*

Proof. We let $A_e=F\times A$ where F is the field of scalars and write $\lambda e+a$ for the ordered pair (λ,a) so that e denotes $(1,0)$ and the set A is identified with $\{0\}\times A$. If we define addition and multiplication in

A_e by the rules $(\lambda_1, a_1)+(\lambda_2, a_2)=(\lambda_1+\lambda_2, a_1+a_2)$ and $(\lambda_1, a_1)(\lambda_2, a_2) =(\lambda_1 \lambda_2, \lambda_1 a_2+\lambda_2 a_1+a_1 a_2)$ then A_e becomes an algebra and $\{0\} \times A$ is clearly a two sided maximal ideal in A_e.

Proposition 4. *If A is a normed algebra, then A_e can be normed by letting $\|\lambda e+a\|=|\lambda|+\|a\|$. The topology induced by this norm is such that $\{0\} \times A$ is a closed subset of A_e. If A is complete so is A_e.*

Proof. The inequality $\|(\lambda_1 e+a_1)(\lambda_2 e+a_2)\| \leqslant \|\lambda_1 e+a_l\| \cdot \|(\lambda_2 e+a_2)\|$ follows from $\|a_1 a_2\| \leqslant \|a_1\| \cdot \|a_2\|$. The other properties of the norm are obvious and one sees that A_e is complete if and only if A is complete.

A normed algebra A is called a *Banach algebra* if A is complete with respect to the uniform structure determined by its norm. If the field of scalars is $\mathbb{C}$ we will speak about a *complex Banach algebra*, so unless the adjective *complex* explicitely appears in the text the discussion concerns both real and complex Banach algebras. If A is a Banach algebra without identity then A_e, the algebra obtained by joining an identity e to A is complete with respect to the norm $\|\lambda e+x\|=|\lambda|+\|x\|$. Thus every Banach algebra can be considered as a closed two sided maximal ideal in a Banach algebra with identity. Of course if A itself has an identity then this embedding is superfluous.

Trivial examples of real Banach algebras are the fields $\mathbb{R}$ and $\mathbb{C}$, the algebra of Hamilton quaternions $\mathbb{H}=\mathbb{R}(-1,-1)$ or any other quaternion algebra $\mathbb{R}(p, q)$ with $p, q<0$. If S is a topological space then $C_b(S)$, the algebra of bounded continuous functions $x: S \rightarrow \mathbb{C}$ normed by $\|x\|=\|x\|_{\infty}=\operatorname{supr}\{|x(s)|: s \in S\}$ is a complex Banach algebra. Since any closed subalgebra A of $C_b(S)$ is complete, it is also a complex Banach algebra, called a *complex function algebra.*

Let A be a real algebra and let $A_{\mathbb{C}}$ be the set of ordered pairs (a_1, a_2) $(a_1, a_2 \in A)$ with the following algebraic structure:

$$\begin{aligned}
(a_1, a_2)+(b_1, b_2) &= (a_1+b_1, a_2+b_2) \\
(\alpha+\beta i)(a_1, a_2) &= (\alpha a_1-\beta a_2, \alpha a_2+\beta a_1) \\
(a_1, a_2)(b_1, b_2) &= (a_1 b_1-a_2 b_2, a_1 b_2+a_2 b_1).
\end{aligned}$$

Then $A_{\mathbb{C}}$ is a complex algebra and $a \mapsto (a, 0)$ is an isomorphism of A into $A_{\mathbb{C}}$. It is called the *complexification* of A. It is known that if A is a normed algebra then its norm can be extended to its complexification and if A is complete then so is $A_{\mathbb{C}}$. However this result will not be used and we omit the proof.

2. The Group of Units and the Quasigroup

Given any ring A with identity we let G or $A^\times$ denote the set of invertible elements of A so that G is a group under the ring multiplication, called the *group of units* or the *group of regular elements* or the *multiplicative group* of A.

Proposition 1. *If A is a Banach algebra with identity e then $e-x\in G$ for every x satisfying $\|x\|<1$ and*

$$(e-x)^{-1}=e+x+x^2+\cdots.$$

Moreover we have

$$\|e-(e-x)^{-1}\|\leqslant\frac{\|x\|}{(1-\|x\|)}.$$

Proof. The convergence of the series follows from $\|x^n\|\leqslant\|x\|^n$ and $\|x\|<1$. Multiplication by $e-x$ shows that the sum of the series is $(e-x)^{-1}$. Since $e-(e-x)^{-1}=-(x+x^2+\cdots)$ the last inequality is a trivial consequence of $\|x^n\|\leqslant\|x\|^n$.

Proposition 2. *If A is a Banach algebra with identity then its group of units G is an open set in A and the map $x\mapsto x^{-1}$ is a homeomorphism of G onto itself.*

A topological algebra A with identity is called an *algebra with continuous inverse* if its group of units G is an open set of A and if $x\mapsto x^{-1}$ is a continuous map of G onto itself. If this is so then $x\mapsto x^{-1}$ is a homeomorphism. Hence we have:

Corollary. *Every Banach algebra with identity is a topological algebra with continuous inverse.*

Proof. It is sufficient to prove the following: If $a\in G$ then for every y satisfying $\|y\|<\alpha=\|a^{-1}\|^{-1}$ we have $a+y\in G$ and

$$\|(a+y)^{-1}-a^{-1}\|\leqslant\frac{\|y\|}{(\alpha-\|y\|)\alpha}.$$

If we let $x=-ya^{-1}$ then Proposition 1 shows that $e+ya^{-1}\in G$ and so $a+y=(e+ya^{-1})a\in G$. Moreover

$$\|(a+y)^{-1}-a^{-1}\|\leqslant\|(e+ya^{-1})^{-1}-e\|\cdot\|a^{-1}\|\leqslant\frac{\|ya^{-1}\|}{(1-\|ya^{-1}\|)\alpha}$$

where $\|ya^{-1}\|\leqslant\|y\|\alpha^{-1}$.

The same reasoning shows that if a has a right (or left) inverse then $a+y$ is right (left) invertible for every y satisfying $\|y\|<\alpha$ and the difference of the right inverses of $a+y$ and a can be estimated in the same manner as above.

In any ring A one can introduce the *circle operation* $x\circ y=x+y-xy$ and verify that $(x\circ y)\circ z=x\circ(y\circ z)$ in the same manner as one shows that forming the symmetric difference leads to an associative operation Δ. If $x\circ y=0$ we say that y is a *right quasi-inverse* of x and x is a *left quasi-inverse* of y. The terminology is justified because if A has an identity e then $x\circ y=0$ is equivalent to $(e-x)(e-y)=e$. If both $x\circ y_1=0$ and $y_2\circ x=0$ hold then

$$y_2(x\circ y_1)=(y_2+x)+y_2y_1-(y_2+x)y_1=y_2-y_1$$

and so by $x\circ y_1=0$ we get $y_1=y_2$. Hence if both right and left quasi-inverses exist for some element x then these coincide and x has a unique *quasi-inverse* y such that $x\circ y=y\circ x=0$. We shall denote this unique element y by x' so that $x\circ x'=x'\circ x=0$.

For any ring A the set of quasi-invertible elements will be denoted by Q and it will be called the *quasigroup* of A. We note that Q is a group under the operation $x\circ y$: For if $x\circ u=0$ and $y\circ v=0$ then one proves by using the associativity of the circle operation that $(x\circ y)\circ(v\circ u)=0$. The neutral element of the group is 0 and the inverse of x is the quasi-inverse x'. The circle operation could be modified to mean $x+y+xy$ and the same concepts could be introduced with respect to this modified circle operator.

Proposition 3. *Let A be an algebra without identity and let A_e be the algebra with identity e adjoined to A. Then $x\in A$ has a right (left) quasi-inverse in A if and only if $e-x$ has a right (left) inverse in A_e.*

Proof. If $x+y=xy$ then $(e-x)(e-y)=e$ and so $e-y$ is the right inverse of $e-x$. Conversely, if $e-x$ has a right inverse then let $y=e-(e-x)^{-1}$ so that $x+y=xy$ and y is a right quasi-inverse of x in A_e. Since $x\in A$ and A is a two sided ideal in A_e we see that $y=xy-x$ belongs to A.

Corollary. *The quasigroup of A and the group of units of A_e are related by $Q=G+e$ and $G=Q+e$.*

It is obvious that the same relation holds between the quasigroup Q and the group of units G of any ring A with identity e. Thus we see that in an algebra with continuous inverse the quasigroup Q is an open set. More precisely we have:

Theorem 4. *A topological algebra A with identity is an algebra with continuous quasi-inverse if and only if A is an algebra with continuous inverse.*

Here of course a topological algebra is called an *algebra with continuous quasi-inverse* if the set Q is open and if $x \mapsto x'$ is a continuous map of Q onto Q. (If $x \mapsto x'$ is continuous then by $x'' = x$ it is a homeomorphism.) The proof follows from $Q = G + e$ and the formulas $x' = e - (e - x)^{-1}$ and $x^{-1} = e - (e - x)'$. One also sees that if A_e is an algebra with continuous inverse then A is an algebra with continuous quasi-inverse.

Proposition 5. *In any Banach algebra A the quasi-inverse x' exists for every $x \in A$ satisfying $\|x\| < 1$, it is given by $x' = -(x + x^2 + \cdots)$ and the quasigroup is an open subset of A.*

Proof. By the corollary of Proposition 3 we have $Q = G + e$. Hence the existence of x' follows from Proposition 1. By Proposition 2 G is an open set in A_e, so its translate Q is open in A_e and by $Q \subseteq A$ it is open also in A. Similarly one can verify immediately the following:

Proposition 6. *If x is a nilpotent element of the ring A, say $x^n = 0$, then $x \in Q$ and $x' = -(x + x^2 + \cdots + x^{n-1})$.*

Proposition 7. *If the ring element $x \in A$ has a right (left) quasi-inverse then $h(x) \neq 1$ for every homomorphism h of A into another ring with identity* 1.

Proof. If $x + y = xy$ then $h(x) + h(y) = h(x)h(y)$ and so $h(x) = 1$ would imply that $1 = 0$.

Theorem 8. *If A is a Banach algebra and if $h: A \to \mathbb{C}$ is a homomorphism then h is continuous and $\|h\| \leqslant 1$.*

Proof. Suppose that $|h(a)| > \|a\|$ for some a in A. Then $x = h(a)^{-1} a$ would be such that $\|x\| < 1$ and $h(x) = 1$. This is a contradiction because by Proposition 5 $x \in Q$ and so by the last proposition $h(x) \neq 1$.

Theorem 9. *If A is a topological algebra with continuous quasi-inverse then every homomorphism $h: A \to \mathbb{C}$ is continuous.*

Proof. If h were not continuous then there would exist a sequence of elements $a_1, a_2, \ldots$ such that $|h(a_n)| \geqslant n \|a_n\|$ $(n = 1, 2, \ldots)$. We let $x_n = h(a_n)^{-1} a_n$ so that $x_n \to 0$ and $h(x_n) = 1$. Since $0 \in Q$ and Q is an open set we have $x_n \in Q$ for all sufficiently large values of n. This contradicts Proposition 7.

An element x of the Banach algebra A is called a *left topological zero divisor* if there is a sequence x_n $(n = 1, 2, \ldots)$ of elements of norm one such that $xx_n \to 0$. Similarly one defines a *right topological zero divisor* and a *two sided topological zero divisor.*

Proposition 10. *If A is a Banach algebra with identity then the boundary points of $A^\times$ are two sided topological zero divisors.*

Proof. Let $x_n \in A^\times$ for $n=1,2,\ldots$ and let $x_n \to x$ where $x \in \partial A^\times$. Then the sequence of norms $\|x_n^{-1}\|$ cannot be bounded because otherwise $x_n^{-1}x - e = x_n^{-1}(x-x_n) \to 0$ or $x_n^{-1}x \to e$ and so $A^\times$ being open for large n we would have $x_n^{-1}x \in A^\times$ and $x = x_n(x_n^{-1}x) \in A^\times$. We replace x_n by a subsequence such that for the new x_n's we have $\|x_n^{-1}\| \to \infty$. If we let $y_n = \|x_n^{-1}\|^{-1}(x_n^{-1})$ then $\|y_n\| = 1$ and

$$xy_n = (x-x_n)y_n + \|x_n^{-1}\|^{-1} e \to 0$$

and similarly $y_n x \to 0$.

Proposition 11. *If x is a left topological zero divisor in a Banach algebra with identity then x has no left inverse.*

Proof. Suppose $\|x_n\| = 1$ for $n=1,2,\ldots$ and $xx_n \to 0$. If we also suppose that x has a left inverse x^{-1} then we arrive at a contradiction because $x^{-1}(xx_n) = x_n \to 0$.

3. The Maximal Ideal Space

The set of all homomorphisms $h: A \to \mathbb{C}$ of a Banach algebra A, including also the *trivial homomorphism* $\theta: A \to 0 \in \mathbb{C}$ will be denoted by $\mathscr{M}_0$ and $\mathscr{M}$ will be the set of non-trivial homomorphisms of A into $\mathbb{C}$. For every $a \in A$ we let $\hat{a}: \mathscr{M}_0 \to \mathbb{C}$ be the function defined by $\hat{a}(h) = h(a)$ and $\hat{a}(\theta) = 0$. The restricted map $\hat{a}: \mathscr{M} \to \mathbb{C}$ is called the *Fourier transform* of a because if $A = L^1(\mathbb{R})$ then as we shall see in a while $\mathscr{M}$ can be identified with the set of all reals and then $\hat{a}$ becomes the classical Fourier transform of $a \in L^1(\mathbb{R})$.
It is clear that $\widehat{a+b} = \hat{a} + \hat{b}$ and $\widehat{ab} = \hat{a}\hat{b}$, similarly $\widehat{\lambda a} = \lambda \hat{a}$ for every scalar λ. Therefore the set of functions $\hat{a}: \mathscr{M} \to \mathbb{C}$ is an algebra over the same field as A which in the case of commutative A we shall denote by $\hat{A}$. By Theorem 2.8 every $h \in \mathscr{M}_0$ is continuous and $\|h\| \leqslant 1$, so $\hat{a}$ is a bounded function and

$$\|\hat{a}\| = \sup_{h \in \mathscr{M}} |\hat{a}(h)| = \sup_{h \in \mathscr{M}_0} |\hat{a}(h)| = \sup_{h \in \mathscr{M}_0} |h(a)| \leqslant \|a\| .$$

$\{\hat{a}: a \in A\}$ with this seminorm $\|\hat{a}\|$ is called the *dual algebra* of A. The Fourier transform map $a \mapsto \hat{a}$ is a norm decreasing homomorphism of A onto $\{\hat{a}: a \in A\}$. The reason to use the notation $\hat{A}$ only in the commutative case will become clear later in Section IV.2.

We topologize $\mathcal{M}_0$ with the weakest topology for which all the functions $\hat{a}: \mathcal{M}_0 \to \mathbb{C}$ are continuous. This will be called the *Gelfand topology* of $\mathcal{M}_0$. A base for the neighborhood filter of some $h \in \mathcal{M}_0$ consists of the sets

$$N_h(\varepsilon; a_1, \ldots, a_n) = \{k: k \in \mathcal{M}_0 \text{ and } |\hat{a}_i(k) - \hat{a}_i(h)| < \varepsilon \text{ for } i = 1, 2, \ldots, n\}$$

where $\varepsilon > 0$ and $a_1, \ldots, a_n \in A$ and $n = 1, 2, \ldots$ are arbitrary. Since $|h(a)| \leqslant \|a\|$ for every $a \in A$ we see that $\mathcal{M}_0 \subseteq \prod \{D(a): a \in A\}$ where $D(a) = \{\lambda: \lambda \in \mathbb{C} \text{ and } |\lambda| \leqslant \|a\|\}$. A fundamental system of neighborhoods in the product space $\prod D(a)$ is formed by the sets

$$N_f(\varepsilon; a_1, \ldots, a_n) = \{g: g \in \prod D(a) \text{ and } |f(a_i) - g(a_i)| < \varepsilon \text{ for } i = 1, \ldots, n\}.$$

Therefore the Gelfand topology of $\mathcal{M}_0$ is identical with the topology induced on $\mathcal{M}_0$ by the product topology of $\prod D(a)$.

Proposition 1. *$\mathcal{M}_0$ is a compact Hausdorff space.*

Proof. By Tihonov's theorem $\prod D(a)$ is a compact Hausdorff space and so it is enough to prove that $\mathcal{M}_0$ is a closed subspace of $\prod D(a)$. Now if $f \in \bar{\mathcal{M}}_0$ then for any $\varepsilon > 0$, any scalar λ and $a, b \in A$ there is an h in $\mathcal{M}_0$ which belongs to the neighborhood $N_f(\varepsilon; a, b, a+b, ab, \lambda a)$. This means that $|f(a) - h(a)| < \varepsilon$, $|f(b) - h(b)| < \varepsilon$, $|f(a+b) - h(a+b)| < \varepsilon$, etc. By $h \in \mathcal{M}_0$ we have $h(a+b) = h(a) + h(b)$ and so $\varepsilon > 0$ being arbitrary we obtain $f(a+b) = f(a) + f(b)$. Similarly we can prove that $f(ab) = f(a) f(b)$ and $f(\lambda a) = \lambda f(a)$. Hence if $f \in \bar{\mathcal{M}}_0$ then $f \in \mathcal{M}_0$.

The set $\mathcal{M}$ of non-trivial homomorphisms $h: A \to \mathbb{C}$ endowed with the Gelfand topology is called the *carrier space* or the *maximal ideal space* of the Banach algebra A. If we associate with each h its kernel I then I is a two sided maximal ideal in A and $I_1 \neq I_2$ if $h_1 \neq h_2$ because $a + I \mapsto h(a)$ is an isomorphism of A/I into $\mathbb{C}$. Hence the set $\mathcal{M}$ can be identified with the set of kernels I and the expression introduced above is justified.

Proposition 2. *The maximal ideal space of a Banach algebra A is a locally compact Hausdorff space and if A has an identity then $\mathcal{M}$ is compact.*

Proof. For any $h \in \mathcal{M}$ we can choose disjoint open sets O_h and O_θ in $\mathcal{M}_0$ containing h and θ, respectively. Then $\bar{O}_h \subseteq c\, O_\theta \subseteq \mathcal{M}$ where $\bar{O}_h$ denotes the closure of O_h in $\mathcal{M}_0$. Since $\mathcal{M}_0$ is compact we see that $\bar{O}_h$ is a compact neighborhood of h in $\mathcal{M}$. Now suppose that A has a unity e. Then for $0<\varepsilon<1$ the neighborhood $N_\theta(\varepsilon; e)$ contains only θ because $h(e)=1$ for $h\neq\theta$. Therefore θ is an isolated point of the compact space $\mathcal{M}_0$ and so $\mathcal{M}$ itself is compact.

If $\mathcal{M}$ is not compact then $\mathcal{M}_0$ is the one point compactification of $\mathcal{M}$. Since $\hat{a}: \mathcal{M}_0 \to \mathbb{C}$ is continuous and $\hat{a}(\theta)=0$ we see that a vanishes at infinity.

4. The Spectrum of an Element

Let A be a complex algebra with continuous inverse and let G be its group of units. If $x\in A$ then $r(x)=\{\lambda: x-\lambda e\in G\}$ is called the *resolvent set* of x and its complement $\sigma(x)$ is called the *spectrum*. If A has no identity but is an algebra with continuous quasi-inverse then $r(x)$ is defined to be the resolvent set of x in A_e and similarly $\sigma(x)$ denotes the spectrum of x in A_e. We let $\sigma(x)^\times = \sigma(x) \cap \mathbb{C}^\times$ so that $\sigma(x)^\times$ is either the spectrum or the spectrum without the characteristic value 0.

Since G is open $r(x)$ contains with each λ also a sufficiently small neighborhood of λ and so $r(x)$ is an open set. Moreover, if A is a complex Banach algebra and if $|\lambda|>\|x\|$ then by Proposition 2.1 $x/\lambda - e\in G$ and so $\lambda\in r(x)$. Hence we have the following:

Proposition 1. *The resolvent set $r(x)$ is open. If A is a complex Banach algebra then $r(x)$ contains the set $\{\lambda: |\lambda|>\|x\|\}$, the spectrum $\sigma(x)$ is compact and it is contained in $\{\lambda: |\lambda|\leqslant\|x\|\}$.*

If $\lambda\neq 0$ then $x-\lambda e\in G$ if and only if $x/\lambda - e\in G$ and so by the corollary of Proposition 2.3 we have:

Proposition 2. *A non-zero complex number λ belongs to $r(x)$ if and only if $x/\lambda\in Q$. If A has no identity then $0\notin r(x)$ and if A has identity then $0\notin r(x)$ if and only if $x\notin G$.*

We can now easily prove the following:

Proposition 3. *If A and B are complex Banach algebras and if $\rho: A\to B$ is a homomorphism of A into B then $\sigma(\rho(x))^\times \subseteq \sigma(x)^\times$ for every $x\in A$.*

Proof. Let $Q(A)$ and $Q(B)$ denote the quasigroups of A and B, respectively. Since ρ is a ring homomorphism $x\in Q(A)$ implies that $\rho(x)\in Q(B)$. Thus if $\lambda\notin\sigma(x)$ and $\lambda\neq 0$ then $x/\lambda\in Q(A)$ so $\rho(x)/\lambda\in Q(B)$ and $\lambda\notin\sigma(\rho(x))$.

Corollary 1. *If ρ is not trivial and B has identity then* $\sigma(\rho(x)) \subseteq \sigma(x)$.

Indeed, if $0 \in \sigma(x)$ then the conclusion is an immediate consequence of the proposition and if $0 \notin \sigma(x)$ then A has an identity e and x is invertible. Since ρ is not trivial $\rho(e)$ is the identity of B and $\rho(x^{-1})$ is the inverse of $\rho(x)$. Hence $0 \notin \sigma(\rho(x))$.

Corollary 2. *If B is a subalgebra of the complex Banach algebra A and $\sigma_A(x)$, $\sigma_B(x)$ denote the spectra of x in A and B, then $\sigma_A(x)^\times \subseteq \sigma_B(x)^\times$. If A has identity and $B \neq 0$ then $\sigma_A(x) \subseteq \sigma_B(x)$.*

Let A be a complex Banach algebra with identity e, let $x \in A$ and for $\lambda \in r(x)$ let $x_\lambda = (x - \lambda e)^{-1}$. The Banach algebra valued map $\lambda \mapsto x_\lambda$ defined on $r(x)$ is called the *resolvent* or *resolvent function* of the element x. Using the formula $\xi' = e + (\xi - e)^{-1}$ we see that

$$x_\lambda = (x - \lambda e)^{-1} = \frac{1}{\lambda}\left(\frac{x}{\lambda}\right)' - \frac{e}{\lambda}$$

for any $\lambda \neq 0$ in $r(x)$. The elements $x - \lambda_1 e$ and $x - \lambda_2 e$ commute for any λ_1, λ_2 and so if $\lambda_1, \lambda_2 \in r(x)$ then $x_{\lambda_1} x_{\lambda_2} = x_{\lambda_2} x_{\lambda_1}$. It follows that

$$x_{\lambda_1} - x_{\lambda_2} = x_{\lambda_1} x_{\lambda_2}(x - \lambda_2 e) - x_{\lambda_2} x_{\lambda_1}(x - \lambda_1 e) = x_{\lambda_1} x_{\lambda_2}(\lambda_1 e - \lambda_2 e).$$

Hence we proved that the resolvent function satisfies the *resolvent equation*

$$x_{\lambda_1} - x_{\lambda_2} = (\lambda_1 - \lambda_2) x_{\lambda_1} x_{\lambda_2}$$

for any pair $\lambda_1, \lambda_2 \in r(x)$.

Proposition 4. *The resolvent function vanishes at infinity, it is strongly differentiable and its derivative is x_λ^2.*

Proof. If $|\lambda| > \|x\|$ then by Proposition 2.1

$$\|x_\lambda\| = \frac{1}{|\lambda|}\left\|\left(\frac{x}{\lambda} - e\right)^{-1}\right\| \leq \frac{1}{|\lambda|}\left(\|e\| + \frac{\|x\|}{|\lambda| - \|x\|}\right)$$

which shows that $\|x_\lambda\| \to 0$ as $\lambda \to \infty$. By Proposition 2.2 $x \mapsto x^{-1}$ is continuous on G and so $x_{\lambda_1} \to x_\lambda$ as $\lambda_1 \to \lambda$, proving incidentally the continuity of the resolvent on $r(x)$. By the resolvent equation for $\lambda_1 \to \lambda$ we have

$$\frac{x_{\lambda_1} - x_\lambda}{\lambda_1 - \lambda} = x_{\lambda_1} x_\lambda \to x_\lambda^2.$$

Proposition 5. *The function $\lambda \to (\lambda x)'$ is strongly differentiable in its domain of definition; at $\lambda=0$ the derivative is $-x$ and if $\lambda \neq 0$ then*

$$\frac{d(\lambda x)'}{d\lambda} = \frac{(\lambda x)' - ((\lambda x)')^2}{\lambda}.$$

Proof. For $|\lambda| < 1/\|x\|$ by Proposition 2.5 we have $(\lambda x)' = -(\lambda x + \lambda^2 x^2 + \cdots)$ and so $(\lambda x)'/\lambda \to -x$ as $\lambda \to 0$. If $\mu \neq 0$ and $\mu \in r(x)$ then $(x/\mu)' = e + \mu x_\mu$ and so by the last proposition

$$\frac{d}{d\mu}\left(\frac{x}{\mu}\right)' = x_\mu + \mu x_\mu^2.$$

Using the substitution $\mu = 1/\lambda$ we obtain

$$\frac{d(\lambda x)'}{d\lambda} = -\lambda^{-2}(x_\mu + \mu x_\mu^2) = -\lambda^{-1}(\mu x_\mu)(e + \mu x_\mu) = \lambda^{-1}(e - (\lambda x)')(\lambda x)'.$$

Theorem 6. *If A is a complex normed algebra and if $x \in A$ then $\sigma(x)$ is not void.*

Proof. We may restrict our attention to normed algebras with identity because if A is without identity then by definition $0 \in \sigma(x)$. Let $\bar{A}$ denote the completion of A. Let us suppose that the spectrum $\sigma(x)$ is void i. e. the residual set $r(x) = \mathbb{C}$. We fix some $\mu \in r(x)$ and choose a continuous linear functional $f: \bar{A} \to \mathbb{C}$ such that $f(x_\mu) \neq 0$. The existence of such functionals follows from the Hahn-Banach theorem. Using f we define $\varphi: \mathbb{C} \to \mathbb{C}$ by the formula $\varphi(\lambda) = f(x_\lambda)$. By Proposition 4 φ is an entire function vanishing at infinity. Hence by Liouville's theorem φ is identically 0 in contradiction to $\varphi(\mu) = f(x_\mu) \neq 0$. Therefore $r(x)$ must be a proper subset of $\mathbb{C}$ and the spectrum is not void.

Theorem 7. *Let A be a complex normed algebra with identity e such that every non-zero element is invertible. Then A is commutative and topologically isomorphic to the field of complex numbers. Namely for each x in A there is a unique λ_x in $\mathbb{C}$ such that $x = \lambda_x e$. If $\|e\| = 1$ then $x \mapsto \lambda_x$ is an isometric isomorphism of A onto $\mathbb{C}$.*

Proof. For any $x \in A$ the spectrum $\sigma(x)$ is not void, so there is some $\lambda_x \in \mathbb{C}$ such that $x - \lambda_x e$ is not invertible. By our hypothesis $x - \lambda_x e = 0$ and so λ_x is uniquely determined by x. The bijective map $x \mapsto \lambda_x$ is the desired isomorphism. If $\|e\| = 1$ then obviously $\|x\| = |\lambda_x|$ for every $x \in A$, and in general $|x| = |\lambda_x|$ where $|x|$ denotes the topologically equivalent norm introduced in Proposition 1.2.

Proposition 8. *If* $x_1,\ldots,x_n$ *are pairwise commutative elements in a semigroup with identity then* $x_1\ldots x_n$ *is invertible if and only if* $x_1,\ldots,x_n$ *are all invertible.*

Proof. If the right inverses $x_1^{-1},\ldots,x_n^{-1}$ exist then $x_n^{-1}\ldots x_1^{-1}$ is a right inverse of $x_1\ldots x_n$. Conversely, if $(x_1\ldots x_n)^{-1}$ is a right inverse of the product then by the commutativity hypothesis $(x_1\ldots x_{k-1}x_{k+1}\ldots x_n)(x_1\ldots x_n)^{-1}$ is a right inverse of x_k.

Proposition 9. *If* A *is a complex Banach algebra with identity,* $x\in A$ *and if* p *is a polynomial with complex coefficients, then* $p(\sigma(x))=\sigma(p(x))$.

Proof. Let $\lambda\in\sigma(p(x))$ and let $\lambda_1,\ldots,\lambda_n\in\mathbb{C}$ be the roots of $p(t)-\lambda$ so that $p(\lambda_k)=\lambda$ for $k=1,\ldots,n$. Then $p(x)-\lambda e=a_n(x-\lambda_1 e)\ldots(x-\lambda_n e)$ and by Proposition 8 $x-\lambda_k e$ is not invertible for at least one of the indices k. This proves that $\sigma(p(x))\subseteq p(\sigma(x))$. Conversely, if $\lambda\in p(\sigma(x))$ then $\lambda=p(\lambda_k)$ for some root λ_k of the polynomial $p(t)-\lambda$ where $x-\lambda_k e$ is not invertible. Hence using Proposition 8 again we see that $p(x)-\lambda e$ is not invertible and so $\lambda\in\sigma(p(x))$.

Proposition 10. *Let* A *be a Banach algebra with identity and let* B *be a Banach subalgebra containing* e. *Then for every* x *in* B *the boundary of* $\sigma_B(x)$ *is contained in the boundary of* $\sigma_A(x)$.

Proof. If λ is in the boundary of $\sigma_B(x)$ then $x-\lambda e$ is in the boundary of $B^\times$ and so by Proposition 2.10 it is a two sided topological zero divisor. Hence by Proposition 2.11 $\lambda\in\sigma_A(x)$. Since λ is a boundary point of $\sigma_B(x)$ and $\sigma_A(x)\subseteq\sigma_B(x)$ we see that λ is on the boundary of $\sigma_A(x)$.

Corollary. *If in addition the resolvent set* $r_A(x)$ *is connected then* $\sigma_A(x)=\sigma_B(x)$.

Proof. Suppose that there is a λ in $\sigma_B(x)\cap r_A(x)$. Let $N>\|x\|$ so that $N\in r_B(x)$. Since $r_A(x)$ is connected and $\lambda, N\in r_A(x)$ there is a polygonal arc π connecting λ with N and lying entirely in $r_A(x)$. We have $\lambda\in\sigma_B(x)$ and $N\notin\sigma_B(x)$ so there is a point μ on π which is a boundary point of $\sigma_B(x)$. Since $\pi\subseteq r_A(x)$ we have $\mu\notin\sigma_A(x)$ where $\sigma_A(x)$ is a closed set. This contradicts Proposition 10.

5. The Spectral Norm Formula

We shall now study the relation between four quantities which can be associated with any element x of a Banach algebra A. These are the norm $\|x\|$, the semi-norm $\|\hat{x}\|$, the *spectral radius* or *spectral norm*

$$\|x\|_{sp} = \text{supr}\,\{|\lambda|: \lambda\in\sigma(x)\}$$

and the limit

$$v(x) = \lim \sqrt[n]{\|x^n\|}\,.$$

We know that $\|\hat{x}\|\leqslant\|x\|$, similarly $\|x\|_{sp}\leqslant\|x\|$ and using $\|x^n\|\leqslant\|x\|^n$ we get $v(x)\leqslant\|x\|$ where $v(x)$ is interpreted as the limit superior of the sequence. The existence of the limit defining $v(x)$ can be seen in any normed algebra by substituting $\alpha_n=\|x^n\|$ in the following:

Lemma 1. *If* $\alpha_n\geqslant 0$ $(n=1,2,\ldots)$ *and* $\alpha_{m+n}\leqslant\alpha_m\alpha_n$ *for all* m,n then $\sqrt[n]{\alpha_n}$ $(n=1,2,\ldots)$ *is a convergent sequence.*

Proof. Fix $k=1,2,\ldots$ and $l=0,1,\ldots,k-1$ and let $n=km+l\,(m=0,1,\ldots)$. Then $\alpha_n\leqslant(\alpha_k)^m\alpha_l$ and so

$$\sqrt[n]{\alpha_n}\leqslant(\alpha_k)^{\frac{1}{k}}\Big((\alpha_k)^{-\frac{l}{k}}\alpha_l\Big)^{\frac{1}{n}}.$$

Therefore $\limsup\sqrt[n]{\alpha_n}\leqslant\sqrt[k]{\alpha_k}$ where $n\equiv l\,(\text{mod}\,k)$ and $n\to\infty$. Thus also $\limsup\sqrt[n]{\alpha_n}\leqslant\sqrt[k]{\alpha_k}$ as $n\to\infty$ and so $\limsup\sqrt[n]{\alpha_n}\leqslant\liminf\sqrt[k]{\alpha_k}$.

Proposition 2. *For any complex Banach algebra* A *and any* $x\in A$ *we have* $\|x\|_{sp}\leqslant\sqrt[n]{\|x^n\|}$ *for* $n=1,2,\ldots$

Proof. If $\lambda\in\sigma(x)$ then by Proposition 4.9 we have $\lambda^n\in\sigma(x^n)$ and so $|\lambda^n|\leqslant\|x^n\|_{sp}\leqslant\|x^n\|$. Hence $|\lambda|\leqslant\sqrt[n]{\|x^n\|}$ for every $\lambda\in\sigma(x)$.

Theorem 3. *For any complex Banach algebra* A *and any* $x\in A$ *we have* $\|x\|_{sp}=v(x)$.

Proof. In view of the last proposition it is sufficient to show that $\|x\|_{sp}\geqslant\limsup\sqrt[n]{\|x^n\|}$. This will both show the existence of the limit and the desired equality. If $\lambda\neq 0$ and $\|x\|_{sp}<|1/\lambda|$ then $1/\lambda\notin\sigma(x)$ and so by Proposition 4.2 $\lambda x\in Q$. Hence we see that $(\lambda x)'$ exists for every λ in the open disk $D=\{\lambda: |\lambda|<1/\|x\|_{sp}\}$.

Let f be any continuous linear functional on the complex vector space A and define $\varphi: D\to\mathbb{C}$ by $\varphi(\lambda)=f((\lambda x)')$. By Proposition 4.5 $(\lambda x)'$ is a strongly differentiable function in its domain of definition and so φ is holomorphic in D. For $\|\lambda x\|<1$ we have $(\lambda x)'=-(\lambda x+\lambda^2x^2+\cdots)$ by Proposition 2.5 and so using the continuity of f we get the Taylor series

$$\varphi(\lambda)=-\sum_1^\infty f(x^n)\,\lambda^n.$$

This will be convergent in D and so we proved that $f((\lambda x)^n)\to 0$ as $n\to\infty$ for every fixed $\lambda\in D$.

We look at $(\lambda x)^n$ as an element of the second conjugate space of A so that $f((\lambda x)^n)$ is the value of the linear functional $(\lambda x)^n$ at $f\in A^*$. Since the sequence of values $(\lambda x)^n(f)$ is bounded for $\lambda\in D$ and any $f\in A^*$ by the principle of uniform boundedness $\|(\lambda x)^n\|\leqslant B(\lambda)$ for some bound $B(\lambda)$. We obtain $\limsup\sqrt[n]{\|x^n\|}\leqslant 1/|\lambda|$ for any $\lambda\in D$ i.e. for any complex λ satisfying $|\lambda|<1/\|x\|_{sp}$. Hence $\limsup\sqrt[n]{\|x^n\|}\leqslant\|x\|_{sp}$ and the proof of the theorem is completed.

Proposition 4. *In any complex Banach algebra we have* $\hat{x}(\mathscr{M})\subseteq\sigma(x)$.

Proof. If $\lambda\in r(x)$ then $(x-\lambda e)y=e$ where e, y belong to A or to A_e according as A is with or without unity. If A is not A_e and if $h\in\mathscr{M}$ then h can be extended to A_e by letting $h(e)=1$. Thus in any case if $\lambda\in r(x)$ and $h\in\mathscr{M}$ then by $(x-\lambda e)y=e$ we have $(h(x)-\lambda)h(y)=1$ and so $h(x)=\hat{x}(h)\neq\lambda$. We proved that $r(x)$ lies in the complement of $\hat{x}(\mathscr{M})$.

Corollary. *In any complex Banach algebra* $\|\hat{x}\|\leqslant\|x\|_{sp}$.

Our next object is to prove that in any commutative, complex Banach algebra we have $\|\hat{x}\|=\|x\|_{sp}$ but this can be proved only after developing some of the ideal theory in Banach algebras. Then we will easily prove that $\hat{x}(\mathscr{M})$ and $\sigma(x)$ differ in at most 0 and this will immediately show that $\|\hat{x}\|=\|x\|_{sp}$.

The *spectrum* $\sigma(x)$ and the *spectral norm* $\|x\|_{sp}$ can be defined in any real or complex algebra A. Namely $\sigma(x)$ denotes the spectrum of $x\in A$ in the complexification $A_{\mathbb{C}}$. Therefore by Proposition 4.2

$$\sigma(x)^{\times}=\left\{\lambda:\lambda\in\mathbb{C}^{\times}\text{ and }\frac{x}{\lambda}\notin Q_{\mathbb{C}}\right\}$$

where $Q_{\mathbb{C}}$ is the quasigroup of $A_{\mathbb{C}}$ and

$$\|x\|_{sp}=\operatorname{supr}\{|\lambda|:\lambda\in\sigma(x)^{\times}\}.$$

If $\sigma(x)^{\times}$ is void then by definition $\|x\|_{sp}=0$.

Proposition 5. *If A is a complex normed algebra then* $\|x\|_{sp}\geqslant\nu(x)$ *for every element x of A.*

Proof. Let $\bar{A}$ denoted the completion of A so that $\bar{A}$ is a complex Banach algebra. We let $\bar{Q}$ denote the quasigroup of $\bar{A}$ and $\overline{\sigma(x)}^{\times}$ be the non-zero spectrum of x in $\bar{A}$. We have $Q\subseteq\bar{Q}$ and so $\sigma(x)^{\times}\supseteq\overline{\sigma(x)}^{\times}$ for any x in

A. Therefore the spectral norm of x as an element of $\bar{A}$ is at most $\|x\|_{sp}$. The conclusion follows from Theorem 3.

If A is a real algebra than $\bar{a}=\overline{(a_1,a_2)}=(a_1,-a_2)$ defines a conjugation $a\mapsto\bar{a}$ in its complexification $A_{\mathbb{C}}$ such that $\overline{a+b}=\bar{a}+\bar{b}$, $\overline{\lambda a}=\bar{\lambda}\bar{a}$ and $\overline{ab}=\bar{a}\bar{b}$. It follows that $x\in Q_{\mathbb{C}}$ if and only if $\bar{x}\in Q_{\mathbb{C}}$. If x is in A then using $\overline{\lambda x}=\bar{\lambda}x$ we see that $\lambda\in\sigma(x)^{\times}$ if and only if $\bar{\lambda}\in\sigma(x)^{\times}$. Therefore if A is a real algebra then $\sigma(x)=\overline{\sigma(x)}$ for every x in A.

Proposition 6. *If A is a real algebra and if $\lambda=\alpha+\beta i$ is in $\mathbb{C}^{\times}$ then $\lambda\in\sigma(x)^{\times}$ if and only if $|\lambda|^{-2}(2\alpha x-x^2)$ is quasisingular in A.*

Proof. One sees immediately that

$$\frac{x}{\lambda}\circ\frac{x}{\bar{\lambda}}=\frac{x}{\bar{\lambda}}\circ\frac{x}{\lambda}=|\lambda|^{-2}(2\alpha x-x^2).$$

An element $a\in A_{\mathbb{C}}$ is quasiregular if and only if $a\circ b=b\circ a=0$ for some element $b\in A_{\mathbb{C}}$. Moreover if $a\in A$ then it is easy to prove that $b\in A$. Hence by the associativity of the circle operation $x/\lambda\in Q_{\mathbb{C}}$ if and only if $|\lambda|^{-2}(2\alpha x-x^2)$ is in Q.

We had seen that the non-negative quantity $v(x)=\lim\sqrt[n]{\|x^n\|}$ is defined for every x in any real or complex normed algebra A. Clearly $0\leqslant v(x)\leqslant\|x\|$ and $v(\lambda x)=|\lambda|\cdot v(x)$ for every scalar λ. Using $xyxy\ldots xy=x(yx)\ldots(yx)y$ we obtain $v(xy)=v(yx)$ for any $x,y\in A$. If $xy=yx$ then $v(xy)\leqslant v(x)v(y)$. For any $m,n=1,2,\ldots$ and $x\in A$ we have

$$\sqrt[mn]{\|x^{mn}\|}\leqslant\sqrt[mn]{\|x^n\|^m}=\sqrt[n]{\|x^n\|}.$$

Using this inequality one sees that $v(x)=\inf\sqrt[n]{\|x^n\|}$. An element x of the normed algebra A is called *topologically nilpotent* if $v(x)=0$.

6. Commutative Banach Algebras and their Ideals

It is natural to ask which ideals I are kernels of homomorphisms $h\in\mathscr{M}$. In the commutative case this question is answered by Gelfand's fundamental theorem on commutative Banach algebras. Although this is a deep result, it can be proved in a sequence of easy steps, most of which are self-contained and interesting results in algebra. One of the main steps is the Gelfand-Mazur theorem which has been stated and proved already earlier (Theorem 4.7). The second half of this section is devoted to the topologies of the maximal ideal space. In addition to the Gelfand topology another important topology called the hull-kernel topology is introduced and the relationship of these two topologies is studied.

A great deal of this material is applicable to ring theory and algebraic geometry where the corresponding topologies are known as the structure space topology and the spectral topology.

If A is a ring or an algebra over some field F then a two sided ideal I of A will be called a *regular ideal* provided the quotient ring A/I has an identity U. As earlier we suppose that U is not the zero element I of A/I and so a regular ideal is necessarily a proper subset of A. The elements u of U are the *relative identities* or *relative unities* of A with respect to I. Hence u is a relative identity if and only if $au-a\in I$ and $ua-a\in I$ for every $a\in A$. If u_1, u_2 are both relative identities modulo the same ideal I then by $u_1u_2-u_1\in I$ and $u_1u_2-u_2\in I$ we have $u_1-u_2\in I$. This shows that if I is regular then its relative identity U is uniquely determined. Of course if A has an identity e then every ideal I properly contained in A is regular and $U=e+I$.

Proposition 1. *If U is the identity of A/I modulo the ideal I then $U\cap Q=\emptyset$.*

Proof. Suppose that u has a quasi-inverse v, so that $u+v=uv$. If $u\in U$ then u being a relative identity $u=uv-v\in I$ and so $U=I$ which implies that $I=A$. Indeed if $u\in U=I$ and $x\in A$ then by $ux-x\in I$ we have $x\in I$.

We notice that if I is a regular ideal and if $I\subseteq J\subset A$ then J is also regular because if $au-a\in I$ then $au-a\in J$, etc. We also see that if I is an ideal in a normed algebra A then $\bar{I}$ is also an ideal in A: For if $\|a-i\|<\varepsilon$ then $\|ax-ix\|<\varepsilon\|x\|$ and so if $a\in\bar{I}$ then $ax\in\bar{I}$ for every $x\in A$ and similarly $xa\in\bar{I}$.

Proposition 2. *If I is a regular maximal ideal of the Banach algebra A then I is a closed set in A.*

Proof. Let u be a relative identity of I so that $U=u+I$. By Proposition 1 we have $U\subseteq cQ$ where cQ is a closed set by Proposition 2.5. Hence cQ contains the closure of $U=u+I$ which is $u+\bar{I}$. Since $u+\bar{I}\subseteq cQ$ and $Q\neq\emptyset$ we see that $\bar{I}$ is properly contained in A. Hence by the maximality of I we have $\bar{I}=I$.

Proposition 3. *If I is a closed ideal of the Banach algebra A then A/I is a Banach algebra under the norm $\|\xi\|=\operatorname{glb}\{\|x\|: x\in\xi\}$.*

Proof. Since I is closed $\|\xi\|=\|x+I\|=\operatorname{glb}\{\|x+i\|: i\in I\}$ is a proper norm and so A/I is a normed vector space which is complete because A is complete: Indeed, from every Cauchy sequence in A/I one can select a subsequence $\alpha_1, \alpha_2, \ldots$ such that $\|\alpha_n-\alpha_{n+1}\|<2^{-n}\,(n=1,2,\ldots)$

and so one can determine by induction a sequence of elements $a_n \in \alpha_n$ such that $\|a_n - a_{n+1}\| < 2^{-n}$. Hence $a_1, a_2, \ldots$ is a Cauchy sequence in A and so it has a limit $a \in A$. The coset $\alpha = a + I$ is then the limit of the original Cauchy sequence given in A/I. One can easily check also that $\|\alpha\beta\| \leqslant \|\alpha\| \cdot \|\beta\|$ for any $\alpha, \beta \in A/I$.

Proposition 4. *If A is a maximal ideal in a ring A then A/I has no proper ideals.*

Proof. For every ideal J in A/I the set $\bigcup\{\eta : \eta \in J\}$ is an ideal in A which includes I and is proper if J is proper. Since I is maximal this new ideal is either I or A and so J is the zero ideal or $J = A/I$.

Proposition 5. *If $\mathscr{A}$ is a commutative ring with identity and without proper ideals then $\mathscr{A}$ is a field.*

Proof. If $\alpha \in \mathscr{A}$ and if $\alpha \neq 0$ then $(\alpha) = \alpha A$, the ideal generated by α is not 0. Therefore $(\alpha) = A$ and $\alpha\beta = \varepsilon$ for some $\beta \in \mathscr{A}$ where ε denotes the identity of $\mathscr{A}$.

We recall that a ring without proper ideals is called a *simple ring*. Hence the above proposition states that the only commutative simple rings having identity are fields.

Proposition 6. *If A is a commutative ring and I is a regular ideal of A then I is maximal if and only if A/I is a field.*

Proof. Let $\mathscr{A} = A/I$ so that $\mathscr{A}$ is a commutative ring with identity. If $\mathscr{J}$ is an ideal in $\mathscr{A}$ then $J = \bigcup \mathscr{J}$ is an ideal in A which contains I and one has $I \subset J \subset A$ if and only if $\{I\} \subset \mathscr{J} \subset \mathscr{A}$ where $\{I\}$ denotes the zero ideal of $\mathscr{A}$. Hence $\mathscr{A}$ has no proper ideals if and only if I is maximal.

Similarly one proves that a regular ideal I of a commutative ring A is a prime ideal if and only if A/I is an integral domain.

Theorem 7. *Let A be a complex, commutative Banach algebra. Then every regular maximal ideal of A is the kernel of some non-trivial homomorphism $h: A \to \mathbb{C}$ and conversely, the kernel of every h is a regular maximal ideal. The relative identities of I are all mapped by h into* 1.

Proof. Let I be a regular maximal ideal in A and let U be the identity of $\mathscr{A} = A/I$. Then by Proposition 3 $\mathscr{A}$ is a complex, commutative Banach algebra. Since I is maximal by Proposition 4 the quotient $\mathscr{A}$ has no proper ideals and so $\mathscr{A}$ being commutative by Proposition 5 every $\alpha \neq 0$ is invertible. Thus by Theorem 4.7 $\mathscr{A}$ is topologically isomorphic to $\mathbb{C}$ under the map $\xi \mapsto \lambda_\xi$ where $\xi = \lambda_\xi \varepsilon$ and ε is the identity of $\mathscr{A}$. We let $\mathscr{A} \to \mathbb{C}$ denote this isomorphism and let $A \to \mathscr{A}$ be the natural homomorphism of A onto A/I. Then the composed map $A \to \mathscr{A} \to \mathbb{C}$ is a homomorphism $h: A \to \mathbb{C}$ whose kernel is the given ideal I. Conversely, if h is a non-trivial homomorphism of A onto $\mathbb{C}$ then its kernel I is a regular ideal in A. The maximality of I can be easily proved by a direct reasoning but it is more elegant to refer to Proposition 6.

Corollary. *In the complex, commutative case the maximal ideal space can be identified with the properly topologized set of regular maximal ideals.*

Proposition 8. *Let A be a ring, let $a \in A$ and let I be the left ideal $I = \{x - xa : x \in A\}$. Then a has a left quasi-inverse if and only if $I = A$.*

Proof. If $I = A$ then $xa = x + a$ for suitable $x \in A$. Conversely, if $x \circ a = 0$ for some $x \in A$ then $a = xa - x \in I$ and so $ya \in I$ for any $y \in A$. Since $y - ya$ is in I we get $y \in I$ where y is an arbitrary element of A.

A *regular left ideal* of a ring A is a left ideal I which has a *right relative identity* i.e. an element $u \in A$ such that $au - a \in I$ for every $a \in A$. The element u in general does not belong to I and the ideal I need not be proper. Similarly I is a *regular right ideal* of A if there exists a *left relative identity* u such that $ua - a \in I$ for every $a \in A$.

Proposition 9. *If I is a regular left (right) ideal and I is properly contained in the ring A then I contains none of its right (left) relative identities.*

Proof. If u is a relative right identity of I and if $u \in I$ then $a - au \in I$ for every $a \in A$ and also $au \in I$. Hence $a \in I$ and $A = I$.

Proposition 10. *If A is a Banach algebra and I is a left (right) regular maximal ideal in A then I is a closed set.*

Proof. Let U denote the set of right relative identities of the regular left ideal I of the ring A and let Q be the quasigroup of A. If $u \in Q \cap U$ and v is a quasi-inverse of u then $u = vu - v \in I$ and so $u \in I$ and $A = I$. Thus if I is properly contained in A then $U \subseteq cQ$. If u is a right relative identity of I then $u + I \subseteq U$ and so $u + I \subseteq cQ$. If A is a Banach algebra then by Proposition 2.5 cQ is closed and so $u + \bar{I} = \overline{u + I} \subseteq cQ$. Since Q is not void it follows that the regular left ideal $\bar{I}$ is properly contained in A. Hence by the maximality of I we have $I = \bar{I}$ and I is closed.

Proposition 11. *Every regular left (right) ideal which is properly contained in the ring A can be extended to a regular maximal left (right) ideal.*

Proof. Let u be a right relative identity of I so that $u \notin I$ by Proposition 9. Consider the set $\mathscr{S}$ of all left ideals J such that $I \subseteq J \subset A$ and partially order $\mathscr{S}$ by inclusion. Since $I \subseteq J$ the element u will be a right relative identity of every J and so by Proposition 9 none of these ideals J can contain u. Thus every linearly ordered subset $\mathscr{C}$ of $\mathscr{S}$ has an upper bound in $\mathscr{S}$, namely its union $\bigcup \{J : J \in \mathscr{C}\}$. Thus by Zorn's lemma I is included in some maximal left ideal $J \subset A$.

Proposition 12. *If A is a ring then $a \in A$ has no left (right) quasi-inverse if and only if a is a right (left) relative identity of some regular maximal left ideal.*

Proof. If we let $I = \{x - xa : x \in A\}$ then I is a left ideal, a is a right relative identity of I and $I \subset A$ by Proposition 8. Now I can be extended to a regular maximal left ideal by Proposition 11 and of course a remains a right relative identity. Conversely, suppose a is a right relative identity of the regular maximal left ideal I. Then $a \notin I$ and $xa - x \in I$ for every x in A and so $xa - x = a$ is not possible.

We can now return to the study of the relationship between the spectrum $\sigma(x)$ and the range of the Fourier transform $\hat{x} : \mathscr{M} \to \mathbb{C}$.

Theorem 13. *In any Banach algebra A we have $\hat{x}(\mathscr{M}) \subseteq \sigma(x)$ for every $x \in A$.*

Proof. Let $h \in \mathscr{M}$ be given. Let $h(x) = \lambda = \alpha + i\beta \neq 0$ and put $y = |\lambda|^{-2}(2\alpha x - x^2)$. Since $h(y) = 1$ by Proposition 2.7 $y \notin Q$. Hence by Proposition 5.6 we get $h(x) \in \sigma(x)$. If $h(x) = 0$ and A has no identity then $h(x) = 0 \in \sigma(x)$ by Proposition 4.2. If A has a unity e then $h(e) = 1$ and so by $h(x) = 0$ we have $x \notin G$. Thus $h(x) = 0 \in \sigma(x)$.

Theorem 14. *If x belongs to the complex, commutative Banach algebra A then $\sigma(x)$ and $\hat{x}(\mathscr{M})$ differ at most in 0 which might belong to $\sigma(x)$ and not to $\hat{x}(\mathscr{M})$. If A has an identity then $\sigma(x) = \hat{x}(\mathscr{M})$.*

Proof. We already proved that in general $\hat{x}(\mathscr{M}) \subseteq \sigma(x)$. Let $\lambda \neq 0$ be in $\sigma(x)$ so that $x/\lambda \notin Q$ by Proposition 4.2. Thus by Proposition 12 x/λ is a relative identity with respect to some regular maximal ideal I and by Theorem 7 I is the kernel of a non-trivial homomorphism $h: A \to \mathbb{C}$. Since x/λ is a relative identity with respect to I by the same theorem we have $h(x/\lambda) = 1$ and so $\hat{x}(h) = h(x) = \lambda$ proving that $\lambda \in \hat{x}(\mathscr{M})$. If A has an identity and $\lambda = 0 \in \sigma(x)$ then x is not invertible and so Ax is a proper ideal of A containing x. By Proposition 11 Ax is included in a maximal ideal I. Since $x \in I$ we have $h_I(x) = \hat{x}(I) = 0$ and $0 \in \hat{x}(\mathscr{M})$.

Corollary. *If A is a complex, commutative Banach algebra then we have $\|\hat{x}\| = \|x\|_{sp}$ for any x in A.*

Part of the reasoning given in the foregoing proof shows that after discarding zero $\sigma(x)$ for any x can be identified with certain regular maximal ideals of the algebra. Therefore in some sense $\mathcal{M}$ can be considered as the spectrum for all elements of the algebra or for the algebra itself.

Given a commutative ring A with identity 1 it is customary to call the set of prime ideals of A the *spectrum* of A and to denote it by $\mathrm{Spec}(A)$, or shortly by X.

Let A be an algebra, let L be a left ideal of A and let

$$I = (L\colon A) = \{a\colon a \in A \text{ and } aA \subseteq L\}.$$

Then I is a two sided ideal called the *quotient* of L. An ideal P of A is called a *primitive ideal* if $P = (L\colon A)$ where L is a left regular maximal ideal. If u is a relative identity for L then $u \notin L$ and so $u \notin P$. Hence every primitive ideal P is properly contained in A. *Every regular maximal ideal I is primitive.* For I is contained in a regular maximal left ideal L. Then $IA \subseteq I \subseteq L$ so $I \subseteq (L\colon A)$ and by the maximality of I we have $I = (L\colon A)$.

Now let A be a Banach algebra and let L be a left regular maximal ideal of A. Then L is closed by Proposition 10. Let $a_n \in (L\colon A)$ i.e. let $a_n x \in L$ for every x in A and suppose that $a_n \to a$. Then $a_n x \to ax$ and so $ax \in L$ for every x in A and $a \in (L\colon A)$. This shows that in a Banach algebra every primitive ideal is closed.

The concept of primitive ideals is meaningful not only in algebras but also in arbitrary rings. The set of primitive ideals P of a ring R is called the *structure space* of R and as the name indicates a topology is also defined on this set. Similarly if A is an algebra then the properly topologized set of primitive ideals of A is called the *structure space* of A.

The same construction can be used to define a topology in at least four situations: If $\mathfrak{P}$ is the set of primitive ideals in a ring or in an algebra then the topology attached to $\mathfrak{P}$ is called the *Jacobson topology* or the *structure space topology*. If $\mathcal{M}$ is the set of regular maximal ideals of a Banach algebra then we speak about the *Stone topology* or the *hull-kernel topology*. Finally if the set is $\mathrm{Spec}(A)$ for some commutative ring A with identity then the corresponding topology is known as the *Zariski topology* or the *spectral topology* of $\mathrm{Spec}(A)$.

We shall consider the case when A is an algebra and the set is $\mathfrak{P}$, the set of primitive ideals or $\mathcal{M}$, the set of regular maximal ideals. We know that $\mathcal{M} \subseteq \mathfrak{P}$ and we shall define the hull-kernel topology as the trace of the structure space topology on $\mathcal{M}$. The following result will concern the definition:

Proposition 15. *If P is a primitive ideal and L_1, L_2 are left ideals in the algebra such that $L_1 L_2 \subseteq P$ then $L_1 \subseteq P$ or $L_2 \subseteq P$.*

Corollary. *If P is primitive then P is prime.*

Proof. Let $P=(L\colon A)$ and let L_1, L_2 be left ideals such that $L_1 L_2 \subseteq P$ but $L_2 \nsubseteq P$. Then $L_2 A \nsubseteq L$ and so $L_2 A + L$ is a left ideal properly containing L. Hence by the maximality of L we have $L_2 A + L = A$. Now we can easily show that $L_1 \subseteq P$: Indeed $L_1 A \subseteq L_1 L_2 A + L_1 L \subseteq P + L \subseteq L$ because L is regular and so $L_1 \subseteq P$.

If J is an ideal in the algebra A we define $h(J)$, the *hull* of J as the set of all primitive ideals P containing J:

$$h(J) = \{P\colon P \in \mathfrak{P} \text{ and } P \supseteq J\}.$$

Proposition 16. *The set of all hulls satisfies the axioms for closed sets and so it defines a topology on $\mathfrak{P}$.*

Proof. Clearly $h(A)$ is empty and $h(0)=\mathfrak{P}$. To prove that the intersection of a family of closed sets $\mathscr{C}_i = h(J_i)\,(i \in \mathscr{I})$ is a closed set notice that

$$\bigcap h(J_i) = \{P\colon P \in \mathfrak{P} \text{ and } P \supseteq J_i \text{ for } i \in \mathscr{I}\} = \{P\colon P \supseteq \bigcup J_i\} = h(J)$$

where $J = \sum J_i$. To prove that the union of two closed sets is closed let $\mathscr{C}_k = h(J_k)\,(k=1,2)$. Then by $J_1 \cap J_2 \subseteq J_k\,(k=1,2)$ one obtains $h(J_1 \cap J_2) \supseteq h(J_1) \cup h(J_2)$. The proof of the opposite inclusion is as follows: Suppose that P is in $h(J_1 \cap J_2)$. Then $P \supseteq J_1 \cap J_2$ where $J_1 \cap J_2 \supseteq J_1 J_2$ because J_1 and J_2 are two sided ideals. Hence $P \supseteq J_1 J_2$ and so $P \supseteq J_1$ or $P \supseteq J_2$ by Proposition 15. We proved that $P \in h(J_1)$ or $P \in h(J_2)$.

For any set of ideals $\mathscr{S}$ of the algebra A we define the *kernel* of the set, $k(\mathscr{S})$ as $\bigcap \{J\colon J \in \mathscr{S}\}$. By definition we let $k(\emptyset)=A$. If $\mathscr{S}$ consists of primitive ideals then $h(k(\mathscr{S})) \supseteq \mathscr{S}$.

Proposition 17. *For any set $\mathscr{S}$ of primitive ideals the closure of $\mathscr{S}$ in the structure space topology is $\bar{\mathscr{S}} = h(k(\mathscr{S}))$.*

Note. In the case of the hull-kernel topology one defines $h(J)=\{I\colon I \supseteq J\}$ where I denotes regular maximal ideals. Since the hull-kernel topology is the trace of the structure space topology on $\mathcal{M}$ for any subset $\mathscr{S}$ of $\mathcal{M}$ one has $\bar{\mathscr{S}} = h(k(\mathscr{S}))$.

Proof. By the definition of the closed sets we have $\bar{\mathscr{S}}=\bigcap\{h(J): h(J)\supseteq\mathscr{S}\}$. If $\mathscr{S}\subseteq h(J)$ then $J\subseteq P$ for every $P\in\mathscr{S}$ and so $J\subseteq\bigcap\{P: P\in\mathscr{S}\}=k(\mathscr{S})$ and conversely if $J\subseteq k(\mathscr{S})$ then $\mathscr{S}\subseteq h(J)$. Thus among all $h(J)$ such that $h(J)\supseteq\mathscr{S}$ there is a minimal one which corresponds to a maximal J, namely $k(\mathscr{S})$. Hence $\bar{\mathscr{S}}=h(k(\mathscr{S}))=hk(\mathscr{S})$.

Proposition 18. *The structure space $\mathfrak{P}$ is a T_0 space and the hull-kernel topology of $\mathscr{M}$ is a T_1 topology.*

Proof. Let $P_k\,(k=1,2)$ be primitive ideals. Then $h(P_k)$ is a closed set containing $P_k\,(k=1,2)$. If $P_2\in h(P_1)$ then $P_2\supseteq P_1$ and similarly if $P_1\in h(P_2)$ then $P_1\supseteq P_2$. Hence $P_2\notin h(P_1)$ or $P_1\notin h(P_2)$ unless $P_1=P_2$. Therefore $\mathfrak{P}$ is a T_0 space. If I is a regular maximal ideal then $h(I)=\{I\}$ and so single points form closed sets in $\mathscr{M}$.

Theorem 19. *If A is an algebra with identity then $\mathscr{M}$ is compact in the hull-kernel topology.*

Note. One can also prove that $\mathfrak{P}$ is compact if A has an identity.

Proof. If $\mathscr{C}$ is a closed set of $\mathscr{M}$ then $\mathscr{C}=h(J)$ where $J=k(\mathscr{C})$. For $a\in J$ let (a) denote the ideal generated by a so that $(a)\subseteq J$ and consequently $h((a))\supseteq h(J)=\mathscr{C}$. This shows that $\mathscr{C}\subseteq\bigcap\{h((a)): a\in J\}$. To prove the opposite inclusion notice that "$I\in h((a))$ for every $a\in J$" means that "$a\in I$ for every $a\in J$". Hence if I belongs to the intersection then $I\supseteq J$ and so $I\in\mathscr{C}$ by $\mathscr{C}=h(J)$. Thus every closed set $\mathscr{C}$ can be written in the form $\mathscr{C}=\bigcap h((a))$ where $a\in J=k(\mathscr{C})$. From this we see that the sets $h((a))\,(a\in A)$ form a base for the closed sets of $\mathscr{T}_{hk}$.

By a well known compactness criterion it is now sufficient to prove the following: If for a family $a_i\,(i\in\mathscr{I})$ of elements in A we have $\bigcap h((a_i))=\emptyset$ then we have the same for a finite subfamily.

Let us suppose that such a family $a_i\,(i\in\mathscr{I})$ is given. Then $\bigcap h((a_i))=h(\sum(a_i))=\emptyset$. But every proper ideal of A can be included in a regular maximal ideal because A has an identity. Therefore we must have $\sum(a_i)=A$. Thus we have $e=b_{i_1}+\cdots+b_{i_n}$ for suitable indices i_k and elements b_{i_k} in (a_{i_k}). This proves that $A=(a_{i_1})+\cdots+(a_{i_n})$ and so $h((a_{i_1}))\cap\cdots\cap h((a_{i_n}))=\emptyset$.

In what follows let S be a non-void set and let A be an algebra of functions $a: S\to\mathbb{C}$. The set $I_s=\{a: a\in A \text{ and } a(s)=0\}$ is a two sided ideal in A called the *fixed ideal* of the point s or the ideal fixed at s.

Proposition 20. *If s is a point in the set S such that some non zero elements of the algebra A vanish at s and others do not then the fixed ideal I_s is a regular maximal ideal in A.*

Proof. By the hypothesis I_s is an ideal such that $\{0\} \subset I_s \subset A$. If $x \notin I_s$ i.e. $x(s) \neq 0$ then for any y in A put

$$y = \left\{y - \frac{y(s)}{x(s)}x\right\} + \frac{y(s)}{x(s)}x.$$

The first term is an element of I_s and the second is in the principal ideal (x). Therefore $y \in I_s + (x)$ and so $I_s + (x) = A$. This proves that I_s is maximal. The map $y \mapsto y(s)$ is a homomorphism of A into $\mathbb{C}$ with kernel I_s. Since $a(s) \neq 0$ for some a in A the homomorphism is surjective and so A/I_s is isomorphic to the field $\mathbb{C}$. Therefore I_s is a regular maximal ideal of A.

We let F denote the set of those s in S for which I_s is a regular maximal ideal. Thus F corresponds to the subspace of fixed ideals of the maximal ideal space $\mathcal{M}$. If for each $s \in S$ there exist a, b in A such that $a(s) = 0$ and $b(s) \neq 0$ then $F = S$ and we can view S as being embedded in $\mathcal{M}$.

Proposition 21. *Let $\mathcal{T}$ be a topology on the set S such that every element a of A is continuous on S. Then the trace of $\mathcal{T}_{hk}$ on the fixed part F of S is not stronger than the trace of $\mathcal{T}$.*

Proof. For $C \subseteq F$ let

$$k(C) = \bigcap \{I_c : c \in C\} = \{a : a \in A \text{ and } a \equiv 0 \text{ on } C\}.$$

For any function $f: S \to \mathbb{C}$ we define the $\mathcal{T}$-closed set

$$Z(f) = \{s : s \in S \text{ and } f(s) = 0\}.$$

If we let

$$Z = F \cap \bigcap \{Z(a) : a \in k(C)\}$$

then Z has the following properties: Z contains C, and Z is a closed set in F relative to the topology $\mathcal{T}$.

The closure of C in $\mathcal{M}$ is $hk(C)$ and therefore C is closed in F relative to the topology induced on F by $\mathcal{T}_{hk}$ if and only if

$$F \cap hk(C) = C \tag{1}$$

where the fixed ideals and the corresponding points in S have been identified. Now we have

$$F \cap hk(C) = \{s : s \in F \text{ and } I_s \supseteq k(C)\} = \{s : s \in F \text{ and } a(s) = 0 \text{ for all } a \in k(C)\}.$$

By the definition of Z the second description shows that $F \cap hk(C) \supseteq Z$. Conversely if $s \in F$ but $s \notin Z$ then there is an a in $k(C)$ such that

$a(s)\neq 0$ and so $F\cap hk(C)=Z$. Hence if C is closed relative to $\mathcal{T}_{hk}$ then by (1) we have $C=Z$ where Z is closed by our earlier conclusion relative to the topology $\mathcal{T}$. Therefore on F we have $\mathcal{T}_{hk}\leqslant\mathcal{T}$.

Proposition 22. *We have* $\mathcal{T}_{hk}=\mathcal{T}$ *on* F *if and only if for every closed set* C *of* F *in the topology* $\mathcal{T}$ *and for every point* $s\in F-C$ *there is an* a *in* A *such that* $a(s)\neq 0$ *but* $a\equiv 0$ *on* C.

Corollary. *If* $\mathcal{T}_{hk}=\mathcal{T}$ *for some topology* $\mathcal{T}$ *then* $\mathcal{T}_{hk}$ *is completely regular.*

Proof. We continue the reasoning developed during the proof of Proposition 21: We already proved that $C\subseteq Z$ always holds. Thus a necessary and sufficient condition for $Z=C$ is that $Z\subseteq C$. In view of the definition of Z this takes place if and only if for every $s\in F-C$ there is an a in $k(C)$ such that $s\notin Z(a)$. In other words there is an a such that $a\equiv 0$ on C and $a(s)\neq 0$. We have $\mathcal{T}_{hk}=\mathcal{T}$ if and only if this holds for every $\mathcal{T}$-closed set C.

If $F=S$ i. e. if for each s in S there exist a, b in A such that $a(s)=0$ and $b(s)\neq 0$ and if $\mathcal{T}_{hk}=\mathcal{T}$ on S then we say that A is a *regular function algebra.*

Let A be a Banach algebra and consider the dual algebra of Fourier transforms $\hat{a}\colon \mathcal{M}\to\mathbb{C}$ where $\hat{a}(I)=h_I(a)$. Thus $\{\hat{a}\colon a\in A\}$ is a function algebra on the set $S=\mathcal{M}$. If a point of S i. e. a regular maximal ideal I is given then $h_I\colon A\to\mathbb{C}$ is a surjective map so there are a, b in A such that $h_I(a)=0$ and $h_I(b)\neq 0$. Consequently $\hat{a}(s)=0$ and $\hat{b}(s)\neq 0$ proving that F is the set $\mathcal{M}$. Let $\mathcal{T}$ denote the Gelfand topology of $\mathcal{M}$ that is to say the weakest topology such that all the Fourier transforms $\hat{a}$ are continuous. If $\{\hat{a}\colon a\in A\}$ is a regular function algebra on $\mathcal{M}$ i. e. if the hull kernel topology of $\mathcal{M}$ coincides with the Gelfand topology then A is called a *regular Banach algebra.*

In order to see an example of a non-regular function algebra consider for S the closed unit disk in the complex plane and the algebra A of complex valued functions $a\colon S\to\mathbb{C}$ which are continuous on S and analytic in the interior of S. If $\mathcal{T}$ denotes the usual topology of S and $\mathcal{T}_w$ the weakest topology which makes every a in A continuous then one can easily verify that $\mathcal{T}_{hk}<\mathcal{T}_w=\mathcal{T}$. We only have to refer to Proposition 22.

Proposition 23. *If* S *is a compact topological space then every maximal ideal of the algebra of complex valued functions* $C(S)$ *is fixed and the maximal ideal space of* S *can be identified with* S *itself.*

Proof. For f in $C(S)$ we let $Z(f)=\{x: f(x)=0\}$ which is a closed set. Let I be an ideal of $C(S)$ and suppose there is an f in I such that $Z(f)=\emptyset$. Then $1/f$ is continuous on S, belongs to $C(S)$ and so $f \cdot 1/f=1$ is in I and $I=C(S)$. Therefore every function chosen from a proper ideal I vanishes somewhere. Now we can prove that the family of sets $Z(f)\,(f\in I)$ have the finite intersection property. Indeed if $f\in I$ then by $\bar{f}\in C(S)$ we have $f\bar{f}=|f|^2\in I$. Therefore given $f_1,\ldots,f_n$ in I we have

$$Z(f_1)\cap\cdots\cap Z(f_n)=Z(|f_1|^2+\cdots+|f_n|^2)\neq\emptyset\,.$$

Since S is compact the total intersection $\bigcap\{Z(f): f\in I\}$ is not void and so the elements of I have a common zero. Therefore every proper ideal I is fixed. It is now easy to show that the usual topology, the Gelfand topology and the hull kernel topology are all the same on S.

7. Radical and Semisimplicity

Most of the material discussed here is pure algebra although some of the concepts such as left quasi-inverses were used or introduced in the context of Banach algebras. This pure algebraic material concerns both arbitrary rings and also algebras over some field F. Actually we can even suppose that F is a not necessarily commutative field i.e. it is a division ring. Thus for every definition or result on rings there is a corresponding one for algebras or for algebras with an identity which in this section will be denoted by 1 instead of e. There are few paragraphs at the end which concern Banach algebras. These are on the radical of a Banach algebra and on the concept of a semisimplicity.

Let A be a ring and let Q_l denote the set of all ring elements a which have left quasi-inverses, so that $a\in Q_l$ if and only if $b\circ a=b+a-ba=0$ for a suitable $b\in A$. Similarly we let Q_r be the set of ring elements having right quasi-inverses and so $Q=Q_l\cap Q_r$ is the set which we called earlier the quasi-group of A.

The *Jacobson radical* of the ring is the set $J(A)=\{x: Ax\subseteq Q_l\}$ i.e. the set of those ring elements for which the left ideal Ax consists entirely of left quasi-invertible elements. The radical can be defined in several different ways, for instance instead of $Ax\subseteq Q_l$ one can require that $xA\subseteq Q_r$, or $Ax\subseteq Q$, or $xA\subseteq Q$. The equivalence of these four definitions becomes obvious in a moment. First, the equivalence of $Ax\subseteq Q_l$ and $Ax\subseteq Q$, and also the equivalence of $xA\subseteq Q_r$ and $xA\subseteq Q$ is obvious from the following:

Proposition 1. *If I is a left (or right) ideal and if $I \subseteq Q_l$ (or $I \subseteq Q_r$) then $I \subseteq Q$.*

Proof. If $x \in I$ then $y \circ x = 0$ for some y and so $y = yx - x \in I \subseteq Q_l$. Hence y has a left quasi-inverse which must be x because x is a right quasi-inverse of y. We proved that $x \circ y = y \circ x = 0$ and so $x \in Q$. The case of right ideals is similar.

The conditions $Ax \subseteq Q$ and $xA \subseteq Q$ are also equivalent because we have the following:

Proposition 2. *If $ab \in Q_l$ then $ba \in Q_l$ and if $ab \in Q_r$ then $ba \in Q_r$, so that $ab \in Q$ implies $ba \in Q$.*

Proof. If $ab \in Q_l$ then $x \circ ab = 0$ for some x and one verifies in a line that $y = bxa - ba$ satisfies $y \circ ba = 0$. Similarly if $ab \in Q_r$, say $ab \circ x = 0$ then $ba \circ y = 0$ for $y = bxa - ba$.

Proposition 3. *The radical $J(A) \subseteq Q$.*

Proof. If $x \in J(A)$ then $Ax \subseteq Q$ so $x^2 \in Q$ and so $x^2 \circ y = y \circ x^2 = 0$ for some $y \in A$. We get $x \circ (-x \circ y) = (x \circ (-x)) \circ y = x^2 \circ y = 0$ and similarly $(y \circ (-x)) \circ x = 0$. Thus x has both right and left quasi-inverses i.e. $x \in Q$.

Theorem 4. *The radical $J(A)$ is a two sided ideal.*

Proof. If $x \in J(A)$ then $Ax \subseteq Q$ and $xA \subseteq Q$ so that for any $a \in A$ we have $A(ax) \subseteq Q$ and $(xa)A \subseteq Q$. This proves that $ax, xa \in J(A)$ if $x \in J(A)$ and $a \in A$. Now let $x, y \in J(A)$ and let $a \in A$. Since $ax \in J(A) \subseteq Q$ we have $u \circ ax = 0$ for some $u \in A$ and similarly by $(a - ua)y \in J(A) \subseteq Q$ we have $v \circ (ay - uay) = 0$ for some $v \in A$. Therefore

$$(v \circ u) \circ (ax + ay) = v \circ (u \circ (ax + ay)) = v \circ (ay - uay) = 0 .$$

This proves that $a(x+y) \in Q_l$ for every $a \in A$ and so $x + y \in J(A)$.

Theorem 5. *The Jacobson radical $J(A)$ is the intersection of all regular left (right) maximal ideals of the ring A.*

Proof. Let J denote the intersection of all regular left maximal ideals of A. If no such ideal exists we let $J = A$. We are going to prove that $J \subseteq J(A)$. The case $J(A) = A$ being trivial let $x \notin J(A)$. Then $ax \notin Q$ for some $a \in A$ and so by Proposition 6.12 ax is a right relative identity of a regular left maximal ideal I of A. By Proposition 6.9 $ax \notin I$ so $x \notin I$ and we proved that $J \subseteq J(A)$.

Next we prove that $J(A) \subseteq J$ by showing that $J(A) \subseteq I$ where I is an arbitrary regular left maximal ideal of A. By Theorem 4 the radical is a two sided ideal so $J'=\{i+j: i\in I \text{ and } j\in J(A)\}$ is a regular left ideal containing I. Since I is maximal $J'=I$ or $J'=A$. Suppose $J'=A$ and denote by u a right relative identity of I. Then $u=i+j$ for some $i\in I$ and $j\in J(A)$. Since $j\in Q$ by Proposition 3 we can apply Proposition 6.8 and determine for any $a\in A$ an $x\in A$ such that $x-xj=a$. Hence $a=x-x(u-i)=(x-xu)+xi$ belongs to I and so $I=A$. We proved that $J'=A$ implies $I=A$. Thus I being a proper subset of A we have $J'=I$ and $J(A)\subseteq I$. The theorem is proved.

We say that a ring A satisfies the *minimum condition for left ideals*, if every non-void set of left ideals contains a minimal ideal. Here minimality is understood with respect to inclusion so that the minimal element does not properly contain any other ideal of the set except the zero ideal. The ring A is said to satisfy the *descending chain condition* for left ideals if for every descending chain of left ideals $I_1\supseteq I_2\supseteq\cdots\supseteq I_n\supseteq\cdots$ there is a natural number n such that $I_n=I_{n+1}=\cdots$. It is easy to see that the minimum condition and the descending chain condition are equivalent. The same concepts can be defined also for right ideals but if nothing specific is stated then we will always consider these notions for left ideals.

If the ring A has an additional multiplication by scalars which makes it into an algebra over some field F then we can speak about the dimension of A over F. One can immediately see that any finite dimensional algebra with identity satisfies the descending chain condition.

An element a of the ring A is called *nilpotent* if $a^n=0$ for some exponent $n=1,2,\ldots$ For example square matrices with entries from some field and with non-zero entries only above the main diagonal are all nilpotent.

An ideal I (left, right or two sided) is called a *nilpotent ideal* if $I^n=0$ for some $n\geqslant 1$. If $a_1\ldots a_n=0$ for any $a_1,\ldots,a_n$ in I then $I^n=0$ and conversely, if $I^n=0$ then I has this property. The ideal (left, right or two sided) is called a *nil ideal* if every element of I is nilpotent. Hence we have:

Lemma 6. *Every nilpotent ideal is a nil ideal.*

We can now return to quasiregularity and the Jacobson radical. An ideal I will be called *quasiregular* if every $a\in I$ has a quasi-inverse a'.

Lemma 7. *Every nil ideal is quasiregular.*

Proof. If $a\in I$ and $a^n=0$ then $a'=-(a+\cdots+a^{n-1})$ satisfies $a\circ a' = a'\circ a=0$.

Theorem 8. *If A is a ring with minimum condition then the Jacobson radical $J(A)$ is nilpotent.*

Proof. For simplicity let $J=J(A)$ and consider $J\supseteq J^2\supseteq\cdots\supseteq J^n\supseteq\cdots$. By the descending chain condition there is an n such that $N=J^n=J^{n+1}=\cdots$. Clearly we have $N^2=N$. The rest of the proof is a reasoning by contradiction: *If N were not* 0 we could take the set $\mathcal{S}$ of left ideals contained in N and satisfying $NI\neq 0$:

$$\mathcal{S}=\{I: I\subseteq N \text{ and } NI\neq 0\}.$$

Then $\mathcal{S}$ is not void because $N\in\mathcal{S}$ and so by the minimum condition $\mathcal{S}$ contains a minimal element I_m. By $NI_m\neq 0$ we can choose an element a in I_m such that $Na\neq 0$. Since Na is a left ideal contained in I_m and $N(Na)=N^2a=Na\neq 0$ we have $Na\in\mathcal{S}$ and so $Na=I_m$. Therefore by $a\in I_m$ there is an x in N such that $a=xa$. Since $N\subseteq J$ and J consists of left quasi-invertible elements there is an x' such that $x'\circ x=0$. Thus $a=a-(x'\circ x)a=(a-xa)-x'(a-xa)=0$ which contradicts $a\neq 0$. Therefore we must have $N=0$ and so J is nilpotent.

Theorem 9. *If A is a ring with minimum condition then $J(A)$ is the union of all nilpotent left ideals.*

Proof. Let J denote the union of all nilpotent left ideals of A. By Theorem 8 the Jacobson radical $J(A)$ is contained in J. To prove the converse choose a nilpotent left ideal I. By Lemma 6 I is a nil ideal and so by Lemma 7 it is quasiregular. Hence for any $x\in I$ we have $Ax\subseteq I\subseteq Q$ and so $x\in J(A)$. This proves that $I\subseteq J(A)$ and so $J(A)=J$.

Note. Of course in the same manner we can prove that $J(A)$ is the union of all nilpotent right ideals.

An element e of a ring A is called *idempotent* if e is not 0 and $e^2=e$. For instance a square matrix one of whose rows is filled with ones and the rest with zeros is idempotent. If the ring A has an identity element which we prefer to denote now by 1 then 1 is an idempotent. Our next object is to show the existence of idempotents in various ideals of A. This will give us a method by which the *Wedderburn-Artin structure theorem* can be proved.

Proposition 10. *If I is a minimal left ideal in the ring A then $I^2=0$ or I contains an idempotent e such that $I=Ae$.*

Proof. Suppose $I^2\neq 0$ and choose a in I such that $Ia\neq 0$. Since Ia is a left ideal by the minimality of I we have $Ia=I$. Let

$$(0:a)=\{x: x\in A \text{ and } xa=0\}.$$

Then $I\cap(0:a)$ is a left ideal and by $Ia\neq 0$ it is properly contained in I, so by the minimality of I it is 0. By $Ia=I$ we can choose e in I such that $ea=a$. Then we have $(e-e^2)a=0$ so $e-e^2$ belongs to $I\cap(0:a)$ which is 0. Therefore $e^2=e$ and $e\neq 0$ because $ea=a\neq 0$. Since Ie is a left ideal contained in I we have $Ie=I$.

If A is a ring satisfying the descending chain condition then the above result can be extended as follows:

Proposition 11. *If A is a ring with identity satisfying the minimum condition for left ideals then every non-nilpotent left ideal contains an idempotent element.*

Proof. The set of those non-nilpotent left ideals which are contained in the given ideal I_0 is not void, so this family of ideals has a minimal element J. Since J^2 is contained in J we have $J^2=J$ and also $J\subseteq I_0$. Let $\mathscr{S}$ be the family of those left ideals I for which $I\subseteq J$ and $JI\neq 0$. Since $J\in\mathscr{S}$ the family has a minimal element I_1. By $JI_1\neq 0$ we can choose x in I_1 such that $Jx\neq 0$. As Jx is a left ideal contained in the ideal I_1 we have $Jx=I_1$. Therefore there is an element e in J such that $ex=x$. By $x\neq 0$ we have $e\neq 0$. Then $e^2x=x$ and so $(e^2-e)x=0$. Let

$$(0:x)=\{y: y\in A \text{ and } yx=0\}.$$

Then $(0:x)$ is a left ideal and so $(0:x)\cap J$ is a left ideal contained in the ideal J which J is minimal with respect to the property $J\subseteq I_0$. Since $Jx\neq 0$ we see that $(0:x)\cap J$ is properly contained in J and so $(0:x)\cap J=0$. Therefore $e^2-e=0$ and e is an idempotent in I_0.

A ring is called *semisimple* if its radical $J(A)=0$. If A is a ring with minimum condition then using Theorem 9 one can immediately see that $A/J(A)$ is a semisimple ring.

In the following discussion we suppose that A is a semisimple algebra with identity 1 which satisfies the left minimum condition. As it was mentioned earlier all the results and proofs discussed here, except those on Banach algebras, hold for rings as well as for algebras. Thus we could consider a semisimple ring having identity 1 with the left minimum condition. The change of language from rings to algebras is more for variety than for any other reason.

Let A_1 be a minimal left ideal of A and let e_1 be an idempotent in A_1 so that $A_1=Ae_1$ by the minimality. Then $A(1-e_1)$ is a left ideal of A and so by the minimality of A_1 we have $A_1\cap A(1-e_1)=0$ or $A=A_1$ and one can choose $e_1=1$. Since every x in A can be represented in the form $x=xe_1+x(1-e_1)$ this shows that $A=A_1=Ae_1$ with $e_1=1$ or

$$A=A_1\oplus A(1-e_1).$$

From here on we reason by induction: Suppose $A_1, \dots, A_k$ are minimal left ideals with certain generating idempotents $e_1, \dots, e_k$ such that

$$A = A_1 \oplus \cdots \oplus A_k + A(1 - e_1 - \cdots - e_k).$$

If $1 \neq e_1 + \cdots + e_k$ then the non-zero ideal $A(1 - e_1 - \cdots - e_k)$ contains a minimal ideal A_{k+1} which is not nilpotent because A is semisimple. Hence by Proposition 11 A_{k+1} has an idempotent e_{k+1} which will be generating i.e. $A_{k+1} = A e_{k+1}$ because A_{k+1} is minimal. Again by the minimality of A_{k+1} we have $A_{k+1} = A(1 - e_1 - \cdots - e_k)$ or

$$A_{k+1} \cap A(1 - e_1 - \cdots - e_k - e_{k+1}) = 0.$$

In the first case replace e_{k+1} by $1 - e_1 - \cdots - e_k$. Since we have

$$x(1 - e_1 - \cdots - e_k) = x e_{k+1} + x(1 - e_1 - \cdots - e_k - e_{k+1})$$

we proved that

$$A(1 - e_1 - \cdots - e_k) = A_{k+1} \oplus A(1 - e_1 - \cdots - e_{k+1})$$

and so

$$A = A_1 \oplus \cdots \oplus A_k \oplus A_{k+1} \oplus A(1 - e_1 - \cdots - e_{k+1}).$$

The algebra A satisfies the minimum condition and the chain $A(1 - e_1 - \cdots - e_k)$ $(k = 1, 2, \dots)$ is descending so after a finite number of steps we arrive at the decomposition

$$A = A_1 \oplus \cdots \oplus A_n$$

where the $A_k (k = 1, \dots, n)$ denote left minimal ideals. Hence we proved the following:

Proposition 12. *Every semisimple algebra with identity* 1 *and satisfying the left minimum condition is a direct sum of minimal left ideals* A_k $(k = 1, \dots, n)$. *Moreover one can find an idempotent* e_k *in each* A_k *such that* $e_1 + \cdots + e_n = 1$, $e_k e_l = 0$ *for* $k \neq l$ *and* $A_k = A e_k$.

Two left ideals I and J of the algebra A will be called *operator isomorphic* if there is a bijective map $\varphi: I \to J$ such that

$$\begin{aligned} &\varphi(a_1 + a_2) = \varphi(a_1) + \varphi(a_2) && \text{for } a_1, a_2 \in I, \\ &\varphi(\lambda a) = \lambda \varphi(a) && \text{for } \lambda \in F \text{ and } a \in I, \\ &\varphi(x a) = x \varphi(a) && \text{for } x \in A \text{ and } a \in I. \end{aligned}$$

Thus I and J are operator isomorphic if there is a map φ which is both an F-module isomorphism and also an A-module isomorphism of I onto J.

We recall that an algebra A is called *simple* if its two sided ideals are only 0 and A itself. If A is simple then its radical is either 0 or A so if A has an identity 1 then by Proposition 3 and $1 \notin Q$ we have $J(A)=0$. Thus a simple algebra with identity is semisimple.

Proposition 13. *If A is an algebra with identity and if A is the direct sum of operator isomorphic minimal left ideals then A is simple.*

Proof. Let $A=A_1 \oplus \cdots \oplus A_n$ and let $\varphi_k: A_k \to A_1$ $(k=1,\dots,n)$ be operator isomorphisms. The object is to show that a non-zero two sided ideal I is A. Since A has an identity 1 we have $I=IA_1+\cdots+IA_n$ and so $IA_k \neq 0$ for some index k. Since A_k is a minimal left ideal $IA_k=A_k$. Thus using the isomorphism $\varphi_l^{-1}\varphi_k$: $A_k \to A_l$ we see that $IA_l \neq \{0\}$ and so $IA_l=A_l$ and $A_l \subseteq I$ $(l=1,\dots,n)$. Therefore $A=A_1+\cdots+A_n \subseteq I$ and so $A=I$.

An *operator homomorphism* φ of the left ideal I into the left ideal J of the algebra A is a map $\varphi: I \to J$ which is both a vector space and an A-module homomorphism.

Theorem 14. *Every semisimple algebra A with identity satisfying the minimum condition is a direct sum of simple algebras,*

$$A=A_1 \oplus \cdots \oplus A_m$$

where $A_1,\dots,A_m$ are ideals of A and $A_kA_l=0$ if $k \neq l$. This decomposition is unique and the simple algebras A_k $(k=1,\dots,n)$ are called the simple components of A.

Proof. By Proposition 12 A is a direct sum of minimal left ideals and so one can write

$$(A_{11} \oplus \cdots \oplus A_{1n_1}) \oplus \cdots \oplus (A_{m1} \oplus \cdots \oplus A_{mn_m})$$

where the left ideals $A_{k1},\dots,A_{kn_k}$ are operator isomorphic for every $k=1,\dots,m$ and the ideals of distinct blocks are not isomorphic. We let

$$A_k=A_{k1} \oplus \cdots \oplus A_{kn_k}.$$

In order to prove that A_k is a right ideal let a non-zero a in A be fixed. We plan to prove that $A_{kl}a \subseteq A_k$ for every $l=1,\dots,n_k$. If this were not the case we would have

$$(1) \qquad A_{kl}a \cap A_{ij} \neq 0 \quad \text{for some } i,j \quad \text{where } i \neq k.$$

Denote by $\varphi: A_{kl} \to A_{ij}$ the map given by $\varphi(x) = \pi_{ij}(xa)$ where π_{ij} is the canonical projection $\pi_{ij}: A \to A_{ij}$. Then it is clear that φ is an F-vector space homomorphism. By (1) φ is not the zero homomorphism. We prove that φ is an operator homomorphism. In fact Proposition 12 also states the existence of generating idempotents $e_{11}, \ldots, e_{mn_k}$ such that

$$1 = e_{11} + \cdots + e_{ij} + \cdots + e_{mn_k}.$$

Multiplying from the left by xa we see that $\varphi(x) = \pi_{ij}(xa) = xa\,e_{ij}$. Similarly we have $\varphi(yx) = yxae_{ij}$ and so

(2) $$\varphi(yx) = y\varphi(x) \quad \text{for any } x \in A_{kl} \quad \text{and } y \in A.$$

From (2) we see that the kernel of φ is a left ideal in A_{kl}. However A_{kl} is a minimal left ideal, so the kernel is $\{0\}$ and φ is an isomorphism. Moreover (2) also shows that $\varphi(A_{kl})$ is a left ideal in A_{ij}. Since A_{ij} is a left minimal ideal and φ is not zero it follows that φ is surjective. Therefore φ is an operator isomorphism of A_{kl} onto A_{ij} which contradicts $i \neq k$. We proved that A_k is a two sided ideal in A.

The remaining statements follow very easily: Since $A_k A_l \subseteq A_k \cap A_l$ and $A_k \cap A_l = \{0\}$ for $k \neq l$ we have $A_k A_l = 0$ $(k \neq l)$. Let $1 = e_1 + \cdots + e_m$. If $x \in A_k$ then $x = xe_1 + \cdots + xe_k + \cdots + xe_m$ where $xe_l \in A_k \cap A_l$ by A_k and A_l being ideals of A. Hence $x = xe_k$ for every $x \in A_k$ and so e_k is a right identity of A_k. Similarly one proves that $x = e_k x$ for every $x \in A_k$. Since A_k has an identity by Proposition 13 A_k is a simple algebra. Finally if I is any ideal in A then by the simplicity of A_k we have $I \cap A_k = \{0\}$ or $I \cap A_k = A_k$. This proves the uniqueness of the decomposition.

Theorem 15. *An algebra A with identity over the skew field F is simple if and only if it is isomorphic to $M_n(D)$ for some skew field D over F where n and D are uniquely determined by A.*

For the case of rings the corresponding result states the following:

Theorem 16. *If A is a simple ring with identity then there is a unique natural number n and a skew field F such that A is isomorphic to $M_n(F)$.*

Proof. The simplicity of $M_n(D)$ was proved in Section 1.

An *operator endomorphism* of a left ideal I of A is an operator homomorphism of I into itself. Under addition, multiplication by elements of the field F and under composition the operator endomorphisms of A form an algebra over F which we denote by $\mathrm{Hom}(A, A)$. We are going to prove that $\mathrm{Hom}(A, A)$ is isomorphic in the algebra sense to $M_n(D)$ for some n and D.

In the beginning of the proof of Theorem 14 we have seen that $A = A_1 \oplus \cdots \oplus A_n$ where $A_1, \ldots, A_n$ are operator isomorphic minimal left ideals of A. Let π_i denote the canonical projection of A onto $A_i (i = 1, \ldots, n)$. Then we have $1 = \pi_1 + \cdots + \pi_n$ and $\pi_i \pi_j = 0$ if $i \neq j$. Here 1 denotes the identity map $1: A \to A$. We choose an operator isomorphism $\tau_i: A_1 \to A_i$ for every i which we shall keep fixed.

Given φ in $\operatorname{Hom}(A, A)$ we define the operator endomorphism $\varphi_{ij}: A_1 \to A_1$ by

(1) $$\varphi_{ij} = \tau_i^{-1} \pi_i \varphi \tau_j \qquad (i, j = 1, \ldots, n).$$

If ψ is an operator endomorphism of the minimal left ideal A_1 into itself then both its kernel and its image are left ideals of A. Therefore by the minimality of A_1 either ψ is 0 or ψ is an operator isomorphism of A_1 onto itself and so it is invertible. It follows that the set of operator endomorphisms ψ of A_1 form a skew field D which is an algebra over F. Therefore using (1) we can associate with each φ of $\operatorname{Hom}(A, A)$ a matrix $(\varphi_{ij}) (i, j = 1, \ldots, n)$ whose entries φ_{ij} belong to the division ring D.

Conversely, if (φ_{ij}) is an arbitrary matrix of elements $\varphi_{ij} \in D$ then (φ_{ij}) defines a φ in $\operatorname{Hom}(A, A)$ such that the matrix associated with φ is the given matrix (φ_{ij}): For we can let

$$\varphi = \sum_{i,k} \tau_i \varphi_{ik} \tau_k^{-1} \pi_k .$$

Hence we constructed a one to one correspondence between $\operatorname{Hom}(A, A)$ and the matrix algebra $M_n(D)$ which is an algebra over F.

It is clear that the correspondence $(\varphi_{ij}) \leftrightarrow \varphi$ is additive and homogeneous over F. To see that it is multiplicative consider the $i-k$ entry of the product matrix $(\varphi_{ij})(\psi_{jk})$: By (1) we have

$$\begin{aligned} \sum \varphi_{ij} \psi_{jk} &= \sum (\tau_i^{-1} \pi_i \varphi \tau_j)(\tau_j^{-1} \pi_j \psi \tau_k) = \tau_i^{-1} \pi_i \varphi \left(\sum \pi_j\right) \psi \tau_k \\ &= \tau_i^{-1} \pi_i (\varphi \psi) \tau_k \end{aligned}$$

which is the $i-k$ entry of the matrix which corresponds to $\varphi \psi$. Thus the one-to-one map between $\operatorname{Hom}(A, A)$ and $M_n(D)$ is an isomorphism.

Now we construct a bijection between the elements of $\operatorname{Hom}(A, A)$ and those of A. Namely, with φ in $\operatorname{Hom}(A, A)$ we associate the algebra element $\varphi(1) = a$. The map is injective because $\varphi(x) = \varphi(x \cdot 1) = x \varphi(1) = x a$ and so φ is uniquely determined by $\varphi(1) = a$. It is also surjective because given a in A an operator homomorphism φ is defined by $\varphi(x) = x a \, (x \in A)$. This bijection is clearly an isomorphism between A and $\operatorname{Hom}(A, A)$. Therefore we proved that A is isomorphic to $M_n(D)$.

The uniqueness of n and D in the foregoing theorem means the following:

$M_{n_1}(D_1)$ and $M_{n_2}(D_2)$ are isomorphic matrix algebras if and only if $n_1 = n_2$ and D_1 is isomorphic to D_2.

The essential part is the proof of the isomorphism $D_1 \approx D_2$ which can be done by using an abstract model of these algebras such as the simple algebra A which is given at present. The critical step is to prove the following:

If e is a generating idempotent of the minimal left ideal I then eAe is a skew field and any two skew fields obtained in this manner are isomorphic.

If I and J are maximal left ideals then either $I = J$ or $I \cap J = 0$ and $I \oplus J \subseteq A$. By the reasoning given in the beginning of the proof of Theorem 14 we know that I and J are operator isomorphic, say $\tau: I \to J$. From the earlier parts of the present proof we also know that $\mathrm{Hom}(I, I)$ and $\mathrm{Hom}(J, J)$ are skew fields. If ψ is in $\mathrm{Hom}(I, I)$ then $\tau\psi\tau^{-1}$ is in $\mathrm{Hom}(J, J)$ and so $\psi \to \tau\psi\tau^{-1}$ is an isomorphism of the first skew field onto the other.

The proof of the lemma will be completed by showing that if e is a generating idempotent of the minimal left ideal I of A i. e. if $I = Ae$ then the skew field $\mathrm{Hom}(I, I)$ is isomorphic to eAe. Given φ in $\mathrm{Hom}(I, I)$ we let $f(\varphi) = \varphi(e)$ so that $f(\varphi) \in Ae$. But $ef(\varphi) = e\varphi(e)$ $= \varphi(ee) = \varphi(e) = f(\varphi)$ and this shows that $f(\varphi) \in eAe$. The map is obviously a homomorphism and since I is minimal it is an isomorphism. f is surjective because a in Ae is the f image of φ where $\varphi(x) = xa$.

Now the proof of $D_1 \approx D_2$ is very easy. The matrix ε_k $(k = 1, 2)$ having 1 in the $1-1$ place and zero everywhere else is a generating idempotent of the minimal ideal $M_{n_k}(D_k)\varepsilon_k$. Under the isomorphism of $M_{n_k}(D_k)\varepsilon_k$ onto A let ε_k be mapped into e_k so that $\varepsilon_k M_{n_k}(D_k)\varepsilon_k$ is isomorphic to $e_k A e_k$ and consequently $\varepsilon_1 M_{n_1}(D_1)\varepsilon_1 \approx \varepsilon_2 M_{n_2}(D_2)\varepsilon_2$. However $\varepsilon_k M_{n_k}(D_k)\varepsilon_k$ consists of those matrices which have an arbitrary element of D_k in the $1-1$ place and zero everywhere else. Thus $\varepsilon_k M_{n_k}(D_k)\varepsilon_k \approx D_k$ and so $D_1 \approx D_2$. The equality of n_1 and n_2 is a simple dimensionality argument, namely the dimension of $M_{n_k}(D_k)$ over $D_1 \approx D_2$ is n_k $(k = 1, 2)$.

Proposition 17. *If A is a complex, commutative Banach algebra then*

$$J(A) = \bigcap \{I : I \in \mathscr{M}\} = \{x : \hat{x} = 0\} = \ker(A \to \hat{A}).$$

Proof. The first statement is obtained by Theorem 5. The second can be seen by noticing that $\hat{x}(h) = h(x) = 0$ if and only if x belongs to the associated ideal I i. e. to the kernel of h. In the last expression for $J(A)$ the map $A \to \hat{A}$ denotes the Fourier transform.

A Banach algebra A is called *semisimple* if its radical is 0. The last proposition shows that a complex, commutative Banach algebra is semisimple if and only if the Fourier transform map $A \to \hat{A}$ is an isomorphism. We see that A is semisimple if and only if $\|\hat{x}\|=0$ implies $x=0$. Using the corollary of Proposition 5.4 we see that if A is a complex, commutative, semisimple Banach algebra then the spectrum $\sigma(x)$ contains some $\lambda \neq 0$ for every $x \neq 0$ in A.

For example let S be a topological space and let $C_b(S)$ be the algebra of bounded continuous functions $x: S \to \mathbb{C}$. Then

$$J(A) = \bigcap \{I: I \in \mathcal{M}\} \subseteq \bigcap \{I_s: s \in S\}$$

where $I_s = \{x: x(s)=0\}$ is the maximal ideal fixed at $s \in S$. (See p. 23 of Section 6.) Hence if $x \in J(A)$ then $x(s)=0$ for every $s \in S$ and so *the function algebra $C_b(S)$ is semisimple.*

By Propositions 6.2 and 17 the radical of a complex, commutative Banach algebra A is a closed subset of A. The same is true for arbitrary Banach algebras by Proposition 6.10 and Theorem 5. Hence we have:

Theorem 18. *If A is a Banach algebra then its radical $J(A)$ is closed.*

Topologically nilpotent elements were introduced at the end of Section 5. An ideal I is called a *topologically nil ideal* if every element of I is topologically nilpotent.

Proposition 19. *The radical of a normed algebra is a topologically nil ideal.*

Proof. Let $a \in J(A)$ where A is a normed algebra over $\mathbb{R}$ or $\mathbb{C}$ and let $\lambda = \alpha + i\beta \neq 0$. Then $|\lambda|^{-2}(2\alpha a - a^2)$ is in $J(A)$. Hence by Propositions 3 and 5.6 it is a quasisingular element in A when A is considered as an algebra over the reals. Thus by Propositions 4.2 and 5.6 we have $\sigma(a)=0$ where A is again considered over $\mathbb{R}$. Hence if A is a real algebra then $v(a)=0$ and a is nilpotent. If A is a complex algebra we have to prove that $\sigma(a)=0$ also when A is considered as an algebra over $\mathbb{C}$. Then the same reasoning will show that a is nilpotent.

By Proposition 4.2 it is sufficient to prove that if $\lambda = \alpha + i\beta \neq 0$ then λa is quasiregular. We know that λa is quasiregular in the real sense, that is to say $(\alpha a, \beta a)$ has a quasi-inverse (b, c) in the complexification of A. Therefore $(\alpha a, \beta a) \circ (b, c) = 0$ and so

$$\alpha a + b - \alpha a b + \beta a c = 0 \quad \text{and} \quad \beta a + c - \beta a b - \alpha a c = 0.$$

These equations show that $(\lambda a) \circ q = 0$ and $q \circ (\lambda a) = 0$ with $q = b + ic$. Hence q is the quasi-inverse of λa and so λa is quasiregular.

8. Involutive Algebras

The basis of the present discussion is

Definition 1. *An algebra A with involution is an algebra over* $\mathbb{R}$ *or* $\mathbb{C}$ *with a map* $x \mapsto x^*$ *defined on A such that*

$$x^{**} = x, \tag{1}$$

$$(x+y)^* = x^* + y^*, \tag{2}$$

$$(\lambda x)^* = \bar{\lambda} x^*, \tag{3}$$

$$(xy)^* = y^* x^* \tag{4}$$

for every $x, y \in A$ *and every scalar* λ.

If A is an algebra over $\mathbb{R}$ then (3) reduces to $(\lambda x)^* = \lambda x^*$. An algebra with involution is often called a *star algebra*, a * *algebra* or an *algebra with adjoint operation*; in the Russian literature the expression "*symmetric ring*" is used. From (1) it is clear that the adjoint map $x \mapsto x^*$ is a bijection.

If $x = x^*$ then x is called a *self-adjoint* or *Hermitian element* and one sees immediately that $x + x^*$ and $x^* x$ are Hermitian elements for every $x \in A$. If A is a complex algebra then every element x of A can be represented uniquely in the form $x = x_1 + i x_2$ where x_1, x_2 are self adjoint: Namely we have $x_1 = \frac{1}{2}(x + x^*)$ and $x_2 = 1/2i\,(x - x^*)$. These elements are sometimes referred to as the *Hermitian components* of x. If $x \circ y = 0$ then $y^* \circ x^* = 0$ and similarly, if A has a unity then $xy = e$ implies $y^* x^* = e$ because $e = e^*$ by $e^* e = e^*$ and $(e^* e)^* = e^{**}$.

A *normed algebra with involution* means *only* that the involutive algebra is normed, so in general the involution need not be continuous. Since $x \mapsto x^*$ is an antilinear map A is a *normed algebra with continuous involution* if and only if $\|x^*\| \leqslant C\|x\|$ for some constant $C > 0$. For several large classes of normed * algebras the involution is actually an isometry. Let us suppose that $\|x^* x\| = \|x\|^2$ for every $x \in A$. Then $\|x\|^2 \leqslant \|x^*\| \cdot \|x\|$, so $\|x\| \leqslant \|x^*\| \leqslant \|x^{**}\| = \|x\|$ and $\|x^*\| = \|x\|$. Hence we have:

Proposition 2. *If the norm satisfies* $\|x^* x\| = \|x\|^2$ *then the involution is an isometry.*

A Banach algebra with involution satisfying $\|x^* x\| = \|x\|^2$ is called a B^* *algebra*. If $\|x^* x\| = \|x\|^2$ then by $e^* = e$ we have $\|e\| = 1$ because $\|e\|^2 = \|e^* e\| = \|e^*\| = \|e\|$. If the involutive algebra A is such that $\|x^*\| = \|x\|$ then A is called *star normed*.

We list a few examples of involutive Banach algebras:

1. The algebra $C(S)$ consisting of the bounded continuous functions on the topological space S is a Banach algebra with involution under the map $x^* = \bar{x}$ and norm $\|x\| = \operatorname{supr} |x(s)|$. Clearly $\|x^* x\| = \operatorname{supr} |x(s)|^2 = \|x\|^2$ and so $C_b(S)$ is a B^* algebra and the involution is an isometry.

2. Let $\mathscr{H}$ be a Hilbert space and let $\mathscr{L}(\mathscr{H})$ be the algebra of continuous operators $A: \mathscr{H} \to \mathscr{H}$ normed by the operator norm $|A|$. The Hilbert space adjoint defined by the formula $(Ax, y) = (x, A^* y)$ is an involution on $\mathscr{L}(\mathscr{H})$. We have

$$\|Ax\|^2 = (Ax, Ax) = (x, A^* A x) \leqslant |A^* A| \cdot \|x\|^2$$

and so $|A|^2 \leqslant |A^* A| \leqslant |A^*| \cdot |A|$. We get $|A| \leqslant |A^*| \leqslant |A^{**}| = |A|$, so $|A^*| = |A|$ and by our earlier inequality $|A^* A| = |A|^2$. We proved that $\mathscr{L}(\mathscr{H})$ is a B^* algebra for any real or complex Hilbert space $\mathscr{H}$.

3. Any self adjoint subalgebra of an involutive algebra is involutive and the properties $\|x^*\| \leqslant C\|x\|$, $\|x^*\| = \|x\|$ and $\|x^* x\| = \|x\|^2$ are preserved under the restriction. We obtain that any closed, self adjoint subalgebra of $\mathscr{L}(\mathscr{H})$ is a B^* algebra. Any Banach algebra obtained in this manner is called a *C^* algebra.*

4. A closed subalgebra of $C_b(S)$ which contains with each x also its complex conjugate $\bar{x}$ is a B^* algebra. Such an algebra is called an *involutive function algebra.*

5. The *convolution algebra* $L^1(\mathbb{R})$ has an involution, namely $f^*(x) = f(-x)$ which is an isometry but $L^1(\mathbb{R})$ is not a B^* algebra. For instance if $f(x)$ is 0 for $|x| > 1$, it is -1 for $-1 \leqslant x < 0$ and 1 for $0 \leqslant x \leqslant 1$ then $f^* * f$ is 0 for $|x| \geqslant 2$, it is $|x| - 2$ for $1 \leqslant |x| \leqslant 2$ and $2 - 3|x|$ for $0 \leqslant |x| \leqslant 1$. Thus $\|f^* * f\| = \frac{8}{3} < 4 = \|f\|^2$.

6. The *convolution algebra* $L^1(G)$ of any locally compact topological group is also involutive and the involution $f^*(x) = \overline{f(x^{-1})}\Delta(x^{-1})$ is an isometry. For the definition of the invariant measure, the modular function Δ and the convolution product $*$ see Sections V.2 and V.4.

In Section 1 we have seen that every algebra A can be embedded in an algebra A_e with identity e and if A is normed then this norm can be extended to a norm of A_e. If A is involutive then its involution can be extended to A_e by defining $(\lambda, a)^* = (\bar{\lambda}, a^*)$ for any scalar λ and $a \in A$. If the involution of A is continuous then so is its extension to A_e. Moreover if $\|a^*\| = \|a\|$ for every a in A then the same relation holds also in A_e. In other words if A is star normed then so is A_e. The algebra A_e is called the *involutive algebra obtained from A by the adjunction of an identity e.*

Proposition 3. *Every B^* algebra A can be embedded isometrically as an ideal into another B^* algebra A_e which has an identity e.*

Note. The enveloping algebra is called the *B^* algebra obtained from A by adjoining the identity element e*. If A has an identity then $A_e = A$.

Proof. Algebraically A_e will be the same as the algebra introduced in Proposition 1.3 but in order to have the B^* property its norm will be different from the norm $\|a+\lambda e\| = \|a\| + |\lambda|$. We consider A as a Banach space and let $\mathcal{L}(A)$ be the Banach algebra of linear operators $T: A \to A$ normed by the operator norm. We associate with each a in the algebra A the operator $A(a)$ mapping x into ax for $x \in A$. Then $a \mapsto A(a)$ is an isomorphism of the algebra A into $\mathcal{L}(A)$. The involutive property of this map will be established in a moment. First let A_e be the subalgebra of $\mathcal{L}(A)$ generated by the operators $A(a)\,(a \in A)$ and the identity I.

We prove that

$$A_e = \{A(a) + \lambda I : a \in A \text{ and } \lambda \in \mathbb{C}\}. \tag{5}$$

Let $A(a_n) + \lambda_n I\ (n = 1, 2, \ldots)$ be a Cauchy sequence with respect to the operator norm. If $\lambda_n\ (n = 1, 2, \ldots)$ were not bounded we could find a subsequence $\lambda_{n_k}\ (k = 1, 2, \ldots)$ such that $|\lambda_{n_k}| \to \infty$. But then

$$|\lambda_{n_k}^{-1} A(a_{n_k}) + I| \leqslant |\lambda_{n_k}|^{-1} |A(a_{n_k}) + \lambda_{n_k} I| \to 0$$

and so $-\lambda_{n_k}^{-1} a_{n_k}$ would have a limit e in A such that $ex = x$ and $xe^* = x$ for every x in A. Thus A would have an identity element. But hypothesis this is not the case and so $\lambda_n\ (n = 1, 2, \ldots)$ is a bounded sequence. We can now suppose that λ_n is convergent, say $\lambda_n \to \lambda$ where $\lambda \in \mathbb{C}$. Then $A(a_n)\ (n = 1, 2, \ldots)$ is also convergent and so $a_n \to a$ for some $a \in A$. It follows that $A(a_n) + \lambda_n I \to A(a) + \lambda I$. This completes the proof of (5). If we define $(A(a) + \lambda I)^* = A(a^*) + \bar{\lambda} I$ then A_e becomes an involutive Banach algebra and $a \mapsto A(a)$ will be an involutive isomorphism of A into A_e. If A is a real algebra then we obtain the same result with $(A(a) + \lambda I)^* = A(a^*) + \lambda I$.

The map $a \mapsto A(a)$ is norm preserving: For it is clear that $|A(a)| \leqslant \|a\|$ and by the B^* property we have

$$\|a^2\| = \|a a^*\| = \|A(a) a^*\| \leqslant |A(a)| \cdot \|a^*\| = |A(a)| \cdot \|a\|.$$

We prove that the operator norm satisfies the identity $|T^* T| = |T|^2$ where $T = A(a) + \lambda I$. In fact if $x \in A$ then by the B^* property of the norm of A we have

$$\begin{aligned}\|Tx\|^2 &= \|ax + \lambda x\|^2 = \|(ax + \lambda x)^* (ax + \lambda x)\| = \|(x^* a^* + \bar{\lambda} x^*)(ax + \lambda x)\| \\ &= \|x^* T^* T x\| \leqslant |T^* T| \cdot \|x\|^2.\end{aligned}$$

Therefore $|T|^2 \leqslant |T^* T| \leqslant |T^*| \cdot |T| = |T|^2$ and so $|T|^2 = |T^* T|$.

Proposition 4. *If A is a complex B^* algebra and if x is a self adjoint element of A then $\sigma(x) \subseteq \mathbb{R}$.*

Note. If $\sigma(x)$ is real for every self adjoint element x in the involutive algebra A then the involution is called *Hermitian.*

Proof. If A does not have an identity we consider an element x of A and compare its spectrum in A with that in A_e. The existence of the enveloping B^* algebra was established above in Proposition 3. If $\lambda \in \mathbb{C}^\times$ and x/λ is quasiregular in A_e then A being an ideal in A_e we see that $(x/\lambda)'$ belongs to A. Hence by Proposition 4.2 we have $\sigma_A(x)^\times \subseteq \sigma_{A_e}(x)^\times$. Conversely if $(x/\lambda)'$ exists in A then x/λ is quasiregular in A_e and so $\sigma_A(x)^\times \supseteq \sigma_{A_e}(x)^\times$. Therefore we may suppose that A is a B^* algebra with identity. Now let $\lambda = a + ib \in \sigma(x)$. For any $\beta \in \mathbb{R}$ we have $\lambda + i\beta \in \sigma(x + i\beta e)$. Then $|\lambda + i\beta| \leqslant \|x + i\beta e\|_{sp} \leqslant \|x + i\beta e\|$ and

$$|\lambda + i\beta|^2 \leqslant \|x + i\beta e\|^2 = \|(x - i\beta e)(x + i\beta e)\| = \|x^2 + \beta^2 e\| \leqslant \|x^2\| + \beta^2 .$$

Hence we proved that $a^2 + b^2 + 2\beta b \leqslant \|x\|^2$ for every $\beta \in \mathbb{R}$. It follows that $b = 0$ and λ is real.

Corollary. *If x is a self adjoint element of the B^* algebra A then $\hat{x}(\mathcal{M}) \subseteq \mathbb{R}$.*

This follows from Proposition 5.4.

Proposition 5. *If A is a complex B^* algebra then $\{\hat{x} : x \in A\}$ is self adjoint and $\bar{\hat{x}} = \widehat{x^*}$ for every x in A.*

Proof. Let $x = x_1 + ix_2$ where $x_1 = x_1^*$ and $x_2 = x_2^*$. Then $x^* = x_1 - ix_2$ and we obtain $\widehat{x^*} = \hat{x}_1 - i\hat{x}_2 = \overline{\hat{x}_1 + i\hat{x}_2} = \bar{\hat{x}}$.

If S is a locally compact space we let $C_o(S)$ denote the algebra of continuous functions $f: S \to \mathbb{C}$ vanishing at infinity normed by $\|f\| = \operatorname{supr} |f(s)|$. The symbol $C_0(S)$ will stand for the algebra of complex valued continuous f on S with compact support. If S is compact then $C(S) = C_o(S) = C_0(S)$.

Theorem 6. *If A is a complex B^* algebra then $\{\hat{x} : x \in A\}$ is dense in $C_o(\mathcal{M})$.*

Proof. Let $\mathcal{M}_0$ be the space of all homomorphisms $h: A \to \mathbb{C}$ including the trivial one θ. By Proposition 3.1 $\mathcal{M}_0$ is compact and by Proposition 3.2 $\mathcal{M}$ is a locally compact subspace of $\mathcal{M}_0$. We also know that $\mathcal{M}$ and $\mathcal{M}_0$ are Hausdorff spaces. If $\mathcal{M}$ is compact it would be sufficient to consider only $\mathcal{M}$ but taking $\mathcal{M}_0$ the case of compact $\mathcal{M}$ is included. We let $\hat{x}$ denote the extended Fourier transform of x so that $\hat{x}: \mathcal{M}_0 \to \mathbb{C}$ and $x(\theta)=0$. Let $\mathcal{A}$ bet the set of real valued $\hat{x}$ $(x \in A)$. By $\widehat{\lambda x} = \lambda \hat{x}$ and by using $\hat{x}+\hat{y}=\widehat{x+y}$ and $\hat{x}\hat{y}=\widehat{xy}$ we see that $\mathcal{A}$ is a real algebra and by the definition of the topology of $\mathcal{M}_0$ we have $\mathcal{A} \subseteq C(\mathcal{M}_0)$. We prove that $\mathcal{A}$ separates the points of $\mathcal{M}$: If $h_1 \neq h_2$ then $h_1(x) \neq h_2(x)$ for some $x \in A$. We let $x=y+iz$ where $y=y^*$ and $z=z^*$. By the corollary of Proposition 4 we have $\hat{y}, \hat{z} \in \mathcal{A}$ and $y(h_1) \neq y(h_2)$ or $z(h_1) \neq z(h_2)$. For each h in $\mathcal{M}$ there is an x in A such that $h(x) \neq 0$ and so setting again $x=y+iz$ we have $\hat{y}(h) \neq 0$ or $\hat{z}(h) \neq 0$ where $y, z \in \mathcal{A}$. Thus at no point of $\mathcal{M}$ do all elements of $\mathcal{A}$ vanish simultaneously. Since $\mathcal{M}_0$ is compact by the Stone-Weierstrass theorem $\mathcal{A}$ is dense in the algebra of real valued continuous functions defined on $\mathcal{M}_0$ and vanishing at θ.

The following important theorem is essentially due to Gelfand and Naimark who proved it in the special case when A has an identity. For this reason one often refers to it as the *Gelfand-Naimark theorem.*

Theorem 7. *If A is a complex, commutative B^* algebra then the Fourier transform map $x \mapsto \hat{x}$ is an isometric isomorphism of A onto $C_o(\mathcal{M})$.*

Proof. If $x=x^*$ then by Proposition 9 and Theorem 5.3 we have $\|x\| = \|x\|_{sp}$ and so $\|\hat{x}\| = \|x\|$ by the corollary of Theorem 6.14. If x is arbitrary then x^*x is self adjoint and hence $\|\widehat{x^*x}\| = \|x^*x\|$. By Proposition 5 we have $\widehat{x^*x} = \widehat{x^*}\hat{x} = |\hat{x}|^2$ and so $\|x^*x\| = \|\hat{x}\|^2$. Finally the B^* property gives $\|x^*x\| = \|x\|^2$ so we conclude that $\|\hat{x}\| = \|x\|$ for any x in A. By the completeness of A and by $\|\hat{x}\| = \|x\|$ we see that $\hat{A}$ is complete with respect to the norm $\|\hat{x}\|$ and so it is a closed subset of $C_o(\mathcal{M})$. By Theorem 6 we know that $\hat{A}$ is dense in $C_o(\mathcal{M})$ and so $\hat{A} = C_o(\mathcal{M})$.

Let A be a complex algebra with involution. By a *representation* of A we mean an algebra homomorphism $\rho: A \to \mathcal{L}(\mathcal{H})$ for some complex Hilbert space $\mathcal{H}$ such that $\rho(x^*) = \rho(x)^*$ for every x in A. These maps ρ are usually called *star representations* but we shall use this expression only in order to stress this additional property. This terminology is in accordance with the one which will be introduced in Section IV.1. A representation of a normed algebra A is called *continuous* provided there is a constant $C>0$ such that $|\rho(x)| \leqslant C\|x\|$ for every x in A. The smallest possible value of C will be called the *norm* of ρ and will be denoted by $|\rho|$.

Theorem 8. *Let A be a Banach algebra with continuous involution and let $\rho: A \to \mathcal{L}(\mathcal{H})$ be a representation. Then ρ is continuous in the norm* *topology of $\mathscr{L}(\mathscr{H})$. If A is star normed then $|\rho(x)| \leqslant \|x\|$ for every x in A i.e. $|\rho| \leqslant 1$.*

Proof. By Proposition 4.3 we have $\sigma(\rho(x))^\times \subseteq \sigma(x)^\times$ and so

$$\|\rho(x)\|_{sp} \leqslant \|x\|_{sp}. \tag{6}$$

By Proposition 9 and Theorem 5.3 we know that $\|T\|_{sp} = |T|$ for every self adjoint operator T in $\mathscr{L}(\mathscr{H})$. Therefore using (6) and the continuity of the involution we obtain

$$\begin{aligned} |\rho(x)|^2 &= |\rho(x)^* \rho(x)| = |\rho(x^* x)| = \|\rho(x^* x)\|_{sp} \\ &\leqslant \|x^* x\|_{sp} \leqslant \|x^* x\| \leqslant \|x^*\| \cdot \|x\| \leqslant C\|x\|^2. \end{aligned}$$

Hence $|\rho(x)| \leqslant \sqrt{C}\,\|x\|$ and so ρ is continuous.

Proposition 9. *If A is a normed involutive algebra such that $\|x^* x\| = \|x\|^2$ for every x in A then $\|x^n\| = \|x\|^n$ for every self adjoint element x and $n = 1, 2, \ldots$*

Proof. For any $n = 1, 2, \ldots$ we have $\|x^{2n}\| = \|(x^n)^* (x^n)\| = \|x^n\|^2$. Therefore $\|x^{2^k}\| = \|x\|^{2^k}$ for $k = 1, 2, \ldots$ If n is arbitrary choose $m \geqslant 0$ such that $m + n = 2^k$ for some $k > 0$. Then $\|x\|^m \cdot \|x^n\| \geqslant \|x^m\| \cdot \|x^n\| \geqslant \|x^{m+n}\| = \|x\|^{m+n}$ and so $\|x^n\| \geqslant \|x\|^n \geqslant \|x^n\|$.

Proposition 10. *For every self adjoint element x of a complex B^* algebra we have $\|x\|_{sp} = \|x\|$.*

Proof. Combine Proposition 9 with Theorem 5.3.

Theorem 11. *Every complex, normed, involutive algebra satisfying $\|x^* x\| = \|x\|^2$ is semisimple.*

Corollary 1. *Every complex B^* algebra is semisimple.*

Proof. If a is a self adjoint element then by Propositions 9 and 5.5 we have $\|a\|_{sp} \geqslant v(a) = \|a\|$. Now let x belong to the radical $J(A)$ and let $a = x^* x$. By Theorem 7.4 λa is in $J(A)$ for every λ in $\mathbb{C}$ and so $\lambda a \in Q$ by Proposition 7.3. This implies that $1/\lambda$ is not in $\sigma(a)^\times$ for every λ in $\mathbb{C}^\times$ and so $\|a\|_{sp} = 0$. Hence using our earlier remark with $a = x^* x$ we obtain $\|x\|^2 = \|x^* x\| = \|a\| \leqslant \|a\|_{sp} = 0$ and so $x = 0$ and $J(A) = 0$.

Corollary 2. *If $\mathscr{H}$ denotes a complex Hilbert space then every self adjoint, not necessarily closed subalgebra $\mathscr{A}$ of $\mathscr{L}(\mathscr{H})$ is semisimple.*

Proof. By Example 2) given near the beginning of this section we have $\|A^* A\| = \|A\|^2$ for every element A in $\mathscr{A}$.

In the following discussion let A be a complex B^* algebra with identity e. We let G be the group of regular elements and S its complement, the set of singular elements of A. By Proposition 2.2 G is open and S is closed. Let Z_0 denote the set of two sided topological zero divisors of A. In terms of these notations Proposition 2.10 states that $\bar{G} \cap S \subseteq Z_0$. Similarly by Proposition 2.11 we have $Z \subseteq S$ where Z is the set of left and right topological zero divisors of A. Since A is star normed we have $Z_0 = Z_0^*$ and $Z = Z^*$.

Following Rickart we define

$$\lambda(x) = \inf \frac{\|xy\|}{\|y\|} \quad \text{and} \quad \rho(x) = \inf \frac{\|yx\|}{\|y\|}.$$

Then A being star normed $\lambda(x^*) = \rho(x)$. It is clear that $x \in Z_0$ if and only if $\lambda(x) = \rho(x) = 0$ and $x \in Z$ if and only if $\lambda(x) = 0$ or $\rho(x) = 0$. We have

$$\lambda(x)\lambda(y) \leqslant \lambda(xy) \quad \text{and} \quad \rho(x)\rho(y) \leqslant \rho(xy) \tag{7}$$

for every x, y in A. In fact

$$\lambda(x)\lambda(y) \leqslant \inf \frac{\|xyz\|}{\|yz\|} \inf \frac{\|yz\|}{\|z\|} \leqslant \inf \frac{\|xyz\|}{\|z\|} = \lambda(xy).$$

The corresponding inequality for ρ is proved similarly. One can also prove that

$$|\lambda(x) - \lambda(y)| \leqslant \|x - y\| \quad \text{and} \quad |\rho(x) - \rho(y)| \leqslant \|x - y\|$$

but these inequalities will not be used. It is also clear that $\lambda(xy) \leqslant \|x\| \lambda(y)$ and $\rho(xy) \leqslant \rho(x)\|y\|$ for every x, y in A. Moreover the proofs of all these results hold in any normed algebra A.

Lemma 12. *If $xy \in Z$ then $x \in Z$ or $y \in Z$.*

Proof. We have $\lambda(xy) = 0$ or $\rho(xy) = 0$. Hence by (7) $\lambda(x) = 0$, $\lambda(y) = 0$, $\rho(x) = 0$ or $\rho(y) = 0$.

Proposition 13. *If A is a complex B^* algebra with identity then $S = Z$.*

Proof. The inclusion $Z \subseteq S$ is very easy to prove: For instance if $xy = e$ and $\|x_n\| = 1$ then $1 = \|x_n xy\| \leqslant \|x_n x\| \cdot \|y\|$ and so $x_n x \to 0$ is not possible. In order to prove that $S \subseteq Z$ let us first suppose that x is a self adjoint element in S. Then by the corollary of Proposition 4 we have $\sigma(e-x) \subseteq \mathbb{R}$ and so $(e-x)-(1+(i/n))e$ is in G for every $n=1, 2, \ldots$ In other words $x+(i/n)e \in G$ for $n=1,2,\ldots$ and so $x \in \bar{G}$. As we have seen $\bar{G} \cap S \subseteq Z$ so we obtain $x \in Z$. Next let x be an arbitrary element in S. Then xx^* or x^*x is in S because otherwise x would be invertible. Therefore xx^* or x^*x belongs to Z. Hence by Lemma 12 x or x^* is in Z. Since $Z=Z^*$ we see that $x \in Z$.

Proposition 14. *Let A be a complex B^* algebra with identity and let B be a closed involutive subalgebra of A. Then $B^\times = A^\times \cap B$.*

Proof. The inclusion $B^\times \subseteq A^\times \cap B$ is obvious. Suppose that x is a singular element of B. Then by Proposition 13 $x \in Z_B$ and so by $Z_B \subseteq Z_A$ we have $x \in Z_A$. Hence by the same proposition $x \in S_A$ i.e. $x \notin A^\times$.

Corollary. *If B is a closed involutive subalgebra of the complex B^* algebra A then $Q_B = Q_A \cap B$.*

This follows by using the corollary of Proposition 2.3.

Proposition 15. *If B is a closed involutive subalgebra of the complex B^* algebra A and $x \in B$ then $\sigma_A(x)^\times = \sigma_B(x)^\times$.*

Proof. Since $A \supseteq B$ we have $\sigma_A(x) \subseteq \sigma_B(x)$. Let $\lambda \in \sigma_B(x)^\times$. Then by Proposition 4.2 $x/\lambda \notin Q_B$. Hence by the above corollary $x/\lambda \notin Q_A$ and so $\lambda \in \sigma_A(x)^\times$.

Proposition 16. *Let A and B be complex B^* algebras and let $\varphi: A \to B$ be an involutive homomorphism of A into B. Then $\|a\| \geqslant \|\varphi(a)\|$ for every a in A.*

Note. If φ is not involutive then the conclusion holds for all self adjoint a in A.

Proof. First let x be a self adjoint element of A. If $\lambda \in \mathbb{C}^\times$ and $x/\lambda \in Q_A$ then $\varphi(x/\lambda) \in Q_B$ and so by Proposition 4.2 we have $\sigma_A(x) \supseteq \sigma_B(\varphi(x))$. Consequently $\|x\| = \|x\|_{sp} \geqslant \|\varphi(x)\|_{sp} = \|\varphi(x)\|$ by Proposition 10. If a is an arbitrary element of A then

$$\|a\|^2 = \|a^* a\| \geqslant \varphi(a^* a)\| = \|\varphi(a)^* \varphi(a)\| = \|\varphi(a)\|^2 .$$

This completes the proof. We also notice that instead of the B^* property it is sufficient to suppose that $\|a\|^2 \geqslant \|a^* a\|$ for every $a \in A$.

9. H* Algebras

These are complex Banach algebras A in which a map $x \mapsto x^*$ can be defined such that x^* acts as an adjoint with respect to an inner product which is compatible with the algebra norm. If A is semisimple then $x \mapsto x^*$ is uniquely determined and it is an involution.

Definition 1. *An H^* algebra A is a complex Banach algebra which is also a complex Hilbert space with an inner product (x, y) such that*

1) *the algebra norm $\|x\|$ and the Hilbert space norm $\sqrt{(x, x)}$ are the same for every x in A,*

2) *for every element x there is at least one element x^*, called an adjoint of x, such that $(xy, z) = (y, x^* z)$ and $(xy, z) = (x, z y^*)$ for every x, y and z in A.*

The importance of H^* algebras is due to the fact that a natural extension of the Wedderburn structure theorems holds for them. Some of these general results will be needed later in the representation theory of compact groups. Ambrose, the discoverer of a somewhat more restricted class of these algebras was guided by the direct sum decomposition of representations of compact groups and in particular by the decomposition of $L^2(G)$ under the left regular representations of G.

Given z in A we have $Az = 0$ if and only if $zA = 0$ because for instance $(az, y) = (z, a^* y) = \overline{(a^* y, z)} = \overline{(a^*, z y^*)} = (z y^*, a^*)$ and so $zA = 0$ implies $az \perp y$ for every y in A. It follows that the set $Z = \{z : Az = 0\} = \{z : zA = 0\}$ is a closed ideal, called the *annihilator ideal* of A.

Theorem 2. *The radical $J(A)$ of the H^* algebra A is its annihilator ideal Z.*

Corollary. *The radical is a closed ideal.*

Proof. By definition $J(A) = \{x : Ax \subseteq Q_l\}$ where Q_l is the set of right quasi inverses. Since $0 \in Q_l$ we see that $Z \subseteq J(A)$. In order to see the opposite inclusion for each fixed a in A consider the map $\sigma(a): A \to A$ defined by $x \mapsto ax$. It is clearly a linear operator on the complex Hilbert space A and it is continuous with $|\sigma(a)| \leqslant \|a\|$. The map $a \mapsto \sigma(a)$ is a homomorphism of the algebra A onto $\sigma(A)$ and its kernel is Z. Therefore A/Z is isomorphic to $\sigma(A)$. By $(ax, y) = (x, a^* y)$ we have $\sigma(a)^* = \sigma(a^*)$ and so the image algebra $\sigma(A)$ is self adjoint. Therefore by the second corollary of Theorem 8.11 $\sigma(A)$ is semisimple. Since semisimplicity is a pure algebraic concept it follows that A/Z is semisimple and so $J(A) \subseteq Z$ because by $J(A/Z) = Z$ the intersection of all regular left maximal ideals containing Z is Z itself.

Corollary. *An H^* algebra A is semisimple if and only if its annihilator ideal $Z=0$.*

If x is an element such that $x^*x=0$ for at least one adjoint x^* then for any a we have $(xa,xa)=(a,x^*xa)=0$. Therefore $xa=0$ for every a in A and $x\in Z$. Conversely if $x\in Z$ then $x^*x=0$ for all adjoints x^* of x. Therefore in view of Theorem 2 we have:

Proposition 3. *The radical of the H^* algebra A consists of those x in A for which $x^*x=0$ for at least one adjoint x^* of x.*

Note. We could require that $x^*x=0$ for *all* adjoints x^*.

Corollary. *The H^* algebra A is semisimple if and only if $x^*x=0$ implies $x=0$.*

Let us suppose that A is semisimple and x_1^*, x_2^* are adjoints of x. Then $((x_1^*-x_2^*)a,b)=0$ for every $a,b\in A$ and so $x_1^*-x_2^*\in Z$ and $x_1^*=x_2^*$. Therefore we have:

Proposition 4. *If the H^* algebra A is semisimple then x^* is uniquely determined by x and the map $x\mapsto x^*$ is an involution.*

We shall now turn to the structure theory of semisimple H^* algebras. As in the Wedderburn structure theory *idempotents* play an important role. A selfadjoint idempotent e is called *reducible* if there exist non-zero, self adjoint orthogonal idempotents e_1, e_2 such that $e=e_1+e_2$. If e_1, e_2 are self adjoint idempotents then $(e_1,e_2)=(e_1,e_1e_2)=(e_1e_2,e_1e_2)$. Therefore $e_1\perp e_2$ implies that $e_1e_2=e_2e_1=0$ and conversely if $e_1e_2=0$ then $e_1\perp e_2$. This explains the following definition: An arbitrary idempotent e is called reducible if $e=e_1+e_2$ where e_1, e_2 are non-zero idempotents such that $e_1e_2=e_2e_1=0$.

A non-zero, self adjoint idempotent can be reduced only a finite number of times: For $\|e\|\geqslant 1$ for any idempotent e and if $e=e_1+e_2$ is a reduction then $\|e\|^2=\|e_1\|^2+\|e_2\|^2$. Therefore the existence of a non-zero, self adjoint idempotent will prove the existence of a non-zero, self adjoint, irreducible idempotent.

Lemma 5. *If $\mathscr{H}$ is a Hilbert space and $a\colon \mathscr{H}\to\mathscr{H}$ is a self adjoint operator then $|a^n|=|a|^n$ for every $n=1,2,\ldots$*

Proof. In Example 2 of the last section we had seen that $\mathscr{L}(\mathscr{H})$ is a B^* algebra. Therefore the lemma follows from Proposition 8.9.

Proposition 6. *Every non-zero left ideal of the semisimple H^* algebra A contains a non-zero self adjoint, irreducible idempotent.*

Corollary. *The algebra A has a non-zero, self adjoint, irreducible idempotent.*

Proof. For any a in A let $|a| = \text{supr}\, \|ax\|/\|x\|$ so that $|a|$ is the norm of the linear operator $x \mapsto ax$. Clearly $|a| \leqslant \|a\|$. If b is a non-zero element of the left ideal I then by the semisimplicity $b^* b \neq 0$ and so a positive scalar multiple of $b^* b$ is an element a in I such that $|a| = 1$. Since the underlying linear space is a Hilbert space $x \mapsto ax$ is a self adjoint operator and by the last lemma we have $|a^n| = 1$ for every $n = 1, 2, \ldots$
We prove that a^{2k} $(k = 1, 2, \ldots)$ is a Cauchy sequence: For any $m \geqslant n$ we have

$$(a^m, a^n) = (a^{m-n}. a^n, a^n) \leqslant |a^{m-n}| (a^n, a^n) = (a^n, a^n).$$

Moreover if $m - n = 2d \geqslant 0$ then

$$(a^m, a^m) = (a^{2d} a^{n+d}, a^{n+d}) \leqslant |a^{2d}| (a^{n+d}, a^{n+d}) = (a^m, a^n).$$

Therefore

$$1 \leqslant (a^m, a^m) \leqslant (a^m, a^n) \leqslant (a^n, a^n)$$

for any two even integers $m \geqslant n$. We see that (a^{2k}, a^{2k}) $(k = 1, 2, \ldots)$ is a decreasing sequence with limit $\lambda \geqslant 1$ and the above inequality also shows that $(a^{2k}, a^{2l}) \to \lambda$ as $k, l \to \infty$. Therefore $\|a^{2k} - a^{2l}\| \to 0$ as $k, l \to \infty$ simultaneously. Let $e = \lim a^{2k}$ so that $\|e\| \geqslant 1$ and $e \neq 0$. The element e is self adjoint and $e^2 = \lim a^{4k} = e$ shows that e is an idempotent. Moreover $e = (\lim a^{2(k+1)}) a^2 = e a^2$ and so by $a \in I$ we have $e \in I$.

The idempotent e constructed above might not be irreducible. However if $e = e_1 + e_2$ is a reduction of e then $e_1 e = e_1^2 + e_1 e_2 = e_1$ and so $e_1 \in I$ by $e \in I$. We had seen already that after finitely many reductions we arrive at an irreducible idempotent. This idempotent belongs to the given left ideal I.

Proposition 7. *A non-zero subset I is a minimal left ideal of the semisimple H^* algebra A if and only if $I = Ae$ where e is a non-zero, self adjoint, irreducible idempotent.*

Corollary. *Every minimal left ideal is closed.*

Proof. Suppose I is a minimal left ideal of A. By the last proposition I has a non-zero, self adjoint idempotent e. Since I is minimal and Ae is a left subideal of I we have $I=Ae$. Suppose that $e=e_1+e_2$ where e_1, e_2 are self adjoint idempotents and $e_1 e_2=e_2 e_1=0$. Then Ae_1 and Ae_2 are orthogonal left subideals of I. Hence one of them, say $Ae_1=I$ and the other, $Ae_2=0$. Hence $e_2^2=e_2=0$ and $e_1=e$ proving that e is not reducible. Thus every self adjoint idempotent e of I is irreducible and $I=Ae$.

Conversely, suppose that $I=Ae$ where e is a non-zero, self adjoint idempotent of A. If I has a non-zero, proper subideal J then by Proposition 6 J contains a self adjoint idempotent j. We let $e_1=ej$ and $e_2=e-e_1$. By $j\in J\subseteq I=Ae$ we have $j=ae$ for some $a\in A$ and so $je_1=jej=(ae)e(ae)=(ae)^2=j^2=j\neq 0$. Therefore $je_1\neq 0$ and $e_1\neq 0$. Also $e_2=e-e_1\neq 0$ because $Ae_1=J$ while $Ae=I$. The proposition is proved.

By Proposition 7 and the corollary of Proposition 6 there are non-zero, minimal left ideals in A. For self adjoint idempotents e_1, e_2 we have $Ae_1\perp Ae_2$ if and only if $e_1\perp e_2$ i.e. $e_1 e_2=e_2 e_1=0$. For $(a_1 e_1, a_2 e_2)=(e_1 e_2, a_1^* a_2)$ for any a_1, a_2 in A and $(e_1 e_2, a)=(a^* e_1, e_1 e_2)$ for any $a\in A$. A *minimal closed ideal* means a non-zero closed ideal which contains no non-zero, proper closed subideals. Similarly one can speak about minimal closed left and right ideals. The algebra A is called *topologically simple* if its only closed ideals are 0 and A itself.

Theorem 8. *A semisimple H^* algebra A has minimal ideals, namely the ideal generated by any minimal left ideal is minimal.*

Any minimal ideal J is self adjoint and $J^2=J$.

Any two distinct minimal ideals are orthogonal.

Any minimal closed ideal is a topologically simple H^* algebra.

Any closed ideal I of A is the direct sum of the minimal closed ideals contained in it.

Proof. Let $I=Ae$ be a minimal left ideal and let J be the ideal generated by I so that J is the intersection of all ideals containing the set AeA. Then J is self adjoint because $(AeA)^*=AeA$ and the adjoint of any ideal is an ideal. In order to prove that J is minimal suppose that J_1 is a subideal of J. We have $J_1 I\subseteq J_1\cap I\subseteq I$ where $J_1 I$ is a left ideal. Hence by the minimality of I either $J_1 I=I$ or $J_1 I=0$. In the first case $J_1\supseteq I$ so $J_1\supseteq J$ and $J_1=J$ by $J_1\subseteq J$. In the second case $J_1 IA=0$ i.e. $J_1 AeA=0$. Since every element of J is expressible as a sum of terms belonging to AeA obtain $J_1 J=0$. Now $J_1 J_1^*\subseteq J_1 J=0$ by $J_1^*\subseteq J^*=J$, so $J_1 J_1^*=0$. By the semisimplicity of A and by the corollary of Proposition 3 $xx^*=0$ implies $x=0$ so we see that $J_1=0$. We proved that J is a minimal ideal.

All minimal ideals J are self adjoint because by Proposition 6 J contains an idempotent e and so it is the smallest ideal containing the self adjoint set AeA which was proved to be self adjoint.

We also notice that $J^2=J$ for any minimal ideal J because J^2 is an ideal contained in J and $J^2=JJ^*\neq 0$ by $J\neq 0$ and the semisimplicity criterion mentioned above.
Let J_1 and J_2 be minimal ideals of A. By $J_1 J_2 \subseteq J_1 \cap J_2 \subseteq J_1, J_2$ we have $J_1 J_2=0$ or $J_1=J_2$. Hence if J_1, J_2 are distinct minimal ideals then $J_1 J_2=J_2 J_1=0$. Thus if $x, y \in J_1$ and $z \in J_2$ then $(xy, z)=(y, x^* z)=0$ by $x^* z \in J_1 J_2$. It follows that $J_1=J_1^2 \perp J_2$.

By definition a minimal closed ideal has no proper closed subideals and so it is topologically simple. Since every such ideal is self adjoint it is an H^* algebra.

If J is a left ideal then so is $J^\perp$ because if $a\in A$, $y\in J^\perp$ and $z\in J$ then $(ay, z)=(y, a^* z)=0$. Let I be a closed, two sided ideal and let J be the subspace spanned by the minimal left ideals contained in I. Then J is a closed left ideal and so $J\subset I$ would imply that $J^\perp \cap I$ would contain a minimal left ideal by Propositions 6 and 7. Therefore $J=I$ and I is spanned by its minimal left ideals. We have seen that every such ideal is included in a minimal closed ideal contained in I. Hence I is spanned by its minimal closed ideals and so by their orthogonality it is their direct sum.

Corollary. *A semisimple H^* algebra can be expressed uniquely as the direct sum of its pairwise orthogonal, minimal closed ideals each of which is a topologically simple H^* algebra.*

By slightly strengthening the reasoning given at the end of the foregoing proof we obtain the following:

Proposition 9. *Every minimal closed ideal I of a semisimple H^* algebra can be expressed as a direct sum of pairwise orthogonal, minimal left ideals.*

Note. Generally the decomposition is not unique.

Corollary. *Every topologically simple H^* algebra is expressible as a direct sum of pairwise orthogonal, minimal left ideals.*

Proof. Let $e_i (i\in\mathscr{I})$ be a maximal family of pairwise orthogonal, self adjoint, irreducible idempotents lying in I. Then $Ae_i (i\in\mathscr{I})$ is a collection of pairwise orthogonal, minimal left ideals. Let J be their direct sum. Then $J\subset I$ would imply that $J^\perp \cap I$ is a non-zero, closed ideal and so by Propositions 6 and 7 it would contain a self adjoint, irreducible idempotent which could be joined to the family $e_i (i \in \mathscr{I})$. Therefore $J = I$.

Proposition 10. *If is a non-zero, irreducible, self adjoint idempotent in the semisimple H^* algebra A then the self adjoint, involutive subalgebra eAe is isomorphic to the complex number field $\mathbb{C}$, namely for each x in eAe there is a unique complex number λ_x such that $x = \lambda_x e$ and $\lambda_x = 0$ only if $x = 0$.*

Proof. It is clear that eAe is a complex, involutive Banach algebra with identity e. By Proposition 7 $I = Ae$ is a minimal left ideal. If $x \neq 0$ and $x \in I$ then $Ax \subseteq I$ and by the semisimplicity $Ax \neq 0$. Therefore by the minimality of I we have $Ax = Ae$. Let a in A be such that $ax = ee = e$. Now if $x \in eAe$ then by $e^2 = e$ we have $x = exe$ and so

$$(eae)(exe) = ea(exe) = (ea)x = e(ax) = e^2 = e\,.$$

This proves that every non-zero x in eAe has a left inverse, namely $(ea)x = e$. If we start with the minimal right ideal eA then by the same reasoning we see that x has a right inverse. Therefore eAe is a division algebra and Theorem 4.7 can be applied. Since $x^* = (\lambda_x e)^* = \bar{\lambda}_x e$ we have $\lambda_{x^*} = \bar{\lambda}_x$ for any x in A.

Lemma 11. *If e is a self adjoint, irreducible idempotent then the projection of any $x \in A$ on the subspace Ae is xe.*

Proof. Let $x_i e (i \in \mathscr{I})$ be a MONS in Ae. Then $(x, x_i e) = (xe, x_i e)$ for every $i \in \mathscr{I}$. Hence the projection Px is

$$Px = \sum (x, x_i e) x_i e = \sum (xe, x_i e) x_i e = xe\,.$$

Theorem 12. *Let A be a topologically simple H^* algebra. Then the following three statements are equivalent:*

1) *A is finite dimensional;*
2) *A has an identity;*
3) *there is a non-zero central element in A.*

Proof. Decompose I into a direct sum of minimal closed left ideals Ae_i $(i \in \mathscr{I})$. If I is finite dimensional let $\mathscr{I} = \{1, \ldots, n\}$ and $e = e_1 + \cdots + e_n$. Then e is a self adjoint idempotent because $e_i e_j = 0$ for $i \neq j$ by the orthogonality of the e_i's. Also if $x \in I$ then by Lemma 11 $x = \sum x e_i = xe$. Similarly we have $x = ex$ and so I is an algebra with identity e. Therefore 1) implies 2). If I has an identity e then e is in the center of I and so 2) implies 3). The following construction needed to prove that 3) implies 1) will also lead to the propositions given below.

Let $e_i (i \in \mathscr{I})$ be a maximal family of pairwise orthogonal, self adjoint, irreducible idempotents of the topologically simple H^* algebra A. While proving Proposition 9 we had seen that $A e_i (i \in \mathscr{I})$ is a collection of pairwise orthogonal minimal left ideals spanning A. Let one of the indices i, say $1 \in I$ be fixed. If i in $\mathscr{I}$ is not 1 then $e_1 A e_i$ is not 0. For A being topologically simple the ideal H generated by $A e_i A$ is dense in A and so $e_1 A e_i A e_1 = 0$ would imply $e_1 x e_1 = 0$ for every $x \in A$. Hence $e_1 a e_i b e_1 \neq 0$ for some $a, b \in A$ and so $e_1 a e_i \neq 0$. Let x be a non-zero element in $e_1 A e_i$. Then $x x^* \in e_1 A e_1$ and so by Proposition 10 $x x^* = \lambda e_1$ where $\lambda > 0$. Therefore $e_{1i} = \lambda^{-\frac{1}{2}} x$ is a non-zero element in $e_1 A e_i$ such that $e_{1i} e_{1i}^* = e_1$. Then we have $e_{1i}^* \in e_i A e_1$ and so $e_{1i}^* e_{1i} \in e_i A e_i$. Since

$$e_{1i}^* e_{1i} e_{1i}^* e_{1i} = e_{1i}^* e_1 e_{1i} = e_{1i}^* e_{1i}$$

we see that $e_{1i}^* e_{1i}$ is a non-zero, self adjoint idempotent in $e_i A e_i$. Hence by Proposition 10 $e_{1i}^* e_{1i} = e_i$ where by hypothesis $i \neq 1$. Now we define $e_{11} = e_{11}^* = e_1$ and also $e_{i1} = e_{1i}^*$ for $i \in I$. Moreover we let $e_{ij} = e_{i1} e_{1j}$ for $i, j \in \mathscr{I}$. If $i = 1$ or $j = 1$ the old and the new definitions coincide because $e_{1j} \in e_1 A e_j$ and $e_{i1} \in e_i A e_1$. We have $e_{ij} e_{jk} = e_{ik}$ and $e_{ij} e_{kl} = 0$ if $j \neq k$. Also $e_{ij}^* = e_{ji}$ and $e_{ii} = e_i$. Moreover $e_{ij} \perp e_{kl}$ except when $i = k$ and $j = l$. Finally we have $(e_{ij}, e_{ij}) = (e_1, e_1)$ because

$$(e_{ij}, e_{ij}) = (e_{i1} e_{1j}, e_{i1} e_{1j}) = (e_{1i} e_{1i}^*, e_{1j} e_{1j}^*) = (e_1, e_1).$$

In particular we have $(e_i, e_i) = (e_{ii}, e_{ii}) = (e_1, e_1)$ for every $i \in \mathscr{I}$. Therefore we proved:

Proposition 13. *If A is a topologically simple H^* algebra then any two orthogonal, non-zero, self adjoint, irreducible idempotents have the same norm.*

Now we are going to prove that the orthogonal system $e_{ij} (i, j \in \mathscr{I})$ is maximal. First we show that $e_i A e_j$ is a one dimensional subspace, namely $e_i A e_j = \mathbb{C} e_{ij}$: For $e_i x e_{ji} = e_i x e_{ji} e_{ii} = e_i x e_{ji} e_i$ belongs to $e_i A e_i$ and so $e_i x e_{ji} = c e_i$ for some $c \in \mathbb{C}$. Multiplying on the right by e_{ij} we obtain $e_i x e_j = c e_{ij}$. The direct sum of the minimal right ideals $e_i A$ is A. This follows from the maximality of the system $e_i (i \in \mathscr{I})$ in the same way as the corresponding result for the left ideals $A e_i$. Therefore given x in A by using both the left and right versions of Lemma 11 we obtain

$$x = \sum_j x e_j = \sum_j \sum_i e_i (x e_j) = \sum_{i,j} c_{ij} e_{ij}.$$

The maximality of the system is proved.

Our results can be summarized as follows:

Proposition 14. *Let e_i $(i \in \mathscr{I})$ be a maximal system of pairwise orthogonal, self adjoint, irreducible idempotents in the topologically simple H^* algebra A. Then the system can be extended to a complete orthogonal system e_{ij} $(i, j \in \mathscr{I})$ of non-zero idempotents such that*

1) $e_{ii} = e_i$ *for every* $i \in \mathscr{I}$,

2) $e_{ij} e_{kl} = 0$ *if* $j \neq k$ *and* $e_{ij} e_{jk} = e_{ik}$ *for any* $i, j, k \in \mathscr{I}$,

3) $e_{ij}^* = e_{ji}$ *for every* $i, j \in \mathscr{I}$,

4) $(e_{ij}, e_{kl}) = 0$ *except when* $i = k$ *and* $j = l$,

5) $\|e_{ij}\|$ *is the same for every* i, j *in* $\mathscr{I}$, *namely it is equal to the common norm of the elements of the original system* e_i $(i \in \mathscr{I})$.

Now we complete the proof of Theorem 12 by showing that 3) implies 1). Suppose x is a non-zero central element in A. We have $x e_i = (x e_i) e_i = e_i x e_i$ and so the projection of x on $A e_i$ is $c_i e_i$ for some $c_i \in \mathbb{C}$ by Proposition 10 and Lemma 11. Therefore we have $x = \sum c_i e_i$. But multiplying $x e_i = c_i e_i$ by e_{ij} from the right we obtain $x e_{ij} = c_i e_{ij}$ and multiplying $e_j x = c_j e_j$ by e_{ij} from the left we get $e_{ij} x = c_j e_{ij}$. Since x is central it follows that $c_i e_{ij} = c_j e_{ij}$. Therefore all the c_i's are equal and so $x = \sum c_i e_i$ implies that A is finite dimensional. Theorem 12 is now proved.

Remarks

The history of abstract harmonic analysis began in 1939 with the publication of Gelfand's preliminary note on normed rings (1). The complete paper entitled "Normierte Ringe" appeared in 1941 (2). The first book on this subject is a classic written by Loomis in 1953 (1). There exist three other important monographs on this topic and related material by Hille and Phillips (1), Naimark (1) and Rickart (1), respectively. A concise summary of the theory of commutative Banach algebras was written by Mackey (1). The same material is discussed to a certain extent in several other books e.g. in the second volume of Dunford and Schwartz (1) and in Simmons' book (1). Harmonic analysis on locally compact groups has a much older history starting with the works of Daniel Bernoulli and Euler in the 18-th century and Fourier's "Théorie analytique de la chaleur" which appeared in 1822.

Normed algebras were considered before Gelfand by Michal and Martin in 1934 (1) and Nagumo in 1936 (1). The operation $(x, y) \mapsto x + y + xy$

was considered by several earlier writers and its systematic use in ring theory and abstract harmonic analysis started around 1945 in the works of Jacobson (1) and Kaplansky (1). The more advantageous circle operation $x \circ y = x + y - xy$ was introduced somewhat later by Hille (1) and Segal (1) who used the term adverse for what is now called quasi-inverse. Topological zero-divisors were introduced by Silov (1).

Topological algebras with continuous inverses were studied in several C.R. notes and papers by Waelbroeck. For references see the Zbl. or the Mathematical Reviews. The Gelfand topology and the maximal ideal space were introduced in Gelfand's famous paper (2). The concept of a spectrum is more than a hundred years old. Characteristic roots of matrices often occur in the works of Frobenius and Sylvester and in some of the papers of Cayley and Hamilton. The name spectrum was introduced by Hilbert. The origin of the resolvent $(x - \lambda e)^{-1}$ is hard to locate for in the case of finite dimensional vector spaces it occurs already in Frobenius' papers. F. Riesz (1) considered in 1911 the resolvent and the resolvent set $r(x)$ when the underlying Hilbert space is finite dimensional. Not too long ago it was still common to deal with $(\lambda x - e)^{-1}$ instead of $(x - \lambda e)^{-1}$ and in some instances the spectrum $\sigma(x)$ was also defined in accordance with this practice as the set of those λ's for which $\lambda x - e$ is not invertible.

Theorem 4.7 is an extension of a result which goes back to Frobenius and states that a finite dimensional skew field over the reals is isomorphic to $\mathbb{R}$, $\mathbb{C}$ or the Hamilton quaternions $\mathbb{H}$. It is often called the Gelfand-Mazur theorem because it was announced in a C.R. note without proof by S. Mazur in 1938 (1) and proved later by Gelfand in his fundamental paper on normed rings (2). Proposition 4.9 is a simple special case of the spectral mapping theorem which is discussed in Chapter II.

The formula $\|x\|_{sp} = \lim \|x^n\|^{1/n}$ was proved by Gelfand in (2) and elements satisfying $\|x^n\|^{1/n} \to 0$ were called *generalized nilpotents* there. The term regular ideal is due to Segal who introduced it in (1). Many writers use the expression *modular ideal* instead of regular ideal. This name was coined by Jacobson in the same paper where the Jacobson radical is introduced (1). The concept of a strong radical is due to Segal (1). Theorems 6.7 and 6.14 are essentially due to Gelfand (2) who proved them in the special case when the commutative Banach algebra A has an identity. The original definition of the hull-kernel topology is due to M. Stone (1) and it dates from 1937. The one for primitive rings which is called the structure space topology is due to Jacobson (2). Theorem 6.19 is also due to Jacobson and it appeared in the same article in 1945. A paper by Gelfand and Silov (1) from 1941 is also relevant. The proof given in the text is based on M. Schreiber's paper (1) dated 1957.

We already mentioned that $J(A)$ was introduced by Jacobson in (2). One can prove that $J(A)$ is the intersection of the kernels of all irreducible morphisms of A. This proof and a great deal of additional information on $J(A)$ can be found in Rickart's book (1) and Jacobson's monograph on the structure of rings (3). The Wedderburn-Artin structure theorem was first proved by Wedderburn in 1908 (1) for finite dimensional algebras over a field. In 1927 Artin (1) proved the corresponding theorem for rings satisfying both the ascending and the descending chain conditions for left ideals. The fact that only the descending chain condition is needed was realized by Hopkins (1).

The theory of involutive algebras started in 1855 with the publication of Hermite's C.R. note (1) on Hermitian matrices. Since the algebra $\mathscr{L}(\mathscr{H})$ is a B^* algebra the early theory of these algebras can be traced back to the beginnings of operator theory in Hilbert spaces. *Every complex B^* algebra is isometrically isomorphic to a subalgebra of $\mathscr{L}(\mathscr{H})$ for a suitable Hilbert space $\mathscr{H}$.* This important result was essentially proved by Gelfand and Naimark in 1943 (1). More precise comments on the proof of this theorem will be given in Chapter III where the symmetry of B^* algebras will be discussed. Proposition 8.4 is due to Gelfand and Naimark (1) and we adopted to the general situation Arens' proof who restricted himself to algebras with an identity (1, 2). Theorem 8.7 was published by Gelfand and Naimark in (1). Propositions 2.10, 8.13, 8.14 and 8.15 are due to Rickart (2) and his paper contains a great deal more on $\lambda(x)$, $\rho(x)$ and the relation between $\sigma_A(x)$ and $\sigma_B(x)$. It has been mentioned already in the text that H^* algebras were introduced by Ambrose (1).

Chapter II

Operators and Operator Algebras

Almost all the material discussed in this chapter will be needed later but a reader who is somewhat familiar with it already might prefer to use it only for reference during the detailed study of some of the later chapters. The topics of this chapter are both important and interesting. The following elementary and concise summary of this part of linear functional analysis can be read without familiarity with the contents of Chapter I. A casual acquaintence with Section I.4. will suffice.

1. Topologies on Vector Spaces and on Operator Algebras

This section contains a brief summary of those basic concepts from the theory of topological vector spaces which are essential for our purposes. It is not intended to be an introduction to this subject. We start with topological vector spaces: These are defined over a field F which also has a topological structure and for simplicity we suppose that F is normed by a norm $|\cdot|$. We shall be concerned only with the case when F is the reals $\mathbb{R}$ or the field of complex numbers $\mathbb{C}$. A vector space $\mathscr{V}$ over F is called a *topological vector* space if $\mathscr{V}$ is a topological space and the two vector space operations are continuous. Hence $(x,y)\mapsto x+y$ and $(\lambda,x)\mapsto\lambda x$ are continuous functions from $\mathscr{V}\times\mathscr{V}$ and $F\times\mathscr{V}$ into $\mathscr{V}$, respectively.

If $\mathscr{V}$ is a topological vector space and $x\in\mathscr{V}$ and O is an open set in $\mathscr{V}$ then $(-x,y)\mapsto y-x$ being a continuous map we see that $x+O$ is also an open set. Similarly one proves that if O is open and $\lambda\in F$, $\lambda\neq 0$ then λO is open. We also see that if C is a closed set in $\mathscr{V}$ then so are $x+C$ and λC for $\lambda\neq 0$. Let $\mathscr{N}(x)$ denote the neighborhood filter of $x\in\mathscr{V}$ and let $\mathscr{B}(x)$ be a basis for $\mathscr{N}(x)$. It follows from above that $\mathscr{N}(x)=\mathscr{N}(0)+x$ and $\mathscr{B}(x)=\mathscr{B}(0)+x$ where 0 is used to denote the zero vector in $\mathscr{V}$. Incidentally we shall use the same symbol to denote the zero element of F.

Let $\mathscr{V}$ be any vector space over $\mathbb{R}$ or $\mathbb{C}$ and let C be a subset of $\mathscr{V}$. The set C is called *convex* if $x_1, x_2\in\mathscr{V}$ implies that $\lambda_1 x_1+\lambda_2 x_2\in C$

for every $\lambda_1, \lambda_2 \geqslant 0$ satisfying $\lambda_1 + \lambda_2 = 1$. It follows that if $x_1, \ldots, x_n \in C$ then $\lambda_1 x_1 + \cdots + \lambda_n x_n$ is in C for every choice of $\lambda_1, \ldots, \lambda_n \geqslant 0$ as long as $\lambda_1 + \cdots + \lambda_n = 1$. We see that convexity corresponds to the intuitive concept of a convex set used in geometry. A *locally convex vector space* is a vector space $\mathscr{V}$ over the reals such that $\mathscr{N}(0)$ has a basis $\mathscr{B}(0)$ which consists of convex, open sets of $\mathscr{V}$.

Let $\mathscr{V}$ be a vector space over a normed field such as for instance $\mathbb{R}$ or $\mathbb{C}$. A function $x \mapsto \|x\|$ from $\mathscr{V}$ into $[0, \infty)$ is called a *norm* if it has the following properties: $\|x\| = 0$ if and only if x is the zero vector; $\|x+y\| \leqslant \|x\| + \|y\|$ for every x, y in $\mathscr{V}$ and $\|\lambda x\| = |\lambda| \cdot \|x\|$ for every $\lambda \in F$ and $x \in \mathscr{V}$. Occasionally one also encounters a *pseudo-norm* which differs from a norm by the property that $\|x\| = 0$ is possible without x being 0. Every *normed vector space* i. e. a vector space with a norm function is locally convex. For the sets $\{x\colon \|x\| < \varepsilon\}$ $(\varepsilon > 0)$ are all convex and their family $\mathscr{B}(0)$ is a convex, open filter base for $\mathscr{N}(0)$.

An *inner product* is a map $\mathscr{V} \times \mathscr{V} \to \mathbb{C}$ with function values (x, y) where $\mathscr{V}$ is a real or complex vector space such that the following conditions are satisfied: $(x, x) \geqslant 0$ for every $x \in \mathscr{V}$ and $(x, x) = 0$ if and only if $x = 0$; $(x+y, z) = (x, z) + (y, z)$ for all $x, y, z \in \mathscr{V}$; $(\lambda x, y) = \lambda(x, y)$ for every $\lambda \in F$ and $x, y \in \mathscr{V}$; $(x, y) = \overline{(y, x)}$ for $x, y \in \mathscr{V}$. If $F = \mathbb{R}$ one also requires that the function values (x, y) be all real. Hence $(\cdot\,, \cdot)$ is a bilinear or hermitian form according as $F = \mathbb{R}$ or $\mathbb{C}$. The function $\|x\| = (x, x)^{\frac{1}{2}}$ is a norm. If the first condition is relaxed so that only $(x,x) \geqslant 0$ is required then we speak about a *pseudo-inner product*. These will come up on a number of occasions. The meaning of an *inner product space* or a *pseudo-inner product space* is self explanatory but we have to point out that sometimes this terminology is used only in the case when $F = \mathbb{R}$. In the complex case then one speaks about a *unitary space*.

A *Banach space* is a normed vector space which is complete with respect to its norm. These spaces will be denoted by $\mathscr{X}, \mathscr{Y}, \mathscr{Z}$ or by special symbols e. g. $L^1(X)$. A *Hilbert space* is an inner product space over $\mathbb{R}$ or $\mathbb{C}$ which is complete with respect to the norm derived from its inner product. Most of the time we shall deal with complex Hilbert spaces and unless there is a special reason not to do so we shall use $\mathscr{H}, \mathscr{K}, \mathscr{L}, \mathscr{M}, \mathscr{N}$ and $\mathscr{P}$ to denote such spaces. The symbols $\mathscr{L}, \ldots, \mathscr{P}$ will usually denote subspaces of $\mathscr{H}$ or $\mathscr{K}$. A *subspace* will always mean a closed linear subset of the topological vector space in question. If this is not the case we shall speak about a *linear manifold*. We know that if $\mathscr{X}$ is a finite dimensional normed vector space over $F = \mathbb{R}$ or $F = \mathbb{C}$ then $\mathscr{X}$ is complete. Hence finite dimensional linear manifolds are always subspaces. A *pre-Hilbert space* means a real or complex inner product space. A subset $\mathscr{S}$ of the topological vector space $\mathscr{V}$ is called *total* if the linear manifold spanned by $\mathscr{S}$ is dense in $\mathscr{V}$.

Let $\mathscr{I}$ be an index set and for each $i \in \mathscr{I}$ let $\mathscr{V}_i$ be a vector space over the same field of scalars F. The *direct sum* of the $\mathscr{V}_i$'s is itself a vector space $\mathscr{V}$ over F which consists of functions $x: \mathscr{I} \to \bigcup \mathscr{V}_i$ such that $x(i) \in \mathscr{V}_i$ for every $i \in \mathscr{I}$. If $\mathscr{I}$ is finite one considers all such functions x and if is infinite then by making various restrictions on x one obtains different kinds of direct sums. We are interested only in the situation when each $\mathscr{V}_i$ is a normed vector space. Then we require that $x(i)$ be different from 0 for at most denumerably many indices and in addition $\|x\| = \sum \|x(i)\|$ be finite. Then $\|\cdot\|$ is a norm on $\mathscr{V}$ and if every $\mathscr{V}_i$ is a Banach space then so is $\mathscr{V}$. If each $\mathscr{V}_i$ is an inner product space and v^1, v^2 are in $\mathscr{V} = \sum \mathscr{V}_i$ then by the BCS inequality the infinite series $\sum (v_i^1, v_i^2)$ is absolutely convergent and it defines an inner product (v^1, v^2) in $\mathscr{V}$. If every $\mathscr{V}_i$ is a Hilbert space then $\mathscr{V}$ is also a Hilbert space. Often it is more convenient to use x_i to denote the values of x instead of $x(i)$.

If $\mathscr{V}$ and $\mathscr{W}$ are topological vector spaces over the same field F then we let $A, B, \ldots, S, T, U, \ldots$ denote *continuous linear operators* such as $A: \mathscr{V} \to \mathscr{W}$. *Continuous linear functionals* will be denoted by various lower case symbols e.g. f, g, x^*, y^*, v^*, etc. We let $\mathscr{V}^*$ be the *dual space* of $\mathscr{V}$ i.e. the vector space of continuous linear functionals $v^*: \mathscr{V} \to F$. We shall need a topology on $\mathscr{V}^*$ only when $\mathscr{V}$ is a Banach space over $\mathbb{R}$ or $\mathbb{C}$ and in that case $\mathscr{V}^*$ is a normed vector space with $\|v^*\| = \text{lub}\,\{|v^*(v)|: v \in \mathscr{V} \text{ and } \|v\| \leqslant 1\}$. We recall that the Banach space $\mathscr{X}$ is called *reflexive* if every element of $\mathscr{X}^{**}$ is of the form $x^* \mapsto x^*(x)$ where x is a fixed element of $\mathscr{X}$ and x^* varies over $\mathscr{X}^*$.

The *adjoint* of an operator will have one of the following two meanings: First if $\mathscr{X}$ and $\mathscr{Y}$ are Banach spaces over the same field and $A: \mathscr{X} \to \mathscr{Y}$ then its adjoint $A^*: \mathscr{Y}^* \to \mathscr{X}^*$ such that $A^* y^*(x) = y^*(Ax)$ for every $x \in \mathscr{X}$ and $y^* \in \mathscr{Y}^*$. Next if $\mathscr{H}$ and $\mathscr{K}$ are Hilbert spaces and $A: \mathscr{H} \to \mathscr{K}$ then $A^*: \mathscr{K} \to \mathscr{H}$ such that $(Ax, y) = (x, A^* y)$ for every $x \in \mathscr{H}$ and $y \in \mathscr{K}$. In the case of Hilbert spaces most of the time we shall use this second, *Hilbert space adjoint*. If the field of scalars is $\mathbb{C}$ then we have $(\lambda A)^* = \lambda A^*$ for the first, general adjoint, while $(\lambda A)^* = \bar{\lambda} A^*$ holds for the Hilbert space adjoint.

A *topological algebra* is an algebra A over a field F such that A is a topological vector space and for every a in A the maps $x \mapsto ax$ and $x \mapsto xa$ are continuous on A. Therefore the multiplication $(x, y) \mapsto xy$ is not supposed to be continuous as a function of two variables. We already know what is meant by a *normed algebra*. We observe that every normed algebra is a topological algebra and the maps $x \mapsto ax$ and $x \mapsto xa$ are actually uniformly continuous because for instance $\|ax-ay\| \; \|a\| \cdot \|x-y\|$. If x_0, y_0, x, y belong to the normed algebra A then we have

$$\|xy - x_0 y_0\| \leqslant \|x_0\| \cdot \|y - y_0\| + \|x - x_0\| \cdot \|y - y_0\| + \|y_0\| \cdot \|x - x_0\|.$$

This inequality shows us two things: Multiplication is continuous as a function of two variables and it is uniformly continuous if the two variables are restricted to the unit ball of A.

Given two topological vector spaces $\mathscr{X}$ and $\mathscr{Y}$ over the same field of scalars we let $\mathscr{L}(\mathscr{X}, \mathscr{Y})$ be the vector space of continuous linear operators $A: \mathscr{X} \to \mathscr{Y}$ and let $\mathscr{L}(\mathscr{X})$ be the operator algebra $\mathscr{L}(\mathscr{X}, \mathscr{X})$. We recall that if $\mathscr{X}$ and $\mathscr{Y}$ are normed then by the continuity of A the *operator norm*

$$|A| = \operatorname*{supr}_{x \neq 0} \frac{\|Ax\|}{\|x\|} = \operatorname*{supr}_{\|x\| \leqslant 1} \|Ax\| = \operatorname*{supr}_{\|x\| = 1} \|Ax\|$$

exists and has the usual properties of a norm. Thus $\mathscr{L}(\mathscr{X}, \mathscr{Y})$ is a normed vector space which is seen to be complete if $\mathscr{Y}$ is a Banach space. We also have the relation $|AB| \leqslant |A| \cdot |B|$ and so $\mathscr{L}(\mathscr{X})$ is a Banach algebra with respect to the operator norm and it has identity $I: \mathscr{X} \to \mathscr{X}$. By an operator algebra we mean any closed subalgebra of $\mathscr{L}(\mathscr{X})$ where $\mathscr{X}$ is a Banach space.

The topology induced on $\mathscr{L}(\mathscr{X}, \mathscr{Y})$ by the operator norm is often called the *uniform topology* of $\mathscr{L}(\mathscr{X}, \mathscr{Y})$. We shall denote this topology by $\mathscr{T}_u$. A base for the neighborhood filter of the zero operator $0: \mathscr{X} \to \mathscr{Y}$ consists of the open sets

$$O(\varepsilon) = O_0(\varepsilon) = \{A: |A| < \varepsilon\}.$$

In addition to the operator norm topology several other important topologies can be introduced on $\mathscr{L}(\mathscr{X}, \mathscr{Y})$, some of which are discussed here. We fix a vector $x \in \mathscr{X}$ and consider the map $\mathscr{L}(\mathscr{X}, \mathscr{Y}) \to \mathscr{Y}$ defined by $A \to Ax$. We topologize $\mathscr{Y}$ by its norm and we take the weakest topology on $\mathscr{L}(\mathscr{X}, \mathscr{Y})$ which makes these maps $A \to Ax$ $(x \in \mathscr{X})$ continuous. A basis for the neighborhood filter of the zero operator 0 consists of the open sets

$$O(\varepsilon; x_1, \ldots, x_n) = \{A: \|Ax_k\| < \varepsilon \text{ for } k = 1, \ldots, n\}$$

where $\varepsilon > 0$, $n = 1, 2, \ldots$ and the vectors $x_1, \ldots, x_n$ are arbitrary. This is called the *strong topology* of $\mathscr{L}(\mathscr{X}, \mathscr{Y})$ and it will be denoted by $\mathscr{T}_s$.

The topology is meaningful also in the more general case when $\mathscr{X}$ and $\mathscr{Y}$ are topological vector spaces.

The *weak topology* $\mathscr{T}_w$ of $\mathscr{L}(\mathscr{X},\mathscr{Y})$ is defined as follows: We fix a continuous linear functional $y^*\in\mathscr{Y}^*$ and a vector $x\in\mathscr{X}$ and consider the map $\mathscr{L}(\mathscr{X},\mathscr{Y})\to\mathbb{C}$ given by $A\to y^*(Ax)$. The weakest topology on $\mathscr{L}(\mathscr{X},\mathscr{Y})$ which makes all these functions continuous is the weak topology $\mathscr{T}_w$. A basis for the neighborhood filter of the zero operator consists of the following kind of open sets: Choose $\varepsilon>0$ and $x_1,\dots,x_n\in\mathscr{X}$ and $y_1^*,\dots,y_n^*\in\mathscr{Y}^*$ where $n=1,2,\dots$ is arbitrary and let

$$O(\varepsilon;x_1,\dots,x_n;y_1^*,\dots,y_n^*)=\{A:|y_k^*(Ax_k)|<\varepsilon \text{ for } k=1,\dots,n\}.$$

Again the topology is meaningful also when $\mathscr{X}$ and $\mathscr{Y}$ are arbitrary, not necessarily normed topological vector spaces.

There are also *ultrastrong* or *strongest* and *ultraweak* or *weakest* topologies on $\mathscr{L}(\mathscr{X},\mathscr{Y})$. In the first a base for the neighborhood filter of 0 consists of the sets

$$O(\varepsilon;x_1,x_2,\dots)=:\{A:\sum\|Ax_k\|^2<\varepsilon^2\}$$

where $\varepsilon>0$ is arbitrary and $x_1,x_2,\dots$ are chosen in $\mathscr{X}$ such that $\sum\|x_k\|^2<\infty$. In the ultraweak topology we consider again an $\varepsilon>0$, vectors $x_1,x_2,\dots\in\mathscr{X}$ such that $\sum\|x_k\|^2<\infty$ and in addition linear functionals $y_1^*,y_2^*,\dots$ such that $\sum\|y_k^*\|^2<\infty$. Then we let

$$O(\varepsilon;x_1,x_2,\dots;y_1^*,y_2^*,\dots)=\{A:\sum|y_k^*(Ax_k)|<\varepsilon^2\}.$$

We have $\mathscr{T}_w\leqslant\mathscr{T}_s\leqslant\mathscr{T}_u$ because

$$O\left(\frac{\varepsilon}{\max\|x_k\|}\right)\subseteq O(\varepsilon;x_1,\dots,x_n)$$

and also

$$O\left(\frac{\varepsilon}{\max\|y_k^*\|;x_1,\dots,x_n}\right)\subseteq O(\varepsilon;x_1,\dots,x_n;y_1^*,\dots,y_n^*).$$

Hence the families of open and closed sets are related in the same manner and so for any set $\mathscr{S}$ of continuous linear operators the various closures of $\mathscr{S}$ satisfy $\bar{\mathscr{S}}_w\supseteq\bar{\mathscr{S}}_s\supseteq\bar{\mathscr{S}}_u$. An immediate corollary is the following:

Proposition 1. *If $\mathscr{S}$ is weakly closed then it is both strongly and uniformly closed.*

In particular we see that every weakly closed subalgebra of $\mathscr{L}(\mathscr{X},\mathscr{Y})$ is an operator algebra.

Although we shall not need it we remark that $\mathscr{L}(\mathscr{X}, \mathscr{Y})$ is a locally convex topological vector space under any of the topologies $\mathscr{T}_w$, $\mathscr{T}_s$ or $\mathscr{T}_u$. Multiplication in $\mathscr{L}(\mathscr{X})$ is continuous only in the uniform topology, but as a function of one variable it is continuous also under $\mathscr{T}_s$ and $\mathscr{T}_w$. If $\mathscr{X}=\mathscr{Y}=\mathscr{H}$ is a Hilbert space then $A \mapsto A^*$ is a continuous map with respect to both $\mathscr{T}_u$ and $\mathscr{T}_w$.

Theorem 2. *Let $\mathscr{V}$ be a vector space over $\mathbb{R}$ and let $p: \mathscr{V} \to \mathbb{R}$ be such that $p(x+y) \leqslant p(x)+p(y)$ and $p(\alpha x)=\alpha p(x)$ for all x, y in $\mathscr{V}$ and $\alpha \geqslant 0$. Suppose that $f: \mathscr{W} \to \mathbb{R}$ is a linear functional on a subspace $\mathscr{W}$ of $\mathscr{V}$ and $f(x) \leqslant p(x)$ for every x in $\mathscr{W}$. Then f can be extended to a linear functional $f: \mathscr{V} \to \mathbb{R}$ such that $f(x) \leqslant p(x)$ for every x in $\mathscr{V}$.*

Note. A function p having the properties stated in the theorem is called a *gauge function* or a *gauge.*

Proof. Let $\mathscr{X}$ be a subspace of $\mathscr{V}$ containing $\mathscr{W}$ and let $f: \mathscr{X} \to \mathbb{R}$ be an extension of $f: \mathscr{W} \to \mathbb{R}$ to $\mathscr{X}$ such that $f(x) \leqslant p(x)$ for all $x \in \mathscr{X}$. We suppose that $\mathscr{X}$ is a proper subspace of $\mathscr{V}$ and we choose an arbitrary vector v in $\mathscr{V}$ such that $v \notin \mathscr{X}$. Let $\mathscr{Y}$ be the subspace spanned by $\mathscr{X}$ and y so that each $y \in \mathscr{Y}$ can be written in a unique way as $y=x+\lambda v$ where $x \in \mathscr{X}$ and $\lambda \in \mathbb{R}$. For every such y we define $f(y)=f(x)+\lambda\alpha$ where α is a fixed real number. This gives a linear extension of f to the enlarged domain $\mathscr{Y}$ for any choice of α. If $x_1, x_2 \in \mathscr{X}$ then

$$f(x_1)-f(x_2)=f(x_1-x_2) \leqslant p(x_1-x_2) \leqslant p(x_1+v)+p(-v-x_2)$$

and so we obtain $-p(-v-x_2)-f(x_2) \leqslant p(x_1+v)-f(x_1)$. Therefore α can be chosen such that

$$\operatorname{lub}\{-p(-v-x_2)-f(x_2): x_2 \in \mathscr{X}\} \leqslant \alpha \leqslant \operatorname{glb}\{p(x_1+v)-f(x_1): x_1 \in \mathscr{X}\}.$$

Then for any $\lambda>0$ we have $f(\lambda^{-1}x)+\alpha \leqslant p(\lambda^{-1}x+v)$ which gives $f(x)+\lambda\alpha \leqslant p(x+\lambda v)$ i.e. $f(y) \leqslant p(y)$. Similar reasoning proves this inequality also when $y=x+\lambda v$ with $\lambda<0$. We proved the following:

If $f: \mathscr{X} \to \mathbb{R}$ is an extension of $f: \mathscr{W} \to \mathbb{R}$ such that $f(x) \leqslant p(x)$ for every $x \in \mathscr{X}$ and $\mathscr{X}$ is not $\mathscr{V}$ then $f: \mathscr{X} \to \mathbb{R}$ has a proper extension $f: \mathscr{Y} \to \mathbb{R}$ such that $f(y) \leqslant p(y)$ for every $y \in \mathscr{Y}$.

We shall now apply Zorn's lemma: We consider the set $\mathscr{F}$ of all extensions $f: \mathscr{X} \to \mathbb{R}$ such that $f(x) \leqslant p(x)$ for all x in $\mathscr{X}$. We define $f_1 \leqslant f_2$ for two such extensions $f_i: \mathscr{X}_i \to \mathbb{R}$ $(i=1,2)$ if $\mathscr{X}_1 \subseteq \mathscr{X}_2$ and the restriction of f_2 to $\mathscr{X}_1$ is f_1. Every linearly ordered subset of $\mathscr{F}$ has an upper bound in $\mathscr{F}$ and so by Zorn's lemma there is an extension f in

$\mathscr{F}$ which is maximal with respect to the partial ordering $\leqslant$. By the foregoing statement given in italics this f is defined on $\mathscr{V}$.

Corollary 1. *Let $\mathscr{Y}$ be a linear manifold in the normed real vector space $\mathscr{X}$ and let $f: \mathscr{Y} \to \mathbb{R}$ be a continuous linear functional with norm*

$$\|f\| = \operatorname{supr}\{|f(y)|: y \in \mathscr{Y} \text{ and } \|y\| \leqslant 1\}.$$

Then f can be extended to a continuous linear functional $f: \mathscr{X} \to \mathbb{R}$ whose norm remains the original $\|f\|$.

In order to prove this corollary we let $p(x) = \|f\| \cdot \|x\|$ so that $p(-x) = p(x)$ for every $x \in \mathscr{X}$. Then $f(y) \leqslant \|f\| \cdot \|y\| = p(y)$ for all y in $\mathscr{Y}$. By the theorem f can be extended to $\mathscr{X}$ such that $f(x) \leqslant p(x)$ for every $x \in \mathscr{X}$. Then we also have $-f(x) = f(-x) \leqslant p(-x) = p(x)$ and so $|f(x)| \leqslant p(x) = \|f\| \cdot \|x\|$ for every $x \in \mathscr{X}$.

Corollary 2. *Let $\mathscr{V}$ be a real vector space, let $\|\cdot\|$ be a semi-norm on $\mathscr{V}$, let $\mathscr{W}$ be a subspace of $\mathscr{V}$ and v a vector in $\mathscr{V}$. We define*

$$\delta = \operatorname{glb}\{\|v + w\|: w \in \mathscr{W}\}$$

so that $\delta \geqslant 0$. Then there exists a linear functional $f: \mathscr{V} \to \mathbb{R}$ such that f is zero on $\mathscr{W}$, $f(v) = \delta$ and $f(x) \leqslant \|x\|$ for every $x \in \mathscr{V}$.

In order to derive this corollary we first define f on the subspace $\mathscr{Y}$ spanned by $\mathscr{W}$ and v. Every element y of $\mathscr{Y}$ can be written uniquely in the form $y = w + \lambda v$ and so we can define $f(y) = \lambda\delta$. Then f vanishes on $\mathscr{W}$ and $f(v) = \delta$. Furthermore by the definition of δ we have

$$f(w + \lambda v) \leqslant |\lambda| \cdot \|v + \lambda^{-1} w\| = \|\lambda v + w\|$$

i.e. $f(y) \leqslant \|y\|$ for every $y \in \mathscr{Y}$. Applying Theorem 2 we obtain the desired linear functional $f: \mathscr{V} \to \mathbb{R}$.

We note that Corollary 1 has an analogue for complex vector spaces and the combination of these two results is usually called the *Hahn-Banach theorem* although this name is frequently given also to Theorem 2.

Theorem 3. *Let $\mathscr{Y}$ be a linear manifold in the real or complex normed vector space $\mathscr{X}$ and let f be a continuous linear functional f on $\mathscr{Y}$ with norm $\|f\|$. Then f can be extended to a continuous linear functional f on $\mathscr{X}$ whose norm remains the original $\|f\|$.*

Proof. We consider only the complex case because the other case is already settled. Using $r(y) = \mathscr{Re}\, f(y)$ we can write $f(y) = r(y) - ir(iy)$ where $r: \mathscr{Y} \to \mathbb{R}$ is an additive functional such that $r(\lambda y) = \lambda r(y)$ for all $\lambda \in \mathbb{R}$ and $y \in \mathscr{Y}$. Since $|f(y)| \leqslant \|f\| \cdot \|y\|$ we have $|r(y)| \leqslant \|f\| \cdot \|y\|$

for every $y \in \mathscr{Y}$. Using Corollary 1 one can extend r to an additive functional $r: \mathscr{X} \to \mathbb{R}$ such that $r(\lambda x) = \lambda r(x)$ and $|r(x)| \leqslant \|f\| \cdot \|x\|$ for all $\lambda \in \mathbb{R}$ and $x \in \mathscr{X}$. Then we define the linear extension of f to $\mathscr{X}$ by letting $f(x) = r(x) - ir(ix)$ for $x \in \mathscr{X}$. Now if $x \in \mathscr{X}$ and $f(x) = e^{it}|f(x)|$ then $|f(x)| = e^{-it} f(x) = f(e^{-it} x) = r(e^{-it} x)$ because $|f(x)|$ is real. Thus we have $|f(x)| \leqslant \|f\| \cdot \|e^{-it} x\| = \|f\| \cdot \|x\|$. Hence the norm of the extended functional is $\|f\|$.

Proposition 4. *Let $\mathscr{V}$ be a real or complex normed vector space, let $\mathscr{W}$ be a subspace of $\mathscr{V}$ and let $v \in \mathscr{V}$ but $v \notin \mathscr{W}$. Then there is a continuous linear functional f on $\mathscr{V}$ such that f is identically zero on $\mathscr{W}$ and $f(v) \neq 0$.*

Note. The proposition can be extended to locally convex vector spaces but we need only the above special case.

Proof. Since $v \in \mathscr{W}$ there is an open neigborhood O_v of v such that $O_v \cap \mathscr{W} = \emptyset$ and so there is a $\delta > 0$ such that $\|v - w\| \geqslant \delta$ for every w in $\mathscr{W}$. Hence by Corollary 2 of Theorem 2 there is a linear functional f satisfying the requirements.

Corollary. *The points of a real or complex normed vector space $\mathscr{V}$ can be separated by continuous linear functionals i. e. if $v_1 \neq v_2$ then there is a continuous linear functional f on $\mathscr{V}$ such that $f(v_1) \neq f(v_2)$.*

Indeed since $\mathscr{V}$ is normed we can use the last proposition with $v = v_1 - v_2$ and $\mathscr{W} = 0$. We note that the corollary can also be extended to locally convex vector spaces $\mathscr{V}$.

Let $\mathscr{V}$ be a real or complex vector space and let $\mathscr{S}$ be a subset of $\mathscr{V}$. A point x is called an *internal point* of $\mathscr{S}$ if for every y in $\mathscr{V}$ there is an $\varepsilon(y) = \varepsilon$ such that $x + \lambda y \in \mathscr{S}$ for every scalar λ satisfying $|\lambda| < \varepsilon$.

Theorem 5. *Let $\mathscr{V}$ be a real or complex vector space and let C_1 and C_2 be disjoint convex subsets of $\mathscr{V}$ such that C_1 has an internal point. Then there is a non-zero linear functional f such that*

$$\text{glb}\{\mathscr{Re}\, f(x_1): x_1 \in C_1\} \geqslant \text{lub}\{\mathscr{Re}\, f(x_2): x \in C_2\}.$$

Note. One says that f *separates* C_1 *and* C_2.

Proof. First we suppose that $\mathscr{V}$ is a real vector space. If f separates C_1 and C_2 then f also separates $C_1 - v$ and $C_2 - v$ for every v in $\mathscr{V}$. Hence one can suppose that 0 is an internal point of C_1. Then for every x in C_2 the zero vector 0 is an internal point of the convex set $C = C_1 - C_2 + x$. Thus given x in C_2 we can define

$$p(v) = \text{glb}\{a: a > 0 \text{ and } a^{-1} v \in C\}$$

for every $v \in \mathscr{V}$. It is clear that $p(0)=0$, $p(v) \geqslant 0$ and $p(\alpha v)=\alpha p(v)$ for every $\alpha \geqslant 0$. We also have $p(v+w) \leqslant p(v)+p(w)$ for every $v, w \in \mathscr{V}$: Indeed given $\varepsilon>0$ we can choose a and b such that $a<p(v)+\varepsilon, b<p(w)+\varepsilon$ and $a^{-1} v$, $b^{-1} w \in C$. Since C is convex it follows that $v+w \in(a+b) C$ and so $p(v+w) \leqslant a+b \leqslant p(v)+p(w)+2 \varepsilon$. Since $C_1 \cap C_2=\emptyset$ we have $x \notin C$ and so by $0 \in C$ and the convexity of C we can have $a^{-1} x \in C$ with $a>0$ only if $a>1$. Therefore $p(x) \geqslant 1$. If $v \in C$ then clearly $p(v) \leqslant 1$.

We define the linear functional f on the subspace $\mathbb{R} x$ by $f(\lambda x)=\lambda p(x)$ where $\lambda \in \mathbb{R}$. By the Hahn-Banach theorem f can be extended to a linear functional on $\mathscr{V}$ such that $f(v) \leqslant p(v)$ for every $v \in \mathscr{V}$. We have $f(v) \leqslant p(v) \leqslant 1 \leqslant p(x)=f(x)$ for every v in C and so f separates C and $\{x\}$. From this one can easily prove that f separates C_1-C_2 and $\{0\}$. By a similar reasoning we see that C_1 and C_2 are separated by f.

We shall now suppose that $\mathscr{V}$ is a complex vector space. Regarding $\mathscr{V}$ as a vector space over $\mathbb{R}$ using the foregoing we can determine a non-zero functional $g: \mathscr{V} \rightarrow \mathbb{R}$ such that g is additive, $g(\lambda x)=\lambda g(x)$ for every $\lambda \in \mathbb{R}$ and

$$\operatorname{glb}\{g(x_1): x_1 \in C_1\} \geqslant \operatorname{lub}\{g(x_2): x \in C_2\}.$$

If we let $f(v)=g(v)-i g(i v)$ for all $v \in \mathscr{V}$ then f is a non-zero linear functional separating C_1 and C_2.

Theorem 6. *Let C_1 and C_2 be disjoint convex sets in the topological vector space $\mathscr{V}$. If one of them has an interior point then C_1 and C_2 can be separated by a non-zero continuous linear functional.*

Proof. By Theorem 5 there is a non-zero linear functional f separating C_1 and C_2. Hence it is sufficient to prove that f is continuous. If $\mathscr{V}$ is a complex vector space then $f(x)=\mathscr{Re} f(x)-i \mathscr{Re} f(i x)$ and so it is sufficient to investigate the continuity of $g=\mathscr{Re} f$. Let x be an interior point of C_1 and let O_0 be an open neighborhood of 0 such that $O_0=-O_0$ and $x+O_0 \subseteq C_1$. Since $g(O_0) \subseteq g(C_1)-g(x)$ where $\operatorname{glb} g(C_1)>-\infty$ by $O_0=-O_0$ there is an $a>0$ such that $g(O_0) \subseteq(-a, a)$. Hence we have $|g(v)| \leqslant \varepsilon$ for every v in $\varepsilon a^{-1} O_0$.

Proposition 7. *Let C_1 and C_2 be disjoint closed convex sets in the locally convex topological vector space $\mathscr{V}$ and let C_1 be compact. Then there is a continuous linear functional f on $\mathscr{V}$ such that*

$$\operatorname{glb}\{\mathscr{Re} f(x_1): x_1 \in C_1\}>\operatorname{lub}\{\mathscr{Re} f(x_2): x_2 \in C_2\}.$$

Proof. The set C_1-C_2 is convex and it is closed because C_1 is compact. Since $0 \notin C_1-C_2$ there is a convex neighborhood C_0 of 0 such that

$C_0 \cap (C_1 - C_2) = \emptyset$. By Theorem 6 there is a non-zero continuous linear functional f which separates C_0 and $C_1 - C_2$. We can find v in $\mathscr{V}$ such that $f(v)=1$ and so $f(\alpha v)=\alpha$ for every $\alpha>0$. If α is sufficiently small then $\alpha v \in C_0$. Using this value of α for any $x_1 \in C_1$ and $x_2 \in C_2$ we have the following:

$$\mathscr{Re}\, f(x_1) - \mathscr{Re}\, f(x_2) = \mathscr{Re}\, f(x_1 - x_2) \geqslant \mathrm{glb}\{\mathscr{Re}\, f(x) : x \in C_1 - C_2\}$$
$$\geqslant \mathrm{lub}\{\mathscr{Re}\, f(y) : y \in C_0\} \geqslant \alpha > 0 .$$

Hence $\mathscr{Re}\, f(x_1) \geqslant \mathscr{Re}\, f(x_2) + \alpha$ for every $x_i \in C_i (i=1,2)$.

On one occasion we shall use the Krein-Millman theorem which we are going to state and prove here. Let K be a convex subset of a vector space $\mathscr{V}$ over $\mathbb{R}$ or $\mathbb{C}$. A point x will be called an *extreme point* of K if $x \in K$ but x is not contained in any line segment belonging to K. This means that if $x = \lambda_1 x_1 + \lambda_2 x_2$ with $\lambda_1, \lambda_2 > 0$ and $\lambda_1 + \lambda_2 = 1$ then $x_1 \notin K$ or $x_2 \notin K$. In other words if $x_1, x_2 \in K$ and $x = \lambda_1 x_1 + \lambda_2 x_2$ with $\lambda_1, \lambda_2 \geqslant 0$ and $\lambda_1 + \lambda_2 = 1$ then $x = x_1$ or $x = x_2$. The closed convex hull of a set S lying in the topological vector space $\mathscr{V}$ is by definition the intersection of all closed convex sets containing S.

Theorem 8. *If K is a compact convex set in a locally convex topological vector space $\mathscr{V}$ then K is the closed convex hull of its extreme points.*

The proof is based on the concept of an *extremal subset* or *supporting set* S of a convex set K. This means a closed convex subset S of K such that if S contains an interior point of a line segment L belonging to K then S contains L. It is clear that a set S consisting entirely of extreme points of K is a supporting set. The intersection of a family of supporting sets of K is itself a supporting set of K.

Proof. First we prove that every non-void supporting set S of K contains a minimal non-void supporting set: We consider the set $\mathscr{S}$ of all non-void supporting sets contained in S and partially order $\mathscr{S}$ such that $S_1 \leqslant S_2$ if and only if $S_1 \supseteq S_2$. The intersection of a linearly ordered subset of $\mathscr{S}$ is a non-void supporting set of K by the compactness of K. Hence by Zorn's lemma $\mathscr{S}$ contains a minimal element.

Suppose that a supporting set S of K contains two distinct points, say x and y. Then there is a continuous linear functional f on $\mathscr{V}$ such that $\mathscr{Re}\, f(x) < \mathscr{Re}\, f(y)$. The set M of those points $x \in K$ at which $\mathscr{Re}\, f$ attains its maximum on K is a supporting set of K. Therefore $M \cap S$ is a supporting set of K which does not contain x. Hence a minimal supporting set consists of a single point which is therefore extremal. Thus our reasoning shows that every non-void supporting set of K contains at least one extreme point of K. Since K is a supporting set of itself we see that every non-void compact set in a locally convex vector space $\mathscr{V}$ contains extreme points.

Now we prove that K is spanned by its extreme points. Let f be any continuous linear functional on $\mathscr{V}$ and let m_K be the maximum of $\mathscr{Re}\, f$ on K. We have seen that $S=\{x: x\in K \text{ and } \mathscr{Re}\, f(x)=m_K\}$ is a non-void supporting set of K and so S contains an extreme point of K. Hence $m_K=m_E$ where m_E denotes the maximum of $\mathscr{Re}\, f$ on the set E of extreme points of K. Now let C denote the closed convex hull of E so that $C\subseteq K$. If $x\notin C$ then by Proposition 7 there is a continuous linear functional such that $\mathscr{Re}\, f(x)>m_C$. Thus using $m_C=m_K$ we see that $x\notin K$ and so $K\subseteq C$ and $K=C$.

2. Compact Operators

In what follows let $\mathscr{X}$, $\mathscr{Y}$, $\mathscr{Z}$ be Banach spaces, let $\mathscr{S}$ be some set in one of them and in particular let $\mathscr{B}$ denote the unit ball. The topologies we consider will be the uniform topologies of these spaces i.e. the topology determined by the norm. A set $\mathscr{S}$ is called *relatively compact* in $\mathscr{Y}$ if its uniform closure $\bar{\mathscr{S}}$ in $\mathscr{Y}$ is a compact space. If $\mathscr{S}$ is relatively compact in $\mathscr{Y}$ then it is bounded and so there is a positive λ such that $\|s\|\leqslant\lambda$ for every s in $\mathscr{S}$.

Definition 1. *A linear operator $T:\mathscr{X}\to\mathscr{Y}$ is compact if $T\mathscr{B}$ is relatively compact in $\mathscr{Y}$.*

In other words T is compact if the uniform closure in $\mathscr{Y}$ of the T-image of the unit ball $\mathscr{B}$ of $\mathscr{X}$ is compact. Every operator whose range $T\mathscr{X}$ is finite dimensional is necessarily compact.

Proposition 2. *The operator $T:\mathscr{X}\to\mathscr{Y}$ is compact if and only if $T\mathscr{S}$ is relatively compact in $\mathscr{Y}$ for every bounded set $\mathscr{S}$ of $\mathscr{X}$.*

Proof. If $\mathscr{S}$ is bounded then $\mathscr{S}\subseteq\lambda\mathscr{B}$ for a sufficiently large $\lambda>0$ and $\overline{T\lambda\mathscr{B}}=\lambda\overline{T\mathscr{B}}$ is compact if $\overline{T\mathscr{B}}$ is a compact set. Therefore $\overline{T\mathscr{S}}$ being a closed subset of $\overline{T\lambda\mathscr{B}}$ is also compact. Since $\mathscr{B}$ is a bounded set the converse statement is obvious.

Every compact operator T is continuous with respect to the norm topologies because $T\mathscr{B}$ being relatively compact is a bounded set.

Theorem 3. *The compact operators form a subspace of $\mathscr{L}(\mathscr{X},\mathscr{Y})$ in its uniform operator topology.*

Proof. If T is compact then clearly λT is compact for any scalar λ because $\lambda T\mathscr{B}=|\lambda|\,T\mathscr{B}$ and so $\overline{\lambda T\mathscr{B}}=|\lambda|\,\overline{T\mathscr{B}}$. Similarly if S and T are compact then so ist $S+T$ because

$$(S+T)\mathscr{B}\subseteq S\mathscr{B}+T\mathscr{B}\subseteq\overline{S\mathscr{B}}+\overline{T\mathscr{B}}$$

and the sum of two compact sets is compact. Thus the compact operators form a linear manifold in $\mathscr{L}(\mathscr{X},\mathscr{Y})$. In order to prove that this is a closed linear manifold it will be sufficient to show the following:

Proposition 4. *If T_n $(n=1,2,\ldots)$ are compact operators and if $T_n\to T$ in the uniform operator topology of $\mathscr{L}(\mathscr{X},\mathscr{Y})$ then T is compact; briefly $|T_n-T|\to 0$ implies that T is also compact.*

Proof. We recall that a set X is precompact with respect to a uniform structure $\mathscr{U}$ if for every U in $\mathscr{U}$ there exist $x_1,\ldots,x_n$ in X such that $U[x_1]\cup\cdots\cup U[x_n]=X$. In particular a metric space X is precompact if for each $\varepsilon>0$ there are points $x_1,\ldots,x_n$ in X such that the balls $\{x\colon d(x,x_n)<\varepsilon\}$ cover X. We also know that a space X is compact if and only if there exists a uniform structure $\mathscr{U}$ which is complete and precompact. Actually $\mathscr{U}$ is uniquely determined by the topology of the compact space X.

Now $\overline{T\mathscr{B}}$ being a closed set in a Banach space is complete. Therefore it is sufficient to show that $\overline{T\mathscr{B}}$ is precompact. Given $\varepsilon>0$ there is an m such that $|T_m-T|<\varepsilon/3$. By the compactness of $\overline{T_m\mathscr{B}}$ there are $x_1,\ldots,x_n$ in $\mathscr{B}$ such that the balls

$$\left\{y\colon \|y-T_m x_k\|<\frac{\varepsilon}{3}\right\}\qquad (k=1,\ldots,n)$$

cover $\overline{T_m\mathscr{B}}$. From this it follows that $\{y\colon \|y-T_m x_k\|<\varepsilon\}$ $(k=1,\ldots,n)$ is a cover of $\overline{T\mathscr{B}}$ and so $\overline{T\mathscr{B}}$ is precompact: Indeed if $z\in\overline{T\mathscr{B}}$ then for a suitable $x\in\mathscr{B}$ we have $\|z-Tx\|<\varepsilon/3$ and so by $|T_m-T|<\varepsilon/3$ we also have $\|z-T_m x\|<2\varepsilon/3$. Thus for some x_k we have $\|z-T_m x_k\|<\varepsilon$.

Theorem 5. *The compact operators form a closed, two sided ideal in $\mathscr{L}(\mathscr{X})$.*

In view of Theorem 3 it will be enough to prove the following:

Proposition 6. *Let $\mathscr{X}\xrightarrow{S}\mathscr{Y}\xrightarrow{T}\mathscr{Z}$ be linear operators. If S is continuous and T is compact, then TS is compact. If S is compact and T is continuous then TS is compact.*

Proof. First we have $(TS)\mathscr{B}\subseteq|S|\,T\mathscr{B}$, so $\overline{(TS)\mathscr{B}}\subseteq|S|\,\overline{T\mathscr{B}}$ which is compact by the compactness of T. Similarly in the second case $T\overline{(S\mathscr{B})}$

being the continuous image of a compact space is compact and so it is a closed subset of $\mathscr{Y}$. Thus by $TS\mathscr{B} \subseteq \overline{TS\mathscr{B}}$ we see that $\overline{TS\mathscr{B}}$ is compact.

We recall that if $\mathscr{H}$ and $\mathscr{K}$ are vector spaces over the same field and if the range of the linear operator $T: \mathscr{H} \to \mathscr{K}$ is finite dimensional then T is called an *operator of finite rank*. From here on let $\mathscr{H}$ and $\mathscr{K}$ be real or complex Hilbert spaces.

Theorem 7. *An operator $T: \mathscr{H} \to \mathscr{K}$ is compact if and only if there exists a directed set N and operators T_n $(n \in N)$ of finite rank such that $T_n \to T$ in norm.*

Note. The same result holds also for operators between Banach spaces. However the proof is more difficult.

Proof. Every operator of finite rank is bounded and consequently it is compact. Hence the fact that the limit of operators T_n having finite dimensional ranges is a compact operator follows immediately form Theorem 5. To prove the converse let N be the set of all those finite dimensional subspaces of $\mathscr{K}$ which are of the form $T\mathscr{X}_n$ for some subspace $\mathscr{X}_n$ of $\mathscr{H}$. We shall use the same symbol n for the index n and the subspace $T\mathscr{X}_n$. Then as $n \subseteq \mathscr{K}$ we see that $T^{-1}n^{\perp}$ is a subspace of $\mathscr{H}$. We let $|T|_n$ denote the norm of the restriction of T to $T^{-1}n^{\perp}$. We order N by inclusion; so $n_1 \leqslant n_2$ if $T\mathscr{X}_{n_1} \subseteq T\mathscr{X}_{n_2}$.

We prove that $|T|_n \to 0$ as $n \to \infty$: Otherwise there is an $\varepsilon > 0$ such that given any $m = T\mathscr{X}_m$ in N we have $|T|_n > \varepsilon$ for some $n = T\mathscr{X}_n$ satisfying $m < n$. Therefore there is an x in $T^{-1}n^{\perp}$ satisfying $\|x\| = 1$ and $\|Tx\| \geqslant \varepsilon$. Hence starting with $m_0 = T0 = 0$ we can construct a sequence $m_0, m_1, \ldots, m_k, \ldots$ such that

$$m_k = T\{x_0, \ldots, x_{k-1}\} \qquad (k = 1, 2, \ldots)$$

and x_k satisfies the conditions $Tx_k \in (m_k)^{\perp}$, $\|x_k\| = 1$ and $\|Tx_k\| \geqslant \varepsilon$. We see that $Tx_i \perp Tx_k$ for $i \neq k$ and so $\|Tx_i - Tx_k\| \geqslant \sqrt{2}\varepsilon$ if $i \neq k$. The negation of $|T|_n \to 0$ led to a sequence $Tx_0, Tx_1, \ldots, Tx_k, \ldots$ with $x_k \in \mathscr{B}$ which has no convergent subsequence. This contradicts the relative compactness of $T\mathscr{B}$.

Now we can define the operators $T_n: \mathscr{H} \to \mathscr{K}$ by letting $T_n = T$ on $\mathscr{X}_n$ and $T_n = 0$ on $T^{-1}n^{\perp}$: Indeed for any $x \in \mathscr{H}$ we have $Tx = a + b$ with $a \in n$ and $b \in n^{\perp}$, so $a = T\xi$ for some $\xi \in \mathscr{X}_n$ and $b = T(x - \xi)$ proving that $x - \xi \in T^{-1}n^{\perp}$. Hence $T_n: \mathscr{H} \to \mathscr{K}$ can be defined by letting $T_n x = a$ where $Tx = a + b$ with $a \in n$ and $b \in T(T^{-1}n^{\perp})$. The range of T_n is finite dimensional, namely $T_n\mathscr{H} = n$ and $(T - T_n)\mathscr{H} \subseteq T(T^{-1}n^{\perp})$, so $|T - T_n| = |T|_n \to 0$ as $n \to \infty$. Q.e.d.

Theorem 8. *The compact operators form a self-adjoint, closed, two sided ideal in* $\mathcal{L}(\mathcal{H})$.

By Theorem 5 it is sufficient to prove the following:

Proposition 9. *If* $T: \mathcal{H} \to \mathcal{H}$ *is compact then so is its adjoint* $T^*: \mathcal{H} \to \mathcal{H}$.

Proof. First we notice that if $T\mathcal{H}$ is finite dimensional then so is $T^*\mathcal{H}$: Let $T\mathcal{H} = \mathcal{K}$ and $\mathcal{H} = \mathcal{K} \oplus \mathcal{K}^{\perp}$; then by $T^*\mathcal{K}^{\perp} = 0$ we have $T^*\mathcal{H} = T^*\mathcal{K}$. Now if T is compact, then by Theorem 7 we have $T_n \to T$ where each T_n has finite dimensional range. Then the range of T_n^* is also finite dimensional and $T_n^* \to T^*$. Thus T^* is compact by Theorem 7.

3. The Spectral Theorem for Compact Operators

Let $\mathcal{H}$ be a complex Hilbert space and let $A: \mathcal{H} \to \mathcal{H}$ be a continuous linear operator. For any $\lambda \in \mathbb{C}$ we let $\mathcal{M}_\lambda = \{x: Ax = \lambda x\}$, so in general $\mathcal{M}_\lambda = 0$ and if $\mathcal{M}_\lambda \neq 0$ then it is the *characteristic subspace* belonging to the *characteristic value* λ. The set of these values is called the *point spectrum* of A and it is denoted by $\sigma_p(A)$. If A is self adjoint then every characteristic value is real because if $Ax = \lambda x$ with $x \neq 0$ then $\lambda(x,x) = (Ax, x) = (x, Ax) = \bar{\lambda}(x,x)$. Similarly we see that for self adjoint A we have $\mathcal{M}_{\lambda_1} \perp \mathcal{M}_{\lambda_2}$ for any $\lambda_1 \neq \lambda_2$: Indeed if $x_k \in \mathcal{M}_{\lambda_k}$ $(k=1,2)$ then $\lambda_1(x_1, x_2) = (Ax_1, x_2) = (x_1, Ax_2) = \lambda_2(x_1, x_2)$ and so $(x_1, x_2) = 0$.

Proposition 1. *If* $A: \mathcal{H} \to \mathcal{H}$ *is continuous and self adjoint then*

$$|A| = \operatorname*{supr}_{x \neq 0} \frac{|(Ax, x)|}{\|x\|^2}.$$

Proof. Introduce the notation $N = \operatorname*{supr}_{x \in \mathcal{B}} |(Ax, x)|$ where $\mathcal{B}$ is the unit ball of $\mathcal{H}$. Since A is self adjoint for any $x \in \mathcal{H}$ and real $\lambda \neq 0$ we have

$$\begin{aligned}\|Ax\|^2 = \frac{1}{4}\Big\{&\Big(A\Big(\lambda x + \frac{1}{\lambda}Ax\Big), \lambda x + \frac{1}{\lambda}Ax\Big)\\ &-\Big(A\Big(\lambda x - \frac{1}{\lambda}Ax\Big), \lambda x - \frac{1}{\lambda}Ax\Big)\Big\}.\end{aligned}$$

Therefore

$$\begin{aligned}\|Ax\|^2 &\leqslant \frac{N}{4}\Big\{\Big\|\lambda x + \frac{1}{\lambda}Ax\Big\|^2 + \Big\|\lambda x - \frac{1}{\lambda}Ax\Big\|^2\Big\}\\ &= \frac{N}{2}\Big\{\lambda^2\|x\|^2 + \frac{1}{\lambda^2}\|Ax\|^2\Big\}.\end{aligned}$$

If $x \neq 0$ then by choosing $\lambda^2 = \|Ax\|/\|x\|$ we get $\|Ax\|^2 \leqslant N\|Ax\| \cdot \|x\|$, and so $|A| \leqslant N$. The opposite inequality is obvious.

Proposition 2. *If $A: \mathscr{H} \to \mathscr{H}$ is compact and self adjoint then there exist a $\lambda \in \mathbb{R}$ and a non-zero x in $\mathscr{H}$ such that $|\lambda| = |A|$ and $Ax = \lambda x$.*

Corollary. *There is an x with $\|x\| = 1$ such that $\|Ax\| = |A|$ and not only $|A|_{sp} = |A|$ but actually there is a $\lambda \in \sigma_p(A)$ such that $|\lambda| = |A|_{sp} = |A|$.*

Proof. By Proposition 1 and by the compactness of A we can find $x_1, x_2, \ldots, x_n, \ldots$ in $\mathscr{H}$ such that $\|x_n\| = 1$, the sequence $Ax_1, Ax_2, \ldots$ is convergent, $\lambda = \lim(Ax_n, x_n)$ exists and either $\lambda = |A|$ or $\lambda = -|A|$. Then we have

$$0 \leqslant \|Ax_n - \lambda x_n\|^2 = \|Ax_n\|^2 + \lambda^2 - 2\lambda(Ax_n, x_n) \leqslant |A|^2 + \lambda^2 - 2\lambda(Ax_n, x_n).$$

Since the right hand side approaches 0 we see that $\|Ax_n - \lambda x_n\| \to 0$ and so by the convergence of $Ax_1, Ax_2, \ldots$ the sequence $x_1, x_2, \ldots$ has limit x. By the continuity of A we get $Ax_n \to Ax$ and so $Ax = \lambda x$ where $|\lambda| = |A|$.

Proposition 3. *For any compact and self adjoint $A: \mathscr{H} \to \mathscr{H}$ there exist a sequence of characteristic values $\lambda_1, \lambda_2, \ldots$ and corresponding characteristic vectors $x_1, x_2, \ldots$ such that we have:*

1) $|\lambda_1| \geqslant |\lambda_2| \geqslant \cdots \geqslant |\lambda_k| \geqslant \cdots > 0$

and if the sequence is infinite then $|\lambda_k| \to 0$.

2) $\|x_k\| = 1$ *for every k and $x_k \perp x_l$ for $k \neq l$.*

3) *For every $x \in \mathscr{H}$ we have*

$$Ax = \sum \lambda_k(x, x_k)x_k = \sum (Ax, x_k)x_k.$$

Proof. We construct the sequences by induction using Proposition 2: Suppose that $x_1, \ldots, x_{n-1}$ are given such that $x_k \in \{0, x_1, \ldots, x_{k-1}\}^\perp = \mathscr{H}_k$ for every $k = 1, 2, \ldots, n-1$ and $Ax_k = \lambda_k x_k$ where $|\lambda_k| = |A_k|$ and $A_k = A | \mathscr{H}_k$ denotes the restriction of A to $\mathscr{H}_k$. Then we let $\mathscr{H}_n = \{0, x_1, \ldots, x_{n-1}\}^\perp$ and notice that $A: \mathscr{H}_n \to \mathscr{H}_n$, because if $x \in \mathscr{H}_n$ then we have

$$(Ax, x_k) = (x, Ax_k) = \lambda_k(x, x_k) = 0 \qquad (k = 1, \ldots, n-1).$$

We let $A_n = A | \mathscr{H}_n$ so that $A_n = A_n^*$ and A_n is compact. If $A_n = 0$ we stop; otherwise by Proposition 2 there is a $\lambda_n \in \mathbb{R}$ and $x_n \in \mathscr{H}_n$ such that $Ax_n = \lambda_n x_n$ and $\|x_n\| = 1$ and $|\lambda_n| = |A_n| > 0$.

Now it is clear that $|\lambda_1| \geqslant |\lambda_2| \geqslant \cdots$ and we can prove that if the sequence is not finite then $|\lambda_n| \to 0$: For the operator A being compact $Ax_1, Ax_2, \ldots$ contains a convergent subsequence and if $k \leqslant l$ then

$$\|Ax_k - Ax_l\|^2 = \|\lambda_k x_k - \lambda_l x_l\|^2 = \lambda_k^2 + \lambda_l^2 > \lambda_k^2 \geqslant \lambda_{k+1}^2 \geqslant \cdots.$$

In order to prove 3) let $x \in \mathscr{H}$ be arbitrary and let

$$y_n = x - \sum_{k=1}^{n-1} (x, x_k) x_k,$$

so that $y_n \in \mathscr{H}_n$. If the construction terminates in n steps then $Ay_n = 0$ and $Ax = \sum \lambda_k (x, x_k) x_k$. If the construction does not terminate in finitely many steps then by $y_n \in \mathscr{H}_n$ we have

$$\|Ay_n\| \leqslant |A_n| \; \|y_n\| = |\lambda_n| \; \|y_n\| \leqslant |\lambda_n| \cdot \|x\| \to 0.$$

Therefore $Ay_n \to 0$, that is

$$Ax = \sum (x, x_k) Ax_k = \sum \lambda_k (x, x_k) x_k = \sum (Ax, x_k) x_k.$$

Theorem 4. *Let $A: \mathscr{H} \to \mathscr{H}$ be a self adjoint and compact operator. Then its point spectrum is countable and each characteristic value occurs with finite multiplicity, that is $\dim \mathscr{M}_\lambda < \infty$ for every λ in $\sigma_p(A)$. Furthermore the only possible accumulation point of $\sigma_p(A)$ is 0.*

If $\mu_1, \mu_2, \ldots$ is the sequence of non-zero, distinct characteristic values and if $\mathscr{M}_\lambda$ denotes the characteristic subspace of $\lambda \in \mathbb{C}$, then

$$\mathscr{H} = \mathscr{M}_0 \oplus \mathscr{M}_{\mu_1} \oplus \cdots \oplus \mathscr{M}_{\mu_n} \oplus \cdots.$$

For every $x \in \mathscr{H}$ we have

$$Ax = \sum \mu_n E(\mu_n) x$$

where $E(\mu_n)$ denotes the projection $\mathscr{H} \to \mathscr{M}_{\mu_n}$.

Proof. First we prove that a complex number $\lambda \neq 0$ belongs to $\sigma_p(A)$ only if $\lambda = \lambda_k$ for some λ_k occurring in Proposition 3: For, if $Ax = \lambda x$ then

$$\lambda x = \sum \lambda (x, x_i) x_i = Ax = \sum \lambda_k (x, x_k) x_k$$

where $\{x_i\}$ $(i \in \mathscr{I})$ denotes some extension of the sequence $x_1, x_2, \ldots$ to a MONS in $\mathscr{H}$. Hence $\lambda = \lambda_k$ for every k satisfying $(x, x_k) \neq 0$. As $|\lambda_k| \to 0$ each value of $|\lambda_k|$ appears only finitely many times in 1) of Proposition 3. This shows that $\sigma_p(A)$ is discrete.

Next we notice that $\mathscr{M}_\lambda$ is the linear span of those x_k's from Proposition 3 for which $\lambda_k=\lambda$: For this span is clearly a subspace of $\mathscr{M}_\lambda$ and if $x\in\mathscr{M}_\lambda$ then from the above identity we obtain

$$\lambda x=\sum_{\lambda_k=\lambda}\lambda_k(x,x_k)x_k\,.$$

Hence each $\mathscr{M}_\lambda$ is finite dimensional. Moreover, if $x\perp\mathscr{M}_\lambda$ for every λ in $\mathbb{C}$ then $x\perp x_k$ for every $k=1,2,\ldots$ and so by 3) we have $Ax=0$. This shows that $\mathscr{H}=\mathscr{M}_0\oplus\mathscr{M}_{\mu_1}\oplus\mathscr{M}_{\mu_2}\oplus\cdots$. The spectral decomposition formula $A=\sum\mu_n E(\mu_n)$ is the same as 3) because

$$\sum_{\lambda_k=\mu_n}\lambda_k(x,x_k)x_k=\mu_n\sum_{\lambda_k=\mu_n}(x,x_k)x_k=\mu_n E(\mu_n)x\,.$$

Note. The spectral decomposition formula can be stated in a slightly different form: Choose an orthogonal basis in each $\mathscr{M}_\lambda$ including $\mathscr{M}_0$ and arrange these basis elements in a single sequence $x_1,x_2,\ldots$ Then $Ax=\sum\lambda_k(x,x_k)x_k$ where λ_k denotes the characteristic value belonging to x_k. Here the x_k's form a MONS in $\mathscr{H}$.

4. Hilbert-Schmidt Operators

Here we summarize the basic facts about certain compact operators, called Hilbert-Schmidt operators in Hilbert spaces. We shall use the abbreviation HS for Hilbert-Schmidt and MONS for a maximal orthonormal system in Hilbert space.

Definition 1. *A linear operator* $T\colon\mathscr{H}\to\mathscr{K}$ *is called Hilbert-Schmidt, or HS, if for some MONS* $\{x_i\}$ $(i\in\mathscr{I})$ *in* $\mathscr{H}$ *one has* $\sum\|Tx_i\|^2<\infty$.

We notice that for any two MONS in $\mathscr{H}$ we have $\sum\|Tx_i\|^2=\sum\|Ty_j\|^2$: In fact, if $\{z_j\}$ $(j\in\mathscr{J})$ is a MONS in $\mathscr{K}$ then

$$\sum_i\|Tx_i\|^2=\sum_{i,j}|(Tx_i,z_j)|^2=\sum_{i,j}|(x_i,T^*z_j)|^2=\sum_j\|T^*z_j\|^2$$

and so applying this relation twice we get

$$\sum\|Tx_i\|^2=\sum\|T^*z_k\|^2=\sum\|Ty_j\|^2\,.$$

The *HS norm* or *double norm* of a HS operator T is defined to be $\|T\|=(\sum\|Tx_i\|^2)^{\frac{1}{2}}$ which is independent of the particular MONS used to define it. We also see that $\|T\|=\|T^*\|$.

The operator norm and the HS norm are related by the inequality $|T|\leqslant\|T\|$: For given $\varepsilon>0$ we can choose an $x\in\mathscr{H}$ such that $\|x\|=1$

and $\|Tx\| > |T| - \varepsilon$ and by Zorn's lemma we can join to x other elements to obtain a MONS in $\mathscr{H}$. Using this system containing x we have $\|T\|^2 \geqslant \|Tx\|^2 > |T|^2 - 2\varepsilon|T|$.

The triangle inequality $\|S+T\| \leqslant \|S\| + \|T\|$ can be proved by using the expression $\|T\|^2 = \sum |(Tx_i, y_j)|^2$ which holds for any two MONS $\{x_i\}$ $(i \in \mathscr{I})$ in $\mathscr{H}$ and $\{y_j\}$ $(j \in \mathscr{J})$ in $\mathscr{K}$. Indeed by Minkowski's inequality we have

$$\|S+T\| = \{\sum |(Sx_i, y_j) + (Tx_i, y_j)|^2\}^{\frac{1}{2}} \leqslant \{\sum |(Sx_i, y_j)|^2\}^{\frac{1}{2}} + \{\sum |(Tx_i, y_j)|^2\}^{\frac{1}{2}} = \|S\| + \|T\|.$$

Theorem 2. *The set of HS operators $T: \mathscr{H} \to \mathscr{H}$ is a two sided ideal in the algebra $\mathscr{L}(\mathscr{H})$ and $\|AT\|, \|TA\| \leqslant |A| \cdot \|T\|$ for every A in $\mathscr{L}(\mathscr{H})$ and T which is HS.*

Proof. We know that the HS operators form a linear submanifold in $\mathscr{L}(\mathscr{H})$. If $A \in \mathscr{L}(\mathscr{H})$ and T is HS then

$$\|AT\|^2 = \sum \|ATx_i\|^2 \leqslant |A|^2 \sum \|Tx_i\|^2 = |A|^2 \, \|T\|^2,$$

so AT is HS and $\|AT\| \leqslant |A| \cdot \|T\|$. Furthermore

$$\|TA\| = \|(TA)^*\| = \|A^* T^*\| \leqslant |A^*| \cdot \|T^*\| = |A| \cdot \|T\|,$$

so TA is also HS and $\|TA\| \leqslant |A| \cdot \|T\|$.

Theorem 3. *The HS operators $T: \mathscr{H} \to \mathscr{H}$ form a Banach algebra under the HS norm which has an identity if and only if the Hilbert space $\mathscr{H}$ is finite dimensional.*

Proof. If S and T are HS operators then by our previous results $\|ST\| \leqslant |S| \cdot \|T\| \leqslant \|S\| \cdot \|T\|$. Thus only the completeness has to be verified. Let $T_1, T_2, \ldots$ be a sequence of HS operators such that $\|T_m - T_n\| \to 0$ as $m, n \to \infty$. Then $|T_m - T_n| \to 0$ and so there is a continuous operator T such that $|T_m - T| \to 0$. Since we are dealing with a Cauchy sequence there is a constant k such that $\|T_n\| \leqslant k$ for every $n = 1, 2, \ldots$

We prove that T is HS and $\|T_m - T\| \to 0$. Let $\{x_i\}$ $(i \in \mathscr{I})$ be a MONS in $\mathscr{H}$ and let $\mathscr{J}$ be any finite subset of $\mathscr{I}$. Then as $T_n \to T$ in the strong operator topology

$$\sum \|Tx_j\|^2 = \lim_{n\to\infty} \sum_{j \in \mathscr{J}} \|T_n x_j\|^2 \leqslant k^2$$

and so T is HS and $\|T\| \leqslant k$. To prove that $\|T_m - T\| \to 0$ let $m(\varepsilon)$ be chosen such that $\|T_m - T_n\| < \varepsilon$ for all $m, n \geqslant m(\varepsilon)$. Then for $m \geqslant m(\varepsilon)$ we have

$$\sum_{j\in\mathscr{J}} \|(T_m - T)x_j\|^2 = \lim_{n\to\infty} \sum_{j\in\mathscr{J}} \|(T_m - T_n)x_j\|^2 \leqslant \limsup_{n\to\infty} \|T_m - T_n\|^2 \leqslant \varepsilon^2 .$$

Hence we obtain $\|T_m - T\|^2 \leqslant \varepsilon^2$ for all $m \geqslant m(\varepsilon)$ and so $\|T_m - T\| \to 0$ as $m \to \infty$. The identity operator I is HS if and only if $\mathscr{H}$ is finite dimensional.

We are now going to prove that in the normed vector space of HS operators $T: \mathscr{H}_1 \to \mathscr{H}_2$ an inner product (S, T) can be introduced such that for any triple of HS operators R, S, T we have $(RS, T) = (S, R^* T)$. Namely we prove that if S, T are HS operators and $\{x_i\}$ $(i \in \mathscr{I})$ is a MONS then the series $\sum (Sx_i, Tx_i)$ is absolutely convergent and its sum is independent of the MONS used. Thus we can define

$$(S, T) = \sum (Sx_i, Tx_i)$$

where $\{x_i\}$ $(i \in \mathscr{I})$ is any MONS in $\mathscr{H}$.

The absolute convergence of the series can be seen as follows: For any two MONS $\{x_i\}$ $(i \in \mathscr{I})$ in $\mathscr{H}_1$ and $\{z_k\}$ $(k \in \mathscr{K})$ in $\mathscr{H}_2$ we have

$$\begin{aligned} \sum |(Sx_i, Tx_i)| &= \sum_i \Big| \sum_k (Sx_i, z_k) \overline{(Tx_i, z_k)} \Big| \leqslant \sum_i \sum_k |(Sx_i, z_k)| \, |(Tx_i, z_k)| \\ &\leqslant \Big\{ \sum_{i,k} |(Sx_i, z_k)|^2 \Big\}^{\frac{1}{2}} \Big\{ \sum_{i,k} |(Tx_i, z_k)|^2 \Big\}^{\frac{1}{2}} = \|S\| \cdot \|T\| . \end{aligned}$$

The independence of the sum from the MONS is proved similarly: Let $\{x_i\}$ $(i \in \mathscr{I})$ and $\{y_j\}$ $(j \in \mathscr{J})$ be MONS in $\mathscr{H}_1$ and let $\{z_k\}$ $(k \in \mathscr{K})$ be a MONS in $\mathscr{H}_2$. By the absolute convergence we have

$$\begin{aligned} \sum (Sx_i, Tx_i) &= \sum_i \sum_k (Sx_i, z_k) \overline{(Tx_i, z_k)} = \sum_k \sum_i (T^* z_k, x_i) \overline{(S^* z_k, x_i)} \\ &= \sum_k (T^* z_k, S^* z_k) . \end{aligned}$$

Therefore using this identity also for the system $\{y_j\}$ $(j \in \mathscr{J})$ we obtain

$$\sum (Sx_i, Tx_i) = \sum (T^* z_k, S^* z_k) = \sum (Sy_j, Ty_j) .$$

The relation $(RS, T) = (S, R^* T)$ and the usual properties of an inner product are obvious from the definition. It is also clear that $(T, T) = \|T\|^2$.

Theorem 4. *Every HS operator is compact.*

Proof. By Theorem 2.7 is sufficient to show that for every HS operator T there exists a sequence $T_1, T_2, \ldots$ of operators of finite rank such that $\|T_n - T\| \to 0$. We fix a MONS $\{x_i\}$ $(i \in \mathscr{I})$ and for any given $n = 1, 2, \ldots$ choose a finite subset $\mathscr{I}_n$ of $\mathscr{I}$ such that

$$\sum_{i \notin \mathscr{I}_n} \|T x_i\|^2 < n^{-2}.$$

We define the operator T_n by letting $T_n x_i = T x_i$ for $i \in \mathscr{I}_n$ and $T_n x_i = 0$ for $i \notin \mathscr{I}_n$. Then the range of T_n is finite dimensional and we have

$$\|T_n - T\| = \sum_{i \notin \mathscr{I}_n} \|T x_i\|^2 < n^{-2}.$$

Thus $|T_n - T| \leqslant \|T_n - T\| < n^{-2}$ for every $n = 1, 2, \ldots$

A combination of Theorems 2, 3 and 4 shows that the HS operators form a self adjoint two sided ideal in the algebra of compact operators.

HS operators acting on the L^2-space $L^2(X) = L^2(X, \mu)$ of an arbitrary measure space (X, μ) can be characterized by the existence of a kernel. Namely, if $K \in L^2(X \times X)$ then one can easily see that

$$(A_K f, g) = \int_X \int_X K(x, y) f(y) \overline{g(x)} \, dy \, dx$$

exists for every $f, g \in L^2(X)$ and thus defines a linear operator $A_K \colon L^2(X) \to L^2(X)$ in the weak sense. The operator A_K turns out to be of the HS class and conversely, if A is a HS operator acting on $L^2(X)$ then one can find a unique kernel $K \in L^2(X \times X)$ such that $A = A_K$. One can prove also that

$$(A_K f)(x) = \int_X K(x, y) f(y) \, dy$$

for almost every x in X.

We give only a short outline of the proofs because later we shall need the generalizations of these results to certain operator valued functions which results will be proved in full detail. (See Sections VI.5, VI.6, VI.7 and in particular Propositions VI.6.10, VI.7.12 and Theorem VI.7.11.)

Let θ_i $(i \in \mathscr{I})$ be a MONS in $L^2(X)$. Then the functions $\phi_{ij}(x, y) = \theta_i(x) \overline{\theta_j(y)}$ form a MONS in $L^2(X \times X)$. The orthogonality of the system is obvious and the maximality can be proved by establishing the validity of Parseval's formula by using the maximality of θ_i $(i \in \mathscr{I})$:

$$\int \int |f(x, y)|^2 \, dx \, dy = \sum \int |g_i(y)|^2 \, dy$$

where $g_i(y) = \int f(x,y)\overline{\theta_i(x)}\,dx$. Here we have

$$\int |g_i(y)|^2\,dy = \sum_{j\in\mathscr{I}} |\int g_i(y)\,\theta_j(y)\,dy|^2$$

and so we get

$$\int\int |f(x,y)|^2\,dx\,dy = \sum |\int\int f(x,y)\,\overline{\phi_{ij}(x,y)}\,dx\,dy|^2.$$

Now if $K \in L^2(X\times X)$ then in the L^2-topology we have $K = \sum (K,\phi_{ij})\phi_{ij}$ where

$$(K,\phi_{ij}) = \int\int K(x,y)\overline{\theta_i(x)}\theta_j(y)\,dx\,dy = (A_K\theta_j,\theta_i).$$

Therefore by the completeness of ϕ_{ij} we have

$$\|K\|^2 = \sum |(K,\phi_{ij})|^2 = \sum |(A_K\theta_j,\theta_i)|^2 = \|A_K\|^2$$

and so $\|A_K\| < \infty$ proving that A_K belongs to the HS class.

Next let A be a HS operator acting on $L^2(X)$. In order to find a kernel K for A we define a linear functional $l: L^2(X\times X)\to\mathbb{C}$ by $l(\phi_{ij}) = (A\theta_i,\theta_j)$ $(i,j\in\mathscr{I})$. Then using the orthogonality of the ϕ_{ij}'s one verifies that $|l| = \|A\|$. Therefore by the Riesz representation theorem there is a K in $L^2(X\times X)$ such that $l(\phi) = (\phi, K)$ for every $\phi \in L^2(X\times X)$. In particular one shows that if $\phi = f(y)\overline{g(x)}$ then

$$l(\phi) = (Af,g) = \int K(x,y)\,f(y)\overline{g(x)}\,dx\,dy$$

so that K is the kernel of A.

An alternate process of determining the kernel K is as follows: Since A is a HS operator we have $\sum |(A\theta_j,\theta_i)|^2 < \infty$ and so $K = \sum (A\theta_j,\theta_i)\phi_{ij}$ exists by the completeness of $L^2(X\times X)$. The function K defines a HS operator A_K and for any f, g in $L^2(X)$ we have

$$(A_K f, g) = \int\int K(x,y)\,f(y)\overline{g(x)}\,dy\,dx.$$

In particular

$$(A_K\theta_j,\theta_i) = \int\int K(x,y)\overline{\phi_{ij}(x,y)}\,dy\,dx = (K,\phi_{ij}) = (A\theta_j,\theta_i).$$

Therefore $(A_K - A)\theta_j \perp \theta_i$ for every θ_i and θ_j. By the maximality of θ_i $(i\in\mathscr{I})$ we see that $A_K = A$ and K is the kernel of A.

Let $\mathscr{H}$ denote the n dimensional vector space over $\mathbb{C}$ with inner product $(x,y) = x_1\bar{y}_1 + \cdots + x_n\bar{y}_n$. Since $\mathscr{H}$ is finite dimensional every linear operator $A: \mathscr{H}\to\mathscr{H}$ is in the HS class and if (a_{ij}) $(i,j = 1,\ldots,n)$ denotes the matrix of A with respect to the basis $(1,0,\ldots,0),\ldots,$ $(0,\ldots,0,1)$ then

$\|A\| = \sum |a_{ij}|^2$. Moreover if A and B are such linear transformations with matrices (a_{ij}) and (b_{ij}) then their HS inner product is

$$(A,B) = \sum a_{ij}\overline{b_{ij}}.$$

One sees that $(A,B) = \operatorname{tr} AB^*$ where tr denotes the trace and B^* is the Hermitian adjoint of B.

5. Trace Class Operators

The purpose of this section is to collect the basic information on trace class operators in complex Hilbert spaces. Most of this material will be used later.

Definition 1. *A linear operator $T \in \mathscr{L}(\mathscr{H})$ is called a trace class operator if it is compact and if for every MONS x_i $(i \in \mathscr{I})$ we have $\sum |(Tx_i, x_i)| < \infty$.*

It is obvious that the trace class, or briefly TC operators form a self adjoint vector space.

Simple examples are the operators $x \otimes y$ $(x, y \in \mathscr{H})$ defined by $z \mapsto (x \otimes y)z = (z,y)x$. One sees immediately that $x \otimes y$ is compact, $|x \otimes y| = \|x\| \cdot \|y\|$ and $(x \otimes y)^* = y \otimes x$. Moreover

$$((x \otimes y)x_i, x_i) = ((x_i, y)x, x_i) = (x_i, y)(x, x_i)$$

so $x \otimes y$ is in the trace class and $\operatorname{tr}(x \otimes y) = (x,y)$ will follow from the definition of the trace. One also sees that the HS norm is $\|x \otimes y\| = \|x\| \cdot \|y\|$.

We note that the symbol $x \otimes y$ will also be used to denote another concept, namely the tensor product of x with y which can also be interpreted as an operator $x \otimes y \colon \mathscr{H} \to \mathscr{H}$. However it is defined by $z \mapsto (x \otimes y)z = (y,z)x$ and so it is a *conjugate linear* or *anti-linear* operator. In general $T \colon \mathscr{X} \to \mathscr{Y}$ is called conjugate linear if it is additive and if $T\lambda x = \bar{\lambda} Tx$ for every λ in $\mathbb{C}$. The trace class and the HS class are meaningful concepts also for these anti-linear operators.

Theorem 2. *If R, S are HS operators in $\mathscr{L}(\mathscr{H})$ then RS is in the trace class. If T is a self adjoint trace class operator then $T = RS$ where R, S are HS operators.*

Proof. If $T = RS$ with R, S in HS then T being HS by Theorem 4.4 it is compact. Moreover for any MONS x_i $(i \in \mathscr{I})$ we have

$$\sum |(Sx_i, R^* x_i)| \leqslant \|S\| \cdot \|R\|$$

so T belongs to TC. Conversely, if $T \in TC$ and $T = T^*$ then by Proposition 3.3

$$Tx = \sum_{k=1}^{\infty} \lambda_k (x, x_k) x_k$$

for any $x \in \mathscr{H}$ where $x_1, x_2, \ldots$ denotes an ONS of characteristic vectors of T. Since $T \in TC$ we obtain by extending $x_1, x_2, \ldots$ to a MONS x_i $(i \in \mathscr{I})$

$$\sum_{k=1}^{\infty} |\lambda_k| = \sum_{k=1}^{\infty} |(T x_k, x_k)| \leqslant \sum_{i \in \mathscr{I}} |(T x_i, x_i)| < \infty .$$

Now we define the linear operators R and S by

$$Rx = \sum_{k=1}^{\infty} \sqrt{|\lambda_k|}\,(x, x_k) x_k \quad \text{and} \quad Sx = \sum_{k=1}^{\infty} \sqrt{|\lambda_k|}\,(\operatorname{sgn} \lambda_k)(x, x_k) x_k .$$

Then using the same MONS we see that

$$\|R\|^2 = \sum_{i \in \mathscr{I}} \|R x_i\|^2 = \sum_{k=1}^{\infty} \|R x_k\|^2 = \sum_{k=1}^{\infty} |\lambda_k| < \infty .$$

Similarly we show that $\|S\| < \infty$ and so R and S are HS operators. For any x_k $(k = 1, 2, \ldots)$ we have

$$(RS) x_k = R(\sqrt{|\lambda_k|}\,(\operatorname{sgn} \lambda_k) x_k) = \lambda_k x_k = T x_k .$$

For the remaining x_i's we have again $RSx_i = R0 = 0 = Tx_i$. Therefore we see that $T = RS$.

Proposition 3. *For any trace class operator T the sum $\sum (Tx_i, x_i)$ is independent of the MONS x_i $(i \in \mathscr{I})$.*

Proof. If $T = T^*$ then by Theorem 2 we have $T = RS$ where R, S are HS operators and so

$$\sum (T x_i, x_i) = \sum (S x_i, R^* x_i) = (S, R^*).$$

Here the right hand side is independent of x_i $(i \in \mathscr{I})$. If T is an arbitrary TC operator then

$$\sum (T x_i, x_i) = \sum \left(\frac{1}{2} (T + T^*) x_i, x_i \right) + i \sum \left(\frac{1}{2i} (T - T^*) x_i, x_i \right).$$

Since $\frac{1}{2}(T + T^*)$ and $1/2i(T - T^*)$ are self adjoint TC operators the right hand side is independent of the choice of the MONS x_i $(i \in \mathscr{I})$.

Definition 4. *For any TC operator T we define* $\operatorname{tr} T = \sum (Tx_i, x_i)$ *where the series is absolutely convergent and its sum is independent of the choice of the MONS used to compute it.*

The trace mapping $T \mapsto \operatorname{tr} T$ is a linear functional on the vector space of TC operators and the foregoing proof shows that the real and imaginary parts of $\operatorname{tr} T$ are $\operatorname{tr} \frac{1}{2}(T+T^*)$ and $\operatorname{tr} 1/2i(T-T^*)$, respectively. If T is self adjoint its trace is real and it is $\sum \lambda_k$ where λ_k $(k=1,2,\ldots)$ are the characteristic values of T each being counted with its finite multiplicity. We also see that $\operatorname{tr} T^* = \overline{\operatorname{tr} T}$. From the proof of Theorem 2 we can also see that if T is self adjoint and is in the trace class then $\sum |\lambda_k|$ is convergent.
The trace class and the trace are unitary invariants i.e. if $T_1: \mathscr{H}_1 \to \mathscr{H}_1$ is a trace class operator and $U: \mathscr{H}_1 \to \mathscr{H}_2$ is a unitary transformation then $T_2 = U T_1 U^{-1}$ is a trace class operator and $\operatorname{tr} T_1 = \operatorname{tr} T_2$.

Theorem 5. *The trace class operators form a two sided ideal both in the HS class and also in* $\mathscr{L}(\mathscr{H})$.

Proof. By applying Theorem 2 we see that any TC operator T can be written in the form $T = R_1 S_1 + R_2 S_2$ where R_1, S_1, R_2, S_2 are HS operators. Now we use Theorems 4.2 and 2 to show that AT and TA are again TC operators for any $A \in \mathscr{L}(\mathscr{H})$.

If instead of applying Theorem 2 we look at its proof we see that every TC operator T can be represented in the form

$$T = \left(+\sqrt{\frac{T+T^*}{2}}\right)\sqrt{\frac{T+T^*}{2}} + i\left(+\sqrt{\frac{T-T^*}{2i}}\right)\sqrt{\frac{T-T^*}{2i}}$$

where $+\sqrt{A}$ is the positive square root of the self adjoint, continuous operator A and the linear operator $\sqrt{A}$ is such that $(+\sqrt{A})(\sqrt{A}) = (\sqrt{A})(+\sqrt{A}) = A$.

Proposition 6. *If* $T: \mathscr{H} \to \mathscr{H}$ *is of finite rank then T belongs to the trace class. If* $T\mathscr{H} = \mathscr{K}$ *and if* $E: \mathscr{H} \to \mathscr{K}$ *is the projection onto* $\mathscr{K}$ *then* $\operatorname{tr} T = \operatorname{tr} ETE = \operatorname{tr} T|\mathscr{L}$ *for any subspace* $\mathscr{L}$ *containing* $\mathscr{K}$.

Proof. Let x_i $(i \in \mathscr{I})$ be a MONS in $\mathscr{H}$, let $y_1, \ldots, y_n$ be a MONS in $\mathscr{K}$ and let y_j $(j \in \mathscr{J})$ be can extension of y_j $(j \leqslant n)$ to a MONS of $\mathscr{H}$. For any $i \in \mathscr{I}$ we have

$$(Tx_i, x_i) = \sum_{j \leqslant n} (Tx_i, y_j)\overline{(x_i, y_j)}.$$

Therefore

$$\sum_{i \in \mathscr{I}} |(Tx_i, x_i)| \leqslant \sum_{j \leqslant n} \sum_{i \in \mathscr{I}} |(Tx_i, y_j)| \cdot |(x_i, y_j)|.$$

By the BCS inequality

$$\sum_{i\in\mathcal{I}} |(Tx_i,y_j)|\cdot|(x_i,y_j)| \leqslant \|T^* y_j\|\cdot\|y_j\|.$$

Therefore $\sum|(Tx_i,x_i)|$ is convergent and T belongs to TC. We compute the trace by using the MONS y_j $(j\in\mathcal{J})$ and obtain

$$\operatorname{tr} T=\sum(Ty_j,y_j)=\sum_{j\leqslant n}(Ty_j,y_j)=\operatorname{tr} T|\mathcal{K}.$$

Similarly if $\mathcal{L}\supseteq\mathcal{K}$ and y_j $(j\in\mathcal{J})$ is an extension of a MONS of $\mathcal{L}$ to $\mathcal{H}$ then we obtain $\operatorname{tr} T=\operatorname{tr} T|\mathcal{L}$.

Proposition 7. *If $\mathcal{H}$ is finite dimensional and $A,B\in\mathcal{L}(\mathcal{H})$ then* $\operatorname{tr} AB=\operatorname{tr} BA$.

Proof. Let $x_1,\dots,x_n$ be a MONS in $\mathcal{H}$. Since $\mathcal{H}$ is finite dimensional A and B are HS operators and so

$$\operatorname{tr} AB=\sum(ABx_i,x_i)=\sum(Bx_i,A^*x_i)=(B,A^*)$$

where (B,A^*) denotes the HS inner product of B and A^*. Hence it is sufficient to prove that $(A,B^*)=(B,A^*)$. By the finiteness of the dimension we have

$$\begin{aligned}(A^*,B)&=\sum_i(A^*x_i,Bx_i)=\sum_i\sum_j(A^*x_i,x_j)\overline{(Bx_i,x_j)}\\&=\sum_i\sum_j\overline{(Ax_j,x_i)}(B^*x_j,x_i)=\sum_j\overline{(Ax_j,B^*x_j)}=\overline{(A,B^*)}.\end{aligned}$$

Proposition 8. *If $T:\mathcal{H}\to\mathcal{H}$ is of finite rank and if $S\in\mathcal{L}(\mathcal{H})$ then* $\operatorname{tr} ST=\operatorname{tr} TS$.

Proof. Both ST and TS are of finite rank so by Proposition 6 their traces exist. Let $\mathcal{L}$ be the finite dimensional subspace spanned by $T\mathcal{H}$ and $ST\mathcal{H}$. Then $ST:\mathcal{L}\to\mathcal{L}$ and $TS:\mathcal{L}\to\mathcal{L}$ and so by Propositions 6 and 7 we have $\operatorname{tr} ST=\operatorname{tr} ST|\mathcal{L}=\operatorname{tr} TS|\mathcal{L}=\operatorname{tr} TS$.

The following lemma will be used later on a number of occasions without any further comment:

Lemma 9. *If $T:\mathcal{H}\to\mathcal{H}$ is a trace class operator and if $U:\mathcal{H}\to\mathcal{H}$ is unitary then UTU^* is in the trace class and* $\operatorname{tr} UTU^*=\operatorname{tr} T$.

Proof. If x_i $(i\in\mathcal{I})$ is a MONS in $\mathcal{H}$ then $\sum|(UTU^*x_i,x_i)|$ is finite because $(UTU^*x_i,x_i)=(TU^*x_i,U^*x_i)$ and U^*x_i $(i\in\mathcal{I})$ is also a MONS in $\mathcal{H}$. Hence UTU^* is in TC and $\operatorname{tr} UTU^*=\sum(TU^*x_i,U^*x_i)=\operatorname{tr} T$.

Proposition 10. *If A, B in $\mathscr{L}(\mathscr{H})$ are self adjoint and AB is in the trace class then* $\operatorname{tr} AB$ *is real,* BA *is in the trace class and* $\operatorname{tr} AB = \operatorname{tr} BA$.

Proof. For any MONS $x_i (i \in \mathscr{I})$ we have

$$\sum |(BAx_i, x_i)| = \sum |(x_i, BAx_i)| = \sum |(ABx_i, x_i)| < \infty$$

because $A = A^*$, $B = B^*$ and AB is in TC. Therefore

$$\operatorname{tr} AB = \sum (ABx_i, x_i) = \sum (x_i, BAx_i) = \overline{\operatorname{tr} BA}.$$

Finally

$$\operatorname{tr} AB = \sum (Bx_i, Ax_i) = \sum_i \sum_j (Bx_i, x_j)(x_j, Ax_i) = \sum_j \sum_i (Ax_j, x_i)(x_i, Bx_j)$$
$$= \sum (Ax_j, Bx_j) = \operatorname{tr} BA.$$

Therefore $\operatorname{tr} AB = \operatorname{tr} BA$ and so by $\operatorname{tr} AB = \overline{\operatorname{tr} BA}$ the trace is real.

Proposition 11. *If A, B in $\mathscr{L}(\mathscr{H})$ are Hilbert-Schmidt operators then* $(A^*, B^*) = (B, A)$.

Proof. Let $x_i (i \in \mathscr{I})$ be a MONS. Then we have

$$(A^*, B^*) = \sum (A^* x_i, B^* x_i) = \sum (A^* x_i, x_j)(x_j, B^* x_i) = \sum (Bx_j, x_i)(x_i, Ax_j)$$
$$= \sum (Bx_j, Ax_j) = (B, A).$$

Proposition 12. *If A and B are Hilbert-Schmidt operators in $\mathscr{L}(\mathscr{H})$ then* $\operatorname{tr} AB = \operatorname{tr} BA$.

Proof. By the reasoning given in the beginning of the proof of Theorem 2 we know that AB and BA are in the trace class. By Proposition 11 we have

$$\operatorname{tr} AB = (B, A^*) = (A, B^*) = \operatorname{tr} BA.$$

On several occasions we shall use the extensions of some of these results to the case when the Hilbert space $\mathscr{H}$ is replaced by a more general vector space $\mathscr{V}$.

Proposition 13. *If $\mathscr{V}$ is a finite dimensional vector space over some field F and A, B are linear transformations on $\mathscr{V}$ then* $\operatorname{tr} AB = \operatorname{tr} BA$.

Proof. Let $v_1, \ldots, v_n$ be an ordered basis of $\mathscr{V}$ and let (a_{ij}) and (b_{ij}) denote the matrices of A and B, respectively. Then

$$\operatorname{tr} AB = \operatorname{tr}(a_{ij})(b_{ij}) = \sum_{i=1}^{n} \sum_{j=1}^{n} a_{ij} b_{ji} = \sum_{j=1}^{n} \sum_{i=1}^{n} b_{ji} a_{ij} = \operatorname{tr}(b_{ij})(a_{ij}) = \operatorname{tr} BA.$$

Proposition 14. *Let $\mathscr{V}$ and $\mathscr{W}$ be finite dimensional vector spaces over the field F. Let $S: \mathscr{V} \to \mathscr{W}$ be an invertible linear transformation of $\mathscr{V}$ onto $\mathscr{W}$ and let $A: \mathscr{V} \to \mathscr{V}$. Then we have* $\operatorname{tr} S A S^{-1} = \operatorname{tr} A$.

Proof. Let $v_1, \ldots, v_n$ be a basis of $\mathscr{V}$ so that $S v_1, \ldots, S v_n$ is a basis of $\mathscr{W}$. Let (a_{ij}) be the matrix of A with respect to $v_1, \ldots, v_n$ i.e. let

$$(A v_1, \ldots, A v_n) = (v_1, \ldots, v_n)(a_{ij}).$$

Then we have

$$(S A S^{-1}(S v_1), \ldots, S A S^{-1}(S v_n)) = (S v_1, \ldots, S v_n)(a_{ij})$$

and so the matrix of $S A S^{-1}$ with respect to $S v_1, \ldots, S v_n$ is also (a_{ij}).

Proposition 15. *Let $\mathscr{V}$ be a vector space over some field F and let $A: \mathscr{V} \to \mathscr{V}$ be an operator of finite rank. Then its range $A\mathscr{V}$ is a finite dimensional A stable subspace of $\mathscr{V}$ and so we can define* $\operatorname{tr} A = \operatorname{tr}(A|A\mathscr{V})$. *Then we have:*

1. *If $\mathscr{W}$ is an A stable subspace of $\mathscr{V}$ such that $\mathscr{W} \supseteq A\mathscr{V}$ then* $\operatorname{tr}(A|\mathscr{W}) = \operatorname{tr} A$.

2. *If $\mathscr{W}$ is another vector space and $S: \mathscr{V} \to \mathscr{W}$ is an invertible linear transformation then $S A S^{-1}$ is of finite rank and* $\operatorname{tr} S A S^{-1} = \operatorname{tr} A$.

3. *If $A: \mathscr{V} \to \mathscr{V}$ and $B: \mathscr{V} \to \mathscr{V}$ are of finite rank then* $\operatorname{tr} A B = \operatorname{tr} B A$.

Proof. In order to prove 1. we choose a basis for $A\mathscr{V}$ and extend it to a basis of $\mathscr{W}$. The matrix of $A|\mathscr{W}$ with respect to this basis is $\begin{pmatrix} M & 0 \\ 0 & 0 \end{pmatrix}$ where M is the matrix of $A|A\mathscr{V}$. Hence $\operatorname{tr} A|\mathscr{W} = \operatorname{tr} A|\mathscr{V} = \operatorname{tr} A$.

Next we consider 2. The space $S A\mathscr{V}$ is the range of $S A S^{-1}$ and $S|A\mathscr{V}$ is an invertible linear map of $A\mathscr{V}$ onto $S A\mathscr{V}$. Hence we can apply Proposition 14 with $A\mathscr{V}$, $S A\mathscr{V}$, $S|A\mathscr{V}$ and $A|\mathscr{V}$ instead of $\mathscr{V}$, $\mathscr{W}$, S and A. We obtain $\operatorname{tr}(S|A\mathscr{V})(A|\mathscr{V})(S|A\mathscr{V})^{-1} = \operatorname{tr} A|\mathscr{V} = \operatorname{tr} A$.

Since the operator standing on the left hand side is $S A S^{-1}|S A\mathscr{V}$ we see that $\operatorname{tr} S A S^{-1} = \operatorname{tr} S A S^{-1}|S A\mathscr{V} = \operatorname{tr} A$.

Finally in order to prove 3. we let $\mathscr{W} = A\mathscr{V} + B\mathscr{V}$. Then $\mathscr{W}$ contains the range of $A B$ and $B A$, moreover $\mathscr{W}$ is stable under A, B, $A B$ and $B A$ because for instance $A B\mathscr{W} \subseteq A B\mathscr{V} \subseteq A\mathscr{V} \subseteq \mathscr{W}$. Hence by 1. we have

$$\operatorname{tr} A B = \operatorname{tr} A B|\mathscr{W} = \operatorname{tr}(A|\mathscr{W})(B|\mathscr{W}) = \operatorname{tr}(B|\mathscr{W})(A|\mathscr{W}) = \operatorname{tr} B A|\mathscr{W} = \operatorname{tr} B A.$$

Lemma 16. *Let $\mathscr{V}$ be a finite dimensional vector space over the field F of characteristic 0 and let $T: \mathscr{V} \to \mathscr{V}$ be a linear operator such that $\operatorname{tr} T^k = 0$ for every $k = 1, 2, \ldots$ Then T is nilpotent.*

Proof. We shall prove the lemma by induction on $n=\dim\mathscr{V}$. If $n=1$ then the matrix of T is 0 and so $T=0$. Let $n\geqslant 2$ and suppose that the lemma holds for vector spaces of dimension $n-1$. Let p be the minimal polynomial of T so that $p(T)=0$. By taking the trace of both sides of this equation we see that the constant term of p is 0. Hence 0 is a characteristic root of T. Since $\mathscr{V}$ is finite dimensional there is a corresponding characteristic vector, say v. Let $v_1,\ldots,v_n$ be any ordered basis of $\mathscr{V}$ such that $v_1=v$. Then the matrix of T with respect to this ordered basis is $M=\begin{pmatrix}0 & * \\ 0 & N\end{pmatrix}$ where N is an $(n-1)\times(n-1)$ matrix. Since $M^k=\begin{pmatrix}0 & * \\ 0 & N^k\end{pmatrix}$ we see that $\operatorname{tr} N^k=0$ for all $k=1,2,\ldots$ Hence by the induction hypothesis N is nilpotent and so $N^k=0$ for a large value of k. Thus M^k is a strictly upper triangular matrix and so M and T are nilpotent.

Lemma 17. *Let $\mathscr{V}$ be a finite dimensional vector space over a field of characteristic* 0 *and let S and T be linear transformations on $\mathscr{V}$ such that $ST-TS$ commutes with S. Then $ST-TS$ is nilpotent.*

Proof. Let $B=(ST-TS)^{k-1}T$ for some $k=1,2,\ldots$ Then

$$(ST-TS)^k=(ST-TS)^{k-1}(ST-TS)=(ST-TS)^{k-1}ST-BS$$

and so by the commutativity property $(ST-TS)^k=SB-BS$. Therefore by Proposition 13 we have $\operatorname{tr}(ST-TS)^k=0$ for every $k=1,2,\ldots$ Hence $ST-TS$ is nilpotent by Lemma 16.

6. Vector Valued Line Integrals

Here we describe a few basic facts about line integrals of vector valued functions. The reason to include this material is that it will be needed in the next section. First we discuss the meaning of $\int_C f(z)dz$ where C is an oriented rectifiable arc or an oriented simple closed curve and $f\colon C\to\mathscr{X}$ is a continuous function along C with values in a complex Banach space $\mathscr{X}$.

Let C be parametrized by $z\colon[0,1]\to C$ so that $z(0)=a$ and $z(1)=b$ are the end points of C, where it is possible that $a=b$. A sequence of points $z_0=z(0)$, $\zeta_1=z(t_1)$, $z_1=z(u_1)$, $\zeta_2=z(t_2)$, $z_2=z(u_2),\ldots,$ $\zeta_n=z(t_n)$, $z_n=z(1)$ with $0=u_0<t_1<u_1<\cdots<t_n<u_n=1$ will be called a partition of C. We let P denote the set of all partitions $p=(z_0,\zeta_1,z_1,\zeta_2,\ldots,\zeta_n,z_n)$. We make a directed set out of P by partially ordering it; we say that $p\leqslant p'=(z'_0,\zeta'_1,z'_1,\zeta'_2,\ldots,\zeta'_{n'},z'_{n'})$ provided

$$\max_{1\leqslant k\leqslant n}|z_k-z_{k-1}|\geqslant\max_{1\leqslant k\leqslant n'}|z'_k-z'_{k-1}|\,.$$

It is clear that $\leqslant$ is a non-strict partial ordering i.e. $p \leqslant p'$ and $p' \leqslant p$ does not necessarily imply that $p=p'$. We also see that the partition p and in general the structure of the directed set P is independent of the parametrization $z: [0,1] \to C$.

Given a Banach space valued continuous function $f: C \to \mathscr{X}$ we can construct a net with index set P by associating with each partition p the Riemann sum

$$S(f,p) = f(\zeta_1)(z_1 - z_0) + \cdots + f(\zeta_k)(z_k - z_{k-1}) + \cdots + f(\zeta_n)(z_n - z_{n-1}).$$

Using the uniform continuity of f we can prove that we have a Cauchy net: This can be done by comparing $S(f,p)$ and $S(f,p')$ with a third sum $S(f,p'')$ where $z_0'', z_1'', \ldots, z_{n''}''$ consist of the points $z_0, z_1, \ldots, z_n$ and z_0', $z_1', \ldots, z_{n'}'$ and $\zeta_1', \ldots, \zeta_{n''}'$ are chosen such that p'' is a partition. Since $\mathscr{X}$ is a complete space the Cauchy net $S(f,p)(p \in P)$ has a limit which is denoted by $\int_C f(z)\,dz$.

The fact that $\int_C f(z)\,dz$ is the limit of the Cauchy net $S(f,p)(p \in P)$ means the following: Given $\varepsilon > 0$ there is a $\delta = \delta(\varepsilon) > 0$ such that

$$\left| \sum_{k=1}^{n} f(\zeta_k)(z_k - z_{k-1}) - \int_C f(z)\,dz \right| < \varepsilon$$

for every partition $p = (z_0, \zeta_1, z_1, \ldots, \zeta_n, z_n)$ satisfying $\max\limits_{1 \leqslant k \leqslant n} |z_k - z_{k-1}| \leqslant \delta$.

If $\mathscr{X}$ is a reflexive Banach space it is possible to give a different definition of the line integral $\int_C f(z)\,dz$ by relying on the existence of ordinary line integrals of complex valued functions. For if $f: C \to \mathscr{X}$ and $x^* \in \mathscr{X}^*$ is a continuous linear functional then $x^* f: C \to \mathbb{C}$ is a continuous function and so $\ell(x^*) = \int_C x^*(f(z))\,dz$ exists. The linear functional $\ell: \mathscr{X}^* \to \mathbb{C}$ in continuous and clearly $|\ell| \leqslant l(C) \|f\|_\infty$ where $l(C)$ denotes the arc length of C and the supremum of f is taken on C. Therefore by the reflexivity of $\mathscr{X}$ there is an element in $\mathscr{X}$, which we call $\int_C f(z)\,dz$ such that $x^*\left(\int_C f(z)\,dz\right) = \ell(x^*)$ for every $x^* \in \mathscr{X}^*$. If the line integral is defined in this sense we speak about *integration in the weak sense* while our earlier definition gives the *integral* $\int_C f(z)\,dz$ *in the strong sense.* As far as values are concerned the two definitions yield the same integral.

The differentiability of a vector valued function can also be interpreted in various ways. Here we shall speak only about differentiability in the complex variable sense. Thus in order to attach a meaning to $f'(z)$ first of all it is required that f be defined in some neighborhood of z. Then we have a choice between *strong* and *weak* definitions of *differentiability*. In the first we ask that

$$\frac{1}{h}\{f(z+h)-f(z)\}\to f'(z)$$

as $h\to 0$ in the topology of the complex plane, while in the second we want that the complex valued x^*f be differentiable at z in the ordinary complex variable sense for every x^* in $\mathscr{X}^*$. A somewhat stronger requirement is that $\ell(x^*)=(x^*f)'(z)$ be a continuous functional on $\mathscr{X}^*$. If $\mathscr{X}$ is reflexive then differentiability in this sense leads to a derivative $f'(z)$ such that $\ell(x^*)=x^*(f'(z))$. Although it is known that these various differentiability concepts coincide we will not rely on this fact. If the concept of differentiability comes up in the conclusion of a theorem it will mean differentiability in the strong sense and if it appears as a hypothesis then it is sufficient to suppose only weak differentiability.

Proposition 1. *If $f: D\to\mathscr{X}$ is a Banach space valued function defined and differentiable in the open domain D lying in $\mathbb{C}$ and if C is an oriented simple closed curve contained in D, then $\int_C f(z)\,dz=0$.*

Proof. By hypothesis x^*f is a holomorphic complex valued function in D for every $x^*\in\mathscr{X}^*$. Therefore by the classical form of the proposition $\int_C x^*(f(z))\,dz=0$. Since the weak and strong definitions of the integral coincide we have $x^*\left(\int_C f(z)\,dz\right)=0$ for every x^* in $\mathscr{X}^*$. By the Hahn-Banach theorem the only element of $\mathscr{X}$ satisfying this condition is 0.

If we supposed strong differentiability throughout D then we could obtain a direct proof of the proposition by repeating the reasoning used in the theory of ordinary complex variables.

An immediate consequence of our proposition is the following:

Corollary. *If $\mathscr{X}$ is a complex Banach space, D is an open set in the complex plane and $f: D\to\mathscr{X}$ is differentiable then*

$$\int_{C_1} f(z)\,dz=\int_{C_2} f(z)\,dz$$

for any two homotopic oriented simple closed curves C_1 and C_2 in D.

Using the same reasoning as in the proof of Proposition 1 we can derive from the corresponding complex variable formula that

$$f(a)=\frac{1}{2\pi i}\int_C \frac{f(z)}{z-a}\,dz$$

provided a is an interior point of the counter-clockwise oriented simple closed curve C lying in the simply connected domain D of $f\colon D\to\mathscr{X}$. The well known formulas

$$f^{(n)}(a)=\frac{n!}{2\pi i}\int_C \frac{f(z)}{(z-a)^{n+1}}\,dz$$

could also be extended to Banach space valued functions f. None of this will be used in the rest of the book.

7. Homomorphisms into A. The Spectral Mapping Theorem

Let A be a Banach algebra, $\hat{A}$ its dual and $x\in A$. We let f denote a complex valued function defined on an open set D_f in the complex plane containing the spectrum $\sigma(x)$. We suppose that f is differentiable at every point of its domain of definition. Thus f is a piecewise holomorphic function on D_f. If the spectrum contains 0 we also require that $f(0)=0$. The object of this section is to prove the existence of another element y in A such that $\hat{y}=f\hat{x}$. This will show that the dual algebra A is closed under piecewise analytic mappings. If the algebra A has an identity then it is not necessary to require that f be 0 at $z=0$. We shall use the notation $y=f(x)$ where x is the given element of A and $f(x)$ denotes the element y corresponding to the function $f\colon D_f\to\mathbb{C}$.

If A is commutative then Theorem I.6.14 is applicable and we see that ignoring the possible characteristic value $\lambda=0$ the formula $\sigma(f(x))=f(\sigma(x))$ holds with $f(x)=y$. This can be considered as a generalization of the *spectral mapping theorem* discussed in Proposition I.4.9. We shall see that it holds also for non-commutative algebras and it is not necessary to treat the value $\lambda=0$ separately.

We note that in general the element y is not uniquely determined by the property $\hat{y}=f\hat{x}$ because if y is a solution then so is $y+ab-ba$ for any a,b in A. We shall however give a simultaneous construction of the elements y for a large algebra of piecewise analytic functions f and later prove that the set of y's obtained in this manner is a homomorphic image of this algebra in A.

Theorem 1. *Let $f: D_f \to \mathbb{C}$ be differentiable in the complex sense and let $\sigma(x) \subseteq D_f$. Let D be an open set contained in D_f and covering $\sigma(x)$ which is bounded by a sequence of rectifiable Jordan curves $\Gamma_0, \Gamma_1, \ldots$ such that $0 \notin D$ if $0 \notin \sigma(x)$ and $f(0) \neq 0$. Then for $\Gamma = \Gamma_0 \cup \Gamma_1 \cup \cdots$*

$$y = -\frac{1}{2\pi i} \int_\Gamma \frac{f(\lambda)}{\lambda} \left(\frac{x}{\lambda}\right)' d\lambda$$

exists, is independent of the choice of D and it is such that $\hat{y} = f\hat{x}$.

Note. If the Banach algebra A containing x has an identity then an alternate solution is

$$y = \frac{1}{2\pi i} \int_\Gamma f(\lambda)(\lambda e - x)^{-1} d\lambda.$$

Before entering into further discussion and starting the proof we mention the following example: If $|\lambda| > \|x\|$ then

$$\lambda^n (\lambda e - x)^{-1} = \lambda^{n-1} \left(e + \frac{x}{\lambda} + \frac{x^2}{\lambda^2} + \cdots\right).$$

Moreover if $\Gamma = \Gamma_0$ is a large circle centered around $\lambda = 0$ then

$$\frac{1}{2\pi i} \int_\Gamma \frac{d\lambda}{\lambda} = 1 \quad \text{and} \quad \int_\Gamma \lambda^k d\lambda = 0 \quad \text{for } k \neq -1, k \in \mathbb{Z}.$$

Thus we obtain

$$\frac{1}{2\pi i} \int_\Gamma \lambda^n (\lambda e - x)^{-1} d\lambda = x^n$$

for $n = 0, 1, 2, \ldots$ Therefore for any complex polynomial p we have

$$\frac{1}{2\pi i} \int_\Gamma p(\lambda)(\lambda e - x)^{-1} d\lambda = p(x).$$

Suppose that f_1 and f_2 both satisfy the requirements. Then one easily sees that Γ can be chosen such that it lies within the domain of both f_1 and f_2. From this it follows that $(f_1 + f_2)(x) = f_1(x) + f_2(x)$. Clearly the Banach algebra element y associated with the function λf is $\lambda f(x)$. We shall prove that in addition we have $(f_1 f_2)(x) = f_1(x) f_2(x)$. This indicates a certain homomorphism property between algebras which can be made precise by introducing the algebra of function germs:

Let C be a closed set in the complex plane $\mathbb{C}$ and let $f, g, \ldots$ be complex valued functions defined on various open neighborhoods $D_f, D_g, \ldots$ of C and such that $f, g, \ldots$ are differentiable at each point of their domains of definitions. We let $f \sim g$ if $f = g$ on $D_f \cap D_g$ and we let $\varphi, \psi, \ldots$ denote the equivalence classes obtained in this manner. Each φ is called a *germ of analytic functions* on C. Addition and multiplication of function germs can be defined by operating on representative functions so that for instance $\varphi\psi$ is the germ containing the functions fg where $f \in \varphi$ and $g \in \psi$. The set of germs endowed with these operations and with the obvious multiplication by complex scalars λ is called the algebra of holomorphic function germs and is denoted by $H(C)$. Since C and D need not be connected sets strictly speaking we are dealing with componentwise analytic function germs. Clearly we can speak about the value of a germ at some point z of C. If $0 \notin C$ we let $H_0(C)$ denote the algebra $H(C)$ and if $0 \in C$ then we let $H_0(C)$ be the subalgebra of function germs φ in $H(C)$ vanishing at 0.

Theorem 2. *Let A be a Banach algebra (with identity), let C be a closed set in $\mathbb{C}$ and let x in A be such that $\sigma(x) \subseteq C$. Then there exists a homomorphism $\tau: H_0(C) \to A$ such that $\widehat{\tau(\varphi)} = \varphi\hat{x}$ for every $\varphi \in H_0(C)$.*

We start the proof by first establishing Theorem 1 under the hypothesis that a system of Jordan curves Γ satisfying the requirements exist. Then we discuss how one constructs Γ when not only f but a pair of such functions f_1, f_2 is given. This will show that τ is a vector space homomorphism. An additional reasoning involving the choice of Γ and integration along these curves will show that $(f_1 f_2)(x) = f_1(x) f_2(x)$.

Proof of Theorem 1. Here we suppose the existence of a domain D bounded by a curve system $\Gamma = \Gamma_1 \cup \cdots \cup \Gamma_n$ such that $\sigma(x) \subset D$ and $\Gamma \subset r(x)$. Then $(x/\lambda)'$ exists and is continuous along Γ so the integrand is continuous along the compact domains of integration Γ_k. By the uniform continuity of the integrand the line integrals $\int_{\Gamma_1}, \int_{\Gamma_2}, \ldots$ exist.

If $\lambda \in \Gamma$ then by the existence of $(x/\lambda)'$ we have $h(x/\lambda) \neq 1$ i.e. $\hat{x}/\lambda(h) \neq 1$ for any $h \in \mathcal{M}$. (See Proposition I.2.7.) Now if $\xi + \xi' = \xi\xi'$ then $\hat{\xi} + \hat{\xi}' = \hat{\xi}\hat{\xi}'$ and so

$$\hat{\xi}'(h) = \frac{\hat{\xi}(h)}{\hat{\xi}(h) - 1} = 1 + \frac{1}{\hat{\xi}(h) - 1}$$

provided $\hat{\xi}(h) \neq 1$ which holds for $\xi = x/\lambda$ when $\lambda \in \Gamma$ by $\Gamma \subset r(x)$.

Using the weak definition of the integral we obtain

$$\hat{y}(h) = -\frac{1}{2\pi i}\int_{\Gamma} \frac{f(\lambda)}{\lambda}\,d\lambda + \frac{1}{2\pi i}\int_{\Gamma} \frac{f(\lambda)}{\lambda - \hat{x}(h)}\,d\lambda.$$

If $\hat{x}(h)=0$ then $\hat{y}(h)=0$ so $\hat{y}(h)=(f\hat{x})(h)$ by $f(0)=0$. If $\hat{x}(h)=h(x)\neq 0$ then by Proposition I.5.4 $h(x)\in\sigma(x)^{\times}$ where $\sigma(x)^{\times}$ is covered by the domain which is bounded by the curves $\Gamma_1, \Gamma_2, \ldots$ Therefore

$$\frac{1}{2\pi i}\int_{\Gamma} \frac{f(\lambda)}{\lambda - \hat{x}(h)}\,d\lambda = f(\hat{x}(h))$$

because $\hat{x}(h)$ lies in the interior of exactly one of the curves $\Gamma_1, \Gamma_2, \ldots$ Thus $\hat{y}=f\hat{x}$ will be proved by showing that $\int_{\Gamma}(f(\lambda)/\lambda)\,d\lambda=0$. But if $0\notin\sigma(x)$ and $f(0)\neq 0$ then 0 lies outside the curves $\Gamma_1, \Gamma_2, \ldots$ so the value of the integral is either $2\pi i f(0)=0$ or it is 0 by the path of integration excluding 0.

If D_1 and D_2 are domains both satisfying the requirements for some function f then $D=D_1\cap D_2$ is also such a domain. Moreover both D_1 and D_2 can be obtained from D by joining to it a sequence of simply connected domains and part of the boundaries of these. The fact that y is independent of the choice of the domain used to compute it can be seen by realizing that $\sigma(x)$ lies outside these simply connected domains and so integration of

$$\frac{f(\lambda)}{\lambda - \hat{x}(h)}$$

along their boundaries yields 0.

The additive nature of τ is now obvious because using the intersection of the domains of f_1 and f_2 we see that $(f_1+f_2)(x)=f_1(x)+f_2(x)$. To prove the corresponding multiplicative result we choose the domains D_1 and D_2 such that the boundaries of D_1 and D_2 are disjoint and $D_1\subset D_2\subseteq D_{f_1}\cap D_{f_2}$ where D_{f_i} is the domain of definition of f_i $(i=1,2)$. Let Γ_i be the boundary of D_i and let $\lambda_i\in\Gamma_i$ $(i=1,2)$. Then

$$(1) \qquad \frac{1}{2\pi i}\int_{\Gamma_2} \frac{f_2(\lambda_2)}{\lambda_2 - \lambda_1}\,d\lambda_2 = f_2(\lambda_1)$$

because λ_1 lies outside of all but one boundary component of Γ_2.

Similarly

$$(2) \qquad \frac{1}{2\pi i}\int_{\Gamma_1} \frac{f_1(\lambda_1)}{\lambda_1-\lambda_2}\,d\lambda_1 = 0$$

because λ_2 lies outside of all the boundary components of Γ_1 or inside precisely two of them oriented in opposite directions.

We now suppose that A has an identity element e so that we may write

$$\frac{1}{\lambda}\left(\frac{x}{\lambda}\right)' = x_\lambda + \frac{e}{\lambda}$$

where $x_\lambda = (x-\lambda e)^{-1}$. (See Section I.4 p. 11.) Since $\int_\Gamma (f(\lambda)/\lambda)\,d\lambda = 0$ we have

$$f(x) = -\frac{1}{2\pi i}\int_\Gamma \frac{f(\lambda)}{\lambda}\left(\frac{x}{\lambda}\right)' d\lambda = -\frac{1}{2\pi i}\int_\Gamma f(\lambda)\,x_\lambda\,d\lambda.$$

Therefore

$$\begin{aligned} f_1(x)\,f_2(x) &= -\frac{1}{4\pi^2}\int_{\Gamma_1} f_1(\lambda_1)\,x_{\lambda_1}\,d\lambda_1 \int_{\Gamma_2} f_2(\lambda_2)\,x_{\lambda_2}\,d\lambda_2 \\ &= -\frac{1}{4\pi^2}\int_{\Gamma_2}\int_{\Gamma_1} f_1(\lambda_1)\,f_2(\lambda_2)\,x_{\lambda_1}x_{\lambda_2}\,d\lambda_1\,d\lambda_2 . \end{aligned}$$

Applying the resolvent equation and splitting the resulting integral by (2) we obtain

$$\begin{aligned} f_1(x)\,f_2(x) &= \frac{1}{4\pi^2}\int_{\Gamma_2}\int_{\Gamma_1} f_1(\lambda_1)\,f_2(\lambda_2)\frac{x_{\lambda_1}}{\lambda_2-\lambda_1}\,d\lambda_1\,d\lambda_2 \\ &= -\frac{1}{4\pi^2}\int_{\Gamma_1} f_1(\lambda_1)\,x_{\lambda_1}\int_{\Gamma_2}\frac{f_2(\lambda_2)}{\lambda_2-\lambda_1}\,d\lambda_2\,d\lambda_1 . \end{aligned}$$

Thus by (1) we have

$$f_1(x)\,f_2(x) = \frac{1}{2\pi i}\int_{\Gamma_1} f_1(\lambda_1)\,f_2(\lambda_1)\,x_{\lambda_1}\,d\lambda_1 = (f_1 f_2)(x).$$

Since every Banach algebra A can be extended to another A_e having identity the theorem is proved in the general case too.

Having established Theorem 2 we can now prove the following general *spectral mapping theorem*:

Theorem 3. *Suppose that the Banach algebra A containing x has an identity. Let f be a piecewise analytic function defined in a neighborhood of* $\sigma(x)$ *and let* $f(x)=\tau(f)$ *where* $\tau\colon H_0(\sigma(x))\to A$. *Then* $\sigma(f(x))=f(\sigma(x))$.

Proof. Given $f\colon D_f\to\mathbb{C}$ define $g\colon D_f\to\mathbb{C}$ by

$$g(\mu)=\frac{f(\lambda)-f(\mu)}{\lambda-\mu}$$

where $\lambda\in\sigma(x)$ is fixed. Since the numerator vanishes with order at least 1 the function g is defined and is differentiable also at λ. Applying the homomorphism τ to the equation

$$(\lambda-\mu)g(\mu)=f(\lambda)-f(\mu)=g(\mu)(\lambda-\mu)$$

we obtain

$$(\lambda e-x)g(x)=f(\lambda)e-f(x)=g(x)(\lambda e-x).$$

This shows that if $(f(x)-f(\lambda)e)^{-1}$ exists then $x-\lambda e$ is invertible in contradiction to $\lambda\in\sigma(x)$. Therefore we have $f(\lambda)\in\sigma(f(x))$ for every $\lambda\in\sigma(x)$.

Next suppose that $\lambda\in\sigma(f(x))$ and $\lambda\notin f(\sigma(x))$. Let $h(\mu)=(f(\mu)-\lambda)^{-1}$ which is differentiable on D_f except at most at those μ for which $f(\mu)=\lambda$, but these $\mu\notin\sigma(x)$. Applying τ we obtain

$$h(x)(f(x)-\lambda e)=e=(f(x)-\lambda e)h(x)$$

in contradiction to $\lambda\in\sigma(f(x))$.

8. Unbounded Operators

Let $\mathscr{H}$ and $\mathscr{K}$ be real or complex Hilbert spaces, let $\mathscr{D}_T$ be a linear manifold in $\mathscr{H}$ and let $T\colon\mathscr{D}_T\to\mathscr{K}$ be a linear map. Then T is called a *linear operator* with domain of definition $\mathscr{D}_T$ and range $T\mathscr{D}_T$ lying in $\mathscr{K}$. It is possible that $\mathscr{D}_T$ consists of the single element 0 and we always have $T0=0$. If $\mathscr{D}_T$ is a dense subset of $\mathscr{H}$ then we say that T is a *densely defined operator*. We note that T can be everywhere defined without being continuous. Two linear operators S and T are called equal if $\mathscr{D}_S=\mathscr{D}_T$ and S and T have the same function values. If $\mathscr{D}_S\subseteq\mathscr{D}_T$ and $Sx=Tx$ for every x in $\mathscr{D}_S$ then we say that T is an extension of S. The *graph* of T is the set

$$\mathscr{G}_T=\{(x,Tx)\colon x\in\mathscr{D}_T\}.$$

The operator $T: \mathscr{D}_T \to \mathscr{K}$ is called *closed* if $\mathscr{G}_T$ is a closed subset of the product space $\mathscr{H} \times \mathscr{K}$. In order that T be a closed operator the following condition is necessary and sufficient: If $x_n \to x$ and $Tx_n \to y$ then $x \in \mathscr{D}_T$ and $Tx = y$. We note that the closure of $\mathscr{G}_T$ is not necessarily the graph of an operator. If $\overline{\mathscr{G}_T}$ is the graph of the extension $\bar{T}$ of T then T is called a *closable operator* and $\bar{T}$ is its *closure*. Some writers call T an operator only if it is closed and densely defined. Any other linear map $T: \mathscr{D}_T \to \mathscr{K}$ is then called a preoperator. We shall not follow this practice.

The *adjoint* T^* of a densely defined operator $T: \mathscr{D}_T \to \mathscr{K}$ is a linear map from $\mathscr{K}$ into $\mathscr{H}$. The vector y belongs to $\mathscr{D}_{T^*}$ if there is a vector called T^*y in $\mathscr{H}$ such that $(Tx, y) = (x, T^*y)$ for every x in $\mathscr{D}_T$. Since $\mathscr{D}_T$ is dense in $\mathscr{H}$ the vector T^*y is uniquely determined by T and y. We have $0 \in \mathscr{D}_{T^*}$ and it is possible that $\mathscr{D}_{T^*} = \{0\}$.

If T_1 and T_2 are operators with domains $\mathscr{D}_{T_1}$ and $\mathscr{D}_{T_2}$ lying in $\mathscr{H}$ then $T_1 + T_2$ is defined as a linear operator with domain $\mathscr{D}_{T_1+T_2} = \mathscr{D}_{T_1} \cap \mathscr{D}_{T_2}$. Similarly if $T: \mathscr{D}_T \to \mathscr{K}$ and $S: \mathscr{D}_S \to \mathscr{L}$ where $\mathscr{D}_T \subseteq \mathscr{H}$ and $\mathscr{D}_S \subseteq \mathscr{K}$ then $ST: \mathscr{D}_{ST} \to \mathscr{L}$ where $\mathscr{D}_{ST} = \mathscr{D}_S \cap T\mathscr{D}_T$. The meaning of the scalar multiple λT of $T: \mathscr{D}_T \to \mathscr{K}$ is selfexplanatory.

It is obvious that $(\lambda T)^* = \bar{\lambda} T^*$. It is also clear that if T is an extension of S then S^* is an extension of T^*. Similarly $(T_1 + T_2)^*$ is an extension of $T_1^* + T_2^*$ and $(ST)^*$ is an extension of $T^* S^*$.

Proposition 1. *The adjoint operator* $T^*: \mathscr{D}_{T^*} \to \mathscr{H}$ *is a closed linear operator.*

This result will become selfevident from the description of the graph of T^* which we are now going to determine: We define $U: \mathscr{H} \oplus \mathscr{K} \to \mathscr{K} \oplus \mathscr{H}$ by $U(x, y) = (iy, -ix)$ where $x \in \mathscr{H}$ and $y \in \mathscr{K}$. Similarly we define $V: \mathscr{K} \oplus \mathscr{H} \to \mathscr{H} \oplus \mathscr{K}$ by $V(y, x) = (ix, -iy)$ where $x \in \mathscr{H}$ and $y \in \mathscr{K}$. It is obvious that U and V are isometric linear bijections and $UV = VU = I$.

Proposition 2. *For any densely defined operator* $T: \mathscr{D}_T \to \mathscr{K}$ *we have* $\mathscr{G}_{T^*} = (U\mathscr{G}_T)^\perp$ *where* $\perp$ *denotes the orthogonal complement in* $\mathscr{K} \oplus \mathscr{H}$.

Proof. We have $(a, b) \in (U\mathscr{G}_T)^\perp$ if and only if $((a, b), (iTx, -ix)) = 0$ for every x in $\mathscr{D}_T$. Therefore (a, b) is in $(U\mathscr{G}_T)^\perp$ if and only if $(Tx, a) = (x, b)$ for all x in $\mathscr{D}_T$. This last statement holds if and only if $a \in \mathscr{D}_{T^*}$ and $T^* a = b$ which is the same as $(a, b) \in \mathscr{G}_{T^*}$.

Now let T be a linear operator in $\mathscr{H}$ so that $\mathscr{D}_T \subseteq \mathscr{H}$ and $T\mathscr{D}_T \subseteq \mathscr{H}$. If T is densely defined and $(Tx, y) = (x, Ty)$ for every x, y in $\mathscr{D}_T$ then T is called a *symmetric operator*. We say that T is *self adjoint* if $T = T^*$.

This can take place only if T is symmetric and closed. The operator T is called *positive* or *positive semi-definite* if it is self adjoint and $(Tx, x) \geqslant 0$ for every x belonging to $\mathscr{D}_T$.

Proposition 3. *If $T: \mathscr{D}_T \to \mathscr{K}$ is a closed, densely defined linear operator from $\mathscr{H}$ into $\mathscr{K}$ then $T^* T$ is densely defined and positive definite.*

Proof. Since T^* is a closed operator its graph $\mathscr{G}_{T^*}$ is a closed subset of $\mathscr{K} \oplus \mathscr{H}$. By Proposition 2 we have $U\mathscr{G}_T + \mathscr{G}_{T^*} = \mathscr{K} \oplus \mathscr{H}$. Thus if $x \in \mathscr{H}$ then there exist u in $\mathscr{D}_T$ and v in $\mathscr{D}_{T^*}$ such that $(0, -ix) = (iTu, -iu) + (v, T^* v)$. This means that $v = -iTu$ and $x = u + iT^* v$ and so $x = u + T^* Tu$. We proved that every x in $\mathscr{H}$ can be written in the form $x = u + T^* Tu$ where $u \in \mathscr{D}_{T^*T}$. We can now prove that $\mathscr{D}_{T^*T}$ is dense in $\mathscr{H}$. For given x in $(\mathscr{D}_{T^*T})^\perp$ we have $x = u + T^* Tu$ for a suitable u in $\mathscr{D}_{T^*T}$. Then by the orthogonality condition we have

$$(u + T^* Tu, u) = (u, u) + (Tu, Tu) = 0$$

and this shows that $u = 0$ and $x = 0$. Thus $T^* T$ is a densely defined operator. The fact that $T^* T$ is self adjoint and positive definite is obvious from the definition of T^*.

Proposition 4. *If T is a closable operator then $\bar{T}^* = T^*$ and $T^{**} = \bar{T}$.*

Proof. Since T is closable we have $\mathscr{G}_{\bar{T}} = \overline{\mathscr{G}_T}$ and so $U\mathscr{G}_{\bar{T}} = U\overline{\mathscr{G}_T} = \overline{U\mathscr{G}_T}$ by the continuity of U. Hence by Proposition 2 $\mathscr{G}_{\bar{T}^*} = (U\mathscr{G}_T)^\perp = \mathscr{G}_{T^*}$. This shows that $\bar{T}^* = T^*$. Since $\overline{U\mathscr{G}_T} = (\mathscr{G}_{T^*})^\perp$ we have $V(\overline{U\mathscr{G}_T}) = V((\mathscr{G}_{T^*})^\perp)$. By the continuity of V and $VU = I$ we have $V(\overline{U\mathscr{G}_T}) \subseteq \overline{VU\mathscr{G}_T} = \overline{\mathscr{G}_T}$. Moreover $V(\overline{U\mathscr{G}_T}) \supseteq VU\mathscr{G}_T = \mathscr{G}_T$ and so $V(\overline{U\mathscr{G}_T}) \supseteq \overline{\mathscr{G}_T}$. Therefore $V(\overline{U\mathscr{G}_T}) = \overline{\mathscr{G}_T}$ and so $\overline{\mathscr{G}_T} = V((\mathscr{G}_{T^*})^\perp)$. Hence by Proposition 2 $\overline{\mathscr{G}_T}$ is the graph of T^{**} and so T being closable $\mathscr{G}_{\bar{T}} = \mathscr{G}_{T^{**}}$. This proves that $\bar{T} = T^{**}$.

Proposition 5. *Let $T: \mathscr{D}_T \to \mathscr{H}$ be a closed linear operator and let $\mathscr{S}$ be a subset of $\mathscr{L}(\mathscr{H})$ such that $TS = ST$ on $\mathscr{D}_T$ for every S in $\mathscr{S}$. Then we have $TA = AT$ on $\mathscr{D}_T$ for every A in $\bar{\mathscr{S}}$ where $\bar{\mathscr{S}}$ denotes the closure of $\mathscr{S}$ in $\mathscr{L}(\mathscr{H})$ under the strong operator topology of $\mathscr{L}(\mathscr{H})$.*

Note. By hypothesis we have $TA = AT$ for every A in the smallest algebra $\tilde{\mathscr{A}}$ containing $\mathscr{S}$ and the identity I. In the next chapter we shall prove that the strong closure of $\tilde{\mathscr{A}}$ is the smallest weakly closed algebra $\mathscr{A}$ containing $\mathscr{S}$ and I. Hence by the above proposition we have $TA = AT$ for every A in $\mathscr{A}$.

Proof. Let $\mathscr{I}$ be a directed set and let $S_i (i \in \mathscr{I})$ be a net with values $S_i \in \mathscr{S}$ such that $S_i \to A$ in the strong operator topology. Hence if $x \in \mathscr{D}_T$ then

$S_i x \to A x$ and $S_i T x \to A T x$. By hypothesis we have $T S_i = S_i T$ for every i in $\mathscr{I}$. Therefore $S_i x \to A x$ and $T S_i x \to A T x$. Since T is a closed operator this implies that $T A x = A T x$. We proved that $T A = A T$ on $\mathscr{D}_T$ for every $A \in \mathscr{L}(\mathscr{H})$ lying in the strong closure of $\mathscr{S}$.

Lemma 6. *Let $\mathscr{H}$ and $\mathscr{K}$ be Hilbert spaces over the same field, let $T: \mathscr{D}_T \to \mathscr{K}$ be a densely defined operator and let $A \in \mathscr{L}(\mathscr{H})$ and $B \in \mathscr{L}(\mathscr{K})$. Then $T A = B T$ on $\mathscr{D}_T$ implies that $A^* T^* = T^* B^*$ on $\mathscr{D}_{T^*}$.*

Proof. We want to prove that if $x \in \mathscr{D}_{T^*}$ then $B^* x \in \mathscr{D}_{T^*}$ and $A^* T^* x = T^* B^* x$. We know that if $x \in \mathscr{D}_{T^*}$ and $y \in \mathscr{D}_T$ then $(Ty, B^* x)$ exists and is equal to $(B T y, x)$. By $y \in \mathscr{D}_T$ and $B T = T A$ we have $A y \in \mathscr{D}_T$ and $B T y = T A y$. Hence

$$(T y, B^* x) = (B T y, x) = (T A y, x).$$

Since $x \in \mathscr{D}_{T^*}$ and $A y \in \mathscr{D}_T$ we have $(T A y, x) = (A y, T^* x)$ and so $(T y, B^* x) = (A y, T^* x) = (y, A^* T^* x)$. This shows that $B^* x \in \mathscr{D}_{T^*}$ and $T^* B^* x = A^* T^* x$. Q.e.d.

Lemma 7. *If $\mathscr{X}$ is a Banach space and $E: \mathscr{X} \to \mathscr{X}$ is a continuous linear operator satisfying $E^2 = E$ then $E\mathscr{X}$ is a subspace of $\mathscr{X}$.*

Proof. It is obvious that $E\mathscr{X}$ is a linear manifold in $\mathscr{X}$. Let $\xi_1, \xi_2, \ldots$ be a convergent sequence of elements $\xi_n \in E\mathscr{X}$ with limit $\xi \in \mathscr{X}$. Since E is continuous and $\|E\xi_n - E\xi\| \leqslant |E| \cdot \|\xi_n - \xi\|$ we see that $E\xi_n \to E\xi$ as $n \to \infty$. On the other hand by $\xi_n \in E\mathscr{X}$ and $E^2 = E$ we have $E\xi_n = \xi_n$ and so $E\xi_n \to \xi$ as $n \to \infty$. Using $E\xi_n \to E\xi$ we obtain $E\xi = \xi$ and so $\xi \in E\mathscr{X}$. This proves that $E\mathscr{X}$ is a closed subset of $\mathscr{X}$.

We are going to discuss now Jacobson's formulation of Schur's lemma and a theorem of Jacobson which will be used in Section VII.4. The original proof of Jacobson's theorem was written with such clarity that it is presented here almost unchanged.

In what follows we let V denote a commutative group written additively with elements $x, y, z, \ldots$ We let $\mathscr{S}$ denote a set of endomorphisms $\alpha: V \to V$. The identity endomorphism will be denoted by 1. For the sake of simplicity we let Z_γ denote the kernel of an endomorphism $\gamma: V \to V$ and we let γV stand for the image of V under γ. Let $\mathscr{S}'$ denote the set of those endomorphisms $\varphi: V \to V$ which commute with every α in $\mathscr{S}$. We call $\mathscr{S}'$ the *commutant* of $\mathscr{S}$. A subgroup W of V will be called *$\mathscr{S}$-invariant* or *$\mathscr{S}$-stable* if $\alpha W \subseteq W$ for every α in $\mathscr{S}$. A set of endomorphisms $\alpha: V \to V$ is *irreducible* if the only $\mathscr{S}$-invariant subgroups are 0 and V. A set $\mathscr{S}$ is said to satisfy the *ascending chain condition* if every chain $0 \subset W_1 \subset W_2 \subset \cdots$ of $\mathscr{S}$-invariant subgroups of V is of finite length. If $\mathscr{S}$ is irreducible then every chain has length at most 1 and the ascending chain condition holds. The subgroup Z_φ is $\mathcal{S}$-invariant for every φ in $\mathcal{S}'$ because if $z \in Z_\varphi$ and $\alpha \in \mathcal{S}$ then $\varphi(\alpha z) = \alpha(\varphi z) = \alpha(0) = 0$ and so $\alpha z \in Z_\varphi$.

Proposition 8. *If $\mathcal{S}$ satisfies the ascending chain condition and φ is an endomorphism in $\mathcal{S}'$ such that $\varphi V = V$ then* $\ker \varphi = 0$.

Proof. It is obvious that $0 \subseteq Z_\varphi \subseteq Z_{\varphi^2} \subseteq Z_{\varphi^3} \subseteq \cdots$ and so we have $Z_{\varphi^k} = Z_{\varphi^{k+1}}$ for a sufficiently large value of k. If $z \in Z_{\varphi^k}$ then by $\varphi V = V$ we have $z = \varphi x$ for some x in V and so $\varphi^k z = \varphi^{k+1} x = 0$. Hence $x \in Z_{\varphi^{k+1}} = Z_{\varphi^k}$ and so $\varphi^{k-1} z = \varphi^k x = 0$ which shows that $z \in Z_{\varphi^{k-1}}$. We proved that $Z_{\varphi^k} = Z_{\varphi^{k+1}}$ implies that $Z_{\varphi^{k-1}} = Z_{\varphi^k}$ and so in finitely many steps we arrive at $Z_{\varphi^0} = Z_\varphi$ where $Z_{\varphi^0} = Z_1 = 0$.

The following theorem is *Jacobson's version of Schur's lemma*:

Theorem 9. *If $\mathcal{S}$ is an irreducible set of endomorphisms then the non-zero elements of $\mathcal{S}'$ are all automorphisms and consequently $\mathcal{S}'$ is a skew field.*

Proof. If $\varphi: V \to V$ belongs to $\mathcal{S}'$ then φV is an $\mathcal{S}$-invariant subgroup of V. Hence if $\varphi \neq 0$ then by the irreducibility of $\mathcal{S}$ we have $\varphi V = V$. Since $\mathcal{S}$ is irreducible the ascending chain condition holds and so by Proposition 8 we obtain $\ker \varphi = 0$. Hence φ is a bijection and so φ^{-1} exists and clearly $\varphi^{-1} \in \mathcal{S}'$. Hence $\mathcal{S}'$ is a skew field.

For comparison's sake we also state and prove the classical *Schur's lemma*:

Proposition 10. *If V is a vector space over a field F and $\mathcal{S}$ is an irreducible set of linear transformations $S: V \to V$ then $\mathcal{S}'$ is a skew field.*

Proof. Let $T: V \to V$ be a non-zero element of $\mathcal{S}'$. Then $\ker T$ is an $\mathcal{S}$-invariant subspace of V and so $\ker T = 0$ by $T \neq 0$ and the irreducibility of $\mathcal{S}$. Similarly TV being an invariant subspace we have $TV = V$.

There is another, more restricted but equally important version of Schur's lemma which will be used on several occasions:

Theorem 11. *Let F be an algebraically closed field, let V be a finite dimensional vector space over F and let $\mathcal{S}$ be an irreducible set of linear transformations. Then every element of $\mathcal{S}'$ is of the form λI where $\lambda \in F$ and $I: V \to V$ is the identity transformation.*

Proof. Let p be the characteristic polynomial of T. Since F is algebraically closed we have

$$p(x) = (x - \lambda_1) \cdots (x - \lambda_n)$$

for suitable $\lambda_1, \ldots, \lambda_n$ in F. Hence we have

$$(T-\lambda_1 I)\cdots(T-\lambda_n I)=0\,.$$

Since $\mathscr{S}'$ is a skew field and $\lambda_1 I, \ldots, \lambda_n I$ and T belong to $\mathscr{S}'$ we see that $T-\lambda_i I=0$ for some i $(1\leqslant i\leqslant n)$.

Again let V be a commutative group written additively with elements $x, y, \ldots$ and let $\mathscr{F}$ denote a skew field of endomorphisms $\alpha\colon V\to V$. The pair $(V, \mathscr{F})$ will be called *a vector space over the skew field of endomorphisms* $\mathscr{F}$. By αx we denote the image of $x\in V$ under the endomorphisms α. We let $\mathscr{L}(V)$ be the vector space of all linear transformations $A\colon V\to V$. Thus every element A of $\mathscr{L}(V)$ is an additive map and satisfies $A\alpha=\alpha A$ for every α in $\mathscr{F}$. A subset $\mathscr{A}$ of $\mathscr{L}(V)$ will be called *k-fold transitive* if for any two ordered set of vectors $x_1, \ldots, x_k$ and $y_1, \ldots, y_k$ such that the x_i's are linearly independent over $\mathscr{F}$, there is an A in $\mathscr{A}$ such that $Ax_i=y_i$ for every $i=1, \ldots, n$.

Lemma 12. *A subring $\mathscr{A}$ of $\mathscr{L}(V)$ is irreducible if and only if $\mathscr{A}$ is 1-fold transitive.*

Proof. Clearly if $\mathscr{A}$ is 1-fold transitive then it is irreducible. Conversely let $\mathscr{A}$ be a subring of $\mathscr{L}(V)$ which is irreducible. If $x\in V$ and $x\neq 0$ then $\mathscr{A}x=\{Ax\colon A\in\mathscr{A}\}$ is a subgroup of V invariant under $\mathscr{A}$ that is to say $A(\mathscr{A}x)\subseteq\mathscr{A}x$ for every A in $\mathscr{A}$. Therefore by the irreducibility of $\mathscr{A}$ we have $\mathscr{A}x=0$ or $\mathscr{A}x=V$. The set $\mathscr{F}x=\{\alpha x\colon \alpha\in\mathscr{F}\}$ is a subgroup, $\mathscr{F}x\neq 0$ and if we suppose that $\mathscr{A}x=0$ then $\mathscr{F}x$ is invariant under $\mathscr{A}$. Hence by the irreducibility of $\mathscr{A}$ it follows that $\mathscr{F}x=V$. Let us suppose that $\mathscr{A}x=0$. Let $A\in\mathscr{A}$ and $y\in V$. Then $y=\alpha x$ for a suitable α in $\mathscr{F}$ by $\mathscr{F}x=V$ and so $Ay=A\alpha x=\alpha Ax=\alpha 0=0$. Therefore $A=0$ and so $\mathscr{A}=0$. Since $\mathscr{A}=0$ would contradict the irreducibility of $\mathscr{A}$ we have $\mathscr{A}x\neq 0$ i.e. $\mathscr{A}x=V$. Therefore every y in V is of the form Ax for a suitable A in $\mathscr{A}$.

If $\mathscr{A}$ is a subset of $\mathscr{L}(V)$ that is k-fold transitive for every $k=1,2,\ldots$ then following Jacobson we say that $\mathscr{A}$ is *dense* in $\mathscr{L}(V)$.

Theorem 13. *Let V be a commutative group, let $\mathscr{A}$ be an irreducible ring of endomorphisms and let $\mathscr{F}$ be the skew field of endomorphisms $\varphi\colon V\to V$ commuting with the elements of $\mathscr{A}$. Then $\mathscr{A}$ is a dense subset of $\mathscr{L}(V)$.*

Proof. Since the elements of $\mathscr{A}$ commute with every φ in $\mathscr{F}$ they are all elements of $\mathscr{L}(V)$ and so it is natural to denote them by $A, B, C, \ldots$ etc. Let $x_1, \ldots, x_k$ be independent elements of V and let $y_1, \ldots, y_k\in V$. We have to show the existence of an A in $\mathscr{A}$ such that $Ax_i=y_i$ for every $i=1,\ldots,k$. By Lemma 12 we know this in the special case when $k=1$.

Let us suppose that $k=2$. First we prove that there is an A_2 in $\mathscr{L}(V)$ such that $A_2 x_1=0$ and $A_2 x_2 \neq 0$. If this is not the case then any equation of the form $X x_1=Y x_1$ where $X, Y \in \mathscr{A}$ implies $X x_2=Y x_2$. Hence the map $y=X x_1 \mapsto X x_2=y'$ is well defined and since every y is representable in the form $X x_1$ its domain of definition is the whole space V. It is obvious that $(y_1+y_2)'=y_1'+y_2'$ and $(By)'=By'$ for every B in $\mathscr{A}$ because if $y=X x_1$ then $By=BXx_1$ and so $(By)'=BXx_2=By'$. Thus the map $y \to y'$ belongs to $\mathscr{F}$ and for this reason it will be denoted by φ. Then we have $\varphi X x_1=X x_2$ and $\varphi X x_1=X \varphi x_1$ for every X in $\mathscr{A}$. Hence $X(x_2-\varphi x_1)=0$ and this shows that $\varphi x_1=x_2$ in contradiction to the linear independence of x_1 and x_2. We proved the existence of an A_2 in $\mathscr{A}$ such that $A_2 x_1=0$ and $A_2 x_2=0$. Similarly there is an A_1 in $\mathscr{A}$ such that $A_1 x_1 \neq 0$ and $A_1 x_2=0$. Now using the result in the special case when $k=1$ we can choose B_1 and B_2 in $\mathscr{A}$ such that $B_1 A_1 x_1=y_1$ and $B_2 A_2 x_2=y_2$. Then $A=B_1 A_1+B_2 A_2$ has the required properties.

Now we let $k>2$ and suppose that the theorem holds for $k-1$. As in the case $k=2$ it is sufficient to prove that there is an A_k in $\mathscr{A}$ such that $A_k x_i=0$ for $i=1,\ldots,k-1$ and $A_k x_k \neq 0$. By the induction hypothesis there is a B in $\mathscr{A}$ such that $Bx_1=\cdots=Bx_{k-2}=0$ and $Bx_k \neq 0$. Let $\mathscr{B}$ denote the set of maps B in $\mathscr{A}$ having this property. Let us first suppose that Bx_{k-1} and Bx_k are independent over $\mathscr{F}$. Then by the case $k=2$ there is a C in $\mathscr{A}$ such that $CBx_{k-1}=0$ and $CBx_k \neq 0$. We can then choose $A_k=CB$. Next let Bx_{k-1} and Bx_k be linearly dependent, say $\varphi Bx_k=Bx_{k-1}$. Then we choose D in $\mathscr{A}$ such that $Dx_1=\cdots=Dx_{k-2}=0$ and $D(\varphi x_k-x_{k-1}) \neq 0$. If $Dx_k=0$ and $Dx_{k-1} \neq 0$ then there is a T in $\mathscr{A}$ such that $Bx_{k-1}=TDx_{k-1}$. Then for $A_k=B-TD$ we have $A_k x_1=\cdots=A_k x_{k-1}=0$ and $A_k x_k=Bx_k \neq 0$. Now we suppose that $Dx_k \neq 0$ and consequently $D \in \mathscr{B}$. If $Dx_{k-1}=0$ then we choose $A_k=D$. If Dx_{k-1} and Dx_k are independent over $\mathscr{F}$ then we find C such that $CDx_{k-1}=0$ but $CDx_k \neq 0$ and let $A_k=CD$. Hence there remains the case when $Dx_{k-1}=\psi Dx_k$ for some $\psi \neq 0$ in $\mathscr{F}$. It is obvious that $\varphi \neq \psi$. By the case $k=1$ there is a T in $\mathscr{A}$ such that $TDx_{k-1}=Bx_{k-1}$. Then

$$TDx_k=T\psi^{-1}(Dx_{k-1})=\psi^{-1}(TDx_{k-1})=\psi^{-1}(Bx_{k-1})=\psi^{-1}\varphi(Bx_k)$$

and so $(\psi^{-1}\varphi B-TD)x_k=0$. Hence if we let $A_k=B-TD$ then by the existence of $(1-\psi^{-1}\varphi)^{-1}$ we have $A_k x_k=(B-TD)x_k \neq 0$. By $TDx_{k-1}=Bx_{k-1}$ we have $A_k x_{k-1}=0$ and obviously $A_k x_i=0$ by

$Bx_i=0$ and $Dx_i=0$ for every $i \leqslant k-2$. We proved the existence of A_k and completed the proof of Jacobson's theorem.

Note. The theorem will be used in Section VII.4 in the special case when V is the additive group of a complex Banach space $\mathscr{X}$ and $\mathscr{A}$ is an irreducible algebra of continuous linear operators $S: \mathscr{X} \to \mathscr{X}$. Since other types of irreducibility will also enter in the discussion we shall say that $\mathscr{A}$ is an *algebraically irreducible* set of operators. The field $\mathscr{F}$ includes all the operators λI where $\lambda \in \mathbb{C}$ and in the application $\mathscr{F}$ will contain no other operators. Therefore the elements of $\mathscr{L}(V)$ will be everywhere defined but not necessarily bounded linear operators $A: \mathscr{X} \to \mathscr{X}$. Hence given a continuous linear operator $A: \mathscr{X} \to \mathscr{X}$ and a neighborhood

$$N_A(\varepsilon; x_1, \dots, x_n) = \{T: \|(T-A)x_k\| < \varepsilon \text{ for } k=1,\dots,n\}$$

in the strong operator topology by Jacobson's theorem there is an operator S in $\mathscr{A}$ such that $S \in N_A(\varepsilon: x_1, \dots, x_n)$.

On one occasion in Chapter VII we shall use Burnside's theorem which will be proved here because it is naturally related to Schur's lemma which is needed in its proof. We start with an elementary result from linear algebra whose proof is left for the reader:

Lemma 14. *Let $\mathscr{A}$ be a proper subspace of the vector space $\mathscr{L}$ of dimension m and let $E_1, \dots, E_m$ be a basis of $\mathscr{L}$. Then there exist scalars $b_1, \dots, b_m$, not all zero and such that for every A in $\mathscr{A}$ we have*

$$\sum_{k=1}^{m} a_k b_k = 0$$

where $A = \sum_{k=1}^{m} a_k E_k$.

Note. The lemma has a converse but that will not be needed.

Theorem 15. *Let $\rho: G \to \mathscr{U}(\mathscr{V})$ be an irreducible representation of the group G in the finite dimensional vector space $\mathscr{V}$ over the algebraically closed field F. Then the algebra generated by the operators $\rho(x)$ $(x \in G)$ is the full operator algebra $\mathscr{L}(\mathscr{V})$.*

Proof. Let $\mathscr{A}$ denote the algebra generated by the operators $\rho(x)$ $(x \in G)$ and let $\mathscr{B}$ be the vector space of operators $B: \mathscr{V} \to \mathscr{V}$ such that $\operatorname{tr} AB = 0$ for every $A \in \mathscr{A}$. We are going to show that the only operator in $\mathscr{B}$ is 0. Let the representation $\sigma: G \to \mathscr{L}(\mathscr{B})$ be defined by $\sigma(x)B = \rho(x)B$ where $x \in G$ and $B \in \mathscr{B}$. Since $\mathscr{B}$ is finite dimensional

σ has an irreducible component $\mathscr{P}$ or $\mathscr{B}$ is $\{0\}$. Let us suppose that $\mathscr{B}\neq\{0\}$ and fix $\mathscr{P}$. Given v in $\mathscr{V}$ we define the linear map $T_v\colon\mathscr{P}\to\mathscr{V}$ by $T_vB=Bv$ where $B\in\mathscr{P}$. Then we have

$$(1)\qquad T_v\sigma(x)B = T_v\rho(x)B=\rho(x)Bv=\rho(x)T_vB$$

for every $B\in\mathscr{P}$ and so T_v intertwines $\sigma|\mathscr{P}$ with ρ. Since $\ker T_v$ is a σ-invariant subspace of $\mathscr{P}$ we see that T_v is injective or $T_v=0$. If $T_v=0$ for every v in $\mathscr{V}$ then $B=0$ and so by $\mathscr{P}\neq 0$ there is a v in $\mathscr{V}$ such that $T_v\neq 0$. Since $\operatorname{im} T_v$ is an invariant subspace of the irreducible representation ρ it follows that T_v is surjective. Thus T_v is an isomorphism of $\mathscr{P}$ onto $\mathscr{V}$.

For every w in $\mathscr{V}$ we have $T_v^{-1}T_w\in(\sigma|\mathscr{P})'$ and so by Schur's lemma $T_v^{-1}T_w=\lambda_w I$ or

$$(2)\qquad T_w=\lambda_w T_v$$

where $\lambda_w\in F$ and in particular $\lambda_v=1$. Let $v_1=v,v_2,\dots,v_n$ be a basis of $\mathscr{V}$ and let $v_1^*,v_2^*,\dots,v_n^*$ be the corresponding dual basis.

We have $\rho(x)Bv_j=\sum_{i=1}^{n}\lambda_{ij}v_i$ for suitable $\lambda_{ij}\in F$. Therefore $v_j^*(\rho(x)Bv_j)=\lambda_{jj}$ and so

$$0=\operatorname{tr}\rho(x)B=\sum_{j=1}^{n}v_j^*(\rho(x)Bv_j).$$

Applying (1) and (2) we obtain

$$(3)\qquad \left(\sum_{j=1}^{n}\lambda_{v_j}v_j^*\right)(T_v\rho(x)B)=0$$

where $x\in G$ and $B\in\mathscr{P}$ are arbitrary. Since $\mathscr{P}$ is an irreducible component of σ and $\rho(x)B=\sigma(x)B$ the elements $\rho(x)B$ ($x\in G$ and $B\in\mathscr{P}$) generate $\mathscr{P}$. But T is an isometry of $\mathscr{P}$ onto $\mathscr{V}$ and so (3) implies that

$$\sum_{j=1}^{n}\lambda_{v_j}v_j^*=0.$$

By the linear independence of $v_1^*,\dots,v_n^*$ we obtain $\lambda_{v_1}=\dots=\lambda_{v_n}=0$. This contradicts $\lambda_{v_1}=\lambda_v=1$ and so the hypothesis $\mathscr{B}\neq\{0\}$ is incorrect.

Now we apply Lemma 14 with $\mathscr{L}=\mathscr{L}(\mathscr{V})$, $m=n^2$ and the basis E_{ij} $(i,j=1,\dots,n)$ where $E_{ij}v_k=\delta_{jk}v_i$ for all $i,j,k=1,\dots,n$. If $\mathscr{A}$ were a proper subspace of $\mathscr{L}(\mathscr{V})$ then we could find scalars b_{ij} $(i,j=1,\dots,n)$, not all zero and such that $\sum_{i,j=1}^{n}a_{ij}b_{ji}=0$ for every $A=\sum a_{ij}E_{ij}$ in $\mathscr{A}$.

But this equation is the same as $\operatorname{tr} AB=0$ where $B=\sum b_{ij}E_{ij}$. Therefore by what have been already proved we must have $\dim \mathscr{A}=n^2$ and so $\mathscr{A}=\mathscr{L}(\mathscr{V})$.

Remarks

There are many books dealing with normed vector spaces and we mention only those written by Banach (1), Day (1), Edwards (1) and Dunford and Schwartz (1). As far as topological vector spaces are concerned the number of choices is considerably more restricted. There are two monographs available on the subject: One is by Köthe (1) and another by H. H. Schaefer (1). A fair amount of material can be found also in the books of J. Horváth (1) and Trèves (1) on topological vector spaces and distributions. Since Bourbaki has a volume on very many worthwhile topics we shall not point out the availability of this series of books in each particular instance.

The credit for the introduction of the weak and strong operator topologies goes to Hilbert, F. Riesz and v. Neumann. Weak and strong convergence occur explicitely in Riesz's book (2) and the various topologies of $\mathscr{L}(\mathscr{X},\mathscr{Y})$ were first described in one of v. Neumann's paper (1). Theorem 1.3, the Hahn-Banach theorem was first proved independently by H. Hahn (1) and Banach (2, 3) in the case of real vector spaces. Its extension to vector spaces over $\mathbb{C}$ is due to Bohnenblust and Sobczyk (1) and also to Soukhomlinov (1). Theorem 1.2 was proved by Banach (2, 3). Additional results related to the Hahn-Banach theorem and more literature can be found in the book of Edwards (1). The Krein-Milman theorem was published in 1940 (1) and Kelley's simple proof dates from 1951 (1).

The concept of a compact operator was introduced by F. Riesz in 1918 (3). Similarly one can define weakly compact operators: A linear operator $T: \mathscr{X} \to \mathscr{Y}$ is weakly compact if the weak closure of $T\mathscr{B}$ in $\mathscr{Y}$ is a compact set. This concept was first used by Kakutani (1) and Yosida (1) in 1938. Schauder (1) proved that if $T: \mathscr{X} \to \mathscr{Y}$ is compact then so is $T^*: \mathscr{Y}^* \to \mathscr{X}^*$. We proved this only in the special case when $\mathscr{X}=\mathscr{Y}=\mathscr{H}$, a Hilbert space. An operator T which is the uniform limit of operators of finite rank is called completely continuous. Theorem 2.7 shows that in the case of Banach spaces complete continuity means compactness.

Replace the field $\mathbb{R}$ or $\mathbb{C}$ in the definition of a Banach space by a field F which is complete with respect to a non-archimedean valuation $|\cdot|$. An *ultra-metric Banach space* over F is a complete normed vector space over

F such that the norm satisfies the inequality $\|x+y\| \leqslant \max\{\|x\|, \|y\|\}$. Such Banach spaces occur implicitely in the works of Dwork (1) on the zeta functions of algebraic varieties over finite fields. The importance of these Banach spaces and their completely continuous operators in the methods devised by Dwork was realized by Serre (1). The related works of M. Krasner are also of great interest. For references see the Zbl. and the Mathematical Reviews.

The spectral theorem of compact operators and the proof given in Section 3 are due to F. Riesz (3). See also the book by F. Riesz and Sz. Nagy (1, 2). The second volume of Dunford and Schwartz's book (1) contains a great wealth of information on HS and TC operators and on the more general C_p classes of linear operators. A book written by Schatten (1) also contains many important results on these classes of linear operators. MONS is of course an abbreviation for "maximal orthonormal system". For the sake of simplicity we shall prefer the notation x_i $(i \in \mathscr{I})$ instead of the more common $\{x_i\}$ $(i \in \mathscr{I})$ which we also used in the beginning.

In the proof of the existence of the HS inner product we used the discrete special case of a well known inequality which we will call most of the time the BCS inequality. According to Bourbaki and Weil the finite, discrete special case of this inequality was known already to the ancient Babylonians. Thus it is senseless to insist that it should be named after one or the other of these three mathematicians. Bunyakowski was a number theorist who published a number of papers in the Bull. Acad. Sci. St. Pétersbourg during the second half of the last century. Among other things he considered polynomials which represent infinitely many primes. For details see V. Bouniakowski in all three volumes of Dickson's History (1).

It is clear that the HS operators form an H^* algebra under the trace inner product (S, T). One can also prove that every TC operator $T: \mathscr{X} \to \mathscr{Y}$ can be represented in the form $T = RS$ where R and S are HS operators. For our purposes the form $T = R_1 S_1 + R_2 S_2$ is sufficient. The identity $\operatorname{tr} ST = \operatorname{tr} TS$ can also be extended to the situation when ST is supposed to belong to the TC. However we shall never use more than the special cases proved in Section 5. Lemma 5.17 is known as "Jacobson's lemma". It will be used in Section VIII.2.

Although we shall not need it we point out that for every Hilbert space $\mathscr{H}$ the trace class TC in naturally isomorphic to the dual of $\mathscr{L}\mathscr{C}(\mathscr{H})$, the space of compact operators $A: \mathscr{H} \to \mathscr{H}$. In fact if $T: \mathscr{H} \to \mathscr{H}$ is a trace class operator then $f_T(A) = \operatorname{tr} AT$ defines a continuous linear functional f_T on $\mathscr{L}\mathscr{C}(\mathscr{H})$ and $T \mapsto f_T$ is a linear bijection of TC into

$\mathscr{LC}(\mathscr{H})^*$. In order to see that $T \mapsto f_T$ is a surjective map suppose that f is a continuous linear functional on $\mathscr{LC}(\mathscr{H})$. Then $f \mid$ HS is continuous with respect to the HS norm and so HS being a Hilbert space there is a HS operator $T: \mathscr{H} \to \mathscr{H}$ such that $f(A) = (A, T^*) = \operatorname{tr} TA = f_T(A)$ for every A in HS. Using the spectral theorem of compact operators one proves that $f(A) = f_T(A)$ for every compact operator A.

The largest proper ideal of $\mathscr{L}(\mathscr{H})$ *is* $\mathscr{LC}(\mathscr{H})$. This important result was proved by Calkin (1) in 1940. Subsequently a number of results had been published on the closed ideals of $\mathscr{L}(\mathscr{H})$. For instance Gramsch (1) gave an explicit description of all these ideals. For additional bibliography on this topic we refer to the recent paper of Porta (1).

The integration theory which is briefly discussed in Section 6 began with a paper of Wiener (1). A more detailed exposition of this theory can be found in the book of Hille and Phillips (1). The spectral mapping theorem given in Section 7 was first proved by Gelfand in (2). More about the one dimensional spectral mapping theorem and on related results can be found in Hille and Phillips (1). The theorem can be extended also to functions of several complex variables but the proof is considerably more difficult. For literature and details we refer the reader to the following: The last section in Chapter I of the book of Gunning and Rossi (1), Chapter III of Hörmander's book on several complex variables (1) and Section III.5 in Rickart's monograph (1).

The material presented in Section 8 was selected because it will be needed later. For further information on unbounded operators the reader should consult Riesz and Sz.-Nagy (1, 2) and Dunford and Schwartz (1). Propositions 8.3 and 8.4 were proved by v. Neumann in (2). Jacobson's version of Schur's lemma and Jacobson's theorem appeared in (4).

Chapter III

The Spectral Theorem, Stable Subspaces and v. Neumann Algebras

The general remarks given in the beginning of the last chapter apply equally well at present. The main object is to prove the spectral theorem for self adjoint continuous operators in Hilbert space and introduce the reader to the elements of the theory of v. Neumann algebras. The contents of Sections 1 and 2 are needed only for the proof of the spectral theorem, which is given in Section 3. Almost all the material presented in Sections 4, 5 and 6 will be used later. Most of it can be read without being familiar with the contents of Chapter I. Similarly an understanding of the statement of the spectral theorem requires no background in Banach algebra theory but the proof relies heavily on the fundamental results discussed in Chapter I.

1. Linear Functionals on Vector Lattices and their Extensions

We recall that an *ordered vector space* V is a vector space over the reals with a partial ordering such that

(a) $x \leqslant y$ and $y \leqslant x$ imply $x = y$,

(b) $x \leqslant y$ implies $x + z \leqslant y + z$ for every $z \in V$, and

(c) $x \geqslant 0$ implies $\lambda x \geqslant 0$ for every $\lambda \geqslant 0$.

A vector lattice or *Riesz space* is an ordered vector space V such that any two elements x, y have a least upper bound $x \cup y$ and consequently a greatest lower bound $x \cap y$. If $\operatorname{lub} A$ exists for all those subsets A of V which are bounded from above then we speak about a *complete vector lattice*. We let $P = \{x : x \geqslant 0\}$ and call P the *positive cone* of V. Every $x \in V$ can be written in the form $x = x_1 - x_2$ where $x_1, x_2 \in P$ because $x = (x \cup 0) + (x \cap 0)$. A linear functional $\mu : V \to \mathbb{R}$ is called *positive*, in symbols $\mu \geqslant 0$, if $\mu(x) \geqslant 0$ for every $x \geqslant 0$.

We fix a sublattice L of the complete vector lattice V and a family $\mathscr{A}$ of non-void subsets A of L satisfying at least the first five of the following conditions:

(1) *each* $A \in \mathscr{A}$ *is bounded from above by some element of* V,

(2) *for every* $a \in L$ *there is an* $A \in \mathscr{A}$ *such that* $a = \operatorname{lub} A$,

(3) $\lambda A \in \mathscr{A}$ *for every* $A \in \mathscr{A}$ *and* $\lambda \geqslant 0$,

(4) $A_1, A_2 \in \mathscr{A}$ *imply* $A_1 + A_2 \in \mathscr{A}$,

(5) $A \cup 0$, $A \cap 0 \in \mathscr{A}$ *for every* $A \in \mathscr{A}$,

(6) $A_1 \cup A_2$, $A_1 \cap A_2 \in \mathscr{A}$ *for every* $A_1, A_2 \in \mathscr{A}$.

Here $A_1 \cup A_2 = \{a_1 \cup a_2 : a_1 \in A_1 \text{ and } a_2 \in A_2\}$ and $A_1 \cap A_2$, $A_1 + A_2$ are defined similarly. Instead of $A \cup \{b\} = \{a \cup b : a \in A\}$ we shall write $A \cup b$. The notations $A \cap b$ and $A + b$ have similar meanings.

We let $U = U(\mathscr{A})$ be the set of least upper bounds $u = \operatorname{lub} A$ where $A \in \mathscr{A}$. One sees immediately that $L \subseteq U$ and U is a cone i.e. $U + U \subseteq U$ and $\lambda U \subseteq U$ for every $\lambda \geqslant 0$: For we can easily prove that

$$\operatorname{lub} A_1 + \operatorname{lub} A_2 = \operatorname{lub}(A_1 + A_2):$$

For simplicity let $u_i = \operatorname{lub} A_i$ $(i = 1, 2)$. If $a_i \in A_i$ $(i = 1, 2)$ then $a_i \leqslant u_i$ and so $a_1 + a_2 \leqslant u_1 + u_2$. This shows that $\operatorname{lub}(A_1 + A_2) \leqslant u_1 + u_2$. Next let v be an upper bound of $A_1 + A_2$. Then given a_2 in A_2 we have $a_1 \leqslant v - a_2$ for every $a_1 \in A_1$ and so $u_1 \leqslant v - a_2$. Thus $a_2 \leqslant v - u_1$ for every a_2 in A_2 and this shows that $u_2 \leqslant v - u_1$ and $u_1 + u_2 \leqslant v$. Similarly we can prove that

$$\operatorname{lub} A_1 \cup \operatorname{lub} A_2 = \operatorname{lub}(A_1 \cup A_2):$$

If $a_i \in A_i$ $(i = 1, 2)$ then $a_i \leqslant u_i$ and so $a_1 \cup a_2 \leqslant u_1 \cup u_2$. Therefore we have $\operatorname{lub}(A_1 \cup A_2) \leqslant u_1 \cup u_2$. Now let v be an upper bound of $A_1 \cup A_2$. If $a_i \in A_i$ $(i = 1, 2)$ then $a_1 \cup a_2 \leqslant v$ and so by $a_i \leqslant a_1 \cup a_2$ we obtain $a_i \leqslant v$ where $i = 1, 2$. Hence $u_i < v$ for $i = 1, 2$ and consequently $u_1 \cup u_2 \leqslant v$. This completes the proof. Hence if (5) holds then $U \cup 0 \subseteq U$ and if (6) holds then $U \cup U \subseteq U$.

From here on we suppose that the complete vector lattice V is *infinitely distributive*: This means that for every x in V and for every subset A of V which is bounded from above we have

$$x \cap \operatorname{lub} A = \operatorname{lub}(x \cap A).$$

Then we can easily prove that

$$\operatorname{lub} A_1 \cap \operatorname{lub} A_2 = \operatorname{lub}(A_1 \cap A_2)$$

for every A_i in $\mathscr{A}$ $(i = 1, 2)$. Indeed if $a_i \in A_i$ $(i = 1, 2)$ then $a_i \leqslant u_i = \operatorname{lub} A_i$ and so $a_1 \cap a_2 \leqslant u_1 \cap u_2$. Therefore $\operatorname{lub}(A_1 \cap A_2) \leqslant u_1 \cap u_2$. Now let v be an upper bound of $A_1 \cap A_2$. Given a_1 in A_1 by the infinite distributivity property we have

$$a_1 \cap u_2 = a_1 \cap \operatorname{lub} A_2 = \operatorname{lub}(a_1 \cap A_2) \leqslant \operatorname{lub}(A_1 \cap A_2) \leqslant v.$$

Therefore $a_1 \cap u_2 \leqslant v$ and so $\operatorname{lub}(A_1 \cap u_2) \leqslant v$. By applying the law of infinite distributivity we obtain $u_1 \cap u_2 \leqslant v$. Now we see that $U \cap 0 \subseteq U$ when (5) holds and $U \cap U \subseteq U$ if we have (6).

Next we fix a vector sublattice of the infinitely distributive complete vector lattice V, say L, and choose a set $\mathscr{A}$ satisfying (1)—(5) and a *positive linear functional* $\mu: L \to \mathbb{R}$ such that

(7) $\quad \operatorname{glb} \mu(a - a \cap A) = 0 \quad$ if $a \in L,\ A \in \mathscr{A}$ and $\operatorname{glb}(a - a \cap A) = 0$.

Proposition 1. *If* $A_1, A_2 \in \mathscr{A}$ *and if* $\operatorname{lub} A_1 \leqslant \operatorname{lub} A_2$ *then* $\operatorname{lub} \mu(A_1) \leqslant \operatorname{lub} \mu(A_2)$.

Proof. We fix $a_1 \in A_1$ and apply the infinite distributivity property: $\operatorname{lub}(a_1 \cap A_2) = a_1 \cap \operatorname{lub} A_2 = a_1$ and so $\operatorname{glb}(a_1 - a_1 \cap A_2) = 0$. By the hypothesis made on μ we get $\operatorname{glb} \mu(a_1 - a_1 \cap A_2) = 0$ and so

$$\operatorname{lub} \mu(A_2) \geqslant \operatorname{lub} \mu(a_1 \cap A_2) \geqslant \mu(a_1)$$

where $a_1 \in A_1$ is arbitrary.

For any $u \in U$ we let $\mu(u) = \operatorname{lub} \mu(A)$ where A is any set in $\mathscr{A}$ satisfying $u = \operatorname{lub} A$. By the foregoing proposition $\mu(u)$ is independent of the set A used to define it. In particular, if $u \in L$ then by (2) we may choose $A \in \mathscr{A}$ such that $u = \operatorname{lub} A$ and using (7) see that the new and the old definitions of μ coincide on L.

Proposition 2. *The functional* $\mu: U \to \mathbb{R}$ *satisfies*

(α) $\mu(u_1) \leqslant \mu(u_2)$ *for* $u_1 \leqslant u_2$,

(β) $\mu(\lambda u) = \lambda \mu(u)$ *for* $\lambda \geqslant 0$, *and*

(γ) $\mu(u_1 + u_2) = \mu(u_1) + \mu(u_2)$ *for any* $u_1, u_2 \in U$.

Proof. Let $u_k = \operatorname{lub} A_k$ $(k = 1, 2)$, let $a_1 \in A_1$ and let $A \in \mathscr{A}$ be such that $a_1 = \operatorname{lub} A$. Then by Proposition 1 we have $\mu(a_1) \leqslant \mu(u_2)$ and so $\mu(u_1) \leqslant \mu(u_2)$. Using the same notations as before let $a_k \in A_k$ $(k = 1, 2)$ so that by (α) we now have $\mu(a_k) \leqslant \mu(u_k)$ and $\mu(a_1 + a_2) \leqslant \mu(u_1) + \mu(u_2)$. Since $u_1 + u_2 = \operatorname{lub}(A_1 + A_2)$ we obtain $\mu(u_1 + u_2) \leqslant \mu(u_1) + \mu(u_2)$. On the other hand $a_1 + a_2 \leqslant u_1 + u_2$ so $\mu(a_1) + \mu(a_2) \leqslant \mu(u_1 + u_2)$. Finally ($\beta$) follows from $\operatorname{lub}(\lambda A) = \lambda \operatorname{lub} A$ where $\lambda \geqslant 0$.

Consider the vector space $U-U=\{u_1-u_2\colon u_1, u_2\in U\}$ so that $L\subseteq U-U\subseteq V$. If $v=u_1-u_2=u_3-u_4$ then by the additivity of μ on U we have $\mu(u_1)-\mu(u_2)=\mu(u_3)-\mu(u_4)$. Hence if we let $\mu(v)=\mu(u_1)-\mu(u_2)$ then the definition is independent of the decomposition used and so μ is extended to a positive linear functional $\mu\colon U-U\to\mathbb{R}$. Occasionally we shall use the notation W or $W(L,\mathscr{A})$ instead of $U-U$.

Let $x\in V$ be such that for every $\varepsilon>0$ there exist $-u_1, u_2\in U$ satisfying $u_1\leqslant x\leqslant u_2$ and $\mu(u_2-u_1)<\varepsilon$. The set of these $x\in V$ will be denoted by $L^1(\mu)=L^1(L,\mathscr{A},\mu)$ so that $L\subseteq U\subseteq L^1(\mu)$ because if $u=\operatorname{lub} A$ then we can choose $u_2=u$ and $u_1\in A\subseteq L\subseteq -U$ such that $\mu(u_2)-\mu(u_1)<\varepsilon$. For any x in V we can define

$$\mu^-(x)=\operatorname{lub}\{\mu(u)\colon -u\in U \text{ and } u\leqslant x\},$$
$$\mu^+(x)=\operatorname{glb}\{\mu(u)\colon u\in U \text{ and } x\leqslant u\}.$$

By definition $x\in L^1(\mu)$ if and only if $\mu^-(x)=\mu^+(x)$ so we can define $\mu\colon L^1(\mu)\to\mathbb{R}$ by $\mu(x)=\mu^-(x)=\mu^+(x)$.

Proposition 3. *$L^1(\mu)$ is a vector space over $\mathbb{R}$ containing W and $x\cup 0$, $x\cap 0\in L^1(\mu)$ for every $x\in L^1(\mu)$. The extended functional $\mu\colon L^1(\mu)\to\mathbb{R}$ is linear, positive and $|\mu(x)|\leqslant\mu(|x|)$ for every $x\in L^1(\mu)$. If $\mathscr{A}$ satisfies condition (6) then $x_1\cup x_2$, $x_1\cap x_2\in L^1(\mu)$ for any $x_1, x_2\in L^1(\mu)$.*

Proof. It is obvious that $x\in L^1(\mu)$ implies $-x\in L^1(\mu)$ and $\lambda x\in L^1(\mu)$ for every $\lambda\geqslant 0$. If $u_{1k}\leqslant x_k\leqslant u_{2k} (k=1,2)$ then $u_{11}+u_{12}\leqslant x_1+x_2\leqslant u_{21}+u_{22}$ and by the additivity and positiveness of μ on $U-U$ we have

$$0\leqslant\mu(u_{21}+u_{22})-\mu(u_{11}+u_{12})=\mu(u_{21})-\mu(u_{11})+\mu(u_{22})-\mu(u_{12}).$$

Hence if $x_1, x_2\in L^1(\mu)$ then $x_1+x_2\in L^1(\mu)$ and it is proved that $L^1(\mu)$ is a vector space over $\mathbb{R}$.
Similarly if $u_{1k}\leqslant x_k\leqslant u_{2k} (k=1,2)$ then

$$u_{11}\cup u_{12}\leqslant x_1\cup x_2\leqslant u_{21}\cup u_{22}$$

where

$$(u_{21}\cup u_{22})-(u_{11}\cup u_{12})\leqslant(u_{21}-u_{11})+(u_{22}-u_{12})$$

because

$$u_{2k}=u_{1k}+(u_{2k}-u_{1k})\leqslant(u_{11}\cup u_{12})+(u_{21}-u_{11})+(u_{22}-u_{12}).$$

We are going to use this pair of inequalities with various choices of x_k and u_{kl}.

First choosing $x=x_1\in L^1(\mu)$ and $x_2=0$ we see that $x\cup 0\in L^1(\mu)$. Hence by the linearity of $L^1(\mu)$ it follows that $L^1(\mu)$ is closed also under the operations $x\cap 0$ and $|x|$. By $0\in U$ we have $\mu^-(x)\geqslant 0$ for $x\geqslant 0$ and so μ is a positive functional on $L^1(\mu)$. Then the inequality $|\mu(x)|\leqslant\mu(|x|)$ follows immediately from $-|x|\leqslant x\cap 0\leqslant x\leqslant x\cup 0\leqslant |x|$.

If $\mathscr{A}$ satisfies (6) then $U\cup U\subseteq U$ and $U\cap U\subseteq U$. Hence in our pair of inequalities $-(u_{11}\cup u_{12})\in U$ and $u_{21}\cup u_{22}\in U$. Moreover we have

$$\mu(u_{21}\cup u_{22})-\mu(u_{11}\cup u_{12})\leqslant\mu(u_{21})-\mu(u_{11})+\mu(u_{22})-\mu(u_{12})\,.$$

We derive that $x_1, x_2\in L^1(\mu)$ implies $x_1\cup x_2\in L^1(\mu)$.

Before specializing these results to vector lattices of functions $x\colon S\to\mathbb{R}$ and applying them to complex valued linear functionals φ we first prove a theorem which permits a decomposition of φ into a linear combination of linear functionals $\mu\geqslant 0$.

Proposition 4. *If L is a vector lattice and if* $a, b, c\geqslant 0$ *then* $a\cap(b+c)\leqslant(a\cap b)+(a\cap c)$.

Proof. Let $m=a\cap(b+c)$ so that $m\leqslant a$ and $m\leqslant b+c$. Thus by $m\leqslant b+a$ we have $m\leqslant(b+a)\cap(b+c)=b+(a\cap c)$. Similarly $m\leqslant a+(a\cap c)$ by $m\leqslant a$ and $a\cap c\geqslant 0$. Therefore

$$m\leqslant(a+(a\cap c))\cap(b+(a\cap c))=(a\cap b)+(a\cap c)\,.$$

Proposition 5. *If L is a vector lattice and if* $x_1, x_2, y_1, y_2\geqslant 0$ *are such that* $x_1+x_2=y_1+y_2$ *then there exist* $z_{11}, z_{12}, z_{21}, z_{22}\geqslant 0$ *such that* $x_1=z_{11}+z_{12}$, $x_2=z_{21}+z_{22}$ *and* $y_1=z_{11}+z_{21}$, $y_2=z_{12}+z_{22}$.

Proof. We let $z_{11}=x_1\cap y_1$, $z_{12}=x_1-z_{11}$ and $z_{21}=y_1-z_{11}$. If we let $z_{22}=x_2-z_{21}$ then by $x_1+x_2=y_1+y_2$ we have $z_{22}=y_2-z_{12}$. Clearly $z_{11}, z_{12}, z_{21}\geqslant 0$ so we only have to prove that $z_{22}\geqslant 0$. We have

$$\begin{aligned}-(z_{12}\cap z_{21})&=(-z_{12})\cup(-z_{21})=(z_{11}-x_1)\cup(z_{11}-y_1)=z_{11}+(-x_1\cup -y_1)\\&=z_{11}-(x_1\cap y_1)=0\end{aligned}$$

and so $z_{12}\cap z_{21}=0$. Next by $z_{21}=z_{12}+x_2-y_2$ we have $z_{21}\leqslant z_{12}+x_2$ and so

$$z_{21}=z_{21}\cap(z_{12}+x_2)\leqslant(z_{21}\cap z_{12})+(z_{21}\cap x_2)=z_{21}\cap x_2\,.$$

Thus $z_{21}\leqslant x_2$ and so by $z_{22}=x_2-z_{21}$ we have $z_{22}\geqslant 0$.

Following others we call a linear functional $\alpha\colon L\to\mathbb{R}$ *relatively bounded* if $\mathrm{lub}\{\alpha(y)\colon |y|\leqslant x\}$ is finite for every $x\geqslant 0$.

Proposition 6. *Let L be a vector lattice and let* $\alpha: L \to \mathbb{R}$ *be a relatively bounded linear functional. Then*

$$\mu(x) = \operatorname{supr}\{\alpha(y): 0 \leqslant y \leqslant x\}, \quad \textit{and}$$
$$\mu(x) = \mu(x \cup 0) - \mu(-(x \cap 0))$$

define a linear functional $\mu: L \to \mathbb{R}$ *such that both* $\mu \geqslant 0$ *and* $\nu = \mu - \alpha \geqslant 0$.

Note. Essentially the proposition states that every relatively bounded linear functional α is a difference of two positive linear functionals μ and ν.

Proof. We prove that on the positive cone $P = \{x: x \geqslant 0\}$ we have $\mu(x_1 + x_2) = \mu(x_1) + \mu(x_2)$. Indeed

$$\begin{aligned}\mu(x_1) + \mu(x_2) &= \operatorname{supr}\{\alpha(y_1 + y_2): 0 \leqslant y_k \leqslant x_k \text{ and } k = 1, 2\}\\ &\leqslant \operatorname{supr}\{\alpha(y): 0 \leqslant y \leqslant x_1 + x_2\} = \mu(x_1 + x_2).\end{aligned}$$

Next if $0 \leqslant y_1 \leqslant x_1 + x_2$ then x_1, x_2, y_1 and $y_2 = x_1 + x_2 - y_1$ satisfy the hypothesis of Proposition 5. Therefore $y_1 = z_{11} + z_{21}$ and $z_{12} = x_1 - z_{11} \geqslant 0$, $z_{22} = x_2 - z_{21} \geqslant 0$. In other words $0 \leqslant z_{11} \leqslant x_1$, $0 \leqslant z_{21} \leqslant x_2$ and so

$$\alpha(y_1) = \alpha(z_{11}) + \alpha(z_{21}) \leqslant \mu(x_1) + \mu(x_2).$$

Since y_1 is an arbitrary element satisfying $0 \leqslant y_1 \leqslant x_1 + x_2$ we get $\mu(x_1 + x_2) \leqslant \mu(x_1) + \mu(x_2)$. Now we can extend the definition of μ to L by $\mu(x) = \mu(x \cup 0) - \mu(-(x \cap 0))$. One can easily verify that $\mu(\lambda x) = \lambda \mu(x)$ by considering first the case $\lambda \geqslant 0$ and then $\lambda \leqslant 0$. We notice that if $x = y - z$ where $y, z \in P$ then $\mu(x) = \mu(y) - \mu(z)$ because if $y_1, y_2, z_1, z_2 \in P$ and $y_1 + z_2 = y_2 + z_1$ then $\mu(y_1) - \mu(z_1) = \mu(y_2) - \mu(z_2)$. Hence the additivity of μ on L follows from

$$\begin{aligned}\mu(x_1 + |x_1| + x_2 + |x_2|) &= \mu(x_1 + |x_1|) + \mu(x_2 + |x_2|),\\ \mu(x_k) &= \mu(x_k + |x_k|) - \mu(|x_k|) \qquad (k = 1, 2),\\ \mu(x_1 + x_2) &= \mu(x_1 + x_2 + |x_1| + |x_2|) - \mu(|x_1| + |x_2|).\end{aligned}$$

Clearly $\mu \geqslant 0$ and as $\alpha(x) \leqslant \mu(x)$ for $x \in P$ we have $\nu = \mu - \alpha \geqslant 0$.

One also verifies that conversely, if $\alpha = \mu - \nu$ where $\mu, \nu \geqslant 0$ then α is relatively bounded: If $|y| \leqslant x$ then $\mu(y) \leqslant \mu(x)$ and $\nu(-y) \leqslant \nu(x)$ so $\alpha(y) \leqslant \mu(x) + \nu(x)$. Therefore we proved:

Theorem 7. *A linear functional* $\alpha: L \to \mathbb{R}$ *defined on the vector lattice* L *is relatively bounded if and only if it is of the form* $\alpha = \mu - \nu$ *where* $\mu, \nu \geqslant 0$.

Applications follow in the next sections.

2. Linear Functionals on Lattices of Functions

A *normed vector lattice* or *Banach lattice* is an infinitely distributive vector lattice V which has a vector space norm such that $|x| \leqslant |y|$ implies $\|x\| \leqslant \|y\|$. Since $0 \leqslant x \cup 0 \leqslant |x|$ it follows that $\|x \cup 0\| \leqslant \|x\|$ for every x in V. Furthermore if $x, y \in V$ are such that $|x| = |y|$ then $\|x\| = \|y\|$. For instance, if S is a set and V consists of bounded functions $x: S \to \mathbb{R}$ then $\|x\| = \|x\|_\infty$ satisfies the requirements.

Proposition 1. *Let V be a normed vector lattice, L a sublattice, $\mathscr{A}$ a family satisfying* (1)—(6) *and μ a positive, continuous linear functional on L. Then the extension of μ to $L^1(\mu)$ is continuous and its norm is the same as that of the original μ.*

Proof. First we prove that the norm of the extension of μ to U is the same as the norm of the original $\mu: L \to \mathbb{R}$. If $u \in U$ then $u \cup 0$ and $u \cap 0$ belong to U and by the positivity of the extended functional $\mu: U \to \mathbb{R}$ we have

$$\mu(u \cap 0) \leqslant \mu(u) = \mu(u \cap 0) + \mu(u \cup 0) \leqslant \mu(u \cup 0)\,.$$

This shows that $|\mu(u)| \leqslant \mu(|u|)$ for any $u \in U$ and so it is sufficient to show that $\mu(u) \leqslant \|\mu\| \cdot \|u\|$ for $u \geqslant 0$ where $\|\mu\|$ denotes the norm of the original μ. Let $u = \operatorname{lub} A$ where $a \geqslant 0$ for every $a \in A$. Then

$$\mu(u) = \operatorname{lub} \mu(a) \leqslant \operatorname{lub} \|\mu\| \cdot \|a\| = \|\mu\| \operatorname{lub} \|a\| \leqslant \|\mu\| \cdot \|u\|$$

because if $a \in A$ then $\|a\| \leqslant \|u\|$ by $0 \leqslant a \leqslant u$.

By Proposition 1.3 the extended functional is positive and $|\mu(x)| \leqslant \mu(|x|)$ for every $x \in L^1(\mu)$. Moreover $\| |x| \| = \|x\|$ for any x in V. Thus it is enough to prove that $\mu(x) \leqslant \|\mu\| \cdot \|x\|$ for every $x \geqslant 0$ in $L^1(\mu)$ where $\|\mu\|$ denotes the norm of the original μ. By $\mu(x) = \mu^-(x)$ for any $\varepsilon > 0$ there is a $u \leqslant x$ such that $-u \in U$ and $\mu(x) \leqslant \mu(u) + \varepsilon$. By $U \cap 0 \subseteq U$ we may suppose that $0 \leqslant u \leqslant x$. Since $\mu(u) = -\mu(-u)$ and $\|-u\| = \|u\|$ for $-u \in U$ we have $|\mu(u)| \leqslant \|\mu\| \cdot \|u\|$ where $\|\mu\|$ denotes the norm of $\mu: L \to \mathbb{R}$. Therefore $\|u\| \leqslant \|x\|$ and $\mu(x) - \varepsilon \leqslant \mu(u) \leqslant \|\mu\| \cdot \|x\|$ where $\varepsilon > 0$ is arbitrary.

Proposition 2. *If L is a normed vector lattice and $\alpha: L \to \mathbb{R}$ is a continuous linear functional then there exist continuous linear functionals $\mu, \nu \geqslant 0$ such that $\|\mu\|, \|\nu\| \leqslant \|\alpha\|$ and $\alpha = \mu - \nu$.*

Proof. Since α is continuous it is relatively bounded and so by Proposition 1.6 we have a decomposition $\alpha = \mu - \nu$ where $\mu, \nu \geqslant 0$. Given

$x \in L$ and $\varepsilon > 0$ there exist $y, z \in L$ such that

$$0 \leqslant \mu(x \cup 0) \leqslant \alpha(y) + \varepsilon \quad \text{and} \quad \alpha(z) - \varepsilon \leqslant \mu(x \cap 0) \leqslant 0$$

where $0 \leqslant y \leqslant x \cup 0$ and $x \cap 0 \leqslant z \leqslant 0$. Hence $\|y\|, \|z\| \leqslant \|x\|$, so $|\alpha(y)|$, $|\alpha(z)| \leqslant \|\alpha\| \cdot \|x\|$ and $\varepsilon > 0$ being arbitrary $|\mu(x)| \leqslant \|\alpha\| \cdot \|x\|$.

Corollary. *If L is a normed vector lattice and $\alpha: L \to \mathbb{C}$ is a continuous linear functional then $\alpha = (\mu_1 - \nu_1) + i(\mu_2 - \nu_2)$ where $\mu_k, \nu_k \geqslant 0$ and $\|\mu_k\|, \|\nu_k\| \leqslant \|\alpha\|$ for $k = 1, 2$.*

Let S be a locally compact space and let $V = V(S)$ be the complete, infinitely distributive vector lattice of all bounded functions $x: S \to \mathbb{R}$. Let L be the sublattice of those x which are continuous and vanish at infinity. V and L become normed vector lattices under the norm $\|x\| = \|x\|_\infty$. Let $\mathscr{A} = \mathscr{A}(L)$ consist of all non-void subsets A of L such that

(a) *A is bounded from above in V,*

(b) *if $a_1, a_2 \in A$ then $a_1, a_2 \leqslant a$ for some $a \in A$.*

It is clear that $\mathscr{A}$ satisfies conditions (1)—(6) stated in the beginning of Section 1. Thus we can speak about the cone $U = U(\mathscr{A}) = U(S, L)$ and the linear manifold $W = W(S, L)$.

Similarly we can consider the sublattice L_0 of those x in V which are continuous and have compact support. We let $\mathscr{A}_0 = \mathscr{A}_0(L) = \mathscr{A}_0(S)$ be the family of all sets $A \subseteq L_0$ such that (a), (b) hold and the functions x in A have a common compact support.

Proposition 3. *If μ is a real valued, continuous linear functional on L or on L_0 then (7) holds with $\mathscr{A} = \mathscr{A}(L)$ or with $\mathscr{A} = \mathscr{A}_0(L)$, respectively.*

Proof. Let $a \in L$, $A \in \mathscr{A}$ and $\operatorname{glb}(a - a \cap A) = 0$. We also choose an element x_0 in A. Since the elements of L vanish at infinity given $\varepsilon > 0$ there is a compact set C in S such that $0 \leqslant (a - a \cap x_0)(s) \leqslant \varepsilon$ for every $s \notin C$. By $\operatorname{glb}(a - a \cap A) = 0$ and by the continuity of the elements of L for every $s \in S$ there is an x in A and an open neighborhood O_s such that $(a - a \cap x)(t) < \varepsilon$ for every $t \in O_s$. Since C is compact it can be covered by a finite family of sets O_s, say belonging to the points $s_1, \ldots, s_n \in C$. Let $x_1, \ldots, x_n$ be the corresponding functions in A and let x in A be such that $x_0, x_1, \ldots, x_n \leqslant x$. Then $0 \leqslant (a - a \cap x)(s) \leqslant \varepsilon$ for every $s \in S$. Hence $0 \leqslant \mu(a - a \cap x) \leqslant \|\mu\| \cdot \|a - a \cap x\| \leqslant \|\mu\| \, \varepsilon$ where x in A depends on ε and $\varepsilon > 0$ is arbitrary. The same proof holds also when L is replaced by L_0 and $\mathscr{A}(L)$ by $\mathscr{A}_0(L)$.

Consider $C_0(S)$, the set of continuous functions $x: S \to \mathbb{C}$ having compact support as a vector space over $\mathbb{R}$ normed by $\|x\| = \|x\|_\infty$. Let L_0 be the subspace formed by the real x and let iL_0 be the subspace of x's with imaginary values. Let $\varphi: C_0(S) \to \mathbb{C}$ be a continuous linear functional. By restricting φ to L_0 and to iL_0 we obtain two continuous functionals $\alpha_k: L_0 \to \mathbb{C}$ satisfying $\|\alpha_k\| \leqslant \|\varphi\|$. Applying Proposition 2 we can find $\mu_{kl}, \nu_{kl} \geqslant 0$ such that

$$\alpha_k = (\mu_{k1} - \nu_{k1}) + i(\mu_{k2} - \nu_{k2}),$$

and $\|\mu_{kl}\|, \|\nu_{kl}\| \leqslant \|\varphi\|$ for $k, l = 1, 2$.

Every x in $C_0(S)$ can be represented in the form $x = x_1 + ix_2$ where $x_1, x_2 \in L_0$. Therefore by

$$\varphi(x) = \varphi(x_1 + ix_2) = \alpha_1(x_1) + \alpha_2(x_2)$$

we have

$$\begin{aligned}\varphi(x_1 + ix_2) = &(\mu_{11} - \nu_{11})(x_1) + i(\mu_{12} - \nu_{12})(x_1) \\ &+ (\mu_{21} - \nu_{21})(x_2) + i(\mu_{22} - \nu_{22})(x_2).\end{aligned}$$

Proposition 3 shows that condition (7) of Section 1 holds for these functionals μ_{kl}, ν_{kl} and so by Proposition 1 they can be extended to continuous, positive functionals on the real vector spaces $L^1(\mu_{kl})$ and $L^1(\nu_{kl})$, respectively. This gives an extension of $\varphi: C_0(S) \to \mathbb{C}$ to a continuous linear functional $\varphi: L^1(\varphi) \to \mathbb{C}$ where $L^1(\varphi)$ denotes the complex vector space obtained from $L^1(\mu_{11}) \cap \cdots \cap L^1(\nu_{22})$ by complexification:

$$L^1(\varphi) = (1 + i)\{L^1(\mu_{11}) \cap \cdots \cap L^1(\nu_{22})\}.$$

According to Proposition 1.3 we have $L^1(\mu_{kl}), L^1(\nu_{kl}) \supseteq W = U - U$ and so $L(\varphi) \supseteq (1 + i)W$. By condition (b) every A in $\mathscr{A}$ is a directed set and $\varphi(a), \mu_{kl}(a), \nu_{kl}(a)\ (a \in A)$ are nets over A. Since $u = \operatorname{lub} A$ is $\lim A$ with respect to the ordering of the lattice we have $\mu_{kl}(u) = \lim \mu_{kl}(a)$ and $\nu_{kl}(u) = \lim \nu_{kl}(a)$ and so $\varphi(u) = \lim \varphi(a)$. By applying Proposition 1 we can also derive an upper estimate for the norm of the extension in terms of the norm of the originally given φ.

We proved the following:

Proposition 4. *If S is a locally compact space then every continuous linear functional $\varphi: C_0(S) \to \mathbb{C}$ has a continuous extension $\varphi: L^1(\varphi) \to \mathbb{C}$ where $L^1(\varphi) \supseteq (1 + i)W$. For any A in $\mathscr{A}_0$ we have $\lim \varphi(a) = \varphi(u)$ where $\varphi(a)$ denotes the net with directed set A and $u = \lim A$. The norm of the extension does not exceed $4\|\varphi\|$ where $\|\varphi\|$ denotes the norm of the original functional.*

3. The Spectral Theorem for Self Adjoint Operators in Hilbert Space

Let S be a compact space and as earlier let $V=V(S)$ be the vector lattice of continuous, real valued functions and let $\|x\|=\|x\|_\infty$. For any x in $C(S)$, the algebra of complex valued, continuous functions on S, define

$$Z(x)=\{s\colon s\in S \text{ and } x(s)=0\}$$

so that $Z(x)$ is a closed subset of S. Let c_x denote the characteristic function of $Z(x)$ and let $\mathscr{A}$ be the family of all sets $A\subseteq L$ satisfying conditions (a) and (b) of Section 2. Since S is compact the algebra of continuous functions vanishing at infinity and $C_0(S)$ are the same as $C(S)$. For the same reason the sublattices L_0 and L coincide and similarly $\mathscr{A}_0=\mathscr{A}$.

Proposition 1. *We have $c_x\in -U$ for every x in L and actually there is a decreasing sequence $a_1, a_2, \ldots$ in L such that c_x is the pointwise limit of $a_1, a_2, \ldots$*

Proof. We define $x_n=1\cap n|x|$ and $a_n=1-x_n$ where $n=1, 2, \ldots$ Then $a_n\in L$, $a_m\geqslant a_n$ for $m\leqslant n$ and $0\leqslant a_n\leqslant 1$. Moreover $a_n(s)=1$ if and only if $s\in Z(x)$ and $a_n(s)=0$ if and only if $|x(s)|\geqslant 1/n$. If $s\notin Z(x)$ then $|x(s)|\geqslant 1/n$ for all sufficiently large n and so $a_n(s)=0$ for $n\geqslant n(s)$. Hence c_x is the pointwise limit of the decreasing sequence $a_1, a_2, \ldots$ If we let $A=\{-a_1, -a_2, \ldots\}$ then A is in $\mathscr{A}$ and $-c_x=\operatorname{lub} A$.

The functions x in $C(S)$ and c_y where y is in $C(S)$ generate a complex algebra which will be denoted by $B(S)$. By $Z(y_1)\cap Z(y_2)=Z(|y_1|+|y_2|)$ we have $c_{y_1} c_{y_2}=c_{|y_1|+|y_2|}$ and so a typical element of $B(S)$ is of the form $a=x_1 c_{y_1}+\cdots+x_n c_{y_n}$ where $x_1, \ldots, x_n$, $y_1, \ldots, y_n\in C(S)$. From this it is plain that $B(S)$ is closed under complex conjugation. Since every a is a bounded function on S we can introduce the norm $\|a\|=\|a\|_\infty$ on $B(S)$.

Proposition 2. *Let $\mathscr{H}$ be a complex Hilbert space and let $\rho\colon C(S)\to\mathscr{L}(\mathscr{H})$ be a continuous star representation of $C(S)$. Then ρ has a continuous extension $\rho\colon B(S)\to\mathscr{L}(\mathscr{H})$ such that if $x\in U$, $x_n\in L$ and $x_n(s)\,(n=1,2,\ldots)$ is an increasing sequence with limit $x(s)$ for every $s\in S$, then for every ξ, η in $\mathscr{H}$ we have*

$$(\rho(x_n)\xi, \eta)\to(\rho(x)\xi, \eta)$$

as $n\to\infty$. The norm of the extension is at most 4 times the original norm $|\rho|$.

Proof. Given ξ, η in $\mathscr{H}$ and x in $C(S)$ let $\varphi(x, \xi, \eta)=(\rho(x)\xi, \eta)$. Then φ is linear in x, ξ and anti-linear in η. Moreover ρ being continuous

$$|\varphi(x, \xi, \eta)|\leqslant|\rho|\cdot\|\xi\|\cdot\|\eta\|\cdot\|x\|$$

where $|\rho| = \operatorname{supr}\{|\rho(x)|: \|x\| \leqslant 1\}$ is the norm of ρ. For fixed ξ, η in $\mathscr{H}$ we obtain a continuous linear functional $\varphi: C(S) \to \mathbb{C}$ which by Proposition 2.4 has a continuous extension $\varphi: L^1(\varphi) \to \mathbb{C}$ where $L^1(\varphi) \supseteq (1+i)W$. By Proposition 1 we have $-c_y \varepsilon U$ for any y in $C(S)$ and so $-c_y = \operatorname{lub} A$ for some A in $\mathscr{A}$. If $x \in L$ and $x \geqslant 0$ then $xA \in \mathscr{A}$ and $\operatorname{lub} xA = -xc_y$ so that $-xc_y \in U$. Using $x = x \cup 0 + x \cap 0$ we see that $xc_y \in W$ for any x in L and y in $C(S)$. Therefore xc_y belongs to $(1+i)W$ for every x, y in $C(S)$. This proves that $B(S) \subseteq L^1(\varphi)$ and so φ has a continuous extension $\varphi: B(S) \to \mathbb{C}$. By Proposition 2.4 we have $\lim \varphi(x_n) = \varphi(u)$ because the natural ordering of $x_1, x_2, \ldots$ is the same as the one derived from the lattice ordering of L.

For any $x \in L^1(\varphi)$ and $\varepsilon > 0$ there is an a in $C(S)$ such that $|\varphi(x) - \varphi(a)| < \varepsilon$: Indeed, if $x = x_1 + ix_2$ then there exist u_{kl}, v_{kl} in U $(k,l = 1,2)$ such that $x_k < u_{kl}, x_k < v_{kl}$ and

$$0 \leqslant \mu_{kl}(u_{kl}) - \mu_{kl}(x_k) < \frac{\varepsilon}{8} \quad \text{and} \quad 0 \leqslant \nu_{kl}(v_{kl}) - \nu_{kl}(x_k) < \frac{\varepsilon}{8}.$$

Next we can choose a_l in L $(l = 1,2)$ such that $a_l \leqslant w_l = u_{1l} \cap u_{2l} \cap v_{1l} \cap v_{2l}$ and

$$|\mu_{kl}(a_l) - \mu_{kl}(w_l)| < \frac{\varepsilon}{8} \quad \text{and} \quad |\nu_{kl}(a_l) - \nu_{kl}(w_l| < \frac{\varepsilon}{8}.$$

Then we have $|\varphi(x) - \varphi(a)| < \varepsilon$ with $a = a_1 + ia_2$. Hence using the linearity of the original φ in ξ and η we see that the extended functional is linear in x, ξ and anti-linear in η. Moreover by Proposition 2.4 we have $|\varphi(x, \xi, \eta)| \leqslant 4|\rho| \cdot \|\xi\| \cdot \|\eta\| \cdot \|x\|$ for every $x \in L^1(\varphi)$ and ξ, η in $\mathscr{H}$.

By the last inequality for fixed x and ξ the anti-linear functional $\eta \mapsto \varphi(x, \xi, \eta)$ is continuous. Therefore there is an element of $\mathscr{H}$, which we call $\rho(x)\xi$ such that $\varphi(x, \xi, \eta) = (\rho(x)\xi, \eta)$ for any η in $\mathscr{H}$. Since φ is linear in ξ we see that $\rho(x)$ is a linear operator on $\mathscr{H}$ and from the above inequality we obtain $|\rho(x)| \leqslant 4|\rho| \cdot \|x\|$. The linearity of ρ in x follows from the linearity of φ in $x \in L^1(\varphi)$.

Now we are ready to prove the *spectral theorem* for self adjoint, continuous, linear operators acting on a complex Hilbert space.

Theorem 3. *For every continuous, self adjoint linear operator* $A: \mathscr{H} \to \mathscr{H}$ *acting on the complex Hilbert space* $\mathscr{H}$ *there exists a function* $\lambda \mapsto E_\lambda$ *associating with each real number* λ *a self adjoint projection* $E_\lambda: \mathscr{H} \to \mathscr{H}$ *such that*

1) *if* $\lambda_1 \leqslant \lambda_2$ *then* $E_{\lambda_1} E_{\lambda_2} = E_{\lambda_1}$ *and* $E_\lambda = 0$ *if* $\lambda < -|A|$ *and* $E_\lambda = I$ *for* $\lambda \geqslant |A|$;
2) *if* $TA = AT$ *then* $TE_\lambda = E_\lambda T$;

3) *if* $f\colon \mathbb{R}\to\mathbb{C}$ *is continuous then* $f(A)=\int_{-\infty}^{+\infty} f(\lambda)\,dE_\lambda$ *exists in the strong sense and in particular* $A=\int_{-\infty}^{\infty} \lambda\, dE_\lambda$;

4) *if* $f, g\colon \mathbb{R}\to\mathbb{C}$ *are continuous functions which coincide on* $\sigma(A)$ *then* $f(A)=g(A)$ *and conversely;*

5) *the map* $f\mapsto f(A)$ *is a complex star-algebra homomorphism of* $C(\mathbb{R})$ *into* $\mathscr{L}(\mathscr{H})$, *in particular* $f(A)g(A)=fg(A)$, *and the image is the involutive Banach algebra generated by* A *in* $\mathscr{L}(\mathscr{H})$;

6) *if* $f\colon \mathbb{R}\to\mathbb{R}$ *is continuous then* $\sigma(f(A))=f(\sigma(A))$.

Note. Since $\mathbb{R}$ is a normal space every continuous $f\colon\sigma(A)\to\mathbb{C}$ has a continuous extension $f\colon\mathbb{R}\to\mathbb{C}$. Thus in view of 4) and 5) we are dealing with an isomorphism of $C(\sigma(A))$ onto the B^* algebra $\mathscr{A}$ generated by A in $\mathscr{L}(\mathscr{H})$. We shall see that this isomorphism is identical with the one given in Theorem I.8.7. The spectrum of A as an element of $\mathscr{A}$ is the same as an element of $\mathscr{L}(\mathscr{H})$. This will be seen during the proof and will be stated also explicitely in Proposition 6 below. One can also show that the function $\lambda\mapsto E_\lambda$ is essentially unique.

Proof. We slightly modify the notation in the proof by letting $a\colon\mathscr{H}\to\mathscr{H}$ denote the given continuous, self adjoint operator. We let A be the smallest self adjoint, uniformly closed algebra of continuous linear operators containing a and the identity operator e. This is the B^* algebra denoted in the above note by $\mathscr{A}$. It is the uniform closure in $\mathscr{L}(\mathscr{H})$ of the algebra $\mathbb{C}[a]$ of complex polynomials in a and so A is commutative. Since A has an identity by Proposition I.3.2 the maximal ideal space $\mathscr{M}$ of A is a compact Hausdorff space. By Theorem I.8.7 the algebra A is isometrically isomorphic to $C(\mathscr{M})$ under the Fourier transform map $x\mapsto\hat{x}$ and so $\hat{A}=C(\mathscr{M})$.

Let $\rho\colon C(\mathscr{M})=\hat{A}\to A$ be the inverse map $\hat{x}\mapsto x$ so that ρ is a continuous, faithful star representation of $\hat{A}$ in $\mathscr{L}(\mathscr{H})$. We let $B(\mathscr{M})$ denote the vector space generated by $C(\mathscr{M})$ and by the characteristic functions c_x of the closed sets $Z(x)$ where $x\in C(\mathscr{M})$. (If $x=x_1+ix_2\in C(\mathscr{M})$ then $Z(x)=Z(|x_1|+|x_2|)$ where $|x_1|+|x_2|\in L(\mathscr{M})$.) By Proposition 2 $\rho\colon\hat{A}\to A$ has an extension $\rho\colon B(\mathscr{M})\to\mathscr{L}(\mathscr{H})$ so that the extended ρ is a star representation of $B(\mathscr{M})$ and $\|\rho(x)\|\leqslant 4\,|\rho|\cdot\|x\|_\infty$ where $|\rho|$ denotes the original norm.

For real λ and x in $L(\mathscr{M})$ let

$$Z(x,\lambda)=Z(x\cup\lambda-\lambda)=\{s\colon (x\cup\lambda-\lambda)(s)=0\}=\{s\colon x(s)\leqslant\lambda\}.$$

From this it is clear that $c_{x,\lambda}$, the characteristic function of the set $Z(x,\lambda)$, belongs to $B(\mathscr{M})$. We let $c_\lambda=c_{\hat{a},\lambda}$ be the characteristic function of

$Z(\hat{a},\lambda)=Z(\hat{a}\cup\lambda-\lambda)$ where a is the given linear operator and so $\hat{a}\in L(\mathcal{M})$. Then we define $E_\lambda=\rho(c_\lambda)$ for $-\infty<\lambda<\infty$. Since ρ is a representation and $c_\lambda^2=c_\lambda$ we have $E_\lambda^2=E_\lambda$. Similarly as $\overline{c_\lambda}=c_\lambda$ it follows that $E_\lambda^*=E_\lambda$. Moreover if $\lambda_1\leqslant\lambda_2$ then $c_{\lambda_1}c_{\lambda_2}=c_{\lambda_1}$ and so $E_{\lambda_1}E_{\lambda_2}=E_{\lambda_1}$ i.e. we have $E_{\lambda_1}\leqslant E_{\lambda_2}$ for $\lambda_1\leqslant\lambda_2$. By Theorem I.8.7 we have $\|\hat{a}\|=\|a\|$. As $c_\lambda=0$ for $\lambda<-\|a\|=-\|\hat{a}\|$ and $c_\lambda=1$ for $\lambda\geqslant\|a\|=\|\hat{a}\|$ we see that $E_\lambda=0$ for $\lambda<-\|a\|$ and $E_\lambda=I=e$ for $\lambda\geqslant\|a\|$.

Given x in $L(\mathcal{M})$ by Proposition 1 there is a decreasing sequence of functions a_n in $L(\mathcal{M})$ such that c_x in the pointwise limit of this sequence. Applying this to $x=\hat{a}\cup\lambda-\lambda$ we find $a_n\in L(\mathcal{M})$ such that a_n decreases to c_λ. Therefore by Proposition 2 $\rho(a_n)\to\rho(c_\lambda)=E_\lambda$ in the weak sense.

Now suppose that T in $\mathcal{L}(\mathcal{H})$ satisfies $Ta=aT$. Then $Tp(a)=p(a)T$ for every complex polynomial in a and A being the uniform closure of polynomials $p(a)$ we also have $Tq=qT$ for every q in A. In other words we have $T\rho(x)=\rho(x)T$ for every x in $C(\mathcal{M})$.

We fix ξ,η in $\mathcal{H}$ and write down $\rho(a_n)\to\rho(c_\lambda)$ in the weak sense:

$$(\rho(a_n)\xi,T^*\eta)\to(\rho(c_\lambda)\xi,T^*\eta).$$

The right hand side is the same as $(T\rho(c_\lambda)\xi,\eta)$ while the left hand side by the commutativity can be written as follows:

$$(\rho(a_n)\xi,T^*\eta)=(T\rho(a_n)\xi,\eta)=(\rho(a_n)T\xi,\eta).$$

Thus we also have

$$(\rho(a_n)\xi,T^*\eta)\to(\rho(c_\lambda)T\xi,\eta)$$

and so $\rho(c_\lambda)T=T\rho(c_\lambda)$. Since $\rho(c_\lambda)=E_\lambda$ we proved that $E_\lambda T=TE_\lambda$ when $aT=Ta$.

We can now turn to the existence of the integral representing $f(a)$ in the strong sense. By hypothesis the restricted function $f\colon[-\|a\|,\|a\|]\to\mathbb{C}$ is uniformly continuous, so given $\varepsilon>0$ there is a $\delta=\delta(\varepsilon)>0$ such that $|\lambda-\lambda'|<\delta$ implies $|f(\lambda)-f(\lambda')|<\varepsilon$. Let

$$\lambda_0<\lambda_1<\cdots<\lambda_k<\cdots<\lambda_n$$

with $\lambda_0\leqslant-\|a\|$ and $\lambda_n\geqslant\|a\|$ be any partition satisfying

$$\max_{1\leqslant k\leqslant n}|\lambda_k-\lambda_{k-1}|\leqslant\delta.$$

For simplicity let c_k denote the characteristic function of $\{\lambda:\lambda\leqslant\lambda_k\}$. Then we have

$$\|f-\{f(\lambda_1)(c_1-c_0)+\cdots+f(\lambda_n)(c_n-c_{n-1})\}\|_\infty<\varepsilon.$$

Therefore we also have

$$\|f(\hat{a}(\cdot))-\{f(\lambda_1)(c_{\lambda_1}-c_{\lambda_0})+\cdots+f(\lambda_n)(c_{\lambda_n}-c_{\lambda_{n-1}})\}\|_\infty<\varepsilon.$$

Since $\|\rho(f)\|\leqslant 4|\rho|\cdot\|f\|$ for every f in $B(\mathcal{M})$ we obtain

$$\|f(a)-\sum f(\lambda_k)(E_{\lambda_k}-E_{\lambda_{k-1}})\|<4|\rho|\varepsilon,$$

where $f(a)=\rho(\hat{a}\circ f)$. This shows that the Riemann-Stieltjes integral $\int_{-\infty}^{+\infty} f(\lambda)dE_\lambda$ exists and it is equal to $f(a)$. Of course then we also have

$$|(f(a)\xi,\eta)-\sum f(\lambda_k)((E_{\lambda_k}-E_{\lambda_{k-1}})\xi,\eta)|<4|\rho|\varepsilon\|\xi\|\cdot\|\eta\|$$

and so

$$(f(a)\xi,\eta)=\int_{-\infty}^{+\infty} f(\lambda)d(E_\lambda\xi,\eta).$$

If we let $f(\lambda)=\lambda$ on $\mathbb{R}$ then $\rho(\hat{a}\circ f)=\rho(\hat{a})=a$ and so $a=\int\lambda dE_\lambda$.

Since A has an identity by Theorem I.6.14 we have $\sigma(x)=\hat{x}(\mathcal{M})$ for any x in A. In particular $\sigma(a)=\hat{a}(\mathcal{M})$ and so if f and g coincide on $\sigma(a)$ then $f(a)=\rho(\hat{a}\circ f)=\rho(\hat{a}\circ f)=g(a)$ proving part 4) of the theorem.

To see part 5) first notice that $\hat{a}\circ\lambda f=\lambda(\hat{a}\circ f)$ and $\hat{a}\circ(f+g)=\hat{a}\circ f+\hat{a}\circ g$ and also $\hat{a}\circ fg=(\hat{a}\circ f)(\hat{a}\circ g)$. Hence the homomorphism property follows from the fact that the original $\rho\colon C(\mathcal{M})\to A$ is an isomorphism. The map $f\mapsto f(a)$ is into A because if $f\colon\mathbb{R}\to\mathbb{C}$ is continuous then so is $\hat{a}\circ f$ and so $\rho(\hat{a}\circ f)=f(a)\in A$. If $f=p$ where p is a complex polynomial then by the homomorphism property $f(a)=p(a)$ is the operator obtained by substituting a in p. Therefore to see that the image algebra under $f\mapsto f(a)$ is A it is sufficient to show that the image is complete. However by Theorem I.8.7 and by $\hat{a}(\mathcal{M})=\sigma(a)$ we have $\|f(a)\|=|\rho(f\circ\hat{a})|=\|f\|$ where $\|f\|=\operatorname{supr}\{|f(t)|\colon t\in\sigma(a)\}$ and so the image algebra is isometrically isomorphic to the function algebra $C(\sigma(a))$.

Part 6) follows from Theorem I.6.14: If $f\colon\mathbb{R}\to\mathbb{R}$ is continuous then the original ρ being the inverse of the Fourier transform map by $\rho(\hat{a}\circ f)=f(a)\in A$ we have $\widehat{f(a)}=\hat{a}\circ f$ and so

$$\sigma(f(a))=\widehat{f(a)}(\mathcal{M})=f\circ\hat{a}(\mathcal{M})=f(\hat{a}(\mathcal{M}))=f(\sigma(a)).$$

Near the end of proving part 5) we incidentally proved also the following corollary of the Gelfand-Naimark theorem:

Theorem 4. *If A is a continuous, self adjoint operator in Hilbert space then the B^* algebra generated by A and I is isometrically isomorphic to the involutive algebra $C(\sigma(A))$ where $\sigma(A)$ denotes the spectrum of A.*

Note. Here $\sigma(A)$ denotes the spectrum of A in the commutative B^* algebra obtained by considering the uniform closure of the algebra of polynomials in A with complex coefficients. Later, at the end of this section it will be proved to be the same as the spectrum of A in $\mathscr{L}(\mathscr{H})$.

Let $\{A\}'$ denote the set of continuous operators commuting with A and let $\{A\}''$ be the set of those T in $\mathscr{L}(\mathscr{H})$ which commute with every operator in $\{A\}'$.

Proposition 5. *If $A: \mathscr{H} \to \mathscr{H}$ is self adjoint then it can be approximated in the operator norm by real linear combinations of projections $P \in \{A\}''$.*

Proof. By part 2) of the spectral theorem we have $E_\lambda \in \{A\}''$ for every λ in $(-\infty, \infty)$. If we let $f(\lambda) = \lambda$ in the proof of part 3) we obtain

$$\left|A - \sum \lambda_k (E_{\lambda_k} - E_{\lambda_{k-1}})\right| < 4|\rho|\varepsilon$$

where $\lambda_0, \ldots, \lambda_n$ in $\mathbb{R}$ depend on the choice of the arbitrarily small $\varepsilon > 0$.

Proposition 6. *If $A: \mathscr{H} \to \mathscr{H}$ is a self adjoint operator and if $\mathscr{B} = \mathscr{B}(A)$ is the smallest uniformly closed subalgebra of $\mathscr{L}(\mathscr{H})$ which contains A and I then $\sigma_{\mathscr{B}}(A) = \sigma_{\mathscr{L}(\mathscr{H})}(A)$.*

Proof. In the beginning of the proof of the spectral theorem we had seen that the B^* algebra $\mathscr{B}$ is isometrically isomorphic to $C(\mathscr{M})$ under the Fourier transform map. Since $A = A^*$ we have $\hat{A} = \bar{\hat{A}}$ and so $\hat{A}(\mathscr{M})$ is a subset of the interval $[-|A|, |A|]$. By Theorem I.6.14 $\hat{A}(\mathscr{M}) = \sigma_{\mathscr{B}}(x) \supseteq \sigma_{\mathscr{L}(\mathscr{H})}(x)$ which shows that the resolvent set $r(A)$ is connected. Now we can apply the corollary of Proposition I.4.10 with $\mathscr{B}$ and $\mathscr{L}(\mathscr{H})$ and obtain the desired conclusion.

4. Normal Elements and Normal Operators

Let A be a complex Banach algebra with identity e and let x be an arbitrary element of A. Then the series $\sum_{n=0}^{\infty} (1/n!)x^n$ with $x^0 = e$ is absolutely convergent and so it defines an element of A which will be denoted by $\exp x$ and called the *exponential* of x. Clearly $\exp 0 = e$ where 0 and e are the zero and identity elements, respectively. For any x and y in A we have

$$(\exp x)(\exp y) = \sum_{n=0}^{\infty} \frac{1}{n!} \sum_{k+l=n} \frac{n!}{k!\,l!} x^k y^l.$$

Thus

$$(1) \qquad (\exp x)(\exp y) = \exp(x+y) \quad \text{if } xy = yx.$$

In particular $(\exp x)(\exp -x) = e$ for any x in A and hence $\exp x \neq 0$ for every x in A. One also sees that if a, x, y are such that $ay = xa$ then $a \exp y = (\exp x)a$.

For fixed x in A and λ in $\mathbb{C}$ let $f(\lambda) = f(\lambda; x) = \exp \lambda x$. Then using the power series defining $f(\lambda)$ one sees that $1/\lambda\{f(\lambda) - e\} \to x$ as $\lambda \to 0$. Hence (1) shows that $\lambda \mapsto \exp \lambda x$ is a differentiable map from $\mathbb{C}$ to A.

If A is a B^* algebra then we have $(\exp x)^* = \exp x^*$ for every x in A. Thus if $x = x^*$ then $(\exp ix)^* = \exp -ix$ and so by (1) $(\exp ix)^{-1} = (\exp ix)^*$. For instance if $\mathscr{H}$ is a complex Hilbert space and $T: \mathscr{H} \to \mathscr{H}$ is a continuous, self adjoint operator then we see that $\exp iT$ is a unitary operator and so its norm is 1. The equation $\|\exp ix\| = 1$ holds for self adjoint x in any B^* algebra because if $x = x^*$ then $yy^* = e$ for $y = \exp ix$ and so $\|y\|^2 = \|yy^*\| = \|e\| = 1$.

An element x of the B^* algebra A is called *normal* if $xx^* = x^*x$ and x is called *unitary* if $xx^* = x^*x = e$. This is an obvious generalization of the terminology used in the case of continuous operators in Hilbert spaces. We notice that $(\exp x)(\exp x^*) = \exp(x + x^*)$ for every normal element x. We also know that if x is unitary then $\|x\| = 1$.

Theorem 1. *Let x and y be normal elements in the complex B^* algebra A. Then $ay = xa$ with a in A implies $ay^* = x^*a$.*

Proof. Since $ay = xa$ we have $a \exp i\bar{\lambda} y = (\exp i\bar{\lambda} x)a$ and

$$a = (\exp i\bar{\lambda} x) a (\exp -i\bar{\lambda} y)$$

for any λ in $\mathbb{C}$. If we multiply this equation from the left by $\exp i\lambda x^*$ and from the right by $\exp -i\lambda y^*$ then by the normality of x and y the resulting equation can be written as

$$f(\lambda) = (\exp i\lambda x^*) a (\exp -i\lambda y^*) = \exp(i\lambda x^* + i\bar{\lambda} x) a \exp(-i\bar{\lambda} y - i\lambda y^*).$$

Here $\lambda x^* + \bar{\lambda} x$ and $-\bar{\lambda} y - \lambda y^*$ are self adjoint and so the exponentials are unitary with norm 1. Therefore $\lambda \mapsto f(\lambda)$ is a bounded analytic function from $\mathbb{C}$ to A. Hence by Liouville's theorem f is constant and consequently its derivative is 0. The product formula for differentiation holds for Banach algebra valued functions of a complex variable. Thus computing the derivative of f at $\lambda = 0$ we obtain $0 = ix^*a - iay^*$.

The following corollary is due to Fuglede and was in fact the result from which the foregoing theorem of Putnam and the proof due to M. Rosenbloom evolved:

Corollary. *If the normal operator* $T\colon \mathscr{H}\to\mathscr{H}$ *commutes with* $S\colon \mathscr{H}\to\mathscr{H}$ *then one also has* $ST^*=T^*S$.

Let $\mathscr{H}$ be a complex Hilbert space and let $T\colon \mathscr{H}\to\mathscr{H}$ be a normal operator i.e. let T be in $\mathscr{L}(\mathscr{H})$ and let $TT^*=T^*T$. The uniform closure in $\mathscr{L}(\mathscr{H})$ of the algebra of polynomials $p(T,T^*)$ in T and T^* with complex coefficients is a B^* algebra with identity which we shall denote by $\mathscr{B}=\mathscr{B}(T)$.

Lemma 2. *We have* $\sigma(p(T,T^*))=\{p(\lambda,\bar{\lambda})\colon \lambda\in\sigma(T)\}$.

Proof. By Theorem I.6.14 $\sigma(p(T,T^*))=\widehat{p(T,T^*)}(\mathscr{M})$ where $\mathscr{M}$ is the maximal ideal space of $\mathscr{B}$. Using the homomorphism property of the Fourier transform we have $\widehat{p(T,T^*)}=p(\hat{T},\widehat{T^*})=p(\hat{T},\bar{\hat{T}})$. Therefore

$$\widehat{p(T,T^*)}(\mathscr{M})=\{p(\hat{T}(I),\overline{\hat{T}(I)})\colon I\in\mathscr{M}\}.$$

The conclusion follows from $\hat{T}(\mathscr{M})=\sigma(T)$.

The following theorem of Gelfand and Naimark is an extension of Theorem 3.4 to normal operators:

Theorem 3. *If* $T\colon \mathscr{H}\to\mathscr{H}$ *is a normal operator then the* B^* *algebra* $\mathscr{B}=\mathscr{B}(T)$ *generated by* T, T^* *and* I *in the uniform operator topology is isometrically isomorphic to* $C(\sigma(T))$ *where* $\sigma(T)$ *is the spectrum of* T *as an element of* $\mathscr{B}(T)$.

Proof. Given $p(x,y)$ in $\mathbb{C}[x,y]$ define p in $C(\sigma(T))$ by $p(\lambda)=p(\lambda,\bar{\lambda})$. Since T is normal $\mathscr{B}(T)$ is commutative and so by Theorem I.8.7, the corollary of Theorem I.6.14 and Lemma 2

$$\begin{aligned}|p(T,T^*)|&=\|p(T,T^*)\|_{sp}=\operatorname{supr}\{|p(\lambda,\bar{\lambda})|\colon \lambda\in\sigma(T)\}\\&=\operatorname{supr}\{|p(\lambda)|\colon \lambda\in\sigma(T)\}=\|p\|.\end{aligned}$$

The map $p\mapsto p(T,T^*)$ is a norm preserving star isomorphism of the algebra of these polynomials onto the corresponding polynomial operators. The algebra is separating and $\overline{p(\lambda)}=\bar{p}(\bar{\lambda},\lambda)$ so the Stone-Weierstrass theorem is applicable and so these polynomials are dense in $C(\sigma(T))$. Similarly the polynomials $p(T,T^*)$ form a dense set in $\mathscr{B}(T)$. Hence by uniform continuity the map $p\mapsto p(T,T^*)$ can be extended to a star isomorphism of $C(\sigma(T))$ onto $\mathscr{B}(T)$.

Proposition 4. *If* $T\colon \mathscr{H}\to\mathscr{H}$ *is a normal operator then the weakly closed algebra* $\mathscr{A}=\mathscr{A}(T)$ *generated by* T, T^* *and* I *is commutative.*

Proof. It is clear that the algebra $\tilde{\mathscr{A}}$ of complex polynomials $p(T,T^*)$ is commutative. Suppose that $A\in\tilde{\mathscr{A}}$ and B belongs to $\mathscr{A}$, the weak closure of $\tilde{\mathscr{A}}$ in $\mathscr{L}(\mathscr{H})$. Given x, y in $\mathscr{H}$ and $\varepsilon > 0$ we can find an operator A' in $\tilde{\mathscr{A}}$ such that

$$A' \in O_B\left(\frac{\varepsilon}{2}; x, Ax; A^* y, y\right).$$

Hence using the identity

$$AB - BA = A(B - A') + (A' - B)A$$

we obtain $|((AB - BA)x, y)| < \varepsilon$. Therefore $(AB - BA)x = 0$ for every x in $\mathscr{H}$ and so $AB = BA$. Now let B_1 and B_2 belong to $\mathscr{A}$, let $\varepsilon > 0$ and $x, y \in \mathscr{H}$. We can choose an operator A in $\tilde{\mathscr{A}}$ such that

$$A \in O_{B_2}\left(\frac{\varepsilon}{2}; x, B_1 x; B_1^* y, y\right).$$

Since we have

$$B_1 B_2 - B_2 B_1 = B_1 (B_2 - A) + (A - B_2) B_1$$

we obtain $|((B_1 B_2 - B_2 B_1)x, y)| < \varepsilon$. This shows that $B_1 B_2 = B_2 B_1$.

In the following sequence of propositions let A be a B^* algebra with identity e, let $a \in A$ and let B be an involutive subalgebra of A containing a and e. In some of the results a will be a normal element and B will be supposed to be commutative. For simplicity the spectrum of a in B will be denoted by $\sigma(a)$ instead of $\sigma_B(a)$. Important special cases are furnished by operator algebras over complex Hilbert spaces $\mathscr{H}$. For instance A can be $\mathscr{L}(\mathscr{H})$, the normal operator a can then be denoted by T and for B one can choose either $\mathscr{A}(T)$ or $\mathscr{B}(T)$.

Proposition 5. *If B is any involutive subalgebra containing the element a then $\sigma(a^*) = \overline{\sigma(a)}$.*

Proof. An element b is invertible if and only if b^* is invertible and $(b^*)^{-1} = (b^{-1})^*$. Hence $\lambda \notin \sigma(a)$ if and only if $(a^* - \bar{\lambda} e)^{-1}$ exists i.e. $\bar{\lambda} \notin \sigma(a^*)$.

Proposition 6. *If a is normal, B is commutative and $\sigma(a) \subset \mathbb{R}$ then $a = a^*$. Conversely, if $a = a^*$ then for any B we have $\sigma(a) \subset \mathbb{R}$.*

Proof. Let $\mathscr{M}$ be the maximal ideal space of B. By Theorem I.6.14 we have $\hat{a}(\mathscr{M}) = \sigma(a)$ and so by hypothesis $\hat{a}(I) = \bar{\hat{a}}(I)$ for every I in $\mathscr{M}$. Since $\hat{a} = \bar{\hat{a}} = \widehat{a^*}$ by Theorem I.8.7 $a = a^*$. The converse statement follows from Proposition 5.

Proposition 7. *If a is normal, B is commutative and if $\sigma(a)$ lies on the unit circle $|\lambda|=1$ then a is unitary. Conversely, if a is unitary and B is commutative then $|\sigma(a)|=1$.*

Proof. Again we use $\hat{a}(\mathcal{M})=\sigma(a)$ and the isometric isomorphism of B onto $C(\mathcal{M})$. If $\sigma(a)$ lies on the unit circle then $\hat{a}\bar{\hat{a}}(I)=1$ for every I in $\mathcal{M}$ so $\hat{a}\bar{\hat{a}}=\hat{e}$ and $aa^*=e$. Conversely, from $aa^*=e$ we get $\hat{a}\bar{\hat{a}}=\hat{e}=1$ and so $|\hat{a}|\equiv 1$ i.e. $|\sigma(a)|=1$.

Since $\mathcal{L}(\mathcal{H})$ for $\dim\mathcal{H}>1$ is not commutative our results are not applicable with this choice of $A=B$. However if T is a normal operator then by Proposition 4 we can let $A=\mathcal{L}(\mathcal{H})$ and $B=\mathcal{A}(T)$ and so in order to obtain results with $B=\mathcal{L}(\mathcal{H})$ it is sufficient to prove that the spectrum of T in $\mathcal{L}(\mathcal{H})$ is the same as that in $\mathcal{A}(T)$. This will be done in Section 6. (See Proposition 6.16 and its corollaries.)

A linear operator $A:\mathcal{H}\to\mathcal{H}$ is called *positive*, and one writes $A\geqslant 0$, if $A=A^*$ and $(Ax,x)\geqslant 0$ for every x in $\mathcal{H}$. If $(Ax,x)>0$ for $x\neq 0$ then A is called *positive definite* and one denotes this fact by $A>0$. For example if E is a projection then $E=E^*$ by convention and $(Ex,x)=(Ex,Ex)\geqslant 0$ by $E^2=E$.

For projections the symbol $E\geqslant 0$ and more generally $E_1\geqslant E_2$ has also another meaning, namely we write $E_1\geqslant E_2$ if $E_1\mathcal{H}\supseteq E_2\mathcal{H}$. Let $x=y+z$ with $y\in E_1\mathcal{H}$ and $z\in(E_1\mathcal{H})^\perp$ and let $y=u+v$ with $u\in E_2\mathcal{H}$ and $v\in(E_2\mathcal{H})^\perp$. Then by $x=y+z=u+(v+z)$ we have $E_1E_2x=E_2E_1x=u=E_2x$. Therefore we see that if $E_1\geqslant E_2$ then $E_1E_2=E_2E_1=E_2$ and so E_1-E_2 is also a projection. Conversely if $E_1E_2=E_2E_1$ and E_1-E_2 is a projection then $E_1E_2=E_2E_1=E_2$ and $E_1\mathcal{H}\supseteq E_1E_2\mathcal{H}=E_2\mathcal{H}$ proving $E_1\geqslant E_2$.

Lemma 8. *If E_1, E_2 are projections then $E_1\geqslant E_2$ if and only if E_1-E_2 is positive.*

Proof. If $E_1\geqslant E_2$ then we proved that E_1-E_2 is a projection and so it is positive. Conversely, if E_1-E_2 is positive then for any x in $E_2\mathcal{H}$ we have

$$0\leqslant((E_1-E_2)x,x)=(E_1x-x,x)=(E_1x,x)-(x,x)$$

and so $\|E_1x\|\cdot\|x\|\geqslant(E_1x,x)\geqslant(x,x)$. But $\|E_1x\|<\|x\|$ unless $E_1x=x$, proving that x is in $E_1\mathcal{H}$.

If A is a positive operator then $A=A^*$ and so by Proposition 3.6 its spectrum is the same with respect to $\mathcal{A}(A)$, $\mathcal{B}(A)$ and $\mathcal{L}(\mathcal{H})$.

Proposition 9. *If A is positive then $\sigma(A)\subset[0,\infty)$ and conversely, if the self adjoint operator A is such that $\sigma(A)\subset[0,\infty)$ then A is a positive operator.*

Note. The second half will be improved in Section 6, namely it is sufficient to suppose that A is normal and $\sigma(A)\subset[0,\infty)$.

Proof. Suppose that $A=A^*$ and $\sigma(A)$ lies in $[0,\infty)$. Define the continuous function $f\colon \mathbb{R}\to[0,\infty)$ by $f(\lambda)=\lambda$ for $\lambda\geqslant 0$ and $f(\lambda)=0$ for $\lambda\leqslant 0$. Since $f(\lambda)=\lambda$ on $\sigma(A)$ by the spectral theorem of self adjoint operators (Theorem 3.3) we have

$$(Ax,x)=\int \lambda\, d(E_\lambda x,x)=\int f(\lambda)\, d(E_\lambda x,x).$$

By Lemma 8 and part 1) of the theorem $(E_\lambda x,x)$ is not decreasing and $f\geqslant 0$. Hence $(Ax,x)\geqslant 0$ for any x in $\mathscr{H}$.

If $A=A^*$ then by Proposition 6 the spectrum is real so it is sufficient to prove that $(-\infty,0)\subseteq r(A)$. For $\lambda<0$ and $x\neq 0$

$$((A-\lambda I)x,x)=(Ax,x)-\lambda(x,x)>0$$

and so $A-\lambda I$ is injective. From $(Tx,y)=(x,T^*y)$ we see that for any operator $T\colon\mathscr{H}\to\mathscr{H}$ we have $(\mathscr{N}_T)^\perp=\overline{T^*\mathscr{H}}$. Hence $((A-\lambda I)\mathscr{H})^\perp=\mathscr{N}_{A-\lambda I}=0$ which proves that $(A-\lambda I)\mathscr{H}$ is dense in $\mathscr{H}$. By

$$\|x\|\cdot\|(A-\lambda I)x\|\geqslant((A-\lambda I)x,x)\geqslant-\lambda\|x\|^2$$

we see that $\|(A-\lambda I)x\|\geqslant-\lambda\|x\|$ or $\|(A-\lambda I)^{-1}y\|\leqslant-1/\lambda\|y\|$ where $y=(A-\lambda I)x$. The first of these inequalities shows that $(A-\lambda I)\mathscr{H}$ is complete and so being dense in $\mathscr{H}$ it is $\mathscr{H}$ itself. By the second $|(A-\lambda I)^{-1}|\leqslant-1/\lambda$ so the operator $(A-\lambda I)^{-1}$ is continuous.

Theorem 10. *If $A\geqslant 0$ then there exists another operator called $+\sqrt{A}$ such that $+\sqrt{A}\geqslant 0$ and $(+\sqrt{A})^2=A$. If T is such that $AT=TA$ then $(+\sqrt{A})T=T(+\sqrt{A})$.*

Proof. Since A is positive $\sigma(A)\subset[0,\infty)$ where by Proposition 3.6 $\sigma(A)$ can be understood with respect to $\mathscr{A}(A)$, $\mathscr{B}(A)$ or $\mathscr{L}(\mathscr{H})$. By applying Theorem 3.3 with $f\colon\mathbb{R}\to[0,\infty)$ where $f(\lambda)=+\sqrt{\lambda}$ for $\lambda\geqslant 0$ and $f(\lambda)=0$ for $\lambda\leqslant 0$ we obtain an operator $f(A)$ such that $f(A)^2=f^2(A)$. Since $f^2(\lambda)=\lambda$ for $\lambda\in\sigma(A)$ by part 4) of the theorem $f^2(A)=A$. Hence $f(A)=+\sqrt{A}$ satisfies the requirement $(+\sqrt{A})^2=A$. The positivity of $+\sqrt{A}$ follows from Lemma 8 and part 1) by the reasoning used in the proof of Proposition 9.

Proposition IV.4.1 which will be introduced in the next chapter is naturally related to the present Theorem 10.

5. Stable Subspaces and Commutants

The purpose of this section is to collect in one place some of the elementary facts on those subspaces of linear topological spaces which are mapped into themselves by a given set of linear transformations. The material included here will be needed in the next section and throughout the chapter on representation theory.

If $\mathscr{X}$ is a topological vector space then an arbitrary, not necessarily closed subspace will be called a *linear manifold*. If the linear manifold is a closed subset of $\mathscr{X}$ then it will be called a *subspace*. Thus a subspace will always mean a closed subspace of $\mathscr{X}$. If no topology is given on $\mathscr{X}$ then it will be viewed as a discrete vector space and so any linear manifold can be called a subspace.

Following our earlier practice by an operator $A: \mathscr{X} \to \mathscr{Y}$ we shall mean a linear operator from $\mathscr{X}$ into $\mathscr{Y}$. Moreover, if $\mathscr{X}$ and $\mathscr{Y}$ are topological vector spaces then unless the contrary is stated we shall assume that A is a continuous map.

Let $\mathscr{S}$ be a set of linear operators $S: \mathscr{X} \to \mathscr{X}$ where $\mathscr{X}$ is some vector space. Then a linear manifold $\mathscr{Y}$ of $\mathscr{X}$ is called *stable*, or *invariant* under $\mathscr{S}$ if $S\mathscr{Y} \subseteq \mathscr{Y}$ for every $S \in \mathscr{S}$. The set $\mathscr{S}$ generates an *algebra* $\tilde{\mathscr{A}}$ consisting of linear combinations of the identity operator I and operators of the form $T = S_1 \ldots S_n$ where $n \geqslant 1$ and $S_1, \ldots, S_n \in \mathscr{S}$. The following is obvious:

Proposition 1. *A linear manifold $\mathscr{Y}$ is stable under $\mathscr{S}$ if and only if it is stable under $\tilde{\mathscr{A}}$, the algebra generated by $\mathscr{S}$.*

Proposition 2. *Let the linear operator $P: \mathscr{X} \to \mathscr{X}$ be idempotent, $P^2 = P$, let $\mathscr{M} = P\mathscr{X}$ and $\mathscr{N} = (I-P)\mathscr{X}$. Then $\mathscr{M}$ and $\mathscr{N}$ are stable under $\mathscr{S}$ if and only if $PS = SP$ for every S in $\mathscr{S}$.*

Proof. If $\mathscr{M}$ is stable then $SP\mathscr{X} = S\mathscr{M} \subseteq \mathscr{M}$. Thus if x is any element of $\mathscr{X}$ then $SPx = Px'$ for some x' in $\mathscr{X}$ and so by $P^2 = P$ we obtain $PSP = SP$. If $\mathscr{N}$ is also stable then similarly we have $(I-P)S(I-P) = S(I-P)$, that is to say $PSP = PS$. Hence if both $\mathscr{M}$ and $\mathscr{N}$ are stable then $PS = PSP = SP$. Conversely, if $PS = SP$ then

$$S\mathscr{M} = SP\mathscr{X} = PS\mathscr{X} \subseteq P\mathscr{X} = \mathscr{M}$$

and in addition $(I-P)S = S(I-P)$. By the same reasoning we also get

$$S\mathscr{N} = S(I-P)\mathscr{X} = (I-P)S\mathscr{X} \subseteq (I-P)\mathscr{X} = \mathscr{N}.$$

Now let $\mathscr{X}$ be a topological vector space and let $\mathscr{L}(\mathscr{X})$ be the algebra of continuous linear operators $A: \mathscr{X} \to \mathscr{X}$. Given a set $\mathscr{S}$ of operators

S in $\mathscr{L}(\mathscr{X})$ we let $\mathscr{A}$ denote the weak closure of the algebra $\tilde{\mathscr{A}}$ in $\mathscr{L}(\mathscr{X})$ and call $\mathscr{A}$ the *weakly closed algebra generated by* $\mathscr{S}$. If $\mathscr{X}=\mathscr{H}$ is a Hilbert space and $\tilde{\mathscr{A}}$ is self adjoint then as we shall see later $\mathscr{A}$ could be interpreted also as the strong or uniform closure of $\tilde{\mathscr{A}}$ in $\mathscr{L}(\mathscr{H})$. (See the remarks after Theorem 6.10.)

If $\mathscr{X}$ is a topological vector space and if the operator P in $\mathscr{L}(\mathscr{X})$ is idempotent then $\mathscr{M}=P\mathscr{X}$ and $\mathscr{N}=(I-P)\mathscr{X}$ are closed because for instance $\mathscr{M}=\{x: x\in\mathscr{X} \text{ and } (I-P)x=0\}$. Therefore every continuous projection P determines a decomposition $\mathscr{X}=\mathscr{M}+\mathscr{N}$ where $\mathscr{M},\mathscr{N}$ are subspaces and $\mathscr{M}\cap\mathscr{N}=0$. Conversely, every such decomposition determines a not necessarily continuous idempotent P: Namely one can define $Px=m$ where $x=m+n$ with $m\in\mathscr{M}$ and $n\in\mathscr{N}$ provided $\mathscr{X}=\mathscr{M}+\mathscr{N}$ where $\mathscr{M}\cap\mathscr{N}=0$ and $\mathscr{M}$ and $\mathscr{N}$ are linear manifolds.

Suppose $\mathscr{X}$ is a Banach space and $\mathscr{M}$ and $\mathscr{N}$ are subspaces of $\mathscr{X}$. Then using the *closed graph theorem* one can see that P is continuous: For if $x_k\to x$ $(k=1,2,\ldots)$ and if $Px_k=m_k\to m$ then $\mathscr{M}$ and $\mathscr{N}$ being closed we have $m\in\mathscr{M}$ and $x-m=\lim(x_k-m_k)\in\mathscr{N}$. Thus $x=m+n$ and so $Px=m$ which is the condition appearing in the closed graph theorem. We shall use the continuity of P most often in the special case when $\mathscr{X}=\mathscr{H}$ is a Hilbert space, $P^2=P$ and $P=P^*$. Then P is seen to be continuous immediately by noticing that $(I-P)x\perp Px$ for every x in $\mathscr{H}$. Namely $\|Px\|^2\leqslant\|Px\|^2+\|(I-P)x\|^2=\|x\|^2$ and so $|P|\leqslant 1$.

Proposition 3. *For any topological vector space $\mathscr{X}$, and for any family $\mathscr{S}$ of continuous operators $S:\mathscr{X}\to\mathscr{X}$ we have: If a linear manifold $\mathscr{Y}$ is stable under $\mathscr{S}$ then so is its closure $\overline{\mathscr{Y}}$.*

Proof. It is known that a function $f: X\to Y$ from a topological space X into another Y is continuous if and only if $f(\overline{S})\subseteq\overline{f(S)}$ for every subset S of X. Thus replacing f by the continuous operator S chosen from the family $\mathscr{S}$ and the subset S by $\mathscr{Y}$ we get $S\overline{\mathscr{Y}}\subseteq\overline{S\mathscr{Y}}\subseteq\overline{\mathscr{Y}}$. This proves the proposition. If $\mathscr{X}$ is a normed vector space one can reason more directly: If $\bar{y}\in\overline{\mathscr{Y}}$ then given $S\neq 0$ in $\mathscr{S}$ and $\varepsilon>0$ there is a y in $\mathscr{Y}$ such that $\|\bar{y}-y\|<\varepsilon/|S|$ and so $\|S\bar{y}-Sy\|<\varepsilon$ where $Sy\in\mathscr{Y}$. Hence $S\bar{y}\in\overline{\mathscr{Y}}$.

Given a vector space $\mathscr{X}$ and a set $\mathscr{S}$ of linear operators $S:\mathscr{X}\to\mathscr{X}$ we let $\mathscr{S}'$ denote the set of all those linear operators $T:\mathscr{X}\to\mathscr{X}$ which commute with every S in $\mathscr{S}$ and call $\mathscr{S}'$ the *commutant* or *commuting algebra* of $\mathscr{S}'$. If $\mathscr{X}$ is a topological vector space and if $\mathscr{S}\subseteq\mathscr{L}(\mathscr{X})$ then by definition the elements of $\mathscr{S}'$ are restricted to continuous operators, so that

$$\mathscr{S}'=\{T: T\in\mathscr{L}(\mathscr{X}) \text{ and } ST=TS \text{ for all } S\in\mathscr{S}\}.$$

It is obvious that the commutant is an algebra with identity I. If $\mathcal{S}_1 \subseteq \mathcal{S}_2$ then $\mathcal{S}_2' \subseteq \mathcal{S}_1'$. Moreover $\mathcal{S} \subseteq \mathcal{S}''$ and so $\mathcal{S}' \subseteq (\mathcal{S}')'' = (\mathcal{S}'')' \subseteq \mathcal{S}'$ proving that $\mathcal{S}' = \mathcal{S}'''$. We also see that if $\mathcal{S}$ is an algebra then $\mathcal{S} \cap \mathcal{S}'$ is the center of both $\mathcal{S}$ and $\mathcal{S}'$.

Proposition 4. *Let $\mathcal{X}$ be a topological vector space, let $\mathcal{S}$ be a family of continuous operators $S: \mathcal{X} \to \mathcal{X}$ and let $\mathcal{Y}$ be an $\mathcal{S}$-stable subspace. Then $T\mathcal{Y}$ and $\overline{T\mathcal{Y}}$ are $\mathcal{S}$-stable for any $T \in \mathcal{S}'$.*

Proof. The stability of $T\mathcal{Y}$ is trivial and it is independent of the topological and algebraic structure of $\mathcal{X}$: If $T \in \mathcal{S}'$ and $S \in \mathcal{S}$ then $S(T\mathcal{Y}) = T(S\mathcal{Y}) \subseteq T\mathcal{Y}$. Since every $S \in \mathcal{S}$ is continuous $\overline{T\mathcal{Y}}$ is stable by Proposition 3.

The next four propositions are stated and proved for locally convex topological vector spaces but they will be used only in the particular case of Banach spaces. The proofs in this special situation are simpler in the sense that they depend only on the ordinary Hahn-Banach theorem instead of its generalization to topological vector spaces. Precisely we need the following corollary of the theorem: If $x \neq 0$ then there exists a continuous linear functional x^* such that $x^*(x) \neq 0$.

Proposition 5. *Let $\mathcal{X}_1, \mathcal{X}_2$ be locally convex topological vector spaces, let $\mathcal{S}$ be a subset of $\mathcal{L}(\mathcal{X}_1)$ and let $f: \mathcal{S} \to \mathcal{L}(\mathcal{X}_2)$. Then*

$$(\mathcal{S}, f) = \{T\colon T \in \mathcal{L}(\mathcal{X}_1, \mathcal{X}_2) \text{ and } TS = f(S)T \text{ for all } S \in \mathcal{S}\}$$

is a weakly closed subspace of $\mathcal{L}(\mathcal{X}_1, \mathcal{X}_2)$.

Proof. Let T belong to the weak closure of $(\mathcal{S}, f)$ and let $S \in \mathcal{S}$. We prove that $TS = f(S)T$ by showing that $(TS - f(S)T)x_1 = 0$ for every fixed $x_1 \in \mathcal{X}_1$. This will be done by verifying that

$$x_2^*((TS - f(S)T)x_1) = 0$$

for every $x_2^* \in \mathcal{X}_2^*$. Fix S in $\mathcal{S}$, x_1 in $\mathcal{X}_1$, x_2^* in $\mathcal{X}_2^*$ and $\varepsilon > 0$, and consider the following weak neighborhood of T:

$$O_T(\tfrac{1}{2}\varepsilon; x_1, Sx_1; f(S)^* x_2^*, x_2^*).$$

Since T belongs to the weak closure of $(\mathcal{S}, f)$ this neighborhood contains an operator V from $(\mathcal{S}, f)$. Thus

$$TS - f(S)T = (T - V)S - f(S)(T - V),$$

and in addition

$$|f(S)^* x_2^*((T - V)x_1)| < \tfrac{1}{2}\varepsilon \quad \text{and} \quad |x_2^*((T - V)Sx_1)| < \tfrac{1}{2}\varepsilon.$$

Therefore by $f(S)^* x_2^* = x_2^* f(S)$ we have

$$|x_2^*((TS - f(S)T)x_1)| < \varepsilon$$

where $\varepsilon > 0$ is arbitrary.

Proposition 6. *If $\mathscr{X}$ is a locally convex topological vector space and $\mathscr{S}$ is a subset of $\mathscr{L}(\mathscr{X})$ then $\mathscr{S}'$ is a weakly closed subspace of $\mathscr{L}(\mathscr{X})$.*

Proof. Apply Proposition 5 with $\mathscr{X}_1 = \mathscr{X}_2 = \mathscr{X}$ and f the identity map.

By applying Proposition II.1.1 we obtain:

Corollary. *If $\mathscr{S} \subseteq \mathscr{L}(\mathscr{X})$ where $\mathscr{X}$ is a Banach space then $\mathscr{S}'$ is a Banach algebra with identity.*

Proposition 7. *Let $\mathscr{X}$ be a locally convex topological vector space and let $\mathscr{S} \subseteq \mathscr{L}(\mathscr{X})$. Then $\mathscr{S}' = (\overline{\mathscr{S}_w})'$ where $\overline{\mathscr{S}_w}$ denotes the weak closure of $\mathscr{S}$ in $\mathscr{L}(\mathscr{X})$.*

Proof. Since the inclusion $(\overline{\mathscr{S}_w})' \subseteq \mathscr{S}'$ is obvious it is sufficient to prove that if $A \in \mathscr{S}'$ then $A \in (\overline{\mathscr{S}_w})'$. Let $T \in \overline{\mathscr{S}_w}$, $x \in \mathscr{X}$, $x^* \in \mathscr{X}^*$ and $\varepsilon > 0$ be given. By $T \in \overline{\mathscr{S}_w}$ the weak neighborhood

$$O_T(\tfrac{1}{2}\varepsilon; x, Ax; x^*A, x^*)$$

contains an operator $S \in \mathscr{S}$. Therefore by $A \in \mathscr{S}'$ we have

$$AT - TA = A(T-S) + (S-T)A$$

and also

$$|x^*(A(S-T)x)| < \tfrac{1}{2}\varepsilon \quad \text{and} \quad |x^*((T-S)Ax)| < \tfrac{1}{2}\varepsilon.$$

Hence $|x^*((AT - TA)x)| < \varepsilon$ where $\varepsilon > 0$, $x \in \mathscr{X}$ and $x^* \in \mathscr{X}^*$ are arbitrary. By the generalized Hahn-Banach theorem this implies $AT = TA$ and so $A \in (\overline{\mathscr{S}_w})'$.

Proposition 8. *Let $\mathscr{X}$ be a locally convex topological vector space and let $\mathscr{S} \subseteq \mathscr{L}(\mathscr{X})$. Then $\mathscr{S}' = \mathscr{A}'$ where $\mathscr{A}$ is the weakly closed algebra generated by $\mathscr{S}$.*

Proof. First of all it is clear that $\mathscr{S}' = \tilde{\mathscr{A}}'$ where $\tilde{\mathscr{A}}$ denotes the algebra generated by $\mathscr{S}$. By Proposition 7 we have $\tilde{\mathscr{A}}' = \mathscr{A}'$ since $\mathscr{A}$ by definition is the weak closure of $\tilde{\mathscr{A}}$ in $\mathscr{L}(\mathscr{X})$.

Proposition 9. *Let $\mathscr{X}$ be a Banach space, let $\mathscr{S} \subseteq \mathscr{L}(\mathscr{X})$ and let $\mathscr{M}, \mathscr{N}$ be stable subspaces under $\mathscr{S}$ such that $\mathscr{X} = \mathscr{M} + \mathscr{N}$ and $\mathscr{M} \cap \mathscr{N} = 0$. Then $\mathscr{M}$ and $\mathscr{N}$ are stable under $\mathscr{A}$, the weakly closed algebra generated by $\mathscr{S}$.*

Proof. Let $P: \mathscr{X} \to \mathscr{M}$ be the projection associated with the decomposition $\mathscr{X} = \mathscr{M} + \mathscr{N}$. Since $\mathscr{X}$ is a Banach space and $\mathscr{M}$ and $\mathscr{N}$ are closed P is continuous. By Proposition 1 $\mathscr{M}$ and $\mathscr{N}$ are stable under $\tilde{\mathscr{A}}$ and so by Proposition 2 P belongs to $(\tilde{\mathscr{A}})'$. Applying Proposition 8 we see that $P \in \mathscr{A}'$ and so again by Proposition 2 $\mathscr{M}$ and $\mathscr{N}$ are stable under $\mathscr{A}$.

Proposition 10. *A subspace $\mathscr{K}$ of a Hilbert space $\mathscr{H}$ is stable under a set of operators $\mathscr{S} \subseteq \mathscr{L}(\mathscr{H})$ if and only if it is stable under the weakly closed algebra $\mathscr{A}$ generated by $\mathscr{S}$.*

Proof. If $\mathscr{K}$ is stable under $\mathscr{S}$ then by Proposition 1 it is stable under the algebra $\tilde{\mathscr{A}}$ generated by $\mathscr{S}$. Now given A in $\mathscr{A}$, let N be a directed set and let A_n $(n \in N)$ be a net with A_n in $\tilde{\mathscr{A}}$ such that $A_n \to A$ in the weak topology of $\mathscr{L}(\mathscr{H})$. For any x in $\mathscr{K}$ and y in $\mathscr{K}^{\perp}$ we have $A_n x \in \mathscr{K}$ and so $0 = (A_n x, y) \to (Ax, y)$. This proves that $Ax \perp \mathscr{K}^{\perp}$ i.e. $A\mathscr{K} \subseteq \mathscr{K}$ and so $\mathscr{K}$ is stable under $\mathscr{A}$. Since $\mathscr{S} \subseteq \mathscr{A}$ the stability of $\mathscr{K}$ under $\mathscr{A}$ clearly implies that $\mathscr{K}$ is invariant under $\mathscr{S}$.

Proposition 11. *Let $\mathscr{H}$ be a Hilbert space and let $\mathscr{S} \subseteq \mathscr{L}(\mathscr{H})$ be such that the weakly closed algebra generated by $\mathscr{S}$ is self adjoint i.e. $\mathscr{A} = \mathscr{A}^*$. Then a subspace $\mathscr{K}$ is stable if and only if $\mathscr{K}^{\perp}$ is stable.*

Proof. In view of Proposition 10 stability with respect to $\mathscr{S}$ and $\mathscr{A}$ mean the same thing. Hence by $\mathscr{K} = \mathscr{K}^{\perp\perp}$ it is sufficient to see that if $\mathscr{K}$ is stable under $\mathscr{A}$ then so is $\mathscr{K}^{\perp}$. Since $\mathscr{A} = \mathscr{A}^*$ this is an immediate consequence of the following simple remark:

Lemma 12. *If $A \in \mathscr{L}(\mathscr{H})$ and if $A\mathscr{K} \subseteq \mathscr{K}$ then $A^* \mathscr{K}^{\perp} \subseteq \mathscr{K}^{\perp}$.*

Proof. If $x \in \mathscr{K}^{\perp}, y \in \mathscr{K}$ then by $A\mathscr{K} \subseteq \mathscr{K}$ we have $(A^* x, y) = (x, Ay) = 0$ and so $A^* x \perp \mathscr{K}$.

Both the foregoing lemma and the last proposition show that if $\mathscr{S}$ is a self adjoint set of operators in Hilbert space then a subspace $\mathscr{K}$ is stable if and only if $\mathscr{K}^{\perp}$ is stable.

Proposition 13. *Let $\mathscr{H}$ be a Hilbert space and let $\mathscr{S} \subseteq \mathscr{L}(\mathscr{H})$ be such that the weakly closed algebra generated by $\mathscr{S}$ is self adjoint. Then a subspace $\mathscr{K}$ is stable under $\mathscr{S}$ if and only if the associated projection $P: \mathscr{H} \to \mathscr{K}$ belongs to $\mathscr{S}'$.*

Proof. Apply Propositions 11 and 2 with $\mathscr{M} = \mathscr{K}, \mathscr{N} = \mathscr{K}^{\perp}$.

Corollary. *The projection of an invariant subspace into any other invariant subspace is again invariant.*

For if $\mathscr{K}$ is an invariant subspace then $P: \mathscr{H} \to \mathscr{K}$ belongs to $\mathscr{S}'$. Hence if $\mathscr{L}$ is another invariant subspace and $S \in \mathscr{S}$ then

$$S(P\mathscr{L}) = (SP)\mathscr{L} = (PS)\mathscr{L} = P(S\mathscr{L}) \subseteq P\mathscr{L} .$$

6. von Neumann Algebras

Throughout this section Hilbert spaces $\mathscr{H}$ will always mean complex Hilbert spaces even if this fact is not explicitely mentioned. The algebras $\mathscr{A}$ considered here are certain subalgebras of $\mathscr{L}(\mathscr{H})$ and they can be defined in two different ways. In both of these definitions one supposes that $\mathscr{A}$ is self adjoint, i.e. $\mathscr{A} = \mathscr{A}^*$. In one of the definitions the additional requirement is that $\mathscr{A} = \mathscr{A}''$. Then by Proposition 5.6 $\mathscr{A}$ is a weakly closed algebra and by its corollary $\mathscr{A}$ is a Banach algebra with identity. The other way of defining v. Neumann algebras is by supposing that the self adjoint algebra $\mathscr{A}$ is weakly closed and contains the identity operator I. The relation $\mathscr{A} = \mathscr{A}''$ is then one of the fundamental results of the theory. Historically this was the way that these algebras were first introduced and we are going to proceed in the same manner:

Definition 1. *A complex algebra $\mathscr{A}$ of continuous operators $A: \mathscr{H} \to \mathscr{H}$ where $\mathscr{H}$ is a complex Hilbert space is called a v. Neumann algebra if $\mathscr{A}$ is self adjoint, contains the identity operator and is weakly closed in $\mathscr{L}(\mathscr{H})$.*

Note. These operator algebras are often called *W^*-algebras.*

For example $\mathscr{L}(\mathscr{H})$ itself and $\mathbb{C}I = \{\lambda I: \lambda \in \mathbb{C}\}$ are v. Neumann algebras. If $\mathscr{S} \subseteq \mathscr{L}(\mathscr{H})$ and if $\mathscr{S} = \mathscr{S}^*$ then by Proposition 5.6 $\mathscr{S}'$ is a v. Neumann algebra. More generally, we may consider $(\mathscr{S} \cup \mathscr{S}^*)'$ for any set of continuous operators in $\mathscr{L}(\mathscr{H})$. The intersection of an arbitrary family of v. Neumann algebras belonging to the same Hilbert space $\mathscr{H}$ is itself a v. Neumann algebra. In particular, if $\mathscr{S}$ is an arbitrary set of operators then the intersection of all v. Neumann algebras containing $\mathscr{S}$, which includes at least $\mathscr{L}(\mathscr{H})$ itself, is a v. Neumann algebra, called the v. Neumann algebra generated by $\mathscr{S}$. It is the smallest one among these algebras containing the set $\mathscr{S}$.

We recall that the center of an algebra $\mathscr{A}$ is the commutative subalgebra

$$c\mathscr{A} = \{C: C \in \mathscr{A} \text{ and } AC = CA \text{ for every } A \in \mathscr{A}\} .$$

If $\mathscr{A}$ is an algebra of continuous operators then its center is $\mathscr{A} \cap \mathscr{A}'$ and so if $\mathscr{A}$ is a v. Neumann algebra then $c\mathscr{A}$ is also a v. Neumann algebra.

More generally one can prove in a few lines that if an algebra $\mathscr{A}$ is weakly closed then so is its center $c\mathscr{A}$ and if $\mathscr{A}=\mathscr{A}^*$ then $c\mathscr{A}=(c\mathscr{A})^*$.

Definition 2. *A v. Neumann algebra $\mathscr{A}$ is called a factor if its center is trivial i.e. $c\mathscr{A}=\mathscr{A}\cap\mathscr{A}'=\mathbb{C}I$.*

Let $\mathscr{H}_i$ $(i\in\mathscr{I})$ be Hilbert spaces, let $\mathscr{H}$ be their direct sum i.e. let $\mathscr{H}=\sum\mathscr{H}_i$ and let $A_i\in\mathscr{L}(\mathscr{H}_i)$ be such that $\operatorname{supr}|A_i|<\infty$. For each $x=(x_i)\in\mathscr{H}$ we let $Ax=(A_ix_i)$ so that $A\colon\mathscr{H}\to\mathscr{H}$ is a continuous linear operator, $|A|=\operatorname{supr}|A_i|$ and each $\mathscr{H}_i$, interpreted as a subspace of $\mathscr{H}$, is invariant under A. If the operators A_i are chosen from given subsets $\mathscr{S}_i$ of $\mathscr{L}(\mathscr{H}_i)$ then the A's form a subset of $\mathscr{L}(\mathscr{H})$, called the *direct sum* of the $\mathscr{S}_i$'s and it is denoted by $\sum\mathscr{S}_i$.

We see immediately that $I\in\sum\mathscr{S}_i$ if and only if $I_i\in\mathscr{S}_i$ for every $i\in\mathscr{I}$ and $\sum\mathscr{S}_i$ is an algebra if and only if each $\mathscr{S}_i$ is an algebra. Moreover we have $(\sum\mathscr{S}_i)^*=\sum\mathscr{S}_i^*$. One can also prove that $\sum\mathscr{S}_i$ is weakly closed in $\mathscr{L}(\mathscr{H})$ if and only if $\mathscr{S}_i$ is weakly closed in $\mathscr{L}(\mathscr{H}_i)$ for every $i\in\mathscr{I}$. Hence $\sum\mathscr{A}_i$ is a v. Neumann algebra if and only if every $\mathscr{A}_i$ is a v. Neumann algebra. One calls $\sum\mathscr{A}_i$ the *direct sum* or *direct product* of the v. Neumann algebras $\mathscr{A}_i$ and one uses also the notation $\prod\mathscr{A}_i$ instead of $\sum\mathscr{A}_i$.

It is natural to ask what can be said about the commuting algebra of a direct sum $\sum\mathscr{S}_i$ but in general these commutants can be characterized only in terms of certain morphisms of the algebra generated by $\sum\mathscr{S}_i$ and their intertwining operators which will be introduced only in the next chapter. Nevertheless several important corollaries can be easily stated and will be discussed here. One of these will be needed in the proof of *v. Neumann's fundamental theorem*, which states that $\mathscr{A}=\mathscr{A}''$ for every one of his algebras.

Proposition 3. *For any choice of $\mathscr{S}_i\subseteq\mathscr{L}(\mathscr{H}_i)$ we have $\sum\mathscr{S}_i'\subseteq(\sum\mathscr{S}_i)'\cap\sum\mathscr{L}(\mathscr{H}_i)$.*

Proof. If $\sum\mathscr{S}_i$ is empty, the statement is trivial. Let E_i denote the projection $E_i\colon\mathscr{H}\to\mathscr{H}_i$ so that $I=\sum E_i$. Then for any T in $\mathscr{L}(\mathscr{H})$we have $T=\sum T_{ij}$ where $T_{ij}=E_iTE_j$ belongs to $\mathscr{L}(\mathscr{H}_j,\mathscr{H}_i)$. If T is in $\sum\mathscr{L}(\mathscr{H}_i)$ then $T=\sum T_{ii}$ and T belongs to $\sum\mathscr{S}_i$ if and only if T is in $\sum\mathscr{L}(\mathscr{H}_i)$ and $T_{ii}\in\mathscr{S}_i$ for every $i\in\mathscr{I}$. Now let A be in $\sum\mathscr{S}_i$ and T in $\sum\mathscr{S}_i'$ so that $A=\sum A_{ii}$ with $A_{ii}\in\mathscr{S}_i$ and $T=\sum T_{jj}$ with $T_{jj}\in\mathscr{S}_j'$. Since $E_iE_j=0_j$ for $i\neq j$ we see that

$$AT=\sum A_{ii}T_{ii}=\sum T_{ii}A_{ii}=TA.$$

This proves that $\sum\mathscr{S}_i'\subseteq(\sum\mathscr{S}_i)'$.

Proposition 4. *If* $\mathcal{S}_i \subseteq \mathcal{L}(\mathcal{H}_i)$ *and if* $\sum \mathcal{S}_i$ *is not void then* $\sum \mathcal{S}_i' = (\sum \mathcal{S}_i)' \cap \sum \mathcal{L}(\mathcal{H}_i)$.

Proof. Since the inclusion $\subseteq$ has been proved in general it is sufficient to notice the following: If $i \in \mathcal{I}$ and $A_{ii} \in \mathcal{S}$ are fixed then there exists an A in $\sum \mathcal{S}_i$ such that $E_i A E_i = A_{ii}$: Indeed if we choose any element of $\sum \mathcal{S}_i$ then its i-th defining component can be replaced by A_{ii} and the new operator will still belong to $\sum \mathcal{S}_i$. Now if $T = \sum T_{ii}$ is in $(\sum \mathcal{S}_i)'$ and A in $\sum \mathcal{S}_i$ then we have $AT = TA$ or $\sum A_{ii} T_{ii} = \sum T_{ii} A_{ii}$. Hence by applying a fixed E_i both from the left and the right we obtain $A_{ii} T_{ii} = T_{ii} A_{ii}$ where as we had seen A_{ii} can be any element of $\mathcal{S}_i$. We obtain $T_{ii} \in \mathcal{S}_i'$ and $T \in \sum \mathcal{S}_i'$.

Proposition 5. *If* $\mathcal{S}_i \subseteq \mathcal{L}(\mathcal{H}_i)$ *and if* $\mathcal{S}_i$ *contains both* 0_i *and* I_i *then* $\sum \mathcal{S}_i' = (\sum \mathcal{S}_i)'$.

Proof. In view of Proposition 4 it is sufficient to prove that $(\sum \mathcal{S}_i)' \subseteq \sum \mathcal{L}(\mathcal{H}_i)$. Since $I_i \in \mathcal{S}_i$ and $0_j \in \mathcal{S}_j$ for every $j \neq i$ we see that $E_i: \mathcal{H} \to \mathcal{H}_i$ belongs to $\sum \mathcal{S}_i$. Hence, if T is in $(\sum \mathcal{S}_i)'$ then $TE_i = E_i T$. We have $T = \sum T_{ij}$ where $T_{ij} = E_i T E_j$ the right hand side being interpreted as an operator from $\mathcal{H}_j$ to $\mathcal{H}_i$. We obtain $T_{ij} = E_i T E_j = T(E_i E_j)$ and so $T_{ij} = 0$ for $i \neq j$. We proved that T belongs to $\sum \mathcal{L}(\mathcal{H}_i)$.

It seems appropriate to include now a few observations which will be used only later and return to the study of v. Neumann algebras after these remarks: Let $\mathcal{I}$ be an index set, let $\mathcal{H} = \sum \{\mathcal{H}_i : i \in \mathcal{I}\}$ and let $\mathcal{S}$ be a family of operators $S: \mathcal{H} \to \mathcal{H}$ such that $S\mathcal{H}_i \subseteq \mathcal{H}_i$ for every i in $\mathcal{I}$. We let E_i denote the projection $E_i: \mathcal{H} \to \mathcal{H}_i$. Since $\mathcal{H}_i^{\perp} = \sum \mathcal{H}_j$ where summation is on all $j \neq i$ we also have $S\mathcal{H}_i^{\perp} \subseteq \mathcal{H}_i^{\perp}$ and so by Proposition 5.2 $SE_i = E_i S$ for every $S \in \mathcal{S}$. For any $T: \mathcal{H} \to \mathcal{H}$ let $T_{ij} = E_i T E_j$ $(i, j = 1, \ldots, n)$. Finally let $(\mathcal{S} \mid \mathcal{H}_j, \mathcal{S} \mid \mathcal{H}_i)$ be the set of those operators T in $\mathcal{L}(\mathcal{H}_j, \mathcal{H}_i)$ for which $TS = ST$ on $\mathcal{H}_j$ or in other words $T(SE_j) = (SE_i) TE_j$ on $\mathcal{H}$.

Proposition 6. *We have* $T \in \mathcal{S}'$ *if and only if* $T_{ij} \in (\mathcal{S} \mid \mathcal{H}_j, \mathcal{S} \mid \mathcal{H}_i)$ *for every* $i, j \in \mathcal{I}$.

Proof. First, if $T \in \mathcal{S}'$ then

$$T_{ij}(SE_j) = E_i T E_j S E_j = S E_i T E_j = (SE_i) T_{ij}$$

and so $T_{ij} \in (\mathcal{S} \mid \mathcal{H}_j, \mathcal{S} \mid \mathcal{H}_i)$. Conversely, if T_{ij} is an element of $(\mathcal{S} \mid \mathcal{H}_j, \mathcal{S} \mid \mathcal{H}_i)$ for every i, j in $\mathcal{I}$ then $T \in \mathcal{S}'$: For $S = \sum S_{kk}$ where $S_{kk} = E_k S E_k$ and so

$$TS = \Big(\sum_{i,j} T_{ij}\Big)\Big(\sum_{k} S_{kk}\Big) = \sum_{i,j} T_{ij} S_{jj}$$

where $T_{ij}S_{jj}=T_{ij}(SE_j)=(SE_i)T_{ij}=S_{ii}T_{ij}$. Thus

$$TS=\sum_{i,j}S_{ii}T_{ij}=\Big(\sum_k S_{kk}\Big)\Big(\sum_{i,j}T_{ij}\Big)=ST.$$

Let $\mathscr{I}$ be an index set and let $\tilde{\mathscr{H}}=\sum\mathscr{H}_i$ where $\mathscr{H}_i$ denotes the same Hilbert space $\mathscr{H}$ for every $i\in\mathscr{I}$. For each A in $\mathscr{L}(\mathscr{H})$ let $\tilde{A}=\sum A_i$ where $A_i=A$ for every i in $\mathscr{I}$ and if $\mathscr{S}\subseteq\mathscr{L}(\mathscr{H})$ let $\tilde{\mathscr{S}}=\{\tilde{A}: A\in\mathscr{S}\}$. As earlier let E_i denote both the projection $E_i\colon\tilde{\mathscr{H}}\to\mathscr{H}_i=\mathscr{H}$ and its restriction $\mathscr{H}_i\to\mathscr{H}_i$. For T in $\mathscr{L}(\tilde{\mathscr{H}})$ let $T_{ik}=E_iTE_k$ so that T_{ik} belongs to $\mathscr{L}(\mathscr{H}_k,\mathscr{H}_i)=\mathscr{L}(\mathscr{H})$.

Proposition 7. *We have $T\in(\tilde{\mathscr{S}})'$ if and only if $T_{ik}\in\mathscr{S}'$ for every $i,k\in\mathscr{I}$.*

Proof. If $\tilde{A}\in\tilde{\mathscr{S}}$ then $\tilde{A}=\sum A_{ii}$ where $A_{ii}=E_i\tilde{A}E_i$ denotes the same operator $A\in\mathscr{S}$ for every $i\in\mathscr{I}$. Furthermore if T is in $\mathscr{L}(\tilde{\mathscr{H}})$ then $T=\sum T_{jk}$ and so we have

$$\tilde{A}T=\sum_{i,k}A_{ii}T_{ik}\quad\text{and}\quad T\tilde{A}=\sum_{i,k}T_{ik}A_{kk}.$$

If $\tilde{A}T=T\tilde{A}$ then by applying a fixed E_i from the left and another E_k from the right we obtain $A_{ii}T_{ik}=T_{ik}A_{kk}$. Hence if $T\in(\tilde{\mathscr{S}})'$ then $T_{ik}\in\mathscr{S}'$. Conversely, if $T_{ik}\in\mathscr{S}'$ for every $i,k\in\mathscr{I}$ then for every A in $\mathscr{S}$ we have $A_{ii}T_{ik}=T_{ik}A_{kk}$ where $A_{ii}=A_{kk}=A$ and so $\tilde{A}T=T\tilde{A}$ proving that $T\in(\tilde{\mathscr{S}})'$.

Proposition 8. *For any $\mathscr{S}\subseteq\mathscr{L}(\mathscr{H})$ we have $(\tilde{\mathscr{S}})''=\widetilde{\mathscr{S}''}$.*

Proof. We notice that $\tilde{I}_{ij}=E_i\tilde{I}E_j$ when interpreted as an operator from $\tilde{\mathscr{H}}$ to $\tilde{\mathscr{H}}$ belongs to $(\tilde{\mathscr{S}})'$ because $(\tilde{I}_{ij})_{kl}$ is 0 unless $i=k$ and $j=l$ in which case it is the identity of $\mathscr{L}(\mathscr{H})$. For any A in $\mathscr{L}(\tilde{\mathscr{H}})$ we have $\tilde{I}_{ij}A=\tilde{I}_{ij}\sum_{k,l}A_{kl}=\sum_l\tilde{I}_{ij}A_{jl}$ and similarly $A\tilde{I}_{ij}=\sum_k A_{ki}\tilde{I}_{ij}$. Hence, if A is in $(\tilde{\mathscr{S}})''$ then $\sum_l\tilde{I}_{ij}A_{jl}=\sum_k A_{ki}\tilde{I}_{ij}$ and so by fixing k and applying E_k from the left and E_j from the right we obtain $E_k\tilde{I}_{ij}A_{jj}=A_{ki}\tilde{I}_{ij}$. By choosing $j=i$ we see that $A_{ki}=0$ for every $i\neq k$. For $k=i$ the last equation is reduced to $\tilde{I}_{ij}A_{jj}=A_{ii}\tilde{I}_{ij}$ from which we derive that $A_{jj}=A_{ii}$ for any two indices $i,j\in\mathscr{I}$. Moreover, if $T\in\mathscr{S}'$ then by the last proposition $\tilde{T}\in(\tilde{\mathscr{S}})'$ and so by $A\in(\tilde{\mathscr{S}})''$ we obtain $\tilde{T}A=A\tilde{T}$ i.e. $TA_{kk}=A_{kk}T$ for every $k\in\mathscr{I}$. Hence we proved that if $A\in(\tilde{\mathscr{S}})''$, then $A\in\widetilde{\mathscr{S}''}$. The converse statement is an easy consequence of Proposition 7: For if $A=\tilde{B}$ where $B\in\mathscr{S}''$ then the object is to prove that $AT=TA$ for every T in $\tilde{\mathscr{S}}'$. But then $T_{ik}\in\mathscr{S}'$ and so the components of A being 0's and B's we have $A_{ij}T_{kl}=T_{kl}A_{ij}$ for any $i,j,k,l\in\mathscr{I}$.

Proposition 9. *Let* $\mathscr{A} \subseteq \mathscr{L}(\mathscr{H})$ *be a self adjoint algebra, let* $\mathscr{A}$ *contain the identity operator and let* $x \in \mathscr{H}$. *Then the linear manifold* $\{Tx: T \in \mathscr{A}''\}$ *is stable under* $\mathscr{A}$ *and it is contained in the closure of the* $\mathscr{A}$*-stable manifold* $\{Ax: A \in \mathscr{A}\}$.

Corollary. *Given* $T \in \mathscr{A}''$, $x \in \mathscr{H}$ *and* $\varepsilon > 0$ *there is an* A *in* $\mathscr{A}$ *such that* $\|(A-T)x\| < \varepsilon$ *i.e.* $A \in O_T(\varepsilon; x) \cap \mathscr{A}$.

Proof. The invariance of $\{Tx: T \in \mathscr{A}''\}$ under $\mathscr{A}$ follows from $\mathscr{A} \subseteq \mathscr{A}''$: Indeed, if A is in $\mathscr{A}$, then $A(Tx) = (AT)x$ where $AT \in \mathscr{A}''$. Denote by $\mathscr{K}$ the closure of $\{Ax: A \in \mathscr{A}\}$ in $\mathscr{H}$ so that $\mathscr{K}$ is invariant under $\mathscr{A}$. By $\mathscr{A} = \mathscr{A}^*$ and Lemma 5.12 $\mathscr{K}^\perp$ is also $\mathscr{A}$-invariant. Hence by Proposition 5.2 the projection $P: \mathscr{H} \to \mathscr{K}$ belongs to $\mathscr{A}'$. Now if $T \in \mathscr{A}''$ then $TP = PT$ and so by $x = Ix \in \mathscr{K}$ we have $Px = x$ and $Tx = TPx = PTx \in P\mathscr{H} = \mathscr{K}$.

Theorem 10. *Let* $\mathscr{A}$ *be a self adjoint algebra of operators in Hilbert space and let* $\mathscr{A}$ *contain the identity operator. Then* $\mathscr{A}''$ *is contained in the strong closure of* $\mathscr{A}$, *i.e.* $\mathscr{A}'' \subseteq \overline{(\mathscr{A})}_s$.

This important result, known as *v. Neumann's density theorem*, can be used as a very sensitive and versatile tool and it has many important consequences. A few immediate corollaries will be pointed out now, before its proof: First $\mathscr{A} \subseteq \mathscr{A}''$ and so if $\mathscr{A}$ is weakly closed then $\mathscr{A} \subseteq \mathscr{A}'' \subseteq \overline{(\mathscr{A})}_s \subseteq \overline{(\mathscr{A})}_w = \mathscr{A}$. Hence we have:

Corollary. *If* $\mathscr{A}$ *is a v. Neumann algebra then* $\mathscr{A} = \mathscr{A}''$.

We also see that if $\mathscr{A}$ is strongly closed then by the density theorem $\mathscr{A}'' \subseteq \overline{(\mathscr{A})}_s = \mathscr{A}$ and so $\mathscr{A} = \mathscr{A}''$ where by Proposition 5.6 $\mathscr{A}''$ is weakly closed. Hence we see that v. Neumann algebras could be defined as strongly closed self adjoint algebras with identity.

Proof. In order to prove Theorem 10 let $S \in \mathscr{A}''$ and a strong neighborhood of S, say $O_S(\varepsilon; x_1, \ldots, x_n)$ ($\varepsilon > 0$ and $x_1, \ldots, x_n \in \mathscr{H}$) be given. It is sufficient to show that O_S contains some operator A from the given algebra $\mathscr{A}$. We form the new Hilbert space $\tilde{\mathscr{H}} = \mathscr{H}_1 \oplus \cdots \oplus \mathscr{H}_n$, where $\mathscr{H}_k$ is the same space $\mathscr{H}$ for every $k = 1, \ldots, n$. For each A in $\mathscr{A}$ we let $\tilde{A} = A \oplus \cdots \oplus A$ so that the set of $\tilde{A}$'s is a self adjoint algebra $\tilde{\mathscr{A}}$ of operators from $\tilde{\mathscr{H}}$ to $\tilde{\mathscr{H}}$ with identity $\tilde{I}$. Since S belongs to $\mathscr{A}''$ by Proposition 8 $\tilde{S}$ is in $(\tilde{\mathscr{A}})''$. We let $x = (x_1, \ldots, x_n)$ where $x_1, \ldots, x_n$ are the elements determining the given strong neighborhood O_S in $\mathscr{L}(\tilde{\mathscr{H}})$ and apply the corollary of Proposition 9 with our Hilbert space $\tilde{\mathscr{H}}$, the algebra $\tilde{\mathscr{A}}$ and the operator $\tilde{S}$ in place of T. We find an $\tilde{A}$ in $\tilde{\mathscr{A}}$ such that $\|(\tilde{A} - \tilde{S})x\| < \varepsilon$, or in other words there exists an A in $\mathscr{A}$

such that $\sum \|(A-S)x_k\|^2 < \varepsilon^2$. Hence we determined an operator A which belongs to $\mathscr{A} \cap O_S(\varepsilon; x_1, \ldots, x_n)$ and so we completed the proof of the density theorem.

Proposition 11. *A self adjoint operator algebra $\mathscr{A}$ is a v. Neumann algebra if and only if $\mathscr{A} = \mathscr{A}''$.*

Proof. If $\mathscr{A} = \mathscr{A}''$ then $\mathscr{A}$ is weakly closed by Proposition 5.6 and obviously it contains the identity. The converse statement is identical with the corollary of Theorem 10.

We return for a moment to the situation which occurs in Propositions 7 and 8. Clearly $\widetilde{\mathscr{S}^*} = (\tilde{\mathscr{S}})^*$ for any set of operators $\mathscr{S} \subseteq \mathscr{L}(\mathscr{H})$ and if $\mathscr{S}$ is an algebra, say $\mathscr{S} = \mathscr{A}$, then $\tilde{\mathscr{A}}$ is isometrically isomorphic to $\mathscr{A}$. Moreover, if $\mathscr{A}$ is a v. Neumann algebra then so is $\tilde{\mathscr{A}}$ because by Proposition 8 we have $(\tilde{\mathscr{A}})'' = \widetilde{\mathscr{A}''} = \tilde{\mathscr{A}}$. This could be proved also without using the density theorem, because more generally a set $\mathscr{S}$ is weakly closed in $\mathscr{L}(\mathscr{H})$ if and only if $\tilde{\mathscr{S}}$ is weakly closed in $\mathscr{L}(\tilde{\mathscr{H}})$: For, by applying the *principle of uniform boundedness* one proves that if $A_k \to A$ weakly, then $\tilde{A}_k \to \tilde{A}$ in the weak topology of $\mathscr{L}(\tilde{\mathscr{H}})$. Indeed let the linear functional $f_k \colon \mathscr{H} \oplus \mathscr{H} \to \mathbb{C}$ be defined by $f_k(x,y) = (A_k x, y)$. Since $f_k(x,y) \to (Ax,y)$ the set $\{f_k(x,y)\}$ is bounded for every fixed (x,y) in $\mathscr{H} \oplus \mathscr{H}$. Hence by the principle there is a $C > 0$ such that $|f_k| \leqslant C$ for every k, that is to say $|(A_k x, y)| \leqslant C\sqrt{\|x\|^2 + \|y\|^2}$ for every k and x, y in $\mathscr{H}$. Now it is easy to show that if $\tilde{x}, \tilde{y}$ are in $\tilde{\mathscr{H}}$ then

$$(\tilde{A}_k \tilde{x}, \tilde{y}) = \sum (A_k x_i, y_i) \to \sum (A x_i, y_i) = (\tilde{A}\tilde{x}, \tilde{y}).$$

Proposition 12. *Every v. Neumann algebra is a B^* algebra with identity with respect to the operator norm.*

Proof. Since $\mathscr{A}$ is weakly closed it is uniformly closed.

Lemma 13. *Let A be a Banach algebra with identity and let a in A satisfy $\|a\| < 1$. Then*

$$\sum_{n=0}^{\infty} \binom{\frac{1}{2}}{n} a^n$$

converges in A to an element called $\sqrt{e+a}$ which commutes with a and for which $(\sqrt{e+a})^2 = e + a$. If A is a B^* algebra and a is self adjoint then so is $\sqrt{e+a}$.

Note. Here $\binom{\frac{1}{2}}{0}=1$ and for $n\geqslant 1$ the symbol $\binom{\frac{1}{2}}{n}$ denotes the value of the polynomial

$$\binom{t}{n}=\frac{t(t-1)\cdots(t-n+1)}{n!}$$

at $t=\frac{1}{2}$ so that $\binom{\frac{1}{2}}{1}=1$ and for $n\geqslant 2$

$$\binom{\frac{1}{2}}{n}=\frac{(-1)^{n-1}}{2n}\cdot\frac{1\cdot 3\cdots(2n-3)}{2\cdot 4\cdots(2n-2)}.$$

Proof. Since $\left|\binom{\frac{1}{2}}{n}\right|\leqslant 1$ for every $n=0,1,\ldots$ the series is absolutely convergent. When computing the square of the limit term by term multiplication gives

$$(\sqrt{e+a})^2=\sum_{n=0}^{\infty} c_n a^n$$

where

$$c_n=\sum_{k+l=n}\binom{\frac{1}{2}}{k}\binom{\frac{1}{2}}{l}.$$

Using Leibnitz's rule one sees that c_n is the n-th derivative of $\sqrt{1+x}\sqrt{1+x}$ at $x=0$. Thus $c_0=c_1=1$ and $c_n=0$ for $n\geqslant 2$. Of course we are dealing with the Taylor series of $(1+x)^{\frac{1}{2}}$ evaluated at the algebra element a. If a is self adjoint so is every term of the series.

Proposition 14. *Every v. Neumann algebra is generated by its unitary operators.*

Proof. It is sufficient to prove that every self adjoint element A satisfying $|A|<1$ can be written in the form $A=\frac{1}{2}(U+U^*)$ where U is a unitary operator in the algebra $\mathscr{A}$. Since $\mathscr{A}$ is a B^* algebra with identity Lemma 13 is applicable. If we let $U=A+i\sqrt{I-A^2}$ then $U^*=A-i\sqrt{I-A^2}$ and by the commutativity of A and $\sqrt{I-A^2}$ we have $UU^*=U^*U=I$. Clearly we have $A=\frac{1}{2}(U+U^*)$.

Corollary. *Every matrix $A\in M_n(\mathbb{C})$ is a linear combination of unitary elements of $M_n(\mathbb{C})$.*

In fact $\mathscr{L}(\mathbb{C}^n)$ is a v. Neumann algebra which is isomorphic to the involutive algebra $M_n(\mathbb{C})$.

Proposition 15. *Every v. Neumann algebra is the uniform closure of the algebra generated by its projections.*

Proof. Every A in $\mathscr{A}$ can be written in the form

$$A=\frac{A+A^*}{2}+i\frac{A-A^*}{2i}=A_1+iA_2$$

where A_1, A_2 are self adjoint and A_1, A_2 belong to $\mathscr{A}$ by $\mathscr{A}=\mathscr{A}^*$. If A is self adjoint by Proposition 3.5 we can find $\lambda_1,\dots,\lambda_n$ in $\mathbb{R}$ and projections $P_1,\dots,P_n$ such that each $P_k\in\{A\}''\subseteq\mathscr{A}''$ and

$$|A-(\lambda_1 P_1+\cdots+\lambda_n P_n)|\leqslant\varepsilon.$$

By $\mathscr{A}=\mathscr{A}''$ the projections P_k all belong to $\mathscr{A}$.

Corollary. *The weak closure of the algebra generated by the projections of $\mathscr{A}$ is $\mathscr{A}$ itself and so the smallest v. Neumann algebra containing all the projection operators of $\mathscr{A}$ is $\mathscr{A}$.*

Proposition 16. *Let $\mathscr{A}$ be a v. Neumann subalgebra of $\mathscr{L}(\mathscr{H})$. Then every element of $\mathscr{A}$ has the same spectrum in $\mathscr{A}$ as in $\mathscr{L}(\mathscr{H})$.*

Proof. By the second corollary of Proposition I.4.3. it is sufficient to prove that if $(A-\lambda I)^{-1}$ exists then it belongs to $\mathscr{A}$ provided A belongs to $\mathscr{A}$. Now if T is in $\mathscr{A}'$ then $T(A-\lambda I)=(A-\lambda I)T$ and so $(A-\lambda I)^{-1}$ commutes with every $T\in\mathscr{A}'$. Hence $(A-\lambda I)^{-1}\in\mathscr{A}''=\mathscr{A}$.

If $T:\mathscr{H}\to\mathscr{H}$ is a normal operator then the weakly closed algebra $\mathscr{A}=\mathscr{A}(T)$ introduced in Proposition 4.4 is a v. Neumann subalgebra of $\mathscr{L}(\mathscr{H})$. Hence in view of Propositions 4.6, 4.7 and 4.9 we obtain the following:

Corollary. *Let $T:\mathscr{H}\to\mathscr{H}$ be a normal operator and let $\sigma(T)$ be the spectrum of T as an element of $\mathscr{L}(\mathscr{H})$. Then $\sigma(T)\subset\mathbb{R}$ implies that T is self adjoint. Similarly if $|\sigma(T)|=1$ then T is unitary and if $\sigma(T)\subset[0,\infty)$ then T is positive.*

By Theorem I.8.7 we have $|T|=\|T\|_{sp}$ where T is normal and the spectral norm is taken with respect to the algebra $\mathscr{A}(T)$. Therefore using Proposition 16 we obtain:

Proposition 17. *If $T:\mathscr{H}\to\mathscr{H}$ is normal then $|T|=\|T\|_{sp}$ where $\|T\|_{sp}$ denotes the spectral norm in $\mathscr{L}(\mathscr{H})$.*

7. Measures on Locally Compact Spaces

In this section we give the fundamentals of Bourbaki's approach to measure theory and a few specific results which will be used later. The basic idea is to regard integrals as primary concepts and define measures in terms of these rather than first introduce measurable sets and their measures and then define integrals.

Let X be a locally compact space and let K be an arbitrary compact subset of X. We let $C_K(X)$ denote the set of all those continuous functions $f: K \to \mathbb{C}$ for which K is a carrier of f. For every such function f the norm

$$\|f\| = \|f\|_\infty = \text{supr}\,\{|f(x)|: x \in X\}$$

is well defined and $C_K(X)$ is a Banach algebra under this norm. However most of the time we shall view $C_K(X)$ only as a complex Banach space. Similarly $C_0(X)$, the vector space of all complex valued functions $f: K \to \mathbb{C}$ having compact support will be considered her as a vector space over $\mathbb{C}$. Clearly we have $C_0(X) = \bigcup C_K(X)$ where the union is formed with respect to all compact subsets of X.

The vector space $C_0(X)$ can be topologized as follows: A subset O of $C_0(X)$ will be called open if $O \cap C_K(X)$ is an open subset of $C_K(X)$ for every compact subset K of X. The empty set $\emptyset$ and $C_0(X)$ are clearly open and one can easily see that the intersection of finitely many open sets is also open. Similarly one shows that the union of an arbitrary family of open sets is again an open set. Under this topology $C_0(X)$ is a locally convex topological vector space. The topology defined in this manner is actually the direct limit of the topologies of the subspaces $C_K(X)$. From here on we shall view $C_0(X)$ as a complex, locally convex topological vector space.

Proposition 1. *A linear functional $\mu: C_0(X) \to \mathbb{C}$ is continuous if and only if the restriction of μ to $C_K(X)$ is continuous for every compact subset K of X.*

Proof. Since μ is linear it is sufficient to investigate its continuity only at 0. First we suppose that μ itself is continuous. Then there is an open set O in $C_0(X)$ such that O contains 0 and $|\mu(f)| < 1$ for every $f \in O$. Let K be a compact subset of X. Then by the definition of the topology of $C_0(X)$ we see that $O \cap C_K(X)$ is an open subset of $C_K(X)$ which contains 0. Hence there is an $\varepsilon > 0$ such that

$$\{f: f \in C_K(X) \text{ and } \|f\| < \varepsilon\} \subseteq O \cap C_K(X) \subseteq O.$$

Thus we have $|\mu(f)|<1$ for every f in $C_K(X)$ satisfying $\|f\|<\varepsilon$. This shows that the restriction of μ to $C_K(X)$ is continuous.

Conversely let μ be such that $\mu|C_K(X)$ is continuous for every compact set K. In order to prove that μ is continuous at 0 it is sufficient to show that

$$\{f: f\in C_0(X) \text{ and } |\mu(f)|<1\}$$

is an open subset of $C_0(X)$. For this purpose it is sufficient to prove that

$$\{f: f\in C_K(X) \text{ and } |\mu(f)|<1\}$$

is an open set in $C_K(X)$ for every K. But this is obvious from the continuity of the restriction of μ to $C_K(X)$.

Definition 2. *A continuous linear functional* $\mu: C_0(X)\to\mathbb{C}$ *is a measure in the Bourbaki-Radon sense.*

Since each $C_K(X)$ is a normed vector space by Proposition 1 we see that a linear functional $\mu: C_0(X)\to\mathbb{C}$ is a measure if and only if for each compact set K there is a positive constant m_K such that $|\mu(f)|\leqslant m_K\|f\|$ for every f in $C_K(X)$.

A not a priori continuous linear functional $\mu: C_0(X)\to\mathbb{C}$ is called *positive* if $\mu(f)\geqslant 0$ for every non-negative f in $C_0(X)$. From this point on we suppose that the locally compact X is a Hausdorff space. The reason for this restriction will be obvious from the proof of the following fundamental result:

Proposition 3. *Every positive linear functional* $\mu: C_0(X)\to\mathbb{C}$ is a *measure on the locally compact Hausdorff space* X.

Proof. Let K be a compact subspace of X. Since X is a locally compact Hausdorff space K has a compact neighborhood N. Since X is completely regular there is a continuous function $g: X\to[0,1]$ such that $g(x)=1$ for every $x\in K$ and $g(y)=0$ for every $y\notin N$. Now if $f\in C_K(X)$ then $-\|f\|g\leqslant f\leqslant\|f\|g$ and so

$$-\|f\|\mu(g)\leqslant\mu(f)\leqslant\|f\|\mu(g).$$

This shows that $|\mu(f)|\leqslant\mu(g)\|f\|$ for every f in $C_K(X)$ and so μ is continuous on $C_0(X)$.

A linear functional $\mu: C_0(X)\to\mathbb{C}$ is called *real* if $\mu(f)$ is real for every real valued f in $C_0(X)$. We will be interested mainly in continuous real functionals.

Proposition 4. *A measure* $\mu\colon C_0(X)\to\mathbb{C}$ *is real if and only if there exist two positive measures* $\mu_k\colon C_0(X)\to\mathbb{C}$ $(k=1,2)$ *such that* $\mu=\mu_1-\mu_2$.

Proof. It is obvious that the difference of two positive measures is a real measure. Conversely suppose that μ is a real measure on the locally compact Hausdorff space X. Let $C_{0r}(X)$ denote the vector space of real valued functions f belonging to $C_0(X)$ and let α denote the restriction of μ to $C_{0r}(X)$. Let g be a non-negative function in $C_{0r}(X)$ and let K be the support of g. If f in $C_{0r}(X)$ is such that $|f|\leqslant g$ then K is a carrier of f and so $|\alpha(f)|=|\mu(f)|\leqslant m_K\|f\|\leqslant m_K\|g\|$. This shows that

$$\operatorname{lub}\{\alpha(f)\colon f\in C_{0r}(X) \text{ and } |f|\leqslant g\}$$

is finite. Therefore α is a relatively bounded, real valued linear functional on $C_{0r}(X)$. By Proposition 1.6 there are two positive linear functionals $\alpha_k\colon C_{0r}(X)\to\mathbb{C}$ such that $\alpha=\alpha_1-\alpha_2$. Let μ_k $(k=1,2)$ denote the extension of α_k to $C_0(X)$. Then μ_k is a positive linear functional and so by Proposition 3 it is a positive measure. Since $\alpha=\alpha_1-\alpha_2$ we have $\mu=\mu_1-\mu_2$.

More careful analysis shows that every real measure μ has a unique *minimal decomposition* $\mu=\mu_1-\mu_2$. This means that if $\mu=\nu_1-\nu_2$ is another decomposition into positive measures then $\nu_1=\mu_1+\pi$ and $\nu_2=\mu_2+\pi$ for some positive measure π. It is clear that every complex measure $\mu\colon C_0(X)\to\mathbb{C}$ can be uniquely decomposed as $\mu=\nu_1+i\nu_2$ where ν_1 and ν_2 are real measures on X. Therefore an arbitrary measure μ is the same as a linear combination of positive linear functionals with complex coefficients.

The simplest example of a measure is the Dirac measure $\varepsilon_a\colon C_0(X)\to\mathbb{C}$ where a is a given point in X. It is given by $\varepsilon_a(f)=f(a)$ for every f in $C_0(X)$. Other measures can be constructed by taking a linear combination of finitely many distinct ε_a measures. Let $\mathscr{S}$ be a σ-algebra containing all the compact subsets of X and let μ be a countably additive positive measure on X such that $\mu(K)$ is finite for every compact subset K of X. Then

$$\mu(f)=\int_G f(x)\,d\mu(x)$$

exists for every f in $C_0(X)$ and the linear functional defined in this manner is a measure in the Bourbaki-Radon sense: For if K is a compact set in X then $|\mu(f)|\leqslant\mu(K)\|f\|$ for every f in $C_0(X)$.

The existence of the support of a measure $\mu\colon C_0(X)\to\mathbb{C}$ can be proved by relying on the following:

Lemma 5. *Let O_1 and O_2 be open sets in the locally compact Hausdorff space X and let K be a compact set contained in $O_1 \cup O_2$. Then there exist continuous functions $g_k: X \to [0,1]$ $(k=1,2)$ such that O_k is a carrier of g_k, for all $x \in K$ one has $(g_1+g_2)(x)=1$ and $0 \leqslant (g_1+g_2)(x) \leqslant 1$ for every x in X.*

Note. One can consider also the situation when a finite number of open sets $O_1, \ldots, O_n$ are given but we need only the above special case.

Proof. Let $\bar{X}$ denote the one-point compactification of X and let O_0 be the complement of K in $\bar{X}$. Then O_0, O_1 and O_2 cover $\bar{X}$ and so there are continuous functions $\bar{g}_k: \bar{X} \to [0,1]$ such that $(\bar{g}_1+\bar{g}_2+\bar{g}_3)(x)=1$ for every $x \in X$ and $\bar{g}_k(x)=0$ for every $x \notin O_k$. The existence of these functions follows from the well known lemma on partition of unity. The restrictions of $\bar{g}_1$ and $\bar{g}_2$ to X are the desired functions.

Given the measure μ let C be a closed subset of X such that if $f \in C_0(X)$ and cC is a carrier of f then $\mu(f)=0$. A set C satisfying these requirements is called a *carrier* of the measure μ. If C_1 and C_2 are carriers then so is $C_1 \cap C_2$: For let $O_k = cC_k$ $(k=1,2)$ and let f in $C_0(X)$ be such that $O_1 \cup O_2$ is a carrier of f. Then K, the support of f, is contained in $O_1 \cup O_2$. Let g_1 and g_2 be the functions whose existence is assured by Lemma 5. Then $f = f(g_1+g_2)$ and O_k is a carrier of fg_k $(k=1,2)$. Hence $\mu(fg_k)=0$ for $k=1,2$ and so $\mu(f)=0$. We proved that the intersection of finitely many carriers is also a carrier.

Now let S denote the intersection of all carriers C of the measure μ. Then S is obviously a closed subset of X. We prove that S is itself a carrier of μ: For let $f \in C_0(X)$, let K be the support of f and suppose that $K \subseteq cS$. Since $S = \bigcap C$ we have $K \subseteq \bigcup cC$ where C varies over all carriers of μ. The sets cC are open and so K being compact there are finitely many among these carriers, say $C_1, \ldots, C_n$ such that $K \subseteq cC_1 \cup \cdots \cup cC_n$. Since $C_1 \cap \cdots \cap C_n$ is a carrier of μ we obtain $\mu(f)=0$. We proved that the measure μ has a smallest carrier S. It is called the *support* of the measure μ.

It is now obvious what is meant by a *measure of compact support*. The set of measures $\mu: C_0(X) \to \mathbb{C}$ having compact support is a complex vector space which is denoted by $M_0(X)$. If $f: X \to \mathbb{C}$ is a continuous function then $fg \in C_0(X)$ for every g in $C_0(X)$ and $|\mu(fg)| \leqslant m_K \|f\| \cdot \|g\|$ for every g in $C_K(X)$. Therefore $g \mapsto \mu(fg)$ defines a measure on X which is denoted by $f\mu$. This operation turns the vector space of all measures on X into a $C(X)$ module. If μ has compact support then so does $f\mu$ and so $M_0(X)$ is a submodule. If X is a compact Hausdorff space then every μ has compact support and $M_0(X)$ is the $C(X)$ module of all measures on X.

Let μ be a measure on X with compact support K and let $\varphi: X \to \mathbb{C}$ be any continuous function such that $\varphi(x)=1$ for every x in K. If $f: X \to \mathbb{C}$ is continuous then we can define $\mu(f)$ by $\mu(f)=\mu(\varphi f)$ because $\mu(\varphi f)$ is independent of the particular choice of φ. Hence $f \mapsto \mu(f)$ defines a linear functional on $C(X)$. The vector space $C(X)$ can be turned into a locally convex topological vector space by introducing the topology of uniform convergence on compacta. The linear functional $\mu: C(X) \to \mathbb{C}$ is continuous with respect to this topology. The restriction of the topology of uniform convergence on compacta to $C_0(X)$ is not stronger than the direct limit topology introduced on $C_0(X)$ in the beginning. Hence if $\mu: C(X) \to \mathbb{C}$ is a continuous linear functional then its restriction to $C_0(X)$ defines a measure which we denote by the same symbol μ. By the continuity of μ there is a neighborhood N_0 of 0 in $C(X)$ such that $|\mu(f)|<1$ for every f in N_0. We may suppose that

$$N_0 = \{f: f \in C(X) \text{ and } |f(x)| < \varepsilon \text{ for all } x \in K\}$$

for some $\varepsilon>0$ and a suitable compact subset K of X. If $f \in C(X)$ and $f(x)=0$ for every $x \in K$ then $\lambda f \in N_0$ for every $\lambda>0$ and so $|\mu(\lambda f)|<1$ i.e. $|\mu(f)|<1/\lambda$. This shows that $\mu(f)=0$ and so K is a compact carrier of μ. We proved the following:

Proposition 6. *A linear functional* $\mu: C_0(X) \to \mathbb{C}$ *is a measure of compact support if and only if it is the restriction of a continuous linear functional* $\mu: C(X) \to \mathbb{C}$. *Hence this correspondence identifies* $M_0(X)$ *with* $C(X)^*$.

For any measure $\mu: C_0(X) \to \mathbb{C}$ one defines the norm $\|\mu\|$ by

$$\|\mu\| = \operatorname{lub}\{|\mu(f)|: f \in C_0(X) \text{ and } \|f\| \leqslant 1\}.$$

Hence $\|\mu\|$ is a non-negative number or the symbol $+\infty$. The measure μ is called *bounded* if $\|\mu\|$ is finite. The set of bounded measures is a $C(X)$ submodule of the module of all measures on X. It will be denoted by the symbol $M(X)$. It is clear that $M_0(X)$ is a submodule of $M(X)$. Hence if X is compact then $M(X)$ is the module of all measures on X. If X is compact and $\mu \geqslant 0$ then $\|\mu\| = \mu(\mathit{1})$ where $\mathit{1}: X \to \mathbb{C}$ is identically 1 on X.

Let $C_o(X)$ denote the vector space of all continuous functions $f: X \to \mathbb{C}$ which vanish at infinity. We know that $C_o(X)$ is a Banach space under the norm $\|f\| = \operatorname{lub}\{|f(x)|: x \in X\}$ and the topology induced on $C_o(X)$ by this norm is the topology of uniform convergence. If μ is a bounded measure on X then it is continuous with respect to this topology and so it can be extended to a continuous linear functional $\mu: C_o(X) \to \mathbb{C}$.

The trace of the topology of uniform convergence is not stronger than the direct limit topology of $C_0(X)$. Hence if a linear functional $\mu: C_o(X) \to \mathbb{C}$ is continuous then its restriction to $C_0(X)$ is a measure which is obviously bounded. Thus we have the following:

Proposition 7. *A linear functional $\mu: C_0(X) \to \mathbb{C}$ is a bounded measure if and only if it is the restriction of a continuous linear functional $\mu: C_o(X) \to \mathbb{C}$. Hence this correspondence identifies $M(X)$ with $C_o(X)^*$.*

As a corollary we see that $M(X)$ is a Banach space under the norm $\|\mu\|$.

We now turn to the definition of the *total variation* $|\mu|$ of a measure μ. We shall see that $|\mu|$ is also a measure on the same space X. First we note that if $\mu \geqslant 0$ then $|\mu(f)| \leqslant \mu(|f|)$ for all f in $C_0(X)$. Indeed by $|f| = f \cup 0 - f \cap 0$ and the additivity of μ we have

$$|\mu(f)| \leqslant |\mu(f \cup 0)| + |\mu(f \cap 0)| = \mu(f \cup 0) - \mu(f \cap 0) = \mu(|f|).$$

Lemma 8. *If $\mu: C_0(X) \to \mathbb{C}$ is a measure and f is a non-negative function in $C_0(X)$ then*

$$|\mu|(f) = \operatorname{lub}\{|\mu(g)| : g \in C_0(X) \text{ and } 0 \leqslant |g| \leqslant f\}$$

is finite.

Proof. Let $g = g_1 + i g_2$ where g_1 and g_2 are real and $|g| \leqslant f$. Furthermore let

$$\mu = \mu_{11} - \mu_{12} + i(\mu_{21} - \mu_{22})$$

where μ_{ij} $(i, j = 1, 2)$ are non-negative measures on X. Then

$$|\mu_{ij}(g_k)| \leqslant \mu_{ij}(|g_k|) \leqslant \mu_{ij}(|g|) \leqslant \mu_{ij}(f)$$

for every $i, j, k = 1, 2$. Therefore

$$|\mu(g)| \leqslant (\mu_{11} + \mu_{12} + \mu_{21} + \mu_{22})(f)$$

for every g in $C_0(X)$ satisfying $|g| \leqslant f$.

Lemma 9. *If μ is a measure on X and f_1, f_2 are non-negative functions in $C_0(X)$ then $|\mu|(f_1 + f_2) = |\mu|(f_1) + |\mu|(f_2)$.*

Proof. Given $\varepsilon > 0$ we can find g_k $(k = 1, 2)$ in $C_0(X)$ such that $|g_k| \leqslant f_k$ and $|\mu(g_k)| + \varepsilon/2 > |\mu|(f_k)$. Then

$$\frac{|\mu(g_2)|}{\mu(g_2)}(\lambda \mu(g_1) + \mu(g_2)) = |\mu(g_1)| + |\mu(g_2)|$$

provided $\lambda = |\mu(g_1)|\,\mu(g_2)/\mu(g_1)\,|\mu(g_2)|$ exists and so

$$|\mu(g_1)| + |\mu(g_2)| = |\mu(\lambda g_1 + g_2)| \leqslant |\mu|\,(f_1 + f_2).$$

The same inequality holds also if $\mu(g_1)$ or $\mu(g_2)$ is zero. This proves that

$$|\mu|\,(f_1) + |\mu|\,(f_2) \leqslant |\mu|\,(f_1 + f_2).$$

In order to prove the opposite inequality we choose any g in $C_0(X)$ such that $|g| \leqslant f_1 + f_2$. We let $h_1 = f_1 \cap |g|$ and $h_2 = |g| - h_1$ so that $h_1 + h_2 = |g|$. It is clear that $0 \leqslant h_k \leqslant f_k$ $(k=1,2)$ and $h_1, h_2 \in C_0(X)$. We let $g_k(x) = (g(x)/|g(x)|)\,h_k$ if $g(x) \neq 0$ and we let $g_k(x) = 0$ if $g(x) = 0$. Then $|g_k| = h_k \leqslant f_k$ and $g_1 + g_2 = g$. Therefore

$$|\mu(g)| = |\mu(g_1 + g_2)| \leqslant |\mu(g_1)| + |\mu(g_2)| \leqslant |\mu|\,(f_1) + |\mu|\,(f_2).$$

This shows that $|\mu|\,(f_1 + f_2) \leqslant |\mu|\,(f_1) + |\mu|\,(f_2)$. The lemma is proved.

By Lemma 9 the function $|\mu|$ defined on the non-negative elements of $C_0(X)$ can be extended to a linear functional $|\mu|\colon C_0(X) \to \mathbb{C}$. Since μ is positive by Proposition 3 $|\mu|$ is a measure in the Bourbaki-Radon sense. It is called the *total variation* of the measure μ.

Lemma 10. *If $\mu \in M(X)$ then $|\mu| \in M(X)$ and the norm of $|\mu|$ is $\|\mu\|$. Conversely if $|\mu|$ is bounded then so is μ.*

Proof. If μ is bounded and f, g in $C_0(X)$ are such that $0 \leqslant |g| \leqslant f$ then $|\mu(g)| \leqslant \|\mu\| \cdot \|g\| \leqslant \|\mu\| \cdot \|f\|$. Therefore $|\mu|(f) \leqslant \|\mu\| \cdot \|f\|$ for every $f \geqslant 0$. If f is an arbitrary element of $C_0(X)$ then by the positivity of $|\mu|$ we have $\big||\mu|(f)\big| \leqslant |\mu|(|f|) \leqslant \|\mu\| \cdot \|f\|$. Thus $|\mu| \in M(X)$ and the norm of $|\mu|$ is at most $\|\mu\|$. In order to see the opposite inequality we note that if $\varepsilon > 0$ then there is an f in $C_0(X)$ such that $\|f\| = 1$ and $|\mu(f)| \geqslant \|\mu\| - \varepsilon$. Then by $\big|\mu(|f|)\big| \geqslant |\mu(f)|$ we see that $\big\||\mu|\big\| \geqslant \|\mu\|$. If $|\mu|$ is bounded then there is an $m > 0$ such that $\big||\mu|(f)\big| \leqslant m\|f\|$ for all f in $C_0(X)$. Hence if $f \in C_0(X)$ then $|\mu(f)| \leqslant |\mu|(|f|) \leqslant m\|f\|$. This shows that μ is also bounded.

Proposition 11. *Let X be a closed subset of the locally compact Hausdorff space $\tilde{X}$. For μ in $C_o(X)^*$ define $\tilde{\mu}\colon C_o(\tilde{X}) \to \mathbb{C}$ by $\tilde{\mu}(\tilde{f}) = \mu(f)$ where f is the restriction of $\tilde{f}$ to X. Then $\mu \mapsto \tilde{\mu}$ is an isometric injection of $C_o(X)^*$ into $C_o(\tilde{X})^*$.*

Proof. The map $\mu \mapsto \tilde{\mu}$ is clearly a homomorphism of $C_o(X)^*$ into $C_o(\tilde{X})^*$. If $\|\tilde{f}\| = 1$ then $\|f\| \leqslant 1$ and so $|\tilde{\mu}(\tilde{f})| \leqslant \|\mu\| \cdot \|f\| \leqslant \|\mu\|$. This proves that $\|\tilde{\mu}\| \leqslant \|\mu\|$. Given $\varepsilon > 0$ there is an f in $C_0(X)$ such that $\|f\| = 1$ and $|\mu(f)| > \|\mu\| - \varepsilon$. Since f has compact support it can be extended to a function $\tilde{f}$ on $\tilde{X}$ such that $\tilde{f} \in C_0(\tilde{X})$ and $\|\tilde{f}\| = \|f\| = 1$. Then we have $\|\tilde{\mu}\| \geqslant |\tilde{\mu}(\tilde{f})| = |\mu(f)| > \|\mu\| - \varepsilon$ and this shows that $\|\tilde{\mu}\| \geqslant \|\mu\|$.

For the sake of completeness we include the proof of the following result which has been used in the proof of the last proposition and also in the proof of Proposition 7:

Proposition 12. *If X is a locally compact Hausdorff space then $C_0(X)$ is dense in $C_o(X)$ in the norm topology.*

Given f in $C_o(X)$ we can choose a compact set C such that $|f(x)| < \varepsilon$ for every $x \notin C$. There is a compact set S in X such that $C \subseteq S^i$. We can also choose $g: X \to [0,1]$ such that g is continuous, it is identically 1 on C and zero on $c\,S^i$. The existence of g follows from the compactness of C and the complete regularity of X. Then fg is in $C_0(X)$ and $fg(x) = f(x)$ for every $x \in C$. If $x \notin C$ then

$$|fg(x) - f(x)| \leqslant |f(x)| \cdot |g(x)| + |f(x)| \leqslant 2\varepsilon.$$

Therefore $\|fg - f\| \leqslant 2\varepsilon$.

Let $\mu: C_0(X) \to \mathbb{C}$ be a measure on the locally compact Hausdorff space X. The integration of scalar functions $f: X \to \mathbb{C}$ with respect to μ is developed as follows: If $f \in C_0(X)$ then by definition

$$\int_X f(x)\, d\mu(x) = \mu(f). \tag{1}$$

For more general complex valued functions we proceed as follows: First of all it is sufficient to define the integral only for real valued functions f because for general f we can define the integral by integrating its real and imaginary parts separately. Moreover μ can be decomposed into a complex linear combination of positive measures. Hence it is enough to define $\int_X f\, d\mu$ in the special case when $f: X \to \mathbb{R}$ and $\mu \geqslant 0$. We consider then the restricted linear functional $\mu: C_{0r}(X) \to \mathbb{R}$ and extend it to a linear functional $\mu: L^1(\mu) \to \mathbb{R}$ by using Proposition 2.1. For every f in $L^1(\mu)$ the integral is then defined by (1). This is the classical Daniell approach to integration theory which was adapted by Bourbaki. Alternatively we can consider the original, not necessarily positive $\mu: C_0(X) \to \mathbb{C}$ and apply Proposition 4. However in order to develop the details of the theory one has to study the special case when $\mu \geqslant 0$. Since the rest is more along the usual lines of measure and integration theory we stop the discussion at this point.

Remarks

The contents of Sections 1 and 2 were needed only to prove the spectral theorem but almost all of this material is fundamental also for the development of the Riesz-Bourbaki integration theory on locally compact spaces. We suppose that the reader is causally acquainted with this theory which will be used in the later parts of this book. If he became familiar with the contents of Sections 1 and 2 he can easily master the Riesz-Bourbaki integration theory by consulting the appropriate volume of Bourbaki's Éléments (1, 2). He will notice that the terminology and several results given in Sections 1 and 2 come from Bourbaki. For instance the "two by two lemma" stated as Proposition 1.5 is the last result in Chapter II. § 1. in Bourbaki.

For more information on infinitely distributive lattices and Banach lattices see Garrett Birkhoff's book on lattice theory (1). Several mathematicians including Luxembourg, Nakano and Zaanen published various results on order completeness, order closed sets, ideals and the Riesz property. We mention only a relatively recent paper by Masterson (1) where further references can be found.

The spectral theorem for continuous self adjoint operators in Hilbert space was first formulated and proved by Hilbert (1). The proof given in Section 3 is a detailed and somewhat modified version of the proof outlined by Loomis on p. 95 of his book (1). The spectral theorem can be extended to normal, not necessarily continuous linear operators in Hilbert space. Sz. Nagy (1) wrote a book on this subject and a large part of Halmos' book (1) is also devoted to the spectral theorem. Similarly the book by Riesz and Sz. Nagy (1, 2) contains much valuable information on this subject. Probably the most detailed account in existence is in the treatise of Dunford and Schwartz (1). One can find there also the references to the various papers written on this subject.

Fuglede's paper (1) appeared in 1950 and the one by Putnam (1) followed it a year later. The proof of Putnam's Theorem 4.1 given in the text is due to M. Rosenblum (1). The reasoning in the proof of Theorem 4.3 can be used to prove a more general theorem on "operational calculus" in an arbitrary B^* algebra which is due to Gelfand and Naimark (1):

If A is a complex B^* algebra with identity e and a is a normal element of A then there is a unique isometric isomorphism $f \mapsto f(a)$ of $C(\sigma(a))$ into A such that $z \mapsto z$ is mapped into a and $z \mapsto 1$ into e. For every f in $C(\sigma(a))$ we have $\sigma(f(a)) = f(\sigma(a))$.

A slightly modified theorem holds for B^* algebras without identities. This result can be found in Rickart's book (1) and the one for algebras A with identity is discussed for instance in Dixmier's important monograph on B^* algebras (1).

At the end of Chapter I we already mentioned the fundamental result which states that every complex B^* algebra is isometrically isomorphic to a C^* algebra. This is the reason why many authors, including Dixmier speak about C^* algebras when they mean a complex B^* algebra in the terminology adapted by us. Gelfand and Naimark (1) could prove this theorem only under the additional hypothesis that every element of the form x^*x is quasiregular. They expressed the belief that this property holds for every complex B^* algebra or in any case there will be a proof which will avoid the use of this property. An involutive algebra such that every element of the form x^*x is quasiregular is called *symmetric*. In 1953 Kaplansky proved that every complex B^* algebra is symmetric and so supplied the last link needed to complete the proof of the Gelfand-Naimark theorem. The papers of Fukamiya (1) and Kelley and Vaught (1) played an important part in Kaplansky's reasoning. For additional information on the symmetry of involutive algebras we refer the reader to Rickart's book (1).

The notion of *positivity* can be extended to elements in a complex Banach algebra: We write $x \geqslant 0$ if $\lambda \geqslant 0$ for every λ in $\sigma(x)$. For example if x is a self adjoint element then by Propositions I.4.9 and I.8.4 we have $\sigma(x^2) = \sigma(x)^2 \subseteq [0, \infty)$ and so $x^2 \geqslant 0$. If A is a complex B^* algebra and if a_1 and a_2 are self adjoint elements of A satisfying $a_1, a_2 \geqslant 0$ then $a_1 + a_2 \geqslant 0$.

The principal results presented in Section 6 were proved by v. Neumann in (2). There is an excellent monograph v. Neumann algebras by Dixmier (2) and a large part of Sakai's book (1) is also devoted to these algebras. In addition the lecture notes of J. T. Schwartz (1) on W^* algebras are also very usable.

The material presented in Section 7 is a variant of Bourbaki's basic measure theory, the main difference being the consideration of complex valued functions. Bourbaki uses the real vector space of continuous functions $f: X \to \mathbb{R}$ with compact support. Daniell's original paper appeared in 1917—1918 and a concise exposition of the details of his approach to integration theory can be found in Loomis' book (1). The same is available in several texts on measure and integration theory. The works of F. Riesz and a paper of Radon (1) were also fundamental.

Chapter IV

Elementary Representation Theory in Hilbert Space

Almost all the material presented in this chapter is essential for the understanding of at least one of the topics treated in the remaining chapters. It is advisable to be familiar with the contents of Sections 1—5 before continuing beyond Section V.4. The first four sections of Chapter V are independent of the material presented in the current chapter.

The prerequisites are as follows: One should be completely familiar with the contents of Sections III.5 and III.6. In order to read Section 6 one has to know all that it is presented in Sections II.4 and II.5. This requires a superficial acquaintance with the contents of Sections II.2 and II.3. We shall not need anything from Chapter I.

As far as the various parts of the current chapter are concerned the reading of any section requires detailed knowledge of all the earlier sections of the chapter. The only exception is Section 6 which is not a prerequisite for the later ones.

1. Representations and Morphisms

The purpose of these sections is to give a useful definition of representations in complex Hilbert spaces and study the basic properties of these, including the existence of a multiplicity for irreducible components. The definition will include as special cases not only representations of groups, rings and algebras but also projective group representations in the unitary case and the representation theory introduced by Mackey in his Chicago lectures in 1955.

The *objects* S of a representation theory in complex Hilbert space are sets with a fixed type of topological and algebraic structure which does not exceed that of some subsets of $\mathscr{L}(\mathscr{H})$ where $\mathscr{H}$ is a complex Hilbert space. For instance S can be a set, a semigroup, a vector space over $\mathbb{R}$ or $\mathbb{C}$, a $\mathbb{Z}$-module, an algebra or any of these with an additional topological structure, like a topological group or ring or a Banach algebra with involution. Both these individual objects S and the categories formed by them will be called *admissible* for representation theory in Hilbert space. Thus an admissible category is distinguished by the fact that it contains objects $\mathscr{S}$ where $\mathscr{S} \subseteq \mathscr{L}(\mathscr{H})$ and the topological and algebraic structure of $\mathscr{S}$ is induced by $\mathscr{L}(\mathscr{H})$. Since there are several topologies attached to $\mathscr{L}(\mathscr{H})$ we have a choice in determining the topology of $\mathscr{S}$. These will be called the *special objects* of the admissible category.

A function $\mu: S \to \mathscr{L}(\mathscr{H})$ will be called a *morphism* of the admissible object S in the Hilbert space $\mathscr{H}$ provided $\mathscr{S} = \mu(S)$ is an object in the category of S and $\mu: S \to \mathscr{S}$ is a morphism in the sense of category theory i. e. μ commutes with the algebraic and topological structures of S and $\mathscr{S}$, respectively. For example, let the admissible objects be topological groups and the special objects be groups of operators endowed with the norm topology derived from $\mathscr{L}(\mathscr{H})$. Then a morphism $\mu: S \to \mathscr{S}$ satisfies $\mu(s\,t) = \mu(s)\,\mu(t)$ and, depending on the choice of the morphisms of the category, μ might be required to be a continuous map or an open map.

For any function $f: S \to \mathscr{L}(\mathscr{H})$ we let $\mathscr{S} = f(S)$ and $\tilde{\mathscr{A}}$ be the algebra of operators generated by $\mathscr{S}$. Thus the elements of $\tilde{\mathscr{A}}$ have the form $A = \lambda_1 X_1 + \cdots + \lambda_m X_m$ where $X_1, \ldots, X_m$ denote products of the type $X = f(s_1) \ldots f(s_n)$ or the identity operator I. Finally we let $\mathscr{A}$ be the closure of $\tilde{\mathscr{A}}$ in $\mathscr{L}(\mathscr{H})$ under the weak operator topology so that $\mathscr{A}$ is the weakly closed algebra generated by the set $\mathscr{S}$ in $\mathscr{L}(\mathscr{H})$. In general $\mathscr{A}$ is not self adjoint and so it is properly contained in the v. Neumann algebra generated by $\mathscr{S}$.

Definition 1. *A function $\rho: S \to \mathscr{L}(\mathscr{H})$ is a representation of the admissible object S in the Hilbert space $\mathscr{H}$ provided*

1) *$\rho: S \to \mathscr{S}$ is a morphism of S onto the special object $\mathscr{S} = \rho(S)$,*

2) *the weakly closed algebra $\mathscr{A}$ generated by $\mathscr{S}$ is self adjoint, $\mathscr{A} = \mathscr{A}^*$.*

Thus a morphism $\rho: S \to \mathscr{L}(\mathscr{H})$ is a representation of the object S in $\mathscr{H}$ if and only if the weakly closed algebra $\mathscr{A}$ is a v. Neumann algebra. It is possible that not only the weak closure $\mathscr{A}$ but the algebra $\tilde{\mathscr{A}}$ is self adjoint. If this is the case we shall speak about a *simple representation* in Hilbert space. This terminology is introduced only for the purpose of brevity.

Important special cases are obtained by considering only those categories in which either one or two functions are attached to each object S. If a single function $S \to S$ mapping s into s^* is given we shall speak about an *object S with a map* and if a pair of functions $S \to S$ and $p: S \to \mathbb{C}$ are associated with S then we call S a *Mackey object.*

The special objects are all Mackey objects and the associated maps are always the operator adjoint * and $p \equiv 1$.

A *Mackey type representation* of the *object S with a map* in the Hilbert space $\mathscr{H}$ is a function $\rho: S \to \mathscr{L}(\mathscr{H})$ such that $\rho: S \to \mathscr{S}$ is a morphism of S onto the special object $\mathscr{S} = \rho(S)$ and $\rho(s)^* = p(s)\rho(s^*)$ for every s in S where $p: S \to \mathbb{C}$ is some function depending not only on S but also on the function ρ. It follows that the algebra $\tilde{\mathscr{A}}$ generated by $\mathscr{S}$ is self adjoint and so $\mathscr{A}$ is a v. Neumann algebra and ρ is a representation. If S is a *Mackey object* then its Mackey type representation is a morphism satisfying $\rho(s^*) = p(s)\rho(s^*)$ where p is the second map attached to S. We can however consider S only as an object with a map and obtain Mackey type representations with multipliers different from the map attached to S.

We list a number of special cases.

1. The *representations of involutive algebras* in Hilbert space are all of Mackey type. Here the object is an algebra A over $\mathbb{C}$ or over one of its self adjoint subrings such as $\mathbb{R}$ with an involution $a \mapsto a^*$ which serves as the first map while the second is $p \equiv 1$. Since $\rho: A \to \mathscr{L}(\mathscr{H})$ is an involutive homomorphism we have $\rho(a)^* = \rho(a^*)$. If A is a topological algebra we can select one of the topologies of $\mathscr{L}(\mathscr{H})$ and require that $\rho: S \to \mathscr{S}$ be continuous. By doing so we obtain for instance a *strongly* or *weakly continuous representation* of A.

2. *Unitary group representations* are also of Mackey type. The objects are groups G and the two functions attached to G are $g^* = g^{-1}$ and $p \equiv 1$. Let $\mathscr{U}(\mathscr{H})$ denote the group formed by the unitary elements of $\mathscr{L}(\mathscr{H})$. The representation $\rho: G \to \mathscr{U}(\mathscr{H})$ is a group homomorphism with unitary values and so $\rho(g)^* = \rho(g)^{-1} = \rho(g^{-1}) = \rho(g^*)$. In the case of topological groups these representations are further classified by various continuity conditions on ρ.

3. A *projective group representation* in Hilbert space is a function $\rho: G \to \mathscr{L}(\mathscr{H})$ such that $\rho(e) = I$, for each $x, y \in G$ there is a $\sigma(x,y) \in \mathbb{C}$ for which

$$\rho(xy) = \sigma(x,y)\rho(x)\rho(y)$$

and the weakly closed algebra generated by $\mathscr{S} = \rho(G)$ is self adjoint. It follows that each $\rho(x)$ is invertible and $\rho(x)^{-1} = \sigma(x, x^{-1})\rho(x^{-1})$. Therefore the multiplier $\sigma: G \times G \to \mathbb{C}^\times$ is uniquely determined by ρ. Using $(xy)z = x(yz)$ we see that

$$\sigma(xy, z)\sigma(x,y) = \sigma(x, yz)\sigma(y,z)$$

and so $\sigma(x,e)=\sigma(e,x)=\sigma(e,e)=1$. If $\sigma\equiv 1$ then ρ is an ordinary group representation in $\mathscr{H}$.

Unitary projective group representations give a simple example of Mackey type representations where the objects are of the Mackey type but they are regarded only as objects with a map. They are projective representations such that every $\rho(x)$ is a unitary operator i.e. $\rho: G\to\mathscr{U}(\mathscr{H})$. In this case $\rho(x)^*=\rho(x)^{-1}=\sigma(x,x^{-1})\rho(x^*)$ where $x^*=x^{-1}$ and so the algebra generated by $\mathscr{S}=\rho(G)$ is automatically self adjoint. We also get $|\sigma|\equiv 1$ but in general σ is not identically 1 and it depends not only on the group G but also on the choice of ρ.

4. If $\mathscr{A}$ is a *v. Neumann algebra*, then its unitary operators form a group $\mathscr{G}=\mathscr{A}\cap\mathscr{U}(\mathscr{H})$ because $I\in\mathscr{A}$ and $\mathscr{A}=\mathscr{A}^*$. By Proposition III.6.14 the algebra $\tilde{\mathscr{A}}$ generated by the operators in $\mathscr{G}$ is $\mathscr{A}$ itself. Hence for every v. Neumann algebra $\mathscr{A}$ there exists a locally compact topological group and a norm continuous unitary representation of this group, namely $\mathscr{G}$ with the discrete topology and its identity representation in the Hilbert space of $\mathscr{A}$, such that the algebra $\tilde{\mathscr{A}}$ associated with this unitary group representation is the given v. Neumann algebra $\mathscr{A}$.

Let $\rho: S\to\mathscr{L}(\mathscr{H})$ be a morphism and let $\mathscr{K}$ be a subspace of $\mathscr{H}$ which is invariant under the operators $\rho(s)\,(s\in S)$. Then we can consider the map $\rho|\mathscr{K}: S\to\mathscr{L}(\mathscr{K})$ where $(\rho|\mathscr{K})(s)=\rho(s)|\mathscr{K}$ for each $s\in S$. It is clear that $\rho|\mathscr{K}$ is a morphism, called the *restriction* of ρ to the stable subspace $\mathscr{K}$. We let $\tilde{\mathscr{A}}_\rho|\mathscr{K}$ denote the algebra of operators $\tilde{A}|\mathscr{K}$ where $\tilde{A}\in\tilde{\mathscr{A}}_\rho$.

Lemma 2. *If $\mathscr{K}$ is an invariant subspace of the morphism $\rho: S\to\mathscr{L}(\mathscr{H})$ then $\tilde{\mathscr{A}}_\rho|\mathscr{K}=\tilde{\mathscr{A}}_{\rho|\mathscr{K}}$.*

Proof. This is clear from the definitions of $\tilde{\mathscr{A}}_\rho$ and $\tilde{\mathscr{A}}_\rho|\mathscr{K}$.

The following result indicates the importance of the concept of simple representations:

Lemma 3. *If $\rho: S\to\mathscr{L}(\mathscr{H})$ is a simple representation and if $\mathscr{K}$ is an invariant subspace of ρ then $\rho|\mathscr{K}$ is a simple representation.*

Proof. This follows from Lemma 1 by $(A|\mathscr{K})^*=A^*|\mathscr{K}$.

If S is an object with a map and if $\rho: S\to\mathscr{L}(\mathscr{H})$ is a Mackey type representation then ρ is simple and our lemma can be applied. We also see that

$$(\rho(s)|\mathscr{K})^* = \rho(s)^*|\mathscr{K} = p(s)\rho(s^*)|\mathscr{K} = p(s)(\rho(s^*)|\mathscr{K})$$

and so we have the following stronger result:

Lemma 4. *If S is an object with a map and $\rho: S \to \mathscr{L}(\mathscr{H})$ is a Mackey type representation then $\rho|\mathscr{K}$ is also a Mackey type representation for every invariant subspace $\mathscr{K}$ of ρ. The same holds if S is a Mackey object.*

The significance of Mackey objects becomes clear when we consider direct sums of representations. If $\rho_i: S \to \mathscr{L}(\mathscr{H}_i)$ for $i \in \mathscr{I}$ are morphisms then using the terminology introduced on p. 128 we can define the *direct sum* $\rho = \sum \rho_i$ to be the morphism $\rho: S \to \mathscr{L}(\mathscr{H})$ where $\mathscr{H} = \sum \mathscr{H}_i$ and $\rho(s) = \sum \rho_i(s)$ for every $s \in S$. If the ρ_i's are representations then in general ρ is only a morphism. This can happen also in the special case when S is an object with a map $s \mapsto s^*$ and each ρ_i is a Mackey type representation of S. However if S is a Mackey object, that is a second map $p: S \to \mathbb{C}$ is attached to S and $\rho_i(s)^* = p(s)\rho_i(s)$ for every $i \in \mathscr{I}$, then ρ is also a Mackey type representation of S. Actually we have the following more general result:

Lemma 5. *If S is an object with a map and if the Mackey type representations $\rho_i: S \to \mathscr{L}(\mathscr{H}_i)$ $(i \in \mathscr{I})$ have a common multiplier p then $\rho = \sum \rho_i$ is a Mackey type representation and p is a multiplier for ρ.*

Proof. We have

$$\rho(s)^* = \left(\sum \rho_i(s)\right)^* = \sum \rho_i(s)^* = \sum p(s)\rho_i(s^*) = p(s)\rho(s^*).$$

Proposition 6. *If $\rho: S \to \mathscr{L}(\mathscr{H})$ is a simple representation then $\mathscr{A}_\rho$ is the strong closure of $\tilde{\mathscr{A}}_\rho$.*

Proof. By v. Neumann's density Theorem III.6.10 we have

$$(\tilde{\mathscr{A}}_\rho)'' \subseteq \overline{(\tilde{\mathscr{A}}_\rho)}_s \subseteq \overline{(\tilde{\mathscr{A}}_\rho)}_w = \mathscr{A}_\rho.$$

Since $(\tilde{\mathscr{A}}_\rho)''$ is weakly closed and contains $\tilde{\mathscr{A}}_\rho$ is follows that $\mathscr{A}_\rho \subseteq (\tilde{\mathscr{A}}_\rho)''$ and so $\mathscr{A}_\rho = \overline{(\tilde{\mathscr{A}}_\rho)}_w = \overline{(\tilde{\mathscr{A}}_\rho)}_s$.

Lemma 7. *If $\rho: S \to \mathscr{L}(\mathscr{H})$ is a representation and $\mathscr{K}$ is an invariant subspace of ρ then $\rho|\mathscr{K}$ is a representation.*

Proof. Let P denote the projection $\mathscr{H} \to \mathscr{K}$ so that $P \in (\rho, \rho)$ by Proposition III.5.13. We have $\tilde{\mathscr{A}}_{\rho|\mathscr{K}} = \tilde{\mathscr{A}}|\mathscr{K}$ by Lemma 2. We prove that $A|\mathscr{K}$ belongs to $\mathscr{A}_{\rho|\mathscr{K}}$ for every A in $\mathscr{A}_\rho$. For given $\varepsilon > 0$ and $x_1, \ldots, x_n$, $y_1, \ldots, y_n$ in $\mathscr{K}$ there is an operator B in $\tilde{\mathscr{A}}$ such that $|((A-B)x_k, y_k)| < \varepsilon$ for every $k = 1, \ldots, n$. Hence $T = B|\mathscr{K}$ belongs to the given weak neighborhood of $A|\mathscr{K}$. Thus $A|\mathscr{K}$ belongs to the weak closure of $\tilde{\mathscr{A}}_{\rho|\mathscr{K}}$ which is $\mathscr{A}_{\rho|\mathscr{K}}$. Now in order to prove that

$(\mathscr{A}_{\rho|\mathscr{K}})^* = \mathscr{A}_{\rho|\mathscr{K}}$ let an operator A in $\mathscr{A}_{\rho|\mathscr{K}}$ be given. Then there exist A_n in $\tilde{\mathscr{A}}_{\rho|\mathscr{K}}$ such that $A_n \to A$ weakly on $\mathscr{K}$. But $A_n = B_n|\mathscr{K}$ where B_n is in $\tilde{\mathscr{A}}_\rho$ and this implies that $A_n^* = B_n^*|\mathscr{K}$ is in $\mathscr{A}_{\rho|\mathscr{K}}$. Since $A_n \to A$ implies $A_n^* \to A^*$ where $A_n^* \in \mathscr{A}_{\rho|\mathscr{K}}$ and $\mathscr{A}_{\rho|\mathscr{K}}$ being weakly closed we obtain $A^* \in \mathscr{A}_{\rho|\mathscr{K}}$.

The v. Neumann algebra $\mathscr{A}_{\rho|\mathscr{K}}$ associated with the *subrepresentation* $\rho|\mathscr{K}$ will be determined later. (See Propositions 8.29 and 8.30.)

2. Irreducible Components, Equivalence

Representations can be considered also in the more general situation in which $\mathscr{X}$ is a topological vector space and $\mathscr{L}(\mathscr{X})$ denotes the algebra of continuous operators $A: \mathscr{X} \to \mathscr{X}$. Since in this generality no restriction is imposed upon the weakly closed algebra $\mathscr{A}$ generated by the operators $\rho(s)(s \in S)$ we prefer to call such a function $\rho: S \to \mathscr{L}(\mathscr{X})$ a *morphism* of the object S in the topological vector space $\mathscr{X}$.

Let $\rho: S \to \mathscr{L}(\mathscr{X})$ be a morphism and let $\mathscr{S} = \rho(S)$ so that $\mathscr{S} \subseteq \mathscr{L}(\mathscr{X})$. If a subspace $\mathscr{Y}$ of $\mathscr{X}$ is stable under the family $\mathscr{S}$ then $\mathscr{Y}$ is called a *stable* or *invariant subspace of the morphism* ρ, or an *invariant component* of ρ. The intersection of an arbitrary family of stable subspaces is again stable. If $\mathscr{Y}$ is invariant we let $\rho|\mathscr{Y}$ be the restriction of ρ to $\mathscr{Y}$ so that $\rho|\mathscr{Y}: S \to \mathscr{L}(\mathscr{Y})$ consists of the operators $\rho(s)|\mathscr{Y}$. We call $\rho|\mathscr{Y}$ a *submorphism* of ρ. If $\mathscr{X} = \mathscr{H}$ is a Hilbert space and $\mathscr{K}$ is a stable subspace of ρ then $\rho|\mathscr{K}$ is essentially the same as ρE where $E: \mathscr{H} \to \mathscr{K}$ is the projection operator onto $\mathscr{K}$.

Proposition 1. *A subspace $\mathscr{K}$ of a Hilbert space $\mathscr{H}$ is a stable subspace of the morphism $\rho: S \to \mathscr{L}(\mathscr{H})$ if and only if $\mathscr{K}$ is stable under the weakly closed algebra generated by ρ.*

Proof. This is a simple reformulation of Proposition III.5.10.

A morphism $\rho: S \to \mathscr{L}(\mathscr{X})$ is called *irreducible* if the family of operators $\mathscr{S} = \rho(S)$ is irreducible i.e. if the only $\mathscr{S}$-stable subspaces are 0 and $\mathscr{X}$ itself. Otherwise ρ is called a *reducible morphism.* If $\mathscr{X} = \mathscr{H}$ is a Hilbert space and if there is a $\mathscr{K} \neq 0, \mathscr{H}$ such that both $\mathscr{K}$ and $\mathscr{K}^\perp$ are invariant subspaces of $\mathscr{H}$ then we say that ρ is *completely reducible.* By Proposition III.5.11 a representation in Hilbert space is either irreducible or it is completely reducible.

We shall also speak about the components of a set of operators $S: \mathscr{X} \to \mathscr{X}$ and about a reducible or irreducible set $\mathscr{S}$ of operators S. For example we shall consider irreducible algebras of operators $A: \mathscr{H} \to \mathscr{H}$ where $\mathscr{H}$ is a complex Hilbert space.

A subspace $\mathscr{P}$ of the Hilbert space $\mathscr{H}$ is called an *irreducible component* of the morphism $\rho: S \to \mathscr{L}(\mathscr{H})$ if $\rho | \mathscr{P}$ is irreducible. If $\mathscr{K}$ is stable and $\mathscr{P}$ is irreducible then $\mathscr{K} \cap \mathscr{P}$ being stable either $\mathscr{P} \subseteq \mathscr{K}$ or $\mathscr{P} \cap \mathscr{K} = 0$. In particular if $\mathscr{P}_1$ and $\mathscr{P}_2$ are irreducible components of ρ then either $\mathscr{P}_1 = \mathscr{P}_2$ or $\mathscr{P}_1 \cap \mathscr{P}_2 = 0$. Later we shall prove that if ρ is a representation then either $\mathscr{P} \subseteq \mathscr{K}$ or $\mathscr{P} \perp \mathscr{K}$.

Theorem 2. *Let $\mathscr{A}$ be a self adjoint algebra of continuous operators in complex Hilbert space. Then $\mathscr{A}$ is irreducible if and only if $\mathscr{A}' = \mathbb{C} I$.*

Proof. If $\mathscr{A}$ is reducible then by $\mathscr{A} = \mathscr{A}^*$ it is completely reducible and so $\mathscr{H} = \mathscr{K} + \mathscr{K}^\perp$ where $\mathscr{K}$ and $\mathscr{K}^\perp$ are non zero stable subspaces. Then by Proposition III.5.13 $\mathscr{A}'$ contains the projection $P: \mathscr{H} \to \mathscr{K}$ and so $\mathscr{A}' \neq \mathbb{C} I$. Conversely, if $\mathscr{A}$ is irreducible then the only projections in $\mathscr{A}'$ are 0 and I. By Proposition III.5.6 $\mathscr{A}'$ is a v. Neumann algebra. Hence by Proposition III.6.15 we conclude that $\mathscr{A}' = \mathbb{C} I$.

If $\rho_k: S \to \mathscr{L}(\mathscr{X}_k)$ $(k = 1, 2)$ we let $(\rho_1, \rho_2) = (\rho_1(S), \rho_2(S))$ so that $T \in (\rho_1, \rho_2)$ if and only if $T \in \mathscr{L}(\mathscr{X}_1, \mathscr{X}_2)$ and $T \rho_1(s) = \rho_2(s) T$ for every s in S. Thus (ρ_1, ρ_2) is a vector space over $\mathbb{C}$.

Proposition 3. *A representation $\rho: S \to \mathscr{L}(\mathscr{H})$ is irreducible if and only if $\dim(\rho, \rho) = 1$.*

Proof. By Proposition III.5.10 we are interested in the irreducibility of the v. Neumann algebra $\mathscr{A}$ generated by ρ. Since $(\rho, \rho) = \mathscr{A}'$ the result is an immediate consequence of the last theorem.

Theorem 2 implies the following:

Proposition 4. *A representation $\rho: S \to \mathscr{L}(\mathscr{H})$ is irreducible if and only if $(\rho, \rho) = \mathbb{C} I$.*

Hence if ρ is irreducible then (ρ, ρ) is a field isomorphic to $\mathbb{C}$.

Theorem 5. *A v. Neumann algebra $\mathscr{A}$ of continuous operators $A: \mathscr{H} \to \mathscr{H}$ is irreducible if and only if $\mathscr{A} = \mathscr{L}(\mathscr{H})$.*

Proof. If $\mathscr{A}$ is irreducible, then by the foregoing theorem $\mathscr{A}' = \mathbb{C} I$ and so $\mathscr{A} = \mathscr{A}'' = \mathscr{L}(\mathscr{H})$. Now suppose that $\mathscr{A} = \mathscr{L}(\mathscr{H})$ so that a T in $\mathscr{A}'$ commutes with every projection $P: \mathscr{H} \to \mathscr{H}$. We fix $x \neq 0$ in $\mathscr{H}$ and let P denote the projection $P: \mathscr{H} \to \mathbb{C} x$. We obtain $Tx = TPx = PTx = \lambda(x) x$ for some $\lambda(x) \in \mathbb{C}$ and so we proved that $Tx = \lambda(x) x$ for every x in $\mathscr{H}$. Using the linearity of T we can show that $Tx = \lambda x$

where λ is a fixed element of $\mathbb{C}$ and so $T=\lambda I$. Hence $\mathscr{A}'=\mathbb{C}I$ and so $\mathscr{A}$ is irreducible by Theorem 2.

Proposition 6. *A representation $\rho: S\to\mathscr{L}(\mathscr{H})$ is irreducible if and only if the algebra generated by I and the operators $\rho(s)(s\in S)$ is dense in $\mathscr{L}(\mathscr{H})$ in the weak operator topology.*

Proof. This is an immediate consequence of Theorem 5 and Proposition III.5.10.

If ρ is a simple, irreducible representation, that is $\tilde{\mathscr{A}}_\rho=(\tilde{\mathscr{A}}_\rho)^*$ then by Proposition 1.6 $\mathscr{A}_\rho$ is the strong closure of $\tilde{\mathscr{A}}_\rho$ and so $\tilde{\mathscr{A}}_\rho$ is dense in $\mathscr{L}(\mathscr{H})$ in the strong operator topology. Of course if ρ is any, not necessarily simple representation for which $\overline{(\tilde{\mathscr{A}}_\rho)}_s=\mathscr{L}(\mathscr{H})$ then by $\overline{(\tilde{\mathscr{A}}_\rho)}_s\subseteq\overline{(\tilde{\mathscr{A}}_\rho)}_w$ and by Proposition 6 ρ is irreducible.

Thus if ρ is irreducible then by the foregoing theorem $\overline{(\tilde{\mathscr{A}})}_s=\overline{(\tilde{\mathscr{A}})}_w=\mathscr{L}(\mathscr{H})$. Conversely if $\overline{(\tilde{\mathscr{A}})}_s$ or $\overline{(\tilde{\mathscr{A}})}_w$ is $\mathscr{L}(\mathscr{H})$ then $\mathscr{A}=\mathscr{L}(\mathscr{H})$ and so ρ is irreducible.

Theorem 5 and Proposition 6 are extensions to complex Hilbert spaces of the classical *theorem of Burnside* which states the following: If A is a non-trivial, irreducible algebra of linear transformations in a finite dimensional vector space V over an algebraically closed field then $A=\mathscr{L}(V)$, the algebra of all linear transformations of V into V.

In other, more general representation theories one calls representations also the morphisms $\rho: S\to\mathscr{L}(\mathscr{X})$ where $\mathscr{X}$ is a Hilbert space, a Banach space or possibly a locally convex topological vector space. Such a generalized representation ρ is called *topologically irreducible* if its only closed invariant subspaces are 0 and $\mathscr{X}$ and ρ is called *completely irreducible* if the strong closure of the algebra generated by I and $\rho(S)$ is $\mathscr{L}(\mathscr{X})$. In this context the above proposition means that a representation in complex Hilbert space in the sense used by us is topologically irreducible if and only if it is completely irreducible. In general complete irreducibility implies topological irreducibility but not conversely.

Proposition 7. *If the v. Neumann algebra of the representation $\rho: S\to\mathscr{L}(\mathscr{H})$ is commutative then ρ is irreducible if and only if $\mathscr{H}$ is one dimensional.*

Proof. If $\dim\mathscr{H}=1$ then ρ is clearly irreducible. Conversely, if ρ is irreducible then by Theorem 5 $\mathscr{A}=\mathscr{L}(\mathscr{H})$ and so by the commutativity of $\mathscr{A}$ we have $\dim\mathscr{H}=1$.

Definition 8. *Two morphisms $\rho_k: S\to\mathscr{L}(\mathscr{H}_k)$ $(k=1,2)$ of the object S are called equivalent if there exists a unitary map U of $\mathscr{H}_1$ onto $\mathscr{H}_2$ such that $\rho_2 U=U\rho_1$ i.e. $\rho_2(s)U=U\rho_1(s)$ for every s in S.*

If ρ_1 and ρ_2 are equivalent we write $\rho_1 \approx \rho_2$. Thus $\rho_1 \approx \rho_2$ if and only if (ρ_1, ρ_2) contains an isometry of $\mathscr{H}_1$ onto $\mathscr{H}_2$. It is clear that $\approx$ is a reflexive, symmetric and transitive relation. We shall use $\rho_1 \leqslant \rho_2$ to denote that ρ_1 is equivalent to a submorphism of ρ_2. The reflexive and transitive nature of the relation $\leqslant$ is obvious. We also note that $\rho_1 \leqslant \rho_2$ and $\rho_2 \leqslant \rho_1$ imply that $\rho_1 \approx \rho_2$ but this result will not be needed until much later. For a formal statement and proof see Theorem 9.19.

Two morphisms $\rho_k: S \to \mathscr{L}(\mathscr{H}_k)$ $(k=1,2)$ are called *disjoint*, or in symbols $\rho_1 \downarrow \rho_2$, if the only invariant subspaces $\mathscr{K}_k$ $(k=1,2)$ such that $\rho_1 | \mathscr{K}_1 \approx \rho_2 | \mathscr{K}_2$ are the zero spaces of $\mathscr{H}_1$ and $\mathscr{H}_2$, respectively. If ρ_1 and ρ_2 are *not disjoint* we write $\rho_1 \uparrow \rho_2$ so that $\rho_1 \uparrow \rho_2$ means the existence of non trivial, equivalent submorphisms. Then we say that ρ_1 and ρ_2 *intertwine*.

If ρ_1 and ρ_2 are morphisms of the same object S and if no submorphism of ρ_1 is disjoint from ρ_2 then we shall write $\rho_1 \preccurlyeq \rho_2$ and we shall say that ρ_1 is *covered* by ρ_2. It is obvious that $\rho \preccurlyeq \rho$ and one can easily see that $\preccurlyeq$ is a transitive relation. If π is irreducible and $\pi \preccurlyeq \rho$ then $\pi \leqslant \rho$.

Definition 9. *Two morphisms $\rho_k: S \to \mathscr{L}(\mathscr{H}_k)$ $(k=1,2)$ are called quasi-equivalent, $\rho_1 \sim \rho_2$, if $\rho_1 \preccurlyeq \rho_2$ and $\rho_2 \preccurlyeq \rho_1$.*

We see that $\sim$ is an equivalence relation and $\approx$ implies $\sim$.

If $\rho_1 \approx \rho_2$ and ρ_1 is irreducible then so is ρ_2. Therefore one can speak about an *irreducible equivalence class* of morphisms of an object S. A typical equivalence class of irreducible representations will be denoted by ω and $\hat{S}$ will stand for the *set of all irreducible equivalence classes* ω of the object S. If $\rho: S \to \mathscr{L}(\mathscr{H})$ is a morphism and $\mathscr{P}$ is a subspace of $\mathscr{H}$ such that $\rho | \mathscr{P}$ belongs to ω then we shall write $\mathscr{P} \in \omega$ instead of the more detailed form $\rho | \mathscr{P} \in \omega$ and we shall say that ω *occurs* in ρ.

The purpose of the following discussion is to clarify the extent to which the invariant subspaces and irreducible components of a representation ρ are determined by the v. Neumann algebra $\mathscr{A} = \mathscr{A}_\rho$ associated with ρ. We start with a concept from operator theory.

Definition 10. *A linear operator $U: \mathscr{H}_1 \to \mathscr{H}_2$ is called a partial isometry if there is a subspace $\mathscr{K}$ of $\mathscr{H}_1$, called the initial domain of U such that $U: \mathscr{K} \to U\mathscr{K}$ is an isometry and U is equal to the zero operator on $\mathscr{K}^{\perp}$. The subspace $U\mathscr{K} = U\mathscr{H}_1$ is called the final or terminal domain of U.*

A projection $P: \mathscr{H} \to \mathscr{K}$ gives a simple example of a partial isometry with identical initial and terminal domains $P\mathscr{H} = \mathscr{K}$.

Lemma 11. *If U is a partial isometry then U^*U is the projection $P: \mathscr{H}_1 \to \mathscr{K}$ and similarly UU^* is the projection $Q: \mathscr{H}_2 \to U\mathscr{K}$. Conversely if*

$U^*U: \mathcal{H}_1 \to \mathcal{K}$ is a projection then $U: \mathcal{H}_1 \to \mathcal{H}_2$ is a partial isometry with initial domain $\mathcal{K}$.

Proof. If $x, y \in \mathcal{K}$ then by $\|Ux\|^2 = \|x\|^2$, $\|Uy\|^2 = \|y\|^2$ and $\|Ux+Uy\|^2 = \|x+y\|^2$ we have $(Ux, Uy) + (Uy, Ux) = (x, y) + (y, x)$ and so $\mathcal{Re}(Ux, Uy) = \mathcal{Re}(x, y)$. Replacing y by iy we get $\mathcal{Im}(Ux, Uy) = \mathcal{Re}(Ux, Uiy) = \mathcal{Re}(x, iy) = \mathcal{Im}(x, y)$. Therefore $(Ux, Uy) = (x, y)$ for every x, y in $\mathcal{K}$. If y is an arbitrary element of $\mathcal{H}_1$ then by writing $y = u + v$ with $u \in \mathcal{K}$ and $v \in \mathcal{K}^\perp$ we see that $(Ux, Uy) = (x, y)$ for any x in $\mathcal{K}$ and y in $\mathcal{H}_1$. Since $(U^*Ux, y) = (Ux, Uy)$ we proved that $U^*Ux = x$ for every $x \in \mathcal{K}$. Moreover $U\mathcal{K}^\perp = 0$ shows that U^*U is zero on $\mathcal{K}^\perp$. Therefore $U^*U = P: \mathcal{H}_1 \to \mathcal{K}$.

Next we prove that UU^* is the projection $\mathcal{H}_2 \to U\mathcal{K}$. Let $T: \mathcal{H}_2 \to \mathcal{H}_1$ be defined by $T = U^{-1}$ on $U\mathcal{K}$ and $T = 0$ on $(U\mathcal{K})^\perp$. For x in $U\mathcal{K}$ we have $Tx \in \mathcal{K}$ and so $Tx = U^*UTx = U^*x$ and $x = UTx = UU^*x$. Moreover if $x \in (U\mathcal{K})^\perp$ i.e. $(U\mathcal{H}_1)^\perp$ then $U^*x = 0$ and $UU^*x = 0$. Finally let U^*U be a projection of $\mathcal{H}_1$ onto $\mathcal{K}$. Then for x in $\mathcal{K}$ we have $\|Ux\| = \|x\|$ because $(Ux, Ux) = (U^*Ux, x) = (x, x)$ and similarly for y in $\mathcal{K}^\perp$ we have $Uy = 0$ because $(Uy, Uy) = (U^*Uy, y) = 0$.

Lemma 12. *If $U: \mathcal{H} \to \mathcal{H}$ is a partial isometry with initial domain $\mathcal{K}_1$ and terminal domain $\mathcal{K}_2$ then $U^*: \mathcal{H} \to \mathcal{H}$ is a partial isometry with initial domain $\mathcal{K}_2$ and terminal domain $\mathcal{K}_1$.*

Proof. By Lemma 11 UU^* is the projection $\mathcal{H} \to \mathcal{K}_2$. Therefore if $x \in \mathcal{K}_2$ then $(U^*x, U^*x) = (UU^*x, x) = (x, x)$ proving that U^* is an isometry on $\mathcal{K}_2$. Moreover if $x \in \mathcal{K}_2^\perp$ then $(U^*x, U^*x) = (UU^*x, x) = (0, x) = 0$ and so U^* is zero on $\mathcal{K}_2^\perp$. We proved that U^* is a partial isometry with initial domain $\mathcal{K}_2$. By Lemma 11 U^*U reduces to the identity on $\mathcal{K}_1$ and so $U^*\mathcal{H} = U^*\mathcal{K}_2 = U^*U\mathcal{K}_1 = \mathcal{K}_1$.

Definition 13. *Two projections P_1 and P_2 belonging to a v. Neumann algebra $\mathcal{A}$ are called equivalent with respect to $\mathcal{A}$, in symbols $P_1 \sim P_2$, if there is an operator V in $\mathcal{A}$ such that $V^*V = P_1$ and $VV^* = P_2$.*

If $\mathcal{H}$ is the Hilbert space of $\mathcal{A}$ and $\mathcal{K}_k = P_k\mathcal{H}$ $(k = 1, 2)$ then by Lemma 11 V and V^* are partial isometries with initial domains $\mathcal{K}_1$ and $\mathcal{K}_2$, respectively. Lemma 12 shows that the terminal domains are $\mathcal{K}_2$ and $\mathcal{K}_1$. Therefore $P_1 \sim P_2$ if and only if a partial isometry V exists in $\mathcal{A}$ with initial domain $P_1\mathcal{H}$ and terminal domain $P_2\mathcal{H}$. The reflexive and symmetric properties of the relation $\sim$ are now obvious. In order to see the transitivity let $\mathcal{K}_k = P_k\mathcal{H}$ $(k = 1, 2, 3)$ where $P_k \in \mathcal{A}$. Suppose V_1 and V_2 are operators demonstrating the relations $P_1 \sim P_2$ and $P_2 \sim P_3$, respectively. Then $V_1 | \mathcal{K}_1$ is an isometry of $\mathcal{K}_1$ onto $\mathcal{K}_2$ and similarly

$V_2|\mathscr{K}_2$ maps $\mathscr{K}_2$ isometrically onto $\mathscr{K}_3$. Thus $V_2 V_1 \in \mathscr{A}$, it maps $\mathscr{K}_1$ isometrically onto $\mathscr{K}_3$ and it is zero on $\mathscr{K}_1^\perp$. By Lemma 11 $(V_2 V_1)^*(V_2 V_1) = P_1$ and $(V_2 V_1)(V_2 V_1)^* = P_3$.

We recall that if $\mathscr{K}$ is an invariant subspace of a representation $\rho: S \to \mathscr{L}\mathscr{H})$ then by Proposition III.5.13 the associated projection $P: \mathscr{H} \to \mathscr{K}$ belongs to $(\rho, \rho) = \mathscr{A}'_\rho$.

Proposition 14. *Let $\rho: S \to \mathscr{L}(\mathscr{H})$ be a representation, let $\mathscr{K}_k$ $(k=1,2)$ be invariant subspaces and let P_k denote the corresponding projections $\mathscr{H} \to \mathscr{K}_k$. Then $\rho|\mathscr{K}_1 \approx \rho|\mathscr{K}_2$ if and only if $P_1 \sim P_2$ in (ρ, ρ).*

Proof. Let $U: \mathscr{K}_1 \to \mathscr{K}_2$ be an isometry such that $\rho_2 U = U \rho_1$ where $\rho_k = \rho|\mathscr{K}_k$ $(k=1,2)$. We define the continuous linear operator $V: \mathscr{H} \to \mathscr{H}$ by letting $V = U$ on $\mathscr{K}_1$ and $V = 0$ on $\mathscr{K}_1^\perp$. We prove that $V \in (\rho, \rho)$: Given x in $\mathscr{H}$ we let $x = y + z$ where $y \in \mathscr{K}_1$ and $z \in \mathscr{K}_1^\perp$. Then

$$\rho V x = \rho V(y+z) = \rho V y = \rho U y = \rho_2 U y = U \rho_1 y,$$

$$V \rho x = V \rho(y+z) = V \rho y + V \rho z = V \rho y = U \rho_1 y$$

and so $\rho V x = V \rho x$ where $x \in \mathscr{H}$ is arbitrary. The initial domain of V is $\mathscr{K}_1$ and its terminal domain is $\mathscr{K}_2$. Therefore by Lemma 11 $V^* V = P_1$ and $V V^* = P_2$. We proved that $P_1 \sim P_2$ in (ρ, ρ).

Conversely if $P_1 \sim P_2$ in (ρ, ρ) then by Definition 13 there is a partial isometry V in (ρ, ρ) such that $V^* V = P_1$ and $V V^* = P_2$. By Lemma 11 the initial domains of V and V^* are $\mathscr{K}_1$ and $\mathscr{K}_2$, respectively. Hence by Lemma 12 V maps $\mathscr{K}_1$ isometrically onto $\mathscr{K}_2$. By $V \in (\rho, \rho)$ we see that $\rho|\mathscr{K}_1 \sim \rho|\mathscr{K}_2$.

By Proposition 1 we know that the invariant subspaces $\mathscr{K}$ of a representation $\rho: S \to \mathscr{L}(\mathscr{H})$ are the same as those of the v. Neumann algebra $\mathscr{A} = \mathscr{A}_\rho$. Since $(\rho, \rho) = \mathscr{A}'$ the foregoing proposition shows that $\mathscr{A}$ also determines which of these subspaces are equivalent i.e. when is $\rho|\mathscr{K}_1 \sim \rho|\mathscr{K}_2$. Therefore given any invariant subspace $\mathscr{P}$ we can decide from $\mathscr{A}$ whether $\rho|\mathscr{P}$ is irreducible. However $\mathscr{A}$ does not contain sufficient information to determine whether or not $\rho|\mathscr{P}$ is equivalent to a given irreducible representation π of S.

Definition 15. *A vector $x \in \mathscr{H}$ is called a cyclic vector of the morphism $\rho: S \to \mathscr{L}(\mathscr{H})$ if $\rho(S)x$ generates a dense linear manifold of $\mathscr{H}$. If a cyclic vector exists then $\rho: S \to \mathscr{L}(\mathscr{H})$ is called a cyclic morphism.*

Thus x is a cyclic vector of the cyclic morphism ρ if and only if the smallest invariant subspace containing x is $\mathscr{H}$ itself. For example if S is the additive group of $\mathbb{Z}$, the Hilbert space is $L^2(T)$ where T is the unit circle and if $\rho(s)f(t)=e^{2\pi i s t}f(t)$ $(-\frac{1}{2}\leqslant t<\frac{1}{2})$ then 1 is a cyclic vector. For by the Weierstrass approximation theorem the linear combinations

$$\lambda_1 e^{2\pi i s_1 t}+\cdots+\lambda_n e^{2\pi i s_n t} \qquad (s_1,\ldots,s_n\in\mathbb{Z})$$

are dense in $C(T)$ which in turn is dense in $L^2(T)$.

Theorem 16. *If the object S has a multiplicative structure and if* $\rho\colon S\to\mathscr{L}(\mathscr{H})$ *is a morphism then* ρ *is the direct sum of cyclic subspaces* $\mathscr{K}_i\,(i\in\mathscr{I})$ *and a subspace* $\mathscr{K}_0$ *such that* $\rho(s)x=0$ *for every* $s\in S$ *and* $x\in\mathscr{K}_0$.

Proof. Let $x\in\mathscr{H}$ and let $\mathscr{K}=\mathscr{K}(x)$ denote the closure of the linear manifold generated by $\rho(S)x$. It is possible that $\mathscr{K}(x)=0$. In any case $\mathscr{K}$ is a stable subspace because

$$\rho(s)(\lambda_1\rho(s_1)x+\cdots+\lambda_n\rho(s_n)x)=\lambda_1\rho(s s_1)x+\cdots+\lambda_n\rho(s s_n)x\,.$$

Clearly $\rho|\mathscr{K}(x)$ is a cyclic submorphism of ρ with cyclic generator x. We are going to apply Zorn's lemma: We consider families consisting of non-zero subspaces $\mathscr{K}_i(i\in\mathscr{I})$ such that $\mathscr{K}_i\perp\mathscr{K}_j$ if $i\neq j$ and $\rho|\mathscr{K}_i$ is a cyclic morphism. We partially order the set of such families $\mathscr{K}_i(i\in\mathscr{I})$ by the inclusion relation. Every linearly ordered subset has an upper bound, namely the union of the individual families in the chain. Thus by Zorn's lemma there exists a family $\mathscr{K}_i(i\in\mathscr{I})$ which is maximal with respect to inclusion. Then $\mathscr{K}_0=(\sum\mathscr{K}_i)^{\perp}$ consists only of such vectors x for which $\rho(S)x=0$ because otherwise $\mathscr{K}(x)$ could be joined to our family in contradiction to its maximality. Thus $\mathscr{K}_0\oplus\sum\mathscr{K}_i$ is the desired cyclic decomposition.

Proposition 17. *If* $\pi\colon S\to\mathscr{L}(\mathscr{H})$ *is an irreducible morphism then* π *is cyclic or* $\pi(S)\mathscr{H}=0$. *If* $\pi(s)x\neq 0$ *for some s in S then x is a cyclic vector.*

Proof. In the last proof we had seen that $\mathscr{K}(x)$, the subspace generated by $\pi(S)x$ is invariant. Hence if $\pi(S)x\neq 0$ then $\mathscr{K}(x)=\mathscr{H}$ by the irreducibility of π and x is a cyclic vector.

Corollary. *Every irreducible, unitary group representation is cyclic and any non-zero vector is a cyclic vector.*

If $\rho\colon S\to\mathscr{K}(\mathscr{H})$ is a reducible morphism and if $\mathscr{K}$ is a proper invariant subspace of ρ then no vector in $\mathscr{K}$ is a cyclic generator of ρ. Therefore a group morphism $\rho\colon G\to Gl(\mathscr{H})$ is irreducible if and only if every non-zero vector in $\mathscr{H}$ is a cyclic vector of ρ.

Lemma 18. *If* $\pi_k\colon S\to\mathscr{L}(\mathscr{H}_k)\,(k=1,\ldots,n)$ *are equivalent, finite dimensional, irreducible representations and if* $n\leqslant d=\dim\mathscr{H}_1=\cdots=\dim\mathscr{H}_n$ *then* $\rho=\pi_1\oplus\cdots\oplus\pi_n$ *is a cyclic morphism.*

Proof. We may suppose that $\mathscr{H}_1 = \cdots = \mathscr{H}_n = \mathscr{H}$ and $\pi_1 = \cdots = \pi_n = \pi$. Given a MONS x_i $(i \in \mathscr{I})$ in $\mathscr{H}$ we choose n indices $i_1, \ldots, i_n$ in $\mathscr{I}$ and let

$$x = (x_{i_1}, \ldots, x_{i_n}).$$

We prove that x is a cyclic vector for the direct sum ρ. First we notice that the vectors

$$x_{i1} = (x_i, 0, \ldots, 0), \ldots, x_{in} = (0, \ldots, 0, x_i) \qquad (i \in \mathscr{I})$$

form a base in $\mathscr{H} \oplus \cdots \oplus \mathscr{H}$. Next we choose the linear operators

$$A_{ik}: \mathscr{H} \to \mathscr{H} \qquad (i \in \mathscr{I} \text{ and } k = 1, \ldots, n)$$

such that $A_{ik} x_{i_k} = x_{i_k}$ and $A_{ik} x_{i_l} = 0$ for $l \neq k$. Clearly we have

(1) $$(A_{ik} \oplus \cdots \oplus A_{ik}) x = (0, \ldots, x_i, \ldots, 0) = x_{ik}.$$

Since π is irreducible by Proposition 2.6 the operator $A_{ik} \oplus \cdots \oplus A_{ik}$ belongs to $\mathscr{A}_\rho$, the v. Neumann algebra associated with $\rho = \pi \oplus \cdots \oplus \pi$. Since π is finite dimensional it is actually in $\tilde{\mathscr{A}}_\rho$ and so by (1) x is a cyclic vector for ρ.

A similar result holds in the infinite dimensional case but then it is necessary to suppose that the π_k's are Mackey type representations of a Mackey object S. In that case ρ is also a simple representation and so by Proposition 1.6 $\mathscr{A}_\rho$ is the strong closure of $\tilde{\mathscr{A}}_\rho$. Hence by (1) each vector x_{ik} can be approximated by vectors of the form Ax where $A \in \tilde{\mathscr{A}}_\rho$. Therefore x is again a cyclic vector for ρ. One can also extend the result to the case of denumerably many infinite dimensional π_k's. Then a cyclic vector is

$$x = (x_{i_1}, \tfrac{1}{2} x_{i_2}, \tfrac{1}{3} x_{i_3}, \ldots).$$

Lemma 19. *If $\rho: S \to \mathscr{L}(\mathscr{H})$ is a representation of the admissible object S in the finite dimensional Hilbert space $\mathscr{H}$ then ρ is a direct sum of irreducible components.*

Proof. Let $\mathscr{P}_1$ be an invariant subspace of positive, minimal dimension. Then $\mathscr{P}_1$ is an irreducible component of ρ. Since ρ is a representation $\mathscr{P}_1^\perp$ is an invariant subspace and so by the same reasoning $\rho | \mathscr{P}_1^\perp$ has an irreducible component $\mathscr{P}_2$. In finitely many steps we obtain a decomposition $\mathscr{H} = \mathscr{P}_1 \oplus \cdots \oplus \mathscr{P}_k$ where each $\mathscr{P}_i$ is an irreducible component of ρ.

3. Intertwining Operators

Let $\mathscr{X}_1$ and $\mathscr{X}_2$ be locally convex topological vector spaces and let $\rho_k: S \to \mathscr{L}(\mathscr{X}_k)$ $(k=1,2)$ be morphisms of the object S. A continuous linear operator $T: \mathscr{X}_1 \to \mathscr{X}_2$ will be called an *intertwining operator* of ρ_1 and ρ_2 in this order provided $T\rho_1(s) = \rho_2(s)T$ for every s in S, or briefly if $T\rho_1 = \rho_2 T$. We let (ρ_1, ρ_2) denote the *vector space of all intertwining operators* T. In particular we let $(\rho, \rho) = \mathscr{S}'$ where $\mathscr{S}'$ denotes the commutant of the set $\mathscr{S} = \rho(S)$ and call (ρ, ρ) the *commuting algebra* or the *commutant* of the morphism $\rho: S \to \mathscr{L}(\mathscr{X})$.

Proposition 1. *The vector space of intertwining operators* (ρ_1, ρ_2) *is a weakly closed subspace of* $\mathscr{L}(\mathscr{X}_1, \mathscr{X}_2)$.

Proof. Apply Proposition III.5.5 with $\mathscr{S} = \rho_1(S)$ and $f: \mathscr{S} \to \mathscr{L}(\mathscr{X}_2)$ given by $f(\rho_1(s)) = \rho_2(s)$.

Similarly, by applying Proposition III.5.8 we see the following:

Proposition 2. *If* $\rho: S \to \mathscr{L}(\mathscr{H})$ *is a representation in Hilbert space, then its commutant* (ρ, ρ) *is a v. Neumann algebra, namely* $(\rho, \rho) = \mathscr{A}'$ *where* $\mathscr{A}$ *is the v. Neumann algebra associated with* ρ.

If ρ is a representation then (ρ, ρ) is a v. Neumann algebra and as such it is generated by its projections. (See Proposition III.6.15 and its corollary.) By Proposition III.5.13 a projection P belongs to (ρ, ρ) if and only if $P\mathscr{H}$ is an invariant subspace of ρ. Moreover by Proposition 2.14 $\mathscr{K}_1$ and $\mathscr{K}_2$ are equivalent invariant subspaces if and only if $P_1 \sim P_2$ in (ρ, ρ). Hence the commutant contains all the necessary information to find out which are the invariant subspaces of ρ and to decide whether the restrictions of ρ to two such subspaces are equivalent or not. For instance we have:

Lemma 3. *If* $\rho_k: S \to \mathscr{L}(\mathscr{H}_k)$ $(k=1,2)$ *are representations then every invariant subspace of* ρ_1 *is invariant under* ρ_2 *if and only if* $(\rho_1, \rho_1) \subseteq (\rho_2, \rho_2)$.

The next result shows that if ρ_1 and ρ_2 are equivalent morphisms then the vector space (ρ_1, ρ_2) is essentially determined by either one of the commutants.

Proposition 4. *If* $\rho_1 \approx \rho_2$, *where* ρ_1 *and* ρ_2 *are morphisms, then* $(\rho_1, \rho_2) = (\rho_2, \rho_2)U$ *and* $(\rho_2, \rho_1) = U^*(\rho_2, \rho_2)$ *for any unitary operator* U *in* (ρ_1, ρ_2).

Proof. Since $\rho_1 = U^* \rho_2 U$ we have $T \in (\rho_1, \rho_2)$ if and only if $TU^* \rho_2 U = \rho_2 T$ i. e. $(TU^*)\rho_2 = \rho_2(TU^*)$ or $TU^* \in (\rho_2, \rho_2)$ i. e. $T \in (\rho_2, \rho_2) U$. The second relation is obtained in a similar manner.

Corollary. *If* $\rho_1 \approx \rho_2$ *then* $\dim(\rho_1, \rho_1) = \dim(\rho_1, \rho_2) = \dim(\rho_2, \rho_1) = \dim(\rho_2, \rho_2)$.

Proposition 5. *Let* $\rho_k : S \to \mathscr{L}(\mathscr{H}_k)$ $(k = 1, 2, 3, 4)$ *be morphisms and let* $\rho_1 \approx \rho_3$ *and* $\rho_2 \approx \rho_4$. *Then for any unitary operators* U_{13} *in* (ρ_1, ρ_3) *and* U_{24} *in* (ρ_2, ρ_4) *we have* $U_{24}(\rho_1, \rho_2) = (\rho_3, \rho_4) U_{13}$.

Corollary. *If* $\rho_1 \approx \rho_3$ *and* $\rho_2 \approx \rho_4$ *then* $(\rho_1, \rho_2) = 0$ *if* $(\rho_3, \rho_4) = 0$.

Proof. The inclusion $U_{24}(\rho_1, \rho_2) U_{13}^* \subseteq (\rho_3, \rho_4)$ can be obtained by inspection from the diagram

$$\begin{array}{ccccccc}
\mathscr{H}_3 & \xrightarrow{U_{13}^*} & \mathscr{H}_1 & \xrightarrow{T} & \mathscr{H}_2 & \xrightarrow{U_{24}} & \mathscr{H}_4 \\
\rho_3 \downarrow & \circlearrowright & \rho_1 \downarrow & \circlearrowright & \downarrow \rho_2 & \circlearrowright & \downarrow \rho_4 \\
\mathscr{H}_3 & \xrightarrow{U_{13}^*} & \mathscr{H}_1 & \xrightarrow{T} & \mathscr{H}_2 & \xrightarrow{U_{24}} & \mathscr{H}_4
\end{array}$$

Proposition 6. *If S is a Mackey object such that the map* $S \to S$ *is surjective and if* $\rho_k : S \to \mathscr{L}(\mathscr{H}_k)$ $(k = 1, 2)$ *are Mackey type representations then* $(\rho_1, \rho_2)^* = (\rho_2, \rho_1)$.

Proof. For any s in S we have $\rho_k(s) = \overline{p(s)}\, \rho_k(s^*)^*$. Hence if $T \in (\rho_1, \rho_2)$ then

$$T^* \rho_2(s) = \overline{p(s)}\, T^* \rho_2(s^*)^* = \overline{p(s)}\, \rho_1(s^*)^* T^* = \rho_1(s) T^*$$

and so $T^* \in (\rho_2, \rho_1)$ and $(\rho_1, \rho_2)^* \subseteq (\rho_2, \rho_1)$.

Proposition 7. *If* $\rho : S \to \mathscr{L}(\mathscr{H})$ *is a representation,* $\mathscr{H}_1$ *and* $\mathscr{H}_2$ *are stable subspaces,* E_k *denotes the projection of* $\mathscr{H}$ *onto* $\mathscr{H}_k$ *and* $\rho_k = \rho \,|\, \mathscr{H}_k = \rho E_k$ *then* $(\rho_1, \rho_2) = E_2(\rho, \rho) E_1$.

Proof. By Proposition III.5.13 $E_k \in (\rho, \rho)$ and so for any $T \in (\rho, \rho)$ we have

$$(E_2 T E_1)\rho_1 = E_2 T E_1 \rho E_1 = \rho E_2 T E_1^2 = \rho_2(E_2 T E_1).$$

Conversely, if $V \in (\rho_1, \rho_2)$ then $T = V E_1$ gives an operator in $\mathscr{L}(\mathscr{H})$ such that $T = V$ on $\mathscr{H}_1$ and

$$\begin{aligned} T\rho x &= T E_1 \rho x = T E_1 \rho(E_1 x) = T \rho_1(E_1 x) = V \rho_1(E_1 x) \\ &= \rho_2 V(E_1 x) = \rho_2 T x = \rho T x \end{aligned}$$

proving that $T \in (\rho, \rho)$.

Proposition 8. *If $\rho: S \to \mathscr{L}(\mathscr{H})$ is a representation and if ρ_1, ρ_2 are representations such that $\rho_1, \rho_2 \leqslant \rho$ then $(\rho_1, \rho_2)^* = (\rho_2, \rho_1)$.*

Proof. By Proposition 5 it is sufficient to consider the special case when ρ_1, ρ_2 are subrepresentations of ρ. Hence let $\mathscr{H}_1$ and $\mathscr{H}_2$ be stable subspaces of ρ, let E_k $(k=1,2)$ be the projection of $\mathscr{H}$ onto $\mathscr{H}_k$ and let $\rho_k = \rho E_k$. Now Proposition 7 can be applied to get $(\rho_1, \rho_2) = E_2(\rho, \rho) E_1$ and $(\rho_2, \rho_1) = E_1(\rho, \rho) E_2$. Thus the conclusion follows from the self adjoint property of (ρ, ρ).

Lemma 9. *Let $\mathscr{H}$ be a Hilbert space and let $A: \mathscr{D}_A \to \mathscr{H}$ be an arbitrary, not necessarily continuous linear operator such that $AP = PA$ for every projection $P: \mathscr{H} \to \mathscr{H}$. Then $A = \lambda I | \mathscr{D}_A$ for some scalar λ.*

Proof. Given x in $\mathscr{D}_A$ let P be the projection of $\mathscr{H}$ onto the subspace generated by x. Then $Ax = APx = PAx$ and so $Ax = \lambda_x x$ for some scalar λ_x. If x and y are independent elements in $\mathscr{D}_A$ then using the additivity of A one sees that $\lambda_x = \lambda_{x+y}$ and $\lambda_y = \lambda_{x+y}$ proving that $\lambda_x = \lambda_y$. The same holds also if x and y are dependent. Hence $Ax = \lambda x$ for every x in $\mathscr{D}_A$ where λ is a constant scalar.

Proposition 10. *Let S be a Mackey object, let $\pi: S \to \mathscr{L}(\mathscr{H})$ be an irreducible, and let $\rho: S \to \mathscr{L}(\mathscr{K})$ be an arbitrary Mackey type representation of S. Let T be a closed, densely defined operator from $\mathscr{H}$ to $\mathscr{K}$ such that its domain $\mathscr{D}_T$ is stable under π and $T\pi = \rho T$ on $\mathscr{D}_T$. Then $\mathscr{D}_T = \mathscr{H}$ and $T = \lambda U$ where $\lambda \geqslant 0$ and $U: \mathscr{H} \to \mathscr{K}$ is an isometry.*

Proof. Since T is closed by Proposition II.8.3 $T^* T$ is a densely defined operator. For any s in S we have $\rho(s^*) T = T\pi(s^*)$ so using $\rho(s)^* = p(s)\rho(s^*)$ and $\pi(s)^* = p(s)\pi(s^*)$ we see that $\rho(s)^* T = T\pi(s)^*$ and so $T^* \rho(s) = \pi(s) T^*$ by Lemma II.8.6. Therefore we have $(T^* T)\pi(s) = \pi(s)(T^* T)$ on $\mathscr{D}_{T^*T}$ for all s in S and applying Proposition II.8.5 we see that $(T^* T)A = A(T^* T)$ for every A in the v. Neumann algebra $\mathscr{A} = \mathscr{A}_\pi$ associated with π. Since π is irreducible by Proposition 2.6 we have $\mathscr{A} = \mathscr{L}(\mathscr{H})$. Therefore applying Lemma 9 we obtain $T^* T = \mu I | \mathscr{D}_{T^*T}$. By $(Tx, Tx) = \mu(x, x)$ we have $\mu = \lambda^2$ for some $\lambda \geqslant 0$. Now if $\lambda > 0$ and $x, y \in \mathscr{D}_{T^*T}$ then

$$\mu(x, y) = (T^* Tx, y) = (Tx, Ty)$$

and so $1/\lambda\, T$ is an isometry from $\mathscr{D}_{T^*T}$ to $\mathscr{K}$. If we let $U: \mathscr{H} \to \mathscr{K}$ be its extension to $\mathscr{H}$ then $T | \mathscr{D}_{T^*T} = \lambda U | \mathscr{D}_{T^*T}$. Since T is closed and it has a continuous extension to $\mathscr{H}$ we have $\mathscr{D}_T = \mathscr{H}$.

Corollary. *If the Mackey type representations* π *and* ρ *are such that a non-zero map* T *exists then* T *can be extended to* $\mathscr{H}$ *and* $T=\lambda U$ *where* $\lambda>0$ *and* $U\in(\pi,\rho)$. *The irreducible representation* π *is equivalent under* U *to a subrepresentation of* ρ.

The proposition and its corollary can be interpreted as generalizations of Schur's lemma.

4. Schur's Lemma

Various closely related statements are known under the name Schur's lemma. First we describe briefly some of the purely algebraic results and then we shall turn to the detailed discussion of the situation in Hilbert space.

The most elementary form concerns sets of $n\times n$ matrices A over a field F. Such a set $\mathscr{A}$ is called *reducible* if there exist integers p and q such that $p+q=n$ and there exists an invertible matrix T such that for every A in $\mathscr{A}$ we have

$$T^{-1}AT=\begin{pmatrix} A_{11} & 0 \\ A_{21} & A_{22} \end{pmatrix}$$

where A_{11} is $p\times p$, A_{21} is $q\times p$ and A_{22} is $q\times q$. If such matrix T does not exist then $\mathscr{A}$ is called an *irreducible set of matrices*. Let $\mathscr{A}$ and $\mathscr{B}$ be irreducible sets of $n\times n$ matrices over a field F, let $f\colon\mathscr{A}\to\mathscr{B}$ be a one-to-one map and let T be an $n\times n$ matrix such that $TA=f(A)T$ for every A in $\mathscr{A}$. Then Schur's lemma states that T is either 0 or a non singular matrix.

A more sophisticated version deals with a set $\mathscr{A}$ of *endomorphisms* of a commutative group G. The set $\mathscr{A}$ is called *irreducible* if the only subgroups which are mapped into themselves by the elements of $\mathscr{A}$ are 0 and G. The lemma states that if $\mathscr{A}$ is an irreducible set of endomorphisms then the ring of those endomorphisms which commute with any A in $\mathscr{A}$ is a skew field. For precise details see Proposition II.8.10.

A form which is closely tied to representation theory concerns a left A-module M where A is a ring or an algebra. We call M irreducible if its only submodules are 0 and M itself. Then the lemma states that the set of morphisms $\mathrm{Hom}(M,M)$ is a skew field. If A is an algebra over an algebraically closed field F then actually $\mathrm{Hom}(M,M)$ is $F\cdot I$ where $I\colon M\to M$ is the identity map. One can also show that if M and N are non isomorphic left A-modules then $\mathrm{Hom}(M,N)=0$. This statement can also be considered as part of Schur's lemma.

Let V_1 and V_2 be vector spaces over the same field F and let $\rho_k: G \to \mathscr{L}(V_k)$ $(k=1,2)$ be representations of the group G on V_k. Let $T: V_1 \to V_2$ be in (ρ_1, ρ_2) and let $N = \{x: Tx = 0\}$. Then it is easy to see that N is an invariant subspace of ρ_1 and TV_1 is an invariant subspace of ρ_2. Thus if ρ_1 is irreducible then $N=0$ or $N=V_1$ and so T is either 0 or $N=0$ and so T is either 0 or it is invertible. In the latter case $\rho_2 = T\rho_1 T^{-1}$ on TV_1 so ρ_2 is the extension to V_2 of a representation of G which is *similar* to ρ_1. If ρ_2 is also irreducible then TV_1 is 0 and T is trivial or $TV_1 = V_2$ and T is a bijection. Therefore if both ρ_1 and ρ_2 are irreducible and $T \neq 0$ then it follows that ρ_1 and ρ_2 are similar representations. If $\rho_1 = \rho_2 = \rho$ is irreducible then (ρ, ρ) is an algebra over F such that every non zero element has an inverse and so (ρ, ρ) is a division ring over F.

In what follows the vector spaces V_1 and V_2 will be complex Hilbert spaces and the operators T of (ρ_1, ρ_2) will always supposed to be continuous. Instead of assuming that ρ_1 and ρ_2 are irreducible we shall look at the restrictions $\rho_1 | N$ and $\rho_2 | \overline{TV_1}$ and compare these two representations. The classical Schur's lemma will appear as a corollary to a more general theorem. The discussion starts with a few lemmas from functional analysis and after that comes the main theorem on Mackey type representations which includes Schur's lemma for the case $F = \mathbb{C}$ as a trivial special case. The rest of the section deals with disjointness of representations. These criteria for disjointness will be needed in the definition of the multiplicity of irreducible components.

If $T: \mathscr{H}_1 \to \mathscr{H}_2$ is a linear operator then we let $\mathscr{N}_T$ denote the *null space* or *kernel* of T. Hence $\mathscr{N}_T = \{x : x \in \mathscr{H}_1 \text{ and } Tx = 0\}$.

Proposition 1. *If $\mathscr{H}_1$ and $\mathscr{H}_2$ are complex Hilbert spaces and if T is in $\mathscr{L}(\mathscr{H}_1, \mathscr{H}_2)$ then there exists an operator $H = \sqrt{T^*T}$ such that*

a) $0 \leq H = H^* \in \mathscr{L}(\mathscr{H}_1)$,

b) $H^2 = T^*T$,

c) $SH^2 = H^2S$ *implies* $SH = HS$ *for any* S *in* $\mathscr{L}(\mathscr{H}_1)$,

d) $\|Hx\| = \|Tx\|$ *for all* x *in* $\mathscr{H}_1$,

e) $\mathscr{N}_H = \mathscr{N}_T$ *and* $\overline{H\mathscr{H}_1} = \overline{T^*\mathscr{H}_2}$.

Proof. Consider the v. Neumann algebra generated by the self adjoint and positive operator $A = T^*T$ in $\mathscr{L}(\mathscr{H}_1)$. Propositions III.4.9 and III.6.16 imply that $\sigma(A) \subseteq [0, \infty)$ where $\sigma(A)$ denotes the spectrum of A in $\mathscr{A}$. Let $\mathscr{M}$ be the maximal ideal space of $\mathscr{A}$ and let

$$\varphi: C(\mathscr{M}) = \hat{\mathscr{A}} \to \mathscr{A}$$

be the inverse Fourier transform map described in Theorem I.8.7. By Theorem I.6.14 $\hat{A}(\mathcal{M})=\sigma(A)\subseteq[0,\infty)$ and so we see that $f=\sqrt{\hat{A}}\in C(\mathcal{M})$ and we can define $H=\varphi(f)\in\mathcal{A}$. Since φ is an involutive isometric isomorphism we have $H^*=\varphi(\bar{f})=\varphi(f)=H$ and $H^2=\varphi(\hat{A})=A=T^*T$. By $\hat{H}=f\geqslant 0$ and by $\sigma(H)=\hat{H}(\mathcal{M})$ we obtain $\sigma(H)\subseteq[0,\infty)$ and so by Propositions III.4.9 and III.6.16 $H\geqslant 0$. This proves a) and b).

In order to prove c) notice that $\mathcal{A}$ is generated by $A=A^*$ and so $SH^2=H^2S$ i.e. $SA=AS$ implies that $S\in\mathcal{A}'$. Since H is in $\mathcal{A}$ we get $SH=HS$.

For any x in $\mathcal{H}_1$ we have

$$(Hx,Hx)=(H^2x,x)=(T^*Tx,x)=(Tx,Tx)$$

and so $\|Hx\|=\|Tx\|$ and $\mathcal{N}_H=\mathcal{N}_T$. The second half of e) follows from

Proposition 2. *If A is in $\mathcal{L}(\mathcal{H}_1,\mathcal{H}_2)$ then $\mathcal{N}_A=(A^*\mathcal{H}_2)^\perp$ and $(\mathcal{N}_A)^\perp=\overline{A^*\mathcal{H}_2}$.*

Proof. It is a trivial consequence of $(Ax,y)=(x,A^*y)$.

Proposition 3. *If T is in $\mathcal{L}(\mathcal{H}_1,\mathcal{H}_2)$ then $T=UH$ where $H=\sqrt{T^*T}$ and U is a partial isometry from $\overline{T^*\mathcal{H}_2}$ to $\overline{T\mathcal{H}_1}$.*

Proof. By part d) of Proposition 1 we have $\|Hx\|=\|Tx\|$ for x in $\mathcal{H}_1$ and so we can define an isometry $U:\overline{H\mathcal{H}_1}\to\overline{T\mathcal{H}_1}$ by letting $U(Hx)=Tx$ for x in $\mathcal{H}_1$. On $(H\mathcal{H}_1)^\perp$ we define $U\equiv 0$ and obtain an operator $U:\mathcal{H}_1\to\mathcal{H}_2$ which clearly satisfies $T=UH$. By part e) we have $\overline{H\mathcal{H}_1}=\overline{T^*\mathcal{H}_2}$.

Proposition 4. *If $\rho_k: S\to\mathcal{L}(\mathcal{H}_k)$ $(k=1,2)$ are irreducible morphisms then either $\rho_1\approx\rho_2$ or $\rho_1\downarrow\rho_2$.*

Proof. This is clear from the definitions of $\approx$ and $\downarrow$.

Theorem 5. *Let S be an object with a map and let $\rho_k: S\to\mathcal{L}(\mathcal{H}_k)$ $(k=1,2)$ be Mackey type representations. Then for any T in (ρ_1,ρ_2)*

$$\mathcal{M}_1=\overline{T^*\mathcal{H}_2}\quad\text{and}\quad\mathcal{M}_2=\overline{T\mathcal{H}_1}$$

are stable subspaces. Moreover $\rho_1|\mathcal{M}_1\approx\rho_2|\mathcal{M}_2$ holds if and only if $\rho_1|\mathcal{M}_1$ and $\rho_2|\mathcal{M}_2$ have a common multiplier p.

Proof. The space $\mathcal{M}_1$ is stable if $(\mathcal{M}_1)^\perp$ is stable and $(\mathcal{M}_1)^\perp=\mathcal{N}_T$ by Proposition 2. The stability relations $\rho_1\mathcal{N}_T\subseteq\mathcal{N}_T$ and $\rho_2\mathcal{M}_2\subseteq\mathcal{M}_2$ follow immediately from $T\rho_1=\rho_2 T$. We let $\tilde{\rho}_1=\rho_1|\mathscr{M}_1$ and $\tilde{\rho}_2=\rho_2|\mathscr{M}_2$ so that by $T\mathscr{H}_1\subseteq\mathscr{M}_2$ and by $T^*\mathscr{H}_2\subseteq\mathscr{M}_1$ we have

$$T\rho_1=\tilde{\rho}_2 T \quad\text{and}\quad \tilde{\rho}_1^* T^* = T^*\rho_2^*.$$

Now suppose that

$$\tilde{\rho}_1(s)^* = p(s)\tilde{\rho}_1(s^*) \quad\text{and}\quad \tilde{\rho}_2(s)^* = p(s)\tilde{\rho}_2(s^*)$$

for every $s\in S$ with the same function $p: S\to\mathbb{C}$. Then we obtain

$$\begin{aligned} T^*T\tilde{\rho}_1(s) &= T^*\tilde{\rho}_2(s)T = T^*\overline{p(s)}\tilde{\rho}_2(s^*)^*T = \overline{p(s)}\,T^*\tilde{\rho}_2(s^*)^*T \\ &= \overline{p(s)}\tilde{\rho}_1(s^*)^*T^*T = \tilde{\rho}_1(s)T^*T \end{aligned}$$

or $H^2\tilde{\rho}_1=\tilde{\rho}_1 H^2$ where $H=\sqrt{T^*T}$. By Proposition 1 we now have $H\tilde{\rho}_1=\tilde{\rho}_1 H$. By Proposition 3 we can write $T=UH$ where U is a partial isometry from $\mathscr{M}_1=\overline{T^*\mathscr{H}_2}=\overline{H\mathscr{H}_1}$ to $\mathscr{M}_2=\overline{T\mathscr{H}_1}$. Then we obtain

$$U\tilde{\rho}_1 H = UH\tilde{\rho}_1 = T\tilde{\rho}_1 = \tilde{\rho}_2 T = \tilde{\rho}_2 UH$$

and so $(U\tilde{\rho}_1-\tilde{\rho}_2 U)H=0$. In other words $U\tilde{\rho}_1=\tilde{\rho}_2 U$ on $H\mathscr{H}_1$ which is dense in $\mathscr{M}_1$. Hence $U:\mathscr{M}_1\to\mathscr{M}_2$ is a unitary map which belongs to $(\tilde{\rho}_1,\tilde{\rho}_2)$. This proves that $\tilde{\rho}_1\approx\tilde{\rho}_2$.

Conversely, if there exists a unitary operator U in $(\tilde{\rho}_1,\tilde{\rho}_2)$ then $\tilde{\rho}_k(s)^*=p_k(s)\tilde{\rho}_k(s^*)$ $(k=1,2)$ and $\tilde{\rho}_2=U\tilde{\rho}_1 U^*$ imply

$$p_2(s)\tilde{\rho}_2(s^*) = \tilde{\rho}_2(s)^* = U\tilde{\rho}_1(s)^*U^* = p_1(s)U\tilde{\rho}_1(s^*)U^* = p_1(s)\tilde{\rho}_2(s^*).$$

Hence we obtain $(p_2(s)-p_1(s))\tilde{\rho}_2(s^*)=0$ for all s in S. Thus for any fixed s either $p_1(s)=p_2(s)$ or $\tilde{\rho}_2(s^*)=0$ in which case we may modify p_1 and p_2 by letting $p_1(s)=p_2(s)=0$. This completes the proof.

An immediate corollary of Theorem 5 is the following generalized version of Schur's lemma which is due to Mackey:

Proposition 6. *Let $\rho_k: S\to\mathscr{L}(\mathscr{H}_k)$ $(k=1,2)$ be Mackey type representations of the Mackey object S. Then for any $T\in(\rho_1,\rho_2)$ the subspaces $\mathscr{M}_1=\overline{T^*\mathscr{H}_2}$ and $\mathscr{M}_2=\overline{T\mathscr{H}_1}$ are stable and $\rho_1|\mathscr{M}_1\approx\rho_2|\mathscr{M}_2$.*

We notice that there are no continuity restrictions imposed upon the representations ρ_1 and ρ_2. Special cases in which for example S is a group and ρ_1, ρ_2 are continuous group representations, or S is an involutive algebra and ρ_1, ρ_2 are star representations were considered earlier. We also introduced already the irreducibility criteria expressed in Theorem 2.2 and Propositions 2.3 and 2.4.

Schur's lemma gives an important disjointness criterion which in view of Proposition 6 can be now stated as follows:

Proposition 7. *Let $\rho_k: S \to \mathscr{L}(\mathscr{H}_k)$ be Mackey type representations of the Mackey object S. Then $\rho_1 \downarrow \rho_2$ if and only if $(\rho_1, \rho_2) = 0$.*

Proof. If $(\rho_1, \rho_2) \neq 0$ then $\rho_1 \uparrow \rho_2$ by Proposition 6. Conversely, let $\rho_1 | \mathscr{K}_1 \approx \rho_2 | \mathscr{K}_2$ where $\mathscr{K}_1, \mathscr{K}_2 \neq 0$ and let $U: \mathscr{K}_1 \to \mathscr{K}_2$ be a unitary map which belongs to $(\rho_1 | \mathscr{K}_1, \rho_2 | \mathscr{K}_2)$. We extend U to a partial isometry $U: \mathscr{H}_1 \to \mathscr{H}_2$ from $\mathscr{K}_1$ to $\mathscr{K}_2$ by letting $U \equiv 0$ on $\mathscr{K}_1^{\perp}$. Then $U \rho_1 = \rho_2 U$ on $\mathscr{K}_1^{\perp}$ and by hypothesis the same holds on $\mathscr{K}_1$. Hence $U \rho_1 = \rho_2 U$ on $\mathscr{H}_1$ where $U \neq 0$ and so $(\rho_1, \rho_2) \neq 0$.

Theorem 8. *Let ρ be a representation and let ρ_1, ρ_2 be representations of the same object S such that $\rho_1 \leqslant \rho$ and $\rho_2 \leqslant \rho$. Then $\rho_1 \downarrow \rho_2$ if and only if $(\rho_1, \rho_2) = 0$.*

Note. Since $\rho_1, \rho_2 \leqslant \rho$ by Proposition 3.8 $(\rho_1, \rho_2)^* = (\rho_2, \rho_1)$ and so $(\rho_1, \rho_2) = 0$ and $(\rho_2, \rho_1) = 0$ fail or hold simultaneously.

Proof. First of all by the Corollary of Proposition 3.5 it is sufficient to consider the special case when ρ_1 and ρ_2 are actually submorphisms of ρ. In order to settle this special case we consider the following:

Proposition 9. *Let $\rho: S \to \mathscr{L}(\mathscr{H})$ be a morphism, let $\mathscr{H}_1, \mathscr{H}_2$ be stable subspaces of ρ and let $\rho_k = \rho | \mathscr{H}_k$ $(k = 1, 2)$. Denote by $\tilde{\mathscr{A}}, \tilde{\mathscr{A}}_1, \tilde{\mathscr{A}}_2$ the algebras generated by these morphisms and let $\mathscr{A}, \mathscr{A}_1, \mathscr{A}_2$ be their weak closures. Let $\sigma: \mathscr{A} \to \mathscr{L}(\mathscr{H})$ be the identity representation and $\sigma_k = \sigma | \mathscr{H}_k$. Then we have the following:*

a) $\sigma_k(\tilde{\mathscr{A}}) = \tilde{\mathscr{A}}_k$ *for* $k = 1, 2$;

b) $\rho_1 \downarrow \rho_2$ *if and only if* $\sigma_1 \downarrow \sigma_2$;

c) $(\rho_1, \rho_2) = (\sigma_1, \sigma_2)$.

If in this proposition we let ρ be a representation then $\mathscr{A}$ is a v. Neumann algebra and so it can be considered as a Mackey object with respect to the operator adjoint and $p \equiv 1$. Then σ being a Mackey representation of a Mackey object by Proposition 7 we have $\sigma_1 \downarrow \sigma_2$ if and only if $(\sigma_1, \sigma_2) = 0$. Hence by b) and c) $\rho_1 \downarrow \rho_2$ if and only if $(\rho_1, \rho_2) = 0$. The proof of Theorem 8 is now reduced to the proof of the above proposition.

Proof of Proposition 9. To prove a) introduce the projections $E_k: \mathscr{H} \to \mathscr{H}_k$ and notice that $\rho E_k = E_k \rho$ by the stability of $\mathscr{H}_k$. (See Proposition III.5.13.) Now if $A \in \tilde{\mathscr{A}}$ then $A = \lambda_1 X_1 + \cdots + \lambda_m X_m$ where each X_k is of the form

$X = \rho(s_1)\dots\rho(s_n)$. Hence by

$$\sigma_k X E_k = X E_k = X E_k^n = \rho(s_1) E_k \dots \rho(s_n) E_k = \rho_k(s_1)\dots\rho_k(s_n)$$

we see that $\sigma_k(A) \in \tilde{\mathscr{A}}_k$. Conversely, if $X^{(k)} = \rho_k(s_1)\dots\rho_k(s_n)$ then by $\rho_k = \rho E_k$ we get $X^{(k)} = \sigma_k(X)$ where $X = \rho(s_1)\dots\rho(s_n)$ and so $X^{(k)} \in \sigma_k(\tilde{\mathscr{A}})$ and $\tilde{\mathscr{A}}_k \subseteq \sigma_k(\tilde{\mathscr{A}})$.

To prove b) it is sufficient to show that $\rho_1 \upharpoonleft \rho_2$ implies $\sigma_1 \upharpoonleft \sigma_2$: Suppose $\mathscr{K}_k \subseteq \mathscr{H}_k$ $(k=1,2)$ are stable subspaces of ρ_k $(k=1,2)$ such that $\rho_1 | \mathscr{K}_1$ and $\rho_2 | \mathscr{K}_2$ are equivalent morphisms. Then there exists a unitary map $U: \mathscr{K}_1 \to \mathscr{K}_2$ such that $(\rho E_2) U = U(\rho E_1)$ where E_k denotes now the projection $E_k: \mathscr{H} \to \mathscr{K}_k$. By the stability of $\mathscr{K}_k$ we have $\rho E_k = E_k \rho$ and so we obtain $(A E_2) U = U(A E_1)$ for every $A \in \tilde{\mathscr{A}}$. Finally let $A_n \in \tilde{\mathscr{A}}$ $(n=1,2,\dots)$ be such that $A_n \to A$ weakly. Then for any x, y in $\mathscr{H}$ on the one hand we have

$$(U A_n E_1 x, y) = (A_n E_2 U x, y) \to (A E_2 U x, y)$$

and on the other hand

$$(U A_n E_1 x, y) = (A_n E_1 x, U^* y) \to (A E_1 x, U^* y) = (U A E_1 x, y).$$

Hence we proved that $(A E_2) U = U(A E_1)$ for every A in $\mathscr{A}$ and so $\sigma_1 | \mathscr{K}_1$ is equivalent to $\sigma_2 | \mathscr{K}_2$.

To prove c) it is sufficient to show that $(\rho_1, \rho_2) \subseteq (\sigma_1, \sigma_2)$. We let $\tilde{\sigma}: \tilde{\mathscr{A}} \to \mathscr{L}(\mathscr{H})$ be the identity morphism and we first verify that $(\rho_1, \rho_2) \subseteq (\tilde{\sigma}_1, \tilde{\sigma}_2)$: Indeed, if $E_k: \mathscr{H} \to \mathscr{H}_k$ then by the stability of $\mathscr{H}_1$ and $\mathscr{H}_2$ under ρ we have $T \rho E_1 = \rho E_2 T$ for any $T \in (\rho_1, \rho_2)$ and so $T A E_1 = A E_2 T$ for every $A \in \tilde{\mathscr{A}}$ i.e. $T \tilde{\sigma}_1 = \tilde{\sigma}_2 T$ or $T \in (\tilde{\sigma}_1, \tilde{\sigma}_2)$. Next we prove that if $T \in (\tilde{\sigma}_1, \tilde{\sigma}_2)$ then $T \in (\sigma_1, \sigma_2)$: Let $A \in \mathscr{A}$ and let $A^{(n)} \in \tilde{\mathscr{A}}$ $(n=1,2,\dots)$ be such that $A^{(n)} \to A$ in the weak sense. Then for any $x \in \mathscr{H}_1$ and $y \in \mathscr{H}_2$ we have

$$(T A^{(n)} x, y) = (A^{(n)} x, T^* y) \to (A x, T^* y) = (T A x, y), \text{ and}$$

$$(T A^{(n)} x, y) = (A^{(n)} T x, y) \to (A\, T x, y).$$

Hence $(T A x, y) = (A\, T x, y)$ for $x \in \mathscr{H}_1$ and $y \in \mathscr{H}_2$, proving that $T(A | \mathscr{H}_1) = (A | \mathscr{H}_2) T$ and so $T \in (\sigma_1, \sigma_2)$.

Proposition 10. *Let $\rho: S \to \mathscr{L}(\mathscr{H})$ be a representation, let $\mathscr{K}$ be a stable subspace and let $E: \mathscr{H} \to \mathscr{K}$. Then $\rho E \downharpoonleft \rho(I-E)$ if and only if $E \in c(\rho, \rho)$ where $c(\rho, \rho)$ denotes the center of the algebra (ρ, ρ).*

Proof. Let ρ_1, ρ_2 denote the submorphisms $\rho_1 = \rho E$ and $\rho_2 = \rho(I-E)$. Since $\mathscr{K}$ is stable by Proposition III.5.13 $E \in (\rho, \rho) = \mathscr{A}'$. By Proposition 3.7 we have

$$(I-E)(\rho,\rho)E = (\rho_1, \rho_2) \text{ and } E(\rho,\rho)(I-E) = (\rho_2, \rho_1).$$

Now if $\rho_1 \downarrow \rho_2$ then by Theorem 8 $(\rho_1, \rho_2) = (\rho_2, \rho_1) = 0$ and so for any $T \in (\rho, \rho)$ we have $(I-E)TE = 0$ and $ET(I-E) = 0$. Hence $ETE = TE = ET$ and E is in $c(\rho, \rho)$. Conversely, if $E \in c(\rho, \rho)$ then for any $T \in (\rho, \rho)$ we have $(I-E)TE = 0$ proving that $(\rho_1, \rho_2) = 0$. Applying the same theorem we see that $\rho_1 \downarrow \rho_2$.

A morphism $\rho: S \to \mathscr{L}(\mathscr{H})$ is called *primary* or *factorial* if it is not the direct sum of two disjoint submorphisms, that is to say if the simultaneous stability of $\mathscr{K}$ and $\mathscr{K}^\perp$ and $\rho|\mathscr{K} \downarrow \rho|\mathscr{K}^\perp$ imply that $\mathscr{K} = 0$ or $\mathscr{K}^\perp = 0$.

Proposition 11. *A representation* $\rho: S \to \mathscr{L}(\mathscr{H})$ *is primary if and only if the v. Neumann algebra* (ρ, ρ) *is a factor.*

Proof. By Proposition III.6.15 a v. Neumann algebra is generated by its projections and so $c(\rho, \rho)$, the center of (ρ, ρ) is equal to $\mathbb{C}I$ if and only if the only projections in $c(\rho, \rho)$ are 0 and I. Therefore the result is an immediate corollary of Proposition 10.

Lemma 12. *Let* $\rho: S \to \mathscr{L}(\mathscr{H})$ *be a representation and let* $\rho_i\ (i \in \mathscr{I})$ *be subrepresentations of* ρ. *Then* $\rho_0 \leqslant \sum \rho_i$ *and* $\rho_0 \neq 0$ *imply that* $\rho_0 \uparrow \rho_i$ *for some* $i \in \mathscr{I}$.

Proof. We may suppose that ρ_0 is a submorphism of $\sum \rho_i$. Let $\mathscr{H}_0$ and $\mathscr{H}_i\ (i \in \mathscr{I})$ be subspaces such that $\rho_0 = \rho|\mathscr{H}_0$ and $\rho_i = \rho|\mathscr{H}_i$. Since $\mathscr{H}_0 \subseteq \sum \mathscr{H}_i$ and $\mathscr{H}_0 \neq \{0\}$ there is at least one index i such that the projection $P_i: \mathscr{H}_0 \to \mathscr{H}_i$ is not zero. If we can show that this particular P_i belongs to (ρ_0, ρ_i) then $\rho_0 \uparrow \rho_i$ will follow by Theorem 8. To see that $P_i \in (\rho_0, \rho_i)$ let $x_0 \in \mathscr{H}_0$ be given and let $x_0 = x_i + y_i$ where $x_i \in \mathscr{H}_i$ and $y_i \in \mathscr{H}_i^\perp$. Then on the one hand

$$\rho_i(s) P_i x_0 = \rho_i(s) x_i = \rho(s) x_i\,.$$

On the other hand we have

$$\rho(s) x_0 = \rho(s) x_i + \rho(s) y_i$$

where $\rho(s) x_i \in \mathscr{H}_i$ and $\rho(s) y_i \in \mathscr{H}_i^\perp$. Therefore

$$P_i \rho_0(s) x_0 = P_i \rho(s) x_0 = \rho(s) x_i$$

and so $\rho_i(s) P_i = P_i \rho_0(s)$.

Corollary. *If* ρ *is a representation and* $\rho_i\,(i\in\mathscr{I})$ *are subrepresentations* ρ *then* $\rho\downarrow\rho_i\,(i\in\mathscr{I})$ *implies* $\rho\downarrow\sum\rho_i$.

For if ρ is a morphism and $\rho\uparrow\sum\rho_i$ then ρ has a submorphism ρ_0 such that $\rho_0\leqslant\sum\rho_i$. Hence by the lemma $\rho_0\uparrow\rho_i$ for some i and so $\rho\uparrow\rho_i$ for the same index i.

5. Multiplicity of Irreducible Components

The purpose of this section is to show that if $\rho\colon S\to\mathscr{L}(\mathscr{H})$ is a representation and ω is an equivalence class of irreducible representations of S then it is meaningful to speak about the multiplicity $n(\omega,\rho)$ of ω in ρ. It will be proved that $n(\omega,\rho)$ can be interpreted as the number of elements in any maximal family of pairwise orthogonal irreducible components $\mathscr{P}$ belonging to the class ω. A stronger result on direct sum decomposition and some related material will also be considered.

Proposition 1. *Let* $\rho\colon S\to\mathscr{L}(\mathscr{H})$ *be a representation and let* $\mathscr{K}_1$ *and* $\mathscr{K}_2$ *be invariant subspaces of* $\mathscr{H}$. *Then* $\rho|\mathscr{K}_1\downarrow\rho|\mathscr{K}_2$ *implies* $\mathscr{K}_1\perp\mathscr{K}_2$.

Proof. Let P_k denote the projection $\mathscr{H}\to\mathscr{K}_k$. Then by Proposition 3.7 we have $(\rho_1,\rho_2)=P_2(\rho,\rho)P_1$. Since $I\in(\rho,\rho)$ we obtain $P_2P_1\in(\rho_1,\rho_2)$. But $(\rho_1,\rho_2)=0$ by Theorem 4.8 and so $P_2P_1=0$.

We add the following related result:

Proposition 2. *Let* $\rho\colon S\to\mathscr{L}(\mathscr{H})$ *be a representation, let* $\mathscr{K}$ *and* $\mathscr{L}$ *be invariant subspaces and let* E *and* F *be the associated projections. Then* $\rho|\overline{E\mathscr{L}}\downarrow\rho|\overline{F\mathscr{K}}$ *implies that* $EFE=FEF=0$.

Proof. By Proposition III.5.13 we know that E and F are in (ρ,ρ). Since $E\mathscr{L}\subseteq\mathscr{K}$ and $F\mathscr{K}\subseteq\mathscr{L}$ we have $\overline{E\mathscr{L}}\subseteq\mathscr{K}$ and $\overline{F\mathscr{K}}\subseteq\mathscr{L}$. Moreover $F(E\mathscr{L})\subseteq F\mathscr{K}$ and F being continuous $F(\overline{E\mathscr{L}})\subseteq\overline{F(E\mathscr{L})}\subseteq\overline{F\mathscr{K}}$ so that $F\colon\overline{E\mathscr{L}}\to\overline{F\mathscr{K}}$. By $E,F\in(\rho,\rho)$ and Proposition III.5.4 we know that $E\mathscr{L}$, $F\mathscr{K}$ and $\overline{E\mathscr{L}}$, $\overline{F\mathscr{K}}$ are invariant subspaces. Since $F\rho=\rho F$ we proved that

$$\begin{array}{ccc}\overline{E\mathscr{L}} & \xrightarrow{F} & \overline{F\mathscr{K}}\\ \Big\downarrow{\scriptstyle\rho} & & \Big\downarrow{\scriptstyle\rho}\\ \overline{E\mathscr{L}} & \xrightarrow{F} & \overline{F\mathscr{K}}\end{array}$$

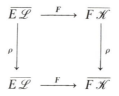

is a commutative diagram. Hence by $F(\overline{E\mathscr{L}}) \subseteq \overline{F\mathscr{K}}$ we see that $F|\overline{E\mathscr{L}}$ belongs to $(\rho|\overline{E\mathscr{L}}, \rho|\overline{F\mathscr{K}})$. Applying Theorem 4.8 we obtain $F|\overline{E\mathscr{L}}=0$. Therefore $F|E\mathscr{L}=0$ or $FEF=0$. Similarly one proves that $EFE=0$.

A special case of Proposition 1 can be derived from Proposition 2 as follows: Let $\mathscr{K}$ be irreducible, let $l \in \mathscr{L}$ and let $l=x+y$ where $x \in \mathscr{K}$ and $y \in \mathscr{K}^{\perp}$. Then we have $Fx=FEl=FEFl=0$ and so $x \in \mathscr{L}^{\perp} \cap \mathscr{K}$. Since $\mathscr{K}$ is irreducible and $\mathscr{L}^{\perp}$ is invariant we have $\mathscr{L}^{\perp} \cap \mathscr{K}=0$ or $\mathscr{L}^{\perp} \cap \mathscr{K}=\mathscr{K}$. In the first case we proved that $x=0$ and so $l=y \in \mathscr{K}^{\perp}$ for every $l \in \mathscr{L}$, i.e. $\mathscr{L} \subseteq \mathscr{K}^{\perp}$. In the second case $\mathscr{K} \subseteq \mathscr{L}^{\perp}$ and so in either case we have $\mathscr{K} \perp \mathscr{L}$.

As earlier let $\hat{S}$ be the set of all equivalence classes of irreducible representations of the object S. We fix a class ω in $\hat{S}$ and a representation $\rho: S \to \mathscr{L}(\mathscr{H})$, and consider irreducible components $\mathscr{P}$ of ρ such that $\mathscr{P}$ belongs to the class ω. If there is at least one such $\mathscr{P}$ then we can form orthogonal families $\{\mathscr{P}_i\}$ of such irreducible subspaces $\mathscr{P}_i \in \omega$ and by Zorn's lemma we see the existence of a family which is maximal with respect to inclusion.

Proposition 3. *Let $\rho: S \to \mathscr{L}(\mathscr{H})$ be a representation, let $\omega \in \hat{S}$ and let $\{\mathscr{P}_i\}$ and $\{\mathscr{Q}_j\}$ be maximal orthogonal families of irreducible components $\mathscr{P}_i, \mathscr{Q}_j \in \omega$. Then we have $\sum \mathscr{P}_i = \sum \mathscr{Q}_j$.*

Proof. We let $\mathscr{K} = \sum \mathscr{P}_i$ so that $\mathscr{K}$ is a subspace of $\mathscr{H}$. Let $\mathscr{Q} \in \omega$ and suppose that $\mathscr{Q} \uparrow \mathscr{K}^{\perp}$ i.e. the restrictions $\rho|\mathscr{Q}$ and $\rho|\mathscr{K}^{\perp}$ are not disjoint. Then by the irreducibility of $\mathscr{Q}$ there would be a $\mathscr{P} \in \omega$ such that $\mathscr{P} \subseteq \mathscr{K}^{\perp}$. However this would contradict the maximality of the family $\{\mathscr{P}_i\}$. Hence if $\mathscr{Q} \in \omega$ then $\mathscr{Q} \downarrow \mathscr{K}^{\perp}$ and so by Proposition 1 $\mathscr{Q} \perp \mathscr{K}^{\perp}$ or in other words $\mathscr{Q} \subseteq \mathscr{K}$. Applying this result to each member of the family $\{\mathscr{Q}_j\}$ we see that $\sum \mathscr{Q}_j \subseteq \mathscr{K} = \sum \mathscr{P}_i$. The conclusion follows from the symmetry of the situation.

In view of the last proposition we can define $\mathscr{H}_\omega = \sum \mathscr{P}_i$ where $\{\mathscr{P}_i\}$ is any maximal orthogonal system of irreducible components $\mathscr{P}_i \in \omega$ of the given representation $\rho: S \to \mathscr{L}(\mathscr{H})$. If ω does not occur in ρ we let $\mathscr{H}_\omega = 0$. Let $\mathscr{P}_1, \mathscr{P}_2$ be inequivalent irreducible components of the representation ρ. Then $\mathscr{P}_1 \downarrow \mathscr{P}_2$ and so by Proposition 1 we have $\mathscr{P}_1 \perp \mathscr{P}_2$. Hence if $\omega_1 \neq \omega_2$ and $\mathscr{P}_k \in \omega_k$ $(k=1, 2)$ then $\mathscr{P}_1$ and $\mathscr{P}_2$ are orthogonal. It follows that $\mathscr{H}_{\omega_1} \perp \mathscr{H}_{\omega_2}$ for distinct $\omega_1, \omega_2 \in \hat{S}$ and we can define $\mathscr{H}_{dI} = \sum \{\mathscr{H}_\omega : \omega \in \hat{S}\}$. Our results can be summarized as follows:

Theorem 4. *One can associate with every representation $\rho: S \to \mathscr{L}(\mathscr{H})$ and any equivalence class $\omega \in \hat{S}$ a uniquely determined subspace $\mathscr{H}_\omega$ of $\mathscr{H}$ such that*

1) *$\mathscr{H}_\omega$ is a direct sum of irreducible components belonging to the class ω;*

2) *if $\mathscr{P}$ is an irreducible component of ρ, then $\mathscr{P}\subseteq\mathscr{H}_\omega$ or $\mathscr{P}\perp\mathscr{H}_\omega$ according as $\mathscr{P}\in\omega$ or $\mathscr{P}\notin\omega$;*

3) *if $\omega_1\neq\omega_2$ then $\mathscr{H}_{\omega_1}\perp\mathscr{H}_{\omega_2}$, and so*

4) *$\mathscr{H}_{dI}=\sum\{\mathscr{H}_\omega:\omega\in\hat{S}\}$ is a subspace of $\mathscr{H}$ which is uniquely determined by ρ, and*

5) *if $\mathscr{P}$ is any irreducible component of ρ then $\mathscr{P}\subseteq\mathscr{H}_{dI}$ and hence*

6) *$\mathscr{H}=\mathscr{H}_{dI}\oplus(\mathscr{H}_{dI})^\perp$ where $\mathscr{H}_{dI}$ is a direct sum of irreducible components while $(\mathscr{H}_{dI})^\perp$ contains no irreducible components of ρ.*

7) *if $\mathscr{K}$ is an invariant subspace of ρ and if $\mathscr{K}\subseteq\mathscr{H}_{dI}$ then $\mathscr{K}$ is a direct sum of irreducible components.*

Proof. Only the last part needs some explanation. Clearly $\rho_0=\rho|\mathscr{K}$ is a subrepresentation of $\rho|\mathscr{H}_{dI}$ where $\mathscr{H}_{dI}$ can be written as a direct sum of irreducible components. If $\mathscr{K}$ is not $\{0\}$ then by Lemma 4.12 ρ_0 is not disjoint from one of these components and so a submorphism of ρ_0 is equivalent to an irreducible component $\rho|\mathscr{P}$. Hence we proved: If $\mathscr{K}\subseteq\mathscr{H}_{dI}$ is not $\{0\}$ and $\mathscr{K}$ is invariant then $\mathscr{K}$ has an irreducible component.

In order to show that $\mathscr{K}$ is a direct sum of irreducible components let us now consider a family $\{\mathscr{P}_i\}$ $(i\in\mathscr{I})$ of pairwise orthogonal irreducible subspaces of $\mathscr{K}$ which is maximal with respect to inclusion. The existence of such a family follows from Zorn's lemma. Then $\mathscr{K}\cap(\sum\mathscr{P}_i)^\perp$ is an invariant subspace of $\mathscr{H}_{dI}$ and so either it is $\{0\}$ or it has an irreducible component. Since the second possibility contradicts the maximality we have $\mathscr{K}=\sum\mathscr{P}_i$.

Theorem 5. *Let $\rho: S\to\mathscr{L}(\mathscr{H})$ be a representation and let $\omega\in\hat{S}$. If a maximal orthogonal system of irreducible components $\mathscr{P}_i\in\omega$ is finite then every other such system is finite and it has the same number of irreducible components.*

Proof. We suppose that ω is an equivalence class of subrepresentations of ρ because otherwise our number is 0. Let $\mathscr{A}$ be the v. Neumann algebra associated with ρ and let $\mathscr{A}_\omega=\mathscr{A}|\mathscr{H}_\omega$. Select any maximal orthogonal system of irreducible subspaces $\mathscr{P}_1,\dots,\mathscr{P}_m\in\omega$ and let $\sigma_i:\mathscr{A}_\omega\to\mathscr{A}_\omega|\mathscr{P}_i=\mathscr{A}|\mathscr{P}_i=\mathscr{A}_i$. Since ω is an equivalence class of irreducible morphisms $\mathscr{A}_i$ is an irreducible v. Neumann algebra and so by Theorem 2.5 $\mathscr{A}_i=\mathscr{L}(\mathscr{P}_i)$. Furthermore σ_i is an irreducible representation of $\mathscr{A}_\omega$ on $\mathscr{A}_i$ and so by

Proposition 2.4 $(\sigma_i, \sigma_i) = \mathscr{A}_i' = \mathbb{C} I$. Using $\mathscr{H}_\omega = \mathscr{P}_1 \oplus \cdots \oplus \mathscr{P}_m$ we get from Proposition III.6.6

$$\mathscr{A}_\omega' = \sum_{i,j=1}^{m} (\sigma_i, \sigma_j)$$

where the right hand side is a direct sum. By Proposition 3.4 $\dim(\sigma_i, \sigma_j) = \dim(\sigma_i, \sigma_i) = 1$ and so we get $\dim \mathscr{A}_\omega' = m^2$. Thus for any finite orthogonal system $\mathscr{P}_1, \ldots, \mathscr{P}_m$ the cardinality m has an invariant meaning which is uniquely determined by ω and ρ.

Now if $\{\mathscr{Q}_j\}$ is any other, possibly infinite orthogonal system of irreducible components $\mathscr{Q}_j \in \omega$ then for any finite subset $\mathscr{Q}_1, \ldots, \mathscr{Q}_n$ we have

$$\mathscr{H}_\omega = \mathscr{Q}_0 \oplus \mathscr{Q}_1 \oplus \cdots \oplus \mathscr{Q}_n$$

where $\mathscr{Q}_0$ is the orthogonal complement of $\mathscr{Q}_1 \oplus \cdots \oplus \mathscr{Q}_n$ in $\mathscr{H}_\omega$. By the same reasoning as above we obtain

$$\mathscr{A}_\omega' = \sum_{i,j=0}^{n} (\sigma_i, \sigma_j)$$

and so $\dim \mathscr{A}_\omega' = n^2 + \cdots \geqslant n^2$. Thus n is bounded and so every orthogonal system $\{\mathscr{Q}_j\}$ is finite provided one of them is finite.

In view of the preceeding theorem we can make the following definition: For any equivalence class ω of irreducible representations of the object S and for every representation $\rho: S \to \mathscr{L}(\mathscr{H})$ we let $n(\omega, \rho) = n(\omega)$ denote the number of components in any maximal orthogonal system of irreducible components $\mathscr{P}_i \in \omega$ provided this number is finite. If any, and thus every such system is infinite we let $n(\omega, \rho) = \infty$. The finite or infinite cardinal $n(\omega)$ will be called the *multiplicity* of ω in ρ. We will say that ω occurs in ρ with *finite* or *infinite multiplicity* according as $n(\omega)$ is finite or not. The function $\omega \mapsto n(\omega, \rho)$ mapping $\hat{S}$ into $\{0, 1, 2, \ldots, \infty\}$ is the *multiplicity* of ρ.

If $\mathscr{P}$ is an irreducible component of the representation $\rho: S \to \mathscr{L}(\mathscr{H})$ we define $n(\mathscr{P}) = n(\mathscr{P}, \rho)$ as the multiplicity of the equivalence class ω defined by $\mathscr{P}$. Thus $n(\mathscr{P}, \rho) = n(\omega, \rho)$ where $\mathscr{P} \in \omega$.

Let $\rho: S \to \mathscr{L}(\mathscr{H})$ be a representation and let $\mathscr{A}$ be the associated v. Neumann algebra. By Proposition 2.1 a subspace $\mathscr{P}$ of $\mathscr{H}$ is an irreducible component of ρ if and only if it is stable and irreducible under $\mathscr{A}$. Hence if $\sigma: \mathscr{A} \to \mathscr{L}(\mathscr{H})$ denotes the identity representation then there is a natural one-to-one correspondence between those elements of $\hat{S}$ and $\hat{\mathscr{A}}$ which occur with positive multiplicities in ρ and σ, respectively. Moreover by Proposition 2.1 and by the definition of the multiplicity we have $n(\mathscr{P},\rho)=n(\mathscr{P},\sigma)=n(\mathscr{P},\mathscr{A})$ for every irreducible component. We arrived at the following result:

Proposition 6. *If* $\rho: S\to\mathscr{L}(\mathscr{H})$ *is a representation and if* $\mathscr{A}$ *is the v. Neumann algebra associated with* ρ *then* $n(\mathscr{P},\rho)=n(\mathscr{P},\mathscr{A})$ *for every irreducible component.*

As a simple application of Theorem 4 we prove:

Proposition 7. *If* $\rho: S\to\mathscr{L}(\mathscr{H})$ *is a primary representation containing an irreducible subrepresentation then all such subrepresentations belong to the same equivalence class* $\omega\in\hat{S}$ *and* $\mathscr{H}=\mathscr{H}_\omega$.

Proof. If π_1 and π_2 are irreducible submorphisms of ρ then $\pi_1\upharpoonright\pi_2$ and so $\pi_1\approx\pi_2$. Therefore by Theorem 4 $\mathscr{H}=\mathscr{H}_\omega\oplus(\mathscr{H}_\omega)^\perp$ where $(\mathscr{H}_\omega)^\perp$ contains no irreducible components. Therefore by part 7) of the theorem we have $\rho|\mathscr{H}_\omega\downharpoonright\rho|(\mathscr{H}_\omega)^\perp$. Since ρ is primary this implies $(\mathscr{H}_\omega)^\perp=0$ and so $\mathscr{H}=\mathscr{H}_\omega$.

The following set of results will lead to Theorem 11 which shows that if ρ is a finite dimensional Mackey type representation of a Mackey object such that $\mathscr{H}=\mathscr{H}_{dI}$ and the multiplicities $n(\omega,\rho)$ are sufficiently small then ρ is a cyclic representation.

Theorem 8. *Let* $\rho_k: S\to\mathscr{L}(\mathscr{H}_k)$ *be Mackey type representations with a common multiplier and let* $a_k\in\mathscr{H}_k$ $(k=1,2)$. *We let* $\mathscr{K}_1$, $\mathscr{K}_2$ *and* $\mathscr{K}$ *denote the smallest invariant subspaces of* ρ_1,ρ_2 *and* $\rho_1\oplus\rho_2$ *containing the vectors* a_1, a_2 *and* (a_1,a_2), *respectively. If we let*

$$\mathscr{L}_1=\{x_1:(x_1,0)\in\mathscr{K}\} \quad \textit{and} \quad \mathscr{L}_2=\{x_2:(0,x_2)\in\mathscr{K}\}$$

then $\rho_1|\mathscr{K}_1\cap\mathscr{L}_1^\perp$ *is equivalent to* $\rho_2|\mathscr{K}_2\cap\mathscr{L}_2^\perp$.

Proof. We see that $\mathscr{L}_k$ is an invariant subspace of ρ_k. Let P_k be the projection operator $\mathscr{H}_k\to\mathscr{L}_k$ so that $P_k\in(\rho_k,\rho_k)$ by Proposition III.5.13. We let

$$\mathscr{L}=\{(x_1,x_2):(x_1,x_2)\in\mathscr{K} \text{ and } x_k\perp\mathscr{L}_k \text{ for } k=1,2\}.$$

Then $\mathscr{L}$ is an invariant subspace of $\rho_1\oplus\rho_2$. We define the continuous linear operator $T_k:\mathscr{L}\to\mathscr{H}_k$ by $T_k(x_1,x_2)=x_k$. These T_k are injections because for instance if $T_1(x_1,x_2)=0$ then $x_1=0$ and $(0,x_2)\in\mathscr{K}$. Thus $x_2\in\mathscr{L}_2\cap\mathscr{L}_2^\perp$ and $x_2=0$.

It is clear since $\mathscr{K}\subseteq\mathscr{K}_1\oplus\mathscr{K}_2$ that the range of T_k lies in $\mathscr{K}_k\cap\mathscr{L}_k^\perp$. We prove that $T_k\mathscr{L}$ is dense in $\mathscr{K}_k\cap\mathscr{L}_k^\perp$. Let x_1 in $\mathscr{K}_1\cap\mathscr{L}_1^\perp$ and $\varepsilon>0$

be given. Then by the definition of $\mathscr{K}_1$ there is a vector $y_1 \in \mathscr{H}_1$ such that $\|x_1 - y_1\| < \varepsilon$ and $y_1 = A_1 a_1$ for some $A_1 \in \tilde{\mathscr{A}}_{\rho_1}$. If

$$A_1 = \sum \lambda_i \rho_1(s_{i1}) \dots \rho_1(s_{ik_i})$$

then we let

$$A_2 = \sum \lambda_i \rho_2(s_{i1}) \dots \rho_2(s_{ik_i})$$

and define $y_2 = A_2 a_2$. We see that $(y_1, y_2) \in \mathscr{K}$. Since $P_k y_k \in \mathscr{L}_k$ we have $(P_1 y_1, 0), (0, P_2 y_2) \in \mathscr{K}$ and so $(z_1, z_2) \in \mathscr{K}$ where $z_k = y_k - P_k y_k$ is in $\mathscr{L}_k^\perp$. Thus $(z_1, z_2) \in \mathscr{L}$ and $T_1(z_1, z_2) = z_1$. We have

$$\|P_1 y_1\| = \|P_1(y_1 - x_1)\| \leqslant \|y_1 - x_1\|$$

and so

$$\|x_1 - z_1\| = \|x_1 - y_1 + P_1 y_1\| \leqslant \|x_1 - y_1\| + \|P_1 y_1\| < 2\varepsilon .$$

We proved that the range of T_1 is dense in $\mathscr{K}_1 \cap \mathscr{L}_1^\perp$.

Finally we have

$$T_k(\rho_1 \oplus \rho_2)(x_1, x_2) = T_k(\rho_1 x_1, \rho_2 x_2) = \rho_k x_k = \rho_k T_k(x_1, x_2)$$

and so T_k is an intertwining operator for $\rho_1 \oplus \rho_2 | \mathscr{L}$ and $\rho_k | \mathscr{K}_k \cap \mathscr{L}_k^\perp$. Since ρ_1, ρ_2 and $\rho_1 \oplus \rho_2$ have the same multiplier we may apply Theorem 4.5 and observing that $\mathscr{L} = \overline{T_k^*(\mathscr{K}_k \cap \mathscr{L}_k^\perp)}$ we obtain

$$(\rho_1 \oplus \rho_2) | \mathscr{L} \approx \rho_k | \mathscr{K}_k \cap \mathscr{L}_k^\perp \qquad (k = 1, 2) .$$

Although it will not be needed, for the sake of completeness we add the following:

Lemma 9. *The vectors* $(a_1, P_2 a_2)$ *and* $(P_1 a_1, a_2)$ *are cyclic vectors for the representations* $(\rho_1 | \mathscr{K}_1) \oplus (\rho_2 | \mathscr{L}_2)$ *and* $(\rho_1 | \mathscr{L}_1) \oplus (\rho_2 | \mathscr{K}_2)$, *respectively.*

Proof. Let $\mathscr{M}_2$ denote the smallest invariant subspace of $\rho_1 \oplus \rho_2$ which contains $(a_1, P_2 a_2)$. If (x_1, x_2) is in $\mathscr{K}_1 \oplus \mathscr{L}_2$ then $P_2 x_2 = x_2$ and $(0, x_2) \in \mathscr{K}$. Since $P_2 \in (\rho_2, \rho_2)$ we obtain $(0, x_2) = (0, P_2 x_2) \in \mathscr{M}_2$. Thus if $(x_1, x_2) \perp \mathscr{M}_2$ then $(x_2, x_2) = 0$ and so $x_2 = 0$. Now suppose that $(x_1, 0) \in \mathscr{K}_1 \oplus \mathscr{L}_2$ and $(x_1, 0) \perp \mathscr{M}_2$. Then

$$(x_1, \rho_1(s_1) \dots \rho_1(s_k) a_1) = 0$$

for all choices of k and $s_1, \dots, s_k$ in S. Therefore using the definition of $\mathscr{K}_1$ we obtain $x_1 \perp \mathscr{K}_1$ and so $x_1 = 0$ by $x_1 \in \mathscr{K}_1$. We proved that $\mathscr{M}_2$ is dense in $\mathscr{K}_1 \oplus \mathscr{L}_2$.

Theorem 10. *Let* $\rho_k: S \to \mathscr{L}(\mathscr{H}_k)$ $(k=1,2)$ *be disjoint, cyclic, Mackey type representations with a common multiplier. Then* $\rho_1 \oplus \rho_2$ *is cyclic and if* a_1, a_2 *are cyclic vectors for* ρ_1 *and* ρ_2 *then* (a_1, a_2) *is a cyclic vector for* $\rho_1 \oplus \rho_2$.

Proof. Since a_k is a cyclic vector for ρ_k by Theorem 8 we obtain $\rho_1 | \mathscr{L}_1^{\perp} \approx \rho_2 \mathscr{L}_2^{\perp}$. Using $\rho_1 \downarrow \rho_2$ we see that $\mathscr{L}_k = \mathscr{H}_k$ for $k=1,2$. Therefore by the definition of $\mathscr{L}_1$ and $\mathscr{L}_2$ we have $(x_1, 0), (0, x_2) \in \mathscr{K}$ for every $x_1 \in \mathscr{H}_1$ and $x_2 \in \mathscr{H}_2$. Hence $\mathscr{K} = \mathscr{H}_1 \oplus \mathscr{H}_2$ and (a_1, a_2) is a cyclic vector.

Theorem 11. *Let* $\pi_1, \ldots, \pi_m$ *be inequivalent, finite dimensional, irreducible Mackey type representations with a common multiplier and let* $d_1, \ldots, d_m$ *be the corresponding degrees. Then*

$$\rho = n_1 \pi_1 \oplus \cdots \oplus n_m \pi_m$$

is a cyclic representation for all choices of $n_1 \leqslant d_1, \ldots, n_m \leqslant d_m$.

Proof. Apply Lemma 2.18 and Theorem 10.

Note. In Section VII.1 we shall see that if the common object S is a compact topological group G and if we restrict our attention to continuous, unitary group representations then the result given in the foregoing theorem is the best possible: If ρ is a cyclic representation of G with finitely many irreducible components then ρ is equivalent to one of the representations given in the theorem.

The following material concerns the extension of Theorem 10 to denumerably many representations ρ_k. The main result is Theorem 15.

Lemma 12. *Every subrepresentation of a cyclic representation is cyclic.*

Proof. Let a be a cyclic vector of $\rho: S \to \mathscr{L}(\mathscr{H})$ and let $\mathscr{K}$ be an invariant subspace of ρ. We let P denote the associated projection $\mathscr{H} \to \mathscr{K}$. Then $c = Pa$ is a cyclic vector for $\rho | \mathscr{K}$: For if $x \in \mathscr{K}$ and $\varepsilon > 0$ are given then there is an A in $\tilde{\mathscr{A}}_\rho$ such that $\|Aa - x\| < \varepsilon$. Hence using $P \in (\rho, \rho)$ we obtain $\|APa - x\| < \varepsilon$ where $A | \mathscr{K}$ is in $\tilde{\mathscr{A}}_{\rho|\mathscr{K}}$ by Lemma 1.1.

Proposition 13. *Let* $\mathscr{K}$ *and* $\mathscr{L}$ *be invariant subspaces of the Mackey type representation* $\rho: S \to \mathscr{L}(\mathscr{H})$ *and let* a *in* $\mathscr{K}$ *be such that its projection into* $\mathscr{L}$ *is a cyclic vector for* $\rho | \mathscr{L}$. *Then* $\rho | \mathscr{L} \leqslant \rho | \mathscr{K}$.

Proof. Let P and Q denote the projections associated with $\mathscr{K}$ and $\mathscr{L}$ so that $P, Q \in (\rho, \rho)$. Then

(1) $$PQ \in (\rho | \mathscr{L}, \rho | \mathscr{K}) \quad \text{and} \quad QP \in (\rho | \mathscr{K}, \rho | \mathscr{L}).$$

We will apply Theorem 4.5 with $\rho_1=\rho|\mathscr{L}$, $\rho_2=\rho|\mathscr{K}$ and $T=PQ$. Suppose that $Tx=0$ for some x in $\mathscr{L}$. Then for any $A\in\tilde{\mathscr{A}}_\rho$ we have by (1)

$$0=(PQx,Aa)=(x,QPAa)=(x,AQPa)=(x,AQa).$$

Since Qa is a cyclic vector for $\rho|\mathscr{L}$ we obtain $x=0$. Hence T is an injection and so by Proposition 4.2 $\overline{T^*\mathscr{K}}=\mathscr{L}$. We obtain $\rho|\mathscr{L}\approx(\rho|\mathscr{K})|T\mathscr{L}$.

Proposition 14. *If $\rho_i\colon S\to\mathscr{L}(\mathscr{H}_i)$ $(i\in\mathscr{I})$ is a family of pairwise disjoint representations and if $\rho\colon S\to\mathscr{L}(\mathscr{H})$ is a representation such that $\rho_i\leqslant\rho$ for every $i\in\mathscr{I}$ then $\sum\rho_i\leqslant\rho$.*

Proof. Let $U_i\colon\mathscr{H}_i\to U_i\mathscr{H}_i=\mathscr{K}_i\subseteq\mathscr{H}$ be a unitary intertwining operator of ρ_i and $\rho|\mathscr{K}_i$. If $i\neq j$ then $\rho|\mathscr{K}_i\downarrow\rho|\mathscr{K}_j$ and so $\mathscr{K}_i\perp\mathscr{K}_j$ by Proposition 1. Hence we can define an isometry $U\colon\sum\mathscr{H}_i\to\sum\mathscr{K}_i=\mathscr{K}\subseteq\mathscr{H}$ by the formula $U(x_i)=\sum U_ix_i$. Since U intertwines $\sum\rho_i$ and $\rho|\mathscr{K}$ we see that $\sum\rho_i$ is equivalent to $\rho|\mathscr{K}$.

Theorem 15. *Let $\rho_k\colon S\to\mathscr{L}(\mathscr{H}_k)$ $(k=1,2,\ldots)$ be a sequence of pairwise disjoint, cyclic Mackey type representations with a common multiplier. Then their direct sum $\rho=\sum\rho_k$ is cyclic.*

Proof. Let a_k be a cyclic vector of ρ_k and let $\|a_k\|=2^{-k}$. Then $a=(a_1,a_2,\ldots)$ is a vector in $\mathscr{H}=\sum\mathscr{H}_k$. Let $\mathscr{K}$ be the smallest invariant subspace of ρ which contains a. Let $\mathscr{L}_k=\{(0,\ldots,x_k,0,\ldots)\colon x_k\in\mathscr{H}_k\}$ and let $P_k\colon\mathscr{H}\to\mathscr{L}_k$ so that P_ka is a cyclic vector for $\rho|\mathscr{L}_k$. By Proposition 13 we have $\rho|\mathscr{L}_k\leqslant\rho|\mathscr{K}$ where $\rho_k\approx\rho|\mathscr{L}_k$. Since the ρ_k's are disjoint by Proposition 14 we obtain $\rho\leqslant\rho|\mathscr{K}$. Now Lemma 12 can be applied because $\rho|\mathscr{K}$ is a cyclic representation.

6. The General Trace Formula

The main results of this section are the general trace formula expressed in Theorem 14 and the closely related Theorem 12 and Proposition 13. They will be used on several occasions in Chapter VII. The earlier parts of the section are the preparations which are necessary for these results.

Proposition 1. *If $\mathscr{A}$ is a v. Neumann algebra, E a projection in $\mathscr{A}$ with range $\mathscr{M}$ and if P is a projection onto some $\mathscr{A}$-stable subspace, then $P\mathscr{M}\subseteq\mathscr{M}$.*

Proof. By Proposition III.5.13 we have $P\in\mathscr{A}'$ so $EP=PE$ and $P\mathscr{M}=PE\mathscr{H}=EP\mathscr{H}\subseteq E\mathscr{H}=\mathscr{M}$.

Proposition 2. *Let $\mathscr{A}$ be a v. Neumann algebra, let $A \in \mathscr{A}$ and let $\mathscr{M}_\lambda = \{x: Ax = \lambda x\}$ where $\lambda \in \mathbb{C}$. Then the projection E_λ having range $\mathscr{M}_\lambda$ belongs to $\mathscr{A}$ for every $\lambda \in \mathbb{C}$.*

Proof. According to Proposition III.5.13 $E_\lambda \in \mathscr{A} = \mathscr{A}''$ if and only if $\mathscr{M}_\lambda$ is stable under $\mathscr{A}'$. Now if $T \in \mathscr{A}'$ and $x \in \mathscr{M}_\lambda$ then $A(Tx) = T(Ax) = \lambda(Tx)$, so $Tx \in \mathscr{M}_\lambda$ and $T\mathscr{M}_\lambda \subseteq \mathscr{M}_\lambda$. Hence $E_\lambda \in \mathscr{A}$.

Proposition 3. *Let $\mathscr{A}$ be a v. Neumann algebra, let E in $\mathscr{A}$ and F in $\mathscr{A}'$ be projections such that E is of finite rank and $EF \neq 0$. Then there exists a projection P in $\mathscr{A}'$ such that $P \leqslant F$, $EP \neq 0$ and $\mathscr{A}P = \{AP: A \in \mathscr{A}\}$ is irreducible.*

Let $\mathscr{H}$ be the Hilbert space of $\mathscr{A}$ so that $\mathscr{A} \subseteq \mathscr{L}(\mathscr{H})$ and introduce the notations $\mathscr{M} = E\mathscr{H}$, $\mathscr{K} = F\mathscr{H}$ and $\mathscr{P} = P\mathscr{H}$ where P is the projection occurring in the conclusion. We shall call $\mathscr{P}$ an irreducible component of the algebra $\mathscr{A}$. In view of Proposition III.5.13 and $EP = PE$ the proposition can be reformulated and proved as follows:

Proposition 4. *Let $\mathscr{A}$ be a v. Neumann algebra, let $\mathscr{A} \subseteq \mathscr{L}(\mathscr{H})$ and let $\mathscr{M}$ be a finite dimensional subspace of $\mathscr{H}$ such that $\mathscr{M} \neq 0$ and the associated projection $E: \mathscr{H} \to \mathscr{M}$ belongs to $\mathscr{A}$. Suppose $\mathscr{K}$ is a subspace of $\mathscr{H}$ which is stable under $\mathscr{A}$ and is such that $\mathscr{K} \cap \mathscr{M} \neq 0$. Then there exists an irreducible component of $\mathscr{A}$ such that $\mathscr{P} \subseteq \mathscr{K}$ and $\mathscr{M} \cap \mathscr{P} \neq 0$.*

Proof. Fix an $\mathscr{A}$-stable subspace $\mathscr{N}$ of $\mathscr{H}$ such that $\mathscr{K} \supseteq \mathscr{N}$ and the dimension of $\mathscr{M} \cap \mathscr{N}$ is minimal positive. The existence of such $\mathscr{N}$ follows from the hypothesis because $\mathscr{N} = \mathscr{K}$ satisfies the inclusion and the positivity requirements. Having fixed such a $\mathscr{N}$ we let

$$\mathscr{P} = \bigcap_{\mathscr{K} \supseteq \mathscr{L}} \{\mathscr{L}: \mathscr{L} \cap \mathscr{M} = \mathscr{M} \cap \mathscr{N}\}$$

where $\mathscr{L}$ varies over all $\mathscr{A}$-stable subspaces contained in $\mathscr{K}$ and satisfying $\mathscr{L} \cap \mathscr{M} = \mathscr{M} \cap \mathscr{N}$. Being an intersection of stable subspaces $\mathscr{P}$ is stable and $\mathscr{M} \cap \mathscr{P} = \mathscr{M} \cap \mathscr{N} \neq 0$. Clearly we have $\mathscr{P} \subseteq \mathscr{K}$ and so it is sufficient to prove that $\mathscr{P}$ is irreducible under $\mathscr{A}$.

Let $\mathscr{K}_1$ be an $\mathscr{A}$-stable subspace of $\mathscr{P}$ and let $\mathscr{K}_2 = \mathscr{K}_1^\perp \cap \mathscr{P}$ so that

(1) $$\mathscr{P} = \mathscr{K}_1 \oplus \mathscr{K}_2 .$$

Since $\mathscr{A}$ is self adjoint by Proposition III.5.11 $\mathscr{K}_2$ is stable under $\mathscr{A}$. By $\mathscr{K}_i \subseteq \mathscr{P}$ $(i=1,2)$ we have $\mathscr{M} \cap \mathscr{K}_i \subseteq \mathscr{M} \cap \mathscr{P} = \mathscr{M} \cap \mathscr{N}$ so if $\mathscr{M} \cap \mathscr{K}_i \neq 0$ then by the minimality of the dimension of $\mathscr{M} \cap \mathscr{N}$ we obtain

$$\dim(\mathscr{M} \cap \mathscr{N}) \leqslant \dim(\mathscr{M} \cap \mathscr{K}_i) \leqslant \dim(\mathscr{M} \cap \mathscr{P}) = \dim(\mathscr{M} \cap \mathscr{N}).$$

Therefore we have:

(2) $$\mathcal{M}\cap\mathcal{K}_i=\mathcal{M}\cap\mathcal{P}=\mathcal{M}\cap\mathcal{N} \quad \text{if } \mathcal{M}\cap\mathcal{K}_i\neq 0.$$

By $\mathcal{M}\cap\mathcal{P}=\mathcal{M}\cap\mathcal{N}\neq 0$ we can find a non-zero x in $\mathcal{M}\cap\mathcal{P}$. Then $x=k_1+k_2$ where $k_i\in\mathcal{K}_i$ $(i=1,2)$ because $x\in\mathcal{P}$ and (1) holds. Moreover since $x\in\mathcal{M}$ and $\mathcal{K}_1$ and $\mathcal{K}_2$ being stable subspaces of $\mathcal{A}$ by Proposition 1 we have $k_1, k_2\in\mathcal{M}$. Thus $0\neq x=k_1+k_2$ where $k_i\in\mathcal{M}\cap\mathcal{K}_i$ for $i=1,2$ and so $\mathcal{M}\cap\mathcal{K}_1\neq 0$ or $\mathcal{M}\cap\mathcal{K}_2\neq 0$.

Therefore by (2) we have $\mathcal{M}\cap\mathcal{K}_i=\mathcal{M}\cap\mathcal{N}$ for $i=1$ or $i=2$. By the definition of $\mathcal{P}$ we obtain $\mathcal{K}_i\supseteq\mathcal{P}$ and so $\mathcal{K}_i=\mathcal{P}$. This shows that $\mathcal{K}_1=0$ or $\mathcal{K}_1=\mathcal{P}$ and so $\mathcal{P}$ is irreducible and Proposition 4 is proved.

By choosing $\mathcal{H}=\mathcal{K}$ we obtain the following corollary:

Proposition 5. *If the v. Neumann algebra $\mathcal{A}$ contains a projection E of finite, positive rank then $\mathcal{A}$ has an irreducible component which intersects the range of E.*

Let $\rho: S\to\mathcal{L}(\mathcal{H})$ be a representation and let $\mathcal{K}_1, \mathcal{K}_2$ be stable subspaces such that $\rho|\mathcal{K}_1\approx\rho|\mathcal{K}_2$. Then we can find a unitary map $U: \mathcal{K}_1\to\mathcal{K}_2$ such that $UA^{(1)}=A^{(2)}U$ for every A in $\mathcal{A}$ where $\mathcal{A}$ is the v. Neumann algebra associated with ρ and for any A in $\mathcal{L}(\mathcal{H})$ the symbol $A^{(i)}$ denotes $A|\mathcal{K}_i$. Since U is unitary it follows that $A^{(1)}$ and $A^{(2)}$ have essentially identical properties: For instance $|A^{(1)}|=|A^{(2)}|$ and $A^{(1)}$ is compact if and only if $A^{(2)}$ is a compact operator. Similarly $A^{(1)}$ and $A^{(2)}$ belong to the HS class or to the trace class simultaneously and in that case $\|A^{(1)}\|=\|A^{(2)}\|$ or $\operatorname{tr} A^{(1)}=\operatorname{tr} A^{(2)}$. Also we have $A^{(1)}=0$ if and only if $A^{(2)}=0$. Thus if ω in $\hat{S}$ is such that $n(\omega,\rho)>0$ then it is meaningful to speak about $|A(\omega)|$, about A being compact on ω, being a HS operator or a trace class operator on ω. We can also define the HS norm $\|A(\omega)\|$ and the trace $\operatorname{tr} A(\omega)$. Although $A(\omega)$ itself is not defined $\operatorname{tr} A(\omega)$ for instance means the trace of the restriction $A|\mathcal{P}$ where $\mathcal{P}$ is any irreducible component of ρ belonging to the class ω.

Proposition 6. *If A is a compact operator in the v. Neumann algebra $\mathcal{A}=\mathcal{A}_\rho$ then $A(\omega)=0$ for every $\omega\in\hat{S}$ satisfying $n(\omega,\rho)=\infty$.*

Proof. Since $\mathcal{A}$ is self adjoint it is sufficient to restrict our attention to the special case when $A=A^*$. Let $\mathcal{B}$ be the unit ball of $\mathcal{H}$ and let $\mathcal{P}=\mathcal{P}_1, \mathcal{P}_2, \mathcal{P}_3, \dots$ be pairwise orthogonal irreducible components of ρ belonging to the class ω. Let $U_k: \mathcal{P}\to\mathcal{P}_k$ be unitary operators from $(\rho|\mathcal{P}, \rho|\mathcal{P}_k)$ so that $AU_k=U_kA$. If $x\in\mathcal{P}\cap\mathcal{B}$ and $x_k=U_kx$ then

$x_k \in \mathscr{R}_k \cap \mathscr{B}$ and $\|Ax_k\| = \|Ax\|$. Since the $\mathscr{R}_k$'s are $\mathscr{A}$-stable and pairwise orthogonal we obtain

$$\|Ax_k - Ax_l\| = \sqrt{2}\,\|Ax\| \quad \text{for } k \neq l.$$

By $x_k \in \mathscr{B}$ and the compactness of the operator A the last relation implies that $\|Ax\| = 0$. Therefore $Ax = 0$ for every $x \in \mathscr{P} \cap \mathscr{B}$ and so $A|\mathscr{P}$ is 0.

If $\mathscr{A}$ is a v. Neumann algebra of operators $A: \mathscr{H} \to \mathscr{H}$ and if $\mathscr{P}$ is an irreducible component of $\mathscr{A}$ then the *multiplicity* of $\mathscr{P}$ in $\mathscr{A}$ can be defined as $n(\mathscr{P}, \mathscr{A}) = n(\mathscr{P}, \sigma)$ where $\sigma: \mathscr{A} \to \mathscr{L}(\mathscr{H})$ is the identity representation of $\mathscr{A}$ so that the algebra $\mathscr{A}_\sigma$ associated with σ is $\mathscr{A}$ itself. Therefore $n(\mathscr{P})$ denotes the finite or infinite cardinality of any maximal orthogonal system of subspaces $\mathscr{R}_k$ of $\mathscr{H}$ such that $(\mathscr{A}|\mathscr{P}, \mathscr{A}|\mathscr{R}_k)$ contains unitary operators.

By Proposition III.5.10 a subspace $\mathscr{K}$ of $\mathscr{H}$ is stable under the v. Neumann algebra $\mathscr{A}$ of a representation $\rho: S \to \mathscr{L}(\mathscr{H})$ if and only if it is stable under ρ and so in particular $\mathscr{P}$ is an irreducible component of $\mathscr{A}$ if and only if it is an irreducible component of ρ. Moreover if $U: \mathscr{P}_1 \to \mathscr{P}_2$ is a unitary operator then $U\rho_1 = \rho_2 U$ if and only if $UA^{(1)} = A^{(2)}U$ for every A in $\mathscr{A}$. Hence we have the following:

Proposition 7. *Let $\mathscr{A}$ be the v. Neumann algebra associated with the representation $\rho: S \to \mathscr{L}(\mathscr{H})$. Then $\mathscr{A}$ and ρ have the same irreducible components $\mathscr{P}$ and $n(\mathscr{P}, \mathscr{A}) = n(\mathscr{P}, \rho)$.*

Proposition 8. *If the v. Neumann algebra $\mathscr{A}$ contains a non-zero, compact operator A then it has an irreducible component $\mathscr{P}$ which occurs with finite multiplicity $n(\mathscr{P}, \mathscr{A}) > 0$ and*

$$n(\mathscr{P}, \mathscr{A}) \leqslant \min\{\dim \mathscr{M} : \mathscr{M} \cap \mathscr{P} \neq 0\}$$

where $\mathscr{M}$ varies over the characteristic subspaces of the operators $\frac{1}{2}(A + A^*)$ and $1/2i(A - A^*)$.

Proof. The two operators given in the proposition are self adjoint, compact elements of $\mathscr{A}$ and at least one of them is not 0. This non-zero operator has a characteristic subspace $\mathscr{M} = \mathscr{M}_\lambda$ $(\lambda \neq 0)$ which by the compactness of the operator is finite dimensional: This fact is part of Theorem II.3.4 but it is much simpler to see it directly: If x_i $(i \in \mathscr{I})$ is an orthonormal family in $\mathscr{M}$ then by $Tx_i = \lambda x_i$ $(i \in \mathscr{I})$ the family Tx_i $(i \in \mathscr{I})$ is orthogonal and every Tx_i has the same positive norm $|\lambda|$. Since $\overline{T\mathscr{B}}$ is compact the family must be finite. According to Proposition 2 the associated projection $E: \mathscr{H} \to \mathscr{M}$ belongs to $\mathscr{A}$. Therefore by Proposition 5 there is an irreducible component $\mathscr{P}$ of $\mathscr{A}$ such that $\mathscr{M}\cap\mathscr{P}\neq 0$. The finiteness of $n(\mathscr{P},\mathscr{A})$ and the upper estimate given in the proposition follow from the following:

Proposition 9. *If $\mathscr{P}$ is an irreducible component of $\mathscr{A}$, and if $\mathscr{M}\cap\mathscr{P}\neq 0$ where $\mathscr{M}$ is a characteristic subspace of an operator $A\in\mathscr{A}$ then $n(\mathscr{P},\mathscr{A})\leqslant \dim\mathscr{M}$.*

Proof. Let $\mathscr{P}=\mathscr{P}_1,\ldots,\mathscr{P}_n$ be pairwise orthogonal, equivalent copies of $\mathscr{P}$ in $\mathscr{H}$. Then there exist unitary maps $U_k:\mathscr{P}\to\mathscr{P}_k$ such that $U_k A^{(1)}=A^{(k)}U_k$ where $A^{(k)}=A\,|\,\mathscr{P}_k$. Let $\mathscr{M}$ belong to the characteristic value λ and let $x\neq 0$ belong to $\mathscr{M}\cap\mathscr{P}$. Then $Ax=\lambda x$ i.e. $A^{(1)}x=\lambda x$ and so $A^{(k)}(U_k x)=\lambda(U_k x)$ or in other words $x_k=U_k x\in\mathscr{M}$. Since $x_k\in\mathscr{P}_k$ the non-zero vectors $x_1,\ldots,x_n$ are pairwise orthogonal and so they are linearly independent elements of $\mathscr{M}$. This proves that $n\leqslant\dim\mathscr{M}$.

Using a simplified version of the above reasoning we obtain the following:

Proposition 10. *If $\mathscr{M}$ is a characteristic subspace of one of the operators $A\in\mathscr{A}$ and if $\mathscr{P}$ and $\mathscr{Q}$ are $\mathscr{A}$-stable subspaces such that $\mathscr{P}\cap\mathscr{M}\neq 0$ while $\mathscr{Q}\cap\mathscr{M}=0$ then $\mathscr{P}$ and $\mathscr{Q}$ are not equivalent.*

In both of these propositions $\mathscr{A}$ could be any subset of $\mathscr{L}(\mathscr{H})$ but they will be used only when $\mathscr{A}$ is a v. Neumann algebra.

Proposition 11. *Let $\mathscr{M}_\lambda$ $(\lambda\in\Lambda_0)$ be a family of subspaces of the Hilbert space $\mathscr{H}$ and let $\mathscr{M}=\sum\mathscr{M}_\lambda$ be the subspace generated by the $\mathscr{M}_\lambda$'s. Suppose $\mathscr{L}$ is a subspace of $\mathscr{H}$ such that the projection of $\mathscr{M}_\lambda$ on $\mathscr{L}$ belongs to $\mathscr{M}_\lambda$ for every $\lambda\in\Lambda_0$. Then $\mathscr{L}\cap\mathscr{M}=\sum(\mathscr{L}\cap\mathscr{M}_\lambda)$.*

Note. Let E_λ and P be the projections associated with the subspaces $\mathscr{M}_\lambda$ and $\mathscr{L}$, respectively. In terms of these the condition relating $\mathscr{M}_\lambda$ and $\mathscr{L}$ states that $PE_\lambda=E_\lambda P$.

Proof. First we prove the remark made in the above note: If $PE_\lambda=E_\lambda P$ then for any $x\in\mathscr{M}_\lambda$ we have $Px=PE_\lambda x=E_\lambda(Px)\in\mathscr{M}_\lambda$. Conversely suppose that $P\mathscr{M}_\lambda\subseteq\mathscr{M}_\lambda$. Since $P=P^*$ by Lemma III.5.12 we have $P\mathscr{M}_\lambda^\perp\subseteq\mathscr{M}_\lambda^\perp$. Now given x in $\mathscr{H}$ let $x=y+z$ with $y\in\mathscr{M}_\lambda$ and $z\in\mathscr{M}_\lambda^\perp$. Then we have $PE_\lambda x=Py$ and $Px=Py+Pz$ where $Py\in\mathscr{M}_\lambda$ and $Pz\in\mathscr{M}_\lambda^\perp$. Hence $E_\lambda Px=E_\lambda Py=Py=PE_\lambda x$.

If the $\mathscr{M}_\lambda$'s form an orthogonal family then $\mathscr{M}=(\sum E_\lambda)\mathscr{H}$ and $\mathscr{L}\cap\mathscr{M}_\lambda=(PE_\lambda)\mathscr{H}$. Hence the conclusion is the geometric analogue of the distributivity formula $P\sum E_\lambda=\sum PE_\lambda$. However it is simpler to apply the following reasoning which covers the general, not necessarily orthogonal case:

The inclusion $\supseteq$ is obvious so we prove that $\subseteq$ holds. Let $x \in \mathscr{L} \cap \mathscr{M}$ be given and let $\varepsilon > 0$. Then there exist $\lambda_1, \ldots, \lambda_n$ and $x_i \in \mathscr{M}_{\lambda_i}$ such that

$$x = x_1 + \cdots + x_n + r \quad \text{where } \|r\| < \varepsilon .$$

Write $x_i = y_i + z_i$ where $y_i \in \mathscr{L}$ and $z_i \in \mathscr{L}^{\perp}$. Since $x_i \in \mathscr{M}_{\lambda_i}$ by hypothesis $y_i \in \mathscr{M}_{\lambda_i}$ and so $y_i \in \mathscr{L} \cap \mathscr{M}_{\lambda_i}$ proving that

$$y = y_1 + \cdots + y_n \in \sum (\mathscr{L} \cap \mathscr{M}_\lambda) .$$

If we let $z = z_1 + \cdots + z_n$ then $x = y + z + r$ and

$$\|r\|^2 = \|x - y - z\|^2 = \|x - y\|^2 + \|z\|^2$$

by $x, y \in \mathscr{L}$ and $z \in \mathscr{L}^{\perp}$. Hence $\|x - y\| < \varepsilon$ where $y = y(\varepsilon)$ belongs to the linear span of the $\mathscr{L} \cap \mathscr{M}_\lambda$ spaces.

Theorem 12. *Let $\mathscr{A}$ be a v. Neumann algebra, let Γ, Λ be index sets and let $\mathscr{M}_\lambda^\gamma$ $(\lambda \in \Lambda)$ for each fixed $\gamma \in \Gamma$ be a family of finite dimensional orthogonal subspaces of the Hilbert space $\mathscr{H}$ of $\mathscr{A}$. Suppose that the projection $E_\lambda^\gamma : \mathscr{H} \to \mathscr{M}_\lambda^\gamma$ belongs to $\mathscr{A}$ for every $\lambda \in \Lambda$ and $\gamma \in \Gamma$. Then $\mathscr{H}$ has an $\mathscr{A}$-stable subspace $\mathscr{K}$ for which the following obtain:*

1) *$\mathscr{K} \supseteq \mathscr{M}_\lambda^\gamma$ for each $\lambda \in \Lambda$ and $\gamma \in \Gamma$;*

2) *the restriction of $\mathscr{A}$ to $\mathscr{K}$ is a direct sum of irreducible components;*

3) *for each of these components $\mathscr{P}$ there is an $\mathscr{M}_\lambda^\gamma$ such that $\mathscr{M}_\lambda^\gamma \cap \mathscr{P} \neq 0$;*

4) *$\mathscr{K}$ is maximal with respect to property 3) i.e. if $\mathscr{Q}$ is an irreducible component and $\mathscr{Q} \subseteq \mathscr{K}^{\perp}$ then $\mathscr{Q} \cap \mathscr{M}_\lambda^\gamma = 0$ for every $\lambda \in \Lambda$ and $\gamma \in \Gamma$.*

Proof. For each $\gamma \in \Gamma$ we let $\mathscr{M}_0^\gamma = (\sum \mathscr{M}_\lambda^\gamma)^{\perp}$. Since the $\mathscr{M}_\lambda^\gamma$'s are orthogonal $\sum E_\lambda^\gamma$ is a projection and by $E_\lambda^\gamma \in \mathscr{A}$ we see that the projection $E_0^\gamma = I - \sum E_\lambda^\gamma$ having range $\mathscr{M}_0^\gamma$ also belongs to $\mathscr{A}$.

If every $\mathscr{M}_\lambda^\gamma$ is 0 then we can choose $\mathscr{K} = 0$ and 1)—4) are satisfied. Excluding this trivial case there is, by Proposition 5, an irreducible component $\mathscr{P}$ such that $\mathscr{P} \cap \mathscr{M}_\lambda^\gamma \neq 0$ for some $\lambda \in \Lambda$ and $\gamma \in \Gamma$. Let $\mathscr{F} = \{\mathscr{P}_i\}$ be a family of pairwise orthogonal irreducible components $\mathscr{P}_i$ of $\mathscr{A}$ such that for each $\mathscr{P}_i$ there is a γ, λ pair with $\mathscr{P}_i \cap \mathscr{M}_\lambda^\gamma \neq 0$. We denote by Φ the set of all non-void families $\mathscr{F}$. Above we proved the existence of an $\mathscr{F}$ consisting of a single component $\mathscr{P}$, so we see that Φ is not void. We partially order Φ so that $\mathscr{F}_1 \leqslant \mathscr{F}_2$ means $\mathscr{F}_1 \subseteq \mathscr{F}_2$ and verify that each chain has an upper bound in Φ. Then by Zorn's lemma there exists a family $\mathscr{F} = \{\mathscr{P}_i\}$ which is maximal with respect to inclusion. Let $\mathscr{K}=\sum \mathscr{P}_i$ where $\mathscr{P}_i$ runs through such a maximal family $\mathscr{F}$. Since each $\mathscr{P}_i$ is $\mathscr{A}$-stable so are $\mathscr{K}$ and $\mathscr{K}^\perp$. By the construction $\mathscr{K}$ satisfies 2), 3) and 4).

In order to prove 1) we apply Proposition 11 to the family consisting of the subspaces $\mathscr{M}_0^\gamma$ and $\mathscr{M}_\lambda^\gamma$ $(\lambda\in\Lambda)$ where γ is fixed and we choose $\mathscr{L}=\mathscr{K}^\perp$. Since $E_\lambda^\gamma\in\mathscr{A}$ for $\lambda=0$ and also for every $\lambda\in\Lambda$ by Proposition 1 the projection of $\mathscr{M}_\lambda^\gamma$ into $\mathscr{L}$ belongs to $\mathscr{M}_\lambda^\gamma$ and so Proposition 11 is applicable. Since

$$\mathscr{M}=\mathscr{M}_0^\gamma+\sum \mathscr{M}_\lambda^\gamma=\mathscr{H}$$

we get

$$\mathscr{L}\cap\mathscr{M}=\mathscr{K}^\perp\cap\mathscr{H}=\mathscr{K}^\perp=(\mathscr{K}^\perp\cap\mathscr{M}_0^\gamma)+\sum(\mathscr{K}^\perp\cap\mathscr{M}_\lambda^\gamma).$$

We have $\mathscr{K}^\perp\cap\mathscr{M}_\lambda^\gamma=0$ for every $\lambda\in\Lambda$, because otherwise by Proposition 4 there would exist a $\mathscr{P}\subseteq\mathscr{K}^\perp$ which could be joined to the system $\mathscr{F}$ in contradiction to its maximality. Therefore we obtain from above $\mathscr{K}^\perp=\mathscr{K}^\perp\cap\mathscr{M}_0^\gamma$, that is to say $\mathscr{K}^\perp\subseteq\mathscr{M}_0^\gamma$ and $\mathscr{K}\supseteq(\mathscr{M}_0^\gamma)^\perp=\sum\mathscr{M}_\lambda^\gamma$. We proved that $\mathscr{K}\supseteq\mathscr{M}_\lambda^\gamma$ for every $\gamma\in\Gamma$ and $\lambda\in\Lambda$, so 1) holds.

Proposition 13. *Let $\mathscr{A}$ be a v. Neumann algebra and let $A^\gamma(\gamma\in\Gamma)$ be compact, self adjoint operators in $\mathscr{A}$. Then there is an $\mathscr{A}$-stable subspace $\mathscr{K}$ such that we have:*

1) $\mathscr{K}^\perp\subseteq\mathscr{M}_0^\gamma=\{x\colon A^\gamma x=0$ for every $\gamma\in\Gamma\}$;

2) *the restriction of $\mathscr{A}$ to $\mathscr{K}$ is a direct sum of irreducible components $\mathscr{P}$;*

3) *the multiplicity of each $\mathscr{P}$ is finite and*

$$n(\mathscr{P},\mathscr{A})\leqslant\min\{\dim\mathscr{M}_\lambda^\gamma\colon \mathscr{M}_\lambda^\gamma\cap\mathscr{P}\neq 0 \text{ and } \gamma\in\Gamma,\ \lambda\in\mathbb{R}^\times\}$$

where $\mathscr{M}_\lambda^\gamma=\{x\colon A^\gamma x=\lambda x\}$;

4) *if $\mathscr{Q}$ is an irreducible component of $\mathscr{A}$ and if $\mathscr{Q}\cap\mathscr{M}_\lambda^\gamma\neq 0$ for some $\gamma\in\Gamma$ and $\lambda\in\mathbb{R}^\times$ then any decomposition of $\mathscr{K}$ into a direct sum of irreducible components contains exactly $n(\mathscr{Q},\mathscr{A})$ equivalents of $\mathscr{Q}$.*

Proof. Since A^γ is self adjoint and compact each $\mathscr{M}_\lambda^\gamma(\lambda\neq 0)$ is finite dimensional and $\mathscr{M}_\lambda^\gamma\perp\mathscr{M}_\mu^\gamma$ if $\lambda\neq\mu$. By Proposition 2 we have $E_\lambda^\gamma\in\mathscr{A}$, so we can apply Theorem 12 with $\Lambda=\mathbb{R}^\times$ and $\mathscr{M}_\lambda^\gamma=\{x\colon A^\gamma x=\lambda x\}$. Conclusions 1) and 2) are the same as the ones given there. The upper estimate for $n(\mathscr{P},\mathscr{A})$ given in 3) is a direct consequence of Proposition 9. The finiteness of $n(\mathscr{P},\mathscr{A})$ follows from this upper estimate and the existence of an $\mathscr{M}_\lambda^\gamma$ intersecting $\mathscr{K}$.

To prove 4) consider a decomposition of $\mathscr{K}$ into irreducible components and select those irreducible parts $\mathscr{P}_1,\ldots,\mathscr{P}_n$ which are equivalent to $\mathscr{Q}$. The finiteness of the number of these components follows from 3). We prove that $\mathscr{P}_1, \ldots, \mathscr{P}_n$ is a maximal orthogonal family of irreducible components equivalent to $\mathscr{Q}$ and so $n = n(\mathscr{Q}, \mathscr{A})$ will follow. Indeed if $\mathscr{P}$ is any component which is equivalent to $\mathscr{Q}$ then by Proposition 10 $\mathscr{P} \cap \mathscr{M}_\lambda^\gamma \neq 0$ and so by 1) we also have $\mathscr{P} \cap \mathscr{K} \neq 0$. However $\mathscr{K}$ is stable and $\mathscr{P}$ is irreducible, so $\mathscr{P} \subseteq \mathscr{K}$. Now

$$(\mathscr{P}_1 \oplus \cdots \oplus \mathscr{P}_n)^\perp \cap \mathscr{K}$$

is a direct sum of irreducible components, none of which is equivalent to $\mathscr{Q}$ and hence to $\mathscr{P}$. Therefore by Proposition 5.1 $\mathscr{P}$ is orthogonal to all of these and so $\mathscr{P} \subseteq \mathscr{P}_1 \oplus \cdots \oplus \mathscr{P}_n$. Q.e.d.

Theorem 14. *Let $\rho: S \to \mathscr{L}(\mathscr{H})$ be a representation of the object S, let $\mathscr{A}$ be the v. Neumann algebra generated by ρ and let $A \in \mathscr{A}$. If A is a trace class operator then*

$$\operatorname{tr} A = \sum_{\omega \in \hat{S}} n(\omega) \operatorname{tr} A(\omega)$$

where $n(\omega) = n(\omega, \rho)$ and the series is absolutely convergent. If A is a Hilbert-Schmidt operator then

$$\|A\|^2 = \sum_{\omega \in \hat{S}} n(\omega) \|A(\omega)\|^2.$$

Proof. It is sufficient to restrict our attention to the special case of self adjoint operators: For trace class operators this follows from the additivity of the trace and for HS operators one uses the identity

$$\|A\|^2 = \left\| \frac{A + A^*}{2} \right\|^2 + \left\| \frac{A - A^*}{2i} \right\|^2.$$

Thus let A be a self adjoint and compact operator. We apply Proposition 13 with this single operator A instead of a whole family A^γ $(\gamma \in \Gamma)$. In each irreducible component of a direct sum decomposition of $\mathscr{K}$ choose an orthonormal basis and complete these systems by selecting an orthonormal basis also in $\mathscr{K}^\perp$. Denote these unit vectors by $\ldots, x_i, \ldots$ Then the contribution to the sum $\sum (A x_i, x_i)$ of the orthonormal vectors lying in $\mathscr{P}$ is $\operatorname{tr} A(\omega)$ where $\mathscr{P} \in \omega \in \hat{S}$. By 4) for each $\mathscr{P}$ occuring in $\mathscr{K}$ there are $n(\mathscr{P}, \mathscr{A})$ pairwise orthogonal components in $\mathscr{K}$ which are equivalent to $\mathscr{P}$. The total contribution of these is $n(\omega) \operatorname{tr} A(\omega)$ because by Proposition 5.5 we have $n(\mathscr{P}, \mathscr{A}) = n(\mathscr{P}, \rho) = n(\omega)$ where $\mathscr{P} \in \omega$. Therefore

$$\sum_{x_i \in \mathscr{K}} (A x_i, x_i) = \sum_{\omega} n(\omega) \operatorname{tr} A(\omega) \tag{3}$$

where each ω is an equivalence class of some $\mathscr{P} \subseteq \mathscr{K}$. If ω is not one of these classes but $n(\omega) > 0$ then $\operatorname{tr} A(\omega) = 0$ because if $\mathscr{Q}$ is an irreducible component of ρ and $\mathscr{Q} \in \omega$ then $n(\omega) = n(\mathscr{Q}, \mathscr{A}) > 0$ and by 4) $\mathscr{Q} \cap \mathscr{M}_\lambda = 0$ for every $\lambda \neq 0$. Hence by the spectral theorem of compact operators $A | \mathscr{Q}$ is 0. Therefore the summation in the right hand side of (3) can be extended to every $\omega \in \hat{S}$ without changing the value of the sum. Similarly if $x_i \in \mathscr{K}^\perp$ then by 1) $A x_i = 0$ and so on the left hand side of (3) the summation can be modified to cover every x_i in $\mathscr{H}$ without affecting the sum.

If A is a HS operator then the same reasoning can be repeated with $\|A x_i\|^2$ instead of $(A x_i, x_i)$.

7. Primary Representations and Factorial v. Neumann Algebras

In this section we will continue the study of invariant subspaces of representations $\rho: S \to \mathscr{L}(\mathscr{H})$ along the same lines as it was done in Section 5. The subspaces $\mathscr{H}_\omega$ introduced in Theorem 5.4 are primary components of ρ i.e. $\rho | \mathscr{H}_\omega$ is a primary representation of S. This is an immediate consequence of Proposition III.6.7. One can also see that each $\mathscr{H}_\omega$ is maximal with respect to inclusion that is to say if $\mathscr{Q}$ is a primary component of ρ and $\mathscr{Q} \supseteq \mathscr{H}_\omega$ then $\mathscr{Q} = \mathscr{H}_\omega$. We know that $\mathscr{H}_\omega$ contains a minimal invariant subspace which is an irreducible component. Our main object is to show that if this last property is dropped as a requirement then ρ might have other maximal primary components.

We start with a sequence of results on primary representations and related topics after which the main result will be stated and proved. We shall then give an explicit example which shows that primary representations and factorial v. Neumann algebras without maximal invariant subspaces indeed exist. Finally we will more closely investigate the case when there is a minimal invariant subspace. Factorial v. Neumann algebras with this property are said to be factorial of type I. We will construct a class of such algebras and will show that every factorial type I v. Neumann algebra is spatially isomorphic to an algebra in the constructed class.

We recall that a representation $\rho: S \to \mathscr{L}(\mathscr{H})$ is *primary* if and only if the associated v. Neumann algebra is a factor i.e. $\mathscr{A} \cap \mathscr{A}' = \mathbb{C} I$. By definition ρ is primary if it is not the direct sum of two disjoint subrepresentations. (See Proposition 4.11 and the paragraph preceding it.)

Proposition 1. *If* $\rho_i: S \to \mathcal{L}(\mathcal{H}_i)$ $(i \in \mathcal{I})$ *is a family of primary, pairwise intertwining Mackey type representations with a common multiplier then their direct sum* $\rho = \sum \rho_i$ *is primary.*

Proof. Let $\mathcal{H} = \sum \mathcal{H}_i$ and let P_i be the projection $P_i: \mathcal{H} \to \mathcal{H}_i$ where $\mathcal{H}_i$ is interpreted as a subspace of $\mathcal{H}$. Since $\mathcal{H}_i$ is an invariant subspace of ρ we have $P_i \in (\rho, \rho)$. Let P be a projection in $c(\rho, \rho)$ so that

$$(1) \qquad \rho | P\mathcal{H} \downarrow \rho | (I-P)\mathcal{H}$$

by Proposition 4.10. We are going to prove that P is 0 or I. Since $PP_i = P_i P$ we see that $\rho | PP_i\mathcal{H}$ is a subrepresentation of $\rho | P\mathcal{H}$ and similarly $\rho | (I-P)P_i\mathcal{H}$ is a subrepresentation of $\rho | (I-P)\mathcal{H}$. Hence by (1) $\rho PP_i \downarrow \rho(I-P)P_i$ or by the commutativity $\rho P_i P \downarrow \rho P_i(I-P)$. Since $\rho | P_i P\mathcal{H}$ and $\rho | P_i(I-P)\mathcal{H}$ are subrepresentations of $\rho | P_i\mathcal{H} \approx \rho_i$ we must have $P_i P = 0$ or $P_i(I-P) = 0$.

Suppose that $P_i P = P_i$ for some index $i \in \mathcal{I}$. Then as we are going to prove $P_j P = P_j$ for every j in $\mathcal{I}$. On the contrary suppose that $P_j P = 0$ for some index j. Then $\rho | (I-P)P_j\mathcal{H} = \rho | P_j\mathcal{H} \approx \rho_j$ and $\rho | PP_i\mathcal{H} = \rho | P_i\mathcal{H} \approx \rho_i$ would be disjoint by (1) contradicting our hypothesis $\rho_i \uparrow \rho_j$. Thus if $P_i P = P_i$ for some $i \in \mathcal{I}$ then the same holds for all i in $\mathcal{I}$. Since $\sum P_i = I$ we obtain $P = I$. The other possibility is that $P_i P = 0$ for every $i \in \mathcal{I}$. Then by the same reasoning $P = 0$. We proved that the only projections in $c(\rho, \rho)$ are 0 and I. Since $c(\rho, \rho)$ is a v. Neumann algebra by the corollary of Proposition III.6.15 $c(\rho, \rho) = \mathbb{C}I$ and so by Proposition 4.11 ρ is a primary representation.

Corollary. *If S is an object with a map and if* π_i $(i \in \mathcal{I})$ *are equivalent irreducible Mackey type representations of S then* $\rho = \sum \pi_i$ *is a primary representation.*

As we saw already this result holds for arbitrary, not necessarily Mackey type representations. In particular if $\pi: S \to \mathcal{L}(\mathcal{H})$ is an irreducible Mackey type representation then $n\pi$ is primary for any finite or infinite cardinal n.

Lemma 2. *Every subrepresentation of a primary representation is primary.*

Proof. Let $\rho | \mathcal{K}$ be a subrepresentation of $\rho: S \to \mathcal{L}(\mathcal{H})$. To prove that $\rho | \mathcal{K}$ is primary suppose that $\mathcal{K} = \mathcal{L} \oplus \mathcal{M}$ where $\mathcal{L}$ and $\mathcal{M}$ are invariant subspaces and $\rho | \mathcal{L} \downarrow \rho | \mathcal{M}$. We are going to show that $\mathcal{L} = 0$ or $\mathcal{M} = 0$. Let $\mathcal{N}_i$ $(i \in \mathcal{I})$ be a maximal orthogonal family of invariant subspaces lying in $\mathcal{K}^\perp$ such that $\rho | \mathcal{L} \downarrow \rho | \mathcal{N}_i$. If we allow one of the $\mathcal{N}_i$ to be the zero space then the existence of such a family follows from

Zorn's lemma. If $\mathcal{N} = \sum \mathcal{N}_i$ then by the corollary of Lemma 4.12 we have $\rho|\mathcal{L} \downarrow \rho|\mathcal{N}$. By $\rho|\mathcal{L} \downarrow \rho|\mathcal{M}$ we obtain

(2) $$\rho|\mathcal{L} \downarrow \rho|\mathcal{M} \oplus \mathcal{N}.$$

By the maximality every non-zero subrepresentation of $\rho|\mathcal{K}^\perp \cap \mathcal{N}^\perp$ intertwines $\rho|\mathcal{L}$. Therefore we must have $\rho|\mathcal{K}^\perp \cap \mathcal{N}^\perp \downarrow \rho|\mathcal{M} \oplus \mathcal{N}$ by (2). Using (2) again we obtain

(3) $$\rho|\mathcal{L} \oplus (\mathcal{K}^\perp \cap \mathcal{N}^\perp) \downarrow \rho|\mathcal{M} \oplus \mathcal{N}.$$

Since ρ is primary and

(4) $$\mathcal{L} \oplus (\mathcal{K}^\perp \cap \mathcal{N}^\perp) \oplus (\mathcal{M} \oplus \mathcal{N}) = \mathcal{H}$$

we must have $\mathcal{L} \oplus (\mathcal{K}^\perp \cap \mathcal{N}^\perp) = 0$ or $\mathcal{M} \oplus \mathcal{N} = 0$. In the first case $\mathcal{L} = 0$ and in the second $\mathcal{M} = 0$.

Corollary. *If $\rho: S \to \mathcal{L}(\mathcal{H})$ is a primary representation then any two non-zero subrepresentations of ρ intertwine.*

This follows from Lemma 2 by using Proposition 5.1: If $\mathcal{K}_1$ and $\mathcal{K}_2$ are invariant subspaces and $\rho|\mathcal{K}_1 \downarrow \rho|\mathcal{K}_2$ then $\mathcal{K}_1 \perp \mathcal{K}_2$ and so $\rho|\mathcal{K}_1 \oplus \mathcal{K}_2$ is not a primary subrepresentation of ρ. Hence ρ is not primary.

Lemma 3. *Let $\mathcal{L}$ and $\mathcal{M}$ be invariant subspaces of the representation ρ and let $\rho|\mathcal{L} \downarrow \rho|\mathcal{M}$. Then there is an invariant subspace $\mathcal{Q}$ such that $\mathcal{L} \subseteq \mathcal{Q}$, $\mathcal{M} \subseteq \mathcal{Q}^\perp$ and the associated projection $Q \in c(\rho, \rho)$.*

Proof. We use essentially the same reasoning as in the proof of Lemma 2. Thus starting with $\rho|\mathcal{L}$ and $\rho|\mathcal{M}$ we construct $\mathcal{N}$ such that (3) and (4) hold. If we let $\mathcal{Q} = \mathcal{L} \oplus (\mathcal{K}^\perp \cap \mathcal{N}^\perp)$ then $\rho|\mathcal{Q} \downarrow \rho|\mathcal{Q}^\perp$ and so by Proposition 4.10 $Q \in c(\rho, \rho)$.

If $\mathcal{A}$ is a v. Neumann algebra and E a projection in $\mathcal{A}$ then there is a smallest projection P in $c\mathcal{A}$ such that $E \leqslant P$. It is called the *central support* of E. Its existence follows from $I \in c\mathcal{A}$. For its range is the intersection of the ranges of all central projections majorizing E.

Proposition 4. *Let $\mathcal{K}$ be an invariant subspace of the representation ρ: $S \to \mathcal{L}(\mathcal{H})$, let E be the associated projection, P the central support of E and $\mathcal{L} = P\mathcal{H}$. Then $\rho|\mathcal{K}$ is quasiequivalent to $\rho|\mathcal{L}$.*

Proof. Suppose that ρE and ρP are not quasiequivalent. Clearly $\rho P \succcurlyeq \rho E$. Then $\rho E \not\succcurlyeq \rho P$ and so there is a projection $F \neq 0$ in $\mathcal{A}_\rho$ such that $F \leqslant P$ and $\rho E \downarrow \rho F$. By Lemma 3 there is a projection Q in $c(\rho, \rho)$ such that $Q \leqslant P, E \leqslant Q$ and $F \leqslant P - Q$. Since $F \neq 0$ we obtain $Q < P$ in contradiction to the minimality of P among the central projections majorizing E.

Proposition 5. *If* $\rho_k : S \to \mathscr{L}(\mathscr{H}_k)\,(k=1,2)$ *are primary representations then* $\rho_1 \leqslant \rho_2$, $\rho_2 \leqslant \rho_1$ *or* $\rho_1 \downarrow \rho_2$.

Proof. Consider a family of pairs $(\mathscr{K}_1^i, \mathscr{K}_2^i)\,(i \in \mathscr{I})$ of invariant subspaces of ρ_1 and ρ_2 such that $\rho_1 | \mathscr{K}_1^i \approx \rho_2 | \mathscr{K}_2^i$ for every $i \in \mathscr{I}$. We also require that $\mathscr{K}_k^i \perp \mathscr{K}_k^j$ for $i \neq j$ and $k=1,2$. If we allow one of the pairs to be $(0,0)$ then by Zorn's lemma there exists a family of pairs which is maximal with respect to the above properties. Having fixed such a maximal family we let $\mathscr{K}_k = \sum \mathscr{K}_k^i$. Then

$$(5) \qquad \rho_1 | \mathscr{K}_1 \approx \rho_2 | \mathscr{K}_2 \quad \text{while } \rho_1 | \mathscr{K}_1^\perp \downarrow \rho_2 | \mathscr{K}_2^\perp .$$

If $\mathscr{K}_1 = 0$ then $\rho_1 \downarrow \rho_2$ and if $\mathscr{K}_1 = \mathscr{H}_1$ then $\rho_1 \leqslant \rho_2$. If $\mathscr{K}_1 \neq 0$, $\mathscr{H}_1$ then ρ_1 being primary $\rho_1 | \mathscr{K}_1$ and $\rho_1 | \mathscr{K}_1^\perp$ intertwine and so by (5) either a submorphism of $\rho_2 | \mathscr{K}_2$ is disjoint from $\rho_2 | \mathscr{K}_2^\perp$ or $\mathscr{K}_2^\perp = 0$. Since ρ_2 is primary by the corollary of Lemma 2 the first case is not possible. Hence $\mathscr{K}_2^\perp = 0$ and $\rho_2 \leqslant \rho_1$.

Proposition 6. *If* $\rho_k : S \to \mathscr{L}(\mathscr{H}_k)$ *are intertwining primary representations then* $\rho_1 \sim \rho_2$.

Proof. Since $\rho_1 \uparrow \rho_2$ by Proposition 5 we have $\rho_1 \leqslant \rho_2$ or $\rho_2 \leqslant \rho_1$. We may suppose that $\rho_1 \leqslant \rho_2$. Then $\rho_1 \preccurlyeq \rho_2$ and so it is sufficient to see that $\rho_2 \preccurlyeq \rho_1$. If ρ_2 had a submorphism which is disjoint from ρ_1 then by $\rho_1 \leqslant \rho_2$ we had found two submorphisms of ρ_2 which are disjoint. But this would contradict the corollary of Lemma 2. Hence $\rho_2 \preccurlyeq \rho_1$ and $\rho_1 \sim \rho_2$.

By applying Proposition 6 we obtain:

Lemma 7. *If* ρ_1, ρ_2 *are primary representations of the same object then* $\rho_1 \sim \rho_2$ *or* $\rho_1 \downarrow \rho_2$.

Corollary. *Two primary representations intertwine if and only if they are quasiequivalent.*

Before we continue further we introduce the concept of a *multiplicity free representation*:

Definition 8. *A representation* $\rho : S \to \mathscr{L}(\mathscr{H})$ *is called multiplicity free if for every invariant subspace* $\mathscr{K}$ *of* ρ *we have* $\rho | \mathscr{K} \downarrow \rho | \mathscr{K}^\perp$.

If ρ is multiplicity free then $n(\omega, \rho) \leqslant 1$ for every $\omega \in \hat{S}$. Let $\pi_1, \ldots, \pi_m$ be equivalent, irreducible representations of S. Then $\rho = n_1 \pi_1 + \cdots + n_m \pi_m$ is multiplicity free if and only if $n_1, \ldots, n_m \leqslant 1$. These facts clarify the origin of the terminology. We note that a multiplicity free representation need not have any irreducible components.

Proposition 9. *A representation* $\rho: S \to \mathscr{L}(\mathscr{H})$ *is multiplicity free if and only if* (ρ, ρ) *is commutative.*

Proof. Let $\mathscr{A} = \mathscr{A}_\rho$ be the v. Neumann algebra associated with ρ so that $\mathscr{A}' = (\rho, \rho)$. We know that $\mathscr{A} \cap \mathscr{A}'$ is the common center of $\mathscr{A}$ and $\mathscr{A}'$. Let $\mathscr{K}$ be an invariant subspace of ρ and let E be the associated projection $\mathscr{H} \to \mathscr{K}$. Then by Proposition 4.10 we have $\rho | \mathscr{K} \downarrow \rho | \mathscr{K}^\perp$ if and only if $E \in c(\rho, \rho)$. Hence if ρ is multiplicity free then $E \in (\rho, \rho)$ implies that $E \in c(\rho, \rho)$. Since $c(\rho, \rho)$ is a v. Neumann algebra by the corollary of Proposition III.6.15 we obtain $(\rho, \rho) \subseteq c(\rho, \rho)$ and so (ρ, ρ) is commutative. Conversely if $c(\rho, \rho) = (\rho, \rho)$ then for any invariant $\mathscr{K}$ the corresponding E lies in $c(\rho, \rho)$ and so $\rho | \mathscr{K} \downarrow \rho | \mathscr{K}^\perp$ by Proposition 4.10.

Lemma 10. *If* ρ *is primary then* ρ *is multiplicity free if and only if it is irreducible.*

Proof. If ρ is multiplicity free then the only invariant subspaces are $\mathscr{K} = 0$ and $\mathscr{K} = \mathscr{H}$ because we must have $\rho \mathscr{K} \downarrow \rho | \mathscr{K}^\perp$ and ρ is primary. The converse implication is obvious. Another line of reasoning is as follows: Since ρ is primary (ρ, ρ) is a factor i.e. $c(\rho, \rho) = \mathbb{C} I$. Thus by Proposition 9 ρ is multiplicity free if and only if $(\rho, \rho) = \mathbb{C} I$. Now apply the irreducibility criterion given in Proposition 2.4.

Given an object S with a map we let $\hat{S}$ denote the set of quasiequivalence classes of primary representations of S. A typical element of $\hat{S}$ will be denoted by φ. For irreducible representations equivalence and quasi-equivalence mean the same thing and irreducible representations are primary. Therefore every equivalence class $\omega \in \hat{S}$ belongs to a unique class φ_ω in $\hat{S}$. Thus $\hat{S}$ can be considered as a subset of $\hat{S}$. Proposition 5.7 gives the structure of the representations ρ belonging to φ_ω.

We can extend the results obtained in Section 5 to arbitrary classes $\varphi \in \hat{S}$ when S is an object with a map. First we generalize Proposition 5.3: We fix a class φ in $\hat{S}$ and consider primary components $\mathfrak{P}$ of ρ such that $\rho | \mathfrak{P}$ belongs to φ. If this is the case then we say that φ occurs in ρ and we write $\mathfrak{P} \in \varphi$ and $\varphi \in \rho$. A primary component $\mathfrak{P}$ of ρ will be called *maximal* if it is not properly contained in any other primary component of ρ. If φ occurs in ρ then by Zorn's lemma there exists a family $\{\mathfrak{P}_i\}$ of pairwise orthogonal primary components $\mathfrak{P}_i \in \varphi$ which is maximal with respect to inclusion. If the empty collection is also permitted then it is not necessary to suppose that φ occurs in ρ.

Proposition 11. *Let S be an object with a map, let $\rho: S \to \mathscr{L}(\mathscr{H})$ be a Mackey type representation and let $\varphi \in \hat{S}$. If $\{\mathfrak{P}_i\}$ is a maximal family of pairwise orthogonal primary components $\mathfrak{P}_i \in \varphi$ then $\mathfrak{P} = \sum \mathfrak{P}_i$ is a maximal primary component of ρ.*

Proof. By Proposition 1 $\rho | \mathfrak{P}$ is a primary representation. Let $\mathfrak{P} \subseteq \mathscr{Q}$ where $\mathscr{Q}$ is a primary component of ρ. Then by Lemma 2 $\mathfrak{P}_\infty = \mathfrak{P}^\perp \cap \mathscr{Q}$ is a primary component of ρ or $\mathfrak{P}_\infty = 0$. But in the first case the components $\mathfrak{P}_\infty$ and $\mathscr{Q}$ would intertwine and so by $\mathscr{Q} \in \varphi$, Lemma 2 and the corollary of Lemma 7 we would have $\mathfrak{P}_\infty \in \varphi$. This would contradict the maximality of the system $\{\mathfrak{P}_i\}$ with respect to inclusion. Hence $\mathfrak{P}_\infty = 0$ and $\mathscr{Q} \subseteq \mathfrak{P}$.

Corollary 1. *If ρ is a Mackey type representation of an object S with a map then every primary component of ρ is included in a unique maximal primary component.*

The uniqueness of the enveloping maximal component follows from Proposition 1: For if $\mathfrak{P}_1$ and $\mathfrak{P}_2$ are maximal primary components of ρ and $\mathfrak{P}_1 \neq \mathfrak{P}_2$ then by Proposition 1 $\mathfrak{P}_1 \downarrow \mathfrak{P}_2$.

Corollary 2. *The space $\mathfrak{P}$ is uniquely determined by the class φ and the representation ρ.*

For let $\mathscr{Q} = \sum \mathscr{Q}_j$ where $\{\mathscr{Q}_j\}$ is a second maximal orthogonal family of primary components $\mathscr{Q}_j \in \varphi$. Then $\mathfrak{P}$ and $\mathscr{Q}$ intertwine and so $\mathfrak{P} + \mathscr{Q}$ is a primary component of ρ by Proposition 1. Hence by the maximality of $\mathfrak{P}$ and $\mathscr{Q}$ we have $\mathfrak{P} = \mathscr{Q}$.

For each $\varphi \in \hat{S}$ we can now define a subspace $\mathscr{H}_\varphi$ of the representation space $\mathscr{H}$: If φ does not occur in $\rho: S \to \mathscr{L}(\mathscr{H})$ then we let $\mathscr{H}_\varphi = 0$. If φ occurs in ρ then we let $\mathscr{H}_\varphi = \sum \mathfrak{P}_i$ where $\{\mathfrak{P}_i\}$ is a maximal family of pairwise orthogonal primary components $\mathfrak{P}_i \in \varphi$. Proposition 5.1 implies that $\mathscr{H}_{\varphi_1} \perp \mathscr{H}_{\varphi_2}$ if $\varphi_1 \neq \varphi_2$. Clearly every maximal primary component $\mathfrak{P}$ of ρ is of the form $\mathscr{H}_\varphi$ where $\mathfrak{P} \in \varphi$.

If $\omega \in \hat{S}$ then according to the definition of $\mathscr{H}_\omega$ in Section 5 we have $\mathscr{H}_\omega = \sum \mathscr{P}_i$ where $\{\mathscr{P}_i\}$ $(i \in \mathscr{I})$ is a maximal family of pairwise orthogonal irreducible components $\mathscr{P}_i \in \omega$. By the maximality there is no primary component $\mathfrak{P}$ such that $\mathfrak{P} \in \varphi_\omega$ and $\mathfrak{P} \perp \mathscr{P}_i$ for every $i \in \mathscr{I}$. Therefore $\{\mathscr{P}_i\}$ $(i \in \mathscr{I})$ is a maximal family of pairwise orthogonal primary components belonging to the class φ. This proves that $\mathscr{H}_\omega = \mathscr{H}_{\varphi_\omega}$ for every ω in $\hat{S}$.

Our discussion can be summarized in the following:

Theorem 12. *If S is an object with a map then one can associate with every Mackey type representation $\rho\colon S \to \mathscr{L}(\mathscr{H})$ and every quasiequivalence class $\varphi \in \hat{S}$ a uniquely determined subspace $\mathscr{H}_\varphi$ of $\mathscr{H}$ such that*

1) *$\mathscr{H}_\varphi$ is the maximal primary component of ρ belonging to the class φ or $\mathscr{H}_\varphi = 0$ according as φ occurs in ρ or not.*

2) *If $\mathfrak{P}$ is a primary component of ρ then $\mathfrak{P} \subseteq \mathscr{H}_\varphi$ or $\mathfrak{P} \perp \mathscr{H}_\varphi$ according as $\mathfrak{P} \in \varphi$ or $\mathfrak{P} \notin \varphi$.*

3) *If $\varphi_1 \neq \varphi_2$ then $\mathscr{H}_{\varphi_1} \perp \mathscr{H}_{\varphi_2}$ and so*

4) *$\mathscr{H}_d = \sum \{\mathscr{H}_\varphi : \varphi \in \hat{S}\}$ is a subspace of $\mathscr{H}$ which is uniquely determined by ρ, and*

5) *if $\mathfrak{P}$ is a primary component of ρ then $\mathfrak{P} \subseteq \mathscr{H}_d$. Hence*

6) *$\mathscr{H} = \mathscr{H}_d \oplus (\mathscr{H}_d)^\perp$ where $\mathscr{H}_d$ is the direct sum of all maximal primary components while $(\mathscr{H}_d)^\perp$ contains no primary component of ρ.*

7) *For each ω in $\hat{S}$ we have $\mathscr{H}_\omega = \mathscr{H}_{\varphi_\omega}$ where φ_ω is the unique class in $\hat{S}$ which contains ω.*

8) *One can write $\mathscr{H}_d = \mathscr{H}_{dI} \oplus \mathscr{H}_{dc}$ where $\mathscr{H}_{dc}$ is the orthogonal complement of $\mathscr{H}_{dI}$ in $\mathscr{H}_d$. The maximal primary components making up $\mathscr{H}_{dc}$ contain no minimal invariant subspaces of ρ.*

The subspace $\mathscr{H}_{dI}$ will be called the *type I discrete part* of ρ and $\mathscr{H}_{dc}$ is the *discrete-continuous part* of ρ. Finally $\mathscr{H}_c = (\mathscr{H}_{dI} \oplus \mathscr{H}_{dc})^\perp$ will be called the *purely continuous* part of ρ. Later we shall see that both $\mathscr{H}_{dc}$ and $\mathscr{H}_c$ can be split in a unique manner into two orthogonal parts called the *type II* and *type III discrete-continuous parts* and *type II* and *type III purely continuous parts* of ρ. The notations will be $\mathscr{H}_{dc} = \mathscr{H}_{dII} \oplus \mathscr{H}_{dIII}$ and $\mathscr{H}_c = \mathscr{H}_{cII} \oplus \mathscr{H}_{cIII}$. Therefore $\mathscr{H}$ is an orthogonal direct sum of five uniquely determined invariant subspaces of ρ.

We shall now turn to the proof of the existence of primary representations of Mackey objects which are not of type I. In Definition 2.13 we introduced an equivalence relation $\sim$ for the projections of a v. Neumann algebra. Using this relation we can now define what is meant by *finite* and *infinite v. Neumann algebras.*

Definition 13. *A v. Neumann algebra $\mathscr{A}$ is called finite if the only projection which is equivalent in $\mathscr{A}$ to the identity I is I itself.*

It is clear that if the underlying Hilbert space is finite dimensional then $\mathscr{A}$ is finite. By definition a *representation* $\rho: S \to \mathscr{L}(\mathscr{H})$ is *finite* if (ρ, ρ) is a finite v. Neumann algebra. In view of Proposition 1.14 this means that ρ is not equivalent to any of its subrepresentations. The primary representations which we are going to construct will be finite and not of type I. *Infinite* means not finite both for algebras and representations.

Lemma 14. *The representation* $\rho: S \to \mathscr{L}(\mathscr{H})$ *is infinite if and only if* ρ *has a non-zero invariant subspace* $\mathscr{M}$ *such that* $\rho|\mathscr{M} \oplus \rho|\mathscr{M} \approx \rho|\mathscr{M}$.

Note. This is the only thing which we need but the following proof actually offers more: If ρ is infinite then there is an $\mathscr{M}$ such that the direct sum of denumerably many copies of $\rho|\mathscr{M}$ is equivalent to $\rho|\mathscr{M}$.

Proof. By hypothesis there is a projection E in (ρ, ρ) such that $E < I$ and $\rho E \approx \rho$. Let $\mathscr{K} = E\mathscr{H}$ and $\mathscr{L} = \mathscr{K}^{\perp}$. By $\rho \approx \rho E$ there is an isometry U mapping $\mathscr{H}$ onto $\mathscr{K}$ which belongs to $(\rho, \rho E)$. Since $\mathscr{L}$ is an invariant subspace of ρ the relation $U\rho = \rho E U$ can be restricted to $\mathscr{L}$ and we obtain

$$(U|\mathscr{L})(\rho|\mathscr{L}) = (\rho|U\mathscr{L})(U|\mathscr{L}).$$

Hence $U|\mathscr{L}$ is a unitary intertwining operator between $\rho|\mathscr{L}$ and $\rho|U\mathscr{L}$. It follows that $\rho|\mathscr{L}$ and $\rho|U\mathscr{L}$ are equivalent subrepresentations of ρ. In general we see that $\rho|\mathscr{L}, \rho|U\mathscr{L}, \rho|U^2\mathscr{L}, \ldots$ are all equivalent representations. For any $m = 1, 2, \ldots$ we have

$$U^m\mathscr{L} = U^{m-1}(U\mathscr{L}) \subseteq U^{m-1}(U\mathscr{H}) = U^{m-1}\mathscr{K} \subseteq \mathscr{K}$$

and so $\mathscr{L} \perp U^m\mathscr{L}$. Therefore $\mathscr{L}, U\mathscr{L}, U^2\mathscr{L}, \ldots, U^m\mathscr{L}, \ldots$ are pairwise orthogonal non-zero subspaces of $\mathscr{H}$. We let

$$\mathscr{M} = \mathscr{L} + U^2\mathscr{L} + U^4\mathscr{L} + \cdots \quad \text{and} \quad \mathscr{M}_1 = U\mathscr{L} + U^3\mathscr{L} + U^5\mathscr{L} + \cdots$$

Then we have

$$\rho|\mathscr{M} \oplus \rho|\mathscr{M}_1 \approx \rho|\mathscr{M} + \mathscr{M}_1 \quad \text{and} \quad \rho|\mathscr{M} \approx \rho|\mathscr{M}_1.$$

Moreover U^m is an isometry of $U^m\mathscr{L}$ onto $U^{2m}\mathscr{L}$ and

$$(U^m|U^m\mathscr{L})(\rho|U^m\mathscr{L}) = (\rho|U^{2m}\mathscr{L})(U^m|U^m\mathscr{L})$$

for every $m = 0, 1, \ldots$ Hence the direct sum of the operators $U^m (m = 0, 1, \ldots)$ is a unitary intertwining operator of $\rho|\mathscr{M} + \mathscr{M}_1$ and $\rho|\mathscr{M}$. Now we see that

$$\rho|\mathscr{M} \oplus \rho|\mathscr{M} \approx \rho|\mathscr{M} \oplus \rho|\mathscr{M}_1 \approx \rho|\mathscr{M} + \mathscr{M}_1 \approx \rho|\mathscr{M}.$$

This proves the lemma. In order to obtain the result mentioned in the note one defines

$$\mathcal{M}_m = U^{\nu_{m1}} \mathcal{L} + U^{\nu_{m2}} \mathcal{L} + U^{\nu_{m3}} \mathcal{L} + \cdots$$

where we can take for example

$$\begin{array}{llllll}
(\nu_{1n}) = (1 & 2 & 4 & 7 & 11 \ldots) \\
(\nu_{2n}) = (3 & 5 & 8 & 12 & 17 \ldots) \\
(\nu_{3n}) = (6 & 9 & 13 & 18 & 24 \ldots) \\
(\nu_{4n}) = (10 & 14 & 19 & 25 & 32 \ldots) \\
\ldots\ldots\ldots\ldots\ldots\ldots\ldots\ldots
\end{array}$$

In order to show that not all primary representations are of type I we are going to study the regular representations of an infinite discrete group G. An invariant measure μ can be introduced on G by assigning to each finite set S the number of elements as $\mu(S)$. The invariance means that $\mu(aS) = \mu(S)$ for every a in G and actually μ is not only left invariant but biinvariant because $\mu(Sa) = \mu(S)$ for all a in G. A function $f: G \to \mathbb{C}$ is integrable with respect to μ if and only if $f(x) \neq 0$ for at most denumerably many x in G and $\sum_{x \in G} |f(x)| < \infty$. The integral of f will be denoted by $\int_G f(x)dx$ or simply by $\sum f(x)$. We let $L^2(G)$ be the space of square integrable complex valued functions.

The underlying Hilbert space of the *left regular representation* $\lambda: G \to \mathcal{U}(\mathcal{H})$ is $\mathcal{H} = L^2(G)$ and $\lambda(a)f(x) = f(a^{-1}x)$, or more briefly $\lambda(a)f = {}^a f$ for every $f \in \mathcal{H}$ and $a \in G$. Since μ is left invariant it is clear that λ is a unitary representation of G in $\mathcal{H}$. The Hilbert space of the *right regular representation* $\rho: G \to \mathcal{U}(\mathcal{H})$ is the same $\mathcal{H} = L^2(G)$ but now it is the right invariance of the integration that matters. For we let $\rho(a)f = f_a$ for all f in $\mathcal{H}$ and $a \in G$.

For every a in G let δ_a denote the function on G with values $\delta_a(a) = 1$ and $\delta_a(x) = 0$ for $x \neq a$. Then $\{\delta_a\}$ $(a \in G)$ is a MONS in $L^2(G)$. We note that $\lambda(a)\delta_b = \delta_{ab}$ for every a, b in G. For any linear operator A in $\mathcal{L}(\mathcal{H})$ we let ω_A denote the function $x \mapsto (A\delta_e, \delta_x)$.

Lemma 15. *If A is in (λ, λ) then*

$$\omega_A(a^{-1}x) = (A\delta_a, \delta_x). \tag{6}$$

Similarly for A and B in (λ, λ) we have

$$\omega_{AB}(x) = \sum_{y \in G} \omega_A(y^{-1}x)\,\omega_B(y) \tag{7}$$

and also

(8) $$\omega_{A^*}(x)=\overline{\omega_A(x^{-1})}\,.$$

If A belongs to (ρ,ρ) *then*

(9) $$\omega_A(xa^{-1})=(A\delta_a,\delta_x)\,.$$

Proof. Using the MONS $\{\delta_x\}(x\in G)$ we have

$$\begin{aligned}A\delta_a=A\lambda(a)\delta_e=\lambda(a)A\delta_e&=\lambda(a)\textstyle\sum\omega_A(x)\delta_x\\&=\textstyle\sum\omega_A(x)\delta_{ax}=\sum\omega_A(a^{-1}x)\delta_x\,.\end{aligned}$$

Therefore $(A\delta_a,\delta_x)=\omega_A(a^{-1}x)$. The proof of (9) is similar. Using (6) we have

$$\sum\omega_A(y^{-1}x)\omega_B(y)=\sum(\delta_y,A^*\delta_x)(B\delta_e,\delta_y)=(B\delta_e,A^*\delta_x)$$

from which we obtain (7). Formula (8) follows immediately from (6) by letting $x=e$, using the definition of $\omega_{A^*}(a)$ and replacing a by x in the resulting equation.

Proposition 16. *The left regular representation of a countable group is finite.*

Note. By a countable group we mean a group G with at most denumerably many elements.

Proof. Let $\mathcal{M}$ be an invariant subspace of λ such that

(10) $$\lambda|\mathcal{M}\oplus\lambda|\mathcal{M}\approx\lambda|\mathcal{M}$$

and let P be the projection $\mathcal{H}\to\mathcal{M}$. Since $\lambda|\mathcal{M}$ is a subrepresentation of λ by (10) $\mathcal{M}$ can be split into the direct sum of two equivalent invariant subspaces. We let E and F denote the projections of $\mathcal{H}$ onto these subspaces. Then $E,F\in(\lambda,\lambda)$ by Proposition III.5.13, and by (10) we have $\lambda E\approx\lambda F\approx\lambda(E+F)$. By Proposition 2.14 there is an operator A in (λ,λ) such that $A^*A=E$ and $AA^*=F$. Therefore by (7) and (8) of Lemma 15 we have

(11) $$\omega_E(e)=\omega_{A^*A}(e)=\sum|\omega_A(y)|^2=\sum|\omega_A(y^{-1})|^2=\omega_{AA^*}(e)=\omega_F(e)\,.$$

Since $\lambda E\approx\lambda(E+F)$ we can perform the same computation with E and $E+F$ in place of E and F and obtain

$$\omega_E(e)=\omega_{E+F}(e)=\omega_E(e)+\omega_F(e)\,.$$

Hence $\omega_F(e)=0$ and so from (11) we get $\omega_A=0$. Thus

$$(A\delta_x,\delta_y)=(\delta_x,A^*\delta_y)=\overline{(A^*\delta_y,\delta_x)}=\overline{\omega_{A^*}(y^{-1}x)}=\omega_A(x^{-1}y)=0$$

for any x, y in G. Since $\{\delta_x\}(x\in G)$ is a MONS we obtain $A\delta_x=0$ and x being arbitrary $A=0$. This shows that $E=A^*A=0$ and consequently $\mathcal{M}=0$. Therefore the only λ-invariant subspace satisfying (10) is $\mathcal{M}=0$. Thus λ is a finite representation by Lemma 14.

Proposition 17. *Let G be a denumerable group such that every element except the identity has infinitely many conjugates. Then the left regular representation of G is primary.*

Proof. As earlier let λ and ρ denote the left and right regular representations of G. For any a and b in G we have $\lambda(a)\rho(b)=\rho(b)\lambda(a)$. Therefore $\lambda(a)\in(\rho,\rho)$ and $\rho(a)\in(\lambda,\lambda)$ for every a in G. Now if A is in $(\lambda,\lambda)'$ then A commutes with every $\rho(a)$ and so $A\in(\rho,\rho)$ proving that $(\lambda,\lambda)'\subseteq(\rho,\rho)$. Similarly we have $(\rho,\rho)'\subseteq(\lambda,\lambda)$.

Now let A belong to $(\lambda,\lambda)\cap(\lambda,\lambda)'$. Then $A\in(\lambda,\lambda)$ and $A\in(\rho,\rho)$ and so by (6) and (9) of Lemma 15 we have $\omega_A(a^{-1}x)=\omega_A(xa^{-1})$. Hence if A is in the center of (λ,λ) then $\omega_A(x)=\omega_A(axa^{-1})$ for any $a, x\in G$ and so ω_A is constant on each conjugacy class of G. Since $A\in(\lambda,\lambda)$ by (7) and (8) of Lemma 15 we have $\sum|\omega_A(y)|^2=\omega_{A^*A}(e)<\infty$ and so ω_A is in $L^2(G)=\mathcal{H}$. Now if every conjugacy class except $\{e\}$ is infinite then by the last inequality $\omega_A(y)=0$ at every $y\neq e$. Using (6) of Lemma 15 we obtain $(A\delta_a,\delta_x)=0$ for $a\neq x$ and by the same equation we have $(A\delta_a,\delta_a)=\omega_A(e)$. Since $\{\delta_x\}(x\in G)$ is a MONS in $\mathcal{H}$ this shows that $A\delta_a=\omega_A(e)\delta_a$ for every a in G and so $A=\omega_A(e)I$. We proved that the center of (λ,λ) is $\mathbb{C}I$ and so λ is a primary representation.

Theorem 18. *Let G be a denumerable group such that all of its conjugacy classes except $\{e\}$ are infinite. Then the left regular representation of G is finite and it has no irreducible components.*

Proof. On the contrary suppose that $\lambda: G\to\mathcal{U}(\mathcal{H})$ has an irreducible component $\mathcal{P}$. We let ω denote the equivalence class containing $\pi=\lambda|\mathcal{P}$ so that $\omega\in\hat{G}$. By Proposition 17 λ is primary and so by Proposition 5.7 we have $\mathcal{H}=\mathcal{H}_\omega$. Therefore by Theorem 5.4 $\lambda\approx n(\omega,\lambda)\pi$ where $n(\omega,\lambda)$ is a finite or infinite cardinal. If $n(\omega,\lambda)$ were infinite then we would have $\lambda\oplus\lambda\approx\lambda$ which by Lemma 14 would contradict Proposition 16. Therefore λ is the direct sum of finitely many equivalent irreducible components or the starting hypothesis is incorrect. In the first case by Proposition III.6.6 the commuting algebra (λ,λ) would consist of all operators of the form $A=(a_{ij}U_{ij})$ where U_{ij} are fixed unitary operators among the finitely many irreducible components and $a_{ij} \in \mathbb{C}$. Therefore (λ, λ) would be finite dimensional which is not the case because $(\lambda, \lambda) \supseteq \{\rho(x) : x \in G\}$ and $\sum \alpha_k \rho(x_k) = 0$ implies $\sum \alpha_k \delta_{x_k} = 0$ where by the orthogonality the δ_{x_k}'s are independent. Hence our starting hypothesis is incorrect and so λ has no irreducible components.

The discussion will be completed by showing the existence of groups satisfying the hypothesis of the theorem. For example we can consider the group of those permutations of the set $\{1, 2, 3, \ldots\}$ which permute only finitely many integers at a time. Let

$$p = \left(\begin{array}{ccc|cc} 1 & 2 & \ldots m & m+1 & m+2 \ldots \\ n_1 & n_2 & \ldots n_m & m+1 & m+2 \ldots \end{array}\right)$$

and let $x = km$ $(k = 1, 2, \ldots)$ and

$$h_k = \left(\begin{array}{c|cc|cc} 1 \ldots m & m+1 \ldots x & x+1 \ldots x+m & x+m+1 & x+m+2 \ldots \\ x+1 \ldots x+m & m+1 \ldots x & 1 \ldots m & x+m+1 & x+m+2 \ldots \end{array}\right)$$

One can easily see that $h_k p h_k^{-1}$ $(k = 1, 2, \ldots)$ are distinct conjugates of p unless p is the identity permutation.

Another simple example is the group of linear transformations $\xi = ax + b$ where $a > 0$ and a, b are rational numbers. This is a semidirect product with group operation $(a, b)(c, d) = (ac, ad + b)$. The conjugates of (a, b) are the pairs $(a, (1-a)y + bx)$ where x, y are rationals.

Definition 19. *A factorial v. Neumann algebra is called a factor of type I if it has an irreducible component.*

Let $\mathscr{H}$ denote the underlying Hilbert space of the type I factor $\mathscr{A}$. If $\mathscr{P}$ is an irreducible component of $\mathscr{A}$ then by Proposition III.5.13 the associated projection $P: \mathscr{H} \to \mathscr{P}$ is a minimal projection in $\mathscr{A}'$. Conversely if P is a minimal projection in $\mathscr{A}'$ then by definition P is not 0 and so $\mathscr{P} = P\mathscr{H}$ is an irreducible component of $\mathscr{A}$. Therefore an alternative definition is as follows: The factorial v. Neumann algebra $\mathscr{A}$ is of type I if its commutant $\mathscr{A}'$ has a minimal projection. A similar sounding but considerably more distant characterization is often taken as the definition: Then it is required that $\mathscr{A}$ itself have a minimal projection or equivalently that $\mathscr{A}'$ have an irreducible component. The reason for this alternative is that if $\mathscr{A}$ is a factor of type I then so is $\mathscr{A}'$. (See Proposition 25.)

When type II and III factors will be introduced we shall see that these last characterizations of type I factors fit more closely into the general pattern. However it is natural to take Definition 19 as the starting point.

For the first basic questions are on the existence of irreducible components and the possibility of decomposing a representation ρ or a v. Neumann algebra $\mathscr{A}$ into a direct sum of such components.

The simplest examples of type I factors are the full operator algebras $\mathscr{L}(\mathscr{H})$ where $\mathscr{H}$ is any complex Hilbert space. For $\mathscr{L}(\mathscr{H})$ is irreducible by Theorem 2.5. The v. Neumann algebras $\mathscr{A}_\lambda$ generated by the representations considered in Theorem 18 give examples of factors which are not of type I.

A large class of factorial type I algebras can be constructed as follows: Let m, n be cardinals and let M and N be sets such that $m = \operatorname{card} M$ and $n = \operatorname{card} N$. Let $x: M \to \mathbb{C}$ be a function such that its value x_μ at $\mu \in M$ differs from 0 at no more than denumerably many places and $\sum |x_\mu|^2 < \infty$. If x and y are such functions we define $(x, y) = \sum x_\mu \overline{y_\mu}$ and let $\mathscr{H}_m$ be the complex Hilbert space of all such functions $x: M \to \mathbb{C}$. Next we define $\mathscr{H}_{mn}$ as the direct sum of n copies of $\mathscr{H}_m$ over the index set N. Thus if x is in $\mathscr{H}_{mn}$ then $x: N \to \mathscr{H}_m$, its value at ν is x_ν and $\sum \|x_\nu\|^2 < \infty$. The inner product of $x = (x_\nu) = ((x_{\mu\nu}))$ and $y = (y_\nu) = ((y_{\mu\nu}))$ is

$$(x, y) = \sum (x_\nu, y_\nu) = \sum x_{\mu\nu} \overline{y_{\mu\nu}}\,.$$

For any A in $\mathscr{L}(\mathscr{H}_m)$ let the symbol A denote also the linear operator $A: \mathscr{H}_{mn} \to \mathscr{H}_{mn}$ given by $Ax = A(x_\nu) = (Ax_\nu)$. The new A is continuous and has the same norm $|A|$ as the original operator acting on $\mathscr{H}_m$: First one proves that $\|Ax\| \leqslant |A| \cdot \|x\|$ for every x in $\mathscr{H}_{mn}$ where $|A|$ is the old norm. Then one proves that the norm is at least as large as the old one by specializing x to elements such that $x_\nu = 0$ for every index ν except a fixed one. The adjoint of a new A is obtained by taking the adjoints of the component operators. Therefore this association defines an injection of $\mathscr{L}(\mathscr{H}_m)$ into $\mathscr{L}(\mathscr{H}_{mn})$ which is an involutive isometric isomorphism between $\mathscr{L}(\mathscr{H}_m)$ and its image which we denote by $\mathscr{A}_{mn}$.

One easily verifies that $\mathscr{A}_{mn}$ is weakly closed so it is a v. Neumann algebra: For if $\sum A_\nu$ is the weak limit of $\sum A_\nu^{(k)}$ where $A_\nu^{(k)}$ is the same operator $A_\nu^{(k)}$ for each fixed k and arbitrary ν then $A_\nu = \lim A_\nu^{(k)}$ and so A_ν is the same for every ν. By Theorem 2.5 the center of $\mathscr{L}(\mathscr{H}_m)$ is $\mathbb{C}I$ and I of $\mathscr{L}(\mathscr{H}_m)$ corresponds under our isomorphism to I of $\mathscr{L}(\mathscr{H}_{mn})$. Therefore $\mathscr{A}_{mn}$ being isomorphic to $\mathscr{L}(\mathscr{H}_m)$ is a factor. For every ν the algebra obtained by restricting the elements A of $\mathscr{A}_{mn}$ to $\mathscr{H}_m^{(\nu)}$ is the complete operator algebra $\mathscr{L}(\mathscr{H}_m^{(\nu)}) = \mathscr{L}(\mathscr{H}_m)$. Therefore by Theorem 2.5 $\mathscr{H}_m^{(\nu)}$ is an irreducible component and so $\mathscr{A}_{mn}$ is a type I factor.

Proposition 20. *Every type I factor is a direct sum of equivalent irreducible components.*

Proof. We look at the representation $\rho: \mathscr{A} \to \mathscr{L}(\mathscr{H})$ of the abstract involutive algebra $\mathscr{A}$ obtained from the given v. Neumann algebra $\mathscr{A}$ by ignoring the additional structure which makes the operator $A: \mathscr{H} \to \mathscr{H}$ out of the abstract element A. Since $\mathscr{A}$ is a factor by Propositions 2.1 and 4.11 ρ is a primary representation with an irreducible component. Therefore by Theorem 5.4 and Proposition 5.7 ρ is a direct sum of equivalent irreducible components. Now the result follows from Proposition 2.1.

Definition 21. *A representation $\rho: S \to \mathscr{L}(\mathscr{H})$ is called a primary representation of type I if the v. Neumann algebra associated with ρ is a type I factor.*

Two operator algebras $\mathscr{A}_1$ and $\mathscr{A}_2$ are called *spatially isomorphic* if there is a unitary map $U: \mathscr{H}_1 \to \mathscr{H}_2$ of the underlying Hilbert spaces such that $U \mathscr{A}_1 U^* = \mathscr{A}_2$. If this is so then $A \mapsto U A U^*$ is an involutive isomorphism of $\mathscr{A}_1$ onto $\mathscr{A}_2$ which preserves the operator norm and also the Hilbert-Schmidt norm and the trace when they exist. The same holds for the full operator algebras $\mathscr{L}(\mathscr{H}_1)$ and $\mathscr{L}(\mathscr{H}_2)$. Moreover by the isomorphism of $\mathscr{L}(\mathscr{H}_1)$ and $\mathscr{L}(\mathscr{H}_2)$ if $A, B \in \mathscr{L}(\mathscr{H}_1)$ and $AB = BA$ then UAU^* and UBU^* commute with each other and conversely. By restricting A to $\mathscr{A}_1$ we obtain:

Proposition 22. *If $U: \mathscr{H}_1 \to \mathscr{H}_2$ determines a spatial isomorphism of $\mathscr{A}_1$ onto $\mathscr{A}_2$ then it gives also a spatial isomorphism of $\mathscr{A}_1'$ onto $\mathscr{A}_2'$.*

We also notice that if $\rho_k: S \to \mathscr{L}(\mathscr{H}_k)$ $(k = 1, 2)$ are equivalent representations then the associated v. Neumann algebras are spatially isomorphic.

Theorem 23. *Every factorial type I v. Neumann algebra is spatially isomorphic to some $\mathscr{A}_{mn}$.*

Proof. Let $\mathscr{H}$ be the underlying Hilbert space of $\mathscr{A}$ and let $\mathscr{A}$ denote also the abstract involutive algebra obtained from $\mathscr{A}$. Let $\rho: \mathscr{A} \to \mathscr{L}(\mathscr{H})$ be the representation associating with each A in $\mathscr{A}$ the operator $A: \mathscr{H} \to \mathscr{H}$. By Proposition 20 $\mathscr{H}$ is an orthogonal direct sum of equivalent irreducible components, say $\mathscr{H} = \sum \{\mathscr{P}_\nu : \nu \in N\}$ where N is our standard index set corresponding to the cardinality n of the set of orthogonal components. We don't have to know whether or not N and n depend on the decomposition because in any case having fixed the irreducible components N and n are uniquely determined. By Proposition 20 the irreducible components $\mathscr{P}_\nu$ are equivalent and so $m = \dim \mathscr{P}_\nu$ is independent of ν.

We prove that $\mathscr{A}$ is spatially isomorphic to $\mathscr{A}_{mn}$. Let $o \in N$ be fixed and let $U_\nu: \mathscr{P}_\nu \to \mathscr{P}_o$ be a unitary map which gives a unitary equivalence between $\mathscr{A} | \mathscr{P}_o$ and $\mathscr{A} | \mathscr{P}_\nu$. Therefore denoting by A_ν the restricted operator $A | \mathscr{P}_\nu$ we have

$$A_o = U_\nu A_\nu U_\nu^* \tag{12}$$

for every A in $\mathscr{A}$. Since $\mathscr{P}_o$ is an irreducible component by Theorem 2.5 $\mathscr{A}_o = \mathscr{A} | \mathscr{P}_o = \mathscr{L}(\mathscr{P}_o)$. By $m = \dim \mathscr{P}_o = \dim \mathscr{H}_m$ there is a unitary map $U: \mathscr{P}_o \to \mathscr{H}_m$ and so $U \mathscr{A}_o U^* = \mathscr{L}(\mathscr{H}_m)$. Having fixed the maps U_ν $(\nu \in N)$ and U we define a unitary transformation $V: \mathscr{H} \to \mathscr{H}_{mn}$ as follows: For any ν we have a unitary map $U U_\nu: \mathscr{P}_\nu \to \mathscr{H}_m$ and $\mathscr{H}_{mn}$ is a direct sum of copies of $\mathscr{H}_m$, namely

$$\mathscr{H}_{mn} = \sum \{\mathscr{H}_m^{(\nu)}: \nu \in N\}.$$

Hence by $\mathscr{H} = \sum \{\mathscr{P}_\nu: \nu \in N\}$ the direct sum $V = \sum U U_\nu$ is a unitary map of $\mathscr{H}$ onto $\mathscr{H}_{mn}$.

The proof will be completed by showing that $A \mapsto V A V^{-1}$ is a spatial isomorphism of $\mathscr{A}$ onto $\mathscr{A}_{mn}$. For this purpose we introduce the projection $P_\nu: \mathscr{H}_{mn} \to \mathscr{H}_m^{(\nu)} = \mathscr{H}_m$ and the injection $I_\nu: \mathscr{H}_m \to \mathscr{H}_{mn}$ of $\mathscr{H}_m$ into the ν-th summand of $\mathscr{H}_{mn}$. Since the restriction of V^* to $\mathscr{H}_m^{(\nu)}$ is $(U U_\nu)^*$ we have $A_\nu V^* P_\nu = A_\nu U_\nu^* U^* P_\nu$. Similarly as V restricted to $\mathscr{P}_\nu$ is $U U_\nu$ we obtain

$$V A V^* P_\nu = I_\nu U U_\nu A U_\nu^* U^* P_\nu.$$

Hence using (12) we have

$$V A V^* P_\nu = I_\nu (U A_o U^*) P_\nu.$$

It follows from the definition of P_ν and I_ν that $P_{\nu_1} P_{\nu_2} = 0$ if $\nu_1 \neq \nu_2$ and $P_\nu I_\nu = I$ where $I: \mathscr{H}_m \to \mathscr{H}_m$ is the identity map. Therefore from above

$$P_{\nu_1}(V A V^*) P_{\nu_2} = 0 \quad \text{if } \nu_1 \neq \nu_2$$

and

$$P_\nu (V A V^*) P_\nu = (U A_o U^*) P_\nu.$$

This shows that $V A V^*$ is an element of $\mathscr{A}_{m,n}$. The map $A \mapsto V A V^*$ is a homomorphism of $\mathscr{A}$ onto $\mathscr{A}_{mn}$ and if A is in its kernel then A_o must be 0. By (12) this implies that $A_\nu = 0$ for every ν and so A is the zero operator in $\mathscr{A}$.

Proposition 24. *For finite m, n, m', n' the algebras $\mathscr{A}_{mn}$ and $\mathscr{A}_{m'n'}$ are spatially isomorphic if and only if $m = m'$ and $n = n'$.*

Proof. As abstract algebras $\mathscr{A}_{mn}$ and $\mathscr{A}_{m'n'}$ are isomorphic to $\mathscr{L}(\mathscr{H}_m)$ and $\mathscr{L}(\mathscr{H}_{m'})$, respectively. Therefore the spatial isomorphism of $\mathscr{A}_{mn}$ and $\mathscr{A}_{m'n'}$ implies that $m=m'$. The dimension of the underlying Hilbert space of $\mathscr{A}_{mn}$ is mn so by $m=m'$ the spatial isomorphism implies that $n=n'$.

Proposition 25. *If $\mathscr{A}$ is a type I factor then so is $\mathscr{A}'$.*

Proof. By Theorem 23 $\mathscr{A}$ is spatially isomorphic to $\mathscr{A}_{mn}$ for some cardinals m and n. Therefore by Proposition 22 $\mathscr{A}'$ is spatially isomorphic to $(\mathscr{A}_{mn})'$. Thus $\mathscr{A}'$ has an irreducible component if and only if $(\mathscr{A}_{mn})'$ has one and so if and only if $(\mathscr{A}_{mn})''$ has a minimal projection. By the corollary of Theorem III.6.10 we have $(\mathscr{A}_{mn})''=\mathscr{A}_{mn}$. Since $\mathscr{A}_{mn}$ is the isomorphic image of $\mathscr{L}(\mathscr{H}_m)$ we see that $\mathscr{A}_{mn}$ has minimal projections and so $\mathscr{A}'$ has irreducible components. The algebras $\mathscr{A}$ and $\mathscr{A}'$ have the same center so $\mathscr{A}'$ is a factor.

8. Algebras and Representations of Type I

Here we shall consider a class of involutive algebras which are especially important in the case of v. Neumann algebras. For factorial v. Neumann algebras this concept was already introduced earlier and such algebras were said to be of type I. After discussing the basic results we will turn to representations such that the associated v. Neumann algebras belong to our class.

Let A be an algebra over a field F and let a be in A. Then aAa is a subalgebra of A and if e is an idempotent then

$$eAe=\{x\colon x\in A \text{ and } ex=xe=x\}.$$

The element a is called *abelian* if aAa is commutative. If e_1, e_2 are idempotents and $e_1e_2=e_2e_1=e_1$ then we shall write $e_1\leqslant e_2$. This is a reflexive and transitive relation satisfying $0\leqslant e$ and also $e\leqslant 1$ if A has an identity element 1. We are interested in the special case when A is an involutive algebra over $\mathbb{C}$ and we will consider only self adjoint idempotents. From here on an idempotent will always supposed to be self adjoint even if this property is not explicitely stated.

Lemma 1. *Suppose that A is an involutive algebra over $\mathbb{C}$ and a is a self adjoint abelian element of A. Let e be a self adjoint idempotent which is a multiple of a. Then e is abelian.*

Proof. Let $e=ab$ and let x, y be in eAe. We have $e^*=b^*a^*=b^*a=e$. Hence

$$xy=(exe)(eye)=(abxb^*a)(abyb^*a)=(abyb^*a)(abxb^*a)$$
$$=(eye)(exe)=yx.$$

The center of A will be denoted by cA and an element a in A will be called *central* if $a \in cA$. We are interested in those involutive algebras A over $\mathbb{C}$ in which every non-zero central idempotent majorizes a non-zero abelian idempotent. These could be called involutive complex algebras of abstract type I. An isomorphism preserves this property.

Definition 2. *A v. Neumann algebra is said to be of type I if every non-zero central projection majorizes a non-zero abelian projection.*

We recall that a projection E in $\mathscr{A}$ is called an *abelian projection* if $E\mathscr{A}E$ is commutative. It is a *central projection* if E is in $c\mathscr{A}=\mathscr{A}\cap\mathscr{A}'$.

Proposition 3. *If $\mathscr{A}_1$ and $\mathscr{A}_2$ are isomorphic v. Neumann algebras and if $\mathscr{A}_1$ is of type I then so is $\mathscr{A}_2$.*

Proof. The property of being of abstract type I is preserved by an involutive isomorphism.

Proposition 4. *Every commutative v. Neumann algebra is of type I.*

Proof. Since $\mathscr{A}$ is commutative every projection is abelian.

Our next object is to prove that our earlier definition of type I factorial v. Neumann algebras is in accordance with the present general definition.

By a *minimal projection* of an operator algebra $\mathscr{A}$ we understand a non-zero projection E in $\mathscr{A}$ such that $P<E$ implies $P=0$ for every projection P in $\mathscr{A}$. More generally we can speak about *minimal idempotents* in involutive algebras A over $\mathbb{C}$.

Proposition 5. *If $\mathscr{A}$ is a v. Neumann algebra then every minimal projection is abelian.*

Note. We already know that a v. Neumann algebra need not have any minimal projections.

Proof. Let E be a minimal projection in $\mathscr{A}$ and let $\mathscr{K}=E\mathscr{H}$. By Proposition III.5.13 a projection P is in $\mathscr{A}$ if and only if $P\mathscr{H}$ is an invariant subspace of $\mathscr{A}'$. Therefore $\mathscr{K}$ is an invariant component of $\mathscr{A}'$ and so $\mathscr{A}'|\mathscr{K}=\mathscr{L}(\mathscr{K})$ and $(\mathscr{A}'|\mathscr{K})'=\mathbb{C}I$ by Propositions 2.4 and 2.5. If

$A \in \mathscr{A}$ and $T \in \mathscr{A}'$ then by $E \in \mathscr{A}$ we have $(EAE)T = T(EAE)$ and so $EAE|\mathscr{K}$ belongs to $(\mathscr{A}'|\mathscr{K})'$. Therefore $EAE|\mathscr{K} = \lambda_A I$ where I is the identity of $\mathscr{L}(\mathscr{K})$. From this it is clear that $E\mathscr{A}E$ is commutative.

Now let $\mathscr{A}$ be a factorial v. Neumann algebra with an irreducible component. Then by Proposition 7.25 $\mathscr{A}'$ is also factorial and it has an irreducible component $\mathscr{P}$. The corresponding projection $E: \mathscr{H} \to \mathscr{P}$ is a minimal projection in $\mathscr{A}$ and $E \neq 0$. Since $\mathscr{A}$ is factorial its only non-zero central projection is I which majorizes E. Hence $\mathscr{A}$ is of type I.

Our reasoning also proves the following:

Lemma 6. *Every factorial v. Neumann algebra containing a minimal projection is of type I.*

Note. See also Proposition 21.

A v. Neumann subalgebra of a type I v. Neumann algebra need not be a type I algebra. It is also possible that $\mathscr{A}_1 \subseteq \mathscr{A}_2$ where $\mathscr{A}_1$ is of type I while $\mathscr{A}_2$ is not. These phenomena occur already in the case of factorial v. Neumann algebras. For instance $\mathscr{L}(\mathscr{H})$ and $\mathbb{C}I$ are always type I factors and the example discussed in Section 7 shows that if $\mathscr{H}$ is suitably chosen then there is a factor between them which is not of type I.

Proposition 7. *The direct sum of type I v. Neumann algebras is a type I algebra.*

Proof. Direct sums are defined in Section III.6. If E is a projection in $\mathscr{A} = \sum \mathscr{A}_i$ then $E = \sum E_i$ where E_i is a projection in $\mathscr{A}_i$. If in addition E is central then E_i is in $\mathscr{A}_i'$ for every i and so E_i is a central projection in $\mathscr{A}_i$. If E is not 0 then there is an index i such that $0 < E_i$. We fix such an index i and choose an abelian projection P_i in $\mathscr{A}_i$ such that $0 < P_i \leq E_i$ and for every other index j we let $P_j = 0$. Then $0 < P = \sum P_i \leq E$ and P is a projection in $\mathscr{A}$. Since $P\mathscr{A}P$ is isomorphic with $\sum P_i \mathscr{A}_i P_i$ we see that P is abelian.

Let E be a projection in the v. Neumann algebra $\mathscr{A}$ and let $\mathscr{K} = E\mathscr{H}$. Consider an A in $\mathscr{A}$ such that $EA = AE = A$ i.e. $A\mathscr{K} \subseteq \mathscr{K}$ and $A\mathscr{K}^\perp = 0$. An operator $A \in \mathscr{A}$ will satisfy these conditions if and only if it is of the form $A = ETE$ where $T \in \mathscr{A}$. Therefore the set of all these operators is the subalgebra $E\mathscr{A}E$ of $\mathscr{A}$. If we let

$$\mathscr{A}_E = \{A|\mathscr{K} : A \in E\mathscr{A}E\}$$

then $\mathscr{A}_E$ is an involutive subalgebra of $\mathscr{L}(\mathscr{K})$ with identity $E|\mathscr{K}$. It is called the *reduced algebra* of $\mathscr{A}$.

Lemma 8. *The map* $A \mapsto A|\mathcal{K}$ *is an isomorphism of the involutive algebra* $E\mathcal{A}E$ *onto* $\mathcal{A}_E$. *The inverse image of an element* T *of* $\mathcal{A}_E$ *under this isomorphism is* TE.

Proof. The map is obviously a homomorphism and it is surjective by the definition of $\mathcal{A}_E$. If A is in $E\mathcal{A}E$ then $A\mathcal{K}^{\perp}=0$, so if $A|\mathcal{K}=0$ then $A=0$. This shows that $A \mapsto A|\mathcal{K}$ is injective.

Lemma 9. *The algebra* $E\mathcal{A}E$ *is strongly closed.*

Proof. By Theorem III.6.10 the v. Neumann algebra $\mathcal{A}$ is strongly closed and the operators satisfying $A\mathcal{K}\subseteq\mathcal{K}$ and $A\mathcal{K}^{\perp}=0$ form a strongly closed subspace of $\mathcal{L}(\mathcal{H})$.

Proposition 10. *The reduced algebra* $\mathcal{A}_E$ *is a v. Neumann algebra.*

Proof. It is sufficient to prove that $\mathcal{A}_E$ is strongly closed. If $T: \mathcal{K}\to\mathcal{K}$ and T is in the strong closure of $\mathcal{A}_E$ then $A=ETE$ is in the strong closure of $E\mathcal{A}E$ and $A|\mathcal{K}=T$. Hence $A\in E\mathcal{A}E$ and $T\in\mathcal{A}_E$.

Lemma 11. *If* $\mathcal{A}$ *is a v. Neumann algebra, E is a projection in* $\mathcal{A}$ *and P a projection in* $\mathcal{A}_E$ *then* $P=A|\mathcal{K}$ *where A is a projection in* $\mathcal{A}$ *and* $\mathcal{K}$ *range of E.*

Proof. By the definition of $\mathcal{A}_E$ we have $P=A|\mathcal{K}$ for some A in $E\mathcal{A}E$. By $EA=A$ we have $A\mathcal{K}\subseteq\mathcal{K}$. Since $(A|\mathcal{K})^2=P^2=P=A|\mathcal{K}$ by $AE=A$ and $A\mathcal{K}\subseteq\mathcal{K}$ we obtain $A^2=A$. If $x, y\in\mathcal{H}$ then by $P=P^*$

$$(Ax,y)=(AEx,Ey)=(PEx,Ey)=(Ex,PEy)=(Ex,AEy)=(x,Ay)$$

proving that $A^*=A$.

Proposition 12. *If the underlying Hilbert space is finite dimensional then the v. Neumann algebra is isomorphic to a direct sum of finitely many factorial v. Neumann algebras whose underlying Hilbert spaces are also finite dimensional.*

Proof. If $\mathcal{A}$ is not factorial then by Proposition III.6.15 its center contains a pair of non-zero orthogonal projections E_1, E_2 such that $E_1+E_2=I$. Then $\mathcal{A}$ is isomorphic to the direct sum of $E_1\mathcal{A}E_1$ and $E_2\mathcal{A}E_2$. We know that $E_k\mathcal{A}E_k$ is isomorphic to the v. Neumann algebra $\mathcal{A}_{E_k}$ with underlying Hilbert space $\mathcal{K}_k=E_k\mathcal{H}$. Since we have $\dim\mathcal{K}_k<\dim\mathcal{H}$ this process can be repeated only finitely many times.

Proposition 13. *If the underlying Hilbert space is finite dimensional then the v. Neumann algebra is of type I.*

Proof. First we suppose that $\mathscr{A}$ is factorial. Consider a non-zero projection E in $\mathscr{A}$ such that the dimension of $\mathscr{K}=E\mathscr{H}$ is as small as possible. Then E is a minimal projection and so $\mathscr{A}$ is of type I by Proposition 6. If $\mathscr{A}$ is not factorial we use Propositions 3, 7 and 12.

Let $\mathscr{A}$ be a v. Neumann algebra of operators $A:\mathscr{H}\to\mathscr{H}$ and let $\mathscr{M}$ be a non-void subset of $\mathscr{H}$. Then we let $(\mathscr{A},\mathscr{M})$ denote the smallest $\mathscr{A}$ invariant subspace containing $\mathscr{M}$. Hence $(\mathscr{A},\mathscr{M})$ is the closed subspace spanned by the vectors Ax where $A\in\mathscr{A}$ and $x\in\mathscr{M}$.

Lemma 14. 1) *If $\mathscr{M}$ is an $\mathscr{A}$ invariant subspace then* $(\mathscr{A},\mathscr{M})=\mathscr{M}$.

2) *For any $\mathscr{A}$ and $\mathscr{M}$ we have* $(\mathscr{A},(\mathscr{A},\mathscr{M}))=(\mathscr{A},\mathscr{M})$.

3) *If* $\mathscr{A}_1\subseteq\mathscr{A}_2$ *and* $\mathscr{M}_1\subseteq\mathscr{M}_2$ *then* $(\mathscr{A}_1,\mathscr{M}_1)\subseteq(\mathscr{A}_2,\mathscr{M}_2)$.

Proof. Parts 1) and 3) are obvious from the definition of $(\mathscr{A},\mathscr{M})$ and 2) follows from 1).

Proposition 15. *If $c\mathscr{A}$ is the center of the v. Neumann algebra $\mathscr{A}$ then* $(\mathscr{A}',(\mathscr{A},\mathscr{M}))=((c\mathscr{A})',\mathscr{M})$.

Proof. We have $\mathscr{A}\subseteq\mathscr{Z}'$ and $\mathscr{A}'\subseteq\mathscr{Z}'$ where for the sake of simplicity $\mathscr{Z}=c\mathscr{A}$. Therefore by Lemma 14 $(\mathscr{A},\mathscr{M})\subseteq(\mathscr{Z}',\mathscr{M})$ and

$$(\mathscr{A}',(\mathscr{A},\mathscr{M}))\subseteq(\mathscr{A}',(\mathscr{Z}',\mathscr{M}))\subseteq(\mathscr{Z}',(\mathscr{Z}',\mathscr{M}))=(\mathscr{Z}',\mathscr{M}).$$

In order to prove the opposite inclusion it is sufficient to shows that $(\mathscr{A}',(\mathscr{A},\mathscr{M}))$ is a $\mathscr{Z}'$-invariant space containing $\mathscr{M}$. By $I\in\mathscr{A},\mathscr{A}'$ we have $(\mathscr{A}',(\mathscr{A},\mathscr{M}))\supseteq\mathscr{M}$ and by Proposition III.5.13 invariance with respect to $\mathscr{Z}'$ means invariance under both $\mathscr{A}$ and $\mathscr{A}'$. The $\mathscr{A}'$-invariance of $(\mathscr{A}',(\mathscr{A},\mathscr{M}))$ is obvious from the definition. The sums of the vectors $Tx\,(T\in\mathscr{A}',x\in(\mathscr{A},\mathscr{M}))$ form a dense subset in $(\mathscr{A}',(\mathscr{A},\mathscr{M}))$ and $ATx=T(Ax)$ is in $(\mathscr{A}',(\mathscr{A},\mathscr{M}))$. Hence $A(\mathscr{A}',(\mathscr{A},\mathscr{M}))$ is in $(\mathscr{A}',(\mathscr{A},\mathscr{M}))$ and so $(\mathscr{A}',(\mathscr{A},\mathscr{M}))$ is invariant under $\mathscr{A}$.

Again let $\mathscr{A}$ be a v. Neumann algebra with underlying Hilbert space $\mathscr{H}$, let A be in $\mathscr{A}$, let $\mathscr{K}=\overline{A\mathscr{H}}$ and let E be the projection $\mathscr{H}\to\mathscr{K}$. Then there is a smallest subspace $\mathscr{L}$ in $\mathscr{H}$ such that $\mathscr{K}\subseteq\mathscr{L}$ and $\mathscr{L}$ is invariant under both $\mathscr{A}$ and $\mathscr{A}'$. It is the intersection of the ranges of those central projections which majorize E. Hence the projection P associated with $\mathscr{L}$ is the smallest central majorant of E. It is called the *central support* of A. If $\mathscr{A}$ is factorial then the only central projections are 0 and I. Hence in that case the central support of any non-zero operator A of $\mathscr{A}$ is I.

Lemma 16. *If* $\mathscr{A}$ *is a v. Neumann algebra and E is a projection in* $\mathscr{A}$ *then the projection associated with* $(\mathscr{A}, E\mathscr{H})$ *is the central support of E.*

Proof. Since $E \in \mathscr{A}$ the associated subspace $\mathscr{K} = E\mathscr{H}$ is $\mathscr{A}'$ invariant and so $(\mathscr{A}', \mathscr{K}) = \mathscr{K}$. By Proposition 15

$$(\mathscr{A}, \mathscr{K}) = (\mathscr{A}, (\mathscr{A}', \mathscr{K})) = (\mathscr{Z}', \mathscr{K})$$

where $\mathscr{Z}$ is the common center of $\mathscr{A}$ and $\mathscr{A}'$. Hence by Proposition III.5.13 $(\mathscr{A}, E\mathscr{H})$ is the smallest subspace containing $\mathscr{K}$ such that the associated projection is in $\mathscr{Z}'' = \mathscr{Z}$.

For a v. Neumann algebra $\mathscr{A}$ and a projection E in $\mathscr{A}'$ let $\mathscr{A}_E = \{A | \mathscr{K} : A \in \mathscr{A}\}$ where $\mathscr{K} = E\mathscr{H}$. This is called the *algebra induced by* $\mathscr{A}$ *on the invariant subspace* $\mathscr{K}$. If $E \in \mathscr{A} \cap \mathscr{A}'$ then $\mathscr{A}_E$ had been defined already as the reduced algebra $\{A | \mathscr{K} : A \in E\mathscr{A}E\}$ and the old and the new definitions coincide: Indeed by $E \in \mathscr{A}'$ we have $A\mathscr{K} \subseteq \mathscr{K}$ for every $A \in \mathscr{A}$, so A is in $E\mathscr{A}E$ if and only if $A\mathscr{K}^{\perp} = 0$. Hence for any A in $\mathscr{A}$ we have $AE \in E\mathscr{A}E$ and $A | \mathscr{K} = AE | \mathscr{K}$.

Proposition 17. *The induced algebra* $\mathscr{A}_E$ *is a v. Neumann algebra with underlying Hilbert space* $\mathscr{K} = E\mathscr{H}$.

Proof. By $(A | \mathscr{K})^* = A^* | \mathscr{K}$ we know that $\mathscr{A}_E$ is self adjoint and it clearly contains the identity of $\mathscr{L}(\mathscr{K})$. To see that $\mathscr{A}_E$ is weakly closed it is sufficient to replace $\mathscr{A}$ by $\mathscr{A}'$ in the next proposition and refer to Proposition III.5.6.

Proposition 18. *If E is a projection in the v. Neumann algebra* $\mathscr{A}$ *then* $(\mathscr{A}_E)' = (\mathscr{A}')_E$.

Proof. Let $T \in \mathscr{A}'$, $x \in \mathscr{K} = E\mathscr{H}$ and let A be in $E\mathscr{A}E$. Then by $T\mathscr{K} \subseteq \mathscr{K}$ we have

$$(T|\mathscr{K})(A|\mathscr{K})x = (T|\mathscr{K})Ax = (TE)Ax = T(EA)x = T(AE)x$$
$$= (AE)Tx = (A|\mathscr{K})(T|\mathscr{K})x.$$

This proves that $(\mathscr{A}')_E \subseteq (\mathscr{A}_E)'$.

In order to prove the opposite inclusion let a unitary operator T in $(\mathscr{A}_E)'$ be given. We are going to find another operator S in $\mathscr{A}'$ such that $S | \mathscr{K} = T$. As earlier we let $(\mathscr{A}, \mathscr{K})$ denote the smallest $\mathscr{A}$ invariant subspace of $\mathscr{H}$ containing $\mathscr{K}$ and let P be the associated projection $\mathscr{H} \to (\mathscr{A}, \mathscr{K})$. By Lemma 16 P is the central support of E and so $P \in \mathscr{A} \cap \mathscr{A}'$.

First we define S on a dense subset of $(\mathscr{A}, \mathscr{K})$ by letting

$$S(\sum A_i x_i) = \sum A_i T x_i \tag{1}$$

for every $A_1, \ldots, A_n \in \mathscr{A}$ and $x_1, \ldots, x_n \in \mathscr{K}$. The fact that S is well defined follows from

$$\|S(\sum A_i x_i)\| = \|\sum A_i x_i\|$$

which can be verified as follows:

By $T \in (\mathscr{A}_E)'$ and $x_i \in \mathscr{K} = E\mathscr{H}$ we have $A_i T x_i = A_i E T x_i$ and so

$$\|\sum A_i T x_i\|^2 = \sum (A_i E T x_i, A_j E T x_j) = \sum ((E A_j^* A_i E) T x_i, T x_j).$$

Since T is a unitary operator in $(\mathscr{A}_E)'$ we obtain by (1)

$$\|S(\sum A_i x_i)\|^2 = \sum (E A_j^* A_i E x_i, x_j) = \sum (A_i x_i, A_j x_j) = \|\sum A_i x_i\|^2.$$

Therefore S is well defined and it has a unique unitary extension $S: (\mathscr{A}, \mathscr{K}) \to (\mathscr{A}, \mathscr{K})$. Finally S can be extended to a continuous linear operator $S: \mathscr{H} \to \mathscr{H}$ by letting $S = PSP$ on $\mathscr{H}$.

For any A in $\mathscr{A}$ we have

$$SA(\sum A_i x_i) = S(\sum A A_i x_i) = \sum A A_i T x_i = AS(\sum A_i x_i)$$

and so we obtain $SA = AS$ on $(\mathscr{A}, \mathscr{K})$. Since P is in $\mathscr{A} \cap \mathscr{A}'$ considering S as an operator on $\mathscr{H}$ we have

$$SA = (PSP)A = P(SA)P = P(AS)P = A(PSP) = AS.$$

Therefore S is in $\mathscr{A}'$. Using (1) we obtain $Sx = Tx$ for every x in $\mathscr{K}$ and so $S|\mathscr{K} = T$. We proved that $(\mathscr{A}')_E$ contains all the unitary elements of $(\mathscr{A}_E)'$. By Proposition III.5.6 $(\mathscr{A}_E)'$ is a v. Neumann algebra so by Proposition III.6.14 we obtain $(\mathscr{A}_E)' \subseteq (\mathscr{A}')_E$.

Proposition 19. *Let E be a projection in $\mathscr{A}'$, let P be its central support and let $\mathscr{K} = E\mathscr{H}$ and $\mathscr{L} = P\mathscr{H}$. Then $A|\mathscr{L} \mapsto A|\mathscr{K}$ is an isomorphism of $\mathscr{A}_P$ onto $\mathscr{A}_E$.*

Proof. Being a restriction map $A|\mathscr{L} \to A|\mathscr{K}$ is a homomorphism and by the definition of $\mathscr{A}_E$ it is clearly surjective. By Lemma 16 we have $(\mathscr{A}', \mathscr{K}) = \mathscr{L}$. If $A \in \mathscr{A}$ then $\ker A$ is invariant under $\mathscr{A}'$, so if $\ker A \supseteq \mathscr{K}$ then $\ker A \supseteq (\mathscr{A}', \mathscr{K}) = \mathscr{L}$. Hence if $A|\mathscr{K}$ is zero then so is $A|\mathscr{L}$.

Corollary. *Let E be a projection in $\mathscr{A}'$ and let P be its central support. Then $P = I$ is a necessary and sufficient condition that $A \mapsto A|\mathscr{K}$ be an isomorphism of $\mathscr{A}$ onto $\mathscr{A}_E$.*

Indeed the sufficiency follows from $\mathscr{A}_I = \mathscr{A}$ and the necessity from $\ker(I-P) \supseteq \mathscr{K} = E\mathscr{H}$.

Proposition 20. *If $\mathscr{A}$ is a v. Neumann algebra and if E is a projection in $\mathscr{A}$ or in $\mathscr{A}'$ then $(c\mathscr{A})_E = c(\mathscr{A}_E)$. If E is a projection in $\mathscr{A}$ and if S is a projection in $(c\mathscr{A})_E$ then there is a projection T in $c\mathscr{A}$ such that $S = T|\mathscr{K}$ where $\mathscr{K}$ is the range of E.*

Proof. Here and in the second and third paragraphs of the proof let $E \in \mathscr{A}$. If $S \in (c\mathscr{A})_E$ then $S = T|\mathscr{K}$ where $T \in c\mathscr{A}$ and $\mathscr{K} = E\mathscr{H}$. Thus $S \in \mathscr{A}_E$ by $c\mathscr{A} \subseteq \mathscr{A}$ and we also have $TA = AT$ for every A in $\mathscr{A}$ because $T \in c\mathscr{A} \subseteq \mathscr{A}'$. Hence if $A \in E\mathscr{A}E$ then $A|\mathscr{K}$ and $T|\mathscr{K}$ commute and so $S = T|\mathscr{K}$ is in $(\mathscr{A}_E)'$. We proved that $S \in \mathscr{A}_E \cap (\mathscr{A}_E)' = c(\mathscr{A}_E)$.

Now we suppose that $S \in c(\mathscr{A}_E)$ and prove that $S \in (c\mathscr{A})_E$: We have $S \in (\mathscr{A}')_E$ and so $S = T|\mathscr{K}$ where $T \in \mathscr{A}'$. Let P denote the central support of E and let $\mathscr{L} = P\mathscr{H}$. By $T\mathscr{K} \subseteq \mathscr{K}$ and $T \in \mathscr{A}'$ we have $PT|\mathscr{K} = T|\mathscr{K}$ and $PT \in \mathscr{A}'$. Hence replacing PT by T the new T satisfies $S = T|\mathscr{K}$ and $T \in \mathscr{A}'$ and also $PT = T$. Applying Proposition 18 we see that $S \in c(\mathscr{A}_E) = c((\mathscr{A}_E)') = c((\mathscr{A}')_E)$. By Proposition 19 $R|\mathscr{L} \to R|\mathscr{K}$ $(R \in \mathscr{A}')$ is an isomorphism of $c((\mathscr{A}')_P)$ onto $c((\mathscr{A}')_E)$. Therefore by $S \in c((\mathscr{A}')_E)$ and Proposition 18 $T|\mathscr{L} \in c((\mathscr{A}')P) = (\mathscr{A}')P \cap ((\mathscr{A}')P)' = (\mathscr{A}')_P \cap (\mathscr{A}_P)'' = (\mathscr{A}')_P \cap \mathscr{A}_P$. This shows that $T|\mathscr{L} = A|\mathscr{L}$ where $A \in \mathscr{A}$. Since we have $T = PT = TP = AP \in \mathscr{A}$ and $T \in \mathscr{A}'$ we see that $T \in c\mathscr{A}$ and so $S = T|\mathscr{K} \in (c\mathscr{A})_E$.

Let us now suppose that $S^2 = S$ and $S^* = S$. Then by $S = T|\mathscr{K}$ we have $T^2 - T|\mathscr{K} = 0$ and $T^* - T|\mathscr{K} = 0$. Therefore $T^2 - T|\mathscr{L}$ and $T^* - T|\mathscr{L}$ belong to the kernel of the isomorphism $R|\mathscr{L} \mapsto R|\mathscr{K}$ $(R \in \mathscr{A}')$ and so $T^2P = TP$ and $T^*P = TP$. Hence by $T = TP = PT$ we have $T^2 = T^2P = TP = T$ and $T^* = T^*P = TP = T$.

Now we suppose that E is a projection in $\mathscr{A}'$. Then by Proposition 18 $((\mathscr{A}')_E)' = (\mathscr{A}'')_E = \mathscr{A}_E$. Hence applying that part of the proposition which has been already proved we have $(c\mathscr{A})_E = (c\mathscr{A}')_E = c((\mathscr{A}')_E) = c(\mathscr{A}_E)$.

Proposition 21. *Two necessary and sufficient conditions that a factorial v. Neumann algebra be of type I are as follows:*

1) *contains a minimal projection.*
2) *has an irreducible component.*

Proof. We already proved that either 1) or 2) implies that $\mathscr{A}$ is of type I. (See Lemma 6 and the paragraph preceding it.) Let $\mathscr{A}$ be of type I and let E be a non-zero abelian projection in $\mathscr{A}$. By Proposition 20 $\mathscr{A}_E$ is factorial and by the commutativity of $\mathscr{A}_E$ we have $c(\mathscr{A}_E)=\mathscr{A}_E\cap(\mathscr{A}_E)'=\mathscr{A}_E$. Thus $\mathscr{A}_E=\mathbb{C}I$ where I is the identity of $\mathscr{L}(\mathscr{K})$ and $\mathscr{K}=E\mathscr{H}$. By Proposition 18 we have $(\mathscr{A}')_E=\mathscr{L}(\mathscr{K})$ and so $\mathscr{K}$ is an irreducible component of $\mathscr{A}'$ by Theorem 2.5. Thus E is a minimal projection in $\mathscr{A}$ and 1) holds. By Proposition 7.25 $\mathscr{A}$ also has an irreducible component so 2) holds too.

Proposition 22. *A v. Neumann algebra $\mathscr{A}$ is of type I if and only if it contains an abelian projection whose central support is the identity.*

Proof. First suppose that $\mathscr{A}$ contains an abelian projection E with central support I. Let P be a central projection. Then by Lemma 1 $F=EP$ is an abelian projection majorized by P. If $F=0$ then $E=E(I-P)$ and so $I-P$ is a central projection majorizing E. Hence $I-P=I$ and $P=0$. For every non-zero central projection P we found a non-zero abelian projection F majorized by it. Hence $\mathscr{A}$ is of type I.

Now suppose that $\mathscr{A}$ is of type I. Consider families P_i $(i\in\mathscr{I})$ of non-zero pairwise orthogonal central projections P_i satisfying the following: For each index i there is an abelian projection E_i in $\mathscr{A}$ such that the central support of E_i is P_i. Since $\mathscr{A}$ is of type I it has a non-zero abelian projection E and so its central support P gives an example of a family consisting of a single projection and satisfying the conditions. Therefore by Zorn's lemma there exists a family P_i $(i\in\mathscr{I})$ which satisfies the conditions and is maximal with respect to inclusion. By the orthogonality $E=\sum E_i$ is a non-zero abelian projection in $\mathscr{A}$.

We will complete the proof by showing that the central support of E is I. Let $P=\sum P_i$ so that P is a central projection. If $I-P$ were not zero then $\mathscr{A}$ being of type I we could find a non-zero abelian projection majorized by $I-P$ and its central support would be majorized by $I-P$. This central support could then be joined to the family P_i $(i\in\mathscr{I})$ in contradiction to its maximality. Therefore we have $P=\sum P_i=I$. Now let Q be the central support of E. Since E_i has central support P_i and QP_i is a central projection majorizing E_i we have $QP_i\geqslant P_i$. Hence $Q=\sum QP_i\geqslant\sum P_i=P=I$ and so $Q=I$.

Proposition 23. *The commutant of a v. Neumann algebra of type I is isomorphic to a v. Neumann algebra whose commutant is abelian.*

Proof. Let E be an abelian projection in $\mathscr{A}$ with central support I. The existence of such a projection follows from the last proposition. Then by Proposition 19 $\mathscr{A}'=(\mathscr{A}')_I$ is isomorphic to the v. Neumann algebra $\mathscr{B}=(\mathscr{A}')_E$. By Proposition 18 $\mathscr{B}'=\mathscr{A}_E$ which is abelian.

Lemma 24. *Let E and F be projections in $\mathscr{A}$ with $E \geqslant F$, let $\mathscr{K} = E\mathscr{H}$ and suppose that the central support of $F|\mathscr{K}$ in $\mathscr{A}_E$ is $E|\mathscr{K}$. Then E and F have the same central support in $\mathscr{A}$.*

Proof. Let P and Q denote the central supports of E and F in $\mathscr{A}$. By $E \geqslant F$ we have $P \geqslant Q$ and so it is sufficient to prove that $P \leqslant Q$. Since Q is a central projection in $\mathscr{A}$ and $Q \geqslant F$ using Proposition 20 we see that $Q|\mathscr{K}$ is a central projection in $\mathscr{A}_E$ majorizing $F|\mathscr{K}$. Hence by hypothesis $Q|\mathscr{K} \geqslant E|\mathscr{K}$ and $Q \geqslant E$.

Lemma 25. *If E is a projection in $\mathscr{A}$ or in $\mathscr{A}'$ such that $\mathscr{A}_E$ is of type I and P is the central support of E then $\mathscr{A}_P$ is also of type I.*

Proof. If E is $\mathscr{A}'$ then the lemma follows from Propositions 3 and 19. Now let E belong to $\mathscr{A}$, let $\mathscr{K} = E\mathscr{H}$ and $\mathscr{L} = P\mathscr{H}$. Since $\mathscr{A}_E$ is of type I there is a projection F in $\mathscr{A}$ such that $F|\mathscr{K}$ is an abelian projection in $\mathscr{A}_E$ and the central support of $F|\mathscr{K}$ in $\mathscr{A}_E$ is $E|\mathscr{K}$. Hence by the last lemma the central support of F in $\mathscr{A}$ is P. Now let $Q|\mathscr{L}$ be the central support of $F|\mathscr{L}$ in $\mathscr{A}_P$. Then by Proposition 20 Q is a central element of $\mathscr{A}$ and $F = FP \leqslant QP$. Hence $QP \geqslant P$ and so $Q|\mathscr{L} = P|\mathscr{L}$. By Lemma 8 $F|\mathscr{L}$ is an abelian projection in $\mathscr{A}_P$.

Lemma 26. *Let $\mathscr{A}$ be a v. Neumann algebra. Let E be a projection in $\mathscr{A}$ or $\mathscr{A}'$ such that $\mathscr{A}_E$ is commutative and let P be its central support. Then $\mathscr{A}_P$ is of type I.*

Proof. Since $\mathscr{A}_E$ is commutative it is of type I by Proposition 4. Therefore $\mathscr{A}_P$ is of type I by Proposition 25.

Theorem 27. *If $\mathscr{A}$ is a v. Neumann algebra then there is a unique central projection P in $\mathscr{A}$ such that $\mathscr{A}_P$ is of type I and if $\mathscr{A}_S$ is of type I for some S in $\mathscr{A}$ or $\mathscr{A}'$ then $P \geqslant S$.*

Proof. If $\mathscr{A}_S$ is of type I only for $S=0$ then we let $P=0$. If we exclude this special case then by Zorn's lemma and Lemma 25 there exists a maximal family P_i $(i \in \mathscr{I})$ of pairwise orthogonal, non-zero, central projections such that $\mathscr{A}_{P_i}$ is of type I for every $i \in \mathscr{I}$. Let $\mathscr{K}_i = P_i\mathscr{H}$ and $\mathscr{K} = \sum \mathscr{K}_i$. Hence $\mathscr{K} = P\mathscr{H}$ where $P = \sum P_i$ is a central projection. Since P and P_i $(i \in \mathscr{I})$ are central $\mathscr{A}_P$ is isomorphic to $\sum \mathscr{A}_{P_i}$ because $P\mathscr{A}P$ is isomorphic to $\sum P_i \mathscr{A} P_i$ under the *surjective* map $PAP \mapsto \sum P_i A P_i$. Hence $\mathscr{A}_P$ is of type I by Propositions 3 and 7.

Now let S denote a projection in $\mathscr{A}$ or in $\mathscr{A}'$ such that $\mathscr{A}_S$ is of type I and let $\mathscr{L} = S\mathscr{H}$. We are going to prove that $\mathscr{K} \supseteq \mathscr{L}$ i.e. $P \geqslant S$. By Propositions 25 we may suppose that S is in the common center of $\mathscr{A}$

and $\mathscr{A}'$. By Proposition 22 there is a projection T in $\mathscr{A}$ such that $T|\mathscr{L}$ is abelian in $\mathscr{A}_S$ and the central support of $T|\mathscr{L}$ in $\mathscr{A}_S$ is $S|\mathscr{L}$. By $P\in\mathscr{A}'$ it follows that $(I-P)T$ is a projection and $(I-P)T|\mathscr{M}$ is in $\mathscr{A}_{(I-P)S}$ where $\mathscr{M}=(I-P)S\mathscr{H}=(I-P)\mathscr{L}$. Let U in $\mathscr{A}$ be such that $U|\mathscr{M}$ is the central support of $(I-P)T|\mathscr{M}$ in $\mathscr{A}_{(I-P)S}$. By Proposition 11 U is a central element in $\mathscr{A}$. Let

$$V=U(I-P)S+(I-(I-P)S).$$

Then V is a central projection, $V|\mathscr{M}=U|\mathscr{M}$ and $V|\mathscr{M}^{\perp}$ is the identity. Hence using $(I-P)T|\mathscr{M}\leqslant U|\mathscr{M}$ we obtain $(I-P)T\leqslant V$. Since $T\leqslant S$ we have $PT\leqslant PS$ and so by the orthogonality

$$T=(I-P)T+PT\leqslant V+PS.$$

Therefore the central support of $T|\mathscr{L}$ being $S|\mathscr{L}$ we obtain $S\leqslant V+PS$ i.e. $(I-P)S\leqslant V$. We proved that $(I-P)S|\mathscr{M}=V|\mathscr{M}=U|\mathscr{M}$ and so the central support of $(I-P)T|\mathscr{M}$ in $\mathscr{A}_{(I-P)S}$ is the identity element. Therefore by Lemma 1 and Proposition 22 $\mathscr{A}_{(I-P)S}$ is a type I algebra. Since $(I-P)S\perp P_i$ for every $i\in\mathscr{I}$ and the family P_i $(i\in\mathscr{I})$ is maximal it follows that $(I-P)S=0$ i.e. $P\geqslant S$.

The uniqueness of P can now be seen from the facts which have been proved already: Namely we see that among the projections S in $\mathscr{A}$ such that $\mathscr{A}_S$ is a type I algebra there exists a largest one P which belongs to the center of $\mathscr{A}$. This projection P is the one whose existence is stated in the theorem.

Definition 28. *A representation $\rho: S\to\mathscr{L}(\mathscr{H})$ is called a representation of type I if the commuting algebra (ρ,ρ) is of type I.*

Every finite dimensional representation is of type I. All trivial representations are of type I. If ρ is irreducible then it is of type I and so by Proposition 7 any direct sum of irreducible representations is of type I. According to Theorem 5.4 there is a largest subspace $\mathscr{H}_{dI}$ of $\mathscr{H}$ on which $\rho: S\to\mathscr{L}(\mathscr{H})$ is a direct sum of irreducible components. Hence $\rho|\mathscr{H}_{dI}$ is a type I representation. It is called the *type I discrete part* of ρ. If $\mathscr{A}=\mathscr{A}_\rho$ and P in $c(\rho,\rho)$ is the projection whose existence is stated in Theorem 27 then $\mathscr{H}_{dI}\subseteq\mathscr{H}_I$ where $\mathscr{H}_I=P\mathscr{H}$. If equality takes place we say that the type I part of ρ is discrete. In general the subrepresentation $\rho|\mathscr{H}_I$ is called the type I part of ρ.

Proposition 29. *Let $\rho: S\to\mathscr{L}(\mathscr{H})$ be a representation, let $\mathscr{K}_1$ and $\mathscr{K}_2$ be invariant subspaces of ρ and let E_1 and E_2 be the corresponding projections. Then $T:\mathscr{K}_1\to\mathscr{K}_2$ is in $(\rho|\mathscr{K}_1,\rho|\mathscr{K}_2)$ if and only if TE_1 is in $\mathscr{A}'_\rho$.*

Proof. First suppose that T is in $(\rho|\mathscr{K}_1, \rho|\mathscr{K}_2)$. Then for any x in $\mathscr{K}_1$ we have

$$(TE_1)\rho x = T(E_1\rho)x = T(\rho E_1)x = T\rho x = \rho T x = \rho(TE_1)x\,.$$

Hence TE_1 belongs to (ρ, ρ). Conversely if $T: \mathscr{K}_1 \to \mathscr{K}_2$ and TE_1 is in (ρ, ρ) then for any x in $\mathscr{K}_1$ we have

$$T(\rho|\mathscr{K}_1)x = TE_1\rho x = \rho TE_1 x = (\rho|\mathscr{K}_2)Tx\,.$$

Thus T is an element of $(\rho|\mathscr{K}_1, \rho|\mathscr{K}_2)$.

Proposition 30. *Let $\mathscr{K}$ be an invariant subspace of the representation $\rho: S \to \mathscr{L}(\mathscr{H})$. Then $\mathscr{A}_{\rho|\mathscr{K}} = (\mathscr{A}_\rho)_E$ where E denotes the projection associated with $\mathscr{K}$.*

Proof. By setting $\mathscr{K}_1 = \mathscr{K}_2 = \mathscr{K}$ and $E_1 = E_2 = E$ in Proposition 29 we obtain:

$$T \in (\rho|\mathscr{K}, \rho|\mathscr{K}) = (\mathscr{A}_{\rho|\mathscr{K}})' \quad \text{if and only if} \quad TE \in \mathscr{A}_\rho'.$$

If a projection $P: \mathscr{K} \to \mathscr{K}$ is such that PE belongs to $\mathscr{A}_\rho'$ then $PE|\mathscr{K} = P \in (\mathscr{A}_\rho')_E$. Conversely if P is in $(\mathscr{A}_\rho')_E$ then $P = Q|\mathscr{K}$ for some Q in $\mathscr{A}_\rho'$ and $PE = QE \in \mathscr{A}_\rho'$. Therefore we see that a projection $P: \mathscr{K} \to \mathscr{K}$ belongs to $(\mathscr{A}_{\rho|\mathscr{K}})'$ if and only if it is in $(\mathscr{A}_\rho')_E$. By Proposition III.6.15 we obtain $(\mathscr{A}_{\rho|\mathscr{K}})' = (\mathscr{A}_\rho')_E$. Finally using Proposition 18 and the corollary of Theorem III.6.10 we get $\mathscr{A}_{\rho|\mathscr{K}} = (\mathscr{A}_\rho)_E$.

Corollary. *Let $\mathscr{A}$ be a factorial v. Neumann algebra with underlying Hilbert space $\mathscr{H}$, let $\mathscr{K}$ be an invariant subspace and let E be the associated projection. Then $\mathscr{A}_E$ is a factorial v. Neumann algebra.*

For let $\rho: \mathscr{A} \to \mathscr{L}(\mathscr{H})$ be the representation of the abstract involutive algebra $\mathscr{A}$ which associates with the element A of $\mathscr{A}$ the operator $A: \mathscr{H} \to \mathscr{H}$. Then ρ is primary and so $\rho|\mathscr{K}$ is a primary representation. By Proposition 30 $\mathscr{A}_{\rho|\mathscr{K}} = \mathscr{A}_\rho|\mathscr{K} = (\mathscr{A}_\rho)_E = \mathscr{A}_E$ and so by Lemma 7.2 and Proposition 4.11 $\mathscr{A}_E$ is a factor. The same result can be obtained by using Proposition 20.

Proposition 31. *If $\mathscr{A}$ is a factor then so are all the v. Neumann algebras obtained from it by induction or reduction.*

Proof. The case of induced algebras was discussed above. In the case of reduced algebras apply Proposition 20.

Proposition 32. *If $\mathscr{A}$ is a type I v. Neumann algebra and if E is a projection in $\mathscr{A}'$ then $\mathscr{A}_E$ is isomorphic to $\mathscr{A}$.*

Proof. Let P denote the central support of E and let $\mathscr{L}=P\mathscr{H}$. Let F be an abelian projection in $\mathscr{A}$ with central support the identity. Then $F|\mathscr{L}$ is an abelian projection in $\mathscr{A}_P$. Let Q be a projection in $\mathscr{A}$ be such that $Q|\mathscr{L}$ is the central support of $F|\mathscr{L}$. Then QP is a projection in $\mathscr{A}$ and $QP|\mathscr{L}$ being a central element of $\mathscr{A}_P$ by Proposition 20 QP is central. Since $QP\geqslant F$ and the central support of F is I we obtain $PQ=QP=I$. Hence $P=I$ and so by Proposition 19 $\mathscr{A}_E\approx\mathscr{A}_P=\mathscr{A}_I=\mathscr{A}$.

Corollary. *If $\mathscr{A}$ is of type I and E is a projection in $\mathscr{A}'$ then $\mathscr{A}_E$ is of type I.*

Proposition 33. *If $\mathscr{A}$ is a type I v. Neumann algebra and if E is a projection in $\mathscr{A}$ then $\mathscr{A}_E$ is of type I.*

Proof. By Proposition 20 every non-zero central projection of $\mathscr{A}_E$ is of the form $P|\mathscr{K}$ where P is a projection in the center of $\mathscr{A}$ and $P\leqslant E$. We can choose a non-zero abelian projection F in $\mathscr{A}$ such that $F\leqslant P$. Since $P\leqslant E$ we have $F\leqslant E$ and so $F|\mathscr{K}$ is a non-zero abelian projection in $\mathscr{A}_E$ and $F|\mathscr{K}\leqslant P|\mathscr{K}$.

Proposition 34. *Every subrepresentation of a type I representation is of type I.*

Proof. Let $\mathscr{K}$ be an invariant subspace of $\rho: S\to\mathscr{L}(\mathscr{H})$ and let E be the associated projection. By Propositions 18 and 30 we have

$$(\rho|\mathscr{K},\rho|\mathscr{K})=(\mathscr{A}_{\rho|\mathscr{K}})'=((\mathscr{A}_\rho)_E)'=((\rho,\rho)'_E)'=(\rho,\rho)_E.$$

Since E is in (ρ,ρ) by Proposition 33 $(\rho|\mathscr{K},\rho|\mathscr{K})$ is of type I.

Proposition 35. *If $\rho_k: S\to\mathscr{L}(\mathscr{H}_k)$ $(k=1,2)$ are quasiequivalent representations and ρ_1 is of type I then so is ρ_2.*

Corollary. *It is meaningful to speak about type I quasiequivalence classes of representations.*

Proof. Suppose that ρ_2 is not of type I and $\rho_2\preccurlyeq\rho_1$. Let P be the projection in the center of (ρ_2,ρ_2) whose existence is stated in Theorem 27 and let $\mathscr{H}_{2I}=P\mathscr{H}$. Then $(\mathscr{H}_{2I})^\perp$ has a non-zero ρ_2 invariant subspace $\mathscr{L}_2$ such that $\rho_2|\mathscr{L}_2\approx\rho_1|\mathscr{L}_1$ for some $\mathscr{L}_1\subseteq\mathscr{H}_1$. Let S be the projection of $\mathscr{H}$ onto $\mathscr{L}_2$ so that $S\in(\rho_2,\rho_2)$. Then by the same reasoning as in the proof of Proposition 34 we have $(\rho_2|\mathscr{L}_2,\rho_2|\mathscr{L}_2)=(\rho_2,\rho_2)_S$. Hence by Theorem 27 $(\rho_2|\mathscr{L}_2,\rho_2|\mathscr{L}_2)$ is not of type I and so ρ_1 is not a type I representation by Proposition 3 and 34.

Theorem 36. *If $\rho_i: S\to\mathscr{L}(\mathscr{H}_i)$ $(i\in\mathscr{I})$ are Mackey type representations with a common multiplier and if every ρ_i is of type I then $\rho=\sum\rho_i$ is a type I representation.*

Proof. Let $\mathscr{A}_i = \mathscr{A}_{\rho_i}$ be the v. Neumann algebra associated with ρ_i and similarly let $\mathscr{A} = \mathscr{A}_\rho$. Let $\mathscr{H} = \sum \mathscr{H}_i$ and let $E_i: \mathscr{H} \to \mathscr{H}_i$ where the same symbol $\mathscr{H}_i$ is used to denote the canonical image of the originally given Hilbert space $\mathscr{H}_i$ in $\mathscr{H}$. The image $\mathscr{H}_i$ is an invariant subspace of ρ because if $(x_i) \in \mathscr{H}$ and $x_j = 0$ for all $j \neq i$ then $\rho((x_i)) = (\rho_i(x_i))$ with $\rho_j(x_j) = 0$ for all $j \neq i$. Therefore E_i is a projection in (ρ, ρ). Using the same reasoning as in the proof of Proposition 34 we see that

$$(\rho \,|\, \mathscr{H}_i, \rho \,|\, \mathscr{H}_i) = (\rho, \rho)_{E_i}.$$

Since $\rho_i \approx \rho \,|\, \mathscr{H}_i$ the canonical injection of $\mathscr{H}_i$ into $\mathscr{H}$ induces a spatial isomorphism of $\mathscr{A}_i$ onto $\mathscr{A}_{E_i}$. Therefore by Proposition 7.22 (ρ_i, ρ_i) and $(\rho, \rho)_{E_i}$ are spatially isomorphic under the same map.

By hypothesis (ρ_i, ρ_i) is of type I and so by Proposition 3 $(\rho, \rho)_{E_i}$ is a type I algebra for every i in $\mathscr{I}$. By Theorem 27 there is a largest projection P in $c(\rho, \rho)$ such that $(\rho, \rho)_P$ is of type I. Since $(\rho, \rho)_{E_i}$ if of type I by the same theorem we have $E_i \leqslant P$ for every i in $\mathscr{I}$. Therefore P is the identity operator on $\mathscr{H}$ and so $(\rho, \rho) = (\rho, \rho)_P$. We proved that (ρ, ρ) is of type I.

9. Type II and III v. Neumann Algebras

The first part of this section deals with finite projections and finite v. Neumann algebras. These concepts are needed in the theory of type II and type III v. Neumann algebras. The middle contains the basic definitions and theorems on these types of algebras. These concepts were first introduced in the case of factorial algebras over separable Hilbert spaces by Murray and v. Neumann. The abstract global theory covering non-factorial types and the decomposition of an arbitrary v. Neumann algebra into such parts is of more recent origin and it is due to Kaplansky. In the last part of the section we discuss some results which are valid only in the case of factorial algebras and present some examples.

In Definition 7.13 we introduced the concept of a finite v. Neumann algebra and in Definition 2.13 we told what is meant by the equivalence of two projections with respect to a v. Neumann algebra $\mathscr{A}$. Now we define a *finite projection*:

Definition 1. *A projection E belonging to the v. Neumann algebra $\mathscr{A}$ is called finite if the only projection in $\mathscr{A}$ which is majorized by and equivalent to it is E itself.*

In order to compare the present definition with Definition 7.13 we prove the following:

Lemma 2. *If E and F are projections in $\mathscr{A}$, E is finite and $F \leqslant E$ then F is finite.*

Proof. We suppose that F is infinite, namely $G<F$ and $G \sim F$. We shall prove that E is also infinite by exhibiting a projection $P<E$ such that $P \sim E$. Let $\mathscr{H}$ be the underlying Hilbert space of $\mathscr{A}$ and let $\mathscr{K}=E\mathscr{H}$, $\mathscr{L}=F\mathscr{H}$ and $\mathscr{M}=G\mathscr{H}$. We define $\mathscr{N}=\mathscr{M}+(\mathscr{K} \cap \mathscr{L}^{\perp})$ and let P be the projection $\mathscr{H} \rightarrow \mathscr{N}$. Since $\mathscr{K}$, $\mathscr{L}^{\perp}$ and $\mathscr{M}$ are invariant under $\mathscr{A}'$ so is $\mathscr{N}$ and consequently $P \in \mathscr{A}$. By $G<F$ we have $\mathscr{M} \subset \mathscr{L}$ and so $\mathscr{N} \subset \mathscr{L}+(\mathscr{K} \cap \mathscr{L}^{\perp})=\mathscr{K}$ and $P<E$. In order to prove that $P \sim E$ let U be a partial isometry in $\mathscr{A}$ with initial domain $\mathscr{L}$ and terminal domain $\mathscr{M}$. Similarly let V be the partial isometry in $\mathscr{A}$ whose initial and terminal domains are $\mathscr{K} \cap \mathscr{L}^{\perp}$. In other words V is the projection $\mathscr{H} \rightarrow (\mathscr{K} \cap \mathscr{L}^{\perp})$. Then one can easily verify that $U+V$ is a partial isometry in $\mathscr{A}$ with initial domain $\mathscr{K}$ and terminal domain $\mathscr{N}$.

Now let $\mathscr{A}$ be a finite v. Neumann algebra i.e. let I be a finite projection. Then by the lemma the projections of $\mathscr{A}$ are all finite. Therefore we see that $\mathscr{A}$ is finite if and only if its projections are all finite. One can easily construct an infinite $\mathscr{A}$ with finite non-zero projections. For instance if $\mathscr{H}$ is infinite dimensional then $\mathscr{A}=\mathscr{L}(\mathscr{H})$ is infinite but every projection whose range is finite dimensional is finite in $\mathscr{A}$.

Proposition 3. *A projection E belonging to the v. Neumann algebra $\mathscr{A}$ is finite if and only if $\mathscr{A}_E$ is finite.*

Proof. As usual we let $\mathscr{K}=E\mathscr{H}$ and we suppose that $\mathscr{A}_E$ is finite. In order to prove that E is finite in $\mathscr{A}$ let a projection F be given in $\mathscr{A}$ such that $F \leqslant E$ and $F \sim E$. Then there is a partial isometry V in $\mathscr{A}$ with initial domain $E\mathscr{H}$ and terminal domain $\mathscr{L}=F\mathscr{H}$. Then $V|\mathscr{K}$ is an isometric mapping of $\mathscr{K}$ onto $\mathscr{L}$ and so $E|\mathscr{K}$ and $F|\mathscr{K}$ are equivalent projections in $\mathscr{A}_E$. Hence $\mathscr{A}_E$ being finite we have $F|\mathscr{K}=E|\mathscr{K}$ and so $F=E$.

Conversely let E be a finite projection in $\mathscr{A}$. Every projection of $\mathscr{A}_E$ is of the form $F|\mathscr{K}$ where F is a projection in $E\mathscr{A}E$ i.e. $F \leqslant E$ and $F \in \mathscr{A}$. Suppose that there is an isometry $U: \mathscr{K} \rightarrow \mathscr{K}$ in $\mathscr{A}_E$ with initial domain $\mathscr{K}$ and terminal domain $\mathscr{L}$. Then $U=V|\mathscr{K}$ for some V in $E\mathscr{A}E$. Since V maps $\mathscr{K}$ isometrically onto $\mathscr{L}$ and V is 0 on $\mathscr{K}^{\perp}$ we see that V is a partial isometry with initial domain $\mathscr{K}$ and terminal domain $\mathscr{L}$. Hence $F \sim E$ and so $F=E$ by $F \leqslant E$ and the finiteness of E. We proved that $F|\mathscr{K}=E|\mathscr{K}$.

Definition 4. *A v. Neumann algebra $\mathscr{A}$ is called purely infinite if every non-zero central projection is infinite.*

We have the following analogue of the last proposition:

Proposition 5. *Let $\mathscr{A}$ be a v. Neumann algebra and let P be a central projection. Then $\mathscr{A}_P$ is purely infinite if and only if the non-zero central projections of $\mathscr{A}$ which are majorized by P are all infinite.*

Proof. By Proposition 8.20 every central projection in $\mathscr{A}_P$ is of the form $C|\mathscr{K}$ where C is a central projection in $P\mathscr{A}P$ that is to say C is a projection in $\mathscr{A}\cap\mathscr{A}'$ and $C\leqslant P$. Thus it will be sufficient to prove that $C|\mathscr{K}$ is finite in $\mathscr{A}_P$ if and only if C is finite in $\mathscr{A}$. By Proposition 3 $C|\mathscr{K}$ is finite in $\mathscr{A}_P$ if and only if $(\mathscr{A}_P)_{C|\mathscr{K}}$ is a finite algebra. Hence if we prove that $(\mathscr{A}_P)_{C|\mathscr{K}}=\mathscr{A}_C$ then by the same proposition $C|\mathscr{K}$ is finite in $\mathscr{A}_P$ if and only if C is finite in $\mathscr{A}$. But given A in $P\mathscr{A}P$ its restriction $A|\mathscr{K}$ lies in $(C|\mathscr{K})\mathscr{A}_P(C|\mathscr{K})$ if and only if A is in $C\mathscr{A}C$. Hence by $(A|\mathscr{K})|C\mathscr{H}=A|C\mathscr{H}$ we have $(\mathscr{A}_P)_{C|\mathscr{K}}=\mathscr{A}_C$.

Proposition 6. *Every v. Neumann algebra $\mathscr{A}$ contains a unique central projection P such that $\mathscr{A}_P$ is finite and $\mathscr{A}_{I-P}$ is purely infinite.*

Note. The algebra $\mathscr{A}_P$ is called the *finite part* of $\mathscr{A}$ and $\mathscr{A}_{I-P}$ is its *infinite part.*

Proof. Let $P_i\,(i\in\mathscr{I})$ be a family of pairwise orthogonal finite central projections which is maximal with respect to inclusion. The existence of such a family follows form Zorn's lemma provided we also admit the empty collection as one of the families. We let $P=\sum P_i$ so that P is a central projection. If the family is empty then we let $P=0$. Let F be a projection in $\mathscr{A}$ such that $F\sim P$ and $F\leqslant P$. Now if V is a partial isometry in $\mathscr{A}$ such that $V^*V=F$ and $VV^*=P$ then $V_i=VP_i$ is a partial isometry in $\mathscr{A}$ with $V_i^*V_i=FP_i$ and $V_iV_i^*=PP_i=P_i$. Hence $FP_i\sim P_i$ where $FP_i\leqslant P_i$ and so by the finiteness of P_i we have $FP_i=P_i$. Here i is arbitrary in $\mathscr{I}$ and so we conclude that $F=P$ and P is finite. Suppose Q is a finite central projection in $\mathscr{A}$. Then $(I-P)Q$ must be 0 because otherwise the family $P_i\,(i\in\mathscr{I})$ would not be maximal. Thus $Q\leqslant P$ and so P is the largest finite central projection in $\mathscr{A}$. We also see that the non-zero central projections majorized by $I-P$ are all infinite. Hence by Propositions 3 and 5 we proved that $\mathscr{A}_P$ is finite and $\mathscr{A}_{I-P}$ is purely infinite.

Lemma 7. *Every abelian v. Neumann algebra is finite.*

Proof. Let E be a projection in $\mathscr{A}$ which is equivalent to the identity I. Then by Definition 2.13 there is a partial isometry V in $\mathscr{A}$ such that $V^*V=I$ and $VV^*=E$. By the commutativity of $\mathscr{A}$ we obtain $I=E$.

Proposition 8. *The direct sum of finite v. Neumann algebras is a finite algebra.*

Proof. Let $\mathscr{A}_i$ $(i \in \mathscr{I})$ be a family of finite v. Neumann algebras and for each i let $\mathscr{H}_i$ denote the underlying Hilbert space of $\mathscr{A}_i$. Let P_i denote the projection of $\mathscr{H} = \sum \mathscr{H}_i$ onto the canonical image of $\mathscr{H}_i$ in $\mathscr{H}$. Then P_i is a central projection with respect to $\mathscr{A} = \sum \mathscr{A}_i$ and $P_i \perp P_j$ for $i \neq j$. It is obvious that $P = \sum P_i$ is the identity of $\mathscr{A}$. Therefore by the first half of the proof of the last proposition $\mathscr{A}_P = \mathscr{A}_I = \mathscr{A}$ is a finite v. Neumann algebra.

The following discussion will lead to a characterization of finite projections. (See Proposition 10.)

Proposition 9. *Let $\mathscr{K}$ be a complex Hilbert space and let $\mathscr{L}$ be a subspace. Let $\mathscr{A}$ be a v. Neumann subalgebra of $\mathscr{L}(\mathscr{K})$ and let $B: \mathscr{K} \to \mathscr{L}$ be a bijection which belongs to $\mathscr{A}$. Then there is a positive operator P in $\mathscr{A}$ and an isometry $U: \mathscr{K} \to \mathscr{L}$ in $\mathscr{A}$ such that $B = UP$.*

Note. This is a close analogue of another result in which $\mathscr{K}$ and $\mathscr{L}$ are arbitrary and $B: \mathscr{K} \to \mathscr{L}$ is any continuous operator. Then again $B = UP$ where $U: \mathscr{K} \to \mathscr{L}$ is a partial isometry, $P \geqslant 0$ and $P \in \mathscr{L}(\mathscr{K})$. If $T \in \mathscr{L}(\mathscr{K})$ and $TB = BT$ then $TP = PT$. UP is called the *polar decomposition* of B. See also Proposition 4.3.

Proof. By Theorem III.4.10 we can define $P = +\sqrt{B^* B}$. Then for any x in $\mathscr{K}$ we have

$$\|Px\|^2 = (Px, Px) = (P^2 x, x) = (B^* B x, x) = (Bx, Bx) = \|Bx\|^2 .$$

From this we see that $\|Px\| = \|Bx\|$ for $x \in \mathscr{K}$. By Proposition 4.2 we have $0 = \ker B = (B^* \mathscr{L})^\perp$ and so $B^* \mathscr{L} = \mathscr{K}$. Therefore $P\mathscr{K} \supseteq P^2 \mathscr{K} = B^* B \mathscr{K} = B^* \mathscr{L} = \mathscr{K}$ proving that P is surjective. Hence if we let $U(Px) = Bx$ then U is a linear norm preserving map of $\mathscr{K}$ onto $\mathscr{L}$. Then by the definition $UP = B$.

Now let $T \in \mathscr{A}'$. Since $B, B^* \in \mathscr{A}$ by Theorem III.4.10 we have $TP = PT$. This shows that $P \in \mathscr{A}'' = \mathscr{A}$. Again let $T \in \mathscr{A}'$ and let $x \in \mathscr{K}$. Then

$$TU(Px) = TBx = BTx = UPTx = UT(Px).$$

Since $P\mathscr{K} = \mathscr{K}$ we have $TU = UT$ and so $U \in \mathscr{A}'' = \mathscr{A}$.

If $\mathscr{H}$ is a complex Hilbert space and $A \in \mathscr{L}(\mathscr{H})$ then it is possible that A is a *left inverse* but not a *right inverse*. For example let $\mathscr{H}$ be the

Hilbert space consisting of the sequences $x=(x_1, x_2, \ldots)$ where $x_n \in \mathbb{C}$ and $\sum |x_n|^2 < \infty$, let

$$A(x_1, x_2, x_3, \ldots) = (x_2, x_3, \ldots) \quad \text{and} \quad B(x_1, x_2, x_3, \ldots) = (0, x_1, x_2, x_3, \ldots).$$

Then A and B are linear operators, $|A|=|B|=1$ and obviously $AB=I$. Now if A had a left inverse then this would be necessarily B, but $BA \neq I$. If we restrict ourselves to operators A belonging to a v. Neumann subalgebra $\mathscr{A}$ of $\mathscr{L}(\mathscr{H})$ then the situation might be different: For instance $\mathscr{A}=\mathbb{C}I$ gives a trivial example for the case when A is a left inverse if and only if it is a right inverse.

Proposition 10. *Let $\mathscr{A}$ be a v. Neumann algebra and let E be a projection in $\mathscr{A}$. Then E is finite if and only if the set of left and right inverses of $E\mathscr{A}E$ coincide.*

Proof. First suppose that E is an infinite projection. Then by definition there is a projection $F<E$ in $\mathscr{A}$ such that $E \sim F$ with respect to $\mathscr{A}$. Then using Definition 2.13 we see the existence of a partial isometry V in $\mathscr{A}$ such that $V^*V=E$ and $VV^*=F$. Let $\mathscr{K}$ be the initial domain of V and let $\mathscr{L}$ be its terminal domain. Since V is 0 on $\mathscr{K}^\perp$ we have $V(I-E)=0$ i.e. $VE=V$. For the same reason $EV=V$ on $\mathscr{K}^\perp$ and by $V\mathscr{K}=\mathscr{L}\subset\mathscr{K}$ we also have $EV=V$ on $\mathscr{K}$. This proves that $EV=V$. Hence V and V^* belong to $E\mathscr{A}E$. In $E\mathscr{A}E$ the identity element is E so V^* is the left inverse of V but V^* is not a right inverse because $VV^*=F\neq E$.

Now let A and B in $E\mathscr{A}E$ be such that $AB=E$ but $BA=Q\neq E$. We are going to prove that E is an infinite projection. For the sake of simplicity let $\mathscr{K}=E\mathscr{H}$ where $\mathscr{H}$ is the Hilbert space of $\mathscr{A}$, let $\mathscr{L}=B\mathscr{K}$ and let F be the projection $\mathscr{H}\to\mathscr{L}$. For any operator S in $E\mathscr{A}E$ let $\tilde{S}=S|\mathscr{K}$ so that $\tilde{S}\in\mathscr{L}(\mathscr{K})$. The set of these operators $\tilde{S}$ is the reduced algebra which we shall denote by $\tilde{\mathscr{A}}$ in this proof. We notice that $\tilde{E}$ is the identity of $\tilde{\mathscr{A}}$ and $\tilde{B}: \mathscr{K}\to\mathscr{L}$ is a bijection because $\tilde{A}\tilde{B}=\tilde{E}$.

By $B\in E\mathscr{A}E$ we have $B=EB$ and so $\mathscr{L}=B\mathscr{K}\subseteq\mathscr{K}$. We prove that $\mathscr{L}$ is a proper subset of $\mathscr{K}$. For let us suppose that $\mathscr{L}=\mathscr{K}$. Then $\tilde{B}$ is invertible and $\tilde{A}=\tilde{B}^{-1}$ by $\tilde{A}\tilde{B}=\tilde{E}$. Thus for x in $\mathscr{H}$ we have

$$BAx = BAEx = \tilde{B}\tilde{A}Ex = \tilde{E}(Ex) = Ex.$$

This shows that $BA=E$ which contradicts the hypothesis $BA=Q\neq E$. Hence $\mathscr{L}\subset\mathscr{K}$ and $F<E$.

Since $\tilde{B}: \mathscr{K}\to\mathscr{L}$ is an bijection by Proposition 9 we have $\tilde{B}=\tilde{U}\tilde{P}$ where $\tilde{P}$ is a positive operator in $\tilde{\mathscr{A}}$ and $\tilde{U}: \mathscr{K}\to\mathscr{L}$ is an isometry which belongs to $\tilde{\mathscr{A}}$. Now $\tilde{U}$ can be extended in a unique manner to a partial isometry $U: \mathscr{H} \to \mathscr{H}$ such that the initial domain of U is $\mathscr{K}$ and its terminal domain is $\mathscr{L}$. Since $\tilde{U} \in \tilde{\mathscr{A}}$ we have $\tilde{U} = \tilde{V}$ for some $V \in E\mathscr{A}E$.

Let $T \in \mathscr{A}'$ so that $T\mathscr{K} \subseteq \mathscr{K}$ and $T\mathscr{K}^{\perp} \subseteq \mathscr{K}^{\perp}$ by $E \in \mathscr{A}$. Then we have $UTE = VTE = TVE = TUE$. Moreover $UT(I-E)\mathscr{H} = UT\mathscr{K}^{\perp} \subseteq U\mathscr{K}^{\perp} = 0$ and similarly $TU(I-E)\mathscr{H} = TU\mathscr{K}^{\perp} = T0 = 0$. Therefore we have

$$UT(I-E) = TU(I-E) \quad \text{and} \quad UTE = TUE$$

which show that $UT = TU$. Since T is an arbitrary element of $\mathscr{A}'$ we obtain $U \in \mathscr{A}'' = \mathscr{A}$. Hence the initial domain $\mathscr{K}$ of U and its terminal domain $\mathscr{L}$ are equivalent with respect to $\mathscr{A}$. By $E\mathscr{H} = \mathscr{K} \supset \mathscr{L} = F\mathscr{H}$ we see that E is an infinite projection.

Using Proposition 10 one can give alternate formulations of the definitions of type II and type III v. Neumann algebras or one can readily state necessary and sufficient conditions that an algebra belong to these types. We will not state these results explicitely.

Definition 11. *A v. Neumann algebra $\mathscr{A}$ is called a type II algebra if it contains no abelian projections except 0 and if every non-zero central projection majorizes a non-zero finite projection.*

In particular a factor $\mathscr{A}$ is of type II if it contains no non-zero abelian projection and if it has at least one non-zero finite projection. We see that the non-zero projections of a type II factor form at least two equivalence classes with respect to $\mathscr{A}$. *A finite type II algebra* is said to be of *type II_1* and similarly an *infinite type II algebra* is called an algebra of *type II_∞*. The factors associated with the primary representations described in Theorem 7.18 are examples of finite type II factors.

Proposition 12. *The direct sum of type II v. Neumann algebras is of type II.*

Proof. The proof of this result is very similar to that of Proposition 8.7. Let E be a central projection in $\mathscr{A}$ and let $E = \sum E_i$ be the decomposition of E into central projections E_i in $\mathscr{A}_i$. If $E \neq 0$ then $E_i \neq 0$ for at least one index i. We fix such an index i and choose a finite projection F_i in $\mathscr{A}_i$ such that $0 < F_i \leqslant E_i$. For every other index j let $F_j = 0$. Then $F = \sum F_i$ is a finite projection in $\mathscr{A}$ satisfying $0 < F < E$. One can easily see that there exists no non-zero abelian projection in $\mathscr{A}$.

Proposition 13. *Let $\mathscr{A}$ be a v. Neumann algebra and let E be a projection in $\mathscr{A}$. Then $\mathscr{A}_E$ is of type II if and only if E majorizes no non-zero abelian projection and $F \leqslant E$ for some non-zero finite projection F in $\mathscr{A}$.*

Proof. We let $\mathscr{K}=E\mathscr{H}$ where $\mathscr{H}$ is the underlying Hilbert space of $\mathscr{A}$. Every projection in $\mathscr{A}_E$ is of the form $F|\mathscr{K}$ where F is a projection in $\mathscr{A}$ and $F\leqslant E$. Moreover $C|\mathscr{K}$ is a central projection in $\mathscr{A}_E$ if and only if C is in the center of $\mathscr{A}$ and $C\leqslant E$. During the proof of Proposition 5 we have seen that $(\mathscr{A}_E)_{F|\mathscr{K}}=\mathscr{A}_F$. Therefore $F|\mathscr{K}$ is abelian in $\mathscr{A}_E$ if and only if F is abelian in $\mathscr{A}$. Similarly by Proposition 3 $F|\mathscr{K}$ is finite in $\mathscr{A}_E$ if and only if F is finite in $\mathscr{A}$. Now the result is a direct consequence of Definition 11.

Definition 14. *A v. Neumann algebra $\mathscr{A}$ is called a type III algebra if $\mathscr{A}$ contains no non-zero abelian projection and all projections in $\mathscr{A}$ are infinite.*

A factor is of type I if and only if it has a non-zero abelian projection. Hence a factor is of type III if and only if it is neither of type I nor of type II. We see that a factorial algebra is of exactly one of the three types. If $\mathscr{A}$ is not a factor then in general $\mathscr{A}$ belongs to none of these types. However it is clear that the three types are mutually exclusive; $\mathscr{A}$ can belong to at most one of them. Moreover we are going to see that every v. Neumann algebra is uniquely expressible as a direct sum of three algebras of types I, II and III, respectively. It is obvious from Definitions 7.13 and 14 that every type III algebra is infinite.

Proposition 15. *The direct sum of type III algebras is of type III.*

Proof. Let E be a non-zero projection in $\mathscr{A}=\sum\mathscr{A}_i$. Then $E=\sum E_i$ where at least one of the projections E_i is not zero. Then by hypothesis E_i is an infinite projection in $\mathscr{A}_i$. Hence if we let $F=\sum F_i$ where $F_i=E_i$ and $F_j=0$ for every $j\neq i$ then F is an infinite projection in $\mathscr{A}$ and $F\leqslant E$. Therefore by Lemma 2 E is an infinite projection and so $\mathscr{A}$ is of type III by Lemma 7.

Proposition 16. *Let $\mathscr{A}$ be a v. Neumann algebra and let E be a projection in $\mathscr{A}$. Then $\mathscr{A}_E$ is of type III if and only if E majorizes neither abelian nor finite non-zero projections.*

Proof. See the facts listed in the proof of Proposition 13. The result is an immediate consequence of these and Definition 14.

Theorem 17. *For every v. Neumann algebra $\mathscr{A}$ there exist uniquely determined central projections such that they are pairwise orthogonal, their sum is the identity of $\mathscr{A}$ and the corresponding reduced algebras $\mathscr{A}_I$, $\mathscr{A}_{II}$ and $\mathscr{A}_{III}$ are of types I, II and III, respectively.*

The projection P_I majorizes all the abelian projections and P_{II} majorizes every finite projection whose central support majorizes no non-zero abelian projection. For every finite projection E we have $E\leqslant P_I+P_{II}$.

Note. The Hilbert spaces $P_I\mathscr{H}$, $P_{II}\mathscr{H}$ and $P_{III}\mathscr{H}$ will be denoted by $\mathscr{H}_I$, $\mathscr{H}_{II}$ and $\mathscr{H}_{III}$, respectively.

Proof. Let $\mathscr{H}_I$ be the type I part of the underlying Hilbert space $\mathscr{H}$ which is described in Theorem 8.27. We consider systems $(P_i, E_i)(i\in\mathscr{I})$ consisting of pairs of projections P_i and E_i in $\mathscr{A}$ such that the P_i's are orthogonal and

1) P_i is the central support of E_i and $P_i\mathscr{H}\subseteq(\mathscr{H}_I)^\perp$.

2) E_i is a non-zero finite projection.

3) P_i does not majorize a non-zero abelian projection.

If we consider the empty family as such a system of pairs then by Zorn's lemma there exists a system $(P_i, E_i)(i\in\mathscr{I})$ which is maximal with respect to inclusion. Let $P_{II}=\sum P_i$ so that $P_{II}\in c\mathscr{A}$ and let $\mathscr{H}_{II}=P_{II}\mathscr{H}$.

By condition 1) we have $\mathscr{H}_{II}\subseteq(\mathscr{H}_I)^\perp$ where $\mathscr{H}_I$ is the underlying Hilbert space of the type I part of $\mathscr{A}$. We let $\mathscr{H}_{III}=(\mathscr{H}_I+\mathscr{H}_{II})^\perp$ and P_{III} be the projection $\mathscr{H}\to\mathscr{H}_{III}$. Then by $P_I, P_{II}\in c\mathscr{A}$ we have $P_{III}\in c\mathscr{A}$. By Theorem 8.27 we know that P_I majorizes the abelian projections of $\mathscr{A}$. By Propositions 12 and 13 $\mathscr{A}_{II}$ is a type II algebra.

Let E be a non-zero finite projection in $\mathscr{A}$ such that its central support P majorizes no non-zero abelian projection. We let $\mathscr{K}=P\mathscr{H}$ and F be the projection $\mathscr{H}\to\mathscr{L}$ where $\mathscr{L}=\mathscr{K}\cap\mathscr{H}_I$. Since $F\leqslant PP_I$ the central support of F is majorized by PP_I and so by Theorem 8.27 it must be zero. This shows that $F=0$ and $\mathscr{K}\subseteq(\mathscr{H}_I)^\perp$ by $P_I\mathscr{K}\subseteq\mathscr{K}$. Now let $\mathscr{L}=\mathscr{K}\cap(\mathscr{H}_{II})^\perp$ and again let F be the projection $\mathscr{H}\to\mathscr{L}$. Then $F\leqslant P$ and $(P_i, E_i)(i\in\mathscr{I})$ being a maximal family it follows that $F=0$. This shows that $E\leqslant P_{II}$ because $P_{II}\mathscr{K}\subseteq\mathscr{K}$ by $P_{II}\in c\mathscr{A}$.

Next let E be an arbitrary finite projection in $\mathscr{A}$, let $\mathscr{K}=E\mathscr{H}$ and $\mathscr{L}=\mathscr{K}\cap\mathscr{H}_{III}$. We let F denote the projection $\mathscr{H}\to\mathscr{L}$ and P its central support. Since $(P_i, E_i)(i\in\mathscr{I})$ is a maximal system and $P\leqslant P_{III}$ we must have $F=P=0$ and so $\mathscr{L}=0$. Hence $\mathscr{K}\subseteq(\mathscr{H}_{III})^\perp=\mathscr{H}_I+\mathscr{H}_{II}$ and $E\leqslant P_I+P_{II}$. This shows that every projection majorized by P_{III} is infinite and so $\mathscr{A}_{III}$ is a type III algebra.

Now it is easy to establish the uniqueness of P_I, P_{II} and P_{III}. By Theorem 8.27 we already know that P_I is uniquely determined by $\mathscr{A}$. Suppose P is a central projection such that $\mathscr{A}_P$ is of type II. Then by Proposition 13 P majorizes no non-zero abelian projection and in addition there exists a non-zero finite projection F such that we have $F\leqslant P$. Therefore by Theorem 8.27 and by what have proved here already we have $P\leqslant P_{II}$. Therefore P_{II} is the largest central projection such that the corresponding reduced algebra is of type II. This already proves that P_{III} is uniquely determined. By Proposition 16 we see that P_{III} is the largest central projection such that the corresponding reduced algebra is of type III.

Now we have at our disposal all the concepts and results for the basic classification and splitting of v. Neumann algebras and representations. Let $\mathscr{A}$ be a v. Neumann algebra with underlying Hilbert space $\mathscr{H}$. By Theorem 17 there is a natural splitting of $\mathscr{H}$ into the pairwise orthogonal subspaces $\mathscr{H}_I$, $\mathscr{H}_{II}$ and $\mathscr{H}_{III}$. Similarly Theorem 6 shows that $\mathscr{H}$ splits into $\mathscr{H}_1$ and its orthogonal complement $\mathscr{H}_\infty$ such that $\mathscr{A}\,|\,\mathscr{H}_1$ is finite and $\mathscr{A}\,|\,\mathscr{H}_\infty$ is purely infinite. Since a type III algebra is always infinite we have $\mathscr{H}_{III} \subseteq \mathscr{H}_\infty$. In general $\mathscr{H}_1$ cuts across both $\mathscr{H}_I$ and $\mathscr{H}_{II}$ and so these subspaces split into $\mathscr{H}_{1I}$ and $\mathscr{H}_{\infty I}$ and similarly $\mathscr{H}_{1II}$ and $\mathscr{H}_{\infty II}$, respectively. They are the *finite* and *infinite type I* and *type II parts of* $\mathscr{A}$.

Further natural splitting of $\mathscr{H}$ can be obtained by considering also the commutant $\mathscr{A}'$ and observing its invariant subspaces. In particular we can apply Theorems 5.4 and 7.12 with ρ being the natural representation of the abstract involutive algebra obtained from $\mathscr{A}'$ on $\mathscr{A}'$. We note that in general $\mathscr{H}_{dI}$ and $\mathscr{H}_d$ are only $\mathscr{A}'$-invariant. Since $\mathscr{H}_{dI}$ is a direct sum of irreducible components by Theorem IV.2.2 and Proposition 8.17 or by Proposition 8.5 the corresponding projections in $\mathscr{A}$ are all abelian. Hence we see that $\mathscr{H}_{dI} \subseteq \mathscr{H}_I$ and so $\mathscr{H}_I$ splits into $\mathscr{H}_{dI}$, the *discrete type I part* and its orthogonal complement in $\mathscr{H}_I$ which will be called the *continuous type I part* of $\mathscr{A}$ and it will be denoted by $\mathscr{H}_{cI}$. The *discrete-continuous part* $\mathscr{H}_{dc}$ was defined just after Theorem 7.12 as the orthogonal complement of $\mathscr{H}_{dI}$ in $\mathscr{H}_d$. The intersection of $\mathscr{H}_d$ with $\mathscr{H}_{II}$ is $\mathscr{H}_{dII}$, the *type II discrete-continuous part* and similarly $\mathscr{H}_d \cap \mathscr{H}_{III}$ is the *type III discrete-continuous part* $\mathscr{H}_{dIII}$. The orthogonal complement of $\mathscr{H}_{dII}$ in $\mathscr{H}_{II}$ is the *purely continuous type II part* $\mathscr{H}_{cII}$. Similarly the *purely continuous type III part* $\mathscr{H}_{cIII}$ which was first mentioned in Section 7 is $(\mathscr{H}_{dIII})^\perp \cap \mathscr{H}_{III}$. The same terminology and notations $\mathscr{A}_d$, $\mathscr{A}_{dc}$, $\mathscr{A}_{dI}$, $\mathscr{A}_{cI}$, etc. can be used for the corresponding algebras.

A representation $\rho: S \to \mathscr{L}(\mathscr{H})$ is called a *type II representation* if its commuting algebra (ρ, ρ) is a type II v. Neumann algebra. Similarly ρ is a *type III representation* if (ρ, ρ) is of type III. Using Propositions 8.30 and 13 and 16 from this section we see that *every subrepresentation of a type II representation is of type II*. Similarly *if* $\rho: S \to \mathscr{L}(\mathscr{H})$ *is of type III then so are all the subrepresentations of* ρ.

Proposition 18. *If* $\rho_i: S \to \mathscr{L}(\mathscr{H}_i)$ $(i \in \mathscr{I})$ *are Mackey type representations with a common multiplier and if every* ρ_i *is of type II then* $\rho = \sum \rho_i$ *is a type II representation. Similarly if every* ρ_i *is of type III then* ρ *is of type III.*

Proof. The reasoning is the same as that which was used to prove Theorem 8.36. The first paragraph of that proof is applicable word by word. In the second paragraph we replace "Theorem 8.27" by "Theorem 17".

By that theorem P_{II} is maximal among those central projections P for which $(\rho, \rho)_P$ is of type II and similarly P_{III} is maximal among those P for which $(\rho, \rho)_P$ is of type III.

In Definition 2.13 we introduced an equivalence relation $\sim$ on the set of projections E of a v. Neumann algebra $\mathscr{A}$. We let (E) denote the equivalence class containing the projection E. We shall write $(E_1) \leqslant (E_2)$ if there is a projection $F \leqslant E_2$ such that $E_1 \sim F$. It is easy to see that the relation $\leqslant$ is well defined, reflexive and transitive. We are going to prove that $(E_1) \leqslant (E_2)$ and $(E_2) \leqslant (E_1)$ imply $(E_1) = (E_2)$ i.e. $E_1 \sim E_2$. This will easily follow from an important fact on the equivalence of representations which was already mentioned in Section 2 but not needed earlier.

Theorem 19. *Let* $\rho_k: S \to \mathscr{L}(\mathscr{H}_k)\,(k=1,2)$ *be representations such that* $\rho_1 \leqslant \rho_2$ *and* $\rho_2 \leqslant \rho_1$. *Then we have* $\rho_1 \approx \rho_2$.

Proof. Let $U: \mathscr{H}_1 \to U\mathscr{H}_1 \subseteq \mathscr{H}_2$ be an isometry such that $U\rho_1 = \rho_2 U$ on $\mathscr{H}_1$ and similarly let $V: \mathscr{H}_2 \to V\mathscr{H}_2 \subseteq \mathscr{H}_1$ be an isometry satisfying $V\rho_2 = \rho_1 V$ on $\mathscr{H}_2$. Using $W = UV$ we define a sequence of subspaces of $\mathscr{H}_2$:

$$\mathscr{K}_n = W^{n-1}\mathscr{H}_2 \cap (W^n \mathscr{H}_2)^\perp \quad \text{for } n=1,2,\ldots \tag{1}$$

Since $W \in (\rho_2, \rho_2)$ we see that $\mathscr{K}_n$ is an invariant subspace of ρ_2. We also define

$$\mathscr{K}_\infty = (W\mathscr{H}_2) \cap (W^2\mathscr{H}_2) \cap (W^3\mathscr{H}_2) \cap \cdots \tag{2}$$

so that $\mathscr{K}_\infty$ is ρ_2-invariant. It is clear that $\mathscr{K}_\infty \perp \mathscr{K}_n$ for every n. Moreover if $m<n$ then $\mathscr{K}_n \subseteq W^{n-1}\mathscr{H}_2 \subseteq W^m \mathscr{H}_2 \perp \mathscr{K}_m$ and so the subspaces $\mathscr{K}_n$ are pairwise orthogonal. We have $W^{n-1}\mathscr{H}_2 = \mathscr{K}_n \oplus W^n\mathscr{H}_2$ for $n=1,2,\ldots$ and so

$$\mathscr{H}_2 = \mathscr{K}_1 \oplus \cdots \oplus \mathscr{K}_n \oplus W^n \mathscr{H}_2$$

where $W\mathscr{H}_2 \supseteq W^2\mathscr{H}_2 \supseteq \cdots$. Hence if $x \in \mathscr{K}_\infty$ then $x \in W^n\mathscr{H}_2$ and so $x \perp \mathscr{K}_n$ for $n=1,2,\ldots$ This shows that $\mathscr{K}_\infty \subseteq (\mathscr{K}_1 \oplus \mathscr{K}_2 \oplus \cdots)^\perp$. We also see that

$$(\mathscr{K}_1 \oplus \mathscr{K}_2 \oplus \cdots)^\perp \subseteq (\mathscr{K}_1 \oplus \cdots \oplus \mathscr{K}_n)^\perp = W^n \mathscr{H}_2$$

and so $(\mathscr{K}_1 \oplus \mathscr{K}_2 \oplus \cdots)^\perp \subseteq \mathscr{K}_\infty$. We proved that

$$\mathscr{H}_2 = \mathscr{K}_\infty \oplus \sum_1^\infty \mathscr{K}_n . \tag{3}$$

The same reasoning also shows that

(4) $$W\mathscr{H}_2 = \mathscr{K}_\infty \oplus \sum_2^\infty \mathscr{K}_n .$$

Next we are going to prove that

(5) $$W\mathscr{K}_\infty = \mathscr{K}_\infty \quad \text{and} \quad W\mathscr{K}_n = \mathscr{K}_{n+1} \quad \text{for } n=1,2,\ldots$$

First $x\in\mathscr{K}_\infty$ implies $x\in W^n\mathscr{H}_2$ and $Wx\in W^{n+1}\mathscr{H}_2$ for $n=1,2,\ldots$ Hence by $W\mathscr{H}_2 \supseteq W^2\mathscr{H}_2$ we obtain $Wx\in\mathscr{K}_\infty$. Now let $x\in\mathscr{K}_\infty$. Then $x\in W^2\mathscr{H}_2$ and so $x=Wy$ for some y in $W\mathscr{H}_2$. Therefore $x=Wy\in W^{n+1}\mathscr{H}_2$ and so $y\in W^n\mathscr{H}_2$ for $n=1,2,\ldots$ This shows that $y\in\mathscr{K}_\infty$ and so $x\in W\mathscr{K}_\infty$. We proved that $\mathscr{K}_\infty = W\mathscr{K}_\infty$. In order to prove the other relation first let $x\in W\mathscr{K}_n$. Then $x=Wy$ with $y\in\mathscr{K}_n$ and so $x=Wy \in W(W^{n-1}\mathscr{H}_2) = W^n\mathscr{H}_2$. Furthermore $y \perp W^n\mathscr{H}_2$ and so W being an isometry $x=Wy \perp W^{n+1}\mathscr{H}_2$. This proves that $x\in\mathscr{K}_{n+1}$. Next let $x\in\mathscr{K}_{n+1}$ so that $x\in W^n\mathscr{H}_2$ and so $x=Wy$ where $y\in W^{n-1}\mathscr{H}_2$. We also have $x=Wy\perp W^{n+1}\mathscr{H}_2$ which implies that $y\perp W^n\mathscr{H}_2$. Therefore we proved that $y\in\mathscr{K}_n$ and $x\in W\mathscr{K}_n$.

Since $W\mathscr{H}_2 \subseteq U\mathscr{H}_1$ we can define $\mathscr{L}_1 = (W\mathscr{H}_2)^\perp \cap U\mathscr{H}_1 = \mathscr{K}_1 \cap U\mathscr{H}_1$. Then we have

(6) $$U\mathscr{H}_1 = W\mathscr{H}_2 \oplus \mathscr{L}_1 .$$

For $n=1,2,\ldots$ we let $\mathscr{L}_{n+1} = W^n\mathscr{L}_1$ so that $\mathscr{L}_n \subseteq \mathscr{H}_2$ for every $n \geqslant 1$. Using (5) we also see that $\mathscr{K}_n \supseteq \mathscr{L}_n$ for $n\geqslant 1$. Therefore we can define $\mathscr{M}_n = \mathscr{K}_n \cap (\mathscr{L}_n)^\perp$ for $n=1,2,\ldots$ Then

(7) $$\mathscr{K}_n = \mathscr{L}_n \oplus \mathscr{M}_n \quad \text{for } n=1,2,\ldots$$

Moreover by (5) and the method used in its proof we can show that

(8) $$\mathscr{M}_{n+1} = W\mathscr{M}_n \quad \text{for } n=1,2,\ldots$$

From (3) and (7) we obtain

(9) $$\mathscr{H}_2 = \sum_1^\infty \mathscr{L}_n \oplus \sum_1^\infty \mathscr{M}_n \oplus \mathscr{K}_\infty$$

where $\sum$ denotes orthogonal summation. Similarly by (4), (6) and (7) we have

(10) $$U\mathscr{H}_1 = \sum_1^\infty \mathscr{L}_n \oplus \sum_2^\infty \mathscr{M}_n \oplus \mathscr{K}_\infty .$$

Let P denote the projection of $\mathscr{H}_2$ onto $\sum_1^\infty \mathscr{L}_n \oplus \mathscr{K}_\infty$ and let

$$S = W(I-P)+P.$$

Using (9) we see that $I-P$ is the projection of $\mathscr{H}_2$ onto $\sum_1^\infty \mathscr{M}_n$. We also notice that S maps each $\mathscr{L}_n$ identically onto itself. Similarly S reduces to the identity on $\mathscr{K}_\infty$. Moreover $S|\mathscr{M}_n = W|\mathscr{M}_n$ and so S maps $\mathscr{M}_n$ isometrically onto $\mathscr{M}_{n+1}$. Therefore by (10) S is an isometry of $\mathscr{H}_2$ onto $U\mathscr{H}_1$ and so $T=U^*S$ is an isometry of $\mathscr{H}_2$ onto $\mathscr{H}_1$.

By $\rho_2 U = U\rho_1$ the subspace $U\mathscr{H}_1$ is invariant under ρ_2 and $W\mathscr{H}_2$ is also ρ_2-invariant by $W\in(\rho_2,\rho_2)$. Hence $\mathscr{L}_1$ is ρ_2-invariant and so $W\in(\rho_2,\rho_2)$ shows that $\mathscr{L}_n$ is invariant under ρ_2 for every $n=1,2,\ldots$ Since we also know that $\mathscr{K}_\infty$ is invariant under ρ_2 we see that $P\in(\rho_2,\rho_2)$ and so $S\in(\rho_2,\rho_2)$. Multiplying $U\rho_1=\rho_2 U$ from the left and the right by U^* we obtain $U^*\rho_2=\rho_1 U^*$. Therefore

$$T\rho_2 = U^*(S\rho_2) = U^*(\rho_2 S) = (U^*\rho_2)S = (\rho_1 U^*)S = \rho_1 T$$

proving that $T\in(\rho_2,\rho_1)$. Hence T is a unitary operator demonstrating the equivalence of ρ_2 and ρ_1.

Proposition 20. *If E_1 and E_2 are projections in a v. Neumann algebra such that $(E_1)\leqslant(E_2)$ and $(E_2)\leqslant(E_1)$ then $(E_1)=(E_2)$ i.e. $E_1\sim E_2$.*

Proof. Let $\mathscr{H}$ be the underlying Hilbert space of the v. Neumann algebra $\mathscr{A}$ and let $E_k\mathscr{H}=\mathscr{H}_k\,(k=1,2)$. In order to apply Theorem 19 we choose for S the abstract involutive algebra obtained from $\mathscr{A}'$ by ignoring its additional structure. For $T\in\mathscr{A}'$ let $\rho_k(T)=T|\mathscr{H}_k$ so that $\rho_k\colon\mathscr{A}'\to\mathscr{L}(\mathscr{H}_k)$. By $(E_1)\leqslant(E_2)$ there is a partial isometry $U\colon\mathscr{H}_1\to\mathscr{H}_2$ with initial domain $\mathscr{H}_1$ such that $U\in\mathscr{A}$. This shows that $\rho_1\leqslant\rho_2$. Similarly by $(E_2)\leqslant(E_1)$ we obtain $\rho_2\leqslant\rho_1$. By Theorem 19 $\rho_1\approx\rho_2$ and so there is an isometry $W\colon\mathscr{H}_1\to\mathscr{H}_2$ which intertwines ρ_1 and ρ_2. If we extend W to a partial isometry $W\colon\mathscr{H}\to\mathscr{H}$ then $W\in\mathscr{A}''=\mathscr{A}$, its initial domain is $E_1\mathscr{H}$ and its terminal domain is $E_2\mathscr{H}$. Hence $E_1\sim E_2$.

Lemma 21. *If $\mathscr{A}$ is a factor and E_1, E_2 are projections in $\mathscr{A}$ then $(E_1)\leqslant(E_2)$ or $(E_1)\geqslant(E_2)$.*

Proof. This result is a simple consequence of the corollary of Lemma 7.2.

Corollary. *If $\mathscr{A}$ is a factor then the set of equivalence classes of its projections is linearly ordered.*

Lemma 22. *Let $\mathscr{A}$ be a v. Neumann algebra and let E_i $(i=1,2,\ldots)$ and F_i $(i=1,2,\ldots)$ be two sequences of pairwise orthogonal projections in $\mathscr{A}$ and let $E_i \sim F_i$ for every $i \geqslant 1$. Then $\sum E_i \sim \sum F_i$.*

Proof. Let $\rho: \mathscr{A}' \to \mathscr{L}(\mathscr{H})$ be the usual representation of $\mathscr{A}'$, the abstract algebra, on $\mathscr{A}'$, the v. Neumann algebra. We let $\rho_i' = \rho \mid \mathscr{K}_i$ and $\rho_i'' = \rho \mid \mathscr{L}_i$ where $\mathscr{K}_i = E_i \mathscr{H}$ and $\mathscr{L}_i = F_i \mathscr{H}$. Then by hypothesis there is an isometry U_i of $\mathscr{K}_i$ onto $\mathscr{L}_i$ which belongs to $\mathscr{A}$ and it intertwines ρ_i' and ρ_i''. Then $U = \sum U_i$ is in $\mathscr{A}$, it maps $\mathscr{K} = \sum \mathscr{K}_i$ isometrically onto $\mathscr{L} = \sum \mathscr{L}_i$ and it intertwines $\rho \mid \mathscr{K}$ and $\rho \mid \mathscr{L}$. Hence $\rho \mid \mathscr{K} \approx \rho \mid \mathscr{L}$ wich means the same as $\sum E_i \sim \sum F_i$ by Proposition 2.14.

Proposition 23. *Let $\mathscr{A}$ be a factor and let E and F be non-zero projections in $\mathscr{A}$. Let E_i $(i \in \mathscr{I})$ be any family of pairwise orthogonal projections $E_i \leqslant E$ such that $E_i \sim F$ and the family is maximal with respect to inclusion. Then there is a projection P in $\mathscr{A}$ such that $(P) < (F)$ and*

$$E = \sum E_i + P. \tag{11}$$

If the index set $\mathscr{I}$ is infinite then we have $E \sim \sum E_i$.

Proof. If we also consider the empty family then by Zorn's lemma there is a family $E_i (i \in \mathscr{I})$ which is maximal with respect to inclusion. For any such family we can write E in the form given in (11). Then by Lemma 21 and the maximality of the family we have $(P) < (F)$.

If $\mathscr{I}$ is an infinite index set then let $\mathscr{I}_o$ be a subset obtained by omitting one index, say i_o. Then $\mathscr{I}_o$ being infinite we have

$$(E) = \left(\sum E_i + P\right) = \left(\sum_{i \neq i_o} E_i + P\right) \leqslant \left(\sum_{i \neq i_o} E_i + E_{i_o}\right) = \left(\sum E_i\right) \leqslant (E).$$

Hence by Proposition 20 it follows that $(E) = (\sum E_i)$.

Proposition 24. *Let $\mathscr{A}$ be a factor without minimal projections and let the underlying Hilbert space $\mathscr{H}$ be separable. Then $\mathscr{A}$ is of type II if and only if its projections form at least three distinct equivalence classes and $\mathscr{A}$ is of type III if and only if there are two equivalence classes of projections.*

Proof. The zero projection forms a class by itself and another class is the one containing I. In the discussion following Definition 7.19 we have seen that $\mathscr{A}$ is a factor without irreducible subspaces and so it is not of type I. If $\mathscr{A}$ is of type II then let E be a non-zero finite projection in $\mathscr{A}$. If I is an infinite projection then the classes containing 0, E and I are distinct. If I is finite then no projection of $\mathscr{A}$ being minimal one can find a non-zero projection satisfying $I > F$. Then the classes 0, F and I are distinct.

Conversely let E and F be inequivalent non-zero projections in $\mathscr{A}$. By Lemma 21 we may suppose that $(E) \leqslant (F)$. We can then also suppose that $E < F$. We are going to prove by contradiction that E is finite. We suppose that E is infinite and more precisely that V is a partial isometry in $\mathscr{A}$ with initial domain $\mathscr{K} = E\mathscr{H}$ whose terminal domain is a proper subspace of $\mathscr{K}$. We are going to proceed in the same manner as in the proof of Lemma 7.14.

Since $\mathscr{K} \supset V\mathscr{K} \supset V^2\mathscr{K} \supset \cdots$ is an infinite sequence of non-zero, $\mathscr{A}'$-invariant subspaces of $\mathscr{H}$

$$\mathscr{L}_n = V^{n-1}\mathscr{K} \cap (V^n\mathscr{K})^{\perp} \qquad (n = 1, 2, \ldots)$$

defines a sequence of pairwise orthogonal, non-zero, $\mathscr{A}'$-invariant subspaces of $\mathscr{H}$. Let P_n be the projection of $\mathscr{H}$ onto $\mathscr{L}_n$. Then VP_n is a partial isometry in $\mathscr{A}$ with initial domain $\mathscr{L}_n$ and terminal domain $\mathscr{L}_{n+1}$. Therefore $P_1, P_2, \ldots$ are equivalent non-zero pairwise orthogonal projections and $P_n \leqslant E < F$ for every $n = 1, 2, \ldots$

Now consider a family $P_i\,(i \in \mathscr{I})$ of pairwise orthogonal projections such that

1) $P_i \leqslant E$ and $P_i \sim P_1$ for every $i \in \mathscr{I}$;
2) $P_n\,(n = 1, 2, \ldots)$ is a subfamily of $P_i\,(i \in \mathscr{I})$;
3) the family $P_i\,(i \in \mathscr{I})$ is maximal with respect to inclusion.

The existence of such a family follows from Zorn's lemma. Applying Proposition 23 with P_1 in place of F we obtain

$$E \sim \sum_{n \in N} P_n + \sum_{i \notin N} P_i \tag{12}$$

where N denotes the set of natural numbers.

Since $E < F$ we can replace E by F in the foregoing construction and obtain

$$F \sim \sum_{n \in N} P_n + \sum_{j \notin N} P_j \tag{13}$$

where $\mathscr{J}$ is an index set containing N, the P_j's are pairwise orthogonal and $P_j \sim P_1$ for every j in $\mathscr{J}$. Since the Hilbert space $\mathscr{H}$ is supposed to be separable $\mathscr{I}$ and $\mathscr{J}$ are denumerable sets. Therefore we can apply Lemma 22 and obtain $E \sim F$ from (12) and (13). This is a contradiction and so E is a finite projection and $\mathscr{A}$ is of type II.

The analogous statement concerning type III factors follows from the fact that a factor $\mathscr{A}$ belongs to exactly one of the three types.

We shall now present a v. Neumann algebra $\mathscr{A}$ such that $\mathscr{A}=\mathscr{A}'$ and $\mathscr{A}$ has no irreducible components. This will give an example for an abelian v. Neumann algebra having neither irreducible components nor minimal projections. This will show that the type I continuous part $\mathscr{H}_{cI}$ can actually exist and so in the general, non-factorial case the type I property is not as strong as the existence of minimal projections or irreducible components. We shall actually find a representation such that the associated v. Neumann algebra will have the above described properties.

The left regular representation λ of the additive group $\mathbb{R}^+$ is a unitary representation acting on the Hilbert space $\mathscr{H}=L^2(\mathbb{R}^+)$ of functions $f\colon \mathbb{R}^+\to\mathbb{C}$ which are square integrable in the Lebesgue sense. The operator $\lambda(x)$ is the translation by $-x$ so that $\lambda(x)f(y)=f(y-x)$ for all y in $\mathbb{R}^+$. It is clear that $\lambda(x)$ is unitary and $\lambda\colon \mathbb{R}^+\to\mathscr{U}(\mathscr{H})$ is a representation of $\mathbb{R}^+$. The right regular representation ρ acts on the same Hilbert space $\mathscr{H}$ and $\rho(x)f(y)=f(x+y)$.

We let $\mathscr{A}_\lambda$ denote the v. Neumann algebra associated with λ. We have $\lambda(x)\lambda(y)=\lambda(y)\lambda(x)$ for all x,y in $\mathbb{R}^+$ and so by Proposition III.5.10 $\lambda(x)A=A\lambda(x)$ for every $x\in\mathbb{R}^+$ and $A\in\mathscr{A}_\lambda$. This shows that $\mathscr{A}_\lambda\subseteq(\mathscr{A}_\lambda)'=(\lambda,\lambda)$. The opposite inclusion is less obvious and its proof requires the use of the convolution product. The relation $\mathscr{A}_\lambda=(\mathscr{A}_\lambda)'$ is a simple special case of a theorem of Segal on the regular representations of an arbitrary unimodular group. Since the proof of this general result can be easily isolated and specialized to the case of the reals we shall not prove that $\mathscr{A}_\lambda\supseteq(\mathscr{A}_\lambda)'$. The reader will find it very easy to follow the general proof which is given in Section V.6. The reading can be made even easier by replacing multiplication in G by addition and G by $\mathbb{R}^+$.

Te fact that λ has no irreducible components is also a special instance of a general result. For it is a consequence of the non-compactness of the group. If, as in the present case, G is commutative then the result is not difficult to prove and its proof can be found in Section V.6. (See Theorem V.6.11.)

Remarks

The definition of a "representation ρ of an admissible object S in Hilbert space" is due to the author. The more restricted "Mackey type representation of a Mackey object S" was introduced by Mackey in 1955 (2).

It seemed appropriate to call the somewhat larger class of maps $\rho: S \to \mathscr{K}(\mathscr{H})$ "Mackey type representations of the object S with a map" and it is hoped that these two expressions will be used also by others. In contrast the term "morphism" was introduced only in order to be able to speak about a "representation ρ in Hilbert space" or a "representation ρ" without stating explicitely on every occasion that $\mathscr{A}_\rho$ is supposed to be a v. Neumann algebra. In the case of finite dimensional vector spaces and in particular in Section VIII.2 the term "morphism" will be replaced by the conventional "representation".

The terms and definitions introduced in Sections 2 and 3 are all standard an can be found for instance in Dixmier (1), Loomis (1), Mackey (2, 3) and Naimark (1). Mackey's paper (3) is a very important summary of the major developments in the theory of infinite dimensional group representations up to 1963. Many of these definitions were in use already for a long time at least in some special situations or in related fields. The equivalence of partial isometries was introduced by Murray and v. Neumann (1). A somewhat more restricted version of Proposition 3.10 can be found in Naimark's book (1) but the appropriate reference is unfortunately missing. (See p. 353 and p. 356.)

Schur's lemma was already mentioned at the end of Chapter II. Propositions 4.1 and 4.3 are due to v. Neumann (3). Theorem 4.5 and Proposition 4.6 first appeared in Mackey's notes (2) but the finite dimensional special cases of these results were known earlier. Theorem 4.8 and Proposition 4.9 were proved by the author. The same holds for all other statements on "representations in Hilbert space" but in many instances the old proofs can be taken over without changes in the reasonings. In a number of cases it is very hard or impossible to decide to whom a basic result is due. This is the situation for instance in the case of Theorems 5.4 and 5.5 which were known to Godement, Harish-Chandra, Mackey and others. Theorem 5.8 is a symmetric formulation of a result of Mackey (2). The material presented in Section 6 was first obtained in 1967 by the author and it is published here for the first time.

The basis of the theory discussed in Sections 7, 8 and 9 was laid down in a series of important papers by Murray and v. Neumann (1, 2, 3) and v. Neumann (4). The last paper on this subject by v. Neumann (5) was written in 1938 but it was published much later. This paper deals with direct integral decompositions of v. Neumann algebras and it will not be discussed in this book. Murray and v. Neumann were originally interested in representation theory but their papers deal almost exclusively with v. Neumann algebras. It was Mackey (2, 6) who adapted the relevant parts of this material to representation theory. The non-factorial, global type theory of v. Neumann algebras is due to Dixmier (3) and Kaplansky (2) and it was also considered by Segal (5). The example of type II factorial representations discussed in Section 7 is the first such example found and it is due to Murray and v. Neumann (3). There is a fair amount of modern literature on this topic and we mention only two writings of Sakai (1, 2) which contains further references. Proposition 8.32 is due to Fakler; I proved only the corollary and he noticed that the same reasoning also gives the proposition. It is known that an arbitrary v. Neumann algebra $\mathscr{A}$ is of type I, II or III if and only if $\mathscr{A}'$ belongs to the same type. (See pp. 112—113 in Sakai's book (1).) Hence we could say that ρ is of type I, II and III, respectively provided $\mathscr{A}_\rho=(\rho,\rho)'$ is that type of algebra.

Preliminary Remarks to Chapter V

In the following study of topological groups and homogeneous spaces we shall *not* assume tacitly that the Hausdorff separation axiom holds. Similarly a locally compact topological group will *not* necessarily be a Hausdorff space. This generality requires the use of slighly finer techniques which are summarized here. If the reader is interested only in Hausdorff groups he can ignore the following remarks. If he is unfamiliar with topological groups then he should first read Section V.1 and return to this before reading Section V.2. If a space X is uniformizable then for every closed set C and point $p\notin C$ there is a continusous function $f: X\to[0,1]$ such that $f\equiv 1$ on C and $f(p)=0$. We shall prove that *every* topological group is uniformizable. Hence the separating functions f exist for every C and $p\notin C$. If G is locally compact and C is a compact but not necessarily closed set then we can construct a continuous function $f: G\to[0,1]$ such that its support is compact and $f\equiv 1$ on C.

Let $\mathcal{O}$ denote the family of open subsets of the topological group G. Then $H=\bigcap\{O: O\in\mathcal{O}$ and $e\in O\}$ is an invariant subgroup which consists of those elements of G which can not be separated from e by open sets. For $O\in\mathcal{O}$ implies $O^{-1}\in\mathcal{O}$ and $a^{-1}Oa\in\mathcal{O}$ for all $a\in G$. Moreover if $e\in O$ but $x\notin O$ then $Q=O^{-1}x$ is in $\mathcal{O}$ and $x\in Q$ but $e\notin Q$. Hence $cH=\bigcup\{Q: Q\in\mathcal{O}$ and $e\notin Q\}$ which proves that H is closed. Therefore G/H is a Hausdorff group. A function $f: G\to\mathbb{C}$ is continuous if and only if f is constant on the cosets xH and the function φ_f defined by it on G/H is continuous there. Clearly $\|f\|_\infty=\|\varphi_f\|_\infty$, the supports of f and φ_f determine each other and $f\in C_0(G)$ if and only if $\varphi_f\in C_0(G)$. Thus if μ is a linear functional which defines a positive measure on G/H then $f\mapsto\mu(\varphi_f)$ is the linear functional of a positive measure on G.

Chapter V

Topological Groups, Invariant Measures, Convolutions and Representations

This chapter is almost independent of the previous ones and it can be read accordingly. For the first four sections only elementary point set theoretical and measure theoretical background is presupposed. Thus Section 1 can be read immediately and before starting Sections 2, 3 or 4 it is sufficient to be familiar with the contents of Section III.7. The remaining two sections require a casual acquaintence with the fundamental concepts of representations in Hilbert spaces. A partial reading of Sections IV.1, 2 and 3 is satisfactory for this purpose. In Section 5 we shall use also some basic facts about v. Neumann algebras $\mathscr{A}$. In particular we need that $\mathscr{A}$ is its second commutant and that $\mathscr{A}$ is weakly closed. All of this is discussed and proved in Section III.6. The text contains detailed references to this background material at the places where it is used. See also the preliminary remarks on the last page.

1. Topological Groups and Homogeneous Spaces

A *topological group* is a set G with the structure of a group and with a topology such that the group operations $x \mapsto x^{-1}$ and $(x,y) \mapsto xy$ are continuous functions on G and $G \times G$, respectively. The additive group of the reals and more generally the additive structure of $\mathbb{R}^n$ give simple examples of topological groups if the usual topology of $\mathbb{R}^n$ is chosen. Other examples of topological groups are the multiplicative groups of $\mathbb{R}$ and $\mathbb{C}$ and the one dimensional torus $T = \{e^{2\pi i\alpha}: 0 \leqslant \alpha < 1\}$.

If A and B are sets in G we let $AB = \{ab: a \in A \text{ and } b \in B\}$ and $A^{-1} = \{a^{-1}: a \in A\}$. We write aB and Ab instead of $\{a\}B$ and $A\{b\}$. If $A = A^{-1}$ then A is called *symmetric*. It is obvious that $(AB)C = A(BC)$ and $(AB)^{-1} = B^{-1}A^{-1}$.

By the continuity of $x \mapsto x^{-1}$ we see that O^{-1} is an open set if O is open. Similarly by the continuity of $x \mapsto a^{-1}x$ and $x \mapsto xa^{-1}$ one sees that aO and Oa are open sets in G for every $a \in G$ provided O is open. From this it follows that if A or B is open then so are AB and BA. If A is closed and B is finite then AB and BA are closed sets. Since

$(x,y) \mapsto xy$ is continuous it follows that AB is compact for every compact A and B because $A \times B$ is compact in the product topology. If A is connected then so is its continuous image A^{-1}. Moreover if A and B are connected sets then AB is connected because if a is in A then aB connects the connected sets $Ab\,(b \in B)$. This can be seen also from the fact that the product $A \times B$ of the connected sets A, B is connected and so its continuous image AB is connected.

The connectedness of A^{-1} and AB for connected A and B implies that the union of all connected sets containing the identity element e is a subgroup. Thus the component containing the identity is a closed normal subgroup of G which is called the *identity component* of G. If H is an open subgroup of G then the cosets xH for $x \notin H$ are all open and so the complement of their union, H is closed.

Since the translates aO and Oa are open sets a set N_x is a neighborhood of x if and only if $x^{-1}(N_x)$ and/or $(N_x)x^{-1}$ are neighborhoods of e. It follows that the neighborhood filter $\mathcal{N}(x)$ of x is $x\mathcal{N}(e) = \mathcal{N}(e)x$ where $\mathcal{N}(e)$ is the filter of neighborhoods of e. Similar statement holds for filter bases. It is easy to find necessary and sufficient conditions that a family of subsets be a base for $\mathcal{N}(e)$. Namely we have:

Theorem 1. *A family $\mathcal{B}(e)$ of sets $A, B, \ldots \subseteq G$ is a base for $\mathcal{N}(e)$ of some topological group G if and only if the following requirements are satisfied:*

1. *For every $x \in G$ and $A \in \mathcal{B}(e)$ there is a B in $\mathcal{B}(e)$ such that $B \subseteq x^{-1}Ax$.*
2. *For every pair of sets A, B in $\mathcal{B}(e)$ there is a C in $\mathcal{B}(e)$ such that $C \subseteq A \cap B$.*
3. *The identity belongs to every A of $\mathcal{B}(e)$.*
4. *For every A in $\mathcal{B}(e)$ there is a $B \in \mathcal{B}(e)$ such that $B^{-1} \subseteq A$.*
5. *For every $A \in \mathcal{B}(e)$ there is a B in $\mathcal{B}(e)$ such that $B^2 \subseteq A$.*

If $\mathcal{B}(e)$ satisfies these requirements then there is exactly one topology on the group G under which G is a topological group and $\mathcal{B}(e)$ is a base for $\mathcal{N}(e)$.

Note. We obtain an axiom system for $\mathcal{N}(e)$ instead of $\mathcal{B}(e)$ by adding to 1.—5. the following:

If $A \in \mathcal{B}(e)$ and $B \supseteq A$ then $B \in \mathcal{B}(e)$.

Proof. First we check that every base $\mathcal{B}(e)$ of the neighborhood filter $\mathcal{N}(e)$ of a topological group satisfies 1.—5. Conditions 1., 2. and 3. are obviously satisfied and 4. follows from the continuity of $x \mapsto x^{-1}$: Since $x \mapsto x^{-1}$ is continuous at e there is an open set O such that $e \in O^{-1} \subseteq A$

and then we can find a $B\in\mathscr{B}(e)$ such that $B\subseteq O$. Therefore $B^{-1}\subseteq O^{-1}\subseteq A$. The last condition can be verified by using the continuity of $(x,y)\mapsto xy$: By the continuity of the product at (e,e) there are neighborhoods $N_e^{(1)}$ and $N_e^{(2)}$ such that $N_e^{(1)}N_e^{(2)}\subseteq A$. We can find a $B\in\mathscr{B}(e)$ contained in $N_e^{(1)}\cap N_e^{(2)}$ and then we have $B^2\subseteq A$.

Next we suppose that G is a group and $\mathscr{B}(e)$ is a family of sets in G which satisfies 1.—5. Then we define $\mathscr{N}(x)$ as the family of sets xN_e where N_e is any subset of G containing some element A of $\mathscr{B}(e)$. We have $N\in\mathscr{N}(x)$ if and only if $xB\subseteq N$ for some $B\in\mathscr{B}(e)$. Using requirement 5. we prove that the sets $\mathscr{N}(x)\,(x\in G)$ are the neighborhood filters of a topology of G: We have to show that given N_x in $\mathscr{N}(x)$ there is an $O_x\in\mathscr{N}(x)$ such that $O_x\subseteq N_x$ and $N_x\in\mathscr{N}(y)$ for every $y\in O_x$. Now given $N_x=xN_e$ we can choose B in $\mathscr{B}(e)$ such that $B^2\subseteq N_e$. We let $O_x=xB$. If $b\in B$ then $xbB\subseteq xB^2\subseteq xN_e=N_x$ and so $N_x\in\mathscr{N}(xb)$ for every $y=xb\in O_x$.

Now we show that under this topology G is a topological group. First, in order to verify the continuity of $x\mapsto x^{-1}$ we fix $N_{x^{-1}}=x^{-1}N_e$. By the definition of $\mathscr{N}(e)$ there is an A in $\mathscr{B}(e)$ such that $A\subseteq N_e$ and by 4. there is a B for which $B^{-1}\subseteq A$. Now if $\xi\in Bx$ then $\xi^{-1}\in x^{-1}B^{-1}\subseteq x^{-1}N_e=N_{x^{-1}}$ and so it is sufficient to show that Bx is a neighborhood of x. Here, for the first time we need condition 1. If C in $\mathscr{B}(e)$ is such that $xCx^{-1}\subseteq B$ then $xC\subseteq Bx$ and as $xC\in\mathscr{N}(x)$ we see that $Bx\in\mathscr{N}(x)$.

Finally we prove that $(x,y)\mapsto xy$ is continuous on $G\times G$: Given x,y and $N_{xy}=xyN_e$ by 5. we can choose a B in $\mathscr{B}(e)$ such that $B^2\subseteq N_e$. By 1. we have $yBy^{-1}\in\mathscr{N}(e)$ and so $xyBy^{-1}\in\mathscr{N}(x)$. Similarly $yB\in\mathscr{N}(y)$. If $\xi\in xyBy^{-1}$ and $\eta\in yB$ then $\xi\eta\in(xyBy^{-1})(yB)=xyB^2\subseteq xyN_e=N_{xy}$. The continuity of the product at (x,y) is proved.

Theorem 2. *Every topological group is uniformizable. A base for the left uniform structure of G consists of the sets $\{(x,y): x^{-1}y\in N_e\}$ where $N_e\in\mathscr{N}(e)$ and the sets $\{(x,y): xy^{-1}\in N_e\}$ $(N_e\in\mathscr{N}(e))$ form a base for the right uniform structure of G.*

Proof. If $U=\{(x,y): x^{-1}y\in N_e\}$ and $B\in\mathscr{N}(e)$ is such that $B^2\subseteq N_e$ then $V=\{(x,y): x^{-1}y\in B\}$ satisfies $V\circ V\subseteq U$. The remaining conditions for a uniform structure are obviously satisfied. Since $U[x]=xN_e$ the neighborhood filter of any x in the uniform topology is the same as in the original topology of G. The right uniform structure can be treated similarly.

The topological groups G_1 and G_2 are called *isomorphic* if a bicontinuous group isomorphism exists between G_1 and G_2. Similarly a *homomorphism* of G_1 into G_2 means a continuous group homomorphism. If G is a topological group and $x \cdot y$ denotes yx then G with the binary operation $\cdot$ and the original topology is a topological group which is isomorphic to the original topological group G under the isomorphism $x \mapsto x^{-1}$. If we introduce the binary operation $x * y = x b^{-1} a^{-1} y$ where a, b are given elements of G we obtain another topological group G^* which is isomorphic to the original one under the map $x \mapsto axb$ of G onto G^*. The maps $T(a,b)$ given by $x \mapsto axb$ themselves form a group if we define $T(a,b)\,T(c,d) = T(ca, bd)$ which is the composition of the two maps in the diagrammatic order. A topology can be introduced on the set of $T(a, b)$ maps via the one-to-one map $(a, b) \mapsto T(a, b)$ where $(a,b) \in G \times G$.

One easily sees that if H is a subgroup of the topological group G then its closure $\bar{H}$ is also a subgroup. If H is a normal subgroup then so is $\bar{H}$. Moreover a subgroup H is open if and only if it contains an interior point.

The following is a list of illustrative examples of topological groups. Except for the first and the last two examples the groups described here are not only topological groups but actually manifolds.

1. An arbitrary group G becomes a topological group if we endow G with the discrete topology.

2. The *general linear group* $\mathrm{Gl}(n,F)$ over the field F consists of those $n \times n$ matrices with entries in F which are invertible. If F is topological e.g. if F is $\mathbb{C}$ or $\mathbb{R}$ then $\mathrm{Gl}(n,F)$ is an open set in the n^2 dimensional product space $F \times \cdots \times F$ because it consists of those matrices g whose determinant $|g| \neq 0$. The group with the subspace topology is a topological group because the matrix product and the inverse depend continuously on the set of matrix entries.

3. Any closed subgroup G of $\mathrm{Gl}(n,\mathbb{C})$ is a locally compact topological group. We will be using the expression *locally compact group* instead of the longer locally compact topological group.

4. The group of those matrices g of $\mathrm{Gl}(n,\mathbb{C})$ which have determinant 1 is the special linear group $\mathrm{Sl}(n,\mathbb{C})$. It is a closed normal subgroup of $\mathrm{Gl}(n,\mathbb{C})$ because it is the inverse image of 1 under the continuous homomorphism $g \mapsto |g|$ of $\mathrm{Gl}(n,\mathbb{C})$ onto $\mathbb{C}^\times$. Thus $\mathrm{Sl}(n,\mathbb{C})$ is a locally compact group. The elements of $\mathrm{Sl}(n,\mathbb{C})$ are often called *unimodular matrices* and occasionally $\mathrm{Sl}(n,\mathbb{C})$ is referred to as the *unimodular group*.

5. We can define similarly the *special linear group* $\mathrm{Sl}(n,F)$ over any field F. If F is a topological field then we obtain a topological group e.g. $\mathrm{Sl}(n,\mathbb{R})$.

6. The group of $n \times n$ orthogonal matrices $\mathrm{O}(n,\mathbb{R})$ is a closed subgroup of $\mathrm{Gl}(n,\mathbb{R})$ because it is the inverse image of the closed set con-

sisting of the identity matrix alone under the map $g \mapsto g g^t$ of $\mathrm{Gl}(n, \mathbb{R})$ into itself. By the orthonormality of the rows $x_{i1}^2 + \cdots + x_{in}^2 = 1$ for $i = 1, \ldots, n$ and so by the Heine-Borel theorem the *orthogonal group* $\mathrm{O}(n, \mathbb{R})$ is a compact topological group.

7. An analogous situation exists for the *unitary group* $\mathrm{U}(n)$ consisting of those complex matrices g for which $g g^* = i$ where $g^* = \overline{g}$ and i is the identity. Thus $\mathrm{U}(n)$ is a compact group.

8. One denotes by $\mathrm{SO}(n, \mathbb{R})$ and calls the *special orthogonal group* the intersection $\mathrm{Sl}(n, \mathbb{R}) \cap \mathrm{O}(n, \mathbb{R})$. Similarly the *special unitary group* $\mathrm{SU}(n)$ is the intersection $\mathrm{Sl}(n, \mathbb{C}) \cap \mathrm{U}(n)$.

9. If $\mathscr{V}$ is a topological vector space over $\mathbb{C}$ or $\mathbb{R}$ we let $\mathscr{L}(V)$ be the algebra of continuous linear operators $T: \mathscr{V} \to \mathscr{V}$. The group of units of $\mathscr{L}(\mathscr{V})$ is called the *general linear group* of $\mathscr{V}$ and is denoted by $\mathrm{Gl}(\mathscr{V})$. It consists of those linear transformations T which have continuous inverse. If $\mathscr{V}$ is a Banach space then $\mathscr{L}(\mathscr{V})$ is a Banach algebra under the operator norm and so by Proposition I.2.2 $\mathrm{Gl}(\mathscr{V})$ is an open set in $\mathscr{L}(\mathscr{V})$ and $T \mapsto T^{-1}$ is continuous on $\mathrm{Gl}(\mathscr{V})$. By $|ST| \leqslant |S| \cdot |T|$ the product map $(S, T) \mapsto ST$ is also continuous and so $\mathrm{Gl}(\mathscr{V})$ is a topological group under the induced topology.

10. If V is a complex Hilbert space $\mathscr{H}$ then we can define the subgroup $\mathscr{U}(\mathscr{H})$ of unitary operators and if V is finite dimensional then it is meaningful to speak about $\mathrm{Sl}(\mathscr{V})$, the special linear group of $\mathscr{V}$. Both $\mathscr{U}(\mathscr{H})$ and $\mathrm{Sl}(\mathscr{V})$ are topological groups under the induced topology.

If G is a topological group and H is a subgroup we let G/H denote the set of left cosets xH and topologize G/H by the quotient topology of the equivalence relation $x \sim y$ meaning $x^{-1} y \in H$. The space G/H is called the (left) *homogeneous space* of G with respect to H. Similarly we let $H \backslash G$ denote the set of right cosets Hx topologized by the quotient topology corresponding to the equivalence relation $x \sim y$ meaning $x y^{-1} \in H$. We will concentrate on G/H but our results apply also to $H \backslash G$ with obvious modifications. The map $p: G \to G/H$ defined by $p(x) = xH$ is called the *projection* or the *natural map* of G onto G/H. It is continuous, in a little while we shall prove that it is open and we can see by looking at examples that in general it is not closed.

Proposition 3. *The homogeneous space G/H is a Hausdorff space if and only if H is a closed subgroup of G.*

Proof. If G/H is a Hausdorff space then $p(e) = \{H\}$ is a closed set in G/H and so $p^{-1}(p(e)) = H$ is a closed set in G. Conversely, let us suppose that H is closed and select two points $\xi \neq \eta$ in G/H. We let

$\xi = xH$ and $\eta = yH$ so that $x^{-1}y \notin H$ and so H being closed there is a symmetric neighborhood N_e such that $x^{-1}yN_e \cap H = \emptyset$. This implies that $x^{-1}y \notin N_e H$ which will be used in a moment.

We now choose a symmetric open set O such that $e \in O \subseteq O^2 \subseteq xN_e x^{-1}$ and prove that OxH and OyH are disjoint neighborhoods of x and y in G/H. In fact if we had $o_1 x h_1 = o_2 y h_2$ for $o_1, o_2 \in O$ and $h_1, h_2 \in H$ then

$$y h_2 h_1^{-1} = o_2^{-1} o_1 x \in O^{-1} O x = O^2 x \subseteq x N_e$$

and so $x^{-1} y \in N_e h_1 h_2^{-1} \subseteq N_e H$ would follow.

Proposition 4. *Every homogeneous space G/H is uniformizable, its right uniform structure has a base consisting of the uniformities*

$$U(N_e) = \{(\xi_1, \xi_2) \colon \xi_1, \xi_2 \in G/H \text{ and } \xi_1 \xi_2^{-1} \cap N_e \neq \emptyset\}$$

where N_e is a neighborhood of e. In the base for the left uniform structure one requires that $\xi_1^{-1} \xi_2 \cap N_e \neq \emptyset$.

Note. Clearly $\xi_1 \xi_2^{-1}$ is to be interpreted as a subset of G. An alternate description of the uniformities is

$$U(N_e) = \{(\xi_1, \xi_2) \colon \xi_k = x_k H \text{ and } x_1 h x_2^{-1} \in N_e \text{ for suitable } h \in H\}.$$

Proof. As $e \in \xi \xi^{-1}$ we have $I \subseteq U(N_e)$ for any N_e. Next we see that $U(N_e^1 \cap N_e^2) \subseteq U(N_e^1) \cap U(N_e^2)$ and $U(N_e^{-1}) = U(N_e)^{-1}$. If O_e is such that $O_e^2 \subseteq N_e$ then we have

$$U(O_e) \circ U(O_e) \subseteq U(N_e) \colon$$

Indeed, if $x_1 h_1 x_2^{-1} \in O_e$ and $x_2 h_2 x_3^{-1} \in O_e$ then $x_1 (h_1 h_2) x_3^{-1} \in O_e^2 \subseteq N_e$. Therefore the $U(N_e)$ sets form a base for a uniform structure $\mathscr{U}_r$.

Let $p \colon G \to G/H$ be the projection of G onto G/H. We suppose that Q in G/H is open with respect to $\mathscr{U}_r$. If $x \in p^{-1}(Q)$ then $\xi = xH \in Q$ and so there is an N_e such that $xy^{-1} \in N_e$ implies $\eta = yH \in U(N_e)[\xi] \subseteq Q$. Thus $y \in p^{-1}(Q)$ if $y \in N_e^{-1} x = N_x$. This proves that $p^{-1}(Q)$ is open in G and so Q is open in the quotient topology.

Next suppose that Q is open in the quotient topology and $\xi = xH \in Q$. Then $x \in p^{-1}(Q)$ so there is an N_x such that $x \in N_x \subseteq p^{-1}(Q)$. We let $N_e = x N_x^{-1}$ and prove that $U(N_e)[\xi] \subseteq Q$: If $\eta \in U(N_e)[\xi]$ then $\xi \eta^{-1} \cap N_e \neq \emptyset$ i.e. $xhy^{-1} \in N_e$ where h is in H and $\eta = yH$. Therefore $xhy^{-1} \in x N_x^{-1}$ i.e. $hy^{-1} \in N_x^{-1}$, that is to say $yh^{-1} \in N_x \subseteq p^{-1}(Q)$. Hence $y \in p^{-1}(Q)$ and so $\eta \in Q$ proving that $U(N_e)[\xi] \subseteq Q$.

Definition 5. *We say that the locally compact group G acts on the Hausdorff space S if a continuous map $G\times S\to S$ is given such that $g_1(g_2 s) = (g_1 g_2)s$ where gs denotes the image of (g,s) under the given map and if $s\mapsto gs$ is a homeomorphism for every $g\in G$.*

If $gs=s$ for every $g\in G$ and $s\in S$ we say that G *acts trivially* on S. If for any ordered pair (s_1, s_2) there is a g in G such that $gs_1=s_2$ then G is said to act *transitively* on S. For example if K is a closed subgroup of the locally compact group G then $S=G/K$ gives rise to the situation where G acts transitively on S, namely $(g, xK)\mapsto(gx)K$. Given $s_k=x_k K$ $(k=1,2)$ we have $gs_1=s_2$ for $g=x_2x_1^{-1}$. The continuity of the map is proved as follows: Given an open set Q in G/K we have $gxK\in Q$ if and only if $gx\in p^{-1}Q$ where $p^{-1}Q$ is an open set. The set of those pairs (g,x) for which $gx\in p^{-1}Q$ is the inverse image of $p^{-1}Q$ under the continuous map $(g,x)\mapsto gx$. Thus the inverse image of the open set Q under the map $(g,xK)\mapsto gxK$ is open and the map itself is continuous. Moreover we also notice that for fixed g and variable xK *the map $xK\mapsto gxK$ of G/K onto itself is a homeomorphism*: The map is obviously bijective. If Q is an open set in G/K then the inverse image of $g^{-1}Q$ under the projection p is $g^{-1}\bigcup Q$ where $\bigcup$ denotes the union of the cosets in Q. The set Q being open so are $p^{-1}(Q)=\bigcup Q$ and $g^{-1}\bigcup Q$. Hence $g^{-1}Q$ is open in G/K and so $xK\mapsto gxK$ is continuous.

Returning to the general situation when G acts on the space S we fix a point s_o in S and consider

$$K=\{g: g\in G \text{ and } gs_o=s_o\}.$$

Then K is clearly a subgroup of G which is closed because it is the inverse image of the single point s_o under the continuous map $g\mapsto gs_o$. It is called the *stability subgroup* of G at s_o or the *isotropy subgroup* of G at s_o.

Let K be the stability subgroup of G at a point s_o of S and let

$$\varphi: G/K\to S$$

defined by $\varphi(gK)=gs_o$. The map is well defined because if g_1, g_2 belong to the same coset i.e. if $g_2^{-1}g_1\in K$ then $g_1s_o=g_2s_o$. Conversely, if $g_1s_o=g_2s_o$ then $g_2^{-1}g_1s_o=s_o$ and so $g_2^{-1}g_1\in K$. Therefore the map is an injection of G/K into S. If G acts transitively on S then φ is surjective and so φ is a one-to-one correspondence of G/K onto S. We plan to show that under mild restrictions φ is a homeomorphism. First we recall the following:

Lemma 6. *If K is a subgroup of G and if G/K is endowed with the quotient topology then the natural map $p: G\to G/K$ is open and continuous.*

Proof. The map p is continuous because the quotient topology is the strongest topology on G/K which makes p continuous. On the other hand if O is an open set in G then

$$p^{-1}\{xK: x\in O\} = \bigcup\{xK: x\in O\} = OK$$

is open in G and so by the definition of the quotient topology $\{xK: x\in O\} = pO$ is open.

Proposition 7. *If K is the stability subgroup of s_o then the map $\varphi: G/K \to S$ defined by $\varphi(gK) = g s_o$ is a one-to-one, continuous map of G/K onto S.*

Proof. We had already seen that φ is a bijection. We notice that $\psi = \varphi p$ maps g in G into $g s_o$ in S and so by the continuity of the action map $G\times S \to S$ the function $\psi: G \to S$ is continuous. Now in order to see that φ is continuous we choose an open set Q in S. Then $\psi^{-1}(Q)$ is open in G and so $p: G\to G/K$ being open $p(\psi^{-1}(Q)) = \varphi^{-1}(Q)$ is an open set in G/K.

Theorem 8. *Let G and S be locally compact, let S be a Hausdorff space and suppose that the Lindelöf property holds in G. If G acts transitively on S and K is the stability subgroup of a point then $\varphi: G/K \to S$ is a homeomorphism.*

In view of the last proposition it is sufficient to prove that φ is an open map. The proof of that is based on the following:

Lemma 9. *If S is a locally compact Hausdorff space and if $S = \bigcup S_n$ $(n = 1, 2, \ldots)$ where each S_n is closed then there exists at least one index n such that $(S_n)^i \neq \emptyset$.*

Proof. Let us suppose that $(S_n)^i = \emptyset$ for every $n = 1, 2, \ldots$ We choose a non-void open set O_1 in S such that $\bar{O}_1$ is compact. Then we can find a point a_1 in $O_1 - S_1$. Since S is regular there is an open set O_2 such that $a_1 \in O_2 \subseteq \bar{O}_2 \subseteq O_1$ and $\bar{O}_2 \cap S_1 = \emptyset$. Next we choose a_2 in $O_2 - S_2$. Again there is an open set O_3 such that $a_2 \in O_3 \subseteq \bar{O}_3 \subseteq O_2$ and $\bar{O}_3 \cap S_2 = \emptyset$. In general $(S_n)^i = \emptyset$ so there is a point a_n in $O_n - S_n$ and then there is an open set O_{n+1} such that $a_n \in O_{n+1} \subseteq \bar{O}_{n+1} \subseteq O_n$ and $\bar{O}_{n+1} \cap S_n = \emptyset$. Now $\bar{O}_1 \supseteq \bar{O}_2 \supseteq \cdots \supseteq \bar{O}_n \supseteq \cdots$ are non-void compact sets, so their intersection is not void. If $s \in \bigcap \bar{O}_n$ then $s \notin S_n$ for every $n = 1, 2, \ldots$ and so $\bigcup S_n$ is not the total space S.

Proof of Theorem 8. It is enough to prove that $\psi = \varphi p: G \to S$ is an open map because if O is an open set in G/K then $p^{-1}(O)$ is open in G and $\varphi(O) = \psi(p^{-1}(O))$. So $\varphi(O)$ is open if ψ is an open map. Let O be an open set in G, let $x \in O$ and let C be a compact neighborhood of e in G such that $C = C^{-1}$ and $xC^2 \subseteq O$. We are going to prove that $\psi(x)$ belongs to the interior of $\psi(O)$.

The sets $(gC)^i (g \in G)$ form an open cover of G so by the Lindelöf property there is a countable subcover, say $G = \bigcup g_n C \ (n=1,2,\ldots)$. Since $Gs_o = S$ we have $S = \bigcup (g_n C \cdot s_o)$ where C is compact and so by the continuity of $(g,s) \mapsto gs_o$ the sets $g_n C \cdot s_o$ are compact and hence closed. By the lemma there is an index n such that $(g_n C \cdot s_o)^i \neq \emptyset$. Therefore $(Cs_o)^i \neq \emptyset$ because $s \mapsto g_n s$ is a homeomorphism.

If $hs_o \in (Cs_o)^i$ then $s_o \in (h^{-1} Cs_o)^i$ and h is an element of C. Hence by $h^{-1} C \subseteq C^2$ and $xC^2 \subseteq O$ we have $s_o \in (h^{-1} C s_o)^i \subseteq C^2 s_o \subseteq x^{-1} O s_o$. Therefore multiplying by x we obtain

$$\psi(x) = x s_o \in x(h^{-1} C \cdot s_o)^i \subseteq O s_o = \psi(O) .$$

We proved that $\psi(x) \in (\psi(O))^i$.

We are now going to describe some important instances of groups acting on a space and the corresponding homogeneous spaces.

1. Let $G = I_o(\mathbb{R}^2)$, the *identity component of the group of isometries* of the plane $\mathbb{R}^2$. In simpler terms G is the *group of motions* of the plane and it can be described as the set of pairs (a, s) where $a = \begin{pmatrix} a_1 & a_2 \\ -a_2 & a_1 \end{pmatrix}$ and $s = \begin{pmatrix} s_1 \\ s_2 \end{pmatrix}$. Here a_1, a_2 and s_1, s_2 are real numbers and $a_1^2 + a_2^2 = 1$. The matrix a corresponds to a rotation through an angle φ where $\tan\varphi = a_2/a_1$ and s means a translation by the vector s. The group operation is $(a,s)(b,t) = (ab, at+s)$. One calls G the *semidirect product* of the group of rotations a and the group of translations s.

The group G acts transitively on $\mathbb{R}^2$, namely if $g = (a, s)$ then $gx = \begin{pmatrix} x_1 \\ x_2 \end{pmatrix} = ax + s$. The stability subgroup K corresponding to the point $x_o = \begin{pmatrix} 0 \\ 0 \end{pmatrix}$ is $SO(2, \mathbb{R})$, the group of matrices a. By Theorem 8 we see that

$$I_o(\mathbb{R}^2)/SO(2, \mathbb{R})$$

is homeomorphic to the plane $\mathbb{R}^2$.

A second identification shows that the topological space of *double cosets*

$$SO(2, \mathbb{R}) \backslash I_o(\mathbb{R}^2) / SO(2, \mathbb{R})$$

is homeomorphic to the half line $[0, \infty)$. For the vectors x which are equivalent under the application of $SO(2, \mathbb{R})$ from the left describe a circle of radius $|x|$ and centered around the origin.

2. Let G be the *special orthogonal group* $SO(3, \mathbb{R})$ consisting of all 3×3 orthogonal matrices with determinant 1. The group acts transitively on the *sphere* S^2 interpreted as unit column vectors x with components x_1, x_2, x_3 so that $x = (x_1, x_2, x_3)^t$. The element gx is the matrix product of $g \in SO(3, \mathbb{R})$ with $x \in S^2$. If we choose $x_o = (0, 0, 1)^t$ then the stability subgroup consists of all orthogonal matrices of determinant 1 which have the structure

$$\begin{pmatrix} * & * & 0 \\ * & * & 0 \\ 0 & 0 & 1 \end{pmatrix}$$

By omitting the third row and third column we get an isomorphism of K onto $SO(2, \mathbb{R})$. Hence by Theorem 8 $SO(3, \mathbb{R})/SO(2, \mathbb{R})$ is homeomorphic to the sphere S^2.

The double coset $SO(2, \mathbb{R}) g\, SO(2, \mathbb{R})$ is the orbit of the point corresponding to $g\, SO(2, \mathbb{R})$ on S^2 around the x_3-axis. These parallel circles form the zones of the sphere S^2. One can see that the interval $[-1, 1]$ is homeomorphic to $SO(2, \mathbb{R}) \backslash SO(2, \mathbb{R})/SO(2, \mathbb{R})$.

3. Let G be the *special linear group* $Sl(2, \mathbb{R})$ consisting of 2×2 real matrices $g = \begin{pmatrix} a_{11} & a_{12} \\ a_{21} & a_{22} \end{pmatrix}$ with $a_{11} a_{22} - a_{12} a_{21} = 1$. We let S be the *upper half plane*

$$\{z : z = x + iy \text{ with } x, y \in \mathbb{R} \text{ and } y > 0\}.$$

The action of the group G consists of mapping z into

$$gz = \frac{a_{11} z + a_{12}}{a_{21} z + a_{22}}.$$

We can easily see that G acts transitively on S because i is mapped into $gi = u + iv$ by the action of that group element g whose entries satisfy

$$a_{11} = a_{21} u + a_{22} v \quad \text{and} \quad a_{12} = a_{22} u - a_{21} v.$$

The stability subgroup corresponding to the point i turns out to be $SO(2, \mathbb{R})$ and so by Theorem 8 we obtain that $Sl(2, \mathbb{R})/SO(2, \mathbb{R})$ is homeomorphic to the upper half plane. The image of the coset $g\, SO(2, \mathbb{R})$ under the homeomorphism is the point gi in the upper half plane.

The image of a double coset KgK under this homeomorphism is the set of all complex numbers kgi where $k\in K$. If we let $gi=z$ and

$$k=\begin{pmatrix} \cos\varphi & \sin\varphi \\ -\sin\varphi & \cos\varphi \end{pmatrix}$$

then this image consists of the complex numbers $(z+\tau)/(1-z\tau)$ where $\tau=\tan\varphi$ and $-\infty<\tau<\infty$. Therefore the point set corresponding to the double coset is the image of the real line $-\infty<\tau<+\infty$ under the linear fractional transformation having matrix $\begin{pmatrix} 1 & z \\ -z & 1 \end{pmatrix}$. Since every such transformation preserves circles the point set is a circle lying in the upper half plane. A short computation shows that the imaginary part of $w=(\tau+z)/(1-\tau z)$ has extremal values at the same points where $\mathcal{R}e\, w=0$. Hence the center of the circle is on the imaginary axis. Now the circle can be easily constructed because z and $-1/z$ lie on it. The double coset space $\mathrm{SO}(2,\mathbb{R})\backslash\mathrm{Sl}(2,\mathbb{R})/\mathrm{SO}(2\,\mathbb{R})$ is homeomorphic to the half line $[0,\infty)$.

4. Let G be a group, K a closed subgroup and N a normal subgroup of G. Every g in G can be written uniquely in the form $g=nk$ if and only if $G=NK$ and $N\cap K=\{e\}$. If this is the case then we say that G is the *semidirect product* of the subgroups N and K. Since N is normal we can define the automorphisms $\alpha_k\colon N\to N$ by $\alpha_k(n)=knk^{-1}$ where $k\in K$. The map $k\mapsto\alpha_k$ is a homomorphism of K into the group of all automorphisms of N because

$$\alpha_{k_1}\alpha_{k_2}(n)=k_1(k_2nk_2^{-1})k_1^{-1}=(k_1k_2)n(k_1k_2)^{-1}=\alpha_{k_1k_2}(n)\,.$$

The structure of G is uniquely determined by the groups N and K and the homomorphism $k\mapsto\alpha_k$: Indeed $g_1g_2=n_1k_1n_2k_2=n_1(k_1n_2k_1^{-1})k_1k_2$ $=n_1\alpha_{k_1}(n_2)k_1k_2$.

We now look at the converse situation where two groups N and K and a homomorphism α of K into the group of automorphisms $Aut\,N$ $\subset\mathrm{Hom}(N,N)$ are given. We define $N\,\textcircled{\alpha}\,K$, the *semidirect product* of N and K with respect to α as follows: $N\,\textcircled{\alpha}\,K$ consists of the pairs (n,k) and

$$(n_1,k_1)(n_2,k_2)=(n_1\alpha_{k_1}(n_2),k_1k_2)\,.$$

One easily verifies that this binary operation defines a group structure on $N\times K$. If N and K are topological groups we consider the product topology on $N\,\textcircled{\alpha}\,K$. If N and K are locally compact we require α to be such that $N\,\textcircled{\alpha}\,K$ is a locally compact group.

5. Let N be the additive group of $\mathbb{R}^n$ the elements being written as column vectors and let $K=\mathrm{SO}(n,\mathbb{R})$, the *special orthogonal group* of $n\times n$ real matrices. We define the automorphism $\alpha_k\colon N\to N$ by $\alpha_k(v)=kv$ where $k\in K, v\in N$ and kv is a matrix product. The elements of $N \circledR K$ are pairs (v,k) where k can be considered a rotation and v a translation in $\mathbb{R}^n$. Thus the semidirect product $G=\mathbb{R}^n \circledR \mathrm{SO}(n,\mathbb{R})$ is isomorphic to the *group of motions* of the n-dimensional Euclidean space. G acts transitively on $S=\mathbb{R}^n$.

If we choose the element $s_0=0$ in the space S then the action of (v,k) on $x\in\mathbb{R}^n$ being $kx+v$ we see that $K=\mathrm{SO}(n,\mathbb{R})$ and Theorem 8 shows that

$$\mathbb{R}^n \circledR \mathrm{SO}(n,\mathbb{R})/\mathrm{SO}(n,\mathbb{R})$$

is homeomorphic to $\mathbb{R}^n$. The double coset space is homeomorphic to the half line $[0,\infty)$.

6. If G is $\mathrm{SO}(n+1,\mathbb{R})$ and S is the n-sphere S^n with points $x=(x_1,\dots,x_{n+1})^t$ then G acts on S^n by mapping x into gx where $g=(a_{ij})$ belongs to G. The action is transitive so Theorem 8 is again applicable. If we choose as fixed point $x_o=(0,\dots,0,1)^t$ then K consists of those orthogonal matrices with determinant 1 whose structure is

$$\begin{pmatrix} * & \dots & * & * & 0 \\ \dots & \dots & \dots & \dots & \dots \\ * & \dots & * & * & 0 \\ 0 & \dots & 0 & 0 & 1 \end{pmatrix},$$

Therefore K is isomorphic to $\mathrm{SO}(n,\mathbb{R})$ and $\mathrm{SO}(n+1,\mathbb{R})/\mathrm{SO}(n,\mathbb{R})$ is homeomorphic to the n-sphere S^n. The double coset space is homeomorphic to the closed interval $[-1,+1]$.

7. We let G be the *general linear group* $\mathrm{Gl}(n,\mathbb{R})$ and denote by P the set of $n\times n$ *symmetric, positive definite matrices* topologized by the usual topology of the $\frac{1}{2}n(n+1)$ dimensional Euclidean space. Positive definite means that $x^t p x>0$ for every $x^t=(x_1,\dots,x_n)$ where t denotes transposing and $x_1,\dots,x_n\neq 0,\dots,0$ in $\mathbb{R}$. The group G acts on P by mapping p into gpg^t. The action is continuous and it is transitive because if $p\in P$ then by the *principal axis theorem* there is an $n\times n$ orthogonal matrix a such that $a^t p a$ is a diagonal matrix. Then the application of a diagonal matrix will further transform p into the $n\times n$ identity matrix i.

A stability subgroup K will be selected by choosing as fixed point the matrix i. Then we have $a\in K$ if and only if $aa^t=i$ and so $K=\mathrm{O}(n,\mathbb{R})$.

Theorem 8 implies that $\mathrm{Gl}(n,\mathbb{R})/\mathrm{O}(n,\mathbb{R})$ is homeomorphic to the space P of $n\times n$ positive definite, symmetric matrices. We see that the double coset space

$$\mathrm{O}(n,\mathbb{R})\backslash\mathrm{Gl}(n,\mathbb{R})/\mathrm{O}(n,\mathbb{R})$$

is homeomorphic to $(0,\infty)\times\cdots\times(0,\infty)$ by the principal axis theorem. Hence it can be considered as $\mathbb{R}^n$.

8. We denote by G the *special linear group* $\mathrm{Sl}(n,\mathbb{C})$ *over the field* $\mathbb{C}$ and by P the set of $n\times n$ *hermitian, positive definite matrices.* These form an open set in the n^2 dimensional Euclidean space. We let G act on P according to the rule $g\cdot p=gpg^*$ where g^* is the hermitian conjugate $\bar{g}^t$. The action is transitive because for every hermitian matrix h there is a unitary matrix u such that uhu^* is a real diagonal matrix and so a further reduction by a real diagonal matrix of determinant 1 leads to the identity matrix i. One can also see that the action is continuous.

As stationary point we choose i so that K consists of those elements of the group for which $gg^*=i$. Thus K is the *special unitary group* $\mathrm{SU}(n,\mathbb{C})$. By Theorem 8 the homogeneous space $\mathrm{Sl}(n,\mathbb{C})/\mathrm{SU}(n,\mathbb{C})$ is homeomorphic to P. By the result quoted on the reduction of hermitian matrices to real diagonal ones we see that in this situation the double coset space

$$\mathrm{SU}(n,\mathbb{C})\backslash\mathrm{Sl}(n,\mathbb{C})/\mathrm{SU}(n,\mathbb{C})$$

is homeomorphic to $(0,\infty)\times\cdots\times(0,\infty)$ i.e. to $\mathbb{R}^n$.

Theorem 10. *Suppose G and H are topological groups, G has the Lindelöf property and H is a Hausdorff space. Then for every continuous homomorphism ρ of G onto H the induced map $\bar{\rho}\colon G/\ker\rho\to H$ is a homeomorphism.*

Proof. In order to apply Theorem 8 we let $S=H$ and $g\cdot h=\rho(g)h$. Since ρ is surjective G acts transitively on S. The stability subgroup of the identity element of H is $K=\ker\rho$. Hence $\varphi\colon G/K\to H$ is a homeomorphism. Clearly $\varphi(gK)=g\cdot e=\rho(g)e=\rho(g)=\bar{\rho}(gK)$ and so $\bar{\rho}$ is a homeomorphism of G/K onto H.

Proposition 11. *If C is compact, O is open and if $C\subset O\subseteq G$ then there is a symmetric neighborhood of the identity N_e such that $CN_e\cup N_eC\subseteq O$.*

Proof. Since C is compact and G is uniformizable we can find a continuous function f with compact support such that $f(x)=1$ for $x\in C$ and $f(x)=0$ for $x\notin O$. Since f is uniformly continuous with respect to both the left and the right uniform structures of G by Theorem 2 we can find a symmetric neighborhood N_e such that $|f(x)-f(y)|<\frac{1}{3}$ if $x^{-1}y\in N_e$ of if $xy^{-1}\in N_e$. Now given x in C and $y\in xN_e$ by $f(x)=1$ we have $f(y)\neq 0$ and so $y\in O$. This proves that $xN_e\subseteq O$ for every x in C and so $CN_e\subseteq O$. Similarly we see that $N_eC\subseteq O$.

Lemma 12. *If H is a closed subgroup of G and if C is compact then CH is a closed set.*

Proof. By Lemma 6 the image of C under the natural map p is a compact set in G/H. Hence by Proposition 3 the image set is closed and so by Lemma 6 the inverse image CH is a closed subset of G.

Lemma 13. *If f is in $C_o(G)$ then f is left and right uniformly continuous.*

Proof. Given $\varepsilon>0$ there is a compact set C such that $|f(x)|\leqslant\varepsilon/2$ for every $x\notin C$. We may suppose that C is a neighborhood of e and $C=C^{-1}$. Since C^2 is compact the restriction of f to C^2 is uniformly continuous and so there is a neighborhood N_e such that $|f(x)-f(y)|<\varepsilon$ for every $x,y\in C^2$ satisfying $xy^{-1}\in N_e$. We can require also that $N_e\subseteq C$. Now let x and y in G be such that $xy^{-1}\in N_e$. If $x,y\notin C$ then $|f(x)-f(y)|\leqslant\varepsilon$ by the choice of C. If x or y is in C then by $N_e\subseteq C$ we have $x,y\in C^2$ and so $|f(x)-f(y)|<\varepsilon$ by $xy^{-1}\in N_e$. Therefore f is right uniformly continuous. The proof of the left uniform continuity is similar.

2. Haar Measure

This section is devoted to the proof of existence and properties of an invariant measure and invariant integration on an arbitrary locally compact topological group G. By a *left invariant measure* we understand a regular Borel measure on G such that

1) $\mu(xE)=\mu(E)$ for every measurable set E;
2) $0\leqslant\mu(C)<\infty$ for every compact set C;
3) $0<\mu(C)$ for every compact set C whose interior is not void.

Similarly μ is called a *right invariant measure* on G if in addition to 2) and 3) we have $\mu(Ex)=\mu(E)$ for every measurable set E. If μ is such that $\mu(xE)=\mu(Ex)=\mu(E)$ for all measurable sets E then μ is called a *biinvariant measure*. In 1933 Alfréd Haar proved that on any locally compact group G there exists a left invariant and also a right invariant measure. For this reason we often speak about left, right and biinvariant Haar measures and Haar integrals. Our first object is to prove the existence of Haar measures.

Somewhat later, in 1936 von Neumann proved that any two left or right invariant measures differ only in a positive factor and so one can say that the left and right Haar measures are *essentially unique*. It will be our second task to prove this uniqueness property. If the group admits a biinvariant measure then it is called a *unimodular group*. For these groups it is very easy to show the essential uniqueness of the invariant measure. We shall begin the uniqueness proof with this special case. The rest of the section deals with the actual computation of the Haar measure and for a number of specific groups the invariant measure will be explicitly given.

The integral associated with the left invariant measure is called the *left invariant integral* or *left Haar integral*. The invariance property means that

$$\int_G f(ax)\,dx = \int_G f(x)\,dx$$

for every integrable function f. The additive group of the reals $\mathbb{R}^+$ together with the Lebesgue measure of the real line furnish a simple, non-trivial example of a biinvariant measure. Another example consists of the multiplicative group of the positive reals and the measure dx/x where dx denotes the Lebesgue measure on the half line $(0, \infty)$. One sees immediately that

$$\int_G f(ax)\frac{dx}{x} = \int_G f(x)\frac{dx}{x}$$

for any function $f\colon (0, \infty) \to \mathbb{C}$ for which $f(x)/x$ is Lebesgue integrable.

Next we give an example of a group with a left invariant integral which is not right invariant. Let

$$G = \{(x, y)\colon x, y \in \mathbb{R} \text{ and } x \neq 0\}$$

where $(a, b)(x, y) = (ax, ay + b/x)$ and consider the measure $x^{-2}dx\,dy$. Then

$$\int_G f\left(ax, ay + \frac{b}{x}\right)\frac{dx\,dy}{x^2} = \int_G f(\xi, \eta)\left|\frac{\partial(x, y)}{\partial(\xi, \eta)}\right| \frac{d\xi\,d\eta}{\left(\frac{\xi}{a}\right)^2} = \int_G f(\xi, \eta)\frac{d\xi\,d\eta}{\xi^2}$$

and so the measure is left invariant.

Similar computation show that $dx\,dy/x^2$ is not a right invariant measure on G. Instead we find that $dx\,dy$ leads to a right invariant integration on G.

We give also an example of a non-commutative group G with biinvariant Haar measure. For interpret $G=\mathbb{R}^3$ as the set of matrices

$$\begin{pmatrix} 1 & x_1 & x_2 \\ 0 & 1 & x_3 \\ 0 & 0 & 1 \end{pmatrix}$$

and define multiplication accordingly. Hence

$$(x_1,x_2,x_3)(y_1,y_2,y_3)=(x_1+y_1,x_2+y_2+x_1y_3,x_3+y_3).$$

The identity element is $(0,0,0)$, the inverse is $((x_1,x_2,x_3)^{-1}=(-x_1,-x_2+x_1x_3,-x_3)$ and the center of the group consists of the elements $(0,x_2,0)$ where $x_2\in\mathbb{R}$. One easily verifies that Lebesgue integration in 3 dimensional space is a biinvariant integration on our group which is therefore unimodular.

If G is a compact group $\mu(G)$ is finite and so it is possible and customary to normalize the Haar measure of G such that $\mu(G)=1$. Similarly if G is a discrete group then we normalize μ by requiring that the measure of each group element be 1. If the group is finite it is usually best to normalize with respect to the discreteness and not the compactness of G.

Proof of the existence. First a few words about Haar's original, intuitive version of the proof. One chooses an open set O_o whose closure is compact. The measure of O_o will be considered unity and one estimates the measure of any other open set O having compact closure $\bar{O}$ by the minimal number of left translates of O_o needed to cover O. It this number is denoted by $(O:O_o)$ then for any three such open sets one has $(A:C)\leqslant(A:B)(B:C)$.

In order to obtain a more precise comparison of the measures of O_o and O one chooses a sequence of open neighborhoods O_n of e and one takes the ratio $(O:O_n)/(O_o:O_n)$ as an approximate measure. By the last inequality one has $1/(O_o:O)\leqslant(O:O_n)/(O_o:O_n)\leqslant(O:O_o)$.

If the bounded sequence of these ratios is convergent then its limit will be the desired measure of O.

The following detailed proof due to André Weil is based on the original ideas of Haar but instead of measures it is formulated in terms of linear functionals and for this reason it is necessary to suppose that G is a Hausdorff group. However as we proved in the Preliminary remarks preceding Chapter V it is sufficient to consider this special case. Also in order to handle non-separable groups it is necessary to use some kind of transfinite argument which is neatly done by Weil via the axiom of choice in the form of Tychonoff's theorem.

Let $C_0(G)$ denote the algebra of complex valued continuous functions on G with compact support. If f, g are non-negative elements of $C_0(G)$ and if g is not identically zero then by the uniformizability of G the function f can be majorized by a positive linear combination of left translates of g i. e. there exist constants $c_1,\dots,c_n>0$ and groups elements $a_1,\dots,a_n$ such that

$$f(x)\leqslant \sum c_i g(a_i x).$$

We define $(f:g)$ as the greatest lower bound of the admissible sums $\sum c_i$. It is clear that

$$\begin{aligned}({}_a f:g)&=(f:g) \quad \text{for every } a \text{ in } G,\\ (f_1+f_2:g)&\leqslant(f_1:g)+(f_2:g),\\ (\lambda f:g)&=\lambda(f:g) \quad \text{for any } \lambda>0,\\ (f_1:g)&\leqslant(f_2:g) \quad \text{if } f_1\leqslant f_2.\end{aligned}$$

We also have $\|f\|_\infty\leqslant(f:g)\|g\|_\infty$ because if x is chosen such that $f(x)=\|f\|_\infty$ then

$$\|f\|_\infty\leqslant\sum c_i g(a_i x)\leqslant\|g\|_\infty\sum c_i.$$

Finally, if f, g, h are non-negative elements of $C_0(G)$ then

$$(f:h)\leqslant(f:g)\,(g:h):$$

Indeed, if $f(x)\leqslant\sum c_i g(a_i x)$ and $g(x)\leqslant\sum d_j h(b_j x)$ then we obtain $f(x)\leqslant\sum c_i d_j h(b_j a_i x)$. Therefore

$$(f:h)\leqslant \operatorname{glb}\sum c_i d_j\leqslant(\operatorname{glb}\sum c_i)\,(\operatorname{glb}\sum d_j)=(f:g)\,(g:h).$$

Now we replace f by a non-negative $f_o\neq 0$ which we keep fixed and let $\alpha_\varphi(f)=(f:\varphi)/(f_o:\varphi)$. Then

$$\frac{1}{(f_o:f)}\leqslant\alpha_\varphi(f)\leqslant(f:f_o) \tag{1}$$

for every non-negative φ in $C_0(G)$. Also we have

$$\alpha_\varphi(f_1+f_2)\leqslant\alpha_\varphi(f_1)+\alpha_\varphi(f_2). \tag{2}$$

We prove that *given* $\varepsilon>0$ *and non-negative* f_1, f_2 in $C_0(G)$ *there is a neighborhood* N_e *such that*

$$\alpha_\varphi(f_1)+\alpha_\varphi(f_2)\leqslant\alpha_\varphi(f_1+f_2)+\varepsilon \tag{3}$$

for all non-negative φ *whose support lies in* N_e.

Indeed let g be a non-negative element of $C_0(G)$ such that $g(x)=1$ for every x in the support of f_1+f_2. Define $f=f_1+f_2+\delta g$ where $\delta>0$ will be determined later. We let $h_k(x)=f_k(x)/f(x)$ $(k=1,2)$ if $f(x)>0$ and let $h_k(x)=0$ if $f(x)=0$. Then h_k is non-negative, belongs to $C_0(G)$ and so it is uniformly continuous. Thus there is a neighborhood of the identity N_e such that

$$|h_k(x)-h_k(y)|<\eta \quad \text{whenever } x^{-1}y\in N_e.$$

The more precise value of $\eta>0$ will be determined later. Now let φ be chosen such that its support lies in N_e and let

$$f(x)\leqslant \sum c_i\varphi(a_i x) \quad \text{where } \sum c_i \leqslant (f:\varphi)(1+2\eta).$$

If $\varphi(a_i x)\neq 0$ then $a_i x\in N_e$ and so $|h_k(a_i^{-1})-h_k(x)|<\eta$. Thus

$$f_k(x)=h_k(x)f(x)\leqslant \sum c_i(h_k(a_i^{-1})+\eta)\varphi(a_i x)$$

and so

$$(f_k:\varphi)\leqslant \sum c_i(h_k(a_i^{-1})+\eta).$$

Since $h_1(x)+h_2(x)\leqslant 1$ for any $x\in G$ we obtain

$$(f_1:\varphi)+(f_2:\varphi)\leqslant \sum c_i(1+2\eta)\leqslant (1+2\eta)^2(f:\varphi).$$

Dividing by $(f_o:\varphi)$ we obtain

$$\alpha_\varphi(f_1)+\alpha_\varphi(f_2)\leqslant (1+2\eta)^2\alpha_\varphi(f)\leqslant (1+2\eta)^2(\alpha_\varphi(f_1+f_2)+\delta\,\alpha_\varphi(g)).$$

Using (1) we see that the desired inequality (3) holds if η and δ are so small that

$$((1+2\eta)^2-1)(f_1+f_2:f_o)+\delta(1+2\eta)^2(g:f_o)<\varepsilon.$$

For every non-negative $f\in C_0(G)$ we consider the closed interval $X_f=[(f:f_o)^{-1},(f:f_o)]$ and let $X=\prod X_f$ be topologized by the product topology so that X is a compact Hausdorff space. By (1) every α_φ is a point in X whose f-coordinate is $\alpha_\varphi(f)$. For every neighborhood N_e we define $C(N_e)$ to be the closure of the set of those α_φ's whose φ has support in N_e. These are compact sets in X and their family has the finite intersection property because $C(N_e^1)\cap\cdots\cap C(N_e^n)\supseteq C(N_e^1\cap\cdots\cap N_e^n)$. By the compactness of X there is a point α which lies in every $C(N_e)$. Thus given $\varepsilon>0$ and finitely many $f_1,\ldots,f_n$ there is a non-negative φ with support in N_e such that

$$|\alpha(f_i)-\alpha_\varphi(f_i)|<\varepsilon \qquad (i=1,\ldots,n). \tag{4}$$

By choosing $n=3$ and $f_1, f_2, f_3 = f_1 + f_2$ and using (2) and (3) we obtain $\alpha(f_1)+\alpha(f_2)=\alpha(f_1+f_2)$. Since each α_φ is left invariant, non-negative and homogeneous and satisfies (1) by (4) the same will hold for α. The measure corresponding to the non-negative linear functional defined on $C_0(C)$ by the map α is the left Haar measure of G.

Proof of the uniqueness. We already mentioned that if the group G admits a biinvariant Haar measure δx then the essential uniqueness of all Haar measures of G can be easily proved. Here are the details: Let dx denote any other *right* invariant measure of G. We choose a continuous function g with compact support such that $\int_G g(x)dx=1$ and let $\lambda=\int_G g(x^{-1})\delta x$. Then for any f in $C_0(G)$ we have

$$\begin{aligned}\int_G f(x)\delta x &= \int_G g(y)dy \int_G f(x)\delta x = \int_G g(y) \int_G f(yx)\delta x\, dy\\ &= \int_G\int_G g(y) f(yx)dy\,\delta x.\end{aligned}$$

Next in the inner integral we let $y \to yx^{-1}$ and obtain

$$\int_G f(x)\delta x = \int_G\int_G g(yx^{-1}) f(y) dy\,\delta x = \int_G f(y)\left(\int_G g(yx^{-1})\delta x\right)dy.$$

Substituting $x \to xy$ and using the right invariance of δx we see that the inner integral is λ. Thus we proved that $\int_G f(x)\delta x = \lambda \int_G f(y)dy$ for every f in $C_0(G)$. The same type of reasoning can be applied also to any left invariant measure dy.

Now we turn to the uniqueness proof for arbitrary locally compact groups. Let $dx, \delta x$ be left invariant measures on G and let α, β denote the corresponding linear functionals in $C_0(G)^*$. It will be sufficient to prove the existence of a constant $c>0$ such that $\beta(f)=c\alpha(f)$ for every non-negative $f \in C_0(G)$.

Let f, f_0 be non-negative and not identically zero elements of $C_0(G)$. We denote by C a common compact carrier of the functions f and f_0. We choose O to be an open set such that $\bar{O}$ is compact and $C \subset O$. Since $\bar{O}$ is compact and G is uniformizable we can choose a function g in $C_0(G)$ such that $g(x)=1$ for every $x \in \bar{O}$.

Next, given $\varepsilon>0$ by Theorem 1.2, Proposition 1.11 and the uniform continuity of f we can find a symmetric neighborhood of the identity N_e such that $CN_e \cup N_e C \subseteq O$ and $|f(yx)-f(x)|<\varepsilon$, $|f(xy)-f(x)|<\varepsilon$ for every $x \in G$ and $y \in N_e$. Hence we have $|f(xy)-f(yx)|<2\varepsilon$ for $x \in G$ and $y \in N_e$. Since g is 1 on $\bar{O}$ and $CN_e \cup N_e C \subseteq O$ we have

$$|f(xy)-f(yx)| < 2\varepsilon g(x) \tag{5}$$

for every $x \in G$, $y \in N_e$.

Let h be a not identically zero continuous function on G with support in N_e and such that $h(x)=h(x^{-1})$ for every $x \in G$. We have

$$\alpha(h)\beta(f) = \int_G \int_G h(y) f(x) \delta x\, dy = \int_G \int_G h(y) f(yx) \delta x\, dy.$$

We also have

$$\beta(h)\alpha(f) = \int_G \int_G h(x) f(y) \delta x\, dy = \int_G \int_G h(y^{-1}x) f(y) \delta x\, dy$$

and so by the symmetry of h

$$\beta(h)\alpha(f) = \int_G \int_G h(x^{-1}y) f(y)\, dy\, \delta x = \int_G \int_G h(y) f(xy)\, dy\, \delta x.$$

Therefore using the fact that the support of h lies in N_e from (5) we obtain

$$\begin{aligned} |\alpha(h)\beta(f)-\beta(h)\alpha(f)| &\leqslant \int_G \int_G h(y) |f(yx)-f(xy)| \delta x\, dy \\ &\leqslant \varepsilon \int_G h(y)\, dy \int_G g(x) \delta x = \varepsilon \alpha(h) \beta(g). \end{aligned}$$

Dividing by $\alpha(h)\alpha(f)$ we obtain

$$\left| \frac{\beta(f)}{\alpha(f)} - \frac{\beta(h)}{\alpha(h)} \right| \leqslant \varepsilon \left| \frac{\beta(g)}{\alpha(f)} \right|.$$

Since exactly the same holds with f_o in place of f we obtain

$$\left| \frac{\beta(f)}{\alpha(f)} - \frac{\beta(f_o)}{\alpha(f_o)} \right| \leqslant \varepsilon \left(\left| \frac{\beta(g)}{\alpha(f)} \right| + \left| \frac{\beta(g)}{\alpha(f_o)} \right| \right).$$

Here $\varepsilon>0$ is independent of f, f_o and g, so by letting $\varepsilon \to 0$ we obtain $\beta(f)=c\alpha(f)$ where $c=\beta(f_o)/\alpha(f_o)$ is independent of f. The essential uniqueness of the left Haar measure is proved.

Proposition 1. *A locally compact group G has finite Haar measure if and only if it is compact.*

Proof. Compact sets have finite Haar measure so if G itself is compact then $\mu(G)$ is finite. If G is not compact then choose an open neighborhood O of e such that $\bar{O}$ is compact. No finite family of left translates of O can cover G because otherwise G would be compact. Therefore one can select a sequence of points $e=x_1, x_2, \ldots$ in G such that x_{n+1} does not belong to $x_1 O \cup \cdots \cup x_n O$ for every $n=1,2,\ldots$ If Q is a symmetric neighborhood of the identity such that $Q^2 \subseteq O$ then $x_1 Q, x_2 Q, \ldots$ are pairwise disjoint and so $\mu(G) \geqslant \mu(x_1 Q) + \cdots + \mu(x_n Q) = n\mu(Q)$ for any $n=1,2,\ldots$ Therefore $\mu(G)$ cannot be finite.

Let α be a left invariant measure on G in the linear functional sense i.e. let $\alpha(f) = \int\limits_G f(x)\,d\alpha(x)$ be considered only for functions $f \in C_0(G)$. As usual let $f_a\colon G \to \mathbb{C}$ denote the right translate mapping $x \mapsto f(xa)$ and similarly let ${}_bf$ be the function $x \mapsto f(bx)$. Keeping a fixed consider the linear functional β defined by $\beta(f) = \alpha(f_a)$ or in terms of integral notation $\beta(f) = \int\limits_G f(xa)\,d\alpha(x)$. It is obvious that $\beta \in C_0(G)^*$ and so β defines a non-trivial measure on G. As ${}_a(f_b) = ({}_af)_b$ ans as α is left invariant we see that

$$\beta({}_bf) = \alpha(({}_bf)_a) = \alpha({}_b(f_a)) = \alpha(f_a) = \beta(f).$$

This shows that β is another left invariant measure on G and so by the uniqueness of invariant measures it differs from α only in a constant factor which we denote by $\Delta(a)$. Therefore $\alpha(f) = \Delta(a)\beta(f)$. In terms of integrals this is the same as

$$\int\limits_G f(x)\,dx = \Delta(a) \int\limits_G f(xa)\,dx$$

where the measure is denoted by dx instead of $d\alpha(x)$. By letting a vary over G we obtain a function $\Delta\colon G \to (0,\infty)$ called the *modular function* of the group G.

By definition $\Delta(ab)\alpha(f_{ab}) = \alpha(f)$ and on the other hand by $(f_b)_a = f_{ab}$ we have

$$\Delta(a)\Delta(b)\alpha(f_{ab}) = \Delta(a)\Delta(b)\alpha((f_b)_a) = \Delta(b)\alpha(f_b) = \alpha(f).$$

Therefore $\Delta(ab) = \Delta(a)\Delta(b)$ for any $a, b \in G$ and so *Δ is a homomorphism of G into the multiplicative group of the positive real numbers.* Sometimes a homomorphism of a group G into $(0,\infty)^\times$ is called a *real character* so we can say that Δ is a real character.

We are now going to prove that *the modular function Δ is continuous.* Since Δ is a homomorphism it is sufficient to prove continuity at e. Let $f\colon G \to (0,\infty)$ be not identically zero, continuous and have compact support S. Let C be a compact neighborhood of e and let $g\colon G \to (0,\infty)$ be continuous, identically one on SC and have compact support. Since

f is uniformly continuous with respect to the left uniform structure of G given $\varepsilon>0$ there is a symmetric neighborhood $N_e \subseteq C$ such that $|f(x_1)-f(x_2)|<\varepsilon$ whenever $x_1^{-1}x_2 \in N_e$. Thus if $a \in N_e$ then we have $|f(xa^{-1})-f(x)|<\varepsilon$ for every $x \in G$. Using the functional notation instead of integrals for $a \in N_e$ we have the estimate

$$|\Delta(a)-1|\alpha(f)=|\alpha(f^a-f)| \leqslant \alpha(|f^a-f|) \leqslant \alpha(\varepsilon g)=\varepsilon\alpha(g).$$

Here f and g are fixed, $\alpha(f)\neq 0$ and $\varepsilon>0$ is arbitrary and N_e depends on ε. Therefore by restricting a to a sufficiently small neighborhood N_e of e one can make $|\Delta(a)-1|$ arbitrarily close to 0. It is proved that Δ is continuous at e and hence everywhere.

Proposition 2. We have the formula

$$\int_G f(x^{-1})dx=\int_G f(x)\Delta(x^{-1})dx$$

where dx is the left Haar measure of G.

Proof. Since dx is left invariant and $f_a(x^{-1})=f((a^{-1}x)^{-1})$ we see that $\alpha(f)=\int_G f(x^{-1})dx$ is a right invariant integral. Similarly, if we let $\beta(f)=\int_G f(x)\Delta(x)^{-1}dx$ then β is a right invariant integral because

$$\beta(f_a)=\int_G f_a(x)\Delta(x)^{-1}dx=\Delta(a)\int_G f(xa)\Delta(xa)^{-1}dx=\int_G f(x)\Delta(x)^{-1}dx.$$

Therefore by the uniqueness of right invariant integrals we have $\beta(f)=c\alpha(f)$ for some constant $c>0$. The fact that $c=1$ follows from the continuity of Δ at e: Let S be a compact, symmetric neighborhood of e and let f be the characteristic function of S. Then by $\beta(f)=c\alpha(f)$ we have

$$\int_G f(x)(c-\Delta(x)^{-1})dx=\int_S (c-\Delta(x)^{-1})dx=0.$$

Since S has finite, positive measure $\mu(S)$ the last relation can be written in the form

$$\mu(S)(c-1)=\int_S (\Delta(x)^{-1}-1)dx.$$

By the continuity of Δ at e we can choose S such that $|\Delta(x)^{-1}-1|<\varepsilon$ for every $x \in S$ and so we obtain $\mu(S)|c-1|<\varepsilon\mu(S)$. Therefore $|c-1|<\varepsilon$ and the proposition is proved.

Proposition 3. *The left and right Haar measures are absolutely continuous with respect to each other and the right measure is $\Delta(x^{-1})dx$ where dx denotes a left measure.*

Proof. Since the right Haar measure is essentially unique it is sufficient to check that $\int_G f(x)\Delta(x^{-1})dx$ is a right invariant integral. This follows immediately from the definition of $\Delta(a)$. The details were part of the proof of Proposition 2.

The group G is unimodular if and only if its modular function Δ is identically 1. This follows from the definition of the modular function.

Every compact group is unimodular. For $\Delta: G \to (0, \infty)^\times$ being a homomorphism the image group $\Delta(G)$ is a compact subgroup of $(0, \infty)^\times$. The only such subgroup is the trivial subgroup $\{1\}$ and so Δ is identically 1.

Commutative groups are unimodular.

If the identity e has a compact neighborhood which is invariant under inner automorphisms then G is unimodular. For if C is such a neighborhood then $0 < \mu(C) < \infty$ and $\mu(xCx^{-1}) = \mu(C)$ for every x in G. However $\mu(xCx^{-1}) = \Delta(x)^{-1}\mu(C)$ and so $\Delta(x) = 1$ for every $x \in G$.

If G coincides with its closed commutator subgroup then G is unimodular. For every commutator we have $\Delta(xyx^{-1}y^{-1}) = 1$ and so by the continuity of Δ we have $\Delta(x) = 1$ for every x in the closed commutator subgroup of G.

Proposition 4. *The Haar measure is inverse invariant i.e. we have* $\int_G f(x^{-1})dx = \int_G f(x)dx$ *if and only if G is unimodular*.

Proof. If Δ is identically 1 then the integral is inverse invariant by Proposition 2. Conversely $\int_G f(x^{-1})dx$ is right invariant so if it is equal to the left invariant integral $\int_G f(x)dx$ then Δ is identically 1.

Proposition 5. *The Haar measure of $G_1 \times G_2$ is the product measure and the modular function of $G_1 \times G_2$ is $\Delta_1 \Delta_2$. The product group is unimodular if and only if G_1 and G_2 are unimodular groups*.

Proof. The product of the left invariant Haar measures dx_1 and dx_2 is a left invariant measure on $G_1 \times G_2$ so by the uniqueness property it is the Haar measure of $G_1 \times G_2$. One can also see that the modular function of the product is $\Delta_1 \Delta_2$ by considering

$$\begin{aligned}\mu(A_1 a_1 \times A_2 a_2) &= \mu_1(A_1 a_1)\mu_2(A_2 a_2) = \Delta_1(a_1)\Delta_2(a_2)\mu_1(A_1)\mu_2(A_2)\\ &= \Delta_1(a_1)\Delta_2(a_2)\mu(A_1 \times A_2).\end{aligned}$$

Proposition 6. *The Haar measure of an open subgroup H of a group G is the measure induced on H by the Haar measure of G*.

Proof. The induced measure is invariant and all compact subsets of H have finite measure. If O is a non-void open set in H then it is open also in G and so its measure is not 0. Thus the induced measure has all the characteristic properties of an invariant measure.

Let G be a Lie group and let $(x_1,\ldots,x_n)$ be a local coordinate system in some coordinate neighborhood (O,φ) containing the identity so that $\varphi: O\to\varphi(O)\subseteq\mathbb{R}^n$ is a homeomorphism and $\varphi(O)$ is homeomorphic to $\mathbb{R}^n$. The Haar integral of functions $f: G\to\mathbb{C}$ whose support lies in N_e can be easily determined by realizing that the Haar measure of G is absolutely continuous with respect to the Lebesgue measure in $\varphi(O)$ and so

$$\int_G f(x)dx = \int_{\varphi(O)} f(\varphi^{-1}(x_1,\ldots,x_n))\omega(x_1,\ldots,x_n)dx_1\ldots dx_n$$

for some function ω where $dx_1\ldots dx_n$ denotes the Lebesgue measure in $\mathbb{R}^n$. We simplify the notation by using the same symbol x to denote the group element x and the corresponding set of coordinates $(x_1,\ldots,x_n)$. We shall also use dx both for the left invariant measure on G and the Lebesgue measure $dx_1\ldots dx_n$. Then we can write

$$\int_G f(x)dx = \int f\varphi^{-1}(x)\omega(x)dx.$$

If a is sufficiently close to e we let $y=ax$ and use y to denote also the n-tuple of functions giving the coordinates of y in terms of those of a and x. Since dx is left invariant we must have

$$\int f\varphi^{-1}(a^{-1}y)\omega(y)dy = \int f\varphi^{-1}(x)\omega(x)dx$$

where in terms of our conventions $a^{-1}y$ denotes $\varphi(a^{-1}y)$ with $a^{-1}y$ in G. By performing the change of variables $y\to ax$ in the Euclidean integral standing on the left hand side we obtain

$$\int f\varphi^{-1}(x)\left\|\frac{\partial y}{\partial x}\right\|\omega(ax)dx = \int f\varphi^{-1}(x)\omega(x)dx.$$

This equation holds for every continuous f whose support lies in O and so we obtain $\|\partial y/\partial x\|\,\omega(ax)=\omega(x)$. From this we obtain $\omega(a)=\omega(e)\|\partial y/\partial x\|^{-1}$. As dx is determined only up to a positive constant factor we can choose $\omega(e)=1$. Therefore we proved that

$$\int_G f(x)dx = \int_{\varphi(O)} f(\varphi^{-1}(x_1,\ldots,x_n))\left\|\frac{\partial(y_1,\ldots,y_n)}{\partial(x_1,\ldots,x_n)}\right\|^{-1} dx_1\ldots dx_n.$$

Hence $y_1,\ldots,y_n$ are the coordinates of $y=xu$ in terms of those of x and u. The Jacobian is taken in absolute value and it is evaluated at $u=e$.

This method was used to determine the invariant integrals in the following list of specific groups. The list is distinguished by the fact that a coordinate neighborhood (O,φ) exists such that the Haar measure of the complement of O is zero. At the end we determine also the Haar measure of the special linear group $\mathrm{Sl}(2,\mathbb{R})$ essentially by the same method although in this case the choice of the coordinates is less obvious.

1. The Haar measure of the *additive group* of $\mathbb{R}^n$ is the n-dimensional Lebesgue measure dx.

2. The Haar measure of the *additive group* of $\mathbb{C}^n$ is the $2n$-dimensional Lebesgue measure $dx_1\,dy_1\ldots dx_n\,dy_n$ where $(x_1+iy_1,\ldots,x_n+iy_n)$ denotes a generic point of $\mathbb{C}^n$.

3. The Haar measure of the *multiplicative group* $\mathbb{R}^\times$ of the field of reals is $dx/|x|$ where dx is the Lebesgue measure on $-\infty<x<+\infty$.

4. The Haar measure of the *multiplicative group* $\mathbb{C}^\times$ is $dx\,dy/(x^2+y^2)$ where $dx\,dy$ is the Lebesgue measure of the plane $-\infty<x,y<+\infty$. Alternatively in terms of polar coordinates the Haar measure is $dr\,d\varphi/r$ where $0<r<\infty$ and $0\leqslant\varphi<2\pi$, dr denotes the Lebesgue measure of the half line and $d\varphi$ is the invariant measure of the circle i.e. Lebesgue integral of the interval $[0,2\pi)$.

5. If $\mathbb{R}^+/\mathbb{Z}^+$ is parametrized by $e^{i\varphi}$ $(0\leqslant\varphi<2\pi)$ then the Haar measure is the measure $d\varphi$ mentioned above.

6. The Haar measure of the *general linear group* $\mathrm{Gl}(2,\mathbb{R})$ is

$$\frac{dx_{11}\,dx_{12}\,dx_{21}\,dx_{22}}{\begin{vmatrix} x_{11} & x_{12} \\ x_{21} & x_{22} \end{vmatrix}^2}$$

where $x=\begin{pmatrix} x_{11} & x_{12} \\ x_{21} & x_{22} \end{pmatrix}$ denotes a general element of $\mathrm{Gl}(2,\mathbb{R})$. The group is unimodular.

7. Let G be the group of all matrices $\begin{pmatrix} x & y \\ 0 & 1 \end{pmatrix}$ where $x\neq 0$ is real and $-\infty<y<+\infty$, the group operation being matrix multiplication. Then the left Haar measure is $dx\,dy/x^2$ and the right Haar measure is $dx\,dy/|x|$. In both cases $dx\,dy$ denotes the planar Lebesgue measure. $\mathrm{Gl}(2,\mathbb{R})$ and its present subgroup G give an example of a *unimodular group with a non-unimodular subgroup*. The modular function of G is $\Delta(x,y)=1/|x|$.

8. The *general linear group* $\mathrm{Gl}(2,\mathbb{C})$ is unimodular with Haar measure

$$\frac{dx_{11}\,dy_{11}\,dx_{12}\,dy_{12}\,dx_{21}\,dy_{21}\,dx_{22}\,dy_{22}}{\left\|\begin{matrix} z_{11} & z_{12} \\ z_{21} & z_{22}\end{matrix}\right\|^4}$$

where $dx_{11}\dots dy_{22}$ is an eight dimensional Lebesgue measure. As earlier the double bar denotes the absolute value of the determinant.

9. The *special unitary group* $\mathrm{SU}(2)$ is unimodular. If the generic element is $\begin{pmatrix} z_1 & z_2 \\ -\bar{z}_2 & \bar{z}_1\end{pmatrix}$ with $z_1=x_1+ix_2$, $z_2=x_3+ix_4$ then the map

$$\begin{pmatrix} z_1 & z_2 \\ -\bar{z}_2 & \bar{z}_1\end{pmatrix} \mapsto (x_1,x_2,x_3,x_4)$$

is a homeomorphism of $\mathrm{SU}(2)$ onto the sphere

$$x_1^2+x_2^2+x_3^2+x_4^2=1.$$

The Haar measure of $\mathrm{SU}(2)$ in terms of this interpretation is the ordinary Lebesgue measure of the sphere S^3.

10. The Haar measure of the *real general linear group* $\mathrm{Gl}(n,\mathbb{R})$ is

$$\frac{dx_{11}\,dx_{12}\dots dx_{n,n-1}\,dx_{nn}}{\left\|\begin{matrix} x_{11} \dots x_{1n} \\ x_{n1} \dots x_{nn}\end{matrix}\right\|^n}$$

where $dx_{11}\dots dx_{nn}$ denotes Lebesgue measure in $\{x\colon x\in\mathbb{R}^{n^2} \text{ and } \det x\neq 0\}$. The group is unimodular.

11. The *complex general linear group* $\mathrm{Gl}(n,\mathbb{C})$ is unimodular and its Haar measure is

$$\frac{dx_{11}\,dy_{11}\dots dx_{nn}\,dy_{nn}}{\|\det z\|^{2n}}$$

where $dx_{11}\dots dy_{nn}$ denotes Lebesgue measure in $2n^2$ dimension and $\det z$ is the determinant of the generic element $z=(z_{kl})=(x_{kl}+iy_{kl})$.

12. Let $x=x_0+x_1 i+x_2 j+x_3 k$ denote a Hamilton quaternion so that $i^2=j^2=k^2=-1$ and $ij=k$, $jk=i$ and $ki=j$. The norm of x is defined to be

$$Nx=x_0^2+x_1^2+x_2^2+x_3^2.$$

This is the square of the linear space norm $\|x\|$ introduced in Section I.1. The *multiplicative group* $\mathbb{H}^\times$ of the algebra $\mathbb{H}$ of Hamilton quaternions is unimodular and its Haar measure is

$$\frac{dx_0\,dx_1\,dx_2\,dx_3}{(N\,x)^2}$$

where $dx_0 \ldots dx_3$ denotes Lebesgue integration.

13. The quaternion algebra $H(p,q)$ for $p,q\in\mathbb{R}$ is the real algebra of general quaternions $x=x_0+x_1 i+x_2 j+x_3 k$ where $i^2=p$, $j^2=q$, $k^2=-pq$, etc. (See Section I.1.) The norm in the ring theoretic sense is

$$N\,x = x_0^2 - p\,x_1^2 - q\,x_2^2 + p\,q\,x_3^2.$$

The multiplicative group $H(p,q)^\times$ consists of those x for which $N\,x\neq 0$. This group is unimodular and the Haar measure is

$$\frac{dx_0\,dx_1\,dx_2\,dx_3}{(N\,x)^2}$$

where again $dx_0 \ldots dx_3$ denotes the four dimensional Lebesgue measure restricted to the set $\{x: x\in\mathbb{R}^4 \text{ and } N\,x\neq 0\}$.

14. The group consisting of the upper triangular matrices

$$\begin{pmatrix} x_{11} & x_{12} & \cdots & x_{1,n-1} & x_{1n} \\ 0 & x_{22} & \cdots & x_{2,n-1} & x_{2n} \\ \cdot & \cdot & \cdot & \cdot & \cdot \\ 0 & 0 & \cdots & x_{n-1,n-1} & x_{n-1,n} \\ 0 & 0 & \cdots & 0 & x_{nn} \end{pmatrix}$$

satisfying $x_{11}x_{22}\ldots x_{n-1,n-1}x_{nn}\neq 0$ is not unimodular for $n>1$. Its left Haar measure is

$$\frac{dx_{11}\,dx_{12}\ldots dx_{n-1,n-1}\,dx_{nn}}{|x_{11}^n\,x_{22}^{n-1}\ldots x_{n-1,n-1}^2\,x_{nn}|}$$

while the right Haar measure is

$$\frac{dx_{11}\,dx_{12}\ldots dx_{n-1,n-1}\,dx_{nn}}{|x_{11}\,x_{22}^2\ldots (x_{n-1,n-1})^{n-1}\,x_{nn}^n|}.$$

15. The group of upper triangular matrices

$$\begin{pmatrix} 1 & x_{12} & x_{13} & \dots & x_{1,n-1} & x_{1n} \\ 0 & 1 & x_{23} & \dots & x_{2,n-1} & x_{2n} \\ \dots & \dots & \dots & \dots & \dots & \dots \\ 0 & 0 & 0 & \dots & 1 & x_{n-1,n} \\ 0 & 0 & 0 & \dots & 0 & 1 \end{pmatrix}$$

is unimodular for every $n=1,2,\dots$ The biinvariant measure is the $\frac{1}{2}(n-1)n$ fold Lebesgue measure $dx_{12}\dots dx_{1n}\dots dx_{n-2,n-1}\,dx_{n-2,n}\,dx_{n-1,n}$.

16. If A and B are compact sets then AB is compact and so its Haar measure $\mu(AB)<\infty$. The following gives an example of *two sets* A, B *of finite Haar measure such that* $\mu(AB)=+\infty$: Let G be the group given in 7. and let

$$A=B=\{(x,y)\colon 1\leqslant x<\infty \text{ and } 0\leqslant y\leqslant 1\}.$$

The left Haar measure of A is

$$\mu(A)=\int_0^1\int_1^\infty \frac{dx\,dy}{x^2}=1.$$

We have

$$\begin{aligned} AB &= \{(x\xi, x\eta+y)\colon 1\leqslant x,\ \xi<\infty \text{ and } 0\leqslant y,\ \eta\leqslant 1\} \\ &= \{(\alpha,\beta)\colon 1\leqslant\alpha<\infty \text{ and } 0\leqslant\beta\leqslant\alpha+1\}. \end{aligned}$$

Therefore

$$\mu(AB)=\int_1^\infty\int_0^{\alpha+1} d\beta\,\frac{d\alpha}{\alpha^2}=+\infty.$$

17. In order to describe the *Haar measure of the group* $\mathrm{Sl}(2,\mathbb{R})$ we first parametrize the elements

$$g=\begin{pmatrix} a_{11} & a_{12} \\ a_{21} & a_{22} \end{pmatrix}$$

by introducing

$$g\cdot i=\frac{a_{11}i+a_{12}}{a_{21}i+a_{22}} \quad\text{and}\quad \theta(g)=\arg(a_{21}i+a_{22})$$

so that $g\cdot i$ is a point in the upper half plane S and θ is on the unit circle T. First we prove:

The map $\mathrm{Sl}(2,\mathbb{R})\to S\times T$ *given by* $g\mapsto(g\cdot i,\theta(g))$ *is a homeomorphism.*

The map is surjective, because given any $z \in S$ there is a g in $\mathrm{Sl}(2,\mathbb{R})$ such that $g \cdot i = z$ and actually $gk \cdot i = g \cdot ki = g \cdot i = z$ for every k in $\mathrm{SO}(2,\mathbb{R})$. Thus by choosing k suitably we can obtain any $\theta = \theta(gk)$ given in advance. Next, if $(g_1 \cdot i, \theta(g_1)) = (g_2 \cdot i, \theta(g_2))$ then by $\theta(g_1) = \theta(g_2)$ the ratio

$$\lambda = \frac{(a_{21}^1\, i + a_{22}^1)}{(a_{21}^2\, i + a_{22}^2)} \tag{5}$$

is real and so $a_{21}^1 = \lambda a_{21}^2$ and $a_{22}^1 = \lambda a_{22}^2$. Hence from $g_1 \cdot i = g_2 \cdot i$ we obtain

$$a_{11}^1\, i + a_{12}^1 = \lambda(a_{11}^2\, i + a_{12}^2).$$

Again using the fact that λ is real we see that $a_{11}^1 = \lambda a_{11}^2$ and $a_{12}^1 = \lambda a_{12}^2$. By g_1 having determinant $|g_1| = \lambda^2$ we obtain $\lambda = \pm 1$. Finally by (5)

$$\theta(g_1) = \arg(a_{21}^1\, i + a_{22}^1) = \arg \lambda + \arg(a_{21}^2\, i + a_{22}^2) = \arg \lambda + \theta(g_2)$$

and so the equality of $\theta(g_1)$ and $\theta(g_2)$ leads to $\lambda = 1$.

From the definitions it is clear that $g \cdot i$ and $\theta(g)$ are continuous functions of g. Conversely we notice that if $r = |a_{21}\, i + a_{22}|$ then by $(x+iy)e^{i\theta} = r^{-1}(a_{11}\, i + a_{12})$ the ratios a_{11}/r and a_{12}/r depend continuously on x, y, θ. Similarly from $re^{i\theta} = a_{21}\, i + a_{22}$ we see that a_{21}/r and a_{22}/r are continuous functions of the data x, y, θ. Finally using

$$r^{-2} = \left(\frac{a_{11}}{r}\right)\left(\frac{a_{22}}{r}\right) - \left(\frac{a_{12}}{r}\right)\left(\frac{a_{21}}{r}\right)$$

we see that a_{11}, a_{12}, a_{21} and a_{22} depend continuously on x, y, θ. Therefore the map $G \to S \times T$ is a homeomorphism.

The invariant integral of $\mathrm{Sl}(2,\mathbb{R})$ in terms of the variables $z = x + iy$ and θ is

$$\int_G f(g)\, dg = \int_0^{\infty} \int_{-\infty}^{+\infty} \int_0^{2\pi} f(x,y,\theta)\, d\theta \frac{dx\, dy}{y^2}.$$

In order to prove that the integral is left invariant first we remark that if $z = g \cdot i$ and $\theta = \theta(g)$ then for any $a \in G$ we have $ag \cdot i = a \cdot gi = a \cdot z$ and $\theta(ag) = \theta(g) + \arg(a_{21} z + a_{22})$ where $a = (a_{ij})$. Next, under the substitution $w = a \cdot z$ and $\psi = \theta + \arg(a_{21} z + a_{22})$ we have

$$\int_S \int_T f(w,\psi)\, d\psi \frac{du\, dv}{v^2} = \int_S \int_T f(z,\theta) \left\| \frac{\partial(u,v,\psi)}{\partial(x,y,\theta)} \right\| \frac{dx\, dy}{v^2}$$

with $v = v(x, y, \theta)$ on the right hand side. If (z, θ) corresponds to g in G then as we saw (w, ψ) corresponds to $h = ag$ of G. Thus the left hand side of the above equation is $\int_G f(ag)\,dg$. The object is to show that the right hand side is $\int_G f(g)\,dg$. By using the Cauchy-Riemann equations we see that

$$\left|\frac{\partial(u, v, \psi)}{\partial(x, y, \theta)}\right| = \begin{vmatrix} u_x & u_y \\ v_x & v_y \end{vmatrix} = u_x^2 + v_x^2 = \left|\frac{dw}{dz}\right|^2 .$$

Since

$$\frac{dw}{dz} = \frac{d(a \cdot z)}{dz} = (a_{21} z + a_{22})^{-2}$$

and

$$v = \mathscr{I}\!m\, w = \frac{\mathscr{I}\!m\, z}{|a_{21} z + a_{22}|^2}$$

we have

$$\left|\frac{\partial(u, v, \psi)}{\partial(x, y, \theta)}\right| = \frac{v^2}{y^2} .$$

Hence the integral on the right hand side is indeed $\int_G f(g)\,dg$.

3. Quasi-Invariant and Relatively Invariant Measures

Let S be a topological space and let G be a topological group acting on S. A measure μ on S is called a *G-invariant measure* if for every g in G and for every measurable subset E of S the set gE is measurable and $\mu(g \cdot E) = \mu(E)$. One is primarily interested in non-trivial invariant Borel measures on a locally compact space S and extensions of these. It is possible that no non-trivial invariant measure exists on S. This can happen already in the special case when G is a locally compact group, H is a closed subgroup of G and S is the homogeneous space G/H.

A measure μ is called *relatively invariant* with respect to the action of G if for each g in G there is a positive constant $D(g)$ such that $D(g)\,\mu(gE) = \mu(E)$ for every measurable set E in S. Still it is possible to find situations where the only relatively invariant measure is the trivial one. We notice that if μ is relatively invariant then for every g in G another relatively invariant measure μ_g is defined by $\mu_g(E) = \mu(gE)$. By the relative invariance of μ we have $\mu_g(E) = 0$ if and only if $\mu(E) = 0$. Hence the measures μ_g $(g \in G)$ are all absolutely continuous with respect to each other, or in other words the measures μ_g $(g \in G)$ are all equivalent. Although in general a homogeneous space G/H does not admit non-trivial relatively invariant measures one can always construct a non-trivial Borel measure μ on G/H such that its zero sets are G invariant i.e. $\mu(E)=0$ if and only if $\mu(gE)=0$ where g in G is arbitrary. A measure having this property is called a *quasi-invariant measure* on G/H. Thus μ is quasi-invariant if and only if $\mu(E)=0$ implies that $\mu(gE)=0$ for every $g\in G$. In other words every translated measure μ_g must be equivalent to μ.

The existence of a non-trivial quasi-invariant measure is important because it can be used in the construction of induced representations of a not necessarily unimodular group G with respect to a closed subgroup H. The purpose of the present section is to prove the existence of such measures and discuss a few fundamental properties. This will be followed by additional remarks on relatively invariant measures. We start with some topological preliminaries needed in the construction of certain special functions $\rho: G\to(0,\infty)$.

The existence of quasi-invariant measures on homogeneous spaces G/H depends on the paracompactness of G/H. A topological space X is called *paracompact* if each open cover of X has a locally finite, open refinement. A cover $Q_j\ (j\in\mathscr{J})$ is called a *refinement* of the cover $O_i\ (i\in\mathscr{I})$ if for each Q_j there is an O_i containing it. *Local finiteness* of $Q_j\ (j\in\mathscr{J})$ means that every point x has a neighborhood N_x intersecting only a finite number of Q_j's.

Proposition 1. *Every locally compact group is paracompact.*

Corollary. *Every locally compact T_0 group is normal.*

Proof. Let O be a symmetric open neighborhood of the identity such that its closure is compact. Then using $O^{-1}=O$ we see that $H=\bigcup\{O^n: 1\leqslant n<\infty\}$ is an open subgroup of G. Since G is the disjoint union of the distinct open subsets $xH\ (x\in G)$ it is sufficient to prove that H is paracompact. The open subgroup H is also closed because its complement in G is $\bigcup\{xH: x\notin H\}$ where each xH is open. Therefore $\overline{O^n}\subseteq H$ for every $n=1,2,\ldots$ and so H is the union of countably many compact sets. Hence H is a uniformizable Lindelöf space and so it is paracompact.

Proposition 2. *Let H be a closed subgroup of the locally compact group G. Then there is a continuous function $f: G\to(0,\infty)$ with support S such that for every compact subset C of G the intersection $CH\cap S$ is compact and $\int_H f(xh)\,dh=1$ for every $x\in G$.*

Proof. Let p denote the canonical projection $p: G\to G/H$. If O is a relatively compact, open neighborhood of the identity then $p(Ox)\ (x\in G)$

is an open cover of G/H. By the paracompactness of G/H we can find a locally finite refinement of this covering, say $Q_i^1\,(i\in\mathscr{I})$ and a paracompact Hausdorff space being normal we can find two other locally finite open covers $Q_i^k\ (i\in\mathscr{I}, k=2,3)$ such that $\overline{Q_i^3}\subseteq Q_i^2\subseteq\overline{Q_i^2}\subseteq Q_i^1$ for every $i\in\mathscr{I}$. Since $Q_i^1\subseteq p(Ox)$ for some x in G where $\overline{Ox}$ is compact there are non-void, relatively compact open sets $O_i^k\ (i\in\mathscr{I}, k=1,2,3)$ in G such that $pO_i^k=Q_i^k$ and $\overline{O_i^3}\subseteq O_i^2\subseteq\overline{O_i^2}\subseteq O_i^1$. For instance having fixed $x=x_i$ we can take $O_i^k=p^{-1}(Q_i^k)\cap Ox$.

Given $x\in G$ by the local finiteness of $Q_i^1\ (i\in\mathscr{I})$ there is a neighborhood N_{px} of px in G/H which intersects only finitely many of the O_i^1 sets. Since $p^{-1}N_{px}$ intersects O_i^1 only if N_{px} intersects Q_i^1 we found a neighborhood $N_x=p^{-1}N_{px}$ of x in G which intersects only finitely many of the O_i^1 sets. Thus if we choose the continuous function $g_i\colon G\to[0,1]$ such that $g_i(x)=1$ for $x\in O_i^3$ and $g_i(x)=0$ for $x\notin O_i^2$ then for each point $x\in G$ there is a neighborhood N_x such that all but finitely many of the g_i's vanish in N_x. Therefore $g=\sum g_i$ defines a non-negative continuous function on G. The existence of g_i follows from the compactness of O_i^3 and the regularity of G.

We prove that if S is the smallest, and hence closed carrier of g and C is a compact set in G then $CH\cap S$ is compact. By Lemma 1.12 CH is a closed set so it will be sufficient to show that $CH\cap S$ is covered by finitely many of the absolutely compact $\overline{O_i^2}$ sets. For each $x\in C$ there is a saturated neighborhood $N_x=p^{-1}N_{px}$ which meets only finitely many of the $\overline{O_i^2}$ sets. By Lemma 1.6 pN_x contains px in its interior. Since pC is compact finitely many of the pN_x sets will cover pC. We select all those $\overline{O_i^2}$ sets which meet any one of these finitely many N_x neighborhoods.

Since $O_i^2\ (i\in\mathscr{I})$ is a locally finite system of sets in G we see that $\bigcup\overline{O_i^2}=\overline{\bigcup O_i^2}$ is a closed set and so

$$O=c\bigcup\overline{O_i^2}=\bigcap c\overline{O_i^2}$$

is open. If $y\in c\overline{O_i^2}$ then $y\notin O_i^2$ and so $g_i(y)=0$. Hence if $y\in O$ then $g_i(y)=0$ for every i in $\mathscr{I}$ and $g(y)=0$. We proved that $g\equiv 0$ on the open set O and so $S\cap O=\emptyset$. If $s\in CH\cap S$ then $s\notin O$ and so $s\in\overline{O_i^2}$ for some index i. By $s\in CH$ we have $ps\in pC$ and so ps is covered by one of the finitely many $pN_x=N_{px}$ selected earlier to cover pC. We see that $s\in\overline{O_i^2}\cap N_x$ and so $\overline{O_i^2}$ is one of the finitely many specially selected sets.

Finally we replace g by $f=g/g_0$ where $g_0(x)=\int_H g(xh)\,dh$. The existence of the integral and the continuity of g_0 are proved by a reasoning which will be needed again. Therefore it will be given in the form of a separate lemma. In order to assure the existence and the continuity of f we still have to show that g_0 is strictly positive. Since $pO_i^3 = Q_i^3$ $(i \in \mathscr{I})$ is a cover of G/H for every x in G there is an index such that xH intersects O_i^3. Hence $xh \in O_i^3$ for some h in H and so $g_i(xh)=1$ and $g(xh) \neq 0$. Since g is non-negative and continuous we obtain $g_0(x)>0$ where $x \in G$ is arbitrary.

Lemma 3. *If* $g: G \to \mathbb{C}$ *is a continuous function with support* S *and if* $CH \cap S$ *is compact for every compact set* C *contained in* G *then* $g_0(x) = \int_H D(h) g(xh)\, dh$ *defines a continuous function on* G *for any locally bounded measurable* $D: H \to \mathbb{C}$.

Proof. Given a point x in G we choose a symmetric, compact neighborhood C of x. If $y \in C$ and $h \in H$ then we can have $g(yh) \neq 0$ only if $yh \in CH \cap S$ and so only if $h \in C(CH \cap S)$. If we let $K = C(CH \cap S) \cap H$ then H being closed K is compact and $g_0(y) = \int_K D(h) g(yh)\, dh = \int_H D(h) g(yh)\, dh$ exists for any $y \in C$.
Therefore we have

$$|g_0(y) - g_0(x)| \leqslant D_K \int_K |g(yh) - g(xh)|\, dh \quad \text{with } D_K = \operatorname{supr}\{|D(h)| : h \in K\}$$

for the fixed x and for any $y \in C$. Since CK is compact g is uniformly continuous there and so given $\varepsilon > 0$ there is a neighborhood N_e with $xN_e \subseteq C$ such that $|g(\xi) - g(\eta)| < \varepsilon / D_K \mu(K)$ for any ξ, η in CK satisfying $\xi \eta^{-1} \in N_e$. In particular if y in C satisfies $(yh)(xh)^{-1} = yx^{-1} \in N_e$ then we have $|g(yh) - g(xh)| < \varepsilon / D_K \mu(K)$ for every h in K. Therefore $|g_0(y) - g_0(x)| < \varepsilon$ for every $y \in N_x = xN_e$.

Proposition 4. *Let* Δ *be the modular function of the locally compact group* G *and let* δ *be the modular function of the closed subgroup* H. *Then there is a continuous function* $\rho: G \to (0, \infty)$ *such that*

$$\rho(xh) = \frac{\delta}{\Delta}(h)\, \rho(x)$$

for every $x \in G$ *and* $h \in H$.

Proof. We let

$$\rho(x) = \int_H \frac{\Delta}{\delta}(h) f(xh)\, dh$$

where f denotes the non-negative, continuous function whose existence is stated in Proposition 2. The quotient Δ/δ is positive and continuous on H and ${}_x f$ is non-negative, not identically zero and continuous there. Hence ρ is strictly positive and by the left invariance of dh it satisfies the functional equation stated in the proposition. The continuity of ρ can be seen from Lemma 3.

Proposition 5. *For any f in $C_0(G)$ let $\tilde{f}$ be defined by $\tilde{f}(x)=\int_H f(xh)\,dh$. Then $\tilde{f}(xh)=\tilde{f}(x)$ for every $x\in G$ and $h\in H$. If $\tilde{f}$ is interpreted as a function on G/H then $\tilde{f}\in C_0(G/H)$, namely if S is a support of f then pS is a support of $\tilde{f}$.*

Proof. The continuity of $\tilde{f}$ follows from the uniform continuity of f or from Lemma 1.12 and Lemma 3. The invariance property $\tilde{f}(xh)=\tilde{f}(x)$ is an immediate consequence of the definition and the left invariance of dh. To prove that $\tilde{f}: G/H\to\mathbb{C}$ has compact support we notice that if S is the support of f then $\tilde{f}(xH)=0$ unless $xh\in S$ for some h in H and so $\tilde{f}(\xi)=0$ unless $\xi=xH\in\{sH: s\in S\}=pS$. Since S is compact so is its continuous image pS.

Lemma 6. *If K is a compact set in G/H then there is a compact set C in G such that $pC=K$.*

Proof. Choose a closed, compact neighborhood of e, say N_e. Then for any $x\in G$ the image set $p(xN_e)$ is a compact neighborhood of px because by Lemma 1.6 p is both continuous and open. Since $p(xN_e^i)\,(x\in K)$ is an open cover of K there exists a finite subcover and so there are points $x_1,\dots,x_n$ in G such that $p(x_1N_e\cup\cdots\cup x_nN_e)\supseteq K$. The set $x_1N_e\cup\cdots\cup x_nN_e$ is compact, so $p^{-1}K$ being closed $C=(x_1N_e\cup\cdots\cup x_nN_e)\cap p^{-1}K$ is also compact. We have $pC=K$.

Theorem 7. *The map $f\mapsto\tilde{f}$ of $C_0(G)$ into $C_0(G/H)$ is surjective. More precisely for every compact set K in G/H there exist a compact set C in G and a positive constant M such that if $\varphi\in C_0(G/H)$ has its support in K then $\varphi=\tilde{f}$ for some f in $C_0(G)$ whose support lies in C and satisfies the inequality $\|f\|_\infty\leqslant M\|\varphi\|_\infty$.*

Proof. Given $\varphi\in C_0(G/H)$ with compact support K, by the last lemma there is a compact set C in G such that $pC=K$. By the compactness of C and the uniformizability of G we can choose a function g in $C_0(G)$ such that it has a positive minimum m on C and non-negative on G. Then we define for every x satisfying $px\in K$ the function values

$$f(x)=\frac{g(x)\varphi(px)}{\int_H g(xh)\,dh}=\frac{g(x)\varphi(px)}{\tilde{g}(px)}.$$

If $px = xH \in K$ then by $pC = K$ there is a c in C such that $px = pc$ and so there is an h in H such that $xh \in C$. Therefore by the positivity of g on C we see that $\tilde{g}(px) \geqslant m$. Thus $\varphi/\tilde{g}$ defines a continuous function on G/H and consequently $f = g\varphi/\tilde{g}$ is continuous on G. We also see that $\|f\|_\infty \leqslant M\|\varphi\|_\infty$ with $M = \|g\|_\infty/m$. If $\tilde{g}(px) \neq 0$ then clearly

$$\tilde{f}(px) = \frac{1}{\tilde{g}(px)} \int_H \varphi(p(xh))g(xh)\,dh = \frac{\varphi(px)}{\tilde{g}(px)} \int_H g(xh)\,dh = \varphi(px).$$

Hence the image of f under our map if $\tilde{f} = \varphi$.

A function $f: G \to \mathbb{C}$ is called *locally integrable* if it is measurable and if it is integrable on every compact set $C \subseteq G$. Since the modular function Δ is continuous we can integrate f on C with respect to both the left and the right Haar measures of G.

Lemma 8. *Let $\rho: G \to (0, \infty)$ be locally integrable and such that $\rho(xh) = (\delta/\Delta)(h)\rho(x)$ for every $x \in G$ and $h \in H$. Then $\tilde{f} = 0$ implies that*

$$\int_G f(x)\rho(x)\,dx = 0.$$

Proof. Since $\tilde{f} = 0$ we have $\int_H f(xh^{-1})\delta(h^{-1})\,dh = 0$ for every x in G. Let $g \in C_0(G)$ be such that $\tilde{g}$ is 1 on pC where C denotes the support of f and g is non-negative on G. Since pC is compact and G/H is regular the existence of such g follows from the last theorem. Then we have

$$\int_G \int_H \rho(x)g(x)f(xh^{-1})\delta(h^{-1})\,dh\,dx = 0.$$

We interchange the order of integration and then in the inner integral we make the substitution $x \geqq xh$. By applying $\rho(xh) = (\delta/\Delta)(h)\rho(x)$ and by reversing again the order of integration we obtain

$$\int_G \int_H \rho(x)f(x)g(xh)\,dh\,dx = \int_G f(x)\rho(x)\tilde{g}(x)\,dx = 0.$$

Since $\tilde{g}(x) = 1$ on C the desired conclusion follows.

Theorem 9. *There exist non-trivial quasi-invariant measures μ on G/H provided there is a locally bounded and measurable $\rho: G \to (0, \infty)$ satisfying $\rho(xh) = (\delta/\Delta)(h)\rho(x)$. Then such a measure is defined by*

$$\mu(\tilde{f}) = \int_{\frac{G}{H}} \tilde{f}(\xi)\,d\mu(\xi) = \int_G f(x)\rho(x)\,dx.$$

Note. The existence of a strictly positive and continuous ρ satisfying the requirements is stated in Proposition 4.

Proof. By Lemma 8 $\mu(\tilde{f})$ does not depend on the particular f used to define it and by Theorem 7 it is defined for every element of $C_0(G/H)$. Theorem 7 also shows that μ is a continuous linear functional on $C_K(G/H)$ for any compact set K in G/H. Therefore μ is a non-trivial measure on G/H. In order to see the quasi-invariance of μ first we notice that ${}^x\tilde{f} = \widetilde{{}^x f}$ for every f in $C_0(G)$ and $x \in G$. Let μ_x denote the translate of μ by $x \in G$ so that $\mu_x(\tilde{f}) = \mu({}^x\tilde{f})$ or in terms of measurable sets $\mu_x(E) = \mu(xE)$. Then

$$\mu_x(\tilde{f}) = \mu(\widetilde{{}^x f}) = \int_G f(x^{-1}y)\rho(y)\,dy = \int_G f(y)\lambda_\rho(x,y)\rho(y)\,dy.$$

Here $\lambda_\rho(x,y) = \rho(xy)/\rho(y)$ depends only on the coset of y and so it defines a function $\lambda_{\rho x}$ on G/H. Clearly $\widetilde{f\lambda_{\rho x}} = \tilde{f}\lambda_{\rho x}$ and so

$$\mu_x(\tilde{f}) = \mu(\tilde{f}\lambda_{\rho x})$$

for every $f \in C_0(G)$. Since the map $f \mapsto \tilde{f}$ is surjective we see that μ_x is absolutely continuous with respect to μ and the Radon-Nikodym derivative is $\lambda_{\rho x}$. This implies that μ is quasi-invariant.

Corollary. *If μ denotes the quasi-invariant measure corresponding to the function ρ then*

$$\int_G f(x)\rho(x)\,dx = \int_{\frac{G}{H}} \int_H f(xh)\,dh\,d\mu(\xi).$$

Lemma 10. *The function $\lambda_\rho(x,y) = \rho(xy)/\rho(y)$ satisfies the following relations:*

1) $\lambda_\rho(x,yh) = \lambda_\rho(x,y)$ *for every* $x, y \in G$ *and* $h \in H$;

2) $\lambda_\rho(xy,\zeta) = \lambda_\rho(x,y\zeta)\lambda_\rho(y,\zeta)$ *for every* $x, y \in G$ *and* $\zeta \in G/H$;

3) $\lambda_\rho(h,k) = (\delta/\Delta)(h)$ *for every* $h, k \in H$.

Proof. We have

$$\lambda_\rho(x,yh) = \frac{\rho(xyh)}{\rho(yh)} = \frac{\rho(xyh)}{\rho(xy)}\,\frac{\rho(y)}{\rho(yh)}\,\frac{\rho(xy)}{\rho(y)}.$$

Therefore using the fundamental relation $\rho(xh) = (\delta/\Delta)(h)\rho(x)$ with xy and y in place of x we obtain 1). Since $\lambda_\rho(h,k) = \lambda_\rho(h,e)$ the same formula implies 3) while 2) is a direct consequence of the definition of λ_ρ.

Proposition 11. *Any two quasi-invariant measures μ_1, μ_2 defined by means of continuous positive functions ρ_1, ρ_2 are equivalent.*

Proof. We have

$$\mu_1(\tilde{f}) = \int_G f(x)\rho_1(x)dx = \int_G f(x)\frac{\rho_1(x)}{\rho_2(x)}\rho_2(x)dx = \mu_2\left(\widetilde{f\frac{\rho_2}{\rho_1}}\right).$$

Since $\rho_2(x)/\rho_1(x)$ depends only on the coset xH it defines a function φ on G/H and $\widetilde{f\rho_2/\rho_1} = \tilde{f}\varphi$. Therefore $\mu_1(\tilde{f}) = \mu_2(\tilde{f}\varphi)$ for every $f \in C_0(G)$ and so μ_1 is absolutely continuous with respect to μ_2.

Proposition 12. *Let G be a locally compact group and let K, N be subgroups of G such that $(k,n) \mapsto kn$ is a homeomorphism of $K \times N$ onto G. Then $kN \mapsto k$ is a homeomorphism of G/N onto K and the left Haar measure of K is an invariant measure on G/N.*

If Δ and δ are the modular functions of G and N then a left Haar measure of G is $dx = (\Delta/\delta)(n)\,dn\,dk$ where dk, dn denote left Haar measures on K and N, respectively.

Proof. Each coset in G/N is of the form kN where k in K is uniquely determined by the coset. Therefore $kN \mapsto k$ is a one-to-one map of G/N onto K. If O is an open set of K then its inverse image under our map is $O \cdot N$. This is an open set in G/N because the inverse image of $O \cdot N$ under the natural map $p: G \to G/N$ is ON which is the same as the image of $O \times N$ under $(k,n) \mapsto kn$. Next let a subset S of K be such that $S \cdot N$ is an open set in G/N. Then SN is open in G so $S \times N$ is open in $K \times N$ and S is open in K. We proved that $kN \mapsto k$ and its inverse are continuous.

Let $\rho: G \to (0, \infty)$ be defined by $\rho(kn) = \delta(n)/\Delta(n)$. Then ρ defines a quasi-invariant measure μ on G/N. If using $kN \mapsto k$ we transfer μ onto K then for any $f \in C_0(K)$ and $a \in K$ we have

$$\int_K f(ak)d\mu(k) = \int_{\frac{G}{N}} f(akN)d\mu(kN) = \int_{\frac{G}{N}} f(kN)\lambda_\rho(a,kN)d\mu(kN)$$

where $\lambda_\rho(a,kN) = \rho(akN)/\rho(kN) = 1$ by the definition of ρ. Therefore the transferred measure is a left Haar measure on K and so by the Corollary to Theorem 9 we have

$$\int_G f(x)\rho(x)dx = \int_K \int_N f(kn)\,dn\,dk$$

for every $f \in C_0(G)$. Replacing $f\rho$ by f we obtain the relation $dx = (\Delta/\delta)(n)\,dn\,dk$.

Proposition 13. *The modular function* $D: G \to (0, \infty)$ *of a relatively invariant measure on* G/H *is a continuous real character such that* $D(h) = (\Delta/\delta)(h)$ *for every* h *in* H.

Proof. The character property $D(ab) = D(a)D(b)$ is an immediate consequence of the definition. Using it we see that it is sufficient to prove continuity at e. The following reasoning is essentially the same which was used to prove the continuity of the modular function Δ of G: We let $X = G/H$ and choose $f: X \to (0, \infty)$ such that it is continuous, not identically zero and has compact support S in X. Having fixed a compact, symmetric neighborhood C of e in G we choose the continuous $g: X \to (0, \infty)$ such that it is identically 1 on CS and its support is compact. Since f is uniformly continuous on X given $\varepsilon > 0$ by Proposition 1.4 we have $|f(a\xi) - f(\xi)| < \varepsilon$ for every ξ in X and a in N_e where N_e is a suitable symmetric neighborhood of e in G. We may also suppose that $N_e \subseteq C$. Then we have

$$|f(a\xi) - f(\xi)| < \varepsilon g(\xi)$$

for every $\xi \in X$ and $a \in N_e$. Therefore

$$|D(a) - 1| \int_X f(\xi)\, d\xi \leqslant \int_X |f(a\xi) - f(\xi)|\, d\xi \leqslant \varepsilon \int_X g(\xi)\, d\xi$$

for every a in N_e. Since g is fixed while $\varepsilon > 0$ can be chosen arbitrarily small we see that D is continuous at e.

In order to show that $\Delta(h) = D(h)\delta(h)$ for all h in H we first prove the following result:

Proposition 14. *If* $d\xi$ *is a non-trivial relatively invariant measure on* G *with modular function* D *then*

$$\int_{\frac{G}{H}} \int_H D(xh) f(xh)\, dh\, d\xi$$

is a left Haar integral of f *on* G *for every* f *in* $C_0(G)$.

Proof. Using the continuity of D and Proposition 5 we see that the above expression defines a positive linear functional α on $C_0(G)$. If a compact set S is given in G we can choose g in $C_0(G)$ such that $g(x) = 1$ for every x in S and g is non-negative on G. Now if S is a support of $f \in C_0(G)$ then $|\alpha(f)| \leqslant M_S \|f\|_\infty$ with

$$M_S = \int_{\frac{G}{H}} \int_H D(xh) g(xh)\, dh\, d\xi.$$

Therefore α defines a measure on G. By the relative invariance of $d\xi$ for any $f \in C_0(G)$ we have

$$\alpha({}_a f) = D(a^{-1}) \int_{\frac{G}{H}} \int_H D(axh) f(axh) dh d\xi = \alpha(f).$$

Therefore the measure defined by α is left invariant. The conclusion follows from the uniqueness of the Haar measure.

Now we can complete the proof of Proposition 13 by showing that $\Delta(h) = D(h)\delta(h)$ as follows: We choose some f in $C_0(G)$ such that its integral is not zero. Then

$$\Delta(a) \int_G f(x) dx = \int_G f(xa^{-1}) dx = \int_{\frac{G}{H}} \int_H D(xh) f(xha^{-1}) dh d\xi$$

$$= D(a)\delta(a) \int_{\frac{G}{H}} \int_H D(xh) f(xh) dh d\xi = D(a)\delta(a) \int_G f(x) dx$$

for any $a \in H$ and so the desired relation follows.

Theorem 15. *A necessary and sufficient condition for the existence of a non-trivial relatively invariant measure on G/H with modular function D is that D be a continuous real character on G such that $D(h) = (\Delta/\delta)(h)$ for every h in H.*

If D satisfies the requirements the relatively invariant measure associated with it is essentially uniquely determined by D.

Proof. If we let $\rho(x) = 1/D(x)$ for $x \in G$ then ρ satisfies the requirements of Theorem 9 and so it defines a quasi-invariant measure μ on G/H. Since the Radon-Nikodym derivative of μ_x with respect to μ is $\lambda_\rho(x,y) = \rho(xy)/\rho(y) = 1/D(x)$ we see that

$$\int_X f(\eta) d\mu_x(\eta) = \int_X f(x^{-1}\eta) d\mu(\eta) = \frac{1}{D(x)} \int_X f(\eta) d\mu(\eta).$$

Applying this with f being the characteristic function of the set E we obtain $\mu(E) = D(x)\mu(xE)$. Therefore μ is relatively invariant with modular function D. The other half of the statement is the content of Proposition 13. If D satisfies the conditions then by Theorem 7, Lemma 8 and Proposition 14 the linear functional μ: $C_0(G/H) \to (0, \infty)$ is uniquely determined by the Haar measure of G. Since the Haar measure is essentially unique so is μ.

Now let $\varphi \in C_0(G/H)$ and let $f \in C_0(G)$ be such that $\tilde{f} = \varphi$ where $\tilde{f}$ is the function obtained from f according to Proposition 5 and Theorem 7.

Since G acts on G/H we can define ${}_a\varphi$ for any $a\in G$ by ${}_a\varphi(\xi)=\varphi(a\xi)$ where $\xi\in G/H$. Using the definition of f we see that ${}_a\tilde{f}=\widetilde{{}_af}$ for every $a\in G$. Now let μ be a quasi-invariant measure determined by the ρ-function $\rho\colon G\to(0,\infty)$. Then by Theorem 9 and Lemma 10 we have

$$\mu({}_a\tilde{f})=\int_G {}_af(x)\rho(x)\,dx=\int_G f(x)\frac{\rho(a^{-1}x)}{\rho(x)}\rho(x)\,dx=\mu(\lambda(a^{-1},\cdot)\tilde{f})$$

and from this we obtain the following:

Proposition 16. *If $\varphi\colon G/H\to\mathbb{C}$ is integrable with respect to the quasi-invariant measure μ then so is ${}_a\varphi$ for every $a\in G$ and*

$$\int_{\frac{G}{H}} {}_a\varphi(\xi)\,d\mu(\xi)=\int_{\frac{G}{H}} \lambda(a^{-1},\xi)\varphi(\xi)\,d\mu(\xi)\,.$$

Corollary. *If μ is a relatively invariant measure with modular function D then*

$$\int_{\frac{G}{H}} {}_a\varphi(\xi)\,d\mu(\xi)=D(a)\int_{\frac{G}{H}} \varphi(\xi)\,d\mu(\xi)\,.$$

If G/H is compact then the measure of G/H is finite for all quasi-invariant measures derived from continuous, positive ρ-functions. This result is formally stated in Proposition VI.6.14 and it is proved there.

Now we give some examples:

1. If G and its subgroup H are both unimodular then $\Delta=\delta=1$ and so G/H admits an invariant measure by the Corollary of Proposition 16.

2. If the subgroup H is unimodular and if Δ denotes the modular function of G then $\Delta/\delta=\Delta$ has a continuous extension from H to G. Therefore G/H admits a relatively invariant measure with modular function Δ.

3. If H is a normal subgroup of G then the left Haar measure of G/H is an invariant measure on G/H.

4. Let H be a normal subgroup of G. Then by 3. there is a non-trivial invariant measure on G/H and so $\Delta=\delta$ on H. In particular if H lies in the kernel of $\Delta\colon G\to(0,\infty)$ then H is unimodular. By Examples 10 and 11 of Section 2 the groups $\mathrm{Sl}(m,\mathbb{R})$ and $\mathrm{Sl}(n,\mathbb{C})$ are unimodular.

5. If G is unimodular while H is not then no invariant measure exits on G/H. An example for this is $G=\mathrm{Gl}(2,\mathbb{R})$ and the subgroup consisting of the matrices $h=\begin{pmatrix}x & y\\ 0 & 1\end{pmatrix}$ where $x,y\in\mathbb{R}$ and $x\neq 0$. By Example 7 of

Section 2 the modular function of H is $\delta(h)=1/|x|=1/|\det h|$. Since $(\Delta/\delta)(g)=|\det g|$ is a continuous extension of Δ/δ from H to G the quotient G/H admits a nontrivial relatively invariant measure with modular function $|\det g|$.

6. Let $G=\{(x,y): 0<x \text{ and } -\infty<y<\infty\}$ and define $(x_1,y_1)(x_2,y_2) =(x_1x_2, x_1y_2+y_1)$ so that $(1,0)$ is the identity. The elements $(1,y)\,(y\in\mathbb{Z})$ form a discrete subgroup of G which we denote by H. Geometrically G is the half plane $\{(s,t): 0<s\}$ and the cosets gH are in one-to-one correspondence with the points of the cylinder obtained from the strip

$$\{(s,t): 0<s \text{ and } -\tfrac{1}{2}\leqslant y\leqslant\tfrac{1}{2}\}$$

by identification of the boundary points $(s,-\frac{1}{2})$ and $(s,\frac{1}{2})$. Since H is a commutative group it is unimodular. By Example 7 of Section 2 G is not unimodular. Nevertheless G/H admits a nontrivial invariant measure because $(\Delta/\delta)(h)=1$ for every $h\in H$. The corresponding invariant integral can be best described by interpreting $f: G/H\to\mathbb{C}$ as a function on G with the property that $f(x,y+z)=f(x,y)$ for every $z\in\mathbb{Z}$. Then we have

$$\int_{\frac{G}{H}} f(\xi)\,d\xi=\int_0^\infty\int_{-\frac{1}{2}}^{\frac{1}{2}} f(x,y)\frac{d\eta\, d\xi}{x}. \tag{1}$$

Indeed

$$\int_0^\infty\int_{-\frac{1}{2}}^{\frac{1}{2}} f(ax,bx+y)\frac{dy\,dx}{x}=\int_0^\infty\int_{\frac{b\xi}{a}-\frac{1}{2}}^{\frac{b\xi}{a}+\frac{1}{2}} f(\xi,\eta)\left|\frac{\partial(x,y)}{\partial(\xi,\eta)}\right|\frac{d\eta\,d\xi}{\frac{\xi}{x}}.$$

Since the Jacobian of the transformation is $1/a$ and f is periodic in η with period 1 the last integral is the same which occurs in (1). Of course G/H has also a relatively invariant measure with modular function $D=\Delta$.

7. Let $G=\mathrm{Sl}(2,\mathbb{R})$ and $H=\mathrm{SO}(2,\mathbb{R})$. By the discussion in Example 3 of the second half of Section 1 we know that G/H is homeomorphic to the upper half plane

$$\{z: z=x+iy \text{ with } x,y\in\mathbb{R} \text{ and } y>0\}.$$

We are going to show that there is an invariant measure on G/H by proving that

(2)
$$\int_{0}^{\infty}\int_{-\infty}^{\infty} f(x,y)\frac{dx\,dy}{y^2}$$

is a G invariant integral in the upper half plane.

We want to prove that

$$\int_{0}^{\infty}\int_{-\infty}^{\infty} f(g^{-1}\cdot(u,v))\frac{du\,dv}{v^2}=\int_{0}^{\infty}\int_{-\infty}^{\infty} f(x,y)\frac{dx\,dy}{y^2}$$

where

$$w=u+vi=g\cdot z=\frac{a_{11}z+a_{12}}{a_{21}z+a_{22}}$$

and $a_{11}a_{22}-a_{12}a_{21}=1$. By performing the substitution $w\mapsto z$ we obtain the integral

(3)
$$\int_{0}^{\infty}\int_{-\infty}^{\infty} f(x,y)\left|\frac{\partial(u,v)}{\partial(x,y)}\right|\frac{dx\,dy}{v^2}.$$

We easily find that

$$\frac{dw}{dz}=(a_{21}z+a_{22})^{-2}$$

and that

$$\mathscr{Im}\,w=\mathscr{Im}\,z|a_{21}z+a_{22}|^{-2}.$$

8. Let G be SO $(n+1,\mathbb{R})$ and $H=\mathrm{SO}(n,\mathbb{R})$ then according to Example 6 in Section 1 the quotient G/H is homeomorphic to S^n. We parametrize S^n by the Euclidean coordinates $x_1,\ldots,x_{n+1}$ subject to $x_1^2+\cdots+x_{n+1}^2=1$. Then a G invariant measure of S^n is

$$dv=\frac{1}{2}\pi^{-\frac{n}{2}}\Gamma\left(\frac{n}{2}\right)\frac{dx_1\ldots dx_n}{x_{n+1}}.$$

The normalizing factor is so chosen that the total measure of S^n is unity.

Proposition 17. *If μ is a σ-finite quasi-invariant measure on the locally compact group G then μ is equivalent to the Haar measure of G.*

Note. The left and right Haar measures being equivalent it makes no difference which one is used.

Proof. Since μ is σ-finite it is equivalent to a finite measure on G so we may suppose that $\mu(G)$ is finite. If E is a Borel set in G then $S=\{(x,y): xy^{-1}\in E\}$ is a Borel set in $G\times G$ and their characteristic functions are related by $c_S(x,y)=c_E(xy^{-1})$. By Fubini's theorem

$$\begin{aligned}\alpha(E^{-1})\mu(G)&=\int_G\int_G c_{E^{-1}}(x)\,d\alpha(x)\,d\mu(y)=\int_G\int_G c_{E^{-1}}(y^{-1}x)\,d\alpha(x)\,d\mu(y)\\&=\int_G\int_G c_{xE}(y)\,d\mu(y)\,d\alpha(x)=\int_G \mu(xE)\,d\alpha(x)\,.\end{aligned}$$

Hence for every Borel set E we have

$$\alpha(E^{-1})\mu(G)=\int_G \mu(xE)\,d\alpha(x)$$

where α is the left Haar measure of G. Now if $\mu(E)=0$ then by the quasi-invariance $\mu(xE)=0$ for every $x\in G$ and so by the above formula $\alpha(E^{-1})=0$ and $\alpha(E)=0$. Conversely if $\alpha(E)=0$ then $\alpha(E^{-1})=0$ and so $\int_G \mu(xE)\,d\alpha(x)=0$ and $\mu(xE)=0$ for almost all $x\in G$. By $\mu(G)>0$ and the quasi-invariance of μ we obtain $\mu(E)=0$.

4. Convolutions of Functions and Measures

Let G be a locally compact group and let dx be a left Haar measure on G. By definition the convolution of f and g at x is

$$f*g(x)=\int_G f(xy)g(y^{-1})\,dy=\int_G f(y)g(y^{-1}x)\,dy$$

for any dx measurable $f: G\to\mathbb{C}$ and $g: G\to\mathbb{C}$ for which the integral is meaningful at $x\in G$. If a function $f*g$ is defined locally almost everywhere by this formula then it will be called the *convolution* or the *convolution product* of f and g. If G is commutative then $f*g=g*f$ but in general $*$ is not a commutative operation. We have

$$f*g(x)=\int_G f(xy^{-1})g(y)\Delta(y^{-1})\,dy$$

so if G is unimodular then

$$f*g(x)=\int_G f(xy^{-1})g(y)\,dy=\int_G f(y^{-1})g(yx)\,dy\,.$$

For $f: G \to \mathbb{C}$ we define $f^*: G \to \mathbb{C}$ by $f^*(x) = \overline{f(x^{-1})}\Delta(x^{-1})$. It is clear that $f^{**} = f$ for any such function f. If $f \in L^1(G)$ then by Proposition 2.2 $f^* \in L^1(G)$ and $\|f^*\|_1 = \|f\|_1$.

Theorem 1. *If G is a locally compact group then under addition, multiplication by complex scalars, the convolution product, the map $f \mapsto f^*$ and the norm $\|f\|_1$ the set $L^1(G)$ is a complex, involutive, star normed Banach algebra.*

We already know that $L^1(G)$ is a Banach space and we just saw that it is star normed. All other statements expressed in the theorem are obvious consequences of the definition of the convolution product except the associativity of $*$ and two other essential points which we state and prove separately:

Proposition 2. *If f and g are in $L^1(G)$ then $f*g$ exists almost everywhere it belongs to $L^1(G)$ and $\|f*g\|_1 \leqslant \|f\|_1 \cdot \|g\|_1$.*

The proof of this is based on the following important observations from which the result will easily follow by Fubini's theorem. The same method will be used on a number of occasions later.

For any locally compact group G the map $h: (x,y) \mapsto (yx,y)$ is a continuous bijection of $G \times G$ onto itself with continuous inverse $(x,y) \mapsto (y^{-1}x,y)$. Given a linear functional μ associated with a measure on $G \times G$ using the homeomorphism h we can construct another measure v on $G \times G$ by letting $v(f) = \mu(f \circ h)$ for every f in $C_0(G \times G)$. In fact if $C \times C$ is a symmetric, compact carrier of f which contains e then $C^2 \times C^2$ is a carrier of $f \circ h: (x,y) \mapsto f(yx,y)$ and so the restriction of v to $C_{C \times C}(G \times G)$ is continuous in the $\|\cdot\|_\infty$ norm.

If μ is the functional associated with the product measure $dx\,dy$ where dx, dy denote Haar measures on G then v is a left invariant measure because $h \circ f(ax,by) = f(byax,by)$ and by Proposition 2.5 $dx\,dy$ is invariant under the transformation $(x,y) \to (ax,by)$. Therefore by the essential uniqueness of the Haar measure we have $v = \lambda\mu$ for some $\lambda > 0$. We proved the following:

If $f: G \times G \to \mathbb{C}$ is integrable with respect to the left Haar measure of $G \times G$ then so is $h \circ f$ where $h(x,y) = (yx,y)$ and if the measures are properly adjusted then their integrals are equal for all choices of f in $L^1(G \times G)$.

As a corollary we see that h maps measurable sets into measurable sets and $h \circ f$ is measurable for every measurable f.

Let f and g be measurable with respect to the left Haar measures dx and dy of G. By the properties of the product measure $(x,y)\mapsto f(x)g(y)$ is a measurable function on $G\times G$, it is integrable if f and g are integrable and in that case

$$\int_{G\times G} f(x)g(y)\,dx\,dy=\int_G\int_G f(x)g(y)\,dx\,dy.$$

In terms of the function φ defined by $(x,y)\mapsto f(y^{-1}x)g(y)$ our function is expressible as $\varphi\circ h$. Therefore we obtain:

If f and g are in $L^1(G)$ then $(x,y)\mapsto f(y^{-1}x)g(y)$ belongs to $L^1(G\times G)$ and if the measures are properly adjusted then

$$\int_{G\times G} f(y^{-1}x)g(y)\,dx\,dy=\int_G\int_G f(x)g(y)\,dx\,dy$$

for every $f,g\in L^1(G)$.

Now it is easy to prove Proposition 2: If $f,g\in L^1(G)$ then by the last result $(x,y)\mapsto g(y^{-1}x)f(y)$ is integrable on $G\times G$ and so by Fubini's theorem $f*g(x)$ exists almost everywhere. Moreover replacing f and g by $|f|$ and $|g|$ in the last equation we obtain

$$\|f*g\|_1=\int_G |f*g(x)|\,dx\leqslant\int_G\int_G |g(y^{-1}x)|\cdot|f(y)|\,dy\,dx\leqslant\|f\|_1\cdot\|g\|_1\,.$$

We shall now discuss a sequence of propositions concerning the convolution of various types of functions and afterwards we will introduce convolutions of measures. The following notations will be used:

$C_o(G)=$ the space of complex valued continuous functions vanishing at infinity;

$C_0(G)=$ the space of complex valued continuous functions with compact support;

$C_u(G)=$ the space of complex valued left uniformly continuous functions;

$Li(G)=$ the space of locally integrable functions;

$L^p(G)=$ the space of p-th power integrable functions;

$L^\infty(G)=$ the space of essentially bounded, measurable functions.

The elements of the last three spaces are also complex valued.

The superscript R will indicate that the property should be considered from the right, e. g. $L^{1R}(G)$ stands for the space of complex valued functions which are integrable with respect to the right Haar measure of G. By space we mean a vector space over $\mathbb{C}$ and in the first, second, fifth and last case the vector space is understood to be normed by $\|f\|_\infty$, $\|f\|_\infty$, $\|f\|_p$ and $\|f\|_\infty$, respectively. The p-th power norm with respect to the right Haar measure will be denoted by $\|f\|_p^R$. The p-th power norms depend on the choice of the left and right Haar measures on G. If p and q both occur in the same proposition then $1<p,\, q<\infty$ and it is tacitly supposed that $1/p+1/q=1$.

One easily verifies that if f and g are integrable functions with compact support then $f*g$ is a function of compact support. It is also easy to see that if $f\in L^1(G)$ and g vanishes at infinity then so does $f*g$.

Proposition 3. *If* $f\in L^1(G)$ *and* $g\in L^\infty(G)$ *then* $f*g\in L^\infty(G)$ and $\|f*g\|_\infty \leqslant \|f\|_1\cdot\|g\|_\infty$.

Proof. One obtains everything by properly reading the following:

$$|(f*g)(x)| \leqslant \int_G |f(xy)|\,|g(y^{-1})|\,dy \leqslant \|g\|_\infty \int_G |f(xy)|\,dy = \|f\|_1\cdot\|g\|_\infty .$$

Proposition 4. *If* $f\in L^\infty(G)$ *and* $g\in L^{1R}(G)$ *then* $f*g\in L^\infty(G)$ *and* $\|f*g\|_\infty \leqslant \|f\|_\infty\cdot\|g\|_1^R$.

Proof. This follows from

$$|(f*g)(x)| \leqslant \int_G |f(xy)|\,|g(y^{-1})|\,dy \leqslant \|f\|_\infty \int_G |g(y^{-1})|\,dy = \|f\|_\infty\,\|g\|_1^R .$$

Proposition 5. *If* $f\in L^p(G)$ *and* $g\in L^{qR}(G)$ *then* $f*g\in L^\infty(G)$ *and* $\|f*g\|_\infty \leqslant \|f\|_p\,\|g\|_q^R$.

Proof. This follows by an application of Hölder's inequality:

$$\begin{aligned}|(f*g)(x)| &\leqslant \int_G |f(xy)|\cdot|g(y^{-1})|\,dy \leqslant \Big(\int_G |f(xy)|^p\,dy\Big)^{\frac{1}{p}}\Big(\int_G |g(y^{-1})|^q\,dy\Big)^{\frac{1}{q}}\\ &= \|f\|_p\,\|g\|_q^R .\end{aligned}$$

Let us suppose that G is compact and its Haar measure is normalized such that $\int_G dx=1$. If $f\in L^2(G)$ then by the BCS inequality $f\in L^1(G)$ and $\|f\|_1\leqslant\|f\|_2$. By the last proposition $\|f*g\|_\infty\leqslant\|f\|_2\,\|g\|_2$ and so it follows that $\|f*g\|_2\leqslant\|f\|_2\,\|g\|_2$. Therefore the inner product

$$(f,g)=\int_G f(x)\overline{g(x)}\,dx$$

and the norm derived from it make $L^2(G)$ into a Hilbert space and a Banach algebra. We prove that $(f*g,h)=g,f^* *h)$: In fact

$$\begin{aligned}(f*g,h) &= \int_G\Big(\int_G f(xy)g(y^{-1})\,dy\Big)\overline{h(x)}\,dx = \int_G\Big(\int_G f(xy^{-1})g(y)\,dy\Big)\overline{h(x)}\,dx\\ &= \int_G g(y)\overline{\Big(\int_G f^*(yx^{-1})h(x)\,dx\Big)}\,dy = (g,f^* *h).\end{aligned}$$

The identity $(f*g,h)=(f,h*g^*)$ can be proved similarly or it can be obtained from the other one by using $\|f\|_2=\|f^*\|_2$ to derive first $(g^*,f^*)=(f,g)$. The H^* algebra obtained in this manner is called the *L^2 algebra of the compact group G*. It will be considered again in Section VII.1 after Theorem 12.

Proposition 6. *If $f\in L^1(G)$ and $g\in L^p(G)$ where $1<p<\infty$ then $f*g\in L^p(G)$ and $\|f*g\|_p\leqslant\|f\|_1\|g\|_p$.*

Note. The case $p=1$ is considered in Proposition 2.

Proof. We prove that the integral defining $f*g(x)$ exists locally almost everywhere and it coincides with a function $f*g$ belonging to $L^p(G)$. We show that

$$\varphi(h)=\int_G (f*g)(x)h(x)\,dx$$

exists for every $h\in L^q(G)$. In fact

$$|\varphi(h)|=\left|\int_G (f*g)(x)h(x)\,dx\right|\leqslant\int_G |f(y)|\int_G |g(y^{-1}x)|\cdot|h(x)|\,dx\,dy$$

$$\leqslant\int_G |f(y)|\left\{\int_G |g(y^{-1}x)|^p\,dx\right\}^{\frac{1}{p}}\left\{\int_G |h(x)|^q\,dx\right\}^{\frac{1}{q}}dy=\|f\|_1\|g\|_p\|h\|_q$$

where the right hand side is known to be finite. Hence the existence of $\varphi(h)$ follows from Fubini's theorem. We also see that $h\mapsto\varphi(h)$ is a continuous linear functional on $L^q(G)$ whose norm is at most $\|f\|_1\|g\|_p$. By choosing for h the characteristic functions of sets of finite measure we see that $f*g(x)$ exists locally almost everywhere on G. The existence of $f*g\in L^p(G)$ coinciding with the locally defined convolution product follows from the Riesz representation theorem.

Note. Some of these propositions could be formulated in a more general setting. For instance Propositions 2 and 6 can be extended in the following manner:

Let H be a closed subgroup of G and let both G and H be unimodular. If $f\in L^1(G)$ and $g\in L^p(G/H)$ where $1\leqslant p<\infty$ then $f*g\in L^p(G/H)$ and $\|f*g\|_p\leqslant\|f\|_1\|g\|_p$.

The proofs concerning the cases $p=1$ and $p>1$ are essentially the same as the ones given for Propositions 2 and 6 respectively.

Proposition 7. *If $f\in C_0(G)$ and $g\in Li(G)$ then $f*g\in C(G)$.*

Proof. Let C be a compact neighborhood of the identity e, let S be a compact carrier of f. We plan to prove that $f*g$ is continuous at a given x. Since f is uniformly continuous with respect to the right uniform structure of G there is a symmetric neighborhood $N_e = N_e(\varepsilon)$ of e such that $N_e \subseteq C$ and $|f(a) - f(b)| < \varepsilon$ for every a, b satisfying $ab^{-1} \in N_e$. Therefore $|f(xy) - f(zy)| < \varepsilon$ for every $y \in G$ and $x, z \in G$ satisfying $xz^{-1} \in N_e$. Thus if $xz^{-1} \in N_e$ then

$$\begin{aligned} |f*g(x) - f*g(z)| &= \left| \int_G (f(xy) - f(zy)) g(y^{-1}) dy \right| \\ &\leqslant \int_{x^{-1}S \cup z^{-1}S} |f(xy) - f(zy)| \cdot |g(y^{-1})| dy \\ &\leqslant \varepsilon \int_{x^{-1}S \cup z^{-1}S} |g(y^{-1})| dy. \end{aligned}$$

By $xz^{-1} \in N_e \subseteq C$ we have $z^{-1} \in x^{-1}C$ and by $e \in C$ we have $x^{-1} \in x^{-1}C$, so $x^{-1}S \cup z^{-1}S \subseteq x^{-1}CS$. Since $S^{-1}C^{-1}x$ is compact and Δ is continuous our last integral is majorized by some upper bound $B(x, S, C) = B(x, f, C)$. Therefore

$$|f*g(x) - f*g(z)| \leqslant \varepsilon B(x, f, C)$$

for every z in $N_x = N_e x$ where $N_e \subseteq C$.

Similar reasoning leads to

Proposition 8. *If $f \in L^1(G)$ and $g \in C_u(G)$ then $f*g \in C_u(G)$.*
If $f \in C_u^R(G)$ and $g \in L^{1R}(G)$ then $f*g \in C_u^R(G)$.

Proof. For instance the proof of the second statement is based on the inequality

$$|f*g(x) - f*g(z)| \leqslant \int_G |f(xy) - f(zy)| \cdot |g(y^{-1})| dy \leqslant \varepsilon \int_G |g(y^{-1})| dy$$

which holds provided $xz^{-1} = (xy)(zy)^{-1} \in N_e(\varepsilon)$.

In what follows we shall use the standard notation ${}_xf$ and f^x to denote the functions obtained from f defined on G by letting ${}_xf(y) = f(xy)$ and $f^x(y) = f(yx^{-1})$. If f is complex valued and bounded then clearly $\|{}_xf\|_\infty = \|f^x\|_\infty = \|f\|_\infty$. We also have $\|{}_xf\|_p = \|f\|_p$ and $\|f^x\|_p = \Delta(x)^{1/p} \|f\|_p$. Similarly we define f_x and ${}^xf$ by $f_x(y) = f(yx)$ and ${}^xf(y) = f(x^{-1}y)$. Then for complex valued f we have $\|f_x\|_\infty = \|{}^xf\|_\infty = \|f\|_\infty$ and $\|f_x\|_p = \Delta(x)^{-1/p} \|f\|_p$.

Proposition 9. *If $f \in L^p(G)$ where $1 \leqslant p < \infty$ then the map $x \mapsto {}_xf$ of G into $L^p(G)$ is uniformly continuous from the right.*

Proof. Given $\varepsilon>0$ choose $g\in C_0(G)$ with compact support S such that $\|f-g\|_p<\varepsilon$. We fix a compact symmetric neighborhood C of the identity e. Next g being right uniformly continuous on G there is a symmetric neighborhood N_e contained in C such that

$$|g(a)-g(b)|\leqslant \varepsilon\,\mu(C\,S)^{-\frac{1}{p}}$$

for all a, b such that $ab^{-1}\in N_e$. Here $\mu(S)$ denotes the left Haar measure of S. In particular for all $x\in N_e$ we have

$$\|g-{}_xg\|_\infty\leqslant \varepsilon\,\mu(C\,S)^{-\frac{1}{p}}.$$

This gives $\|g-{}_xg\|_p<\varepsilon$ for every $x\in N_e$. Now

$$\|f-{}_xf\|_p\leqslant\|f-g\|_p+\|g-{}_xg\|_p+\|{}_xg-{}_xf\|_p<3\varepsilon$$

for x in N_e. Thus, if $ab^{-1}\in N_e$ then with $x=ab^{-1}$ we have

$$\|{}_af-{}_bf\|_p=\|{}_xf-f\|_p<3\varepsilon.$$

Proposition 10. *If $f\in L^{pR}(G)$ where $1\leqslant p<\infty$ then the map $x\mapsto f^x$ of G into $L^{pR}(G)$ is uniformly continuous from the right.*

Proof. This is essentially the left—right dual of Proposition 9. The seeming asymmetry is due to the exponent -1 in the definition of f^x. Left uniform continuity would hold for the function f_x.

Proposition 11. *If $f\in L^1(G)$ and $g\in L^\infty(G)$ then $f*g$ is in $C_u^R(G)\cap L^\infty(G)$ and $\|f*g\|_\infty\leqslant\|f\|_1\,\|g\|_\infty$.*

Proof. The existence and boundedness of $f*g$ comes from the estimate

$$|f*g(x)|\leqslant\int_G|f(xy)|\cdot|g(y^{-1})|\,dy\leqslant\|g\|_\infty\int_G|f(xy)|\,dy=\|g\|_\infty\,\|f\|_1.$$

To prove the uniform continuity from the right we notice that

$$\begin{aligned}|f*g(x)-f*g(z)|&\leqslant\int_G|f(xy)-f(zy)|\cdot|g(y^{-1})|\,dy\\&=\int_G|{}_xf(y)-{}_zf(y)|\cdot|g(y^{-1})|\,dy\leqslant\|g\|_\infty\,\|{}_xf-{}_zf\|_1.\end{aligned}$$

Thus we can apply Proposition 9 and obtain the uniform continuity of $f*g$ from the right.

Proposition 12. *If $f\in L^\infty(G)$ and $g\in L^{1R}(G)$ then $f*g$ is in $C_u(G)\cap L^\infty(G)$ and $\|f*g\|_\infty\leqslant\|f\|_\infty\,\|g\|_1^R$.*

Proof. This is the exact left—right dual of Proposition 11.

Proposition 13. *If* $1/p+1/q=1$ *and* $f\in L^p(G)$ *and* $g\in L^{qR}(G)$ *then* $f*g$ *exists and it is continuous on* G.

Proof. The existence of $f*g$ follows from Hölder's inequality. For any $x, z\in G$ we have

$$|f*g(x)-f*g(z)| \leq \int_G |{}_xf(y)-{}_zf(y)|\cdot|g(y^{-1})|\,dy \leq \|{}_xf-{}_zf\|_p\|g\|_q^R.$$

Therefore the continuity of $f*g$ at x follows from Proposition 9.

Proposition 14. *For measurable* f *and* g *and for* a, x *in* G *we have:*

1) $({}_af*g)(x) = {}_a(f*g)(x)$.
2) $(f*g_a)(x) = (f*g)_a(x)$.
3) $(f_a*g)(x) = \Delta(a^{-1})(f*{}_ag)(x)$.

If one of the expressions exists and is finite then the other one occurring on the same line exists and is equal to it.

Proof. 1) is obvious from the definition of the convolution product. 2) is obtained by using the substitutions $y\to ay$ and $y\to a^{-1}y$, respectively. Similarly 3) is proved by using the substitutions $y\to ya^{-1}$ and $y\to ya$.

Corollary. *If* G *is commutative then*

$${}_af*g(x) = f_a*g(x) = f*{}_ag(x) = f*g_a(x).$$

We also mention the following:

If G *is commutative,* $f=f^*$, $g=g^*$ *and* $f, g\in L^1(G)$ *then* $f*g\in L^\infty(G)$.

The proof of this depends on the concept of positive definite functions. (See Section VII.2 and in particular Proposition VII.2.14.) Since $f=f^*$ the function $f*f$ is positive definite and so it is bounded. Hence $f*g$ is also bounded by

$$f*g = \tfrac{1}{2}((f+g)*(f+g)-f*f-g*g).$$

The definition of convolution product can be extended to measures. Let X be a locally compact space and let μ and ν be complex measures on X. We shall use the same symbol μ to denote the linear functional $\mu: C_0(X)\to\mathbb{C}$ associated with the measure μ. Given a compact set C in X we let $C_C(X)$ be the space of those f in $C_0(X)$ whose support lies in C. We know that a linear functional $\mu: C_0(X)\to\mathbb{C}$ corresponds to a measure if and only if for every compact set C the restriction of μ to $C_C(X)$ is a continuous linear functional i. e. there is a constant m_C such that $|\mu(f)| \leqslant m_C \|f\|_\infty$ for every f in $C_C(X)$. It is clear that if μ is an element of $C_o(X)^*$ then μ satisfies this requirement and so it corresponds to some measure on X. The convolution $\mu * v$ will be defined only if G is a locally compact group and the linear functionals μ and v are continuous on $C_o(G)$ i. e. $\mu, v \in C_o(G)^*$.

Lemma 15. *If* $f \in C_o(G)$ *and* μ *is in* $C_o(G)^*$ *then* $x \mapsto \mu({}_x f)$ *defines a function* $\mu f : G \to \mathbb{C}$ *which belongs to* $C_o(G)$.

Proof. We have ${}_x f \in C_o(G)$ and so $\mu({}_x f) = \mu f(x)$ is defined for every x in G. By Lemma 1.13 f is right uniformly continuous and so given $\varepsilon > 0$ there is a neighborhood N_e such that $|f(u) - f(v)| < \varepsilon$ for every u and v satisfying $uv^{-1} \in N_e$. It follows that $\|{}_x f - {}_y f\|_\infty \leqslant \varepsilon$ if $xy^{-1} \in N_e$. Therefore

$$|\mu f(x) - \mu f(y)| = |\mu({}_x f - {}_y f)| \leqslant \|\mu\| \cdot \|{}_x f - {}_y f\|_\infty \leqslant \|\mu\| \varepsilon$$

provided $xy^{-1} \in N_e$. This shows that μf is right uniformly continuous on G.

We still have to prove that $\mu f(x) \to 0$ as $x \to \infty$. It will be sufficient to do this in the special case when f is non-negative. We may also suppose that μ is a non-negative measure. Since μ corresponds to a continuous linear functional on $C_o(G)$ we have $\mu(G) < \infty$. Therefore there is a symmetric compact set S in G such that $0 < \mu(S)$ and $\mu(cS) \leqslant \varepsilon$. Having fixed S we choose the compact set C such that $|f(x)| \leqslant \varepsilon / \mu(S)$ for every $x \notin C$. Then we have

$$0 \leqslant \mu f(x) = \mu({}_x f) = \int_S f(xy)\,dy + \int_{cS} f(xy)\,dy \leqslant \int_S f(xy)\,dy + \varepsilon \|f\|_\infty .$$

Suppose that $f(xy) > \varepsilon/\mu(S)$ and $y \in S$. Then $xy \in C$ and so $x \in CS$. Hence if $x \notin CS$ and $y \in S$ then $f(xy) \leqslant \varepsilon/\mu(S)$. This shows that if $x \notin CS$ then $\int_S f(xy)\,dy \leqslant \varepsilon$. We proved that $0 \leqslant \mu f(x) \leqslant \varepsilon + \varepsilon \|f\|_\infty$ for every x in the complement of the compact set CS.

Let $M(G)$ denote the vector space of those measures μ on the locally compact group G for which the associated linear functional belongs to $C_o(G)^*$. Let μ and v be measures in $M(G)$ and let $f \in C_o(G)$. Then by Lemma 15 $\mu f \in C_o(G)$ and so we can define $\mu * v(f) = v(\mu f)$. This gives a linear functional $\mu * v : C_o(G) \to \mathbb{C}$. Since $|\mu f(x)| = |\mu({}_x f)| \leqslant \|\mu\| \cdot \|f\|_\infty$ we have $\|\mu f\|_\infty \leqslant \|\mu\| \cdot \|f\|_\infty$. Therefore

$$|\mu * v(f)| \leqslant \|v\| \cdot \|\mu f\|_\infty \leqslant \|\mu\| \cdot \|v\| \cdot \|f\|_\infty$$

and so $\mu * \nu$ is a continuous linear functional on $C_o(G)$. We also see that $\|\mu * \nu\| \leqslant \|\mu\| \cdot \|\nu\|$. The measure associated with the functional $\mu * \nu$ is called the *convolution of the measures* μ *and* ν. It will be denoted by the same symbol $\mu * \nu$. We have $\mu * \nu \in M(G)$ or in terms of functionals $\mu * \nu \in C_o(G)^*$. Hence $*$ is a binary operation on $M(G)$ and $C_o(G)^*$, respectively.

Proposition 16. *The convolution of measures is an associative operation.*

Proof. Let $\lambda, \mu, \nu \in C_o(G)^*$ and $f \in C_o(G)$. Then

$$(\lambda * \mu) * \nu(f) = (\lambda * \mu)(\nu f) = \lambda(\mu(\nu f))$$
$$= \lambda(\mu * \nu(f)) = \lambda * (\mu * \nu)(f).$$

We see that both the measures and the corresponding functionals form complex algebras called the *measure algebra* $M(G)$ and the convolution algebra of $C_o(G)^*$. They are isometrically isomorphic Banach algebras under the norm $\|\mu\|$ by Proposition III.7.7.

The algebras $M(G)$ *and* $C_o(G)^*$ *have identities.*

For given a in G we can define δ_a on $C_o(G)$ by $\delta_a(f) = f(a)$. Then δ_a belongs to $C_o(G)^*$ and $\|\delta_a\| = 1$. We have

$$\delta_a f(x) = \delta_a({}_x f) = {}_x f(a) = f(x a) = f_a(x)$$

and so $\delta_a f = f_a$. Using this we obtain $\delta_a * \delta_b = \delta_{ab}$ for every $a, b \in G$. We shall write ε instead of δ_e. Since $\mu(\varepsilon f) = \mu(f)$ and $\varepsilon(\mu f) = \mu f$ we have $\mu * \varepsilon = \mu$ and $\varepsilon * \mu = \mu$ for every μ in $C_o(G)^*$. Hence ε and the associated Dirac measure are identities in $C_o(G)^*$ and $M(G)$, respectively.

For any function f on G let f' be defined by $f'(x) = f(x^{-1})$. It is clear that $f'' = f$. If the ranges of f and g lie in a vector space then $(f+g)' = f' + g'$ and $(\lambda f)' = \lambda f'$. We see that $f' \in C_o(G)$ for every f in $C_o(G)$ and $\|f'\| = \|f\|$. For μ in $C_o(G)^*$ we define $\bar{\mu}$ by $\bar{\mu}(f) = \overline{\mu(\bar{f})}$ so that $\bar{\mu}$ is in $C_o(G)^*$ and $\|\bar{\mu}\| = \|\mu\|$. The measure corresponding to $\bar{\mu}$ is the complex conjugate of the measure μ. For f in $C_o(G)$ and μ in $C_o(G)^*$ let $\mu^*(f) = \overline{\mu(\bar{f}')} = \bar{\mu}(f')$. Then μ^* is in $C_o(G)^*$ and $\|\mu^*\| = \|\mu\|$. The measure corresponding to μ^* will be denoted by the same symbol μ^*. It is clear that $\mu^{**} = \mu$ and $(\mu + \nu)^* = \mu^* + \nu^*$. We also have $(\lambda \mu)^* = \bar{\lambda} \mu^*$ for any $\lambda \in \mathbb{C}$.

Proposition 17. *If* $\mu, \nu \in M(G)$ *and* $f \in C_o(G)$ *then*

$$\mu * \nu(f) = \int_G \int_G f(x y) \, d\nu(y) \, d\mu(x).$$

Proof. We have $\nu f(x)=\nu({}_x f)=\int_G f(xy)\,d\nu(y)$ and $\mu*\nu(f)=\int_G \nu f(x)\,d\mu(x)$.

Corollary. *If G is commutative then so are $M(G)$ and $C_o(G)^*$.*

The corollary is obtained by using Fubini's theorem.

Proposition 18. *$M(G)$ and $C_o(G)^*$ are isometrically isomorphic star normed complex involutive Banach algebras.*

Proof. We already proved everything except the identity $(\mu*\nu)^*=\nu^**\mu^*$. Now if $\mu,\nu\in M(G)$ then by Proposition 17 and Fubini's theorem

$$\begin{aligned}\overline{(\mu*\nu)^*(f)}=\mu*\nu(\overline{f}')&=\int_G\int_G \overline{f'(xy)}\,d\nu(y)\,d\mu(x)\\&=\int_G\int_G \overline{f'(xy)}\,d\mu(x)\,d\nu(y).\end{aligned}$$

On the other hand $\overline{\nu^**\mu^*(f)}=\overline{\nu^*(\mu^*f)}=\nu(\overline{(\mu^*f)'})$ where

$$\overline{(\mu^*f)'(y)}=\overline{\mu^*f(y^{-1})}=\overline{\mu^*({}^yf)}=\mu(\overline{({}^yf)'})=\int_G \overline{({}^yf)'(x)}\,d\mu(x)=\int_G \overline{f'(xy)}\,d\mu(x).$$

Therefore we have

$$\overline{\nu^**\mu^*(f)}=\int_G\int_G \overline{f'(xy)}\,d\mu(x)\,d\nu(y).$$

We proved that $(\mu*\nu)^*(f)=\nu^**\mu^*(f)$ for every f in $C_o(G)$.

Proposition 19. *For any μ,ν in either $M(G)$ or $C_o(G)^*$ we have $\overline{\mu*\nu}=\overline{\mu}*\overline{\nu}$.*

Proof. First of all $\overline{\nu}f=\overline{\nu\overline{f}}$ for any f in $C_o(G)$ because $\overline{\nu}f(x)=\overline{\nu}({}_xf)=\overline{\nu({}_x\overline{f})}$ for all x in G. Now we have

$$\overline{\mu}*\overline{\nu}(f)=\overline{\mu}(\overline{\nu}f)=\overline{\mu(\overline{\overline{\nu}f})}=\overline{\mu(\nu\overline{f})}=\overline{\mu*\nu(\overline{f})}=\overline{\mu*\nu}(f).$$

For μ in $C_o(G)^*$ we define μ' by $\mu'(f)=\mu(f')$ where $f\in C_o(G)$. This defines μ' also for measures $\mu\in M(G)$. It is clear that $\mu''=\mu$. We also have

$$\mu^*=(\overline{\mu})'=\overline{\mu'} \tag{1}$$

because $\mu^*(f)=\overline{\mu}(f')=(\overline{\mu})'(f)$ and $\mu^*(f)=\overline{\mu((\overline{f})')}=\overline{\mu'(\overline{f})}$. Similarly we have

$$\mu'=(\overline{\mu})^*=\overline{\mu^*} \tag{2}$$

because replacing μ by $\overline{\mu}$ in (1) we obtain $(\overline{\mu})^*=\mu'$ and taking the conjugate in (1) we have $\overline{\mu^*}=\mu'$.

Proposition 20. *For any μ, ν in either $M(G)$ or $C_o(G)^*$ we have $(\mu * \nu)' = \nu' * \mu'$.*

Proof. By Proposition 18 we have

$$(\overline{\mu} * \overline{\nu})^* = (\overline{\nu})^* * (\overline{\mu})^* .$$

By (2) and Proposition 19 the left hand side is $(\mu * \nu)'$ and by (2) the right hand side is $\nu' * \mu'$.

Now we change notation and let φ denote an arbitrary element of $C_o(G)$. If $f \in L^1(G)$ and dx is a left Haar measure of G then

$$\mu_f(\varphi) = \int_G \varphi(x) f(x) dx$$

defines a linear functional μ in $C_o(G)^*$ and $\|\mu_f\| = \|f\|_1$. The corresponding measure $f(x)dx$ is absolutely continuous with respect to dx. We have $\overline{\mu_f} = \mu_{\bar{f}}$ for every f in $L^1(G)$.

We recall that for $f: G \to \mathbb{C}$ we defined in the beginning of this section $f^*: G \to \mathbb{C}$ by $f^*(x) = \overline{f(x^{-1})} \Delta(x^{-1})$ where Δ is the modular function of G.

Proposition 21. *For every f in $L^1(G)$ we have $(\mu_f)^* = \mu_{f^*}$.*

Proof. If φ is in $C_o(G)$ then we have

$$\begin{aligned}(\mu_f)^*(\varphi) &= \int_G \varphi'(x) d\overline{\mu}_f(x) = \int_G \varphi(x^{-1}) \overline{f(x)} dx \\ &= \int_G \varphi(x) \overline{f(x^{-1})} \Delta(x^{-1}) dx = \int_G \varphi(x) f^*(x) dx = \mu_{f^*}(\varphi) .\end{aligned}$$

Proposition 22. *For any f and g in $L^1(G)$ we have $\mu_f * \mu_g = \mu_{f*g}$.*

Proof. By Proposition 17 for any φ in $C_o(G)$ we have

$$\mu_f * \mu_g(\varphi) = \int_G \int_G \varphi(xy) d\mu_g(y) d\mu_f(x) = \int_G \int_G \varphi(xy) g(y) f(x) dx dy .$$

Using the subsitutions $x \to xy^{-1}$ and $y \to y^{-1}$ we obtain

$$\begin{aligned}\mu_f * \mu_g(\varphi) &= \int_G \varphi(x) \int_G f(xy^{-1}) g(y) \Delta(y^{-1}) dy dx \\ &= \int_G \varphi(x) \int_G f(xy) g(y^{-1}) dy dx = \int_G \varphi(x)(f * g)(x) dx .\end{aligned}$$

Hence $\mu_f * \mu_g(\varphi) = \mu_{f*g}(\varphi)$ for every φ in $C_o(G)$.

Proposition 23. *Let G be a closed subgroup of the locally compact Hausdorff group $\tilde{G}$. For μ in $C_o(G)^*$ define $\tilde{\mu}: C_o(\tilde{G}) \to \mathbb{C}$ by $\tilde{\mu}(\tilde{\varphi}) = \mu(\varphi)$ where $\varphi = \tilde{\varphi} | G$. Then $\mu \mapsto \tilde{\mu}$ is an isometric injection of the convolution algebra of $C_o(G)^*$ into that of $C_o(G)^*$.*

Proof. By Proposition III.7.11 we already know that $\mu \mapsto \tilde{\mu}$ is an isometry and a vector space isomorphism. We are going to prove that $\widetilde{\mu_1 * \mu_2}(\tilde{\varphi}) = \tilde{\mu}_1 * \tilde{\mu}_2(\tilde{\varphi})$ for every μ_1, μ_2 in $C_o(G)^*$ and $\tilde{\varphi}$ in $C_o(G)$. Given $\tilde{\varphi}$ we let $\varphi = \tilde{\varphi} | G$ so that $\varphi \in C_o(G)$ and ${}_x(\tilde{\varphi}) | G = {}_x\varphi$ for every $x \in G$. Hence with x in G we have

$$\tilde{\mu}_2 \tilde{\varphi}(x) = \tilde{\mu}_2({}_x(\tilde{\varphi})) = \mu_2({}_x(\tilde{\varphi}) | G) = \mu_2({}_x\varphi) = \mu_2 \varphi(x)$$

and so $\tilde{\mu}_2 \tilde{\varphi} | G = \mu_2 \varphi$. Therefore

$$\begin{aligned} \tilde{\mu}_1 * \tilde{\mu}_2(\tilde{\varphi}) &= \tilde{\mu}_1(\tilde{\mu}_2 \tilde{\varphi}) = \mu_1(\tilde{\mu}_2 \tilde{\varphi} | G) = \mu_1(\mu_2 \varphi) \\ &= \mu_1 * \mu_2(\varphi) = \widetilde{\mu_1 * \mu_2}(\tilde{\varphi}) . \end{aligned}$$

Corollary. *There is a natural injection of the measure algebra $M(G)$ into $M(\tilde{G})$.*

In Sections VII.3 and VII.4 we shall use convolutions of measures extensively. The following material is developed with these applications in mind.

Let K be a closed subgroup of the locally compact Hausdorff group G. Then by Proposition 23 we can identify a measure in $M(K)$ with a measure in $M(G)$. In particular every function a in $L^1(K)$ can be identified with an element μ_a of $M(G)$. If $K = G$ then μ_a has already been defined as $a(x)dx$ and this coincides with the present definition. We see that we can identify $f * a$ and $a * f$ for a in $L^1(K)$ and f in $L^1(G)$ as the two convolution products of the corresponding measures in $M(G)$.

Proposition 24. *If K is a closed subgroup of the locally compact Hausdorff group G and $a \in L^1(K)$, $f \in L^1(G)$ then $a * f$ and $f * a$ are locally integrable on G and locally almost everywhere we have*

$$a * f(x) = \int_K a(k) f(k^{-1} x) dk , \tag{3}$$

$$f * a(x) = \int_K f(x k^{-1}) a(k) dk . \tag{4}$$

We have $\mu_a * \mu_f(\varphi) = \mu_{a*f}(\varphi)$ and $\mu_f * \mu_a(\varphi) = \mu_{f*a}(\varphi)$ for all $\varphi \in C_o(G)$.

Proof. If φ is in $C_o(G)$ then

$$\mu_f \varphi(k) = \mu_f({}_k f) = \int_G \varphi(k x) f(x) dx .$$

Therefore

$$\mu_a * \mu_f(\varphi) = \mu_a(\mu_f\,\varphi) = \int_K \Big(\int_G \varphi(kx) f(x)\,dx\Big)\, a(k)\,dk$$
$$= \int_K \Big(\int_G \varphi(x) f(k^{-1}x)\,dx\Big)\, a(k)\,dk\,.$$

We have

$$\int_K \int_G |\varphi(x)|\cdot|f(k^{-1}x)|\cdot|a(k)|\,dx\,dk \leqslant \|\varphi\|_\infty \cdot \|f\|_1 \cdot \|a\|_1$$

so Fubini's theorem can be applied and we obtain

$$\mu_a * \mu_f(\varphi) = \int_G \Big(\int_K a(k) f(k^{-1}x)\,dk\Big)\, \varphi(x)\,dx$$

the inner integral being defined almost everywhere on the set where $\varphi(x) \neq 0$. This shows that $a*f(x)$ is defined as a function on G by (3) locally almost everywhere with respect to the left Haar measure of G. At the exceptional values of x we define $a*f(x)=0$. Then $a*f$ is locally integrable and $\mu_a * \mu_f(\varphi) = \mu_{a*f}(\varphi)$.

Similarly we have

$$\mu_f * \mu_a(\varphi) = \int_G (\mu_a\,\varphi)(x) f(x)\,dx = \int_G \mu_a({}_x\varphi) f(x)\,dx$$
$$= \int_G \Big(\int_K \varphi(xk) a(k)\,dk\Big) f(x)\,dx\,.$$

Again we have an upper estimate

$$\int_G \int_K |\varphi(xk)|\cdot|a(k)|\cdot|f(x)|\,dx\,dk \leqslant \|\varphi\|_\infty \cdot \|a\|_\infty \cdot \|f\|_1$$

and so Fubini's theorem is applicable: Since $\Delta \equiv 1$ on K we obtain

$$\mu_f * \mu_a(\varphi) = \int_K \int_G a(k)\varphi(xk) f(x)\,dx\,dk = \int_K \int_G a(k)\varphi(x) f(xk^{-1})\,dx\,dk$$
$$= \int_G \Big(\int_K f(xk^{-1}) a(k)\,dk\Big)\varphi(x)\,dx\,.$$

Thus $f*a(x)$ is defined locally almost everywhere on G.

Proposition 25. *If K is a compact subgroup of G and $1 \leqslant p < \infty$ then for a in $L^\infty(K)$ and f in $L^p(G)$ the integrals* (3) *and* (4) *exist almost everywhere and $a*f$, $f*a$ belong to $L^p(G)$.*

Proof. This is an easy consequence of Fubini's theorem: Consider $g\colon G \times K \to [0, \infty)$ where $g(x,k) = |f(xk^{-1})|$. Then $g \in L^p(G \times K)$ and if

K is normalized then $\|g\|_p = \|f\|_p$. Hence by the same theorem $\int_K g(x,k)\,dk$ exists for almost all x in G. Therefore by $|f(xk^{-1})a(k)| \leqslant \|a\|_\infty g(x,k)$ we see that $a*f(x)$ exists almost everywhere.

Proposition 26. *If K is a compact subgroup of G, the function $f: G \to \mathbb{C}$ is continuous and a belongs to $L^\infty(K)$ then $a*f$ and $f*a$ are continuous on the group G.*

Proof. Let C be a compact, symmetric neighborhood of e in G and let $x \in G$ be such that $a*f(x)$ is defined by (3). The restriction of f to $K(Cx \cup xC)$ is left uniformly continuous. Hence given $\varepsilon > 0$ there is a neighborhood N_e such that $N_e \subseteq C$ and $|f(a) - f(b)| \leqslant \varepsilon$ for every a, b in $K(Cx \cup xC)$ satisfying $a^{-1}b \in N_e$. Suppose that $y \in xN_e = N_x$. Then $k^{-1}x \in KCx$ and $k^{-1}y \in KxC$ for every k in K. Moreover $(k^{-1}x)^{-1}(k^{-1}y) = x^{-1}y \in N_e$. Therefore

$$|a*f(y) - a*f(x)| \leqslant \|a\|_\infty \int_K |f(k^{-1}y) - f(k^{-1}x)|\,dk \leqslant \varepsilon \|a\|_\infty$$

for every $y \in N_x$.

5. The Algebra Representation Associated with $\rho: S \to \mathscr{L}(\mathscr{H})$

In what follows we shall consider representations of objects which in addition to their algebraic structure also have certain topological and measure theoretic properties: We suppose that the object S is a locally compact topological space. The elements of the σ-algebra generated by the compact subsets of S will be called the *Borel sets* of S. We suppose that a countably additive measure $d\mu$ is defined on the algebra of these Borel sets. If the object S has all these properties then it will be called a *Borel object*. A simple example is a locally compact topological group together with its left or right Haar measure.

Simple and important examples of representations of a Borel object are the regular representations of a locally compact group G. Here $S = G$, the measure is the left Haar measure dx and $\mathscr{H} = L^2(G)$. The *left regular representation* of G is defined by

$$(\lambda(x)f)(y) = f(x^{-1}y) = {}^x f(y)$$

where $x, y \in G$. Since

$$\|\lambda(x)f\|^2 = \int_G f(x^{-1}y)\overline{f(x^{-1}y)}\,dy = \int_G |f(y)|^2\,dy = \|f\|_2^2$$

we see that λ is a unitary group representation. In Section 6 it will be proved that in a certain sense λ is continuous, namely $x \mapsto \lambda(x)f$ is a continuous map from G into $\mathscr{H}$ for each fixed $f \in \mathscr{H}$.

The *right regular representation* ρ of G is defined similarly: Here the measure is the right Haar of G, the Hilbert space is $\mathscr{H} = L^{2R}(G)$ and $(\rho(x)f)(y) = f(yx) = f_x(y)$.

Let S be a Borel object and let $\rho: S \to \mathscr{L}(\mathscr{X})$ be a morphism of S in a reflexive Banach space $\mathscr{X}$. We suppose that ρ is bounded and we let $|\rho| = \text{lub}\{|\rho(s)| : s \in S\}$. Tacitly we shall also suppose that ρ is weakly measurable i.e. $s \mapsto \xi^*(\rho(s)\xi)$ is a measurable function for every $\xi \in \mathscr{X}$ and $\xi^* \in \mathscr{X}^*$. Our task is to associate with ρ a morphism $\sigma: L^1(S) \to \mathscr{L}(\mathscr{X})$ in a natural way and study the relationship between ρ and σ. Formally σ is given by the expression

$$\sigma(f) = \int_S f(s)\rho(s)\,ds$$

and the precise definition is based on the weak interpretation of this integral. Namely for any $f \in L^1(G)$ and $x \in \mathscr{X}$, $x^* \in \mathscr{X}^*$ we let

$$\ell(x^*) = \int_S f(s)x^*(\rho(s)x)\,ds\,.$$

The right hand side is an ordinary Lebesgue integral so $\ell(x^*)$ is meaningful and actually

$$|\ell(x^*)| \leqslant \|f\|_1\,\|x^*\| \cdot |\rho| \cdot \|x\|\,.$$

Therefore $x^* \to \ell(x^*)$ is a continuous linear functional on $\mathscr{X}^*$. By the reflexivity of $\mathscr{X}$ the functional ℓ defines an element $\sigma(f)x$ of $\mathscr{X}$ such that $\ell(x^*) = x^*(\sigma(f)x)$. The map $x \mapsto \sigma(f)x$ is a linear operator $\sigma(f)$ on $\mathscr{X}$ and we see from above that $|\sigma(f)x| = \|\ell\| \leqslant \|f\|_1|\rho| \cdot \|x\|$ and so $|\sigma(f)| \leqslant |\rho| \cdot \|f\|_1$. The morphism $\sigma: L^1(G) \to \mathscr{L}(\mathscr{X})$ is defined by $f \mapsto \sigma(f)$. It bounded and

$$|\sigma| = \text{lub}\{|\sigma(f)| : \|f\|_1 \leqslant 1\} \leqslant |\rho|\,.$$

Definition 1. *A Borel object S is called a Borel object with involution if a continuous involution $s \mapsto s^*$ is given on S such that the measure ds^* defined by*

$$\int_S \varphi(s)\,ds^* = \int_S \varphi(s^*)\,ds \qquad (\varphi \in C_0(G))$$

is absolutely continuous with respect to ds.

Given a Borel object with involution S and a function $f: S\to\mathbb{C}$ we define f^* by $f^*(s)=\overline{f(s^*)}(ds^*/ds)(s)$. If $f\in L^1(S)$ then $f^*\in L^1(S)$ and $\|f\|_1=\|f^*\|_1$ because

$$\|f\|_1=\int_S |f(s^{**})|ds=\int_S |f(s^*)|ds^*=\int_S |\overline{f(s^*)}|\frac{ds^*}{ds}(s)\,ds=\int_S |f^*(s)|ds=\|f^*\|_1.$$

Therefore $f\mapsto f^*$ is a norm preserving involution of the vector space $L^1(S)$.

Next we are going to prove that if $\mathscr{X}=\mathscr{H}$ is a Hilbert space then σ associated with a bounded representation ρ of S is itself a representation and actually $\sigma(f^*)=\sigma(f)^*$: In fact given $\alpha,\beta\in\mathscr{H}$ and $f\in L^1(S)$ we have

$$\begin{aligned}(\sigma(f)^*\alpha,\beta)=\overline{(\sigma(f)\beta,\alpha)}&=\int_S \overline{f(s)}(\alpha,\rho(s)\beta)\,ds\\ &=\int_S \tilde f(s^*)(\rho(s^*)\alpha,\beta)\,ds\end{aligned}$$

where $\tilde f(s)=\overline{f(s^*)}$. Therefore by the definition of ds^*

$$(\sigma(f)^*\alpha,\beta)=\int_S \tilde f(s)(\rho(s)\alpha,\beta)\,ds^*=\int_S f^*(s)(\rho(s)\alpha,\beta)\,ds=(\sigma(f^*)\alpha,\beta)\,.$$

This proves that $\sigma(f)^*=\sigma(f^*)$ which implies that the algebra $\tilde{\mathscr{A}}$ generated by the identity and the operators $\sigma(f)\,(f\in L^1(S))$ is self adjoint.

Our results can be summarized as follows:

Theorem 2. *Let S be a Borel object and let $\rho: S\to\mathscr{L}(\mathscr{X})$ be a bounded morphism of S into the reflexive Banach space $\mathscr{X}$. Then there exists an associated vector space morphism $\sigma: L^1(S)\to\mathscr{L}(\mathscr{X})$ which is also bounded and $|\sigma|\leqslant|\rho|$. It is given by*

$$\sigma(f)=\int_S f(s)\rho(s)\,ds$$

which is interpreted in the weak sense. If S is a Borel object with an involution and $\mathscr{X}=\mathscr{H}$ is a Hilbert space then $\sigma(f)^*=\sigma(f^*)$ where $f^*(s)=\overline{f(s^*)}(ds^*/ds)(s)$ and consequently σ is a simple representation in Hilbert space.

The following observation will be used on several occasions:

Proposition 3. *Let X be a measure space, let $\mathscr{H}=L^2(X)$ and let $\rho: S\to\mathscr{L}(\mathscr{H})$ be a bounded morphism. Then for any α in $L^2(X)$ the integral*

$$(A_f\alpha)(x)=\int_S f(s)(\rho(s)\alpha)(x)\,ds$$

exists for locally almost all $x\in X$, the function $A_f\alpha$ belongs to $L^2(X)$ and $A_f=\sigma(f)$.

Proof. The existence of $(A_f\alpha)(x)$ for almost all x is a consequence of Fubini's theorem: For by using the σ-finite measures $|f(s)|\,ds$ and $|\beta(x)|\,dx$ for $\alpha, \beta \in \mathscr{H}$ we have

$$\begin{aligned}(\sigma(f)\alpha, \beta) &= \int_S f(s)(\rho(s)\alpha, \beta)\,ds = \int_S f(s) \int_X \rho(s)\alpha(x)\overline{\beta(x)}\,dx\,ds \\ &= \int_G \left(\int_S f(s)\rho(s)\alpha(x)\,ds\right)\overline{\beta(x)}\,dx = \int_X (A_f\alpha)(x)\overline{\beta(x)}\,dx \\ &= (A_f\alpha, \beta)\,.\end{aligned}$$

On several occasions we shall need the existence of the morphism $\sigma: L^1(S) \to \mathscr{L}(\mathscr{X})$ when the Banach space $\mathscr{X}$ is not necessarily reflexive but the bounded morphism $\rho: S \to \mathscr{L}(\mathscr{X})$ is strongly continuous. Then $\sigma(f)$ is defined as the integral of a vector valued function. We suppose that the reader is to some extent familiar with the integration theory of Banach space valued functions $f: S \to \mathscr{X}$. This is an easy supplement to the theory of integration of scalar valued functions and can be found for instance in Bourbaki's Intégration or in Part I of Linear Operators by Dunford and Schwartz. The basic idea is to approximate f by simpler vector valued functions $g: S \to \mathscr{X}$ whose integrals are already defined. For this purpose Bourbaki uses continuous functions g with compact support and approximation is understood in terms of the L^1-norm of $s \mapsto \|f(s) - g(s)\|$. Dunford and Schwartz approximate f by simple functions g and use convergence in measure. In the first theory f is integrable if and only if for every $\varepsilon > 0$ there is a continuous g with compact support such that $\| \|f - g(\cdot)\| \|_1 < \varepsilon$. In the second a necessary and sufficient condition for the integrability of f is the existence of a sequence of simple functions g_n such that

$$\lim_{n\to\infty} \mu\{s: s \in S \text{ and } \|f(s) - g_n(s)\| > \varepsilon\} = 0\,.$$

Of course in both cases it is assumed that g is measurable in a certain sense.

Now let $\rho: S \to \mathscr{L}(\mathscr{X})$ be a bounded and strongly continuous morphism in the Banach space $\mathscr{X}$ and let $f \in L^1(S)$. Given any ξ in $\mathscr{X}$ we consider the vector valued function $s \mapsto f(s)\rho(s)\xi$. If g is another scalar valued function on S then for every s in S we have

$$\|f(s)\rho(s)\xi - g(s)\rho(s)\xi\| \leqslant |\rho| \cdot \|\xi\| \cdot \|f(s) - g(s)\|\,.$$

Thus if f is approximated by g then $s \mapsto g(s)\rho(s)\xi$ will approximate $s \mapsto f(s)\rho(s)\xi$. Hence this function is integrable and so we can define

$$\sigma(f)\xi = \int_S f(s)\rho(s)\xi\,ds\,.$$

We omitted the detail of checking the measurability of the integrand.

Another way to define $\sigma(f)$ is by considering the Banach space $\mathscr{L}(\mathscr{X})$ instead of ξ and $\mathscr{X}$. Then we take the function $s \mapsto f(s)\rho(s)$ and note that

$$|f(s)\rho(s) - g(s)\rho(s)| \leqslant |\rho| \cdot \|f(s) - g(s)\|$$

for every s in S. One also verifies that $s \mapsto f(s)\rho(s)$ is measurable. Hence we can define the operator valued integral

$$\sigma(f) = \int_S f(s)\rho(s)\,ds.$$

By using simple functions to approximate f we see that $\xi \mapsto \sigma(f)\xi$ and $\sigma(f)$ are the same operator. If $\varphi: S \to \mathscr{X}$ is integrable then so is $s \mapsto \|\varphi(s)\|$ and $\left\|\int_S \varphi(s)\,ds\right\| \leqslant \int_S \|\varphi(s)\|\,ds$. Using this inequality we obtain $\|\sigma(f)\xi\| \leqslant |\rho| \cdot \|f\|_1 \|\xi\|$ and $|\sigma(f)| \leqslant |\rho| \cdot \|f\|_1$.

Now we turn to the special case when the Borel object is a locally compact group G and $d\mu = dx$ is its left Haar measure. We let $x^* = x^{-1}$ so that $\underline{dx^*}$ is its right Haar measure, $(dx^*/dx)(x) = \Delta(x^{-1})$ and $f^*(x) = \overline{f(x^{-1})}\Delta(x^{-1})$. Thus G is a Borel object with involution and so if a unitary group representation $\rho: G \to \mathscr{U}(\mathscr{H})$ is given the associated $\sigma: L^1(G) \to \mathscr{L}(\mathscr{H})$ is a vector space representation satisfying $\sigma(f)^* = \sigma(f^*)$. We are going to prove that in this case

$$\sigma(f_1)\sigma(f_2) = \sigma(f_1 * f_2).$$

For, if S is a semigroup, then a straight forward and short computation shows that

$$(\sigma(f_1)\sigma(f_2)\alpha, \beta) = \int_S \int_S f_1(x) f_2(y) (\rho(xy)\alpha, \beta)\,dx\,dy$$

where $\alpha, \beta \in \mathscr{H}$. Thus if $S = G$ is a group then

$$\begin{aligned}(\sigma(f_1)\sigma(f_2)\alpha, \beta) &= \int_G \int_G f_1(x) f_2(x^{-1}y)(\rho(y)\alpha, \beta)\,dy\,dx \\ &= \int_G (f_1 * f_2)(y)(\rho(y)\alpha, \beta)\,dy = (\sigma(f_1 * f_2)\alpha, \beta).\end{aligned}$$

We proved the following:

Theorem 4. *If $\rho: G \to \mathscr{U}(\mathscr{H})$ is a measurable representation of the locally compact group G then*

$$\sigma(f) = \int_G f(x)\rho(x)\,dx$$

defines in the weak sense a bounded representation $\sigma: L^1(G) \to \mathscr{L}(\mathscr{H})$ of the involutive algebra $L^1(G)$ and $|\sigma| \leqslant |\rho|$.

We have the analogous result also when $\rho: G \to \mathscr{L}(\mathscr{X})$ is strongly continuous but the Banach space $\mathscr{X}$ is not necessarily reflexive. The only difference is that σ is then defined in the strong sense. If g_1 and g_2 are simple functions in $L^1(G)$ then $\sigma(g_1 * g_2) = \sigma(g_1)\sigma(g_2)$ is an easy consequence of the left invariance of the Haar measure of G. If f_1 and f_2 are arbitrary functions in $L^1(G)$ then we can choose simple functions g_k in $L^1(G)$ such that $\|f_k - g_k\|_1 < \varepsilon$ for $k = 1, 2$. Then

$$\|f_1 * f_2 - g_1 * g_2\|_1 \leqslant \varepsilon(\|f_1\|_1 + \|f_2\|_1 + \varepsilon).$$

Using the inequality $|\sigma(f)| \leqslant |\rho| \cdot \|f\|_1$ and the identity

$$\begin{aligned}\sigma(f_1)\sigma(f_2) - \sigma(f_1 * f_2) = \sigma(f_1 - g_1)\sigma(f_2) + \sigma(g_1)\sigma(f_2 - g_2) \\ - \sigma(f_1 * f_2 - g_1 * g_2)\end{aligned}$$

we obtain $\sigma(f_1 * f_2) = \sigma(f_1)\sigma(f_2)$.

As a simple example we determine the representation $\sigma: L^1(G) \to \mathscr{L}(\mathscr{H})$ associated with the left regular representation of G. Since $\mathscr{H} = L^2(G)$ and $\lambda(x)\alpha(y) = \alpha(x^{-1}y)$ we have

$$\begin{aligned}(\sigma(f)\alpha, \beta) &= \int_G f(x)(\rho(x)\alpha, \beta)\,dx = \int_G f(x) \int_G \alpha(x^{-1}y)\overline{\beta(y)}\,dy\,dx \\ &= \int_G \Big(\int_G f(x)\alpha(x^{-1}y)\,dx\Big)\overline{\beta(y)}\,dy = \int_G (f * \alpha)(y)\overline{\beta(y)}\,dy = (f * \alpha, \beta).\end{aligned}$$

Theorem 5. *The involutive algebra representation $\sigma: L^1(G) \to \mathscr{L}(\mathscr{H})$ associated with the left regular representation λ of G is given by $\sigma(f)\alpha = f * \alpha$ where $f \in L^1(G)$ and $\alpha \in L^2(G)$.*

Proof. Proposition 4.6 shows that $f * \alpha \in L^2(G)$ for every $f \in L^1(G)$ and $\alpha \in L^2(G)$.

Proposition 6. *Let $\rho: S \to \mathscr{L}(\mathscr{H})$ be a representation of the Borel object S and let $\mathscr{A} = \mathscr{A}_\rho$ be the associated v. Neumann algebra. Then $\sigma(f) \in \mathscr{A}$ for every $f \in L^1(S)$.*

Proof. Let $T \in \mathscr{A}'$ so that $T\rho(s) = \rho(s)T$ for every $s \in S$. Now given $\alpha, \beta \in \mathscr{H}$ we have

$$((\sigma(f)T - T\sigma(f))\alpha, \beta) = \int_S f(s)((\rho(s)T - T\rho(s))\alpha, \beta)\,ds = 0.$$

Therefore α, β being arbitrary $\sigma(f)T = T\sigma(f)$ for every $T \in \mathscr{A}'$. We proved that $\sigma(f) \in \mathscr{A}'' = \mathscr{A}$.

Proposition 7. *Let $\rho: S \to \mathscr{L}(\mathscr{H})$ be a weakly continuous representation of the Borel object S and let $\sigma: L^1(S) \to \mathscr{L}(\mathscr{H})$ be the associated vector* *space morphism. Then the v. Neumann algebra $\mathscr{A}_\rho$ is generated by the operators $\sigma(f)$ and their adjoints where f varies over all real valued continuous functions with compact support.*

Note. We can let f vary over all simple functions having compact support.

Proof. We split the proof into three parts, the first of which consists of the proof of Lemma 8 stated below.

Since ρ is weakly continuous given $s \in S$, finitely many elements α_i $(i=1,\ldots,m)$, β_j $(j=1,\ldots,n)$ in $\mathscr{H}$ and $\varepsilon>0$ there is a compact neighborhood N_s such that

$$|(\rho(t)\alpha_i - \rho(s)\alpha_i, \beta_j)| < \varepsilon \tag{1}$$

for every i,j and $t \in N_s$. Since N_s is compact we have $0<\mu(N_s)<\infty$. We define $f: S \to \mathbb{R}$ by

$$f(t) = \begin{cases} c=\mu(N_s)^{-1} & \text{if } t \in N_s \\ 0 & \text{if } t \notin N_s \end{cases}. \tag{2}$$

We have

$$(\sigma(f)\alpha_i, \beta_j) = \int_S f(t)\,(\rho(t)\alpha_i, \beta_j)\,dt$$

and by (1) we may write

$$(\rho(t)\alpha_i, \beta_j) = (\rho(s)\alpha_i, \beta_j) + \varepsilon(t)$$

for every $t \in N_s$ where $|\varepsilon(t)| \leqslant \varepsilon$. Thus by (2) we have

$$\begin{aligned}\int_S f(t)\,(\rho(t)\alpha_i, \beta_j)\,dt &= c\int_{N_s} (\rho(t)\alpha_i, \beta_j)\,dt \\ &= c(\rho(s)\alpha_i, \beta_j)\int_{N_s} dt + c\int_{N_s} \varepsilon(t)\,dt.\end{aligned}$$

Hence

$$(\sigma(f)\alpha_i, \beta_j) = (\rho(s)\alpha_i, \beta_j) + r$$

where

$$|r| \leqslant c\left|\int_{N_s} \varepsilon(t)\,dt\right| < c\varepsilon \int_{N_s} dt = \varepsilon.$$

We proved the following:

Lemma 8. *Given $s \in S$, finitely many α_i, β_j in $\mathscr{H}$ and $\varepsilon>0$ there is a simple function f with compact support such that*

$$|(\sigma(f)\alpha_i, \beta_j) - (\rho(s)\alpha_i, \beta_j)| < \varepsilon$$

for every α_i and β_j.

Now let $\mathscr{T}$ be the family of all operators $\sigma(f)$ where f varies over all simple functions with compact support. If we wish we can also suppose that f assumes only one non-zero value. Moreover we let $\mathscr{B}$ denote the v. Neumann algebra generated by $\mathscr{T}$.

We choose a $T \in \mathscr{B}'$, fix $s \in S$, also $\alpha, \beta \in \mathscr{H}$ and $\varepsilon > 0$. We let in the preceeding lemma $\alpha_1 = \alpha$, $\alpha_2 = T\alpha$ and $\beta_1 = \beta$, $\beta_2 = T^*\beta$. By the lemma there is a simple function f with compact support such that four inequalities hold. Among these we have

$$|(\sigma(f)\alpha, T^*\beta) - (\rho(s)\alpha, T^*\beta)| < \varepsilon$$

$$|(\sigma(f)T\alpha, \beta) - (\rho(s)T\alpha, \beta)| < \varepsilon.$$

Therefore

$$|(T\sigma(f)\alpha, \beta) - (T\rho(s)\alpha, \beta)| < \varepsilon$$

$$|(\sigma(f)T\alpha, \beta) - (\rho(s)T\alpha, \beta)| < \varepsilon$$

and so

$$|((T\sigma(f) - \sigma(f)T)\alpha, \beta) - ((T\rho(s) - \rho(s)T)\alpha, \beta)| < 2\varepsilon.$$

By our hypothesis we have $T\sigma(f) = \sigma(f)T$, so

$$|((T\rho(s) - \rho(s)T)\alpha, \beta)| < 2\varepsilon.$$

By letting $\varepsilon \to 0$ we obtain

$$((T\rho(s) - \rho(s)T)\alpha, \beta) = 0$$

and so β and α being arbitrary $T\rho(s) = \rho(s)T$. Applying Proposition III.5.8 we see that T commutes with every element of $\mathscr{A}_\rho$. Since T in $\mathscr{B}'$ is arbitrary we have $\mathscr{B}' \subseteq \mathscr{A}_\rho'$ and so $\mathscr{B} = \mathscr{B}'' \supseteq \mathscr{A}_\rho'' = \mathscr{A}_\rho$. By Proposition 6 we have $\mathscr{B} \subseteq \mathscr{A}_\rho$ so we proved that $\mathscr{A}_\rho = \mathscr{B}$.

The last step of the proof of Proposition 7 consists of showing that the simple functions can be replaced by continuous ones. Let $\mathscr{S}$ denote the set of operators $\sigma(f)$ where $f: S \to \mathbb{R}$ varies over all continuous functions with compact support. We let $\mathscr{C}$ be the v. Neumann algebra generated by $\mathscr{S}$. We notice that $\mathscr{S}$ and its uniform closure $\overline{\mathscr{S}_u}$ generate the same weakly closed algebra $\mathscr{C}$. For if $\mathscr{S}$ generates $\mathscr{C}$ then so does its weak closure $\overline{\mathscr{S}_w}$ but $\mathscr{S} \subseteq \overline{\mathscr{S}_u} \subseteq \overline{\mathscr{S}_w}$ so $\overline{\mathscr{S}_u}$ generates also $\mathscr{C}$. Similarly $\mathscr{T}$ and $\overline{\mathscr{T}_u}$ generate the same algebra $\mathscr{B} = \mathscr{A}_\rho$. Now we know that

$$|\sigma(f) - \sigma(g)| \leqslant |\sigma| \cdot \|f - g\|_1$$

and any continuous function with compact support can be approximated in the $\|\cdot\|_1$ norm by simple functions having compact support and conversely. Therefore $\mathscr{S} \subseteq \overline{\mathscr{T}_u}$ and $\mathscr{T} \subseteq \overline{\mathscr{S}_u}$ proving that $\mathscr{A}_\rho = \mathscr{B} = \mathscr{C}$. This completes the proof of Proposition 7.

Theorem 9. *A weakly continuous representation of the Borel object S and the associated vector space morphism $\sigma: L^1(S) \to \mathscr{L}(\mathscr{H})$ generate the same v. Neumann algebra.*

This is an immediate consequence of Propositions 6 and 7. The reason to state Proposition 7 explicitely is its usefulness. We also note that Proposition 7 has a generalization to reflexive Banach spaces, namely one proves that $\mathscr{A}_\rho \subseteq \mathscr{T}''$.

Proposition 10. *Let $\rho: S \to \mathscr{L}(\mathscr{H})$ be a weakly continuous and bounded morphism of the Borel object S and let $\sigma: L^1(S) \to \mathscr{L}(\mathscr{H})$ be the vector space morphism derived from ρ. Then for every x in $\mathscr{H}$ we have: If $\sigma(f)x=0$ for every non-negative $f \in C_0(S)$ satisfying $f=f^*$ then $\rho(s)x=0$ for every s in S.*

Corollary. *If $\rho: G \to \mathrm{Gl}(\mathscr{H})$ is a weakly continuous and bounded group representation and $\sigma: L^1(G) \to \mathscr{L}(\mathscr{H})$ is the associated algebra representation then no $x \in \mathscr{H}$, except $x=0$ is annihilated by every $\sigma(f)$ where $f=f^*$ is in $C_0(G)$ and $f \geqslant 0$.*

Proof. Let $x \in \mathscr{H}$, $s \in S$, $y \in \mathscr{H}$ and $\varepsilon > 0$ be given in this order. By the weak continuity of ρ there is a symmetric neighborhood N_s such that

$$|(\rho(t)x,y)-(\rho(s)x,y)| < \varepsilon \quad \text{for every } t \in N_s. \tag{3}$$

Since S is a locally compact Hausdorff space there is a g in $C_0(S)$ such that $g \geqslant 0$, the set N_s is a carrier of g and $\int_S g(t)\,dt > 0$. If we let $f=\lambda(g+g^*)$ where λ is a suitable positive constant then $f \geqslant 0$, $f \in C_0(S)$, $f=f^*$ and $\int_S f(t)\,dt=1$. Using the definition of the operator $\sigma(f)$ we obtain

$$(\sigma(f)x-\rho(s)x,y)=\int_S f(t)\,(\rho(t)x-\rho(s)x,y)\,dt$$

and so by (3) and by the other properties of f

$$|(\sigma(f)x-\rho(s)x,y)| < \varepsilon.$$

Now suppose that $\sigma(f)x=0$ for every f in $C_0(S)$. Then we obtain $|(\rho(s)x,y)| \leqslant \varepsilon$ where x, s, y are fixed and $\varepsilon > 0$ is arbitrary. It follows that $(\rho(s)x,y)=0$ for every $y \in \mathscr{H}$ and $\rho(s)x=0$ for every $s \in S$.

Lemma 11. *If $f \in L^p(G)$ $(1 \leqslant p < \infty)$ and $\varepsilon > 0$ are given then there is a neighborhood N_e such that $\|u*f-f\|_p < \varepsilon$ and $\|f*u-f\|_p < \varepsilon$ for every non-negative u in $L^1(G)$ satisfying $\int_G u(x)\,dx=1$ and $u(x)=0$ for $x \notin N_e$.*

Proof. If h is in $L^q(G)$ where $1/p+1/q=1$ then by Proposition 4.6 for any non-negative, integrable u we have

$$(u*f-f,h)=\int_G\int_G (u(y)(f(y^{-1}x))-f(x))\overline{h(x)}\,dy\,dx.$$

Interchanging the order of integration and applying Hölder's inequality we obtain

$$|(u*f-f,h)| \leqslant \|h\|_q \int_G \|{}^y f-f\|_p u(y)\,dy.$$

Therefore by looking at the norm of the linear functional $h \mapsto (u*f-f,h)$ mapping $L^q(G)$ into $\mathbb{C}$ we obtain

$$\|u*f-f\|_p \leqslant \int_G \|{}^y f-f\|_p u(y)\,dy.$$

By Proposition 4.9 we can choose N_e such that $\|{}^y f-f\|_p<\varepsilon$ for every $y \in N_e$. Thus if $\int_G u(x)\,dx=1$ and $u(y)=0$ for $y \notin N_e$ then $\|u*f-f\|_p<\varepsilon$.

The proof of the other inequality proceeds exactly the same way except if G is not unimodular then a factor $m=\int_G u(x^{-1})\,dx$ appears as follows:

$$\|f*u-f\|_p \leqslant \int_G \|m f_y - f\|_p \frac{u(y^{-1})}{m}\,dy.$$

However if N_e is sufficiently small then m is arbitrarily close to 1 because Δ being continuous we can approximate 1 by $\Delta(x^{-1})$ in N_e and obtain approximately $\int_G u(x^{-1})\Delta(x^{-1})\,dx=\int_G u(x)\,dx=1$.

Let $\mathscr{N}$ denote the set of all neighborhhods N of the identity e of G. Ordering $\mathscr{N}$ by inverse inclusion i.e. letting $N_1 \leqslant N_2$ if $N_1 \supseteq N_2$ we obtain a directed set. We construct a net u_N $(N \in \mathscr{N})$ by selecting a non-negative function u_N from $L^1(G)$ such that $\int_G u_N(x)\,dx=1$ and $u_N(x)=0$ for $x \notin N$. By Lemma 11 we have $u_N * f \to f$ and $f * u_N \to f$ for every $f \in L^p(G)$. Due to this fact any such net u_N $(N \in \mathscr{N})$ is called an *approximate identity* for $L^1(G)$. By the lemma we may also suppose that $u_N \in C_0(G)$ for every $N \in \mathscr{N}$.

Let G be compact and for simplicity let $A=L^2(G)$ in the following proof. By Proposition 4.5 A is a Banach algebra.

Proposition 12. *If G is compact then a closed subset of $L^2(G)$ is a left ideal of $L^2(G)$ if and only if it is a left invariant subspace.*

Note. Similar statement holds from the right.

Proof. First suppose that I is a closed left ideal of A. Let u_N $(N \in \mathcal{N})$ be an approximate identity of functions $u_N \in C_0(G)$, let $x \in G$ and $f \in I$ be fixed. Then $u_N * f \to f$ and so ${}_x(u_N * f) = {}_x u_N * f \to {}_x f$. Since ${}_x u_N * f \in I$ it follows that ${}_x f \in I$ and so I is left invariant.

Next suppose that I is a left invariant subspace of A. The algebra A is a Hilbert space. If $f \in I$, $g \in I^\perp$ and $h \in A$ then

$$(h * f, g) = \int_G h(x) \int_G f(x^{-1} y) \overline{g(y)} \, dy \, dx = \int_G h(x) ({}^x f, g) \, dx = 0.$$

Hence $h * f$ is orthogonal to g for every g in $I^\perp$ and so $h * f \in I^{\perp\perp} = I$. It follows that I is a left ideal of A.

Theorem 13. *A closed subset of $L^1(G)$ is a left ideal of $L^1(G)$ if and only if it is a left invariant subspace.*

The proof is essentially the same as before if we replace the inner product (f, g) by $(f, g) = \int_G f(x) \overline{g(x)} \, dx$ where $g \in L^\infty(G)$. By the Riesz representation theorem the linear functionals $f \mapsto (f, g)$ exhaust all elements of $L^1(G)^*$ as g varies over $L^\infty(G)$. Hence $I^\perp$ can be defined as a subset of $L^1(G)^*$ and then $I^{\perp\perp}$ as a subset of $L^1(G)$. We only have to prove that $I^{\perp\perp} = I$.

More generally let $\mathcal{X}$ be a Banach space, let $\mathcal{X}^*$ be the space of continuous linear functionals on $\mathcal{X}$ and let x^* denote a typical element of $\mathcal{X}^*$. If $\mathcal{S}$ is a subset of $\mathcal{X}$ then we define $\mathcal{S}^\perp$, the *annihilator* of $\mathcal{S}$ by

$$\mathcal{S}^\perp = \{x^* \colon x^*(s) = 0 \text{ for every } s \in \mathcal{S}\}.$$

Also we define

$$\mathcal{S}^{\perp\perp} = \{x \colon x \in \mathcal{X} \text{ and } x^*(x) = 0 \text{ for all } x^* \in \mathcal{S}^\perp\}.$$

Then we have the following:

Lemma 14. *If $\mathcal{Y}$ is a subspace of the Banach space $\mathcal{X}$ then $\mathcal{Y}^{\perp\perp} = \mathcal{Y}$.*

Proof. If $y \in \mathcal{Y}$ and $y^* \in \mathcal{Y}^\perp$ then $y^*(y) = 0$ and so $y \in \mathcal{Y}$ implies $y \in \mathcal{Y}^{\perp\perp}$. To prove the converse it is sufficient to show that for any $x \notin \mathcal{Y}$ there is a $y^* \in \mathcal{Y}^\perp$ such that $y^*(x) \neq 0$. This fact is an immediate consequence of the following:

Lemma 15. *Let $\mathcal{Y}$ be a closed subspace of the Banach space $\mathcal{X}$ and let $x \notin \mathcal{Y}$. Then there is a continuous linear functional $f \in \mathcal{X}^*$ such that $f \in \mathcal{Y}^\perp$ and $f(x) \neq 0$.*

Proof. Let d denote the distance of x from $\mathscr{Y}$ and define f on the subspace spanned by x and $\mathscr{Y}$ by $f(\lambda x + y) = \lambda d$. Then f is linear, $f(x) = d$ and f is zero on $\mathscr{Y}$. Moreover

$$|f| = \underset{\lambda \in \mathbb{C},\, y \in \mathscr{Y}}{\text{supremum}} \frac{|\lambda| d}{\|\lambda x - y\|} = \underset{y}{\text{lub}} \frac{d}{\|x-y\|} = \frac{d}{\text{glb}\|x-y\|} = 1$$

and so f is continuous. By the Hahn-Banach theorem f can be extended to a continuous linear functional having norm 1 and satisfying the requirements.

Note. We see that f can be chosen such that $|f| = 1$ and $f(x) = d = \text{glb}\|x-y\|$.

Let $\sigma: A \to \mathscr{L}(\mathscr{X})$ be a morphism of the algebra A on the topological space $\mathscr{X}$. Then σ is called *non-degenerate* provided the union $\mathscr{R}$ of the ranges of the operators $\sigma(a)$ $(a \in A)$ is dense in $\mathscr{X}$. One sees that

$$\mathscr{N} = \{x: \sigma(a)x = 0 \text{ for every } a \in A\} = \bigcap \ker \sigma(a)$$

is an invariant subspace. If $\mathscr{X}$ is a Hilbert space $\mathscr{H}$ then we can also define $\tilde{\mathscr{R}}$, the union of the ranges of the operators $A \in \mathscr{A}_\sigma$ and $\tilde{\mathscr{N}} = \bigcap \ker A$ where $A \in \mathscr{A}_\sigma$. Now if σ is a representation then using $\mathscr{R} \subseteq \tilde{\mathscr{R}}$ and $\tilde{\mathscr{N}} \subseteq \mathscr{N}$ one proves that $\mathscr{R}^\perp = \mathscr{N}$. Hence the representation $\sigma: A \to \mathscr{L}(\mathscr{H})$ is non-degenerate if and only if $\mathscr{N} = 0$.

Theorem 16. *If $\sigma: L^1(G) \to \mathscr{L}(\mathscr{H})$ is a non-degenerate, bounded representation of $L^1(G)$ then there is a strongly continuous representation $\rho: G \to \mathscr{U}(\mathscr{H})$ such that σ is the algebra representation associated with ρ.*

Note. The same result holds if G is compact and σ is a non-degenerate, bounded representation of the H^* algebra $L^2(G)$. Similarly $\mathscr{H}$ can be replaced by a reflexive Banach space $\mathscr{X}$ but then ρ is only a bounded morphism.

Proof. By Lemma 11 we can choose a net u_N $(N \in \mathscr{N})$ of approximate identities where $\mathscr{N}$ is the directed set of neighborhood of e. For any a in G we have ${}^a u_N * f = {}^a(u_N * f) \to {}^a f$ where ${}^a f(x) = f(a^{-1}x)$, the function f is in $L^1(G)$ and convergence is with respect to the norm. Since σ is bounded it follows that

$$\|\sigma({}^a u_N)\sigma(f) - \sigma({}^a f)\| \to 0 \quad \text{as } N \to \infty. \tag{4}$$

Hence if α belongs to the union of the ranges $\mathscr{R}$, say $\alpha = \sigma(f)\beta$ then $\sigma({}^a u_N)\alpha$ is a convergent sequence and so we can define $\rho(a)\alpha$ as the limit of this sequence. Of course we have $\rho(a)\alpha = \sigma({}^a f)\beta$ and in particular $\rho(e)\alpha = \alpha$. By the boundedness of σ we have $\|\sigma({}^a u_N)\alpha\|$

$\leqslant B\|^a u_N\|_1\|\alpha\| = B\|\alpha\|$ and so $\|\rho(a)\alpha\| \leqslant B\|\alpha\|$. Consequently $\mathscr{R}$ being dense $\rho(a)$ can be extended to a continuous linear operator satisfying $|\rho(a)| \leqslant B$. The extended operator $\rho(e)$ is the identity map. By (4) we have $\rho(a)\sigma(f)\beta = \sigma(^a f)\beta$ for every β and so $\rho(a)\sigma(f) = \sigma(^a f)$. Hence

$$\rho(ab)\sigma(f) = \sigma(^{ab}f) = \sigma(^a(^b(f))) = \rho(a)\sigma(^b f) = \rho(a)\rho(b)\sigma(f).$$

$\mathscr{R}$ being dense it follows that $\rho(ab) = \rho(a)\rho(b)$ for any $a, b \in G$. Since $\rho(e)$ is the identity we see that every $\rho(a)$ is invertible. The morphism ρ is strongly continuous because by Proposition 4.9 $^a f$ as an element of $L^1(G)$ depends continuously on a. Hence $\rho(a)\sigma(f) = \sigma(^a f)$ is a continuous function of a in the strong operator topology and so $\rho(a)$ is strongly continuous as a function of a on the union of the ranges.

In order to prove that $\rho(a)^* = \rho(a^{-1})$ it is sufficient to show that $\rho(a)^*\sigma(f) = \rho(a^{-1})\sigma(f) = \sigma(_a f)$ for every f in $L^1(G)$. This in turn will follow if we have $(^a u_N)^* * f \to {}_a f$ as $N \to \infty$ or in other form $f^* * {}^a u_N \to ({}_a f)^*$. Replacing f^* by g we obtain $g * {}^a u_N \to ({}_a(g^*))^*$ where $^a u_N$ is bounded and so the convolution is meaningful. Now one finds that

$$\Delta(a)\,(g * {}^a u_N) = g^a * u_N \quad \text{and} \quad \Delta(a)\,({}_a(g^*))^* = g^a.$$

Thus the question is reduced to the validity of $g^a * u_N \to g^a$ which holds because u_N $(N \in \mathscr{N})$ is an approximate identity.

Finally to prove that the representation associated with ρ is σ it will be sufficient to prove

$$(\sigma(f)\alpha, \beta) = \int_G f(x)\,(\rho(x)\alpha, \beta)\,dx$$

for every $\beta \in \mathscr{H}$, for every α in a dense subset of $\mathscr{H}$ and every f in $L^1(G)$. Thus it will be enough to show that

$$(\sigma(f)\sigma(g)\alpha, \beta) = \int_G f(x)\,(\rho(x)\sigma(g)\alpha, \beta)\,dx$$

for $\alpha, \beta \in \mathscr{H}$ and $f, g \in L^1(G)$. For fixed α, β the linear functional $a\colon L^1(G) \to \mathbb{C}$ defined by $a(f) = (\sigma(f)\alpha, \beta)$ is continuous and so

$$a(f) = (\sigma(f)\alpha, \beta) = \int_G f(x)\,w(x)\,dx$$

for some bounded, measurable function $w\colon G \to \mathbb{C}$. We obtain

$$\begin{aligned}(\sigma(f*g)\alpha, \beta) &= \int_G\int_G f(y)g(y^{-1}x)w(x)\,dy\,dx = \int_G f(y)\int_G g(y^{-1}x)w(x)\,dx\,dy \\ &= \int_G f(y)\,(\sigma(^y g)\alpha, \beta)\,dy = \int_G f(y)\,(\rho(y)\sigma(g)\alpha, \beta)\,dy.\end{aligned}$$

Since $\sigma(f*g) = \sigma(f)\sigma(g)$ the proof is finished.

Theorem 17. *For every locally compact group G the map $\rho \mapsto \sigma$ is a bijection between the strongly continuous elements of $\hat{G}$ and those of $\widehat{L^1(G)}$. A class σ in $\widehat{L^1(G)}$ is finite dimensional if and only if the corresponding ρ is finite dimensional and we have $d_\sigma = d_\rho$.*

Proof. If σ is irreducible then the union of the ranges $\sigma(f)\mathcal{H}$ is dense in $\mathcal{H}$. Hence for every irreducible σ there is a ρ such that $\rho \mapsto \sigma$. By Propositions 6 and 7 the v. Neumann algebras generated by ρ and σ coincide, so ρ is irreducible if and only if the associated σ is irreducible.

Let $\mu: G \to \mathrm{Gl}(\mathcal{X})$ be a bounded, weakly measurable morphism of the locally compact group G in the reflexive complex Banach space $\mathcal{X}$. The bounded vector space morphism $\sigma: L^1(G) \to \mathcal{L}(\mathcal{X})$ introduced in Theorem 2 satisfies $\sigma(f_1)\sigma(f_2) = \sigma(f_1 * f_2)$. This follows by the same reasoning which was used before Theorem 4 in the special case when $\mu = \rho$ is a unitary representation. Therefore σ is a bounded morphism of the algebra $L^1(G)$ in $\mathcal{X}$.

We let (μ, μ) be the algebra of those continuous linear operators $T: \mathcal{X} \to \mathcal{X}$ which commute with every $\mu(x)(x \in G)$ and similarly we let (σ, σ) be the algebra of those T which commute with every $\sigma(f)(f \in L^1(G))$.

Proposition 18. *If $\mu: G \to \mathcal{L}(\mathcal{X})$ is a bounded strongly continuous morphism of the locally compact group in the reflexive Banach space $\mathcal{X}$ and if $\sigma: L^1(G) \to \mathcal{L}(\mathcal{X})$ is the associated algebra morphism then $(\mu, \mu) = (\sigma, \sigma)$.*

Proof. Let $\xi \in \mathcal{X}$, $\xi^* \in \mathcal{X}^*$ and $\varepsilon > 0$ be given. First we suppose that $T \in (\sigma, \sigma)$ and prove that T commutes with every $\mu(a)(a \in G)$. Since μ is strongly continuous there is a compact neighborhood C of a such that

$$\|\mu(x)T\xi - \mu(a)T\xi\| < \varepsilon \quad \text{and} \quad \|\mu(x)\xi - \mu(a)\xi\| < \varepsilon$$

for every x in C. Let $f \in C_0(G)$ be such that C is a carrier of f and $\int_G |f(x)|\,dx = 1$. Then by $T \in (\sigma, \sigma)$ we have

$$|\xi^*((\mu(a)T - T\mu(a))\xi)| \leq |\xi^*((\mu(a) - \sigma(f))T\xi)| + |\xi^*(T(\sigma(f) - \mu(a))\xi)| .$$

Using the properties of f we obtain

$$|\xi^*((\mu(a) - \sigma(f))T\xi)| = \left| \int_C f(x)\xi^*((\mu(a) - \mu(x))T\xi)\,dx \right| \leq \|\xi^*\|\,\varepsilon .$$

Similarly

$$|\xi^*(T(\sigma(f)-\mu(a))\xi)|=\left|\int_C f(x)\xi^* T((\mu(x)-\mu(a))\xi)dx\right|\leqslant\|\xi^* T\|\varepsilon.$$

Therefore

$$|\xi^*((\mu(a)T-T\mu(a))\xi|\leqslant\|\xi^*\|(1+|T|)\varepsilon$$

and this shows that $\mu(a)T=T\mu(a)$. We proved that $(\sigma,\sigma)\subseteq(\mu,\mu)$.

Now we suppose that $T\in(\mu,\mu)$ and prove that $T\sigma(f)=\sigma(f)T$ for every $f\in L^1(G)$. By the definition of $\sigma(f)$ we have

$$\xi^*(\sigma(f)T\xi)=\int_G f(x)\xi^*(\mu(x)T\xi)dx=\int_G f(x)\xi^*(T\mu(x)\xi)dx.$$

Thus using the definition of T^* we obtain

$$\xi^*(\sigma(f)T\xi)=\int_G f(x)(T^*\xi^*)(\mu(x)\xi)dx=T^*\xi^*(\sigma(f)\xi).$$

Therefore $\xi^*(\sigma(f)T\xi)=\xi^*(T\sigma(f)\xi)$ and so $\sigma(f)T=T\sigma(f)$. This proves that $(\mu,\mu)\subseteq(\sigma,\sigma)$.

Proposition 19. *Let $\mu\colon G\to\mathscr{L}(\mathscr{X})$ be a bounded, strongly continuous morphism and let σ be the associated algebra representation. Then the strongly closed algebra generated by the operators $\mu(x)(x\in G)$ is contained in the strong closure of $\sigma(C_0(G))$.*

Corollary. *If μ is completely irreducible then so is σ.*

Proof. It is sufficient to prove the following: Given $a\in G$, a positive ε and $\xi_1,\ldots,\xi_n$ in $\mathscr{X}$ there is an f in $C_0(G)$ such that

$$\|(\mu(a)-\sigma(f))\xi_k\|<\varepsilon \tag{5}$$

for every $k=1,\ldots,n$. Since μ is strongly continuous a has a compact neighborhood C such that

$$\|\mu(x)\xi_k-\mu(a)\xi_k\|<\varepsilon$$

for every $x\in C$ and $k=1,\ldots,n$. We choose f in $C_0(G)$ such that C is a carrier of f and $\int_G |f(x)|dx=1$. Then for any $\xi^*\in\mathscr{X}^*$ we have

$$|\xi^*((\mu(a)-\sigma(f))\xi_k|=\left|\int_C f(x)\xi^*(\mu(a)\xi_k-\mu(x)\xi_k)dx\right|\leqslant\|\xi^*\|\varepsilon.$$

Hence by the reflexivity of $\mathscr{X}$ we obtain (5).

6. The Regular Representations of Locally Compact Groups

The purpose of this section is to develop some of the fundamental properties of the regular representations of a locally compact group G. We recall that the left regular representation λ which was introduced in the beginning of the last section acts on the Hilbert space $L^2(G)$ while the right regular representation ρ acts on $L^{2R}(G)$. If G is unimodular λ and ρ have the same underlying Hilbert space. By definition $(\lambda(x)f)(y) = {}^x f(y)$ and $(\rho(x)f)(y) = f_x(y)$ for every $x, y \in G$. Therefore

$$(1) \qquad \lambda(a)\rho(b) = \rho(b)\lambda(a) \quad \text{for all } a, b \text{ in } G.$$

As usual we let $\mathscr{A}_\lambda$ and $\mathscr{A}_\rho$ denote the v. Neumann algebras associated with the representations λ and ρ. Thus $\mathscr{A}_\lambda = (\lambda, \lambda)'$ and $\mathscr{A}_\rho = (\rho, \rho)'$ by the corollary of Theorem III.6.10.

Proposition 1. *For any locally compact group G we have* $\mathscr{A}_\lambda = (\lambda, \lambda)' \subseteq (\rho, \rho)$ *and* $\mathscr{A}_\rho = (\rho, \rho)' \subseteq (\lambda, \lambda)$.

Proof. This follows immediately from the above remarks including (1) in particular.

Theorem 2. *The regular representations of a locally compact group are strongly continuous.*

Note. The strong continuity of $\lambda: G \to \mathscr{U}(\mathscr{H})$ means that $x \mapsto \lambda(x)\,(x \in G)$ is continuous with respect to the locally compact topology of G and the strong operator topology of $\mathscr{L}(\mathscr{H})$.

Proof. If f is in $C_0(G)$ then f is uniformly continuous and so given $\varepsilon > 0$ we can find a symmetric neighborhood N_e such that $x^{-1}y \in N_e$ implies $|f(x) - f(y)| < \varepsilon$. In particular we have $|f(xu) - f(x)| < \varepsilon$ if $u \in N_e$ and $x \in G$ is arbitrary. Therefore $\|f_u - f\|_\infty < \varepsilon$ if $u \in N_e$. Since f has a compact carrier S by requiring that N_e be a subneighborhood of a compact neighborhood C of e we obtain $\|f_u - f\|_2 \leqslant \sqrt{\mu(CS)} \cdot \|f_u - f\|_\infty$. This shows that $\|f_u - f\|_2 < \varepsilon$ provided u is restricted to a sufficiently small neighborhood N_e. Finally, if g is an arbitrary function in $L^{2R}(G)$ then there is an f in $C_0(G)$ such that $\|f - g\|_2 < \varepsilon$ and so $\|f_u - g_u\|_2 < \varepsilon$ for every $u \in G$. Thus $\|g_u - g\|_2 < 3\varepsilon$ for $u \in N_e = N_e(\varepsilon)$ and so the right regular representation is strongly continuous.

If $\rho_1 \approx \rho_2$ and ρ_1 is strongly continuous then so is ρ_2. Hence the left regular representation of G is also continuous in the strong sense.

A group representation $\rho: G \to \mathrm{Gl}(\mathscr{H})$ is called *faithful* if ρ is a group isomorphism so if $\rho(a) = I$ only for $a = e$.

Proposition 3. *The regular representations are faithful.*

Proof. If $a \neq e$ then by the uniformizability of G there is a continuous function with compact support such that $f(a) \neq f(e)$. Hence $\rho(a) f(e) = f(a) \neq f(e)$ and so $\rho(a) f \neq f$ and $\rho(a) \neq I$. The left regular representation λ can be treated similarly.

Proposition 4. *The regular representations λ and ρ are equivalent.*

Proof. For $f: G \to \mathbb{C}$ let $U f(x) = f(x^{-1})$ for every x in G. By Propositions 2.2 and 2.3 we have $f \in L^2(G)$ if and only if $U f \in L^{2R}(G)$ and $\|f\|_2 = \|U f\|_2^R$. Therefore U is an isometry of $L^2(G)$ onto $L^{2R}(G)$. Since

$$U\lambda(x) f(y) = \lambda(x) f(y^{-1}) = f(x^{-1} y^{-1}) = U f(yx) = \rho(x) U f(y)$$

we see that $U \in (\lambda, \rho)$ and $\lambda \approx \rho$.

Since λ and ρ are continuous by Theorem 5.4 one can define the representations σ_λ and σ_ρ of the involutive algebra $L^1(G)$ which are associated with λ and ρ, respectively. If G is unimodular then $L^2(G) = L^{2R}(G)$ and so by Theorem 5.5 we have

$$\sigma_\lambda(f) g = f * g = \sigma_\rho(f) g \quad \text{for all } f, g \text{ in } L^2(G)\,. \tag{2}$$

This formula will be used in the proof of a very interesting result due to Segal:

Theorem 5. *Let λ and ρ denote the regular representations of the unimodular group G. Then $(\lambda, \lambda)' = (\rho, \rho)$ and $(\rho, \rho)' = (\lambda, \lambda)$.*

We shall closely follow the original proof which is built up from a sequence of lemmas. The actual proof at the end will take only a few lines. We let $\mathscr{D}$ be the dense linear manifold in $L^2(G)$ consisting of those functions f which have compact support i.e. vanish almost everywhere outside some compact subset of G. The following lemma shows that $f * g$ and $g * f$ are in $L^2(G)$ for every f in $\mathscr{D}$ and g in $L^2(G)$:

Lemma 6. *Let G be unimodular, let f and g belong to $L^2(G)$ and suppose that at least one of them has compact support. Then $f * g(x)$ exists almost everywhere on G and $f * g$ is in $L^2(G)$.*

Proof. We give the proof for the case when g has a symmetric compact carrier S. The other case follows by trivial modifications. It is sufficient to prove the existence of

$$\int_G \left(\int_G |f(xy^{-1})| \cdot |g(y)| \, dy \right)^2 dx\,.$$

The existence of the inner integral together with a suitable upper estimate follows from the BCS inequality: Since the range of integration can be restricted to S we obtain

$$\left(\int_G |f(xy^{-1})| \cdot |g(y)| dy\right)^2 \leqslant \|g\|_2^2 \int_S |f(xy)|^2 dy .$$

Therefore the problem is reduced to proving that the function h defined by the last integral belongs to $L^1(G)$. For this purpose we use Fubini's theorem and the right invariance of the measure dx: If $\mu(S)$ denotes the Haar measure of the compact set then

$$\int_G \int_S |f(xy)|^2 dy dx = \int_S \int_G |f(xy)|^2 dx dy = \|f\|_2^2 \int_S dy = \|f\|_2^2 \mu(S) < \infty .$$

Lemma 7. *Let $p, q \in L^2(G)$ and let $C_1 \subseteq C_2 \subseteq \cdots$ be compact sets in G such that p is almost everywhere zero outside of $C = \bigcup C_n$. Define $p_n(a) = p(a)$ if $a \in C_n$ and $p_n(a) = 0$ if $a \in C_n$. Then $p_n * q \to r$ in the $\|\cdot\|_2$ topology implies $r(a) = p * q(a)$ almost everywhere. Similarly if $q * p_n \to r$ then $r = q * p$.*

Proof. We can restrict our attention to the case of $p_n * q$ because the other case requires only trivial changes in the reasoning. Since $p_n * q \to r$ in the norm topology there is a subsequence $n_1, n_2, \ldots$ such that $p_{n_i} * q \to r$ almost everywhere on G. This is well known and follows for example from Proposition VI.1.2 as a special case. Since $|p_{n_i}(a) q(a^{-1} b)| \leqslant |p(a) q(a^{-1} b)|$ for each fixed b in G the sequence of functions $a \mapsto |p_{n_i}(a) q(a^{-1} b)|$ is dominated by a function belonging to $L^1(G)$. Therefore by Lebesgue's convergence theorem

$$p_{n_i} * q(b) = \int_G p_{n_i}(a) q(a^{-1} b) da \to \int_G p(a) q(a^{-1} b) da = (p * q)(b) .$$

Lemma 8. *It is sufficient to prove that for f in $\mathscr{D}$ and $T \in (\rho, \rho)$ we have $T \sigma_\lambda(f) \in (\lambda, \lambda)'$.*

Proof. By (1) we have $(\rho, \rho)' \subseteq (\lambda, \lambda)$ and so we must prove that $(\lambda, \lambda) \subseteq (\rho, \rho)'$. By Lemma 5.11 and the discussion following it there is an approximate identity u_N $(N \in \mathscr{N}c)$ where $\mathscr{N}c$ is the directed set of compact neighborhoods of e and $u_N \in \mathscr{D}$ for every N in $\mathscr{N}c$. Hence if f is in $L^2(G)$ then by (2) we have $f * u_N = \sigma_\lambda(f) u_N \to f$ and $u_N * f = \sigma_\lambda(u_N) f \to f$ in the norm topology of $L^2(G)$. Thus if $g, h \in L^2(G)$ and $T \in (\rho, \rho)$ then

$$|(T \sigma_\lambda(u_N) g, h) - (Tg, h)| \leqslant \|T \sigma_\lambda(u_N) g - Tg\| \cdot \|h\| \leqslant |T| \cdot \|h\| \cdot \|\sigma_\lambda(u_N) g - g\|_2$$

where the right hand side tends to 0 as $N \to \infty$. This proves that $T \sigma_\lambda(u_N) \to T$ in the weak operator topology. By hypothesis $T \sigma_\lambda(u_N) \in (\lambda, \lambda)'$

and so we obtain $T\in(\lambda,\lambda)'$. Therefore $(\rho,\rho)\subseteq(\lambda,\lambda)'$ and the theorem will follow.

Lemma 9. *Let h be such an element of $L^2(G)$ that $h*:\mathscr{D}\to L^2(G)$ is a continuous operator on $\mathscr{D}$. Then the unique extension of $h*$ to $L^2(G)$ belongs to $(\lambda,\lambda)'$.*

Proof. Let H denote the extended operator $H: L^2(G)\to L^2(G)$. In order to prove that $H\in(\lambda,\lambda)'$ it is sufficient to show that $THg=HTg$ for every T in (λ,λ) and every g in $\mathscr{D}$. Let $C_1\subseteq C_2\subseteq\cdots$ be a sequence of compact sets in G such that if $C=\bigcup C_n$ then h is zero almost everywhere on the complement of C. We define $h_n(x)=h(x)$ for $x\in C_n$ and $h_n(x)=0$ for $x\notin C_n$. Then we have $h_n\to h$ in the norm topology of $L^2(G)$. By Theorem 5.9 $\sigma_\lambda(h_n)\in\mathscr{A}_\lambda=(\lambda,\lambda)'$ and so $T\sigma_\lambda(h_n)g=\sigma_\lambda(h_n)Tg$ for each g in $\mathscr{D}$. By Proposition 4.6 we have

$$\|\sigma_\lambda(h_n)g-\sigma_\lambda(h)g\|_2=\|(h_n-h)*g\|_2\leqslant\|h_n-h\|_2\cdot\|g\|_1\to 0$$

as $n\to\infty$. Therefore $\sigma_\lambda(h_n)g\to\sigma_\lambda(h)g$ and so $T\sigma_\lambda(h_n)\to T\sigma_\lambda(h)g$. We can now apply Lemma 7 with $p_n=h_n$, $q=Tg$ and $r=THg$ because $p_n*q=h_n*Tg=\sigma_\lambda(h_n)Tg\to T\sigma_\lambda(h)g=Th*g=r$. We obtain $h*Tg=THg$.

We complete the proof by showing that $h*Tg=HTg$. For this purpose we apply Lemma 7 again with $p=Tg$, $q=h$ and $r=HTg$. Then $q*p_n=h*p_n=Hp_n$ by the hypothesis of the lemma and by the definition of H. Since $p_n\to p$ in norm by the continuity of H we have $Hp_n\to Hp=HTg=r$. Hence $q*p_n\to r$ and Lemma 7 is applicable. We obtain $q*p=h*Tg=r=HTg$. Thus $THg=h*Tg=HTg$ and so $TH=HT$ where T is an arbitrary element of (λ,λ).

Proof of Theorem 5. Let f,g in $\mathscr{D}$ and T in (ρ,ρ) be given. By Theorem 5.9 $\sigma_\rho(g)\in\mathscr{A}_\rho=(\rho,\rho)'$ and so by (2)

$$T\sigma_\lambda(f)g=T(f*g)=T\sigma_\rho(g)f=\sigma_\rho(g)Tf=Tf*g\,.$$

Therefore $T\sigma_\lambda(f)$ is a continuous extension of $Tf*:\mathscr{D}\to\mathscr{D}$ to an operator in $\mathscr{L}(L^2(G))$. Now we can apply Lemma 9 and obtain $T\sigma_\lambda(f)\in(\lambda,\lambda)'$. The theorem now follows from Lemma 8.

Lemma 10. *Let G be an arbitrary locally compact group and let $\mathscr{K}$ be a non-zero subspace of $L^2(G)$ which is invariant under one of the regular representations of G. Then $\mathscr{K}$ contains a non-zero continuous function.*

Proof. Let g be a non-zero element of $\mathscr{K}$. Then $\lambda(a)$ being a left translation operator $\lambda(a)g\neq 0$ for each a in G. Hence by Proposition 5.10

there is an f in $C_0(G)$ such that $\sigma_\lambda(f)g \neq 0$. By Theorem 5.5 we have $\sigma_\lambda(f)g = f * g$ and so $\sigma_\lambda(f)g$ is continuous by Proposition 4.7. By Theorem 5.9 and Proposition III.5.10 $\mathscr{K}$ is invariant under $\sigma_\lambda(f)$ and so the continuous, non-zero function $\sigma_\lambda(f)g$ belongs to $\mathscr{K}$.

Theorem 11. *The regular representations of a commutative, non-compact group have no irreducible components.*

Proof. Let G be a commutative topological group and let $\mathscr{K}$ be an irreducible component of the left regular representation λ of G. By Lemma 10 $\mathscr{K}$ contains a non-zero function g which is continuous on G and belongs to $L^2(G)$. By Proposition IV.2.7 we have $\mathscr{K} = \mathbb{C} g$. Since $\mathscr{K}$ is invariant under $\lambda(a^{-1})$ we see that $g(ax) = c(a)g(x)$ for some complex factor $c(a)$ and every x in G. From this we obtain $c(ab) = c(a)c(b)$ for all a, b in G. Therefore $g(a) = c(a)g(e)$ for every $a \in G$ where $g(e) \neq 0$. This shows that c is a continuous function on G and c belongs to $L^2(G)$. Let A, B and C denote those measurable subsets of G on which $|c|$ is $=1$, >1 and <1 respectively. Then $\mu(A) + \mu(B) \leqslant \|c\|_2^2 < \infty$ and by the inverse invariance of the Haar measure of G we also have $\mu(C) = \mu(B)$. This shows that $\mu(G)$ is finite and so G is compact by Proposition 2.1.

Note. If G is compact then its regular representations are direct sums of irreducible components. This follows from one of the fundamental theorems in the representation theory of compact groups. (See Theorems VII.1.1 and VII.1.36.) The theorem should be compared also with Theorem VII.1.5 which concerns the non-commutative case.

7. Continuity of Group Representations and the Gelfand-Raikov Theorem

A morphism $\rho: S \to \mathscr{L}(\mathscr{H})$ of the Borel object S is called a *weakly measurable morphism* if $s \mapsto (\rho(s)\alpha, \beta)$ is a measurable function for every α, β in $\mathscr{H}$. Similarly ρ is called *weakly continuous* if $s \mapsto (\rho(s)\alpha, \beta)$ is a continuous function on S for every fixed pair α, β in $\mathscr{H}$. If $s \mapsto \rho(s)$ is a continuous function with respect to the locally compact topology of S and the strong operator topology of $\mathscr{L}(\mathscr{H})$ then ρ is called a *strongly continuous morphism.* Therefore if G is a locally compact group then it is meaningful to speak about the weak measurability and the weak or strong continuity of a morphism $\rho: G \to \mathrm{Gl}(\mathscr{H})$. In the first part of this section we are going to prove three results on the continuity of unitary representations ρ of such groups G.

Theorem 1. *Every weakly continuous unitary representation of a locally compact group is strongly continuous.*

Proof. Let $\rho: G \to \mathscr{U}(\mathscr{H})$ be a weakly measurable unitary representation and let $\sigma: L^1(G) \to \mathscr{L}(\mathscr{H})$ be the associated group algebra representation. First we prove the theorem in the special case when σ is cyclic. The weak continuity will be used only later. We define the linear manifold

$$\mathscr{K} = \{\sigma(f)\alpha : f \in C_0(G)\}$$

where α is a cyclic vector of σ. Then $\mathscr{K}$ is dense in $\mathscr{H}$ because for every g in $L^1(G)$ there is an f in $C_0(G)$ such that $\|f - g\|_1 < \varepsilon$.

Next we shall define a function $\tau: G \to \mathscr{L}(\mathscr{H})$ which will turn out to be a strongly continuous unitary representation of G. Finally we shall prove that $\tau = \rho$ and so ρ is strongly continuous. Given u in G first we define $\tau(u)$ only on $\mathscr{K}$ by

$$\tau(u)\xi = \tau(u)\sigma(f)\alpha = \sigma({}^u f)\alpha$$

where $\xi = \sigma(f)\alpha$ and as usual ${}^u f(v) = f(u^{-1}v)$. We show that $\tau(u)$ is and isometry on $\mathscr{K}$:

$$\begin{aligned}
\|\tau(u)\xi\|^2 = (\sigma({}^u f)\alpha, \sigma({}^u f)\alpha) &= \int_G {}^u f(x)(\rho(x)\alpha, \sigma({}^u f)\alpha)\,dx \\
&= \int_G \int_G {}^u f(x)\overline{{}^u f(y)}(\rho(x)\alpha, \rho(y)\alpha)\,dx\,dy \\
&= \int_G \int_G f(x)\overline{f(y)}(\rho(x)\alpha, \rho(y)\alpha)\,dx\,dy\,.
\end{aligned}$$

In the last step we used the fact that $\rho(u)$ is a unitary operator. Therefore we proved that

$$\|\tau(u)\xi\|^2 = (\sigma(f)\alpha, \sigma(f)\alpha) = \|\xi\|^2\,.$$

Since $\mathscr{K}$ is dense in $\mathscr{H}$ one can extend $\tau(u)$ to a unitary transformation $\tau(u): \mathscr{H} \to \mathscr{H}$. We have

$$\tau(u)\tau(v)\xi = \tau(u)\sigma({}^v f)\alpha = \sigma({}^{uv} f)\alpha = \tau(uv)\alpha$$

so τ is a unitary group representation.

We shall now prove that $\tau: G \to \mathscr{U}(\mathscr{H})$ is a strongly continuous representation. Given $\xi = \sigma(f)\alpha$ we have

$$\begin{aligned}
\|\tau(u)\xi - \xi\| = \|\sigma({}^u f)\alpha - \sigma(f)\alpha\| &= \|\sigma({}^u f - f)\alpha\| \\
&\leqslant \|\alpha\| \cdot |\sigma({}^u f - f)| \leqslant \|\alpha\| \cdot \|{}^u f - f\|_1\,.
\end{aligned}$$

Since $f \in C_0(G)$ by the right uniform continuity of f we have $\|{}^u f - f\|_\infty \to 0$ as $u \to e$. Thus by the same reasoning as in the proof of the continuity of the left regular representation of G we see that $\|{}^u f - f\|_1 \to 0$ as $u \to e$. Hence $\tau(u)\xi \to \xi$ as $u \to e$ for every fixed $\xi \in \mathscr{K}$. Since $\mathscr{K}$ is dense in $\mathscr{H}$ for any given $\eta \in \mathscr{H}$ and $\varepsilon > 0$ we can find a ξ in $\mathscr{K}$ such that $\|\xi - \eta\| < \varepsilon/3$. Since τ is unitary $\|\tau(u)(\xi - \eta)\| < \varepsilon/3$ for every $u \in G$. We know the existence of a neighborhood N_e such that $\|\tau(u)\xi - \xi\| < \varepsilon/3$ for every $u \in N_e$. Thus given $\varepsilon > 0$ we found N_e such that

$$\|\tau(u)\eta - \eta\| \leqslant \|\tau(u)(\eta - \xi)\| + \|\tau(u)\xi - \xi\| + \|\xi - \eta\| < \varepsilon$$

for every $u \in N_e$. We proved that τ is strongly continuous at e.

We will prove now that $\tau = \rho$: For any $\xi \in \mathscr{K}$ and $\beta \in \mathscr{H}$ we have

$$\begin{aligned}(\tau(u)\xi, \beta) &= (\sigma({}^u f)\alpha, \beta) = \int_G f(u^{-1}x)(\rho(x)\alpha, \beta)\,dx \\ &= \int_G f(x)(\rho(ux)\alpha, \beta)\,dx = (\sigma(f)\alpha, \rho(u)^* \beta) \\ &= (\xi, \rho(u)^* \beta).\end{aligned}$$

Therefore $(\tau(u)\xi, \beta) = (\rho(u)\xi, \beta)$ for every fixed $\xi \in \mathscr{K}$ and for every $\beta \in \mathscr{H}$. We have $\tau(u) = \rho(u)$ on $\mathscr{K}$ and consequently also on $\mathscr{H}$.

Finally we can turn to the general case when σ is not cyclic but weakly continuous. By Theorem IV.2.16 and the corollary of Proposition 5.10 σ is a direct sum of cyclic submorphisms of $L^1(G)$. Thus $\mathscr{H} = \sum \mathscr{H}_i$ where each $\mathscr{H}_i$ is such that $\sigma | \mathscr{H}_i \, (i \in \mathscr{I})$ is cyclic. Now in order to prove that ρ is strongly continuous at e let $\xi \in \mathscr{H}$ and $\varepsilon > 0$ be given. We can write

$$\xi = \eta + \xi_{i_1} + \cdots + \xi_{i_n}$$

where $\xi_{i_k} \in \mathscr{H}_{i_k}$ and $\|\eta\| < \varepsilon/4$. By what had been already proved we can choose neighborhoods $N_e^{(k)} \, (k = 1, \ldots, n)$ such that

$$\|\rho(u)\xi_{i_k} - \xi_{i_k}\| < \frac{\varepsilon}{2n}$$

for every $u \in N_e^{(k)}$. Therefore if $u \in N_e = \bigcap N_e^{(k)}$ then $\rho(u)$ being unitary $\|\rho(u)\xi - \xi\| < \varepsilon$.

We also proved the following:

Proposition 2. *If the weakly measurable representation $\rho: G \to \mathscr{U}(\mathscr{H})$ is such that the associated group algebra representation is cyclic then ρ is strongly continuous.*

By Theorem 1 for unitary resentations of a locally compact group weak and strong continuity mean the same thing and so we will simply speak about the *continuity* of such representations.

Theorem 3. *If* $\rho: G \to \mathcal{U}(\mathcal{H})$ *is weakly measurable and if* $\mathcal{H}$ *is separable then* ρ *is continuous.*

Proof. Let σ be the algebra representation associated with ρ. By Theorem IV.2.16 we have $\mathcal{H} = \mathcal{H}_0 \oplus \sum \mathcal{H}_i$ where $\sigma | \mathcal{H}_0$ is zero and each $\sigma | \mathcal{H}_i$ is cyclic. If we can prove that $\mathcal{H}_0 = 0$ then the conclusion follows immediately from the reasoning given in the proof of Theorem 1.

Let α_n $(n=1,2,\ldots)$ be a MONS in $\mathcal{H}$ and let $\alpha \in \mathcal{H}_0$. Then we have

$$(\sigma(f)\alpha, \alpha_n) = \int_G f(x)(\rho(x)\alpha, \alpha_n)\,dx = 0$$

and so $(\rho(x)\alpha, \alpha_n) = 0$ for every $x \notin S_n$ where S_n is a set of measure zero. Consequently $(\rho(x)\alpha, \alpha_n) = 0$ for every $n = 1, 2, \ldots$ and every $x \notin S = \bigcup S_n$. By the maximality of the system α_n $(n=1,2,\ldots)$ we obtain $\rho(x)\alpha = 0$ for $x \notin S$. Since G is of positive or infinite measure there is an x in G such that $x \notin S$ and so $\rho(x)$ being invertible we obtain $\alpha = 0$. We proved that the only element of $\mathcal{H}_0$ is $\alpha = 0$.

We shall now turn to the main object of this section which is the Gelfand-Raikov theorem. A family of representations of a group G is said to *contain sufficiently many representations* if for every x in G there is a representation ρ in the family such that $\rho(x) \neq I$. This happens if and only if for each pair of distinct group elements x_1 and x_2 there is a ρ such that $\rho(x_1) \neq \rho(x_2)$. The Gelfand-Raikov theorem states that every locally compact group has sufficiently many continuous, irreducible unitary representations. In other terminology the set of such representations is *complete* or *total*.

The key to the proof is the Krein-Milman theorem together with an important fact about positive, Hermitian linear functionals on $L^1(G)$ or positive definite functions on G. Since functionals offer generalizations to arbitrary involutive algebras we shall base the proof on the use of positive functionals. The relevant definitions and results make up the first stage towards the Gelfand-Raikov theorem. (Starting before Lemma 4 and including Proposition 12.) This is followed by the material which we described before as an important fact. Namely if ρ is a representation of a complex involutive algebra A then $f(x) = (\rho(x)\alpha, \alpha)$ is a positive Hermitian functional and if A is complete then every such functional comes from some representation ρ. One can find a useful necessary and sufficient condition for the irreducibility of ρ in terms of f. (See from Lemma 13 to Theorem 17 including its proof.) The final part of the section contains the actual proof and some related material such as the concept of the star radical.

A not necessarily continuous linear functional f defined on a complex, involutive algebra A is called *Hermitian* or *symmetric* if $f(x^*)=\overline{f(x)}$ for every x in A. One proves that f is symmetric if and only if it assumes real values on the Hermitian elements of A. For if this condition is satisfied and if $x=x_1+ix_2$ where x_1 and x_2 are Hermitian elements then

$$f(x^*)=f(x_1-ix_2)=\overline{f(x_1)+if(x_2)}=\overline{f(x)}.$$

The functional f is called positive if $f(x^*x)\geq 0$ for every x in the involutive algebra A.

Lemma 4. *If f is a positive functional then $f(y^*x)=\overline{f(x^*y)}$ and*

$$|f(y^*x)|^2\leq f(x^*x)\,f(y^*y)$$

for every x, y in A.

Proof. Given x, y in A and λ in $\mathbb{C}$ we have

$$0\leq f((x+\lambda y)^*(x+\lambda y))=f(x^*x)+\lambda f(x^*y)+\bar{\lambda} f(y^*x)+|\lambda|^2 f(y^*y).$$

Using $\lambda=1$ and $\lambda=i$ we obtain

$$\mathscr{Im} f(x^*y)+\mathscr{Im} f(y^*x)=0 \quad \text{and} \quad \mathscr{Re}\, f(x^*y)=\mathscr{Re}\, f(y^*x),$$

respectively. The stated inequality is the BCS inequality for the bilinear functional $(x,y)=f(y^*x)$.

Lemma 5. *If A is a complex, involutive algebra with identity and if f is a positive functional on A then f is symmetric and it satisfies $|f(x)|^2\leq f(e)f(x^*x)$ for every x in A.*

Proof. Let $y=e$ in the relations given in the previous lemma.

If f is a positive Hermitian linear functional then we let

$$v(f)=\operatorname{supr}\frac{|f(x)|^2}{f(x^*x)}$$

where the supremum is taken for every x satisfying $f(x^*x)\neq 0$. One calls $v(f)$ the *total variation* of f and if $v(f)$ is finite one says that f is of *finite variation.* If A has an identity then choosing $x=e$ we see that $v(f)\geq f(e)$ and so by the last lemma we obtain $v(f)=f(e)$.

On page 38 in Section I.8 we have seen that every involutive algebra A having no identity can be embedded in an involutive algebra A_e with identity e.

Proposition 6. *A positive functional $f\colon A \to \mathbb{C}$ can be extended to A_e if and only if it is Hermitian and its total variation is finite.*

If f is Hermitian and $v(f)$ is finite then there is an extension whose total variation is $v(f)$.

Proof. If f is extendable to A_e then f is Hermitian and of finite variation by Lemma 5. Conversely, if $f\colon A \to \mathbb{C}$ satisfies these requirements then we can define its extension to A_e by

$$f((\lambda, x)) = f(\lambda e + x) = \lambda k + f(x)$$

where k is any positive constant satisfying $k \geqslant v(f)$. To see that the extended functional is positive we first notice that

$$f((\lambda, x)^* (\lambda, x)) = |\lambda|^2 k + 2 \mathscr{Re}\{\lambda f(x^*)\} + f(x^* x).$$

Since

$$\mathscr{Re}\{\lambda f(x^*)\} \geqslant -|\lambda| \cdot |f(x^*)| = -|\lambda| \cdot |f(x)| \geqslant -|\lambda| \cdot \sqrt{k f(x^* x)}$$

we obtain

$$f((\lambda, x)^* (\lambda, x)) \geqslant (|\lambda| \sqrt{k} - \sqrt{f(x^* x)})^2 \geqslant 0.$$

If we choose $k = v(f)$ then $f(e) = k = v(f)$ and so the total variation of the extended functional is $v(f)$.

Lemma 7. *If the positive f_1, f_2 and f are Hermitian then we have $v(f_1 + f_2) \leqslant v(f_1) + v(f_2)$ and $v(\lambda f) = \lambda v(f)$ for any $\lambda \geqslant 0$.*

Proof. Let the same symbol f_k $(k = 1, 2)$ denote an extension of f_k to A_e such that the total variation of the extended functional is the same as that of the original f_k. Then

$$(f_1 + f_2)(e) = f_1(e) + f_2(e) = v(f_1) + v(f_2).$$

Since $(f_1 + f_2)(e)$ is the total variation of $f_1 + f_2$ on A_e we have $v(f_1 + f_2) \leqslant (f_1 + f_2)(e)$. The second relation is trivial.

Lemma 8. *If f is Hermitian then we have $|f(x)|^2 \leqslant v(f) f(x^* x)$ for every x in A.*

Proof. Extend f to A_e without increasing its total variation. Then $v(f) = f(e)$ and so our inequality follows from Lemma 5.

Lemma 9. *Let A be an involutive algebra, let f be a positive functional on A and let $x \in A$. Then $g(y) = f(x^* y x)$ defines a positive Hermitian functional satisfying $v(g) \leq f(x^* x)$.*

Proof. We have

$$g(y^* y) = f(x^* y^* y x) = f((y x)^* (y x)) \geq 0$$

proving the positivity of g. Since

$$g(y^*) = f(x^* y^* x) = f((y x)^* x) = \overline{f(x^* y x)} = \overline{g(y)}$$

the functional g is Hermitian. Finally by Lemma 4

$$|g(y)|^2 = |g(y^*)|^2 = |f(x^* y^* x)|^2 \leq f(x^* x) f((y x)^* (y x)) = f(x^* x) g(y^* y)$$

and so $v(g) \leq f(x^* x)$.

Lemma 10. *Let $\rho_k: A \to \mathscr{L}(\mathscr{H}_k)$ $(k=1,2)$ be cyclic representations of the involutive algebra A with cyclic vectors α_k and let $f_k(x) = (\rho_k(x)\alpha_k, \alpha_k)$. Then $f_1 = f_2$ implies that $\rho_1 \approx \rho_2$.*

Proof. Since $f_1(a^* x) = f_2(a^* x)$ and $\rho_k(A)\alpha_k$ is dense in $\mathscr{H}_k$ we see that $\rho_1(x)\alpha_1 = 0$ implies $\rho_2(x)\alpha_2 = 0$ and conversely. Therefore we can define $T: \rho_1(A)\alpha_1 \to \rho_2(A)\alpha_2$ by $T\rho_1(x)\alpha_1 = \rho_2(x)\alpha_2$. Since

$$(\rho_1(x)\alpha_1, \rho_1(a)\alpha_1) = f_1(a^* x) = f_2(a^* x) = (\rho_2(x)\alpha_2, \rho_2(a)\alpha_2)$$

the map T is an isometry and so it is extendable to an isometry of $\mathscr{H}_1$ onto $\mathscr{H}_2$. We have

$$T\rho_1(x)\{\rho_1(a)\alpha_1\} = T\rho_1(x a)\alpha_1 = \rho_2(x a)\alpha_2 = \rho_2(x) T\{\rho_1(a)\alpha_1\}$$

for every a, x in A and so $T\rho_1(x) = \rho_2(x) T$ for all x in A. We proved that $T: \mathscr{H}_1 \to \mathscr{H}_2$ is a unitary map in (ρ_1, ρ_2).

An analogue of the above lemma will be considered in Section VII.2. (See Proposition VII.2.17.)

Proposition 11. *Let A be a complex Banach algebra with continuous involution and let $f: A \to \mathbb{C}$ be a positive, Hermitian linear functional of finite variation. Then f is continuous and*

$$|f| \leq v(f) \sqrt{\operatorname{supr} \frac{\|x^*\|}{\|x\|}} . \tag{1}$$

Note. If A is star normed i.e. $\|x^*\| = \|x\|$ and if $\|e\| = 1$ then we have $|f| = f(e)$.

Proof. If A has no identity then by Proposition 6 f can be extended to A_e so we may suppose without loss of generality that the original algebra A has an identity element e. Now if $\|x\|<1$ then by Lemma III.6.13 there is an element y in A such that $y^2=e-x$. Moreover if x is Hermitian then so is y and so

$$f(e-x)=f(y^2)=f(y^*y)\geqslant 0$$

proving that $f(x)\leqslant f(e)$. Since x can be replaced by $-x$ we obtain $|f(x)|\leqslant f(e)$ for every Hermitian element x satisfying $\|x\|<1$. If x is an arbitrary, non-zero Hermitian element then we can use this inequality with λx $(0<\lambda\|x\|<1)$ in place of x and obtain $|f(x)|\leqslant f(e)\|x\|$. Finally if x is an arbitrary element of A then by Lemma 5

$$|f(x)|^2\leqslant f(e)f(x^*x)\leqslant f(e)^2\|x^*x\|\leqslant C(f(e)\|x\|)^2$$

where C is such that $\|x^*\|\leqslant C\|x\|$. Therefore f is continuous and $|f|\leqslant f(e)\sqrt{C}$. Since $v(f)=f(e)$ we obtain (1).

A Banach algebra A is called an *algebra with approximate identity* if there is a net e_i $(i\in\mathscr{I})$ with values $e_i\in A$ such that $\|e_i\|\leqslant 1$ and $\lim e_i x=\lim x e_i=x$ for every x in A. In Section 5 we proved that every $L^1(G)$ algebra is a complex, involutive Banach algebra with approximate identity. We also know that every B^* algebra has approximate identity. If the involution of A is continuous and e_i $(i\in\mathscr{I})$ is an approximate identity then clearly $e_i^*x\to x$ and $xe_i^*\to x$ for every $x\in A$.

Proposition 12. *Let A be a complex Banach algebra with continuous involution and with an approximate identity. Then a positive linear functional $f\colon A\to\mathbb{C}$ is continuous if and only if it is Hermitian and its total variation is finite.*

Proof. Let f be continuous and let e_i $(i\in\mathscr{I})$ be an approximate identity. Then by Lemma 4

$$f(x^*)=\lim f(x^*e_i)=\lim\overline{f(e_i^*x)}=\overline{f(x)}.$$

Similarly using Lemma 4 we have

$$|f(x)|^2=\lim|f(e_i^*x)|^2\leqslant\lim f(e_i^*e_i)f(x^*x)\leqslant C|f|f(x^*x).$$

Here C denotes any constant such that $\|x^*\|\leqslant C\|x\|$ for every $x\in A$. The converse statement follows from Proposition 11.

Note. We see that if A is star normed and if $\|e\|=1$ then $|f|>v(f)$ for every continuous f.

Lemma 13. *If $\rho\colon A\to\mathscr{L}(\mathscr{H})$ is a representation of the complex, involutive algebra A then for any vector $\alpha\in\mathscr{H}$ the equation $f(x)=(\rho(x)\alpha,\alpha)$ defines a Hermitian positive functional of finite total variation at most (α,α).*

Proof. The linear functional f is positive because

(2) $$f(x^* x) = (\rho(x^* x)\alpha, \alpha) = (\rho(x)^* \rho(x)\alpha, \alpha) = (\rho(x)\alpha, \rho(x)\alpha) \geqslant 0.$$

The Hermitian property of f follows from

$$f(x^*) = (\rho(x^*)\alpha, \alpha) = (\rho(x)^* \alpha, \alpha) = (\alpha, \rho(x)\alpha) = \overline{f(x)}.$$

Finally by (2) we have

$$|f(x)|^2 = |(\rho(x)\alpha, \alpha)|^2 \leqslant (\rho(x)\alpha, \rho(x)\alpha)(\alpha, \alpha) = f(x^* x)(\alpha, \alpha)$$

and so the total variation of f is at most (α, α).

Lemma 14. *If $\rho: A \to \mathscr{L}(\mathscr{H})$ is a cyclic representation and $f(x) = (\rho(x)\alpha, \alpha)$ where α is a cyclic vector then $v(f) = \|\alpha\|^2$.*

Proof. By Lemma 13 we have $v(f) \leqslant \|\alpha\|^2$. To prove the opposite inequality notice that ρ being cyclic x can be chosen such that $\|\rho(x)\alpha - \alpha\|$ is arbitrarily small. Therefore using

$$|(\rho(x)\alpha, \alpha) - (\alpha, \alpha)| \leqslant \|\rho(x)\alpha - \alpha\| \cdot \|\alpha\|$$

and

$$\frac{|f(x)|^2}{f(x^* x)} = \frac{|(\rho(x)\alpha, \alpha)|^2}{\|\rho(x)\alpha\|^2}$$

one sees that $|f(x)|^2 / f(x^* x)$ can be brought arbitrarily close to $\|\alpha\|^4 / \|\alpha\|^2$.

Let f and g be positive linear functionals on the involutive algebra A. If $f(x^* x) \geqslant g(x^* x)$ for every $x \in A$ then one says that f *dominates* g and one writes $f \geqslant g$. This concept is used in the following lemma which will be needed only on one occasion to prove an important irreducibility criterion.

Lemma 15. *Let A be a complex, involutive algebra, let $\rho: A \to \mathscr{L}(\mathscr{H})$ be a cyclic representation of A with cyclic vector α and let $f(x) = (\rho(x)\alpha, \alpha)$. Then for every Hermitian positive functional of finite variation $g: A \to \mathbb{C}$ dominated by f there is a unique positive operator $T \in (\rho, \rho)$ such that $|T| \leqslant 1$ and $g(x) = (T\rho(x)\alpha, \alpha)$ for every $x \in A$. Conversely, if T is such an operator then the functional g defined by T is positive, Hermitian, of finite total variation and $g \leqslant f$.*

Proof. Let g be a Hermitian, positive linear functional of finite variation such that $g \leqslant f$. Since ρ is cyclic the linear manifold $\mathscr{L} = \rho(A)\alpha$ is dense in $\mathscr{H}$. For x and y in A we let

$$Q(\rho(x)\alpha, \rho(y)\alpha) = g(y^* x).$$

By Lemma 4 and (2) we have

$$|g(y^* x)|^2 \leqslant g(x^* x)\, g(y^* y) \leqslant f(x^* x)\, f(y^* y) = \|\rho(x)\alpha\|^2\, \|\rho(y)\alpha\|^2 .$$

Therefore Q can be extended to a bilinear form $Q: \mathscr{H} \times \mathscr{H} \to \mathbb{C}$ satisfying

$$|Q(\xi,\eta)| \leqslant \|\xi\| \cdot \|\eta\| . \tag{3}$$

Since $Q(\rho(x)\alpha, \rho(x)\alpha) = g(x^* x) \geqslant 0$ and $\mathscr{L}$ is dense in $\mathscr{H}$ we see that $Q(\xi,\xi) \geqslant 0$ for every $\xi \in \mathscr{H}$. Using $g(y^* x) = \overline{g(x^* y)}$ we see that

$$Q(\xi,\eta) = \overline{Q(\eta,\xi)} \tag{4}$$

for all ξ, η in $\mathscr{H}$.

We define $T: \mathscr{H} \to \mathscr{H}$ by $(T\xi,\eta) = Q(\xi,\eta)$ where ξ is fixed so that $Q(\cdot,\eta)$ is a continuous linear functional. By (3) we have $\|T\xi\| \leqslant \|\xi\|$ and so $|T| \leqslant 1$. The positivity of T follows from $(T\xi,\xi) = Q(\xi,\xi) \geqslant 0$. Similarly $T = T^*$ is an immediate consequence of (4). The intertwining nature of T can be seen from

$$\begin{aligned}(T\rho(x)\rho(y)\alpha, \rho(z)\alpha) &= g(z^*(xy)) = g((x^* z)^* y) = (T\rho(y)\alpha, \rho(x^*)\rho(z)\alpha)\\ &= (\rho(x)\, T\rho(y)\alpha, \rho(z)\alpha).\end{aligned}$$

In order to prove that $g(x) = (T\rho(x)\alpha, \alpha)$ we first notice that by (2)

$$|g(x)|^2 \leqslant v(g)\, g(x^* x) \leqslant v(g)\, f(x^* x) = v(g)\, \|\rho(x)\alpha\|^2$$

and so $\rho(x)\alpha \mapsto g(x)$ defines a continuous linear functional on $\mathscr{H}$. Hence there is a vector β in $\mathscr{H}$ such that $g(x) = (\rho(x)\alpha, \beta)$ for every x in A. Since

$$\begin{aligned}(\rho(x)\alpha, T\rho(y)\alpha) &= (\rho(y^*)\, T\rho(x)\alpha, \alpha) = (T\rho(y^* x)\alpha, \alpha) = g(y^* x)\\ &= (\rho(y^* x)\alpha, \beta) = (\rho(x)\alpha, \rho(y)\beta)\end{aligned}$$

and $\mathscr{L}$ is dense in $\mathscr{H}$ we see that $T\rho(y)\alpha = \rho(y)\beta$ for $y \in A$. Therefore

$$(T\rho(x)\alpha, \alpha) = (\rho(x)\beta, \alpha) = \overline{(\rho(x^*)\alpha, \beta)} = \overline{g(x^*)} = g(x).$$

The uniqueness of T follows from the identity

$$g(y^* x) = (T\rho(y^* x)\alpha, \alpha) = (T\rho(x)\alpha, \rho(y)\alpha)$$

and the fact that $\mathscr{L} = \rho(A)\alpha$ is dense in $\mathscr{H}$.

Let us now suppose that $g(x) = (T\rho(x)\alpha, \alpha)$ where T is a positive operator in (ρ, ρ) and $|T| \leqslant 1$. Since

$$g(x^* x) = (T\rho(x^* x)\alpha, \alpha) = (\rho(x^*)\, T\rho(x)\alpha, \alpha) = (T\rho(x)\alpha, \rho(x)\alpha) \geqslant 0$$

the functional g is positive and by

$$g(x^*) = (T\rho(x^*)\alpha, \alpha) = (\rho(x^*)T\alpha, \alpha) = (\alpha, T\rho(x)\alpha) = \overline{g(x)}$$

it is Hermitian. By Theorem III.4.10 the positive operator T has a positive squareroot S which also belongs to (ρ, ρ). Then

$$\begin{aligned} |g(x)|^2 &= |(S^2\rho(x)\alpha, \alpha)|^2 = |(\rho(x)S\alpha, S\alpha)|^2 \\ &\leqslant \|\rho(x)S\alpha\|^2 \|S\alpha\|^2 = g(x^*x)\|S\alpha\|^2. \end{aligned}$$

This proves that $v(g) \leqslant \|S\alpha\|^2$. Using $|T| \leqslant 1$ and (2) we obtain

$$g(x^*x) = (T\rho(x)\alpha, \rho(x)\alpha) \leqslant |T| \cdot \|\rho(x)\alpha\|^2 \leqslant f(x^*x).$$

Therefore f dominates g.

Following Naimark we say that a positive, Hermitian functional of finite variation f is *indecomposable* if every other such functional g dominated by f is of the form $g = \lambda f$. Since $0 \leqslant g(x^*x) \leqslant f(x^*x)$ it follows that $0 \leqslant \lambda \leqslant 1$.

Theorem 16. *Let $\rho: A \to \mathscr{L}(\mathscr{H})$ be a cyclic representation of the complex involutive algebra A and let α be a cyclic vector. Then ρ is irreducible if and only if $f(x) = (\rho(x)\alpha, \alpha)$ is indecomposable.*

Proof. Let ρ be irreducible and let g be dominated by f. By Lemma 15 we have $g(x) = (T\rho(x)\alpha, \alpha)$ where $T \geqslant 0$ and $|T| \leqslant 1$. Since ρ is irreducible by Proposition IV.2.4 we have $T = \lambda I$ and so $g(x) = \lambda(\rho(x)\alpha, \alpha) = \lambda f(x)$. Conversely let f be indecomposable and let E be a projection in (ρ, ρ). Then by Lemma 15 $g(x) = (E\rho(x)\alpha, \alpha)$ is dominated by f and so $g = \lambda f$ where $0 \leqslant \lambda \leqslant 1$. Since T is uniquely determined by g it follows that $E = \lambda I$. However E is a projection and so $\lambda = 0$ or $\lambda = 1$. Applying Propositions III.6.15 and IV.3.2 we obtain $(\rho, \rho) = \mathbb{C}I$ and so ρ is irreducible.

Theorem 17. *Let A be a complex Banach algebra with continuous involution and let f be a positive Hermitian functional of finite variation. Then there is a strongly continuous cyclic representation $\rho: A \to \mathscr{L}(\mathscr{H})$ and a cyclic vector α such that $f(x) = (\rho(x)\alpha, \alpha)$ for every $x \in A$.*

Proof. Let $I = \{x : x \in A \text{ and } f(x^*x) = 0\}$. If $a \in A$ and $x \in I$ then by Lemma 4 $f(ax) = 0$ and so replacing a by x^*a^*a we see that $f((ax)^*(ax)) = 0$. This proves that I is a left ideal of A. Therefore A/I is a pre-Hilbert space under the inner product $(x+I, y+I) = f(y^*x)$. Let $\mathscr{H}$ be the complex Hilbert space obtained by completing A/I.

For every a in A we define $\rho(a): A/I \to A/I$ by $\rho(a)(x+I) = ax+I$. Then $\rho(a)$ is a linear operator, $\rho(ab) = \rho(a)\rho(b)$ and $(\rho(a^*)(x+I), y+I) = (x+I, \rho(a)(y+I))$ by $f(y^*(a^*x)) = f((ay)^*x)$. By Lemma 9 and Proposition 11 we have

$$\begin{aligned}\|\rho(a)(x+I)\|^2 &= f((ax)^*(ax)) = g(a^*a) \leqslant \|g\| \cdot \|a^*a\| \leqslant v(g)\sqrt{C}\,\|a^*a\| \\ &\leqslant f(x^*x)\,C\sqrt{C}\,\|a\|^2 = C\sqrt{C}\,\|a\|^2\,\|x+I\|^2\end{aligned}$$

where C is such that $\|x^*\| \leqslant C\|x\|$. Therefore $\rho(a)$ can be extended to a continuous operator $\rho(a): \mathscr{H} \to \mathscr{H}$ and these operators $\rho(a)$ $(a \in A)$ define a representation $\rho: A \to \mathscr{L}(\mathscr{H})$. Since $|\rho(a)| \leqslant C^{\frac{3}{4}}\,\|a\|$ we see that ρ is strongly continuous. (The same follows from Theorem I.8.8.)

Using $|f(x)|^2 \leqslant v(f)\,f(x^*x)$ we see that $f(x)=0$ for every x in I and so f defines a linear functional $\tilde{f}: A/I \to \mathbb{C}$. Since $f(x^*x) = \|x+I\|^2$ we have $|\tilde{f}| \leqslant v(f)$ and so $\tilde{f}$ can be extended to a continuous linear functional $\tilde{f}: \mathscr{H} \to \mathbb{C}$. Let α be the vector in $\mathscr{H}$ such that $\tilde{f}(x+I) = f(x) = (x+I, \alpha)$ for every $x \in A$. By the definition of ρ we have

$$(\rho(a)\alpha, x+I) = (\alpha, \rho(a^*)(x+I)) = (\alpha, a^*x+I) = \overline{f(a^*x)} = (a+I, x+I)$$

for every a, $x \in A$. Therefore A/I being dense in $\mathscr{H}$ we obtain $\rho(a)\alpha = a+I$ for every a in A. This shows that α is a cyclic vector and $f(x) = (x+I, \alpha) = (\rho(x)\alpha, \alpha)$ for all x in A.

Note. Let $\sigma: A \to \mathscr{L}(\mathscr{K})$ be a cyclic representation with cyclic vector γ and let $f(x) = (\sigma(x)\gamma, \gamma)$. Then by Lemma 10 the representation associated with f is equivalent to σ. The lemma also shows that up to unitary equivalence ρ is uniquely determined by f.

We let $K = K(A)$ denote the set of positive, Hermitian functionals $f: A \to \mathbb{C}$ satisfying $v(f) \leqslant 1$.

Proposition 18. *If A is a complex Banach algebra with continuous involution then K is a convex set in the dual space of A and K is compact in the weak star topology.*

Proof. We have $f \in K$ if and only if f is a linear functional satisfying $f(x^*x) \geqslant 0$, $f(x^*) = \overline{f(x)}$ and $|f(x)|^2 \leqslant f(x^*x)$. Therefore K is a closed subset of the dual space. By Proposition 11 we have

$$|f| \leqslant \sqrt{\operatorname{supr} \frac{\|x^*\|}{\|x\|}}$$

for every f in K and so by Alaoglu's theorem K is compact. In order to prove that K is convex let f_1, f_2 in K and $\lambda_1, \lambda_2 > 0$ with $\lambda_1 + \lambda_2 = 1$

be given. By Proposition 6 f_k $(k=1,2)$ can be extended to A_e without increasing $v(f_k)$. Then $\lambda_1 f_1+\lambda_2 f_2$ is of finite variation and

$$v(\lambda_1 f_1+\lambda_2 f_2)=(\lambda_1 f_1+\lambda_2 f_2)(e)=\lambda_1 v(f_1)+\lambda_2 v(f_2)\leqslant\lambda_1+\lambda_2=1.$$

Theorem 19. *The zero functional is an extreme point of K. An arbitrary $f\in K$ is an extreme point if and only if it is indecomposable and $v(f)=1$.*

Proof. Suppose that $\lambda_1 f_1+\lambda_2 f_2=0$ where $\lambda_1,\lambda_2>0$. Then by $f_k(x^*x)\geqslant 0$ we obtain $f_k(x^*x)=0$ for every x in A. Hence using Lemma 8 we see that $f_1=f_2=0$. Therefore 0 is an extreme point of K.

Now let f be an indecomposable element of K satisfying $v(f)=1$ and let $f=\lambda_1 f_1+\lambda_2 f_2$ where λ_1,λ_2 are positive, $\lambda_1+\lambda_2=1$ and $f_1,f_2\in K$. Since f dominates $\lambda_k f_k$ we have $\lambda_k f_k=\mu_k f$ for some μ_k satisfying $0\leqslant\mu_k\leqslant 1$. By Lemma 7 we have

$$1=v(f)\leqslant\lambda_1 v(f_1)+\lambda_2 v(f_2)$$

and so using $v(f_k)\leqslant 1$ we obtain $v(f_k)=1$. Hence

$$\lambda_k=v(\lambda_k f_k)=v(\mu_k f)=\mu_k$$

and so $f_k=f$ for $k=1,2$. Therefore f is an extreme point of K.

If $f\in K$ and $0<v(f)^{-1}<1$ then $v(f)f$ is in K and so f is not an extreme point because

$$f=v(f)\frac{f}{v(f)}+(1-v(f))0.$$

In order to prove that every extreme point f is an indecomposable functional in view of Theorem 16 it is sufficient to prove that the cyclic representation defined by f is irreducible. Let $\mathscr{H}_1$ be an invariant subspace of ρ, let $\mathscr{H}_2=\mathscr{H}_1^{\perp}$ and let $\alpha=\alpha_1+\alpha_2$ be the decomposition of the cyclic vector with $\alpha_k\in\mathscr{H}_k$. Then α_k is a cyclic vector for the representation $\rho|\mathscr{H}_k$. Thus by Lemma 14

$$1=v(f)=\|\alpha\|^2=\|\alpha_1\|^2+\|\alpha_2\|^2=v(f_1)+v(f_2)$$

where $f_k(x)=(\rho(x)\alpha_k,\alpha_k)$. Due to the invariance and orthogonality of $\mathscr{H}_1,\mathscr{H}_2$ we have $f=f_1+f_2$ and so

$$f=v(f_1)\frac{f_1}{v(f_1)}+v(f_2)\frac{f_2}{v(f_2)}.$$

Here $v(f_1)>0$ by $\mathscr{H}_1\neq 0$ and in case $v(f_2)=0$ the second term means 0. Since f is an extreme point of K by $v(f_1)+v(f_2)=1$ we must indeed have $v(f_2)=0$. By Lemma 8 we obtain $f_2=0$ and so $\rho|\mathscr{H}_2=0$. Therefore $\rho(x)\alpha=\rho(x)\alpha_1\in\mathscr{H}_1$ for every x in A and so α being a cyclic vector $\mathscr{H}_1=\mathscr{H}$. We proved that ρ is irreducible.

Theorem 20. *The set K is the weak star closed convex hull of its extreme points.*

Proof. The set H of Hermitian elements of A is a real Banach space. Let $\tilde{f}$ denote the restriction of any $f\in K$ to H. Then $\tilde{f}$ is a real valued functional on H. The map $f\mapsto\tilde{f}$ preserves linear combinations and it is continuous in the weak star topology. It is a one-to-one map because if $x\in A$ then $x=x_1+ix_2$ for suitable $x_1,x_2\in H$ and so $\tilde{f}=0$ implies $f=0$. By Proposition 18 K is compact so $\tilde{K}$ being the continuous linear image of K is a compact, convex set. Now the conclusion follows by applying the Krein-Milman theorem to the image of K under $f\mapsto\tilde{f}$.

Theorem 21. *Let A be a complex Banach algebra with continuous involution and let $x\in A$. If A has a representation ρ such that $\rho(x)\neq 0$ then it has a strongly continuous irreducible representation π satisfying $\pi(x)\neq 0$.*

Proof. First we notice that $\rho(x^*x)\neq 0$ if and only if $\rho(x)\neq 0$ because $(\rho(x^*x)\alpha,\alpha)=(\rho(x)\alpha,\rho(x)\alpha)$ for every α in the Hilbert space $\mathscr{H}$ of ρ. Hence by hypothesis we have $(\rho(x^*x)\alpha,\alpha)\neq 0$ for some α satisfying $\|\alpha\|\leqslant 1$. Therefore by Lemma 13 $y\mapsto g(y)=(\rho(y)\alpha,\alpha)$ belongs to the set K of positive Hermitian functionals f satisfying both $v(f)\leqslant 1$ and $f(x^*x)>0$. By Theorem 20 K is the convex hull of its extreme points so there is an extreme point f in K such that $f(x^*x)>0$. By Theorem 17 there is a cyclic representation π associated with f and by Theorems 16 and 19 π is irreducible. Moreover if α is the cyclic vector associated with f then using $v(f)=1$ and (2) we see that

$$\|\pi(x^*x)\alpha\|^2=f((x^*x)^*(x^*x))\geqslant|f(x^*x)|^2>0.$$

Obviously this implies that $\pi(x)\neq 0$. The continuity of π follows from Theorem I.8.8.

Part of the foregoing reasoning also proves the following:

Proposition 22. *Let A be a complex Banach algebra with continuous involution and let $x\in A$. Then an irreducible representation π satisfying $\pi(x)\neq 0$ exists if and only if there is a positive Hermitian functional f of finite variation such that $f(x^*x)>0$.*

Lemma 23. *Let A be a normed involutive algebra and let $\rho\colon A\to\mathscr{L}(\mathscr{H})$ be a continuous representation. Then for any Hermitian element x we have $|\rho(x)|\leqslant v(x)$.*

Note. Here $v(x)=\lim\sqrt[n]{\|x^n\|}$. The existence of the limit and properties of v are discussed in Section I.5.

Proof. By the continuity of ρ we have $|\rho(a)|\leqslant C\|a\|$ for some constant C and for every $a\in A$. Therefore by Lemma I.9.5 we have

$$|\rho(x)|^n=|\rho(x^n)|\leqslant C\|x^n\|$$

for $n=1,2,\ldots$ Hence $|\rho(x)|\leqslant\sqrt[n]{C}\sqrt[n]{\|x^n\|}$ and so $|\rho(x)|\leqslant v(x)$.

The *star radical* or *reducing ideal* $R^*(A)$ of an involutive algebra A is defined to be the intersection of the kernels of all irreducible representations of A. If A is a Banach algebra with continuous involution then by Theorem I.8.8 every representation of A is continuous and so $R^*(A)$ is a closed ideal. It is clear that $R^*(A)$ is self adjoint that is $x\in R^*(A)$ implies $x^*\in R^*(A)$. If $R^*(A)=0$ then A is called *star semisimple.*

Proposition 24. *If A is a complex, involutive Banach algebra then its Jacobson radical $J(A)$ is contained in $R^*(A)$.*

Corollary. *If A is star semisimple then it is semisimple.*

Proof. If $x\in J(A)$ then by the ideal property $x^*x\in J(A)$ and so $v(x^*x)=0$ by Proposition I.7.19. Thus by Lemma 23 x^*x is in the kernel of every representation $\rho\colon A\to\mathscr{L}(\mathscr{H})$. Since $\rho(x^*x)=0$ we have $\rho(x)=0$ and so $x\in R^*(A)$.

Proposition 25. *The intersection of the kernels of all representations of a complex Banach algebra with continuous involution is its star radical.*

Proof. Let R be the intersection of the kernels of the representations of A. This concept is meaningful because equivalent representations have a common kernel. By the definition of the star radical we have $R\subseteq R^*(A)$. Now if $x\in R^*(A)$ then $\pi(x)=0$ for every irreducible representation of A and so by Theorem 21 we have $\rho(x)=0$ for every representation $\rho\colon A\to\mathscr{L}(\mathscr{H})$. Hence if $x\in R^*(A)$ then $x\in R$ proving that $R=R^*(A)$.

Proposition 26. *If G is a locally compact group, $f\in L^1(G)$ and $f\neq 0$ then there is an irreducible representation π of the involutive algebra $L^1(G)$ such that $\pi(f)\neq 0$.*

Proof. In view of Theorem 21 it is sufficient to verify the following:

Proposition 27. *The algebra representation associated with the left regular representation of a locally compact group is faithful.*

Proof. Let $\mathscr{H}=L^2(G)$ and let $\sigma: L^1(G)\to\mathscr{L}(\mathscr{H})$ be the involutive algebra representation associated with the left regular representation $\lambda: G\to\mathscr{U}(\mathscr{H})$. By Theorem 5.5 we have $\sigma(f)g=f*g$ for every $g\in\mathscr{H}$. If $f\neq 0$ then there is a g in $C_0(G)$ such that

$$f*g(e)=\int_G f(x^{-1})g(x)\,dx\neq 0.$$

By Proposition 4.8 $f*g$ is continuous and so $\sigma(f)\neq 0$.

Corollary. *The involutive algebra $L^1(G)$ is star semisimple.*

This follows immediately from Proposition 26 and the definition of star semisimplicity. It can be obtained also from Propositions 25 and 27.

Theorem 28. *If G is a locally compact group and the element x is different from the identity of G then there is a continuous, irreducible, unitary representation π of G such that $\pi(x)$ is not the identity operator.*

Note. The conclusion is often stated in one of the following forms: The locally compact group G has *sufficiently many* continuous, irreducible, unitary representations.—The set of continuous, irreducible, unitary representations of a locally compact group G is *complete* or *total.*

Proof. Let $\pi: G\to\mathscr{U}(\mathscr{H})$ be a continuous irreducible representation such that $\pi(x)$ is the identity operator. Then for any $\alpha,\beta\in\mathscr{H}$ and $f\in L^1(G)$ we have

$$\begin{aligned}(\sigma({}^xf)\alpha,\beta)&=\int_G f(x^{-1}y)(\pi(y)\alpha,\beta)\,dy=\int_G f(y)(\pi(x)\pi(y)\alpha,\beta)\,dy\\&=\int_G f(y)(\pi(y)\alpha,\beta)\,dy=(\sigma(f)\alpha,\beta).\end{aligned}$$

Therefore $\pi(x)=I$ implies that $\sigma({}^xf-f)=0$ for every f in $L^1(G)$. Now choose f such that ${}^xf-f\neq 0$. Then by Proposition 26 we have $\sigma({}^xf-f)\neq 0$ for some irreducible representation σ of $L^1(G)$. By Theorem 5.16 and 5.17 σ comes from some continuous, irreducible representation π of G. For this π we have $\pi(x)\neq I$.

Remarks

Special classes of topological groups such as Lie groups and topological transformation groups appeared in the mathematical literature in the later parts of the last century. Axiomatization began around 1925 in the papers of F. Leja (1), Baer (1), Schreier (1) and others. The English translation of Pontrjagin's book appeared in 1939. The second edition is available in both English and German (1,2). A strong motivating force for research in this area was Hilbert's famous fifth problem which was settled by the works of Gleason, Montgomery and Zippin and the efforts of many others. The relevant literature and results can be found in a monograph by Montgomery and Zippin (1). Two other factors influencing the development of the theory of topological groups were classical harmonic analysis and group representations. An important book summarizing these aspects of the theory was written in 1938 by Weil (1). The most up to date and extensive work covering many aspects of this area is a two volume monograph by Hewitt and Ross (1).

A systematic study of the subgroups of a topological group was started by Freudenthal in 1936 (1). The uniformizability of topological groups in the form of complete regularity was proved by Pontrjagin and made known in a letter written to Weil. (See p. 13 of Weil (2).) The left and right uniform structures for such a group were given by Weil (2). The uniformizability of the homogeneous spaces and the uniform structure described in the text are due to the author. If G satisfies the first axiom of countability then G is a metrizable space and there exists a metric which is invariant under one sided translations. This fact was discovered by Garrett Birkhoff (2) and Kakutani (2) and with our present knowledge on the metrizability of uniform spaces it can be easily proved.

The existence of a Haar measure on compact Lie groups was known to Péter and Weyl (1) already. Haar (1) gave his construction only for groups G satisfying the second axiom of countability. The fact that his reasoning can be extended to arbitrary G was pointed out by Kakutani (2) and Weil (3). The essential uniqueness of the Haar measure was proved by v. Neumann in (6) and (7). The second paper is important also for another reason, namely the possibility of defining a measure by a linear functional under suitable conditions is stated there explicitely. For additional information on Haar measure see Hewitt and Ross (1).

The existence of quasi-invariant measures was first proved by Mackey in 1952 (4) for groups G satisfying the second axiom of countability. His proof uses a result of Federer and A. P. Morse (1). The fact that a continuous ρ-function exists in general was first proved by F. Bruhat in 1956 (1,2). The proof presented in Section 3 is due to him. In 1960 independently of Bruhat another proof of the existence of a quasi-invariant measure with continuous ρ-function was published by Loomis (2). In (1) Bruhat also proved that if G is a Lie group then there exist infinitely differentiable ρ-functions on G. Theorem 3.15 is due to Weil (1).

Convolutions first appeared in the theory of finite groups and in various special instances in classical analysis on $\mathbb{R}$. The first uses of convolutions in the case of Lie groups were made by Weyl in 1925 (1). The convolution of measures was introduced in (1) by Weil. Several of the propositions listed in Section 4 were first pointed out in Weil's book. Of course some of these results were already proved earlier in the special case when the underlying group is $\mathbb{R}^+$ or $(\mathbb{R}^+)^n$. A few of the results stated in the text might appear for the first time. They will be needed in Sections VII.3 and VII.4 and they were probably known to Godement. For a more detailed study of convolutions see Hewitt and Ross (1) and W. Rudin (1).

The algebra representation σ associated with the group representation ρ is approximately as old as the representation theory of locally compact groups. The earlier writers used to speak about "the extension of the representation ρ to $L^1(G)$". A true extension is obtained if we replace $L^1(G)$ by $M(G)$. The definition of σ in the case of a reflexive Banach space can be found in Loomis' book (1) and for Hilbert spaces Godement gave the definition already in 1948 (1). The contents of Proposition 5.7 in the special case when S is a group was probably known to others although I have not seen it in print and the proof given in the text is my own. Lemma 11 is in Loomis (1) and probably it originates from a result which can be found in a paper of Segal (2). The group algebra $L^2(G)$ of a compact group was studied in great detail by Kaplansky (3) before the introduction of H^* algebras by Ambrose in (1). Proposition 5.10 and Theorems 5.16 and 5.17 can be found in slightly different forms in Loomis' book (1). Theorem 6.5 is due to Segal and it was published in 1950 (3) but special cases of this result were known already earlier. Theorem 6.11 is well known although I was not able to locate it in the literature. It is possible that there exists a simpler proof but the present reasoning came naturally.

Theorem 6.1 is due to Godement (1). Other results on the continuity of representations and morphisms can be found in the papers by R. T. Moore (1) and Aarnes (1). The Gelfand-Raikov theorem was published in 1943 (1) in a paper which became very important and had a great influence on others. The theorem can be proved also by using positive definite functions instead of positive functionals. A proof along these lines was given in 1948 by Godement (1). The proof given by me is essentially the same as Segal's (2) but I had an opportunity to read also an elegant exposition of this proof by Argabright and I also made some improvements e.g. it is now required only that the involution $x \mapsto x^*$ be continuous. The proof of Theorem 17 uses some simplifications which were introduced by Gil de Lamdarid. The Gelfand-Raikov theorem is discussed also in Rickart's book (1) and in Hewitt and Ross (1). The semisimplicity of $L^1(G)$ was proved in 1947 by Segal (4).

Chapter VI

Induced Representations

This chapter contains the basic theory of induced representations of locally compact groups and some more advanced material on this subject. The theory of induced representations of finite groups was extended to locally compact groups by a single person, George Mackey. His basic definition is given in Section 2 and his main results are proved in Sections 8, 9, 10 and 11. Mackey had to restrict himself to separable Hilbert spaces $\mathscr{H}$ and groups G satisfying the second axiom of countability. For the sake of clarity we shall define induced representations first in this special case. Later in Section 4 we extend Mackey's definition to arbitrary G and $\mathscr{H}$. The theorems given in Sections 8, 9, 10 and 11 are proved in full generality.

If the reader is mainly interested in the basic theory of induced representations then he should read Sections 1, 2, 4, 8, 9, 10 and possibly consult some of the later publications by Mackey. If his interest lies in the material treated in the second half of Chapter VII then it is best to read Sections 1, 2, 5, 6 and 7. The relevant material on tensor products is discussed in Section 3.

The results treated in Section 12 naturally belong to the present chapter. However they involve some of the results given in Section VII.1 the proofs of which do not depend on induced representations. The minimal prerequisites for the rest of this chapter consist of Sections IV.1—IV.5 and Sections V.1—V.3. In order to read Sections 5, 6 and 7 one also needs some of the material treated in Section V.5.

1. The Riesz-Fischer Theorem

The purpose of this section is to recall the Riesz-Fischer reasoning and instead of going all the way to proving the completeness of L^2 spaces stop at the next to the last step and state the conclusion as a theorem. The Riesz-Fischer theorem follows from there by a very simple argument. The reason for this approach is that the revised version of the theorem is applicable to a number of instances where one wishes to prove the completeness of certain function spaces. In particular this device will save us from including repetitious arguments in Sections 2 and 4. We shall also generalize the classical situation by considering Hilbert space valued functions instead of complex valued ones. The results will be needed in this generalized form.

Let (X, μ) be a measure space, let S be a set, $p: S \to X$ surjective and let $\mathscr{H}$ be a Hilbert space. If $\theta: S \to \mathscr{H}$ is a function such that $ps_1 = ps_2$ implies $\|\theta(s_1)\| = \|\theta(s_2)\|$ then a non-negative function is defined on X by the rule $ps \mapsto \|\theta(s)\|$. If this function is measurable we call θ measurable and define $\|\theta\|^2 = \int_X \|\theta(s)\|^2 d\mu$ and if the integral is finite we call θ a *normed function*. As usual two normed functions which differ in norm only on a set of measure zero will be called equivalent and considered identical. We let $\mathscr{L}$ be a vector space of equivalence classes of normed functions $\theta: S \to \mathscr{H}$. The symbol θ will be used both for the individual functions and the equivalence class containing them. The condition that $\|\theta(\cdot)\|$ be defined on X for every $\theta \in \mathscr{L}$ is a considerable restriction on the choice of the normed vector space $\mathscr{L}$. Let θ_1 and θ_2 belong to such a vector space $\mathscr{L}$. In $\mathscr{H}$ we have the identity

$$(\alpha, \beta) = \tfrac{1}{4}\{\|\alpha+\beta\|^2 + i\|\alpha+i\beta\|^2 - \|\alpha-\beta\|^2 - i\|\alpha-i\beta\|^2\}.$$

Hence we see that $(\theta_1(x), \theta_2(x))$ can be defined for every x in X as $(\theta_1(s), \theta_2(s))$ where s denotes any element of S such that $ps = x$. Moreover it follows that $x \mapsto (\theta_1(x), \theta_2(x))$ is a measurable function, it is integrable and

$$(\theta_1, \theta_2) = \int_X (\theta_1(x), \theta_2(x))\, d\mu(x)$$

satisfies $|(\theta_1, \theta_2)| \leqslant \|\theta_1\| \cdot \|\theta_2\|$. Therefore $\mathscr{L}$ is an inner product space.

For example, if $S = X$, p is the identity map and $\mathscr{H}$ is a separable Hilbert space then a vector space $\mathscr{L}$ can be constructed as follows: We consider all functions $\theta: X \to \mathscr{H}$ such that $x \mapsto (\theta(x), \alpha)$ is a measurable function for every α in $\mathscr{H}$. Let α_i $(i \in \mathscr{I})$ be a MONS in $\mathscr{H}$. Then we have

$$\|\theta(x)\|^2 = \sum (\theta(x), \alpha_i)(\alpha_i, \theta(x))$$

and so $\mathscr{I}$ being countable $x \mapsto \|\theta(x)\|^2$ is a measurable function. The functions θ for which $\|\theta\| < \infty$ form a vector space and the set of those θ for which $\|\theta\| = 0$ is a subspace. Identifying equivalent functions we arrive at an inner product space $L^2(X, \mu, \mathscr{H})$ which will be called the *L^2-space of X, μ and $\mathscr{H}$.*

Proposition 1. *If θ_i $(i \in \mathscr{I})$ is a Cauchy net in norm then it is a Cauchy net in measure i.e. for every $\delta, \varepsilon > 0$ there is an $i(\delta, \varepsilon)$ in $\mathscr{I}$ such that*

$$\mu\{ps: \|\theta_i(s) - \theta_j(s)\| \geqslant \delta\} < \varepsilon$$

for every i, j satisfying $i(\delta, \varepsilon) \leqslant i, j$.

Proof. Let $f_{ij}(x) = \|\theta_i(s) - \theta_j(s)\|^2$ where $ps = x$. Since we have a Cauchy net in norm given $\delta, \varepsilon > 0$ there is an $i(\delta, \varepsilon)$ in $\mathscr{I}$ such that

$$\int_X f_{ij}(x)\, d\mu(x) < \delta^2 \varepsilon$$

for every $i, j \geqslant i(\delta, \varepsilon)$. Since $\mu\{x: f_{ij}(x) \geqslant \delta^2\} \geqslant \varepsilon$ would imply the opposite inequality we have $\mu\{x: f_{ij}(x) \geqslant \delta^2\} < \varepsilon$ for every $i, j \geqslant i(\delta, \varepsilon)$.

Proposition 2. *Every Cauchy net in measure θ_i $(i \in \mathscr{I})$ contains for every i_0 in $\mathscr{I}$ a subsequence θ_{i_k} $(i_0 < i_1 < i_2 < \cdots)$ which is convergent almost everywhere.*

Note. By almost everywhere convergence we mean that the projection of the set of divergence points has zero measure in X.

Proof. By hypothesis we can select a subsequence $i_1 < i_2 < \cdots$ of $\mathscr{I}$ such that $i_0 < i_1$ and

$$\mu\{x: f_{ij}(x) \geqslant 2^{-k}\} < 2^{-k}$$

for every $i, j \geqslant i_k$. We let

$$E_k = \{x: f_{i_{k+1}, i_k}(x) \geqslant 2^{-k}\}$$

so that $\mu E_k < 2^{-k}$. Furthermore we let

$$A_l = \bigcup_{k \geqslant l} E_k \quad \text{and} \quad B = \bigcap_{l \geqslant 1} A_l = \lim A_l.$$

By $\mu E_k < 2^{-k}$ we have $\mu A_l < 2^{1-l}$ $(l = 1, 2, \ldots)$ and so $\mu B = 0$.

If $x \notin B$ then $x \notin A_l$ for some $l = l(x) \geqslant 1$. Hence $x \notin E_k$ for every $k \geqslant l(x)$ and so

$$f_{i_{k+1}, i_k}(x) < 2^{-k}$$

for every $k \geqslant l(x)$. Therefore given $\varepsilon > 0$ we can choose k such that $2^{-k/2} < \varepsilon^2(\sqrt{2} - 1)$ and obtain for any s with $ps = x$ and for any $r > q > k$ the following estimate:

$$\begin{aligned}\|\theta_{i_r}(s) - \theta_{i_q}(s)\| &\leqslant \sum_{q \leqslant l < r} \|\theta_{i_{l+1}}(s) - \theta_{i_l}(s)\| = \sum_{q \leqslant l < r} f_{i_{l+1}, i_l}(ps)^{\frac{1}{2}} \\ &< \sum_{q \leqslant l < r} 2^{-\frac{l}{2}} < \varepsilon^2.\end{aligned}$$

We proved that $\theta_{i_1}(s), \theta_{i_2}(s), \ldots$ is a Cauchy sequence for every s such that $ps = x \notin B$ where $\mu B = 0$.

We recall that Fatou's lemma concerns integrable functions on an arbitrary measure space (X, μ) and it states the following:

Proposition 3. *If $f_1, f_2, \ldots$ is a sequence of non-negative integrable functions such that*

$$\liminf \int_X f_n(x)\, d\mu(x) < \infty$$

then $f = \liminf f_n$ is integrable and

$$\int_X f(x)\, d\mu(x) \leqslant \liminf \int_X f_n(x)\, d\mu(x).$$

As an illustrative application of Proposition 1, 2 and 3 we prove the generalized form of the Riesz-Fischer theorem:

Theorem 4. *For every measure space (X, μ) and separable Hilbert space $\mathscr{H}$ the inner product space $L^2(X, \mu, \mathscr{H})$ is complete in its norm.*

Proof. Since $L^2(X, \mu, \mathscr{H})$ is normed by $\|\theta\| = \|\theta\|_2$ it is sufficient to consider only Cauchy sequences instead of Cauchy nets. Let $\theta_1, \theta_2, \ldots$ be a Cauchy sequence in norm. Then it is a Cauchy sequence in measure and so it contains an almost everywhere convergent subsequence $\theta_{i_1}, \theta_{i_2}, \ldots$ We let $\theta = \lim \theta_{i_k}$ so that θ is a Hilbert space valued measurable function on X.

Next given $\varepsilon > 0$ we determine $k(\varepsilon)$ such that $\|\theta_{i_k} - \theta_{i_l}\|_2 \leqslant \varepsilon$ for all $k, l > k(\varepsilon)$. We fix $k \geqslant k(\varepsilon)$ and apply Fatou's lemma to the sequence f_n where $f_n(x) = \|\theta_{i_k}(x) - \theta_{i_{k+n}}(x)\|^2$ and $n = 1, 2, \ldots$ We see that $f(x) = \lim f_n(x) = \|\theta_{i_k}(x) - \theta(x)\|^2$ is integrable and

$$\int_X \|\theta_{i_k}(x) - \theta(x)\|^2\, d\mu(x) \leqslant \varepsilon^2.$$

Thus given $\varepsilon > 0$ we found $k(\varepsilon)$ such that $\|\theta_{i_k} - \theta\| \leqslant \varepsilon$ for every $k \geqslant k(\varepsilon)$. Since $\theta_1, \theta_2, \ldots$ is a Cauchy sequence it follows that $\|\theta_i - \theta\| \to 0$ as $i \to \infty$. Also θ_{i_k} and $\theta_{i_k} - \theta$ being square integrable we see that θ belongs to $L^2(X, \mu, \mathscr{H})$.

As an example we can consider a locally compact group G, a closed subgroup H and a quasi-invariant measure μ on $G/H = X$. The situation is particularly simple if G and H are such that G/H admits an invariant measure $d\xi$. In this case $L^2(G/H, \mathscr{H})$ will always mean the L^2-space of $X = G/H$ with respect to the measure $d\xi$.

In the general theory of induced representation of a locally compact group G we shall deal with Hilbert spaces $\mathscr{H}$ which are not necessarily separable and on several occasions it will be necessary to use the L^2-space of $\mathscr{H}$-valued functions defined on G. The theory of these is somewhat more advanced than the ones discussed above. A brief summary will be given at the end of Section 4.

2. Induced Representations when G/H has an Invariant Measure

Let G be a locally compact group, H a closed subgroup and $\chi: H \to \mathscr{U}(\mathscr{H})$ a unitary representation of H in the complex Hilbert space $\mathscr{H}$. There is a procedure by which a unitary representation of G is constructed from χ which is called the *representation induced on G by* χ and is denoted by ${}_G\rho^\chi$. If no confusion can arise one uses the notations ρ^χ, ${}_G\rho$ or simply ρ.

First we introduce induced representations in the special case when the homogeneous space $X = G/H$ admits an invariant measure and $\mathscr{H}$ is separable. If $\mathscr{H}$ is such then G and H for instance can be unimodular groups. In the general case the construction involves a quasi-invariant measure μ on X and the computations become more involved. After some parts of the theory are developed in the special case of invariant measures it will be easier to discuss the modifications which are necessary due to the presence of the quasi-invariant factor and the non-separability of $\mathscr{H}$.

We start by constructing the underlying Hilbert space $\mathscr{K}$ of the induced representation $\rho^\chi: G \to \mathscr{U}(\mathscr{K})$. This will be followed by the definition of ρ^χ and some properties including the fact that up to equivalence ρ^χ is uniquely determined by the equivalence class of χ. This is of special interest when χ is irreducible. The notation χ is used because in the special case when the Hilbert space $\mathscr{H}$ of χ is one dimensional χ is a group character.

Consider functions $\theta: G \to \mathscr{H}$ such that $x \mapsto (\theta(x), \alpha)$ is a measurable function on G for every fixed $\alpha \in \mathscr{H}$. If θ_1, θ_2 are such functions then $x \mapsto (\theta_1(x), \theta_2(x))$ is a measurable function on G because if α_i $(i = 1, 2, \ldots)$ is a MONS in $\mathscr{H}$ then

$$(\theta_1(x), \theta_2(x)) = \sum (\theta_1(x), \alpha_i)(\alpha_i, \theta_2(x)).$$

Suppose also that $\theta(xh) = \chi(h)^* \theta(x)$ for all $x \in G$ and $h \in H$. Then $x \mapsto (\theta_1(x), \theta_2(x))$ can be interpreted as a function on $X = G/H$ because $\chi(h)^*$ being unitary

$$(\theta_1(xh), \theta_2(xh)) = (\chi(h)^* \theta_1(x), \chi(h)^* \theta_2(x)) = (\theta_1(x), \theta_2(x)).$$

In particular $\|\theta(xh)\|$ is defined on X and we can require that

$$\|\theta\| = \left(\int_X \|\theta(xh)\|^2 \, d\xi\right)^{\frac{1}{2}}$$

be finite. We shall often use the notation $\|\theta(xh)\|^2$ although the integrand is the function $\xi \mapsto \|\theta(p^{-1}x)\|^2$ where x is an arbitrary element of ξ. We see that if $\|\theta_1\|, \|\theta_2\|$ are finite then

$$(\theta_1, \theta_2) = \int_X (\theta_1(x), \theta_2(x)) \, d\xi$$

exists and $|(\theta_1, \theta_2)| \leqslant \|\theta_1\| \cdot \|\theta_2\|$.

Theorem 1. *Let the closed subgroup H of the locally compact group G be such that G/H has an invariant measure and let χ be a unitary representation of H in the separable Hilbert space $\mathscr{H}$. Denote by $\mathscr{K} = \mathscr{K}(G, H, \chi)$ the vector space of equivalence classes of those functions $\theta\colon G \to \mathscr{H}$ for which*

1) *$x \mapsto (\theta(x), \alpha)$ is measurable for every $\alpha \in \mathscr{H}$;*

2) *$\theta(xh) = \chi(h)^* \theta(x)$ for $x \in G$ and $h \in H$ with the possible exception of a set pairs (x, h) such that the corresponding xh's belong to a set of measure zero lying in G/H.*

3) *$\|\theta\|$ is finite.*

Then $\mathscr{K}$ is a Hilbert space with respect to the inner product (θ_1, θ_2).

Proof. If we call $\theta_1 \sim \theta_2$ when $\|\theta_1 - \theta_2\| = 0$ then it is trivial to see that $\sim$ is an equivalence relation and the equivalence classes form an inner product space $\mathscr{K}$. Thus only the completeness of $\mathscr{K}$ has to be verified. For this purpose we use Propositions 1.1, 1.2 and 1.3.

We let $S = G$, $X = G/H$ and $p\colon S \to X$ be the natural map $g \mapsto gH$. We let $d\mu$ be a G—invariant measure $d\xi$ and choose $\mathscr{L} = \mathscr{K}$. Given a Cauchy sequence $\theta_1, \theta_2, \ldots$ in norm by repeating the reasoning given in the proof of Theorem 1.4 we find a function $\theta\colon G \to \mathscr{H}$ such that $\|\theta_i - \theta\| \to 0$ and a subsequence $\theta_{i_1}, \theta_{i_2}, \ldots$ such that $\|\theta_{i_k}(x) - \theta(x)\| \to 0$ almost everywhere on G/H. The completeness of θ will be established by proving that $\theta \in \mathscr{K}$.

Since $(\theta(x), \alpha) = \lim(\theta_{i_k}(x), \alpha)$ where $(\theta_{i_k}(x), \alpha)$ is measurable we see that $(\theta(x), \alpha)$ itself is a measurable function. Moreover we have $\theta_{i_k}(xh) = \chi(h)^* \theta_{i_k}(x)$ with the exception of a set of pairs (x, h) such that the corresponding xh points are covered by a null set lying in G/H. Thus if x, h are not in a null set then by $|\chi(h)^*| = 1$ we have

$$\|\theta(xh) - \chi(h)^* \theta(x)\| \leqslant \|\theta(xh) - \theta_{i_k}(xh)\| + \|\theta_{i_k}(x) - \theta(x)\|$$

with the possible exception of a similar set of points (x,h), due to the countability of the index set. Since the right hand side approaches zero we obtain 2). We also have 3) because $\|\theta_{i_k}\|$ and $\|\theta_{i_k}-\theta\|$ are finite.

Lemma 2. *Let $f: G\to\mathscr{H}$ be a continuous function with compact support where $\mathscr{H}$ is an arbitrary, not necessarily separable Hilbert space. Then the formula*

$$\theta_f(x)=\int_H \chi(h)f(xh)\,dh$$

defines in the weak sense a continuous function $\theta_f: G\to\mathscr{H}$ such that

$$\theta_f(xh)=\chi(h)^*\,\theta_f(x)$$

for every $x\in G$, $h\in H$ and $\|\theta_f(\cdot)\|$ has compact support in X.

Proof. Let S be the support of f. We have $\theta_f(x)\neq 0$ only if $xh\in S$ for some $h\in H$ and so $S\cdot H=pS$ is a support of $\|\theta_f(\cdot)\|$ in X. In order to prove the continuity of θ_f we start with the inequality

$$|(\theta_f(x_1)-\theta_f(x_2),\eta)|\leqslant\int_H \|f(x_1h)-f(x_2h)\|\cdot\|\eta\|\,dh$$

where η is an arbitrary vector in $\mathscr{H}$. Since f is continuous and has compact support it is uniformly continuous and so given $\varepsilon>0$ there is a symmetric neighborhood N_e of the identity e in G such that $|f(x)-f(y)|<\varepsilon$ whenever $xy^{-1}\in N_e$. Therefore if $x_1x_2^{-1}\in N_e$ then the integrand is at most $\varepsilon\|\eta\|$ for every $h\in H$. Now we fix x_1 and a compact symmetric neighborhood C of x_1. Then we have $f(xh)\neq 0$ for some x in C only if $h\in CS$ and so if $x_2\in C\cap N_e x_1$ then

$$|(\theta_f(x_1)-\theta_f(x_2),\eta)|\leqslant\varepsilon\|\eta\|\,\nu(CS)$$

where $\nu(CS)$ denotes the Haar measure of $CS\cap H$ in H. By choosing $\eta=\theta_f(x_1)-\theta_f(x_2)$ we obtain $\|\theta_f(x_1)-\theta_f(x_2)\|\leqslant\varepsilon\,\nu(CS)$ for every $x_2\in C\cap N_e x_1$. Thus θ_f is continuous at x_1.

An easy computation shows that $(\theta_f(xa),\eta)=(\chi(a)^*\theta_f(x),\eta)$ for every η in $\mathscr{H}$ and $a\in H$, $x\in G$ and so θ_f satisfies condition 2) of Theorem 1. Condition 1) holds because θ_f is continuous and 3) is valid because $\|\theta_f(\cdot)\|$ is continuous and has compact support in X.

Corollary. *If $\chi: H\to\mathscr{U}(\mathscr{H})$ is continuous then there are non-zero continuous functions $\theta: G\to\mathscr{H}$ in $\mathscr{K}$ such that $\|\theta(\cdot)\|$ has compact support in X.*

In order to prove this corollary we consider a non-zero vector α in $\mathscr{H}$ and let $f(x)=\varphi(x)\alpha$ for $x\in G$ where $\varphi\in C_0(G)$. Then by Lemma 2 θ_f

is continuous, belongs to $\mathscr{K}$ and $\|\theta_f(\cdot)\|$ has compact support in X. Moreover

$$\theta_f(e) = \int_H \varphi(h)\chi(h)\alpha\, dh = \sigma(\varphi)\alpha$$

where σ is the algebra representation associated with χ. By Proposition V.5.10 we have $\sigma(\varphi)\alpha \neq 0$ for some $\varphi \in C_0(G)$. Hence $\theta_f(e) \neq 0$ for a suitable f of the form $f = \varphi\alpha$.

We emphasize that Lemma 2 and its corollary hold not only for separable but for arbitrary Hilbert spaces $\mathscr{H}$. This fact will be used when the definition of induced representations will be extended to arbitrary G, H and χ. If $\mathscr{H}$ is separable then by Theorem V.7.3 χ is continuous. Hence we see that $\mathscr{K}$ contains non-zero continuous functions $\theta: G \to \mathscr{H}$ such that $\|\theta(\cdot)\|$ has compact support in X.

We let $C_0(G, \mathscr{H})$ denote the vector space of continuous functions $f: G \to \mathscr{H}$ with compact support and let

$$\mathscr{M} = \{\theta_f : f \in C_0(G, \mathscr{H})\}.$$

Then (θ_f, θ_g) is a pseudo-inner product on $\mathscr{M}$ which becomes a proper inner product if equivalent elements of $\mathscr{M}$ are identified. We use the same symbol $\mathscr{M}$ to denote the vector space of these equivalence classes. By Theorem 1 the completion of this inner product space can be regarded as a subspace of $\mathscr{K}$. If $\mathscr{H}$ is separable then one can prove that this subspace is $\mathscr{K}$ itself. Since this fact is not needed at present its proof will be given later in Section 4.

On one occasion, in the proof of the continuity of induced representations we shall use the following result due to Kraljević:

Proposition 3. *Every θ_f is right uniformly continuous on G.*

Proof. Let S be the compact support of the continuous $f: G \to \mathscr{H}$. Then for any x and y in G we have

$$\|\theta_f(x) - \theta_f(y)\|^2 = \left(\int_H \chi(h_1)^*(f(xh_1) - f(yh_1))\, dh_1, \int_H \chi(h_2)^*(f(xh_2) - f(yh_2))\, dh_2\right).$$

Since $\chi(h_1)$ and $\chi(h_2)$ are unitary operators the right hand side can be majorized by

$$\int_H \|f(xh_1) - f(yh_1)\|\, dh_1 \int_H \|f(xh_2) - f(yh_2)\|\, dh_2.$$

Hence we have

$$(1) \qquad \|\theta_f(x) - \theta_f(y)\| \leq \int_H \|f(xh) - f(yh)\|\, dh.$$

In order to estimate the measure of the set where the integrand does not vanish we proceed as follows: If $a \in G$ but $a \notin SH$ then $a^{-1} \notin HS^{-1}$ and so $a^{-1}S \cap H = \emptyset$. Hence if $a \notin SH$ then $m_H(a^{-1}S \cap H) = 0$ where m_H denotes the left Haar measure on H. If $a \in SH$ then $a^{-1} \in hS^{-1}$ for some h in H and so $a^{-1}S \subseteq hS^{-1}S$. Hence

$$m_H(a^{-1}S \cap H) \leqslant m_H(hS^{-1}S \cap H) = m_H(S^{-1}S \cap H).$$

We proved that for every a in G we have $m_H(a^{-1}S \cap H) \leqslant m_H(S^{-1}S \cap H) < \infty$. Since f has compact support it is right uniformly continuous and so given $\varepsilon > 0$ there is a neighborhood N_e of e in G such that $\|f(xh) - f(yh)\| < \varepsilon$ for every h in H provided $xy^{-1} \in N_e$. Thus by (1) we have

$$\|\theta_f(x) - \theta_f(y)\| \leqslant \varepsilon\, m_H(S^{-1}S \cap H)$$

for every x and y in G satisfying $xy^{-1} \in N_e$.

We shall now return to the definition of the induced representation when $X = G/H$ has an invariant measure μ and the Hilbert space $\mathscr{H}$ of the weakly measurable unitary representation χ of the closed subgroup H is separable. If $a \in G$ and $\theta \in \mathscr{K}$ then $x \mapsto \theta(a^{-1}x)$ defines a function on G with values in $\mathscr{H}$ which we shall denoted by $\rho(a)\theta$. Since $x \mapsto (\theta(x), \alpha)$ is measurable and the measure in question is left invariant it follows that $x \mapsto (\rho(a)\theta(x), \alpha)$ is measurable with respect to the left Haar measure of G for every α in $\mathscr{H}$. This shows that condition 1) of Theorem 1 is satisfied by $\rho(a)\theta$. Moreover we have

$$\rho(a)\theta(xh) = \theta(a^{-1}xh) = \chi(h)^*\,\theta(a^{-1}x) = \chi(h)^*\,\rho(a)\theta(x)$$

for every x, h pair such that $a^{-1}xh$ does not belong to a zero set of X. Since μ is invariant under the action of G we see that $\rho(a)\theta$ satisfies condition 2) of Theorem 1. Condition 3) is also satisfied because $\|\rho(a)\theta\| = \|\theta\|$ by the left invariance of μ. Therefore $\rho(a)\theta$ belongs to $\mathscr{K}$ and so $\theta \mapsto \rho(a)\theta$ defines a unitary operator $\rho(a)\colon \mathscr{K} \to \mathscr{K}$. Finally we note that $\rho(ab) = \rho(a)\rho(b)$ for every a and b in G because

$$\rho(ab)\theta(x) = \theta((ab)^{-1}x) = \theta(b^{-1}a^{-1}x) = \rho(b)\theta(a^{-1}x) = \rho(a)\rho(b)\theta(x).$$

We proved that the operators $\rho(a)$ $(a \in G)$ define a unitary representation of G in the Hilbert space $\mathscr{K}$.

Definition 4. *The unitary representation* $\rho = \rho^\chi = {}_G\rho^\chi\colon G \to \mathscr{U}(\mathscr{K})$ *defined by* $\rho(a)\theta(x) = \theta(a^{-1}x)$ *for a and x in G is called the representation induced on G by the unitary representation* $\chi\colon H \to \mathscr{U}(\mathscr{H})$.

Note. Although the definition is meaningful when $\mathscr{H}$ and $\mathscr{K}$ are real Hilbert spaces we tacitly suppose that $\mathscr{H}$ and $\mathscr{K}$ are Hilbert spaces over $\mathbb{C}$.

Proposition 5. *If* $\chi_k: H \to \mathscr{U}(\mathscr{H}_k)$ $(k=1,2)$ *are equivalent representations then* ${}_G\rho^{\chi_1}$ *and* ${}_G\rho^{\chi_2}$ *are also equivalent.*

Proof. Let $U: \mathscr{H}_1 \to \mathscr{H}_2$ be a unitary transformation such that $U\chi_1 = \chi_2 U$. We define the invertible linear operator $U: \mathscr{K}_1 \to \mathscr{K}_2$ by $(U\theta_1)(x) = U(\theta_1(x))$ where $x \in G$. In order to see that U maps $\mathscr{K}_1$ into $\mathscr{K}_2$ we write with x in G and h in H

$$\begin{aligned}(U\theta_1)(xh) &= U(\theta_1(xh)) = U(\chi_1(h)^* \theta_1(x)) = (U\chi_1(h)^* U^*)\, U(\theta_1(x)) \\ &= \chi_2(h)^* (U\theta_1)(x).\end{aligned}$$

Moreover $\|(U\theta_1)(x)\| = \|U(\theta_1(x))\| = \|\theta_1(x)\|$ for every $x \in G$ and so $\|U\theta_1\| = \|\theta_1\|$ proving that $U: \mathscr{K}_1 \to \mathscr{K}_2$ is a unitary map. It is surjective because if $\theta_2 \in \mathscr{K}_2$ then $x \mapsto U^{-1}\theta_2(x)$ defines a function $U^{-1}\theta_2$ in $\mathscr{K}_1$ and $U^{-1}\theta_2$ is mapped into θ_2 by U. Finally we have

$$(U\rho^{\chi_1}(a)\theta_1)(x) = U((\rho^{\chi_1}(a)\theta_1)(x)) = U(\theta_1(a^{-1}x))$$

and

$$(\rho^{\chi_2}(a)\, U\theta_1)(x) = (U\theta_1)(a^{-1}x) = U(\theta_1(a^{-1}x)).$$

Therefore $U\rho^{\chi_1} = \rho^{\chi_2} U$ and $\rho^{\chi_1} \approx \rho^{\chi_2}$.

Proposition 6. *If H is a closed subgroup of G and χ_i $(i=1,2)$ are unitary representations of H then $\rho^{\chi_1 \oplus \chi_2}$ and $\rho^{\chi_1} \oplus \rho^{\chi_2}$ are equivalent.*

Corollary. *The induced representation ρ^χ is irreducible only if χ is irreducible.*

Proof. Let $\mathscr{H}_1, \mathscr{H}_2$ and $\mathscr{H} = \mathscr{H}_1 \oplus \mathscr{H}_2$ be the Hilbert spaces of χ_1, χ_2 and $\chi_1 \oplus \chi_2$, respectively. If θ belongs to $\mathscr{K}$, the space of $\rho^{\chi_1 \oplus \chi_2}$, then $\theta(x) = (\theta(x)_1, \theta(x)_2)$ with $\theta(x)_i$ in $\mathscr{H}_i$ $(i=1,2)$. We let $\theta \mapsto (\theta_1, \theta_2)$ where $\theta_i(x) = \theta(x)_i$. One easily verifies that θ_i belongs to $\mathscr{K}_i$, the space of ρ^{χ_i}. This defines a linear map $U: \mathscr{K} \to \mathscr{K}_1 \oplus \mathscr{K}_2$ which is easily seen to be a bijective isometry. We have

$$(U\rho^{\chi_1 \oplus \chi_2}(a)\theta)(x) = (\rho^{\chi_1 \oplus \chi_2}(a)\theta(x)_1, \rho^{\chi_1 \oplus \chi_2}(a)\theta(x)_2) = (\theta(a^{-1}x)_1, \theta(a^{-1}x)_2)$$

and

$$(\rho^{\chi_1} \oplus \rho^{\chi_2})(a)\, U\theta = \rho^{\chi_1}(a) \oplus \rho^{\chi_2}(a)\, (\theta_1, \theta_2).$$

Therefore we see that U intertwines $\rho^{\chi_1 \oplus \chi_2}$ and $\rho^{\chi_1} \oplus \rho^{\chi_2}$.

If $\chi: H \to \mathscr{U}(\mathscr{H})$ is one dimensional i.e. if $\mathscr{H} = \mathbb{C}$ and χ is a group character then the induced representation ρ^χ is called *monomial*. In particular if χ is the one dimensional identity then the monomial representation ρ^χ is called a *permutation representation*. Thus the left regular representation $\lambda: G \to \mathscr{U}(L^2(G))$ is a permutation representation.

Proposition 7. *If H has finite index $[G:H]$ in G then* $\deg \rho^\chi = [G:H] \deg \chi$.

Corollary. *If ρ^χ is monomial then* $\deg \rho^\chi = [G:H]$.

Proof. Let $\mathscr{H}$ be the Hilbert space of χ, let $v = [G:H]$ and let $x_1, \ldots, x_v$ be representatives from each of the cosets xH. Then we can define the linear map $\varphi: \mathscr{K} \to \mathscr{H}^v$ by $\theta \mapsto (\theta(x_1), \ldots, \theta(x_v))$. By $\theta(xh) = \chi(h)^* \theta(x)$ the values $\theta(x_1), \ldots, \theta(x_v)$ uniquely determine θ, so φ is injective. Conversely, if $(\theta_1, \ldots, \theta_v) \in \mathscr{H}^v$ is given and if x is in G then $x = x_i h$ for a unique $i = i(x)$ $(1 \leqslant i \leqslant v)$ and $h = h(x)$ so we can define $\theta(x) = \chi(h)^* \theta_i$. The function $\theta: G \to \mathscr{H}$ defined in this manner satisfies the measurability condition 1) because $x \mapsto (\theta(x), \alpha)$ is continuous: Indeed H being a closed subgroup of finite index the cosets $x_i H$ $(i = 1, \ldots, v)$ are open sets in G. Then if y is restricted to a sufficiently small neighborhood of x then $i(y) = i(x)$ and so $h(y)^{-1} h(x) = y^{-1} x$. Since $\|\theta\|^2 = \|\theta_1\|^2 + \cdots + \|\theta_v\|^2$ we proved that $\theta \in \mathscr{K}$. It is clear that φ maps θ back to $(\theta_1, \ldots, \theta_v)$ and so φ is continuous. Since G/H is finite we have

$$\|\theta\|^2 = \int_X \|\theta(x)\|^2 \, d\xi = \|\theta(x_1)\|^2 + \cdots + \|\theta(x_v)\|^2$$

and so we see that φ is an isometry between $\mathscr{K}$ and $\mathscr{H}^v$.

Proposition 8. *The left regular representation λ of a locally compact group is the representation induced on G by the one dimensional identity representation of $H = \{e\}$.*

Proof. More generally let H be a closed subgroup of G such that G/H has an invariant measure μ. Let $\mathscr{H}$ be a Hilbert space and let $\chi(h) = I$ for every $h \in H$ where $I: \mathscr{H} \to \mathscr{H}$ is the identity operator. We construct $\mathscr{K} = \mathscr{K}(G, H, \mathscr{H}) = L^2(G/H, \mathscr{H})$ by using μ and let $\rho^\chi = {}_G\rho^\chi$. Then ρ^χ is a generalization of the left regular representation of G. If $H = \{e\}$ then ρ^χ is the left regular representation of G in $L^2(G, \mathscr{H})$ and by choosing $\mathscr{H} = \mathbb{C}$ we obtain the ordinary left regular representation λ of G.

3. Tensor Products

First we review the concept of an *algebraic tensor product* of vector spaces X, Y over the same field of scalars F. One is interested in a vector space $X \otimes Y$ containing among other elements the symbols $x \otimes y$ such that

$$(x_1 + x_2) \otimes y = x_1 \otimes y + x_2 \otimes y$$
$$x \otimes (y_1 + y_2) = x \otimes y_1 + x \otimes y_2$$
$$\lambda(x \otimes y) = \lambda x \otimes y = x \otimes \lambda y$$

where $\lambda \in F$. Moreover one would like that $X \otimes Y$ contain no superfluous elements besides the tensor products $x \otimes y$ and sums of these products.

One can give a precise formulation of this extremal property and prove the existence and essential uniqueness of $X \otimes Y$: One starts with the Cartesian product $X \times Y$ consisting of the ordered pairs (x, y) whose addition is defined by the rule $(x_1, y_1) + (x_2, y_2) = (x_1 + x_2, y_1 + y_2)$ and $\lambda(x, y) = (\lambda x, \lambda y)$. A *product* of X and Y, in this order, is defined to be a pair (V, φ) where V is a vector space over F, $\varphi: X \times Y \to V$ is a bilinear map and the elements $\varphi(x, y)$ $(x \in X, y \in Y)$ generate V. The image vector $\varphi(x, y)$ is to be interpreted as the product $x \otimes y$. If (V, φ) and (W, ψ) are products and if there is a linear map $f: V \to W$ such that $\psi = f\varphi$ then we say that ψ can be factored through V and write $(V, \varphi) \geqslant (W, \psi)$. Two products (V, φ) and (W, ψ) are called *equivalent* if there is an isomorphism $f: V \to W$ such that $\psi = f\varphi$. One can prove that $(V, \varphi) \leqslant (W, \psi)$ and $(W, \psi) \leqslant (V, \varphi)$ together imply that (V, φ) and (W, ψ) are equivalent products. The proof is a simple application of the hypothesis that $\varphi(x, y)(x \in X, y \in Y)$ and $\psi(x, y)(x \in X, y \in Y)$ are generating V and W, respectively. We shall use the symbol $\cong$ to denote the equivalence of products.

The fundamental result of the theory states the existence of maximal elements $(X \otimes Y, \varphi)$ such that $(X \otimes Y, \varphi) \leqslant (W, \psi)$ for every product (W, ψ). In this case the map φ is usually denoted by the symbol $\otimes$, its value at (x, y) is $x \otimes y$ and one writes $(X \otimes Y, \otimes)$ or $X \otimes Y$ instead of $(X \otimes Y, \varphi)$. It follows that any two maximal elements are equivalent. This equivalence class and/or its individual elements $(X \otimes Y, \otimes)$ are called the algebraic tensor product of X and Y. The image of the pair (x, y) under the map $\otimes: X \times Y \to X \otimes Y$ is denoted by $x \otimes y$.

In one process by which one can construct a maximal element $(X \otimes Y, \otimes)$ one starts by forming the vector space $\mathscr{V}$ of all formal finite sums

$$v = \sum c_i(x_i, y_i)$$

where $c_i \in F$, $x_i \in X$ and $y_i \in Y$. More precisely $\mathscr{V}$ is the vector space of all functions $v: X \times Y \to F$ such that $v(x,y) \neq 0$ only for finitely many pairs (x,y). The space $\mathscr{V}$ contains the subspace $\mathscr{W}$ generated by the functions associated with the formal sums

$$\begin{gathered}(x_1+x_2,y)-(x_1,y)-(x_2,y)\\(x,y_1+y_2)-(x,y_1)-(x,y_2)\\(x,\lambda y)-\lambda(x,y)\\(\lambda x,y)-\lambda(x,y)\,.\end{gathered}$$

The quotient space $\mathscr{V}/\mathscr{W}$ together with the map $X \times Y \to \mathscr{V}/\mathscr{W}$ defined by $(x,y) \mapsto (x,y)+\mathscr{W} = x \otimes y$ is a product $(X \otimes Y, \otimes)$. The maximality is proved by introducing a linear map $f: \mathscr{V} \to V$ where $f(c(x,y)) = c\,\varphi(x,y)$ and (V,φ) is the product which we wish to compare with $(X \otimes Y, \otimes)$. Since $f(\mathscr{W})=0$ one can interpret f as a linear map $f: \mathscr{V}/\mathscr{W} \to V$. Then we have $f(x \otimes y) = \varphi(x,y)$ proving that φ factors through $\mathscr{V}/\mathscr{W}$ and so $(\mathscr{V}/\mathscr{W}, \otimes) \geq (V,\varphi)$. Therefore $(X \otimes Y, \otimes) = (\mathscr{V}/\mathscr{W}, \otimes)$ is a tensor product of X and Y.

One proves that $X \otimes Y \cong Y \otimes X$ and $(X \otimes Y) \otimes Z \cong X \otimes (Y \otimes Z)$ for any X, Y, Z over F. One of the simplest constructions concerns the tensor product of X with F interpreted as a vector space over itself: For the underlying vector space of $X \otimes F$ one can choose X itself and define $x \otimes \lambda = \lambda x$ for any $x \in X$ and $\lambda \in F$. It follows that $F \otimes F \cong F$ and in general $F \otimes \cdots \otimes F \cong F$ with $\lambda_1 \otimes \cdots \otimes \lambda_n = \lambda_1 \cdots \lambda_n$.

If X is a vector space over F one defines $X_o^o = F \otimes F$ and $X_o^n = X \otimes X_o^{n-1}$ for $n=1,2,\ldots$ so that $X_o^1 = X$ as a vector space. Using $X_o^m \otimes X_o^n \cong X_o^{m+n}$ one can define an algebra structure on the direct sum of the vector spaces of the products X_o^n $(n=0,1,\ldots)$. This algebra is called the *contravariant tensor algebra* of X and is denoted by some symbol like $T_o(X)$. The *covariant tensor algebra* $T^o(X)$ is defined similarly but one starts with $X_n^o = X^* \otimes X_{n-1}^o$ where X^* is the dual of X.

Let X_1, X_2, Y_1, Y_2 be vector spaces over F and let $A: X_1 \to X_2$ and $B: Y_1 \to Y_2$ be linear operators. A product $(X_2 \otimes Y_2, \varphi)$ is defined by the bilinear map $\varphi(x_1,y_1) = Ax_1 \otimes By_1$ mapping $X_1 \otimes Y_1$ into $X_2 \otimes Y_2$. Let $A \otimes B$ denote the linear map from $X_1 \otimes Y_1$ into $X_2 \otimes Y_2$ which factors φ through $X_1 \otimes Y_1$. This linear operator $A \otimes B$ is called the *tensor product* of A and B.

Now we turn our attention to the algebraic tensor product of Hilbert spaces. If X, Y are complex Hilbert spaces then by an *antilinear map* T we understand a continuous additive operator such that $T(\lambda y) = \bar{\lambda} Ty$. A simple example is the continuous map $x \otimes y: Y \to X$ where $x \in X$,

$y \in Y$ and $(x \otimes y)z = (y, z)x$ for $z \in Y$. Any linear combination of these gives an antilinear map of finite rank.

Theorem 1. *An algebraic tensor product of the Hilbert spaces X, Y is the vector space $X \otimes Y$ of all antilinear maps $T: Y \to X$ of finite rank together with the product operation $(x, y) \mapsto x \otimes y$ where $x \otimes y$ maps z into $(y, z)x$.*

Proof. Let $T: Y \to X$ be an antilinear map of finite rank n. We show that T can be expressed in the form

$$T = (x_1 \otimes y_1) + \cdots + (x_n \otimes y_n)$$

where $x_1, \ldots, x_n \in X$ and $y_1, \ldots, y_n \in Y$. Let the non-zero vectors $x_i = T u_i$ $(i = 1, \ldots, n)$ generate the range of T. For every $i = 1, \ldots, n$ let Y_i be the non-zero subspace

$$Y_i = \{u : u \in Y \text{ and } Tu = \lambda_u x_i\}$$

and define $f_i: Y_i \to \mathbb{C}$ by $f_i(u) = \lambda_u$. Since f_i is a continuous antilinear functional there is a y_i in Y_i such that $f_i(u) = \lambda_u = (y_i, u)$ for every $u \in Y$. Now if z is an arbitrary element of Y and if $Tz = \lambda_1 x_1 + \cdots + \lambda_n x_n$ then we can choose $z_i \in Y_i$ such that $T z_i = \lambda_i x_i$ $(i = 1, \ldots, n)$. We have $f_i(z_i) = \lambda_i$ and

$$T z_i = \lambda_i x_i = f_i(z_i) x_i = (y_i, z_i) x_i = (x_i \otimes y_i) z_i .$$

Moreover $Tz = Tz_1 + \cdots + Tz_n$. Therefore we proved that

$$T = (x_1 \otimes y_1) + \cdots + (x_n \otimes y_n)$$

and so $(X \otimes Y, \otimes)$ is a product. We note that the above decomposition of T is not unique.

We complete the proof by showing that $(X \otimes Y, \otimes)$ and $(\mathscr{V}/\mathscr{W}, \otimes)$ are isomorphic products. First we define a homomorphism $\mathscr{V} \to X \otimes Y$ by mapping $\sum c_i(x_i, y_i)$ into the antilinear operator $\sum c_i(x_i \otimes y_i)$. The kernel contains all formal sums of the type $(x_1 + x_2, y) - (x_1, y) - (x_2, y)$, etc. and so the kernel contains $\mathscr{W}$. We prove that if a sum $\sum_{i=1}^{m} c_i(x_i, y_i)$ belongs to the kernel then it is in $\mathscr{W}$. This is certainly true for sums of length 0 and 1. We suppose that it holds for all sums with $m < n$ and we prove that it is true also for formal sums of length n.

Let $\sum_{i=1}^{n} c_i(x_i, y_i)$ be such that

$$\sum_{i=1}^{n} c_i(y_i, z) x_i = 0 \tag{1}$$

for every $z \in Y$. We suppose that $c_1, \ldots, c_n \neq 0$. If the x_i's are linearly dependent, say $x_n = \lambda_1 x_1 + \cdots + \lambda_{n-1} x_{n-1}$ then by

$$(x_n, y_n) - \sum_{i=1}^{n-1} (\lambda_i x, y_n) \in \mathscr{W}$$
$$(\lambda_i x_i, y_n) - (x_i, \lambda_i y_n) \in \mathscr{W}$$
$$c_i(x_i, y_i) - (x_i, c_i y_i) \in \mathscr{W}$$

we obtain

$$\sum_{i=1}^{n} c_i(x_i, y_i) = \sum_{i=1}^{n-1} (x_i, c_i y_i + c_n \lambda_i y_n) + w$$

where $w \in \mathscr{W}$ and so w belongs to the kernel. Using (1) it is easy to see that the sum on the right hand side belongs to the kernel. Since its length is less than n by the induction hypothesis it belongs to $\mathscr{W}$. Therefore the sum on the left hand side is in $\mathscr{W}$. The other possibility is that the x_i's are linearly independent. Then from (1) we obtain $(y_i, z) = 0$ for every $z \in Y$ and so $y_i = 0 \, (i = 1, \ldots, n)$. Again we see that $\sum c_i(x_i, y_i)$ belongs to $\mathscr{W}$.

We proved that the kernel of the map $\mathscr{V} \to X \otimes Y$ is $\mathscr{W}$ and so $\mathscr{V}/\mathscr{W} \to X \otimes Y$ is an isomorphism between the products $(\mathscr{V}/\mathscr{W}, \otimes)$ and $(X \otimes Y, \otimes)$.

In full analogy to the HS operators one can define the HS norm $\|T\|$ for antilinear maps $T\colon Y \to X$ and for any pair of antilinear maps satisfying $\|S\|, \|T\| < \infty$ one can also introduce the analogue (S, T) of the HS inner product. Namely if $y_i (i \in \mathscr{I})$ is a MONS in Y then

$$\|T\|^2 = \sum \|T y_i\|^2$$

is independent of the choice of the MONS. Similarly, if $\|S\|, \|T\| < \infty$ then

$$(S, T) = \sum (S y_i, T y_i)$$

is an absolutely convergent series and its sum is independent of the MONS $y_i \, (i \in \mathscr{I})$.

The antilinear HS operators form a Hilbert space under the inner product (S, T). The set of antilinear operators of finite rank is dense in this Hilbert space. For if $\sum \|T y_i\|^2 < \infty$ then there is an index n such that $\sum_{n+1}^{\infty} \|T y_i\|^2 < \varepsilon$. Hence if $S\colon Y \to X$ is defined to be T on the span of $y_1, \ldots, y_n$ and zero on its orthogonal complement then S is of finite rank and $\|T - S\|^2 < \varepsilon$.

On the algebraic tensor product $X \otimes Y$ of the Hilbert spaces X, Y one can introduce an inner product such that the topology determined by it is the strongest topology on $X \otimes Y$ which makes the functions $(x, y) \mapsto x \otimes y$ continuous. Namely if the elements of $X \otimes Y$ are antilinear operators $T: Y \to X$ of finite rank then the HS inner product (S, T) satisfies the requirements. The *topological tensor product* of the Hilbert spaces X and Y is defined to be the completion of the algebraic tensor product with respect to the norm derived from this inner product. Since the antilinear HS operators form a Hilbert space and the operators of finite rank are dense in it we have:

Theorem 2. *The topological tensor product $X \otimes Y$ can be interpreted as the space of antilinear HS operators $T: Y \to X$ with the inner product (S, T).*

Corollary. *The topological tensor product of two Hilbert spaces is a Hilbert space.*

Since $\|x \otimes y\| = \|x\| \cdot \|y\|$ the operation $(x, y) \mapsto x \otimes y$ is obviously continuous with respect to the topology derived from the HS inner product. We shall not prove the extremal property of this product which was mentioned earlier. We agree that from here on the tensor product of two Hilbert spaces will always mean the Hilbert space of antilinear HS operators.

Note. One can give the following alternate description of the tensor product $\mathscr{H}_1 \otimes \mathscr{H}_2$: We consider formal expressions

$$\sum_{k=1}^{p} \lambda_k [x_1^k, x_2^k] \quad \text{where } \lambda_k \in \mathbb{C}, \quad x_1^k \in \mathscr{H}_1, \quad x_2^k \in \mathscr{H}_2$$

and add them formally and multiply by scalars in the same way. The vector space obtained in this manner can be identified with the space $\mathscr{V}$ considered in the beginning of this section. We introduce the inner product

$$\left(\sum \lambda_k [x_1^k, x_2^k], \sum \mu_l [y_1^l, y_2^l]\right) = \sum \lambda_k \bar{\mu}_l (x_1^k, y_1^l)(x_2^k, y_2^l).$$

This definition corresponds to the rule

$$(x_1 \otimes x_2, y_1 \otimes y_2) = (x_1, y_1)(x_2, y_2).$$

We identify all expressions equal modulo $\mathscr{N}$ where $\mathscr{N}$ is the set of formal expressions with zero pseudo norm. It is easy to see that the subspace $\mathscr{W}$ considered in the beginning is contained in $\mathscr{N}$. The advantage of this method is that it works in more general situations than Hilbert spaces. For instance in the defintition of the tensor product of two Banach spaces one can start with expressions $\sum \lambda_k[x_k, x_k^*]$ where $x_k \in \mathcal{X}_1$ and $x_k^* \in \mathcal{X}_2^*$.

Proposition 3. *If x_i $(i \in \mathcal{I})$ is a MONS in X and y_j $(j \in \mathcal{J})$ is a MONS in Y then their tensor products $x_i \otimes y_j$ $(i \in \mathcal{I}, j \in \mathcal{J})$ form a MONS in $X \otimes Y$.*

Corollary. *If the Hilbert spaces X, Y are finite dimensional then* $\dim(X \otimes Y) = \dim X \dim Y$.

Using any MONS in Y we find that the HS inner product of $x_1 \otimes y_1$ and $x_2 \otimes y_2$ is $(x_1, x_2)(y_1, y_2)$. From this it follows that $x_i \otimes y_j$ $(i \in \mathcal{I}, j \in \mathcal{J})$ is an *ONS* in $X \otimes Y$. Its maximality can be proved by showing its completeness: A simple computation or direct verification shows that for any $x \in X$ and $y \in Y$ we have

$$x \otimes y = \sum_{i,k} (x, x_i)(y, y_k)(x_i \otimes y_k) . \tag{2}$$

Now suppose that an antilinear operator $T: Y \to X$ is orthogonal to every $x_i \otimes y_j$. Then by (2) T is orthogonal to $x \otimes y$ for every $x \in X$, $y \in Y$. In the proof of Theorem 1 we saw that every antilinear operator of finite rank is a linear combination of operators of the type $x \otimes y$. Thus T is orthogonal to every operator of finite rank. The latter form a dense set in $X \otimes Y$, so T is orthogonal to every element of $X \otimes Y$ and so $T=0$.

Proposition 4. *For any two Hilbert spaces X and Y we have* $X \otimes Y \cong Y \otimes X$.

Note. The $\cong$ sign means that $(X \otimes Y, \otimes)$ and $(Y \otimes X, \otimes)$ are isometrically isomorphic topological tensor products.

Proof. Associate with the antilinear operator $T: Y \to X$ the antilinear operator $T^*: X \to Y$ defined by $(y, T^*x) = (x, Ty)$ where y varies over Y. Then $T \to T^*$ gives an isometric isomorphism of $X \otimes Y$ onto $Y \otimes X$ such that $x \otimes y \mapsto y \otimes x$. Therefore $(X \otimes Y, \otimes)$ and $(Y \otimes X, \otimes)$ are isomorphic topological tensor products.

Proposition 5. *For any three Hilbert spaces X, Y and Z over the same field of scalars there is a unique unitary isomorphism $U: X \otimes (Y \otimes Z) \to (X \otimes Y) \otimes Z$ such that $U(x \otimes (y \otimes z)) = (x \otimes y) \otimes z$ for any $x \in X$, $y \in Y$ and $z \in Z$.*

Proof. Let T belong to $X \otimes (Y \otimes Z)$ so that $T: Y \otimes Z \to X$. In order to define $U(T): Z \to X \otimes Y$ we let $U(T)z$ be the antilinear map $U(T)z: Y \to X$ given by $y \to T(y \otimes z)$. It is clear that $T \to U(T)$ defines a linear map U

of $X \otimes (Y \otimes Z)$ into $(X \otimes Y) \otimes Z$. The map U is surjective because $V: Z \to (X \otimes Y)$ is the image of the map $T: Y \otimes Z \to X$ defined by $T(y \otimes z) = Vz(y)$ for $y \in Y$, $z \in Z$. The fact that U is one-to-one will follow immediately if we prove that U preserves the HS norm i.e. it is unitary. Now if $y_i\ (i \in \mathscr{I})$ and $z_j\ (j \in \mathscr{J})$ are MONS in Y and Z then by Proposition 3

$$\|T\|^2 = \sum_{i,j} \|T(y_i \otimes z_j)\|^2 = \sum_{i,j} \|U(T) z_j(y_i)\|^2 .$$

We have $\|U(T)\|^2 = \sum_j \|U(T) z_j\|^2$ where $\|U(T) z_j\|^2 = \sum_i \|U(T) z_j(y_i)\|^2$. Hence we see that $\|T\|^2 = \|U(T)\|^2$.

The proof of the additional property which uniquely determines the map is a routine computation: One shows that both $U(x \otimes (y \otimes z))\zeta(\eta)$ and $((x \otimes y) \otimes z)\zeta(\eta)$ are equal to $(y, \eta)(z, \zeta)x$ for every ζ, η and x, y, z.

Proposition 6. *If X_1, X_2 are measure spaces then $L^2(X_1) \otimes L^2(X_2) \cong L^2(X_1 \times X_2)$ where $f_1 \otimes f_2$ corresponds to $(x_1, x_2) \mapsto f_1(x_1) f_2(x_2)$.*

Proof. In Section II.4 we proved that $T: L^2(X) \to L^2(X)$ is a HS operator if and only if it can be represented by a kernel belonging to $L^2(X \times X)$. By essentially the same reasoning we can prove that $T: L^2(X_2) \to L^2(X_1)$ is an antilinear HS operator if and only if there exists a kernel K_T in $L^2(X_1 \times X_2)$ such that

$$(Tg)(x_1) = \int_G K_T(x_1, x_2) \overline{g(x_2)}\, dx_2 .$$

Therefore we have a natural one-to-one correspondence between operators T which make up $L^2(X_1) \otimes L^2(X_2)$ and the elements K of $L^2(X_1 \times X_2)$. In the first let $f_1 \otimes f_2$ be the antilinear map $(f_1 \otimes f_2) g = (f_2, g) f_1$ and in the second we let $f_1 \otimes f_2$ denote the kernel $f_1(x) f_2(y)$. Then it is clear that our one-to-one correspondence commutes with the operation $\otimes$. Moreover one can easily show that the linear combinations of the products $f_1 \otimes f_2$ are dense in $L^2(X_1 \times X_2)$. For instance we can apply the following:

Proposition 7. *Let X, Y be measure spaces and let $f_i\ (i \in \mathscr{I})$ and $g_j\ (j \in \mathscr{J})$ be ONS in $L^2(X)$ and $L^2(Y)$, respectively. Then the functions $(x, y) \mapsto f_i(x) g_j(y)$ form an ONS in $L^2(X \times Y)$. If $f_i\ (i \in \mathscr{I})$ and $g_j\ (j \in \mathscr{J})$ are maximal then so is the system of the products $f_i g_j\ (i \in \mathscr{I}, j \in \mathscr{J})$.*

Proof. The proof of the orthogonality is straight forward and it can be omitted. The maximality is proved by showing that Parseval's formula holds: Given h in $L^2(X\times Y)$ by the maximality of the g_j's we have

$$\begin{aligned}\sum_{i,j}|a_{ij}|^2 &= \Big|\sum_{i,j}\int_X\int_Y f_i(x)g_j(y)h(x,y)\,dx\,dy\Big|^2\\ &= \sum_i\int_X\int_X f_i(x)\overline{f_i(\xi)}\Big(\sum_j\int_Y\int_Y g_j(y)\overline{g_j(\eta)}h(x,y)\overline{h(\xi,\eta)}\,dy\,d\eta\Big)\,dx\,d\xi\\ &= \sum_i\int_X\int_X f_i(x)\overline{f_i(\xi)}\Big(\int_Y h(x,y)\overline{h(\xi,y)}\,dy\Big)\,dx\,d\xi\,.\end{aligned}$$

Hence by the maximality of the f_i's we obtain

$$\begin{aligned}\sum_{i,j}|a_{ij}|^2 &= \sum_i\int_Y\Big(\int_X f_i(x)h(x,y)\,dx\Big)\Big(\overline{\int_X f_i(\xi)h(\xi,y)\,d\xi}\Big)\,dy\\ &= \int_Y\int_X h(x,y)\overline{h(x,y)}\,dx\,dy = \|h\|_2^2\,.\end{aligned}$$

Let $\mathcal{H}_k$ $(k=1,2)$ be Hilbert spaces and let $A_k\colon\mathcal{H}_k\to\mathcal{H}_k$ be continuous linear operators. If $T\colon\mathcal{H}_2\to\mathcal{H}_1$ is an antilinear HS operator then $A_1TA_2^*\colon\mathcal{H}_2\to\mathcal{H}_1$ is antilinear and by the antilinear version of Theorem II.4.2 it is a HS operator. Thus we can define the linear operator

$$A_1\otimes A_2\colon\mathcal{H}_1\otimes\mathcal{H}_2\to\mathcal{H}_1\otimes\mathcal{H}_2$$

by the rule $A_1\otimes A_2(T)=A_1TA_2^*$.

Proposition 8. *The operator* $A_1\otimes A_2$ *satisfies* $|A_1\otimes A_2|=|A_1|\cdot|A_2|$ *and* $\|A_1\otimes A_2\|=\|A_1\|\cdot\|A_2\|$ *provided* A_1, A_2 *are in HS class.*

Proof. We have

$$\|(A_1\otimes A_2)(T)\|=\|A_1TA_2^*\|\leqslant|A_1|\cdot\|TA_2^*\|\leqslant|A_1|\cdot|A_2^*|\cdot\|T\|$$

and so $|A_1\otimes A_2|\leqslant|A_1|\cdot|A_2|$. On the other hand one can easily see that

$$(A_1\otimes A_2)(x_1\otimes x_2)=A_1x_1\otimes A_2x_2\,.$$

Therefore

$$\|(A_1\otimes A_2)(x_1\otimes x_2)\|=\|A_1x_1\otimes A_2x_2\|=\|A_1x_1\|\cdot\|A_2x_2\|\,.$$

For suitable choices of x_1 and x_2 the right hand side exceeds $|A_1|\cdot|A_2|\cdot\|x_1\|\cdot\|x_2\|-\varepsilon$. Since $\|x_1\|\cdot\|x_2\|=\|x_1\otimes x_2\|$ we conclude that $|A_1\otimes A_2|=|A_1|\cdot|A_2|$. The HS norm is computed by using Proposition 3:

$$\sum_{i,j}\|(A_1\otimes A_2)(x_i\otimes y_j)\|^2=\sum_{i,j}\|A_1x_i\|^2\,\|A_2y_j\|^2=\|A_1\|^2\,\|A_2\|^2\,.$$

Proposition 9. *For any* $A_k, B_k\in\mathcal{L}(\mathcal{H}_k)$ $(k=1,2)$ *we have*

$$(A_1\otimes A_2)(B_1\otimes B_2)=A_1B_1\otimes A_2B_2\,.$$

Corollary. *If* A_1, A_2 *are invertible, then* $A_1 \otimes A_2$ *is invertible and* $(A_1 \otimes A_2)^{-1} = A_1^{-1} \otimes A_2^{-1}$.

Proof. For any T we have

$$(A_1 \otimes A_2)(B_1 \otimes B_2)T = A_1(B_1 T B_2^*) A_2^* = (A_1 B_1) T (A_2 B_2)^* \\ = (A_1 B_1 \otimes A_2 B_2)(T).$$

Proposition 10. *If* $A_k \in \mathscr{L}(\mathscr{H}_k)$ $(k=1,2)$ *then* $(A_1 \otimes A_2)^* = A_1^* \otimes A_2^*$.

Proof. Let S, T be antilinear operators mapping $\mathscr{H}_2$ into $\mathscr{H}_1$. Then

$$(S,(A_1 \otimes A_2)^* T) = ((A_1 \otimes A_2)S, T) = (A_1 S A_2^*, T) = \operatorname{tr} T^* A_1 S A_2^* \\ = \overline{\operatorname{tr} A_2 S^* A_1^* T}$$

Thus by Theorem II.4.2 and Proposition II.5.12

$$(S,(A_1 \otimes A_2)^* T) = \overline{\operatorname{tr} A_1^* T A_2 S^*} = \overline{(A_2 S^*, T^* A_1)} = \overline{(S^*, A_2^* T^* A_1)} \\ = (S,(A_1^* \otimes A_2^*) T).$$

As a corollary of the last two propositions we have:

Proposition 11. *If* $A_k \in \mathscr{L}(\mathscr{H}_k)$ $(k=1,2)$ *are unitary operators then so is* $A_1 \otimes A_2$.

Proof. If $A_k^{-1} = A_k^*$ $(k=1,2)$ then

$$(A_1 \otimes A_2)^{-1} = A_1^{-1} \otimes A_2^{-1} = A_1^* \otimes A_2^* = (A_1 \otimes A_2)^*.$$

Note. Although it will not be needed we mention that the products $A_1 \otimes A_2$ can be defined in the more general situation when $A_k: \mathscr{H}_k \to \mathscr{K}_k$ $(k=1,2)$. Then $A_1 \otimes A_2: \mathscr{H}_1 \otimes \mathscr{H}_2 \to \mathscr{K}_1 \otimes \mathscr{K}_2$ and it is defined by $(A_1 \otimes A_2)T = A_1 T A_2^*$ where $T: \mathscr{H}_2 \to \mathscr{H}_1$ is an antilinear HS operator. Again $A_1 \otimes A_2$ is a continuous linear operator with $|A_1 \otimes A_2| = |A_1| \cdot |A_2|$.

Theorem 12. *If* $\rho_k: G_k \to \mathscr{U}(\mathscr{H}_k)$ $(k=1,2)$ *are unitary group representations then*

$$(\rho_1 \otimes \rho_2)(x_1, x_2) = \rho_1(x_1) \otimes \rho_2(x)$$

defines a unitary representation $\rho_1 \otimes \rho_2: G_1 \times G_2 \to \mathscr{U}(\mathscr{H}_1 \otimes \mathscr{H}_2)$.

Proof. By the last proposition the operators $(\rho_1 \otimes \rho_2)(x_1, x_2)$ are all unitary. The representation property is an easy exercise:

$$(\rho_1\otimes\rho_2)(x_1y_1,x_2y_2)=\rho_1(x_1y_1)\otimes\rho_2(x_2y_2)=\rho_1(x_1)\rho_1(y_1)\otimes\rho_2(x_2)\rho_2(y_2)$$
$$=(\rho_1(x_1)\otimes\rho_2(x_2))(\rho_1(y_1)\otimes\rho_2(y_2))$$
$$=(\rho_1\otimes\rho_2)(x_1,x_2)(\rho_1\otimes\rho_2(y_1,y_2).$$

We now add that one can define by the same method a second product $\mathscr{H}_1\bar{\otimes}\mathscr{H}_2$ which will be extensively used in Sections 4 and 5. Namely instead of antilinear HS operators one uses linear HS operators as elements of $\mathscr{H}_1\bar{\otimes}\mathscr{H}_2$. The fundamental operator $x\bar{\otimes}y$ is defined by $(x\bar{\otimes}y)z=(z,y)x$ and by choosing a MONS in $\mathscr{H}_2$ one proves that

$$(x_1\bar{\otimes}y_1,x_2\bar{\otimes}y_2)=(x_1,x_2)(y_2,y_1).$$

Therefore one can define a second product $\rho_1\bar{\otimes}\rho_2$ of the unitary group representations $\rho_k\colon G_k\to\mathscr{U}(\mathscr{H}_k)$ $(k=1,2)$ which will have somewhat different properties from the original $\rho_1\otimes\rho_2$.

The product $\rho_1\bar{\otimes}\rho_2$ however is not a genuinely new concept. For if $\mathscr{H}$ is a Hilbert space then we can introduce the *conjugate Hilbert space* or *complex conjugate* $\bar{\mathscr{H}}$ as follows: $\bar{\mathscr{H}}$ has the same additive structure as $\mathscr{H}$, the scalar multiple $\lambda\cdot x$ in $\bar{\mathscr{H}}$ is $\lambda\cdot x=\bar{\lambda}x$ where $\bar{\lambda}x$ denotes the multiple of x by $\bar{\lambda}$ in $\mathscr{H}$. The inner product of x and y in $\bar{\mathscr{H}}$ is (y,x) where $(\cdot,\cdot)$ denotes the inner product in $\mathscr{H}$. Now if $\mathscr{H}_1$, $\mathscr{H}_2$ are Hilbert spaces and if $A\colon\mathscr{H}_1\to\mathscr{H}_2$ is a linear operator, then the same function A can be viewed as an antilinear operator $A\colon\bar{\mathscr{H}}_1\to\mathscr{H}_2$. Moreover every linear operator $T\colon\mathscr{H}\to\mathscr{H}$ can be considered as a linear operator from $\bar{\mathscr{H}}$ to $\bar{\mathscr{H}}$ in which case it will be denoted by $\bar{T}$ instead of T. One also sees that the same function T^* is the adjoint of T in both $\mathscr{H}$ and in $\bar{\mathscr{H}}$ because $(Tx,y)=(x,T^*y)$ is the same as $(y,Tx)=(T^*y,x)$. From all this it follows that $\rho_1\bar{\otimes}\rho_2$ is the tensor product of the representations $\rho_1\colon G_1\to\mathscr{U}(\mathscr{H}_1)$ and $\bar{\rho}_2\colon G_2\to\mathscr{U}(\bar{\mathscr{H}}_2)$ where $\bar{\rho}_2(x_2)=\overline{\rho_2(x_2)}$ is the same function as $\rho_2(x_2)$. In symbols we have $\rho_1\bar{\otimes}\rho_2=\rho_1\otimes\bar{\rho}_2$. The representation $\bar{\rho}\colon G\to\mathscr{U}(\bar{\mathscr{H}})$ obtained from $\rho\colon G\to\mathscr{U}(\mathscr{H})$ is called the *conjugate representation* associated with ρ.

Lemma 13. *Let $\mathscr{H}$, $\mathscr{H}_2$ be the Hilbert spaces, let β_j $(j\in\mathscr{J})$ be a MONS in $\mathscr{H}_2$ and let T belong to $\mathscr{H}\otimes\mathscr{H}_2$. Then $T=\sum T\beta_j\otimes\beta_j$ where $\sum\|T\beta_j\|^2=\|T\|^2$. Conversely, if $\sum\|t_j\|^2<\infty$ with t_j in $\mathscr{H}$ then $T=\sum t_j\otimes\beta_j$ exists and belongs to $\mathscr{H}\otimes\mathscr{H}_2$.*

Note. A similar result holds with $\bar{\otimes}$ instead of $\otimes$.

Proof. It is obvious from the definition of $t\otimes\beta$ that the value of $\sum T\beta_j\otimes\beta_j$ at β_k is $T\beta_k$. Since T and the operator given by the series coincide on the MONS β_j $(j\in\mathscr{J})$ they are identical. For the second half of the lemma notice that

$$\|(\sum t_j \otimes \beta_j)\beta_k\|^2 = \|t_k\|^2$$

where the summation can be taken on any subset $\mathcal{S}$ of $\mathcal{J}$. Thus the series $\sum t_j \otimes \beta_j$ is convergent in the topology determined by the Hilbert-Schmidt norm. By Theorem II.4.3 the sum is an antilinear HS operator.

Proposition 14. *Let $\pi: G \to \mathcal{U}(\mathcal{H})$ be an irreducible representation of the group G with identity e and let $i_k: \{e\} \to \mathcal{U}(\mathcal{H}_k)$ $(k=1,2)$ be the trivial representations. Then we have $(\pi \otimes i_1, \pi \otimes i_2) = I \otimes \mathcal{L}(\mathcal{H}_1, \mathcal{H}_2)$.*

Proof. Let α_i $(i \in \mathcal{I})$ and β_j $(j \in \mathcal{J})$ be MONS in $\mathcal{H}_1$ and $\mathcal{H}_2$, respectively. Let us first suppose that T belongs to $(\pi \otimes i_1, \pi \otimes i_2)$. Since $T: \mathcal{H} \otimes \mathcal{H}_1 \to \mathcal{H} \otimes \mathcal{H}_2$ it follows that $T(h \otimes \alpha_i)$ belongs to $\mathcal{H} \otimes \mathcal{H}_2$. Thus by the last lemma for any $h \in \mathcal{H}$ we have

$$T(h \otimes \alpha_i) = \sum T_{ij} h \otimes \beta_j \tag{3}$$

where $T_{ij} h = (T(h \otimes \alpha_i))\beta_j$. Hence T_{ij} is a continuous linear operator $T_{ij}: \mathcal{H} \to \mathcal{H}$ and $|T_{ij}| \leqslant |T|$. By hypothesis we have $T(\pi \otimes i_1) = (\pi \otimes i_2)T$. Since

$$T(\pi \otimes i_1)(x,e)(h \otimes \alpha_i) = T(\pi(x) \otimes i_1(e))(h \otimes \alpha_i) = T(\pi(x)h \otimes \alpha_i)$$

and

$$(\pi \otimes i_2(x,e))\,T(h \otimes \alpha_i) = (\pi(x) \otimes i_2(e))\,T(h \otimes \alpha_i) = \pi(x)\,T(h \otimes \alpha_i)$$

we obtain

$$\begin{aligned} T_{ij}\pi(x)h &= (T(\pi(x)h \otimes \alpha_i))\beta_j = (\pi \otimes i_2)(x,e)(h \otimes \alpha_i)\beta_j \\ &= \pi(x)\,T(h \otimes \alpha_i)\beta_j = \pi(x)\,T_{ij}h. \end{aligned}$$

Therefore we proved that T_{ij} is in (π, π) and so π being irreducible by Proposition IV.2.4 we have $T_{ij} = t_{ij} I$ for some $t_{ij} \in \mathbb{C}$.

Let h be any element in $\mathcal{H}$ such that $(h,h)=1$. Then

$$|t_{ij}|^2 = (T_{ij}h, T_{ij}h) = (T(h \otimes \alpha_i)\beta_j, T(h \otimes \alpha_i)\beta_j)$$

where $T(h \otimes \alpha_i): \mathcal{H}_2 \to \mathcal{H}$ is a HS operator. Therefore

$$\sum_j |t_{ij}|^2 = \sum_j \|T(h \otimes \alpha_i)\beta_j\|^2 = \|T(h \otimes \alpha_i)\|^2 < \infty.$$

Now we can define the linear operator $S: \mathcal{H}_1 \to \mathcal{H}_2$ by $S\alpha_i = \sum t_{ij}\beta_j$. Then one can easily see that $S^*\beta_k = \sum \overline{t_{jk}}\,\alpha_j$ and so

$$(I \otimes S)(h \otimes \alpha_i)\beta_k = (h \otimes \alpha_i)S^*\beta_k = (h \otimes \alpha_i)\sum \overline{t_{jk}}\,\alpha_j = t_{ik}h.$$

By (3) we have

$$T(h\otimes\alpha_i)\beta_k=\sum_j(t_{ij}h\otimes\beta_j)\beta_k=t_{ik}h$$

and so we see that $I\otimes S=T$. The rest is trivial.

Theorem 15. *If* $\pi_k\colon G_k\to\mathscr{U}(\mathscr{H}_k)$ $(k=1,2)$ *are irreducible representations then so is* $\pi_1\otimes\pi_2\colon G_1\times G_2\to\mathscr{U}(\mathscr{H}_1\otimes\mathscr{H}_2)$.

Proof. If T is in $(\pi_1\otimes\pi_2,\pi_1\otimes\pi_2)$ then T commutes with every operator of the form $\pi_1(x)\otimes\pi_2(e)$ and so by the foregoing proposition $T=I_1\otimes S_2$. Similarly we obtain $T=S_1\otimes I_2$. Now if $h_1\in\mathscr{H}_1$ and $h,h_2\in\mathscr{H}_2$ then

$$(I_1\otimes S_2)(h_1\otimes h_2)h=(h_1\otimes S_2h_2)=(S_2h_2,h)h_1$$

and

$$(S_1\otimes I_2)(h_1\otimes h_2)h=(S_1h_1\otimes h_2)h=(h_2,h)S_1h_1.$$

Therefore by choosing $h=h_2\neq 0$ we shall obtain $S_1=\lambda I_1$ where $\lambda=(S_2h_2,h_2)/(h_2,h_2)$. This gives $T=\lambda I_1\otimes I_2=\lambda(I_1\otimes I_2)$ and so the commuting algebra of $\pi_1\otimes\pi_2$ is $\mathbb{C}I$. Hence by Theorem IV.2.2 $\pi_1\otimes\pi_2$ is irreducible.

For any group G the map $x\mapsto(x,x)$ gives an isomorphic embedding of G in $G\times G$. Thus if $\rho_k\colon G\to\mathscr{U}(\mathscr{H}_k)$ $(k=1,2)$ are representations of G then by restricting $\rho_1\otimes\rho_2$ to this diagonal subgroup of $G\times G$ one obtains a representation of G which is called the *inner tensor product* of the representations ρ_1 and ρ_2 and can be denoted by $\rho_1\otimes\rho_2|G$. We point out that $\pi_1\otimes\pi_2|G$ can be reducible even if both π_1 and π_2 are irreducible representations. We also see that $\pi_1\otimes\pi_2|G_1\times\{e_2\}$ is a primary representation of type I. Namely it is a multiple of π_1 and π_1 occurs in it with multiplicity equal to the dimension of $\mathscr{H}_2$. Similarly $\pi_1\otimes\pi_2|\{e_1\}\times G_2$ is a multiple of π_2. Therefore π_1 and π_2 are determined up to equivalence by $\pi_1\otimes\pi_2$.

Proposition 16. *For* $k=1,2$ *let* $\rho_k\colon G_k\to\mathscr{U}(\mathscr{H}_k)$ *and* $\tilde{\rho}_k\colon G_k\to\mathscr{U}(\tilde{\mathscr{H}}_k)$ *be such that* $\rho_1\approx\tilde{\rho}_1$ *and* $\rho_2\approx\tilde{\rho}_2$. *Then* $\rho_1\otimes\rho_2\approx\tilde{\rho}_1\otimes\tilde{\rho}_2$.

Proof. Let $U_k\in(\rho_k,\tilde{\rho}_k)$ $(k=1,2)$ be unitary operators realizing the equivalence $\rho_k\approx\tilde{\rho}_k$. We define $U\colon\mathscr{H}_1\otimes\mathscr{H}_2\to\tilde{\mathscr{H}}_1\otimes\tilde{\mathscr{H}}_2$ by $U(T)=U_1TU_2^{-1}$ where $T\colon\mathscr{H}_2\to\mathscr{H}_1$ belongs to $\mathscr{H}_1\otimes\mathscr{H}_2$. We prove that U is unitary by showing that $\|U(T)\|=\|T\|$ for every T where $\|\cdot\|$ denotes the HS norm. For let α_i $(i\in\mathscr{I})$ be a MONS in $\tilde{\mathscr{H}}_2$ so that $U_2^{-1}\alpha_i$ is a MONS in $\mathscr{H}_2$. We have

$$\|U(T)\|^2=\sum\|U_1TU_2^{-1}\alpha_i\|^2$$

where $\|U_1 T U_2^{-1} \alpha_i\| = \|T U_2^{-1} \alpha_i\|$ and so the desired conclusion follows. The proof will be completed by showing that U intertwines $\rho_1 \otimes \rho_2$ and $\tilde{\rho}_1 \otimes \tilde{\rho}_2$: Now if $T \in \mathscr{H}_1 \otimes \mathscr{H}_2$ then

$$(U(\rho_1 \otimes \rho_2)(x_1, x_2))T = U(\rho_1(x_1) T \rho_2(x_2)^*) = U_1 \rho_1(x_1) T \rho_2(x_2^{-1}) U_2^{-1}.$$

By the intertwining nature of U_1 and U_2 the right hand side is $(\tilde{\rho}_1 \otimes \tilde{\rho}_2(x_1, x_2) U) T$.

If $\rho: G \to \mathscr{U}(\mathscr{H})$ is a finite dimensional representation then the *character* of ρ is the function $\chi_\rho: G \to \mathbb{C}$ given by $\chi_\rho(x) = \operatorname{tr} \rho(x)$ for $x \in G$.

Proposition 17. *For any two finite dimensional representations* $\rho_k: G_k \to \mathscr{U}(\mathscr{H}_k)$ $(k=1,2)$ *we have*

$$\chi_{\rho_1 \otimes \rho_2}(x_1, x_2) = \chi_{\rho_1}(x_1) \chi_{\rho_2}(x_2)$$

for any $x_1 \in G_1$ *and* $x_2 \in G_2$.

Proof. Since in

$$\operatorname{tr} \rho_1 \otimes \rho_2(x_1, x_2) = \sum (\rho_1 \otimes \rho_2(x_1, x_2) T_s, T_s)$$

the parenthesis denotes a HS inner product we need a MONS, say y_j $(j \in \mathscr{J})$ in $\mathscr{H}_2$. If we choose a similar system x_i $(i \in \mathscr{I})$ in $\mathscr{H}_1$ then by Proposition 3 for T_s $(s \in S)$ we can choose $x_i \otimes y_j$ $(i \in \mathscr{I}, j \in \mathscr{J})$. Then a routine computation shows that

$$(\rho_1 \otimes \rho_2(x_1, x_2)(x_i \otimes y_j), x_i \otimes y_j) = (\rho_1(x_1) x_i, x_i)(\rho_2(x_2) y_j, y_j).$$

Hence the result follows by summing over i and j.

Proposition 18. *For any three representations* $\rho_k: G_k \to \mathscr{U}(\mathscr{H}_k)$ $(k=1,2,3)$ *we have* $\rho_1 \otimes (\rho_2 \otimes \rho_3) \approx (\rho_1 \otimes \rho_2) \otimes \rho_3$.

Note. As usual $G_1 \times (G_2 \times G_3)$ and $(G_1 \times G_2) \times G_3$ are identified via $(x_1, (x_2, x_3)) \mapsto ((x_1, x_2), x_3)$.

Proof. We show that the unitary map appearing in Proposition 5 is an intertwining operator for the representations $\rho_1 \otimes (\rho_2 \otimes \rho_3)$ and $(\rho_1 \otimes \rho_2) \otimes \rho_3$. In fact if $x_k \in G_k$ and $\alpha_k \in \mathscr{H}_k$ $(k=1,2,3)$ then using

$$\begin{aligned} \rho_1 \otimes (\rho_2 \otimes \rho_3)(x_1, (x_2, x_3)) &= \rho_1(x_1) \otimes (\rho_2 \otimes \rho_3)(x_2, x_3) \\ &= \rho_1(x_1) \otimes (\rho_2(x_2) \otimes \rho_3(x_3)) \end{aligned}$$

we obtain

$$\begin{aligned}&U\rho_1\otimes(\rho_2\otimes\rho_3)(x_1,(x_2,x_3))\alpha_1\otimes(\alpha_2\otimes\alpha_3)\\&=U\rho_1(x_1)\alpha_1\otimes(\rho_2(x_2)\alpha_2\otimes\rho_3(x_3)\alpha_3)\\&=(\rho_1(x_1)\alpha_1\otimes\rho_2(x_2)\alpha_2)\otimes\rho_3(x_3)\alpha_3.\end{aligned}$$

Similarly we have

$$\begin{aligned}&(\rho_1\otimes\rho_2)\otimes\rho_3((x_1,x_2),x_3)U\alpha_1\otimes(\alpha_2\otimes\alpha_3)\\&=(\rho_1(x_1)\otimes\rho_2(x_2))\otimes\rho_3(x_3)(\alpha_1\otimes\alpha_2)\otimes\alpha_3\\&=(\rho_1(x_1)\alpha_1\otimes\rho_2(x_2)\alpha_2)\otimes\rho_3(x_3)\alpha_3.\end{aligned}$$

Hence the intertwining nature of U is proved.

If $\rho_k: G\to\mathscr{U}(\mathscr{H}_k)$ $(k=1,2)$ are representations of G then we define $\rho_1\oplus\rho_2$, the *direct sum* of ρ_1 and ρ_2 by $(\rho_1\oplus\rho_2)(x)=\rho_1(x)\oplus\rho_2(x)$ so that $\rho_1\oplus\rho_2: G\to\mathscr{U}(\mathscr{H}_1\oplus\mathscr{H}_2)$. One can easily verify that if $\rho_k: G\to\mathscr{U}(\mathscr{H}_k)$ and $\tilde{\rho}_k: G\to\mathscr{U}(\tilde{\mathscr{H}}_k)$ $(k=1,2)$ are representations and $\rho_k\approx\tilde{\rho}_k$ for $k=1,2$ then $\rho_1\oplus\rho_2\approx\tilde{\rho}_1\oplus\tilde{\rho}_2$. Hence *the direct sum defines an operation* $\oplus$ *on the class* $P(G)$ *of equivalence classes* ρ *of representations of* G.

The linear map $(\alpha_1,\alpha_2)\to(\alpha_2,\alpha_1)$ is a unitary transformation of $\mathscr{H}_1\oplus\mathscr{H}_2$ onto $\mathscr{H}_2\oplus\mathscr{H}_1$ which we denote by U. One can easily verify that U is an intertwining operator of $\rho_1\oplus\rho_2$ and $\rho_2\oplus\rho_1$. This shows that the above defined binary operation $\oplus$ is commutative. Similarly if $\rho_k: G\to\mathscr{U}(\mathscr{H}_k)$ $(k=1,2,3)$ are given then one can immediately see that $(\rho_1\oplus\rho_2)\oplus\rho_3$ and $\rho_1\oplus(\rho_2\oplus\rho_3)$ are equivalent representations of G. Therefore $\oplus$ is an associative operation on $P(G)$. The zero dimensional representation $G\to\mathscr{L}(0)$ can be introduced as the neutral element of $P(G)$ with respect to $\oplus$. Propositions 16 and 18 show that $P(G)$ admits an associative multiplication. We are now going to show that $\oplus$ and $\otimes$ satisfy the distributive law.

Proposition 19. *If* $\rho_k: G\to\mathscr{U}(\mathscr{H}_k)$ $(k=1,2,3)$ *then*

$$(\rho_1\oplus\rho_2)\otimes\rho_3\approx\rho_1\otimes\rho_3\oplus\rho_2\otimes\rho_3.$$

Proof. If $T:\mathscr{H}_3\to\mathscr{H}_1\oplus\mathscr{H}_2$ and $\alpha_3\in\mathscr{H}_3$ then we can define $T_i\alpha_3=(T\alpha_3)_i$ $(i=1,2)$ where $(T\alpha_3)_i$ is the projection of $T\alpha_3$ on $\mathscr{H}_i$. If T is antilinear then $T_i:\mathscr{H}_3\to\mathscr{H}_i$ $(i=1,2)$ are also antilinear operators. Moreover it is obvious that if T is a HS operator then so are T_1 and T_2. Conversely, if $T_i:\mathscr{H}_3\to\mathscr{H}_i$ $(i=1,2)$ are given then $T=T_1\oplus T_2$ defines an operator $T:\mathscr{H}_3\to\mathscr{H}_1\oplus\mathscr{H}_2$ such that its T_1 and T_2 are the operators given in advance. Again, if T_1 and T_2 are antilinear HS operators then so is T.

By choosing a MONS in $\mathscr{H}_3$ we can immediately see that for any two operators S and T we have $(S, T)=(S_1, T_1)+(S_2, T_2)$ where in each instance $(\cdot, \cdot)$ denotes the HS inner product in the corresponding spaces $(\mathscr{H}_1 \oplus \mathscr{H}_2) \oplus \mathscr{H}_3$ and $\mathscr{H}_i \otimes \mathscr{H}_3$ $(i=1, 2)$. Hence if we let

$$U: \mathscr{H}_1 \otimes \mathscr{H}_3 \oplus \mathscr{H}_2 \otimes \mathscr{H}_3 \rightarrow (\mathscr{H}_1 \oplus \mathscr{H}_2) \otimes \mathscr{H}_3$$

be the map defined by $(T_1, T_2) \mapsto T_1 \oplus T_2$ then U is a unitary isomorphism of these two Hilbert spaces.

We claim that U is an intertwining operator of the representations $(\rho_1 \oplus \rho_2) \otimes \rho_3$ and $\rho_1 \otimes \rho_3 \oplus \rho_2 \otimes \rho_3$. Indeed

$$\begin{aligned}(\rho_1 \oplus \rho_2) \otimes \rho_3(T_1, T_2) &= (\rho_1 \oplus \rho_2)(T_1, T_2)\rho_3^* \\ &= (\rho_1 \oplus \rho_2)(T_1 \rho_3^*, T_2 \rho_3^*) = (\rho_1 T_1 \rho_3^*, \rho_2 T_2 \rho_3^*)\end{aligned}$$

and so

$$(U(\rho_1 \oplus \rho_2) \otimes \rho_3)(T_1, T_2) = \rho_1 T_1 \rho_3^* \oplus \rho_2 T_2 \rho_3^* .$$

On the other hand

$$\begin{aligned}(\rho_1 \otimes \rho_3 \oplus \rho_2 \otimes \rho_3) U(T_1, T_2) &= (\rho_1 \otimes \rho_3 \oplus \rho_2 \otimes \rho_3)(T_1 \oplus T_2) \\ &= (\rho_1 \otimes \rho_3) T_1 \oplus (\rho_2 \otimes \rho_3) T_2 \\ &= \rho_1 T_1 \rho_3^* \oplus \rho_2 T_2 \rho_3^* .\end{aligned}$$

This completes the proof.

Proposition 19 implies that $(\rho_1 \oplus \rho_2) \otimes \rho_3 = \rho_1 \otimes \rho_3 \oplus \rho_1 \otimes \rho_3$ where ρ_1, ρ_2, ρ_3 denote equivalence classes of representations of the same group G i.e. $\rho_1, \rho_2, \rho_3 \in P(G)$.

We already mentioned that if $\rho: G \rightarrow \mathscr{U}(\mathscr{H})$ is a representation then another representation $\bar{\rho}$, called the *complex conjugate representation* can be defined by viewing each operator $\rho(x)$ as a linear operator $\bar{\rho}(x): \overline{\mathscr{H}} \rightarrow \overline{\mathscr{H}}$ where $\overline{\mathscr{H}}$ is the conjugate Hilbert space of $\mathscr{H}$. Clearly $\bar{\bar{\rho}} = \rho$ and $\overline{\rho_1 \oplus \rho_2} = \overline{\rho_1} \oplus \overline{\rho_2}$. One can also show that $\overline{\rho_1 \otimes \rho_2} = \overline{\rho_1} \otimes \overline{\rho_2}$. If $\rho_1 \approx \rho_2$ then $\overline{\rho_1} \approx \overline{\rho_2}$ and so for every class ρ in $P(G)$ there is a complex conjugate class $\bar{\rho}$. We see that ρ and $\bar{\rho}$ have the same degree. If ρ is finite dimensional then it has a character $\chi_\rho: G \rightarrow \mathbb{C}$ given by $\chi_\rho(x) = \sum (\rho(x)\alpha_i, \alpha_i)$ where α_i is any MONS in the Hilbert space of ρ. Since in $\overline{\mathscr{H}}$ the inner product is the complex conjugate of the inner product of $\mathscr{H}$ it follows that the character of the class $\bar{\rho}$ is $\bar{\chi}_\rho$ where χ_ρ is the character of ρ. A representation ρ and its complex conjugate $\bar{\rho}$ have the same invariant subspaces. Hence $\bar{\pi}$ is irreducible if π is irreducible and so the conjugation is an involution on $\hat{G}$.

Proposition 20. *The inner tensor product is commutative i.e. if* $\rho_i\colon G \to \mathscr{U}(\mathscr{H}_i)$ $(i=1,2)$ *are unitary representations then* $\rho_1 \otimes \rho_2 \approx \rho_2 \otimes \rho_1$.

Proof. If $T \in \mathscr{H}_1 \otimes \mathscr{H}_2$ then $T^* \in \mathscr{H}_2 \otimes \mathscr{H}_1$ and $T \mapsto T^*$ defines an isometry of $\mathscr{H}_1 \otimes \mathscr{H}_2$ onto $\mathscr{H}_2 \otimes \mathscr{H}_1$. Since we have

$$\begin{aligned} U(\rho_1 \otimes \rho_2)(x)\,T &= U\,\rho_1(x)\,T\,\rho_2(x)^* = \rho_2(x)\,T^*\,\rho_1(x)^* = \rho_2 \otimes \rho_1(x)\,T^* \\ &= \rho_2 \otimes \rho_1(x)\,U\,T \end{aligned}$$

for every $x \in G$ and $T \in \mathscr{H}_1 \otimes \mathscr{H}_2$. Hence U is an intertwining operator and so $\rho_1 \otimes \rho_2$ and $\rho_2 \otimes \rho_1$ are equivalent.

4. Induced Representations for Arbitrary G and H

The material presented in this section starts with the discussion of the modifications needed in the definition of induced representations when G/H admits no invariant measure and the Hilbert space $\mathscr{H}$ is not necessarily separable. We will continue by proving a number of fundamental lemmas which will be needed in the proof of some basic results on induced representations such as the theorem on induction in stages and the tensor product theorem. We shall then prove that the definition given in Section 2 is indeed a special case of the general definition described in the sequel. The section ends with the proof of the continuity of induced representations.

We let G be a locally compact group and H a closed subgroup of G. We suppose that a weakly measurable unitary representation $\chi\colon H \to \mathscr{U}(\mathscr{H})$ is given. In order to define the representation $\rho^\chi\colon G \to \mathscr{U}(\mathscr{K})$ induced on G by χ we introduce a quasi-invariant measure μ on $X = G/H$ which belongs to a continuous ρ-function $\rho\colon G \to (0, \infty)$. We let $\mathscr{M}$ be the complex vector space of those continuous functions $\theta\colon G \to \mathscr{H}$ which satisfy $\theta(xh) = \chi(h)^*\,\theta(x)$ for every $x \in G$ and $h \in H$ and are such that $\|\theta(\cdot)\|$ has compact support in X. We can then introduce the pseudo-inner product

$$(\theta_1, \theta_2) = \int_X (\theta_1(x), \theta_2(x))\,d\mu(\xi)$$

and identify two elements θ and ψ of $\mathscr{M}$ if $\|\theta - \psi\| = 0$ where $\|\cdot\|$ denotes the pseudo-norm derived from $(\cdot\,,\cdot)$. The same symbol θ will be used to denote both equivalence classes and individual functions. The inner product space of equivalence classes will also be denoted by $\mathscr{M}$.

We let $\mathscr{K}$ be the complex Hilbert space obtained by completing the inner product space $\mathscr{M}$. A typical element of $\mathscr{K}$ will be denoted by θ

or ψ. Hence unless otherwise specified θ can denote not only a function $\theta: G \to \mathscr{H}$ or an equivalence class of such functions but also any abstract vector lying in $\mathscr{K}$.

We shall use $C_0(G, \mathscr{H})$ to denote the vector space of continuous functions $f: G \to \mathscr{H}$ with compact support and we let $C_{01}(G, \mathscr{H})$ be the set of those f which are of the form $f = \varphi\alpha$ where $\varphi \in C_0(G)$ and $\alpha \in \mathscr{H}$. By Lemma 2.2 every f in $C_0(G, \mathscr{H})$ determines an element θ_f of $\mathscr{M}$. By the corollary of this lemma if χ is continuous and $\mathscr{H} \neq 0$ then $\mathscr{M}$ and $\mathscr{K}$ are not zero dimensional. Lemma 8 below will show that in general $\dim \mathscr{K} \geqslant \dim \mathscr{H}$.

We can now turn to the definition of the operators $\rho^\chi(a) (a \in G)$ of the induced representation ρ^χ: Following the notations introduced earlier we let $\lambda(x, y)$ denote the Radon-Nikodym derivative of the translated quasi-invariant measure μ_x with respect to the original μ. The basic properties of the λ-function λ are summarized in Lemma V.3.10. For $\theta \in \mathscr{M}$ we let

$$\rho(a)\theta(x) = \theta(a^{-1}x)\sqrt{\lambda(a^{-1}, \xi)} = {}^a\theta(x)\sqrt{\lambda(a^{-1}, \xi)}$$

where $\xi = px \in X$. Since $\xi \mapsto \lambda(a^{-1}, \xi)$ is continuous and $\xi \mapsto a\xi$ is a homeomorphism of X onto itself one can immediately see that $\rho(a)\theta$ belongs to $\mathscr{M}$. We show that $\rho(a)$ preserves the norm: Indeed

$$\begin{aligned}\|\rho(a)\theta\|^2 &= \int_X \|\rho(a)\theta(x)\|^2 d\mu(\xi) = \int_X \|\theta(a^{-1}x)\|^2 \lambda(a^{-1}, \xi) d\mu(\xi) \\ &= \int_X \|\theta(x)\|^2 \lambda(a^{-1}, a\xi) d\mu(a\xi).\end{aligned}$$

Since $\lambda(a, \xi) = d\mu(a\xi)/d\mu(\xi)$ and by Lemma V.3.10 $\lambda(a^{-1}, a\xi)\lambda(a, \xi) = \lambda(e, \xi) = 1$ we obtain

$$\|\rho(a)\theta\|^2 = \int_X \|\theta(x)\|^2 d\mu(\xi) = \|\theta\|^2.$$

It follows that $\rho(a): \mathscr{M} \to \mathscr{M}$ can be extended in a unique manner to a unitary transformation $\rho(a): \mathscr{K} \to \mathscr{K}$.

We prove that $\rho(a_1)\rho(a_2) = \rho(a_1 a_2)$ on $\mathscr{M}$ for every $a_1, a_2 \in G$: For

$$\begin{aligned}\rho(a_1)\rho(a_2)\theta(x) &= \rho(a_1)\theta(a_2^{-1}x)\sqrt{\lambda(a_2^{-1}, px)} \\ &= \theta(a_2^{-1}a_1^{-1}x)\sqrt{\lambda(a_2^{-1}, px)\lambda(a_1^{-1}, pa_2^{-1}x)}.\end{aligned}$$

By $(a_2^{-1}x)H = a_2^{-1}(xH)$ we have $p(a_2^{-1}x) = a_2^{-1}px$ and so by Lemma V.3.10 the expression under the square root sign is $\lambda(a_1 a_2, px)$. Therefore the right hand side of the equation is the same as $\rho(a_1 a_2)\theta(x)$. Hence we have $\rho(a_1)\rho(a_2) = \rho(a_1 a_2)$ on $\mathscr{M}$ and consequently on $\mathscr{K}$.

Definition 1. *The unitary representation* $\rho^{\chi}\colon G\to\mathcal{U}(\mathcal{K})$ *defined by the formula*

$$\rho(a)\theta(x)=\theta(a^{-1}x)\sqrt{\lambda(a^{-1},px)} \qquad (a,x\in G \text{ and } \theta\in\mathcal{M})$$

is called the representation induced on the locally compact group G by the weakly measurable unitary representation $\chi\colon H\to\mathcal{U}(\mathcal{H})$ *of the closed subgroup H.*

The induced representation ρ^{χ} depends on the quasi-invariant measure μ but we are going to prove now that if μ_1 and μ_2 are such measures then the corresponding representations ρ_1^{χ} and ρ_2^{χ} are equivalent. By the corollary of Theorem V.3.9 we have

$$\int_X \tilde{f}(\xi)\,d\mu_k(\xi)=\int_X f(x)\,\rho_k(x)\,dx \qquad (k=1,2)$$

for every $f\in C_0(C)$ where $\rho_k\colon G\to(0,\infty)$ is the positive continous ρ-function determining the measure μ_k. By Theorem V.3.7 $f\mapsto\tilde{f}$ maps $C_0(G)$ onto $C_0(X)$ and so we see from the above equations that

$$\frac{d\mu_1}{d\mu_2}(\xi)=\frac{\rho_1(x)}{\rho_2(x)}$$

where ρ_1 and ρ_2 are strictly positive continuous functions. Let $\mathcal{M}_1$ and $\mathcal{M}_2$ denote the inner product spaces obtained by using in the definition of the inner product (θ,ψ) the measures μ_1 and μ_2 respectively. If $\theta\in\mathcal{M}_1$ then $\theta\sqrt{\rho_2/\rho_1}\in\mathcal{M}_2$ because $\rho_k(xh)=(\delta/\Delta)(h)\,\rho_k(x)$ $(k=1,2)$ for every $x\in G$ and $h\in H$. The map $\theta\mapsto\theta\sqrt{\rho_2/\rho_1}$ is a bijection and it is linear. Thus it can be extended to a unitary operator $U\colon\mathcal{K}_1\to\mathcal{K}_2$ where $\mathcal{K}_k$ denotes the completion of $\mathcal{M}_k$ $(k=1,2)$.

We prove that U intertwines the representation ρ^{χ_1} and ρ^{χ_2}. We have

$$\rho_2(a)\,U\theta(x)=U\theta(a^{-1}x)\sqrt{\lambda_2(a^{-1},x)}={}^{a}\theta(x)\sqrt{\lambda_2(a^{-1},x)}\sqrt{\frac{\rho_1}{\rho_2}(a^{-1}x)}$$

and similarly

$$U\rho_1(a)\theta(x)=(\rho_1(a)\theta)(x)\sqrt{\frac{\rho_1}{\rho_2}(x)}={}^{a}\theta(x)\sqrt{\lambda_1(a^{-1},x)}\sqrt{\frac{\rho_1}{\rho_2}(x)}.$$

Therefore the equivalence of ρ_1 and ρ_2 will be proved by showing that

$$\lambda_1(a,x)\frac{\rho_1}{\rho_2}(x)=\lambda_2(a,x)\frac{\rho_1}{\rho_2}(ax)$$

for every $a,x\in G$. By introducing $\lambda_k(a,x)=\rho_k(ax)/\rho_k(x)$ this relation becomes a trivial identity. Therefore we proved:

Proposition 2. *The induced representations corresponding to the various quasi-invariant measures derived from strictly positive, continuous ρ-functions $\rho: G \to (0, \infty)$ are unitarily equivalent. Therefore up to a unitary equivalence ${}_G\rho^\chi$ is uniquely determined by G, H and $\chi: \to \mathscr{U}(\mathscr{H})$.*

The remaining material discussed in Section 2 incluhing proofs is word by word applicable to the present, general situation. In particular Propositions 2.5, 2.6 and 2.7 hold for arbitrary groups G, H and Hilbert spaces $\mathscr{H}$. Hence ρ^χ depends only on G and the equivalence class containing χ.

Proposition 3. *Every function $\theta \in \mathscr{M}$ is of the form $\theta = \theta_f$ for a suitable $f \in C_0(G, \mathscr{H})$.*

Proof. Let S denote the compact support of $\|\theta(\cdot)\|$ in X. By Propositions V.1.3 and V.1.4 X is a uniform space and so we can choose a continuous function $\varphi: X \to [0,1]$ with compact support such that φ is identically 1 on S. By Theorem V.3.7 there is a continuous function $\psi: G \to \mathbb{R}$ with compact support such that $\varphi(x) = \int_H \psi(xh)\,dh$ for every $x \in G$. If we let $f = \psi\theta$ then $f \in C_0(G, \mathscr{H})$. Moreover if $x \in G$ then

$$\begin{aligned}\theta_f(x) &= \int_H \chi(h) f(xh)\,dh = \int_H \psi(xh)\chi(h)\theta(xh)\,dh \\ &= \int_H \psi(xh)\theta(x)\,dh = \theta(x)\varphi(x) = \theta(x).\end{aligned}$$

The last equation holds because $\varphi(x) = 1$ if $\theta(x) \neq 0$. Hence we proved that $\theta = \theta_f$ with $f = \psi\theta \in C_0(G, \mathscr{H})$.

Lemma 4. *If $f \in C_0(G, \mathscr{H})$ then there is a compact set C in G which contains the support of f and is such that we have: For each $\varepsilon > 0$ there is a $g = \sum_1^n f_i$ with f_i in $C_{01}(G, \mathscr{H})$ such that C is a carrier of g and $\|f - g\|_\infty < \varepsilon$.*

Proof. Let N_e be a compact neighborhood of the identity and let $C = N_e S$ where S is the support of f. Since f is continuous and has compact support it is uniformly continuous. Hence there is an open neighborhood O_e of the identity in G such that

$$\|f(x) - f(y)\| < \varepsilon \quad \text{whenever } xy^{-1} \in O_e. \tag{1}$$

We may also suppose that $O_e \subseteq N_e$. Using the compactness of S we can find $x_1, \ldots, x_n$ in G such that the open sets $O_e x_1, \ldots, O_e x_n$ cover S. We choose a partition of unity for these open sets i. e. we select a family of continuous functions $\varphi_i: G \to [0,1]$ such that

(2) $$\varphi_i(x) \neq 0 \quad \text{only if } x \in O_e x_i \quad (i=1,\dots,n),$$

(3) $$\varphi_1(x) + \dots + \varphi_n(x) = 1 \quad \text{for every } x \in G.$$

We define g for every x in G by

$$g(x) = \varphi_1(x) f(x_1) + \dots + \varphi_n(x) f(x_n).$$

Then $g \in C_{01}(G, \mathscr{H})$ and if $x \in G$ then by (3)

$$\|f(x) - g(x)\| = \|\sum \varphi_i(x)(f(x) - f(x_i))\| \leqslant \sum \varphi_i(x) \|f(x) - f(x_i)\|.$$

If $\varphi_i(x) \neq 0$ then by (1) and (2) we have $\|f(x) - f(x_i)\| < \varepsilon$ and so $\|f(x) - g(x)\| \leqslant \varepsilon \sum \varphi_i(x) = \varepsilon$. Since x is arbitrary we obtain $\|f - g\|_\infty \leqslant \varepsilon$.

Lemma 5. *If $f \in C_0(G, \mathscr{H})$ and the symmetric compact set S in G is a carrier of f then*

$$\|\theta_f\| \leqslant \|f\|_\infty \nu(H \cap S^2) \sqrt{\mu(pS)}$$

where ν denotes the Haar measure of H and μ is the quasi-invariant measure used to define $\mathscr{K}$.

Proof. If $\alpha \in \mathscr{H}$ then using the definition of $\theta_f(x)$ in the weak sense we obtain

$$|(\theta_f(x), \alpha)| \leqslant \|f\|_\infty \|\alpha\| \nu(H \cap x^{-1} S).$$

We substitute $\alpha = \theta_f(x)$ and have

$$\|\theta_f(x)\| \leqslant \|f\|_\infty \nu(H \cap x^{-1} S).$$

We need this inequality only if $x \in SH$ because otherwise $\theta_f(x) = 0$. If $x = sh$ with $s \in S$ and $h \in H$ then $\|\theta_f(x)\| = \|\theta_f(s)\|$ and so S being symmetric we have $\|\theta_f(x)\| \leqslant \|f\|_\infty \cdot \nu(H \cap S^2)$ for every $x \in G$. Now the inequality in question can be obtained by integrating $\|\theta_f(\cdot)\|^2$ over X.

Lemma 6. *For every $f \in C_0(G, \mathscr{H})$ and $a \in G$ we have ${}_a(\theta_f) = \theta_{({}_a f)}$.*

Proof. If $x \in G$ then

$${}_a(\theta_f)(x) = \theta_f(ax) = \int_H \chi(h) f(axh)\, dh = \int_H \chi(h)\, {}_a f(xh)\, dh = \theta_{({}_a f)}(x).$$

Corollary. *The linear manifold $\{\theta_f(x): f \in C_0(G, \mathscr{H})\}$ is independent of $x \in G$.*

If $\varphi \in C_0(X)$ then φ defines a continuous function $\varphi: G \to \mathbb{C}$ such that $\varphi(xh) = \varphi(x)$ for every $x \in G$ and $h \in H$. Given any f in $C_0(G, \mathscr{H})$ and

$x \in G$ we have $\theta_{\varphi f}(x) = \varphi(x)\theta_f(x)$ and so $\varphi\theta_f$ is an element of $\mathcal{M}$. Hence we proved the following:

Lemma 7. *If* $\varphi \in C_0(X)$ *and* $\theta_f \in \mathcal{M}$ *then we have* $\varphi\theta_f = \theta_{\varphi f} \in \mathcal{M}$.

We recall that earlier we defined $C_{01}(G, \mathcal{H})$ as the set of those $f: G \to \mathcal{H}$ which are of the form $f = \varphi\alpha$ where $\alpha \in \mathcal{H}$ and $\varphi \in C_0(G)$. Hence $C_{01}(G, \mathcal{H})$ is a subset of $C_0(G, \mathcal{H})$.

Lemma 8. *If* $\chi: H \to \mathcal{U}(\mathcal{H})$ *is weakly continuous then* $\{\theta_f(x): f \in C_{01}(G, \mathcal{H})\}$ *is total in* $\mathcal{H}$ *for every* x *in* G.

Corollary. *If* χ *is continuous and* $x \in G$ *then the closure of the linear manifold* $\{\theta(x): \theta \in \mathcal{M}\}$ *is* $\mathcal{H}$ *and so* $\dim \mathcal{K} \geqslant \dim \mathcal{H}$.

Proof. Let $\mathcal{H}_1$ be the subspace of those $\beta \in \mathcal{H}$ for which $\beta \perp \theta_f(e)$ for every $f \in C_{01}(G, \mathcal{H})$. If $f = \varphi\alpha$ where $\varphi \in C_0(G)$ and $\alpha \in \mathcal{H}$ then ${}_af = ({}_a\varphi)\alpha$. Hence using Lemma 6 we see that $\mathcal{H}_1$ is the same as the subspace of those $\beta \in \mathcal{H}$ which are orthogonal to $\theta_f(y)$ for every $f \in C_{01}(G, \mathcal{H})$ and $y \in G$. If $\beta \in \mathcal{H}_1$ then $(\theta_f(x), \chi(h)\beta) = (\theta_f(xh), \beta) = 0$ for every $f \in C_{01}(G, \mathcal{H})$ and $h \in H$. Thus $\chi(h)\beta \in \mathcal{H}_1$ and this shows that $\mathcal{H}_1$ is an invariant subspace of the representation χ. Our object is to prove that $\mathcal{H}_1$ is 0. Given β in $\mathcal{H}_1$ we let $g = \varphi\beta$ where $\varphi \in C_0(G)$. Then due to the invariance of $\mathcal{H}_1$ under χ we have $\theta_g(e) \in \mathcal{H}_1$. Hence by the definition of $\mathcal{H}_1$ we obtain $\theta_g(e) \perp \theta_f(y)$ for all $f \in C_{01}(G, \mathcal{H})$ and $y \in G$. By choosing $f = g$ and $y = e$ we see that $\theta_g(e) = 0$. However $\theta_g(e) = \sigma(\varphi)\beta$ where σ is the algebra representation associated with χ. Hence $\beta = 0$ by Proposition V.5.10.

We can now prove that if G/H admits an invariant measure μ and $\mathcal{H}$ is separable then the Hilbert space $\mathcal{K} = \mathcal{K}(G, H, \chi, \mu)$ obtained by completing $\mathcal{M} = \mathcal{M}(G, H, \chi)$ with respect to the inner product determined by μ can be naturally identified with the Hilbert space which was introduced in Theorem 2.1. More generally let G and H be arbitrary, let μ be a quasi-invariant measure on $X = G/H$ and let $\mathcal{H}$ be separable. Let $\theta_k: G \to \mathcal{H}$ $(k = 1, 2)$ be such that $x \mapsto (\theta(x), \alpha)$ is a measurable function on G for every $\alpha \in \mathcal{H}$. In the beginning of Section 2 we proved that $x \mapsto (\theta_1(x), \theta_2(x))$ is then a measurable function on G. We also saw that if $\theta_k(xh) = \chi(h)^* \theta_k(x)$ $(k = 1, 2)$ holds for $x \in G$ and $h \in H$ then $x \mapsto (\theta_1(x), \theta_2(x))$ is defined on X. Moreover if $\|\theta_k\| = \int\limits_X \|\theta_k(x)\|^2 \, d\mu(\xi)$ is finite for $k = 1, 2$ then

$$(\theta_1, \theta_2) = \int\limits_X (\theta_1(x), \theta_2(x)) \, d\mu(\xi)$$

exists. Two functions θ and ψ satisfying these conditions will be called equivalent if $\|\theta - \psi\| = 0$.

Theorem 9. *Let $\mathscr{H}$ be a separable Hilbert space. Denote by $\mathscr{K}=\mathscr{K}(G,H,\chi,\mu)$ the vector space of equivalence classes of those functions $\theta: G\to\mathscr{H}$ for which*

1) $x\mapsto(\theta(x),\alpha)$ is measurable for every $\alpha\in\mathscr{H}$;

2) $\theta(xh)=\chi(h)^*\theta(x)$ for $x\in G$ and $h\in H$ with the possible exception of a set of pairs (x,h) such that the corresponding xh's belong to a set of measure zero lying in G/H.

3) $\|\theta\|$ is finite.

Then $\mathscr{K}$ is a Hilbert space with respect to the inner product (θ_1,θ_2).

Proof. The present theorem is a straightforward generalization of Theorem 2.1 and its proof is essentially identical with the proof given there. The only difference is that the invariant measure $d\xi$ is now replaced by the quasi-invariant measure $d\mu(\xi)$.

In order to distinguish the Hilbert space obtained by completing $\mathscr{M}$ from the one which occurs in the above theorem in the following reasoning we shall denote the first one by $\mathscr{K}'$. If $\mathscr{H}$ is separable and $f\in C_0(G,\mathscr{H})$ then θ_f satisfies conditions 1), 2) and 3). The definition of the inner product (θ_f,θ_g) being the same each equivalence class in $\{\theta_f: f\in C_0(G,\mathscr{H})\}$ is contained in the corresponding class of functions $\theta: G\to\mathscr{H}$ satisfying 1), 2) and 3). Therefore we have a natural inner product preserving embedding of $\mathscr{M}$ in $\mathscr{K}$ and so $\mathscr{K}'$ is naturally isomorphic to a subspace of $\mathscr{K}$. In order to see that this isomorphism is surjective it is sufficient to prove that if $\mathscr{H}$ is separable then $\mathscr{M}$ is dense in $\mathscr{K}$. This can be seen immediately from the following:

Proposition 10. *Let $\mathscr{H}$ be separable. Let $\mathscr{C}$ be a class of functions in $\mathscr{M}$ such that we have:*

1) If $\theta\in\mathscr{C}$ then ${}_a\theta\in\mathscr{C}$ for every a in G.

2) If $\theta\in\mathscr{C}$ then $\varphi\theta\in\mathscr{C}$ for every $\varphi\in C_0(X)$.

3) For some x in G the set $\{\theta(x):\theta\in\mathscr{C}\}$ is total in $\mathscr{H}$.

Then for every choice of the quasi-invariant measure μ the class $\mathscr{C}$ is total in $\mathscr{K}$.

Note. A set $\mathscr{S}$ is called *total* in the Hilbert space $\mathscr{H}$ if the linear span of $\mathscr{S}$ is dense in $\mathscr{H}$.

Proof. Since $\mathscr{H}$ is separable χ is continuous by Theorem V.7.3. We are going to prove that if ψ in $\mathscr{K}$ is orthogonal to $\mathscr{C}$ then $\psi=0$. By Proposition 3 there is a family $\mathscr{F}$ of functions f in $C_0(G,\mathscr{H})$ such that $\{\theta_f: f\in\mathscr{F}\}=\mathscr{C}$. By hypothesis 1) the family $\mathscr{F}$ is invariant under left translations and so by Lemma 6 the set $\{\theta(a): \theta \in \mathscr{C}\}$ is total in $\mathscr{H}$ for every $a \in G$. If we choose $a=e$ then by the separability of $\mathscr{H}$ there is a sequence of functions $f_1, f_2, \ldots$ in $C_0(G, \mathscr{H})$ such that the set of values $\theta_{f_i}(e)$ $(i=1,2,\ldots)$ is total in $\mathscr{H}$. For the sake of simplicity we shall write θ_i instead of θ_{f_i}.

We choose a function φ in $C_0(X)$ and use φ to denote also the corresponding function $\varphi: G \to \mathbb{C}$. By Lemma 7 and 1) we have $\varphi({}^x\theta_i) \in \mathscr{C}$ for every x in G and so

$$(\varphi\,\theta_i(x^{-1}\cdot), \psi) = \int_X \varphi(\eta)\,(\theta_i(x^{-1}y), \psi(y))\,d\mu(\eta)$$

for every $x \in G$ and $i=1,2,\ldots$ We multiply the last equation by $\varphi(\xi)$ where ξ is an arbitrary element of X and integrate on X. Since φ has compact support in X we can apply Fubini's theorem. Using the fact that φ is an arbitrary function in $C_0(X)$ we obtain the following:

If C is a compact subset of X then for μ-almost all η in C we have $(\theta_i(x^{-1}y), \psi(y)) = 0$ for almost all $\xi \in C$. There are only denumerably many choices for i and so there is a μ-null set N in C such that *if* $\eta \notin N$ *then for every* $i=1,2,\ldots$ *and* $y \in \eta$ *we have*

$$(\theta_i(x^{-1}y), \psi(y)) = 0 \tag{4}$$

for almost all $\xi \in C$.

Using this information we are going to prove that ψ is equal to zero locally almost everywhere in X. Let a compact subset K of X be given. Since G is locally compact by using Lemma V.1.6 we see that X is also a locally compact space. Hence there is a compact set C in X such that K lies in the interior of C. Let N denote the exceptional zero set associated with C. Now suppose that $\psi(y) \neq 0$ for some $y \in G$ where $\eta = yH$ belongs to K. Then by the choice of the functions f_i $(i=1,2,\ldots)$ there is an index i such that $(\theta_i(e), \psi(y)) \neq 0$. Since θ_i is continuous this implies the existence of a neighborhood N_e of the identity e such that $(\theta_i(a), \psi(y)) \neq 0$ for all $a \in N_e$. This shows the existence of a neighborhood N_y of y in G such that $(\theta_i(x^{-1}y), \psi(y)) \neq 0$ for every $x \in N_y$. We can chosse N_y so small that N_η, its canonical image in X lies entirely within C. By Theorems V.3.7 and V.3.9 the μ-measure of N_η is positive and so in view of (4) we must have $\eta \in N$. We proved that $\psi(\eta) = 0$ for all $\eta \notin N$ and so ψ is zero almost everywhere in K. Therefore we proved that $\psi(y) = 0$ for all $y \in \eta$ and locally almost all $\eta \in X$. Since $\|\psi\|$ is finite ψ is equal to zero almost everywhere.

Theorem 11. *The induced representation* ρ^χ *is continuous.*

Proof. By Theorem V.7.1 weak and strong continuity are equivalent notions for unitary group representations. Thus it will be sufficient to prove that $a \mapsto \rho^{\chi}(a)\theta$ is continuous at e for every $\theta \in \mathscr{K}$. First we suppose that $\theta \in \mathscr{M}$ and let S denote the compact support of $\|\theta(\cdot)\|$ in X. We choose a compact, symmetric neighborhood C of e in G. Since $\rho: G \to (0, \infty)$ is a strictly positive continuous function so is $\lambda = \lambda_\rho: G \times X \to (0, \infty)$. Hence there is a positive constant M such that

$$(5) \qquad |\lambda(a, \xi)| \leqslant M \quad \text{for all } a \in C \text{ and } \xi \in S.$$

By Proposition V.1.4 X is uniformizable and so the restriction of λ to $C \times S$ is uniformly continuous. Therefore given $\varepsilon > 0$ we can choose a symmetric neighborhood N_e of e in G such that $N_e \subseteq C$ and $|\lambda(a,x) - \lambda(b,y)| < \varepsilon$ whenever $x h y^{-1} \in N_e$ and $a b^{-1} \in N_e$ for some $h \in H$. By choosing $b = e$ and $y = x$ we obtain

$$(6) \qquad |\lambda(a,x) - 1| < \varepsilon \quad \text{for all } a \in N_e \text{ and } x \in G.$$

Similarly by $\theta \in \mathscr{M}$ the function θ is uniformly continuous on G and so the above N_e can be chosen such that we also have

$$(7) \qquad \|\theta(ax) - \theta(x)\| < \varepsilon \quad \text{for all } a \in N_e \text{ and } x \in G.$$

Since $\|\theta(\cdot)\|$ has compact support the foregoing M can be chosen so large that $\|\theta(x)\| \leqslant M$ for every $x \in G$. For all $a, x \in G$ we have

$$\begin{aligned}\|\rho^{\chi}(a)\theta(x) - \theta(x)\| \leqslant{}& \|\theta(a^{-1}x)\sqrt{\lambda(a^{-1},x)} - \theta(x)\sqrt{\lambda(a^{-1},x)}\| \\ &+ \|\theta(x)\,(\sqrt{\lambda(a^{-1},x)} - 1)\|.\end{aligned}$$

Since $\lambda(a,x) \geqslant 0$ and (5) holds, the right hand side can be majorized for $a \in C$ and $x \in G$ by

$$\sqrt{M}\,\|\theta(a^{-1}x) - \theta(x)\| + M\,|\lambda(a^{-1},x) - 1|.$$

Now if $a \in N_e$ then N_e being symmetric by (6) and (7) we obtain

$$\|\rho^{\chi}(a)\theta(x) - \theta(x)\| < \varepsilon(\sqrt{M} + M).$$

The left hand side of the last inequality can differ from zero only if $\xi = px \in C \cdot S$ because $a \in N_e \subseteq C$ and so $p(a^{-1}x) = a^{-1}px \in S$ or $px \in a \cdot S \subseteq C \cdot S$ if x is such a point in G. Hence if we integrate in order to estimate $\|\rho^{\chi}(a)\theta - \theta\|$ it is sufficient to extend the range of integration only to the compact set $C \cdot S$ lying in X. We obtain

$$\|\rho^{\chi}(a)\theta - \theta\| < \varepsilon(\sqrt{M} + M)\mu(C \cdot S)$$

for every $a \in N_e$. Since $\varepsilon > 0$ is arbitrary we see that $a \mapsto \rho^\chi(a)\theta$ is continuous at e. Since ρ^χ is a unitary representation and is dense in $\mathscr{K}$ it follows for general reasons that ρ^χ is strongly continuous at every $\theta \in \mathscr{K}$. For we have the following:

Proposition 12. *If the unitary representation $\rho: G \to \mathscr{U}(\mathscr{K})$ is strongly continuous for every vector in a dense subset $\mathscr{M}$ of $\mathscr{K}$ then ρ is strongly continuous for every element of $\mathscr{K}$.*

Proof. It is sufficient to prove that $a \mapsto \rho(a)\theta$ is continuous at $a = e$ for every fixed $\theta \in \mathscr{K}$. Given $\varepsilon > 0$ we can choose ψ in $\mathscr{M}$ such that $\|\theta - \psi\| < \varepsilon/4$. Since $a \mapsto \rho(a)\psi$ is continuous at $a = e$ there is a neighborhood N_e of e such that $\|\rho(a)\psi - \psi\| < \varepsilon/2$ for every $a \in N_e$. For any a in G we have

$$\|\rho(a)\theta - \theta\| \leqslant \|\rho(a)(\theta - \psi)\| + \|\rho(a)\psi - \psi\| + \|\psi - \theta\|.$$

Hence if $a \in N_e$ then $\rho(a)$ being a unitary operator we obtain $\|\rho(a)\theta - \theta\| < \varepsilon$.

Proposition 13. *If H is a closed subgroup of G and $\bar{\chi}$ denotes the conjugate of the representation $\chi: H \to \mathscr{U}(\mathscr{H})$ then the representation $\rho^{\bar{\chi}}$ induced on G by $\bar{\chi}$ is equivalent to the conjugate representation $\overline{\rho^\chi}$.*

Proof. Let $\mathscr{M}$ be the inner product space defining $\mathscr{K}$, the Hilbert space of ρ^χ. We know that $\mathscr{H}$ and $\bar{\mathscr{H}}$ have the same underlying set and the same additive structure. Thus every $\theta \in \mathscr{M}$ can be interpreted as a function $\tilde{\theta}: G \to \bar{\mathscr{H}}$. Since the norm in $\bar{\mathscr{H}}$ is the same as in $\mathscr{H}$ the function $\tilde{\theta}$ is continuous. Moreover if $x \in G$ and $h \in H$ then

$$\tilde{\theta}(xh) = \theta(xh) = \chi(h^{-1})\theta(x) = \bar{\chi}(h^{-1}) \cdot \theta(x) = \bar{\chi}(h)^* \cdot \theta(x) = \bar{\chi}(h)^* \cdot \tilde{\theta}(x).$$

Therefore $\tilde{\theta} \in \tilde{\mathscr{M}}$ where $\tilde{\mathscr{M}}$ is the linear manifold used to define $\tilde{\mathscr{K}}$, the Hilbert space of $\rho^{\bar{\chi}}$. If we use the same quasi-invariant measure to define the inner product in $\mathscr{M}$ and $\tilde{\mathscr{M}}$ then

$$(\theta_1, \theta_2) = \int_X (\theta_1(x), \theta_2(x))\, d\mu(\xi) = \int_X [\tilde{\theta}_2(x), \tilde{\theta}_1(x)]\, d\mu(\xi) = [\tilde{\theta}_2, \tilde{\theta}_1]$$

where $[\cdot, \cdot]$ is used to denote the inner product in $\bar{\mathscr{H}}$ and $\tilde{\mathscr{K}}$. Since $\overline{(\theta_1, \theta_2)}$ is the inner product of θ_1 and θ_2 in $\bar{\mathscr{K}}$, the conjugate of $\mathscr{K}$ we see that $\theta \mapsto \tilde{\theta}$ can be extended uniquely to an inner product preserving linear map U of $\bar{\mathscr{K}}$ onto $\tilde{\mathscr{K}}$. One can easily see that U intertwines $\overline{\rho^\chi}$ with $\rho^{\bar{\chi}}$.

Lemma 14. *If $\chi: G \to \mathscr{U}(\mathscr{H})$ is continuous then ${}_G\rho^\chi$ is equivalent to χ.*

Proof. The homogeneous space $X=G/H$ consists of the single point $\xi\in G$ and we can let $\mu(X)=\mu(\xi)=1$. For every $\theta\in\mathcal{M}$ let $U\theta=\theta(e)$ so that U maps $\mathcal{M}$ linearly into $\mathcal{H}$. It is clear that $\|U\theta\|=\|\theta\|$ for every $\theta\in\mathcal{M}$. By the corollary of Lemma 8 $U\mathcal{M}$ is dense in $\mathcal{H}$ and so U extends to a linear isometry of $\mathcal{K}$ onto $\mathcal{H}$. Since

$$U_G\rho^\chi(a)\theta = U^a\theta = {}^a\theta(e) = \theta(a^{-1}) = \chi(a)\theta(e) = \chi(a)U\theta$$

we see that U intertwines ${}_G\rho^\chi$ with χ.

There will be a few occasions when it will be necessary to know that the elements of the Hilbert space $\mathcal{K}$ of the induced representations $\rho^\chi\colon G\to\mathcal{U}(\mathcal{K})$ can be identified with functions $\theta\colon G\to\mathcal{H}$. This fact is an easy consequence of the theory of integration of Hilbert space valued functions. This theory can be found for instance in Bourbaki's Intégration which we are going to follow.

Let E be a locally compact space, μ a positive measure on E and F a topological space. A function $\theta\colon E\to F$ is called measurable in the Bourbaki sense if for every compact set K lying in E and for every $\varepsilon>0$ there is a compact subset C of K such that $\mu(K-C)<\varepsilon$ and $\theta|C$ is continuous. If F is a Banach space then the following two conditions are necessary and sufficient for the measurability of θ:

a) If f^* is a continuous linear functional on F then $x\mapsto f^*(\theta(x))$ is a measurable function.

b) For every compact set K lying in E there is a countable set C in F such that $\theta(x)\in\bar{C}$ for almost all x in K.

These conditions show the connection between the general theory and the one discussed by us in the special case when F is a separable complex Hilbert space $\mathcal{H}$.

If θ is measurable then the non-negative function $x\mapsto\|\theta(x)\|$ is also measurable. Let $1\leqslant p<\infty$ and define $\|\theta\|_p=\mu^*(\|\theta(\cdot)\|^p)$ where μ^* denotes the upper integral with respect to the measure μ given on E. One proves that the set of those $\theta\colon G\to F$ for which $\|\theta\|_p<\infty$ is a complete space with respect to the norm $\|\theta\|_p$. The closure of $C_0(E,F)$ in this complete space is denoted by $\mathcal{L}_F^p(E,\mu)$. By identifying the equivalent elements of this space we obtain $L_F^p(E,\mu)$, the *L^p-space of p-th power integrable functions θ with values in the Banach space F*. Hence $L_F^p(E,\mu)$ is a Banach space. If $F=\mathcal{H}$ is a Hilbert space then by using the polarization identity one sees that $L_{\mathcal{H}}^2(E,\mu)$ is a Hilbert space.

Theorem 15. *Let G be a locally compact group, H a closed subgroup, $\mathcal{H}$ an arbitrary complex Hilbert space, $\chi\colon H\to\mathcal{U}(\mathcal{H})$ a weakly measurable unitary representation and μ a quasi-invariant measure on $X=G/H$.*

Then the Hilbert space $\mathscr{K}$ of the induced representation ρ^χ can be identified with a subspace of $L^2(G, \mu, \mathscr{H})$, namely $\mathscr{K}$ is the closure of the linear manifold $\mathscr{M}$ in $L^2(G, \mu, \mathscr{H})$. The elements of $\mathscr{K}$ are represented by functions $\theta: G \to \mathscr{H}$ which are measurable in the Bourbaki sense and satisfy $\theta(xh) = \chi(h)^* \theta(x)$ for every $x \in G$ and $h \in H$. If $\theta: G \to \mathscr{H}$ is continuous and satisfies this identity and $\|\theta\|_2 < \infty$ then θ belongs to $\mathscr{L}^2_{\mathscr{H}}(G, \mu)$ and so θ defines an element of $\mathscr{K}$.

Note. We use $L^2(G, \mu, \mathscr{H})$ to denote the space $L^2_{\mathscr{H}}(G, \mu)$.

Proof. The elements of $\mathscr{M}$ are all continuous functions $\theta: G \to \mathscr{H}$ satisfying $\theta(xh) = \chi(h)^* \theta(x)$. By the definition of μ in Theorem V.3.9 the norm $\|\theta\|$ used on $\mathscr{M}$ in the definition of $\mathscr{K}$ is the same as the L^2-norm $\|\theta\|_2$ used in $\mathscr{L}^2_{\mathscr{H}}(G, \mu)$. Hence $\mathscr{M}$ lies in $\mathscr{L}^2_{\mathscr{H}}(G, \mu)$ and so $\mathscr{K}$ can be considered as a subspace of $L^2(G, \mu, \mathscr{H})$. If θ in $\mathscr{L}^2_{\mathscr{H}}(G, \mu)$ belongs to the equivalence class of an element of $\mathscr{K}$ then by Corollary 2 of Theorem 2 in Chapter IV, § 3, No. 4 in Bourbaki there is a sequence of functions $\theta_n \in \mathscr{M}$ such that $\theta_n \to \theta$ in norm and $\theta_n(x) \to \theta(x)$ for almost every x in G. Hence we have $\theta(xh) = \chi(h)^* \theta(x)$ for almost all $x \in G$ and all $h \in H$.

The *kernel of a representation* $\rho: G \to \mathscr{U}(\mathscr{H})$ is the kernel of the group homomorphism ρ i.e. it is the normal subgroup consisting of those x in G for which $\rho(x) = I$, the identity operator. We are going to prove a theorem which gives a description of the kernel of an arbitrary induced representation $\rho^\chi: G \to \mathscr{U}(\mathscr{K})$. As usual we let Δ and δ denote the modular functions of G and H, respectively. Since $\delta/\Delta: H \to (0, \infty)$ is a group homomorphism we can speak about its kernel.

Theorem 16. *Let H be a closed subgroup of the locally compact group G and let $\chi: H \to \mathscr{U}(\mathscr{H})$ be a continuous unitary representation. Then we have*

$$\ker \rho^\chi = \bigcap_{a \in G} a^{-1} \left(\ker \chi \cap \ker \frac{\delta}{\Delta} \right) a .$$

We start the proof by first establishing the following:

Lemma 17. *If $x, y \in G$ and $x \notin yH$ then there is a θ in $\mathscr{M}$ such that $\theta(x) \neq 0$ and $\theta(yh) = 0$ for every $h \in H$.*

Proof. Let C be a compact neighborhood of e in G such that $y^{-1}xC \cap H = \emptyset$. Then of course $y^{-1}xCH \cap H$ is also empty. Let α be a non-zero vector in $\mathscr{H}$. Since χ is continuous there is a closed neighborhood C_e of e in G such that $\mathscr{Re}(\chi(h)\alpha, \alpha) > 0$ for every h in C_e. We let $K = C_e \cap C$ and choose a non-negative, not identically zero function φ in $C_0(G)$

such that its support S lies in K. Then

$$\mathcal{R}e \int_H \varphi(h)(\chi(h)\alpha, \alpha)\,dh = \int_K \varphi(h)\,\mathcal{R}e(\chi(h)\alpha, \alpha)\,dh > 0$$

by $K \subseteq C_e$. Therefore

$$\int_H \varphi(h)(\chi(h)\alpha, \alpha)\,dh \neq 0\,. \tag{8}$$

Now we define θ by

$$\theta(z) = \int_H \varphi(x^{-1}zh)\chi(h)\alpha\,dh$$

where $z \in G$. Then by (8) we have $\theta(x) \neq 0$. Moreover if $\theta(z) \neq 0$ then $x^{-1}zh \in S$ for some h in H. Hence $z \in xSH$ and $y^{-1}z \in y^{-1}xSH \subseteq y^{-1}xCH$. Therefore if $\theta(z) \neq 0$ then $y^{-1}z \notin H$. It follows that $\theta(yh) = 0$ for every $h \in H$.

Proof of Theorem 16. Let $\mathcal{M}$ be the inner product space whose completion is $\mathcal{K}$. Clearly we have $\rho^\chi(x) = I$ if and only if $\rho^\chi(x)|\mathcal{M}$ is the identity operator on $\mathcal{M}$. If $x \in \ker \rho^\chi$ then for every θ in $\mathcal{M}$ we have

$$\rho^\chi(x)\theta(y) = \theta(x^{-1}y)\lambda(x^{-1}, py)^{\frac{1}{2}} = \theta(y)$$

for all $y \in G$. Choosing $y = e$ we obtain

$$\theta(x^{-1})\lambda(x^{-1}, pe)^{\frac{1}{2}} = \theta(e) \tag{9}$$

and so using $\lambda(x^{-1}, pe) \neq 0$ and Lemma 17 we see that $x \in H$. Then we have

$$\theta(x^{-1}) = \chi(x)\theta(e) \tag{10}$$

and so $\chi(x)$ being a unitary operator (9) and Lemma 8 give $\lambda(x^{-1}, pe) = 1$. Since $\lambda(x^{-1}, pe) = \rho(x^{-1}e)/\rho(e) = (\delta/\Delta)(x^{-1}) = 1$ we proved that $x \in \ker \delta/\Delta$. By (9) and (10) we have $\chi(x)\theta(e) = \theta(e)$ for every θ in $\mathcal{M}$ and so by the corollary of Lemma 8 we obtain $\chi(x)\alpha = \alpha$ for every $\alpha \in \mathcal{H}$. Hence we also proved that $x \in \ker \chi$. Since $\ker \rho^\chi$ is a normal subgroup of G from $x \in \ker \rho^\chi$ we obtain $axa^{-1} \in \ker \rho^\chi$ for every a in G. Thus

$$axa^{-1} \in \ker \chi \cap \ker \frac{\delta}{\Delta} \tag{11}$$

and this proves that

$$\ker \rho^\chi \subseteq a^{-1}\left(\ker \chi \cap \ker \frac{\delta}{\Delta}\right)a$$

for every a in C.

Conversely let x be an element of G such that (11) holds for every $a \in G$. Then for any y in G we have $y^{-1}x^{-1}y \in \ker \chi \subseteq H$ and so by $y^{-1}x^{-1}y \in \ker \delta/\Delta$ we have

$$\rho(x^{-1}y) = \rho(y y^{-1} x^{-1} y) = \frac{\delta}{\Delta}(y^{-1}x^{-1}y)\rho(y) = \rho(y).$$

This gives $\lambda(x^{-1}, y) = \rho(x^{-1}y)/\rho(y) = 1$ for every y in G. Hence for any $\theta \in \mathcal{M}$ and $y \in G$ we have

$$\rho^{\chi}(x)\theta(y) = \theta(x^{-1}y)\lambda(x^{-1}, y)^{\frac{1}{2}} = \theta(x^{-1}y) = \chi(x)\theta(y) = \theta(y).$$

Therefore $\rho^{\chi}(x)\theta = \theta$ for every θ in $\mathcal{M}$ and so $x \in \ker \rho^{\chi}$.

Proposition 18. *If $\chi: H \to \mathcal{U}(\mathcal{H})$ is a faithful continuous representation then $\rho^{\chi}: G \to \mathcal{U}(\mathcal{K})$ is also faithful.*

Proof. By Theorem 16 we have in general $\ker \rho^{\chi} \subseteq \ker \chi$ and so the result is obvious.

Proposition 19. *If no subgroup of H except $\{e\}$ is a normal subgroup of G then ρ^{χ} is a faithful representation.*

Proof. The kernel of ρ^{χ} is a normal subgroup of G and it lies in $\ker \chi$ which is a subgroup of H.

By choosing $H = \{e\}$ and $\mathcal{H} = \mathbb{C}$ in the last proposition we see that the left regular representation λ of G is faithful. Of course this could be also proved directly.

5. The Existence of a Kernel for $\sigma: L^1(G) \to \mathcal{L}(\mathcal{K})$

Let G be a locally compact group and let $\lambda: G \to \mathcal{U}(L^2(G))$ be the left regular representation of G. By Theorem V.5.4 λ determines a representation $\sigma: L^1(G) \to \mathcal{L}(L^2(G))$. By Theorem V.5.5 and Propositions V.4.6 and V.5.3 we have $(\sigma(f)g, h) = (f * g, h)$ for every $g, h \in L^2(G)$ and if G is σ-finite then

$$\sigma(f)g(x) = \int_G f(xy^{-1})g(y)\,dy$$

for almost all $x \in G$. Therefore the function $(x, y) \mapsto f(xy^{-1})$ is a *kernel* for the operator $\sigma(f)$. The purpose of this section is the extension of this result to arbitrary induced representations. For the sake of simplicity we will develop the details only in the special case when the group G and the closed subgroup H are unimodular. The result will be used later only in this special case. The general situation will be described at the end of the section.

Let G be a locally compact group, let H be a closed subgroup of G and let $X=G/H$. As usual we let dx be a left Haar measure of G and dh a left Haar measure of H. We choose a measurable unitary representation $\chi: H\to\mathscr{U}(\mathscr{H})$ and a function $f: G\to\mathbb{C}$. Various conditions will be imposed on f as we progress.

Proposition 1. *If $x, y\in G$ are such that ${}_xf^y\in L^1(H)$ then*

$$K(x,y)=\int_H f(xhy^{-1})\chi(h)\,dh$$

defines a continuous linear operator $K(x,y)\colon\mathscr{H}\to\mathscr{H}$ in the weak sense and $|K(x,y)|\leqslant\|{}_xf^y\|_1$.

Note. Similar result holds when $\chi: H\to\mathscr{L}(\mathscr{H})$ is an operator valued function and if $h\mapsto{}_xf^y(h)|\chi(h)|$ belongs to $L^1(H)$.

Proof. For α, β in $\mathscr{H}$ the formula

$$\varphi(\alpha)=\int_H f(xhy^{-1})(\alpha,\chi(h)\beta)\,dh$$

defines a continuous linear functional $\varphi\colon\mathscr{H}\to\mathbb{C}$. Hence there is an element $K(x,y)\beta$ in $\mathscr{H}$ such that $\varphi(\alpha)=(\alpha, K(x,y)\beta)$. We have

$$|(\alpha, K(x,y)\beta)|\leqslant\|\alpha\|\cdot\|\beta\|\cdot\|{}_xf^y\|_1 \tag{1}$$

Proposition 2. *If $x, y\in G$ and $K(x,y)$ exists in the weak sense then $K(xa, yb)$ exists for every $a, b\in H$ and*

$$K(xa,yb)=\delta(b)\chi(a)^*K(x,y)\chi(b)\,.$$

Corollary. *Given $\xi,\eta\in X$ the operator $K(x,y)$ exists for all or none of the elements $x\in\xi, y\in\eta$.*

Proof. Let $\alpha,\beta\in\mathscr{H}$ be given. Then

$$\begin{aligned}(\chi(a)\alpha, K(x,y)\beta)&=\int_H f(xhy^{-1})(\chi(a)\alpha,\chi(h)\beta)\,dh\\ &=\int_H f(xhy^{-1})(\alpha,\chi(a^{-1}h)\beta)\,dh=\int_H f(xahy^{-1})(\alpha,\chi(h)\beta)\,dh\\ &=(\alpha, K(xa,y)\beta)\end{aligned}$$

and so $(\alpha,\chi(a)^*K(x,y)\beta)=(\alpha,K(xa,y)\beta)$. Similarly we have

$$\begin{aligned}(\alpha,\delta(b)K(x,y)\chi(b)\beta)&=\delta(b)\int_H f(xhy^{-1})(\alpha,\chi(h)\chi(b)\beta)\,dh\\ &=\int_H f(xhb^{-1}y^{-1})(\alpha,\chi(h)\beta)\,dh=(\alpha,K(x,yb)\beta)\,.\end{aligned}$$

From this point on we suppose that G and H are unimodular and the Hilbert space $\mathscr{H}$ of the representation χ is separable. If f is such that ${}_xf^y \in L^1(H)$ then $\|{}_{xa}f^{yb}\|_1 = \|{}_xf^y\|_1$ for $a, b \in H$ and so one can define $\|{}_\xi f^\eta\|_1$ where $\xi = xH$ and $\eta = yH$. We fix an invariant measure μ on X and in integrals we will write $d\xi$ instead of $d\mu(\xi)$. We let $\mathscr{K}$ denote the Hilbert space of the representation ρ induced on G by $\chi: H \to \mathscr{U}(\mathscr{H})$. As earlier we let $\theta, \psi, \ldots$ denote the elements of $\mathscr{K}$.

Proposition 3. *Let ξ in X be such that $K(x,y)$ exists for every $x \in \xi$ and $y \in \eta$ where η belongs to the complement of a zero set of X. Then for every x in ξ*

$$A(x)\theta = \int_X K(x,y)\theta(y)\,d\eta$$

defines in the weak sense a linear operator $A(x): \mathscr{K} \to \mathscr{H}$ which is continuous and satisfies

$$|A(x)| \leqslant \sqrt{\int_X (\|{}_\xi f^\eta\|_1)^2\,d\eta}\,. \tag{2}$$

For any h in H we have

$$A(xh) = \chi(h)^* A(x)\,. \tag{3}$$

Proof. First of all the integrand depends only on η, not on the particular y chosen in η: For by Proposition 2 we have for any h in H

$$K(x, yh)\theta(yh) = K(x,y)\chi(h)\theta(yh) = K(x,y)\theta(y)\,.$$

Next we define the continuous linear functional $\varphi: \mathscr{H} \to \mathbb{C}$ by

$$\varphi(\alpha) = \int_X (\alpha, K(x,y)\theta(y))\,d\eta$$

where $\theta \in \mathscr{K}$. The existence and continuity of φ follows from (1) because by the BCS inequality

$$|\varphi(\alpha)| \leqslant \|\alpha\| \cdot \|\theta\| \sqrt{\int_X \|{}_\xi f^\eta\|_1^2\,d\eta}\,. \tag{4}$$

Hence there is an element $A(x)\theta$ in $\mathscr{H}$ such that $\varphi(\alpha) = (\alpha, A(x)\theta)$ for every $\alpha \in \mathscr{H}$. By letting $\alpha = A(x)\theta$ and using (4) we obtain $\|A(x)\theta\| \leqslant \|\theta\| \sqrt{\int_X \|{}_\xi f^\eta\|_1^2\,d\eta}$.

Note. In the beginning of the proof we have also shown the following: If ξ, η in X are such that $K(x,y)$ exists then for any θ in $\mathscr{K}$ the value of $K(x,y)\theta(y)$ depends only on x and the equivalence class η.

Proposition 4. *If*

$$\|f\| = \left(\int_X \int_X \|{}_\xi f^\eta\|^2 \, d\xi \, d\eta\right)^{\frac{1}{2}}$$

is finite and θ is such that $A(x)\theta$ exists for all x in almost every ξ than $A_f\theta(x) = A(x)\theta$ defines an element $A_f\theta$ of $\mathscr{H}$ and $\|A_f\theta\| \leqslant \|f\|\cdot\|\theta\|$.

Proof. Let S be the set of those ξ in X for which $A(x)\theta$ exists for every x in ξ. For all $\xi \notin S$ and $x \in \xi$ we let $A(x)\theta = 0$. We define $A_f\theta(x) = A(x)\theta$ for every ξ in X and every x in ξ. Then by (3) we have $A_f\theta(xh) = \chi(h)^*(A_f\theta)(x)$ for every $x\in G$. Moreover using (2) we obtain

$$\begin{aligned}\|A_f\theta\|^2 &= \int_X \|A_f\theta(x)\|^2 d\xi = \int_X \|A(x)\theta\|^2 d\xi \leqslant \|\theta\|^2 \int_X |A(x)|^2 d\xi \\ &\leqslant \|\theta\|^2 \int_X \int_X (\|{}_\xi f^\eta\|)^2 d\eta \, d\xi = \|f\|^2\|\theta\|^2 .\end{aligned}$$

This proves the inequality $\|A_f\theta\| \leqslant \|f\|\cdot\|\theta\|$.

Theorem 5. *Let the measurable $f: G\to\mathbb{C}$ be such that $\|f\| < +\infty$. Then $A_f\theta$ exists for every θ in $\mathscr{H}$. The map $\theta\mapsto A_f\theta$ is a continuous linear operator $A_f: \mathscr{H}\to\mathscr{H}$ and its norm satisfies $|A_f| \leqslant \|f\|$.*

Proof. Let $S(\xi)$ be the set of those pairs (ξ,η) in $X\times X$ for which $\|{}_\xi f^\eta\|_1 = +\infty$ and let

$$S_n = \left\{\xi : \mu S(\xi) \geqslant \frac{1}{n}\right\}.$$

By $\|f\| < +\infty$ we have $\mu S_n = 0$ for every $n = 1, 2, \ldots$ and so $S = S_1 \cup S_2 \cup \cdots$ is a set of measure zero. For $\xi\notin S$ we have $\mu S(\xi) = 0$. Hence if $\xi \notin S$ then for every η outside a set of measure zero $\|{}_\xi f^\eta\|_1 < +\infty$ and so by Propositions 1 and 3 $A(x)\theta$ is defined for every $x\in\xi$. Therefore by Proposition 4 $A_f\theta$ is defined for every $\theta\in\mathscr{H}$, it belongs to $\mathscr{H}$ and $\|A_f\theta\| \leqslant \|f\|\cdot\|\theta\|$.

Let $\rho: G\to\mathscr{U}(\mathscr{H})$ be the representation induced on G by χ. If $f: G\to\mathbb{C}$ is a measurable function we have

$$\sigma(f)\theta = \int_G f(x)\rho(x)\,dx$$

in the weak sense for every θ in $\mathscr{H}$ for which the definition is meaningful.

Proposition 6. *If $f: G\to\mathbb{C}$ is measurable and $\theta\in\mathscr{H}$ then $\sigma(f)\theta$ exists if and only if $A_f\theta$ exists and the two are equal.*

Corollary. *We have* $\sigma(f)=A_f$ *both for domain and values and in particular* $\sigma(f)=A_f$ *for every* f *in* $L^1(G)$.

Proof. Let θ and ψ belong to $\mathscr{K}$. If f is such that $\sigma(f)\theta$ exists then

$$(\psi,\sigma(f)\theta)=\int_G f(y)(\psi,\rho(y)\theta)\,dy=\int_G f(y)\int_X(\psi(x),\rho(y)\theta(x))\,d\xi\,dy\,.$$

Using the definition of ρ we obtain

$$(\psi,\sigma(f)\theta)=\int_X\int_G f(y)(\psi(x),\theta(y^{-1}x)\,dy\,d\xi$$

and so by the unimodularity of G

$$(\psi,\sigma(f)\theta)=\int_X\int_G f(xy^{-1})(\psi(x),\theta(y))\,dy\,d\xi\,.$$

Applying Propositions V.3.14 and the unimodularity of H we have

$$(\psi,\sigma(f)\theta)=\int_X\int_X\int_H f(xhy^{-1})(\psi(x),\theta(yh^{-1})\,dh\,d\eta\,d\xi\,.$$

Hence using $\theta\in\mathscr{K}$ we obtain

$$(\psi,\sigma(f)\theta)=\int_X\int_X\int_H f(xhy^{-1})(\psi(x),\chi(h)\theta(y))\,dh\,d\eta\,d\xi\,.$$

Now we can apply Proposition 1 and replace the inner integral by $(\psi(x),K(x,y)\theta(y))$ and obtain

$$(\psi,\sigma(f)\theta)=\int_X\int_X(\psi(x),K(x,y)\theta(y))\,d\eta\,d\xi\,. \tag{5}$$

In view of the definition of $A(x)\theta$ given in Proposition 3 we proved that if $\sigma(f)\theta$ exists then $A(x)\theta$ is defined and

$$(\psi,\sigma(f)\theta)=\int_X(\psi(x),A(x)\theta)\,d\xi\,.$$

Since $A(x)\theta$ is defined for every x in almost all ξ in X we see that $A_f\theta$ is defined and $(\psi,\sigma(f)\theta)=(\psi,A_f\theta)$. Thus the existence of $\sigma(f)\theta$ implies that $A_f\theta$ exists and $\sigma(f)\theta=A_f\theta$.

Conversely suppose that f and θ are such that $A_f\theta$ exists. Then by Proposition 3 we have for any $x\in G$ at which $A_f\theta(x)=A(x)\theta$ exists

$$(\psi(x),A_f\theta(x))=\int_X(\psi(x),K(x,y)\theta(y))\,d\eta\,.$$

Since $A_f\theta$ exists this formula holds for every x in almost all ξ in X and so

$$(\psi, A_f\theta) = \int_X (\psi(x), A_f\theta(x))\,d\xi = \int_X \int_X (\psi(x), K(x,y)\theta(y))\,d\eta\,d\xi\,.$$

Starting from the right hand side of this equation and proceeding backwards in the reasoning given in the first half of the current proof we see that $(\psi, \sigma(f)\theta)$ exists and $(\psi, A_f\theta) = (\psi, \sigma(f)\theta)$. Therefore the existence of $A_f\theta$ implies that $\sigma(f)\theta$ exists and is equal to $A_f\theta$.

Our results can be summarized in the following:

Theorem 7. *Let H be a closed unimodular subgroup of the unimodular group G and let $\chi: H \to \mathscr{U}(\mathscr{H})$ be a representation of H in the separable Hilbert space $\mathscr{H}$. Let $\rho: G \to \mathscr{U}(\mathscr{K})$ denote the representation induced on G by χ and let $\sigma: L^1(G) \to \mathscr{L}(\mathscr{K})$ be the algebra representation associated with ρ. Then for every f in $L^1(G)$ the operator valued integral*

$$K_f(x,y) = \int_H f(xhy^{-1})\chi(h)\,dh$$

exists in the weak sense for every $x \in \xi$, $y \in \eta$ and almost all ξ, η in $X = G/H$. In the weak sense we have

$$\sigma(f)\theta(x) = \int_X K_f(x,y)\theta(y)\,d\eta$$

for every $x \in \xi$ and almost all ξ in X.

Now we turn to the general situation when G is a locally compact group and H is a closed subgroup. Let dx and dh be left Haar measures on G and H and let Δ and δ be the modular functions of dx and dh respectively. Given a representation $\chi: H \to \mathscr{U}(\mathscr{H})$ in the separable Hilbert space $\mathscr{H}$ we let $\sigma: L^1(G) \to \mathscr{L}(\mathscr{K})$ denote the algebra representation associated with the representation induced on G by χ. We shall use the symbol ρ to denote the function $\rho: G \to (0, \infty)$ which determines the quasi-invariant measure μ used in the construction of the Hilbert space $\mathscr{K}$.

Proposition 8. *For every f in $L^1(G)$ and θ in $\mathscr{K}$ we have*

$$\sigma(f)\theta(x) = \int_{\frac{G}{H}} K_f(x,y)\theta(y)\,d\mu(\eta)$$

in the weak sense where

$$K_f(x,y) = \int_H f(xhy^{-1})\frac{D(h)}{\Delta(y)}\frac{1}{\sqrt{\rho(x)\rho(yh^{-1})}}\chi(h)\,dh\,.$$

Note. If G/H has a relatively invariant measure with modular function then D then

$$K_f(x,y)=\int_H f(xhy^{-1})\frac{\sqrt{D(xhy)}}{\Delta(y)}\chi(h)dh.$$

Proof. Instead of giving a detailed proof we will only show that $K_f(x,y)$ is the operator valued kernel given in the proposition and

$$\sigma(f)\theta=\int_{\frac{G}{H}} K_f(\cdot\,,y)\theta\, d\mu(\eta)$$

in the weak sense in $\mathscr{K}$. In the proposition this formula is stated in the weak pointwise sense in $\mathscr{H}$.

Given θ and ψ in $\mathscr{K}$ by definition we have

$$\begin{aligned}(\sigma(f)\theta,\psi)&=\int_G f(y)\,({}_G\rho^\chi(y)\theta,\psi)\,d\eta\\&=\int_G f(y)\int_X(\theta(y^{-1}x)\sqrt{\lambda(y^{-1},\xi)},\psi(x))\,d\xi\,dy\end{aligned}$$

where λ is the function associated with $\rho\colon G\to(0,\infty)$. We make the substitution $y\to xy^{-1}$ and apply Lemma V.3.10:

$$(\rho(f)\theta,\psi)=\int_G\int_X f(xy^{-1})\,(\theta(y),\psi(x))\sqrt{\frac{\rho(y)}{\rho(x)}}\,\Delta(y^{-1})\,d\xi\,dy.$$

Next we use the corollary of Theorem V.3.9 to obtain

$$(\sigma(f)\theta,\psi)$$
$$=\int_X\int_H\int_X f(xh^{-1}y^{-1})\,(\theta(yh),\psi(x))\frac{1}{\sqrt{\rho(x)\rho(yh)}}\Delta(h^{-1}y^{-1})\,d\xi\,dh\,d\eta.$$

Therefore by the definition of θ and using the substitution $h\to h^{-1}$ we obtain

$$(\sigma(f)\theta,\psi)$$
$$=\int_X\int_X\int_H f(xhy^{-1})\,(\chi(h)\theta(y),\psi(x))\frac{1}{\sqrt{\rho(x)\rho(yh^{-1})}}\Delta(hy^{-1})\delta(h^{-1})\,dh\,d\xi\,d\eta.$$

Finally we have

$$(\sigma(f)\theta,\psi)$$

$$=\int_X\left(\int_X\left(\int_H f(xhy^{-1})\frac{D(h)}{\Delta(y)}\frac{1}{\sqrt{\rho(x)\rho(yh^{-1})}}\chi(h)\,dh\right)\theta(y)\,d\eta,\psi(x)\right)d\xi.$$

Proposition 9. *The operator valued kernel $K_f(x,y)$ satisfies*

$$K_f(xa,yb)=\chi(a)^*K_f(x,y)\chi(b)$$

for every $x,y\in G$ and $a,b\in H$.

Proof. First consider $K_f(xa,y)$, perform the substitution $h\to a^{-1}h$ and apply $D(a)\rho(xa)=\rho(x)$. (See Theorem V.3.9.) Next consider $K_f(x,yb)$, perform the substitution $h\to hb$ and apply the same formula.

Note. The difference between the properties expressed in Proposition 2 and the present proposition is due to the fact that the kernel considered there is different from the present one. It is actually the specialization of the present, general kernel to the special case of unimodular group and subgroups.

6. The Direct Sum Decomposition of the Induced Representation $\rho^\chi: G\to\mathscr{U}(\mathscr{K})$

In this section let G be a locally compact unimodular group and let H be unimodular subgroup of G. We let $\chi: H\to\mathscr{U}(\mathscr{H})$ be a weakly measurable unitary representation of H in the separable Hilbert space $\mathscr{H}$. We use $\rho^\chi: G\to\mathscr{U}(\mathscr{K})$ to denote the representation induced on G by χ. Our object is to prove the existence of a direct sum decomposition of $\rho^\chi: G\to\mathscr{U}(\mathscr{K})$ when H is closed, the homogeneous space $X=G/H$ is compact and $\mathscr{H}$ is finite dimensional. The concepts and techniques used in the proof will be further developed and applied again later.

We shall study operator valued functions $K: G\times G\to\mathscr{L}(\mathscr{H})$ such that

$$K(xa,yb)=\chi(a)^*K(x,y)\chi(b) \tag{1}$$

for $x,y\in G$ and $a,b\in H$. The kernels K_f considered in Section 5 belong to this family of functions.

Proposition 1. *If $\theta\in\mathscr{K}$ then $K(x,\cdot)\theta(\cdot)$ is a function on X that is to say $K(x,ya)\theta(ya)=K(x,y)\theta(y)$ for every $x,y\in G$ and $a\in H$.*

Proof. This follows from the definitions of the function K and the Hilbert space $\mathscr{K}$: For

$$K(x,ya)\theta(ya) = K(x,y)\chi(a)\chi(a)^*\theta(y) = K(x,y)\theta(y).$$

Proposition 2. *For every θ and ψ in $\mathscr{K}$ the inner product $(K(x,y)\theta(y),\psi(x))$ is defined on $X \times X$.*

Proof. Given $x,y \in G$ and $a,b \in H$ we have

$$(K(xa,yb)\theta(yb),\psi(xa)) = (\chi(a)^* K(x,y)\chi(b)\chi(b)^*\theta(y),\chi(a)^*\psi(x)).$$

Since $\chi(a)^*$ is unitary the right hand side is equal to

$$(K(x,y)\chi(b)\chi(b)^*\theta(y),\psi(x)) = (K(x,y)\theta(y),\psi(x)).$$

Proposition 3. *The HS norm $\|K(x,y)\|$ is defined on $X \times X$.*

Proof. Let α_i $(i \in \mathscr{I})$ be a MONS in $\mathscr{H}$. Then $\|K(x,y)\|^2 = \sum \|K(x,y)\alpha_i\|^2$ Therefore $\|K(xa,y)\|^2 = \sum \|K(xa,y)\alpha_i\|^2$ and $\chi(a)^*$ being unitary we obtain

$$\|K(xa,y)\alpha_i\|^2 = \|\chi(a)^* K(x,y)\alpha_i\|^2 = \|K(x,y)\alpha_i\|^2.$$

We also have

$$\|K(x,yb)\|^2 = \sum \|K(x,yb)\alpha_i\|^2 = \sum \|K(x,y)\chi(b)\alpha_i\|^2 = \|K(x,y)\|^2$$

because $\chi(b)$ being unitary $\chi(b)\alpha_i$ $(i \in \mathscr{I})$ is a MONS in $\mathscr{H}$.

Proposition 4. *For any $K_k\colon G \times G \to \mathscr{L}(\mathscr{H})$ $(k=1,2)$ satisfying* (1)

$$(K_1(x,y),K_2(x,y)) = \operatorname{tr} K_2(x,y)^* K_1(x,y)$$

is defined on $X \times X$.

Proof. If $x,y \in G$ and $a,b \in H$ then we have

$$\begin{aligned}\operatorname{tr} K_2(xa,y)^* K_1(xa,y) &= \operatorname{tr}(\chi(a)^* K_2(x,y))^*(\chi(a)^* K_1(x,y))\\ &= \operatorname{tr} K_2(x,y)^*\chi(a)\chi(a)^* K_1(x,y)\\ &= \operatorname{tr} K_2(x,y)^* K_1(x,y)\end{aligned}$$

because $\chi(a)$ is unitary. Similarly

$$\begin{aligned}\operatorname{tr} K_2(x,yb)^* K_1(x,yb) &= \operatorname{tr}(K_2(x,y)\chi(b))^*(K_1(x,y)\chi(b))\\ &= \operatorname{tr}\chi(b)^* K_2(x,y)^* K_1(x,y)\chi(b).\end{aligned}$$

Since $\chi(b)^*$ is unitary the right hand side is equal to $\operatorname{tr} K_2(x,y)^* K_1(x,y)$ by Lemma II.5.9.

Let $\mathscr{L}$ be the linear space of those functions $K: G \times G \to \mathscr{L}(\mathscr{H})$ satisfying (1) for which $(x,y) \mapsto (K(x,y)\alpha, \beta)$ is a measurable function on $G \times G$ for every fixed $\alpha, \beta \in \mathscr{H}$. If K_1 and K_2 are in $\mathscr{L}$ then $(x,y) \mapsto (K_1(x,y), K_2(x,y))$ is a measurable function on $G \times G$. For if α_i $(i \in \mathscr{I})$ is a MONS in $\mathscr{H}$ then

$$(K_1(x,y), K_2(x,y)) = \sum (K_1(x,y)\alpha_i, \alpha_j)(\alpha_j, K_2(x,y)\alpha_i)$$

and the terms of the series are measurable functions. Hence for K in $\mathscr{L}$ we can define

$$\|K\|^2 = \int_X \int_X \|K(x,y)\|^2 \, d\xi \, d\eta .$$

If K_k $(k=1,2)$ are such that $\|K_k\| < \infty$ then by the BCS inequality the pseudo-inner product

$$(K_1, K_2) = \int_X \int_X (K_1(x,y), K_2(x,y)) \, d\xi \, d\eta$$

is defined and in particular $(K,K) = \|K\|^2$. As usual we identify K_1 and K_2 if $\|K_1 - K_2\| = 0$.

Definition 5. *We let $\mathscr{L}^2 = \mathscr{L}^2(G,H,\chi)$ denote the inner product space of equivalence classes of those functions $K: G \times G \to HS(\mathscr{H})$ for which*

1) *$(K(\cdot,\cdot)\alpha, \beta)$ is a measurable function on $G \times G$ for every $\alpha, \beta \in \mathscr{H}$;*
2) *$K(xa, yb) = \chi(a)^* K(x,y) \chi(b)$ for almost every $x, y \in G$ and all $a, b \in H$;*
3) *$\|K\|$ is finite.*

If $K \in \mathscr{L}^2$ then we let K^* be the function defined by $K^*(x,y) = K(y,x)^*$. We see that $K^* \in \mathscr{L}^2$ and $\|K^*\| = \|K\|$.

Given θ and ψ in $\mathscr{K}$ we define $\theta \overline{\otimes} \psi: G \times G \to \mathscr{L}(\mathscr{H})$ by $(\theta \overline{\otimes} \psi)(x,y) = \theta(x) \overline{\otimes} \psi(y)$ where $x, y \in G$. Hence if $\alpha \in \mathscr{H}$ then $(\theta \overline{\otimes} \psi)(x,y)\alpha = (\alpha, \psi(y))\theta(x)$. Thus $\theta \overline{\otimes} \psi(x,y)$ is a continuous linear operator on $\mathscr{H}$. We shall prove that $\theta \overline{\otimes} \psi$ belongs to $\mathscr{L}^2$.

Proposition 6. *If θ and ψ are in $\mathscr{K}$ then*

$$(\theta \overline{\otimes} \psi)(xa, yb) = \chi(a)^* (\theta \overline{\otimes} \psi)(x,y) \chi(b)$$

for every $x, y \in G$ and $a, b \in H$.

Proof. Given α in $\mathscr{H}$ we have

$$\begin{aligned}(\theta\bar{\otimes}\psi)(xa,yb)\alpha &= (\alpha,\psi(yb))\theta(xa) = (\alpha,\chi(b)^*\psi(y))\chi(a)^*\theta(x)\\ &= (\chi(b)\alpha,\psi(y))\chi(a)^*\theta(x) = \chi(a)^*\{(\chi(b)\alpha,\psi(y))\theta(x)\}\\ &= \chi(a)^*\{(\theta(x)\bar{\otimes}\psi(y))\chi(b)\alpha\} = \chi(a)^*(\theta\bar{\otimes}\psi)(x,y)\chi(b)\alpha.\end{aligned}$$

Proposition 7. *If* $\theta_1,\theta_2,\psi_1,\psi_2\in\mathscr{K}$ *and* $x,y\in G$ *then we have*

$$((\theta_1\bar{\otimes}\psi_1)(x,y),(\theta_2\bar{\otimes}\psi_2)(x,y)) = (\theta_1(x),\theta_2(x))(\psi_2(y),\psi_1(y)).$$

Proof. In general $(a\bar{\otimes}b,c\bar{\otimes}d)=(a,c)(d,b)$ so using $(\theta_k\bar{\otimes}\psi_k)(x,y)=\theta_k(x)\bar{\otimes}\psi_k(y)$ with $k=1,2$ we obtain the desired result.

Proposition 8. *If* $\theta_1,\theta_2,\psi_1,\psi_2\in\mathscr{K}$ *then*

$$(\theta_1\bar{\otimes}\psi_1,\theta_2\bar{\otimes}\psi_2)=(\theta_1,\theta_2)(\psi_2,\psi_1).$$

In particular for any $\theta,\psi\in\mathscr{K}$ *we have* $\theta\bar{\otimes}\psi\in\mathscr{L}^2$ *and* $\|\theta\bar{\otimes}\psi\|=\|\theta\|\cdot\|\psi\|$.

Proof. Apply Proposition 7 and integrate over $X\times X$.

Corollary 1. *If* $\theta_1\perp\theta_2$ *or* $\psi_1\perp\psi_2$ *then* $\theta_1\bar{\otimes}\psi_1\perp\theta_2\bar{\otimes}\psi_2$.

Corollary 2. *If* $\theta_i\,(i\in\mathscr{I})$ *is a MONS in* $\mathscr{K}$ *then the operator valued functions* $\phi_{ij}=\theta_i\bar{\otimes}\theta_j$ *form an ONS in* $\mathscr{L}^2$.

Given $K\in\mathscr{L}^2$, $\theta\in\mathscr{K}$ and $\beta\in\mathscr{H}$ the inner product $(K(x,\cdot)\theta(\cdot),\beta)$ is a measurable function on X because given a MONS $\alpha_i\,(i\in\mathscr{I})$ in $\mathscr{H}$ we have

$$(\theta(y),K(x,y)^*\beta)=\sum(\theta(y),\alpha_i)(\alpha_i,K(x,y)^*\beta)$$

where $y\mapsto(\theta(y),\alpha_i)$ and $y\mapsto(\alpha_i,K(x,y)^*\beta)$ are measurable functions on G. Given $x\in G$ and $\beta\in\mathscr{H}$ the function $y\mapsto(K(x,y)\theta(y),\beta)$ belongs to $L^2(X)$. For

$$|(K(x,y)\theta(y),\beta)|\leqslant\|K(x,y)\theta(y)\|\cdot\|\beta\|\leqslant\|K(x,y)\|\cdot\|\theta(y)\|\cdot\|\beta\|$$

where $\|K(x,y)\|$ denotes the HS norm. Hence using the BCS inequality we obtain the estimate $\|(K(x,\cdot)\theta(\cdot),\beta)\|_2\leqslant\|K\|(\xi)\|\theta\|\cdot\|\beta\|$ for every $x\in\xi$ and for almost all $\xi\in X$ where $\|K\|(\xi)^2=\int_X\|K(x,y)\|^2\,d\eta$. It follows that

$$l_x(\beta)=\int_X(K(x,y)\theta(y),\beta)\,d\eta$$

is a continuous anti-linear functional on $\mathscr{H}$ and $|l_x|\leqslant\|K\|(\xi)\|\theta\|$. Therefore a vector $A_K(x)$ can be defined in the weak sense for every $x\in\xi$ and almost all $\xi \in X$ by the formula $l_x(\beta) = (A_K \theta(x), \beta)$. Hence we obtain a Hilbert space valued function $x \mapsto A_K \theta(x)$ and formally we can write

$$A_K \theta(x) = \int_X K(x,y)\theta(y)\,d\eta\,.$$

Similarly given $K \in \mathscr{L}^2$ and $\theta, \psi \in \mathscr{K}$ we see that $(x,y) \mapsto (K(x,y)\theta, \psi)$ defines a measurable function on $X \times X$. Moreover

$$|(K(x,y)\theta(y), \psi(x))| \leqslant \|K(x,y)\| \cdot \|\theta(y)\| \cdot \|\psi(x)\|$$

and so by the BCS inequality our function is in $L^2(X \times X)$ and its norm is at most $\|K\| \cdot \|\theta\| \cdot \|\psi\|$. Hence using the continuous anti-linear functional

$$l(\psi) = \int_X \int_X (K(x,y)\theta(y), \psi(x))\,d\xi\,d\eta$$

we can define $A_K \theta \in \mathscr{K}$ by the equation $l(\psi) = (A_K \theta, \psi)$. Since we have $\|A_K \theta\| \leqslant \|K\| \cdot \|\theta\|$ we see that $A_K : \mathscr{K} \to \mathscr{K}$ is a continuous linear operator and $|A_K| \leqslant \|K\|$.

Proposition 9. *For any HS operator* $K \in \mathscr{L}(\mathscr{H})$ *and every* $\alpha, \beta \in \mathscr{H}$ *we have* $(\alpha \overline{\otimes} \beta, K^*) = (K\alpha, \beta)$.

Proof. Let α_i $(i \in \mathscr{I})$ be a MONS in $\mathscr{H}$. Then

$$\begin{aligned}(\alpha \overline{\otimes} \beta, K^*) &= \sum ((\alpha \overline{\otimes} \beta)\alpha_i, K^* \alpha_i) = \sum ((\alpha_i, \beta)\alpha, K^* \alpha_i) \\ &= \sum (\alpha_i, \beta)(K\alpha, \alpha_i) = (K\alpha, \beta)\,.\end{aligned}$$

Given $K \in \mathscr{L}^2$, vectors $\theta, \psi \in \mathscr{K}$ and $x, y \in G$ we can apply the last proposition with $\alpha = \theta(x)$, $\beta = \psi(y)$ and $K = K(y,x)$. We obtain $(\theta(x) \overline{\otimes} \psi(y), K(y,x)^*) = (K(y,x)\theta(x), \psi(y))$ and so

$$(A_K \theta, \psi) = \int_X \int_X (\theta(x) \overline{\otimes} \psi(y), K^*(x,y))\,d\xi\,d\eta\,.$$

Hence we proved the following:

Corollary. *For every* $K \in \mathscr{L}^2$ *and* $\theta, \psi \in \mathscr{K}$ *we have* $(A_K \theta, \psi) = (\theta \overline{\otimes} \psi, K^*)$.

Proposition 10. *If* $K \in \mathscr{L}^2$ *and* θ, ψ *vary over* $\mathscr{K}$ *then the formula*

$$(A_K \theta, \psi) = \int_X \int_X (K(x,y)\theta(y), \psi(x))\,d\eta\,d\xi$$

defines in the weak sense a HS operator $A_K : \mathscr{K} \to \mathscr{K}$.

Note. In the next section we shall see that conversely every HS operator $A: \mathscr{K} \to \mathscr{K}$ can be defined by a suitable kernel $K \in \mathscr{L}^2$.

Proof. We already proved that the above expression defines a continuous linear operator $A_K: \mathscr{K} \to \mathscr{K}$ and $|A_K| \leqslant \|K\|$. In order to prove that A_K belongs to the HS class we choose a MONS $\theta_i\,(i \in \mathscr{I})$ in $\mathscr{K}$ and form the ONS $\phi_{ij}\,(i,j \in \mathscr{I})$ introduced in Corollary 2 of Proposition 8. Then by the corollary of Proposition 9

$$(A_K \theta_i, \theta_j) = (\theta_i \bar{\otimes} \theta_j, K^*) = (\phi_{ij}, K^*).$$

Hence by Bessel's inequality it follows that

$$\|A_K\|^2 = \sum_{i,j} |(A_K \theta_i, \theta_j)|^2 = \sum_{i,j} |(\phi_{ij}, K^*)|^2 \leqslant \|K^*\|^2 = \|K\|^2$$

where the right hand side in finite by $K \in \mathscr{L}^2$.

Proposition 11. *For any Borel subset S of the locally compact group G define $\varphi: G \times G \to [0, \infty]$ by*

$$\varphi(x,y) = \int_H c_S(xhy^{-1})\,dh = \int_H c_{x^{-1}Sy}(h)\,dh$$

where c_S denotes the characteristic function of S. Then for unimodular H, compact G/H and compact S the function φ is bounded.

Proof. Let O be an open neighborhood of the identity element e such that $O^{-1} = O$ and $\bar{O}$ is compact. Then OH is an open subset of G in the G/H-topology of G, or equivalently $\{oH: o \in O\}$ is an open subset of G/H in the quotient topology. By $e \in O$ we have $H \subseteq OH$. For each x in G we consider the set xOH which is open in the G/H-topology of G and covers x. The family $xOH\,(x \in G)$ covers G so there exists a finite subcover, say $x_1 OH, \ldots, x_n OH$.

We notice that by the unimodularity of H we have $\varphi(xh,y) = \varphi(x,yh) = \varphi(x)$ for every $x, y \in G$ and $h \in H$. Now let a and b in G be given. Then $a \in x_i OH$ and $b \in x_j OH$ for some indices $i,j\ (1 \leqslant i,j \leqslant n)$ and so $a = x_i o_a h_a$ and $b = x_j o_b h_b$ for suitable $o_a, o_b \in O$ and $h_a, h_b \in H$. Hence

$$x = a h_a^{-1} = x_i o_a \in x_i O \quad \text{and} \quad y = b h_b^{-1} = x_j o_b \in x_j O$$

and so by the right invariance of φ we have $\varphi(x,y) = \varphi(a,b)$. Since $y \in x_j O$ and $x^{-1} \in O^{-1} x_i^{-1} = O x_i^{-1}$ we see that $x^{-1} S y \subseteq O x_i^{-1} S x_j O$. Therefore

$$\varphi(a,b) = \varphi(x,y) = \int_H c_{x^{-1}Sy}(h)\,dh = \int_H c_{Ox_i^{-1}Sx_jO}(h)\,dh.$$

Thus for all $a, b \in G$ we have

$$0 \leqslant \varphi(a,b) \leqslant \int_H c_A(h)\,dh = m_H(A)$$

where

$$A = \bigcup_{i,j=1}^{n} \bar{O} x_i^{-1} S x_j \bar{O}.$$

Here the closure $\bar{O}$ and the given set S are compact, so A is a compact set in the original, locally compact topology of G and so $m_H(A)$ is finite.

The same reasoning can be used to prove:

Lemma 12. *If G/H is compact then for any compact subset S of G*

$$\varphi(x) = \int_H c_S(xh)\,dh = \int_H c_{x^{-1}S}(h)\,dh$$

defines a bounded function on G.

Note. This is a stronger result then the one obtained from Proposition 11 by letting $y=e$ because here the unimodularity of H is not required.

If G/H is compact and has an invariant measure then the boundedness of $\varphi(\cdot)$ implies:

Lemma 13. *If $g: G/H \to \mathbb{C}$ and $g \in L^1(G/H)$ then g is locally integrable on G.*

Proof. If S is a compact set in G then

$$\int_S |g(x)|\,dx = \int_{\frac{G}{H}} \int_H c_S(xh)|g(xh)|\,dh\,d\xi = \int_{\frac{G}{H}} |g(x)|\,\varphi(x)\,d\xi\,.$$

A second application of the function appearing in Lemma 12 concerns the quasiinvariant measure μ derived from a positive ρ-function defined on G.

Proposition 14. *If G is a locally compact group and H is a subgroup such that G/H is compact then $\mu(G/H)$ is finite for every quasi-invariant measure μ derived from a positive continuous ρ-function.*

Proof. By Lemma V.3.6 there is a compact set C in G such that its image under the natural map $p: G \to G/H$ is G/H itself. Let S be another compact set in G such that C lies in its interior. Let $f: G \to [0,1]$ be a continuous function with compact support such that f is identically 1 on S.

Then given any ξ in $X=G/H$ there is an x in C such that $p(x)=x+H=\xi$ and so there is an open set O_x in G for which $x\in O_x\subseteq S$. Thus

$$\varphi(\xi)=\int_H c_S(xh)\,dh\geqslant\int_H c_{O_x}(xh)\,dh=\mu_H(H\cap x^{-1}O_x) \tag{2}$$

where μ_H denotes the left Haar measure on H. Since $H\cap x^{-1}O_x$ is an open subset of H and it contains e we see that the right hand side of (2) is positive. Hence $\varphi(\xi)>0$ for every $\xi\in X$ and so by the compactness of X there is a constant $m>0$ such that $\varphi(\xi)\geqslant m$ for every $\xi\in X$. Now we use the corollary of Theorem V.3.9 with our function f:

$$\begin{aligned}\int_G f(x)\rho(x)\,dx&=\int_{\frac{G}{H}}\int_H f(xh)\,dh\,d\mu(\xi)\geqslant\int_{\frac{G}{H}}\int_H c_S(xh)\,dh\,d\mu(\xi)\\&=\int_{\frac{G}{H}}\varphi(\xi)\,d\mu(\xi)\geqslant m\int_{\frac{G}{H}}d\mu(\xi)=m\mu\left(\frac{G}{H}\right).\end{aligned}$$

Since $m>0$ and the left hand side is finite we see that $\mu(G/H)<\infty$.

Now we turn our attention to the representation $\rho\colon G\to\mathscr{U}(\mathscr{K})$ induced on G by $\chi\colon H\to\mathscr{U}(\mathscr{H})$. If $\sigma\colon L^1(G)\to\mathscr{L}(\mathscr{K})$ denotes the algebra representation associated with ρ then by Theorem 5.7 the operator valued integral

$$K_f(x,y)=\int_H f(xhy^{-1})\chi(h)\,dh$$

exists in the weak sense for every f in $L^1(G)$, for every $x\in\xi$, $y\in\eta$ and almost all $\xi,\eta\in X$. At the exceptional points ξ,η we can define K_f by supposing for instance that $K_f(x,y)=0$ for every $x\in\xi$ and $y\in\eta$. Then by Proposition 5.9 K_f satisfies (1) for every $x,y\in G$ and $a,b\in H$.

Proposition 15. *If $f\colon G\to\mathbb{C}$ is a continuous function with compact support, then the operator norm $|K_f(x,y)|$ is bounded on $G\times G$.*

Proof. First let α and then another vector β be given in $\mathscr{H}$. Then by definition

$$(K_f(x,y)\alpha,\beta)=\int_H f(xhy^{-1})(\chi(h)\alpha,\beta)\,dh.$$

Therefore $\chi(h)$ being unitary

$$|(K_f(x,y)\alpha,\beta)|\leqslant\|\alpha\|\cdot\|\beta\|\int_H|f(xhy^{-1})|\,dh. \tag{3}$$

If S denotes the compact support of f then

$$\int_H|f(xhy^{-1})|\,dh=\int_{H\cap x^{-1}Sy}|f(xhy^{-1})|\,dh\leqslant\|f\|_\infty\varphi(x,y)$$

where φ denotes the function appearing in Proposition 11. Therefore choosing $\beta = K_f(x,y)\alpha$ from (3) we obtain

$$\|K_f(x,y)\alpha\| \leqslant \|f\|_\infty \varphi(x,y)\|\alpha\| .$$

Hence by Proposition 11 $|K_f(x,y)| \leqslant \|f\|_\infty \|\varphi\|_\infty$ for every $x, y \in G$.

Corollary. *If $\mathscr{H}$ is finite dimensional then the HS norm $\|K_f(x,y)\|$ is bounded on $G \times G$.*

In fact we have

$$\|K_f(x,y)\|^2 = \sum \|K_f(x,y)\alpha_i\|^2 \leqslant |K_f(x,y)|^2 \sum \|\alpha_i\|^2$$

where α_i $(i \in \mathscr{I})$ is a MONS in $\mathscr{H}$. Therefore $\|K_f(x,y)\| \leqslant |K_f(x,y)|\sqrt{\dim \mathscr{H}}$.

Theorem 16. *Let G be a unimodular group and H a closed, unimodular subgroup such that $X = G/H$ is compact. Let $\chi: H \to \mathscr{U}(\mathscr{H})$ be a finite dimensional unitary representation. Then the representation $\rho^\chi: G \to \mathscr{U}(\mathscr{K})$ induced on G by χ is a direct sum of irreducible components, each of which occurs in ρ with finite multiplicity.*

Proof. Let $f: G \to \mathbb{C}$ be any continuous function with compact support. By Proposition 3 the HS norm $\|K_f(x,y)\|$ is defined on $X \times X$ and by the corollary of Proposition 15 $\|K_f(\xi,\eta)\|$ is bounded on $X \times X$. By Proposition 14 we also know that $\mu(G/H)$ is finite. Therefore $\|K_f\|^2 = \int_X \int_X \|K_f(x,y)\|^2 d\xi\, d\eta$ is finite and so $K_f \in \mathscr{L}^2$. Thus by Proposition 10 and Theorem II.4.4 $A_f = A_{K_f}$ is a compact operator for every continuous f with compact support.

Now we turn to Proposition V.5.10 and apply its corollary with the induced representation $\rho^\chi: G \to \mathscr{U}(\mathscr{K})$ and the associated representation $\sigma: L^1(G) \to \mathscr{L}(\mathscr{K})$. We see that $\sigma(f)x \neq 0$ for a suitable $f \in C_0(G)$ satisfying $f = f^*$. In other words $x \notin \mathscr{M}_0(\sigma(f))$ where $\mathscr{M}_0(A)$ denotes the kernel of the operator A.

Let $\mathscr{A}$ be the v. Neumann algebra generated by the representation ρ^χ and let $\mathscr{C}$ be the set of compact, self adjoint operators $A \in \mathscr{A}$. By Proposition V.5.6 every $\sigma(f)$ belongs to $\mathscr{A}$ and so by its compactness $\sigma(f) \in \mathscr{C}$ for every $f \in C_0(G)$ satisfying $f = f^*$. Thus we have

$$\bigcap_{A \in \mathscr{C}} \mathscr{M}_0(A) \subseteq \bigcap_{f \in C_0(G)} \mathscr{M}_0(\sigma(f)) = 0.$$

Therefore Proposition IV.6.13 is applicable with the family of compact operators $\mathscr{C}$ and with $\mathscr{K}^\perp = 0$.

Corollary. *If* $A: \mathscr{K} \to \mathscr{K}$ *belongs to the v. Neumann algebra generated by the operators* $\rho^{\chi}(x)(x \in G)$ *then*

(4) $$\operatorname{tr} A = \sum_{\omega \in \hat{G}} n(\omega, \rho^{\chi}) \operatorname{tr} A(\omega).$$

Similarly for any operator A *in* $\mathscr{L}(\mathscr{K})$ *we have*

(5) $$\|A\|^2 = \sum_{\omega \in \hat{G}} n(\omega, \rho^{\chi}) \|A(\omega)\|^2$$

The meaning of $\operatorname{tr} A(\omega)$ and $\|A(\omega)\|$ is explained in Section IV.6 after Proposition IV.6.5. In order to obtain (4) and (5) from Theorem 16 we fix a direct sum decomposition of ρ^{χ}, say $\mathscr{K} = \sum \mathscr{P}_i$, and choose a MONS α_{ij} $(j \in \mathscr{J}_i)$ in every irreducible component $\mathscr{P}_i$ $(i \in \mathscr{I})$. Then the conclusion follows right away from the definitions of the concepts involved in (4) and (5). We see that the corollary is obtained from Theorem 16 without using Proposition IV.6.13 or Theorem IV.6.14.

7. The Isometric Isomorphism between $\mathscr{L}^2$ and $HS(\mathscr{K}_2, \mathscr{K}_1)$. The Computation of the Trace in Terms of the Associated Kernel

Here we continue the study of the vector spaces of operator valued functions $K: G \times G \to \mathscr{L}(\mathscr{H})$ and their connection with induced representations. Our immediate object is to prove a formula for the trace of a TC operator which will be stated in Theorem 16.

We start with an obvious generalization of the concepts introduced in the last section. We let G_1 and G_2 be arbitrary locally compact groups and in both of them we choose a closed subgroup $H_k (k=1,2)$. We fix two continuous unitary representations $\chi_k: H_k \to \mathscr{U}(\mathscr{H}_k)$ $(k=1,2)$ where $\mathscr{H}_1$ and $\mathscr{H}_2$ are separable complex Hilbert spaces. We shall use K to denote operator valued functions $K: G_1 \times G_2 \to \mathscr{L}(\mathscr{H}_2, \mathscr{H}_1)$ such that

(1) $$K(x_1 a_1, x_2 a_2) = \chi_1(a)^* K(x_1, x_2) \chi_2(a_2)$$

for every $x_k \in G_k$ and $a_k \in H_k$ $(k=1,2)$. The results stated in Propositions 6.1 to 6.4 can be generalized right away and the proofs are essentially the same as the original ones. Hence θ^k denotes a typical element of the Hilbert space $\mathscr{K}_k$ of the representation ρ^{χ_k} induced on G_k by $\chi_k (k=1,2)$. Then we have the following:

Proposition 1. 1. *If* $\theta^2 \in \mathscr{K}_2$ *then*

$$K(x_1, x_2 a_2)\theta^2(x_2 a_2) = K(x_1, x_2)\theta^2(x_2)$$

for every $x_k \in G_k$ $(k=1,2)$ *and* $a_2 \in H_2$.

2. *For every* $\theta^k \in \mathscr{K}_k$ $(k=1, 2)$ *the inner product*

$$(K(x_1, x_2)\theta^2(x_2), \theta^1(x_1))$$

is defined on $X_1 \times X_2$.

3. *For all* K_1, K_2 *the HS inner product*

$$(K_1(x_1, x_2), K_2(x_1, x_2)) = \operatorname{tr} K_2(x_1, x_2)^* K(x_1, x_2)$$

is defined on $X_1 \times X_2$.

As a corollary we see that the HS norm $\|K(x_1, x_2)\|$ is defined on $X_1 \times X_2$ for every function K satisfying (1). Here the value $+\infty$ is accepted as a possibility for the HS norm at some point $(x_1, x_2) \in G_1 \times G_2$.

Now we extend the definition of the linear spaces $\mathscr{L}$ and $\mathscr{L}^2$ to the present general situation:

Definition 2. *We let* $\mathscr{L}$ *denote the linear space of those functions* $K: G_1 \times G_2 : \to \mathscr{L}(\mathscr{H}_2, \mathscr{H}_1)$ *which have the following properties:*

1. *For every* $x_k \in G_k$ *and* $a_k \in H_k$ $(k=1, 2)$ *one has*

$$K(x_1 a_1, x_2 a_2) = \chi_1(a_1)^* K(x_1, x_2) \chi_2(a_2).$$

2. *For every fixed pair of vectors* $\alpha_k \in \mathscr{H}_k$ $(k=1, 2)$

$$(x_1, x_2) \mapsto (K(x_1, x_2)\alpha_2, \alpha_1)$$

is measurable on $G_1 \times G_2$.

In order to define the space $\mathscr{L}^2$ we introduce a quasi-invariant measure μ_k on X_k which is determined by a continuous, positive ρ-function $\rho_k: G_k \to (0, \infty)$ $(k=1, 2)$. Since quasi-invariant measures of X_k form a single equivalence class the following result is independent of the choice of μ_1 and μ_2:

Proposition 3. *If* $K_1, K_2 \in \mathscr{L}$ *and* $\|K_k(x_1, x_2)\|$ $(k=1, 2)$ *is finite for almost all* $(\xi_1, \xi_2) \in X_1 \times X_2$ *then*

$$(x_1, x_2) \mapsto (K_1(x_1, x_2), K_2(x_1, x_2))$$

is a measurable function, it is defined on $X_1 \times X_2$ *and*

$$|(K_1(x_1, x_2), K_2(x_1, x_2))| \leqslant \|K_1(x_1, x_2)\| \cdot \|K_2(x_1, x_2)\|$$

for all $x_1 \in \xi_1$ $x_2 \in \xi_2$ *and almost all* $(\xi_1, \xi_2) \in X_1 \times X_2$.

Proof. Let $\alpha_i^1\,(i\in\mathscr{I})$ and $\alpha_j^2\,(j\in\mathscr{J})$ be MONS in $\mathscr{H}_1$ and $\mathscr{H}_2$, respectively. Then by hypothesis $\mathscr{I}$ and $\mathscr{J}$ are countable index sets. The infinite series

$$\sum_i (K_1(x_1,x_2)\alpha_j^2,\alpha_i^1)(K_2(x_1,x_2)\alpha_j^2,\alpha_i^1) \tag{2}$$

is absolutely convergent and by the BCS inequality the absolute value of its sum does not exceed

$$\|K_1(x_1,x_2)\alpha_j^2\|\cdot\|K_2(x_1,x_2)\alpha_j^2\|$$

for almost all $(\xi_1,\xi_2)\in X_1\times X_2$ and all $x_1\in\xi_1$ and $x_2\in\xi_2$. By hypothesis the terms of (2) are measurable functions and so (2) defines a measurable function on $G_1\times G_2$. Therefore

$$(x_1,x_2)\mapsto(K_1(x_1,x_2)\alpha_j^2,K_2(x_1,x_2)\alpha_j^2)$$

is a measurable function. If we now sum on $j\in\mathscr{J}$ then applying the BCS inequality we obtain

$$\sum_j |(K_1(x_1,x_2)\alpha_j^2,K_2(x_1,x_2)\alpha_j^2)|\leqslant\|K_1(x_1,x_2)\|\cdot\|K_2(x_1,x_2)\|$$

Therefore the infinite series

$$\sum_j (K_1(x_1,x_2)\alpha_j^2,K_2(x_1,x_2)\alpha_j^2)$$

is absolutely convergent for almost every $(\xi_1,\xi_2)\in X_1\times X_2$ and every $x_1\in\xi_1$, $x_2\in\xi_2$. Since the terms are measurable so is the sum of the series which is $(K_1(x_1,x_2),K_2(x_1,x_2))$. The fact that $(K_1(x_1,x_2),K_2(x_1,x_2))$ is defined on $X_1\times X_2$ follows from (1).

We shall now explicitely use the quasi-invariant measures μ_1 and μ_2 introduced on X_1 and X_2 before stating Definition 4: If $K\in\mathscr{L}$ then by Proposition 3 we can define $\|K\|$ by

$$\|K\|^2=\int_{X_1}\int_{X_2}\|K(x_1,x_2)\|^2\,d\mu_1(\xi_1)\,d\mu_2(\xi_2)\,.$$

Moreover if K_1 and K_2 are in $\mathscr{L}$ and $\|K_1\|$, $\|K_2\|$ are finite then we can introduce the pseudo-inner product

$$(K_1,K_2)=\int_{X_1}\int_{X_2}(K_1(x_1,x_2),K_2(x_1,x_2))\,d\mu_1(\xi_1)\,d\mu_2(\xi_2)\,.$$

Definition 4. *We let*

$$\mathscr{L}^2=\mathscr{L}^2(G_1,H_1,\chi_1,\mu_1;G_2,H_2,\chi_2,\mu_2)$$

be the inner product space of equivalence classes of functions K in $\mathscr{L}$ for which $\|K\|$ is finite.

The space $\mathscr{L}^2 = \mathscr{L}^2(G, H, \chi)$ considered in Section 6 is a special case with $G_1 = G_2 = G$, $H_1 = H_2 = H$, $\chi_1 = \chi_2 = \chi$ and $\mu_1 = \mu_2 = \mu$ being an invariant measure on $X = G/H$.

Theorem 5. *The inner product space $\mathscr{L}^2$ is a complex Hilbert space.*

Proof. Since $\mathscr{L}^2$ is known to be an inner product space only its completeness has to be proved. For this purpose we use Propositions 1.1, 1.2, 1.3 and the reasoning applied in the proof of Theorem 1.4. Thus we let $S = G_1 \times G_2, X = X_1 \times X_2 = G_1/H_1 \times G_2/H_2$ and $p: S \to X$ be the map obtained from the natural maps $p_k: G_k \to G_k/H_k = X_k$ $(k=1, 2)$. For $d\mu$ we choose the product measure $d\mu_1(\xi_1) d\mu_2(\xi_2)$ where μ_1, μ_2 denote the quasi-invariant measures chosen in X_1 and X_2, respectively.

Given a Cauchy sequence $K_1, K_2, \ldots$ in $\mathscr{L}^2$ thus we find an element $K: G_1 \times G_2 \to \mathscr{L}(\mathscr{H}_2, \mathscr{H}_1)$ and a sequence $K_{i_1}, K_{i_2}, \ldots$ such that

$$\|K_{i_k}(x) - K(x)\| \to 0$$

for almost every $\xi \in X$ and every $x \in \xi$. We also know that $K(x)$ is a HS operator for almost every ξ and every $x \in \xi$. We wish to prove that $K \in \mathscr{L}^2$. First of all $(K(\cdot)\alpha_2, \alpha_1)$ is μ-measurable for every $\alpha_l \in \mathscr{H}_l$ $(l=1, 2)$ because almost everywhere it is the limit of the sequence of measurable functions $K_{i_k}((\cdot)\alpha_2, \alpha_1)$. This shows that part 1. of Definition 2 is fulfilled. Next by the unitary nature of χ_1 and χ_2 we have

$$|\chi_1(a_1)^* (K_{i_k}(x_1, x_2) - K(x_1, x_2)) \chi_2(a_2)| \leqslant |K_{i_k}(x_1, x_2) - K(x_1, x_2)|$$

for any $x_l \in G_l$ and $a_l \in H_l$ $(l=1, 2)$. Therefore (1) follows from the similar relations holding for the elements K_{i_k} $(k=1, 2, \ldots)$. Hence K satisfies part 2. of Definition 2. By the finiteness of the norms $\|K_{i_k}\|$ and $\|K_{i_k} - K\|$ we obtain $\|K\| < \infty$ and this implies that $\|K(x_1, x_2)\|$ is finite almost everywhere on $X_1 \times X_2$.

Given $\theta^k \in \mathscr{K}_k$ $(k=1, 2)$ we define $\theta^1 \mathbin{\bar{\otimes}} \theta^2: G_1 \times G_2 \to \mathscr{L}(\mathscr{H}_2, \mathscr{H}_1)$ by $(\theta^1 \mathbin{\bar{\otimes}} \theta^2)(x_1, x_2) = \theta^1(x_1) \mathbin{\bar{\otimes}} \theta^2(x_2)$. Thus if $\alpha^2 \in \mathscr{H}_2$ then $(\theta^1 \mathbin{\bar{\otimes}} \theta^2)(x_1, x_2)\alpha^2 = (\alpha^2, \theta^2(x_2))\theta^1(x_1)$. Propositions 6.6, 6.7 and 6.8 can be obviously generalized as follows:

Proposition 6. 1. *If $\theta^k \in \mathscr{K}_k$ $(k=1, 2)$ then*

$$\theta^1 \mathbin{\bar{\otimes}} \theta^2(x_1 a_1, x_2 a_2) = \chi_1(a_1)^* (\theta^1 \mathbin{\bar{\otimes}} \theta^2)(x_1, x_2) \chi_2(a_2)$$

for every $x_k \in G_k$ and $a_k \in H_k$ $(k=1, 2)$.

2. *If $\theta_i^k \in \mathscr{K}_k$ $(i=1, 2)$ and $x_k \in G_k$ $(k=1, 2)$ then*

$$((\theta_1^1 \mathbin{\bar{\otimes}} \theta_1^2)(x_1, x_2), (\theta_2^1 \mathbin{\bar{\otimes}} \theta_2^2)(x_1, x_2)) = (\theta_1^1(x_1), \theta_2^1(x_1))(\theta_1^2(x_2), \theta_2^2(x_2)).$$

3. *If* $\theta_i^k \in \mathcal{K}_k$ $(i=1, 2$ *and* $k=1, 2)$ *then*

$$(\theta_1^1 \overline{\otimes} \theta_1^2, \theta_2^1 \overline{\otimes} \theta_1^2) = (\theta_1^1, \theta_2^1)(\theta_2^2, \theta_1^2).$$

In particular if $\theta^k \in \mathcal{K}_k$ $(k=1, 2)$ *then* $\theta^1 \overline{\otimes} \theta^2 \in \mathcal{L}^2$ *and* $\|\theta^1 \overline{\otimes} \theta^2\|$ $= \|\theta^1\| \cdot \|\theta^2\|$.

Hence if $\theta_i^k \in \mathcal{K}_k$ $(i=1, 2$ and $k=1, 2)$ are such that $\theta_1^1 \perp \theta_2^1$ or $\theta_1^2 \perp \theta_2^2$ then $\theta_1^1 \overline{\otimes} \theta_1^2 \perp \theta_2^1 \overline{\otimes} \theta_2^2$. This implies the following:

Lemma 7. *Let* $\theta_i^1 (i \in \mathcal{I})$ *and* $\theta_j^2 (j \in \mathcal{J})$ *be ONS in* $\mathcal{K}_1$ *and* $\mathcal{K}_2$, *respectively. Then* $\phi_{ij} = \theta_i^1 \overline{\otimes} \theta_j^2$ $(i \in \mathcal{I}$ *and* $j \in \mathcal{J})$ *is an ONS in* $\mathcal{L}^2$.

We can also extend Proposition 6.9:

Lemma 8. *If* $K: \mathcal{H}_2 \to \mathcal{H}_1$ *is a HS operator and* $\alpha^k \in \mathcal{H}_k$ $(k=1, 2)$ *then* $(\alpha^2 \overline{\otimes} \alpha^1, K^*) = (K\alpha^2, \alpha^1)$.

We are going to prove that if the ONS occurring in Lemma 7 are maximal then $\phi_{ij} (i \in \mathcal{I}$ and $j \in \mathcal{J})$ is a MONS. We let $\mathcal{M}_k = \{\theta_f : f \in C_0(G_k, \mathcal{H}_k)\}$ $(k=1, 2)$ where $C_0(G_k, \mathcal{H}_k)$ is the vector space of continuous functions $f_k: G_k \to \mathcal{H}_k$ with compact support. Then we have:

Proposition 9. *The set* $\mathcal{M}_1 \overline{\otimes} \mathcal{M}_2$ *is a total set in the Hilbert space* $\mathcal{L}^2$.

Proof. Let ψ in $\mathcal{L}^2$ be such that $\psi \perp \mathcal{M}_1 \overline{\otimes} \mathcal{M}_2$. It is our object to prove that $\psi = 0$. Since $\mathcal{H}_k$ is separable by Lemma 4.8 there is a sequence of functions $f_1^k, f_2^k, \ldots$ in $C_0(G_k, \mathcal{H}_k)$ such that the vectors $\theta_{f_i^k}(e_k)$ $(i=1, 2, \ldots)$ span the space $\mathcal{H}_k (k=1, 2)$. For the sake of simplicity in notations we shall write θ_i^k instead of $\theta_{f_i^k}$. Let $\varphi_k \in C_0(X_k)$ and $x_k \in G_k$ $(k=1, 2)$ be given. We shall use same symbol φ_k to denote the continuous function $x \mapsto \varphi_k(p_k x) = \varphi_k(\xi_k)$ defined on G_k. Then by Lemmas 4.6 and 4.7 the function $\varphi_k \theta_i^k(x_k \cdot): G_k \to \mathcal{H}_k$ belongs to $\mathcal{M}_k$ for every $i=1, 2, \ldots$ and so

$$(\varphi_1 \theta_i^1(x_1 \cdot) \overline{\otimes} \varphi_2 \theta_j^2(x_2 \cdot), \psi) = 0$$

for all $i, j = 1, 2, \ldots$ Using Lemma 8 and the definition of the inner product in $\mathcal{L}^2$ we obtain

$$\int_{X_1} \int_{X_2} (\varphi_1(y)\theta_i^1(x_1 y_1) \overline{\otimes} \varphi_2(y_2)\theta_j^2(x_2 y_2), \psi(y_1, y_2)) d\mu_1(\eta_1) d\mu_2(\eta_2)$$
$$= \int_{X_1} \int_{X_2} \varphi_1(\eta_1)\overline{\varphi_2(\eta_2)}(\theta_i^1(x_1 y_1), \psi(y_1, y_2)\theta_j^2(x_2 y_2)) d\mu_1(\eta_1) d\mu_2(\eta_2) = 0.$$

We multiply the last equation by $\varphi_1(\xi_1)\overline{\varphi_2(\xi_2)}$ where $\xi_1 \in X_1$ and $\xi_2 \in X_2$ are arbitrary. We can then integrate on $X_1 \times X_2$ and apply Fubini's theorem. Since φ_k is an arbitrary function in $C_0(X_k)$ $(k=1, 2)$ we obtain the following:

For any pair of compact sets C_k in X_k $(k=1,2)$ and for almost all $(\eta_1,\eta_2)\in C_1\times C_2$ we have

$$(3) \qquad (\theta_i^1(x_1 y_1), \psi(y_1,y_2)\,\theta_j^2(x_2 y_2)) = 0$$

for almost all $(\xi_1,\xi_2)\in C_1\times C_2$. Since there are only denumerably many choices for i and j we can find a null set N in $C_1\times C_2$ such that *if* $(\eta_1,\eta_2)\notin N$ *then for every* $i,j=1,2,\ldots$ (3) *holds for almost all* $(\xi_1,\xi_2)\in C_1\times C_2$.

Let us now suppose that $\psi(y_1,y_2)\neq 0$ for some $(y_1,y_2)\in G_1\times G_2$ where $(y_1 H_1, y_2 H_2)=(\eta_1,\eta_2)$ belongs to $C_1\times C_2$ but $(\eta_1,\eta_2)\notin N$. Then by the corollary of Lemma 4.8 and the choice of the functions f_i^k there is a pair of indices i,j such that

$$(\theta_i^1((y_1)^2), \psi(y_1, y_2)\,\theta_j^2((y_2)^2))\neq 0\,.$$

Since θ_i^1 and θ_j^2 are continuous functions we obtain

$$(\theta_i^1(x_1 y_1), \psi(y_1,y_2)\,\theta_j^2(x_2 y_2)) \neq 0$$

for every (x_1,x_2) in a sufficiently small neighborhood of (y_1,y_2). Since we have (3) for almost all $(\xi_1,\xi_2)\in C_1\times C_2$ it follows that (η_1,η_2) is an isolated point of $C_1\times C_2$. Therefore we proved that $\psi(y_1,y_2)=0$ for all $y_k\in\eta_k$ and locally almost all $(\eta_1,\eta_2)\in X_1\times X_2$. Since $\|\psi\|$ is finite we see that ψ is equivalent to the zero element in $\mathcal{L}^2$.

Proposition 10. *Let* θ_i^1 $(i\in\mathcal{I})$ *and* θ_j^2 $(j\in\mathcal{J})$ *be MONS in* $\mathcal{K}_1$ *and* $\mathcal{K}_2$ *respectively. Then* ϕ_{ij} $(i\in\mathcal{I}$ *and* $j\in\mathcal{J})$ *is a MONS in* $\mathcal{L}^2$.

Proof. By Lemma 7 ϕ_{ij} $(i\in\mathcal{I}$ and $j\in\mathcal{J})$ is an ONS in $\mathcal{L}^2$. Let $K\in\mathcal{L}^2$ be orthogonal to every ϕ_{ij}. Since θ_i^1 $(i\in\mathcal{I})$ spans $\mathcal{K}_1$ using part 3 of Proposition 6 we obtain $\psi^1\bar{\otimes}\theta_j^2\perp K$ for every ψ^1 in $\mathcal{K}_1$. Similarly using the fact that θ_j^2 $(j\in\mathcal{J})$ spans $\mathcal{K}_2$ we see that $\psi^1\bar{\otimes}\psi^2\perp K$ for every $\psi^1\in\mathcal{K}_1$ and $\psi^2\in\mathcal{K}_2$. Hence by Proposition 9 $K=0$.

If $K\in\mathcal{L}^2$, the vector θ^2 is fixed in $\mathcal{K}_2$ and θ^1 varies over $\mathcal{K}_1$ then

$$l(\theta^1) = l_{\theta_2}(\theta^1) = \int_{X_1}\int_{X_2} (K(x_1,x_2)\theta^2(x_2),\theta^1(x_1))\,d\mu_1(\xi_1)\,d\mu_2(\xi_2)$$

defines an element $A_K\theta^2\in\mathcal{K}_1$ such that $|l|=\|A_K\theta^2\|$ and $l(\theta^1)=(A_K\theta^2,\theta^1)$ for every $\theta^1\in\mathcal{K}_1$. Since $|l|\leqslant\|K\|\cdot\|\theta^2\|$ we see that $\theta^2\mapsto A_K\theta^2$ is a continuous linear operator belonging to $\mathcal{L}(\mathcal{K}_2,\mathcal{K}_1)$ and $|A_K|\leqslant\|K\|$. Using the same reasoning as in the proof of Propositions 6.9 and 6.10 we can easily prove that A_K belongs to $HS(\mathcal{K}_2,\mathcal{K}_1)$: If K is in $\mathcal{L}^2(G_1,\ldots;G_2,\ldots)$ then $K^*(x_2,x_1)=K(x_1,x_2)^*$ defines a vector in $\mathcal{L}^2(G_2,\ldots;G_1,\ldots)$ which we shall denote by K^*. If $K \in \mathscr{L}^2$ and $\theta^k \in \mathscr{K}_k$ $(k=1,2)$ then by Lemma 8 we have $(A_K\theta^2, \theta^1) = (\theta^2 \overline{\otimes} \theta^1, K^*)$. Now if θ_i^1 $(i \in \mathscr{I})$ and θ_j^2 $(j \in \mathscr{J})$ are MONS in $\mathscr{K}_1$ and $\mathscr{K}_2$ then by Lemma 7 and Proposition 10

$$\sum_j \|A_K \theta_j^2\|^2 = \sum_{i,j} |(A_K \theta_j^2, \theta_i^1)|^2 = \sum_{i,j} |(\phi_{ij}, K^*)|^2 = \|K^*\|^2 = \|K\|^2.$$

Therefore A_K is in the HS class and $\|A_K\|^2 = \|K\|^2$.

Theorem 11. *A linear operator* $A: \mathscr{K}_2 \to \mathscr{K}_1$ *belongs to the HS class* $HS(\mathscr{K}_2, \mathscr{K}_1)$ *if and only if there is a K in* $\mathscr{L}^2$ *such that* $A = A_K$ *i.e.*

$$(4) \qquad (A\theta^2, \theta^1) = \int_{X_1} \int_{X_2} (K(x_1, x_2)\theta^2(x_2), \theta^1(x_1))\, d\mu_1(\xi_1)\, d\mu_2(\xi_2)$$

for every $\theta^k \in \mathscr{K}_k$ $(k=1,2)$. *The kernel* $K = K_A$ *is uniquely determined by A and the map* $A \mapsto K$ *is a norm preserving isomorphism of* $HS(\mathscr{K}_2, \mathscr{K}_1)$ *onto* $\mathscr{L}^2$.

Note. The HS norm $\|A\|$ is used in $HS(\mathscr{K}_2, \mathscr{K}_1)$ so that $\|A\| = \|K_A\|$.

Proof. Let θ_i^1 $(i \in \mathscr{I})$ and θ_j^2 $(j \in \mathscr{J})$ be MONS in $\mathscr{K}_1$ and $\mathscr{K}_2$, respectively. We let $\phi_{ij} = \theta_i^1 \overline{\otimes} \theta_j^2$ $(i \in \mathscr{I}$ and $j \in \mathscr{J})$ be the associated MONS in $\mathscr{L}^2$. If $A: \mathscr{K}_2 \to \mathscr{K}_1$ is a HS operator then

$$\|A\|^2 = \sum_{i,j} |(A\theta_j^2, \theta_i^1)|^2$$

is finite and so

$$(5) \qquad K^* = \sum_{i,j} (\theta_i^1, A\theta_j^2)\, \phi_{ij}$$

exists by the completeness of $\mathscr{L}^2(G_2, \ldots; G_1, \ldots)$. The element $K = K^{**}$ belongs to $\mathscr{L}^2(G_1, \ldots; G_2, \ldots) = \mathscr{L}^2$. Therefore it defines in the weak sense a HS operator A_K. Namely if $\theta^k \in \mathscr{K}_k$ $(k=1,2)$ then $(A_K\theta^2, \theta^1)$ is given by the right hand side of (4). Thus by $\phi_{ji}(x_2, x_1) = \theta_j^2(x_2) \overline{\otimes} \theta_i^1(x_1)$ and Lemma 8 we obtain $(A_K\theta_j^2, \theta_i^1) = (\phi_{ji}, K^*)$. Hence using (5) it follows that $(A_K\theta_j^2, \theta_i^1) = (A\theta_j^2, \theta_i^1)$. This shows that $(A_K - A)\theta_j^2 \perp \theta_i^1$ for every $\theta_i^1 \in \mathscr{K}_1$ and $\theta_j^2 \in \mathscr{K}_2$. By the maximality of these ON systems we see that $A_K = A$ and so the given operator A has kernel K. We have seen that for any K in $\mathscr{L}^2$ we have the relation $\|A_K\| = \|K\|$ and this shows that for each operator A there exists only one kernel K such that $A = A_K$.

Proposition 12. *If* $K \in \mathscr{L}^2$ *then we can express* $A_K\theta^2(x_1)$ *in the weak sense by*

$$A_K\theta^2(x_1) = \int_{X_2} K(x_1, x_2)\theta^2(x_2)\, d\mu_2(\xi_2)$$

for every $\theta^2 \in \mathscr{K}_2$, *all* $x_1 \in \xi_1$ *and almost all* $\xi_1 \in X_1$.

Proof. Given $\theta^2 \in \mathscr{H}_2$ and $x_1 \in G_1$ we let

$$l_{x_1}(\alpha^1) = \int_{X_2} (K(x_1,x_2)\theta^2(x_2),\alpha^1)\, d\mu_2(\xi_2)$$

where α^1 will vary over $\mathscr{H}_1$. By the BCS inequality the integral exists for all $x_1 \in \xi_1$ and almost all $\xi_1 \in X_1$ and

$$|l_{x_1}(\alpha^1)| \leqslant \|K\|(\xi_1) \cdot \|\theta^2\| \cdot \|\alpha^1\|$$

where

$$\|K(\xi_1)\|^2 = \int_{X_2} \|K(x_1,x_2)\|^2 \, d\mu_2(\xi_2).$$

Therefore if ξ_1 is not in the exceptional set then l_{x_1} is a continuous anti-linear functional on $\mathscr{H}_1$ and $|l_{x_1}| \leqslant \|K\|(\xi_1) \cdot \|\theta^2\|$. Let $B\theta^2(x_1)$ denote the unique vector in $\mathscr{H}_1$ such that $l_{x_1}(\alpha^1) = (B\theta^2(x_1), \alpha^1)$ for every $\alpha^1 \in \mathscr{H}_1$. Our object is to show that $A_K\theta^2(x_1) = B\theta^2(x_1)$ for almost all $\xi_1 \in X_1$ and all $x_1 \in \xi_1$.

If ξ_1 belongs to the exceptional set then we define $B\theta^2(x_1) = 0$ for every $x_1 \in \xi_1$. Then it is clear that $B\theta^2 : G_1 \to \mathscr{H}_1$ satisfies

$$B\theta^2(x_1 h_1) = \chi_1(h_1)^* B\theta^2(x_1)$$

for every $x_1 \in G_1$ and $h_1 \in H_1$. If ξ_1 is not exceptional then by choosing $\alpha^1 = B\theta^2(x_1)$ we obtain

$$(B\theta^2(x_1), B\theta^2(x_1)) = \int_{X_2} (K(x_1,x_2)\theta^2(x_2), B\theta^2(x_1))\, d\mu_2(\xi_2).$$

Hence we have

$$\|B\theta^2(x_1)\| \leqslant \int_{X_2} \|K(x_1,x_2)\theta^2(x_2)\| \, d\mu_2(\xi_2).$$

We majorize the integrand by $\|K(x_1,x_2)\| \cdot \|\theta^2(x_2)\|$ and apply the BCS inequality to obtain

$$\|B\theta^2(x_1)\|^2 \leqslant \|\theta^2\|^2 \int_{X_2} \|K(x_1,x_2)\|^2 \, d\mu_2(\xi_2).$$

By integrating on X_1 we arrive at the estimate $\|B\theta^2\| \leqslant \|K\| \cdot \|\theta^2\| < \infty$. We proved that $B\theta^2 \in \mathscr{H}_1$.

Now it is easy to prove that $A_K\theta^2 = B\theta^2$. For let any θ^1 be given in $\mathscr{H}_1$. If ξ_1 is not exceptional and $x_1 \in \xi_1$ then by choosing $\alpha^1 = \theta^1(x_1)$ in the weak definition of $B\theta^2(x_1)$ we have

$$(B\theta^2(x_1), \theta^1(x_1)) = \int_{X_2} (K(x_1,x_2)\theta^2(x_2), \theta^1(x_1))\, d\mu_2(\xi_2).$$

If we integrate both sides of this equation on X_1 then we obtain $(B\theta^2, \theta^1) = (A_K\theta^2, \theta^1)$. Since θ^1 is an arbitrary element of the Hilbert space $\mathscr{K}_1$ we see that $B\theta^2 = A_K\theta^2$.

Now we are going to study the special case when only one set of data G, H, χ and μ is given so that $\mathscr{L}^2 = \mathscr{L}^2(G,H,\chi,\mu)$. If $K, L \in \mathscr{L}^2$ and $x, y, z \in G$ then by (1) we have

$$K(x,yh)L(yh,z) = K(x,y)L(y,z)$$

for every $h \in H$. Therefore we can define $K * L$ as follows:

Definition 13. *For K and L in $\mathscr{L}^2(G,H,\chi,\mu)$ we define $K*L$ in the weak sense by*

$$(K*L(x,z)\alpha,\beta) = \int_X (K(x,y)L(y,z)\alpha,\beta)\,d\mu(\eta)$$

where $\alpha, \beta \in \mathscr{H}$ and $x, y, z \in G$.

One can immediately see that $K*L$ satisfies the functional equation given in (1). Furthermore we can also prove that $\|K*L\| \leqslant \|K\| \cdot \|L\|$ and so $\|K*L\|$ is finite and $K*L \in \mathscr{L}^2$: For if α_i $(i \in \mathscr{I})$ is a MONS in $\mathscr{H}$ then we have

$$\begin{aligned}|(K*L(x,z)\alpha_i,\alpha_j)| &\leqslant \int_X |(L(y,z)\alpha_i, K(x,y)^*\alpha_j)|\,d\mu(\eta)\\ &\leqslant \Big(\int_X \|L(y,z)\alpha_i\|^2\,d\mu(\eta)\Big)^{\frac12}\Big(\int_X \|K(x,y)^*\alpha_j\|^2\,d\mu(\eta)\Big)^{\frac12}.\end{aligned}$$

Therefore

$$\sum_{i,j} |(K*L(x,z)\alpha_i,\alpha_j)|^2 \leqslant \int_X \sum_i \|L(y,z)\alpha_i\|^2\,d\mu(\eta) \int_X \sum_j \|K(x,y)\alpha_j\|^2\,d\mu(\eta)$$

and so

$$\|(K*L(x,z))\|^2 \leqslant \int_X \|L(y,z)\|^2\,d\mu(\eta) \int_X \|K(x,y)\|^2\,d\mu(\eta).$$

The inequality $\|K*L\| \leqslant \|K\| \cdot \|L\|$ follows by integrating on both sides of the last inequality with respect to x and z.

It is clear that the kernels $K \in \mathscr{L}^2$ form a complex vector space and $(K+L)^* = K^* + L^*$ where $K^*(x,y) = K(y,x)^*$ for all $x, y \in G$. Moreover one can easily see from the formula defining $K*L$ in the weak sense that $(K*L)^* = L^* * K^*$. Similarly one can easily verify that the operation $*$ is associative. It turns out that $\mathscr{L}^2$ is an involutive algebra under the operations $K+L$, λK, $K*L$ and K^*. By Theorem 5 and the elementary properties of the norm $\mathscr{L}^2$ *is a star-normed, complex, involutive Banach algebra.* One could also prove directly that $\mathscr{L}^2$ has the H^* property but this follows also from the following:

Theorem 14. *The one-to-one map* $K \mapsto A_K$ *is an involutive, isometric isomorphism between the involutive Banach algebras* $\mathscr{L}^2$ *and* $HS(\mathscr{K})$.

Corollary. $\mathscr{L}^2$ *is an* H^**-algebra.*

Proof. By Theorem 11 $K \mapsto A_K$ is an isometric one-to-one map between $\mathscr{L}^2$ and $HS(\mathscr{K})$. Since the norms of these spaces are derived from inner products it follows that the map $K \mapsto A_K$ preserves inner products. The properties $A_{K+L} = A_K + A_L$ and $A_{\lambda K} = \lambda A_K$ are obvious from the definition of the operators associated with the kernels K, L and λK.

The relation $A_{K^*} = (A_K)^*$ is obtained as follows:

$$((A_K)^* \psi, \theta) = \overline{(A_K \theta, \psi)} = \int_X \int_X (\psi(x), K(x,y)\theta(y))\, d\mu(\eta)\, d\mu(\xi)$$

$$= \int_X \int_X (K^*(y,x)\psi(x), \theta(y))\, d\mu(\xi)\, d\mu(\eta) = (A_{K^*}\psi, \theta).$$

The only other property which needs proving is the identity $A_K A_L = A_{K*L}$. We have

$$(A_K A_L \theta, \psi) = (A_L \theta, A_K^* \psi) = \int_X \int_X (L(x,y)\theta(y), A_K^* \psi(x))\, d\mu(\xi)\, d\mu(\eta)$$

$$= \int_X \int_X (L(x,y)\theta(y), A_{K^*}\psi(x))\, d\mu(\eta)\, d\mu(\xi).$$

Now we use the relation

$$(A_{K^*}\psi(x), \alpha) = \int_X (K^*(x,z)\psi(z), \alpha)\, d\mu(\zeta)$$

with $\alpha = L(x,y)\theta(y)$ which holds for every $x \in \xi$ and almost every $\xi \in X$ by Proposition 12. We obtain

$$(\psi, A_K A_L \theta) = \int_X \int_X \int_X (K^*(x,z)\psi(z), L(x,y)\theta(y))\, d\mu(\zeta)\, d\mu(\eta)\, d\mu(\xi).$$

By changing the order of integration to $d\mu(\xi)\, d\mu(\eta)\, d\mu(\zeta)$ and using $K^*(x,z) = K(z,x)^*$ we arrive at

$$(\psi, A_K A_L \theta) = \int_X \int_X (\psi(z), K*L(z,y)\theta(y))\, d\mu(\eta)\, d\mu(\zeta) = (\psi, A_{K*L}\theta).$$

Hence $A_K A_L \theta = A_{K*L} \theta$ for every $\theta \in \mathscr{K}$.

Now let $A_K: \mathscr{K} \to \mathscr{K}$ be a HS operator such that the associated kernel is of the form $K = L^* * L$ where $L \in \mathscr{L}^2$. Then by Theorem 14 $A_K = (A_L)^* A_L$ and so by the definitions of the trace and the HS norm and by Theorem 14 we have

$$\operatorname{tr} A_K = \|A_L\|^2 = \|L\|^2. \tag{6}$$

We are going to compute this trace in a different manner: Let α_i $(i \in \mathscr{I})$ be a MONS in $\mathscr{H}$. Then by Definition 13

$$(L^* * L(x,x)\alpha_i, \alpha_i) = \int_X (L^*(x,y)L(y,x)\alpha_i, \alpha_i)\, d\mu(\eta)$$

for every $i \in \mathscr{I}$. If we use $L^*(x,y) = L(y,x)^*$ and sum over $i \in \mathscr{I}$ then we obtain

$$\operatorname{tr}(L^* * L)(x,x) = \int_X \|L(y,x)\|^2\, d\mu(\eta).$$

On the left hand side we substitute K for $L^* * L$ and integrate both sides of this equation on X. The right hand side becomes $\|L\|^2$ and so we obtain by (6):

Lemma 15. *If $K \in \mathscr{L}^2$ is of the form $K = L^* * L$ where $L \in \mathscr{K}$ then*

$$\operatorname{tr} A_K = \int_X \operatorname{tr} K(x,x)\, d\mu(\xi). \tag{7}$$

Our object is to extend this formula to arbitrary trace class operators $A: \mathscr{K} \to \mathscr{K}$. The next step in this direction consists of proving it for operators A_K whose kernel is of the form $K = L * M$ where L and M belong to $\mathscr{L}^2$. For this purpose we need the identity

$$\begin{aligned} L * M = \tfrac{1}{4}((L^* + M)^* * (L^* + M) + i(L^* - iM)^* * (L^* - iM) \\ -(L^* - M)^* * (L^* - M) - i(L^* + iM)^* * (L^* + iM)) \end{aligned} \tag{8}$$

which can be verified by performing the various operations indicated on the right hand side. Turning our attention to the corresponding operators and using Lemma 14 we obtain

$$\begin{aligned} \operatorname{tr} A_{L*M} &= \tfrac{1}{4}(\operatorname{tr} A_{(L^*+M)^**(L^*+M)} + \cdots) \\ &= \tfrac{1}{4}\Big(\int_X \operatorname{tr}(L^* + M)^* * (L^* + M)(x,x)\, d\mu(\xi) + \cdots\Big). \end{aligned}$$

On the right hand side we have

$$\int_X \operatorname{tr} \tfrac{1}{4}((L^* + M)^* * (L^* + M)(x,x) + \cdots)\, d\mu(\xi).$$

Hence using (8) once more we obtain

$$\operatorname{tr} A_{L*M} = \int_X \operatorname{tr}(L*M)(x,x)\,d\mu(\xi).$$

Therefore we proved the following:

Lemma 16. *If* $K \in \mathscr{L}^2$ *is of the form* $K = L*M$ *where* $L, M \in \mathscr{K}$ *then* (7) *holds.*

Let us now suppose that $A = A_K: \mathscr{K} \to \mathscr{K}$ is an arbitrary trace class operator. Then by Theorem II.5.2 we have $A = R_1 S_1 + R_2 S_2$ where R_1, S_1 and R_2, S_2 are HS operators. If L_1, M_1 and L_2, M_2 denote the corresponding kernels then by Theorem 14 we have $K = L_1 * M_1 + L_2 * M_2$ and $A_K = A_{L_1*M_1} + A_{L_2*M_2}$. Therefore by applying Lemma 16 twice we obtain the desired formula for $A = A_K$. Thus we proved:

Theorem 17. *If* $A: \mathscr{K} \to \mathscr{K}$ *is a trace class operator then*

$$\operatorname{tr} A = \int_X \operatorname{tr} K(x,x)\,d\mu(\xi)$$

where $K \in \mathscr{L}^2$ *denotes the kernel associated with* $A = A_K$.

This theorem will be used several times in Section VII.6 to prove various trace formulas.

8. The Tensor Product of Induced Representations

The purpose of this section is to establish the following fundamental result concerning induced representations:

Theorem 1. *Let* G_k $(k=1,2)$ *be locally compact groups, let* H_k *be a closed subgroup of* G_k *and let* $\chi_k: H_k \to \mathscr{U}(\mathscr{H}_k)$ *be unitary representations* $(k=1,2)$. *Then the tensor product of the induced representations* $\rho^{\chi_k}: G_k \to \mathscr{U}(\mathscr{K}_k)$ *is equivalent to the representation* ρ^χ *induced on* $G_1 \times G_2$ *by the tensor product* $\chi = \chi_1 \otimes \chi_2$, *in symbols* $\rho^{\chi_1} \otimes \rho^{\chi_2} \approx \rho^{\chi_1 \otimes \chi_2} = \rho^\chi$.

Before giving the proof we shall first consider the special case when the Hilbert spaces $\mathscr{H}_1$ and $\mathscr{H}_2$ are separable because this will show the connection between the above tensor product theorem und the Hilbert spaces $\mathscr{L}^2 = \mathscr{L}^2(G_1, \ldots; G_2, \ldots)$ studied in the last section. If the reader omitted Section 7 he can also omit the coming discussion and start with the paragraph preceeding Lemma 2 which is the beginning of the proof.

By definition the Hilbert space of the tensor product $\rho^{\chi_1} \otimes \rho^{\chi_2}$ is $HS(\overline{\mathscr{K}_2}, \mathscr{K}_1)$ where $\overline{\mathscr{K}_2}$ is the Hilbert space conjugate to $\mathscr{K}_2$. We define $\overline{\mathscr{L}^2}$ by

$$\overline{\mathscr{L}^2} = \overline{\mathscr{L}^2}(G_1, H_1, \chi_1, \mu_1; G_2, H_2, \chi_2, \mu_2) = \mathscr{L}^2(G_1, H_1, \chi_1, \mu_1; G_2, H_2, \overline{\chi_2}, \mu_2).$$

Then $\overline{\mathscr{L}^2}$ consists of functions $K: G_1 \times G_2 \to \mathscr{H}$ where $\mathscr{H} = HS(\overline{\mathscr{H}_2}, \mathscr{H}_1)$ is the Hilbert space of $\chi = \chi_1 \otimes \chi_2 : H_1 \times H_2 \to \mathscr{U}(\mathscr{H})$. Condition 1. of Definition 7.2 can be written in the form

$$K(x_1 h_1, x_2 h_2) = \chi_1 \otimes \chi_2 (h_1, h_2)^* K(x_1, x_2)$$

where $x_k \in G_k$ and $h_k \in H_k$ $(k=1,2)$ are arbitrary. This is the same as condition 2) in Theorem 4.9. Similarly the other two conditions of Definition 7.2 coincide with those of Theorem 4.9. Therefore $\overline{\mathscr{L}^2}$ is identical with the Hilbert space of the representation $\rho = \rho^{\chi} = \rho^{\chi_1 \otimes \chi_2}$ induced on $G = G_1 \times G_2$ by the representation $\chi = \chi_1 \otimes \chi_2$ of the subgroup $H = H_1 \times H_2$.

By Theorem 7.11 $K \mapsto A_K$ is a unitary operator U from $\mathscr{K}$, the Hilbert space of $\rho^{\chi} = \rho^{\chi_1 \otimes \chi_2}$ to $HS(\overline{\mathscr{K}_2}, \mathscr{K}_1)$ whose inverse was denoted there by $A \mapsto K_A$. We are going to prove that U intertwines ρ^{χ} and $\rho^{\chi_1} \otimes \rho^{\chi_2}$. For let $a_k \in G_k$ $(k=1,2)$ and $K \in \mathscr{K}$ be given. Then by $UK = A_K$ we have

$$\rho^{\chi_1} \otimes \rho^{\chi_2}(a_1, a_2) U K = \rho^{\chi_1}(a_1) A_K \rho^{\chi_2}(a_2)^* .$$

In order to determine the image of this HS operator under U^{-1} we choose an arbitrary θ^k in $\mathscr{K}_k$ for $k=1,2$. Then

$$\begin{aligned} &\int_{X_1} \int_{X_2} (\rho^{\chi_1}(a_1) A_K \rho^{\chi_2}(a_2)^* \theta^2(x_2), \theta^1(x_1)) \, d\mu_1(\xi_1) \, d\mu_2(\xi_2) \\ &= \int_{X_1} \int_{X_2} (K(x_1, x_2) \theta^2(a_2 x_2), \theta^1(a_1 x_1)) \sqrt{\lambda_1(a_1, \xi_1) \lambda_2(a_2, \xi_2)} \, d\mu_1(\xi_1) \, d\mu_2(\xi_2). \end{aligned}$$

Hence if we perform the substitutions $x_1 \to a_1^{-1} x_1$ and $x_2 \to a_2^{-1} x_2$ and use Proposition V.3.16 and Lemma V.3.10 then we see that

$$\begin{aligned} U^{-1} \rho^{\chi_1} \otimes \rho^{\chi_2}(a_1, a_2) U K(x_1, x_2) &= K(a_1^{-1} x_1, a_2^{-1} x_2) \sqrt{\lambda_1(a_1^{-1}, \xi_1) \lambda_2(a_2^{-1}, \xi_2)} \\ &= \rho^{\chi}(a_1, a_2) K(x_1, x_2). \end{aligned}$$

We proved that $(\rho^{\chi_1} \otimes \rho^{\chi_2}) U = U \rho^{\chi}$ and so $\rho^{\chi_1} \otimes \rho^{\chi_2} \approx \rho^{\chi}$.

We shall now prove Theorem 1 in the general case when the Hilbert spaces $\mathscr{H}_1$ and $\mathscr{H}_2$ are not supposed to be separable. We let $\mathscr{M}_k$ be the linear manifold introduced in Section 4 whose completion is the Hilbert space $\mathscr{K}_k$ of the induced representation ρ^{χ_k} $(k=1,2)$. Similarly we let $\mathscr{M}$ denote the linear manifold defining the Hilbert space $\mathscr{K}$ of $\rho = \rho^{\chi_1 \otimes \chi_2}$.

If $\theta^k \in \mathcal{M}_k$ and $x_k, y_k \in G_k$ then we let

$$\theta^1 \times \theta^2(x_1, x_2) = \theta^1(x_1) \otimes \theta^2(x_2)$$

so that $\theta^1 \times \theta^2 \colon G_1 \times G_2 \to \mathcal{H}_1 \otimes \mathcal{H}_2 = HS(\bar{\mathcal{H}}_2, \mathcal{H}_1)$. Then we have:

Lemma 2. *In terms of the foregoing notations we have* $\mathcal{M}_1 \times \mathcal{M}_2 \subseteq \mathcal{M}$.

Proof. The elements of $\mathcal{M}_k$ are continuous functions $\theta^k \colon G_k \to \mathcal{H}_k$ such that $\theta^k(x_k h_k) = \chi_k(h_k)^* \theta^k(x_k)$ for every $x_k \in G_k$ and $h_k \in H_k$ and $\|\theta^k(\cdot)\|$ has compact support in X_k. Given $\theta^k \in \mathcal{M}_k$ and $x_k, y_k \in G_k$ $(k=1,2)$ we have

$$\|\theta^1 \times \theta^2(x_1, x_2) - \theta^1 \times \theta^2(y_1, y_2)\|^2$$
$$= \sum_{i,j} |(\theta^1(x_1) \otimes \theta^2(x_2)\alpha_i, \alpha_j) - (\theta^1(y_1) \otimes \theta^2(y_2)\alpha_i, \alpha_j)|^2$$

where $\alpha_i (i \in \mathcal{I})$ is a MONS in $\mathcal{H}_2$. The right hand side can be written as

$$\sum_{i,j \in \mathcal{I}} |(\theta^2(x_2) - \theta^2(y_2), \alpha_i)(\theta^1(x_1), \alpha_j) + (\theta^1(x_1) - \theta^1(y_1), \alpha_j)(\theta^2(y_2), \alpha_i)|.$$

Therefore using the BCS inequality we obtain

$$\|\theta^1 \times \theta^2(x_1, x_2) - \theta^1 \times \theta^2(y_1, y_2)\|^2$$
$$\leqslant \|\theta^2(x_2) - \theta^2(y_2)\|^2 \cdot \|\theta^1(x_1)\|^2 + \|\theta^1(x_1) - \theta^1(y_1)\|^2 \cdot \|\theta^2(y_2)\|^2.$$

This shows that $\theta^1 \times \theta^2 \colon G_1 \times G_2 \to \mathcal{H}_1 \otimes \mathcal{H}_2$ is a continuous function. We can easily see that

$$\|\theta^1(x_1) \otimes \theta^2(x_2)\| = \|\theta^1(x_1)\| \cdot \|\theta^2(x_2)\|$$

and so $\|\theta^1 \times \theta^2\|$ has compact support in $X_1 \times X_2$. Using $\theta^k(x_k h_k) = \chi(h_k)^* \theta^k(x_k)$ we obtain

$$\theta^1 \times \theta^2(x_1 h_1, x_2 h_2) = \chi_1 \otimes \chi_2(h_1, h_2)^* \theta^1 \times \theta^2(x_1, x_2)$$

for all $x_k \in G_k$ and $h_k \in H_k$ $(k=1,2)$. Therefore we proved that if $\theta^k \in \mathcal{M}_k$ then the function $\theta^1 \times \theta^2$ belongs to $\mathcal{M}$. Lemma 2 is proved.

We know that the Hilbert space of $\rho^{\chi_1} \otimes \rho^{\chi_2}$ is $\mathcal{K}_1 \otimes \mathcal{K}_2 = HS(\bar{\mathcal{K}}_2, \mathcal{K}_1)$. If $\theta_i^k \in \mathcal{K}$ and $x_k \in G_k$ $(i=1,2$ and $k=1,2)$ then

$$(\theta_1^1 \times \theta_1^2(x_1, x_2), \theta_2^1 \times \theta_2^2(x_1, x_2)) = (\theta_1^1(x_1), \theta_2^1(x_1))(\theta_1^2(x_2), \theta_2^2(x_2)).$$

Hence by integrating on $X_1 \times X_2$ we obtain

$$(1) \qquad (\theta_1^1 \times \theta_1^2, \theta_2^1 \times \theta_2^2) = (\theta_1^1, \theta_2^1)(\theta_1^2, \theta_2^2) = (\theta_1^1 \otimes \theta_1^2, \theta_2^1 \otimes {}_2^2).$$

The inner product on the left hand side is taken in $\mathcal{K}$ while on the right hand side it is in $\mathcal{K}_1 \otimes \mathcal{K}_2$.

The set $\mathscr{M}_1 \otimes \mathscr{M}_2$ spans a linear manifold $\mathscr{L}$ in $\mathscr{K}_1 \otimes \mathscr{K}_2$ whose elements are finite sums of the form $\sum \theta_i^1 \otimes \theta_i^2$. By (1) for any two sums of this kind

$$(\sum \theta_i^1 \times \theta_i^2, \sum \theta_j^1 \times \theta_j^2) = (\sum \theta_i^1 \otimes \theta_i^2, \sum \theta_j^1 \otimes \theta_j^2).$$

Since this equality implies that

$$\|\sum \theta_i^1 \times \theta_i^2 - \sum \theta_j^1 \times \theta_j^2\| = \|\sum \theta_i^1 \otimes \theta_i^2 - \sum \theta_j^1 \otimes \theta_j^2\|$$

using Lemma 2 we see that

$$\sum \theta_i^1 \otimes \theta_i^2 \mapsto \sum \theta_i^1 \times \theta_i^2 \tag{2}$$

is a well defined map of $\mathscr{L}$ into $\mathscr{M}$ which preserves inner products. The set $\mathscr{M}_k$ is dense in the Hilbert space $\mathscr{K}_k$ and so we see that the linear manifold $\mathscr{L}$ spanned by $\mathscr{M}_1 \otimes \mathscr{M}_2$ is dense in $\mathscr{K}_1 \otimes \mathscr{K}_2$. Therefore the inner product preserving map given in (2) can be extended in a unique manner to an inner product preserving map U of $\mathscr{K}_1 \otimes \mathscr{K}_2$ into $\mathscr{K}$, the completion of $\mathscr{M}$.

We are going to prove that the range of the map U is dense in $\mathscr{K}$ and so U is a unitary transformation of $\mathscr{K}_1 \otimes \mathscr{K}_2$ onto $\mathscr{K}$. After that we shall prove that U intertwines $\rho^{\chi_1} \otimes \rho^{\chi_2}$ with $\rho^{\chi_1 \otimes \chi_2}$. By Proposition 4.3 every $\theta^k \in \mathscr{M}_k$ is of the form $\theta^k = \theta^k_{f_k}$ where $f_k \in C_0(G_k, \mathscr{H}_k)$ $(k=1,2)$. For such f_1 and f_2 we define $f_1 \times f_2 \colon G_1 \times G_2 \to \mathscr{H}_1 \otimes \mathscr{H}_2$ by

$$f_1 \times f_2(x_1, x_2) = f_1(x_1) \otimes f_2(x_2).$$

Then we have:

Lemma 3. *For every* $f_k \in C_0(G_k, \mathscr{H}_k)$ $(k=1,2)$ *we have* $\theta_{f_1} \times \theta_{f_2} = \theta_{f_1 \times f_2}$.

Proof. It is easy to see that $f_1 \times f_2$ is continuous and if S_k is the support of f_k $(k=1,2)$ then $S_1 \times S_2$ is a compact carrier of $f_1 \times f_2$. Let $A \in \mathscr{H}_1 \otimes \mathscr{H}_2$ and $x_1 \in G_1, x_2 \in G_2$. Then

$$\begin{aligned}(\theta_{f_1 \times f_2}(x_1, x_2), A) &= \int_{H_1} \int_{H_2} (\chi_1 \otimes \chi_2(h_1, h_2) f_1 \times f_2(x_1 h_1, x_2 h_2), A)\, dh_1\, dh_2 \\ &= \int_{H_1} \int_{H_2} (\chi_1(h_1)(f_1(x_1 h_1) \otimes f_2(x_2 h_2) \chi_2(h_2)^*, A)\, dh_1\, dh_2.\end{aligned}$$

In order to evaluate the integrand, which is a HS inner product, we choose a MONS α_i $(i \in \mathscr{I})$ in $\mathscr{H}_1$ and obtain

$$\begin{aligned}(\theta_{f_1 \times f_2}(x_1, x_2), A) &= \sum \int_{H_1} (\chi_1(h_1) f_1(x_1 h_1), A\alpha_i)\, dh_1 \int_{H_2} (\chi_2(h_2) f_2(x_2 h_2), \alpha_i)\, dh_2 \\ &= \sum (\theta_{f_2}(x_2), \alpha_i)(\theta_{f_1}(x_1), A\alpha_i) = (\theta_{f_1}(x_1) \otimes \theta_{f_2}(x_2), A) \\ &= (\theta_{f_1} \times \theta_{f_2}(x_1, x_2), A).\end{aligned}$$

Since A is arbitrary in $\mathscr{H}_1 \otimes \mathscr{H}_2$ we obtain $\theta_{f_1 \times f_2}(x_1, x_2) = \theta_{f_1} \times \theta_{f_2}(x_1, x_2)$. In order to prove that the range of U is dense in $\mathscr{K}$ it is sufficient to show that $U\mathscr{L}$ is dense in $\mathscr{K}$. The elements of $U\mathscr{L}$ are functions $\theta = \sum \theta_i^1 \times \theta_i^2$ where the index set $\mathscr{I}$ is finite and $\theta_i^k \in \mathscr{M}_k$ ($k=1,2$ and $i \in \mathscr{I}$). By Proposition 4.3 every $\theta_i^k \in \mathscr{M}_k$ is of the form $\theta^k = \theta^k_{f_i^k}$ for some $f_i^k \in C_0(G_k, \mathscr{H}_k)$ and $U\mathscr{L}$ consists of the functions θ_f where $f = \sum f_i^1 \times f_i^2$ with $f_i^k \in C_0(G_k, \mathscr{H}_k)$ ($k=1,2$ and $i \in \mathscr{I}$). By definition $\mathscr{K}$ is the completion of $\mathscr{M}$ and by Proposition 4.3 every θ in $\mathscr{M}$ is of the form $\theta = \theta_g$ where $g \in C_0(G_1 \times G_2, \mathscr{H}_1 \otimes \mathscr{H}_2)$. Hence in order to prove that $U\mathscr{L}$ is dense in $\mathscr{K}$ it is enough to prove the following: Given $g \in C_0(G_1 \times G_2, \mathscr{H}_1 \otimes \mathscr{H}_2)$ and $\varepsilon > 0$ there is an $f = \sum f_i^1 \times f_i^2$ such that $\|\theta_f - \theta_g\| = \|\theta_{f-g}\| < \varepsilon$. Therefore in view of Lemma 4.5 the problem of showing that U is surjective is reduced to proving the following:

Lemma 4. *If g in $C_0(G_1 \times G_2, \mathscr{H}_1 \otimes \mathscr{H}_2)$ has support S then there is a compact set C in $G_1 \times G_2$ such that $C \supseteq S$ and the following holds: If $\varepsilon > 0$ is given then there is a function $f = \sum f_i^1 \times f_i^2$ such that $f_i^k \in C_0(G_k, \mathscr{H}_k)$ ($k=1,2$ and $i \in \mathscr{I}$), the set C is a carrier of f and $\|f - g\|_\infty \leqslant \varepsilon$.*

Proof. Given $\varepsilon > 0$ by Lemma 4.4 there are finitely many continuous functions $\varphi_h: G \to \mathbb{C}$ with a common compact carrier C and vectors $A_h \in \mathscr{H}_1 \otimes \mathscr{H}_2$ such that $C \supseteq S$ and the least upper bound of the HS norms satisfies $\|g - \sum \varphi_h A_h\| \leqslant \varepsilon/2$. We may suppose that $C = C_1 \times C_2$ where C_k is a compact set in $G_k (k=1,2)$. By the Stone-Weierstrass theorem we can find finitely many functions $\varphi_{hi}^k: G_k \to \mathbb{C}$ such that C_k is a carrier of every φ_{hi}^k and

$$\|\varphi_h - \sum \varphi_{hi}^1 \varphi_{hi}^2\|_\infty \leqslant \frac{1}{H} \min\left(1, \frac{\varepsilon}{4\|A_h\|}\right) \tag{3}$$

where $\|A_h\|$ denotes the HS norm of A_h and H is the number of indices h. By Proposition VI.3.3 there are finitely many vectors α_{hj}^k in $\mathscr{H}_k$ such that

$$\|A_h - \sum \alpha_{hj}^1 \otimes \alpha_{hj}^2\|_\infty \leqslant \frac{\varepsilon}{4(1 + H\|\varphi_h\|_\infty)}. \tag{4}$$

We define $f_{hij}^k = \varphi_{hi}^k \alpha_{hj}^k$. Then by (3) and (4) we have

$$\|g - \sum_{h,i,j} f_{hij}^1 \times f_{hij}^2\|_\infty \leqslant \frac{\varepsilon}{2} + \sum_h \|\varphi_h A_h - \sum_{h,i,j} \varphi_{hi}^1 \varphi_{hi}^2 \alpha_{hj}^1 \otimes a_{hj}^2\|_\infty$$

$$\leqslant \frac{\varepsilon}{2} + \sum_h \left\| \varphi_h - \sum_{i,j} \varphi_{hi}^1 \varphi_{hi}^2 \right\|_\infty \cdot \|A_h\| + \sum_h \left\| \sum_i \varphi_{hi}^1 \varphi_{hi}^2 \right\|_\infty \cdot \left\| A_h - \sum_j \alpha_{hj}^1 \otimes \alpha_{hj}^2 \right\|$$

$$\leqslant \varepsilon .$$

Since U is an isometry of $\mathscr{K}_1 \otimes \mathscr{K}_2$ onto $\mathscr{K}$ we can now complete the proof of Theorem 1 by showing that the restriction of U to the dense subspace $\mathscr{L}$ of $\mathscr{K}_1 \otimes \mathscr{K}_2$ intertwines $\rho^{\chi_1} \otimes \rho^{\chi_2}$ with $\rho^{\chi_1 \otimes \chi_2}$: We have

$$\begin{aligned}\rho^{\chi_1} \otimes \rho^{\chi_2}(a_1, a_2) \sum \theta_i^1 \otimes \theta_i^2 &= \sum \rho^{\chi_1}(a_1)(\theta_i^1 \otimes \theta_i^2)\rho^{\chi_2}(a_2)^* \\ &= \sum \rho^{\chi_1}(a_1)\theta_i^1 \otimes \rho^{\chi_2}(a_2)\theta_i^2 \\ &= \sum {}^{a_1}\theta_i^1 \otimes {}^{a_2}\theta_i^2 \, \lambda_1(a_1^{-1}, \cdot)^{\frac{1}{2}} \lambda_2(a_2^{-1}, \cdot)^{\frac{1}{2}}.\end{aligned}$$

Therefore by (2) we have

$$\begin{aligned}U \rho^{\chi_1} \otimes \rho^{\chi_2}(a_1, a_2) \sum \theta_i^1 \otimes \theta_i^2 &= \sum {}^{a_1}\theta_i^1 \times {}^{a_2}\theta_i^2 \, \lambda_1(a_1^{-1}, \cdot)^{\frac{1}{2}} \lambda_2(a_2^{-1}, \cdot)^{\frac{1}{2}} \\ &= {}^{(a_1, a_2)}\sum \theta_i^1 \times \theta_i^2 \, \lambda_1(a_1^{-1}, \cdot)^{\frac{1}{2}} \lambda_2(a_2^{-1}, \cdot)^{\frac{1}{2}}.\end{aligned}$$

On the other hand

$$\begin{aligned}\rho^{\chi_1 \otimes \chi_2}(a_1, a_2) U \sum \theta_i^1 \otimes \theta_i^2 &= \rho^{\chi_1 \otimes \chi_2}(a_1, a_2) \sum \theta_i^1 \times \theta_i^2 \\ &= {}^{(a_1, a_2)}\sum \theta_i^1 \times \theta_i^2 \, \lambda((a_1, \alpha_2)^{-1}, \cdot)^{\frac{1}{2}}.\end{aligned}$$

We suppose that the quasi-invariant measure μ used on $G_1 \times G_2/H_1 \times H_2$ is defined by the positive ρ-function $(x_1, x_2) \mapsto \rho_1(x_1)\rho_2(x_2)$ where ρ_k is the ρ-function of the measure μ_k used on G_k/H_k $(k=1,2)$. Then

$$\lambda_1(a_1, \cdot)\lambda_2(a_2, \cdot) = \lambda((a_1, a_2), \cdot).$$

This proves that $U\rho^{\chi_1} \otimes \rho^{\chi_2} = \rho^{\chi_1 \otimes \chi_2} U$ on $\mathscr{L}$ and so $\mathscr{L}$ being dense in $\mathscr{K}_1 \otimes \mathscr{K}_2$ we have the same relation on $\mathscr{K}_1 \otimes \mathscr{K}_2$. Theorem 1 is proved.

Proposition 5. *Let H and K be closed subgroups of the locally compact group G such that $G = H \times K$, a direct product. Then for every continuous $\chi: H \to \mathscr{U}(\mathscr{H})$ we have ${}_G\rho^\chi \approx \chi \otimes \lambda$ where λ is the regular representation of K.*

Proof. We apply Theorem 1 with $G_1 = H$, $G_2 = K$, $H_1 = H$ and $H_2 = \{e\}$. For χ_1 we choose the given representation χ and for $\chi_2 = \mathbf{1}$, the one dimensional identity representation. By Proposition 2.8 ${}_K\rho^{\mathbf{1}}$ is λ and by Lemma 2.8 ${}_H\rho^\chi \approx \chi$. Hence by Theorem 1 and Proposition 3.15 we obtain ${}_G\rho^{\chi \otimes \mathbf{1}} \approx {}_H\rho^\chi \otimes {}_K\rho^{\mathbf{1}} \approx \chi \otimes \lambda$. It is clear that $\mathscr{H} \otimes \mathbb{C}$ is isometrically isomorphic to $\mathscr{H}$ under the map $A \mapsto A1$ where $A \in \mathscr{H} \otimes \mathbb{C} = HS(\overline{\mathbb{C}}, \mathscr{H})$ and 1 is the identity of $\mathbb{C}$. Under this map $\chi \otimes \mathbf{1} \approx \chi$ and so by the general form of Proposition 2.5 ${}_G\rho^\chi \approx \chi \otimes \lambda$.

9. The Theorem on Induction in Stages

Here we let G denote a locally compact group with two closed subgroups H_1 and H_2 such that $H_1 \subseteq H_2$. Then H_1 is a closed subgroup of H_2. We let $\chi: H_1 \to \mathcal{U}(\mathcal{H})$ be a weakly measurable unitary representation. We can use χ to induce a representation on G which we denote by $\rho_1^\chi: G \to \mathcal{U}(\mathcal{K}_1)$. We can also induce a representation $\rho_{12}^\chi: H_2 \to \mathcal{U}(\mathcal{K}_{12})$ on the group H_2 and then we can induce a representation $\rho_2^\chi: G \to \mathcal{U}(\mathcal{K}_2)$ on G. One says that ρ_2^χ is the representation induced on G in stages. Our object is to prove that ρ_1^χ and ρ_2^χ are equivalent representations. Hence we are going to establish the following:

Theorem 1. *Let H_1 and H_2 be closed subgroups of the locally compact group G such that $H_1 \subseteq H_2$ and let $\chi: H_1 \to \mathcal{U}(\mathcal{H})$ be a unitary representation. Then the representation induced on G by χ is equivalent to the one obtained by first inducing from H_1 to H_2 and then from there to G.*

By Proposition 4.2 we know that the quasi-invariant measures used on the homogeneous spaces $X_1 = G/H_1, X_2 = G/H_2$ and $X_{12} = H_2/H_1$ can be chosen arbitrarily. The first step in the proof is to adjust these measures such that the rest of the proof becomes as simple as possible.

Lemma 2. *If $\rho_k: G \to (0, \infty)$ is a strictly positive, continuous ρ-function for the space X_k $(k=1,2)$ then $\rho_{12} = (\rho_1/\rho_2)|H_2$ is a strictly positive, continuous ρ-function for X_{12}.*

Proof. Let Δ be the modular function of G and let δ_k be the modular function of H_k $(k=1,2)$. Then by $H_1 \subseteq H_2$ for every $h_1 \in H_1$ and $x \in G$ we have $\rho_k(x h_1) = (\delta_k/\Delta)(h_1)\rho_k(x)$ where $k=1,2$. Therefore $\rho_{12}(x h_1) = (\delta_1/\delta_2)(h_1)\rho_{12}(x)$ for $h_1 \in H_1$ and $x \in H_2$ and so ρ_{12} is a ρ-function for H_2/H_1. The additional properties of ρ_{12} are obvious.

In order to proceed with the proof of Theorem 1 we choose a strictly positive continuous ρ-function $\rho_k: G \to (0, \infty)$ for X_k $(k=1,2)$ and let $\rho_{12}: H_2 \to (0, \infty)$ be the restriction of ρ_1/ρ_2 to H_2. We let μ_1, μ_2 and μ_{12} denote the corresponding quasi-invariant measures on X_1, X_2 and X_{12}, respectively. We define $\lambda_{12}(x,y) = \rho_1(xy)\rho_2(y)/\rho_1(y)\rho_2(xy)$ for $x, y \in G$.

We let $\mathcal{M}_1$ denote the inner product space consisting of equivalence classes of functions $\theta: G \to \mathcal{H}$ whose completion is $\mathcal{K}_1$, the Hilbert space of the induced representation ρ_1^χ. Similarly $\mathcal{M}_{12}$ will denote the inner product space used to define $\mathcal{K}_{12}$, the Hilbert space of ρ_{12}^χ. Hence $\mathcal{M}_{12}$ consists of continuous functions $\phi: H_2 \to \mathcal{H}$ such that $\phi(h_2 h_1) = \chi(h_1)^* \phi(h_2)$ for every $h_k \in H_k$ $(k=1,2)$ and $\|\phi(\cdot)\|$ has compact support in X_{12}.

For $\theta\in\mathcal{M}_1$, $x\in G$ and $h_2\in H_2$ let $\theta(x,h_2)=\theta(xh_2)\sqrt{\lambda_{12}(x,h_2)}$ so that $h_2\mapsto\theta(x,h_2)$ defines a function mapping H_2 into $\mathcal{H}$. By the continuity of θ and the group product $\theta(x,\cdot)$ is continuous. If $h_k\in H_k$ $(k=1,2)$ then

$$\theta(x,h_2h_1)=\theta(xh_2h_1)\sqrt{\lambda_{12}(x,h_2h_1)}=\chi(h_1)^*\theta(xh_2)\sqrt{\lambda_{12}(x,h_2)}$$
$$=\chi(h_1)^*\theta(x,h_2).$$

Therefore $\|\theta(x,\cdot)\|$ is defined on X_{12}. Let S denote the compact support of $\|\theta(\cdot)\|$ in X_1. Since we have $\|\theta(x,h_2)\|\neq 0$ only if $p_1(xh_2)=x(h_2H_1)\in S$ we see that the compact set $x^{-1}S\cap X_{12}$ is a carrier of $\|\theta(x,\cdot)\|$ in X_{12}. Therefore we proved that $\theta(x,\cdot)$ belongs to $\mathcal{M}_{12}$ and hence to $\mathcal{K}_{12}$. Let $\mathcal{M}_2$ be the linear manifold defining $\mathcal{K}_2$, the Hilbert space of the two stage induction. Thus the elements of $\mathcal{M}_2$ are functions $\psi\colon G\to\mathcal{K}_{12}$ satisfying two additional conditions: Namely $\psi(xh_2)=\rho_{12}^{\chi}(h_2)\psi(x)$ for every $x\in G$, $h_2\in H_2$ and $\|\psi(\cdot)\|$ has compact support in X_2.

Lemma 3. *If* $\theta\in\mathcal{M}_1, x\in G$ *and* $\theta(x,h_2)=\theta(xh_2)\sqrt{\lambda_{12}(x,h_2)}$ *then the function* $\psi_\theta\colon G\to\mathcal{K}_{12}$ *defined by* $\psi_\theta(x)=\theta(x,\cdot)$ *belongs to* $\mathcal{M}_2$.

Proof. First let $x\in G$, $h_2\in H_2$ and $y\in H_2$ be given. Then $\psi_\theta(xh_2)(y)=\theta(xh_2,y)=\theta(xh_2y)\sqrt{\lambda_{12}(xh_2,y)}$. Since we have $\lambda_{12}(xh_2,y)=\lambda_{12}(x,h_2y)\cdot\lambda_{12}(h_2,y)$ we obtain

$$\psi_\theta(xh_2)(y)=\theta(x,h_2y)\sqrt{\lambda_{12}(h_2,y)}=\psi_\theta(x)(h_2y)\sqrt{\lambda_{12}(h_2,y)}$$
$$=(\rho_{12}\chi(h^2)^*\psi_\theta(x))(y).$$

Therefore $\psi_\theta(xh_2)=\rho_{12}(h_2)^*\psi_\theta(x)$ for every $x\in G$ and $h_2\in H_2$.

Next we prove that ψ_θ is a continuous function: Let S denote the compact support of $\|\theta(\cdot)\|$ in X_1. If $x,y\in G$ then

$$\|\psi_\theta(x)-\psi_\theta(y)\|^2=\int_{X_{12}}\|\theta(xz)-\theta(yz)\|^2\lambda_{12}(x,z)\,d\mu_{12}(\xi) \tag{1}$$

where z denotes an arbitrary element in ζ for each $\zeta\in X_{12}$. We fix x and choose a compact symmetric neigborhood C of e in G. Let y be an arbitrary element in Cx. Given ζ in X_{12} we have $\|\theta(xz)-\theta(yz)\|\neq 0$ either for every $z\in\zeta$ or for no $z\in\zeta$. The first alternative takes place only if $\|\theta(xz)\|\neq 0$ or $\|\theta(yz)\|\neq 0$ for all $z\in\zeta$. This holds only if $x\cdot\zeta\in S$ or $y\cdot\zeta\in S$ and so only if $\zeta\in x^{-1}\cdot S\subseteq x^{-1}C\cdot S$ or $\zeta\in y^{-1}\cdot S\subseteq x^{-1}C\cdot S$. Hence if $y\in C$ then $x^{-1}C\cdot S$ is a carrier of $\|\theta(x\cdot)-\theta(y\cdot)\|$ in X_{12}. Thus by (1) we have

$$\|\psi_\theta(x)-\psi_\theta(y)\|^2=\int_{x^{-1}C\cdot S}\|\theta(xz)-\theta(yz)\|^2\lambda_{12}(x,z)\,d\mu_{12}(\zeta) \tag{2}$$

for every $y\in Cx$.

By Lemma V.3.6 there is a compact set K in G such that $p_1 K \supseteq x^{-1} C \cdot S$. Since CxK is compact θ is uniformly continuous with respect to the unique uniform structure of CxK. Hence using the trace of the right uniform structure of G on CxK we obtain the following: Given $\varepsilon > 0$ there is a symmetric neighborhood N_e of the identity e in G such that $N_e \subseteq C$ and

(3) $\qquad \|\theta(a)-\theta(b)\| < \tilde{\varepsilon} \quad$ with $\varepsilon = \tilde{\varepsilon} \operatorname{supr}\{\lambda_{12}(x, \zeta) : \zeta \in x^{-1} C \cdot S\}$

for every $a, b \in CxK$ satisfying $ab^{-1} \in N_e$. Let y be an arbitrary point in $N_e x$. If $\zeta \in x^{-1} C \cdot S$ then one can choose z in ζ such that $z \in K$. Hence $xz \in xK \subseteq CxK$ and $yz \in N_e xK \subseteq CxK$ where $(xz)(yz)^{-1} = xy^{-1} \in N_e$. Thus by (3) we have $\|\theta(xz)-\theta(yz)\| < \tilde{\varepsilon}$. Therefore by (2) we proved that $\|\psi_\theta(x)-\psi_\theta(y)\| < \varepsilon \mu_{12}(x^{-1} C \cdot S)$ for every $y \in N_e x$. Here $\mu_{12}(x^{-1} C \cdot S)$ is finite by Proposition 6.14.

Finally we prove that $\|\psi_\theta(\cdot)\|$ has a compact support in the quotient space $X_2 = G/H_2$. Since $\psi_\theta(x)$ is continuous we have $\psi_\theta(x) \neq 0$ if and only if $\psi_\theta(x)(y) \neq 0$ for some $y \in H_2$. Thus if $\psi_\theta(x) \neq 0$ then $\theta(xy) \neq 0$ for some $y \in H_2$ and so $p_1(xy) = xyH_1 \in S$ where S is the compact support of $\|\theta(\cdot)\|$ in X_1. Hence xH_2 belongs to the image of S under the continuous map $xH_1 \mapsto xH_2$. This shows that the image of S in X_2 is a compact carrier of ψ_θ.

Lemma 4. *The map $\theta \mapsto \psi_\theta$ is an isometry.*

Proof. Let θ in $\mathcal{M}$ be given. Then $\|\theta(\cdot)\|^2$ is a continuous function with compact support in X_1 and so by Theorem V.3.7 there is a φ in $C_0(G)$ such that

(4) $$\tilde{\varphi}(x) = \int_{H_1} \varphi(xh_1)\, dh_1 = \|\theta(x)\|^2$$

for every $x \in G$. Therefore by

$$\|\psi_\theta(x)\|^2 = \int_{X_{12}} \|\psi_\theta(x)(y)\|^2 \, d\mu_{12}(\eta) = \int_{X_{12}} \|\theta(xy)\|^2 \lambda_{12}(x, y)\, d\mu_{12}(\eta)$$

we obtain

$$\|\psi_\theta(x)\|^2 = \int_{X_{12}} \tilde{\varphi}(xy) \lambda_{12}(x, y)\, d\mu_{12}(\eta) .$$

Since $\lambda_{12}(x, yh_1) = \lambda_{12}(x, y)$ by (4) we have

$$\|\psi_\theta(x)\|^2 = \int_{X_{12}} \int_{H_1} \varphi(xyh_1) \lambda_{12}(x, yh_1)\, dh_1\, d\mu_{12}(\eta) .$$

By definition $\lambda_{12}(x, yh_1) = \rho_{12}(xyh_1)/\rho_{12}(yh_1)$ and so applying the corollary of Theorem V.3.9 we obtain

$$\|\psi_\theta(x)\|^2 = \int_{H_2} \rho_{12}(xy) \varphi(xy)\, dy .$$

Hence using $\rho_{12}(xy)=\rho_1(xy)/\rho_2(xy)$ the same corollary implies that

$$\|\psi_\theta\|^2=\int_{X_2}\|\psi_\theta(x)\|^2\,d\mu_2(\xi)=\int_G\varphi(x)\rho_1(x)\,dx\,.$$

Similarly by (4) we have

$$\|\theta\|^2=\int_{X_1}\|\theta(x)\|^2\,d\mu_1(\xi)=\int_{X_1}\tilde{\varphi}(x)\,d\mu_1(\xi)=\int_G\varphi(x)\rho_1(x)\,dx$$

and this shows that $\|\psi_\theta\|=\|\theta\|$.

Lemma 5. *If* $\varphi\in C_0(G)$ *and* $\Phi\in\mathscr{M}_{12}$ *then there is a* θ *in* $\mathscr{M}_1$ *such that* $\psi_{\varphi\Phi}=\psi_\theta$ *where* $\varphi\Phi\in C_{01}(G,\mathscr{K}_{12})$.

Proof. By Proposition 4.3 we have $\Phi=\Phi_f$ where $f\in C_0(H_2,\mathscr{H})$. Let $g\in C_0(G,\mathscr{H})$ be defined by

$$g(x)\sqrt{\rho_{12}(x)}=\int_{H_2}\varphi(xh_2)\,f(h_2^{-1})\sqrt{\rho_{12}(h_2^{-1})}\,dh_2$$

where $x\in G$. We replace in this equation x by xy where $y\in H_2$, perform the substitution $h_2\to y^{-1}h_2$, divide both sides by $\sqrt{\rho_{12}(y)}$ and replace y by yh_1 where $h_1\in H_1$. Then we obtain

$$g(xyh_1)\sqrt{\lambda_{12}(x,y)}=\int_{H_2}\varphi(xh_2)\,f(h_2^{-1}yh_1)\sqrt{\lambda_{12}(h_2^{-1},y)}\,dh_2\,.$$

Next we apply the operator $\chi(h_1)$ and integrate the resulting equation in the weak sense on H_1. After interchanging the order of integration on the right hand side we shall thus arrive at the equation

$$(\theta_g(x,y),\alpha)=\int_{H_2}\varphi(xh_2)\sqrt{\lambda_{12}(h_2^{-1},y)}\int_{H_1}(\chi(h_1)\,f(h_2^{-1}yh_1),\alpha)\,dh_1\,dh_2$$

where $\alpha\in\mathscr{H}$. Since $f\in C_0(H_2,\mathscr{H})$ the right hand side can be written in the form

$$\int_{H_2}\varphi(xh_2)\sqrt{\lambda_{12}(h_2^{-1},y)}(\Phi_f(h_2^{-1}y),\alpha)\,dh_2\,.$$

Since α is arbitrary and the functions which occur are continuous we obtain

$$\begin{aligned}\theta_g(x,y)&=\int_{H_2}\varphi(xh_2)\,\phi_f(h_2^{-1}y)\sqrt{\lambda_{12}(h_2^{-1},y)}\,dh_2\\&=\int_{H_2}\varphi(xh_2)\,\rho_{12}^{\chi}(h_2)\,\phi_f(y)\,dh_2=\psi_{\varphi\Phi_f}(x)(y)\,.\end{aligned}$$

Hence $\psi_{\varphi\Phi}=\psi_\theta$ where $\theta=\theta_g$.

Lemma 6. *The set* $\{\psi_\theta\colon\theta\in\mathscr{M}_1\}$ *is dense in* $\mathscr{K}_2$.

Proof. Since $\mathscr{K}_2$ is the completion of $\mathscr{M}_2$ it will be sufficient to prove that every element of $\mathscr{M}_2$ can be approximated arbitrarily closely by elements of the form ψ_θ where $\theta \in \mathscr{M}_1$. If ψ is an element of $\mathscr{M}_2$ then by Proposition 4.3 we have $\psi = \psi_f$ for some f in $C_0(G, \mathscr{K}_{12})$. Hence by Lemmas 4.4 and 4.5 ψ can be approximated as closely as we wish by linear combinations of vectors of the form $\psi_{\varphi\Phi}$ where $\varphi \in C_0(G)$ and $\Phi \in \mathscr{K}_{12}$. Therefore it will be enough to show that each $\psi_{\varphi\Phi}$ can be approximated by the elements of $\{\psi_\theta : \theta \in \mathscr{M}_1\}$. We can further reduce the problem by using a similar reasoning for $\mathscr{K}_{12}$: Since $\mathscr{M}_{12}$ is dense in $\mathscr{K}_{12}$ by Lemma 4.5 we may suppose that $\Phi \in \mathscr{M}_{12}$. By Lemma 5 these elements of $\mathscr{K}_2$ can not only be approximated but are actually of the form $\psi_{\varphi\Phi} = \psi_\theta$ where $\theta \in \mathscr{M}_1$.

The map $\theta \mapsto \psi_\theta$ of $\mathscr{M}_1$ into $\mathscr{M}_2$ which is defined in Lemma 3 is an isometry by Lemma 4. Hence it can be extended in a unique way to a unitary map U of $\mathscr{K}_1$ into $\mathscr{K}_2$. By Lemma 6 the image space of U is dense in $\mathscr{K}_2$ and so U is a unitary map of $\mathscr{K}_1$ onto $\mathscr{K}_2$. We prove that ρ_1^χ and ρ_2^χ are intertwined by $U : \mathscr{K}_1 \to \mathscr{K}_2$: If $\theta \in \mathscr{M}_1$ and $x \in G$ then

$$(\rho_2^\chi(a)\, U\, \theta)(x) = (\rho_2^\chi(a)\, \psi_\theta)(x) = \psi_\theta(a^{-1}x)\sqrt{\lambda_2(a^{-1}, x)}$$

and so

$$(\rho_2^\chi(a)\, U\, \theta)(x)(y) = \theta(a^{-1}xy)\sqrt{\lambda_{12}(a^{-1}x, y)\,\lambda_2(a^{-1}, x)}\,.$$

On the other hand

$$\begin{aligned}(U\rho_1^\chi(a)\theta)(x)(y) &= (U^a\theta\sqrt{\lambda_1(a^{-1}, \cdot)})(x)(y) = \psi_{{}^a\theta\sqrt{\lambda_1(a^{-1},\cdot)}}(x)(y)\\ &= \theta(a^{-1}xy)\sqrt{\lambda_1(a^{-1}, xy)\,\lambda_{12}(x, y)}\,.\end{aligned}$$

Hence in order to prove that $\rho_2^\chi U = U\rho_1^\chi$ it is sufficient to check that

$$\lambda_{12}(ax, y)\,\lambda_2(a, x) = \lambda_1(a, xy)\,\lambda_{12}(x, y)$$

for every $a, x \in G$ and $y \in H_2$. If we express λ_1, λ_2 and λ_{12} in terms of ρ_1 and ρ_2 then in view of $\rho_2(axy) = \rho_2(ax)$ and $\rho_2(xy) = \rho_2(x)$ this becomes a trivial identity. This completes the proof of Theorem 1.

Proposition 7. *Let H be a closed subgroup of the locally compact group G and let λ and Λ denote the left regular representations of H and G, respectively. Then the representation induced on G by λ is equivalent to Λ.*

Proof. Apply Theorem 1 with the following choices: $H_1 = \{e\}$, $H_2 = H$ and $\chi = 1$, the one dimensional identity representation of H_1. By the general version of Proposition 2.7 the representations induced on H_2 and G by χ are λ and Λ, respectively. Hence the conclusion follows right away from Theorem 1.

10. Representations Induced by Representations of Conjugate Subgroups

We are going to prove the following general theorem and give a few corollaries such as the subgroup lemma and the tensor product lemma of induced representations.

Theorem 1. *Let G be a locally compact group and let H_1 and H_2 be closed conjugate subgroups of G under an inner automorphism $\sigma: G \to G$. Let $\chi_i: H_i \to \mathscr{U}(\mathscr{H}_i)$ $(i=1,2)$ be representations such that $U\chi_1 U^{-1} = \chi_2 \sigma$ for a unitary operator $U: \mathscr{H}_1 \to \mathscr{H}_2$. Then the representations ${}_G\rho^{\chi_i}$ $(i=1,2)$ are equivalent.*

The proof will be based on two simple lemmas. We let X_i denote the homogeneous space G/H_i $(i=1,2)$. The modular function of H_i will be denoted by $\delta_i (i=1,2)$. Since σ is a topological isomorphism of H_1 onto H_2 we have $\delta_1 = \delta_2 \sigma$. We let δ be the modular function of G.

Lemma 2. *Let s in G be such that $\sigma = s(\cdot)s^{-1}$ and for every $\xi \in X_1$ let $\xi s^{-1} = (xs^{-1})H_2$ where x is an arbitrary element of ξ. Then $\xi \mapsto \xi s^{-1}$ is a homeomorphism of X_1 onto X_2.*

Proof. The image ξs^{-1} and the inverse image ηs where $\eta \in X_2$ are well defined: For instance if x, x' belong to ξ then $(xs^{-1})^{-1}(x's^{-1}) = sx^{-1}x's^{-1} \in sH_1 s^{-1} = H_2$ and so $xs^{-1}, x's^{-1}$ belong to the same left H_2 coset. It is easy to see that $\xi \mapsto \xi s^{-1}$ is a bijection. In order to prove the continuity of $\eta s \mapsto \eta$ where $\eta \in X_2$ let an open set Q be given in X_2. We prove that $\{yH_1 : ys \in p_2^{-1}Q\}$, the image of Q under $\eta \mapsto \eta s$ is an open set in X_1. Indeed the inverse image of this set under p_1 is

$$\{yh_1 : h_1 \in H_1 \text{ and } ys \in p_2^{-1}Q\} = (p_2^{-1}Qs^{-1})H_1$$

where $p_2^{-1}Qs^{-1}$ is an open set in G. Since the roles of H_1 and H_2 can be interchanged we see that the map $\xi \mapsto \xi s^{-1}$ is also continuous.

Lemma 3. *Let s in G be such that $\sigma = s(\cdot)s^{-1}$ and let $\rho_1: G \to (0, \infty)$ be a positive, continuous ρ-function for the homogeneous space X_1. Then $\rho_2(x) = \rho_1(xs)\delta(s) (x \in G)$ defines a positive, continuous ρ-function for X_2.*

Proof. We have to prove that $\rho_2(xh_2) = (\delta_2/\delta)(h_2)\rho_2(x)$ for every $x \in G$ and $h_2 \in H_2$. Since $h_2 \in H_2$ we have $h_2 = \sigma h_1$ for some $h_1 \in H_1$. Thus $\rho_2(xh_2) = \rho_1(xh_2 s)\delta(s) = \rho_1(xsh_1)\delta(s) = (\delta_1/\delta)(h_1)\rho_1(xs)\delta(s)$ and so by the homomorphism property of δ we obtain

$$\rho_2(xh_2) = \frac{\delta_1(h_1)}{\delta(sh_1 s^{-1})}\rho_2(x) = \frac{\delta_2}{\delta}(h_2)\rho_2(x).$$

Hence ρ_2 is a ρ-function for X_2. The positivity and continuity of ρ_2 are obvious facts.

Proof of Theorem 1. Let ρ_1 be a positive, continuous ρ-function for X_1. We let ρ_2 be related to ρ_1 as is described in Lemma 3 and let μ_1 and μ_2 be the corresponding quasi-invariant measures on X_1 and X_2, respectively. Furthermore we let $\mathscr{M}_i$ denote the inner product space whose completion is $\mathscr{H}_i$, the Hilbert space of the induced representation ${}_G\rho^{\chi_i}(i=1,2)$. We define a linear map $V\colon \mathscr{M}_1 \to \mathscr{M}_2$ which will turn out to be a bijection and we will show that V is norm preserving. Namely if $\theta^1 \in \mathscr{M}_1$ then we let $(V\theta^1)(x) = U\theta^1(xs)$ for all $x \in G$. As earlier s denotes here an element $s \in G$ such that $\sigma(h_1) = s h_1 s^{-1}$ for every h_1 in H_1. Then it is clear that $V\theta^1\colon G \to \mathscr{H}_2$ is continuous map. We prove that

$$V\theta^1(x h_2) = \chi_2(h_2)^* \, V\theta^1(x)$$

for every $x \in G$ and $h_2 \in H_2$. Indeed we have $h_2 = \sigma h_1 = s h_1 s^{-1}$ for a suitable $h_1 \in H_1$ and so

$$\begin{aligned} V\theta^1(x h_2) &= V\theta^1(x s h_1 s^{-1}) = U\theta^1(x s h_1) = U\chi_1(h_1)^* \theta^1(x s) \\ &= (U\chi_1(h_1)U^*)^* \, U\theta^1(x s) = \chi_2(h_2)^* \, V\theta^1(x). \end{aligned}$$

We have $V\theta^1(x) \neq 0$ only if $U\theta^1(xs) \neq 0$. Hence xsH_1 is in S_1, the support of θ^1 in X_1. Thus $\xi = xH_2$ belongs to the image of S_1 under the homeomorphism $\xi \mapsto \xi s^{-1}$ defined in Lemma 2. Hence $S_1 \cdot s^{-1}$ is a compact carrier of $V\theta^1$ in X_2. We proved that $V\theta^1$ is an element of $\mathscr{M}_2$ for every θ^1 in $\mathscr{M}_1$. It is now easy to verify that the map $V\colon \mathscr{M}_1 \to \mathscr{M}_2$ is surjective. Actually it is sufficient to refer to the essentially symmetric roles of H_1 and H_2 in this discussion.

Next we prove that V is an isometry: Given θ^1 in $\mathscr{M}_1$ by Theorem V.3.7 there is a function f in $C_0(G)$ such that $\tilde{f}(\xi) = \int_{H_1} f(x h_1)\, dh_1 = 1$ for every ξ in S_1. Since the map $x \mapsto s x s^{-1}$ is an isomorphism of H_1 onto H_2 we see that

$$\int_{H_2} f(x s^{-1} h_2 s)\, dh_2 = 1$$

for every $\xi \in S_1$ and $x \in \xi$. Since $S_1 \cdot s^{-1}$ is a carrier of $\|V\theta^1(\cdot)\|$ and $\xi s^{-1} = (x s^{-1}) H_2$ we see that

$$\tilde{f}(\eta) = \int_{H_2} f_s(y h_2)\, dh_2 = 1$$

for every η in the support of $\|V\theta^1(\cdot)\|$ in X_2. Therefore

$$\|V\theta^1\|^2 = \int_{X_2} \|V\theta^1(y)\|^2 \tilde{f}(\eta)\, d\mu_2(\eta) = \int_{X_2} \int_{H_2} \|V\theta^1(y h_2)\|^2 f_s(y h_2)\, dh_2\, d\mu_2(\eta).$$

By applying the corollary of Theorem V.3.9 we obtain

$$\|V\theta^1\|^2=\int_G \|V\theta^1(y)\|^2 f(ys)\rho_2(y)\,dy=\int_G \|\theta^1(ys)\|^2 f(ys)\rho_1(ys)\delta(y)\,dy.$$

Hence using the substitution $y\to ys^{-1}$ we see that

$$\begin{aligned}\|V\theta^1\|^2&=\int_G \|\theta^1(y)\|^2 f(y)\rho_1(y)\,dy=\int_{X_1}\int_{H_1} \|\theta^1(yh_1)\|^2 f(yh_1)\,dh_1\,d\mu_1(\eta)\\ &=\int_{X_1}\|\theta^1(y)\|^2 \tilde{f}(\eta)\,d\mu_1(\eta)=\int_{X_1}\|\theta^1(y)\|^2\,d\mu_1(\eta)=\|\theta^1\|^2.\end{aligned}$$

We proved that V maps $\mathcal{M}_1$ isometrically onto $\mathcal{M}_2$.

We can now extend $V\colon \mathcal{M}_1\to\mathcal{M}_2$ to a unitary operator $V\colon \mathcal{K}_1\to\mathcal{K}_2$. It is simple to prove that V intertwines ${}_G\rho^{\chi_1}$ with ${}_G\rho^{\chi_2}$: For let λ_i denote the λ-function associated with ρ_i $(i=1,2)$ and let $a,x\in G$. Then for any θ^1 in $\mathcal{M}_1$ we have

$$\begin{aligned}(V({}_G\rho^{\chi_1}(a))\theta^1)(x)&=(V^a\theta^1\sqrt{\lambda_1(a^{-1},\cdot)})(x)=U^a\theta^1(xs)\sqrt{\lambda_1(a^{-1},xs)}\\ &=U\theta^1(a^{-1}xs)\sqrt{\lambda_1(a^{-1},xs)}.\end{aligned}$$

On the other hand

$$\begin{aligned}(({}_G\rho^{\chi_2}(a))V\theta^1)(x)&=({}_G\rho^{\chi_2}(a)U(\theta^1)_s)(x)={}^aU(\theta^1)_s(x)\sqrt{\lambda_2(a^{-1},x)}\\ &=U\theta^1(a^{-1}xs)\sqrt{\lambda_2(a^{-1},x)}.\end{aligned}$$

Therefore in order to prove the intertwining nature of V it is sufficient to prove that $\lambda_2(a,x)=\lambda_1(a,xs)$ for every $a,x\in G$. By definition $\rho_2(ax)=\rho_1(axs)\delta(s)$ and $\rho_2(x)=\rho_1(xs)\delta(s)$ and so the required identity is a triviality. This completes the proof of Theorem 1.

Let G be any group and let H_1 and H_2 be subgroups of G. Two elements x,y of G will be considered equivalent with respect to the ordered pair H_1,H_2 if $y=h_1xh_2$ for suitable $h_i\in H_i$ $(i=1,2)$. It is easily verified that an equivalence relation is defined in this manner on G. An equivalence class will be called an H_1,H_2 *double coset* of G and will be denoted by D. It is obvious that each double coset is a subset of G of the form $D=H_1xH_2$ where $x\in G$. The set of H_1,H_2 double cosets will be denoted by $H_1\backslash G/H_2$. Two double cosets are either disjoint or they coincide. Every D can be represented as a union of left H_2 cosets and also as a union of right H_1 cosets. Examples for double cosets in topological groups can be found in Section V.1.

In what follows we let G be a locally compact group and let K and H be closed subgroups of G. For any x in G we define $H(x)=K\cap xHx^{-1}$. If x_1,x_2 belong to the same K,H double coset D then $H_1=H(x_1)$ and $H_2=H(x_2)$ are topologically isomorphic groups: If $x_2=kx_1h$ with

$k \in K$ and $h \in H$ then $y \mapsto kyk^{-1}$ is a bicontinuous isomorphism of H_1 onto H_2. Let a representation $\chi: H \to \mathcal{U}(\mathcal{H})$ be given. If $x \in G$ and $y \in H(x)$ then $x^{-1}yx \in H$ and so we can define $\chi_x(y) = \chi(x^{-1}yx)$. Then it is clear that $\chi_x: H(x) \to \mathcal{U}(\mathcal{H})$ is a unitary representation. For the sake of simplicity we shall use the abbreviation $\chi_i = \chi_{x_i}$ $(i=1,2)$. If $x_2 = kx_1h$ and $y \in H_1$ then

$$\chi_2(kyk^{-1}) = \chi(x_2^{-1}kyk^{-1}x_2) = \chi(h^{-1}x_1^{-1}yx_1h) = \chi(h)^{-1}\chi_1(y)\chi(h).$$

Therefore if we identify H_1 with H_2 under the topological isomorphism $y \mapsto kyk^{-1}$ then we see that $\chi(h)$ intertwines χ_2 with χ_1. Hence the pairs H_1, χ_1 and H_2, χ_2 in a certain sense can be considered the same. Using Theorem 1 we can prove that χ_1 and χ_2 induce equivalent representations on K. Thus we shall obtain:

Theorem 4. *Let H and K be closed subgroups of the locally compact group G and let $\chi: H \to \mathcal{U}(\mathcal{H})$ be a unitary representation. For x in G let $H(x) = K \cap xHx^{-1}$ and let $\chi_x: H(x) \to \mathcal{U}(\mathcal{H})$ be the representation defined by $\chi_x(y) = \chi(x^{-1}yx)$ where $y \in H(x)$. Then the representation ${}_K\rho^{\chi_x}$ induced on K by χ_x is determined within equivalence by the double coset $D = KxH$ containing x.*

Note. In view of this result we can write ${}_K\rho^{\chi_D}$ instead of ${}_K\rho^{\chi_x}$ and more precisely we use this symbol to denote the equivalence class containing all the representations ρ^{χ_x} $(x \in D)$.

By applying Theorem 9.1 we see that ${}_G\rho^{\chi_i}$ is equivalent to the representation induced on G by ${}_K\rho^{\chi_i}$ $(i=1,2)$. Hence by the general version of Proposition 2.5 we obtain the following:

Corollary. *The representation induced on G by χ_x depends only on the K, H double coset containing x.*

Another consequence of Theorem 1 will be derived by using the following:

Lemma 5. *If G is a group and $\tilde{G}$ is the diagonal subgroup of $G \times G$ then $(x, y) \mapsto xy^{-1}$ induces a bijection between the $G_1 \times G_2, \tilde{G}$ double cosets of $G \times G$ and the G_1, G_2 double cosets of G.*

Proof. If $g_i \in G_i$ $(i=1,2)$ and $g \in G$ then (g_1xg, g_2yg) is mapped into $g_1(xy^{-1})g_2^{-1}$ and this shows that the image of $(G_1 \times G_2)(x, y)\tilde{G}$ lies in $G_1(xy^{-1})G_2$. It also shows that the image is the entire double coset $G_1(xy^{-1})G_2$. It is obvious that every coset $G_1 z G_2$ $(z \in G)$ can be obtained in this manner. We prove that distinct double cosets are mapped into distinct ones: For if $(G_1 \times G_2)(x_1, y_1)\tilde{G}$ and $(G_1 \times G_2)(x_2, y_2)\tilde{G}$ are both mapped onto the same double coset $G_1 x_1 y_1^{-1} G_2 = G_1 x_2 y_2^{-1} G_2$

then $x_1 y_1^{-1} = g_1 x_2 y_2^{-1} g_2$ for suitable $g_i \in G_i$ $(i=1,2)$ and so we have $y_1 = g_2^{-1} y_2 (x_2^{-1} g_1^{-1} x_1)$. Since $x_1 = g_1 x_2 (x_2^{-1} g_1^{-1} x_1)$ we see that $(x_1, y_1) \in (G_1 \times G_2)(x_2, y_2)\tilde{G}$.

Theorem 6. *Let H_1 and H_2 be closed subgroups of the locally compact group G and let $\chi_i: H_i \to \mathscr{U}(\mathscr{H}_i)$ $(i=1,2)$ be unitary representations. Given (x_1, x_2) in $G \times G$ consider the representations $\chi_{x_i}(\cdot) = \chi_i(x_i^{-1} \cdot x_i)$ of the subgroup $(x_1 H_1 x_1^{-1}) \cap (x_2 H_2 x_2^{-1})$ and let $\chi_{(x_1, x_2)}$ be the restriction of $\chi_{x_1} \otimes \chi_{x_2}$ to the diagonal subgroup. Then the representation induced on $G \times G$ by $\chi_{(x_1, x_2)}$ is determined within equivalence by the H_1, H_2 double coset containing $x_1 x_2^{-1}$.*

Proof. We use Theorem 4 with the following choices: We replace G by $G \times G$ and H by $H_1 \times H_2$. For the subgroup K we choose the diagonal subgroup $\tilde{G}$ of $G \times G$ and we let $\chi = \chi_1 \otimes \chi_2$. Now given $x = (x_1, x_2)$ one can immediately see that the subgroup $H(x) = K \cap x H x^{-1}$ becomes the diagonal subgroup corresponding to $x_1 H_1 x_1^{-1} \cap x_2 H_2 x_2^{-1}$ and so χ_x is the representation $\chi_{(x_1, x_2)}$ described in the above theorem. Therefore by the corollary of Theorem 4 ${}_{G \times G}\rho^{\chi(x_1, x_2)}$ depends only on the double coset $D(x_1, x_2) = H(x_1, x_2) K$. Hence by Lemma 5 it depends only on $H_1 x_1 x_2^{-1} H_2$.

11. Mackey's Theorem on Strong Intertwining Numbers and Some of its Consequences

Let $\rho_i: G \to \mathscr{U}(\mathscr{H}_i)$ $(i=1,2)$ be arbitrary unitary representations of the abstract group G and let (ρ_1, ρ_2) be the vector space of intertwining operators $T \in \mathscr{L}(\mathscr{H}_1, \mathscr{H}_2)$. If the dimension of (ρ_1, ρ_2) is finite then this integer will be called the *intertwining number* of ρ_1 and ρ_2 and it will be denoted by $i(\rho_1, \rho_2)$. If (ρ_1, ρ_2) is infinite dimensional then by definition $i(\rho_1, \rho_2)$ is the symbol ∞. By Proposition IV.3.6 we have $i(\rho_1, \rho_2) = i(\rho_2, \rho_1)$. If T is a HS operator in (ρ_1, ρ_2) then it will be called a *strong intertwining operator* and the vector space of these operators will be denoted by $(\rho_1, \rho_2)_{HS}$. If this space is finite dimensional then its dimension is called the *strong intertwining number* of ρ_1 and ρ_2 and it is denoted by $j(\rho_1, \rho_2)$. Thus i and j are finite cardinals or the symbol ∞ and $j(\rho_1, \rho_2) \leqslant i(\rho_1, \rho_2)$. It is clear that $j(\rho_1, \rho_2) = j(\rho_2, \rho_1)$ for any two representations ρ_1 and ρ_2 of G. It is also obvious that all these concepts are meaningful also when G is replaced by an arbitrary admissible object S. By Proposition IV.3.6 the equations $i(\rho_1, \rho_2) = i(\rho_2, \rho_1)$ and $j(\rho_1, \rho_2) = j(\rho_2, \rho_1)$ will hold if S is a Mackey object and the map $S \to S$ is surjective.

If $\rho: S \to \mathscr{L}(\mathscr{H})$ is a representation in Hilbert space then we let $\mathscr{H}_{dI}$ denote the largest subspace of $\mathscr{H}$ on which ρ is a direct sum of irreducible components. The existence and properties of $\mathscr{H}_{dI}$ are stated in Theorem IV.5.4. Let ${}^0\hat{S}$ denote the set of those equivalence classes $\omega \in \hat{S}$ whose degrees d_ω are finite. Then by the same theorem the direct sum of the subspaces $\mathscr{H}_\omega$ $(\omega \in {}^0\hat{S})$ is a subspace $\mathscr{H}_{fdI}$ of $\mathscr{H}_{dI}$. It is the largest ρ-invariant subspace of $\mathscr{H}$ on which ρ is the direct sum of finite dimensional irreducible components. The restriction of ρ to $\mathscr{H}_{fdI}$ will be called the *strictly finite discrete part* of ρ and it will be denoted by ${}^0\rho$.

We let $\boldsymbol{1}_G$ denote the equivalence class of the one dimensional identity representation $G \to \mathscr{U}(\mathbb{C})$ so that $\boldsymbol{1}_G$ is in ${}^0\hat{G}$. If it causes no confusion we shall write $\boldsymbol{1}$ instead of $\boldsymbol{1}_G$. We shall use $\boldsymbol{1}_G$ or $\boldsymbol{1}$ to denote also a concrete representation belonging to the class $\boldsymbol{1}_G$. Following the terminology introduced in Section IV.5 we let $n(\omega, \rho)$ denote the multiplicity of the class $\omega \in \hat{G}$ in the representation $\rho: G \to \mathscr{U}(\mathscr{H})$. Therefore we have

$$(1) \qquad d_\omega n(\omega, \rho) = \dim \mathscr{H}_\omega .$$

We let $\bar{\rho}$ stand for the complex conjugate of the representation ρ.

Theorem 1. *Let $\rho_i: G \to \mathscr{U}(\mathscr{H}_i)$ $(i=1,2)$ be arbitrary unitary representations of the abstract group G. Then*

$$j(\rho_1, \rho_2) = i({}^0\rho_1, {}^0\rho_2) = n(\boldsymbol{1}_G, \rho_2 \otimes \bar{\rho}_1) = \sum_{\omega \in \hat{G}} n(\omega, {}^0\rho_1) n(\omega, {}^0\rho_2).$$

Note. Since we use $\boldsymbol{1}_G$ it is clear that $\rho_2 \otimes \bar{\rho}_1$ denotes above the inner tensor product of ρ_2 and $\bar{\rho}_1$.

Proof. If T belongs to $HS(\mathscr{H}_1, \mathscr{H}_2)$ then every one of the statements listed below is equivalent to the one following it:

$$\begin{gathered} T \text{ is in the vector space } (\rho_1, \rho_2), \\ \begin{aligned} T\rho_1(x) &= \rho_2(x)\,T && \text{for every } x \in G, \\ \rho_2(x)\,T\rho_1(x)^* &= T && \text{for every } x \in G, \\ \rho_2(x)\,T\bar{\rho}_1(x)^* &= T && \text{for every } x \in G, \\ \rho_2 \otimes \bar{\rho}_1(x)\,T &= T && \text{for every } x \in G. \end{aligned} \\ T \in \mathscr{K}_{1_G} \text{ where } \mathscr{K} = HS(\mathscr{H}_1, \mathscr{H}_2) \text{ is the Hilbert space of } \rho_2 \otimes \bar{\rho}_1. \end{gathered}$$

Therefore we proved that

$$(2) \qquad (\rho_1, \rho_2)_{HS} = \mathscr{K}_{1_G} \quad \text{where } \mathscr{K} \text{ belongs to } \rho_2 \otimes \bar{\rho}_1 .$$

Hence by (1) we obtain

$$(3) \qquad j(\rho_1, \rho_2) = n(\boldsymbol{1}_G, \rho_2 \otimes \bar{\rho}_1) .$$

Our next object is to prove that

$$j(\rho_1, \rho_2) = j({}^0\rho_1, {}^0\rho_2) \tag{4}$$

where ${}^0\rho_i$ is the finite discrete part of ρ_i $(i=1,2)$. For let T be a HS operator in (ρ_1, ρ_2). Then $\mathcal{M}_1 = (\ker T)^\perp$ and $\mathcal{M}_2 = \overline{T\mathcal{H}_1}$ are invariant subspaces of ρ_1 and ρ_2 respectively. This follows for example from Theorem IV.4.5. If we let $A = T^*T$ then $A: \mathcal{H}_1 \to \mathcal{H}_1$ is self adjoint and it is compact because being a HS operator T is compact and T^* is continuous. Therefore by Theorem II.3.4 each characteristic subspace $\mathcal{M}(\lambda)(\lambda \neq 0)$ of A is finite dimensional. Since T is in (ρ_1, ρ_2) every $\mathcal{M}(\lambda)$ is an invariant subspace of ρ_1: Indeed if $\alpha \in \mathcal{M}(\lambda)$ and $x \in G$ then

$$A\rho_1(x)\alpha = T^*T\rho_1(x)\alpha = T^*\rho_2(x)T\alpha = \rho_1(x)T^*T\alpha = \rho_1(x)A\alpha = \lambda\rho_1(x)\alpha$$

and so $\rho_1(x)\alpha \in \mathcal{M}(\lambda)$. Since $\ker T = \ker A$, by Theorem II.3.4 this implies that $\mathcal{M}_1 \subseteq \mathcal{H}_{1f}$ where for the sake of simplicity $\mathcal{H}_{1f} = (\mathcal{H}_1)_{fdI}$. By interchanging the roles of $\mathcal{H}_1$ and $\mathcal{H}_2$ and applying Proposition IV.4.2 we also see that $\mathcal{M}_2 \subseteq \mathcal{H}_{2f} = (\mathcal{H}_2)_{fdI}$. Since $T \in (\rho_1, \rho_2)$ the T-image of each invariant subspace of ρ_1 is an invariant subspace of ρ_2. Therefore $T\mathcal{H}_{1f} \subseteq \mathcal{H}_{2f}$. We have $\mathcal{M}_1 = (\ker T)^\perp \subseteq \mathcal{H}_{1f}$ and so we see that T vanishes on $(\mathcal{H}_{1f})^\perp$. Now we associate with each T in (ρ_1, ρ_2) the operator $\tilde{T}: \mathcal{H}_{1f} \to \mathcal{H}_{2f}$ obtained by restricting T to $\mathcal{H}_{1f}$. Then $T \to \tilde{T}$ is an isomorphism of $(\rho_1, \rho_2)_{HS}$ into $({}^0\rho_1, {}^0\rho_2)_{HS}$. If $\tilde{T}$ is an arbitrary element of $({}^0\rho_1, {}^0\rho_2)_{HS}$ then we define $T: \mathcal{H}_1 \to \mathcal{H}_2$ by letting $T = \tilde{T}$ on $\mathcal{H}_{1f}$ and $T = 0$ on $(\mathcal{H}_{1f})^\perp$. If $\alpha \in \mathcal{H}_1$ decomposes into $\beta + \gamma$ with $\beta \in \mathcal{H}_{1f}$ and $\gamma \in (\mathcal{H}_{1f})^\perp$ then

$$T\rho_1(x)\alpha = T\rho_1(x)\beta = \tilde{T}\rho_1(x)\beta = \rho_2(x)\tilde{T}\beta = \rho_2(x)T(\beta+\gamma) = \rho_2(x)T\alpha$$

and so $T \in (\rho_1, \rho_2)_{HS}$. Hence $T \mapsto \tilde{T}$ is an isomorphism of $(\rho_1, \rho_2)_{HS}$ onto $({}^0\rho_1, {}^0\rho_2)_{HS}$. Therefore we proved (4).

Now we prove that

$$i({}^0\rho_1, {}^0\rho_2) = j({}^0\rho_1, {}^0\rho_2) = \sum_{\omega \in \hat{G}} n(\omega, {}^0\rho_1)\, n(\omega, {}^0\rho_2). \tag{5}$$

For let $\mathcal{H}_{1f} = \sum \mathcal{P}_i^1$ and $\mathcal{H}_{2f} = \sum \mathcal{P}_j^2$ be decompositions of $\mathcal{H}_{1f}$ and $\mathcal{H}_{2f}$ into irreducible components $\mathcal{P}_i^1 (i \in \mathcal{I})$ and $\mathcal{P}_j^2 (j \in \mathcal{J})$ of ${}^0\rho_1$ and ${}^0\rho_2$, respectively. We let E_i^1 denote the projection of $\mathcal{H}_{1f}$ onto $\mathcal{P}_i^1$ and similarly E_j^2 stands for the projection $\mathcal{H}_{2f} \to \mathcal{P}_j^2$. If T is in $\mathcal{L}(\mathcal{H}_{1f}, \mathcal{H}_{2f})$ then we let $T_{ij} = E_j T E_i$ for all $i \in \mathcal{I}$ and $j \in \mathcal{J}$. By the same reasoning as that used to prove Proposition III.6.6 we see that the continuous operator T belongs to $({}^0\rho_1, {}^0\rho_2)$ if and only if $T_{ij} \in ({}^0\rho_1|\mathcal{P}_i^1, {}^0\rho_2|\mathcal{P}_j^2)$ for every $i \in \mathcal{I}$ and $j \in \mathcal{J}$. Hence by Propositions IV.2.4, IV.3.4 and IV.4.7

$({}^0\rho_1|\mathscr{P}_i^1, {}^0\rho_2|\mathscr{P}_j^2)\neq 0$ if and only if ${}^0\rho_1|\mathscr{P}_i^1\approx{}^0\rho_2|\mathscr{P}_j^2$ in which case this vector space is of dimension one. We let B_{ij} denote a basis element if the vector space of intertwining operators is one dimensional and otherwise we let B_{ij} be the zero operator in $\mathscr{L}(\mathscr{P}_i^1, \mathscr{P}_j^2)$. If there are infinitely many pairs $(i,j)\in\mathscr{I}\times\mathscr{J}$ such that $B_{ij}\neq 0$ then $i({}^0\rho_1, {}^0\rho_2)$ and $j({}^0\rho_1, {}^0\rho_2)$ are both ∞. If this is not the case then any operator of the form

$$T=\sum\lambda_{ij}B_{ij}\qquad(\lambda_{ij}\in\mathbb{C}\text{ for all } i\in\mathscr{I}\text{ and } j\in\mathscr{J})$$

is continuous and is in the HS class. Thus in the second case we have $({}^0\rho_1, {}^0\rho_2)=({}^0\rho_1, {}^0\rho_2)_{HS}$ and so $i({}^0\rho_1, {}^0\rho_2)=j({}^0\rho_1, {}^0\rho_2)$. If $B_{ij}\neq 0$ then λ_{ij} varies freely over $\mathbb{C}$ and if $B_{ij}=0$ then we can choose $\lambda_{ij}=0$. Hence $i({}^0\rho_1, {}^0\rho_2)$ is the number of ordered pairs (i,j) with $B_{ij}\neq 0$. Since $\mathscr{P}_i^1$ and $\mathscr{P}_j^2$ are irreducible components, $B_{ij}\neq 0$ if and only if ${}^0\rho_1|\mathscr{P}_i^1$ and ${}^0\rho_2|\mathscr{P}_j^2$ intertwine. Thus we see that (5) holds. Theorem 1 follows at once from (3), (4) and (5).

Corollary. *If π is an irreducible unitary representation of the abstract group G and ρ is a unitary representation of G then $j(\pi,\rho)=n(\pi, {}^0\rho)$.*

In fact if we let $\rho_1=\pi$ and $\rho_2=\rho$ then the sum occuring in Theorem 1 reduces to the single term $n(\omega_\pi, {}^0\pi)n(\omega_\pi, {}^0\rho)$. If d_π is finite then $n(\omega_\pi, {}^0\pi)=n(\omega_\pi,\pi)=1$ and we obtain $j(\pi,\rho)=n(\omega_\pi, {}^0\rho)$. If π is infinite dimensional then $n(\omega_\pi, {}^0\pi)=0$ and so $j(\pi,\rho)=0$. Since ${}^0\rho$ is a direct sum of finite dimensional irreducible components we also have $n(\omega_\pi, {}^0\rho)=0$. Hence again $j(\pi,\rho)=n(\omega_\pi, {}^0\rho)$.

Let us consider from this point on a locally compact group G, a closed subgroup H and a weakly measurable unitary representation $\chi: H\to\mathscr{U}(\mathscr{H})$. We construct the induced representation $\rho^\chi: G\to\mathscr{U}(\mathscr{K})$ by choosing some quasi-invariant measure μ which is derived from a positive ρ-function $\rho: G\to(0,\infty)$. The Hilbert space $\mathscr{K}$ depends on the choice of μ but we know that ρ^χ is uniquely determined within unitary equivalence.

Proposition 2. *The elements of $\mathscr{K}_{1_G}$ are continuous functions $\theta: G\to\mathscr{H}$ and for every x in G we have $\theta(x)=\theta(e)\lambda(x,e)^{-\frac{1}{2}}$ where $\theta(e)\in\mathscr{H}_{1_H}$.*

Proof. We consider the realization of $\mathscr{K}$ as a Hilbert space of equivalence classes of functions $\theta: G\to\mathscr{H}$ as described in Theorem 4.15. If θ is in $\mathscr{K}_{1_G}$ then for almost all $\xi,\eta\in X$ and for all $x\in\xi$, $y\in\eta$ we have

$$\theta(y)=\rho^\chi(x^{-1})\theta(y)=\theta(xy)\lambda(x,y)^{\frac{1}{2}}.$$

Thus if y is not in the exceptional set then

$$\theta(x)=\theta(y)\lambda(xy^{-1},y)^{-\frac{1}{2}}\tag{6}$$

for all $x \in \xi$ and almost all $\xi \in X$. Since λ is continuous we see that θ can be replaced by an equivalent continuous function which we also denote by θ. Then (6) will hold for every x and y in G. Choosing $y=e$ we obtain the expression given in the proposition.

Now for every h in H we have

$$\chi(h)^* \theta(e) = \theta(h) = \theta(e)\lambda(h, e)^{-\frac{1}{2}}.$$

Hence $\chi(h)^*$ being unitary we obtain $\lambda(h, e)=1$ and so $\chi(h)^* \theta(e) = \theta(e)$ for every $h \in H$. Therefore $\theta(e)$ is an invariant vector of the representation χ and so $\theta(e) \in \mathscr{H}_{1_H}$. If $\theta(e)=0$ then we can not conclude that $\lambda(h, e)=1$ but $\theta(e)$ is in $\mathscr{H}_{1_H}$ also in this case.

Proposition 3. *If the identity class 1_G occurs in ρ^χ with positive multiplicity then G/H admits a finite invariant measure. Conversely if G/H has a finite invariant measure then the representation induced on G by the one dimensional identity representation of H contains 1_G.*

Proof. Near the end of the last proof we obtained the following: If there is a $\theta \neq 0$ in $\mathscr{K}_{1_G}$ then the quasi-invariant measure used in the definition of $\mathscr{K}$ is such that $\lambda(h, e)=1$ for every $h \in H$. Hence by Lemma V.3.10 we have $\lambda(h, e) = \rho(he)/\rho(e) = (\delta/\Delta)(h) = 1$ for all $h \in H$. Therefore by Theorem V.3.15 X admits a non-trivial invariant measure $d\xi$. We consider the representation $\tilde{\rho}^\chi$ whose Hilbert space is constructed by using $d\xi$. By Proposition 4.2 it is equivalent to the original ρ^χ and so its Hilbert space $\tilde{\mathscr{K}}$ contains a non-zero vector $\tilde{\theta} \in \tilde{\mathscr{K}}_{1_G}$. Using Proposition 2 we obtain

$$0 < \|\tilde{\theta}(e)\|^2 \mu(X) = \int_X \|\tilde{\theta}(x)\|^2 \, d\xi = \|\tilde{\theta}\|^2 < \infty$$

which shows that $\mu(X)$ is finite. Conversely suppose that X admits a finite invariant measure. Let $\chi: H \to \mathscr{U}(\mathbb{C})$ be the one dimensional identity representation of H and construct the Hilbert space of ρ^χ by using the finite invariant measure. Then those functions $\theta \in \mathscr{K}$ which are constant on G all belong to $\mathscr{K}_{1_G}$.

Lemma 4. *Suppose that the identity class 1_G occurs with positive multiplicity in an induced representation $\rho^\chi: G \to \mathscr{U}(\mathscr{K})$ where $\chi: H \to \mathscr{U}(\mathscr{H})$. Then for every positive, continuous ρ-function $\rho: G \to (0, \infty)$ we have $\lambda(\cdot, e)^{-1} \in L^1(X, \mu)$ where λ is the λ-function and μ is the quasi-invariant measure associated with ρ.*

Proof. Let θ be a non-zero element of $\mathscr{K}_{1_G}$. According to Proposition 2 we have

$$\|\theta\|^2 = \int_X \|\theta(x)\|^2 \, d\mu(\xi) = \|\theta(e)\|^2 \int_X \lambda(\xi, e)^{-1} \, d\mu(\xi).$$

Since $\|\theta(e)\| \neq 0$ we see that $\lambda^{-1}(\cdot, e)^{-1}$ is integrable.

Theorem 5. *Let $\rho^\chi: G \to \mathscr{U}(\mathscr{K})$ be the representation induced on G by the weakly measurable unitary representation $\chi: H \to \mathscr{U}(\mathscr{H})$. Then the elements of $\mathscr{K}_{I_G}$ are continuous functions $\theta: G \to \mathscr{H}$ and $\theta \mapsto \theta(e)$ is an isomorphism of $\mathscr{K}_{I_G}$ into $\mathscr{H}_{I_H}$. If G/H has a finite invariant measure μ then $\theta \mapsto \theta(e)$ is surjective.*

Proof. By Proposition 2 the elements of $\mathscr{K}_{I_G}$ are all continuous functions and $\theta \mapsto \theta(e)$ is an isomorphism of $\mathscr{K}_{I_G}$ into $\mathscr{H}_{I_H}$. If $\mathscr{H}_{I_H}$ is 0 then $\theta \mapsto \theta(e)$ is trivially surjective. If there is a non-zero vector α in $\mathscr{H}_{I_H}$ then we consider the induced representation $\tilde{\rho}^\chi: G \to \mathscr{U}(\tilde{\mathscr{K}})$ where $\tilde{\mathscr{K}}$ is constructed by using the measure μ. If we let $\theta(x) = \alpha$ for every $x \in G$ then $\theta \neq 0$ and $\theta \in \tilde{\mathscr{K}}_{I_G}$. Since $\tilde{\rho}^\chi$ and ρ^χ are equivalent it follows that $\mathscr{K}_{I_G} \neq 0$. Thus by Lemma 4 $\lambda(\cdot, e)^{-1}$ is integrable on X. Thus given any $\beta \in \mathscr{H}_{I_H}$ we can let $\theta(x) = \beta\lambda(x, e)^{-\frac{1}{2}}$ and obtain a continuous function $\theta: G \to \mathscr{H}$ with finite $\|\theta\|$. By the existence of μ and Theorem V.3.15 we have $(\delta/\Delta)(h) = 1$ for every $h \in H$ and so $\rho(xh) = \rho(x)$ for all $x \in G$ and $h \in H$. Hence by $\beta \in \mathscr{H}_{I_H}$ we have

$$\theta(xh) = \beta\lambda(xh, e)^{-\frac{1}{2}} = \chi(h)^* \beta\lambda(x, e)^{-\frac{1}{2}} = \chi(h)^* \theta(x)$$

for all $x \in G$, $h \in H$. Therefore $\theta \in \mathscr{K}_{I_G}$ by Theorem 4.15 and clearly $\theta \mapsto \theta(e) = \beta$. We proved that if G/H has a finite invariant measure then the isomorphism $\theta \mapsto \theta(e)$ is surjective.

Corollary. *If G/H admits a finite invariant measure then* $\dim \mathscr{K}_{I_G} = \dim \mathscr{H}_{I_H}$ *i.e.* $n(\boldsymbol{1}_G, \rho^\chi) = n(\boldsymbol{1}_H, \chi)$.

If $\mathscr{K}_{I_G} \neq 0$ then in view of Proposition 3 the corollary can be applied and so $\dim \mathscr{K}_{I_G} = \dim \mathscr{H}_{I_H}$ holds. We note that if G/H has no finite invariant measure then by Proposition 3 we have $\mathscr{K}_{I_G} = 0$.

Theorem 6. *Let $\rho: G \to \mathscr{U}(\mathscr{L})$ and $\chi: H \to \mathscr{U}(\mathscr{H})$ be continuous representations of the locally compact group G and its closed subgroup H in the separable Hilbert spaces $\mathscr{L}$ and $\mathscr{H}$, respectively. Then we have $\rho \otimes \rho^\chi \approx \rho^{(\rho|H) \otimes \chi}$ where $\otimes$ denotes inner tensor product.*

Proof. Let λ_i $(i \in \mathscr{I})$ be a MONS in $\mathscr{L}$ and let $\mathscr{M}$ denote the linear manifold defining the Hilbert space of $\rho^\chi: G \to \mathscr{U}(\mathscr{K})$. Let $\tilde{\mathscr{K}}$ be the Hilbert space of $\rho^{(\rho|H) \otimes \chi}$ and $\tilde{\mathscr{M}}$ the linear manifold used to define $\tilde{\mathscr{K}}$. We let $\mathscr{A}$ be the linear manifold formed by those elements of $\mathscr{L} \otimes \mathscr{K}$ which are finite sums of the form $\lambda_{i_1} \otimes \theta_1 + \cdots + \lambda_{i_n} \otimes \theta_n$ with distinct $i_1, \ldots, i_n \in \mathscr{I}$ and arbitrary $\theta_1, \ldots, \theta_n \in \mathscr{M}$. It is clear that every element of $\mathscr{A}$ has only one representation of this type.

For every $\lambda \in \mathscr{L}$ and $\theta \in \mathscr{M}$ we define $U(\lambda \otimes \theta)$ to be a function on G with values in $\mathscr{L} \otimes \mathscr{H}$. Namely we let

$$U(\lambda \otimes \theta)(x) = \rho(x)^* \lambda \otimes \theta(x)$$

for every x in G. Since ρ and θ are continuous we see that $U(\lambda \otimes \theta)$ is a continuous function with the same support as θ. If $x \in G$ and $h \in H$ then

$$\begin{aligned} U(\lambda \otimes \theta)(xh) &= \rho(xh)^* \lambda \otimes \theta(xh) = \rho(h)^* \rho(x)^* \lambda \otimes \chi(h)^* \theta(x) \\ &= (\rho(h)^* \otimes \chi(h)^*)(\rho(x)^* \lambda \otimes \theta(x)) \\ &= (\rho(h) \otimes \chi(h))^* U(\lambda \otimes \theta)(x). \end{aligned}$$

Hence we have

$$U(\lambda \otimes \theta)(xh) = (\rho \otimes \chi(h))^* U(\lambda \otimes \theta)(x)$$

for all $x \in G$ and $h \in H$. Since $U(\lambda \otimes \theta)$ is continuous and $\|U(\lambda \otimes \theta)\|$ has compact support in $X = G/H$, by Theorem 4.15 we see that $U(\lambda \otimes \theta)$ belongs to $\tilde{\mathscr{K}}$. Therefore

$$U(\sum \lambda_{i_k} \otimes \theta_k) = \sum U(\lambda_{i_k} \otimes \theta_k)$$

defines a linear map $U: \mathscr{A} \to \tilde{\mathscr{K}}$.

We can easily prove that U is an isometry. We have

$$\begin{aligned} \Big\| \sum_k \rho(x)^* \lambda_{i_k} \otimes \theta_k(x) \Big\|^2 &= \sum_{k,l} (\rho(x)^* \lambda_{i_k} \otimes \theta_k(x), \rho(x)^* \lambda_{i_l} \otimes \theta_l(x)) \\ &= \sum_{k,l} (\lambda_{i_k}, \lambda_{i_l})(\theta_k(x), \theta_l(x)) = \sum_k \|\theta_k(x)\|^2 \end{aligned}$$

for all $x \in G$. Similarly we have

$$\Big\| \sum_k \lambda_{i_k} \otimes \theta_k(x) \Big\|^2 = \sum_k \|\theta_k(x)\|^2$$

for every x in G. Hence

$$\begin{aligned} \Big\| U\Big(\sum_k \lambda_{i_k} \otimes \theta_k\Big) \Big\|^2 &= \int_X \Big\| \sum_k \rho(x)^* \lambda_{i_k} \otimes \theta_k(x) \Big\|^2 d\mu(\xi) = \sum_k \int_X \|\theta_k(x)\|^2 d\mu(\xi) \\ &= \int_X \Big\| \sum_k \lambda_{i_k} \otimes \theta_k(x) \Big\|^2 d\mu(\xi) = \Big\| \sum_k \lambda_{i_k} \otimes \theta_k \Big\|^2. \end{aligned}$$

Next we prove that $U\mathscr{A}$ is dense in $\tilde{\mathscr{K}}$. For this purpose we can use Proposition 4.10. First we note that U being unitary we can extend it in a unique manner to a unitary map U of $\mathscr{K}$ into $\tilde{\mathscr{K}}$. It is obvious that if $\lambda \in \mathscr{L}$ and $\theta \in \mathscr{M}$ then the extended U maps $\lambda \otimes \theta$ into the function $x \mapsto \rho(x)^* \lambda \otimes \theta(x)$. We let $\mathscr{C}$ be the class of all these functions $U(\lambda \otimes \theta)$

where $\lambda \in \mathscr{L}$ and $\theta \in \mathscr{M}$. We recall that by Proposition 4.3 every θ is of the form $\theta = \theta_f$ for some f in $C_0(G, \mathscr{L} \otimes \mathscr{H})$. By Lemma 4.6 we have

$$\rho(ax)^* \lambda \otimes \theta_f(ax) = \rho(x)^* (\rho(a^{-1})\lambda) \otimes \theta_{(af)}(x)$$

and so we see that condition 1) of Proposition 4.10 is satisfied by the class $\mathscr{C}$. Similarly by Lemma 4.7 we have

$$\varphi(x)(\rho(x)^* \lambda \otimes \theta_f(x)) = \rho(x)^* \lambda \otimes \theta_{\varphi f}(x)$$

for every φ in $C_0(X)$ and so condition 2) also holds.

Since $\rho(x)$ is a unitary map, for every x in G the set of vectors $\rho(x)\lambda \otimes \theta(x)$ ($\lambda \in \mathscr{L}$ and $\theta \in \mathscr{M}$) is the same as the set of vectors $\lambda \otimes \theta(x)$ ($\lambda \in \mathscr{L}$ and $\theta \in \mathscr{M}$). Hence by Lemma 4.8 the class $\mathscr{C}$ also satisfies condition 3). By Proposition 4.10 it follows that the extended U is an isometric mapping of $\mathscr{K}$ onto $\tilde{\mathscr{K}}$.

We complete the proof by showing that U intertwines the representations $\rho^{(\rho|H)\otimes\chi}$ and $\rho \otimes \rho^\chi$. Of course it is sufficient to prove that the intertwining relation holds at every $\lambda \otimes \theta$. But then on the one hand

$$\begin{aligned} U(\rho \otimes \rho^\chi)(a)(\lambda \otimes \theta)(x) &= U(\rho(a)\lambda \otimes \rho^\chi(a)\theta)(x) \\ &= \rho(x)^* \rho(a)\lambda \otimes \theta(a^{-1}x)\lambda(a^{-1}, x)^{\frac{1}{2}} \end{aligned}$$

and on the other hand

$$\begin{aligned} \rho^{(\rho|H)\otimes\chi}(a)\, U(\lambda \otimes \theta)(x) &= \rho^{(\rho|H)\otimes\chi}(a)(\rho(\cdot)^* \lambda \otimes \theta)(x) \\ &= \rho(a^{-1}x)^* \lambda \otimes \theta(a^{-1}x)\lambda(a^{-1}, x)^{\frac{1}{2}}. \end{aligned}$$

Hence U is in $(\rho \otimes \rho^\chi, \rho^{(\rho|H)\otimes\chi})$ and the theorem is proved.

We can now combine the foregoing results to derive the following extension of the *Frobenius reciprocity theorem* which generalizes a somewhat more restricted version stated by Mackey:

Theorem 7. *Let G be a locally compact group and let H be a closed subgroup such that G/H has a finite invariant measure. Let $\pi: H \to \mathscr{U}(\mathscr{H})$ and $\rho: G \to \mathscr{U}(\mathscr{L})$ be continuous irreducible representations in the separable Hilbert spaces $\mathscr{H}$ and $\mathscr{L}$ respectively. Then $n(\rho, {}^0\rho^\pi) = n(\pi, {}^0(\rho|H))$.*

Proof. By the corollary of Theorem 1 we have

(7) $$n(\rho, {}^0\rho^\pi) = j(\rho, \rho^\pi) \quad \text{and} \quad n(\pi, {}^0(\rho|H)) = j(\pi, \rho|H).$$

By Theorem 1 and Proposition 4.13 we also have

(8) $$j(\rho, \rho^\pi) = n(\mathbf{1}_G, \rho^{\bar{\pi}} \otimes \rho) \quad \text{and} \quad j(\rho|H, \pi) = n(\mathbf{1}_H, \bar{\pi} \otimes (\rho|H)).$$

Next by the corollary of Theorem 5 we can write

(9) $$n(\mathit{1}_G, \rho^{(\rho|H)\otimes\bar{\pi}}) = n(\mathit{1}_H, (\rho|H)\otimes\bar{\pi}).$$

Finally by Theorem 6 we have

(10) $$\rho\otimes\rho^{\bar{\pi}} \approx \rho^{(\rho|H)\otimes\bar{\pi}}.$$

Now we apply (7), (8), (10), (9), (8) and (7) in this order to obtain

$$\begin{aligned} n(\rho, {}^0\rho^\pi) &= j(\rho, \rho^\pi) = n(1_G, \rho^{\bar{\pi}}\otimes\rho) = n(1_G, \rho\otimes\rho^{\bar{\pi}}) = n(\mathit{1}_G, \rho^{(\rho|H)\otimes\bar{\pi}}) \\ &= n(\mathit{1}_H, (\rho|H)\otimes\bar{\pi}) = n(\mathit{1}_H, \bar{\pi}\otimes(\rho|H)) = j(\rho|H, \pi) \\ &= j(\pi, \rho|H) = n(\pi, {}^0(\rho|H)). \end{aligned}$$

On two occasions we used the fact that within equivalence the inner tensor product of two representations is commutative.

Corollary. *If ρ is finite dimensional then we have* $n(\rho, \rho^\pi) = n(\pi, \rho|H)$.

This follows from the theorem by observing that if ρ is finite dimensional then $n(\pi, {}^0\rho^\pi) = n(\pi, \rho^\pi)$ and similarly $n(\pi, {}^0(\rho|H)) = n(\pi, \rho|H)$. Mackey's extension of the Frobenius reciprocity theorem is the above corollary in the special case when G satisfies the second axiom of countability. The special case of arbitrary compact groups G is due to Weil.

The condition that G/H admits a finite invariant measure is to a certain extent essential for the validity of Theorem 7. For if $d_\rho = \infty$ then $n(\rho, {}^0\rho^\pi) = 0$ for any π and so if the Frobenius reciprocity relation holds then $n(\pi, {}^0(\rho|H)) = 0$ follows. Thus if $d_\rho = \infty$ then for example we can not choose $H = \{e\}$.

We shall now prove a theorem which is a simple consequence of Theorems 1, 5, 6 and is due to Fakler. In order to put his result in true perspective we note that Theorem 6 was first explicitely formulated and proved also by Fakler. It is very likely that Mackey knew about Theorem 6 in the case of separable groups but as far as I know he never explicitely mentioned this fact.

Theorem 8. *Let G be a locally compact group, let H_1, H_2 be closed subgroups of G and let $\chi_i\colon H_i \to \mathscr{U}(\mathscr{H}_i)$ $(i=1,2)$ be continuous unitary representations in separable Hilbert spaces. If G/H_2 has a finite invariant measure then*

$$j(\rho^{\chi_1}|H_2, \chi_2) = j(\rho^{\chi_1}, \rho^{\chi_2}).$$

Proof. By Theorem 1 we have

(11) $$j(\rho^{\chi_1}|H_2, \chi_2) = n(\mathit{1}_{H_2}, \chi_2\otimes\rho^{\bar{\chi}_1}|H_2).$$

Next by the corollary of Theorem 5

(12) $$n(1_{H_2}, (\rho^{\bar{\chi}_1}|H_2)\otimes\chi_2)=n(1_G, \rho^{(\rho^{\bar{\chi}_1}|H_2)\otimes\chi_2}).$$

Then by Theorem 6 we obtain

(13) $$n(1_G, \rho^{(\rho^{\bar{\chi}_1}|H_2)\otimes\chi_2})=n(1_G, \rho^{\bar{\chi}_1}\otimes\rho^{\chi_2}).$$

We apply Theorem 1 once more to get

(14) $$j(\rho^{\chi_1}, \rho^{\chi_2})=n(1_G, \rho^{\chi_2}\otimes\rho^{\bar{\chi}_1}).$$

Now the conclusion follows from (11), (12), (13), (14) and the commutativity of the inner tensor product.

Corollary. *If both G/H_1 and G/H_2 have finite invariant measures then*

$$j(\rho^{\chi_1}|H_2, \chi_2)=j(\rho^{\chi_2}|H_1, \chi_1).$$

12. Isomorphism Theorems Implying the Frobenius Reciprocity Relation

The classical Frobenius reciprocity theorem states that if ρ is a finite dimensional irreducible representation of the finite group G and π is an irreducible representation of the subgroup H then $n(\pi, \rho|H)=n(\rho, \rho^\pi)$. Here we shall consider locally compact groups G and closed subgroups H and prove several theorems on isomorphisms of the vector space of intertwining operators $(\pi, \rho|H)$. Two of these will imply that the Frobenius reciprocity relation $n(\pi, \rho|H)=n(\rho, \rho^\pi)$ remains valid under the hypotheses of these theorems.

Theorem 1. *Let G be a locally compact group and let H be a closed subgroup such that G/H admits a finite invariant measure. Let $\pi: H\to\mathscr{U}(\mathscr{H})$ be a unitary representation and let $\rho: G\to\mathscr{U}(\mathscr{L})$ be a continuous unitary representation. Given an operator A in $(\pi, \rho|H)$ for every λ in $\mathscr{L}$ and x in G we define*

$$(T_A\lambda)(x)=A^*\rho(x)^*\lambda$$

so that $T_A\lambda: G\to\mathscr{H}$. Then $A\mapsto T_A$ is an isomorphism of $(\pi, \rho|H)$ into (ρ, ρ^π) where the Hilbert space $\mathscr{K}$ of ρ^π is constructed by using a finite invariant measure on G/H. If $\mathscr{L}$ is finite dimensional and ρ is irreducible then the map $A\mapsto T_A$ is surjective.

Note. If the finite invariant measure μ is so normalized that $\mu(X)=1$ then the isomorphism $A\mapsto T_A$ is norm decreasing.

Proof. Since ρ is continuous $T_A\lambda$ is a continuous function on G with values in $\mathscr{H}$. If $x\in G$ and $h\in H$ then

$$(T_A\lambda)(xh)=A^*\rho(xh)^*\lambda=A^*\rho(h)^*\rho(x)^*\lambda=\pi(h)^*A^*\rho(x)^*\lambda$$
$$=\pi(h)^*(T_A\lambda)(x).$$

Hence $\|T_A\lambda(\cdot)\|$ is defined on X and so

$$\|T_A\lambda\|^2=\int_X\|T_A\lambda(x)\|^2\,d\mu(\xi)=\int_X\|A^*\rho(x)^*\lambda\|^2\,d\mu(\xi)\leqslant(|A|\cdot\|\lambda\|)^2\mu(X)$$

because ρ is unitary. Therefore by Theorem 4.15 we see that $T_A\lambda$ is an element of the Hilbert space $\mathscr{K}$ of the induced representation ρ^π and $\|T_A\lambda\|\leqslant|A|\cdot\|\lambda\|\mu(X)^{\frac{1}{2}}$. Thus it is obvious that $T_A\colon\mathscr{L}\to\mathscr{K}$ is a continuous linear map and $|T_A|\leqslant|A|\mu(X)^{\frac{1}{2}}$. We prove that T_A intertwines ρ with ρ^π: In fact if $a,x\in G$ and $\lambda\in\mathscr{L}$ then

$$(\rho^\pi(a)T_A\lambda)(x)=(T_A\lambda)(a^{-1}x)=A^*\rho(a^{-1}x)^*\lambda=A^*\rho(x)^*\rho(a)\lambda$$
$$=(T_A\rho(a)\lambda)(x).$$

Now it is clear that $A\mapsto T_A$ is a linear map of $(\pi,\rho|H)$ into (ρ,ρ^π). Since $T_A\lambda$ is continuous and $T_A\lambda(e)=A^*\lambda$ we see that $A\mapsto T_A$ is an isomorphism.

Now let an operator $T\colon\mathscr{L}\to\mathscr{K}$ be given in (ρ,ρ^π). We are going to prove that under the additional conditions stated in the theorem there is an other operator A in $(\pi,\rho|H)$ such that $T_A=T$. Given $\alpha\in\mathscr{H}$, $\lambda\in\mathscr{L}$ and $x\in G$ we let

$$\varphi_\alpha(\rho(x)^*\lambda)=(T\lambda(x),\alpha). \tag{1}$$

Since ρ is irreducible λ is a cyclic vector and this indicates that (1) might define a linear functional $\varphi_\alpha\colon\mathscr{L}\to\mathbb{C}$. If $\lambda_1,\lambda_2\in\mathscr{L}$ and $x_1,x_2\in G$ are such that $\rho(x_1)^*\lambda_1=\rho(x_2)^*\lambda_2$ then $\lambda_2=\rho(x_1x_2^{-1})^*\lambda_1$ and so $T\lambda_2(x_2)=(T\rho(x_1x_2^{-1})^*\lambda_1)(x_2)=(\rho^\pi(x_1x_2^{-1})^*T\lambda_1)(x_2)=(T\lambda_1)(x_1)$. This shows that the function values defined by (1) are independent of the choice of λ. Let $\mathscr{M}$ denote the inner product space whose completion is $\mathscr{K}$ and let θ be a non-zero element of $\mathscr{M}$. We shall use (1) with $\lambda=T^*\theta$ so that $T\lambda=\theta$ is a continuous function on G with values in $\mathscr{H}$.

We extend the definition of φ_α by linearity: In order to see that in this manner we obtain a well defined function let $a_i\in\mathbb{C}$ and $x_i\in G$ $(i=1,\dots,n)$ be given and define

$$v=a_1\rho(x_1)^*\lambda+\cdots+a_n\rho(x_n)^*\lambda. \tag{2}$$

Since T is an intertwining operator we have

$$T\rho(x_i^{-1})\lambda=\rho^\pi(x_i^{-1})T\lambda$$

and so

$$\theta(y)=T\lambda(y)=\rho^\pi(x_i)T\rho(x_i)^*\lambda(y)$$

for every y in G. Using this relation with $y=x_i$ for every $i=1,\ldots,n$ we obtain

$$\begin{aligned}\sum a_i(T\lambda(x_i),\alpha)&=\sum a_i(\rho^\pi(x_i)T\rho(x_i)^*\lambda(x_i),\alpha)\\&=\sum a_i(T\rho(x_i)^*\lambda(e),\alpha)=(T\sum a_i\rho(x_i)^*\lambda(e),\alpha)=(Tv(e),\alpha)\end{aligned}$$

by (2). Therefore $v=0$ implies that $\sum a_i(T\lambda(x_i),\alpha)=0$ and so (1) defines a linear functional φ_α on the dense linear manifold of $\mathscr{L}$ which is spanned by the vectors $\rho(x)^*\lambda$ $(x\in G)$. Moreover the last identity shows that $\varphi_\alpha(v)=(Tv(e),\alpha)$ for every v in the domain of definition of φ_α. Since $\mathscr{L}$ is finite dimensional φ_α can be extended to a continuous linear functional $\varphi_\alpha\colon\mathscr{L}\to\mathbb{C}$. Let $A\alpha$ be the unique vector in $\mathscr{L}$ such that $\varphi_\alpha(\xi)=(\xi,A\alpha)$ for every ξ in $\mathscr{L}$. Then $\alpha\mapsto A\alpha$ is a continuous linear operator $A\colon\mathscr{H}\to\mathscr{L}$. We prove that A is in $(\pi,\rho|H)$: Given $\alpha\in\mathscr{H}$, $x\in G$ and $h\in H$ we have

$$\begin{aligned}(\rho(x)^*\lambda,A\pi(h)\alpha)&=\varphi_{\pi(h)\alpha}(\rho(x)^*\lambda)=(T\lambda(x),\pi(h)\alpha)\\&=(\pi(h)^*T\lambda(x),\alpha)=(T\lambda(xh),\alpha)=\varphi_\alpha(\rho(xh)^*\lambda)\\&=(\rho(xh)^*\lambda,A\alpha)=(\rho(x)^*\lambda,\rho(h)A\alpha).\end{aligned}$$

Since the vectors $\rho(x)^*\lambda$ $(x\in G)$ span $\mathscr{L}$ we obtain $A\pi(h)\alpha=\rho(h)A\alpha$ and so $A\pi(h)=\rho(h)A$ for every h in H. Therefore we see that A is in $(\pi,\rho|H)$. Finally $T_A=T$ because if $\lambda\in\mathscr{L}$ and $\alpha\in\mathscr{H}$ then

$$(T_A\lambda(x),\alpha)=(A^*\rho(x)^*\lambda,\alpha)=(\rho(x)^*\lambda,A\alpha)=\varphi_\alpha(\rho(x)^*\lambda)=(T\lambda(x),\alpha).$$

This completes the proof.

Corollary 1. *If G/H admits a finite invariant measure, $\pi\in\hat{H}$ and $\rho\in\hat{G}$ then $n(\pi,\rho|H)\leqslant n(\rho,\rho^\pi)$.*

Corollary 2. *If G/H admits a finite invariant measure, $\pi\in\hat{H}$ and ρ is a finite dimensional, continuous, irreducible unitary representation of G then $n(\pi,\rho|H)=n(\rho,\rho^\pi)$.*

In Section VII.1 we shall prove without using induced representations that if G is compact then every continuous irreducible unitary representation of G is finite dimensional. Since G and its closed subgroup H

are compact we also know by Theorem V.3.15 and Proposition VI.6.14 that G/H admits a finite invariant measure. Therefore we shall establish the following:

Proposition 2. *If G is compact, H is a closed subgroup, $\pi\in\hat{H}$ and $\rho\in\hat{G}$ then $n(\pi,\rho|H)=n(\rho,\rho^{\pi})$.*

The following theorem and Theorem 4 below should be considered as a pair:

Theorem 3. *Let G be a locally compact group and H a closed subgroup such that the homogeneous space $X=G/H$ has a finite quasi-invariant measure μ. Let $\pi\colon H\to\mathscr{U}(\mathscr{H})$ be a weakly measurable unitary representation and let $\rho\colon G\to\mathscr{U}(\mathscr{L})$ be a continuous unitary representation. Let the Hilbert space $\mathscr{K}$ of the induced representation ρ^{π} be constructed by using μ. Then for any unit vector λ in $\mathscr{L}$*

$$A\mapsto A^*\rho(\cdot)^*\lambda \tag{3}$$

defines a linear map of $(\pi,\rho|H)$ into $\mathscr{K}$.

If ρ is irreducible, X admits a finite invariant measure μ and $\mathscr{K}$ is constructed by using μ then this map is an isomorphism of $(\pi,\rho|H)$ into $\mathscr{K}_{\rho}$, the maximal primary component of ρ^{π} corresponding to the equivalence class containing ρ.

Note. If the finite quasi-invariant measure μ of X is so normalized that $\mu(X)=1$ then this linear map is norm decreasing.

Proof. It is clear that $x\mapsto A^*\rho(x)^*\lambda$ is a continuous function on G with values in $\mathscr{H}$. For any $x\in G$ and $h\in H$ we have

$$A^*\rho(xh)^*\lambda=A^*\rho(h)^*\rho(x)^*\lambda=\pi(h)^*A^*\rho(x)^*\lambda.$$

Therefore $\|A^*\rho(\cdot)^*\lambda\|$ is defined on X and so

$$\int_X\|A^*\rho(x)^*\lambda\|^2\,d\mu(\xi)\leqslant\int_X|A|^2\,\|\rho(x)^*\lambda\|^2\,d\mu(\xi)=|A|^2\int_X\|\lambda\|^2\,d\mu(\xi).$$

Hence $\|A^*\rho(\cdot)^*\lambda\|\leqslant|A|\sqrt{\mu(X)}<\infty$ by $\|\lambda\|=1$ and $\mu(X)<\infty$. By Theorem 4.15 this shows that $A^*\rho(\cdot)^*\lambda$ is an element of $\mathscr{K}$. It is plain that the map described in (3) is linear in A. If the image of A under (3) is zero then for every $\alpha\in\mathscr{H}$ and $x\in G$ we have $(\rho(x)^*\lambda,A\alpha)=0$. Hence the smallest invariant subspace of ρ containing λ is orthogonal to the range of A. Consequently if ρ is irreducible then (3) defines a vector space isomorphism.

From this point on we suppose that X admits a finite invariant measure μ which is so normalized that $\mu(X)=1$. We also suppose that ρ is irreducible. These hypotheses will enter step by step. Let θ denote the image of a given operator A belonging to $(\pi,\rho|H)$ under the map given in (3). We shall prove that $\theta\in\mathscr{K}_\rho$ provided $\mathscr{K}$ is constructed by using the measure μ.

We let $\mathscr{V}$ denote the linear manifold spanned by the vectors $\rho(x)\lambda$ $(x\in G)$ and let $\mathscr{K}_\theta$ be the subspace of $\mathscr{K}$ generated by the vectors $\rho^\pi(x)\theta$ $(x\in G)$. We define $T(\rho(x)\lambda)=\rho^\pi(x)\theta$ for every x in G. If x_1, x_2 and y are in G then $T\rho(x_i)\lambda(y)=\rho^\pi(x_i)\theta(y)=A^*\rho(x_i^{-1}y)^*\lambda=A^*\rho(y)^*\rho(x_i)\lambda$ for $i=1,2$ and so $\rho(x_1)\lambda=\rho(x_2)\lambda$ implies that $T\rho(x_1)\lambda=T\rho(x_2)\lambda$. Hence T is a well defined function on the set of vectors $\rho(x)\lambda$ $(x\in G)$. We extend the definition of T by linearity to $\mathscr{V}$: This is possible because for every y in $G, a_1,\ldots,a_n\in\mathbb{C}$ and $x_1,\ldots,x_n\in G$ we have

$$\begin{aligned}(\textstyle\sum a_i\rho^\pi(x_i)\theta)(y)&=\textstyle\sum a_i\theta(x_i^{-1}y)=\sum a_i A^*\rho(x_i^{-1}y)^*\lambda\\ &=\textstyle\sum A^*\rho(y)^*a_i\rho(x_i)\lambda\end{aligned}\tag{4}$$

and so $\sum a_i\rho(x_i)\lambda=0$ implies that $\sum a_i\rho^\pi(x_i)\theta=0$. Hence $T:\mathscr{V}\to\mathscr{K}_\theta$ is a well defined linear map. By (4) we have

$$\begin{aligned}\|\textstyle\sum a_i\rho^\pi(x_i)\theta\|^2&=\displaystyle\int_X\|A^*\rho(y)^*(\textstyle\sum a_i\rho(x_i)\lambda)\|^2\,d\mu(\eta)\\ &\leqslant|A|^2\,\|\textstyle\sum a_i\rho(x_i)\lambda\|^2\,.\end{aligned}$$

This shows that T is continuous on $\mathscr{V}$. If ρ is irreducible then $\mathscr{V}$ is dense in $\mathscr{L}$ and so T can be extended in a unique manner to a continuous linear operator $T:\mathscr{L}\to\mathscr{K}_\theta$. Since $\mathscr{K}_\theta$ is generated by the vectors $\rho^\pi(x)\theta=T\rho(x)\lambda$ we see that T maps $\mathscr{L}$ onto $\mathscr{K}_\theta$.

We prove that T intertwines ρ with $\rho^\pi|\mathscr{K}_\theta$. Given an element $a\in G$, which we keep fixed, and $x,y\in G$, which we shall vary, we have

$$\begin{aligned}(T\rho(a)(\rho(x)\lambda))(y)&=T\rho(ax)\lambda(y)=\rho^\pi(ax)\theta(y)=\rho^\pi(a)\rho^\pi(x)\theta(y)\\ &=(\rho^\pi(a)T\rho(x)\lambda)(y)\,.\end{aligned}$$

Hence $T\rho(a)(\rho(x)\lambda)=\rho^\pi(a)T(\rho(x)\lambda)$ for every $x\in G$. Since ρ is irreducible λ is a cyclic vector of ρ and so we obtain $T\rho(a)=\rho^\pi(a)T$ where $a\in G$ is arbitrary. Now let $\mathscr{X}$ be an invariant subspace of $\rho^\pi|\mathscr{K}_\theta$. Then due to the intertwining property of T the subspace $T^{-1}\mathscr{X}$ of $\mathscr{L}$ is invariant under the irreducible representation ρ. Hence $T^{-1}\mathscr{X}=0$ or $T^{-1}\mathscr{X}=\mathscr{L}$ and so $\mathscr{X}=0$ or $\mathscr{X}=\mathscr{K}_\theta$. We proved that $\mathscr{K}_\theta$ is either 0 or an irreducible component of the induced representation ρ^π. We have seen already that $\theta=0$ only if $A=0$. Hence if $A\neq 0$ then $\mathscr{K}_\theta$ is an irreducible component of ρ^π which contains θ, the image of A under the map described in (3).

We apply now Proposition IV.4.6 with T and the following choices of the remaining data: We let $\rho_1=\rho$ and $\rho_2=\rho^\pi|\mathscr{K}_\theta$ so that $\mathscr{M}_1=\overline{T^*\mathscr{K}_\theta}$ and $\mathscr{M}_2=T\mathscr{L}=\mathscr{K}_\theta$. We obtain $\rho|\mathscr{M}_1\approx\rho^\pi|\mathscr{K}_\theta$ and so by the irreducibility of ρ we have $\rho\approx\rho^\pi|\mathscr{K}_\theta$. Therefore by Theorem IV.5.4 we arrive at the inclusion $\mathscr{K}_\theta\subseteq\mathscr{K}_\rho$. We proved that the image of $(\pi,\rho|H)$ under the map given in (3) lies $\mathscr{K}_\rho$, the maximal primary component of ρ^π corresponding to the class of the irreducible representation ρ.

Theorem 4. *Let G be a unimodular group and H a unimodular subgroup so that the homogeneous space $X=G/H$ has a finite invariant measure μ. Let $\rho\colon G\to\mathscr{U}(\mathscr{L})$ be a square integrable, continuous, irreducible representation and let $\chi\colon H\to\mathscr{U}(\mathscr{H})$ be a weakly measurable representation. Suppose that the Hilbert space $\mathscr{K}$ of ρ^π is constructed by using μ.*

Let $\lambda_1,\dots,\lambda_m$ be any ON system of vectors in $\mathscr{L}$. For $A_1,\dots,A_m$ in $(\pi,\rho|H)$ let

$$(A_1,\dots,A_m)\mapsto\sum_1^m A_i^*\,\rho(\cdot)^*\,\lambda_i\,. \tag{5}$$

Then this map is a linear isomorphism of $(\pi,\rho|H)\times\cdots\times(\pi,\rho|H)$ into $\mathscr{K}_\rho$. If G is compact and $m=d_\rho$ then the isomorphism is surjective.

The dimension of $\mathscr{K}_\rho$ is $d_\rho n(\rho,\rho^\pi)$ where d_ρ denotes the degree of ρ which we suppose to be finite. Hence choosing $m=d_\rho$ we obtain the following:

Corollary. *If G/H admits a finite invariant measure, $\rho\colon G\to\mathscr{U}(\mathscr{L})$ is square integrable, continuous, irreducible and of finite degree and $\chi\colon H\to\mathscr{U}(\mathscr{H})$ is weakly measurable then $n(\pi,\rho|H)\leqslant n(\rho,\rho^\pi)$.*

If G is compact then our map being surjective we obtain $n(\pi,\rho|H)=n(\rho,\rho^\pi)$ which is Weil's version of the Frobenius reciprocity theorem. Of course this is also follows from Corollary 2 of Theorem 1 and Theorem VII.1.10.

Proof. By Theorem 3 $A_i\mapsto A_i^*\,\rho(\cdot)^*\,\lambda_i$ is an isomorphism of $(\pi,\rho|H)$ into $\mathscr{K}_\rho$. Hence it is clear that (5) defines a linear map into $\mathscr{K}_\rho$. Suppose that the ordered m-tuple $(A_1,\dots,A_m)$ belongs to the kernel of this map. Then for any α in $\mathscr{H}$ we have

$$\int_G\Big(\sum_i A_i\rho(x)^*\lambda_i,\alpha\Big)\overline{\Big(\sum_j A_j\rho(x)^*\lambda_j,\alpha\Big)}\,dx=0\,.$$

Since ρ is square integrable we can multiply out and obtain

$$\sum_{i,j}\int_G(\rho(x)^*\lambda_i,A_i^*\alpha)\overline{(\rho(x)^*\lambda_j,A_j^*\alpha)}\,dx=0\,.$$

Applying Proposition VII.1.29 we get

$$\sum_{i,j} (\lambda_i, \lambda_j)(A_j^* \alpha, A_i^* \alpha) = 0 .$$

Since $\lambda_1, \ldots, \lambda_m$ is an ON system we see that

$$(A_1^* \alpha, A_1^* \alpha) + \cdots + (A_m^* \alpha, A_m^* \alpha) = 0 .$$

Therefore $(A_i \alpha, A_i \alpha) = 0$ for every $i = 1, \ldots, m$ and so $A_1 = \cdots = A_m = 0$.

Now we suppose that G is compact and the Haar measures of the compact groups G and H are normalized. By Theorem VII.1.10 ρ is of finite degree d_ρ. We let $m = d_\rho$ and choose a vector θ in $\mathscr{K}_\rho$. By Theorem 4.15 θ has a concrete realization as a function $\theta: G \to \mathscr{H}$. In order to determine $(A_1, \ldots, A_{d_\rho})$ such that its image is θ we choose α in $\mathscr{H}$ and μ in $\mathscr{L}$ which we shall vary. Then

$$\varphi_\alpha^i(\mu) = d_\rho \int_G (\theta(x), \alpha)(\rho(x)\mu, \lambda_i)\, dx$$

exists and by the BCS inequality

$$|\varphi_\alpha^i(\mu)| \leqslant d_\rho \|\mu\| \cdot \|\theta\| \cdot \|\alpha\|$$

because by $\|\theta(xh)\| = \|\theta(x)\|$ we have

$$\int_G \|\theta(x)\|^2\, dx = \int_X \int_H \|\theta(xh)\|^2\, dh\, d\mu(\xi) = \int_X \|\theta(x)\|^2\, d\mu(\xi) = \|\theta\|^2 .$$

Therefore $\varphi_\alpha^i: \mathscr{L} \to \mathbb{C}$ is a continuous linear functional and $\|\varphi_\alpha^i\| \leqslant d_\rho \|\theta\| \cdot \|\alpha\|$. Hence there is a unique vector in $\mathscr{L}$, which we call $A_i \alpha$ such that $\varphi_\alpha^i(\mu) = (\mu, A_i \alpha)$ for every $\mu \in \mathscr{L}$. We also know that $\|A_i \alpha\| = \|\varphi_\alpha^i\|$. Since α is arbitrary $\alpha \mapsto A_i \alpha$ defines a linear operator $A_i: \mathscr{H} \to \mathscr{L}$ which is continuous and $|A_i| \leqslant d_\rho \|\theta\|$.

We prove that A_i intertwines π and $\rho|H$. Indeed

$$\begin{aligned}(\mu, \rho(h) A_i \alpha) = \varphi_\alpha^i(\rho(h)^* \mu) &= d_\rho \int_G (\theta(x), \alpha)(\rho(x)\rho(h)^* \mu, \lambda_i)\, dx \\ &= d_\rho \int_G (\theta(xh), \alpha)(\rho(x)\mu, \lambda_i)\, dx\end{aligned}$$

by the substitution $x \to xh$. Therefore by $\theta(xh) = \pi(h)^* \theta(x)$ we obtain

$$(\mu, \rho(h) A_i \alpha) = \varphi_{\pi(h)\alpha}^i(\mu) = (\mu, A_i \pi(h) \alpha) .$$

Since μ is arbitrary this shows that $\rho(h) A_i \alpha = A_i \pi(h) \alpha$ for every $\alpha \in \mathscr{H}$ and $h \in H$.

Let $\psi_i = A_i^* \rho(\cdot)^* \lambda_i$ so that $\psi_i \in \mathcal{K}_\rho$ by Theorem 3. It is our object to show that $\psi_1 + \cdots + \psi_{d_\rho} = \theta$. If $\phi \in \mathcal{K}_\rho$ then

$$(\psi_i, \phi) = \int_X (\rho(x)^* \lambda_i, A_i \phi(x))\, d\mu(\xi) = \int_X \varphi^i_{\phi(x)}(\rho(x)^* \lambda_i)\, d\mu(\xi).$$

Hence by the definition of φ^i_α we have

$$(\psi_i, \phi) = d_\rho \int_X \int_G (\theta(y), \phi(x))(\rho(y)\rho(x)^* \lambda_i, \lambda_i)\, dy\, d\mu(\xi).$$

If we perform the substitution $y \to yx$ and change the order of integration then we obtain

$$(\psi_i, \phi) = d_\rho \int_G (\rho(y)\lambda_i, \lambda_i) \int_X (\theta(x), \rho^\pi(y)\phi(x))\, d\mu(\xi)\, dy.$$

Therefore by introducing the inner product of $\mathcal{K}$ we arrive at

$$(\psi_i, \phi) = d_\rho \int_G (\rho(y)\lambda_i, \lambda_i)\overline{(\rho^\pi(y)\phi, \theta)}\, dy.$$

By Theorem IV.5.4 $\mathcal{K}_\rho$ is a direct sum of irreducible components $\mathcal{P}_j$ $(j \in \mathcal{J})$ such that $\rho^\pi | \mathcal{P}_j \approx \rho$ for every $j \in \mathcal{J}$. Let $U_j \in (\rho^\pi | \mathcal{P}_j, \rho)$ be a unitary operator and let $\phi = \sum \phi_j$ with $\phi_j \in \mathcal{P}_j$ for each $j \in \mathcal{J}$. If we replace in the last equation ϕ by ϕ_j and apply Proposition VII.1.24 then we obtain

$$(\psi_i, \phi_j) = (U^{-1}\lambda_i, \phi_j)(\theta, U^{-1}\lambda_i).$$

Since $U^{-1}\lambda_i$ $(i = 1, \ldots, n)$ is a MONS in $\mathcal{P}_j$ we see that

$$(\psi, \phi_j) = \sum_i (\psi_i, \phi_j) = \sum_i (\theta, U^{-1}\lambda_i)(U^{-1}\lambda_i, \phi_j) = (\theta, \phi_j)$$

where $j \in \mathcal{J}$. Thus $(\psi, \phi) = (\theta, \phi)$ where $\phi \in \mathcal{K}_\rho$ is arbitrary. This proves that $\psi = \theta$ and so the map described in (5) is surjective in the special case when G is compact and $m = d_\rho$.

Theorem 5. *Let G be a compact group, H a closed subgroup and let $\rho: G \to \mathcal{U}(\mathcal{L})$ and $\pi: H \to \mathcal{U}(\mathcal{H})$ be continuous irreducible representations. Suppose that the Hilbert space $\mathcal{K}$ of the induced representation ρ^π is constructed by using an invariant measure on G/H. Let $\sigma: G \to \mathcal{U}(\mathcal{M})$ be another continuous irreducible representation and let $\theta \in \mathcal{K}_\rho$. Then for every $\alpha \in \mathcal{H}$ and $\mu_1, \mu_2 \in \mathcal{M}$ we have: If σ is not equivalent to ρ then*

$$\int_G (\theta(x), \alpha)(\sigma(x)\mu_1, \mu_2)\, dx = 0.$$

If σ is equivalent to ρ then

$$\int_G (\theta(x), \alpha)(\sigma(x)\mu_1, \mu_2)\,dx = \int_G (\theta(y), \alpha)(\rho(y)\mu_1, \mu_2)\,dy\,.$$

Proof. We use the concepts and facts developed in the last proof: Let $\psi_1, \ldots, \psi_{d_\rho}$ be the vectors in $\mathscr{K}_\rho$ such that $\psi = \psi_1 + \cdots + \psi_{d_\rho} = \theta$. Then

$$\begin{aligned}\int_G (\psi_i(x), \alpha)(\sigma(x)\mu_1, \mu_2)\,dx &= \int_G (\rho(x)^* \lambda_i, A_i\alpha)(\sigma(x)\mu_1, \mu_2)\,dx \\ &= \int_G \varphi^i_\alpha(\rho(x)^* \lambda_i)(\sigma(x)\mu_1, \mu_2)\,dx\,.\end{aligned}$$

Thus by the definition of φ^i_α the right hand side is equal to

$$\begin{aligned}&d_\rho \int_G \int_G (\theta(y), \alpha)(\rho(y)\rho(x)^* \lambda_i, \lambda_i)(\sigma(x)\mu_1, \mu_2)\,dy\,dx \\ &\qquad = d_\rho \int_G (\theta(y), \alpha) \int_G (\sigma(x)\mu_1, \mu_2)\overline{(\rho(x)\rho(y)^* \lambda_i, \lambda_i)}\,dx\,dy\,.\end{aligned}$$

If σ is not equivalent to ρ then by applying Proposition VII.1.22 we see that the inner integral is zero. Therefore

$$\int_G (\psi_i(x), \alpha)(\sigma(x)\mu_1, \mu_2)\,dx = 0$$

for every $i = 1, \ldots, n$. By summing on i and using $\psi = \theta$ we obtain the desired relation. Similarly if σ and ρ are equivalent then we can use Proposition VII.1.24 and obtain the second identity because in that case the inner integral is

$$d_\rho^{-1}(U^{-1}\mu_1, \rho(y)^* \lambda_i)(\lambda_i, U^{-1}\mu_2)$$

where $U: \mathscr{L} \to \mathscr{M}$ is a unitary operator intertwining ρ with σ.

Remarks

Induced representations in the case of finite groups G were introduced in 1898 by Frobenius (1, 2). The first of these papers contains also the Frobenius reciprocity relation for finite groups G. Induced representations of finite groups were used by a number of leading algebraist including Artin, Blichfeldt, Richard Brauer, Burnside, Speiser and Shoda. For details we refer the reader to the book by Curtis and Reiner (1). It was pointed out already in the beginning of the present chapter that induced representations of locally compact groups were introduced by Mackey. He restricted himself to groups G satisfying the second axiom of countability. The Hilbert spaces $\mathscr{H}$ and $\mathscr{K}$ entering the definition are both separable. He announced his basic definition and results in a note which dates from 1949 (5) and a detailed exposition appeared in 1952—53 (4, 6). Some of his results can be found also in an earlier paper (7) which primarily deals with induced representations of finite groups. Before Mackey's papers appeared induced representations of the inhomogeneous Lorentz group were used by Wigner (1) and the Frobenius reciprocity relation was formulated and proved by Weil (1) for the case of compact groups. The regular representations of locally compact groups G in $L^2(G)$ were also considered earlier. The first uses of induced representations were made by the leading researchers in the field of representation theory of locally compact groups: Gelfand and Naumark, Godement, Harish-Chandra and Mautner.

If the existence of a positive measurable ρ-function is known and $\mathscr{H}$ is separable then $\mathscr{K}$ and ρ^χ can be defined in the same way as was done by Mackey in the case of separable groups. Some work in this direction was published by Greenleaf (1). For arbitrary G and $\mathscr{H}$ induced representations were first defined by Blattner (1) who also proved the theorem on induction in stages in full generality. An improved version of Blattner's definition is due to Argabright: Given the unitary representation $\chi: H \to \mathscr{U}(\mathscr{H})$ let $\mathscr{H}_0(\rho^\chi)$ be the vector space of those continuous functions $f: G \to \mathscr{H}$ which satisfy $f(hx) = (\delta/\Delta)(h)^{\frac{1}{2}} \chi(h) f(x)$ for all $x \in G$, $h \in H$ and have compact support modulo H: This means the existence of a compact set C in G such that HC is a carrier of f. If $f, g \in \mathscr{H}_0(\rho^\chi)$ and K is a compact set in G such that HK is a carrier of both f and g then we can choose a non-negative w in $C_0(G)$ such that $\int_H w(hx)\,dh = 1$ for every x in K. One proves that

$$(f, g) = \int_G (f(x), g(x))\, w(x)\, dx$$

is independent of the choice of w and defines an inner product on $\mathscr{H}_0(\rho^\chi)$. The completion of $\mathscr{H}_0(\rho^\chi)$ is the Hilbert space $\mathscr{H}(\rho^\chi)$ of the induced representation ρ^χ. The operators $\rho^\chi(a)$ $(a \in G)$ are first defined on $\mathscr{H}_0(\rho^\chi)$ by $\rho^\chi(a) f(x) = f(xa)$ $(x \in G)$ and then extended to $\mathscr{H}(\rho^\chi)$ by using their uniform continuity.

A number of basic results such as the theorem on the tensor product of induced representations were extended to arbitrary G and $\mathscr{H}$ by Fakler (1) in his thesis. He uses Argabright's definition of an induced representation. One of the principal reasons for choosing the definition given in Section 4 is that it is well adapted to the material discussed in Sections 6 and 7. It is also closer to Mackey's original definition of $\mathscr{K}$ and ρ^χ.

Proposition 2.3 was published in 1969 by Kraljevic (1). We note concerning Proposition 2.7 and its generalization to arbitrary G and H that $\deg \rho^{\chi} = [G: H] \deg \chi$ always holds i.e. ρ^{χ} is finite dimensional if and only if d_{χ} and $[G: H]$ are both finite. This result was first formulated in the case of separable groups by Schochetman (1). It has some very useful corollaries:

If G is a locally compact group such that every proper open subgroup is of finite index then no finite dimensional unitary representation of G is induced.

From this corollary one obtains a result due to Langworthy (1):

If G is a connected compact group then no irreducible unitary representation of G is induced.

The functions θ_f defined in Lemma 2.2 are all of the special form given in Proposition 4.3. It was pointed out to me that this fact was independently and earlier proved by Kraljevic (1). Lemmas 4.6, 4.7, 4.8, Theorem 4.9 and Proposition 10 were proved by Mackey in (5). I have not seen Theorem 4.11 in print but the result is well known. The proof given in the text is my own. Theorem 4.16 on the kernel of an induced representation is due to Schochetman (1) who proved it in the separable case.

The results stated in Section 5 were believed to be true earlier but actual proofs are given here for the first time. Most of the material presented in Sections 6 and 7 is new. One exception is the overlap with Mackey's treatment of the theorem on the tensor product of induced representations. Theorem 6.16 in the special case when $\mathscr{H} = \mathbb{C}$ and χ is the one dimensional identity representation was stated without proof in 1962 by Godement (2). There is no doubt that Godement knew how to prove his result. The case when H is a discrete subgroup and $\mathscr{H}$ is an arbitrary Hilbert space was stated in the book by Gelfand, Graev and Pyatetskii–Shapiro (1). The proof outlined there is correct only when $\mathscr{H} = \mathbb{C}$ and χ is the one dimensional identity.

Theorem 8.1 in the separable case was proved in (5) by Mackey and it was first extended to the general case by Fakler in (1). The proof given in the text is similar to Fakler's proof but it fits our definition of an induced representation. The separable case of Theorem 9.1 is due to Mackey (5) and the general case was first treated by Blattner in (1). Theorems 10.1 and 10.4 are due to Mackey (5) who considered them only in the separable case. The same holds for Theorem 10.6. As we pointed out already in the text Theorem 11.1 and its corollary are due to Mackey. The same is true for the corollary of Proposition 3 which is stated here for the first time but might have been known to Mackey for a long time. Proposition 11.2 and Theorem 11.5 are probably new but the corollary was already proved by Mackey in (5). The results presented in Section 12 are due to the author.

Induced representations of finite groups are discussed in the introductory book of Burrow (1) and in the monograph of Curtis and Reiner (1). One can also consult Mackey's notes (2) where several examples are worked out. There are multiple reasons why I did not include more material on induced representations. Time and space limitations prevented me from including all the general results known about induced representations of locally compact groups. Since no direct integral decompositions are considered in this book a number of interesting results are automatically excluded. The best place to obtain orientation and references on this additional material is Mackey's expository paper (3). The reader might find a more recent publication of Mackey (8) also helpful. There is a fair amount of literature also on the induced representation of more special groups e.g. Lie groups. Induced representations in Banach spaces and the corresponding Frobenius reciprocity relation were considered by C. Moore (1).

Chapter VII

Square Integrable Representations, Spherical Functions and Trace Formulas

This chapter deals with one of the central topics of this book. The first section contains important basic facts from the theory of representations of compact groups and it should be read in full even if some of the later sections are omitted. Section 2 is essential for the formulation and proof of the Selberg–Tamagawa trace formula which is given in Section 6. Its proof is independent of the material discussed in Sections 3, 4 and 5. However the last section contains the statement and proof of a class of trace formulas which can be appreciated only by knowing about spherical functions as defined by Godement and Harish–Chandra. The theory of these is developed in Section 3. Moreover it is necessary to know from Section 4 and 5 at least the statements of Theorems 4.1 and 5.10.

The first four sections can be read without being familiar with the contents of Chapter VI. In Section 5 we use in a proof induced representations so one should at least superficially read before this proof Section VI.2. In Section 6 induced representations play a fundamental role. Thus before reading it one should be well acquainted with the contents of Sections VI.1, 2, 5, 6, and 7.

1. Square Integrable Representations and the Representation Theory of Compact Groups

The first third of this section is centered around the result expressed in Theorem 1 which will be viewed as a special case of other, more general theorems. In the middle of the section we develop the fundamental results on characters of unitary representations of compact groups. Some of these will be used in the third part where the Péter–Weyl theorem is stated and proved, others will be needed later.

Theorem 1. *Every continuous unitary representation of a compact group is a direct sum of finite dimensional irreducible components.*

As usual in all what follows the measurability of the representation $\rho: G \to \mathscr{U}(\mathscr{H})$ is tacitly assumed unless some stronger condition is explicitely stated. If $\rho: S \to \mathscr{L}(\mathscr{H})$ is a morphism of an arbitrary object S and if $\alpha, \beta \in \mathscr{H}$ then the function $s \mapsto (\rho(s)\alpha, \beta)$ is called a *coefficient* of the morphism ρ. Coefficients will be denoted by $\varphi^{\rho}_{\alpha,\beta}$ or $\varphi_{\alpha,\beta}$ and if β is some specially selected element of $\mathscr{H}$ then the notation φ_{α} will be used. A morphism $\rho: S \to \mathscr{L}(\mathscr{H})$ is called *square integrable* with respect to a measure $d\mu$ given on S if its coefficients are square integrable for every $\alpha, \beta \in \mathscr{H}$ with respect to $d\mu$. If G is a locally compact group then the square integrability of a unitary representation of G will always be understood with respect to its Haar measure. Since $\varphi_{\alpha,\beta}(x^{-1}) = \overline{\varphi_{\beta,\alpha}(x)}$ it makes no difference which one of the two Haar measures is used. If $\rho: G \to \mathscr{U}(\mathscr{H})$ is a unitary representation then its coefficients $\varphi_{\alpha,\beta}$ are bounded functions on G. Hence if G is compact and $\rho: G \to \mathscr{U}(\mathscr{H})$ is weakly measurable then ρ is square integrable.

Lemma 2. *If $\rho_i: S \to \mathscr{L}(\mathscr{H}_i)$ $(i=1,2)$ are equivalent morphisms and if ρ_1 is square integrable then so is ρ_2.*

Proof. Let $\varphi^i_{\alpha,\beta}$ denote the coefficients of ρ_i $(i=1,2)$. If $U: \mathscr{H}_2 \to \mathscr{H}_1$ is a unitary map and $\rho_1 U = U \rho_2$ then $\varphi^2_{\alpha,\beta} = \varphi^1_{U\alpha, U\beta}$ for any $\alpha, \beta \in \mathscr{H}_2$.

Lemma 3. *Every square integrable, irreducible representation of a unimodular group is equivalent to a subrepresentation of the right regular representation of G.*

Proof. Let $\pi: G \to \mathscr{U}(\mathscr{H})$ be a square integrable, irreducible representation of G. We fix a non-zero vector β in $\mathscr{H}$. For every $\alpha \in \mathscr{H}$ we consider $\varphi_\alpha: G \to \mathbb{C}$ given by $\varphi_\alpha(x) = (\pi(x)\alpha, \beta)$. We notice that the functions φ_α $(\alpha \in \mathscr{H})$ form a linear manifold in $L^2(G)$ so π being square integrable their set can be considered as a pre-Hilbert space under

$$(\varphi_{\alpha_1}, \varphi_{\alpha_2}) = \int_G \varphi_{\alpha_1}(x) \overline{\varphi_{\alpha_2}(x)}\, dx\,.$$

By the corollary of Proposition IV.2.17 we have $\|\varphi_\alpha\| > 0$ for every $\alpha \neq 0$. The completion of this manifold we denote by $\mathscr{K}$ so that $\mathscr{K}$ is a subspace of $L^2(G)$. Later we shall see that $\mathscr{K} = \{\varphi_\alpha : \alpha \in \mathscr{H}\}$ and $(\varphi_{\alpha_1}, \varphi_{\alpha_2}) = \lambda^2(\alpha_1, \alpha_2)$ for some fixed $\lambda > 0$.

We let $\rho: G \to \mathscr{U}(\mathscr{K})$ be the unitary representation for which $\rho(x)\varphi_\alpha = (\varphi_\alpha)_x$ where $\varphi_x(y) = \varphi(yx)$ The unitary nature of $\rho(x)$ follows from the unimodularity of G. Clearly

$$\rho(x)\varphi_\alpha = (\varphi_\alpha)_x = \varphi_{\pi(x)\alpha} \tag{1}$$

for every $\alpha \in \mathscr{H}$ and $x \in G$. Let $T: \mathscr{H} \to \mathscr{K}$ be defined by $\alpha \mapsto \varphi_\alpha$. Then T is linear and it is closed because if $\alpha_n \to \alpha$ then $|\varphi_{\alpha_n}(x) - \varphi_\alpha(x)| \to 0$ uniformly on G and so $\|\varphi_{\alpha_n} - \varphi_\alpha\| = \|\varphi_{\alpha_n} - \varphi_\alpha\|_2 \to 0$ and this means that $T\alpha_n = \varphi_{\alpha_n} \to \varphi_\alpha = T\alpha$. By (1) we have

$$T\pi(x)\alpha = \varphi_{\pi(x)\alpha} = \rho(x)\varphi_\alpha = \rho(x)T\alpha$$

and so $T\pi = \rho T$ on $\mathscr{H}$. Therefore Proposition IV.3.10 is applicable and so $T = \lambda U$ where $\lambda > 0$ and $U: \mathscr{H} \to \mathscr{K}$ is an isometry. We have

$$(\varphi_{\alpha_1}, \varphi_{\alpha_2}) = (T\alpha_1, T\alpha_2) = \lambda^2(U\alpha_1, U\alpha_2) = \lambda^2(\alpha_1, \alpha_2).$$

This shows that the pre-Hilbert space $T\mathscr{H}$ is complete and so $U\mathscr{H} = T\mathscr{H} = \mathscr{K}$. Now it is clear that ρ is the restriction of the right regular representation of G to $\mathscr{K}$. Since $U\pi = \rho U$ we see that $\pi \approx \rho$.

Proposition 4. *If G is unimodular and the irreducible representation $\pi: G \to \mathscr{U}(\mathscr{H})$ has a non-zero square integrable coefficient then π is a square integrable representation.*

Proof. Let us suppose that the non-zero coefficient $\varphi_{\alpha,\beta}$ belongs to $L^2(G)$. We keep α fixed and define φ_η for $\eta \in \mathscr{H}$ by $\varphi_\eta = \varphi_{\alpha,\eta}$. Let $\mathscr{D}$ be the set of those η in $\mathscr{H}$ for which $\varphi_\eta \in L^2(G)$. Then $\mathscr{D}$ is a linear manifold in $\mathscr{H}$ and $\mathscr{D}$ is invariant under π because $\varphi_{\pi(a)\eta}(x) = \varphi_\eta(a^{-1}x)$. Since $\varphi_\beta \in \mathscr{D}$ and π is irreducible it follows that $\mathscr{D}$ is dense in $\mathscr{H}$. If we define $T: \mathscr{D} \to L^2(G)$ by $T\eta = \varphi_\eta$ then T is a closed operator: In fact, if $\eta_n \to \eta$ then $\varphi_{\eta_n}(x) \to \varphi_\eta(x)$ for every $x \in G$. Hence if in addition $\|T\eta_n - \varphi\| \to 0$ then $T\eta_n \to \varphi$ in measure and so $\varphi = \varphi_\eta = T\eta$ where $\eta \in \mathscr{D}$. Applying Proposition IV.3.10 with ρ being the left regular representation λ we see that $\mathscr{D} = \mathscr{H}$. It follows that $\varphi_{\alpha,\eta}$ is square integrable for every $\eta \in \mathscr{H}$.

We now fix a non-zero $\eta \in \mathscr{H}$ and consider the new set $\mathscr{D}_\eta = \{\xi : \varphi_{\xi,\eta} \in L^2(G)\}$. Again $\mathscr{D}_\eta$ is a linear manifold and by the unimodularity of the group $\mathscr{D}_\eta$ is invariant under π. Since $\varphi_{\alpha,\eta}$ is in $\mathscr{D}_\eta$ and π is irreducible we see that $\mathscr{D}_\eta$ is dense in $\mathscr{H}$. We define $S_\eta: \mathscr{D}_\eta \to L^2(G)$ by $S_\eta\xi = \varphi_{\xi,\eta}$. Then S_η is a closed operator and so by Proposition IV.3.10 we have $\mathscr{D}_\eta = \mathscr{H}$. This shows that $\varphi_{\xi,\eta}$ is square integrable for every $\xi \in \mathscr{H}$ and $\eta \neq 0$. The case $\eta = 0$ is trivial because $\varphi_{\xi,0} \equiv 0$.

Theorem 5. *An irreducible representation $\pi: G \to \mathscr{U}(\mathscr{H})$ of the unimodular group G is square integrable if and only if it is equivalent to a subrepresentation of the left regular representation of G.*

Proof. Due to Lemma 3 and the equivalence of the left and right regular representations it is sufficient to prove that if the irreducible representation π is equivalent to a subrepresentation of the left regular representation λ then π is square integrable. By Lemma 2 we may suppose that π is actually an irreducible component of λ, say $\pi = \lambda|\mathscr{P}$ where $\lambda: G \to \mathscr{U}(\mathscr{H})$ and $\mathscr{H} = L^2(G)$. If P denotes the projection of $\mathscr{H}$ onto $\mathscr{P}$ then $C_0(G)$ being dense in $L^2(G)$ there is an f in $C_0(G)$ such that $Pf \neq 0$. By ${}^x Pf = P\,{}^x f$ we have

$$\varphi_{Pf,Pf}(x) = (P\,{}^x f, Pf) = (P\,{}^x f, f) = \int_G \overline{f(y)}\, Pf(x^{-1}y)\,dy = \bar{f} * \widetilde{Pf}(x)$$

where $\widetilde{Pf}(z) = Pf(z^{-1})$. Applying Proposition V.4.6 we see that $\|\varphi_{Pf,Pf}\|_2 \leqslant \|f\|_1 \|Pf\|_2 < \infty$. Therefore π has a square integrable coefficient and so π is square integrable by Proposition 4.

A topological group G is called a *group with small invariant neighborhoods* if any given neighborhood N_e of the identity element e contains another neighborhood N of e such that N is invariant under inner automorphisms of G i.e. $x^{-1}Nx = N$ for every x in G.

Proposition 6. *Every compact group is a group with small invariant neighborhoods.*

Proof. Given a neighborhood N_e by parts 2), 4) and 5) of Theorem V.1.1 there is a neighborhood B of e such that $B = B^{-1}$ and $B^3 \subseteq N_e$. The family of interiors xB^i $(x \in G)$ cover G so by the compactness of G we can find $x_1, \ldots, x_n$ in G such that $x_1 B, \ldots, x_n B$ cover G. By parts 1) and 2) of the same theorem we can find a neighborhood C of e such that $x_k^{-1} C x_k \subseteq B$ for every $k = 1, \ldots, n$. Given any g in G we have $g \in x_k B$ for some k and so

$$g^{-1} C g \subseteq B x_k^{-1} C x_k B \subseteq B^3 \subseteq N_e.$$

Therefore N, the union of the sets $g^{-1}Cg$ $(g \in G)$ is an invariant neighborhood of e which is contained in N_e.

A function f defined on the group G is called *central* if $f(xy) = f(yx)$ for every $x, y \in G$. The condition is clearly equivalent to $f(xyx^{-1}) = f(y)$ for $x, y \in G$. The following lemma explains the reason why such functions are called central.

Lemma 7. *If G is unimodular then $f: G \to \mathbb{C}$ belongs to the center of the algebra $L^1(G)$ if and only if f is in $L^1(G)$ and it is a central function.*

Proof. Since G is unimodular we have

$$f*g(x)-g*f(x)=\int_G \{f(xy)-f(yx)\}g(y^{-1})dy$$

for every $g\in L^1(G)$. Hence if f is in the center of $L^1(G)$ then ${}_xf=f_x$ and conversely if f is a central element of $L^1(G)$ then $f*g=g*f$.

Proposition 8. *If G is a unimodular group with small invariant neighborhoods then every irreducible component of the regular representation of G is finite dimensional.*

Proof. Since the two regular representations are equivalent we may restrict our attention to the left regular representation $\lambda: G\to\mathscr{U}(\mathscr{H})$ where $\mathscr{H}=L^2(G)$. Let $\pi=\lambda|\mathscr{P}$ be an irreducible component and let $\sigma: L^1(G)\to\mathscr{L}(\mathscr{P})$ be the algebra representation associated with π. We can immediately see that σ has a kernel, namely

$$(\sigma(f)g)(y)=\int_G f(x)\pi(x)g(y)dx=\int_G f(yx^{-1})g(x)dx.$$

If we restrict our attention to continuous f's with compact support or to characteristic functions of compact sets then the kernel $f(yx^{-1})$ clearly belongs to $L^2(G\times G)$ and so $\sigma(f)$ is a HS operator. (See the remarks at the end of Section II.4 and in the beginning of Section VI.5.)

Let N be an invariant neighborhood of the identity e of G with compact closure and let $c_N: G\to\mathbb{C}$ be the characteristic function of N. Then c_N is a central function in $L^1(G)$ and so by Lemma 7 $c_N*f=f*c_N$ for every f in $L^1(G)$. Since $\sigma: L^1(G)\to\mathscr{L}(\mathscr{P})$ is a representation we obtain

$$\sigma(c_N)\sigma(f)=\sigma(f)\sigma(c_N)$$

for any f in $L^1(G)$. By Proposition III.5.7 and V.5.7 and by Theorem V.5.9 we have $\sigma(c_N)A=A\sigma(c_N)$ for every A in $\mathscr{A}$ where $\mathscr{A}=\mathscr{A}_\sigma=\mathscr{A}_\pi$ is the v. Neumann algebra generated by σ and π. We know that π is irreducible and so by Propositions IV.2.4 and IV.3.2 we have $(\pi,\pi)=\mathscr{A}'=\mathbb{C}I$ and $\sigma(c_N)=k_N I$.

Let g in $\mathscr{P}$ be an element with $\|g\|=\|g\|_2=1$. Then

$$k_N=(\sigma(c_N)g,g)=\int_N(\pi(x)g,g)dx.$$

By Theorem V.6.2 the coefficient $x\to(\pi(x)g,g)$ is continuous at e so by choosing N to be sufficiently small we see that $k_N\neq 0$. Since $k_N\neq 0$ and $k_N I$ is a HS operator the underlying Hilbert space $\mathscr{P}$ is finite dimensional.

Theorem 9. *Every square integrable, irreducible representation of a unimodular group with small invariant neighborhoods is finite dimensional.*

Proof. This follows from Theorem 5 and Proposition 8.

Theorem 10. *Every irreducible representation of a compact group is finite dimensional.*

Proof. By Proposition 6 and by the unimodularity of the group the hypotheses of Theorem 9 are satisfied.

Theorem 11. *If G is a compact group then every irreducible representation σ of the algebra $L^1(G)$ is finite dimensional.*

If ρ_1 and ρ_2 are equivalent representations of the locally compact group G and if ρ_1 is weakly measurable then so is ρ_2. Thus we can speak about a class ω being measurable. We let $\hat{G}_m$ denote the set of measurable equivalence classes of irreducible representations of G.

Theorem 12. *If G is compact then $\hat{G}_m$ consists of the equivalence classes of the irreducible subrepresentations of the regular representation of G.*

Proof. Since G is compact every irreducible, weakly measurable representation of G is square integrable. Hence the conclusion follows from Theorem 5.

Let G be a locally compact group and let $\omega \in \hat{G}$. Since the elements of ω are equivalent the Hilbert spaces of the representations $\pi \in \omega$ all have the same dimension d_ω. In representation theory it is customary to call d_ω the *degree of the class* ω. The set of those equivalence classes ω for which d_ω is finite will be denoted by $\hat{G}_0$ or $\Omega_0(G)$ if the notation $\Omega(G)$ is used instead of $\hat{G}$. By Lemma 2 we can speak about a *square integrable class* ω. We let $\hat{G}_d$ or $\Omega_d(G)$ denote the set of these square integrable classes $\omega \in \hat{G}$. The set $\hat{G}_d$ is called the *discrete series* of G. Some of our results can be summarized by using these notations. For instance if G is compact then $\hat{G} = \hat{G}_0$ and if G is unimodular with small invariant neighborhoods then $\hat{G}_d \subseteq \hat{G}_0$.

As far as Theorem 1 is concerned so far we proved only that the irreducible components, if they exist, are finite dimensional. In order to complete the proof we must show that every continuous representation $\rho: G \to \mathscr{U}(\mathscr{H})$ of the compact group G is a direct sum of irreducible components. We suppose that the Haar measure of G is normalized i.e. $\mu(G)=1$. If $f \in L^2(G)$ then $f \in L^1(G)$ and $\|f\|_1 \leqslant \|f\|_2$. Moreover if $f, g \in L^2(G)$ then by Proposition V.4.5 $f * g \in L^\infty(G)$ and $\|f*g\|_\infty \leqslant \|f\|_2 \|g\|_2$. Since $f*g$ is bounded we have $\|f*g\|_2 \leqslant \|f*g\|_\infty$.

Therefore $f*g\in L^2(G)$ for any $f,g\in L^2(G)$ and $A=L^2(G)$ is a Banach algebra. In Section V.4 it was shown that the inner product of the complex Hilbert space $L^2(G)$ satisfies the identities $(f*g,h)=(g,f^**h)$ and $(f*g,h)=(f,h*g^*)$. Therefore A is an H^* algebra. If $f\in A$ then by Proposition V.4.13 $f*f^*$ is continuous and clearly $f*f^*(e)=\|f\|_2^2$. Therefore $f\neq 0$ implies $f*f^*\neq 0$ and so A is semisimple by the corollary of Proposition I.9.3. It is called the *L^2 algebra* or the *H^* algebra* of the compact group G.

Lemma 13. *If G is compact and I is a non-zero, closed ideal in $L^2(G)$ then I contains a non-zero element which is in the center of $L^2(G)$ and also in $C(G)$.*

Corollary. *If I is minimal then it is finite dimensional and it has an identity element.*

Proof. We choose a non-zero element h in I and let $g=h*h^*$. Then $g\in C(G)$ by Proposition V.4.13. We define f by $f(x)=\int_G g(xyx^{-1})dy$. Then f is continuous and

$$f(e)=g(e)=h*h^*(e)=\|h\|_2^2>0.$$

Since f is continuous and central by Lemma 7 f belongs to the center of $L^2(G)$ because $L^2(G)\subseteq L^1(G)$. By $h\in I$ we have $g\in I$ and so ${}_xg_{x^{-1}}$ is in I for every $x\in G$ by Proposition V.5.12. Thus I being closed we have $f\in I$. If I is minimal then we can apply Theorems I.9.8 and I.9.12 and derive the corollary stated above.

Now let $\rho: G\to\mathcal{U}(\mathcal{H})$ be the continuous representation whose direct sum decomposition is to be proved. We let $\sigma: L^1(G)\to\mathcal{L}(\mathcal{H})$ denote the algebra representation associated with ρ and described in Theorem V.5.4. Let τ be the restriction of σ to $A=L^2(G)$. By Theorems V.5.7 and V.5.9 ρ, σ and τ generate the same v. Neumann algebra and so by Proposition III.5.10 they have identical invariant subspaces.

By Theorem I.9.8, Proposition I.9.9 and their corollaries A can be expressed as an orthogonal direct sum of closed, minimal ideals I. Given a non-zero α in $\mathcal{H}$ by Proposition V.5.10 we have $\tau(f)\alpha\neq 0$ for some f in A and so $\tau(f)\alpha\neq 0$ for at least one minimal ideal I. We fix α and the ideal I and consider the linear manifold

$$\mathcal{K}=\{\tau(f)\alpha: f\in I\}.$$

Using the ideal property we see that $\mathcal{K}$ is invariant under τ. By the corollary of Lemma 13 I is finite dimensional so $\mathcal{K}$ is a finite dimensional subspace of $\mathcal{H}$ which is invariant under τ. Therefore a suitable subspace $\mathscr{P}$ of $\mathscr{K}$ is an irreducible component of τ and hence of ρ. We proved the following:

Lemma 14. *Every continuous unitary representation of a compact group has an irreducible component.*

The fact that τ comes from ρ was used only to assert the existence of an α such that $\tau(f)\alpha \neq 0$ for some f in A and to conclude that $\mathscr{P}$ is an irreducible component of ρ. Now if $\tau: A \rightarrow \mathscr{L}(\mathscr{H})$ is an arbitrary representation of A and if $\tau(A)\mathscr{H} = 0$ then any one dimensional subspace of $\mathscr{H}$ is an irreducible component of τ. Therefore we have:

Lemma 15. *Every representation of the L^2-algebra of a compact group has an irreducible component.*

The proof of Theorem 1 will be completed by proving the following:

Proposition 16. *If every representation of the object S has an irreducible component then every representation $\rho: S \rightarrow \mathscr{L}(\mathscr{H})$ is a direct sum of irreducible components.*

Proof. We consider non-void systems $\{\mathscr{P}_i\}$ $(i \in \mathscr{I})$ of pairwise orthogonal, non-zero, irreducible components $\mathscr{P}_i$ of the given representation ρ. By hypotheses such systems exist. If we partially order the set of all these systems using inclusion as the partial ordering then the chain condition is satisfied and Zorn's lemma is applicable. Let $\{\mathscr{P}_i\}$ $(i \in \mathscr{I})$ be a system which is maximal with respect to inclusion and let $\mathscr{K} = \sum \mathscr{P}_i$. Since $\mathscr{K}$ is invariant under ρ so is $\mathscr{K}^{\perp}$ and $\rho | \mathscr{K}^{\perp}$ is a representation of S. If $\mathscr{K}^{\perp}$ were not zero then by hypothesis $\rho | \mathscr{K}^{\perp}$ would have an irreducible component in contradiction to the maximality of the system $\{\mathscr{P}_i\}$ $(i \in \mathscr{I})$. Hence $\mathscr{K} = \mathscr{H}$ and ρ is the direct sum of the irreducible components $\rho | \mathscr{P}_i$.

If G is compact and $\tau: L^2(G) \rightarrow \mathscr{L}(\mathscr{H})$ is a bounded non-degenerate representation then by $\|f\|_2 \leqslant \|f\|_1$ one sees that τ can be extended to a bounded representation $\sigma: L^1(G) \rightarrow \mathscr{L}(\mathscr{H})$. By Theorem V.5.16 there is a representation $\rho: G \rightarrow \mathscr{U}(\mathscr{H})$ such that σ is the algebra representation associated with ρ. If τ is irreducible then so is ρ and hence $\dim \mathscr{H} < \infty$. Applying Lemma 15 and the foregoing proposition to the H^* algebra of G we obtain:

Theorem 17. *Every bounded representation of the L^2-algebra of a compact group is the direct sum of finite dimensional irreducible components.*

Actually we can first decompose τ into a direct sum of invariant subspaces which give faithful representations of some of the minimal two-sided ideals I of $L^2(G)$ and then decompose these invariant subspaces. The first step can be done for any semisimple H^* algebra.

Theorem 18. *Every bounded representation* $\tau: A \to \mathscr{L}(\mathscr{H})$ *of a semisimple H^* algebra A can be expressed as a direct sum of faithful representations of some of its closed, minimal ideals.*

Proof. By the corollary of Theorem I.9.8 A is the orthogonal direct sum of its minimal ideals I. For each I we define $\mathscr{H}_I$ to be the subspace spanned by the vectors $\tau(a)\alpha$ where $\alpha \in \mathscr{H}$ and $a \in I$. By the ideal property and by Proposition III.5.3 $\mathscr{H}_I$ is an invariant subspace of τ. If I_1, I_2 are distinct closed minimal ideals and $a_k \in I_k$ $(k=1,2)$ then $a_1 a_2 = a_2 a_1 = 0$ because $a_1 a_2 - a_2 a_1 \in I_1 \cap I_2$ which is zero by the minimality. If I_1 and I_2 are as before and $a_k \in I_k$ and $\alpha_k \in \mathscr{H}$ then

$$(\tau(a_1)\alpha_1, \tau(a_2)\alpha_2) = (\alpha_1, \tau(a_1^* * a_2)\alpha_2) = 0$$

because $a_1^* \in I_1$ and so $a_1^* * a_2 = 0$. Therefore the subspaces $\mathscr{H}_I$ are pairwise orthogonal. The restricted map $\tau_I: I \to \mathscr{L}(\mathscr{H}_I)$ is a bounded representation of the topologically simple H^* algebra I and so its kernel is either 0 or I itself. Now τ is the direct sum of those τ_I's which are faithful.

We recall that a morphism $\rho: G \to \mathrm{Gl}(\mathscr{H})$ of a group G is called *bounded* if there is a positive constant B such that $|\rho(x)| \leqslant B$ for every $x \in G$. As earlier we let $\mathrm{Gl}(\mathscr{H})$ stand for the group of units of $\mathscr{L}(\mathscr{H})$ that is to say $\mathrm{Gl}(\mathscr{H})$ is the group of continuous invertible linear transformations $A: \mathscr{H} \to \mathscr{H}$.

Proposition 19. *If* $\rho: G \to \mathrm{Gl}(\mathscr{H})$ *is a bounded, strongly continuous morphism of the compact group G in the Hilbert space $\mathscr{H}$ then*

$$\langle \alpha, \beta \rangle = \int_G (\rho(x)\alpha, \rho(x)\beta)\,dx$$

is an inner product on $\mathscr{H}$ which is equivalent to the original one and under which ρ is a unitary representation.

Proof. If $\alpha = \beta \neq 0$ then the integrand is a non-negative, continuous function which is strictly positive at $x = e$. Hence $\langle \alpha, \alpha \rangle > 0$ for $\alpha \neq 0$. The rest of the properties of an inner product are obviously shared by $\langle \alpha, \beta \rangle$ the continuity of $x \mapsto (\rho(x)\alpha, \rho(x)\beta)$ being a consequence of the continuity of $x \mapsto \|\rho(x)\alpha\|^2$. We have

$$\langle \rho(a)\alpha, \rho(a)\alpha \rangle = \int_G (\rho(xa)\alpha, \rho(xa)\alpha)\,dx = \int_G (\rho(x)\,\alpha, \rho(x)\alpha)\,dx = \langle \alpha, \alpha \rangle$$

and so $\rho(a)$ is a unitary operator for every a in G. To see that the norm $|\alpha|$ corresponding to $\langle\alpha,\beta\rangle$ is equivalent to the original norm $\|\alpha\|$ let B be a bound for the norms $|\rho(x)|$ $(x\in G)$. Clearly we have $\langle\alpha,\alpha\rangle\leqslant B^2(\alpha,\alpha)$ and so $|\alpha|\leqslant B\|\alpha\|$. Given α for any $x\in G$ we have $\|\alpha\|=\|\rho(x^{-1})\rho(x)\alpha\|$ $\leqslant B\|\rho(x)\alpha\|$ and so integrating we obtain $\|\alpha\|^2\leqslant B^2\langle\alpha,\alpha\rangle$ i.e. $\|\alpha\|\leqslant B|\alpha|$. Therefore the two norms $|\alpha|$ and $\|\alpha\|$ are equivalent.

Corollary 1. *If $\mathscr{V}$ is a finite dimensional vector space over $\mathbb{C}$ and if $\rho: G\to \mathrm{Gl}(\mathscr{V})$ is a morphism then there is an inner product on $\mathscr{V}$ such that ρ is a unitary representation of G.*

Applying Theorem 10 we obtain:

Corollary 2. *Every bounded, strongly continuous, irreducible morphism $\rho: G\to \mathrm{Gl}(\mathscr{H})$ of the compact group G is finite dimensional.*

From here on coefficients will play an important role in the present chapter. We shall see that certain special coefficients called zonal spherical functions have very interesting and important properties. Sums of certain coefficients will be called spherical functions of height h and studied later. At present we are going to introduce group characters which being special cases of spherical functions again turn out to be sums of coefficients.

If the morphism $\rho: S\to\mathscr{L}(\mathscr{H})$ is of finite degree then every $\rho(x)$ is a trace class operator and we can define $\chi_\rho: S\to\mathbb{C}$ by $\chi(x)=\operatorname{tr}\rho(x)$. It is called the *character* of the morphism ρ. If $\rho_1\approx\rho_2$ then $\rho_1=U^{-1}\rho_2 U$ for some unitary operator U and so $\operatorname{tr}\rho_1(x)=\operatorname{tr}\rho_2(x)$. This shows that χ_ρ, the character of an irreducible morphism depends only on the equivalence class ω containing ρ and so we can speak about the *character of the class* ω. By Definition II.5.4 if $\alpha_1,\dots,\alpha_d$ is a MONS in $\mathscr{H}$ then the character of $\rho: S\to\mathscr{L}(\mathscr{H})$ is

$$\chi_\rho(x)=\sum(\rho(x)\alpha_i,\alpha_i).$$

There are two important facts about convolutions of characters of irreducible unitary group representations known as the *orthogonality relations* (Propositions 30 and 31). Our immediate object is to develop these and some related material.

First we remark that if ρ is a unitary representation then we have

$$\overline{\chi_\rho(x)}=\chi_\rho(x^{-1})$$

for every x in G. For if α_i $(i\in\mathscr{I})$ is a MONS in the Hilbert space of ρ then $(\rho(x^{-1})\alpha_i,\alpha_i)=(\alpha_i,\rho(x)\alpha_i)$. Furthermore if ρ is unitary then χ_ρ is a central function i. e. $f(ax)=f(xa)$ for every x and a in G. For

$$\chi_\rho(a^{-1}xa)=\sum(\rho(a^{-1}xa)\alpha_i,\alpha_i)=\sum(\rho(x)\rho(a)\alpha_i,\rho(a)\alpha_i)$$

where ρ being unitary $\rho(a)\alpha_i$ $(i\in\mathscr{I})$ is a MONS in $\mathscr{H}$. Therefore the right hand side is $\chi_\rho(x)$ and so $\chi_\rho(a^{-1}xa)=\chi_\rho(x)$ for every $a, x\in G$.

Proposition 20. *For any two group representations* $\rho_k: G\to\mathscr{U}(\mathscr{H})$ *we have* $\chi_{\rho_1\oplus\rho_2}=\chi_{\rho_1}+\chi_{\rho_2}$ *and* $\chi_{\rho_1\otimes\rho_2}=\chi_{\rho_1}\chi_{\rho_2}$.

Proof. The first equation becomes obvious by choosing a MONS in $\mathscr{H}_1$ and $\mathscr{H}_2$ and by using their union as a basis for $\mathscr{H}=\mathscr{H}_1+\mathscr{H}_2$. As far as the second is concerned if α_i and β_j are elements of MONS in $\mathscr{H}_1$ and $\mathscr{H}_2$ then by Proposition VI.3.3 the products $\alpha_i\otimes\beta_j$ form a basis in $\mathscr{H}_1\otimes\mathscr{H}_2$. Now if $A: \mathscr{H}_1\to\mathscr{H}_1$ and $B: \mathscr{H}_2\to\mathscr{H}_2$ we let

$$A\alpha_i=\sum_k a_{ik}\alpha_k \quad\text{and}\quad B\beta_j=\sum_l b_{jl}\beta_l$$

so that $(A\alpha_i\otimes B\beta_j,\alpha_i\otimes\beta_j)=(A\alpha_i,\alpha_i)(B\beta_j,\beta_j)=a_{ii}b_{jj}$. Hence

$$\sum_{i,j}(A\alpha_i\otimes B\beta_j,\alpha_i\otimes\beta_j)=\sum_{i,j}a_{ii}b_{jj}=\operatorname{tr}A\operatorname{tr}B.$$

Thus in general we have $\operatorname{tr}A\otimes B=\operatorname{tr}A\operatorname{tr}B$ and our result is a simple application of this to $A=\rho_1(x)$, $B=\rho_2(x)$.

Proposition 21. *If* G *is locally compact,* $\rho_k: G\to\mathscr{U}(\mathscr{H}_k)$ $(k=1,2)$ *are representations,* $A\in\mathscr{L}(\mathscr{H}_2,\mathscr{H}_1)$ *and if the linear operator*

$$T=\int_G\rho_1(x)A\rho_2(x^{-1})dx$$

exists in the weak sense and is continuous then $T\in(\rho_2,\rho_1)$.

Proof. The intertwining property is proved by writing out the weak version of the following reasoning:

$$\rho_1(a)T=\int_G\rho_1(ax)A\rho_2(x^{-1})dx=\int_G\rho_1(x)A\rho_2(x^{-1}a)dx=T\rho_2(a).$$

Corollary. *If* G *is compact then* T *exists for every* $A\in\mathscr{L}(\mathscr{H}_2,\mathscr{H}_1)$ *and* $T\in(\rho_2,\rho_1)$.

Proposition 22. *If* G *is a compact group,* $\rho_k: G\to\mathscr{U}(\mathscr{H}_k)$ $(k=1,2)$ *are disjoint representations and if* $\alpha_k,\beta_k\in\mathscr{H}_k$ *for* $k=1,2$ *then*

$$\int_G(\rho_1(x)\alpha_1,\beta_1)\overline{(\rho_2(x)\alpha_2,\beta_2)}dx=0.$$

Proof. By the corollary of Proposition 21 the linear operator

$$T = \int_G \rho_1(x)(\alpha_1 \overline{\otimes} \alpha_2)\rho_2(x^{-1})dx$$

exists and belongs to (ρ_2, ρ_1) which is 0 because $\rho_1 \downharpoonright \rho_2$. Therefore $T=0$ and in particular $(T\beta_2, \beta_1)=0$. We note that $\alpha_1 \overline{\otimes} \alpha_2$ denotes the linear operator such that $(\alpha_1 \overline{\otimes} \alpha_2)\xi = (\xi, \alpha_2)\alpha_1$. Therefore

$$(\alpha_1 \overline{\otimes} \alpha_2)\rho_2(x^{-1})\beta_2 = (\rho_2(x^{-1})\beta_2, \alpha_2)\alpha_1$$

and so $(T\beta_2, \beta_1)$ is the integral occurring in the proposition.

Proposition 23. *Let* $\rho: G \to \mathcal{U}(\mathcal{H})$ *be a square integrable representation and let* $\alpha \in \mathcal{H}$. *Then* $T_\alpha \xi(x) = (\xi, \rho(x)\alpha)$ *with* $\xi \in \mathcal{H}$ *and* $x \in G$ *defines a continuous linear operator* $T_\alpha: \mathcal{H} \to L^2(G)$ *and* $T_\alpha \in (\rho, \lambda)$ *where* λ *is the left regular representation of* G.

Proof. Since ρ is square integrable $T_\alpha \xi = \overline{\varphi_{\alpha,\xi}}$ is in $L^2(G)$. Hence T_α is an everywhere defined linear map of $\mathcal{H}$ into $L^2(G)$. If $\eta_n \to \eta$ in $\mathcal{H}$ then $T_\alpha \eta_n(x) \to T_\alpha \eta(x)$ for every x in G. Hence if $T_\alpha \eta_n \to f$ in $L^2(G)$ then $T_\alpha \eta = f$ and so T_α is a closed operator. Therefore by the closed graph theorem $T_\alpha \in \mathcal{L}(\mathcal{H}, L^2(G))$. If $a, x \in G$ and $\eta \in \mathcal{H}$ then

$$T_\alpha \rho(a)\eta(x) = (\rho(a)\eta, \rho(x)\alpha) = (\eta, \rho(a^{-1}x)\alpha) = T_\alpha \eta(a^{-1}x) = \lambda(a) T_\alpha \eta(x).$$

Hence T_α is in (ρ, λ).

The proposition will be used in the proof of three integral formulas involving the coefficients of square integrable representations of a locally compact group G. First we discuss the special case when G is compact.

Proposition 24. *If* G *is compact,* $\pi_k: G \to \mathcal{U}(\mathcal{H}_k)$ $(k=1,2)$ *are equivalent irreducible representations and if* $U \in (\pi_1, \pi_2)$ *then*

$$\int_G (\pi_1(x)\alpha, \beta)\overline{(\pi_2(x)U\gamma, U\delta)}dx = \frac{(\alpha,\gamma)(\delta,\beta)}{d}$$

where d *denotes the common degree of* π_1 *and* π_2.

Proof. Consider

$$T = \int_G \pi_1(x)(\alpha \overline{\otimes} U\gamma)\pi_2(x^{-1})dx.$$

By Proposition 21 $T \in (\pi_2, \pi_1)$ and so by Propositions IV.2.4 and IV.3.4 we have $T = kU^*$ for some $k \in \mathbb{C}$. We compute $(TU\delta, \beta)$ two ways and obtain on the one hand $k(\delta,\beta)$ and on the other hand the integral appearing in the proposition. The constant k can be determined by computing the trace of UT in two different ways: Using $T=kU^*$ we obtain $\operatorname{tr} UT=kd$ and on the other hand

$$\operatorname{tr} UT=\int_G \operatorname{tr} U\pi_1(x)(\alpha\bar{\otimes}U\gamma)\pi_2(x)^{-1}dx.$$

Since $U\pi_1(x)=\pi_2(x)U$ and $\int_G dx=1$ we obtain $\operatorname{tr} UT=\operatorname{tr} U(\alpha\bar{\otimes}U\gamma)$ which can be computed by using a MONS which begins with $U\gamma/\|U\gamma\|$. One easily obtains (α,γ). Hence our integral is $k(\delta,\beta)$ where $kd=(\alpha,\gamma)$.

Corollary. *If G is compact and $\pi\colon G\to\mathscr{U}(\mathscr{H})$ is irreducible then*

$$\int_G (\pi(x)\alpha,\beta)\overline{(\pi(x)\gamma,\delta)}\,dx=\frac{(\alpha,\gamma)(\delta,\beta)}{d}$$

where d is the degree of π.

The same reasoning leads to the following:

Proposition 25. *Let $\rho\colon G\to\mathscr{U}(\mathscr{H})$ be a representation of the compact group G, let $\mathscr{P}_1,\mathscr{P}_2$ be equivalent irreducible components of ρ and let $U\colon\mathscr{P}_1\to\mathscr{P}_2$ be a unitary element of $(\rho|\mathscr{P}_1,\rho|\mathscr{P}_2)$. Then for any β in $\mathscr{H}$ and α,γ,δ in $\mathscr{P}_1$ we have*

$$\int_G (\rho(x)\alpha,\beta)\overline{(\rho(x)U\gamma,\delta)}\,dx=\frac{(\alpha,\gamma)(U^*\delta,\beta)}{d}$$

where d denotes the degree of $\rho|\mathscr{P}_1$.

Proof. The linear operator $T\colon\mathscr{P}_2\to\mathscr{P}_1$ defined by

$$T=\int_G \rho(x)(\alpha\bar{\otimes}U\gamma)\rho(x)^{-1}dx$$

belongs to $(\rho|\mathscr{P}_2,\rho|\mathscr{P}_1)$ and so $T=kU^*$. The rest of the reasoning is identical with that given in the proof of the last proposition.

The result will be used in the proof of a property of spherical functions.

Proposition 26. *Let G be compact, let $\rho\colon G\to\mathscr{U}(\mathscr{H})$ be a representation and let $\mathscr{P}$ be an irreducible component which occurs in ρ with multiplicity 1. Then for any $\alpha,\beta\in\mathscr{H}$ and $\gamma,\delta\in\mathscr{P}$ we have*

$$\int_G (\rho(x)\alpha,\beta)\overline{(\rho(x)\gamma,\delta)}\,dx=\frac{(\alpha,\gamma)(\delta,\beta)}{d}$$

where d denotes the degree of $\rho|\mathscr{P}$.

Proof. First suppose that $\alpha \in \mathscr{P}$. Then the formula holds by Proposition 25. Next suppose that $\alpha \in \mathscr{P}^{\perp}$. If $\beta \in \mathscr{P}$ then by $\rho(x)\alpha \in \mathscr{P}^{\perp}$ the integrand is zero and so the formula is valid. If $\beta \in \mathscr{P}^{\perp}$ then $\rho|\mathscr{P}$ and $\rho|\mathscr{P}^{\perp}$ being disjoint we can apply Proposition 22 with $\mathscr{H}_1 = \mathscr{P}^{\perp}$, $\mathscr{H}_2 = \mathscr{P}$ and $\alpha_1 = \alpha$, $\beta_1 = \beta$, $\alpha_2 = \gamma$, $\beta_2 = \delta$. The integral appearing in Proposition 22 is the present integral and so it is 0 by $\beta \in \mathscr{P}^{\perp}$ i.e. the formula holds again. For arbitrary $\alpha \in \mathscr{H}$ we let $\alpha = \alpha_1 + \alpha_2$ where $\alpha_1 \in \mathscr{P}$ and $\alpha_2 \in \mathscr{P}^{\perp}$ and obtain the desired conclusion in general.

Proposition 27. *Let G be a locally compact group and let* $\rho_k: G \to \mathscr{U}(\mathscr{H}_k)$ $(k=1,2)$ *be disjoint square integrable representations. Then for any* $\alpha_k, \beta_k \in \mathscr{H}_k$ $(k=1,2)$ *we have*

$$\int_G (\rho_1(x)\alpha_1, \beta_1)\overline{(\rho_2(x)\alpha_2, \beta_2)}\,dx = 0 .$$

Proof. Let $T_{\alpha_k}: \mathscr{H}_k \to L^2(G)$ $(k=1,2)$ denote the linear operator which was introduced in Proposition 23. Then $T_{\alpha_k}\beta_k(x) = (\beta_k, \rho_k(x)\alpha_k)$ for $k=1,2$ and so $(T_{\alpha_1}^* T_{\alpha_2}\beta_2, \beta_1) = (T_{\alpha_2}\beta_2, T_{\alpha_1}\beta_1)$ is our integrand. Since $T_{\alpha_k} \in (\rho_k, \lambda)$ we see that $T_{\alpha_1}^* T_{\alpha_2} \in (\rho_2, \rho_1)$ and so $T_{\alpha_1}^* T_{\alpha_2} = 0$ by $\rho_1 \,\rangle\!\!\!\langle\, \rho_2$.

Proposition 28. *Let* $\rho: G \to \mathscr{U}(\mathscr{H})$ *be a square integrable representation and let* $\mathscr{H}_1$ *and* $\mathscr{H}_2$ *be stable subspaces of* ρ *such that* $\rho|\mathscr{H}_1$ *and* $\rho|\mathscr{H}_2$ *are disjoint. Then for any* $\alpha_1 \in \mathscr{H}_1, \alpha_2, \beta_2 \in \mathscr{H}_2$ *and* $\beta_1 \in \mathscr{H}$ *we have*

$$\int_G (\rho(x)\alpha_1, \beta_1)\overline{(\rho(x)\alpha_2, \beta_2)}\,dx = 0 .$$

Proof. Let $T_{\alpha_k}: \mathscr{H}_k \to L^2(G)$ $(k=1,2)$ be the operators introduced in Proposition 23. Then the integrand can be expressed in the form $(T_{\alpha_2}\beta_2, T_{\alpha_1}\beta_1) = (T_{\alpha_1}^* T_{\alpha_2}\beta_2, \beta_1)$. By $T_{\alpha_k} \in (\rho|\mathscr{H}_k, \lambda)$ we have $T_{\alpha_1}^* T_{\alpha_2} \in (\rho|\mathscr{H}_2, \rho|\mathscr{H}_1)$. Hence $\rho|\mathscr{H}_1$ and $\rho|\mathscr{H}_2$ being disjoint the integral is 0.

Another way of proving the proposition is as follows: If $\beta_1 \in \mathscr{H}_1$ then the formula is valid by Proposition 27. If $\beta_1 \in \mathscr{H}_1^{\perp}$ then by $\rho(x)\alpha_1 \in \mathscr{H}_1$ the integrand is zero and the formula holds. Now if β_1 is arbitrary then we can use the result with the projections of β_1 in $\mathscr{H}_1$ and $\mathscr{H}_1^{\perp}$ in place of β_1. By adding the resulting formulas we obtain the general result.

Proposition 29. *Let* $\pi: G \to \mathscr{U}(\mathscr{H})$ *be a square integrable irreducible representation of the unimodular group G. Then there is a positive constant d such that*

$$\int_G (\pi(x)\alpha, \beta)\overline{(\pi(x)\gamma, \delta)}\,dx = \frac{(\alpha,\gamma)(\delta,\beta)}{d}$$

for every α, β, γ *and* δ *in* $\mathscr{H}$.

Note. The constant d is called the *formal degree* of π.

Proof. Let $T_\alpha: \mathscr{H} \to L^2(G)$ and $T_\gamma: \mathscr{H} \to L^2(G)$ be the operators introduced in Proposition 23. Since $T_\alpha \beta(x) = (\beta, \pi(x)\alpha)$ and $T_\gamma \delta(x) = (\delta, \pi(x)\gamma)$ the integrand is $\overline{T_\alpha \beta} \cdot T_\gamma \delta$ and the integral in question is $(T_\gamma \delta, T_\alpha \beta) = (T_\alpha^* T_\gamma \delta, \beta)$. By Proposition 23 we have $T_\alpha, T_\gamma \in (\pi, \lambda)$ so that $T_\alpha^* T_\gamma \in (\pi, \pi)$. Hence by Proposition IV.2.4

$$T_\alpha^* T_\gamma = k(\alpha, \gamma) I$$

where $k(\alpha, \gamma)$ is a constant depending only on α, γ, the representation π and the choice of the Haar measure dx. Therefore the value of the integral is $k(\alpha, \gamma)(\delta, \beta)$. Since G is unimodular by performing the substitution $x \to x^{-1}$ the integral becomes

$$\int_G (\pi(x)\delta, \gamma)\overline{(\pi(x)\beta, \alpha)}\, dx$$

which is $k(\delta, \beta)(\alpha, \gamma)$. Therefore $k(\alpha, \gamma)(\delta, \beta) = k(\delta, \beta)(\alpha, \gamma)$ and so $k(\alpha, \gamma)/\alpha, \gamma)$ is a constant k for every $\alpha, \gamma \in \mathscr{H}$ such that $(\alpha, \gamma) \neq 0$. By choosing $\alpha = \gamma \neq 0$ and $\beta = \delta \neq 0$ we see that $k > 0$. The constant $d = d_\pi = 1/k$ is the formal degree of π.

Proposition 30. *Let G be compact and let χ be the character of the irreducible representation $\pi: G \to \mathscr{U}(\mathscr{H})$. Then $\chi * \chi = (1/d)\chi$ where $d = \chi(e)$ is the degree of π.*

Proof. We choose a MONS, say $\alpha_1, \ldots, \alpha_n$ in $\mathscr{H}$. Then $\chi(x) = \sum (\pi(x)\alpha_i, \alpha_i)$ and

$$\chi * \chi(x) = \sum_{i,j} \int_G (\pi(xy)\alpha_i, \alpha_i)(\pi(y^{-1})\alpha_j, \alpha_j)\, dy\,.$$

The integrand can be written in the form $(\pi(y)\alpha_i, \pi(x)^* \alpha_i)(\alpha_j, \pi(y)\alpha_j)$ and so applying the corollary of Proposition 24 the integral becomes

$$\frac{1}{d}(\alpha_i, \alpha_j)(\alpha_j, \pi(x)^* \alpha_i)\,.$$

Using the orthogonality of the system and summing over $i = j = 1, \ldots, n$ we obtain $(1/d)\chi$.

Proposition 31. *If G is compact and $\rho_k: G \to \mathscr{U}(\mathscr{H}_k)$ $(k = 1, 2)$ are disjoint representations then their characters χ_1, χ_2 satisfy $\chi_1 * \chi_2 = 0$.*

Corollary. *If G is compact and $\omega_1, \omega_2 \in \hat{G}$ are distinct equivalence classes then the corresponding characters χ_1, χ_2 satisfy $\chi_1 * \chi_2 = 0$.*

Proof. Choose a MONS $\alpha_1, \ldots, \alpha_{d_1}$ in $\mathscr{H}_1$ and another $\beta_1, \ldots, \beta_{d_2}$ in $\mathscr{H}_2$. Then

$$\chi_1 * \chi_2(x) = \sum_{i,j} \int_G (\rho_1(xy)\alpha_i, \alpha_i)(\rho_2(y^{-1})\beta_j, \beta_j)\,dy\,.$$

The integrands can be written in the form

$$(\rho_1(y)\alpha_i, \rho_1(x)^*\alpha_i)\overline{(\rho_2(y)\beta_j, \beta_j)}\,.$$

so by Proposition 22 each individual integral is zero.

Theorem 32. *Two irreducible unitary representations of a compact group are equivalent if and only if they have the same character.*

Proof. Let $\rho_k: G \to \mathscr{U}(\mathscr{H}_k)$ $(k=1,2)$ be irreducible representations with characters χ_1 and χ_2, respectively. We know that $\rho_1 \approx \rho_2$ implies $\chi_1 = \chi_2$. Now if ρ_1 and ρ_2 are not equivalent then by the last proposition $\chi_1 * \chi_2 = 0$ while by Proposition 30 $\chi_k * \chi_k = (1/d_k)\chi_k$ $(k=1,2)$. Therefore $\chi_1 = \chi_2$ is not possible.

Let $\rho: S \to \mathscr{L}(\mathscr{H})$ be a representation of the object S and let $\omega \in \hat{S}$. Theorem IV.5.4 states the existence of a uniquely determined subspace $\mathscr{H}_\omega$ of $\mathscr{H}$ which envelops all irreducible components $\mathscr{P}$ of ρ belonging to the class ω. In the special case when S is a compact group G the orthogonal projection $E(\omega): \mathscr{H} \to \mathscr{H}_\omega$ admits an analytic expression. Namely we have:

Theorem 33. *If G is a compact group, $\rho: G \to \mathscr{U}(\mathscr{H})$ is a representation and $\omega \in \hat{G}$ then*

$$E(\omega) = d_\omega \sigma(\overline{\chi}_\omega) = d_\omega \int_G \overline{\chi}_\omega(x)\rho(x)\,dx$$

where χ_ω is the character of ω and d_ω is its degree.

Proof. The operator $E(\omega)$ defined by the integral formula is continuous and using the substitution $x \to x^{-1}$ one can immediately see that it is self adjoint. Since $\chi_\omega * \chi_\omega = (1/d_\omega)\chi_\omega$ we have

$$E(\omega)^2 = d_\omega^2 \sigma(\overline{\chi}_\omega * \overline{\chi}_\omega) = d_\omega \sigma(\overline{\chi}_\omega) = E(\omega)\,.$$

Therefore $E(\omega)$ is a projection. Since $E(\omega)$ is an idempotent its range is a subspace of $\mathscr{H}$ and our object is to show that $E(\omega)\mathscr{H} = \mathscr{H}_\omega$.

Let $\pi: G \to \mathscr{U}(\mathscr{K})$ be an irreducible representation belonging to the class ω and let $\alpha_1, \ldots, \alpha_{d_\omega}$ be a MONS in $\mathscr{K}$. Then the character of the class ω is

$$\chi(x) = \sum (\pi(x)\alpha_i, \alpha_i)\,.$$

Let us suppose that the vector α is in an irreducible component $\mathscr{P}$ of ρ such that $\rho|\mathscr{P}$ belongs to the class ω. We are going to show that $E(\omega)\alpha=\alpha$. First of all $E(\omega)$ belongs to the v. Neumann algebra $\mathscr{A}_\rho$ generated by ρ because by Proposition V.5.6 $\sigma(\overline{\chi}_\omega)\in\mathscr{A}_\rho$. Therefore $E(\omega)\mathscr{P}\subseteq\mathscr{P}$ and in particular $E(\omega)\alpha\in\mathscr{P}$. Hence it is sufficient to prove that $(E(\omega)\alpha,\beta)=(\alpha,\beta)$ for every $\beta\in\mathscr{P}$. Let $U:\mathscr{K}\to\mathscr{P}$ be a unitary intertwining operator in $(\pi,\rho|\mathscr{P})$. By Proposition 24 we have

$$\begin{aligned}(E(\omega)\alpha,\beta)&=d_\omega\sum\int_G \overline{(\pi(x)\alpha_i,\alpha_i)}(\rho(x)\alpha,\beta)\,dx\\&=\sum(\alpha,U\alpha_i)(U\alpha_i,\beta)=(U\alpha,U\beta)=(\alpha,\beta).\end{aligned}$$

This proves that $E(\omega)\mathscr{H}\supseteq\mathscr{H}_\omega$.

By Theorem IV.5.4 we know that $\mathscr{H}$ is the direct sum of the various $\mathscr{H}_\omega$'s and so the opposite inclusion will follow if we prove that $E(\omega_1)E(\omega_2)=0$ for $\omega_1\neq\omega_2$. By the remarks made at the end of Section VI.3 we have

$$\chi_{\overline{\omega}_1}*\chi_{\overline{\omega}_2}=\overline{\chi}_{\omega_1}*\overline{\chi}_{\omega_2}.$$

Hence by Proposition 31

$$\sigma(\overline{\chi}_{\omega_1})\sigma(\overline{\chi}_{\omega_2})=\sigma(\overline{\chi}_{\omega_1}*\overline{\chi}_{\omega_2})=\sigma(0)=0.$$

This shows that $E(\omega_1)E(\omega_2)=0$ and completes the proof.

Let G be a compact group and let $\pi: G\to\mathscr{U}(\mathscr{P})$ be an irreducible representation belonging to the class $\omega\in\hat{G}$. Since π is supposed to be measurable by Theorem V.7.3 its coefficients are continuous on G. We choose a MONS $\alpha_1,\ldots,\alpha_{d_\omega}$ in $\mathscr{P}$ and consider the continuous functions $\varphi_{ij}: G\to\mathbb{C}$ where

$$\varphi^\omega_{i,j}(x)=(\pi(x)\alpha_i,\alpha_j)\qquad (i,j=1,\ldots,d_\omega).$$

They are the entries of the matrix of the linear transformation $\pi(x)$ with respect to the orthonormal basis $\alpha_1,\ldots,\alpha_{d_\omega}$. This explains why the functions $x\mapsto(\pi(x)\alpha,\beta)$ are called the *coefficients* of the representation π.

By the corollary of Proposition 24 the coefficients φ^ω_{ij} form an orthogonal system in $L^2(G)$ and

$$\int_G |\varphi^\omega_{ij}(x)|^2\,dx=\frac{1}{d_\omega}.$$

If ω_1,ω_2 are distinct equivalence classes then by Proposition 22 every $\varphi^{\omega_1}_{ij}$ is orthogonal to every $\varphi^{\omega_2}_{kl}$. Therefore

$$\{\sqrt{d_\omega}\,\varphi^\omega_{ij}\}\qquad(\omega\in\hat{G},\ i,j=1,\ldots,d_\omega)$$

is an ONS of continuous functions in $L^2(G)$.

We are going to prove that the above ONS is maximal. Suppose $f \in L^2(G)$ is orthogonal to φ_{ij}^{ω} for a fixed ω and for every $i, j = 1, \ldots, d_\omega$. Then

$$(\varphi_{ij}^{\omega}, f) = \int_G \overline{f(x)} (\pi(x)\alpha_i, \alpha_j)\, dx = (\sigma(\overline{f})\alpha_i, \alpha_j) = 0 .$$

Since $\sigma(\overline{f})\alpha_i$ is in $\mathscr{P}$ and the α_i's form a MONS in $\mathscr{P}$ it follows that $\sigma(\overline{f})$ is 0 on $\mathscr{P}$. By Theorem 5 $\mathscr{P}$ occurs as an irreducible component in the left regular representation $\lambda\colon G \to \mathscr{U}(L^2(G))$. Restricting λ to $\mathscr{P}$ we obtain a representation equivalent to π and restricting the associated algebra representation σ_λ we obtain an equivalent of our σ. Therefore, if f is orthogonal to every φ_{ij}^{ω} ($\omega \in \hat{G}$ and $i, j = 1, \ldots, d_\omega$) then f is such that $\sigma_\lambda(\overline{f}) = 0$. By Theorem V.5.5 we obtain $\overline{f} * g = 0$ for every $g \in L^2(G)$. Thus the completeness of the ONS will be proved if we establish the following:

Lemma 34. *If G is compact, $f \in L^2(G)$ and if $f * g = 0$ for every $g \in L^2(G)$ then $f = 0$.*

Proof. We may suppose that f is real valued. Then it is sufficient to choose for g the characteristic function of the set $S^{-1} = \{y : f(y^{-1}) > 0\}$. Then by Proposition V.4.13 $f * g$ is continuous and we obtain

$$f * g(e) = \int_G f(y) g(y^{-1})\, dy = \int_S f(y)\, dy .$$

Hence if the hypothesis holds then $f(y) > 0$ only on a set of measure zero. The same type of reasoning shows that $f(y) < 0$ holds only on a null set.

Our conclusions can be summarized as follows:

Theorem 35. *Let G be a compact group and let an irreducible representation $\pi^\omega\colon G \to \mathscr{U}(\mathscr{P}_\omega)$ belonging to the class ω be given for every $\omega \in \hat{G}$. Choose a MONS $\alpha_1^\omega, \ldots, \alpha_{d_\omega}^\omega$ in each $\mathscr{P}_\omega$ and let φ_{ij}^ω denote the coefficients corresponding to this system: $\varphi_{ij}^\omega(x) = (\pi^\omega(x)\alpha_i, \alpha_j)$. Then the functions*

$$\sqrt{d_\omega}\, \varphi_{ij}^\omega \qquad (\omega \in \hat{G} \text{ and } i, j = 1, \ldots, d_\omega)$$

are continuous and they form a MONS in $L^2(G)$.

This important result is usually called the *Péter-Weyl theorem.* It was proved by them in 1927. It can be immediately applied to determine the spectral decomposition of the left regular representation of the compact group G.

Theorem 36. *Every class $\omega \in \hat{G}$ occurs in the left regular representation λ of the compact group G with multiplicity $n(\omega, \lambda) = d_\omega$.*

Proof. One can give an explicit direct sum decomposition of λ in terms of the coefficients φ_{ij}^{ω}, namely one can easily verify that for each fixed i the coefficients $\varphi_{ij}^{\omega}(1 \leqslant j \leqslant d_{\omega})$ generate an irreducible component belonging to the class ω: The invariance of this subspace follows from the identity

$$(1) \qquad \lambda(x)\varphi_{ik}^{\omega} = \sum_j \overline{\varphi_{kj}^{\omega}(x)}\,\varphi_{ij}^{\omega}$$

which holds because

$$\lambda(x)\varphi_{ik}^{\omega}(y) = (\pi^{\omega}(y)\alpha_i, \pi^{\omega}(x)\alpha_k) = \sum_j (\pi^{\omega}(y)\alpha_i, \alpha_j)(\alpha_j, \pi^{\omega}(x)\alpha_k).$$

Instead of showing that the restriction of λ is irreducible and belongs to the class ω we give a second proof of the theorem based on the projection $E(\omega)$ given in Theorem 33. We are going to prove that the dimension of $\mathscr{H}_{\omega} = L^2(G)_{\omega} = E(\omega)L^2(G)$ is d_{ω}^2.

First if $f \in L^2(G)$ then by Theorems 33 and V.5.5

$$(2) \qquad E(\omega)f = d_{\omega}\sigma(\overline{\chi}_{\omega})f = d_{\omega}(\overline{\chi}_{\omega} * f).$$

Next, using (1) with $x=e$, the orthogonality relations and

$$(3) \qquad \chi_{\omega}(x) = \sum \varphi_{rr}^{\omega}(x)$$

one shows in a line or two that

$$(4) \qquad d_{\omega}\overline{\chi_{\omega}} * \overline{\varphi_{ik}^{\omega}} = \overline{\varphi_{ik}^{\omega}} \qquad (i, k = 1, \ldots, d_{\omega}).$$

Therefore $E(\omega)\varphi_{ik}^{\omega} = \varphi_{ik}^{\omega}$ and so φ_{ik}^{ω} is in $\mathscr{H}_{\omega}$ for every $i, k = 1, \ldots, d_{\omega}$. This implies that $\dim \mathscr{H}_{\omega} \geqslant d_{\omega}^2$.

Now if f is an arbitrary element of $L^2(G)$ then by the completeness of the system given in the Péter–Weyl theorem we have

$$f = f_{\omega} + g_n + h_n$$

where

$$f_{\omega} = \sum_{i,k} c_{ik}^{\omega}\overline{\varphi_{ik}^{\omega}},$$

g_n is a finite linear combination of the form

$$g_n = \sum_{\psi \neq \omega}\sum_{i,k} c_{ik}^{\psi}\overline{\varphi_{ik}^{\psi}}$$

and h_n is such that $\|h_n\|_2 \to 0$ as $n \to \infty$. Then by (3), (4) and the orthogonality of the system we have

$$\overline{\chi_\omega} * f = \frac{1}{d_\omega} \sum_{i,k} c_{ik}^\omega \overline{\varphi_{ik}^\omega} + \overline{\chi_\omega} * h_n .$$

Since $\int_G dx = 1$ and $\|\overline{\chi_\omega}\|_\infty \leqslant d_\omega$ we have

$$\|\overline{\chi_\omega} * h_n\|_2 \leqslant d_\omega \|h_n\|_2$$

where $\|h_n\|_2 \to 0$. Therefore

$$\overline{\chi_\omega} * f = \frac{1}{d_\omega} \sum c_{ik}^\omega \overline{\varphi_{ik}^\omega} .$$

Now if $f \in \mathscr{H}_\omega$ then by (2) we see that f is a linear combination of the $\overline{\varphi_{ik}^\omega}$ functions and so $\dim \mathscr{H}_\omega \leqslant d_\omega^2$. We proved that $n(\omega, \lambda) = d_\omega$.

It is worth pointing out that the last theorem is a simple example of the Frobenius reciprocity theorem which was discussed earlier. At present we consider the special case when G is a finite group of say, $n = \operatorname{ord} G$ elements. Then $L^2(G)$ is the vector space of all complex valued functions on G and so its dimension is n. By Theorem 1 the left regular representation λ of G is a direct sum of irreducible components. (Since λ is the induced representation of G when the subgroup H is $\{e\}$ and the representation χ of H is the one dimensional identity we could also apply Theorem VI.6.16.) Hence using the notation of Theorem IV.5.4 we have

$$L^2(G) = \sum_{\omega \in \hat{G}} \oplus L^2(G)_\omega .$$

The dimension of $L^2(G)_\omega$ is $n(\omega) d_\omega$ where $n(\omega)$ is the multiplicity of the class ω in λ and d_ω is its degree. Thus for any finite group G we have

$$\operatorname{ord} G = \sum_{\omega \in \hat{G}} n(\omega) d_\omega .$$

This is actually the simplest, non-trivial example for the general trace formula given in Theorem IV.6.14. If we substitute in accordance with the last theorem d_ω for $n(\omega)$ we obtain

$$\operatorname{ord} G = \sum_{\omega \in \hat{G}} d_\omega^2 .$$

This is a familiar and important relation from the representation theory of finite groups. Here we proved it only in the special case when the field of scalars is $\mathbb{C}$.

As a further application of Theorem 33 we prove the following:

Proposition 37. *If* $\rho: G \to \mathscr{U}(\mathscr{H})$ *is a measurable cyclic representation of the compact group G then for every* $\omega \in \hat{G}$ *we have* $n(\omega, \rho) \leqslant d_\omega$.

Proof. Let α be a cyclic vector of ρ, let $g \in G$ and let $\xi = \xi_g = \rho(g)\alpha$. We use the same symbol ω to denote a representation of G belonging to the class ω. If $\alpha_1, \ldots, \alpha_{d_\omega}$ is a MONS in the Hilbert space of ω then for any x in G we have

$$\begin{aligned}\bar{\chi}_\omega(x g^{-1}) = \chi_\omega(g x^{-1}) = \operatorname{tr} \omega(g)\omega(x)^* &= \sum_i (\omega(g)\omega(x)^* \alpha_i, \alpha_i) \\ &= \sum_{i,j} (\alpha_i, \omega(x)\alpha_j)(\omega(g)\alpha_j, \alpha_i).\end{aligned}$$

Therefore by Theorem 33

$$\begin{aligned}E(\omega)\xi &= d_\omega \int_G \bar{\chi}_\omega(x)\rho(xg)\alpha\, dx = d_\omega \int_G \bar{\chi}_\omega(x g^{-1})\rho(x)\alpha\, dx \\ &= d_\omega \sum_{i,j} (\omega(g)\alpha_j, \alpha_i) \int_G (\alpha_i, \omega(x)\alpha_j)\rho(x)\alpha\, dx.\end{aligned}$$

This shows that $E(\omega)\xi$ is a linear combination of the d_ω^2 vectors occurring on the right hand side. Since they are independent of g we see that the dimension of $E(\omega)\mathscr{H}$ is at most d_ω^2.

A cyclic representation $\sigma: S \to \mathscr{L}(\mathscr{H})$ is called *maximal cyclic* if every other cyclic representation ρ is equivalent to a subrepresentation of σ i.e. $\rho \leqslant \sigma$.

Theorem 38. *Let G be a compact group such that* $L^2(G)$ *is separable. Then the left and right regular representations of G are maximal cyclic.*

Note. The separability hypothesis is satisfied for instance if G satisfies the second axiom of countability.

Proof. If ρ is a cyclic representation of G then $\rho \leqslant \lambda$ by Proposition 37 and Theorems 1 and 36. By Theorem IV.5.11 we know that $d_\omega \pi_\omega$ is cyclic for every ω in $\hat{G}$. Since these primary representations are disjoint λ is cyclic by Theorem IV.5.15.

It is obvious that if G is a locally compact group then the central elements of $L^2(G)$ form a subspace which will be denoted by $L^{o2}(G)$. We have seen that the character χ_ρ of a finite dimensional unitary representation ρ of G is a central function. Hence if G is compact then χ_ρ is in $L^{o2}(G)$. By Theorem 35 the continuous functions χ_ω $(\omega \in \hat{G})$ form an orthonormal system in $L^{o2}(G)$.

Theorem 39. *If G is compact then* χ_ω $(\omega \in \hat{G})$ *is a maximal orthonormal system in* $L^{o2}(G)$.

Proof. Let $f \in L^{o2}(G)$, let $\pi \in \omega \in \hat{G}$ and let α_i $(i \in \mathscr{I})$ be a MONS in the Hilbert space of π. Then

$$(f, \varphi_{ij}^{\omega}) = \int_G f(x)\overline{(\pi(x)\alpha_i, \alpha_j)}\,dx = \int_G f(x)\overline{(\pi(a^{-1}xa)\alpha_i, \alpha_j)}\,dx.$$

Moreover π being unitary

$$\begin{aligned}(\pi(a^{-1}xa)\alpha_i, \alpha_j) &= (\pi(x)\pi(a)\alpha_i, \pi(a)\alpha_j)\\ &= \sum_k (\pi(x)\pi(a)\alpha_i, \alpha_k)(\alpha_k, \pi(a)\alpha_j) = \sum_k \overline{\varphi_{jk}^{\omega}(a)}(\pi(a)\alpha_i, \pi(x)^*\alpha_k).\end{aligned}$$

Since we have

$$(\pi(a)\alpha_i, \pi(x)^*\alpha_k) = \sum_l (\pi(a)\alpha_i, \alpha_l)(\alpha_l, \pi(x)^*\alpha_k) = \sum_l \varphi_{il}^{\omega}(a)\varphi_{lk}^{\omega}(x)$$

we see that

$$(f, \varphi_{ij}^{\omega}) = \sum_{k,l} \varphi_{jk}^{\omega}(a)\overline{\varphi_{il}^{\omega}(a)}(f, \varphi_{lk}^{\omega})$$

where $i,j,k,l = 1, \ldots, d_\omega$. By Theorem 35 the functions $\sqrt{d_\omega}\,\varphi_{jk}^{\omega}$ form a MONS in $L^2(G)$. Hence by integrating the last formula with respect to a we obtain $(f, \varphi_{ij}^{\omega}) = 0$ for every $i \neq j$. Moreover if $i = j$ then

$$d_\omega(f, \varphi_{ii}^{\omega}) = \sum (f, \varphi_{kk}^{\omega}) = (f, \chi_\omega).$$

Hence if $f \perp \chi_\omega$ for every $\omega \in \hat{G}$ then f is orthogonal to every φ_{ij}^{ω} and so $f = 0$ by Theorem 35.

2. Zonal Spherical Functions

In this section we let G be a unimodular group and K a compact subgroup of G. Our main object is to give four definitions of zonal spherical functions and show the equivalence of these. The last theorem will show the connection between zonal spherical functions and representation theory. The unimodularity restriction will be lifted between Definitions 13 and 18.

As usual $C(G)$ denotes the algebra of complex valued, continuous functions on G and $C_0(G)$ is the subalgebra of those functions in $C(G)$ which have compact support. We let $C(K \backslash G)$ denote the set of continuous functions $f: G \to \mathbb{C}$ satisfying $f(kx) = f(x)$ for every $k \in K$ and $x \in G$. Similarly we define $C(G/K)$ as the vector space of right K-invariant, continuous functions f. We let $C(G, K)$ stand for the set of bi-invariant f in $C(G)$ i.e. those f for which $f(k_1 x k_2) = f(x)$ for every $x \in G$ and

$k_1, k_2 \in K$. By Proposition V.4.7 it is clear that if $f \in C(K \backslash G)$ and $g \in C_0(G)$ then $f * g \in C(K \backslash G)$ and similarly $f \in C_0(G)$ and $g \in C(G/K)$ implies that $f * g \in C(G/K)$. Thus $C_0(G, K)$ is a complex convolution algebra.

A function $\omega \in C(K \backslash G)$ is called *right spherical* with respect to the compact subgroup K if $f * \omega = \lambda_f \omega$ for every $f \in C_0(G, K)$. Similarly $\omega \in C(G/K)$ is called a *left spherical function* if $\omega * f = \lambda_f \omega$ for every $f \in C_0(G, K)$.

Definition 1. *A function $\omega: G \to \mathbb{C}$ is called a zonal spherical function with respect to the compact subgroup K if ω belongs to $C(G, K)$, it is right spherical and $\omega(e) = 1$.*

We notice that the normalization $\omega(e) = 1$ is not required for one sided spherical functions and so for instance 0 is both right and left spherical but it is not a zonal spherical function. In the present definition and also in the equivalent characterizations which follow we suppose explicitely that ω is a continuous function. This condition can be relaxed when checking whether a given function $\omega \in Li(G, K)$ is zonal spherical. For in $f * \omega$ the function f can be chosen such that $f \in C_0(G, K)$ and so the continuity of ω follows from Proposition V.4.7. We shall see that a zonal spherical function is necessarily left spherical so we could have required in the definition that ω is both left and right spherical. One of these conditions is superfluous if one knows that $\omega \in C(G, K)$.

If $\omega: G \to \mathbb{C}^\times$ is a continuous group character which takes the value 1 on the compact subgroup K then ω is a zonal spherical function and

$$f * \omega = \omega * f = \omega \int_G f(y^{-1}) \omega(y) \, dy .$$

Therefore zonal spherical functions can be interpreted as generalizations of group characters. By a *quasicharacter* ω we understand a homomorphism of the group G into the multiplicative group $\mathbb{C}^\times$. We see immediately that they are also zonal spherical functions with respect to the subgroup K on which ω is 1.

Lemma 2. *If K is a compact subgroup of G then the neighborhood filter of the identity e has a base consisting of neighborhoods N_e such that $k N_e k^{-1} = N_e$ for every k in K.*

Proof. The reasoning is very similar to that given in the proof of Proposition VII.1.6. Let N be a given neighborhood of e. Choose a symmetric, open neighborhood O of e such that $O^3 \subseteq N$. Then the family Ok $(k \in K)$ covers K and so one can choose $k_1, \ldots, k_n$ in K such that the union of the sets $Ok_1, \ldots, Ok_n$ covers K. By Theorem V.1.1 we can choose a neighborhood C of e such that $k_i C k_i^{-1} \subseteq O$ for every $i=1, \ldots, n$. If k is in K then $k \in O k_i$ for some index i and so

$$k C k^{-1} \subseteq (O k_i) C (O k_i)^{-1} = O(k_i C k_i^{-1}) O \subseteq O^3 \subseteq N .$$

Therefore the union of the sets $k C k^{-1}$ $(k \in K)$ gives a neighborhood N_e satisfying $k N_e k^{-1} = N_e$ and $N_e \subseteq N$.

Lemma 3. *If f is in $C_o(G)$ then*

$$^o f^o(x) = \int_K \int_K f(k_1 x k_2) \, dk_1 \, dk_2$$

defines a function $^o f^o \colon G \to \mathbb{C}$ which belongs to $C_o(G, K)$. If f has compact support C then the support of $^o f^o$ lies in KCK.

Note. It is supposed that the Haar measure of the compact group K is normalized.

Proof. By Lemma V.1.13 f is right uniformly continuous on G so there is a neighborhood N_e such that $|f(u) - f(v)| < \varepsilon$ for every u, v in G satisfying $uv^{-1} \in N_e$. By Lemma 2 we may suppose that $k N_e k^{-1} = N_e$ for every k in K. Then by choosing $u = k_1 x k_2$ and $v = k_1 y k_2$ we obtain $|f(k_1 x k_2) - f(k_1 y k_2)| < \varepsilon$ for any k_1, $k_2 \in K$ and any x, y satisfying $xy^{-1} \in N_e$. The Haar measure of K being normalized we see that $|^o f^o(x) - {}^o f^o(y)| \leq \varepsilon$ for any x and y satisfying $xy^{-1} \in N_e$. We proved that $^o f^o$ is right uniformly continuous on G. We have $f(x) \to 0$ as $x \to \infty$ because given $\varepsilon > 0$ there is a compact set C such that $|f(x)| \leq \varepsilon$ for every $x \notin C$. Hence if $x \notin KCK$ then $|^o f^o(x)| \leq \varepsilon$. If $f \in C_0(G)$ and C is the compact support of f then $f(k_1 x k_2) = 0$ for every $x \notin KCK$ and $k_1, k_2 \in K$. Hence $^o f^o(x) = 0$ if $x \notin KCK$.

Let $Li(G, K)$ be the set of $f \in Li(G)$ satisfying $f(k_1 x k_2) = f(x)$ for $k_1, k_2 \in K$ and $x \in G$. Then $Li(G, K)$ is a linear manifold containing $L^1(G, K)$. If f is in $L^1(G, K)$ then given $\varepsilon > 0$ there is a g in $C_0(G)$ such that $\|f - g\|_1 < \varepsilon$. Then $^o g^o$ is in $C_0(G, K)$ and using the bi-invariance of f we see that $\|f - {}^o g^o\|_1 < \varepsilon$. This proves that $C_0(G, K)$ is a dense linear manifold in $L^1(G, K)$.

Lemma 4. *If K is compact, φ is in $Li(G, K)$ and $\int_G f(x) \varphi(x) \, dx = 0$ for every f in $C_0(G, K)$ then $\varphi = 0$.*

Proof. If the Haar measure of K is normalized then for any f in $C_0(G)$ we have

$$\int_G f(x) \varphi(x) \, dx = \int_G f(x) \Big(\int_K \int_K \varphi(k_1 x k_2) \, dk_1 \, dk_2 \Big) \, dx .$$

By interchanging the order of integration and by letting $k_1 x k_2 \to x$ we obtain

$$\int_G f(x)\varphi(x)\,dx = \int_K \int_K \int_G f(k_1^{-1} x k_2^{-1})\varphi(x)\,dx\,dk_1\,dk_2 = \int_G {}^o f^o(x)\varphi(x)\,dx\,.$$

By Lemma 3 we have ${}^o f^o \in C_0(G, K)$ and so we obtain $\int_G f(x)\varphi(x)\,dx = 0$. Since f is arbitrary in $C_0(G)$ we see that $\varphi = 0$.

For any ω in $Li(G)$ and f in $C_0(G)$ we shall use the notation

$$\hat{f}(\omega) = \hat{\omega}(f) = \int_G f(x^{-1})\omega(x)\,dx\,.$$

By evaluating $f * \omega = \lambda_f \omega$ at e we notice that $\lambda_f = \hat{\omega}(f)$ *for any zonal spherical function* ω. Next by using associativity in the product $f * g * \omega$ we obtain $\hat{\omega}(f * g) = \hat{\omega}(f)\hat{\omega}(g)$. Therefore if ω is a zonal spherical function then $f \mapsto \hat{\omega}(f)$ is a homomorphism of the algebra $C_0(G, K)$ onto $\mathbb{C}$. We shall prove that this is a characteristic property.

Lemma 5. *If K is a compact subgroup of G and if φ and ω in $Li(G, K)$ are such that $\hat{\varphi}(f) = c\hat{\omega}(f)$ for every $f \in C_0(G, K)$ then $\varphi = c\omega$ with the same constant c.*

Proof. By hypothesis

$$\int_G f(x^{-1})(\varphi(x) - c\,\omega(x))\,dx = 0$$

for every f in $C_0(G, K)$. Since $\varphi - c\omega$ is in $Li(G, K)$ we can apply the preceeding lemma and obtain $\varphi - c\omega = 0$.

If the domain of the function φ is G we define φ' on G by $\varphi'(x) = \varphi(x^{-1})$.

Proposition 6. *A function φ is right spherical with respect to K if and only if φ' is left spherical with respect to the same subgroup K. The characteristic values of the operators $*\varphi$ and $\varphi' *$ are related by $\lambda_f(\varphi) = \lambda_{f'}(\varphi')$.*

Proof. We have $\varphi \in C(K \backslash G)$ if and only if $\varphi' \in C(G/K)$. Moreover if $f \in C_0(G, K)$ then $f' \in C_0(G, K)$ and $f' * \varphi = \lambda_{f'}\varphi$. Using the substitution $x \to x^{-1}$ we obtain $\varphi' * f = \lambda_{f'}\varphi'$ and conversely.

Proposition 7. *A function ω satisfying $\omega(e) = 1$ is zonal spherical with respect to K if and only if it is both left and right spherical with respect to the same K. If ω is a zonal spherical function then $f * \omega = \omega * f = \hat{\omega}(f)\omega$.*

Proof. If ω is both left and right spherical then $\omega \in C(G,K)$ and $f*\omega=\lambda_f\omega$ for every f in $C_0(G,K)$ and so ω is zonal spherical. Conversely if we suppose that ω is zonal spherical then ω is by definition right spherical and $\omega \in C(G/K)$. Thus we only have to prove that $\omega * f=\lambda_f\omega$ for f in $C_0(G,K)$. By Proposition V.4.7 we have $\omega * f \in C(G)$. Furthermore using the bi-invariance of ω and f under K and the unimodularity of G we obtain

$$(\omega * f)(k_1 x k_2) = \int_G \omega(x k_2 y^{-1}) f(y) dy = \int_G \omega(x y^{-1}) f(y k_2) dy = \omega * f(x).$$

Hence $\omega * f \in C(G,K)$.

By the unimodularity of G we have for any g in $C_0(G,K)$

$$\begin{aligned}\widehat{(\omega * f)}(g) &= \int_G g(x^{-1}) \int_G \omega(y^{-1}) f(yx) dy\, dx = \int_G (f*g)(y)\omega(y^{-1}) dy \\ &= \hat{\omega}(f*g) = \hat{\omega}(f)\hat{\omega}(g).\end{aligned}$$

Therefore by Lemma 5 we have $\omega * f = \hat{\omega}(f)\omega$. This proves that ω is right spherical with the same characteristic value $\hat{\omega}(f)$ as that for the left convolution operator $f*$.

Theorem 8. *If ω is a zonal spherical function with respect to the compact subgroup K of G then $f \mapsto \hat{\omega}(f)$ is a homomorphism of $C_0(G,K)$ onto $\mathbb{C}$. Conversely, if ω is in $Li(G,K)$ and $f \mapsto \hat{\omega}(f)$ is a homomorphism of $C_0(G,K)$ onto $\mathbb{C}$ then ω is almost everywhere equal to a zonal spherical function or $\omega=0$.*

Proof. We have seen already that if ω is a zonal spherical function then $f \mapsto \hat{\omega}(f)$ is a homomorphism.

For any f, g in $C_0(G,K)$ and ω in $Li(G,K)$ we have

$$\hat{g}(f*\omega) = \int_G g(x^{-1}) \int_G f(xy^{-1})\omega(y) dy\, dx = \int_G \left(\int_G g(x^{-1}) f(xy^{-1}) dx\right)\omega(y) dy.$$

The inner integral is $(g*f)(y^{-1})$. Thus if ω has the homomorphism property then $\hat{g}(f*\omega)=\hat{\omega}(g*f)=\hat{\omega}(f)\hat{\omega}(g)$. Keeping f fixed and letting g vary over $C_0(G,K)$ Lemma 5 becomes applicable and we obtain $f*\omega=\hat{\omega}(f)\omega$ almost everywhere. By Proposition V.4.7 $f*\omega$ is continuous and so ω is almost everywhere equal to a continuous function which is right spherical.

We let ω denote this continuous function belonging to $C(G,K)$. Let us first suppose that $\omega(e)=0$. Then by $f*\omega=\lambda_f\omega$ we have

$$\int_G f(x)\omega(x) dx = f' * \omega(e) = \lambda_{f'}\omega(e) = 0$$

for every f in $C_0(G,K)$. Therefore by Lemma 4 we see that ω is identically 0 on G.

Now we suppose that $\omega(e)\neq 0$ and we are going to prove that $\omega(e)=1$ and consequently ω is a zonal spherical function. The first step is the proof of the following statement:

I. *Given a neighborhood N_e of e in G there is a neighborhood C of e in G such that $CkCk^{-1}\subseteq N_e$ for every $k\in K$.*

The proof is similar to the proof of Proposition VII.1.6. By parts 2), 4) and 5) of Theorem V.1.1 there is a neighborhood B of e in G such that B is symmetric and $B^4\subseteq N_e$. The sets $B^i k$ $(k\in K)$ cover K and so K being compact we can choose $k_1,\dots,k_n$ in K such that the union of $Bk_1,\dots,Bk_n$ covers K. By parts 1) and 2) of the same theorem we can find a neighborhood C of e such that $C\subseteq B$ and $k_iCk_i^{-1}\subseteq B$ for every $i=1,\dots,n$. Now given k in K we have $k\in Bk_i$ for some i $(1\leqslant i\leqslant n)$ and so

$$C(kCk^{-1})\subseteq C(Bk_i)C(Bk_i)^{-1}=CB(k_iCk_i^{-1})B\subseteq CB^3\subseteq B^4\subseteq N_e.$$

Hence C satisfies the requirements and I. is proved.

II. *Given $\varepsilon>0$ there is a compact, symmetric neighborhood C of e such that $|\omega(x)-\omega(e)|\leqslant\varepsilon$ for every $k\in K$ and $x\in CkC$.*

To prove this we choose a compact neighborhood N of e in G. Then NKN is compact and so ω is right uniformly continuous on it. Therefore there is a neighborhood N_e of e in G such that

$$|\omega(x)-\omega(y)|\leqslant\varepsilon \quad \text{if } xy^{-1}\in N_e \quad \text{and } x,y\in NKN.$$

Having determined N_e we now choose C according to I. Now if $k\in K$ and $x\in CkC$ then $xk^{-1}\in CkCk^{-1}\subseteq N_e$ and so $|\omega(x)-\omega(k)|\leqslant\varepsilon$. But $\omega(k)=\omega(e)$ by $\omega\in C(G,K)$ and so $|\omega(x)-\omega(e)|\leqslant\varepsilon$.

III. *Given $\varepsilon>0$ there is an f in $C_0(G,K)$ such that*

$$|\hat{\omega}(f)-\omega(e)|\leqslant\varepsilon \quad \text{and } |\hat{\omega}(f*f)-\omega(e)|\leqslant\varepsilon.$$

In order to prove this we first determine the compact neighborhood C of e in G according to II. Then we choose a not identically zero, non-negative, continuous $g\colon G\to[0,\infty)$ with compact support S such that $S\subseteq C$. Then we define $f={}^og^o/\|{}^og^o\|_1$. Then by Lemma 3 f is in $C_0(G,K)$, its support is in KSK and $\|f\|_1=1$.

If y is such that $f(y^{-1})\neq 0$ by the symmetry of C we have $y\in KS^{-1}K\subseteq KCK$ and so $y=k_1xk_2$ where $x\in C$. Applying II. with x and $k=e$ we obtain $|\omega(x)-\omega(e)|\leqslant\varepsilon$. But $\omega(x)=\omega(y)$ by $\omega\in C_0(G,K)$

and so $|\omega(y)-\omega(e)| \leqslant \varepsilon$. Therefore if y is such that $f(y^{-1}) \neq 0$ then

$$\omega(y)=\omega(e)+\varepsilon(y) \quad \text{where } |\varepsilon(y)| \leqslant \varepsilon.$$

If $f(y^{-1})=0$ we let $\varepsilon(y)=\varepsilon$. Then we have

$$\begin{aligned}\hat{\omega}(f) &= \int_G f(y^{-1})\omega(y)\,dy = \int_G f(y^{-1})(\omega(e)+\varepsilon(y))\,dy \\ &= \omega(e) + \int_G f(y^{-1})\varepsilon(y)\,dy.\end{aligned}$$

Here we used $\|f\|_1=1$ and the unimodularity of G. Consequently $|\hat{\omega}(f)-\omega(e)| \leqslant \varepsilon$ by $\|f\|_1=1$.

The proof of the other inequality is similar: First of all $f*f$ is non-negative and so $\|f*f\|_1=1$ by $\|f\|_1=1$. If x and y are such that $f(x^{-1}y)f(y^{-1}) \neq 0$ then by $S \subseteq C$ we have $x^{-1}y \in KCK$ and $y^{-1} \in KCK$. Therefore by the symmetry of C we have $x=k_1c_1kc_2k_2$ where $c_1, c_2 \in C$ and $k_1, k, k_2 \in K$. By $\omega \in C_0(G,K)$ we have $\omega(x)=\omega(c_1kc_2)$ and so $|\omega(x)-\omega(e)| \leqslant \varepsilon$ by II. Therefore if $f(x^{-1}y)f(y^{-1}) \neq 0$ then

$$\omega(x)=\omega(e)+\varepsilon(x) \quad \text{where } |\varepsilon(x)| \leqslant \varepsilon.$$

If x and y are such that $f(x^{-1}y)f(y^{-1})=0$ then we define $\varepsilon(x)=\varepsilon$. Then

$$\begin{aligned}\hat{\omega}(f*f) &= \int_G\int_G f(x^{-1}y)f(y^{-1})\omega(x)\,dx\,dy \\ &= \int_G\int_G f(x^{-1}y)f(y^{-1})(\omega(e)+\varepsilon(x))\,dx\,dy \\ &= \omega(e) + \int_G (f*f)(x^{-1})\varepsilon(x)\,dx.\end{aligned}$$

We used $\|f*f\|_1=1$ and the unimodularity of G. Hence using $\|f*f\|_1=1$ once more we obtain $|\hat{\omega}(f*f)-\omega(e)| \leqslant \varepsilon$. We proved III.

Now it is easy to see that $\omega(e)=1$: In fact by the homomorphism property $\hat{\omega}(f*f)=\hat{\omega}(f)^2$ for every f in $C_0(G,K)$. Thus by III. we obtain $\omega(e)^2=\omega(e)$. Since we supposed that ω is not identically zero $\omega(e) \neq 0$ and so $\omega(e)=1$. The proof of Theorem 8 is completed.

We also proved the following proposition which will be needed later in Section 6.

Proposition 9. *Given a compact neighborhood C of e in G there exists a not identically zero, non-negative function f in $C_0(G,K)$ such that its support lies in KCK.*

Note. One can also suppose that $f(e)>0$ and $\|f\|_1=1$.

Proof. Start with a continuous function $g: G\to[0,\infty)$ such that $g(e)>0$ and the support of g lies in C. Construct f as is described in III. Then f is bi-invariant under K, its support lies in KCK and $\|f\|_1=1$.

Theorem 10. *A function $\omega\in C(G,K)$ is zonal spherical with respect to the compact subgroup K if and only if with normalized dk we have*

$$\int_K \omega(xky)\,dk=\omega(x)\omega(y)$$

for every x,y in G.

Note. The above integral formula is called the *functional equation* of the zonal spherical function ω.

Proof. If $f\in C_0(G,K)$ then by the substitution $x\to xk$ we have

$$\int_G f(x^{-1})\omega(xy)\,dx=\int_G f(k^{-1}x^{-1})\omega(xky)\,dx=\int_G f(x^{-1})\omega(xky)\,dx.$$

Therefore by integrating this equation over K we obtain

$$\begin{aligned}\int_G f(x^{-1})\int_K \omega(xky)\,dk\,dx&=\int_K\int_G f(x^{-1})\omega(xy)\,dx\,dk\\&=\int_G f(x^{-1})\omega(xy)\,dx=(f*\omega)(y).\end{aligned}$$

Hence using the definition of the zonal spherical function ω we have

$$\int_G f(x^{-1})\int_K \omega(xky)\,dk\,dx=\omega(y)\hat{\omega}(f).$$

We keep y fixed, let f vary over $C_0(G,K)$ and apply Lemma 5. We obtain

$$\int_K \omega(xky)\,dk=\omega(y)\omega(x).$$

Conversely, let us now suppose that ω in $C(G,K)$ satisfies the functional equation. Then for any $f\in C_0(G,K)$ we have

$$\begin{aligned}(f*\omega)(x)&=\int_G f(y^{-1})\omega(yx)\,dy=\int_K\int_G f(y^{-1})\omega(yx)\,dy\,dk\\&=\int_K\int_G f(y^{-1})\omega(ykx)\,dy\,dk.\end{aligned}$$

The last equality was obtained by the substitution $y\to yk$ in the inner integral. Thus by interchanging the order of integration and by applying the functional equation we obtain

$$(f*\omega)(x)=\int_G f(y^{-1})\omega(y)\omega(x)\,dy=\hat{\omega}(f)\omega(x).$$

As an application of the functional equation we prove:

Proposition 11. *If ω is a zonal spherical function with respect to the compact subgroup K and if χ is a quasicharacter which is trivial on K then $\chi\omega$ is a zonal spherical function.*

Proof. Clearly $\chi\omega$ is in $C(G,K)$ and

$$\begin{aligned}\int_K \chi\omega(xky)\,dk &= \int_K \chi(x)\chi(k)\chi(y)\omega(xky)\,dk=\chi(x)\chi(y)\int_K \omega(xky)\,dk\\ &=\chi(x)\chi(y)\omega(x)\omega(y).\end{aligned}$$

Thus $\chi\omega$ satisfies the functional equation.

Proposition 12. *If K is a compact normal subgroup of G then ω is a zonal spherical function on G if and only if ω is a continuous quasicharacter of G which is trivial on K.*

Proof. We know that all continuous quasicharacters trivial on K are zonal spherical functions. Conversely let ω belong to $C(G,K)$ and let

$$\int_K \omega(xky)\,dk=\omega(x)\omega(y).$$

Since K is normal $ky=yl$ for some l in K and so $\omega(xky)=\omega(xy)$. Thus $\omega(xy)=\omega(x)\omega(y)$ for every $x,y\in G$.

There is an important connection between unitary group representations $\rho: G\to\mathcal{U}(\mathcal{H})$ and those zonal spherical functions which are coefficients belonging to a special element of the underlying Hilbert space. The representations themselves are also restricted, only the so called class *1* representations define zonal spherical functions. We start by defining positive definite functions and their connection with unitary group representations.

Definition 13. *A not necessarily continuous, complex valued function φ defined on a group G is called positive definite if for every choice of $c_1,\dots,c_n\in\mathbb{C}$ and $x_1,\dots,x_n\in G$ we have*

$$\sum_{i,j} c_i\bar{c}_j\varphi(x_i x_j^{-1})\geqslant 0.$$

We suppose in the applications in this section that G is unimodular and for $f: G\to\mathbb{C}$ we let $f^*(x)=\overline{f(x^{-1})}$. Then $\varphi=f*f^*$ is a positive definite function on G for every $f \in L^2(G)$ because

$$f * f^*(x_i x_j^{-1}) = \int_G f(x_i x_j^{-1} y) \overline{f(y)} dy = \int_G f(x_i y) \overline{f(x_j y)} dy.$$

The unimodularity of G did not enter and the following sequence of results on positive definite functions holds also for arbitrary G.

Proposition 14. *If* $\varphi: G \to \mathbb{C}$ *is a positive definite function then*

1) $\varphi(x) = \overline{\varphi(x^{-1})}$;

2) $|\varphi(x)| \leqslant \varphi(e)$;

3) $|\varphi(x) - \varphi(y)|^2 \leqslant 2\varphi(e) \mathcal{R}e\{\varphi(e) - \varphi(xy^{-1})\}$.

Corollary. *If φ is continuous at e then it is uniformly continuous on G.*

Proof. By choosing a single $c=1$ and a single point $x=e$ in the characterizing inequality we obtain $\varphi(e) \geqslant 0$. To prove 1) choose two points $x_1 = e$ and $x_2 = x$ and two scalars $c_1 = 1$ and $c_2 = c$. Then the inequality becomes

$$(1 + |c|^2)\varphi(e) + c\varphi(x) + \bar{c}\varphi(x^{-1}) \geqslant 0.$$

If we let $c=1$ then by $\varphi(e) \geqslant 0$ we see that $\varphi(x) + \varphi(x^{-1})$ is real. Similarly substituting $c=i$ we obtain $i\{\varphi(x) - \varphi(x^{-1})\} \in \mathbb{R}$. Therefore $\mathcal{I}m\,\varphi(x) + \mathcal{I}m\,\varphi(x^{-1}) = 0$ and $\mathcal{R}e\,\varphi(x) - \mathcal{R}e\,\varphi(x^{-1}) = 0$.

In order to prove 2) given x we choose c in the above inequality such that $c\varphi(x) = -|\varphi(x)|$. By 1) we have $\bar{c}\varphi(x) = \overline{\bar{c}\varphi(x^{-1})} = -|\varphi(x)|$ and so the inequality becomes $(1 + |c|^2)\varphi(e) \geqslant 2|\varphi(x)|$. By $|c| = 1$ we obtain 2).

Finally 3) can be proved by choosing three points $x_1 = e$, $x_2 = x$, $x_3 = y$ and three scalars, namely $c_1 = 1$,

$$c_2 = \frac{\lambda|\varphi(x) - \varphi(y)|}{\varphi(x) - \varphi(y)}$$

where λ is real and $c_3 = -c_2$. Straightforward elementary computation leads from the characterizing inequality to

$$\varphi(e) + 2\lambda|\varphi(x) - \varphi(y)| + 2\lambda^2(\varphi(e) - \mathcal{R}e\,\varphi(xy^{-1})) \geqslant 0.$$

Since this holds for arbitrary real λ the discriminant of the quadratic polynomial in λ must be non-positive. This will give the inequality expressed in 3).

Proposition 15. *If $\rho: G \to \mathscr{U}(\mathscr{H})$ is a unitary representation and α is an arbitrary element of $\mathscr{H}$ then $\varphi(x) = (\rho(x)\alpha, \alpha)$ defines a positive definite function.*

Note. The conclusion is correct also in the case when ρ is not continuous but we are primarily interested in the continuous case. Then by Proposition 14 φ is uniformly continuous.

Proof. Choose $c_1, \ldots, c_n \in \mathbb{C}$ and $x_1, \ldots, x_n \in G$. Then

$$\begin{aligned} \sum c_i \bar{c}_j \varphi(x_i x_j^{-1}) &= \sum c_i \bar{c}_j (\rho(x_i x_j^{-1})\alpha, \alpha) \\ &= (\sum \bar{c}_j \rho(x_j^{-1})\alpha, \sum \bar{c}_i \rho(x_i^{-1})\alpha) \geqslant 0. \end{aligned}$$

Proposition 16. *If $\varphi: G \to \mathbb{C}$ is a positive definite function then there is a cyclic unitary representation $\rho: G \to \mathscr{U}(\mathscr{H})$ such that $\varphi(x) = (\rho(x)\alpha, \alpha)$ where α is a cyclic generator of ρ. If φ is continuous then ρ is weakly continuous.*

Proof. Let $\mathscr{L}$ be the complex linear space of those $f: G \to \mathbb{C}$ for which $f(y) = \sum_x a(x)\varphi(xy)$ and $a(x) \neq 0$ only for finitely many points x in G. In other words let each f be a linear combination of left translates of φ. For f, g in $\mathscr{L}$ we define

$$(f, g) = \sum_{x,y} \varphi(y^{-1}x) a(x) \overline{b(y)}.$$

Then (f,g) is bilinear and $\overline{(f,g)} = (g,f)$ by part 1) of Proposition 12. Since φ is positive definite we have $(f,f) \geqslant 0$. The BCS inequality follows by the usual reasoning.

Let $\mathscr{N}$ consist of those f in $\mathscr{L}$ for which $(f,f) = 0$. Then $\mathscr{N}$ is a linear manifold and by the BCS inequality $(f,n) = (n,f) = 0$ for f in $\mathscr{L}$ and n in $\mathscr{N}$. It follows that $(f+m, g+n) = (f,g)$ for every $f, g \in \mathscr{L}$ and $m, n \in \mathscr{N}$. Therefore $\mathscr{L}/\mathscr{N}$ is an inner product space with respect to the inner product

$$(\alpha, \beta) = (a + \mathscr{N}, b + \mathscr{N}) = (a, b)$$

where $a \in \alpha$, $b \in \beta$ and $\alpha, \beta \in \mathscr{L}/\mathscr{N}$. We let $\mathscr{H}$ denote the completion of $\mathscr{L}/\mathscr{N}$.

For any a in G define

$$\rho(t) f(y) = \sum_x a(x) \varphi(txy) = \sum_x a(t^{-1}x) \varphi(xy)$$

so that $\rho(t)\colon \mathscr{L}\to\mathscr{L}$. Then ρ is an isometry because

$$(\rho(t)f,\rho(t)g)=\sum_{x,y}\varphi(y^{-1}x)a(t^{-1}x)b\overline{(t^{-1}y)}=\sum_{x,y}\varphi(y^{-1}x)a(x)\overline{b(y)}=(f,g).$$

Therefore $\rho(t)$ can be extended to a unitary operator $\rho(t)\colon \mathscr{H}\to\mathscr{H}$. One sees that the operators $\rho(t)$ $(t\in G)$ form a unitary representation $\rho\colon G\to\mathscr{U}(\mathscr{H})$.

Let $\alpha=d+\mathscr{N}$ where $d\colon G\to\mathbb{C}$ denotes the function defined by the rule $d(y)=\sum a(x)\varphi(xy)$ with $a(x)=0$ for $x\neq e$ and $a(e)=1$. Then α is a cyclic vector for the representation ρ because

$$\sum b(x)\varphi(xy)=\sum b(x)\rho(x)\varphi(y).$$

Moreover by $a(e)=1$ and $a(x)=0$ for $x\neq e$ we have

$$(\rho(t)d,d)=\sum_{x,y}\varphi(y^{-1}x)a(t^{-1}x)\overline{a(y)}=\sum_{x}\varphi(x)a(t^{-1}x)=\varphi(t).$$

Therefore ρ is a unitary cyclic representation such that $\varphi(x)=(\rho(x)\alpha,\alpha)$. It is clear that $(\rho(x)f,g)$ is a finite combination of two-sided translates of φ. Hence if φ is continuous then so is $(\rho(x)f,g)$ for any $f,g\in\mathscr{L}$ and so ρ is a weakly continuous representation.

Note. The proof can be arranged somewhat differently by starting with the linear space of all those functions $f\colon G\to\mathbb{C}$ which satisfy $f(x)\neq 0$ only at finitely many points $x\in G$. The inner product is defined by

$$(f,g)=\sum_{x,y}\varphi(y^{-1}x)f(x)\overline{g(y)}$$

and the representation ρ is given by $\rho(t)f(x)=f(t^{-1}x)$. The cyclic vector is $\alpha=f+\mathscr{N}$ where $f(e)=1$ and $f(x)=0$ for all $x\neq e$.

Proposition 17. *Let* $\rho_k\colon G\to\mathscr{U}(\mathscr{H}_k)$ $(k=1,2)$ *be cyclic representations such that*

(1) $$(\rho_1(x)\alpha_1,\alpha_1)=(\rho_2(x)\alpha_2,\alpha_2)$$

for every $x\in G$ *where* α_k *is a cyclic generator of* ρ_k. *Then* ρ_1 *and* ρ_2 *are equivalent representations.*

Note. This result is often stated in the following somewhat less precise form: *If two cyclic representations have a common coefficient then they are unitarily equivalent.* A similar result can be proved for involutive algebras.

Proof. Let $\varphi(x)=(\rho_k(x)\alpha_k,\alpha_k)$ $(k=1,2)$. Consider the set of all elements of the form

$$\xi_k=\sum a(x)\,\rho_k(x)\,\alpha_k$$

where as in the last proof $a(x)\neq 0$ only at finitely many $x\in G$. It is obvious that the set is a dense linear manifold in $\mathscr{H}_k$ $(k=1,2)$. We define a linear map between these dense manifolds by letting $U\xi_1=\xi_2$ where ξ_1 and ξ_2 have the same coefficients. Then using (1) one sees that U is well defined. Moreover

$$(\xi_k,\eta_k)=\sum_{x,y}a(x)\overline{b(y)}(\rho_k(x)\alpha_k,\rho_k(y)\alpha_k)=\sum_{x,y}a(x)\overline{b(y)}\,\varphi(y^{-1}x)$$

and so $(\xi_1,\eta_1)=(U\xi_1,U\eta_1)$. Therefore U can be extended to a unitary operator $U:\mathscr{H}_1\to\mathscr{H}_2$. We have

$$U\rho_1(a)\xi_1=U\big(\sum a(x)\,\rho_1(ax)\,\alpha_1\big)=U\big(\sum a(a^{-1}x)\,\rho_1(x)\,\alpha_1\big)=\sum a(a^{-1}x)\,\rho_2(x)\,\alpha_2.$$

Similarly

$$\rho_2(a)\,U\xi_1=\rho_2(a)\big(\sum a(x)\,\rho_2(x)\,\alpha_2\big)=\sum a(x)\,\rho_2(ax)\,\alpha_2=\sum a(a^{-1}x)\,\rho_2(x)\,\alpha_2\,.$$

Therefore U intertwines ρ_1 and ρ_2 and so $\rho_1\approx\rho_2$.

Definition 18. *A representation $\rho: G\to\mathscr{U}(\mathscr{H})$ is called a class $\mathit{1}$ representation with respect to the closed subgroup K if $\rho|K$ contains the one dimensional identity representation.*

Thus if ρ is a representation of class $\mathit{1}$ then there exists a non-zero vector α in $\mathscr{H}$ such that $\rho(k)\alpha=\alpha$ for every $k\in K$. If we consider $\mathit{1}$ as the equivalence class in $\hat{K}$ containing the one dimensional identity representation $\pi: K\to\mathscr{U}(\mathbb{C})$ then ρ is a class $\mathit{1}$ representation if and only if $n(\mathit{1},\rho|K)>0$. This form of the definition readily suggests generalizations to other classes than the identity class $\mathit{1}$. These will be considered in the next section on spherical functions.

From here on in this section we suppose again that the group G is unimodular. We let $\mathscr{H}_{\mathit{1}}$ denote the invariant subspace of $\rho|K$ associated with the irreducible class $\mathit{1}\in\hat{K}$. The existence and properties of $\mathscr{H}_{\mathit{1}}$ are stated in Theorem IV.5.4.

Proposition 19. *If φ is a positive definite zonal spherical function then the representation constructed in the proof of Proposition* 16 *is of class $\mathit{1}$, the cyclic generator α of $\rho: G\to\mathscr{U}(\mathscr{H})$ belongs to $\mathscr{H}_{\mathit{1}}$ and $\varphi(x)=(\rho(x)\alpha,\alpha)$.*

Proof. We let $E=E(\mathbf{1})$ be the projection operator $\mathscr{H}\to\mathscr{H}_{\mathbf{1}}$, fix a in G and compute $(E\rho(t)f,g)$ where f and g are arbitrary. Since $(\rho(kt)f,g)=(\rho(t)f,\rho(k^{-1})g)$ and

$$\rho(t)f(y)=\sum a(t^{-1}x)\varphi(xy) \quad\text{and}\quad \rho(k^{-1})g(y)=\sum b(kx)\varphi(xy)$$

we obtain

$$(\rho(kt)f,g)=\sum_{x,y}\varphi(y^{-1}x)a(t^{-1}x)\overline{b(ky)}=\sum_{x,y}a(x)\overline{b(y)}\varphi(y^{-1}ktx).$$

Now we apply Theorem 1.33 with the group K and $\omega=\mathbf{1}$ so that $d_\omega=1$ and $\chi_\omega=1$. By the functional equation of the zonal spherical function ω we obtain

$$\begin{aligned}(E(\mathbf{1})\rho(t)f,g)=\int_K(\rho(kt)f,g)\,dk&=\sum_{x,y}a(x)\overline{b(y)}\int_K\varphi(y^{-1}ktx)\,dx\\&=\sum_{x,y}a(x)\overline{b(y)}\varphi(y^{-1})\varphi(tx).\end{aligned}$$

Therefore if we let $E=E(\mathbf{1})$ then we proved the following formula:

$$(E\rho(t)f,g)=\sum_{x,y}a(x)\overline{b(y)}\varphi(y^{-1})\varphi(tx).$$

We specialize it by substituting for f the function d determining the cyclic generator α so that $d(y)=\sum a(x)\varphi(xy)$ with $a(e)=1$ and $a(x)=0$ for $x\neq e$. Then we obtain $(E\rho(t)d,g)=\varphi(t)(d,g)$.

An arbitrary $f\in\mathscr{L}$ can be written as $f=\sum a(x)\rho(x)d$ and so by the last formula

$$(Ef,g)=\sum a(x)(E\rho(x)d,g)=(d,g)\sum a(x)\varphi(x)=(d,g)(f,d).$$

Since we now have $(Ef,g)=(f,d)(d,g)$ for every $f,g\in\mathscr{L}$ we proved that $Ef=(f,d)d$ for all $f\in\mathscr{L}$. By $(n,d)=0$ we have $E(f+\mathscr{N})=(f+\mathscr{N},d+\mathscr{N})(d+\mathscr{N})$ i.e. the same relation holds in $\mathscr{L}/\mathscr{N}$ which is dense in $\mathscr{H}$. Substituting $f=d$ we obtain $E\alpha=\alpha$ proving that $n(\mathbf{1},\rho|K)>0$. Thus ρ is a class $\mathbf{1}$ representation.

Proposition 20. *The only K-stable vectors of the representation used in the last proposition are those of the form $c\alpha$ where $c\in\mathbb{C}$. In other words we have $n(\mathbf{1},\rho|K)=1$.*

Proof. We use the notations and the facts established in the proof of Proposition 19. If $\rho(k)\gamma=\gamma$ for every $k\in K$ then

$$(E\gamma,\gamma)=\int_K(\rho(k)\gamma,\gamma)\,dk=\int_K(\gamma,\gamma)\,dk=(\gamma,\gamma)$$

and so $(E\gamma, \gamma)=(\gamma, \gamma)(\alpha, \alpha)$ by $d(e)=1$ where $\alpha=d+\mathcal{N}$ is the cyclic generator. On the other hand we saw that $E\gamma=(\gamma, \alpha)\alpha$ and so $(E\gamma, \gamma) =(\gamma, \alpha)(\alpha, \gamma)=(\gamma, \gamma)(\alpha, \alpha)$. The strong form of the BCS inequality gives $\gamma=c\alpha$. Hence $E\mathcal{H}=\mathcal{H}_1=\mathbb{C}\alpha$ and $\dim \mathcal{H}_1=n(1, \rho|K)=1$.

Proposition 21. *The representation ρ considered in the last two propositions is irreducible.*

Proof. Let $\mathcal{U}$ be a ρ-stable subspace. If $\alpha \in \mathcal{U}$ where $\alpha=d+\mathcal{N}$ is the cyclic generator then $\mathcal{U}=\mathcal{H}$. If $\mathcal{U}$ is a proper subspace then $\mathcal{U} \cap \mathbb{C}\alpha=0$ and by $n(1, \rho|K)=1$ we see that $\rho|\mathcal{U}$ and $\rho|\mathbb{C}\alpha$ are disjoint representations of the subgroup K. Since $\rho|\mathbb{C}\alpha$ is one dimensional it is an irreducible representation of K and so by Proposition IV.5.1 $\mathcal{U} \perp \mathbb{C}\alpha$. We proved that if $\mathcal{U}$ is a stable subspace of ρ and if $\mathcal{U} \neq \mathcal{H}$ then $\mathcal{U} \perp \alpha$.

Now let $\mathcal{K}$ be the subspace generated by the set of all stable subspaces $\mathcal{U} \neq \mathcal{H}$. Hence $\mathcal{K}=\overline{\sum \mathcal{U}}$ where each $\mathcal{U} \neq \mathcal{H}$ is a stable subspace of ρ. Then $\mathcal{K}$ and $\mathcal{K}^{\perp}$ are stable. Since $\mathcal{U} \perp \alpha$ for every $\mathcal{U} \neq \mathcal{H}$ we have $\mathcal{K} \perp \alpha$ and so $\alpha \in \mathcal{K}^{\perp}$. But in the beginning of the proof we saw that $\alpha \in \mathcal{K}^{\perp}$ implies $\mathcal{K}^{\perp}=\mathcal{H}$. Therefore $\mathcal{K}=0$ proving that 0 and $\mathcal{H}$ are the only stable subspaces of ρ. Thus ρ is irreducible.

The last set of propositions and Proposition 1.26 give the following:

Theorem 22. *A function ω in $C(G, K)$ is a positive definite zonal spherical function if and only if $\omega(x)=(\rho(x)\alpha, \alpha)$ where $\rho: G \to \mathcal{U}(\mathcal{H})$ is a weakly continuous irreducible representation of class 1 satisfying $n(1, \rho|K)=1$ and α is any unit vector of $\mathcal{H}$ such that $\alpha \in \mathcal{H}_1$ i.e. $\rho(k)\alpha=\alpha$ for every $k \in K$.*

Note. It is plain that ω depends only on the class to which the irreducible representation ρ belongs. Therefore we constructed a natural one-to-one correspondence between positive definite zonal spherical functions and equivalence classes of weakly continuous, irreducible unitary representations of class 1 satisfying $n(1, \omega|K)=1$. The zonal spherical function and the class corresponding to it can be denoted by the same symbol ω.

Proof. First let ω be a positive definite zonal spherical function with respect to the compact subgroup K. Then by Propositions 16 and 19 there is a weakly continuous class 1 irreducible representation $\rho: G \to \mathcal{U}(\mathcal{H})$ and a vector α in the subspace $\mathcal{H}_1$ associated with the class 1 in $\hat{K}$ such that $\omega(x)=(\rho(x)\alpha, \alpha)$ for every $x \in G$. Since $\omega(e)=1$ and $\rho(e)=I$ we see that α is a unit vector. By Proposition 20 every other unit vector of $\mathcal{H}_1$ is of the form $\lambda\alpha$ where $\lambda \in \mathbb{C}$ and $|\lambda|=1$. Hence $n(1, \rho|K)=1$ and we have $\omega(x)=(\dot{\rho}(x)\alpha, \alpha)$ for every unit vector α in $\mathcal{H}_1$.

Next we suppose that $\rho: G \to \mathscr{U}(\mathscr{H})$ is a weakly continuous class $\mathit{1}$ irreducible representation such that $n(\mathit{1}, \rho|K)=1$. Let α be a unit vector in $\mathscr{H}_{\mathit{1}}$. Then $\omega(x)=(\rho(x)\alpha, \alpha)$ defines a continuous function on G which is bi-invariant under K because

$$\omega(k_1 x k_2)=(\rho(k_1 x k_2)\alpha, \alpha)=(\rho(x)\rho(k_2)\alpha, \rho(k_1)^* \alpha)=(\rho(x)\alpha, \alpha)=\omega(x)$$

by $\rho(k_2)\alpha=\rho(k_1^{-1})\alpha=\alpha$. By Proposition 15 ω is positive definite. We prove that ω satisfies the functional equation which is stated in Theorem 10. Let x and y in G be given. We are going to apply Proposition 1.26 with the following choices of the data given there: For the compact group we choose our K and we let $\rho|K$ be the unitary representation occurring in the proposition. By hypothesis $\mathscr{H}_{\mathit{1}}$ is an irreducible component of $\rho|K$ which we let to be the component $\mathscr{P}$ of the proposition. Then we can choose for γ and δ the unit vector α belonging to $\mathscr{H}_{\mathit{1}}$. Finally for the vectors α and β of the proposition we pick $\rho(y)\alpha$ and $\rho(x)^*\alpha$ respectively. The degree of $\mathscr{P}=\mathscr{H}_{\mathit{1}}$ being 1 we obtain

$$\int_K (\rho(k)\rho(y)\alpha, \rho(x)^*\alpha)\overline{(\rho(k)\alpha, \alpha)}\,dk=(\rho(y)\alpha, \alpha)(\alpha, \rho(x)^*\alpha).$$

Using $\rho(k)\alpha=\alpha$ and $\|\alpha\|=1$ we get

$$\int_K (\rho(xky)\alpha, \alpha)\,dk=\omega(y)\omega(x)$$

where $(\rho(xky)\alpha, \alpha)=\omega(xky)$. Hence ω satisfies the functional equation and so it is a zonal spherical function by Theorem 10.

Let $\rho: G \to \mathscr{U}(\mathscr{H})$ and let $\mathscr{H}_{\mathit{1}}$ be the subspace associated with the class $\mathit{1} \in \hat{K}$ and the restricted representation $\rho|K$. Then by Theorem IV.5.4 we have $\alpha \in \mathscr{H}_{\mathit{1}}$ for every α satisfying $\rho(k)\alpha=\alpha$ for all $k \in K$. Conversely if such an α is given then using the definition of $\mathscr{H}_{\mathit{1}}$ one sees immediately that $\alpha \in \mathscr{H}_{\mathit{1}}$. Therefore $\mathscr{H}_{\mathit{1}}=\{\alpha: \rho(k)\alpha=\alpha \text{ for every } k \in K\}$.

Lemma 23. *For every $f \in L^1(K \backslash G)$ and every $\rho: G \to \mathscr{U}(\mathscr{H})$ we have $\sigma(f)\mathscr{H} \subseteq \mathscr{H}_{\mathit{1}}$ where $\mathscr{H}_{\mathit{1}}$ is the subspace associated with $\rho|K$ and the class $\mathit{1} \in \hat{K}$.*

Proof. Let α and β in $\mathscr{H}$ be given. Then for any k in K we have

$$\begin{aligned}(\rho(k)\sigma(f)\alpha, \beta)&=(\sigma(f)\alpha, \rho(k)^*\beta)=\int_G f(x)(\rho(x)\alpha, \rho(k)^*\beta)\,dx\\ &=\int_G f(x)(\rho(kx)\alpha, \beta)\,dx=\int_G f(x)(\rho(x)\alpha, \beta)\,dx=(\sigma(f)\alpha, \beta).\end{aligned}$$

Therefore $\rho(k)\sigma(f)\alpha=\sigma(f)\alpha$ and so $\sigma(f)\alpha \in \mathscr{H}_{\mathit{1}}$.

Theorem 24. *If* $\rho: G \to \mathscr{U}(\mathscr{H})$ *is irreducible and* $L^1(G, K)$ *is commutative then* $\dim \mathscr{H}_1 \leqslant 1$.

Corollary 1. *If* $L^1(G, K)$ *is commutative and* ρ *is a class* ***1*** *irreducible unitary representation of G then* $n(\boldsymbol{1}, \rho|K) = 1$.

Corollary 2. *If* $L^1(G, K)$ *is commutative then there is a natural one-to-one correspondence between the positive definite zonal spherical functions with respect to the compact subgroup K and those weakly continuous* ω *in* $\hat{G}$ *which are of class* ***1***.

Proof. We are going to prove that if $\dim \mathscr{H}_1 > 1$ then ρ is reducible. By Lemma 23 $\mathscr{H}_1$ is an invariant subspace of the representation $\tau: L^1(G, K) \to \mathscr{L}(\mathscr{H})$ where $\tau = \sigma | L^1(G, K)$ and σ is the algebra representation associated with ρ. Since $L^1(G, K)$ is commutative by $\dim \mathscr{H}_1 > 1$ and Proposition IV.2.7 $\tau | \mathscr{H}_1$ is reducible, say $\mathscr{H}_1 = \mathscr{K} \oplus \mathscr{L}$, where $\mathscr{K}$ and $\mathscr{L}$ are stable under τ.

We choose an $\alpha \neq 0$ in $\mathscr{K}$ and let $\mathscr{M}$ denote the subspace generated by the vectors $\sigma(f)\alpha$ where $f \in L^1(G)$. Then $\mathscr{M}$ is stable under σ and so by Theorem V.5.9 and Proposition III.5.10 it is a non-zero invariant subspace of ρ. We prove that $\mathscr{L} \perp \mathscr{M}$. This will show that $\mathscr{M}$ is a non-zero, proper ρ-invariant subspace of $\mathscr{H}$ and ρ is reducible. Let $\beta \in \mathscr{L}$ and let $f \in L^1(G)$. Then

$$(\sigma(f)\alpha, \beta) = \int_G f(x)(\rho(x)\alpha, \beta)\,dx = \int_G f(k_1 x k_2)(\rho(k_1 x k_2)\alpha, \beta)\,dx$$

where k_1, k_2 are arbitrary elements of K. By $\alpha, \beta \in \mathscr{H}_1$ we have

$$(\rho(k_1 x k_2)\alpha, \beta) = (\rho(x)\rho(k_2)\alpha, \rho(k_1)^* \beta) = (\rho(x)\alpha, \beta)\,.$$

Therefore from above

$$(\sigma(f)\alpha, \beta) = \int_G f(k_1 x k_2)(\rho(x)\alpha, \beta)\,dx\,.$$

If we integrate with respect to k_1 and k_2 on $K \times K$ then the Haar measure of K being normalized we obtain $(\sigma(f)\alpha, \beta) = (\sigma({}^o f^o)\alpha, \beta)$. Here ${}^o f^o$ is in $L^1(G, K)$ and so by $\alpha \in \mathscr{K}$ and $\beta \in \mathscr{L}$ we obtain $(\sigma(f)\alpha, \beta) = 0$. We proved that $\sigma(f)\alpha \perp \mathscr{L}$ and this implies that $\mathscr{M} \perp \mathscr{L}$.

3. Spherical Functions of Arbitrary Type and Height

An important generalization of positive definite zonal spherical functions was given by Godement and Harish-Chandra. The starting point is the formula $\omega(x) = (\rho(x)\alpha, \alpha)$ described in Theorem 2.22. Thus $\rho: G$

$\to \mathscr{U}(\mathscr{H})$ is an irreducible representation of class $\mathbf{1}$ and α is a unit vector in $\mathscr{H}$ such that $\alpha \in \mathscr{H}_{\mathbf{1}}$ where $\mathscr{H}_{\mathbf{1}}$ is the subspace associated with $\rho|K$ and the class $\mathbf{1} \in \hat{K}$ according to Theorem IV.5.4. The basic idea which leads to the generalization is the realization that $(\rho(x)\alpha, \alpha)$ is the trace of $\rho(x)$ on the subspace $\mathbb{C}\alpha$. We proved that $\mathscr{H}_{\mathbf{1}}$ is one dimensional and so $\mathscr{H}_{\mathbf{1}} = \mathbb{C}\alpha$. Therefore we can say that the trace is taken on a maximal orthogonal direct sum, or what is the same on the union of all subspaces on which $\rho|K$ reduces to the identity representation.

The generalization consists of two stages in both of which $\mathbf{1}$ is replaced by an arbitrary class $\kappa \in \hat{K}$. In the first stage we consider an irreducible unitary representation $\rho: G \to \mathscr{U}(\mathscr{H})$ such that the subspace $\mathscr{H}_\kappa$ of the restricted representation $\rho|K$ is of positive dimension. If ρ satisfies this condition then we say that *ρ is a class κ representation with respect to the closed subgroup K*. Hence ρ is of class κ if and only if $n(\kappa, \rho|K) > 0$. Thus we are dealing with a simple generalization of the concept of class $\mathbf{1}$ representations. It is easy to see that if ρ_1 is of class κ and if $\rho_2 \approx \rho_1$ then ρ_2 is also a class κ representation. Hence it is meaningful to say that a class ω in $\hat{G}$ is of class κ.

Now let G be a locally compact group, K a compact subgroup, $\kappa \in \hat{K}$, and $\omega \in \hat{G}$. We suppose that ω is of class κ and $\mathscr{H}_\kappa$ is finite dimensional. Since d_κ, the degree of κ is finite the last two conditions are equivalent to the inequalities $0 < (\kappa, \omega|K) < \infty$. Let $\rho: G \to \mathscr{U}(\mathscr{H})$ belong to the class ω. Then by hypothesis $\mathscr{H}_\kappa$ is finite dimensional and $\dim \mathscr{H}_\kappa = d_\kappa n(\kappa, \omega|K)$ Following our earlier practice we let $E(\kappa)$ denote the projection $\mathscr{H} \to \mathscr{H}_\kappa$. For every $x \in G$ the operator $E(\kappa)\rho(x)$ has finite rank and so by Proposition II.5.6 it belongs to the trace class. We define

$$\varphi(x) = \operatorname{tr} E(\kappa)\rho(x) \quad (x \in G)$$

and call $\varphi: G \to \mathbb{C}$ the *spherical function associated with the triple* (K, κ, ω).

It is evident that φ depends only on ω and not the individual representation ρ used to determine it. The multiplicity $n(\kappa, \omega|K) = n(\kappa, \rho|K)$ will be called the *height* of φ. If $\kappa = \mathbf{1}$ then by Theorem 2.20 φ has height one so $\mathscr{H}_\kappa$ is one dimensional, say $\mathscr{H}_\kappa = \mathbb{C}\alpha$ where $\|\alpha\| = 1$ and the trace is $\varphi(x) = (\rho(x)\alpha, \alpha)$. Therefore the spherical functions φ associated with the triples (K, κ, ω) are natural generalizations of the positive definite zonal spherical functions studied in the last section.

The trace is a sum of terms $(\rho(x)\alpha_i, \alpha_i)$ which are positive definite by Proposition 2.15. Thus we see that *these spherical functions φ are also positive definite.* Their boundedness can be seen also directly from the definition and we have

$$\|\varphi\|_\infty = d_\kappa n(\kappa, \omega|K).$$

If ρ is a weakly continuous representation then the terms $(\rho(x)\alpha_i, \alpha_i)$ are continuous on G and so φ is a continuous function. Since it is tacitly assumed that ρ is weakly measurable by Theorem V.7.3 φ is also continuous if $\mathscr{H}$ is separable. Using the same theorem we see that the restriction of φ to K is always continuous.

A spherical function φ is not bi-invariant under K unless $\kappa = 1$. If $d_\kappa = 1$ so that κ is a character of K then

$$\varphi(k_1 x k_2) = \kappa(k_1)\varphi(x)\kappa(k_2)$$

for any $k_1, k_2 \in K$ and $x \in G$. We can easily see that every spherical function satisfies the identity $\varphi(kxk^{-1}) = \varphi(x)$ where k and x are arbitrary elements of K and G, respectively. In fact $E(\kappa)$ being the projection on the $\rho|K$-stable subspace $\mathscr{H}_\kappa$ by Proposition III.5.13 commutes with $\rho(k)$ and so $\rho(k)$ being unitary

$$\begin{aligned}\varphi(kxk^{-1}) &= \operatorname{tr} E(\kappa)\rho(kxk^{-1}) = \operatorname{tr} \rho(k)E(\kappa)\rho(x)\rho(k^{-1}) \\ &= \operatorname{tr} E(\kappa)\rho(x) = \varphi(x).\end{aligned}$$

Let $\mathscr{X}$ be a complex Banach space and let $\mathscr{L}(\mathscr{X})$ be the algebra of continuous linear operators together with its various topologies. A set S with an algebraic and topological structure will be called an *admissible object* if these structures do not exceed those of $\mathscr{L}(\mathscr{X})$. For our purposes it is sufficient to consider the special case when S is a locally compact group G. By a morphism $\mu: S \to \mathscr{L}(\mathscr{X})$ we understand a map which commutes with the algebraic structure of S e.g. if $S = G$ then μ is a homomorphism of G into $\mathrm{Gl}(\mathscr{X})$. We can also require that μ be continuous with respect to the topology of S and one of the topologies of $\mathscr{L}(\mathscr{X})$.

A linear manifold or a subspace $\mathscr{Y}$ of $\mathscr{X}$ is called *invariant* if $\mu(s)\mathscr{Y} \subseteq \mathscr{Y}$ for every s in S. A morphism $\mu: S \to \mathscr{L}(\mathscr{X})$ will be called *topologically irreducible* if the only μ-invariant subspaces are 0 and $\mathscr{X}$ itself.

In order to further extend the definition of spherical functions we shall replace the irreducible unitary representation $\rho: G \to \mathscr{U}(\mathscr{H})$ by a *bounded morphism* $\mu: G \to \mathrm{Gl}(\mathscr{X})$ where $\mathscr{X}$ is an arbitrary complex Banach space. The boundedness of μ means the existence of a positive constant M such that $|\mu(x)| \leqslant M$ for every x in G. Using the inequalities $|\mu(x^{-n})| \cdot |\mu(x^n)| \geqslant 1$ and $|\mu(x^n)| \leqslant |\mu(x)|^n$ where $n = 1, 2, \ldots$ we can easily see that if μ bounded then one has $|\mu(x)| \geqslant 1$ for every x in G.

A morphism $\mu: S \to \mathscr{L}(\mathscr{X})$ is said to be of *finite degree* d_μ if $\mathscr{X}$ is of finite dimension. Two morphisms $\mu_i: S \to \mathscr{L}(\mathscr{X}_i)$ $(i = 1, 2)$ of finite degree are called *similar*, in symbols $\mu_1 \simeq \mu_2$ if there exists a linear *bijection*

$U: \mathscr{X}_1 \to \mathscr{X}_2$ such that $\mu_1 U = U \mu_2$. The vector spaces of similar morphisms have the same dimension and so we can define the *degree* of the similarity class. If $\mu_1 \simeq \mu_2$ and μ_1 is irreducible then so is μ_2. Hence it is meaningful to say that a class is irreducible. We shall use ω to denote a similarity class of irreducible morphisms of finite degree.

If G is a group and $\mu_i: G \to \mathrm{Gl}(\mathscr{X}_i)$ $(i=1,2)$ are similar morphisms of finite degree and μ_1 is bounded then so is μ_2. Hence it makes sense to speak about a bounded similarity class of finite dimensional morphisms of a group G.

Let us suppose that S is not only an algebraic object but also a topological space. Let $\mu: S \to \mathscr{L}(\mathscr{X})$ be a morphism of S in the complex Banach space $\mathscr{X}$. Then μ is defined to be strongly continuous provided the function $s \mapsto \mu(s)\xi$ is continuous for every fixed $\xi \in \mathscr{X}$ with respect to the topology given on S and the topology derived from the norm of $\mathscr{X}$. Let $\mu_1 \simeq \mu_2$ and let U be a linear bijection intertwining μ_1 and μ_2. If μ_1 is strongly continuous then $s \mapsto \mu_2(s)\alpha$ is a continuous function because $\mu_2(s)\alpha = U \mu_1(s)(U^{-1}\alpha)$. Therefore a continuous similarity class of finite dimensional morphisms is a meaningful concept when the admissible object S has a topology. We shall use $\Omega_0(S)$ to denote the *set of bounded, continuous and irreducible similarity classes* of morphisms of finite degree of the topological-algebraic object S. A typical element of $\Omega_0(S)$ will be denoted by κ.

If $\mu: S \to \mathscr{L}(\mathscr{X})$ is a morphism of finite degree then its *character* $\chi_\mu: S \to \mathbb{C}$ can be defined by $\chi(s) = \mathrm{tr}\, \mu(s)$ where the trace depends only on the algebraic structure of $\mathscr{X}$. Since similar matrices have the same trace we see that $\mu_1 \simeq \mu_2$ implies $\chi_{\mu_1} = \chi_{\mu_2}$. Thus we can define the *character* χ_κ *of the similarity class* κ.

Every finite dimensional vector space over the complex numbers can be viewed as a Hilbert space and so if $\mu: G \to \mathrm{Gl}(\mathscr{X})$ is a bounded morphism of the compact group G then μ can be considered as a bounded morphism of G into a Hilbert space. Hence by Corollary 1 of Theorem 1.19 we can view μ as a unitary group representation. If $\kappa \in \Omega_0(G)$ then by Theorem 1.32 the unitary irreducible representations contained in κ are all equivalent and if their unitary equivalence class is denoted by ω_κ then obviously $\omega_\kappa \subseteq \kappa$. Since we are dealing with morphisms of finite degree by Theorem V.7.3 the elements of $\hat{G}$ are all continuous. Thus the inclusion relation $\omega_\kappa \subseteq \kappa$ gives a natural one-to-one correspondence between the elements of $\hat{G}$ and those of $\Omega_0(G)$ where G is a compact group.

A morphism $\mu: G \to \mathrm{Gl}(\mathscr{X})$ of a locally compact group G in a complex Banach space $\mathscr{X}$ will be called *weakly measurable* if for every $\xi \in \mathscr{X}$ and every continuous linear functional $\xi^* \in \mathscr{X}^*$ the function $x \mapsto \xi^*(\mu(x)\xi)$ is measurable with respect to the Haar measures of G.

We will need the analogues of Theorems IV.5.4 and IV.5.5 for morphisms $\mu: G \to \mathrm{Gl}(\mathscr{X})$ in the special case when G is a compact group and the unitary equivalence class ω is replaced by a class $\kappa \in \Omega_0(G)$.

Proposition 1. *Let $\mu: G \to Gl(\mathscr{X})$ be a bounded, weakly measurable morphism of the compact group G in the arbitrary complex Banach space $\mathscr{X}$. Let μ be strongly continuous or let $\mathscr{X}$ be reflexive. For κ in $\Omega_0(G)$ define $\mathscr{X}_\kappa$ as the set theoretical union of those μ-invariant finite dimensional irreducible subspaces $\mathscr{P}$ of $\mathscr{X}$ for which $\mu|\mathscr{P}$ belongs to the class κ. Then we have:*

1. *$\mathscr{X}_\kappa$ is a subspace of $\mathscr{X}$.*

2. *$\mathscr{X}_\kappa$ contains every irreducible component $\mathscr{P}$ such that $\mu|\mathscr{P} \in \kappa$.*

3. *If $\xi \in \mathscr{X}_\kappa$ then there is an irreducible component $\mathscr{P}$ such that $\xi \in \mathscr{P}$ and $\mu|\mathscr{P} \in \kappa$.*

4. *If the Haar measure of G is normalized then*

$$E(\kappa) = d_\kappa \sigma(\overline{\chi}_\kappa) = d_\kappa \int_G \overline{\chi}_\kappa(x) \mu(x) dx$$

defines a continuous linear operator which is idempotent and maps $\mathscr{X}$ onto $\mathscr{X}_\kappa$.

5. *We have $\mu(x)E(\kappa) = E(\kappa)\mu(x)$ for every x in G.*

Note. The integral defining $E(\kappa)$ is taken in the strong or weak sense according as μ is strongly continuous or not. Part 5. and Proposition III.5.2 imply that the range of $I - E(\kappa)$ is also μ-stable.

Proof. First we prove without using the reflexivity of $\mathscr{X}$ that $\mathscr{X}_\kappa$ is a linear manifold. The same result will automatically follow also from the later parts of the proof. It is obvious that $\xi \in \mathscr{X}_\kappa$ implies $\lambda\xi \in \mathscr{X}_\kappa$ for every $\lambda \in \mathbb{C}$. If $\xi_1, \xi_2 \in \mathscr{X}_\kappa$ then there are invariant subspaces $\mathscr{P}_1$ and $\mathscr{P}_2$ such that $\xi_i \in \mathscr{P}_i$ and $\mu|\mathscr{P}_i \in \kappa$. Then $\mathscr{P}_1 + \mathscr{P}_2$ is an invariant subspace of μ and so $\mu|\mathscr{P}_1 + \mathscr{P}_2$ is a bounded morphism. Since $\mathscr{P}_1 + \mathscr{P}_2$ is finite dimensional it can be viewed as a complex Hilbert space and so Corollary 1 of Proposition 1.19 can be applied. Hence $\mu|\mathscr{P}_1 + \mathscr{P}_2$ is a unitary representation in Hilbert space under a suitable inner product on $\mathscr{P}_1 + \mathscr{P}_2 = \mathscr{H}$. The restrictions $\mu|\mathscr{P}_i$ $(i=1,2)$ of the original μ are similar morphisms and so they have the same character. Therefore by Theorem 1.32 the restrictions of the representation $\mu: G \to \mathscr{U}(\mathscr{H})$ to the subspaces $\mathscr{P}_1$ and $\mathscr{P}_2$ are equivalent and they belong to the unitary equivalence class contained in κ. Hence by Theorem IV.5.4 we have $\mathscr{P}_1, \mathscr{P}_2 \subseteq \mathscr{H}_{\omega_\kappa}$ where $\mathscr{H}_{\omega_\kappa}$ is the maximal primary component of

$\mu: G \to \mathscr{U}(\mathscr{H})$ associated with ω_κ. It follows that $\mathscr{P}_1 + \mathscr{P}_2 = \mathscr{H} = \mathscr{H}_{\omega_\kappa}$ and so μ is a primary representation. Let $\mathscr{P}$ be an invariant subspace of $\mathscr{H}$ containing the vector $\xi_1 + \xi_2$ of lowest positive dimension. Then $\mathscr{P}$ is an irreducible component containing $\xi_1 + \xi_2$ and $\mu|\mathscr{P} \in \omega_\kappa$ by $\mathscr{H} = \mathscr{H}_{\omega_\kappa}$. The relation $\mu|\mathscr{P} \in \kappa$ remains valid also when we return to the original norm of the Banach space $\mathscr{X}$. Therefore we found an irreducible component $\mathscr{P}$ of the original morphism μ such that $\xi_1 + \xi_2 \in \mathscr{P}$ and $\mu|\mathscr{P} \in \kappa$ and so $\xi_1 + \xi_2 \in \mathscr{X}_\kappa$. We proved that $\mathscr{X}_\kappa$ is a linear manifold.

The class κ contains a strongly continuous unitary representation and so χ_κ is continuous and $|\chi_\kappa(x)| \leqslant d_\kappa$ for every x in G. Since G is compact it follows that $\bar{\chi}_\kappa \in L^1(G)$ and $\|\bar{\chi}_\kappa\|_1 \leqslant d_\kappa$. Hence $\sigma(\bar{\chi}_\kappa)$ is defined in the weak or strong sense by Theorem V.5.2, it belongs to $\mathscr{L}(\mathscr{X})$ and $|\sigma(\bar{\chi}_\kappa)| \leqslant d_\kappa |\mu|$ where $|\mu|$ denotes the norm of the morphism μ. Using Proposition 1.30 we also see that

$$E(\kappa)^2 = d_\kappa^2 \sigma(\bar{\chi}_\kappa)^2 = d_\kappa^2 \sigma(\bar{\chi}_\kappa * \bar{\chi}_\kappa) = d_\kappa \sigma(\bar{\chi}_\kappa) = E(\kappa).$$

Hence by Lemma II.8.7 the range of $E(\kappa)$ is a closed subset of $\mathscr{X}$.

Let $\xi \in \mathscr{X}_\kappa$ be given. Then there is an irreducible component $\mathscr{P}$ of μ such that $\xi \in \mathscr{P}$ and $\mu|\mathscr{P} \in \kappa$. We renorm $\mathscr{P}$ so that $\mu|\mathscr{P}$ becomes an irreducible unitary representation ρ of class ω_κ. By applying Theorem 1.33 we see that the operator

$$E(\omega_\kappa) = d_{\omega_\kappa} \sigma(\bar{\chi}_{\omega_\kappa})$$

obtained from ρ is the identity map $\mathscr{P} \to \mathscr{P}$. Since

$$E(\omega_\kappa) = d_\kappa \sigma(\bar{\chi}_\kappa)|\mathscr{P} = E(\kappa)|\mathscr{P}$$

it follows that $E(\kappa)\xi = \xi$. Hence we proved that $E(\kappa)\mathscr{X} \supseteq \mathscr{X}_\kappa$.

The next essential step is to prove that $\mathscr{X}_\kappa \supseteq E(\kappa)\mathscr{X}$. We start from the identities

$$\text{(1)} \qquad \mu(a)\sigma(\bar{\chi}_\kappa) = \sigma({}^a\bar{\chi}_\kappa) \quad \text{and} \quad \sigma(\bar{\chi}_\kappa)\mu(a) = \sigma(\bar{\chi}_\kappa^a)$$

which hold for every $a \in G$. These relations are obvious when the integral defining σ is taken in the strong sense. Let σ be defined in the weak sense. Then for any $\xi \in \mathscr{X}$ and $\xi^* \in \mathscr{X}^*$ we have

$$\begin{aligned} \xi^*(\sigma({}^a\bar{\chi}_\kappa)\xi) &= \int_G \bar{\chi}_\kappa(a^{-1}x)\xi^*(\mu(x)\xi)\,dx = \int_G \bar{\chi}_\kappa(x)\xi^*(\mu(a)\mu(x)\xi)\,dx \\ &= \int_G \bar{\chi}_\kappa(x)\mu(a)^*\xi^*(\mu(x)\xi)\,dx = \mu(a)^*\xi^*(\sigma(\bar{\chi}_\kappa)\xi) \\ &= \xi^*(\mu(a)\sigma(\bar{\chi}_\kappa)\xi) \end{aligned}$$

and this implies the first identity. The second can be proved similarly.

Given $\xi \in E(\kappa)\mathscr{X}$ by $E(\kappa)^2 = E(\kappa)$ we have $E(\kappa)\xi = \xi$. Let $\pi: G \to \mathscr{U}(\mathscr{H})$ be an irreducible representation belonging to the class κ and let $\alpha_1, \ldots, \alpha_{d_\kappa}$ be a MONS in $\mathscr{H}$. Then for any a in G we have

$$\begin{aligned}\mu(a)\xi &= \mu(a)E(\kappa)\xi = d_\kappa \mu(a)\sigma(\overline{\chi}_\kappa)\xi = d_\kappa \sigma({}^a\overline{\chi}_\kappa)\xi \\ &= d_\kappa \int_G \sum_i (\pi(a^{-1}x)\alpha_i, \alpha_i)\mu(x)\xi\, dx\,.\end{aligned}$$

Since the α_i's form a base in $\mathscr{H}$ we have

$$(\pi(a^{-1}x)\alpha_i, \alpha_i) = (\pi(x)\alpha_i, \pi(a)\alpha_i) = \sum_j c_{ij}(\pi(x)\alpha_i, \alpha_j).$$

Therefore $\mu(a)\xi$ is a linear combination of the d_κ^2 vectors

$$\xi_{ij} = \int_G (\pi(x)\alpha_i, \alpha_j)\mu(x)\xi\, dx$$

where $i, j = 1, \ldots, d_\kappa$ and the integral is interpreted in the weak sense. We proved that the vectors $\mu(a)\xi$ $(a \in G)$ generate a finite dimensional subspace $\mathscr{Y}$ of $\mathscr{X}$.

The subspace $\mathscr{Y}$ is μ-invariant so we can consider the restricted morphism $\mu|\mathscr{Y}$. Since μ is bounded by Corollary 1 of Theorem 1.19 we can introduce an inner product on $\mathscr{Y}$ such that $\mu|\mathscr{Y}$ becomes a unitary representation $\rho: G \to \mathscr{U}(\mathscr{Y})$. By Theorem V.7.3 ρ is continuous. As earlier let ω_κ denote the unitary equivalence class contained in κ. By Theorem 1.33 we have

$$E(\omega_\kappa)\xi = d_{\omega_\kappa}\sigma(\overline{\chi}_{\omega_\kappa})\xi = d_\kappa \sigma(\overline{\chi}_\kappa)\xi = E(\kappa)\xi = \xi\,.$$

Therefore $\xi \in \mathscr{Y}_{\omega_\kappa}$ and so there is a ρ-invariant subspace $\mathscr{P}$ of $\mathscr{Y}$ such that $\xi \in \mathscr{P}$ and $\rho|\mathscr{P} \in \omega_\kappa$. Since ρ and $\mu|\mathscr{Y}$ differ only in the choice of the norm on $\mathscr{Y}$ we see that $\mu|\mathscr{P} \in \omega_\kappa$ and so $\xi \in \mathscr{X}_\kappa$. Since ξ is an arbitrary element of $E(\kappa)\mathscr{X}$ we proved that $E(\kappa)\mathscr{X} \subseteq \mathscr{X}_\kappa$. Part 5. follows from (1) and the fact that χ_κ is a central function. For we have ${}^a\chi_\kappa = \chi_\kappa^a$ for every $a \in G$.

The range of $E(\kappa)$ is already known to be a closed subset of $\mathscr{X}$. Hence having proved $E(\kappa)\mathscr{X} = \mathscr{X}_\kappa$ we see that $\mathscr{X}_\kappa$ is a subspace of $\mathscr{X}$. This proves part 1. of the proposition. Parts 2. and 3. are obvious from the definition of $\mathscr{X}_\kappa$ and are stated explicitely only for the sake of completeness. The proof of Proposition 1 is finished.

Let again $\mu: G \to \mathrm{Gl}(\mathscr{X})$ be a bounded, weakly measurable morphism of the compact group G in the complex Banach space $\mathscr{X}$ and let μ be strongly continuous or let $\mathscr{X}$ be reflexive. Let $\kappa \in \Omega_0(G)$ and let $d_\kappa n(\kappa, \mu)$ be the dimension of the subspace $\mathscr{X}_\kappa$ introduced in Proposition 1. If $n(\kappa,\mu)>0$ then we say that κ *occurs in* μ *with positive multiplicity* and if $n(\kappa,\mu)$ is finite then κ is said to occur in μ with *finite multiplicity*. The non-negative integer $n(\kappa,\mu)$, or the symbol ∞ if $n(\kappa,\mu)$ is an infinite cardinal, will be called the multiplicity of the class κ in μ. If μ_1 and μ_2 are *similar morphisms* i. e. if $U\mu_1=\mu_2 U$ for an invertible bijection $U\in\mathscr{L}(\mathscr{X}_1,\mathscr{X}_2)$ then $(\mathscr{X}_2)_\kappa=U(\mathscr{X}_1)_\kappa$ and so $n(\kappa,\mu_1)=n(\kappa,\mu_2)$. Hence $n(\kappa,\mu)$ depends only on κ and the similarity class to which μ belongs. If $\mathscr{X}$ is a Hilbert space $\mathscr{H}$ and $\mu=\rho$ where ρ is a unitary representation of G in $\mathscr{H}$ then $n(\kappa,\mu)$ coincides with the multiplicity $n(\omega_\kappa,\rho)$ introduced after the proof of Theorem IV.5.5. If κ occurs in μ with positive multiplicity then one says that μ *is a class* κ *morphism*.

We can now return to the definition of spherical functions. We let G be a locally compact group, K a compact subgroup of G and κ a class in $\Omega_0(K)$. We suppose that $\mathscr{X}$ is an arbitrary complex Banach space and $\mu\colon G\to \mathrm{Gl}(\mathscr{X})$ is a bounded, weakly continuous morphism. We also suppose that μ is strongly continuous or $\mathscr{X}$ is reflexive. These notations and hypotheses will be in effect in the remainder of this section. We suppose that κ occurs in the morphism $\mu|K$ with finite multiplicity. Then $\mathscr{X}_\kappa$ is finite dimensional and so by Proposition II.5.15 the trace of $E(\kappa)\mu(x)$ is defined for every $x\in G$. We let

$$\varphi(x)=\operatorname{tr} E(\kappa)\mu(x)\quad (x\in G)$$

and call $\varphi\colon G\to\mathbb{C}$ the *spherical function associated with* K, κ *and* μ. By Proposition II.5.15 the function φ depends only on the similarity class ω containing μ. Therefore φ is uniquely determined by the triple (K,κ,ω) where $n(\kappa,\omega|K)<\infty$.

In the remainder of this section we describe some of the basic properties of these general spherical functions.

Proposition 2. *If* φ *is a spherical function then* $\varphi(kxk^{-1})=\varphi(x)$ *for every* $k\in K$ *and* $x\in G$.

Proof. By part 5. of Proposition 1 $\mu(k)$ commutes with $E(\kappa)$ and so by Proposition II.5.15 we have

$$\varphi(kxk^{-1})=\operatorname{tr}\mu(k)E(\kappa)\mu(x)\mu(k)^{-1}=\operatorname{tr}E(\kappa)\mu(x)=\varphi(x).$$

Although χ_κ, the character of κ is defined only on K nevertheless we can define $f*\chi_\kappa$ and $\chi_\kappa*f$ where $f\colon G\to\mathbb{C}$. For χ_κ defines a measure on K which can be injected into the measure algebra $M(G)$ according to Proposition V.4.23 and its corollary. We can form the convolution of the injected measure with the function f. If $f\in L^p(G)$ where $1\leqslant p<\infty$

then by Proposition V.4.25 we obtain

$$f*\chi_\kappa(x)=\int_K f(xk)\chi_\kappa(k^{-1})\,dk \quad \text{and} \quad \chi_\kappa*f(x)=\int_K \chi_\kappa(k^{-1})f(kx)\,dk$$

almost everywhere.

Proposition 3. *If φ is a spherical function associated with κ then*

$$d_\kappa \chi_\kappa*\varphi = d_\kappa \varphi*\chi_\kappa=\varphi.$$

Proof. For almost every x in G we have

$$\begin{aligned}\chi_\kappa*\varphi(x)&=\int_K \chi_\kappa(k^{-1})\varphi(kx)\,dk=\int_K \operatorname{tr} E(\kappa)\mu(kx)\chi_\kappa(k^{-1})\,dk\\ &=\operatorname{tr}\Big\{E(\kappa)\Big(\int_K \overline{\chi}_\kappa(k)\mu(k)\,dk\Big)\mu(x)\Big\}=d_\kappa \operatorname{tr}\sigma(\overline{\chi}_\kappa)\sigma(\overline{\chi}_\kappa)\mu(x).\end{aligned}$$

Hence by Proposition 1.30 we obtain

$$\chi_\kappa*\varphi(x)=\operatorname{tr}\sigma(\overline{\chi}_\kappa)\mu(x)=d_\kappa^{-1}\operatorname{tr}E(\kappa)\mu(x)=d_\kappa^{-1}\varphi.$$

The other identity is proved similarly.

Proposition 4. *If φ is any spherical function associated with the class κ then*

$$\int_K \varphi(xky)\overline{\chi}_\kappa(k)\,dk=\int_K \varphi(ykx)\overline{\chi}_\kappa(k)\,dk.$$

Proof. We have

$$\begin{aligned}\int_K \varphi(xky)\overline{\chi}_\kappa(k)\,dk&=\int_K \operatorname{tr}E(\kappa)\mu(xky)\overline{\chi}_\kappa(k)\,dk\\ &=\operatorname{tr}\Big\{E(\kappa)\mu(x)\Big(\int_K \mu(k)\overline{\chi}_\kappa(k)\,dk\Big)\mu(y)\Big\}\\ &=d_\kappa^{-1}\operatorname{tr}E(\kappa)\mu(x)E(\kappa)\mu(y).\end{aligned}$$

Similarly one sees that the other integral is equal to

$$d_\kappa^{-1}\operatorname{tr}E(\kappa)\mu(y)E(\kappa)\mu(x).$$

Hence by using Proposition II.5.15 we see that the two integrals are equal.

Proposition 5. *For every spherical function φ we have*

$$\int_K \varphi(kxk^{-1}y)\,dk=\int_K \varphi(kyk^{-1}x)\,dk.$$

Proof. We define

$$\mu^o(x) = \int_K \mu(kxk^{-1})\,dk$$

in the weak sense. Using a substitution one immediately sees that $\mu^o(x)$ commutes with every $\mu(k)$. We have

$$\begin{aligned} I(x,y) &= \int_K \varphi(kxk^{-1}y)\,dk = \int_K \operatorname{tr} E(\kappa)\,\mu(kxk^{-1}y)\,dk \\ &= \operatorname{tr}\left\{E(\kappa)\left(\int_K \mu(k)\mu(x)\mu(k^{-1})\,dk\right)\mu(y)\right\} = \operatorname{tr} E(\kappa)\,\mu^o(x)\,\mu(y). \end{aligned}$$

By part 5. of Proposition 1 $\mu(k)$ commutes with $E(\kappa)$. Therefore

$$\begin{aligned} \operatorname{tr} E(\kappa)\,\mu^o(x)\,\mu(y) &= \operatorname{tr} \mu(k)\,E(\kappa)\,\mu^o(x)\,\mu(y)\,\mu(k^{-1}) \\ &= \operatorname{tr} E(\kappa)\,\mu^o(x)\,\mu(k)\,\mu(y)\,\mu(k^{-1}). \end{aligned}$$

By integrating on K we obtain

$$\operatorname{tr} E(\kappa)\,\mu^o(x)\,\mu(y) = \operatorname{tr} E(\kappa)\,\mu^o(x)\,\mu^o(y).$$

Hence we proved that $I(x,y) = \operatorname{tr} E(\kappa)\,\mu^o(x)\,\mu^o(y)$. Using Proposition II.5.15 we obtain $I(x,y) = \operatorname{tr} \mu^o(y)\,E(\kappa)\,\mu^o(x)$ and so $I(x,y) = I(y,x)$ will follow by proving that $\mu^o(y)$ commutes with $E(\kappa)$. In fact using the substitution $k \to uk$ we obtain

$$\mu^o(y)\,E(\kappa) = d_\kappa \int_K \overline{\chi}_\kappa(u) \int_K \mu(k)\,\mu(y)\,\mu(k^{-1}u)\,dk\,du = E(\kappa)\,\mu^o(y).$$

Let $C_0^o(G)$ denote the convolution algebra of those f in $C_0(G)$ which satisfy $f(kxk^{-1}) = f(x)$ for every $x \in G$ and $k \in K$. Then we have:

Proposition 6. *If G is unimodular and $f \in C_0^o(G)$ then $f * \varphi = \varphi * f$ for every spherical function φ.*

Proof. We have

$$\begin{aligned} f*\varphi(x) &= \int_G f(y^{-1})\,\varphi(yx)\,dy = \int_G \int_K f(k^{-1}y^{-1}k)\,\varphi(yx)\,dk\,dy \\ &= \int_K \int_G f(y^{-1})\,\varphi(kyk^{-1}x)\,dy\,dk. \end{aligned}$$

Hence using Proposition 5 we obtain

$$f*\varphi(x) = \int_G f(y^{-1}) \int_K \varphi(kxk^{-1}y)\,dk\,dy.$$

We interchange the order of integration, then we let $y \to kyk^{-1}$ and use $f(ky^{-1}k^{-1}) = f(y^{-1})$ and $\varphi(kxyk^{-1}) = \varphi(xy)$ to obtain

$$f * \varphi(x) = \int_K \int_G f(y^{-1})\varphi(xy)\,dy\,dk = \int_G f(y^{-1})\varphi(xy)\,dy = \varphi * f(x).$$

In view of Proposition 1 the projection $E(\kappa)$ needed in the definition of the spherical function φ of type κ can be written in the analytic form

$$E(\kappa) = d_\kappa \int_K \overline{\chi}_\kappa(k)\mu(k)\,dk = d_\kappa \sigma(\overline{\chi}_\kappa)$$

where χ_κ is the character of κ and $\sigma(f) = \int_K f(k)\mu(k)\,dk$ for any integrable $f: K \to \mathbb{C}$. The same symbol σ will be used also to denote the operator $\sigma(f) = \int_G f(x)\mu(x)\,dx$ where $f: G \to \mathbb{C}$.

Lemma 7. *For every integrable* $f: G \to \mathbb{C}$ *we have* $E(\kappa)\sigma(f)E(\kappa) = \sigma(g)$ *where* $g = d_\kappa^2(\overline{\chi}_\kappa * f * \overline{\chi}_\kappa)$.

Proof. A somewhat long but completely straight forward computation shows that if $\xi^* \in \mathscr{X}^*$ then both

$$\xi^*(E(\kappa)\sigma(f)E(\kappa)\xi) \quad \text{and} \quad d_\kappa^2\, \xi^*(\sigma(\overline{\chi}_\kappa * f * \overline{\chi}_\kappa)\xi)$$

are equal to

$$d_\kappa^2 \int_G \int_K \int_K f(x)\overline{\chi}_\kappa(k_1)\overline{\chi}_\kappa(k_2)\xi^*(\mu(k_1 x k_2)\xi)\,dk_1\,dk_2\,dx.$$

The following notations will be used: We let $L^1(\kappa)$ denote the convolution algebra of those integrable $f: G \to \mathbb{C}$ which satisfy the identity

$$d_\kappa \overline{\chi}_\kappa * f = d_\kappa\, f * \overline{\chi}_\kappa = f$$

and we let $L^{o1}(\kappa)$ be the subalgebra of those f's which also satisfy the relation $f(kxk^{-1}) = f(x)$ for $k \in K$ and $x \in G$. Similarly $C_0(\kappa)$ and $C_0^o(\kappa)$ will denote the subalgebras of continuous functions with compact supports.

Proposition 8. *If* G *is unimodular then* $L^1(\kappa)$ *is an involutive algebra.*

Proof. The substitution $k \to k^{-1}$ in the integral defining $\overline{\chi}_\kappa * f^*$ shows that

$$d_\kappa \overline{\chi}_\kappa * f^*(x) = d_\kappa \overline{f * \overline{\chi}_\kappa(x^{-1})} = \overline{f(x^{-1})} = f^*.$$

The relation $d_\kappa f^* * \overline{\chi}_\kappa = f^*$ is proved similarly.

If $f \in L^1(\kappa)$ then $d_\kappa^2 \bar{\chi}_\kappa * f * \bar{\chi}_\kappa = f$ and so by Lemma 7

$$E(\kappa)\sigma(f)E(\kappa) = \sigma(f). \tag{2}$$

Since $E(\kappa)$ maps $\mathscr{X}$ onto $\mathscr{X}_\kappa$ we obtain:

Proposition 9. *The subspace $\mathscr{X}_\kappa$ is stable under all the operators $\sigma(f)$ where $f \in L^1(\kappa)$.*

We let $\tau: L^1(\kappa) \to \mathscr{L}(\mathscr{X}_\kappa)$ denote the morphism obtained by first restricting $\sigma: L^1(G) \to \mathscr{L}(\mathscr{X})$ to $L^1(\kappa)$ and then to its invariant subspace $\mathscr{X}_\kappa$. The morphism τ_0 is defined by further restriction to $C_0(\kappa)$.

Lemma 10. *For any f in $L^1(\kappa)$ we have $\tau(f)=0$ if and only if $\sigma(f)=0$.*

Proof. This follows from (2).

Lemma 11. *If $f \in L^1(G)$ then $g = d_\kappa^2(\bar{\chi}_\kappa * f * \bar{\chi}_\kappa)$ belongs to $C(\kappa)$.*

Proof. The continuity of g follows from Proposition V.4.26. The relations $d_\kappa \bar{\chi}_\kappa * g = d_\kappa g * \bar{\chi}_\kappa = g$ can be obtained from Propositions V.4.16 and 1.30.

We recall that a morphism $\mu: S \to \mathscr{L}(\mathscr{X})$ is called completely irreducible if the algebra generated by the operators $\mu(s)$ $(s \in S)$ is dense in $\mathscr{L}(\mathscr{X})$ in the strong sense.

Proposition 12. *If $\mu: G \to \mathrm{Gl}(\mathscr{X})$ is bounded, strongly continuous and completely irreducible or if μ is bounded $\mathscr{X}$ is reflexive and σ is completely irreducible then $\tau_0: C_0(\kappa) \to \mathscr{L}(\mathscr{X}_\kappa)$ is also completely irreducible.*

Proof. Given a continuous linear operator $A: \mathscr{X}_\kappa \to \mathscr{X}_\kappa$ we extend A to $\mathscr{X}$ by defining $A\xi = A E(\kappa)\xi$ for every $\xi \in \mathscr{X}$. Since μ is completely irreducible by Proposition V.5.19 we can find a generalized sequence of continuous functions f_i with compact supports such that $\sigma(f_i) \to A$ in the strong sense. If we let $g_i = d_\kappa^2(\bar{\chi}_\kappa * f_i * \bar{\chi}_\kappa)$ then by Lemma 11 $g_i \in C_0(\kappa)$ and by Proposition 7 $\tau_0(g_i)$ converges strongly to the original A. Thus τ_0 is completely irreducible.

The next important theorem on spherical functions derived from irreducible unitary representations states that up to a unitary equivalence the functions uniquely determine the irreducible representation ρ to which they belong. The result is due to Harish-Chandra and the proof which we present was given by Godement. The following sequence of results is an algebraic preparation of the proof and it leads to the theorem which states that the finite dimensional irreducible representations of an algebra with identity are determined by their traces.

Proposition 13. *If A is an algebra over $\mathbb{C}$ and if $\pi: A \to \mathscr{L}(\mathscr{H})$ is a finite dimensional irreducible representation of A then* $\ker \pi$ *is a maximal ideal of A.*

Note. More generally A can be an algebra over an arbitrary field F and $\mathscr{H}$ can be any finite dimensional vector space over F. However the result will be needed only if $F = \mathbb{C}$.

Proof. The quotient $A/\ker \pi$ is isomorphic to $\pi(A)$. If $\ker \pi$ is not maximal, say $\ker \pi \subset I \subset A$ then $I/\ker \pi$ is a proper ideal in $A/\ker \pi$. Hence $\pi(A)$ has a proper ideal and so $\pi(A) \neq \mathscr{L}(\mathscr{H})$ by the simplicity of $\mathscr{L}(\mathscr{H})$. (See the corollary of Proposition I.1.1.) By Proposition IV.2.6 π is not irreducible.

Theorem 14. *If $\mathscr{H}_1$ and $\mathscr{H}_2$ are vector spaces and $\varphi: \mathscr{L}(\mathscr{H}_1) \to \mathscr{L}(\mathscr{H}_2)$ is a surjective isomorphism then $\mathscr{H}_1$ and $\mathscr{H}_2$ are isomorphic and there is an isomorphism $u: \mathscr{H}_2 \to \mathscr{H}_1$ such that $\varphi(a) = u^{-1} a u$.*

Note. This will be used only in the case when the field of scalars is $\mathbb{C}$ and $\mathscr{H}_1$ and $\mathscr{H}_2$ are finite dimensional. There is a somewhat deeper result concerning the case when φ is not an algebra isomorphism but only a ring isomorphism.

Proof. First we prove:

If I is a non-zero left ideal in $\mathscr{L}(\mathscr{H})$ and if ξ in $\mathscr{H}$ is such that $I\xi \neq 0$ then $I\xi = \mathscr{H}$.

Indeed we can look at the identity mapping $\pi: \mathscr{L}(\mathscr{H}) \to \mathscr{L}(\mathscr{H})$ as a representation of the abstract algebra $\mathscr{L}(\mathscr{H})$ on $\mathscr{H}$, the image of the abstract element a being the linear operator a. The image algebra is the operator algebra $\mathscr{L}(\mathscr{H})$ so π is irreducible by Proposition IV.2.6. By $aI = I$ we see that $I\xi$ is an invariant subspace of π; since $I\xi \neq 0$ it must be $\mathscr{H}$.

For fixed I and ξ let $\chi: I \to \mathscr{H}$ be defined by $\chi(x) = x\xi$. Then I being a left ideal $\ker \chi$ is also a left ideal. Thus if I is a minimal left ideal then the kernel is zero and χ is invertible. For a in $\mathscr{L}(\mathscr{H})$ let $a_r: I \to I$ be defined by $a_r(b) = ab$ so that $a\chi = \chi a_r: I \to \mathscr{H}$ and consequently if I is minimal then $a_r = \chi^{-1} a \chi$.

Now let $\varphi: \mathscr{L}(\mathscr{H}_1) \to \mathscr{L}(\mathscr{H}_2)$ be an isomorphism, let I_1 be a minimal left ideal in $\mathscr{L}(\mathscr{H}_1)$, let $I_2 = \varphi(I_1)$ and let $\chi_k: I_k \to \mathscr{H}_k$ $(k=1,2)$ be the corresponding maps. Given a_k in $\mathscr{L}(\mathscr{H}_k)$ we have

$$a_{kr} = \chi_k^{-1} a_k \chi_k \qquad (k=1,2). \tag{3}$$

If $b_1 \in I_1$ then

$$\varphi(a_{1r}(b_1)) = \varphi(a_1 b_1) = \varphi(a_1)\varphi(b_1) = \varphi(a_1)_r(\varphi(b_1)),$$

and so setting $\varphi(b_1)=x$ we see that

$$\varphi(a_1)_r = \varphi\, a_{1r}\, \varphi^{-1}. \tag{4}$$

From (3) and (4) we obtain

$$\varphi(a_1) = \chi_2\, \varphi(a_1)_r\, \chi_2^{-1} = \chi_2\, \varphi\, a_{1r}\, \varphi^{-1}\chi_2^{-1} = (\chi_2\, \varphi\, \chi_1^{-1})\, a_1 (\chi_1\, \varphi^{-1}\chi_2^{-1}).$$

Here $u=\chi_1\varphi^{-1}\chi_2^{-1}$ is an isomorphism of $\mathscr{H}_2$ onto $\mathscr{H}_1$ which depends only on the minimal ideals I_1 and I_2. Hence we proved that $\varphi(a_1)=u^{-1}a_1 u$ for every $a_1 \in \mathscr{L}(\mathscr{H}_1)$.

Proposition 15. *If* $\pi_k\colon A \to \mathscr{L}(\mathscr{H}_k)$ $(k=1,2)$ *are finite dimensional irreducible representations of the algebra A over* $\mathbb{C}$ *then* π_1 *and* π_2 *are similar if and only if* $\ker \pi_1 = \ker \pi_2$.

Note. The field of scalars F could be arbitrary but the proposition is used only in the case when $F=\mathbb{C}$.

Proof. It is sufficient to see what happens if $\mathscr{H}_1=\mathscr{H}_2=\mathscr{H}$ and $\ker \pi_1 = \ker \pi_2$. Since $A/\ker \pi_k \approx \pi_k(A)$ under the isomorphism $a+\ker \pi_k \mapsto \pi_k(a)$ the equality of the kernels gives an isomorphism of $\pi_1(A)$ onto $\pi_2(A)$ via

$$\pi_1(a) \mapsto a + \ker \pi_1 = a + \ker \pi_2 \mapsto \pi_2(a).$$

Since $\mathscr{H}$ is finite dimensional the weakly closed algebra generated by $\pi_k(A)$ is $\pi_k(A)$ itself which is $\mathscr{L}(\mathscr{H})$ by Proposition IV.2.6 and the irreducibility of π_k. Therefore $\pi_1(A) \to \pi_2(A)$ constructed above is an automorphism of $\mathscr{L}(\mathscr{H})$ onto itself. But by Theorem 14 every such automorphism is an inner one and so $\mathscr{H}$ being finite dimensional $\pi_1 \simeq \pi_2$.

Theorem 16. *Let A be an algebra over* $\mathbb{C}$ *with identity and let* $\pi_1, \ldots, \pi_r$ *be non-zero, finite dimensional, irreducible representations of A no two of which are equivalent. Then the characters* $\chi_1, \ldots, \chi_r$ *of* $\pi_1, \ldots, \pi_r$ *are linearly independent over* $\mathbb{C}$.

Note. By essentially the same method of proof the result can be extended to fields of characteristic zero.

Corollary. *If A is a complex algebra and its irreducible representations* π_1 *and* π_2 *have the same trace then* $\pi_1 \approx \pi_2$.

Proof. Let $\pi_k: A \to \mathscr{L}(\mathscr{H}_k)$, let $\mathscr{H} = \mathscr{H}_1 \oplus \cdots \oplus \mathscr{H}_r$ and let $\pi = \pi_1 \oplus \cdots \oplus \pi_r$ so that $\pi: A \to \mathscr{L}(\mathscr{H})$. Denote by $\mathscr{A} = \pi(A)$, the algebra formed by the operators $\pi(a)$ $(a \in A)$ and let $\rho_k: \mathscr{A} \to \mathscr{L}(\mathscr{H}_k)$ be defined by $\rho_k(\pi(a)) = \pi_k(a)$. The set of operators is the same for both ρ_k and π_k so ρ_k is an irreducible representation of $\mathscr{A}$ on $\mathscr{H}_k$. The kernel of ρ_k is $\pi(\ker \pi_k) = \mathscr{N}_k$ so $\mathscr{N}_k$ is a maximal ideal of $\mathscr{A}$ by Proposition 13. Since $\rho_1, \ldots, \rho_r$ are not equivalent representations by Proposition 15 the kernels $\mathscr{N}_k$ are all distinct.

Since π_k is irreducible $\pi_k(A)$ is the full matrix algebra and so $\pi_k(A)$ is simple. (See Propositions IV.2.6 and I.1.1.) The homomorphism $\rho_k: \mathscr{A} \to \pi_k(\mathscr{A})$ is surjective so the image of a nilpotent left ideal $\mathscr{I}$ under ρ_k is a nilpotent left ideal in the simple algebra $\pi_k(A)$ and so it is 0. Thus if $\pi(a) \in \mathscr{I}$ then $\pi_k(a) = 0$ for $k = 1, \ldots, r$ and so $\pi(a) = 0$. We proved that the only nilpotent left ideal of $\mathscr{A}$ is 0. Hence by Theorem I.7.8 $\mathscr{A}$ is semisimple. Therefore by Theorem I.7.14 $\mathscr{A}$ is a direct sum of simple rings which are two sided ideals and this decomposition is unique:

$$\mathscr{A} = \mathscr{A}_1 \oplus \cdots \oplus \mathscr{A}_s .$$

The trace of an ideal $\mathscr{I}$ of $\mathscr{A}$ on $\mathscr{A}_k$ is either $\mathscr{A}_k$ or 0. Hence the maximal ideals $\mathscr{N}_k$ are each direct sums of all but one of the components $\mathscr{A}_l$ $(l = 1, \ldots, s)$. We arrange the notation such that the summand which is missing from $\mathscr{N}_k$ is $\mathscr{A}_k$ $(k = 1, \ldots, r)$. This is possible because as we remarked already $\mathscr{N}_1, \ldots, \mathscr{N}_r$ are distinct. We also see from this construction that $\mathscr{A}_k \subset \mathscr{N}_l$ if $k \leqslant r$ and $k \neq l$.

Now we choose e_k in A such that $\pi(e_k)$ is the identity of the simple algebra $\mathscr{A}_k$. Since $\pi(e_k) \in \mathscr{A}_k \subset \mathscr{N}_l$ we have $\pi_l(e_k) = 0$ for $k \leqslant r$ and $k \neq l$. On the other hand due to the choice of notation $\pi(1 - e_k) \in \mathscr{N}_k$ and so $\pi_k(1 - e_k) = 0$ for $k \leqslant r$. It follows that $\chi_l(e_k) = 0$ if $k, l \leqslant r$ and $k \neq l$ and also that $\chi_k(e_k) = d_k$ where d_k is the degree of π_k. Now the linear independence of $\chi_1, \ldots, \chi_r$ follows trivially.

Proposition 17. *Let A be an algebra with identity over $\mathbb{C}$, let π and ρ be finite dimensional representations of A and let π be irreducible. Then $\chi_\pi = \chi_\rho$ implies that* $\ker \pi \supseteq \ker \rho$ *i.e. $\pi(x) = 0$ whenever $\rho(x) = 0$.*

Proof. Since ρ is finite dimensional we have $\rho = n_1 \pi_1 + \cdots + n_k \pi_k$ where $\pi_1, \ldots, \pi_k$ are inequivalent irreducible representations and $n_1, \ldots, n_k$ are positive integer multiplicities. By hypothesis

$$\chi_\pi = n_1 \chi_{\pi_1} + \cdots + n_k \chi_{\pi_k}$$

and so by Theorem 16 π is equivalent to π_i for some i.

Theorem 18. *If* $\rho_k: G \to \mathscr{U}(\mathscr{H}_k)$ $(k=1,2)$ *are irreducible representations of the unimodular group G and for some* $\kappa \in \hat{K}$ *the spherical functions corresponding to* ρ_1 *and* ρ_2 *exist and are equal then* $\rho_1 \approx \rho_2$. *Conversely, if* $\rho_1 \approx \rho_2$ *then for every* κ *the associated spherical functions* φ_1 *and* φ_2 *are equal.*

Proof. Let ρ_k $(k=1,2)$ be irreducible representations of G, let $\kappa \in \hat{K}$ and let $\mathscr{H}_\kappa^k$ $(k=1,2)$ be finite dimensional. We let $\varphi_k(x) = \operatorname{tr} E(\kappa)\rho_k(x)$ and suppose that $\varphi_1 = \varphi_2$. By Proposition 12 the representation $\tau(k): C(\kappa) \to \mathscr{L}(\mathscr{H}_k)$ associated with ρ_k is irreducible. For any f in $C(\kappa)$ we have

$$\operatorname{tr}\tau(k)(f) = \operatorname{tr}\sigma_k(f)|\mathscr{H}_k = \operatorname{tr}\int_G f(x)E(\kappa)\rho_k(x)\,dx = \int_G f(x)\varphi_k(x)\,dx\,.$$

Hence by $\varphi_1 = \varphi_2$ we obtain $\operatorname{tr}\tau(1) = \operatorname{tr}\tau(2)$. Since $\tau(k)$ is irreducible by the corollary of Theorem 16 this implies that $\tau(1) \approx \tau(2)$. Let $\mathscr{R}_k$ $(k=1,2)$ be irreducible components of class κ and let α_k $(k=1,2)$ be non-zero vectors in $\mathscr{R}_k$ $(k=1,2)$ which correspond to each other under some unitary equivalence of $\tau(1)$ and $\tau(2)$. We define

$$\theta_k(x) = (\rho_k(x)\alpha_k, \alpha_k) \quad (k=1,2)\,.$$

Then we have

$$\int_G f(x)\theta_k(x)\,dx = \int_G f(x)(\rho_k(x)\alpha_k, \alpha_k)\,dx = (\sigma_k(f)\alpha_k, \alpha_k) = (\tau(k)(f)\alpha_k, \alpha_k)\,.$$

Since $\alpha_2 = U\alpha_1$ where $U\tau(1) = \tau(2)U$ we have

$$(\tau(1)(f)\alpha_1, \alpha_1) = (\tau(2)(f)\alpha_2, \alpha_2)$$

and so we obtain from above

$$\int_G f(x)\theta_1(x)\,dx = \int_G f(x)\theta_2(x)\,dx \tag{5}$$

where $f \in L(\kappa)$ is arbitrary.

Suppose we can prove that

$$d_\kappa \chi_\kappa * \theta_k = d_\kappa \theta_k * \chi_\kappa = \theta_k\,. \tag{6}$$

Then given any f in $C_0(G)$ by Propositions 3 and V.4.7 we have

$$\begin{aligned}\int_G f(x)\theta_k(x)\,dx &= f' * \theta_k(e) = d_\kappa(f' * \chi_\kappa * \theta_k)(e) = d_\kappa(\theta_k * f' * \chi_k)(e)\\ &= d_\kappa^2(\theta_k * \chi_k * f' * \chi_k)(e) = d_\kappa^2(\theta_k * (\chi_k' * f * \chi_k')')(e)\,.\end{aligned}$$

In these last steps we used the unimodularity of G. Since $\chi'_\kappa = \overline{\chi}_\kappa$ we see that $\chi'_\kappa * f * \chi'_\kappa$ belongs to $C(\kappa)$ and so using (5) we obtain

$$\int_G f(x)\theta_1(x)dx = \int_G f(x)\theta_2(x)dx$$

for any $f \in C_0(G)$. Therefore $\theta_1 = \theta_2$ and so $\rho_k : G \to \mathscr{U}(\mathscr{H}_k)$ being irreducible by Proposition 2.17 we have $\rho_1 \approx \rho_2$.

The proof of (6) is based on Theorem 1.35. For the sake of simplicity we drop the subscript k. Let $\beta_1, \ldots, \beta_d$ be a MONS in $\mathscr{P}$ and let $\gamma_1, \ldots, \gamma_e$ be a MONS in any other irreducible representation belonging to a class different from κ. We fix $x \in G$ and let $f(k) = (\rho(kx)\alpha, \alpha)$ for $k \in K$. Then the Fourier coefficients of f corresponding to the base $\psi_{ij}(x) = \sqrt{e}(\rho(x)\gamma_r, \gamma_s)$ $(r, s = 1, \ldots, e)$ are all zero because by Proposition 1.22

$$\int_K (\rho(k)\rho(x)\alpha, \alpha)\overline{(\rho(k)\gamma_r, \gamma_s)}dk = 0\,.$$

Therefore f has a finite Fourier series which converges to $f(e)$ at e because it is convergent in norm and the functions f and $\varphi_{ij}(k) = \sqrt{d}(\rho(k)\beta_i, \beta_j)$ $(i, j = 1, \ldots, d)$ are continuous. We obtain

$$f(e) = (\rho(x)\alpha, \alpha) = \theta(x) = \sum_{i,j=1}^{d} \varphi_{ij}(e) \int_K f(k)\overline{\varphi_{ij}(k)}dk\,.$$

Since $\varphi_{ij}(e) = \sqrt{d}(\beta_i, \beta_j)$ the sum reduces to

$$\theta(x) = d \sum_{i=1}^{d} \int_K (\rho(kx)\alpha, \alpha)\overline{(\rho(k)\beta_i, \beta_i)}dk\,.$$

By introducing θ and χ_κ we obtain

$$\theta(x) = d \int_K \theta(kx)\chi_\kappa(k^{-1})dk = d(\chi_\kappa * \theta)(x)\,.$$

The other relation involving $\theta * \chi_\kappa$ can be proved similarly. Theorem 18 is proved.

Theorem 19. *Let φ be a spherical function of height one derived from an irreducible unitary representation and belonging to the class $\kappa \in \hat{K}$. Then*

$$d_\kappa \int_K \varphi(kxk^{-1}y)dk = \varphi(x)\varphi(y)$$

for any $x, y \in G$ where d_κ denotes the degree of the class κ.

Note. This is called the functional equation of φ. If $\kappa=1$ then $d_\kappa=1$ and φ being left invariant under K the above relation reduces to the functional equation of the zonal spherical function belonging to the compact subgroup K.

Proof. Let $\alpha_i\ (i\in\mathscr{I})$ be a MONS in $\mathscr{H}_\kappa$ and let $\alpha_j\ (j\in\mathscr{J})$ be an extension of α_i to a MONS in $\mathscr{H}$. We fix x, y in G and take a k in K which is momentarily kept fixed. We have

$$(\rho(kxk^{-1}y)\alpha_i, \alpha_i)=\sum_j (\rho(k^{-1})\rho(y)\alpha_i, \rho(x)^*\alpha_j)(\alpha_j, \rho(k)^*\alpha_i).$$

Hence

$$\int_K (\rho(kxk^{-1}y)\alpha_i, \alpha_i)dk=\sum_j \int_K (\rho(k)\alpha_j, \alpha_i)\overline{(\rho(k)\rho(x)^*\alpha_j, \rho(y)\alpha_i)}dk.$$

We notice that the summation can be restricted to those j's which belong to $\mathscr{I}$ because if $j\notin\mathscr{I}$ then $\mathscr{P}^\perp$ being an invariant subspace of $\rho|K$ we have $(\rho(k)\alpha_j, \alpha_i)=0$ for every $k\in K$.

Since φ is of height one $\mathscr{H}_\kappa$ is an irreducible component and it occurs with multiplicity 1. Therefore we can apply Proposition 1.26 with $\mathscr{P}=\mathscr{H}_\kappa$ and $\alpha=\rho(x)^*\alpha_j$, $\beta=\rho(y)\alpha_i$, $\gamma=\alpha_j$, $\delta=\alpha_i$. We obtain

$$\begin{aligned}\int_K (\rho(kxk^{-1}y)\alpha_i, \alpha_i)dk &= d_\kappa^{-1}(\rho(y)\alpha_i, \alpha_i)\sum_{j\in\mathscr{I}}(\rho(x)\alpha_j, \alpha_j)\\ &= \frac{1}{d_\kappa}(\rho(y)\alpha_i, \alpha_i)\varphi(x).\end{aligned}$$

By summing over $i\in\mathscr{I}$ we obtain the functional equation stated above.

In the next section we shall prove that spherical functions of height one derived from certain morphisms $\mu: G\to \mathrm{Gl}(\mathscr{X})$ in arbitrary complex Banach spaces also satisfy the functional equation given in the last theorem. We shall also prove that the functional equation characterizes these functions.

4. Godement's Theorem on the Characterization of Spherical Functions

The purpose of this section is to prove the theorem which is stated below and point out a few consequences including a proof of the functional equation of spherical functions of height one. These results and the proofs presented below are all due to Godement. Doubtless they are the deepest and most important parts of Godement's work on the theory of spherical functions.

As in the last section G *denotes a unimodular group,* K *is a compact subgroup,* $\kappa \in \hat{K}$ *and* d_κ *is the degree of the class* κ. For simplicity we shall use χ to denote the character of the class κ. The height h is the multiplicity $n(\kappa, \mu|K)$ where μ is the morphism determining the spherical function φ. Again we let $L^1(\kappa)$ and $L^{o1}(\kappa)$ denote the involutive subalgebras of $L^1(G)$ which were introduced in Section 3. We let $C_0^o(\kappa) = L^{o1}(\kappa) \cap C_0(G)$.

Let M be a regular maximal left ideal of $L^1(G)$ and let $\mathscr{X}$ be the complex Banach space $L^1(G)/M$ where $\|f+M\| = \text{glb}\{\|f+m\|_1 : m \in M\}$. Since by Theorem V.5.13 M is invariant under left translations a morphism $\mu: G \to \text{Gl}(\mathscr{X})$ can be defined by letting $\mu(a)(f+M) = {}^a f + M$. By Proposition V.4.9 μ is strongly continuous. As we shall later see μ is topologically irreducible and $|\mu(a)| = 1$ for every a in G. A morphism obtained in this manner is called primitive. The reason for this terminology is the fact that the kernel of the algebra morphism associated with μ is a primitive ideal of $L^1(G)$.

Theorem 1. *Let φ be a spherical function of type κ and height h determined by the strongly continuous and completely irreducible morphism $\mu: G \to \text{Gl}(\mathscr{X})$ where $\mathscr{X}$ is a complex Banach space. Then there is an h dimensional irreducible morphism $v^o: L^{o1}(\kappa) \to \mathscr{L}(\mathscr{H})$ such that $\varphi(f) = d_\kappa \operatorname{tr} v^o(f)$ for every f in $L^{o1}(\kappa)$.*

Conversely suppose that $n(\kappa, \mu|K)$ is finite for every primitive morphism μ. Let $\varphi: G \to \mathbb{C}$ be a bounded, measurable function satisfying $d_\kappa \chi * \varphi = \varphi$ and $\varphi(kxk^{-1}) = \varphi(x)$ for every $k \in K$ and $x \in G$. Suppose that there is an h dimensional irreducible morphism $v^o: L^{o1}(\kappa) \to \mathscr{L}(\mathscr{H})$ such that $\varphi(f) = d_\kappa \operatorname{tr} v^o(f)$ for all $f \in L^{o1}(\kappa)$. Then there is a primitive, topologically irreducible morphism $\mu: G \to Gl(\mathscr{X})$ such that the algebra morphism σ is completely irreducible and φ is locally almost everywhere equal to the spherical function of type κ and height h determined by μ.

First we prove the necessity of the condition: We suppose that φ is a spherical function of type κ and height h. Then for every f in $L^{o1}(\kappa)$ we have

$$\sigma(f)\mu(k) = \int_G f(x)\mu(xk)\,dx = \int_G f(kxk^{-1})\mu(kx)\,dx$$
$$= \mu(k)\int_G f(x)\mu(x)\,dx = \mu(k)\sigma(f).$$

Therefore $\sigma(f)$ belongs to $(\mu|K, \mu|K)$ for every f in $L^{o1}(\kappa)$. In the remainder of this necessity proof we shall write $\mu(k)$ and $\mu|K$ instead of $\mu(k)|\mathscr{X}_\kappa$

and $\mu|K|\mathscr{X}_\kappa$, respectively. By hypothesis $\mathscr{X}_\kappa$ is finite dimensional and by Proposition 3.12 $\tau=\sigma|L^1(\kappa)|\mathscr{X}_\kappa$ is irreducible. Hence if $A\in\mathscr{L}(\mathscr{X}_\kappa)$ then $A=\tau(f)$ for some f in $L^1(\kappa)$. Thus if A is in $(\mu|K, \mu|K)$ then

$$A=\int_K \mu(k)A\mu(k)^{-1}dk=\int_K \mu(k)\tau(f)\mu(k)^{-1}dk=\tau(f^o).$$

Hence $A\in\tau(L^{o1}(\kappa))$ and we proved that $(\mu|K, \mu|K)=\tau(L^{o1}(\kappa))$.

Since $\mu|K$ is a direct sum of h copies of a morphism of type κ by Proposition III.6.7 the algebra $(\mu|K, \mu|K)$ consists of all the composite matrices $(c_{ij}I)$ where $1\leqslant i,j\leqslant h$, $c_{ij}\in\mathbb{C}$ and I is the $d_\kappa\times d_\kappa$ identity matrix. Hence the map γ given by $(c_{ij}I)\mapsto(c_{ij})$ is an isomorphism of $(\mu|K, \mu|K)$ onto the algebra $M_h(\mathbb{C})$ of all $h\times h$ complex matrices. Thus the function $v^o=\gamma\tau^o$ is a morphism of $L^{o1}(\kappa)$ onto $M_h(\mathbb{C})$. Therefore v^o is an h dimensional irreducible morphism of $L^{o1}(\kappa)$. For any composite matrix $(c_{ij}I)$ we have

$$\operatorname{tr}(c_{ij}I)=d_\kappa\sum c_{ii}=d_\kappa\operatorname{tr}(c_{ij})$$

and so

$$\varphi(f)=\operatorname{tr}\tau^o(f)=d_\kappa\operatorname{tr}v^o(f).$$

This shows that the condition stated in the theorem is necessary.

We note that the morphism v^o constructed above is such that $\ker v^o=\ker\tau^o$. This fact will be used later, near the end of the proof.

The proof of the sufficiency of the conditions is considerably more involved and for this reason it will be split into a sequence of lemmas and propositions. We shall use the following notations some of which have been introduced earlier:

$$\begin{aligned} f'(x)&=f(x^{-1})\\ f^o(x)&=\int_K f(kxk^{-1})dk\\ f(g)&=\int_G f(x)g(x)dx\\ f*\chi(x)&=\int_K f(xk)\bar{\chi}(k)dk\\ \chi*f(x)&=\int_K f(kx)\bar{\chi}(k)dk. \end{aligned}$$

In the last two expressions χ is the character of the class $\kappa\in\hat{K}$ and $f\colon G\to\mathbb{C}$ and $x\in G$ are such that the integrals in question exist. It is possible to go through the proof by using only these formulas as the definitions of $\chi*f$ and $f*\chi$. However there is a deeper meaning for these expressions, namely by Proposition V.4.24 they give the convolutions of f with the measure obtained from χ by injecting $M(K)$ into $M(G)$. The existence of these convolutions almost everywhere under suitable conditions is stated in Propositions V.4.24 and V.4.25.

We shall also use a number of elementary facts which are collected here:

1) *We have* $f^o = f$ *if and only if* $f(kxk^{-1}) = f(x)$ *for every* x *in* G *and* k *in* K.
2) *If* $f \in L^2(G)$ *and* $g \in L^2(G)$ *or* $f \in Li(G)$ *and* $g \in C_0(G)$ *then* $f(g^o) = f^o(g)$.
3) *If* $f = f^o$ *and if one side of the following equation exists then so does the other side and* $(f * g)^o(x) = f * g^o(x)$. *Similarly if* $g = g^o$ *then* $(f * g)^o(x) = f^o * g(x)$.
4) *If* $f^o = f$ *and either side exists then* $\chi * f(x) = f * \chi(x)$.
5) *If* f *is in* $L^p(G)$ $(1 \leqslant p < \infty)$ *then* $\chi * f = d_\kappa \chi * \chi * f$ *and* $f * \chi = d_\kappa f * \chi * \chi$.
6) *If* $p = q = 1$ *or if the positive exponents* p, q *are such that* $1/p + 1/q = 1$ *then for any* f *in* $L^p(G)$ *and* g *in* $L^q(G)$ *we have* $f(\chi * g) = (\bar{\chi} * f)(g)$ *and* $f(g * \chi) = (f * \bar{\chi})(g)$.
7) *If* f *is in* $L^p(G)$ *and* $d_\kappa(\chi * f) = f$ *then* $d_\kappa(\chi * f * g) = f * g$.
8) *If* f *is in* $L^p(G)$ *and* $d_\kappa(f * \chi) = f$ *then* $d_\kappa(g * f * \chi) = g * f$.
9) *If either side is meaningful then* $(f * g)'(x) = g' * f'(x)$.

We shall now prove these facts:

If $f = f^o$ then given a in K and x in G we have

$$f(axa^{-1}) = \int_K f(kaxa^{-1}k^{-1})\,dk = \int_K f(kxk^{-1})\,dk = f^o(x) = f(x).$$

This proves the first half of 1). The second half is trivial.

In order to prove 2) interchange the order of integration and perform the substitution $x \to k^{-1}xk$.

The proof of 3) is as follows: If $x \in G$ then

$$\int_K f * g(kxk^{-1})\,dk = \int_K \int_G f(kxk^{-1}y)g(y^{-1})\,dy = \int_K \int_G f(kxyk^{-1})g(ky^{-1}k^{-1})\,dy.$$

By interchanging the order of integration and using 1) we see that $(f * g)^o = f * g^o$. Similarly if $g = g^o$ then $(f * g)^o = f^o * g$.

To prove 4) it is sufficient to notice that $f^o = f$ implies $f(kx) = f(xk)$ for every x in G and k in K.

Statement 5) follows from Proposition 1.30 and the associativity of the convolution which can be verified directly. One can also refer to the associativity of the convolution of measures which is the content of Proposition V.4.16.

The proof of 6) is as follows: We have

$$\int_G f(x)(\chi * g)(x)\,dx = \int_G f(x) \int_K g(kx)\overline{\chi}(k)\,dk\,dx = \int_K \int_G f(x) g(kx)\overline{\chi}(k)\,dx\,dk\,.$$

Hence by using the substitution $x \to k^{-1}x$ and interchanging the order of integration we obtain

$$f(\chi * g) = \int_K \int_G f(k^{-1}x) g(x) \chi(k^{-1})\,dx\,dk = \int_G \left(\int_K f(kx)\chi(k)\,dk\right) g(x)\,dx\,.$$

Therefore

$$f(\chi * g) = \int_G (\overline{\chi} * f)(x) g(x)\,dx = \overline{\chi} * f(g)\,.$$

The other identity given in 6) is proved similarly.

Statements 7) and 8) can be immediately derived by using the associativity of the convolution of measures.

Finally 9) is obtained by using the substitutions $y \to xy$ and $y \to x^{-1}y$ in the integrals defining $(f * g)'(x)$ and $g' * f'(x)$, respectively.

Proposition 2. *If* $\theta: G \to \mathbb{C}$ *is in* $L^p(\kappa)$ *for some* p *satisfying* $1 \leqslant p < \infty$ *and* κ *in* $\hat{K}$ *and if* $({}_h\theta_k)^o = 0$ *for every* h *and* k *in* K *then* $\theta = 0$.

Proof. If α and β are in $L^\infty(K)$ then by Proposition V.4.25 $\alpha * \theta * \beta$ exists almost everywhere and it belongs to $L^p(G)$. Let $\mathscr{V}$ be the vector space spanned by the functions $\alpha * \theta * \beta$ where α and β belong to $L^\infty(K)$. By the first hypothesis we have $d_\kappa \overline{\chi} * \theta = d_\kappa \theta * \overline{\chi} = \theta$. Hence we may restrict ourselves to functions α, β satisfying $d_\kappa \alpha * \overline{\chi} = \alpha$ and $d_\kappa \overline{\chi} * \beta = \beta$. Since $\overline{\chi}$ is measurable and bounded by Proposition V.4.6 $\alpha \mapsto \alpha * d_\kappa \overline{\chi}$ defines a self adjoint linear operator $* d_\kappa \overline{\chi}: L^2(K) \to L^2(K)$. This operator has the kernel $d_\kappa \overline{\chi}(y^{-1}x)$ which belongs to $L^2(K \times K)$ and so it is in the H.S. class. Therefore using Theorems II.3.4 and II.4.4 we see that the characteristic subspace corresponding to the characteristic value 1 is finite dimensional. Similarly we can show that the vector space of those β's which satisfy $d_\kappa \overline{\chi} * \beta = \beta$ is finite dimensional. Therefore $\mathscr{V}$ is a finite dimensional vector space over $\mathbb{C}$.

By the first corollary of Proposition 1.19 we can introduce an inner product on $\mathscr{V}$ such that $\rho(h,k)\varphi(x) = \varphi(h^{-1}xk)$ defines a unitary representation $\rho: K \times K \to \mathscr{U}(\mathscr{V})$. Since $(\overline{\chi})^o = \overline{\chi}$ we have $\overline{\chi} * \alpha = \alpha * \overline{\chi}$ and so by the first hypothesis imposed on θ it follows that $d_\kappa^2 \overline{\chi} * \varphi * \overline{\chi} = \varphi$

for every φ in $\mathscr{V}$. By writing out this equation explicitely we obtain

$$(1) \qquad d_\kappa^2 \int_K \int_K \bar{\chi}(h)\chi(k)\rho(h,k)\,dh\,dk = I$$

where $I: \mathscr{V} \to \mathscr{V}$ is the identity operator.

By Proposition VI.3.16 and the remarks made at the end of Section VI.3 $\bar{\chi}(h)\chi(k)$ is the character of the class $\bar{\kappa} \otimes \kappa$ where $\bar{\kappa}$ denotes the class conjugate to κ. By Theorem VI.3.15 $\bar{\kappa} \otimes \kappa$ is an irreducible class and its degree is obviously $d_{\bar{\kappa}\otimes\kappa} = d_\kappa^2$. Therefore the left hand side of (1) is the integral occurring in Theorem 1.33 and so we obtain $E(\bar{\kappa} \otimes \kappa) = I$. Therefore this theorem shows that ρ is a primary representation of the form $\bar{\kappa} \otimes \kappa \oplus \cdots \oplus \bar{\kappa} \otimes \kappa$ and its character is $n(\bar{\kappa} \otimes \kappa, \rho)\bar{\chi}(h)\chi(k)$. Our object is to show that $n(\bar{\kappa} \otimes \kappa, \rho) = 0$.

Let $\pi: K \to \mathscr{U}(\mathscr{H})$ be an irreducible representation belonging to the class κ. Since K is compact $\mathscr{H}$ is finite dimensional by Theorem 1.10. If we let $\bar{\mathscr{H}}$ be the Hilbert space conjugate to $\mathscr{H}$ then $\bar{\pi}: K \to \mathscr{U}(\bar{\mathscr{H}})$ consists of the same operators $\pi(k): \mathscr{H} \to \mathscr{H}$ and when we write $\bar{\pi}(k): \bar{\mathscr{H}} \to \bar{\mathscr{H}}$ the only difference is that we have another multiplication by scalars and another inner product. Obviously we have $\pi(k) I \pi(k)^* = I$ for every k in K. However if we interpret I as a map of $\mathscr{H}$ onto $\bar{\mathscr{H}}$ then the identity I becomes an element of $\bar{\mathscr{H}} \otimes \mathscr{H}$ and our relation will become $(\bar{\pi} \otimes \pi)(k,k) I = I$ which shows that the restriction of $\bar{\pi} \otimes \pi$ to $\{(k,k): k \in K\}$ has a non-zero invariant vector I.

Applying this result with $\mathscr{H} = \mathscr{V}$ and $\pi \in \kappa$ we obtain the following: If $n(\bar{\kappa} \otimes \kappa, \rho) > 0$ then there is a non-zero element φ in $\mathscr{V}$ such that $\rho(k,k)\varphi = \varphi$ for every k in K. In other words we have $\varphi(k^{-1}xk) = \varphi(x)$ for all $k \in K$, $x \in G$ i.e. $\varphi^o = \varphi$. By the hypothesis of the proposition we have $({}_h\theta_k)^o = 0$ for every $h, k \in K$. Using this one can easily verify that $(\alpha * \theta * \beta)^o = 0$ for every α, β in $L^2(K)$: In fact we have

$$(\alpha * \theta * \beta)^o(x) = \int_K \int_K \int_K \alpha(k_1^{-1})\theta(k_1 k^{-1} x k k_2)\beta(k_2^{-1})\,dk_1\,dk_2\,dk$$

for any x in G. In particular we have $\varphi^o = 0$ and this contradicts $\varphi^o = \varphi \neq 0$. Therefore $n(\bar{\kappa} \otimes \kappa, \rho) = 0$ and so $\mathscr{V} = 0$ i.e. $\alpha * \theta * \beta = 0$ for every α, β in $L^2(K)$. By choosing $\alpha = \beta = d_\kappa \bar{\chi}$ we obtain $\alpha * \theta * \beta = \theta = 0$.

Lemma 3. *If $f \in C_o(G)$ then $f^o \in C_o(G)$ and if $f \in C_0(G)$ then $f^o \in C_0(G)$.*

Proof. By Lemma V.1.13 f is left uniformly continuous so given $\varepsilon > 0$ by Lemma 2.2 there is a neighborhood N_e such that $k N_e k^{-1} = N_e$ for every $k \in K$ and $|f(a) - f(b)| < \varepsilon$ for any a, b satisfying $a^{-1}b \in N_e$. Hence $|f(kxk^{-1}) - f(kyk^{-1})| < \varepsilon$ for every x and y in G such that

$x^{-1}y \in N_e$. Therefore f^o is left uniformly continuous. By hypothesis there is a compact set C such that $|f(x)| \leqslant \varepsilon$ for every $x \notin C$. This shows that $|f^o(x)| \leqslant \varepsilon$ for $x \notin KCK$. Moreover if f has the compact support S then KSK is a compact carrier of f^o.

For $1 \leqslant p < \infty$ we let $L^p(\kappa)$ denote the set of those f in $L^p(G)$ which satisfy $\overline{\chi} * f = f * \overline{\chi} = d_\kappa^{-1} f$. One can easily verify the $L^p(\kappa)$ is a subspace of $L^p(G)$.

Proposition 4. *The linear combinations of the left translates under K of the elements of $C_0^o(\kappa)$ are everywhere dense in $L^p(\kappa)$ for every p satisfying $1 \leqslant p < \infty$.*

Proof. First we show that if f is in $C_0^o(\kappa)$ then its left translates $_a f$ $(a \in K)$ are in $L^p(\kappa)$. For using $\chi(a^{-1}ka) = \chi(k)$ we have

$$\begin{aligned}\overline{\chi} * {_a f}(x) &= \int_K f(akx)\chi(k)\,dk = \int_K f(kax)\chi(k)\,dk \\ &= \overline{\chi} * f(ax) = d_\kappa^{-1} f(ax) = d_\kappa^{-1}\, {_a f}(x).\end{aligned}$$

Similarly we can prove that $_a f * \overline{\chi}(x) = d_\kappa^{-1}\, {_a f}(x)$ for every x in K. Since $_a f \in L^p(G)$ we obtain $_a f \in L^p(\kappa)$.

We shall prove that the set of left translates is dense by showing the following: If a continuous linear functional $\alpha: L^p(\kappa) \to \mathbb{C}$ vanishes on the set of left translates $_a f$ then $\alpha = 0$. Since $L^p(\kappa)$ is a subspace of $L^p(G)$ by the Hahn-Banach theorem α can be extended to a continuous linear functional $\alpha: L^p(G) \to \mathbb{C}$. By the Riesz representation theorem

$$\alpha(f) = \int_G f(x)g(x)\,dx = f(g)$$

where $g \in L^q(G)$ and $1/p + 1/q = 1$. We let $\theta = d_\kappa^2(\chi * g * \chi)$ so that $\theta \in L^q(G)$ by Proposition V.4.25. By 6) we have for any f in $L^p(\kappa)$

$$f(\theta) = f(d_\kappa^2 \chi * g * \chi) = d_\kappa^2 \overline{\chi} * f * \overline{\chi}(g) = f(g)$$

and so $\alpha(f) = f(\theta)$. Therefore in the definition of the extended functional α we may replace g by θ. Applying 6) once more we see that

$$\alpha(f) = d_\kappa^2\, \alpha(\overline{\chi} * f * \overline{\chi}) \tag{2}$$

for every f in $C_0(G)$ because

$$\alpha(f) = f(\theta) = f(d_\kappa^2 \chi * \theta * \chi) = d_\kappa^2 \overline{\chi} * f(\theta * \chi) = d_\kappa^2 \overline{\chi} * f * \overline{\chi}(\theta) = d_\kappa^2\, \alpha(\overline{\chi} * f * \overline{\chi}).$$

The remaining reasoning involves integrations which are most easily described by interpreting them as convolutions of measures. In particular it will be helpful to write $\varepsilon_k * f$ instead of $^k f$ where ε_k is the Dirac measure concentrated at the point k of K.

Given f in $C_0(G)$ by (2) we have

$$\alpha(\varepsilon_k * f^o) = d_\kappa^2 \alpha(\overline{\chi} * \varepsilon_k * f^o * \overline{\chi}).$$

We are going to express the integrand in a different form which will show that the integral vanishes. First of all

$$\varepsilon_k * \overline{\chi} = \overline{\chi} * \varepsilon_k$$

because $(\overline{\chi})^o = \overline{\chi}$. Moreover

$$\overline{\chi} * f^o * \overline{\chi} = d_\kappa^{-1} \overline{\chi} * f^o$$

because $\overline{\chi} * f^o \in C_0^o(\kappa)$ by Lemma 3 and Proposition V.4.26 and by $\overline{\chi} * f^o = f^o * \overline{\chi}$ and Proposition 1.30. Therefore

$$\overline{\chi} * \varepsilon_k * f^o * \overline{\chi} = \varepsilon_k * \overline{\chi} * f^o * \overline{\chi} = d_\kappa^{-1} \varepsilon_k * \overline{\chi} * f^o.$$

Since $\overline{\chi} * f^o$ is in $C_0^o(\kappa)$ by the initial hypothesis on α we obtained $\alpha(\varepsilon_k * f^o) = 0$. By $f^{oo} = f^o$ we have $\varepsilon_k * f^o = f^o * \varepsilon_k$ and so

$$\varepsilon_h * f^o * \varepsilon_k = \varepsilon_h * \varepsilon_k * f^o = \varepsilon_{hk} * f^o.$$

This shows that

$$\alpha(\varepsilon_h * f^o * \varepsilon_k) = 0 \tag{3}$$

for every f in $C_0(G)$ and $h, k \in K$.

For any f and g in $C(G)$ and k in K we have $\varepsilon_k * f(g) = f(\varepsilon_{k^{-1}} * g)$ and $f * \varepsilon_k(g) = f(g * \varepsilon_{k^{-1}})$. Now let

$$\beta = (\varepsilon_h * \theta * \varepsilon_k)^o = ({}_h\theta_k)^o$$

where $h, k \in K$. Then by 2) for any f in $C_0(G)$ we have

$$\begin{aligned}\beta(f) &= (\varepsilon_h * \theta * \varepsilon_k)(f^o) = \theta * \varepsilon_k(\varepsilon_{h^{-1}} * f^o) \\ &= \varepsilon_{h^{-1}} * f^o(\theta * \varepsilon_k) = \varepsilon_{h^{-1}} * f^o * \varepsilon_{k^{-1}}(\theta).\end{aligned}$$

Hence by (3)

$$\beta(f) = \alpha(\varepsilon_{h^{-1}} * f^o * \varepsilon_{k^{-1}}) = 0.$$

Since we proved that $\beta(f) = 0$ for every f in $C_0(G)$ we have $\beta = 0$. Therefore Proposition 2 can be applied and we obtain $\theta = 0$. Hence the original linear functional $\alpha: L^p(\kappa) \to \mathbb{C}$ is zero and Proposition 4 is proved.

Lemma 5. *If* $f: G \to \mathbb{C}$ *is a locally integrable function,* $f^o = f$ *and* $d_\kappa \chi * f = d_\kappa f * \chi = f$ *and if* $f(g) = 0$ *for every* g *in* $C_0^o(\kappa)$ *then* $f = 0$ *locally almost everywhere.*

Proof. If $g \in C_0(\kappa)$ then by 2) we have $f(g) = f^o(g) = f(g^o) = 0$ and so we proved that $f(g) = 0$ for every k in $C_0(\kappa)$. Now let $g \in C_0(G)$. Then by 6)

$$f(g) = d_\kappa^2(\chi * f * \chi)(g) = d_\kappa^2 f(\overline{\chi} * g * \overline{\chi})$$

where $\overline{\chi} * g * \overline{\chi} \in C_0(\kappa)$ by Proposition V.4.26. Therefore $f(g) = 0$ for every g in $C_0(G)$ and so $f = 0$ locally almost everywhere.

Corollary. *If f is continuous then the conditions imply that $f = 0$.*

We shall now actually start the proof of the sufficiency of the condition stated in the theorem by supposing that $\varphi: G \to \mathbb{C}$ is a function which is bounded, measurable, satisfies $\varphi(k x k^{-1}) = \varphi(x)$ and $d_\kappa \chi * \varphi = d_\kappa \varphi * \chi = \varphi$. In addition we make the following *hypothesis*: There is an h dimensional irreducible morphism v^o of $L^{o1}(\kappa)$ such that

$$\varphi(f) = d_\kappa \operatorname{tr} v^o(f) \quad \text{for every } f \in L^{o1}(\kappa). \tag{4}$$

The irreducibility of v^o will be first used only in Lemma 8 and then after the proof of Proposition 9.

Lemma 6. *If φ satisfies the hypothesis then $f' * \varphi = \varphi * f'$ for every f in $C_0^o(\kappa)$.*

Proof. In general we have

$$g(f) = f' * g(e) = g * f'(e). \tag{5}$$

Since φ is a constant multiple of the trace of v^o by Proposition II.5.7 $\varphi(f * g) = \varphi(g * f)$ where f and g are arbitrary elements of $C_0^o(\kappa)$. Hence by (5)

$$\begin{aligned}(g' * f') * \varphi(e) &= (f * g)' * \varphi(e) = \varphi(f * g) = \varphi(g * f) \\ &= \varphi * (g * f)'(e) = \varphi * (f' * g')(e).\end{aligned}$$

Again by (5) $g' * (f' * \varphi)(e) = f' * \varphi(g)$ and $(\varphi * f') * g'(e) = \varphi * f'(g)$. Hence we obtain

$$f' * \varphi(g) = \varphi * f'(g) \tag{6}$$

for every f, g in $C_0^o(\kappa)$. Now the conclusion follows from Lemma 5.

Lemma 7. *If φ satisfies the hypothesis then $f' * \varphi = \varphi * f'$ for every f in $L^p(\kappa)$ where $1 \leqslant p < \infty$.*

Proof. The existence of $f' * \varphi$ and $\varphi * f'$ follows from Proposition V.4.25. Since $\varphi = \varphi^o$ we have $\varepsilon_k * \varphi = \varphi * \varepsilon_k$ and so by the preceeding lemma $f' * \varphi = \varphi * f'$ for every $f = \varepsilon_k * g$ where $k \in K$ and $g \in C_0^o(\kappa)$. By applying Proposition 4 we obtain the desired result.

Lemma 8. *If* φ *satisfies the hypothesis then*

$$J = \{f\colon f \in L^1(\kappa) \text{ and } f' * \varphi = 0\}$$

is a regular ideal in $L^1(\kappa)$ *and* $J \cap L^{o1}(G)$ *is the kernel of* $v^o\colon L^{o1}(\kappa) \to \mathscr{L}(\mathscr{H})$. *One of the relative identities of* J *belongs to* $C_0^o(\kappa)$.

Proof. The commutativity relation expressed in Lemma 7 and 9) imply that J is a two sided ideal in $L^1(\kappa)$.

If f and g belong to $L^1(G)$ then by the unimodularity of G and Proposition V.4.11 $f' * \varphi$ and $(f * g)' * \varphi$ are bounded, continuous functions. We also have

$$\varphi(f * g) = (f * g)' * \varphi(e) = g' * f' * \varphi(e). \tag{7}$$

Thus $f \in J$ implies that $\varphi(f * g) = 0$ for every $g \in L^1(G)$.

Now we suppose that f in $L^{o1}(\kappa)$ is such that $\varphi(f * g) = 0$ for every g in $L^{o1}(\kappa)$. Let g be an arbitrary element of $L^1(G)$. Then by 3) we have

$$\varphi(f * g) = \varphi^o(f * g) = \varphi((f * g)^o) = \varphi(f * g^o).$$

Therefore using $d_\kappa \varphi * \chi = \varphi$ and $d_\kappa f * \bar{\chi} = f$ we obtain

$$\varphi(f * g) = d_\kappa \varphi * \chi(d_\kappa f * \bar{\chi} * g^o).$$

Hence by 6) we have

$$\varphi(f * g) = d_\kappa^2 \varphi(f * \bar{\chi} * g^o * \bar{\chi}) = 0$$

because $\bar{\chi} * g^o * \bar{\chi}$ belongs to $L^{o1}(\kappa)$ by Proposition V.4.25. Therefore by (7) we have

$$\int_G g(x) f' * \varphi(x) dx = 0$$

for every g in $L^1(G)$ and so $f' * \varphi = 0$ and $f \in J$.

We proved that a function f from $L^{o1}(\kappa)$ belongs to J if and only if $\varphi(f * g) = 0$ for every g in $L^{o1}(\kappa)$. Using this characterization of J we can prove that $J \cap L^{o1}(G)$ is the kernel of v^o: By hypothesis

$$\varphi(f * g) = d_\kappa \operatorname{tr} v^o(f * g) = d_\kappa \operatorname{tr} v^o(f) v^o(g) \tag{8}$$

for every f and g in $L^{o1}(\kappa)$. Thus if $v^o(f) = 0$ then $\varphi(f * g) = 0$ for every $g \in L^{o1}(\kappa)$ and so $f \in J$. Conversely if $f \in J \cap L^{o1}(G)$ then $\varphi(f * g) = 0$ for every g in $L^{o1}(\kappa)$ and so choosing $g = f^*$ in (8) we obtain $\operatorname{tr} v^o(f) v^o(f)^* = 0$.

Since $v^o(f)v^o(f)^*$ is a positive operator and the underlying Hilbert space is finite dimensional this implies that all the characteristic values of $v(f)v(f)^*$ are zero. Hence $v^o(f)v^o(f)^*=0$ and $v^o(f)=0$.

In order to see that J is regular we notice that v^o being irreducible by Proposition IV.2.6 $v^o(g)$ assumes as values every element in the full matrix algebra $M_h(\mathbb{C})$ as g varies over $L^{o1}(\kappa)$. Therefore we can find a u in $L^{o1}(\kappa)$ such that $v^o(u)=I$, the identity element of $\mathscr{L}(\mathscr{H})$. Since v^o is a continuous representation every invariant subspace of $v^o|C_0^o(\kappa)$ is invariant under v^o. Therefore $v^o|C_0^o(\kappa)$ is irreducible and so the element u satisfying $v^o(u)=I$ can be chosen in $C_0^o(\kappa)$.

For g in $L^{o1}(\kappa)$ we have by (4) and (5)

$$\begin{aligned} u' * \varphi(g) &= g(u' * \varphi) = g' * u' * \varphi(e) = (u * g)' * \varphi(e) \\ &= \varphi(u * g) = d_\kappa \operatorname{tr} v^o(u) v^o(g) = d_\kappa \operatorname{tr} v^o(g) = \varphi(g). \end{aligned}$$

Therefore $(u' * \varphi - \varphi)(g)=0$ for every g in $L^{o1}(\kappa)$. Hence by Proposition V.4.7 and the corollary of Lemma 5 $u' * \varphi = \varphi$. By Proposition V.4.11 this shows that φ is locally almost everywhere equal to a continuous function which we identify with φ. We have $u \notin J$ because $\varphi(u)=d_\kappa \operatorname{tr} v^o(u)=d_\kappa^2>0$ and so $u' * \varphi \neq 0$. Now we prove that u is a relative identity for J. If $f \in L^1(\kappa)$ then by Lemma 7 and Proposition V.4.11

$$(f * u - f)' * \varphi = u' * f' * \varphi - f' * \varphi = u' * \varphi * f' - f' * \varphi = 0$$

and similarly

$$(u * f - f)' * \varphi = f' * u' * \varphi - f' * \varphi = 0.$$

Therefore we see that $f * u - f$ and $u * f - f$ belong to J for every f in $L^1(\kappa)$.

Proposition 9. *Let A be a regular maximal left ideal in $L^1(\kappa)$ such that A has a right relative identity u satisfying $u=u^o$. Let M be the set of those f in $L^1(G)$ for which*

(9) $$\bar{\chi} * g * f * \bar{\chi} \quad \text{is in } A \text{ for all } g \text{ in } L^1(G).$$

Then M is a regular maximal left ideal in $L^1(G)$ and $A = M \cap L^1(\kappa)$. We also have

(10) $$d_\kappa f * \bar{\chi} - f \in M$$

for every f in $L^1(G)$.

Proof. From the definition it is clear that M is a left ideal. In order to prove that M is regular we must find a right relative identity for M. Consider a relative identity u satisfying $u=u^o$ for A. Then we have $u*\bar{\chi}=\bar{\chi}*u$ and so for any f and g in $L^1(G)$

$$\bar{\chi}*g*(f*u-f)*\bar{\chi}=h*u-h$$

where $h=\bar{\chi}*g*f*\bar{\chi}$ is in $L^1(\kappa)$ by Proposition V.4.25. Since $h*u-h\in A$, by (9) this shows that $f*u-f\in M$ i. e. u is a right relative identity for M.

Since A is maximal in order to prove that $A=M\cap L^1(\kappa)$ it is sufficient to show that $M\cap L^1(\kappa)$ is a proper left ideal containing A. We prove that it is proper by contradiction: Thus we suppose that $u\in M$. This implies that $\bar{\chi}*f*u*\bar{\chi}$ is in A for every f in $L^1(G)$. Hence if $f\in L^1(\kappa)$ then by $u\in L^1(\kappa)$ we have $f*u\in A$. Since u is a relative identity of A we obtain $f\in A$ where f is an arbitrary element of $L^1(\kappa)$. This is a contradiction and so $u\notin M$ and $M\cap L^1(\kappa)$ is a proper left ideal.

In order to prove that A is contained in $M\cap L^1(\kappa)$ let f in A be given. If $g\in L^1(G)$ then $\bar{\chi}*g*\bar{\chi}\in L^1(\kappa)$. Hence A being a left ideal in $L^1(\kappa)$ have

$$\bar{\chi}*g*f*\bar{\chi}=(\bar{\chi}*g*\bar{\chi})*f\in A$$

for every g in $L^1(G)$ and so $f\in M$ by (9).

The fact that $d_\kappa f*\bar{\chi}-f$ belongs to M for every f in $L^1(G)$ follows from Proposition 1.30 because for every g in $L^1(G)$ we have

$$\bar{\chi}*g*(d_\kappa f*\bar{\chi}-f)*\bar{\chi}=d_\kappa(\bar{\chi}*g*f)*(\bar{\chi}*\bar{\chi})-\bar{\chi}*g*f*\bar{\chi}=0.$$

In order to prove that M is maximal suppose that N is a proper left ideal in $L^1(G)$ containing M. Since N is proper it follows that $u\notin N$. We have $A=M\cap L^1(\kappa)\subseteq N\cap L^1(\kappa)$ and $u\notin N\cap L^1(\kappa)$. Therefore by the maximality of A we have $N\cap L^1(\kappa)=A$. If $f\in N$ then by (10) $d_\kappa f*\bar{\chi}-f\in M\subseteq N$ and so $f*\bar{\chi}\in N$. Hence N being a left ideal $\bar{\chi}*g*f*\bar{\chi}\in N$ for every g in $L^1(G)$. Consequently

$$\bar{\chi}*g*f*\bar{\chi}\in N\cap L^1(\kappa)=A$$

for every g in $L^1(G)$ which implies that $f\in M$. We proved that $M=N$ and so M is maximal.

The ideal J whose existence is stated in Lemma 8 is regular and so we can find a regular maximal left ideal A in $L^1(\kappa)$ which contains J. Let M denote the regular left maximal ideal constructed from A according to Proposition 9. Since M is a closed left ideal in $L^1(G)$ by Theorem V.5.13 it is a left invariant subspace of $L^1(G)$ i. e. ${}_af\in M$ for every

$a \in G$ and $f \in M$. Therefore by letting

$$\mu(x)(f+M) = {}^x f + M \tag{11}$$

for $x \in G$ and $f \in L^1(G)$ we defined a morphism $\mu: G \to \mathscr{L}(\mathscr{X})$ where $\mathscr{X} = L^1(G)/M$.

We know that $\mathscr{X}$ is a Banach space under the norm $\|f+M\| = \text{glb}\|f+m\|_1$. If $\mathscr{Y}$ is a subspace of $\mathscr{X}$ which is invariant under μ then $\mathscr{Y} = Y/M$ where Y is a left invariant subspace of $L^1(G)$. Hence by Theorem V.5.13 it is a regular left ideal containing M and so by the maximality of M we have $Y = M$ or $Y = L^1(G)$. Therefore $\mathscr{Y} = \mathscr{X}$ or $\mathscr{Y} = 0$. This shows that $\mu: G \to \mathscr{L}(\mathscr{X})$ is a topologically irreducible morphism. Since ${}^x M = M$ and $\|{}^x f + {}^x m\| = \|f+m\|$ for every $x \in G$ we see that $|\mu(x)| = 1$ for every a in G and so μ is bounded. Since

$$\|({}^x f + M) - (f+M)\| = \|({}^x f - {}^a f) + M\| \leqslant \|{}^x f - {}^a f\|_1$$

by Proposition V.4.9 the function $x \mapsto {}^x f + M$ is continuous for every $f \in L^1(G)$. Hence μ is strongly continuous.

We let $\tilde{\sigma}: L^1(G) \to \mathscr{L}(\mathscr{X})$ be the algebra morphism associated with $\mu: G \to \mathrm{Gl}(\mathscr{X})$. The morphisms $\tilde{\tau}$ and $\tilde{\tau}^o$ are defined by $\tilde{\tau} = \tilde{\sigma}|L^1(\kappa)|\mathscr{X}_\kappa$ and $\tilde{\tau}^o = \tilde{\sigma}|L^{o1}(\kappa)|\mathscr{X}_\kappa$, respectively.

Lemma 10. *For every f and g in $L^1(G)$ we have $\tilde{\sigma}(f)(g+M) = f * g + M$.*

Proof. If $\xi^* \in \mathscr{X}^*$ then ξ^* defines a continuous linear functional $L^1(G) \to \mathbb{C}$ which we denote by the same symbol ξ^*: Namely we let $\xi^*(f) = \xi^*(f+M)$ for every $f \in L^1(G)$. Then we have

$$\begin{aligned}\xi^*(\tilde{\sigma}(f)(g+M)) &= \int_G f(x)\xi^*(\mu(x)(g+M))\,dx = \int_G f(x)\xi^*({}^x g + M)\,dx \\ &= \int_G f(x)\xi^*({}^x g)\,dx\end{aligned}$$

for every $g \in L^1(G)$. If $g \in C_0(G)$ then the right hand side is $\xi^*(f * g)$ and so one can easily see that the same holds for every g in $L^1(G)$. Hence we obtain

$$\xi^*(\tilde{\sigma}(f)(g+M)) = \xi^*(f * g) = \xi^*(f * g + M).$$

Since this holds for every ξ^* in $\mathscr{X}^*$ we obtain the result stated in the lemma.

By the same reasoning we have

$$\xi^*(\tilde{\sigma}(\overline{\chi})(g+M)) = \int_K \overline{\chi}(k)\xi^*({}^k g)\,dk$$

for all $g \in L^1(G)$: If h in $L^\infty(G)$ is the function corresponding to ξ^* then the right hand side can be written as

$$\int_K \overline{\chi}(k) \int_G g(k^{-1}x) h(x) \, dx \, dk = \int_G h(x) \int_K \overline{\chi}(k) g(k^{-1}x) \, dk \, dx .$$

Hence by Proposition V.4.25 we obtain

$$\xi^*(\tilde{\sigma}(\overline{\chi})(g+M)) = \int_G h(x)(\overline{\chi} * g)(x) \, dx = \xi^*(\overline{\chi} * g).$$

We proved that $\tilde{\sigma}(\overline{\chi})(g+M) = \overline{\chi} * g + M$ for every g in $L^1(G)$. Therefore by part 4 of Proposition 3.1 we have

$$E(\kappa)(g+M) = d_\kappa \overline{\chi} * g + M$$

and this shows that

$$\mathscr{X}_\kappa = \frac{\overline{\mathscr{X}} * L^1(G)}{M}.$$

We have a right relative identity for M in $C_0(\kappa)$ and so $u + M \in \mathscr{X}_\kappa$ and $u + M \neq M$ by $u \notin M$. Thus $\mathscr{X}_\kappa$ has positive dimension and μ is a class κ morphism.

Next we shall prove that $\tilde{\sigma}$, the algebra representation associated with μ is completely irreducible. The proof of this fact is based on the following general remarks. Let A be an algebra over $\mathbb{C}$ and let $\sigma: A \to \mathscr{L}(\mathscr{X})$ be a morphism in a complex topological vector space $\mathscr{X}$. Then σ is called *algebraically irreducible* provided the only σ-invariant linear manifolds are 0 and $\mathscr{X}$ itself. It is obvious that algebraic irreducibility implies topological irreducibility.

Proposition 11. *If $\sigma: A \to \mathscr{L}(\mathscr{X})$ is an algebraically irreducible morphism of the complex algebra A in the complex Banach space $\mathscr{X}$ then σ is completely irreducible.*

Proof. Let $\mathscr{A}$ be the algebra of not necessarily continuous linear transformations $T: \mathscr{X} \to \mathscr{X}$ which commute with every $\sigma(x)$ $(x \in A)$. Let ξ be a unit vector in $\mathscr{X}$. Since σ is algebraically irreducible the σ-invariant linear manifold $\{\sigma(x)\xi : x \in A\}$ is $\mathscr{X}$ and so we can define

$$|T| = \operatorname{glb}\{|\sigma(x)| : x \in A \text{ and } \sigma(x)\xi = T\xi\}.$$

One can easily see that $|T_1 + T_2| \leqslant |T_1| + |T_2|$ and $|T_1 T_2| \leqslant |T_1| \cdot |T_2|$ and also $|\lambda T| \leqslant |\lambda| \cdot |T|$ for every $\lambda \in \mathbb{C}$. By Theorem II.9.9 every $T \neq 0$ is invertible and so $|T| \geqslant |T\xi| > 0$. Therefore by the same theorem $\mathscr{A}$ is a normed division ring over $\mathbb{C}$. By Theorem I.4.7 $\mathscr{A}$ is naturally isomorphic to $\mathbb{C}$, namely $\mathscr{A}=\mathbb{C}I$. Now the proposition follows from Theorem II.8.13.

Let $\tilde{\sigma}$ be the algebra representation associated with the morphism μ. Let $\mathscr{Y}$ be a $\tilde{\sigma}$-invariant linear manifold in $\mathscr{X}$ and let $Y=\{g: g+M\in\mathscr{Y}\}$. Then by Lemma 10 Y is a left ideal in $L^1(G)$ and Y contains M. Hence by the maximality of M we have $Y=M$ or $Y=L^1(G)$. This shows that $\tilde{\sigma}$ is algebraically irreducible and so it is completely irreducible by Proposition 11. Thus by Proposition 3.12 $\tilde{\tau}_0$ and $\tilde{\tau}$ are also completely irreducible.

Lemma 12. *We have* $\mathscr{X}_\kappa=L^1(\kappa)/M$.

Proof. If $f\in L^1(\kappa)$ then $d_\kappa\bar{\chi}*f=f$ and so $E(\kappa)(f+M)=f+M$ proving that $f+M\in\mathscr{X}_\kappa$. Conversely if $f+M\in\mathscr{X}_\kappa$ then $d_\kappa\bar{\chi}*f-f\in M$. Hence by formula (10) of Proposition 9 we have

$$d_\kappa^2\bar{\chi}*f*\bar{\chi}-f\in M$$

and this shows that $f+M\in L^1(\kappa)/M$.

A part of the hypothesis of Theorem 1 is that G, K and κ are such that $n(\kappa,\mu|K)$ is finite for every primitive, strongly continuous topologically irreducible morphism μ such that the associated $L^1(G)$ algebra morphism σ is completely irreducible. Therefore by what has been already proved we see that χ_κ is finite dimensional.

Since J is a two sided ideal in $L^1(\kappa)$ if $f\in J$ then $f*g\in J$ for any g in $L^1(\kappa)$. Thus by Lemma 10 if $f\in J$ then $\tilde{\tau}(f)=0$. By Lemma 8 $v^o(f)=0$ implies that $f\in J\cap L^{o1}(\kappa)$ and so $\tilde{\tau}(f)=0$. Thus $v^o(f)=0$ implies that $\tilde{\tau}^o(f)=0$.

Let $\tilde{\varphi}$ be the spherical function associated with the triple (K,κ,μ). Let h be the height of $\tilde{\varphi}$ i.e. let $n(\kappa,\mu|K)=\tilde{h}$. Then by the first part of Theorem 1 there is an $\tilde{h}$-dimensional irreducible morphism $\tilde{v}^o: L^{o1}(\kappa)\to\mathscr{L}(\tilde{\mathscr{H}})$ such that $\tilde{\varphi}(f)=d_\kappa\operatorname{tr}\tilde{v}(f)$ for every f in $L^{o1}(\kappa)$. Moreover $\tilde{v}^o$ is such that $\tilde{v}^o(f)=0$ if and only if $\tilde{\tau}^o(f)=0$. Hence by what has been proved above $v^o(f)=0$ implies that $\tilde{v}(f)=0$ or in other words $\ker v^o\subseteq\ker\tilde{v}^o$. Since v^o and $\tilde{v}^o$ are finite dimensional, non-trivial irreducible morphisms of the complex algebra $L^{o1}(\kappa)$ by Proposition 3.13 we obtain $\ker v^o=\ker\tilde{v}^o$. Thus by Proposition 3.15 v° and $\tilde{v}^o$ are similar. Hence $h=\tilde{h}$ and

$$\varphi(f)=d_\kappa\operatorname{tr}v^o(f)=d_\kappa\operatorname{tr}\tilde{v}^o(f)=\tilde{\varphi}(f).$$

This proves that the function φ is spherical and it is associated with the triple (K,κ,μ). This completes the proof of Theorem 1.

Theorem 13. *Let G, K and $\kappa \in \hat{K}$ be such that $n(\kappa, \mu|K)$ is finite for every primitive morphism $\mu: G \to \mathrm{Gl}(\mathscr{X})$. Then the following three properties are equivalent:*

1. *$L^{o1}(\kappa)$ is commutative.*
2. *$n(\kappa, \rho|K)$ is at most one for every continuous irreducible representation $\rho: G \to \mathscr{U}(\mathscr{H})$.*
3. *$L^{o1}(\kappa)$ is the center of $L^1(\kappa)$.*

Note. Both 1. and 3. imply that $n(\kappa, \mu|K)$ is at most one for every morphism $\mu: G \to \mathrm{Gl}(\mathscr{X})$ where μ is strongly continuous and completely irreducible or $\mathscr{X}$ is a reflexive Banach space and μ is such that the associated algebra morphism σ is completely irreducible.

Proof. If 1. holds then every finite dimensional irreducible morphism of $L^{o1}(\kappa)$ is one dimensional. Hence by Theorem 1 every spherical function determined by a strongly continuous and completely irreducible morphism μ of type κ is of height one. This shows that $n(\kappa, \mu|K)=1$ for every morphism μ satisfying these requirements. If μ is not of class κ then by definition κ does not occur in $\mu|K$. Hence 1. implies 2. and part of the stronger statement given in the note.

By Proposition V.7.26 for every $h \neq 0$ in $L^{o1}(\kappa)$ there is an irreducible representation $\pi: L^{o1}(\kappa) \to \mathscr{L}(\mathscr{H})$ such that $\pi(h) \neq 0$. If 2. holds then by Theorem 1 π is one dimensional and so $\pi(f)\pi(g)=\pi(g)\pi(f)$ for every $f, g \in L^{o1}(\kappa)$. Hence $h = f*g - g*f$ must be 0 and so $L^{o1}(\kappa)$ is commutative.

It is clear that 3. implies 1.

Now we suppose that 1. holds. Then 2. holds and so $\rho|K|\mathscr{H}_\kappa$ is irreducible. If $f \in L^{o1}(\kappa)$ and $k \in K$ then

$$\begin{aligned}\rho(k)\sigma(f) &= \int_G f(x)\rho(kx)\,dx = \int_G f(k^{-1}x)\rho(x)\,dx = \int_G f(xk^{-1})\rho(x)\,dx \\ &= \int_G f(x)\rho(xk)\,dx = \sigma(f)\rho(k)\,.\end{aligned}$$

Hence $\tau(f)=\sigma(f)|\mathscr{H}_\kappa$ is in $(\rho|K|\mathscr{H}_\kappa, \rho|K|\mathscr{H}_\kappa)=\mathbb{C}I$ and $\tau(f)=\lambda_f I$ where I is the identity map of $\mathscr{H}_\kappa$. Therefore $\tau(f*g-g*f)=0$ for every $f \in L^{o1}(\kappa)$ and $g \in L^1(\kappa)$ and applying Lemma 3.10 we see that $\sigma(f*g-g*f)=0$. From this we can prove that $f*g=g*f$ i.e. $L^{o1}(\kappa)$ lies in the center of $L^1(\kappa)$: For if h is a non-zero element of $L^1(\kappa)$ then by Proposition V.7.26 there is an irreducible representation $\pi: L^1(G) \to \mathscr{L}(\mathscr{H})$ such that $\pi(h) \neq 0$ and by Theorem V.5.16 there is a representation $\rho: G \to \mathscr{U}(\mathscr{H})$ such that the algebra representation σ associated with ρ is π.

We still have to prove that every central element f of $L^1(\kappa)$ lies in $L^{o1}(\kappa)$. This amounts to showing that $f(kxk^{-1})=f(x)$ for every $x\in G$ and $k\in K$ or in other words $f=f^o$. Let $\rho: G\to\mathcal{U}(\mathcal{H})$ be any weakly continuous irreducible representation of G and let σ and τ be defined as usual. Then for any g in $L^1(\kappa)$ we have $\sigma(f)\sigma(g)=\sigma(g)\sigma(f)$ and so $\sigma(f)$ is in (σ,σ). By Theorem V.5.9 $(\sigma,\sigma)=(\rho,\rho)$ and by the irreducibility of ρ we have $(\rho,\rho)=\mathbb{C}I$. Hence $\sigma(f)$ is a scalar and so

$$\begin{aligned}\sigma(f)&=\int_K \rho(k)\sigma(f)\rho(k^{-1})\,dk=\int_K\int_G f(x)\rho(kxk^{-1})\,dx\,dk\\&=\int_G\left(\int_K f(kxk^{-1})\,dk\right)\rho(x)\,dx=\sigma(f^o).\end{aligned}$$

Since this holds for every ρ we obtain $f=f^o$ by the reasoning which was also used at the end of the last paragraph.

Theorem 14. *If G, K and κ satisfy the conditions stated in the last theorem then a bounded, continuous function $\psi: G\to\mathbb{C}$ is proportional to a spherical function of height one if and only if*

$$\psi(e)\int_K \psi(kxk^{-1}y)\,dk=\psi(x)\psi(y)\qquad (x,y\in G)\,. \tag{12}$$

Proof. First let φ be a spherical function of type κ and height one and let $\psi=d_\kappa^{-1}\varphi$. Then by Theorem 1

$$\psi(f*g)=\psi(f)\psi(g) \tag{13}$$

for all f,g in $L^{o1}(\kappa)$. If $f,g\in L^1(G)$ then by 2) and 3) we have

$$\psi(f^o*g)=\psi^o(f^o*g)=\psi((f^o*g)^o)=\psi(f^o*g^o)\,.$$

Hence using 6), 7) and 8) we obtain

$$\psi(f^o*g)=d_\kappa^2\,\chi*\psi*\chi(f^o*g^o)=d_\kappa^2\,\psi(\bar{\chi}*f^o*g^o*\bar{\chi})\,.$$

Since $\bar{\chi}*f^o$ and $g^o*\bar{\chi}$ are in $L^{o1}(\kappa)$ by (13) it follows that

$$\psi(f^o*g)=d_\kappa^2\,\psi(\bar{\chi}*f^o)\psi(g^o*\bar{\chi})\,.$$

Therefore applying 2), 3), 6), 7) and 8) we obtain

$$\psi(f^o*g)=\psi(f)\psi(g) \tag{14}$$

where f and g are arbitrary elements of $L^1(G)$.

Every measure $\alpha\in M(G)$ can be approximated in the weak sense by linear combinations of Dirac measures δ_a $(a\in G)$ and every δ_a can be approximated by measures of the form $f(x)dx$. Hence from (14) we obtain

$$\psi(\alpha^o * \beta) = \psi(\alpha)\psi(\beta) \tag{15}$$

for every α, β in $M(G)$. By choosing $\alpha = \delta_x$ and $\beta = \delta_y$ we arrive at (12).

Conversely let us suppose that ψ satisfies (12). This means that (15) holds with $\alpha = \delta_x$ and $\beta = \delta_y$ and so approximating arbitrary α, β by linear combinations of Dirac measures we obtain (15) for arbitrary α, β in $M(G)$. If we specialize (15) to measures of the form $f(x)dx$ where $f \in L^{o1}(\kappa)$ for some $\kappa \in \hat{K}$ then we see that $f \mapsto \psi(f)$ is the trace of a one or zero dimensional morphism of $L^{o1}(\kappa)$. In order to prove ψ is proportional to a spherical function of height one and type κ for some κ in $\hat{K}$ it remains to prove that $d_\kappa \chi_\kappa * \psi = \psi$.

First suppose that $\chi_\kappa * \psi(e) = 0$ for every $\kappa \in \hat{K}$. Then we have

$$\chi_\kappa * \psi(e) = \int_K \overline{\chi}_\kappa(k)\psi(k)\,dk = (\psi|K, \chi_\kappa) = 0\,.$$

Hence by Theorem 1.39 $\psi|K$ is orthogonal to all $f \in L^{o2}(G)$ and so $\psi|K$ is identically zero. In particular $\psi(e) = 0$ and by (12) this implies that $\psi(x) = 0$ for every x in G. Thus by $\psi = 0$ we have $d_\kappa \chi_\kappa * \psi = \psi$.

Now suppose that $\chi_\kappa * \psi(e) \neq 0$ for some κ in $\hat{K}$. Since χ_κ is in $M(G)$ and $\chi_\kappa = \chi_\kappa^o$ by (15) we have

$$\psi(\overline{\chi}_\kappa * \alpha) = \psi(\overline{\chi}_\kappa)\psi(\alpha)$$

for every α in $M(G)$. This shows that

$$\chi_\kappa * \psi = \psi(\overline{\chi}_\kappa)\psi\,. \tag{16}$$

From this relation we obtain

$$\psi(\overline{\chi}_\kappa)\psi = \chi_\kappa * \psi = d_\kappa \chi_\kappa * \chi_\kappa * \psi = d_\kappa \psi(\overline{\chi}_\kappa)\chi_\kappa * \psi = d_\kappa \psi(\overline{\chi}_\kappa)^2 \psi$$

and so $d_\kappa \psi(\overline{\chi_\kappa}) = 1$. Substituting into (16) we see that $d_\kappa \chi_\kappa * \psi = \psi$. This completes the proof.

5. Representations of Groups with an Iwasawa Decomposition

We shall say that the locally compact group G has an *Iwasawa decomposition* $G = KS$ if there is a compact subgroup K and a solvable, connected subgroup S such that $(k, s) \mapsto ks$ is a homeomorphism of $K \times S$ onto G. Other than the connectedness we make no further restriction on the topological nature of S. Solvability means that a sufficiently high commutator subgroup of S consists only of the identity element. A locally compact group G is called a *CCR group* if for every irreducible representation $\pi: G \to \mathscr{U}(\mathscr{H})$ the operators $\sigma(f)$ are compact for every f in $C_0(G)$. The importance of this concept is due to a theorem of Kaplansky which says that *every CCR group is of type I* i.e. the v. Neumann algebra $\mathscr{A}_\rho$ associated with every continuous representation $\rho: G \to \mathscr{U}(\mathscr{H})$ is of type I.

The main object of this section is to prove the following:

Theorem 1. *If G is a locally compact group with sufficiently many finite dimensional representations and an Iwasawa decomposition then G is a CCR group.*

The result will immediately follow after two other theorems are proved. The first of these is Godement's improved version and generalization of an inequality due to Harish-Chandra. In the special case of Hilbert space representations it states that $n(\kappa, \omega|K) \leqslant d_\kappa$ where $\kappa \in \hat{K}$ and $\omega \in \hat{G}$. The other theorem which is due to Godement states that for certain representations $\rho: G \to \mathscr{U}(\mathscr{H})$ the associated algebra representations σ contain sufficiently many compact operators.

We start with a few algebraic preliminaries due to Kaplansky which are followed by some results on σ-complete sets of representations and some others on the representations of solvable groups.

Let $x_1, \ldots, x_r$ be indeterminates and let p be the polynomial

$$p(x_1, \ldots, x_r) = \sum \operatorname{sign}(p_1, \ldots, p_r) x_{p_1} \ldots x_{p_r}$$

where $(p_1, \ldots, p_r)$ is any permutation of $1, \ldots, r$ and sign is ± 1 according as $(p_1, \ldots, p_r)$ is even or odd.

If A is an algebra of dimension d over some field F and if $d < r$ then $p(a_1, \ldots, a_r) = 0$ for every $a_1, \ldots, a_r \in A$.

Since p is multilinear it is sufficient to prove this fact only in the special case when $a_1, \ldots, a_r$ are elements of a basis. Then by $d < r$ at least one repetion occurs among these elements and hence we can perform a transposition which leaves $p(a_1, \ldots., a_r)$ unchanged. On the other hand such transposition changes the sign of each term and so $p(a_1, \ldots, a_r) = 0$.

Now let a natural number $n \geqslant 1$ be fixed. We let $r = r(n)$ denote the smallest value of r for which $p(a_1, \ldots, a_r) = 0$ holds for all complex matrices $a_1, \ldots, a_r \in M_n(\mathbb{C})$. By the foregoing reasoning we know that such r exists and actually $r(n) \leqslant n^2 + 1$. One has $r(1) = 2$, $r(2) = 4$ and

$r(3)=6$, but these numerical values are not important. Instead we need the following:

Lemma 2. *For every* $n=1,2,\ldots$ *we have* $r(n)\geqq r(n-1)+2$.

Proof. Let e_{ij} denote that square matrix of some unspecified order which contains 1 in its i,j-th place and zero everywhere else. If we let $t=r(n-1)-1$ then there exists $(n-1)\times(n-1)$ complex matrices $a_1,\ldots,a_t$ such that $p(a_1,\ldots,a_t)\neq 0$. Let us suppose that the k-th column of $p(a_1,\ldots,a_t)$ contains a non-zero entry. We let $b_1,\ldots,b_t$ be the $n\times n$ matrices obtained from $a_1,\ldots,a_t$ by adding a last row and a last column of zeros. We claim that

(1) $$p(b_1,\ldots,b_t,e_{kn},e_{nn})\neq 0\,.$$

This will imply that $r(n)>t+2=r(n-1)+1$. Indeed if b is any $n\times n$ matrix then $e_{jn}b$ has the n-th row of b in its j-th row and zeros everywhere else. Hence the only possible non-zero terms of (1) are of the form $b_{p_1}\ldots b_{p_t}e_{kn}e_{nn}$ or $b_{p_1}\ldots b_{p_t}e_{nn}e_{kn}$ where $k<n$. But $e_{kn}e_{nn}=e_{kn}$ and $e_{nn}e_{kn}=0$. Hence the left hand side of (1) is equal to $p(b_1,\ldots,b_t)e_{kn}$ and its last column is the k-th column of $p(a_1,\ldots,a_t)$ augmented by a zero.

Since we proved that $r(n)$ is strictly increasing we have the following:

Lemma 3. *There exists a non-commutative polynomial* $p(x_1,\ldots,x_r)$ *with coefficients* ± 1 *such that*

1. *p is multilinear.*
2. $p(a_1,\ldots,a_r)=0$ *for every* $a_1,\ldots,a_r\in M_n(\mathbb{C})$.
3. *There exist* $a_1,\ldots,a_r\in M_{n+1}(\mathbb{C})$ *such that* $p(a_1,\ldots,a_r)\neq 0$.

Using this lemma we are going to prove:

Proposition 4. *Let A be an algebra over $\mathbb{C}$. Suppose that for every non-zero x in A there is a morphism $\mu\colon A\to\mathcal{L}(\mathcal{X})$ of degree at most n such that $\mu(x)\neq 0$. Then every completely irreducible morphism of A has degree at most n.*

Proof. Let p denote the polynomial whose existence and properties are stated in the preceeding lemma. If $x_1,\ldots,x_r\in A$ then $p(x_1,\ldots,x_r)=0$ because otherwise there would be a μ such that $\mu(p(x_1,\ldots,x_r))=p(a_1,\ldots,a_r)\neq 0$ although $a_k=\mu(x_k)\in M_n(\mathbb{C})$. Thus if $\sigma\colon A\to\mathcal{L}(\mathcal{X})$ is an arbitrary morphism of A then

(1) $$p(\sigma(x_1),\ldots,\sigma(x_r))=0$$

for every $x_1,\ldots,x_r\in A$.

Since σ is completely irreducible $\sigma(A)$ is dense in $\mathscr{L}(\mathscr{X})$ in the strong topology of $\mathscr{L}(\mathscr{X})$. Thus given $A_1, \ldots, A_r \in \mathscr{L}(\mathscr{X})$ we can find operators $\sigma(a)$ approximating any one of them in the weak operator topology. Approximating A_1 by the multilinearity of p we obtain

$$p(A_1, \sigma(x_2), \ldots, \sigma(x_r)) = 0 \tag{2}$$

where $x_2, \ldots, x_r$ are arbitrary elements in A. Next approximating A_2 we derive from (2)

$$p(A_1, A_2, \sigma(x_3), \ldots, \sigma(x_r)) = 0 .$$

Finally in n steps we see that

$$p(A_1, \ldots, A_r) = 0 \tag{3}$$

where $A_1, \ldots, A_r$ are arbitrary operators in $\mathscr{L}(\mathscr{X})$.

Now if $\mathscr{X}$ contained an $n+1$ dimensional subspace $\mathscr{Y}$ then we could choose a basis in $\mathscr{Y}$ and find linear operators $A_1, \ldots, A_r$ in $\mathscr{L}(\mathscr{X})$ such that in the given basis the restriction of A_k to $\mathscr{Y}$ is given by the matrix a_k specified in part 3. of Lemma 3. Hence (3) and part 3. of the lemma would lead to a contradiction. Therefore $\mathscr{X}$ is at most n dimensional.

Let Ω be a set of bounded, weakly measurable morphisms μ of a locally compact group G in reflexive complex Banach spaces and for each μ in Ω let σ_μ denote the algebra representation associated with μ in accordance with Theorem V.5.2. The set Ω is called *σ-complete* if for every non-vanishing f in $C_0(G)$ there is a μ in Ω such that $\sigma_\mu(f) \neq 0$. Similarly a set Ω of morphisms $\sigma: A \to \mathscr{L}(\mathscr{H})$ of a subalgebra of $L^1(G)$ is called *complete* if for each $f \neq 0$ in $C_0(G)$ there is a σ in Ω such that $\sigma(f) \neq 0$.

Proposition 5. *Let Ω be a σ-complete set of bounded, weakly measurable morphisms of the locally compact group G in reflexive complex Banach spaces and let K be a compact subgroup of G. Suppose that $n(\kappa, \mu|K) \leqslant h$ for a given κ in $\Omega_0(K)$ and for every $\mu \in \Omega$. Then we have $n(\kappa, \mu|K) \leqslant h$ for every bounded weakly measurable morphism $\mu: G \to \mathrm{Gl}(\mathscr{X})$ where $\mathscr{X}$ is an arbitrary complex Banach space, μ is strongly continuous and completely irreducible or $\mathscr{X}$ is reflexive and μ is completely irreducible.*

Proof. If $\mu \in \Omega$ let σ_μ denote the associated algebra morphism and let $\tau: L^1(\kappa) \to \mathscr{L}(\mathscr{X}_\kappa)$ be obtained by first restricting σ_μ to $L^1(\kappa)$ and then to $\mathscr{X}_\kappa$. By Lemma 3.10 we have $\tau(f) = 0$ if and only if $\sigma_\mu(f) = 0$. Therefore the morphisms τ associated with the various members of Ω form a complete set of morphisms of $L^1(\kappa)$ whose degrees are all at most $n = h d_\kappa$. Therefore by Proposition 4 every completely irreducible morphism $\tau: L^1(\kappa)\to\mathscr{L}(\mathscr{H})$ has degree at most n whether it is associated with an element of Ω or not. By Proposition 3.12 if $\mu: G\to\mathscr{L}(\mathscr{X})$ satisfies the hypothesis then the associated τ is completely irreducible. Hence the degree of μ is at most n and so $n(\kappa, \mu|K)\leqslant h$.

Lemma 6. *If* $\varphi_k: S\to\mathbb{C}$ $(k=1,2)$ *are coefficients of some representations of the object S then* $\varphi_1\varphi_2$ *and* $\varphi_1\bar{\varphi}_2$ *are also coefficients.*

Proof. Let $\varphi_k(s)=(\rho_k(s)\alpha_k^1, \alpha_k^2)$. Then $\varphi_1\varphi_2$ is a coefficient of $\rho_1\otimes\rho_2$, namely

$$(\rho_1\otimes\rho_2(s)(\alpha_1^1\otimes\alpha_2^1), \alpha_1^2\otimes\alpha_2^2)=(\rho_1(s)\alpha_1^1, \alpha_1^2)(\rho_2(s)\alpha_2^1, \alpha_2^2).$$

Similarly $\varphi_1\bar{\varphi}_2$ is a coefficient of $\rho_1\bar{\otimes}\rho_2$ because

$$(\rho_1\bar{\otimes}\rho_2(s)(\alpha_1^1\bar{\otimes}\alpha_2^1), \alpha_1^2\bar{\otimes}\alpha_2^2)=(\rho_1(s)\alpha_1^1, \alpha_1^2)(\alpha_2^2, \rho_2(s)\alpha_2^1).$$

Corollary. *If* $\varphi: S\to\mathbb{C}$ *is a coefficient then so is* $\bar{\varphi}$.

Proof. The identically 1 function is a coefficient of the trivial representation $\rho: S\to\mathscr{L}(\mathscr{H})$, namely if $\alpha_1=\alpha_2=\alpha$ and $(\alpha,\alpha)=1$ then $(\rho(s)\alpha,\alpha)=1$. Hence by Lemma 6 $\bar{\varphi}=1\cdot\bar{\varphi}$ is a coefficient. One can also reason directly as follows: If φ is given by the representation ρ and the vectors α_1,α_2 then $\bar{\varphi}$ is determined by the conjugate representation $\bar{\rho}$ and the same vectors α_1,α_2.

A representation $\rho: G\to\mathscr{U}(\mathscr{H})$ is called *faithful* if its kernel is trivial. A group G is said to have *sufficiently many finite dimensional representations* if for every $x\neq e$ in G there is a finite dimensional continuous representation $\rho: G\to\mathscr{U}(\mathscr{H})$ such that $\rho(x)\neq I$. For example G is such a group if it has a finite dimensional continuous faithful representation. Suppose that G has sufficiently many finite dimensional representations and g_1, g_2 are distinct elements of G. Then $\rho(g_1^{-1}g_2)\neq I$ for some $\rho: G\to\mathscr{U}(\mathscr{H})$ and so

$$\theta(g_1)=(\rho(g_1)\alpha,\beta)\neq(\rho(g_2)\alpha,\beta)=\theta(g_2)$$

for suitable $\alpha,\beta\in\mathscr{H}$. Therefore the coefficients θ of such representations separate the elements of G.

Proposition 7. *If G is a locally compact group with sufficiently many finite dimensional representations then the set of finite dimensional irreducible representations of G is* σ*-complete.*

Note. Since we consider only weakly measurable representations by Theorem V.7.3 the representations considered here are all strongly continuous.

Proof. Let A denote the set of linear combinations of coefficients of finite dimensional irreducible representations of G. Since every finite dimensional representation of G is a direct sum of irreducible components A contains the coefficients of all finite dimensional representations of G. By hypothesis G has sufficiently many of these and so if $\theta(g_1)=\theta(g_2)$ for every θ in A then $g_1=g_2$. Thus in view of Lemma 6 and its corollary A is a separating algebra of continuous functions $\theta: G\to\mathbb{C}$ which contains with each θ also its conjugate $\bar{\theta}$. By the Stone-Weierstrass theorem we see that every continuous function $f: G\to\mathbb{C}$ can be approximated on any compact subset of G by elements of A. Therefore if f is an non-vanishing function in $C_0(G)$ then there is a θ in A such that $\int_G f(x)\theta(x)dx\neq 0$. Since θ is a finite sum of coefficients of irreducible representations there is a finite dimensional irreducible representation $\rho: G\to\mathscr{U}(\mathscr{H})$ such that

$$(\sigma(f)\alpha,\beta)=\int_G f(x)(\rho(x)\alpha,\beta)\,dx$$

does not vanish for some α and β in $\mathscr{H}$. Thus $\sigma(f)\neq 0$ and this shows that the set of finite dimensional irreducible representations is σ-complete.

Theorem 8. *If S is a solvable, connected topological group then every finite dimensional irreducible representation $\pi: S\to\mathscr{U}(\mathscr{H})$ is one dimensional.*

Proof. Let S' be the commutator subgroup of $S=S^{(0)}$ and let $S^{(k)}=(S^{(k-1)})'$ for $k=1,2,\dots$ Since S is solvable $S^{(h)}=\{e\}$ for some $h\geqslant 1$ and the smallest integer $h\geqslant 1$ having this property is called the *height* of S. The theorem will be proved by induction on the height h. If $h=1$ then the group is commutative and so by Proposition IV.2.7 the irreducible representation π is one dimensional.

Now we suppose that the theorem is proved for all groups of height $h-1$ where $h>1$. Then S' having height $h-1$ the theorem is applicable to the commutator subgroup S'. Given a finite dimensional irreducible representation $\pi: S\to\mathscr{U}(\mathscr{H})$ its restriction $\pi|S'$ has an irreducible component which is known to be one dimensional. In other words there is a continuous group character $\lambda: S'\to T$ and a non-zero element a_λ of $\mathscr{H}$ such that

$$\pi(x)\alpha_\lambda=\lambda(x)a_\lambda \quad \text{for } x\in S'. \tag{4}$$

Let Λ denote the set of characters λ for which (4) has a non-zero solution a_λ. Since $\mathscr{H}$ is finite dimensional Λ is a finite set: For if $\lambda_1,\dots,\lambda_m$ are distinct characters then the corresponding characteristic vectors $a_{\lambda_1}, \ldots, a_{\lambda_m}$ are linearly independent. This can be proved by contradiction starting from a shortest possible linear dependence relation and by constructing a shorter one by applying $\pi(x)$ to it and eliminating one of the a_{λ_k}'s. By the invariance of S' in S the element $s^{-1}xs$ belongs to S' for every s in S and x in S'. Hence if λ is a character of S' then for every s in S the formula

$$\lambda_s(x) = \lambda(s^{-1}xs) \tag{5}$$

defines a character of S'. Using (4) we have

$$\pi(x)\pi(s)a_\lambda = \pi(s)\pi(s^{-1}xs)a_\lambda = \lambda_s(x)\pi(s)a_\lambda \tag{6}$$

for every $x \in S'$ and $s \in S$. Thus $\lambda \in \Lambda$ implies that $\lambda_s \in \Lambda$ for every $s \in S$. In the topology of pointwise convergence on S' the finite set Λ is a discrete space of functions. For fixed, continuous λ the map $s \mapsto \lambda_s$ is a continuous function from S to Λ and so S being connected the discrete space $\{\lambda_s : s \in S\}$ contains only one element. We proved that $\lambda_s = \lambda$ for every $s \in S$ and $\lambda \in \Lambda$.

Now (6) reduces to the equation

$$\pi(x)(\pi(s)a_\lambda) = \lambda(x)(\pi(s)a_\lambda)$$

where $x \in S'$ and $s \in S$. This shows that if a_λ satisfies (4) then so does $\pi(s)a_\lambda$. Therefore the space of vectors a_λ satisfying (4) is a π-invariant subspace of $\mathscr{H}$. By the definition of Λ non-zero a_λ's exist and so by the irreducibility of π this subspace must be $\mathscr{H}$ itself. Hence we have

$$\pi(x) = \lambda(x)I \tag{7}$$

for every x in S' where I is the identity operator on $\mathscr{H}$.

Let s_0 be a fixed element of S. Then $\mathscr{H}$ being finite dimensional there is a characteristic vector a_0 and a complex scalar c such that

$$\pi(s_0)a_0 = c a_0 . \tag{8}$$

By (7) for any $s \in S$ we have

$$\pi(s_0)\pi(s) = \lambda(s_0^{-1}s^{-1}s_0 s)\pi(s)\pi(s_0) . \tag{9}$$

Combining this with (8) we see that for every $s \in S$ the vector $\pi(s)a_0$ is a characteristic vector of $\pi(s_0)$ with characteristic value $c\lambda(s_0^{-1}s^{-1}s_0 s) = c\mu_s$. Now we can repeat the reasoning which we applied to obtain (7): Since $\mathscr{H}$ is finite dimensional μ_s can assume only finitely many distinct values, but S is connected and μ_s depends continuously on s

and so using $s \mapsto \mu_s$ we see that $\mu_s = \lambda(s_0^{-1} s^{-1} s_0 s)$ must be a constant. By choosing $s = s_0$ we see that this constant is 1. Thus by (9) the space of characteristic vectors belonging to c and satisfying (8) is invariant under every $\pi(s)$ where $s \in S$. However π is irreducible and so this space must be $\mathscr{H}$ and $\pi(s_0)$ must be the operator $\pi(s_0) = cI$. Since $\pi: S \to \mathscr{U}(\mathscr{H})$ is irreducible it follows that $\mathscr{H}$ is one dimensional.

For any closed subgroup S of G let Ω_S denote the set of representations $\rho^\chi: G \to \mathscr{U}(\mathscr{K})$ induced by one dimensional representations $\chi: S \to \mathscr{U}(\mathbb{C})$.

Proposition 9. *If S is a closed, connected and solvable subgroup of the locally compact group G then every finite dimensional irreducible unitary representation of G is similar to a component of some member of Ω_S.*

Proof. Let $\pi: G \to \mathscr{U}(\mathscr{H})$ be a finite dimensional irreducible representation. Since S is solvable and connected by Theorem 8 the restriction $\pi|S$ has a one dimensional subrepresentation and so there is a non-zero vector β in $\mathscr{H}$ and a group character $\chi: S \to T$ such that $\pi(s)\beta = \chi(s)\beta$ for every $s \in S$.

For every α in $\mathscr{H}$ we define $\theta_\alpha: G \to \mathbb{C}$ by $\theta_\alpha(x) = (\pi(x)^* \alpha, \beta)$, so θ_α is continuous by Theorem V.7.3. We have $\theta_\alpha(xs) = \overline{\chi(s)}\theta_\alpha(x)$ and so θ_α belongs to the Hilbert space $\mathscr{K}^\chi$ of the induced representation ρ^χ. The map $\alpha \mapsto \theta_\alpha$ is a linear operator $A: \mathscr{H} \to \mathscr{K}^\chi$. If α belongs to its kernel then $(\alpha, \pi(x)\beta) = 0$ for every x in G, but π being irreducible β is a cyclic vector and so we obtain $\alpha = 0$. Hence A is an injection. For any a in G we have

$$\theta_{\pi(a)\alpha}(x) = (\pi(x)^* \pi(a)\alpha, \beta) = (\pi(a^{-1}x)^* \alpha, \beta) = \theta_\alpha(a^{-1}x).$$

This equation means that $A\pi = \rho^\chi A$ and so $\mathscr{H}$ being finite dimensional π and $\rho^\chi|A\mathscr{H}$ are similar.

Theorem 10. *Let G be a locally compact group with sufficiently many finite dimensional representations and an Iwasawa decomposition $G = KS$. Then $n(\kappa, \mu|K) \leqslant d_\kappa$ for every $\kappa \in \Omega_0(K)$ any every bounded, weakly measurable morphism $\mu: G \to Gl(\mathscr{X})$ where $\mathscr{X}$ is a complex Banach space, μ is strongly continuous and completely irreducible or $\mathscr{X}$ is reflexive and the associated algebra morphism σ is completely irreducible.*

Note. This theorem shows that for such groups G and morphisms μ the spherical functions associated with (K, κ, μ) exist for every choice of κ.

Proof. Let S be the solvable group appearing in the Iwasawa decomposition $G = KS$. First we prove that Ω_S is a σ-complete set of representations of G. For let any non-zero f in $C_0(G)$ be given. Our object is to find ρ^χ in Ω_S such that the associated $L^1(G)$ algebra representation σ^χ satisfies $\sigma^\chi(f)\neq 0$. By Proposition 7 there is a finite dimensional irreducible representation $\rho: G\to\mathscr{U}(\mathscr{H})$ such that $\sigma_\rho(f)\neq 0$. By Proposition 9 there is an induced representation $\rho^\chi: G\to\mathscr{U}(\mathscr{K}^\chi)$ and a linear injection $A:\mathscr{H}\to\mathscr{K}^\chi$ such that $A\rho=\rho^\chi A$. Then for any $\alpha\in\mathscr{H}$ and $\beta\in A\mathscr{H}$ we have

$$(A\sigma(f)\alpha,\beta)=\int_G f(x)(\rho(x)\alpha, A^*\beta)\,dx=\int_G f(x)(\rho^\chi(x)A\alpha,\beta)\,dx$$
$$=(\sigma^\chi(f)A\alpha,\beta)$$

Thus $A\sigma_\rho(f)A^{-1}=\sigma^\chi(f)|A\mathscr{H}$ and so $\sigma^\chi(f)\neq 0$.

Our next step is to prove that $\rho^\chi|K\approx\lambda$ for every χ where λ is the left regular representation of K. Given f in $L^2(K)$ we define $Uf: G\to\mathbb{C}$ by

$$(Uf)(ks)=\overline{\chi(s)}\,f(k).$$

Then $(Uf)(gs)=\overline{\chi(s)}\,(Uf)(g)$ for every $g\in G$ and $s\in S$ because if $g=ks_0$ then $gs=ks_0s$ gives

$$(Uf)(gs)=\overline{\chi(s_0 s)}\,f(k)=\overline{\chi(s)}\,(\overline{\chi(s_0)}\,f(k))=\overline{\chi(s)}\,(Uf)(g).$$

By Proposition V.3.12 the invariant measure of G/S is the Haar measure of K and so we have

$$\|Uf\|^2=\int_{\frac{G}{S}}|Uf(\xi)|^2\,d\xi=\int_K |f(k)|^2\,dk=\|f\|^2$$

where $\|f\|=\|f\|_2$ and $\|Uf\|$ denotes the norm in $\mathscr{K}^\chi$. Moreover U is surjective because if $\theta\in\mathscr{K}^\chi$ then $\theta=Uf$ where f is the restriction of θ to K. Therefore $U: L^2(K)\to\mathscr{K}^\chi$ is a unitary operator. If λ is the left regular representation of G and $a\in K$ then $U\lambda(a)f=U{}^af$ where ${}^af(x)=f(a^{-1}x)$. Hence for $g=ks$ we have

$$U\lambda(a)f(g)=\overline{\chi(s)}\,f(a^{-1}k).$$

On the other hand by the definition of the induced representation

$$(\rho^\chi(a)Uf)(g)=Uf(a^{-1}g)=\overline{\chi(s)}\,f(a^{-1}k).$$

This shows that $U\in(\lambda,\rho^\chi)$. We proved that $\rho^\chi|K\approx\lambda$.

By Theorem 1.36 we have $n(\kappa,\lambda)=d_\kappa$, hence $n(\kappa,\rho^\chi|K)=d_\kappa$. Since Ω_S is a σ-complete set we can apply Proposition 5 and obtain $n(\kappa,\mu|K)\leqslant d_\kappa$ for every morphism $\mu: G\to\mathrm{Gl}(\mathscr{X})$ satisfying the conditions stated in the proposition.

Theorem 11. *Let G be a locally compact group, K a compact subgroup and $\rho: G \to \mathcal{U}(\mathcal{H})$ a representation of G such that $\mathcal{H}_\kappa$ is finite dimensional for every $\kappa \in \hat{K}$. Then $\sigma(f) = \int_G f(x)\rho(x)dx$ is a compact operator for every f in $C_0(G)$.*

Proof. First we consider a special case, namely we suppose that f satisfies $f = d_\kappa \overline{\chi}_\kappa * f$ for some κ in $\hat{K}$ where d_κ is the degree of κ and χ_κ is its character. Then by Theorem 1.33 we have

$$\begin{aligned} E(\kappa)\sigma(f) &= d_\kappa \int_K \overline{\chi}_\kappa(k)\rho(k)dk \int_G f(x)\rho(x)dx \\ &= d_\kappa \int_G \int_K \overline{\chi}_\kappa(k) f(k^{-1}x)\rho(x)dk\,dx \\ &= \int_G d_\kappa \overline{\chi}_\kappa * f(x)\rho(x)dx = \int_G f(x)\rho(x)dx = \sigma(f) \end{aligned}$$

and so $\sigma(f) = E(\kappa)\sigma(f)$. We see that $\sigma(f)$ is an operator of finite rank and so it is compact by Theorem II.2.7.

Next let $\tau: K \to \mathcal{U}(L^2(G))$ be the restriction of the left regular representation of G to the subgroup K i.e. let $\tau(k)g(x) = g(k^{-1}x)$ for every g in $L^2(G)$. By Theorem 1.1 τ is a direct sum of irreducible components. For our purposes it is sufficient to know that $L^2(G)$ is the direct sum of the subspaces $L^2(G)_\kappa$. Thus if f is in $L^2(G)$ then we can write

$$f = \sum_{\kappa \in \hat{K}} E(\kappa) f \tag{10}$$

where convergence is understood in the $L^2(G)$ norm.

Now we return for a moment to the algebra representation σ associated with ρ. Let C be a compact carrier of $h \in C_0(G)$ and let $\alpha, \beta \in \mathcal{H}$. Then using the BCS inequality we obtain

$$|(\sigma(h)\alpha, \beta)|^2 \leqslant \int_C |h(x)|^2 dx \int_C |(\rho(x)\alpha, \beta)|^2 dx \leqslant \|h\|_2^2 \|\alpha\|^2 \|\beta\|^2 \mu(C)$$

where μ denotes the left Haar measure of G. By choosing $\beta = \sigma(h)\alpha$ we see that $\|\sigma(h)\alpha\| \leqslant \|h\|_2 \|\alpha\| \mu(C)^{\frac{1}{2}}$. Hence we proved: If $h \in C_0(G)$ and C is a compact carrier of h then

$$|\sigma(h)| \leqslant \mu(C)^{\frac{1}{2}} \|h\|_2 . \tag{11}$$

By Theorem 1.33 and Proposition V.4.25 the projection

$$E(\kappa): L^2(G) \to L^2(G)_\kappa$$

is given by

$$E(\kappa) f(x) = d_\kappa \int_K \overline{\chi}_\kappa(k) f(k^{-1}x)dk = d_\kappa \overline{\chi}_\kappa * f(x). \tag{12}$$

If f has compact support B then this formula shows that the functions $E(\kappa)f$ $(\kappa \in \hat{K})$ and f all have the common compact carrier KB. Therefore if $\sum E(\kappa)f$ denotes an arbitrary partial sum of the series occurring in (10) then by (11) we have

$$|\sigma(f) - \sum \sigma(E(\kappa)f)| \leqslant \mu(KB)^{\frac{1}{2}} \|f - \sum E(\kappa)f\|_2 .$$

Hence by (12) we see that $\sigma(f)$ is the limit in the operator norm of a linear combination of operators of the type $\sigma(\overline{\chi}_\kappa * f)$. As the latter are already known to be of finite rank by Theorem II.2.7 it follows that $\sigma(f)$ is compact.

Corollary. *If $\rho: G \to \mathscr{U}(\mathscr{H})$ and K are such that $\mathscr{H}_\kappa$ is finite dimensional for every $\kappa \in \hat{K}$ then ρ is a direct sum of irreducible components each of which occurs in ρ with finite multiplicity.*

The corollary follows from the theorem by the same reasoning which was used at the end of Section VI.6: Namely by Proposition IV.6.13 σ is a direct sum of irreducible components each occurring with finite multiplicity on a subspace $\mathscr{K}$ of $\mathscr{H}$ such that

$$\mathscr{K}^\perp \subseteq \{\alpha : \sigma(f)\alpha = 0 \text{ for all } f = f^* \text{ in } C_0(G)\}.$$

By the corollary of Proposition V.5.10 the right hand side is 0 and so $\mathscr{K} = \mathscr{H}$. Now the result follows from the fact that ρ and σ have identical invariant subspaces.

Now we see that Theorem 1 is an immediate consequence of Theorems 10 and 11.

Theorem 12. *Let G be a locally compact group, K a compact subgroup of G and $\rho: G \to \mathscr{U}(\mathscr{H})$ a continuous representation of G such that $n(\kappa, \rho|K) \leqslant M d_\kappa$ for some $M > 0$ and all $\kappa \in \hat{K}$. Then $\sigma(f) = \int_G f(x)\rho(x)\,dx$ is a HS operator for every f in $L_0^2(G)$. More precisely if C is a compact set in G then there is a constant $M_C > 0$ such that $\|\sigma(f)\| \leqslant M_C \|f\|_2$ for every f in $L^2(G)$ with compact carrier C.*

Proof. We suppose that the Haar measure of K is normalized. Then for any f in $L_0^2(G)$ we have

$$\sigma(f) = \int_G f(x)\rho(x)\,dx = \int_K \int_G f(kx)\rho(kx)\,dx\,dk$$

and so we have

$$\sigma(f) = \int_G \left(\int_K f(kx)\rho(k)\,dk \right) \rho(x)\,dx. \tag{13}$$

By Theorem 1.1 $\rho|K$ is a direct sum of irreducible components and so it is equivalent to the representation

$$\sum_{\kappa\in\hat{K}} n(\kappa,\rho|K)\pi_\kappa$$

where π_κ is a fixed irreducible representation belonging to the class κ. Thus for any x in G we have

$$\left\|\int_K f(kx)\rho(k)dk\right\|^2 = \sum_{\kappa\in\hat{K}} n(\kappa,\rho|K)\left\|\int_K f(kx)\pi_\kappa(k)dk\right\|^2$$

where $\|\cdot\|$ denotes the HS norm. Hence by the inequality $n(\kappa,\rho|K)\leqslant M\,d_\kappa$ we get

$$\left\|\int_K f(kx)\rho(k)dk\right\|^2 \leqslant M\sum_{\kappa\in\hat{K}} d_\kappa\left\|\int_K f(kx)\pi_\kappa(k)dk\right\|^2.$$

By Theorem 1.1 and 1.36 the left regular representation λ of K is equivalent to $\sum_{\kappa\in\hat{K}} d_\kappa\pi_\kappa$ and so by the same reasoning as above we have

$$\left\|\int_K f(kx)\lambda(k)dk\right\|^2 = \sum_{\kappa\in\hat{K}} d_\kappa\left\|\int_K f(kx)\pi_\kappa(k)dk\right\|^2.$$

Hence we obtain

$$(14)\qquad \left\|\int_K f(kx)\rho(k)dk\right\| \leqslant M\left\|\int_K f_x(k)\lambda(k)dk\right\|.$$

Now we use some of the results developed in Sections VI.5, 6 and 7. In Definition VI.6.5 we replace G by our compact group K and we let $H=\{e\}$ and $\mathscr{H}=\mathbb{C}$. Then ρ^χ is λ and

$$\sigma_\lambda(f)=\int_K f(kx)\lambda(k)dk.$$

In the beginning of Section VI.5 it was pointed out that the operator $\sigma_\lambda(f_x)$ has a kernel which is $(k,l)\mapsto f_x(kl^{-1})$ where $(k,l)\in K\times K$. Hence by Theorem VI.7.14 we have

$$\left\|\int_K f(kx)\lambda(k)dk\right\|^2=\|\sigma_\lambda(f)\|^2=\int_K\int_K |f_x(kl^{-1})|^2\,dk\,dl=\|f_x\|_2^2.$$

Thus by (14) we obtain

$$(15)\qquad \left\|\int_K f(kx)\rho(k)dk\right\|^2 \leqslant M\int_K |f(kx)|^2\,dk$$

where x is an arbitrary element of G.

Now we suppose that the compact set C is a carrier of f. Then by (13) we have

$$\|\sigma(f)\| \leqslant \int_{KC} \left\| \int_K f(kx)\rho(k)\,dk \right\| dx$$

and so by the BCS inequality

$$\|\sigma(f)\|^2 \leqslant \mu(KC) \int_G \left\| \int_K f(kx)\rho(k)\,dk \right\|^2 dx$$

where $\mu(KC)$ denotes the left Haar measure of KC. Applying (15) we obtain

$$\|\sigma(f)\|^2 \leqslant \mu(KC) M \int_G \int_K |f(kx)|^2\,dk\,dx = \mu(KC)\|f\|_2^2 .$$

Therefore $\sigma(f)$ is a HS operator and $\|\sigma(f)\| \leqslant M_C \|f\|_2$ where $M_C = \mu(KC)^{\frac{1}{2}}$.

Let $\mu: G \to \mathrm{Gl}(\mathscr{X})$ be a bounded weakly measurable morphism of the locally compact group G in the Banach space $\mathscr{X}$. As usual we suppose that μ is strongly continuous or $\mathscr{X}$ is reflexive so that $\sigma_\mu(f) = \int_G f(x)\mu(x)\,dx$ is defined for every f in $L^1(G)$. Our object is to define a generalization of the concept of the character. If $\mathscr{X}$ is finite dimensional then χ_μ is defined and

$$(16) \qquad \chi_\mu(x) = \operatorname{tr} \mu(x) = \sum \alpha_i^*(\mu(x)\alpha_i)$$

for every x in G where α_i $(i \in \mathscr{I})$ is a basis in $\mathscr{X}$ and α_i^* $(i \in \mathscr{I})$ is the corresponding dual basis. Hence by the boundedness and weak measurability of μ the character χ_μ belongs to $L^\infty(G)$. If f is in $L^1(G)$ then

$$\alpha_i^*(\sigma_\mu(f)\alpha_i) = \int_G f(x)\alpha_i^*(\mu(x)\alpha_i)\,dx$$

and so by (16)

$$(17) \qquad \operatorname{tr} \sigma_\mu(f) = \int_G f(x)\chi_\mu(x)\,dx .$$

We shall see in a moment that $\mathscr{X}$ can be an infinite dimensional Hilbert space $\mathscr{H}$ and χ_μ is meaningless but nevertheless $\sigma_\mu(f)$ is a TC operator with non-zero trace for some f in $L^1(G)$. For any finite or infinite dimensional $\mathscr{X}$ and μ satisfying the foregoing conditions we let $\mathscr{I}_\mu$ be the set of those f in $L^1(G)$ for which $\sigma_\mu(f)$ is a TC operator. Since σ_μ is an algebra homomorphism $\mathscr{I}_\mu$ is an ideal in $\mathscr{L}(\mathscr{X})$ cause if $\mathscr{X} = \mathscr{H}$ is a Hilbert space then we can use Theorem II.5.5. We let $\theta_\mu(f) = \operatorname{tr} \sigma_\mu(f)$ and call $\theta_\mu: \mathscr{I}_\mu \to \mathbb{C}$ the *character* or the *distribution character* of μ. By definition μ has a character if $\sigma_\mu(f)$ is a TC operator for some $f \neq 0$ belonging to $L^1(G)$. The restriction of θ_μ to any non-zero ideal $\mathscr{I}$ is called a character of μ. If μ is not only bounded and weakly measurable but also continuous then $\sigma_\mu(\alpha) = \int_G \mu(x)\,d\alpha(x)$ is defined for every measure α in $M(G)$. We can then consider the ideal $\mathscr{I}_\mu$ of those α in $M(G)$ for which $\sigma_\mu(\alpha)$ is a TC operator. If $\mathscr{I}_\mu$ is not the zero ideal then one says that μ has a generalized character or a generalized distribution character. The significance of the distribution character is clear from the following:

Proposition 13. *If the weakly measurable irreducible unitary representations $\pi_k: G \to \mathscr{U}(\mathscr{H}_k)$ $(k=1,2)$ have a common character which is not identically zero then π_1 and π_2 are equivalent.*

Note. A similar result holds for continuous π_1 and π_2 with a common non-zero character defined on an ideal of $M(G)$.

The proposition is an immediate consequence of the following result:

Proposition 14. *If the weakly measurable irreducible representation $\pi: G \to \mathscr{U}(\mathscr{H})$ has a character then π is determined within equivalence by the restriction of θ_π to any ideal $\mathscr{I}$ on which θ_π is defined and does not vanish identically.*

Proof. For f and g in $\mathscr{I}_\pi$ we define

$$(f,g) = \theta_\pi(f^* * g) = \operatorname{tr} \sigma_\pi(f^*)\sigma_\pi(g).$$

We have $(f,f)=0$ if and only if the HS norm $\|\sigma_\pi(f)\|$ is 0 i.e. if $\sigma_\pi(f)=0$. Thus

$$\mathscr{N} = \{f : f \in \mathscr{I}_\pi \text{ and } (f,f)=0\}$$

is an ideal in $L^1(G)$ and so $(f+\mathscr{N}, g+\mathscr{N}) = (f,g)$ defines an inner product on $\mathscr{I}/\mathscr{N}$. The completion of $\mathscr{I}/\mathscr{N}$ with respect to the norm derived from this inner product will be denoted by $\mathscr{K}$.

For every a in G we have

$$\sigma_\pi({}_a f) = \pi(a^{-1})\sigma_\pi(f) \quad \text{and} \quad \sigma_\pi(f_a) = \sigma_\pi(f)\pi(a^{-1}) \tag{18}$$

and so we see that $\mathscr{N}$ is invariant under left and right translations i.e. ${}_a\mathscr{N} = \mathscr{N}_a = \mathscr{N}$ for every a in G. Therefore given x and y in G the correspondence $f + \mathscr{N} \mapsto {}^x f_y + \mathscr{N}$ defines a linear map on $\mathscr{I}/\mathscr{N}$. By (18) we have

$$\sigma_\pi(({}^x f_y)^* * {}^x f_y) = \sigma_\pi({}^x f_y)^* \sigma_\pi({}^x f_y) = \pi(y)\sigma_\pi(f^*)\sigma_\pi(f)\pi(y)^{-1}.$$

Hence we obtain

$$({}^xf_y, {}^xf_y) = \operatorname{tr} \pi(y)\sigma_\pi(f^* * f)\pi(y)^{-1} = \operatorname{tr} \sigma_\pi(f^* * f) = (f, f).$$

Therefore the map $f + \mathcal{N} \mapsto {}^xf_y + \mathcal{N}$ is an isometry and so it can be extended in a unique manner to a unitary operator $v(x,y): \mathcal{K} \to \mathcal{K}$. It is clear that $(x,y) \mapsto v(x,y)$ is a unitary representation of $G \times G$ in $\mathcal{K}$ which is determined by the restriction of θ_π to $\mathcal{I}$.

Let $\mathrm{HS}(\mathcal{H})$ be the Hilbert space of HS operators $T: \mathcal{H} \to \mathcal{H}$ under the HS inner product and let $U: \mathcal{I}/\mathcal{N} \to \mathrm{HS}(\mathcal{H})$ be defined by $f + \mathcal{N} \mapsto \sigma_\pi(f)$. Since

$$(f + \mathcal{N}, f + \mathcal{N}) = \operatorname{tr} \sigma_\pi(f^*)\sigma_\pi(f) = (\sigma_\pi(f), \sigma_\pi(f))$$

the linear map U is an isometry and so it can be extended to a unitary operator U from $\mathcal{K}$ into $\mathrm{HS}(\mathcal{H})$. The operator $\bar{\pi} \otimes \pi(x, y)$ acts on the antilinear HS operators $T: \bar{\mathcal{H}} \to \mathcal{H}$ i.e. on the elements of $\mathrm{HS}(\mathcal{H})$. We have

$$\begin{aligned} \bar{\pi} \otimes \pi(x, y) U(f + \mathcal{N}) &= \bar{\pi}(x) \otimes \pi(y)\, \sigma_\pi(f) = \pi(x)\sigma_\pi(f)\pi(y)^* \\ &= \sigma_\pi({}^xf_y) = U({}^xf_y + \mathcal{N}) = U\, v(x, y)(f + \mathcal{N}). \end{aligned}$$

Therefore U intertwines v and $\bar{\pi} \otimes \pi | U\mathcal{K}$. By the condition imposed on $\theta_\pi | \mathcal{I}$ we see that $\mathcal{K}$ has positive dimension. Since π is irreducible by Theorem VI.3.15 so is $\bar{\pi} \otimes \pi$ and so it follows that $U\mathcal{K} = \mathrm{HS}(\mathcal{H})$. We proved that $\bar{\pi} \otimes \pi$ is equivalent to v where v is uniquely determined by the ideal $\mathcal{I}$ and the non-trivial restriction $\theta_\pi | \mathcal{I}$.

Theorem 15. *If G is a locally compact group, K is a compact subgroup of G and $\rho: G \to \mathcal{U}(\mathcal{H})$ is a continuous representation such that $n(\kappa, \rho | K) \leq M d_\kappa$ for some $M > 0$ and all $\kappa \in \hat{K}$ then ρ has a non-zero character θ_ρ.*

Proof. If f is in $L_0^2(G)$ then by Theorem 12 $\sigma(f)$ is a HS operator. Hence if $f, g \in L_0^2(G)$ then by Theorem II.5.2 $\sigma(f * g)$ is in the trace class. It follows that the ideal $\mathcal{I}_\rho$ contains the linear manifold generated by the functions $f * g$ where $f, g \in L_0^2(G)$. Therefore ρ has a character θ_ρ. By the corollary of Proposition V.5.10 there is an f in $C_0(G)$ satisfying $f = f^*$ such that $\sigma(f) \neq 0$. Then $\sigma(f * f)$ is a strictly positive trace class operator and so $\operatorname{tr} \sigma(f * f) > 0$ by Theorem II.3.4. Hence θ_ρ is not identically zero on $\mathcal{I}_\rho$.

Theorem 16. *If G is a locally compact group with sufficiently many finite dimensional representations and Iwasawa decomposition $G = KS$ then every continuous irreducible representation $\pi: G \to \mathcal{U}(\mathcal{H})$ has a non-zero character θ_π.*

Proof. By Theorem 10 we have $n(\kappa, \rho|K) \leqslant d_\kappa$ for every $\kappa \in \hat{K}$ and so we can apply Theorem 15.

Harish-Chandra proved that every continuous irreducible representation π of a semi-simple connected Lie group has a non-zero character θ_π and $\mathscr{I}_\pi$ includes every C^∞-function $f: G \to \mathbb{C}$ having compact support. He also proved that the linear functional $\theta_\pi: C_0^\infty \to \mathbb{C}$ is a distribution. This is the reason that θ_ρ is often called a distribution character.

We finish this section by explaining the terminology "*group G with an Iwasawa decomposition*". In Chapter VIII we shall define what is meant by a semisimple Lie group G and its various Lie subgroups. Iwasawa proved that for every connected semisimple Lie group G there exist Lie subgroups K, A and N of G such that the following holds: The group K is in a certain sense "essentially compact", A is abelian, N is nilpotent and A and N are both simply connected. Moreover $(k, a, n) \mapsto kan$ is an analytic diffeomorphism of the product manifold $K \times A \times N$ onto G. Briefly we have the Iwasawa decomposition $G = KAN$. If we let $S = AN$ then S is a solvable Lie subgroup of G and $G = KS$. This is the decomposition whose existence is required when we say that the locally compact group G has an Iwasawa decomposition.

As an illustration we discuss the case of the group $G = \mathrm{Sl}(2, \mathbb{R})$. Then K is the compact group $\mathrm{SO}(2, \mathbb{R})$, the group A consists of the diagonal matrices $\begin{pmatrix} a & 0 \\ 0 & 1/a \end{pmatrix}$ where $a > 0$ and N is the subgroup of parabolic elements $\begin{pmatrix} 1 & b \\ 0 & 1 \end{pmatrix}$ where $b \in \mathbb{R}$. Both A and N are homeomorphic and diffeomorphic to the real line. If

$$g = \begin{pmatrix} g_{11} & g_{12} \\ g_{21} & g_{22} \end{pmatrix} \quad \text{and} \quad k = \begin{pmatrix} \cos\varphi & \sin\varphi \\ -\sin\varphi & \cos\varphi \end{pmatrix}$$

then

$$k^{-1} g = \begin{pmatrix} g_{11}\cos\varphi - g_{21}\sin\varphi & * \\ g_{11}\sin\varphi + g_{21}\cos\varphi & * \end{pmatrix}.$$

The equation $g_{11}\sin\varphi + g_{21}\cos\varphi = 0$ has exactly one solution φ in $0 \leqslant \varphi < 2\pi$ such that $g_{11}\cos\varphi - g_{21}\sin\varphi = 0$. Then by $g_{11}\tan\varphi + g_{21} = 0$ we have $\sin\varphi = -g_{21}/a$ and $\cos\varphi = g_{11}/a$ where $a^2 = g_{11}^2 + g_{21}^2$. Hence we obtain

$$\begin{pmatrix} g_{11} & g_{12} \\ g_{21} & g_{22} \end{pmatrix} = \begin{pmatrix} g_{11}a^{-1} & -g_{21}a^{-1} \\ g_{21}a^{-1} & g_{11}a^{-1} \end{pmatrix} \begin{pmatrix} a & 0 \\ 0 & a^{-1} \end{pmatrix} \begin{pmatrix} 1 & (g_{11}g_{12} + g_{21}g_{22})a^{-2} \\ 0 & 1 \end{pmatrix}.$$

This is the Iwasawa decomposition of the arbitrary group element $g \in G$. The elements of the group $S = AN$ are the matrices $\begin{pmatrix} a & b \\ 0 & a^{-1} \end{pmatrix}$ where $a > 0$ and $b \in \mathbb{R}$. The space of S is homeomorphic to the plane and since $S' = N$ is commutative the group S is solvable.

We also note that the Haar measure of $G = \mathrm{Sl}(2, \mathbb{R})$ can be expressed in terms of the Haar measures of the subgroups K, A and N. Namely we have

$$\int_G f(g)\,dg = \int_K \int_A \int_N f(kan)\,a^{-2}\,dk\,da\,dn = \int_0^{2\pi} \int_0^{\infty} \int_{-\infty}^{\infty} f(\varphi, a, b)\,a^{-2}\,d\varphi\,da\,db\,.$$

6. Trace Formulas

Here we shall derive various consequences of the general trace formula obtained in Section IV.6 by using the theory of spherical functions. The proof of the Selberg-Tamagawa trace formula uses only zonal spherical functions. This is a special case of a family of trace formulas which are considerably deeper for their proof depends among other things on Theorem 4.1.

Theorem 1. *Let G and its closed subgroup H be unimodular, let $\chi: H \to \mathcal{U}(\mathcal{H})$ be a unitary representation of H in the separable Hilbert space $\mathcal{H}$ and let ρ^χ be the representation induced on G by χ. Let f in $L^1(G)$ be such that*

$$\sigma(f) = \int_G f(x)\,\rho^\chi(x)\,dx$$

is a trace class operator. Then

$$\sum_{\omega \in \hat{G}} n(\omega)\,\mathrm{tr}\,\sigma(f)(\omega) = \int_{\frac{G}{H}} \int_H f(xhx^{-1})\,\mathrm{tr}\,\chi(h)\,dh\,d\xi$$

where the series is absolutely convergent and $d\xi$ denotes the invariant measure of G/H.

Notes. 1. In view of Theorem V.5.17 the summation could also be considered over $\omega \in \widehat{L^1(G)}$.

2. The symbol $n(\omega)$ denotes the multiplicity of ω in ρ^χ or in σ. We know that if ω denotes both a class ω in $\hat{G}$ and the class corresponding to it in $\widehat{L^1(G)}$ in accordance with Theorem V.5.17 then $n(\omega, \rho^\chi) = n(\omega, \sigma)$.

3. The expression $\mathrm{tr}\,\sigma(f)(\omega)$ denotes the trace of $\sigma(f)$ on any irreducible component $\mathcal{P}$ such that $\rho^\chi|\mathcal{P}$ or $\sigma|\mathcal{P}$ belongs to the class ω.

4. By Proposition IV.6.6 we know that if $n(\omega)=\infty$ then $\operatorname{tr}\sigma(f)(\omega)$ is 0 and the case $n(\omega)=0$ is of no interest.

5. The invariant measure $d\xi$ and the Haar measures dx and dh are so chosen that $dx=dh\,d\xi$. For the details see Proposition V.3.14.

6. If $\mathscr{H}$ is infinite dimensional then the only f satisfying the requirement is 0. For if $\chi(x)$ is a TC operator then so is $\chi(x)^*$ and $\chi(x)\chi(x)^*=I$.

Proof. We let $\mathscr{A}$ denote the v. Neumann algebra associated with the induced representation ρ^χ. We let $\sigma: L^1(G)\to\mathscr{L}(\mathscr{K})$ be the algebra representation associated with ρ^χ which is described in Theorem V.5.4. By Theorem V.5.9 the v. Neumann algebra associated with σ is $\mathscr{A}$. Since ρ^χ and σ generate the same algebra by Proposition III.5.10 the operator families $\rho^\chi(G)$ and $\sigma(L^1(G))$ have identical invariant subspaces.

The general trace formula stated in Theorem IV.6.14 expresses the trace of an operator $A\in\mathscr{A}$ in the form of an infinite series which for simplicity will be abbreviated by LHS meaning the left hand side of the trace formula. Similarly we shall use RHS to denote the right hand side. If f is in $L^1(G)$ then the operator $\sigma(f)$ is in $\mathscr{A}$. Hence if $\sigma(f)$ is in the trace class then we can apply Theorem IV.6.14 either with $S=G$ or with $S=L^1(G)$. Since ρ^χ is a strongly continuous representation in the first case the summation in LHS is on $\omega\in\hat{G}$ and in the second on $\omega\in\widehat{L^1(G)}$. In either case we see that LHS is equal to RHS which is $\operatorname{tr}\sigma(f)$. The operator $\sigma(f)$ has a kernel K_f which is given in Proposition VI.5.8. By Theorem VI.7.17 the trace of $\sigma(f)$ can be written as

$$\operatorname{tr}\sigma(f)=\int_{\frac{G}{H}}\operatorname{tr}K_f(\xi,\xi)\,d\xi\,.$$

By substituting for K_f the expression given in Proposition VI.5.8 we obtain the expression which appears on RHS in the theorem.

We note that the unimodularity of G and H is not essential. For instance if G and H are such that G/H has a relatively invariant measure $d\xi$ with modular function D then the RHS becomes

$$\int_{\frac{G}{H}}\int_H f(xhx^{-1})\frac{D(x)}{\Delta(x)}\sqrt{D(h)}\operatorname{tr}\chi(h)\,dh\,d\mu(\xi)$$

where Δ is the modular function of G. More generally if $d\mu$ is the quasi-invariant measure defined by a continuous positive function $\rho: G\to(0,\infty)$

then the RHS is

$$\int_{G/H}\int_H f(xhx^{-1})\frac{\Delta(h)}{\delta(h)\Delta(x)\sqrt{\rho(xh^{-1})\rho(x)}}\operatorname{tr}\chi(h)\,dh\,d\mu(\xi)\,.$$

If the measures dx, dh and $d\mu(\xi)$ are so normalized that

$$\int_G g(x)\rho(x)dx = \int_{G/H}\int_H g(xh)\,dh\,d\mu(\xi)$$

for every g in $C_0(G)$ then LHS and RHS are equal. They are different expressions for $\operatorname{tr}\sigma(f)$. If $\chi(h)=I$ for every h in H then the result is interesting only in the special case when $\dim\mathscr{H}<\infty$.

Let G be a unimodular group and let K be a compact subgroup of G. As earlier we let $L^1(G,K)$ denote the set of those f in $L^1(G)$ which are bi-invariant under K. Then $L^1(G,K)$ is an involutive subalgebra of $L^1(G)$. We let $\widehat{L^1(G,K)}$ denote the set of equivalence classes of irreducible representations of $L^1(G,K)$.

Theorem 2. *If G is unimodular and $L^1(G,K)$ is commutative then there is a natural one-to-one correspondence between $\widehat{L^1(G,K)}$ and those zonal spherical functions corresponding to the subgroup K which satisfy $\|\omega\|_\infty=1$. Namely if ω is such a zonal spherical function then the corresponding element of $\widehat{L^1(G,K)}$ is the equivalence class containing the homomorphism $\hat{\omega}: L^1(G,K)\to\mathbb{C}$ where*

(1) $$\hat{\omega}(f)=\int_G f(x^{-1})\omega(x)\,dx\,.$$

Note. The exponent -1 could be omitted. The present formulation is in accordance with the familiar Fourier transform formula valid in the case of a commutative group G. The equivalence class and the corresponding zonal spherical function will be denoted by the same symbol ω.

Proof. 1. Let a zonal spherical function ω satisfying $\|\omega\|_\infty=1$ be given. Then (1) defines a continuous linear functional $\hat{\omega}: L^1(G)\to\mathbb{C}$ and $\|\hat{\omega}\|=1$. By Theorem 2.8 the restriction of $\hat{\omega}$ to $C_0(G,K)$ is a homomorphism of the involutive algebra $C_0(G,K)$ onto $\mathbb{C}$. Using

$$f_n * g_n - f*g = (f_n - f)*g_n + f*(g_n - g)$$

we see that $\hat{\omega}(f*g)=\hat{\omega}(f)\hat{\omega}(g)$ also on $L^1(G,K)$. Then $z\to\hat{\omega}(f)z$ defines a linear operator on $\mathscr{H}=\mathbb{C}$ which we can denote by $\hat{\omega}(f)$ and $f\to\hat{\omega}(f)$ is a bounded irreducible representation of $L^1(G,K)$. The class containing this representation is by definition the image of the zonal spherical function ω satisfying $\|\omega\|_\infty = 1$ in $\widehat{L^1(G, K)}$.

2. Now let a class ω be given in $\widehat{L^1(G, K)}$. By Proposition IV.2.7 we have $d_\omega = 1$. Let $\pi: L^1(G, K) \to \mathscr{L}(\mathscr{H})$ be an irreducible representation such that $\pi \in \omega$. Choose α in $\mathscr{H}$ such that $\|\alpha\| = 1$ and define the linear functional $\hat{\omega}: L^1(G, K) \to \mathbb{C}$ by $\hat{\omega}(f) = (\pi(f)\alpha, \alpha)$. Since $\mathscr{H}$ is one dimensional $\hat{\omega}$ depends only on the class ω which contains π. It is clear that $\hat{\omega}$ is a homomorphism of the involutive algebra $L^1(G, K)$ into $\mathbb{C}$. By using part of the proof of Theorem I.8.8 $\hat{\omega}$ is continuous and $|\hat{\omega}(f)| \leqslant \|f\|_1$ for every f in $L^1(G, K)$. Since $\pi \neq 0$ we have $\hat{\omega}(f * f) > 0$ for some f and so $\|\hat{\omega}\| > 0$. Using the Hahn-Banach theorem $\hat{\omega}$ can be extended to a linear functional $\hat{\omega}: L^1(G) \to \mathbb{C}$ such that its norm $\|\hat{\omega}\|$ coincides with the original norm and so $0 < \|\hat{\omega}\| \leqslant 1$.

By the Riesz representation theorem we can find a function φ in $L^\infty(G)$ such that $\|\varphi\|_\infty = \|\hat{\omega}\|$ and

$$\hat{\omega}(f) = \int_G f(x^{-1})\varphi(x)\,dx.$$

Here we used the unimodularity of G.

We prove that the function $\omega: G \to \mathbb{C}$ obtained from φ by the formula

$$\omega(x) = {}^o\varphi^o(x) = \int_K \int_K \varphi(k_1 x k_2)\,dk_1\,dk_2$$

is almost everywhere equal to a zonal spherical function with respect to the compact subgroup K. It is clear that ω is bi-invariant with respect to K and φ being bounded $\omega \in Li(G)$. We also see that $\|\omega\|_\infty \leqslant \|\varphi\|_\infty \leqslant 1$. If $f \in L^1(G, K)$ then for any k_1, k_2 in K we have

$$\hat{\omega}(f) = \int_G f(x^{-1})\varphi(x)\,dx = \int_G f(k_2 x^{-1} k_1)\varphi(x)\,dx = \int_G f(x^{-1})\varphi(k_1 x k_2)\,dx.$$

By integrating with respect to k_1 and k_2 we obtain

$$\hat{\omega}(f) = \int_G f(x^{-1})\omega(x)\,dx. \tag{2}$$

Since $\hat{\omega}(f) \neq 0$ for some f in $L^1(G, K)$ this shows that the function ω is not identically zero. We know that $\hat{\omega}$ is a homomorphism of the involutive algebra $L^1(G, K)$ onto $\mathbb{C}$. Hence by (2) we can apply Theorem 2.8. It follows that ω is a zonal spherical function. We know already that $\|\omega\|_\infty \leqslant 1$ and so using $\omega(e) = 1$ we obtain $\|\omega\|_\infty = 1$.

3. Let Ω denote the set of zonal spherical functions satisfying $\|\omega\|_\infty = 1$. In 1. we defined a map $\Omega \to \widehat{L^1(G, K)}$ and in 2. we constructed a map

$\widehat{L^1(G,K)} \to \Omega$. We now prove that the composition of these two maps in either order is the identity.

First let ω in Ω be given. In 1. we determined its image in $\widehat{L^1(G,K)}$ as the class containing the representation $f \mapsto \hat{\omega}(f)$ of $L^1(G,K)$. In order to find the image of this class under the map described in 2. we choose $\alpha = 1$ in $\mathscr{H} = \mathbb{C}$ and consider the linear functional

$$f \mapsto (\pi(f)\alpha, \alpha) = (\pi(f)1, 1) = (\hat{\omega}(f)1, 1) = \hat{\omega}(f).$$

Here $\hat{\omega}(f)$ denotes the complex number obtained from the original zonal spherical function ω by the formula given in (2). Therefore we can choose $\varphi = \omega$ when the Riesz representation theorem is used in the next step. Then ${}^o\varphi^o(x)$ is the original zonal spherical function ω and so the composite map $\Omega \to \widehat{L^1(G,K)} \to \Omega$ is the identity.

Next let ω be an element of $\widehat{L^1(G,K)}$. In 2. we associate with this class ω the zonal spherical function ${}^o\varphi^o$ where φ is such that

$$(\pi(f)\alpha, \alpha) = \int_G f(x^{-1})\,{}^o\varphi^o(x)\,dx \tag{3}$$

for every f in $L^1(G,K)$ with some irreducible representation $\pi: L^1(G,K) \to \mathscr{L}(\mathscr{H})$ belonging to the class ω and some α in $\mathscr{H}$ satisfying $\|\alpha\| = 1$. To obtain the image of ${}^o\varphi^o$ under the map $\widehat{L^1(G,K)} \to \Omega$ described in 1. we consider the linear functional ${}^o\varphi^o$ given by the right hand side of (3). Therefore the linear operator $\widehat{{}^o\varphi^o(f)}$ defined in the next step becomes $z \mapsto (\pi(f)\alpha, \alpha)z$ where $z \in \mathbb{C}$. The Hilbert spaces of the representations $\widehat{{}^o\varphi^o}$ and π are $\mathbb{C}$ and $\mathscr{H} = \mathbb{C}\alpha$, respectively. By $\|\alpha\| = 1$ the map $z \mapsto z\alpha$ is a unitary operator $U: \mathbb{C} \to \mathbb{C}\alpha$. We have

$$\pi(f)Uz = \pi(f)z\alpha = z(\pi(f)\alpha) = z(\pi(f)\alpha, \alpha)\alpha = U(\pi(f)\alpha, \alpha)z = U\widehat{{}^o\varphi^o(f)}z$$

because $\mathscr{H}$ being one dimensional $\pi(f)\alpha = (\pi(f)\alpha, \alpha)\alpha$. Hence $\widehat{{}^o\varphi^o}$ and π are equivalent and the class associated with $\widehat{{}^o\varphi^o}$ is the original class ω containing π. We proved that the composite map $\widehat{L^1(G,K)} \to \Omega \to \widehat{L^1(G,K)}$ is the identity.

These facts show that the map $\Omega \to \widehat{L^1(G,K)}$ described in 1. and stated also in the theorem is a bijection. Theorem 2 is proved.

Proposition 3. *Let K be a compact subgroup of the unimodular group G, let $L^1(G,K)$ be commutative and let $\tau: L^1(G,K) \to \mathscr{L}(\mathscr{K})$ be a representation. Then for every class ω in $\widehat{L^1(G,K)}$ the maximal primary component $\mathscr{K}_\omega$ consists of those θ in $\mathscr{K}$ for which $\tau(f)\theta = \hat{\theta}(f)\theta$ for every f in $L^1(G,K)$. Here $\hat{\theta}$ denotes the linear functional $\hat{\theta}: L^1(G,K) \to \mathbb{C}$ determined by the zonal spherical function associated with the class ω.*

Note. The zonal spherical function ω and the functional $\hat{\theta}$ are given in Theorem 2. The subspace $\mathscr{K}_\omega$ is described in Theorem IV.5.4 and in the corollary of Proposition IV.7.11. We say that the elements of $\mathscr{K}_\omega$ belong to the spherical function ω.

Proof. By definition $\mathscr{K}_\omega$ is spanned by a maximal orthogonal system of irreducible components belonging to the class ω. Since $L^1(G,K)$ is commutative each of these components is one dimensional by Proposition IV.2.7. Therefore $\mathscr{K}_\omega$ is spanned by an orthonormal family of vectors θ_i $(i\in\mathscr{I})$ such that $\tau(f)\theta_i=\lambda_{if}\theta_i$ for some $\lambda_{if}\in\mathbb{C}$. In part 1. of the proof of Theorem 2 we have seen that if $\pi\colon L^1(G,K)\to\mathscr{L}(\mathscr{H})$ is in the class ω and α in $\mathscr{H}$ is such that $\|\alpha\|=1$ then

$$(\pi(f)\alpha,\alpha)=\hat{\omega}(f)=\int_G f(x^{-1})\omega(x)\,dx$$

where ω is the zonal spherical function associated with the class ω. If we choose $\mathscr{H}=\mathbb{C}\theta_i$ and $\pi=\tau|\mathscr{H}$ then we obtain

$$\hat{\omega}(f)=(\pi(f)\alpha,\alpha)=(\tau(f)\theta_i,\theta_i)=\lambda_{if}(\theta_i,\theta_i)=\lambda_{if}\,.$$

Therefore we have $\tau(f)\theta_i=\hat{\omega}(f)\theta_i$ for every $i\in\mathscr{I}$.

If θ in $\mathscr{K}$ is such that $\tau(f)\theta=\hat{\theta}(f)\theta$ for every f in $L^1(G,K)$ then $\mathbb{C}\theta$ is an invariant subspace of τ and one can easily see that $\tau|\mathbb{C}\theta$ belongs to the class ω unless $\theta=0$: In fact for any choice of i the irreducible representation $\tau|\mathbb{C}\theta_i$ is known to belong to the class ω. The map $z\theta\mapsto z\|\theta\|\theta_i$ is a unitary transformation U of $\mathbb{C}\theta$ onto $\mathbb{C}\theta_i$ and U intertwines $\tau|\mathbb{C}\theta$ and $\tau|\mathbb{C}\theta_i$. Hence $\tau|\mathbb{C}\theta$ is in ω and by Theorem IV.5.4 belongs to $\mathscr{K}_\omega$.

In what follows we let G be a unimodular group, K a compact subgroup and H a closed unimodular subgroup of G. Let $\chi\colon H\to\mathscr{U}(\mathscr{H})$ be a unitary representation of H and let $\rho^\chi\colon G\to\mathscr{U}(\mathscr{K})$ be the representation induced on G by χ. This determines a representation $\sigma\colon L^1(G)\to\mathscr{L}(\mathscr{K})$. The restriction of σ to $L^1(G,K)$ is a representation $\tau\colon L^1(G,K)\to\mathscr{L}(\mathscr{K})$.

Let ω be in $\widehat{L^1(G,K)}$ and let $\mathscr{K}_\omega$ be the maximal primary component of τ determined by the class ω. (See Theorems IV.5.4 and IV.7.12.) Hence $\mathscr{K}_\omega$ is a subspace of the Hilbert space $\mathscr{K}$ of the induced representation ρ^χ and $\mathscr{K}_\omega$ is invariant under every $\tau(f)$ where $f\in L^1(G,K)$.

As usual for $\theta\in\mathscr{K}$ and $k\in K$ let ${}_k\theta(x)=\theta(kx)$. Since ${}_k\theta=\rho^\chi(k^{-1})\theta$ we have ${}_k\theta\in\mathscr{K}$. We let $\mathscr{K}_I$ be the subspace of $\mathscr{K}$ on which $\rho^\chi|K$ is the identity. Hence $\mathscr{K}_I$ consists of those elements θ in $\mathscr{K}_I$ which satisfy ${}_k\theta=\theta$ for every $k\in K$.

Proposition 4. *If G and H are unimodular and $L^1(G,K)$ is commutative then $\mathscr{K}_\omega \subseteq \mathscr{K}_I$ for every ω in $\widehat{L^1(G,K)}$.*

Proof. Let f be in $L^1(G,K)$ and let $\theta, \psi \in \mathscr{K}$. By Definition VI.2.4 we have

$$(\sigma(f)\theta,\psi) = \int_G f(y)(\rho^\chi(y)\theta,\psi)\,dy = \int_G f(yk) \int_{\frac{G}{H}} (\theta(y^{-1}x),\psi(x))\,d\xi\,dy.$$

Hence using the substitution $yk \to y$ we obtain

$$(\sigma(f)\theta,\psi) = \int_G f(y) \int_{\frac{G}{H}} (\theta(ky^{-1}x),\psi(x))\,d\xi\,dy = (\sigma(f)_k\theta,\psi).$$

This shows that $\sigma(f)\theta = \sigma(f)_k\theta$ for every f in $L^1(G,K)$, $k \in K$ and $\theta \in \mathscr{K}$. If $\theta \in \mathscr{K}_\omega$ for some ω then by Proposition 3

$$\hat{\omega}(f)\theta = \sigma(f)\theta = \sigma(f)_k\theta = \hat{\omega}(f)_k\theta.$$

By Lemma 2.4 we know that $\hat{\omega}$ is not the zero functional and so we obtain $\theta = {}_k\theta$. Since $k \in K$ is arbitrary $\theta \in \mathscr{K}_I$ and we proved that $\mathscr{K}_\omega \subseteq \mathscr{K}_I$.

Definition 5. *The elements of $\mathscr{K}_\omega$ are called automorphic functions belonging to the zonal spherical function ω and the associated class of irreducible representations of $L^1(G,K)$.*

In more detail an automorphic function θ is determined by the unimodular group G, the compact subgroup K and the closed subgroup H, by the representation $\chi: H \to \mathscr{U}(\mathscr{H})$ and the class ω in $\widehat{L^1(G,K)}$. The zonal spherical function associated with ω is given in Theorem 2. It is supposed that the involutive algebra $L^1(G,K)$ is commutative. The automorphic function θ is a function $\theta: G \to \mathscr{H}$ such that

1. the function $x \mapsto (\theta(x),\alpha)$ is measurable for every $\alpha \in \mathscr{H}$;
2. $\theta(xh) = \chi(h)^*\theta(x)$ for every $x \in G$ and $h \in H$;
3. $\int_{\frac{G}{H}} \|\theta(x)\|^2\,d\xi$ is finite;
4. $\sigma(f)\theta = \hat{\theta}(f)\theta$ for every $f \in L^1(G,K)$;
5. $\theta(kx) = \theta(x)$ for every $k \in K$ and $x \in G$.

We also mention that $\sigma(f)\theta$ could also be denoted by $f*\theta$ because due to $\rho^\chi(y)\theta(x) = \theta(y^{-1}x)$ we have $\sigma(f)\theta(x) = \int_G f(y)\theta(y^{-1}x)\,dy$.

Proposition 6. *The multiplicity* $n(\omega,\tau)$ *of a class* $\omega\in\widehat{L^1(G,K)}$ *in* $\tau=\sigma\,|\,L^1(G,K)$ *is equal to the dimension of the vector space of automorphic functions belonging to* ω.

Proof. By the definition of the multiplicity of an irreducible component in Section IV.5 we have $n(\omega,\tau)=\dim\mathscr{K}_\omega$. Therefore the result is an immediate consequence of the definition of automorphic functions.

Proposition 7. *If* $n(\omega,\tau)>0$ *then the zonal spherical function associated with the class* $\omega\in\widehat{L^1(G,K)}$ *is positive definite.*

Proof. Let θ be a non-zero automorphic function in $\mathscr{K}_\omega$. For any h in H and $s,x\in G$ we have

$$(\theta(sxh),\theta(xh))=(\theta(sx),\theta(x))$$

so we can define $\psi\colon G\to\mathbb{C}$ by

$$\psi(s)=\int_{\frac{G}{H}}(\theta(sx),\theta(x))\,d\xi$$

where $d\xi$ is an invariant measure on G/H. The existence of the integral defining ψ follows from the fact that $\|\theta(x)\|$ is square integrable over G/H. By applying the BCS inequality we see not only the existence of $\psi(s)$ but we also obtain $|\psi(s)|\leqslant\psi(e)$. Clearly $\psi(s)=(\rho^\chi(s^{-1})\theta,\theta)$ for all $s\in G$.

We are going to prove that $\psi/\psi(e)$ is a positive definite zonal spherical function and at the end we shall see that this is the function which is associated with the class ω according to Theorem 2.

First of all ψ is not identically zero because $\psi(e)=\|\theta\|^2>0$ by $\theta\neq0$. By Proposition 4 we have $\theta(kx)=\theta(x)$ and so using the invariance of $d\xi$ under the action of G we have

$$\psi(k_1sk_2)=\int_{\frac{G}{H}}(\theta(sk_2x),\theta(x))\,d\xi=\int_{\frac{G}{H}}(\theta(sx),\theta(x))\,d\xi=\psi(s).$$

This shows that ψ is bi-invariant under K.

Next we prove that $f*\psi=\hat{\omega}(f)\psi$ for every f in $L^1(G,K)$: If $s\in G$ then using the substitution $tx\to t$ by the unimodularity of G we have

$$\begin{aligned}f*\psi(s)&=\int_G f(st^{-1})\int_{\frac{G}{H}}(\theta(tx),\theta(x))\,d\xi\,dt\\&=\int_{\frac{G}{H}}\int_G f(sxt^{-1})(\theta(t),\theta(x))\,dt\,d\xi=\int_{\frac{G}{H}}\Big(\int_G f(sxt^{-1})\theta(t)\,dt,\,\theta(x)\Big)d\xi.\end{aligned}$$

By the definition of ρ^{χ} and by Theorem VI.5.7 and the corollary of Theorem V.3.9 the inner integral is $\tau(f)\theta(s\,x)$. Hence by Proposition 3

$$f*\psi(s)=\int_{\frac{G}{H}}(\hat{\omega}(f)\theta(s\,x),\theta(x))\,d\xi=\hat{\omega}(f)\psi(s).$$

Now we prove that ψ is positive definite: Let $c_1,\ldots,c_n\in\mathbb{C}$ and $s_1,\ldots,s_n\in G$ be given. Then

$$\sum c_k\bar{c}_l\psi(s_k s_l^{-1})=\sum c_k\bar{c}_l\int_{\frac{G}{H}}(\theta(s_k s_l^{-1}x),\theta(x))\,d\xi.$$

We make the substitution $x\to s_l x$ and interchange the summation with the integration to obtain

$$\int_{\frac{G}{H}}\sum(c_k\theta(s_k x),c_l\theta(s_l x))\,d\xi=\int_{\frac{G}{H}}\|\sum c_k\theta(s_k x)\|^2\,d\xi.$$

The last integral is non-negative and so ψ is positive definite.

Next we prove that ψ is a continuous function. We already know that ψ is bounded and $|\psi(s)|\leqslant\psi(e)$ for every s in G. We have also shown that $f*\psi=\hat{\omega}(f)\psi$ for every f in $L^1(G,K)$ and so it is sufficient to prove that $f*\psi$ is continuous for some f satisfying $\hat{\omega}(f)\neq 0$. By Lemma 2.3 the space $L^1(G,K)$ contains a non-zero continuous function f with compact support and by Lemma 2.4 f can be chosen such that $\hat{\omega}(f)\neq 0$. Let S denote the compact support of f. We can now reason the same way as in the proof of Proposition V.4.7: We fix x in G, a compact neighborhood C of the identity in G and choose an $\varepsilon>0$. By the right uniform continuity of f there is a symmetric neighborhood N_e in G such that $|f(xy)-f(zy)|<\varepsilon$ for every y in G and for every z satisfying $xz^{-1}\in N_e$. We may suppose that $N_e\subseteq C$. Hence if $xz^{-1}\in N_e$ then

$$|f*\psi(x)-f*\psi(z)|\leqslant\int|f(xy)-f(zy)|\cdot\|\psi\|_\infty\,dy$$

where the integration can be restricted to the set $x^{-1}S\cup z^{-1}S$. By $e\in C$ and $xz^{-1}\in N_e\subseteq C$ we have $x^{-1}S\cup z^{-1}S\subseteq x^{-1}CS$ and so we obtain

$$|f*\psi(x)-f*\psi(z)|\leqslant\varepsilon\|\psi\|_\infty\,\mu(x^{-1}CS).$$

Therefore $f*\psi$ is continuous at x. We could have also applied Theorem V.7.3.

We complete the proof by showing that ψ is a constant multiple of ω, the zonal spherical function associated with the class $\omega\in\widehat{L^1(G,K)}$. Both $\psi/\psi(e)$ and ω are zonal spherical functions and we proved that

$f*\psi=\hat{\omega}(f)\psi$ for every f in $L^1(G,K)$. Therefore by evaluating $f*\psi$ at e we see that $\hat{\psi}(f)=\psi(e)\hat{\omega}(f)$ and so by Lemma 2.5 we have $\psi=\psi(e)\omega$. Proposition 7 is proved.

We are lead to the following:

Theorem 8. *Let G be a unimodular group, K a compact subgroup such that $L^1(G,K)$ is commutative, H a closed, unimodular subgroup of G and $\chi: H\to\mathscr{U}(\mathscr{H})$ a unitary representation with separable $\mathscr{H}$. Then*

$$\sum n(\omega)\hat{f}(\omega)=\int_{\frac{G}{H}}\int_H f(xhx^{-1})\operatorname{tr}\chi(h)\,dh\,d\xi$$

for every f in $L^1(G,K)$ such that $\sigma(f)=\int_G f(x)\rho^\chi(x)dx$ is a trace class operator. Here $n(\omega)$ denotes the dimension of the vector space $\mathscr{K}_\omega$ of automorphic functions belonging to the class ω in $\widehat{L^1(G,K)}$ and $\hat{f}(\omega)=\hat{\omega}(f)$ $=\int_G f(x^{-1})\omega(x)dx$ where ω is the zonal spherical function associated with the class ω. Summation is extended over all positive definite zonal spherical functions with respect to K or what is the same over those classes ω in $\widehat{L^1(G,K)}$ for which the associated zonal spherical function is positive definite and $0<n(\omega)<\infty$. The series is absolutely convergent.

Note. 1. We emphasize that there is no restriction on the closed unimodular subgroup H nor on the homogeneous space G/H. In particular H need not be discrete and G/H need not be compact.

2. The notation $\hat{f}(\omega)$ will be used in the trace formula where f is fixed and ω varies while $\hat{\omega}(f)$ is reserved for the linear functional obtained from ω via the integral given in (1).

Proof. We shall use the notations introduced in the discussion preceding Proposition 4 and in Definition 5. We consider the representation $\tau: L^1(G,K)\to\mathscr{L}(\mathscr{K})$ obtained from $\rho^\chi: G\to\mathscr{U}(\mathscr{K})$. We apply Theorem IV.6.14 with $S=L^1(G,K)$ and $\rho=\tau$. By choosing $A=\tau(f)$ we obtain

$$\sum n(\omega)\operatorname{tr}\tau(f)(\omega)=\operatorname{tr}\tau(f) \tag{4}$$

where the summation is on $\omega\in\widehat{L^1(G,K)}$, $n(\omega)=n(\omega,\tau)$ and the series is absolutely convergent.

By Proposition IV.6.6 we can restrict summation to those classes ω for which $n(\omega)$ is finite. Clearly we can also require that $n(\omega)>0$. By Theorem 2 there is a zonal spherical function ω associated with the class $\omega\in\widehat{L^1(G,K)}$. In the beginning of part 2. of the proof of Theorem 2 we have seen that

$$\hat{\omega}(f)=\int_G f(x^{-1})\omega(x)dx=(\pi(f)\alpha,\alpha)$$

where π is any irreducible representation of $L^1(G,K)$ such that $\pi\in\omega$ and α is any vector in the Hilbert space of π such that $\|\alpha\|=1$. Since $L^1(G,K)$ is commutative the degree of π is one and so

$$\hat{\omega}(f)=(\pi(f),\alpha,\alpha)=\operatorname{tr}\pi(f).$$

By definition $\operatorname{tr}\tau(f)(\omega)=\operatorname{tr}\pi(f)$ and $\hat{f}(\omega)=\hat{\omega}(f)$. Therefore (4) becomes

$$\sum_{0<n(\omega)<\infty} n(\omega)\hat{f}(\omega)=\operatorname{tr}\tau(f). \tag{5}$$

By Proposition 7 the zonal spherical functions associated with the classes ω occurring on LHS are all positive definite. By Definition 5, Theorem 2 and Proposition 6 we may consider summation over all positive definite spherical functions ω with respect to the compact subgroup K. By the definition of τ we have $\tau(f)=\sigma(f)$ for any f in $L^1(G,K)$ where $\sigma\colon L^1(G)\to\mathscr{L}(\mathscr{H})$ is the algebra representation associated with ρ^χ. By Proposition VI.5.8 $\sigma(f)$ has a kernel K_f and by Theorem VI.7.17 we have

$$\operatorname{tr}\sigma(f)=\int_{\frac{G}{H}}\operatorname{tr}K_f(\xi,\xi)d\xi=\int_{\frac{G}{H}}\int_H f(xhx^{-1})\operatorname{tr}\chi(h)d\xi.$$

If we substitute this expression for the RHS of (5) we obtain the result stated in Theorem 8.

The foregoing theorem contains as a special case what we can call the *Selberg-Tamagawa trace formula.* In Selberg's original formulation H is a discrete subgroup of G, the quotient G/H is compact and $\mathscr{H}$ is finite dimensional. It is also required that the operator $\sigma(f)$ be in the trace class. The finite dimensionality of $\mathscr{H}$ and the compactness of G/H were needed in the proof of the spectral decomposition of ρ^χ which is given in Theorem VI.6.16 but as we have just seen the trace formula can be proved without the existence of a decomposition into irreducible components. The spectral decomposition in the special case when $\mathscr{H}$ is one dimensional and χ is the identity representation of the discrete subgroup H was first proved by Tamagawa who also introduced the automorphic functions corresponding to this special case.

We can extend Selberg's trace formula to the general situation where $L^1(G,K)$ is not necessarily commutative and we can prove a trace formula for each class κ in $\hat{K}$ under suitable conditions. In the formula corresponding to the class κ the spherical functions of the class κ of arbitrary heights will occur. As earlier we let $L^1(\kappa)$ denote the involutive algebra of those functions f in $L^1(G)$ which satisfy

$$d_\kappa \overline{\chi}_\kappa * f = d_\kappa f * \overline{\chi}_\kappa = f \tag{6}$$

where χ_κ denotes the character of κ and d_κ is its degree. If $\kappa = \boldsymbol{1}$ then $d_\kappa = 1$ and the integral formulas describing (6) become

$$\int_K f(kx)\,dx = f(x) \quad \text{and} \quad \int_K f(xk)\,dx = f(x).$$

These are equivalent to the left and right invariance of f under K. Therefore $L^1(\boldsymbol{1}) = L^1(G, K)$.

The general setup for the trace formula is essentially the same as before: We let $\rho: G \to \mathscr{U}(\mathscr{K})$ be the representation induced on the unimodular group G by the unitary representation $\chi: H \to \mathscr{U}(\mathscr{H})$ of the closed subgroup H and we let σ denote the associated representation of the algebra $L^1(G)$. For f we choose an element of $L^{o1}(\kappa)$ and we suppose that $\sigma(f)$ is a trace class operator. In the general trace formula expressed in Theorem IV.6.14 we let the object S be the involutive algebra $L^{o1}(\kappa)$ so that the summation on the RHS is over the equivalence classes φ belonging to $\widehat{L^{o1}(\kappa)}$. We let $\tau = \sigma | L^{o1}(\kappa)$.

The principal difference between the cases $\kappa = \boldsymbol{1}$ and $\kappa \neq \boldsymbol{1}$ is that in the general case we have to make two additional hypotheses on the triple (G, K, κ) in order to be able to use the second half of Theorem 4.1: First we suppose that the classes φ are all finite dimensional and second we require that $n(\kappa, \mu)$ is finite for every primitive morphism μ of G. If we suppose that $L^{o1}(\kappa)$ is commutative than $d_\varphi = 1$ for every φ in $\widehat{L^{o1}(\kappa)}$ and the first hypothesis becomes superfluous.

Lemma 9. *If ψ in $Li^o(G)$ is such that $d_\kappa \chi_\kappa * \psi = \psi$ and $\psi(f) = 0$ for every f in $C_0^o(\kappa)$ then $\psi = 0$.*

Proof. Let f be an arbitrary element of $C_0(\kappa)$. Then by $\psi^o = \psi$ and $f^o \in C_0^o(\kappa)$ we have $\psi(f) = \psi^o(f) = \psi(f^o) = 0$. Hence $\psi(f) = 0$ for every f in $C_0(\kappa)$. Now let $f \in C_0(G)$. We have $d_\kappa^{-1} \psi = \chi_\kappa * \psi = \psi * \chi_\kappa$ and so

$$d_\kappa^{-2} \psi(f) = \chi_\kappa * \psi * \chi_\kappa(f) = \psi(\overline{\chi}_\kappa * f * \overline{\chi}_\kappa) = 0.$$

Since $\psi(f) = 0$ for every f in $C_0(G)$ we obtain $\psi = 0$.

Lemma 10. *Let φ in $\widehat{L^{o1}(\kappa)}$ be such that d_φ is finite. Then there is a unique function in $L^{o\infty}(G)$ which we denote also by φ such that $d_\kappa \chi_\kappa * \overline{\varphi} = \overline{\varphi}$ and*

$$\overline{\varphi}(f) = d_\kappa \operatorname{tr} \pi(f) \tag{7}$$

for every $\pi \in \varphi$ and $f \in L^{o1}(\kappa)$.

Proof. We let $l(f) = d_\kappa \operatorname{tr} \pi(f)$ so that l is a linear functional on $L^{o1}(\kappa)$ which is independent of the choice of π in φ. By Theorem I.8.8 for every unit vector α in the Hilbert space of π we have $|(\pi(f)\alpha, \alpha)| \leqslant \|f\|_1$. Hence $|l(f)| \leqslant (d_\kappa d_\varphi)\|f\|_1$ and so l is continuous. By the Riesz representation theorem there is a function φ in $L^\infty(G)$ such that

$$l(f) = \int_G f(x)\overline{\varphi(x)}\,dx = \overline{\varphi}(f)$$

for every f in $L^{o1}(\kappa)$. Since $\overline{\varphi}^o(f) = \overline{\varphi}(f^o) = \overline{\varphi}(f)$ for $f \in L^{o1}(\kappa)$ we may replace φ by φ^o. Then using the new φ we have

$$d_\kappa \chi_\kappa * \overline{\varphi}(f) = \overline{\varphi}(d_\kappa \overline{\chi}_\kappa * f) = \overline{\varphi}(f)$$

for all f in $L^{o1}(\kappa)$. Therefore φ can be replaced by $d_\kappa \overline{\chi}_\kappa * \varphi$. Then the new φ will belong to $L^{o\infty}(\kappa)$ and $l(f) = \overline{\varphi}(f)$ will hold for every f in $L^{o1}(\kappa)$. The uniqueness of φ follows from Lemma 9.

Let $\mathscr{X}$ be a complex Banach space. Then $\overline{\mathscr{X}}$, the *conjugate* of $\mathscr{X}$ is the same set as $\mathscr{X}$, it has the same additive structure and the same norm. The only difference is that scalar multiplication in $\overline{\mathscr{X}}$ is defined by $(\lambda, x) \mapsto \overline{\lambda} x$ where $\overline{\lambda} x$ denotes multiplication in $\mathscr{X}$. One can easily see that if $\mathscr{X}$ is reflexive then so is $\overline{\mathscr{X}}$. If $A: \mathscr{X} \to \mathscr{X}$ is a continuous linear operator then the same function A defines a linear operator $A: \overline{\mathscr{X}} \to \overline{\mathscr{X}}$, for if $\xi \in \overline{\mathscr{X}}$ and $\lambda \in \mathbb{C}$ then $\overline{A}\lambda \cdot \xi = A\overline{\lambda}\xi = \overline{\lambda} A\xi = \lambda \cdot A\xi = \lambda \cdot \overline{A}\xi$. It is clear that $\overline{A}$ is continuous and $|\overline{A}| = |A|$. Thus if $\mu: G \to \mathrm{Gl}(\mathscr{X})$ is a morphism then another morphism $\overline{\mu}: G \to \mathrm{Gl}(\overline{\mathscr{X}})$ is defined by $\overline{\mu}(x) = \overline{\mu(x)}$. If μ is bounded then the same holds for μ. We call $\overline{\mu}$ the *conjugate morphism* of μ. If $\mu = \rho$ is a representation in Hilbert space then $\overline{\mu}$ is the conjugate representation introduced in Section VI.3.

Proposition 11. *If φ is the spherical function determined by G, K, κ and $\mu: G \to \mathrm{Gl}(\mathscr{X})$ then $\overline{\varphi}$ is also a spherical function and it is given by $G, K, \overline{\kappa}$ and $\overline{\mu}: G \to \mathrm{Gl}(\overline{\mathscr{X}})$.*

Proof. According to part 4. of Proposition 3.1 $\mathscr{X}_\kappa = \sigma(\overline{\chi}_\kappa)\mathscr{X}$ and $(\overline{\mathscr{X}})_\kappa = \overline{\sigma}(\overline{\chi_\kappa})\overline{\mathscr{X}}$. Since $\mu(x)$ and $\overline{\mu}(x)$ are identical the same is true for $\sigma(\overline{\chi}_\kappa)$ and $\overline{\sigma}(\chi_\kappa)$. Therefore $\mathscr{X}_\kappa$ and $(\overline{\mathscr{X}})_{\overline{\kappa}}$ have the same underlying set and $(\overline{\mathscr{X}})_{\overline{\kappa}} = \overline{\mathscr{X}_\kappa}$. Let α_i $(i \in \mathscr{I})$ be a basis in $\mathscr{X}_\kappa$ so that the same set of vectors is a basis in $\overline{\mathscr{X}}_\kappa$. If $A\alpha_i=\sum a_{ji}\alpha_i$ then by $\underline{A\alpha_i}=\bar{A}\alpha_i$ we have $\bar{A}\alpha_i=\sum \overline{a_{ji}}\cdot\alpha_i$. Hence if A is of finite rank then $\overline{\operatorname{tr} A}=\operatorname{tr}\bar{A}$. Applying this result with $A=\mu(x)|\mathscr{X}_\kappa$ we obtain

$$\bar{\varphi}(x)=\operatorname{tr}\overline{\mu(x)|\mathscr{X}_\kappa}=\operatorname{tr}\bar{\mu}(x)\,(\overline{\mathscr{X}})_{\bar{\kappa}}.$$

Lemma 12. *Let the unimodular group G, the compact subgroup K and $\kappa\in\hat{K}$ be such that $n(\kappa,\mu|K)$ is finite for every primitive morphism μ of G with completely irreducible σ_μ. Then there is a natural one-to-one correspondence between those elements of $\widehat{L^{o1}(\kappa)}$ for which d_φ is finite and a set of spherical functions of type $\bar{\kappa}$. Namely if φ is in $\widehat{L^{o1}(\kappa)}$ and $d_\varphi<\infty$ then the corresponding spherical function is the unique function $\varphi\colon G\to\mathbb{C}$ such that $\bar{\varphi}(f)=d_\kappa\operatorname{tr}\pi(f)$ for every $\pi\in\varphi$ and $f\in L^{o1}(\kappa)$.*

Proof. Let φ in $\widehat{L^{o1}(\kappa)}$ satisfying $d_\varphi<\infty$ be given and let $\pi\in\varphi$. Then by Lemma 10 there is a function φ in $L^{o\infty}(G)$ such that $d_\kappa\chi_\kappa*\bar{\varphi}=\bar{\varphi}$ and (7) holds. Hence by Theorem 4.1 and Proposition 11 $\bar{\varphi}$ and φ are spherical functions of types κ and $\bar{\kappa}$ resp. By Lemma 9 φ is uniquely determined by the class φ. Now let φ_1,φ_2 be classes of finite degree such that the corresponding spherical functions are equal. If $\pi_i\in\varphi_i$ $(i=1,2)$ then by (7) we have $\operatorname{tr}\pi_1(f)=\operatorname{tr}\pi_2(f)$ and so $\pi_1\approx\pi_2$ and $\varphi_1=\varphi_2$.

Theorem 13. *Let G be a unimodular group and K a compact subgroup of G. Suppose that G, K and a class κ in $\hat{K}$ are such that d_φ is finite for every φ in $\widehat{L^{o1}(\kappa)}$ and $n(\kappa,\mu|K)$ is finite for every primitive morphism μ of G with completely irreducible σ_μ. Let H be a closed, unimodular subgroup of G and let $\chi\colon H\to\mathscr{U}(\mathscr{H})$ be a unitary representation with separable $\mathscr{H}$. Then*

$$\sum n(\varphi)\,f(\bar{\varphi})=d_\kappa\int_{\frac{G}{H}}\int_H f(xhx^{-1})\operatorname{tr}\chi(h)\,dh\,d\xi$$

for every $f\in L^{o1}(\kappa)$ such that $\sigma(f)=\int_G f(x)\rho^\chi(x)\,dx$ is a trace class operator. The summation is restricted to those φ in $\widehat{L^{o1}(\kappa)}$ for which $0<n(\varphi)<\infty$. The series on the left hand side is absolutely convergent.

Note. The symbol $n(\varphi)$ denotes the multiplicity of φ in $\tau=\sigma|L^{o1}(\kappa)$.

Proof. The reasoning used in the proof of Theorem 8 can be applied also at present: By Theorem IV.6.14 we have

$$\sum n(\varphi)\operatorname{tr}\tau(f)(\varphi)=\operatorname{tr}\tau(f) \tag{8}$$

where the summation is extended over every φ in $\widehat{L^{o1}(\kappa)}$, the series is absolutely convergent and $n(\varphi)$ is an abbreviation for the multiplicity

$n(\varphi, \tau)$. By Proposition IV.6.6 the summation can be restricted to those classes φ for which $n(\varphi)$ is finite and positive.

By Lemma 12 there is a spherical function φ associated with each class φ such that $f(\overline{\varphi}) = \overline{\varphi}(f) = d_\kappa \operatorname{tr} \pi(f)$ for every $f \in L^{o1}(\kappa)$ and every irreducible representation $\pi \in \varphi$. We choose for π the restriction of τ to any irreducible component $\mathscr{P}$ of τ such that $\tau|\mathscr{P}$ is in the class φ. Then we obtain

(9) $$\overline{\varphi}(f) = d_\kappa \operatorname{tr} \tau(f)(\varphi).$$

Therefore (8) gives

(10) $$\sum_{0 < n(\varphi) < \infty} n(\varphi) f(\overline{\varphi}) = d_\kappa \operatorname{tr} \tau(f).$$

The RHS of this equation has been determined already at the end of the proof of Theorem 8. By substituting the expression obtained there into (10) we arrive at the trace formula given in Theorem 13.

By definition $\mathscr{K}_\varphi$ is spanned by a maximal orthogonal system of irreducible components of τ belonging to the class φ in $\widehat{L^{o1}(\kappa)}$. If $L^{o1}(\kappa)$ is supposed to be commutative then these components are one dimensional and so $\mathscr{K}_\varphi$ is spanned by an orthonormal family of vectors θ_i $(i \in \mathscr{I})$ such that $\sigma(f)\theta_i = \lambda_f^i \theta_i$ for a suitable λ_f^i in $\mathbb{C}$. Using (9) we see that $\lambda_f^i = d_\kappa^{-1} \overline{\varphi}(f)$ for every i in $\mathscr{I}$. Hence by the maximality of the system and by using Theorem IV.5.4 we conclude that *$\mathscr{K}_\varphi$ consists of those θ in $\mathscr{K}$ for which*

$$\sigma(f)\theta = d_\kappa^{-1} \overline{\varphi}(f)\theta$$

for every f in $L^{o1}(\kappa)$. Since positive definite zonal spherical functions satisfy the identity $\omega^* = \omega$ for those we have $\overline{\omega}(f) = f(\omega)$. Generalizing from the case $\kappa = 1$ we say that *the elements of $\mathscr{K}_\varphi$ belong to the spherical function φ*.

Since the spherical function φ is determined by a class $\varphi \in \widehat{L^{o1}(\kappa)}$ where $d_\varphi = 1$ by the commutativity of $L^{o1}(\kappa)$ it is of height one by Theorem 4.1. Conversely if a spherical function φ of height one and type κ is given then the representation v^o occurring in Theorem 4.1 is of degree one and so one sees that any two such representations corresponding to φ are unitarily equivalent. Therefore in the case of commutative $L^{o1}(\kappa)$ Theorem 4.1 and Lemma 12 give a natural one-to-one correspondence between the elements of $\widehat{L^{o1}(\kappa)}$ and the set of all spherical functions of height one and type $\overline{\kappa}$.

Definition 14. *If $L^{o1}(\kappa)$ is commutative then the elements of $\mathscr{K}_\varphi$ are called automorphic functions belonging to the classes $\kappa \in \hat{K}$ and $\varphi \in \overline{L^{o1}(\kappa)}$ or to the class κ and the associated spherical function φ.*

An important difference between the case $\kappa = \mathit{1}$ considered in Definition 5 and the general situation is that the special automorphic functions belonging to $\kappa = \mathit{1}$ can be interpreted as functions on the homogeneous space $K \backslash G$ while the general ones have to be viewed as functions on G.

Theorem 15. *If G is a group with an Iwasawa decomposition $G = KS$ and if $L^{o1}(\kappa)$ is commutative for a given $\kappa \in \hat{K}$ then the trace formula*

$$\sum n(\varphi) f(\overline{\varphi}) = d_\kappa \int_{\frac{G}{H}} \int_H f(xhx^{-1}) \operatorname{tr} \chi(h) \, dh \, d\xi$$

holds for every f in $L^{o1}(\kappa)$ such that $\sigma(f)$ is in the trace class. Here $n(\varphi)$ denotes the dimension of the vector space $\mathscr{K}_\varphi$ of automorphic functions belonging to the class κ and the spherical function φ. The summation is extended over those spherical functions of type κ and height one for which $0 < n(\varphi) < \infty$.

Proof. By the commutativity of $L^{o1}(\kappa)$ we have $d_\varphi = 1$. By Theorem 5.10 we know that $n(\kappa, \mu | K)$ is finite and so Theorem 13 is applicable. We have seen above that the summation on LHS can be considered on the index set consisting of all spherical functions φ of type κ and height one such that the corresponding class φ satisfies $0 < n(\varphi) < \infty$.

An example of the situation described in Theorem 13 arises when G is a compact group, $K = G$ and H is an arbitrary closed subgroup of G. For if v^o is an irreducible representation of $L^{o1}(\kappa)$ in Hilbert space then v^o can be extended to a representation of $L^1(G)$ by defining $v(f) = d_\kappa^2 v^o(\overline{\chi}_\kappa * f * \overline{\chi}_\kappa)$ for every f in $L^1(G)$. The extension v consists of the same operators as the original v^o and so it is irreducible. According to Theorem 1.11 v is finite dimensional and so the same holds for v^o. Therefore the first hypothesis given in Theorem 13 is satisfied. The second hypothesis on the finiteness of $n(\kappa, \mu | K)$ is also satisfied by Theorems 5.10, 1.10 and V.7.28. Since G is compact every spherical function φ is determined by an irreducible representation π of G in Hilbert space. Hence if α_i ($i \in \mathscr{I}$) is a MONS in this space then $\varphi(x)$ is expressible as a sum of terms $(\pi(x)\alpha_i, \alpha_i)$. Spherical functions are continuous and so G being compact every φ is square integrable. Therefore by Theorem 3.18 and Proposition 1.22 we see that any two distinct spherical functions corresponding to the same class $\overline{\kappa}$ are orthogonal. Thus if we substitute a spherical function φ for f in the LHS of the trace formula given in Theorem 13 then the infinite series reduces to a single term. Since φ is of type $\bar{\kappa}$ the substitution $f=\varphi$ is permissible because Propositions 3.2 and 3.3 show that φ belongs to $L^{o1}(\kappa)$. We also have to verify that $\sigma(\varphi)$ is a TC operator. Since $K=G$ is compact we know that χ_κ and φ are continuous and belong to $L^1(G)$. Hence by Proposition 3.3 $\sigma(\varphi)=d_\kappa\sigma(\chi_\kappa)\sigma(\varphi)$. According to Proposition VI.6.10 $\sigma(\chi_\kappa)$ and $\sigma(\varphi)$ are HS operators and so their product and $\sigma(\varphi)$ are in the TC by Theorem II.5.2. We proved the following:

Theorem 16. *Let G be compact, let $\kappa\in\hat{G}$ and suppose that φ is in $\widetilde{L^{o1}(\kappa)}$. Then for any closed subgroup H and $\chi: H\to\mathscr{U}(\mathscr{H})$ we have*

$$n(\varphi,\tau)=d_\kappa\|\varphi\|_2^{-2}\int_{\frac{G}{H}}\int_H \varphi(xhx^{-1})\operatorname{tr}\chi(h)\,dh\,d\xi$$

where φ on the RHS denotes the spherical function of type $\bar{\kappa}$ associated with the class φ.

If $H=\{e\}$ then we obtain $n(\varphi,\sigma_\lambda)=\|\varphi\|_2^{-2}$ where σ_λ is the algebra representation associated with the left regular representation λ of the compact group G.

Remarks

The fundamental results on the representation theory of compact groups such as Theorem 1.1, Proposition 1.19 and Theorem 1.35 were proved in the joint paper of Péter and Weyl (1). Square integrable representations were studied by Godement (3, 4) and Harish-Chandra (5). The concept of a formal degree was introduced by Harish-Chandra. Groups with small invariant neighborhoods and those with compact invariant neighborhoods has been considered already in the early thirties by van Dantzig (1), Freudenthal (2) and Pontrjagin. (See for example p. 129 in Weil's book.) Instead of giving detailed references on this subject we mention two papers by Grosser and Moskowitz (1, 2) where a survey of the existing literature and various comments can be found.

Characters of finite dimensional representations were used for a long time. The characters of the unitary representations of compact groups were discussed by Péter and Weyl in 1927 (1). A systematic study of characters and their generalizations was undertaken by Godement in 1954 (3, 4). The orthogonality relations expressed in Proposition 22 were proved by Péter and Weyl (1). Propositions 22 and 24 were extended to square integrable representations of $Sl(2,\mathbb{R})$ by Bargmann (1) and then for the case of arbitrary locally compact groups by Godement (5, 6). Propositions 1.25, 1.26 and 1.28 were found by me while I was proving other results such as the functional equation of spherical functions of height 1. I have not seen them elsewhere in the literature. Proposition 1.29 is Harish-Chandra's result on the formal degree (5). The second proof of Proposition 1.28 is due to Fakler.

The result expressed in Theorem 1.33 originates from Harish-Chandra (2) who considered the case when G is a semisimple Lie group, ρ is a morphism in a Banach space and κ is of finite degree. His result is essentially the same as our Proposition 3.1. The theorem is also mentioned without proof by Godement (7). A detailed and satisfactory proof of these results is published here for the first time. (Although I do not doubt that Harish-Chandra could have supplied a detailed proof if he were not pressed by other matters.) The applications of Theorem 1.33 which can be found in the text are probably new.

Péter and Weyl proved not only the result expressed in Theorem 1.35 but they also proved that every continuous function on the compact group G with complex values can be uniformly approximated by linear combinations of coefficients of irreducible representations. This result is often called also the Péter-Weyl theorem. It can be easily derived from the Gelfand-Raikov theorem and the Stone-Weierstrass theorem.

Every square integrable representation of a locally compact group is a direct sum of irreducible square integrable representations.

This interesting theorem was obtained by Kunze (1) in 1970. Not knowing about Kunze's result I conjectured the same proposition and gave it to a younger collegue to work on. Fortunately I came across Kunze's paper a few weeks later. Kunze also uses the same definition of square integrability which can be found in the text.

It can be proved that the zonal spherical functions introduced in Section 2 are straight forward generalizations of the classical zonal spherical functions. This and some other examples can be found in the last chapter of Helgason's book (1). He considers zonal spherical functions only when the underlying group G is a Lie group but many of the proofs and the results stated there hold with very slight modifications for arbitrary locally compact groups. General zonal spherical functions were introduced by Gelfand (3) in 1956. We note that they were considered by Atle Selberg approximately at the same time and possibly actually earlier (1). He does not develop their theory however; for instance the functional equation of zonal spherical functions was proved only in Gelfand's paper. Both Gelfand and Selberg were influenced by the work of Elie Cartan who introduced zonal spherical functions when G/K is a compact, irreducible symmetric space and explicitely described them for instance for the case $SO(n+1, \mathbb{R})/SO(n, \mathbb{R})$.

The general theory of positive definite functions was developed by Godement (1). Among other things he proved that every continuous function $f: G \to \mathbb{C}$ can be uniformly approximated on compact sets by linear combinations of coefficients of continuous, irreducible unitary representations of the underlying group G. Positive definite functions were considered earlier by Gelfand and Raikov (1). For instance Proposition 2.14 was proved by them. The converse part of Theorem 2.8 is new.

The definition of spherical functions of type κ and height h is due to Harish-Chandra (1, 6, 7) who considered them mainly in the special case when G is a semisimple Lie group. His note was soon followed by Godement's paper (7) were much of the theory is fully developed. We point out that the papers of Godement and Harish-Chandra contain much important material which is not covered in the present book. Some of these results are discussed in Helgason's monograph (1) but most of this material is available at present only in the original articles. Theorem 3.18 appears explicitely in Harish-Chandra's note (1). The proof given in the text is based on the outline which is in Godement's paper. The proof of the functional equation expressed in Theorem 3.19 is new and it is of some significance. For the original proof which is due to Godement is based on Godement's Theorem 4.1 which is very deep and difficult to prove.

Proposition 4.11 on complete irreducibility is due to Dieudonné. The rest of Section 4 belongs to Godement. I did rework his original proof, changed the notations, eliminated some minor inaccuracies and added some details. But it is still Godement's original proof. A genuinly new proof, especially if it were shorter and easier would be of considerable interest.

Iwasawa's theorem on the decomposition $G = KAN$ was proved in 1949 (1) and a detailed proof can be found in Helgason's book (1). The proof is based on the Cartan decomposition and the root space decomposition of the complexification of the Lie algebra $\mathfrak{g}$ of G. The decomposition KAN can be explicitely described in many instances and thus one can avoid the use of Iwasawa's theorem. The CCR property was introduced by Kaplansky in 1951 (2). In the same year he proved (4) that the structure space of a CCR algebra is a Baire space and that "CCR implies type I" (5). The fact that every connected semisimple Lie group G is of type I was proved by Harish-Chandra (2) in the following form: Every primary representation $\rho: G \to \mathscr{U}(\mathscr{H})$ has an irreducible component. Lemma 5.2 and the material preceding it is due to Kaplansky (6) and Kolchin. (See the footnote on p. 105 in Kaplansky's article.)

The main results in the first two third of Section 5 were proved in Godement's paper on spherical functions (7) or I formulated them under the influence of the material presented in that paper. Recently I noticed that Godement himself returned to this subject in (4). The results near the end of Section 5 belong to Harish-Chandra (4) or are variants of the results obtained by him.

A closed subgroup K of a locally compact group G is called large or massive if for every continuous irreducible representation $\rho: G \to \mathscr{U}(\mathscr{H})$ and for every class $\kappa \in \hat{K}$ we have $n(\kappa, \rho|K) < \infty$. Harish-Chandra (3) proved that the subgroup K occurring in the Iwasawa decomposition of a semisimple Lie group is large. Hence if K is also compact then one can apply Theorem 5.15 and derive the existence of a distribution character θ_π. Harish-Chandra proved this also in the case when K is not necessarily compact. He also proved that the ideal $\mathscr{I}_\pi$ contains $C_0^\infty(G)$ and

$$\theta_\pi(f) = \int_G f(x)\varphi(x)\,dx$$

for a suitable locally integrable function $\varphi: G \to \mathbb{C}$ and for all f in $C_0^\infty(G)$. Furthermore he proved the existence of a dense open subset O of G such that the complement of O has Haar measure zero and $\varphi|O$ is an analytic function.

Selberg's trace formula was discovered some time between 1952 and 1956 and it was published in a very important paper (1) in 1956. Selberg, with a very good philosophical reason, is reluctant to publish and his writings are very condensed reports. His 1956 paper contains much valuable material including the "trace formula". The significance of this result was immediately recognized in France, Japan and USSR. The first detailed proof was given by Tamagawa in 1960 (1). Its connection with representation theory was emphasized by Godement (2) who outlined a proof of the direct sum decomposition in the case when G/H is compact. The significance of induced representations for the trace formula was clearly recognized by Gelfand probably in 1964 or 1965. There was no reference to induced representations in his 1962 in absentia adress to the International Congress in Stockholm (4).

The present formulation given in Theorem 6.8 is made possible by the general trace formula given in Section IV.6. Similarly Theorem 1 is a new version of the result announced by Gelfand in 1962 (4). The class of trace formulas given in Theorems 6.13 and 6.15 were discovered by myself in 1970 and they are published here for the first time.

The material discussed in Section 6 covers only one aspect of the trace formulas. The next obvious question which one should answer is the following: Given a concrete class of groups G and subgroups H and K, for which functions $f: G \to \mathbb{C}$ is $\sigma(f)$ a trace class operator? There are at present approximately thirty papers dealing with Selberg's trace formula, most of which are applications to automorphic functions and arithmetic questions. (The word "arithmetic" means "general number theoretic and/or algebraic geometric".) Originally I planned to cover a large part of this material in the present book but by now I realize that this will not be possible.

During the last few years some papers appeared also on another topic introduced in Selberg's famous paper. These articles deal with the Selberg zeta function, its interpretation in terms of topological dynamics and its generalizations. I would like to point out that Selberg's paper contains the germs of two other important subjects related to the Selberg zeta function: The first of these is Selberg's explicit formula which is obviously very resemblent to Riemann's classical explicit formulae and their generalization due to Weil (4). The second subject which is mentioned only in one or two sentences is the analogue of the classical Riemann hypothesis. The intensive study of these topics might eventually lead to a better understanding or to the actual solution of this difficult mathematical problem.

Chapter VIII

Lie Algebras, Manifolds and Lie Groups

This chapter gives a concise introduction to the fundamentals of the general theory of Lie algebras and Lie groups. The only prerequisites are the elements of classical advanced calculus, some general topology, linear algebra and a few facts about finite dimensional representations. Therefore the material presented here is almost independent from the earlier chapters: Sections 1, 2, 3 and 4 can be read without being familiar with the rest of the book and the remaining sections depend only on these ones. It is advisable to get acquainted with Section V.1 before reading Sections 5 and 6. This can be easily done because there are no prerequisites for Section V.1.

The content of this chapter together with the preceding material on general representation theory in Hilbert space and the representation theory of locally compact groups and their group algebras lead naturally to the study of the representations of Lie groups. In the present chapter we collected most of the material which one should know before beginning the study of this very interesting and important part of the theory of group representations.

1. Lie Algebras

We recall that a *non-associative algebra* A over a field F is a vector space A over F with a multiplication which satisfies the usual distributive laws, $\lambda(ab)=(\lambda a)b=a(\lambda b)$ for all $\lambda\in F$ and $a,b\in A$ but multiplication need not be associative. In what follows we suppose that A is finite dimensional over F. One simple way to describe the multiplication in such an algebra A is by choosing a basis $a_1,\ldots,a_n$ and expressing the products $a_i a_j$ $(i,j=1,\ldots,n)$ as linear combinations of these basic elements, say $a_i a_j=\sum_{k=1}^{n}\lambda_{ijk}a_k$. The coefficients λ_{ijk} are called the *structural constants* or *multiplication constants* of A with respect to the given basis. One can easily see that A is associative if and only if the multiplication of the basic elements is associative.

In the theory of Lie groups and Lie algebras it is customary to use lower case gothic letters to denote such algebras while capital latin letters are used for the various elements. Multiplication will be denoted by the symbol $[A, B]$ instead of the usual notation AB which will be reserved for the operator product or matrix product in case A and B denote linear transformations or matrices, respectively.

Definition 1. *A non-associative algebra* $\mathfrak{a}$ *over a field* F *is called a Lie algebra if* $[X, X]=0$ *for every* X *in* $\mathfrak{a}$ *and*

$$[X,[Y,Z]]+[Y,[Z,X]]+[Z,[X,Y]]=0 \tag{1}$$

for every X, Y, Z *in* $\mathfrak{a}$.

We shall most of the time tacitly suppose also that $\mathfrak{a}$ is a finite dimensional vector space over F but we note that some of the results can be proved without this additional restriction. Since

$$0=[X+Y,X+Y]=[X,X]+[X,Y]+[Y,X]+[Y,Y]=[X,Y]+[Y,X]$$

we see that

$$[X,Y]+[Y,X]=0 \tag{2}$$

for every X, Y in the Lie algebra $\mathfrak{a}$. If F is not a field of characteristic 2 then (2) could be used in the definition of $\mathfrak{a}$ instead of the condition $[X,X]=0$. The relation stated in (1) is called the *Jacobi identity*. It is simple to prove that if $A_1, \ldots, A_n$ is a basis of the non-associative algebra $\mathfrak{a}$ then $\mathfrak{a}$ is a Lie algebra if and only if $[A_i, A_i]=0$, $[A_i, A_j]+[A_j, A_i]=0$ and

$$[A_i,[A_j,A_k]]+[A_j,[A_k,A_i]]+[A_k,[A_i,A_j]]=0$$

for every $i, j, k=1, \ldots, n$.

Example 2. Let $\mathfrak{a}$ be the vector space $\mathbb{R}^3$ over the reals $\mathbb{R}$, let i, j and k denote the canonical basic vectors $(1,0,0)$, $(0,1,0)$ and $(0,0,1)$, respectively. We define the products of these vectors by the multiplication table

	i	j	k
i	0	k	$-j$
j	$-k$	0	i
k	j	$-i$	0

so that $[X, Y]$ for general X, Y in $\mathfrak{a}$ denotes the *vector product* or *outer product* used by physicists.

Example 3. Let again $\mathfrak{a}$ be $\mathbb{R}^3$ over the field $\mathbb{R}$ with basic vectors i, j and k and let multiplication be defined by the table

	i	j	k
i	0	0	$-j$
j	0	0	i
k	j	$-i$	0

so that for general X, Y in $\mathfrak{a}$ we have

$$[X, Y] = X \times Y - (X \times Y, k)k$$

where $\times$ stands for the vector product and $(\cdot, \cdot)$ denotes the scalar product in $\mathbb{R}^3$.

Example 4. Let $\mathfrak{a}$ be the vector space of all $n \times n$ matrices whose entries belong to a given field F and let $[X, Y] = XY - YX$ where XY and YX denote ordinary matrix products. The Lie algebra obtained in this manner will be denoted by $\mathfrak{gl}(n, F)$.

Example 5. Let $\mathscr{V}$ be a finite dimensional vector space over the field F and let $\mathfrak{a}$ be the vector space $\mathscr{L}(\mathscr{V})$ with multiplication $[A, B] = AB - BA$ where AB and BA denote operator products. This Lie algebra will be denoted by $\mathfrak{gl}(\mathscr{V})$.

Example 6. Let $\mathfrak{a}$ be any associative algebra with elements $X, Y, \ldots$ We define a Lie algebra on the set $\mathfrak{a}$ by retaining its vector space structure and defining $[X, Y] = XY - YX$ for all X, Y in $\mathfrak{a}$. The product $[X, Y]$ is called the *Lie product* of X and Y and $\mathfrak{a}$ with this multiplication is the *Lie algebra of the associative algebra* $\mathfrak{a}$. The Lie algebras described in the last two examples are concrete instances of this general construction.

A subset $\mathfrak{b}$ of the Lie algebra $\mathfrak{a}$ is called a *subalgebra* of $\mathfrak{a}$ if $X + Y, \lambda X$ and $[X, Y]$ belong to $\mathfrak{b}$ for every X, Y in $\mathfrak{b}$ and λ in F. One can also speak about an *ideal* $\mathfrak{i}$ in the Lie algebra $\mathfrak{a}$. By (2) the concept of right, left and two-sided ideals coincide. The subset $\mathfrak{i}$ is an ideal if it is a sub-vector space of $\mathfrak{a}$ and in addition $[X, Y]$ belongs to $\mathfrak{i}$ for every X in $\mathfrak{a}$ and Y in $\mathfrak{i}$. The set

$$\mathfrak{c} = \{C : C \in \mathfrak{a} \text{ and } [C, X] = 0 \text{ for all } X \in \mathfrak{a}\}$$

is an ideal, called the *center* of $\mathfrak{a}$. The Lie algebra $\mathfrak{a}$ is said to be abelian if $\mathfrak{c} = \mathfrak{a}$. We see that if $\mathfrak{a}$ is the Lie algebra of an associative algebra then the center is the set of those C in $\mathfrak{a}$ for which $CX = XC$ for every X in

$\mathfrak{a}$ and the Lie algebra $\mathfrak{a}$ is abelian if and only if its associative algebra is commutative. If $\mathfrak{a}$ is a Lie algebra then the set of all elements of the form $\sum [A_i, B_i]$ is an ideal of $\mathfrak{a}$ which is called the *derived ideal* or *derived algebra* and is usually denoted by $\mathfrak{a}'$ or $\mathscr{D}\mathfrak{a}$.

We can define the *quotient algebra* $\mathfrak{a}/\mathfrak{i}$ when $\mathfrak{i}$ is an ideal in the Lie algebra $\mathfrak{a}$: The elements of $\mathfrak{a}/\mathfrak{i}$ are the additive cosets $X+\mathfrak{i}$ $(X \in \mathfrak{a})$ and the operations are

$$(X+\mathfrak{i})+(Y+\mathfrak{i})=(X+Y)+\mathfrak{i},$$
$$\lambda(X+\mathfrak{i})=\lambda X+\mathfrak{i},$$
$$[X+\mathfrak{i}, Y+\mathfrak{i}]=[X, Y]+\mathfrak{i}.$$

The quotient algebra is again a Lie algebra. Of course the same concepts are meaningful in any non-associative algebra but we are interested in the special case of Lie algebras.

A map $\sigma: \mathfrak{a} \to \mathfrak{b}$ of a Lie algebra $\mathfrak{a}$ into another $\mathfrak{b}$ is called a *homomorphism* if $\mathfrak{a}$ and $\mathfrak{b}$ are defined over the same field F and σ commutes with the operations of $\mathfrak{a}$ and $\mathfrak{b}$ i.e. $\sigma(X+Y)=\sigma X+\sigma Y, \sigma(\lambda X)=\lambda\sigma X$ and $\sigma[X, Y]=[\sigma X, \sigma Y]$ for all $X, Y \in \mathfrak{a}$ and $\lambda \in F$. The set of homomorphisms of $\mathfrak{a}$ into $\mathfrak{b}$ is denoted by $\mathrm{Hom}(\mathfrak{a}, \mathfrak{b})$. If the fields of $\mathfrak{a}$ and $\mathfrak{b}$ are different then $\mathrm{Hom}(\mathfrak{a}, \mathfrak{b})$ is empty. The class of all Lie algebras together with the sets $\mathrm{Hom}(\mathfrak{a}, \mathfrak{b})$ form the *category of Lie algebras*. If $\sigma: \mathfrak{a} \to \mathfrak{b}$ is a Lie algebra homomorphism then the image $\sigma\mathfrak{a}$ is a subalgebra of $\mathfrak{b}$ and the *kernel* $\sigma^{-1}\{0\}$ is an ideal in $\mathfrak{a}$. If σ is injective it is called an *isomorphism* of $\mathfrak{a}$ into $\mathfrak{b}$. This is the case if and only if the kernel of σ is the zero ideal of $\mathfrak{a}$. If σ is a bijection then $\mathfrak{a}$ and $\mathfrak{b}$ are called *isomorphic* Lie algebras. If $\mathfrak{a}=\mathfrak{b}$ then the homomorphism σ is called an *endomorphism*. A bijective endomorphism is an *automorphism*.

Example 7. Consider the complex matrices

$$A_1=\frac{i}{2}\begin{pmatrix}0 & 1\\ 1 & 0\end{pmatrix}, \quad A_2=\frac{1}{2}\begin{pmatrix}0 & -1\\ 1 & 0\end{pmatrix} \quad \text{and } A_3=\frac{i}{2}\begin{pmatrix}1 & 0\\ 0 & -1\end{pmatrix}$$

and let $\mathfrak{a}$ be the set of all linear combinations $z_1A_1+z_2A_2+z_3A_3$ where $z_1, z_2, z_3 \in \mathbb{C}$. Then $\mathfrak{a}$ is a subalgebra of $\mathfrak{gl}(2, \mathbb{C})$ with basis A_1, A_2, A_3 and the multiplication table of the basic vectors is

	A_1	A_2	A_3
A_1	0	A_3	$-A_2$
A_2	$-A_3$	0	A_1
A_3	A_2	$-A_1$	0 .

This Lie algebra is called $\mathfrak{su}(2)$ because it is associated with the group SU(2).

Example 8. Let A_1, A_2, A_3 denote the same vectors as before but consider only real linear combinations $x_1 A_1 + x_2 A_2 + x_3 A_3$. We use the same multiplication as above so that $[X, Y] = XY - YX$ where XY and YX are matrix products. The real Lie algebra $\mathfrak{a}$ obtained in this manner is isomorphic to the one given in Example 2 and $x_1 A_1 + x_2 A_2 + x_3 A_3 \mapsto (x_1, x_2, x_3)$ is an isomorphism.

Example 9. We can form a Lie algebra $\mathfrak{m}(2)$ by taking all real linear combinations of the matrices

$$A_1 = \begin{pmatrix} 0 & 0 & 1 \\ 0 & 0 & 0 \\ 0 & 0 & 0 \end{pmatrix}, \quad A_2 = \begin{pmatrix} 0 & 0 & 0 \\ 0 & 0 & 1 \\ 0 & 0 & 0 \end{pmatrix}, \quad A_3 = \begin{pmatrix} 0 & -1 & 0 \\ 1 & 0 & 0 \\ 0 & 0 & 0 \end{pmatrix}.$$

The multiplication table of the basic vectors is

	A_1	A_2	A_3
A_1	0	0	$-A_2$
A_2	0	0	A_1
A_3	A_2	$-A_1$	0 .

Then $[X, Y] = XY - YX$ for any two vectors X, Y in $\mathfrak{m}(2)$. This Lie algebra and the one described in Example 3 are isomorphic and $x_1 A_1 + x_2 A_2 + x_3 A_3 \mapsto (x_1, x_2, x_3)$ is an isomorphism of $\mathfrak{m}(2)$ onto $\mathfrak{a}$. The reason to call this algebra $\mathfrak{m}(2)$ is that it is associated with the motion group $M(2)$ of the Euclidean plane.

Consider again the Lie algebra $\mathfrak{a}$ of Example 2 and let $\sigma(x_1, x_2, x_3) = (x_2, x_3, x_1)$ for every (x_1, x_2, x_3) in $\mathfrak{a}$. One can easily verify that σ is an automorphism of $\mathfrak{a}$. If $\mathfrak{a}$ denotes the Lie algebra given in Example 3 then the map σ is not an automorphism of $\mathfrak{a}$. One sees that in this case $(x_1, x_2, x_3) \mapsto (-x_2, x_1, x_3)$ is a Lie algebra automorphism.

Definition 10. *A linear map $D: \mathfrak{a} \to \mathfrak{a}$ of a non-associative algebra $\mathfrak{a}$ into itself is called a derivation of $\mathfrak{a}$ if $D(xy) = (Dx)y + x(Dy)$ for every x, y in $\mathfrak{a}$.*

Note. We recall that a non-associative algebra means a not necessarily associative one and so derivations are defined also for ordinary, associative algebras $\mathfrak{a}$.

Formal derivations in algebras of polynomials, partial and directional derivatives in algebras of C^∞ functions are well known examples of derivations. If D_1, D_2 and D are derivations of $\mathfrak{a}$ and λ is in F then $D_1+D_2, \lambda D$ and $[D_1, D_2]=D_1 D_2 - D_2 D_1$ are also derivations. A simple computation shows that the product $[D_1, D_2]$ satisfies the Jacobi identity and so one obtains:

Proposition 11. *The set* $\mathfrak{d}(\mathfrak{a})$ *of all derivations of a non-associative algebra* $\mathfrak{a}$ *is a Lie algebra under operator addition, multiplication of operators by scalars and the product* $[D_1, D_2]=D_1 D_2 - D_2 D_1$.

Note. The Lie algebra $\mathfrak{d}(\mathfrak{a})$ is called the *derivation algebra* of $\mathfrak{a}$.

If $\mathfrak{a}$ is an associative algebra and $a\in\mathfrak{a}$ then $x\mapsto xa-ax$ is a derivation, called the *inner derivation of* $\mathfrak{a}$ *determined by the element a.* By considering the Lie algebra of the associative algebra $\mathfrak{a}$ we are led to the concept of *inner derivation* or *Lie derivative* associated with an element A of the arbitrary Lie algebra $\mathfrak{a}$: This is the linear map $\theta(A)\colon \mathfrak{a}\to\mathfrak{a}$ such that $\theta(A)X=[A, X]$ for every X in $\mathfrak{a}$. By the Jacobi identity we have

$$\begin{aligned}\theta(A)[X, Y]&=[A,[X, Y]]=[[A, X], Y]+[X,[A, Y]]\\&=[\theta(a)X, Y]+[X, \theta(A)Y]\end{aligned}$$

and so $\theta(A)$ is a derivation of $\mathfrak{a}$. It is customary to call $\theta(A)$ the *adjoint mapping* of A and use the notation ad A instead of $\theta(A)$.

Proposition 12. *The Lie derivatives of the Lie algebra* $\mathfrak{a}$ *form an ideal in the derivation algebra* $\mathfrak{d}(\mathfrak{a})$.

The proof of this proposition will at the same time show another result:

Proposition 13. *The map* $\mathrm{ad}\colon\mathfrak{a}\to\mathfrak{gl}(\mathfrak{a})$ *is a homomorphism of the Lie algebra* $\mathfrak{a}$ *into* $\mathfrak{gl}(\mathfrak{a})$.

Note. Here $\mathfrak{gl}(\mathfrak{a})$ denotes the Lie algebra associated with the vector space of $\mathfrak{a}$ as is described in Example 5.

Proof. It is clear that $\mathrm{ad}(A+B)=\mathrm{ad}\,A+\mathrm{ad}\,B$ and $\mathrm{ad}\,\lambda A=\lambda\,\mathrm{ad}\,A$ for every $A, B\in\mathfrak{a}$ and $\lambda\in F$. Now let $D\in\mathfrak{d}(\mathfrak{a})$ and $A, X\in\mathfrak{a}$. Then $[\mathrm{ad}\,A, D]X$ $=(\mathrm{ad}\,A)DX-D(\mathrm{ad}\,A)X=[A, DX]-D[A, X]=-[DA, X]=-(\mathrm{ad}\,DA)X$. Therefore $[\mathrm{ad}\,A, D]=-\mathrm{ad}\,DA$ and this shows that $\mathrm{ad}\,\mathfrak{a}$ is an ideal in $\mathfrak{d}(\mathfrak{a})$. Let us suppose that D is an inner derivation of $\mathfrak{a}$, say $D=\mathrm{ad}\,B$. Then

$$[\mathrm{ad}\,A, \mathrm{ad}\,B]X=-(\mathrm{ad}[B, A])X=(\mathrm{ad}[A, B])X$$

for every $X \in \mathfrak{a}$. Thus $\operatorname{ad}[A, B] = [\operatorname{ad} A, \operatorname{ad} B]$ and so ad is a Lie algebra homomorphism.

The ideal $\operatorname{ad}\mathfrak{a}$ of Lie derivatives is called the *adjoint algebra* of $\mathfrak{a}$. Thus the symbol ad associates with each object $\mathfrak{a}$ of the category of Lie algebras an object $\operatorname{ad}\mathfrak{a}$ of the subcategory whose objects are Lie algebras of the type $\mathfrak{gl}(\mathfrak{a})$ and their subalgebras. We can define $\operatorname{ad}\varphi$ for any Lie algebra homomorphism $\varphi: \mathfrak{a} \to \mathfrak{b}$ by the formula

$$(\operatorname{ad}\varphi)(\operatorname{ad} X) = \operatorname{ad}\varphi(X)$$

where $X \in \mathfrak{a}$. Hence $\operatorname{ad}\varphi$ maps $\operatorname{ad}\mathfrak{a}$ into $\operatorname{ad}\mathfrak{b}$. A routine computation shows that $\operatorname{ad}\varphi$ is actually a homomorphism of $\operatorname{ad}\mathfrak{a}$ into $\operatorname{ad}\mathfrak{b}$. Therefore the new symbol ad maps $\operatorname{Hom}(\mathfrak{a}, \mathfrak{b})$ into $\operatorname{Hom}(\operatorname{ad}\mathfrak{a}, \operatorname{ad}\mathfrak{b})$. One sees that ad is a covariant functor from the category of Lie algebras into the subcategory described above. By the definition of $\operatorname{ad}\varphi$ we also see that

$$\begin{array}{ccc} \mathfrak{a} & \xrightarrow{\operatorname{ad}} & \operatorname{ad}\mathfrak{a} \\ {\scriptstyle\varphi}\downarrow & & \downarrow{\scriptstyle\operatorname{ad}\varphi} \\ \mathfrak{b} & \xrightarrow{\operatorname{ad}} & \operatorname{ad}\mathfrak{b} \end{array}$$

is a commutative diagram.

If $A_1, \ldots, A_n$ is a basis for the Lie algebra $\mathfrak{a}$ then $\operatorname{ad} A_1, \ldots, \operatorname{ad} A_n$ is a set of generators for $\operatorname{ad}\mathfrak{a}$. Thus if $\mathfrak{a}$ and $\operatorname{ad}\mathfrak{a}$ are isomorphic then $\mathfrak{a}$ being finite dimensional $\operatorname{ad} A_1, \ldots, \operatorname{ad} A_n$ is a basis for the adjoint algebra of $\mathfrak{a}$. Conversely if $\operatorname{ad} A_1, \ldots, \operatorname{ad} A_n$ are linearly independent then $\mathfrak{a}$ and $\operatorname{ad}\mathfrak{a}$ are isomorphic Lie algebras. For instance if i, j, k denote the basic vectors of Example 2 then the matrices of the linear transformations $\operatorname{ad} i$, $\operatorname{ad} j$, $\operatorname{ad} k$ are

$$\begin{pmatrix} 0 & 0 & 0 \\ 0 & 0 & -1 \\ 0 & 1 & 0 \end{pmatrix} \quad \begin{pmatrix} 0 & 0 & 1 \\ 0 & 0 & 0 \\ -1 & 0 & 0 \end{pmatrix} \quad \begin{pmatrix} 0 & -1 & 0 \\ 1 & 0 & 0 \\ 0 & 0 & 0 \end{pmatrix}.$$

Hence $\mathfrak{a}$ and $\operatorname{ad}\mathfrak{a}$ are isomorphic. Similarly one can show that the Lie algebra considered in Example 3 is isomorphic to its adjoint.

We can associate with every finite dimensional Lie algebra $\mathfrak{a}$ a bilinear form $B: \mathfrak{a} \times \mathfrak{a} \to F$ by defining

$$B(X, Y) = \operatorname{tr}(\operatorname{ad} X)(\operatorname{ad} Y)$$

for every X, Y in $\mathfrak{a}$. The form B is called the *Killing form* or *natural bilinear form* of $\mathfrak{a}$. It is symmetric i.e. $B(X, Y) = B(Y, X)$ for all $X, Y \in \mathfrak{a}$

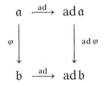

because by Proposition II.5.13 $\operatorname{ad} X \operatorname{ad} Y$ and $\operatorname{ad} Y \operatorname{ad} X$ have identical traces. For instance the Killing form of the Lie algebra of Example 2 is

$$B(X, Y) = -2(X, Y) \tag{3}$$

where (X, Y) is the inner product of X and Y. Similarly if $\mathfrak{a}$ denotes the algebra given in Example 3 then

$$B(X, Y) = -2x_3 y_3 \tag{4}$$

where $X = (x_1, x_2, x_3)$ and $Y = (y_1, y_2, y_3)$. It takes somewhat more computation to determine B when $\mathfrak{a}$ is the Lie algebra $\mathfrak{gl}(n, F)$ of Example 4. The result is

$$B(X, Y) = 2 \operatorname{tr} X \operatorname{tr} Y + 2n \sum_{i,j=1}^{n} X_{ij} Y_{ji} \tag{5}$$

where $X = (X_{ij})$ and $Y = (Y_{ij})$ are $n \times n$ matrices with entries in the field F. This determines also the Killing form of the algebra $\mathfrak{gl}(\mathscr{V})$ discussed in Example 5 because we have the following:

Proposition 14. *If $\mathfrak{a}$ and $\mathfrak{b}$ are Lie algebras and σ is an isomorphism of $\mathfrak{a}$ onto $\mathfrak{b}$ then $B(X, Y) = B(\sigma X, \sigma Y)$ for every X, Y in $\mathfrak{a}$.*

Proof. If $X, Y \in \mathfrak{a}$ and $Z \in \mathfrak{b}$ then

$$\begin{aligned} \sigma(\operatorname{ad} X)(\operatorname{ad} Y)\sigma^{-1} Z &= \sigma(\operatorname{ad} X)[Y, \sigma^{-1} Z] = \sigma[X, [Y, \sigma^{-1} Z]] \\ &= [\sigma X, [\sigma Y, Z]] = (\operatorname{ad} \sigma X)(\operatorname{ad} \sigma Y) Z. \end{aligned}$$

Hence we have

$$\sigma(\operatorname{ad} X)(\operatorname{ad} Y)\sigma^{-1} = \operatorname{ad} \sigma X \operatorname{ad} \sigma Y.$$

By Proposition II.5.14 the trace of the left hand side is $\operatorname{tr}(\operatorname{ad} X)(\operatorname{ad} Y) = B(X, Y)$ while the trace of the right hand side is $B(\sigma X, \sigma Y)$.

Proposition 15. *For every X, Y, Z in the Lie algebra $\mathfrak{a}$ we have*

$$B([X, Y], Z) = B([Y, Z], X) = B([Z, X], Y).$$

Note. Due to (2) a similar identity holds with expressions such as $B(X, [Y, Z])$.

Proof. We have

$$\operatorname{ad}[X, Y] = [\operatorname{ad} X, \operatorname{ad} Y] = (\operatorname{ad} X)(\operatorname{ad} Y) - (\operatorname{ad} Y)(\operatorname{ad} X)$$

and so

$$B([X,Y],Z) = \operatorname{tr}(\operatorname{ad}[X,Y])(\operatorname{ad}Z)$$
$$= \operatorname{tr}(\operatorname{ad}X)(\operatorname{ad}Y)(\operatorname{ad}Z) - \operatorname{tr}(\operatorname{ad}Y)(\operatorname{ad}X)(\operatorname{ad}Z).$$

Hence by Proposition II.5.13 we obtain

$$B([X,Y],Z) = \operatorname{tr}(\operatorname{ad}Z)(\operatorname{ad}X)(\operatorname{ad}Y) - \operatorname{tr}(\operatorname{ad}X)(\operatorname{ad}Z)(\operatorname{ad}Y)$$
$$= \operatorname{tr}[\operatorname{ad}Z, \operatorname{ad}X](\operatorname{ad}Y).$$

Since $[\operatorname{ad}Z, \operatorname{ad}X] = \operatorname{ad}[Z,X]$ the right hand side is $B([Z,X],Y)$.

Let $\mathfrak{a}_1, \mathfrak{a}_2, \mathfrak{a}_3$ be subsets of the Lie algebra $\mathfrak{a}$. Denote by $[\mathfrak{a}_1, \mathfrak{a}_2]$ the subspace spanned by all the products $[A_1, A_2]$ where $A_i \in \mathfrak{a}_i$ $(i=1,2)$. The by the Jacobi identity we have

$$[[\mathfrak{a}_1,\mathfrak{a}_2],\mathfrak{a}_3] \subseteq [[\mathfrak{a}_1,\mathfrak{a}_3],\mathfrak{a}_2] + [[\mathfrak{a}_3,\mathfrak{a}_2],\mathfrak{a}_1]. \tag{6}$$

Suppose that $\mathfrak{a}_1$ and $\mathfrak{a}_2$ are ideals in $\mathfrak{a}$ and $\mathfrak{a}_3 = \mathfrak{a}$. Then this inclusion shows that $[\mathfrak{a}_1, \mathfrak{a}_2]$, the *product of the ideals* $\mathfrak{a}_1$ and $\mathfrak{a}_2$ is itself an ideal. The derived algebra $\mathfrak{a}'$ was defined as $\mathfrak{a}' = [\mathfrak{a}, \mathfrak{a}]$ and so $\mathfrak{a}'$ is an ideal of $\mathfrak{a}$. More generally we define $\mathfrak{a}^{(k)} = [\mathfrak{a}^{(k-1)}, \mathfrak{a}^{(k-1)}]$ for every $k = 1, 2, \ldots$ where $\mathfrak{a}^0 = \mathfrak{a}$. Then $\mathfrak{a}^{(k)}$ is an ideal, $\mathfrak{a}^{(k)} = (\mathfrak{a}^{(k-1)})'$ and clearly $\mathfrak{a} \supseteq \mathfrak{a}' \supseteq \mathfrak{a}^{(2)} \supseteq \mathfrak{a}^{(3)} \supseteq \cdots$. The sequence of ideals $\mathfrak{a}, \mathfrak{a}', \mathfrak{a}^{(2)}, \ldots$ is called the *derived series* of $\mathfrak{a}$. Since the definition of the derived algebra is independent of its possible embedding in a larger Lie algebra we see that if $\mathfrak{i}$ is an ideal in $\mathfrak{a}$ then we have a sequence of ideals $\mathfrak{i} \supseteq \mathfrak{i}' \supseteq \mathfrak{i}^{(2)} \supseteq \mathfrak{i}^{(3)} \supseteq \cdots$ associated with $\mathfrak{i}$.

Similarly if $\mathfrak{i}$ is an ideal in the Lie algebra $\mathfrak{a}$ then we let $\mathfrak{i}^0 = \mathfrak{a}$, $\mathfrak{i}^1 = \mathfrak{i}$, $\mathfrak{i}^2 = [\mathfrak{i}, \mathfrak{i}] = \mathfrak{i}'$ and in general $\mathfrak{i}^k = [\mathfrak{i}^{k-1}, \mathfrak{i}]$ for all $k = 1, 2, \ldots$ Then by (6) $\mathfrak{i}^0, \mathfrak{i}, \mathfrak{i}^2, \mathfrak{i}^3, \ldots$ are ideals and $\mathfrak{i}^0 \supseteq \mathfrak{i} \supseteq \mathfrak{i}^2 \supseteq \cdots$. If $\mathfrak{i}$ is $\mathfrak{a}$ itself the sequence of ideals $\mathfrak{a}, \mathfrak{a}^2, \mathfrak{a}^3, \ldots$ is called the *lower central series* or *central descending series* of $\mathfrak{a}$.

Definition 16. *A Lie algebra* $\mathfrak{a}$ *is called solvable if its derived series is finite i.e. if* $\mathfrak{a}^{(k)} = 0$ *for k sufficiently large.*

For instance the Lie algebra considered in Example 3 is solvable because $\mathfrak{a}^{(2)}$ is generated by the basic vectors i and j and so $\mathfrak{a}^{(3)} = 0$. All one and two dimensional Lie algebras are solvable. For if $\dim \mathfrak{a} = 1$ and A is a basic vector of $\mathfrak{a}$ then $[A, A] = 0$ and so $\mathfrak{a}' = 0$ and $\mathfrak{a}$ is abelian. Moreover if $\dim \mathfrak{a} = 2$ then one possibility is that $\mathfrak{a}' = 0$ which uniquely determines $\mathfrak{a}$ and also shows that $\mathfrak{a}$ is solvable. The other possibility is that $\mathfrak{a}' \neq 0$. Then $\mathfrak{a}'$ is one dimensional and so the basic vectors A_1

and A_2 of $\mathfrak{a}$ can be chosen such that A_1 generates $\mathfrak{a}'$ and $[A_1, A_2] = A_1$. Now by $[A_1, A_1] = 0$ it is clear that $\mathfrak{a}^{(2)} = 0$ and so $\mathfrak{a}$ is solvable. Our reasoning also shows the following:

Proposition 17. *There are two non-isomorphic Lie algebras of dimension 2 over the field F. One of them is the abelian Lie algebra* $\mathfrak{a} = \{x_1 A_1 + x_2 A_2 : x_1, x_2 \in F\}$ *with* $[A_1, A_2] = 0$ *and the other one is characterized by* $[A_1, A_2] = A_1$.

We shall now define another type of Lie algebras which will turn out to be a subclass of the solvable ones.

Definition 18. *The Lie algebra* $\mathfrak{a}$ *is called nilpotent if its lower central series is finite i.e.* $\mathfrak{a}^k = 0$ *for k sufficiently large.*

For example the one dimensional Lie algebra over F is nilpotent. More generally consider the set $\mathfrak{a}$ of those upper triangular matrices in $M_n(F)$ whose diagonal entries are all zeros. Then $\mathfrak{a}$ is a nilpotent subalgebra of $\mathfrak{gl}(n, F)$. This will become clear as soon as one multiplies together two such matrices. One also sees then that $\mathfrak{a}^n = 0$.

Lemma 19. *If* $\mathfrak{i}$ *is an ideal in the Lie algebra* $\mathfrak{a}$ *then* $[\mathfrak{i}^j, \mathfrak{i}^k] \subseteq \mathfrak{i}^{j+k}$ *for every* $j, k = 1, 2, \ldots$

Proof. We have $[\mathfrak{i}^j, \mathfrak{i}] = \mathfrak{i}^{j+1}$ for every $j = 1, 2, \ldots$ Now we can proceed by induction on k: If $[\mathfrak{i}^j, \mathfrak{i}^k] \subseteq \mathfrak{i}^{j+k}$ for every $j = 1, 2, \ldots$ then by (6) we have

$$[\mathfrak{i}^j, \mathfrak{i}^{k+1}] = [\mathfrak{i}^j, [\mathfrak{i}^k, \mathfrak{i}]] = [[\mathfrak{i}^k, \mathfrak{i}], \mathfrak{i}^j] \subseteq [[\mathfrak{i}^k, \mathfrak{i}^j], \mathfrak{i}] + [[\mathfrak{i}^j, \mathfrak{i}], \mathfrak{i}^k].$$

Since $[\mathfrak{i}^k, \mathfrak{i}^j] = [\mathfrak{i}^j, \mathfrak{i}^k] = \mathfrak{i}^{j+k}$ we obtain

$$[\mathfrak{i}^j, \mathfrak{i}^{k+1}] \subseteq [\mathfrak{i}^{j+k}, \mathfrak{i}] + [\mathfrak{i}^{j+1}, \mathfrak{i}^k] = \mathfrak{i}^{j+k+1} + \mathfrak{i}^{j+1+k} = \mathfrak{i}^{j+(k+1)}.$$

Proposition 20. *Every nilpotent Lie algebra is solvable.*

Proof. By the last lemma $\mathfrak{a}' = [\mathfrak{a}, \mathfrak{a}] \subseteq \mathfrak{a}^2$ and if $\mathfrak{a}^{(k)} \subseteq \mathfrak{a}^{2^k}$ for some $k = 1, 2, \ldots$ then by the same lemma $\mathfrak{a}^{(k+1)} = [\mathfrak{a}^{(k)}, \mathfrak{a}^{(k)}] \subseteq [\mathfrak{a}^{2^k}, \mathfrak{a}^{2^k}] \subseteq \mathfrak{a}^{2^{k+1}}$. Hence we have $\mathfrak{a}^{(k)} \subseteq \mathfrak{a}^{2^k}$ for every $k = 1, 2, \ldots$

We note that there exist solvable Lie algebras which are not nilpotent. For example the non-abelian Lie algebra introduced in Proposition 17 is solvable but it is not nilpotent because $\mathfrak{a}^n = \mathfrak{a}^2 = \{x A_1 : x \in F\}$ for every $n \geq 2$.

We shall now prepare the way for the definition of a third important class of Lie algebras. It is obvious from the definitions involved that any homomorphic image of a solvable Lie algebra $\mathfrak{a}$ is solvable. Thus in particular solvability depends only on the isomorphism class containing $\mathfrak{a}$. Similarly if $\mathfrak{a}$ is solvable then so is every subalgebra of $\mathfrak{a}$.

Lemma 21. *If* $\mathfrak{i}$ *is a solvable ideal of* $\mathfrak{a}$ *and* $\mathfrak{a}/\mathfrak{i}$ *is solvable then* $\mathfrak{a}$ *is a solvable algebra.*

Proof. Since $\mathfrak{a}/\mathfrak{i}$ is solvable we have $(\mathfrak{a}/\mathfrak{i})^{(k)}=\mathfrak{i}$ for every sufficiently large $k\in\{1,2,\ldots\}$. Hence if $X\in\mathfrak{a}^{(k)}$ then by $X+\mathfrak{i}\in(\mathfrak{a}/\mathfrak{i})^{(k)}$ we obtain $X\in\mathfrak{i}$ and so $\mathfrak{a}^{(k)}\subseteq\mathfrak{i}$. If k is large enough then by the solvability of $\mathfrak{i}$ we also have $\mathfrak{i}^{(k)}=0$. Hence $\mathfrak{a}^{(2k)}=(\mathfrak{a}^{(k)})^{(k)}\subseteq\mathfrak{i}^{(k)}=0$ and so $\mathfrak{a}$ is solvable.

Lemma 22. *If* $\mathfrak{i}_1$ *and* $\mathfrak{i}_2$ *are solvable ideals in the Lie algebra* $\mathfrak{a}$ *then* $\mathfrak{i}_1+\mathfrak{i}_2$ *is also a solvable ideal of* $\mathfrak{a}$.

Proof. It is obvious that $\mathfrak{i}_2$ is an ideal in $\mathfrak{i}_1+\mathfrak{i}_2$. The map $X\mapsto X+\mathfrak{i}_2$ is a Lie algebra homomorphism of $\mathfrak{i}_1$ into $\mathfrak{a}/\mathfrak{i}_2$, the image of $\mathfrak{i}_1$ being $(\mathfrak{i}_1+\mathfrak{i}_2)/\mathfrak{i}_2$. Therefore the quotient algebra $(\mathfrak{i}_1+\mathfrak{i}_2)/\mathfrak{i}_2$ is solvable. Since $\mathfrak{i}_2$ is also solvable $\mathfrak{i}_1+\mathfrak{i}_2$ is solvable by Lemma 21.

If $\mathfrak{a}$ is finite dimensional then there exists a solvable ideal $\mathfrak{r}$ whose dimension is maximal. If $\mathfrak{i}$ is another solvable ideal of $\mathfrak{a}$ then by Lemma 22 $\mathfrak{r}+\mathfrak{i}$ is a solvable ideal. Therefore by the maximality of the dimension of $\mathfrak{r}$ we have $\mathfrak{r}\supseteq\mathfrak{i}$. This shows that every finite dimensional Lie algebra admits a solvable ideal $\mathfrak{r}$ which contains all solvable ideals of $\mathfrak{a}$. It is called the *solvable radical* of $\mathfrak{a}$.

Definition 23. *The finite dimensional Lie algebra* $\mathfrak{a}$ *is called semisimple if its solvable radical is* 0 *i.e. if the only solvable ideal of* $\mathfrak{a}$ *is* 0.

We shall see that the Lie algebra introduced in Example 2 is semisimple because it has a stronger property called simplicity.

Definition 24. *A Lie algebra* $\mathfrak{a}$ *is called simple if* $\mathfrak{a}'\neq 0$ *and the only ideals of* $\mathfrak{a}$ *are* 0 *and* $\mathfrak{a}$.

Suppose that $\mathfrak{a}$ is a simple algebra. Then its solvable radical $\mathfrak{r}$ is $\mathfrak{a}$ or 0. Let us suppose that $\mathfrak{r}=\mathfrak{a}$ so that $\mathfrak{r}'=\mathfrak{a}'\neq 0$. We have $\mathfrak{r}'\subseteq\mathfrak{r}$ and $\mathfrak{r}'\neq\mathfrak{r}$ because $\mathfrak{r}^{(k)}=0$ for large k. Thus $\mathfrak{r}=\mathfrak{a}$ implies that $\mathfrak{r}'$ is an ideal of $\mathfrak{a}$ which is neither 0 nor $\mathfrak{a}$. Hence $\mathfrak{a}$ is not simple. Hence we proved:

Lemma 25. *Every simple Lie algebra is semisimple.*

Now consider the Lie algebra of Example 2. Since $i=[j,k]$, $j=[k,i]$, $k=[i,j]$ we see that $\mathfrak{a}'=\mathfrak{a}\neq 0$. If $\mathfrak{i}$ is an ideal and $\mathfrak{i}\neq 0$ then $\dim\mathfrak{a}/\mathfrak{i}\leqslant 2$ and so $\mathfrak{a}/\mathfrak{i}$ is solvable. Hence by Lemma 21 and $\mathfrak{a}'=\mathfrak{a}$ the ideal $\mathfrak{i}$ cannot be solvable and so $\dim\mathfrak{i}\geqslant 3$ i.e. $\mathfrak{i}=\mathfrak{a}$. This shows that

$\mathfrak{a}$ is a simple Lie algebra. The same reasoning shows that if $\dim \mathfrak{a} = 3$ and $\mathfrak{a}' = \mathfrak{a}$ then $\mathfrak{a}$ is simple.

In the next section we shall see that simple Lie algebras are the building blocks of semisimple ones in the sense of direct sums. Therefore the study of simple Lie algebras is an important subject. If $F = \mathbb{C}$ then due to the efforts of Elie Cartan and Killing we know the complete list of finite dimensional simple Lie algebras. Naturally in this classification isomorphic algebras are considered to be the same. They come in four sequences, $\mathfrak{a}_n$ $(n \geq 1)$, $\mathfrak{b}_n$ $(n \geq 2)$, $\mathfrak{c}_n$ $(n \geq 3)$, $\mathfrak{d}_n$ $(n \geq 4)$ and in addition there are *five exceptional simple Lie algebras* called $\mathfrak{g}_2, \mathfrak{f}_4, \mathfrak{e}_6, \mathfrak{e}_7$, and $\mathfrak{e}_8$. The last ones are arranged according to an increasing order of dimensions. The proof of these facts has been considerably simplified by later workers in the field, especially by Dynkin and it is also known that $\mathbb{C}$ can be replaced by an algebraically closed field of characteristic zero but the actual proof is still far from being simple or short. We will be content by describing the representative elements of the four principal types and by telling what is meant by the Cartan subalgebra and the roots.

The elements of the series $\mathfrak{a}_n, \mathfrak{b}_n, \mathfrak{c}_n$ and $\mathfrak{d}_n$ are meaningful over an arbitrary field F and they are Lie subalgebras of $\mathfrak{gl}(m, F)$ for suitable values of $m = 2, 3, \ldots$ Namely

$$\mathfrak{a}_n = \{X : X \in \mathfrak{gl}(n+1, F) \text{ and } \operatorname{tr} X = 0\},$$

$$\mathfrak{b}_n = \{X : X \in \mathfrak{gl}(2n+1, F) \text{ and } X + X^t = 0\},$$

$$\mathfrak{c}_n = \{X : X \in \mathfrak{gl}(2n, F) \text{ and } JX + X^t J = 0\},$$

$$\mathfrak{d}_n = \{X : X \in \mathfrak{gl}(2n, F) \text{ and } X + X^t = 0\}.$$

In the third definition J denotes the $2n \times 2n$ matrix $J = \begin{pmatrix} 0 & I \\ -I & 0 \end{pmatrix}$.

If $\mathfrak{a} \neq 0$ is solvable and k is the first index such that $\mathfrak{a}^{(k)} = 0$ then $\mathfrak{i} = \mathfrak{a}^{(k-1)}$ is a non-zero abelian ideal in $\mathfrak{a}$ i.e. $\mathfrak{i}' = 0$. Hence $\mathfrak{i}$ is solvable and so $\mathfrak{a}$ is not semisimple. Therefore *solvability and semisimplicity are incompatible except when* $\mathfrak{a} = 0$. Similar reasoning shows that if $\mathfrak{a} \neq 0$ is nilpotent then its center is not zero: For if $\mathfrak{a}^k = 0$ then $\mathfrak{a}^{k-1}$ is in the center.

Lemma 26. *If* $\mathfrak{i}$ *is an ideal in the Lie algebra* $\mathfrak{a}$ *then so is* $\mathfrak{i}'$.

Proof. By definition $\mathfrak{i}' = [\mathfrak{i}, \mathfrak{i}]$ and using (6) we already proved that the product of two ideals is an ideal. Hence $\mathfrak{i}'$ is an ideal in $\mathfrak{a}$.

Proposition 27. *The Lie algebra* $\mathfrak{a}$ *is semisimple if and only if it has no non-zero abelian ideals.*

Proof. If $\mathfrak{a}$ is not semisimple then by definition it has a non-zero solvable ideal $\mathfrak{i}$. Let k be the first natural number such that $\mathfrak{i}^{(k)}=0$. Then $\mathfrak{i}^{(k-1)}$ is an abelian algebra wich is an ideal in $\mathfrak{a}$ by the last lemma. Since every abelian ideal is solvable the converse statement is trivial.

If $\mathfrak{a}$ is a Lie algebra and $\mathfrak{b}$ is a subalgebra of $\mathfrak{a}$ then

$$\mathfrak{n} = \{X: X\in\mathfrak{a} \text{ and } [X,B]\in\mathfrak{b} \text{ for all } B \text{ in } \mathfrak{b}\}$$

is a subalgebra of $\mathfrak{a}$, called the *normalizer* of $\mathfrak{b}$ in $\mathfrak{a}$: In fact if $X, Y\in\mathfrak{n}$ and $\lambda\in F$ then clearly $X+Y\in\mathfrak{n}$ and $\lambda X\in\mathfrak{n}$. Moreover using

$$[[X,Y],B]=[X,[Y,B]]-[Y,[X,B]]$$

we see that $[X,Y]\in\mathfrak{n}$. Since $X\in\mathfrak{n}$ and $B\in\mathfrak{b}$ imply $[X,B]\in\mathfrak{n}$ the subalgebra $\mathfrak{b}$ is an ideal in $\mathfrak{n}$. If $\mathfrak{c}$ is a subalgebra of $\mathfrak{a}$ and $\mathfrak{b}$ is an ideal in $\mathfrak{c}$ then $\mathfrak{c}\subseteq\mathfrak{n}$. Hence we proved:

Lemma 28. *The normalizer* $\mathfrak{n}$ *of the subalgebra* $\mathfrak{b}$ *of* $\mathfrak{a}$ *is the unique largest subalgebra of* $\mathfrak{a}$ *such that* $\mathfrak{b}$ *is an ideal in* $\mathfrak{n}$.

A subalgebra $\mathfrak{H}$ of a Lie algebra $\mathfrak{a}$ is called a *Cartan subalgebra* if it is nilpotent and it coincides with its own normalizer. There is a different, perhaps somewhat more involved description of these algebras which is important because it is used to prove the existence of Cartan subalgebras of certain Lie algebras. Let $\mathfrak{H}$ be a subalgebra of $\mathfrak{a}$ and consider $\operatorname{ad}: \mathfrak{H}\to\mathfrak{gl}(\mathfrak{a})$, the restriction of the adjoint representation of $\mathfrak{a}$ to $\mathfrak{H}$. The set

$$\mathfrak{a}_0(\mathfrak{H}) = \{X: X\in\mathfrak{a} \text{ and } (\operatorname{ad} H)^k X=0 \text{ for some } k \text{ and all } H\in\mathfrak{H}\}$$

is a subspace of $\mathfrak{a}$ which is called the *root space of* $\operatorname{ad}|\mathfrak{H}$ *belonging to the zero root*. Since $\operatorname{ad} H: \mathfrak{a}\to\mathfrak{a}$ is a derivation we have

$$(\operatorname{ad} H)^m [X,Y] = \sum_{i=0}^{m} [(\operatorname{ad} H)^i X, (\operatorname{ad} H)^{m-i} Y]$$

for all $m=1,2,\ldots$ This shows that $\mathfrak{a}_0(\mathfrak{H})$ is a subalgebra.

Lemma 29. *Let* $\mathfrak{H}$ *be a nilpotent subalgebra of* $\mathfrak{a}$ *such that* $\mathfrak{H}=\mathfrak{a}_0(\mathfrak{H})$. *Then* $\mathfrak{H}$ *is a Cartan subalgebra of* $\mathfrak{a}$.

Note. Using Engel's theorem which will be considered in the next section one can also prove that if $\mathfrak{H}$ is a Cartan subalgebra of $\mathfrak{a}$ then $\mathfrak{H}=\mathfrak{a}_0(\mathfrak{H})$.

Proof. Since $\mathfrak{H}$ is nilpotent we have $\mathfrak{H}^k=0$ for a large positive integer k. If $X\in\mathfrak{n}$, the normalizer of $\mathfrak{H}$, then $[H,X]\in\mathfrak{H}$ and so $(\operatorname{ad}H)^k X=0$ for every H in $\mathfrak{H}$. This shows that $\mathfrak{n}\subseteq\mathfrak{a}_0(\mathfrak{H})$. Hence if $\mathfrak{a}_0(\mathfrak{H})=\mathfrak{H}$ then $\mathfrak{n}\subseteq\mathfrak{H}\subseteq\mathfrak{n}$ and so $\mathfrak{H}=\mathfrak{n}$.

Let A be an arbitrary element of the finite dimensional Lie algebra $\mathfrak{a}$ and let $\mathfrak{H}_A$ be the subspace generated by A. Then $\mathfrak{H}_A$ is an abelian subalgebra of $\mathfrak{a}$ and so we can take $\mathfrak{a}_0(\mathfrak{H}_A)$ and denote its dimension by $n(A)$. If $A\neq 0$ then by $[A,A]=0$ we have $A\in\mathfrak{a}_0(\mathfrak{H}_A)$ and so $n(A)\geqslant 1$. If $A=0$ but $\mathfrak{a}\neq 0$ then $\mathfrak{a}_0(\mathfrak{H}_A)=\mathfrak{a}$ and again $n(A)\geqslant 1$. An element A of the non-zero Lie algebra $\mathfrak{a}$ is called a *regular element* if $n(A)$ is as small as possible. We see that every $\mathfrak{a}\neq 0$ has at least one regular element. The importance of these elements is obvious from the following:

Theorem 30. *If $\mathfrak{a}$ is a finite dimensional Lie algebra over the complex field and A is a regular element then the root space $\mathfrak{a}_0(\mathfrak{H}_A)$ is a Cartan subalgebra of $\mathfrak{a}$.*

Corollary. *Every finite dimensional non-zero complex Lie algebra has a Cartan subalgebra.*

Note. The theorem can be extended to finite dimensional Lie algebras over any infinite field.

Proof. In view of Lemma 29 it is sufficient to prove that $\mathfrak{H}=\mathfrak{a}_0(\mathfrak{H}_A)$ is nilpotent and $\mathfrak{a}_0(\mathfrak{H})=\mathfrak{H}$. Since $A\in\mathfrak{H}$ we have $\mathfrak{H}_A\subseteq\mathfrak{H}$ and so it follows that $X\in\mathfrak{a}_0(\mathfrak{H})$ implies $X\in\mathfrak{H}$. Hence $\mathfrak{a}_0(\mathfrak{H})\subseteq\mathfrak{H}$. Suppose we already know that $\mathfrak{H}$ is nilpotent, say $\mathfrak{H}^k=0$. Then $(\operatorname{ad}H)^k X=[H,[H,\ldots[H,X]\ldots]=0$ for every X in $\mathfrak{H}$ and so we have $X\in\mathfrak{a}_0(\mathfrak{H})$ for every $X\in\mathfrak{H}$ that is to say $\mathfrak{H}\subseteq\mathfrak{a}_0(\mathfrak{H})$. Therefore the problem is reduced to proving that $\mathfrak{H}$ is nilpotent. This is the non-trivial part of the proof and besides Engel's theorem it involves all the root spaces of $\mathfrak{H}_A$. We start by introducing the latter:

For any $\lambda\in\mathbb{C}$ we define

$$\mathfrak{a}_\lambda(\mathfrak{H}_A)=\{X: X\in\mathfrak{a} \text{ and } (\operatorname{ad}A-\lambda I)^k X=0 \text{ for some } k>0\}.$$

The dimension of $\mathfrak{a}_\lambda(\mathfrak{H}_A)$ is positive if and only if λ is a characteristic value of the linear transformation $\operatorname{ad}A:\mathfrak{a}\to\mathfrak{a}$. If this is the case one calls $\mathfrak{a}_\lambda(\mathfrak{H}_A)$ the root space belonging to the characteristic root λ. Thus if $A\neq 0$ then the subalgebra $\mathfrak{a}_0(\mathfrak{H}_A)$ introduced earlier is the root space belonging to 0.

Lemma 31. *For every $A\in\mathfrak{a}$ and $\lambda,\mu\in\mathbb{C}$ we have*

$$[\mathfrak{a}_\lambda(\mathfrak{H}_A),\mathfrak{a}_\lambda(\mathfrak{H}_A)]\subseteq\mathfrak{a}_{\lambda+\mu}(\mathfrak{H}_A).$$

Proof. Since $\operatorname{ad} A$ is a derivation one can easily prove by induction on $m=1,2,\ldots$ that

$$(\operatorname{ad} A-(\lambda+\mu) I)^m [X, Y]=\sum_{i=0}^m \binom{m}{i} [(\operatorname{ad} A-\lambda I)^i X,(\operatorname{ad} A-\mu I)^{m-i} Y].$$

Using this identity we see right away that $X \in \mathfrak{a}_\lambda(\mathfrak{H}_A)$ and $Y \in \mathfrak{a}_\mu(\mathfrak{H}_A)$ imply $[X, Y] \in \mathfrak{a}_{\lambda+\mu}(\mathfrak{H}_A)$.

Lemma 32. *If A is a regular element of $\mathfrak{a}$ then $\operatorname{ad} H$ is nilpotent for all $H \in \mathfrak{a}_0(\mathfrak{H}_A)$.*

Proof. Let $\lambda_0=0, \lambda_1, \ldots, \lambda_r$ denote the characteristic roots of $\operatorname{ad} A: \mathfrak{a} \rightarrow \mathfrak{a}$ and let

$$\tilde{\mathfrak{a}}=\mathfrak{a}_{\lambda_1}(\mathfrak{H}_A) \oplus \cdots \oplus \mathfrak{a}_{\lambda_r}(\mathfrak{H}_A).$$

Then we have $\mathfrak{a}=\mathfrak{H} \oplus \tilde{\mathfrak{a}}$ where $\mathfrak{H}=\mathfrak{a}_0(\mathfrak{H}_A)$, a notation introduced earlier. We know that $[\mathfrak{H}, \mathfrak{H}] \subseteq \mathfrak{H}$ and by Lemma 31 we have $[\mathfrak{H}, \tilde{\mathfrak{a}}] \subseteq \tilde{\mathfrak{a}}$. Therefore $\mathfrak{H}$ and $\tilde{\mathfrak{a}}$ are invariant subspaces of $\operatorname{ad} H$ for every $H \in \mathfrak{H}$. We let $d(H)$ denote the determinant of the linear transformation $\operatorname{ad} H \mid \tilde{\mathfrak{a}}$. Let $H_1, H_2, \ldots$ and $A_1, \ldots, A_s$ denote ordered bases for $\mathfrak{H}$ and $\tilde{\mathfrak{a}}$, respectively. Let M_i be the matrix of $\operatorname{ad} H_i$ with respect to $A_1, \ldots, A_s$. Then for any $H=\sum z_i H_i$ the matrix of $\operatorname{ad} H \mid \tilde{\mathfrak{a}}$ is $\sum z_i M_i$. Hence $d(H)=d(\sum z_i H_i)=p(z_1, \ldots, z_s)$ where p is a polynomial in $z_1, \ldots, z_s \in \mathbb{C}$ with complex coefficients. We have $A \in \mathfrak{H}$ and $d(A) \neq 0$ because $\operatorname{ad} A \mid \tilde{\mathfrak{a}}$ has only non-zero characteristic roots and $d(A)$ is the product of these roots, each counted with its multiplicity. It follows that p can not identically vanish in any non-void open subset of $\mathbb{C}^s$. Therefore the set S of those $z=(z_1, \ldots, z_s)$ for which $p(z_1, \ldots, z_s) \neq 0$ is dense in $\mathbb{C}^s$. If $z \in S$ and $H=\sum z_i H_i$ then by $d(H)=p(z_1, \ldots, z_s) \neq 0$ the characteristic roots of $\operatorname{ad} H \mid \tilde{\mathfrak{a}}$ are all different from 0. Since $\operatorname{ad} H=\operatorname{ad} H\mid\mathfrak{H}+\operatorname{ad} H \mid \tilde{\mathfrak{a}}$ this implies that $\mathfrak{a}_0(\mathfrak{H}_H) \subseteq \mathfrak{H}$: Indeed let $X \in \mathfrak{a}$ be such that $T^k X=0$ and let $T^{k-1} X=X_1+X_2$ with $X_1 \in \mathfrak{H}$ and $X_2 \in \tilde{\mathfrak{a}}$. Then $T X_1+T X_2=0$ and so $T X_2=0$ which implies that $X_2=0$ and $X=X_1 \in \mathfrak{H}$. Then we let $T^{k-2} X=X_1+X_2$ and obtain $T X_1+T X_2=0$ so that $X_2=0$ and $X=X_1 \in \mathfrak{H}$. In a finite number of steps we arrive at $X \in \mathfrak{H}$. Hence if $z \in S$ then $\mathfrak{a}_0(\mathfrak{H}_H) \subseteq \mathfrak{H}=\mathfrak{a}_0(\mathfrak{H}_A)$ and by the regularity of A we have $\mathfrak{a}_0(\mathfrak{H}_H)=\mathfrak{H}$. In other words $\operatorname{ad} H \mid \mathfrak{H}$ is nilpotent. Since $\operatorname{ad} H \mid \mathfrak{H}$ is the same as $\operatorname{ad} H: \mathfrak{H} \rightarrow \mathfrak{H}$ and $\mathfrak{H}$ is finite dimensional there is a common exponent k such that $(\operatorname{ad} H)^k=0$ on $\mathfrak{H}$ for every $H=\sum z_i H_i$ where $z \in S$. Since S is dense in $\mathbb{C}^s$ it follows that $\operatorname{ad} H: \mathfrak{H} \rightarrow \mathfrak{H}$ is nilpotent for every $H \in \mathfrak{H}$.

A theorem due to Engel states that if $\operatorname{ad} H: \mathfrak{H} \rightarrow \mathfrak{H}$ is nilpotent for every H in $\mathfrak{H}$ then the finite dimensional Lie algebra $\mathfrak{H}$ is nilpotent.

Therefore the proof of the existence of a Cartan subalgebra as is stated in Theorem 30 is now reduced to the proof of Engel's criterion for nilpotency. This will be done in the beginning of the next section.

Proposition 33. *If* $\mathfrak{a}$ *is a semisimple Lie algebra then every derivation* D *of* $\mathfrak{a}$ *is a Lie derivative i.e.* $DX=[A,X]$ *for some* $A\in\mathfrak{a}$ *and all* $X\in\mathfrak{a}$.

Proof. The kernel of the adjoint representation is the center $\mathfrak{c}$ which is an abelian ideal of $\mathfrak{a}$. Since $\mathfrak{a}$ is semisimple $\mathfrak{c}$ is 0 and so ad is an isomorphism of $\mathfrak{a}$ onto $\operatorname{ad}\mathfrak{a}$. Hence $\operatorname{ad}\mathfrak{a}$ is semisimple. We have $[D,\operatorname{ad}X]=\operatorname{ad}DX$ for every $X\in\mathfrak{a}$ and every derivation $D:\mathfrak{a}\to\mathfrak{a}$ because

$$[D,\operatorname{ad}X]\,Y=D(\operatorname{ad}X)\,Y-(\operatorname{ad}X)\,D\,Y=D[X,Y]-[X,DY]$$
$$=[DX,Y]=(\operatorname{ad}DX)\,Y.$$

Therefore $\operatorname{ad}\mathfrak{a}$ is an ideal in the algebra $\mathfrak{d}(\mathfrak{a})$ of all derivations of $\mathfrak{a}$. Let B denote the Killing form of $\mathfrak{d}(\mathfrak{a})$ and define

$$(\operatorname{ad}\mathfrak{a})^{\perp}=\{D\colon D\in\mathfrak{d}(\mathfrak{a}) \text{ and } B(D,\operatorname{ad}X)=0 \text{ for all } X\in\mathfrak{a}\}.$$

Proposition 15 shows that $(\operatorname{ad}\mathfrak{a})^{\perp}$ is an ideal and so $\operatorname{ad}\mathfrak{a}\cap(\operatorname{ad}\mathfrak{a})^{\perp}$ is also an ideal in $\mathfrak{d}(\mathfrak{a})$. Since $\operatorname{ad}\mathfrak{a}$ is an ideal in $\mathfrak{d}(\mathfrak{a})$ the Killing form of $\operatorname{ad}\mathfrak{a}$ is the restriction of B to $(\operatorname{ad}\mathfrak{a})\times(\operatorname{ad}\mathfrak{a})$. Hence if $\operatorname{ad}Y$ is in $(\operatorname{ad}\mathfrak{a})^{\perp}\cap\operatorname{ad}\mathfrak{a}$ then by $B(\operatorname{ad}X,\operatorname{ad}Y)=0$ $(X\in\mathfrak{a})$ and the semisimplicity of $\operatorname{ad}\mathfrak{a}$ we have $\operatorname{ad}Y=0$. Therefore $(\operatorname{ad}\mathfrak{a})^{\perp}\cap\operatorname{ad}\mathfrak{a}=0$. Now if $D\in(\operatorname{ad}\mathfrak{a})^{\perp}$ then by the ideal properties $[D,\operatorname{ad}X]\in(\operatorname{ad}\mathfrak{a})^{\perp}\cap\operatorname{ad}\mathfrak{a}=0$ and so $\operatorname{ad}DX=0$ for all $X\in\mathfrak{a}$. Since ad is an isomorphism we obtain $D=0$. We proved that $(\operatorname{ad}\mathfrak{a})^{\perp}=0$ which implies that $\operatorname{ad}\mathfrak{a}$ is the entire space $\mathfrak{d}(\mathfrak{a})$: The proof of this fact is the same as the one used in the case of real and complex finite dimensional inner product spaces.

2. Finite Dimensional Representations of Lie Algebras. Cartan's Criteria and the Theorems of Engel and Lie

By a *representation* ρ of a Lie algebra $\mathfrak{a}$ in a vector space $\mathscr{V}$ we mean a Lie algebra homomorphism $\rho:\mathfrak{a}\to\mathrm{gl}(\mathscr{V})$. The existence of ρ implies that $\mathfrak{a}$ and $\mathscr{V}$ are defined over the same field F. The adjoint representation $\operatorname{ad}:\mathfrak{a}\to\mathrm{gl}(\mathfrak{a})$ gives an example of a representation with $\mathscr{V}=\mathfrak{a}$ being the underlying vector space of $\mathfrak{a}$. The space $\mathscr{V}$ need not be finite dimensional but then we usually suppose that $\mathscr{V}$ is a topological vector space and $\mathrm{gl}(\mathscr{V})$ consists of all the continuous linear operators $A:\mathscr{V}\to\mathscr{V}$.

By a *weight vector* of the representation $\rho: \mathfrak{a} \to \mathfrak{gl}(\mathscr{V})$ we mean a non-zero vector $\boldsymbol{v}$ in $\mathscr{V}$ which is characteristic vector for all the operators $\rho(X)$ $(X \in \mathfrak{a})$. Hence $\rho(X)\boldsymbol{v} = \lambda(X)\boldsymbol{v}$ for every $X \in \mathfrak{a}$ and $\lambda: \mathfrak{a} \to F$ is a linear functional called the weight associated with the weight vector $\boldsymbol{v}$. For every linear functional $\lambda: \mathfrak{a} \to F$ we define

$$\mathscr{V}_\lambda = \{\boldsymbol{v}: \boldsymbol{v} \in \mathscr{V} \text{ and } (\rho(X) - \lambda(X)I)^N \boldsymbol{v} = 0 \text{ for } N \text{ large and all } X \in \mathfrak{a}\}.$$

Then $\mathscr{V}_\lambda$ is a subspace of $\mathscr{V}$ which is called the *weight space associated with the linear functional* λ. If $\mathscr{V}_\lambda$ is of positive dimension then λ is called a *weight* of the representation ρ. The elements of $\mathscr{V}_\lambda$ are called *generalized weight vectors* and the expression weight vector is also used when no confusion can arise. If the field F is not of zero characteristic or is not algebraically closed then one calls weights also some other objects besides the linear functionals. Here we are primarily interested only in these classical weights even if F is an arbitrary field. In addition the vector space $\mathscr{V}$ of the representation ρ will usually be finite dimensional in this section. By a *root* of the Lie algebra $\mathfrak{a}$ we understand a weight of the adjoint representation of $\mathfrak{a}$. The weight space of a root of $\mathfrak{a}$ is called a *root space* of $\mathfrak{a}$.

The following result and its consequences are generalizations and variants of a classical theorem due to Engel and are frequently called *Engel's theorem.* They concern the question of nilpotency of a finite dimensional Lie algebra.

Theorem 1. *Let $\mathfrak{a}$ be a finite dimensional Lie algebra such that* $\operatorname{ad} X$ *is a nilpotent operator for every X in $\mathfrak{a}$. Let $\rho: \mathfrak{a} \to \mathfrak{gl}(\mathscr{V})$ be a finite dimensional representation such that $\rho(X)$ is a nilpotent operator for every $X \in \mathfrak{a}$. Then there is a non-zero vector $\boldsymbol{v}$ in $\mathscr{V}$ such that $\rho(X)\boldsymbol{v} = 0$ for every X in $\mathfrak{a}$.*

Note. In other words the conclusion is that the zero linear functional is a weight of ρ and the weight space $\mathscr{V}_0$ contains weight vectors in the restricted sense.

Proof. Let $n = \dim \mathfrak{a}$. If $n = 1$ then we can choose for $\boldsymbol{v}$ any non-zero vector in $\mathscr{V}$: For $\rho(X)\boldsymbol{v} = \lambda \boldsymbol{v}$ for some $\lambda \in F$ and so $\rho(X)$ being nilpotent $\lambda = 0$ and so $\rho(X)\boldsymbol{v} = 0$ for every $X \in \mathfrak{a}$. Now we suppose that $n \geqslant 2$ and the proposition holds for every dimension smaller than n. Let $\mathfrak{m}$ be a proper subalgebra of $\mathfrak{a}$ such that its dimension is maximal. Since every one dimensional subspace is a subalgebra we have $\dim \mathfrak{m} \geqslant 1$. If $A \in \mathfrak{m}$ then $\mathfrak{m}$ being a subalgebra $X + \mathfrak{m} \mapsto [A, X] + \mathfrak{m}$ defines a linear map $\sigma(A): \mathfrak{a}/\mathfrak{m} \to \mathfrak{a}/\mathfrak{m}$.

Since $[A,[B,X]]-[B,[A,X]]=[[A,B],X]$ we see that σ is a representation of $\mathfrak{m}$ in the vector space $\mathfrak{a}/\mathfrak{m}$. Moreover $\operatorname{ad}A$ being nilpotent we see that $\sigma(A)^k X=(\operatorname{ad}A)^k X+\mathfrak{m}=\mathfrak{m}$ for k sufficiently large. Hence every $\sigma(A)$ is nilpotent and so $\dim\mathfrak{a}/\mathfrak{m}$ being less than n the proposition can be applied with $\mathfrak{m}$ and σ instead of $\mathfrak{a}$ and ρ. Thus we can find a non-zero vector B in $\mathfrak{a}$ such that $B\notin\mathfrak{m}$ and $[A,B]\in\mathfrak{m}$ for every $A\in\mathfrak{m}$.

By $B\notin\mathfrak{m}$ and the maximality of the dimension of $\mathfrak{m}$ we see that B and $\mathfrak{m}$ generate $\mathfrak{a}$. Consider the subspace

$$\mathscr{V}_0=\{\boldsymbol{x}:\boldsymbol{x}\in\mathscr{V} \text{ and } \rho(A)\boldsymbol{x}=0 \text{ for all } A\in\mathfrak{m}\}.$$

We have $\dim\mathfrak{m}<n$ and so by the induction hypothesis we can apply the theorem with $\mathfrak{m}$ and $\rho|\mathfrak{m}$ instead of $\mathfrak{a}$ and ρ. It follows that $\mathscr{V}_0\neq 0$. If $\boldsymbol{x}\in\mathscr{V}_0$ and $A\in\mathfrak{m}$ then we have $[A,B]\in\mathfrak{m}$ and so

$$\rho(A)\rho(B)\boldsymbol{x}=\rho(B)\rho(A)\boldsymbol{x}+\rho([A,B])\boldsymbol{x}=0.$$

This proves that $\mathscr{V}_0$ is invariant under the nilpotent transformation $\rho(B)$. Hence by the nilpotency of $\rho(B)|\mathscr{V}_0$ there is a non-zero vector $\boldsymbol{v}$ in $\mathscr{V}_0$ such that $\rho(B)\boldsymbol{v}=0$. By $\boldsymbol{v}\in\mathscr{V}_0$ we have $\rho(A)\boldsymbol{v}=0$ for every $A\in\mathfrak{m}$ and so $\rho(X)\boldsymbol{v}=0$ for every $X\in\mathfrak{a}$ because B and $\mathfrak{m}$ generate $\mathfrak{a}$.

Corollary. *There is a strictly decreasing sequence of subspaces* $\mathscr{V}=\mathscr{V}_0\supset\mathscr{V}_1\supset\cdots\supset\mathscr{V}_k=0$ *such that* $\rho(X)\mathscr{V}_{i-1}\subseteq\mathscr{V}_i$ *for every* $X\in\mathfrak{a}$ *and* $i=1,\ldots,k$.

In order to prove the corollary consider $\mathscr{V}^*$, the dual of $\mathscr{V}$ and define $\rho^*:\mathfrak{a}\to\mathfrak{gl}(\mathscr{V}^*)$ by $(\rho^*(X)\boldsymbol{v}^*)(\boldsymbol{v})=\boldsymbol{v}^*(\rho(X)\boldsymbol{v})$. Then ρ^* is a representation of $\mathfrak{a}$ in $\mathscr{V}^*$ and it satisfies the conditions of Theorem 1. Hence there is a non-zero linear functional $\boldsymbol{v}^*$ in $\mathscr{V}^*$ such that $\rho(X)^*\boldsymbol{v}^*=0$ for all $X\in\mathfrak{a}$. The kernel of $\boldsymbol{v}^*$ is a proper subspace $\mathscr{V}_1$ of $\mathscr{V}$ such that $\rho(X)\mathscr{V}\subseteq\mathscr{V}_1$. Using the same process we can now construct $\mathscr{V}_2,\mathscr{V}_3,\ldots$

Theorem 2. *A finite dimensional Lie algebra* $\mathfrak{a}$ *is nilpotent if and only if* $\operatorname{ad}X$ *is a nilpotent operator for every* X *in* $\mathfrak{a}$.

Note. This result is called the *Engel criterion.*

Proof. First we suppose that $\operatorname{ad}X$ is nilpotent for all $X\in\mathfrak{a}$. We consider Theorem 1 and its corollary with $\mathfrak{a}$ and the adjoint representation $\operatorname{ad}:\mathfrak{a}\to\mathfrak{gl}(\mathfrak{a})$. The corollary implies the existence of a strictly decreasing sequence $\mathfrak{a}=\mathfrak{a}_0\supset\mathfrak{a}_1\supset\cdots\supset\mathfrak{a}_k=0$ of subspaces of $\mathfrak{a}$ such that $\rho(X)Y=[X,Y]\in\mathfrak{a}_i$ for every $X\in\mathfrak{a}$ and $Y\in\mathfrak{a}_{i-1}$. In other words $[\mathfrak{a},\mathfrak{a}_{i-1}]\subseteq\mathfrak{a}_i$ and so $\mathfrak{a}^i\subseteq\mathfrak{a}_i$ for all $i=1,\ldots,k$. In particular $\mathfrak{a}^k\subseteq\mathfrak{a}_k=0$ and so $\mathfrak{a}$ is nilpotent. The converse statement is trivial because $(\operatorname{ad}X)^k Y$

$=[X,[X,\ldots,[X,Y]\ldots]\in\mathfrak{a}^{k+1}$ for every $k=1,2,\ldots$ and so the nilpotency of $\mathfrak{a}$ implies that of $\operatorname{ad}X$ for all X in $\mathfrak{a}$.

Proposition 3. *Let $\mathscr{V}$ be a finite dimensional vector space and let $\mathfrak{a}$ be a subalgebra of $\mathfrak{gl}(\mathscr{V})$ such that every element of $\mathfrak{a}$ is a nilpotent linear transformation. Then $\operatorname{ad}X$ is nilpotent for every X in $\mathfrak{a}$.*

Proof. For any X in $\mathfrak{gl}(\mathscr{V})$ let $L_X Y=XY$ and $R_X Y=YX$ for all $Y\in\mathfrak{gl}(\mathscr{V})$. Then L_X and R_X are commuting linear operators and $\operatorname{ad}X=L_X-R_X$. Hence by the binomial theorem

$$(\operatorname{ad}X)^m Y=\sum_{i=0}^{m}(-1)^i\binom{m}{i}X^i Y X^{m-i}$$

where $Y\in\mathfrak{gl}(\mathscr{V})$. If $X\in\mathfrak{a}$ then $X^k=0$ for a large value of k and so for $m\geqslant 2k$ we obtain $(\operatorname{ad}X)^m=0$.

Corollary 1. *There is a non-zero vector $\boldsymbol{v}$ in $\mathscr{V}$ such that $X\boldsymbol{v}=0$ for all X in $\mathfrak{a}$.*

In fact we can apply Theorem 1 with the representation $\rho:\mathfrak{a}\to\mathfrak{gl}(\mathscr{V})$ where $\rho(X)=X$ for every X in $\mathfrak{a}$.

Corollary 2. *The vector space $\mathscr{V}$ has a basis $\boldsymbol{v}_1,\ldots,\boldsymbol{v}_n$ such that for every X in $\mathfrak{a}$ and every $i=0,1,\ldots,n$ the image $X\boldsymbol{v}_i$ belongs to the subspace spanned by $\boldsymbol{v}_0,\boldsymbol{v}_1,\ldots,\boldsymbol{v}_{i-1}$ where $\boldsymbol{v}_0=0$. In other words the matrix of each X with respect to the ordered basis $\boldsymbol{v}_1,\ldots,\boldsymbol{v}_n$ is upper triangular and has zeros on its main diagonal.*

Let $\boldsymbol{v}_1\neq 0$ be such that $X\boldsymbol{v}_1=0$ for all $X\in\mathfrak{a}$. Then X defines a linear transformation $\tilde{X}:\mathscr{V}/\mathscr{V}_1\to\mathscr{V}/\mathscr{V}_1$ where $\mathscr{V}_1$ is the one dimensional subspace generated by $\boldsymbol{v}_1$: Namely $\tilde{X}(\boldsymbol{v}+\mathscr{V}_1)=X\boldsymbol{v}+\mathscr{V}_1$ for all $\boldsymbol{v}\in\mathscr{V}$. The set of operators $\tilde{X}$ is a subalgebra of $\mathfrak{gl}(\mathscr{V}/\mathscr{V}_1)$ and every $\tilde{X}$ is nilpotent. Hence if $\mathscr{V}\neq\mathscr{V}_1$ then by Corollary 1 there is a $\boldsymbol{v}_2$ in $\mathscr{V}$ such that $\boldsymbol{v}_2\notin\mathscr{V}_1$ and $\boldsymbol{v}_2+\mathscr{V}_1$ is annihilated by every $\tilde{X}$. We continue this construction by replacing $\mathscr{V}_1$ by the subspace $\mathscr{V}_2$ spanned by $\boldsymbol{v}_1$ and $\boldsymbol{v}_2$, etc.

Corollary 3. *Let $\rho:\mathfrak{a}\to\mathfrak{gl}(\mathscr{V})$ be a finite dimensional faithful representation of the Lie algebra $\mathfrak{a}$ such that every $\rho(X)$ $(X\in\mathfrak{a})$ is a nilpotent operator. Then $\mathfrak{a}$ is nilpotent and $\mathscr{V}$ has an ordered basis such that the matrix of every $\rho(X)$ $(X\in\mathfrak{a})$ is upper triangular and has zeros on its main diagonal.*

For $\rho(\mathfrak{a})$ is a Lie subalgebra of $\mathfrak{gl}(\mathscr{V})$ and so $\operatorname{ad}\rho(X)$ is nilpotent for every $X\in\mathfrak{a}$ by Proposition 3. Hence by Theorem 2 $\rho(\mathfrak{a})$ is nilpotent. Since ρ is faithful we see that $\mathfrak{a}$ is nilpotent. By Corollary 2 there is an ordered basis such that the matrices of the operators $\rho(X)$ $(X \in \mathfrak{a})$ are strictly upper triangular. Corollary 3 is also called Engel's criterion.

We shall now leave nilpotent Lie algebras and turn to the study of the larger category consisting of solvable Lie algebras.

Proposition 4. *Let* $\rho: \mathfrak{a} \to \mathfrak{gl}(\mathscr{V})$ *be a finite dimensional representation of the Lie algebra* $\mathfrak{a}$, *let* $\mathfrak{i}$ *be an ideal in* $\mathfrak{a}$ *and let* $\lambda: \mathfrak{i} \to F$ *be a weight of* $\rho|\mathfrak{i}$ *in the restricted sense. Then* $\lambda(X)=0$ *for every* X *in* $[\mathfrak{a},\mathfrak{i}]$.

Proof. Let $\boldsymbol{v}_1$ be a weight vector in the restricted sense associated with λ i.e. let $\rho(X)\boldsymbol{v}_1 = \lambda(x)\boldsymbol{v}_1$ for all X in $\mathfrak{i}$. We let $\boldsymbol{v}_{i+1} = \rho(X)\boldsymbol{v}_i$ for $i=1,2,\ldots$ where X is a fixed element of $\mathfrak{a}$. Let $\mathscr{V}_i$ denote the subspace of $\mathscr{V}$ spanned by the vectors $\boldsymbol{v}_0, \ldots, \boldsymbol{v}_i$ where $\boldsymbol{v}_0 = 0$. Using induction on i we prove that

$$\rho(Y)\boldsymbol{v}_i - \lambda(Y)\boldsymbol{v}_i \in \mathscr{V}_{i-1} \tag{1}$$

for every $Y \in \mathfrak{i}$ and $i=1,2,\ldots$ This is clear if $i=1$ by the definitions of $\boldsymbol{v}_0$ and $\boldsymbol{v}_1$. We suppose that it holds for the index $i \geqslant 1$. For any Y in $\mathfrak{i}$ we have

$$\rho(Y)\boldsymbol{v}_{i+1} = \rho(Y)\rho(X)\boldsymbol{v}_i = \rho(X)\rho(Y)\boldsymbol{v}_i + \rho([Y,X])\boldsymbol{v}_i$$

and so

$$\rho(Y)\boldsymbol{v}_{i+1} - \lambda(Y)\boldsymbol{v}_{i+1} = \lambda([Y,X])\boldsymbol{v}_i + \rho(X)(\rho(Y)\boldsymbol{v}_i - \lambda(Y)\boldsymbol{v}_i) + (\rho([Y,X])\boldsymbol{v}_i - \lambda([Y,X])\boldsymbol{v}_i).$$

The second term on the right hand side belongs to $\rho(X)\mathscr{V}_{i-1}$ which lies in $\mathscr{V}_i$ by the definition of the vectors $\boldsymbol{v}_0, \ldots, \boldsymbol{v}_{i-1}$. Similarly by (1) the last term is in $\mathscr{V}_{i-1}$. Hence the right hand side belongs to $\mathscr{V}_i$.

Let $\mathscr{W}$ denote the subspace of $\mathscr{V}$ spanned by the vectors $\boldsymbol{v}_1, \boldsymbol{v}_2, \ldots$ By (1) we see that $\rho(Y)\mathscr{W} \subseteq \mathscr{W}$ for all Y in $\mathfrak{i}$ and so ρ can be restricted to the invariant subspace $\mathscr{W}$. If we select a basis from the sequence $\boldsymbol{v}_1, \boldsymbol{v}_2, \ldots$ then (1) also shows that the diagonal entries of the matrix of $\rho(Y)|\mathscr{W}$ with respect to this ordered basis are all equal to $\lambda(Y)$. Hence $\operatorname{tr}\rho(Y)|\mathscr{W} = \lambda(Y)\dim\mathscr{W}$ for all $Y \in \mathfrak{i}$. In particular

$$\lambda([X,Y])\dim\mathscr{W} = \operatorname{tr}\rho([X,Y])|\mathscr{W} = 0$$

because $\rho([X,Y]) = \rho(X)\rho(Y) - \rho(Y)\rho(X)$ and so Proposition II.5.13 can be applied. We proved that $\lambda([X,Y])=0$ for all $X \in \mathfrak{a}$ and $Y \in \mathfrak{i}$. Hence $\lambda(X)=0$ for every X in $[\mathfrak{a},\mathfrak{i}]$. Q.e.d.

Proposition 5. *Let* $\rho: \mathfrak{a} \to \mathfrak{gl}(\mathscr{V})$ *be a finite dimensional representation of the Lie algebra* $\mathfrak{a}$. *If* $\lambda: \mathfrak{a} \to F$ *is a linear functional and* $\mathfrak{i}$ *is an ideal of*

$\mathfrak{a}$ *then*

$$\mathscr{M}_\lambda = \{\boldsymbol{v} : \boldsymbol{v} \in \mathscr{V} \text{ and } \rho(X)\boldsymbol{v} = \lambda(X)\boldsymbol{v} \text{ for all } X \in \mathfrak{i}\}$$

is an invariant subspace of ρ.

Proof. If $\mathscr{M}_\lambda = 0$ then the result is trivial. If $\mathscr{M}_\lambda \neq 0$ then for every $\boldsymbol{v} \in \mathscr{M}_\lambda$, $X \in \mathfrak{i}$ and $Y \in \mathfrak{a}$ we have

$$\rho(X)\rho(Y)\boldsymbol{v} = \rho([X, Y])\boldsymbol{v} + \rho(Y)\rho(X)\boldsymbol{v} = \lambda(X)\rho(Y)\boldsymbol{v} + \lambda([X, Y])\boldsymbol{v}.$$

Since $\mathscr{M}_\lambda \neq 0$ we can apply Proposition 4 and obtain $\rho(X)\rho(Y)\boldsymbol{v} = \lambda(X)\rho(Y)\boldsymbol{v}$ for every $X \in \mathfrak{i}$ and $Y \in \mathfrak{a}$. This shows that $\rho(Y)\boldsymbol{v}$ is in $\mathscr{M}_\lambda$ for every $Y \in \mathfrak{a}$ where $\boldsymbol{v} \in \mathscr{M}_\lambda$ is arbitrary.

Lemma 6. *Every finite dimensional, non-zero, solvable Lie algebra* $\mathfrak{a}$ *has an ideal* $\mathfrak{m}$ *of codimension* 1 *i. e. such that* $1 + \dim \mathfrak{m} = \dim \mathfrak{a}$.

Proof. Since $\mathfrak{a}$ is solvable and $\mathfrak{a} \neq 0$ we have $\mathfrak{a}' \subset \mathfrak{a}$ and so $\mathfrak{a}$ has a subspace $\mathfrak{m}$ such that $\mathfrak{a}' \subseteq \mathfrak{m}$ and $1 + \dim \mathfrak{m} = \dim \mathfrak{a}$. By $[\mathfrak{a}, \mathfrak{m}] \subseteq [\mathfrak{a}, \mathfrak{a}] = \mathfrak{a}' \subseteq \mathfrak{m}$ the set $\mathfrak{m}$ is an ideal of $\mathfrak{a}$.

The following is one of a number of results which are called *Lie's theorems*:

Theorem 7. *If* $\mathfrak{a}$ *is a finite dimensional, solvable Lie algebra over an algebraically closed field of characteristic* 0 *then the irreducible representations of* $\mathfrak{a}$ *are all one dimensional.*

Corollary. *Every finite dimensional representation has a weight vector in the restricted sense.*

Proof. We shall prove the theorem by induction on $n = \dim \mathfrak{a}$. First we note that if the theorem holds for all dimensions less than n then so does its corollary: For the representation ρ of $\mathfrak{a}$ has an irreducible component $\mathscr{P}$ which is then one dimensional and so any generator $\mathscr{P}$ is a weight vector for ρ. Let $\mathfrak{a}$ be n dimensional and let $\pi : \mathfrak{a} \to \mathfrak{gl}(\mathscr{V})$ be irreducible. By Lemma 6 $\mathfrak{a}$ has an ideal $\mathfrak{m}$ of codimension 1 and so by the induction hypothesis $\pi|\mathfrak{m}$ has a weight vector in the restricted sense. Let λ denote the corresponding weight. By Proposition 5 and the irreducibility of π we have $\mathscr{M}_\lambda = \mathscr{V}$. If $X \in \mathfrak{a}$ and $X \notin \mathfrak{m}$ then the field F of $\mathfrak{a}$ being algebraically closed the operator $\pi(X)$ has a characteristic vector $\boldsymbol{v}$. The one dimensional subspace $F\boldsymbol{v}$ is invariant under $\pi(X)$ and by $\boldsymbol{v} \in \mathscr{V} = \mathscr{M}_\lambda$ also under every $\pi(Y)$ $(Y \in \mathfrak{m})$. Since X and $\mathfrak{m}$ generate $\mathfrak{a}$ we see that $F\boldsymbol{v}$ is an invariant subspace of π. Since π is irreducible we obtain $\mathscr{V} = F\boldsymbol{v}$.

Proposition 8. *Let $\mathfrak{a}$ be a finite dimensional solvable Lie algebra over an algebraically closed field of characteristic zero and let $\rho: \mathfrak{a} \to \mathfrak{gl}(\mathscr{V})$ be an irreducible representation. Then $\rho(X) = 0$ for every X in $\mathfrak{a}'$.*

Proof. This is an immediate consequence of Theorem 7. For $\mathscr{V}$ is one dimensional and so any two operators in $\mathscr{L}(\mathscr{V})$ commute with each other. Thus if $X, Y \in \mathfrak{a}$ then $\rho([X, Y]) = \rho(X)\rho(Y) - \rho(Y)\rho(X) = 0$.

Theorem 9. *Let $\mathfrak{a}$ be a finite dimensional solvable Lie algebra over an algebraically closed field of characteristic 0 and let $\rho: \mathfrak{a} \to \mathfrak{gl}(\mathscr{V})$ be a representation of $\mathfrak{a}$ in the finite dimensional vector space $\mathscr{V}$. Then $\mathscr{V}$ has an ordered basis such that the matrix of every $\rho(X)$ $(X \in \mathfrak{a})$ is upper triangular.*

Proof. If $\mathscr{V} \neq 0$ then by the corollary of Theorem 7 there is a common characteristic vector $\boldsymbol{v}_1$ for all the operators $\rho(X)$ $(X \in \mathfrak{a})$. We let $\mathscr{V}_1$ be the one dimensional invariant subspace generated by $\boldsymbol{v}_1$ and ρ_1 the representation subduced on $\mathscr{V}/\mathscr{V}_1$ by ρ. If $\dim \mathscr{V}/\mathscr{V}_1$ is positive then by the same corollary there is a vector $\boldsymbol{v}_2$ in $\mathscr{V}$ such that $\boldsymbol{v}_2 \notin \mathscr{V}_1$ and $\boldsymbol{v}_2 + \mathscr{V}_1$ is a characteristic vector of every $\rho_1(X)$ $(X \in \mathfrak{a})$. We let $\mathscr{V}_2$ be the ρ-invariant subspace of $\mathscr{V}$ generated by $\boldsymbol{v}_1, \boldsymbol{v}_2$ and let ρ_2 be the representation subduced on $\mathscr{V}/\mathscr{V}_2$ by ρ, etc. Hence in a finite number of steps we obtain a basis $\boldsymbol{v}_1, \boldsymbol{v}_2, \ldots, \boldsymbol{v}_m$ such that $\rho(X)\boldsymbol{v}_i \in \mathscr{V}_i$ for every $i = 1, 2, \ldots, m$.

Proposition 10. *Let $\mathfrak{a}$ be a finite dimensional solvable Lie algebra over an algebraically closed field of characteristic zero. Then $B(X, Y) = 0$ for every $X \in \mathfrak{a}'$ and $Y \in \mathfrak{a}$ where B is the Killing form of $\mathfrak{a}$.*

Proof. By Theorem 9 we can choose a basis for $\mathfrak{a}$ such that the matrices of the adjoint representation are all upper triangular. Hence if $X_1, X_2 \in \mathfrak{a}$ then the corresponding diagonal entries of $\operatorname{ad} X_1 \operatorname{ad} X_2$ and $\operatorname{ad} X_2 \operatorname{ad} X_1$ are the same. Consequently the matrix of $\operatorname{ad}[X_1, X_2]$ is strictly upper triangular. Thus $\operatorname{ad}[X_1, X_2] \operatorname{ad} Y$ has a strictly upper triangular matrix for every X_1, X_2, Y in $\mathfrak{a}$. This implies that $B([X_1, X_2], Y) = 0$ and so $B(X, Y) = 0$ for every $X \in \mathfrak{a}'$ and $Y \in \mathfrak{a}$.

Since every nilpotent operator has zero trace we see that the nilpotency of each operator $\rho(X)$ implies that $\operatorname{tr} \rho(X)^2 = 0$ for every $X \in \mathfrak{a}$ where $\rho: \mathfrak{a} \to \mathfrak{gl}(\mathscr{V})$ denotes a finite dimensional representation. Hence the hypothesis $\operatorname{tr} \rho(X)^2 = 0$ is formally weaker than Engel's condition and so it can be expected that it has weaker consequences than nilpotency. The following important theorem known as *Cartan's solvability criterion* is due to Elie Cartan:

Theorem 11. *Let $\mathfrak{a}$ be a finite dimensional Lie algebra over a field of characteristic* 0. *Let $\rho: \mathfrak{a} \to \mathfrak{gl}(\mathscr{V})$ be a faithful representation of $\mathfrak{a}$ in the finite dimensional vector space $\mathscr{V}$ such that $\operatorname{tr} \rho(X)^2 = 0$ for every X in $\mathfrak{a}$. Then $\mathfrak{a}$ is solvable.*

Proof. First we suppose that F is algebraically closed. We shall prove the theorem in this case by induction on $n = \dim \mathfrak{a}$. If $n=0$ or $n=1$ then $\mathfrak{a}$ is trivially solvable. Now let $n \geqslant 2$ and suppose that the theorem holds for every dimension less than n. If $\mathfrak{a}' \subset \mathfrak{a}$ then $\mathfrak{a}'$ is solvable because the hypothesis of the theorem applies also to $\mathfrak{a}'$. Therefore $\mathfrak{a}$ is solvable when $\mathfrak{a}' \subset \mathfrak{a}$ and so we may restrict our attention to the case $\mathfrak{a}' = \mathfrak{a}$. Let $\mathfrak{m}$ be a proper subalgebra of $\mathfrak{a}$ of maximal dimension so that $1 \leqslant \dim \mathfrak{m} < n$. Hence our induction hypothesis implies that $\mathfrak{m}$ is solvable. In the proof of Theorem 1 we have seen that if $A \in \mathfrak{m}$ then $X + \mathfrak{m} \mapsto [A, X] + \mathfrak{m}$ defines a linear operator $\sigma(A): \mathfrak{a}/\mathfrak{m} \to \mathfrak{a}/\mathfrak{m}$ and the set of operators $\sigma(A)$ $(A \in \mathfrak{m})$ is a representation of $\mathfrak{m}$ in the finite dimensional vector space $\mathfrak{a}/\mathfrak{m}$. Since $\mathfrak{m}$ is solvable and F is algebraically closed and of zero characteristic by the corollary of Theorem 7 σ has a weight vector Y in the restricted sense. Let $\lambda: \mathfrak{m} \to F$ be the associated weight so that

$$(2) \qquad [A, Y] - \lambda(A) Y \in \mathfrak{m}$$

for every $A \in \mathfrak{m}$ by the definition of σ. By the maximality of the dimension of $\mathfrak{m}$ we see that Y and $\mathfrak{m}$ generate $\mathfrak{a}$. Therefore $\mathfrak{a}' \subseteq \mathfrak{m}' + [\mathfrak{m}, Y]$. Since $\mathfrak{a}' = \mathfrak{a}$ and $\mathfrak{m} \subset \mathfrak{a}$ this inclusion relation together with (2) shows that we can not have $\lambda(A) = 0$ for every A in $\mathfrak{m}$. Hence there is a vector A in $\mathfrak{m}$ such that $\lambda(A) \neq 0$.

Let $\tau: \mathfrak{a} \to \mathfrak{gl}(\mathscr{W})$ be any finite dimensional representation of $\mathfrak{a}$ which later we shall suppose to be irreducible. Since $\mathfrak{m}$ is solvable by the corollary of Theorem 7 $\tau|\mathfrak{m}$ has a weight vector $\boldsymbol{w}_0$ in the restricted sense. We define $\boldsymbol{w}_{i+1} = \tau(Y)\boldsymbol{w}_i$ for $i = 0, 1, 2, \ldots$ and let $\mathscr{W}_i$ denote the subspace spanned by the vectors $\boldsymbol{w}_0, \ldots, \boldsymbol{w}_i$. If $X \in \mathfrak{m}$ and $i \geqslant 0$ then

$$\begin{aligned} \tau(X)\boldsymbol{w}_{i+1} &= \tau(X)\tau(Y)\boldsymbol{w}_i = \tau(Y)\tau(X)\boldsymbol{w}_i + \tau([X, Y])\boldsymbol{w}_i \\ &= \tau(Y)(\tau(X)\boldsymbol{w}_i + \lambda(X)\boldsymbol{w}_i) + \tau([X, Y] - \lambda(X)Y)\boldsymbol{w}_i . \end{aligned}$$

Since $[X, Y] - \lambda(X) Y$ is in $\mathfrak{m}$ using $i=0$ we see that $\tau(X)\boldsymbol{w}_1 \in \mathscr{W}_1$ and so using $i=1$ it follows that $\tau(X)\boldsymbol{w}_2 \in \mathscr{W}_2$ and in general $\tau(X)\boldsymbol{w}_i \in \mathscr{W}_i$ for all $i = 1, 2, \ldots$ and $X \in \mathfrak{m}$. Hence each *$\mathscr{W}_i$ is an invariant subspace of* $\tau|\mathfrak{m}$. We let $\mu: \mathfrak{m} \to F$ be the weight associated with the weight vector $\boldsymbol{w}_0$ of $\tau|\mathfrak{m}$. Using induction on $i = 0, 1, 2, \ldots$ we can prove that

$$(3) \qquad \tau(X)\boldsymbol{w}_i - (\mu(X) + i\lambda(X))\boldsymbol{w}_i \in \mathscr{W}_{i-1}$$

for all $X \in \mathfrak{m}$ where $\mathscr{W}_{-1}=0$: This is clear for $i=0$ by the definition of the weight μ. Moreover if $i \geqslant 0$ then using $w_{i+1}=\tau(Y)w_i$ we obtain

$$\begin{aligned}\tau(X)w_{i+1}-(\mu(X)+(i+1)\lambda(X))w_{i+1} &= \tau([X,Y]-\lambda(X)Y)w_i \\ &\quad +\tau(Y)(\tau(X)w_i+\lambda(X)w_i-(\mu(X)+(i+1)\lambda(X)w_i)\,.\end{aligned}$$

By (2) and the invariance of $\mathscr{W}_i$ under $\tau|\mathfrak{m}$ the first term on the right hand side belongs to $\mathscr{W}_i$. By the induction hypothesis the second term is in $\tau(Y)\mathscr{W}_{i-1}\subseteq\mathscr{W}_i$. Hence (3) holds for $i+1$ and so for all $i=0,1,2,\ldots$

Denote by q the largest value of i such that $w_0,\ldots,w_i$ are linearly independent. Since $\mathfrak{m}$ and Y span $\mathfrak{a}$ by $w_{i+1}=\tau(Y)w_i$ it follows that $\mathscr{W}_q$ is an invariant subspace of τ and so if τ is irreducible then we have $\mathscr{W}_q=\mathscr{W}$. Therefore in this case $w_0,\ldots,w_q$ is a basis of $\mathscr{W}$ which we keep ordered in this manner. If $X\in\mathfrak{m}$ then using this ordered basis the trace of $\tau(X)$ can be computed from (3) and we obtain

(4) $$\operatorname{tr}\tau(X)=(q+1)\mu(X)+\tfrac{1}{2}q(q+1)\lambda(X)\,.$$

By Proposition II.5.14 we have $\operatorname{tr}\tau([X_1,X_2])=0$ for every $X_1,X_2\in\mathfrak{a}$. Hence using $\mathfrak{a}=\mathfrak{a}'$ we see that $\operatorname{tr}\tau(X)=0$ for every $X\in\mathfrak{a}$. Thus (4) gives $\mu(X)=-\tfrac{1}{2}q\lambda(X)$ and so (3) becomes

$$\tau(X)w_i-(i-\tfrac{1}{2}q)\lambda(X)w_i\in\mathscr{W}_{i-1}$$

where $i=0,1,\ldots,q$. Using this information we can determine the diagonal entries of the matrix of $\tau(X)^2$ with respect to the ordered basis $w_0,\ldots,w_q$ and we obtain

(5) $$\operatorname{tr}\tau(X)^2=\sum_{i=0}^{q}(i-\tfrac{1}{2}q)^2\lambda(X)^2\,.$$

Now we turn to the representation $\rho:\mathfrak{a}\to\mathfrak{gl}(\mathscr{V})$ which occurs in the statement of the theorem and choose a series of invariant subspaces $\mathscr{V}=\mathscr{V}_0\supset\mathscr{V}_1\supset\cdots\supset\mathscr{V}_k=0$ such that the representation subduced on $\mathscr{V}_j/\mathscr{V}_{j+1}$ by ρ is irreducible for every $j=0,1,\ldots,k-1$. We let $q_j=\dim(\mathscr{V}_j/\mathscr{V}_{j+1})-1$ and use (5) to obtain

(6) $$\operatorname{tr}\rho(X)^2=\sum_{j=0}^{k-1}\sum_{i=0}^{q_j}(i-\tfrac{1}{2}q_j)^2\lambda(X)^2$$

for every $X\in\mathfrak{m}$. We pointed out already near the beginning of the proof the existence of an X in $\mathfrak{m}$ such that $\lambda(X)\neq 0$. By the hypothesis of the theorem $\operatorname{tr}\rho(x)^2=0$ and so (6) implies that $q_j=2i$ for all i $(0\leqslant i\leqslant q_j)$. Hence $q_j=0$ and $\dim(\mathscr{V}_j/\mathscr{V}_{j+1})=1$ for $j=0,1,\ldots,k-1$.

Since a one dimensional Lie algebra is solvable using $\mathfrak{a}=\mathfrak{a}'$ and Proposition 8 we see that every $\rho(X)$ $(X\in\mathfrak{a})$ defines the zero operator on each $\mathscr{V}_j/\mathscr{V}_{j+1}$. Thus $\rho(X)$ is nilpotent on $\mathscr{V}$ for every $X\in\mathfrak{a}$. Hence by Corollary 3 of Proposition 3 $\mathfrak{a}$ is nilpotent and consequently solvable.

We shall consider the case when F is not algebraically closed but ρ is faithful. Let $\tilde{F}$ be an algebraically closed field containing F and let $\mathbf{v}_1,\ldots,\mathbf{v}_m$ be an ordered basis of $\mathscr{V}$. Then $\mathbf{v}_1,\ldots,\mathbf{v}_m$ defines an injection of $\mathscr{V}$ into the vector space $\tilde{\mathscr{V}}=\tilde{F}^m$. Similarly using a basis $A_1,\ldots,A_n$ of $\mathfrak{a}$ over F and the multiplication of these basic vectors we can inject $\mathfrak{a}$ into a Lie algebra $\tilde{\mathfrak{a}}$ over $\tilde{F}$. The matrix of each $\rho(A_i)\colon\mathscr{V}\to\mathscr{V}$ defines a linear operator $\tilde{\rho}(A_i)\colon\tilde{\mathscr{V}}\to\tilde{\mathscr{V}}$ $(i=1,2,\ldots,n)$. If we let $\tilde{\rho}(\sum\tilde{\lambda}_i A_i)=\sum\tilde{\lambda}_i\tilde{\rho}(A_i)$ then $\tilde{\rho}\colon\mathfrak{a}\to\mathfrak{gl}(\tilde{\mathscr{V}})$ is a finite dimensional representation. Since $\tilde{\rho}(A_i)$ and $\rho(A_i)$ have the same matrix we have $\operatorname{tr}\tilde{\rho}(A_i)^2=0$ $(i=1,2,\ldots,n)$. Similarly $\operatorname{tr}\tilde{\rho}(A_i+A_j)^2=0$ for all $i<j$. Hence

$$\operatorname{tr}\tilde{\rho}(\sum\tilde{\lambda}_i A_i)^2=\sum_{i<j}\tilde{\lambda}_i\tilde{\lambda}_j\operatorname{tr}(\tilde{\rho}(A_i)\tilde{\rho}(A_j)+\tilde{\rho}(A_j)\tilde{\rho}(A_i)).$$

If we let $\tilde{\lambda}_i=\tilde{\lambda}_j=1$ and we set zero for all the other coefficients then we obtain from above

$$\operatorname{tr}(\tilde{\rho}(A_i)\tilde{\rho}(A_j)+\tilde{\rho}(A_j)\tilde{\rho}(A_i))=0.$$

Therefore $\operatorname{tr}\tilde{\rho}(\tilde{X})^2=0$ for every $\tilde{X}\in\tilde{\mathfrak{a}}$. Since $\tilde{F}$ is algebraically closed we see that $\tilde{\mathfrak{a}}$ is solvable. Hence the subalgebra $\mathfrak{a}$ of $\tilde{\mathfrak{a}}$ is also solvable. Of course by $\mathfrak{a}=\mathfrak{a}'$ actually we have $\mathfrak{a}=0$.

Theorem 12. *Let $\mathfrak{a}$ be a finite dimensional Lie algebra over a field of characteristic* 0. *If $B(X,X)=0$ for every X in $\mathfrak{a}'$ then $\mathfrak{a}$ is solvable. If F is algebraically closed and $\mathfrak{a}$ is solvable then conversely $B(X,X)=0$ for every X in $\mathfrak{a}'$.*

Proof. We apply Theorem 11 with the following data: For $\mathfrak{a}$ we choose $\operatorname{ad}(\mathfrak{a}')$, the image of $\mathfrak{a}'$ under the adjoint map $\operatorname{ad}\colon\mathfrak{a}\to\operatorname{ad}\mathfrak{a}$. For ρ we choose the representation $\rho\colon\operatorname{ad}(\mathfrak{a}')\to\mathfrak{gl}(\mathfrak{a})$ where $\rho(\operatorname{ad}X)=\operatorname{ad}X$ for every $X\in\mathfrak{a}'$. Then ρ is faithful and by hypothesis

$$\operatorname{tr}\rho(\operatorname{ad}X)^2=\operatorname{tr}(\operatorname{ad}X)^2=B(X,X)=0$$

for every X in $\mathfrak{a}'$. Hence by Theorem 11 $\operatorname{ad}(\mathfrak{a}')$ is solvable. The kernel of the adjoint representation $\operatorname{ad}\colon\mathfrak{a}\to\operatorname{ad}\mathfrak{a}$ is the center $\mathfrak{c}$ of $\mathfrak{a}$ and so $\operatorname{ad}(\mathfrak{a}')$ is isomorphic to $\mathfrak{a}'/\mathfrak{c}\cap\mathfrak{a}'$. Hence $\mathfrak{a}'/\mathfrak{c}\cap\mathfrak{a}'$ is solvable. Since $\mathfrak{c}$ is solvable so is its subalgebra $\mathfrak{c}\cap\mathfrak{a}'$ and so $\mathfrak{a}'$ is solvable by Lemma 1.21. Therefore $\mathfrak{a}$ is also solvable. The converse follows from Proposition 10.

Let $B: \mathscr{V} \times \mathscr{V} \to F$ be a symmetric bilinear form on the vector space $\mathscr{V}$ over the field F. If $\mathscr{S}$ is a subset of $\mathscr{V}$ then we let

$$\mathscr{S}^{\perp} = \{v : v \in \mathscr{V} \text{ and } B(v, s) = 0 \text{ for all } s \in \mathscr{S}\}.$$

The form B is called non-degenerate if $\mathscr{V}^{\perp} = 0$. If F is the field of reals then this definition coincides with the usual one. If $B: \mathfrak{a} \times \mathfrak{a} \to F$ is a symmetric bilinear form on the non-associative algebra $\mathfrak{a}$ and $B([X, Y], Z) = B(X, [Y, Z])$ for all X, Y, Z in $\mathfrak{a}$ then B is called *associative*. By Proposition 1.15 we know that the Killing form of a finite dimensional Lie algebra $\mathfrak{a}$ is associative.

We can now turn to a second important result due to Elie Cartan which is called *Cartan's criterion for semisimplicity*:

Theorem 13. *Let $\mathfrak{a}$ be a finite dimensional Lie algebra over a field of characteristic* 0. *If $\mathfrak{a}$ is semisimple and $\rho: \mathfrak{a} \to \mathfrak{gl}(\mathscr{V})$ is a finite dimensional faithful representation then the symmetric bilinear form*

$$B_\rho(X, Y) = \operatorname{tr} \rho(X)\rho(Y) \qquad (X, Y \in \mathfrak{a})$$

is non-degenerate.

If the Killing form of $\mathfrak{a}$ is non-degenerate then $\mathfrak{a}$ is semisimple.

Note. The symmetric bilinear form B_ρ is called the *trace form* of ρ. The *Killing form* is the trace form of the adjoint representation of $\mathfrak{a}$.

Proof. Consider the set

$$\mathfrak{i} = \{X : B_\rho(X, Y) = 0 \text{ for all } Y \in \mathfrak{a}\}.$$

Then it is clear that $\mathfrak{i}$ is a subspace of $\mathfrak{a}$. Using reasoning similar to that in the proof of Proposition 1.15 we see that

$$B_\rho([A, X], Y) = B_\rho([Y, A], X) + B_\rho([X, Y], A)$$

for all A, X, Y in $\mathfrak{a}$. This identity shows that $[A, X] \in \mathfrak{i}$ for every $A \in \mathfrak{a}$ and $X \in \mathfrak{i}$ and so $\mathfrak{i}$ is an ideal in $\mathfrak{a}$. We have $B_\rho(X, X) = 0$ for every X in $\mathfrak{i}$ and so $\mathfrak{i}$ is solvable by Theorem 12. Since $\mathfrak{a}$ is semisimple so is $\mathfrak{i}$ and in Section 1 we have seen that the only solvable, semisimple Lie algebra is 0. Hence $\mathfrak{i} = 0$ and so B_ρ is non-degenerate by definition.

Let us now suppose that $\mathfrak{a}$ is not semisimple. By Proposition 27 there exists then a non-zero abelian $\mathfrak{i}$ in $\mathfrak{a}$. We select an ordered basis for $\mathfrak{a}$ such that a basis of $\mathfrak{i}$ stands at the beginning. Then the matrix of $\operatorname{ad} Y$ for any Y in $\mathfrak{a}$ and the matrix of $\operatorname{ad} X$ for any X in the abelian ideal $\mathfrak{i}$ are of the forms $\begin{pmatrix} * & * \\ 0 & * \end{pmatrix}$ and $\begin{pmatrix} 0 & * \\ 0 & 0 \end{pmatrix}$, respectively. Therefore $B(X, Y) = \operatorname{tr}(\operatorname{ad} X)(\operatorname{ad} Y) = 0$ for every $X \in \mathfrak{i}$ and $Y \in \mathfrak{a}$. Since $\mathfrak{i} \neq 0$ it follows that B is degenerate.

Corollary. *A finite dimensional Lie algebra over a field of characteristic* 0 *is semisimple if and only if its Killing form is non-degenerate.*

The following fundamental theorem on the structure of finite dimensional semisimple Lie algebras over a field of characteristic zero is due to Elie Cartan:

Theorem 14. *A finite dimensional Lie algebra over a field of characteristic* 0 *is semisimple if and only if it is a direct sum of finitely many ideals which are simple algebras.*

Proof. The first half of the theorem could be put in a more general setting as it depends only on the existence of a symmetric bilinear form $B: \mathfrak{a} \times \mathfrak{a} \to F$ which is associative and the fact that no non-zero nilpotent ideals can be found in $\mathfrak{a}$. We suppose that $\mathfrak{a}$ is semisimple and prove the existence of a decomposition

$$\mathfrak{a} = \mathfrak{a}_1 \oplus \cdots \oplus \mathfrak{a}_m \tag{7}$$

by induction on $n = \dim \mathfrak{a}$. If $n = 1$ then $\mathfrak{a}$ is simple. We suppose that a decomposition exists for every dimension less than n. Let $\mathfrak{a}_1$ be a minimal ideal of $\mathfrak{a}$ so that $\mathfrak{a}_1$ is a simple Lie algebra. In the beginning of the proof of Theorem 13 we have seen that

$$\mathfrak{a}_1^{\perp} = \{X : X \in \mathfrak{a} \text{ and } B(X, Y) = 0 \text{ for all } Y \in \mathfrak{a}_1\}$$

is an ideal in $\mathfrak{a}$. We prove that $\mathfrak{a}_1 \cap \mathfrak{a}_1^{\perp} = 0$: If $X, Y \in \mathfrak{a}_1 \cap \mathfrak{a}_1^{\perp}$ and $Z \in \mathfrak{a}$ then $B([X, Y], Z) = B(X, [Y, Z]) = 0$. Since $\mathfrak{a}$ is a semisimple Lie algebra by the corollary of Theorem 13 we see that $[X, Y] = 0$. Therefore $(\mathfrak{a}_1 \cap \mathfrak{a}_1^{\perp})' = 0$ and so $\mathfrak{a}_1 \cap \mathfrak{a}_1^{\perp}$ is an abelian ideal of $\mathfrak{a}$. Hence by Proposition 27 $\mathfrak{a}_1 \cap \mathfrak{a}_1^{\perp} = 0$ and so $\mathfrak{a} = \mathfrak{a}_1 \oplus \mathfrak{a}_1^{\perp}$. An ideal $\mathfrak{i}$ of $\mathfrak{a}_1^{\perp}$ is necessarily an ideal of $\mathfrak{a}$ because if $X \in \mathfrak{a}$, $Y \in \mathfrak{i}$ and $X = X_1 + X_2$ with $X_1 \in \mathfrak{a}_1$, $X_2 \in \mathfrak{a}_1^{\perp}$ then

$$[X, Y] = [X_1, Y] + [X_2, Y] = [X_2, Y] \in \mathfrak{i}.$$

Hence $\mathfrak{a}_1^{\perp}$ is semisimple and so by the induction hypothesis we have $\mathfrak{a}_1^{\perp} = \mathfrak{a}_2 \oplus \cdots \oplus \mathfrak{a}_m$ where $\mathfrak{a}_2, \ldots, \mathfrak{a}_m$ are simple algebras and ideals in $\mathfrak{a}_1^{\perp}$. Hence $\mathfrak{a}_1, \ldots, \mathfrak{a}_m$ are simple ideals of $\mathfrak{a}$ and (7) holds.

Conversely let us suppose that $\mathfrak{a}_1, \ldots, \mathfrak{a}_m$ are simple Lie algebras and $\mathfrak{a}$ is the algebra given by (7). If we choose a basis in each $\mathfrak{a}_i$ then we can easily prove that

$$B(X, Y) = \sum B_i(X_i, Y_i) \tag{8}$$

where $X = X_1 \oplus \cdots \oplus X_m$, $Y = Y_1 \oplus \cdots \oplus Y_m$ and B and B_i denote the Killing forms of $\mathfrak{a}$ and $\mathfrak{a}_i$ $(i = 1, \ldots, m)$, respectively. Since $\mathfrak{a}_i$ is simple it is semisimple and so according to the corollary of Theorem 13 B_i is not degenerate. Hence B is non-degenerate by (8) and so $\mathfrak{a}$ is semisimple.

Let $\mathscr{S}$ be a set of operators $S: \mathscr{V} \to \mathscr{V}$ where $\mathscr{V}$ is a finite dimensional vector space. We let $\mathscr{A}$ denote the associative algebra of operators $A: \mathscr{V} \to \mathscr{V}$ generated by $\mathscr{S}$ and the identity operator I. Let $\mathscr{A}'$ be the commuting algebra of $\mathscr{A}$ so that $T \in \mathscr{A}'$ provided $AT = TA$ for every $S \in \mathscr{S}$. The set $\mathscr{S}$ is called *semisimple* if every $\mathscr{S}$-invariant subspace of $\mathscr{V}$ has a complementary invariant subspace. A finite dimensional representation $\rho: \mathfrak{a} \to \mathfrak{gl}(\mathscr{V})$ of the Lie algebra $\mathfrak{a}$ is called semisimple if the associative algebra $\mathscr{A}$ generated by I and the operators $\rho(X)$ $(X \in \mathfrak{a})$ is semisimple.

Lemma 15. *A semisimple associative algebra of operators has no non-zero nilpotent left ideal.*

Proof. This has to be proved by contradiction: Let $\mathscr{I}$ be a non-zero nilpotent left ideal and let $\mathscr{I}^k \neq 0$ but $\mathscr{I}^{k+1} = 0$. Then $\mathscr{B} = \mathscr{I}^k$ is a non-zero left ideal such that $\mathscr{B}^2 = 0$. Let $\mathscr{V}_1$ be the subspace of $\mathscr{V}$ spanned by the subspaces $B\mathscr{V}$ $(B \in \mathscr{B})$. Then $\mathscr{V}_1$ is $\mathscr{A}$-invariant and so there is an $\mathscr{A}$-invariant complementary subspace $\mathscr{V}_2$. If $B \in \mathscr{B}$ then $B\mathscr{V}_2 \subseteq B\mathscr{V} \cap \mathscr{V}_2 \subseteq \mathscr{V}_1 \cap \mathscr{V}_2 = 0$ and so $B\mathscr{V}_2 = 0$. Since $\mathscr{B}^2 = 0$ we have $B\mathscr{V}_1 = 0$ and so $B\mathscr{V} = B\mathscr{V}_1 + B\mathscr{V}_2 = 0$. Thus $B = 0$ for every B in $\mathscr{B}$ which contradicts $\mathscr{B} \neq 0$.

Lemma 16. *If $\mathscr{A}$ is a semisimple associative algebra of operators over an algebraically closed field then the elements of the center of $\mathscr{A}$ are all semisimple. The underlying vector space of $\mathscr{A}$ admits an ordered basis such that the matrix of each central operator is a diagonal matrix.*

Proof. By hypothesis the underlying vector space $\mathscr{V}$ of $\mathscr{A}$ is a direct sum of irreducible components $\mathscr{P}_i$. We choose a basis in every $\mathscr{V}_i$ and use these vectors as the elements of an ordered basis for $\mathscr{V}$. The matrix of every $A \in \mathscr{A}$ is a diagonal block matrix with blocks A_i corresponding to each $\mathscr{P}_i$. By Burnside's theorem for each i the set of the blocks A_i $(A \in \mathscr{A})$ is the full matrix algebra. Hence if $T \in \mathscr{A} \cap \mathscr{A}'$ then each block $T_i = \lambda_i I_i$ where λ_i is a scalar and I_i is the corresponding identity matrix. Hence T is semisimple.

Theorem 17. *Let $\mathfrak{a}$ be a Lie algebra over an algebraically closed field F of characteristic 0 and let $\rho: \mathfrak{a} \to \mathfrak{gl}(\mathscr{V})$ be a finite dimensional semisimple representation. Let $\mathfrak{c}$ be the center of $\rho(\mathfrak{a})$. Then $\rho(\mathfrak{a})' \cap \mathfrak{c} = 0$ and $\rho(\mathfrak{a})/\mathfrak{c}$ is semisimple.*

Proof. Let $\mathscr{A}$ denote the associative algebra generated by the operators $\rho(X)$ $(X \in \mathfrak{a})$ and I. If $A = [\rho(X), \rho(Y)]$ is in $\mathfrak{c}$ then A is in $\mathscr{A} \cap \mathscr{A}'$ which is the center of $\mathscr{A}$. Thus $A\rho(X) = \rho(X)A$ and so by Lemma II.5.17 A is nilpotent. Hence by Lemma 16 $[\rho(X), \rho(Y)] = 0$ and this shows that $\rho(\mathfrak{a})' \cap \mathfrak{c} = 0$.

Every abelian ideal of $\rho(\mathfrak{a})/\mathfrak{c}$ is the image under the map $\rho(X) \mapsto \rho(X) + \mathfrak{c}$ of an ideal $\mathfrak{j}$ of $\rho(\mathfrak{a})$ such that $[\mathfrak{j}, \mathfrak{j}] \subseteq \mathfrak{c}$. By $\rho(\mathfrak{a})' \cap \mathfrak{c} = 0$ we have $[\mathfrak{j}, \mathfrak{j}] = 0$ and so $\mathfrak{b} = [\rho(\mathfrak{a}), \mathfrak{j}]$ is an abelian ideal of $\rho(\mathfrak{a})$. Let $\mathscr{B}$ be the associative algebra generated by the operators of $\mathfrak{b}$. If $B = [\rho(X), J]$ with $X \in \mathfrak{a}$ and $J \in \mathfrak{j}$ then by $B \in \mathfrak{j}$ the element J commutes with $B = \rho(X)J - J\rho(X)$ and so B is nilpotent by Lemma II.5.17. Since the sum of nilpotent operators is nilpotent it follows that every element of $\mathfrak{b}$ is a nilpotent operator. Since $\mathfrak{b}$ is abelian the elements of $\mathfrak{b}$ commute with each other and so we see that $\mathscr{B}$ is a nilpotent operator algebra.

If $J \in \mathfrak{j}$ then by the definitions of $\mathfrak{b}$ and $\mathscr{B}$ for every X in $\mathfrak{a}$ there is a B in $\mathscr{B}$ such that $\rho(X)J = J\rho(X) + B$. Thus if $J_1, J_2 \in \mathfrak{j}$ then

$$\rho(X)J_1J_2 = J_1J_2\rho(X) + B$$

for a suitable B in $\mathscr{B}$. By induction the same holds for an arbitrary number of factors $J_1, J_2, \ldots, J_k$. Hence by considering sums of such equations we see that

$$\rho(X)B - B\rho(X) \in \mathscr{B} \tag{9}$$

for every X in $\mathfrak{a}$ and B in $\mathscr{B}$.

Let $(\mathscr{A}\mathscr{B})$ and $(\mathscr{B}\mathscr{A})$ denote the operator algebras generated by the products AB and BA $(A \in \mathscr{A}, B \in \mathscr{B})$, respectively. For any subset $\mathscr{S}$ of $\mathscr{L}(\mathscr{V})$ let $\mathscr{S}^k$ be the set of products $S_1, \ldots, S_k$ with $S_1, \ldots, S_k$ in $\mathscr{S}$. Using induction on k from (9) we obtain: If $X_1, \ldots, X_k \in \mathfrak{a}$ and $B \in \mathscr{B}$ then there exist $B_0 \in \mathscr{B}$ and $B_{i_1 \ldots i_j} \in \mathscr{B}$ for every $1 \leqslant i_1 \leqslant i_2 \leqslant \cdots \leqslant i_j \leqslant k$ such that

$$\rho(X_1)\ldots\rho(X_k)B = B\rho(X_1)\ldots\rho(X_k) + \sum_{j=1}^{k-1} B_{i_1 \ldots i_j}\rho(X_{i_1})\ldots\rho(X_{i_j}) + B_0.$$

Consequently for every $A \in \mathscr{A}$ and $B \in \mathscr{B}$ we have

$$AB = \sum B_i A_i \tag{10}$$

with suitable $A_i \in \mathscr{A}$ and $B_i \in \mathscr{B}$. From this we see that $(\mathscr{A}\,\mathscr{B})$ is generated by the operators BA where $A \in \mathscr{A}$ and $B \in \mathscr{B}$. Using induction on $k=1,2,\ldots$ from (10) we see that if $A \in \mathscr{A}$ and $B \in \mathscr{B}$ then

$$(BA)^k = \sum B_i A_i$$

where $A_i \in \mathscr{A}$ and $B_i \in \mathscr{B}^k$. Since $\mathscr{B}$ is a finite dimensional, commutative and nilpotent algebra $\mathscr{B}^k = 0$ for a sufficiently large k and so $(BA)^k = 0$ where $A \in \mathscr{A}$ and $B \in \mathscr{B}$ are arbitrary. Since the elements BA generate $(\mathscr{A}\,\mathscr{B})$ we proved that $(\mathscr{A}\,\mathscr{B})$ is a nilpotent left ideal in $\mathscr{A}$. Hence by Lemma 15 we have $(\mathscr{A}\,\mathscr{B}) = 0$ and so in particular $\mathfrak{b} = [\rho(\mathfrak{a}), \mathfrak{j}] = 0$. This proves that $\mathfrak{j} \subseteq \mathfrak{c}$ and $\rho(\mathfrak{a})/\mathfrak{c}$ is semisimple.

The following result is a generalization of Theorem 7 and Proposition 8 and for that reason it is also known as *Lie's theorem*:

Theorem 18. *Let $\mathfrak{a}$ be a solvable Lie algebra over an algebraically closed field of characteristic 0 and let $\rho\colon \mathfrak{a} \to \mathfrak{gl}(\mathscr{V})$ be a finite dimensional semisimple representation. Then $\rho(X) = 0$ for every X in $\mathfrak{a}'$.*

Proof. Let $\mathfrak{c}$ be the center of $\rho(\mathfrak{a})$. Since $\rho(\mathfrak{a})$ is solvable so is $\rho(\mathfrak{a})/\mathfrak{c}$. But by Theorem 17 $\rho(\mathfrak{a})/\mathfrak{i}$ is also semisimple and so $\rho(\mathfrak{a}) = \mathfrak{c}$. Thus $\rho([X, Y]) = [\rho(X), \rho(Y)] = 0$ for all X, Y in $\mathfrak{a}$.

3. Presheaves and Sheaves

These concepts were introduced by Leray around 1948 and their importance was immediately recognized by Henri Cartan, Weil and other members of the Bourbaki group. These mathematicians not only extended and deepened the theory associated with sheaves but also made an effort to make it available to a larger circle of mathematicians. Sheaves will be used here to obtain a better insight when discussing differentiable manifolds and various concepts associated with them.

Definition 1. *A presheaf of sets $\mathscr{S} = (X, S, \rho)$ is a triple consisting of a topological space X, a family S of non-void sets $S(O)$ associated with each open set O of X and a set ρ of maps $\rho(O,Q)\colon S(O) \to S(Q)$ associated with every ordered pair of open sets $O \supseteq Q$ such that $\rho(O,O)$ is the identity map for every open set O of X and if $O \supseteq Q \supseteq R$ then $\rho(O,R) = \rho(Q,R)\,\rho(O,Q)$.*

Note. The topological space X is called the *base space* and the maps $\rho(O,Q)$ are the *restriction maps*. One can define $\rho(O,Q)$ for arbitrary open sets O, Q in X; if $O \not\supseteq Q$ then by definition $\rho(O,Q)$ is the empty set $\emptyset$.

For example let X be a topological space and let A be a non-void set. If O is a non-void open set in X we define $S(O)=A$ and if $O=\emptyset$ then we let $S(O)$ consist of the empty set i. e. $S(\emptyset)=\{\emptyset\}$. For non-void Q and $O\supseteq Q$ let $\rho(O,Q): S(O)\to S(Q)$ be the identity map of A into itself. Since $S(\emptyset)$ has only one element $\rho(O,\emptyset)$ is uniquely determined by the open set O. It is clear that $\rho(O,O)$ is always the identity and the maps $\rho(O,Q)$ satisfy the transitivity condition described above. Thus $\mathscr{S}=(X,S,\rho)$ is a presheaf called the *presheaf of the simple sheaf associated with the topological space X and the non-void set A*.

The simple sheaf is a special case of the *sheaf of continuous functions with values in a topological space A* which is defined as follows: Since X and A are topological spaces we can speak about the continuous functions $s: O\to A$ where O is a non-void open set in X. The family of these functions is a set which we call $S(O)$. It is convenient to define $S(\emptyset)=\{\emptyset\}$. If $O\supseteq Q\neq\emptyset$ we let $\rho(O,Q)s$ be the restriction of $s: O\to A$ to the subset Q. As earlier the maps $\rho(O,\emptyset): S(O)\to S(\emptyset)$ are uniquely determined by the fact that $S(\emptyset)$ has only one element. Since $\rho(O,O)$ is the identity and the transitivity property obviously holds the triple $\mathscr{S}(X,S,\rho)$ defined by the topological spaces X and A is a presheaf.

Definition 2. *A presheaf $\mathscr{S}=(X,S,\rho)$ is called a sheaf if it has the following properties: Let O_i $(i\in\mathscr{I})$ be open sets in X and let $O=\bigcup O_i$. Then we have:*

(S.1) *If $s', s''\in S(O)$ and $\rho(O,O_i)s'=\rho(O,O_i)s''$ for every $i\in\mathscr{I}$ then $s'=s''$.*

(S.2) *Suppose an element s_i is chosen from each $S(O_i)$ such that $\rho(O_i,O_i\cap O_j)s_i=\rho(O_j,O_i\cap O_j)s_j$ for every $i,j\in\mathscr{I}$. Then there is an s in $S(O)$ such that $\rho(O,O_i)s=s_i$ for every $i\in\mathscr{I}$.*

Note. Statements (S.1) and (S.2) are called the *sheaf axioms*. In view of (S.1) the element s whose existence is stated in (S.2) is uniquely determined by the elements s_i $(i\in\mathscr{I})$.

The presheaf of continuous functions has the additional properties expressed in (S.1) and (S.2): For if $s', s'': O\to A$ and $x\in O$ then $x\in O_i$ for some $i\in\mathscr{I}$ and so using the hypothesis of (S.1) we obtain

$$s'(x)=\rho(O,O_i)s'(x)=\rho(O,O_i)s''(x)=s''(x)$$

which shows that $s'=s''$. Te same reasoning holds even if the maps s' and s'' are not continuous. In order to see (S.2) we construct a map $s: O\to A$ by using the given family of maps s_i $(i\in\mathscr{I})$: To each $x\in O$ there corresponds at least one O_i containing x and so we can define $s(x)=s_i(x)$. The definition is independent of the choice of the index i

because if $x \in O_i \cap O_j$ then by the hypothesis of (S.2) we have

$$s_i(x) = \rho(O_i, O_i \cap O_j) s_i(x) = \rho(O_j, O_i \cap O_j) s_j(x) = s_j(x).$$

Thus $s: O \to A$ is well defined on O and actually s and s_i coincide on the open set O_i. Since s_i is continuous on O_i and the O_i sets cover O it follows that s is continuous.

As a second illustration we show that the presheaf $\mathscr{S}$ associated with the topological space X and the set A is in general not a sheaf: Let $s', s'' \in S(O) = A$ and let $i \in \mathscr{I}$ be such that $O_i \neq \emptyset$. Then $\rho(O, O_i) S(O) = A \to S(O_i) = A$ is the identity map and so by the hypothesis of (S.1) we obtain $s' = s''$. Therefore axiom (S.1) is always satisfied. In order to disprove (S.2) we suppose that X contains a pair of non-void disjoint open sets O_1, O_2 and A consists of at least two elements a_1, a_2.

Consider the sheaf of continuous functions with values in the space of reals $\mathbb{R}$. Then $S(O)$ for every non-void open set O consists of real valued continuous functions $s: O \to \mathbb{R}$ and so $S(O)$ can be given an algebraic structure: Pointwise addition and multiplication defines a ring structure and under multiplication by real scalars $S(O)$ is a commutative algebra with identity. Similarly if A is the field of complex numbers $\mathbb{C}$ with its usual topology then $S(O)$ carries the structure of a commutative complex algebra with identity.

If A is a group or a ring we introduce the discrete topology on A and construct the sheaf of continuous functions with values in A. Then for each open set O the set $S(O)$ is a group or a ring, respectively. We shall consider $S(\emptyset) = \{\emptyset\}$ as a group or ring consisting of a single element. More generally we can start with a topological group or ring A and obtain a corresponding structure on the sets $S(O)$. If A is commutative then so is $S(O)$ and if it is a ring with identity then $S(O)$ has an identity for every $O \neq \emptyset$.

We notice that in every one of the above situations $\rho(O, Q)$ is a homomorphism: If A is a group then it is a group homomorphism, if $A = \mathbb{R}$ or $\mathbb{C}$ then it is an A-algebra homomorphism, etc. The sheaves constructed in this manner are simple examples of presheaves and sheaves of algebraic systems. The most important of these are defined as follows:

Definition 3. *A presheaf of groups (rings or algebras over a fixed ring R) is a presheaf* $\mathscr{S} = (X, S, \rho)$ *such that* $S(O)$ *is a group (a ring or an R-algebra) and every* $\rho(O, Q): S(O) \to S(Q)$ *is a group (ring or R-algebra) homomorphism.*

Note. If $\mathscr{S} = (X, S, \rho)$ is a sheaf then it is called a *sheaf of groups, rings or R-algebras*, respectively.

A triple (X, π, Y) consisting of a pair of topological spaces X, Y and a map $\pi\colon Y \to X$ will be called a *generalized covering space.* The space X is the base space and the map π is the projection. If y is in Y and $\pi(y) = x$ then y is said to be over the base point x. The set $\pi^{-1}(x)$ of points lying over x is called the *fiber* or *fibre* over x. The fibers give a decomposition of Y into disjoint sets. If X is a T_1 space and π is continuous then each fiber is a closed subset of Y. The covering spaces introduced here are generalizations of the classical covering spaces which will be described below. In order to stress the difference between these two concepts sometimes one speaks about "the space Y over the base space X with projection map π" instead of the generalizing covering space (X, π, Y).

A trivial example of a covering space in the classical sense is the triple (X, π, X) where X is an arbitrary topological space and $\pi\colon X \to X$ is the identity map. A slightly less trivial example is obtained by letting $X = \mathbb{Z}_m$, the set of congruence classes of $\mathbb{Z}$ modulo m and Y the set of integers $\mathbb{Z}$. We let π map the integer y into the congruence class containing y. If X and Y are endowed with the discrete topology then $\pi\colon \mathbb{Z} \to \mathbb{Z}_m$ is continuous.

Next let $\mathbb{R}_m$ denote the space obtained from the space of the reals $\mathbb{R}$ by identifying any two real numbers which are congruent modulo m. Let $X = \mathbb{R}_1$, $Y = \mathbb{R}_m$ and let π map the class $y \pmod{m}$ containing the real number y into the class $y \pmod{1}$. Then π is continuous relative to the quotient topologies of $\mathbb{R}_m$ and $\mathbb{R}_1$. Each fiber $\pi^{-1}(x)$ consists of n points of Y which are represented by the reals $x, x+1, \ldots, x+(m-1)$.

Another classical covering space is obtained by considering the spaces $X = \mathbb{R}_1$ and $Y = \mathbb{R}$ and the map $\pi\colon y \mapsto y \pmod{1}$. This is a natural covering of the circle X by the line Y. Every fiber is denumerable. If $X = \mathbb{R}_1 \times \mathbb{R}_1$ and $Y = \mathbb{R}_{m_1} \times \mathbb{R}_{m_2}$ and π maps y_i into $y_i \pmod{1}$ $(i = 1, 2)$ then we obtain an $m_1 m_2$-fold continuous covering of the torus X by the torus Y. If Y is replaced by $Y = \mathbb{R} \times \mathbb{R}$ and by the map $(y_1, y_2) \mapsto (y_1, y_2) \pmod{1}$ then we obtain a covering of the torus X by the plane Y.

The continuous map π occurring in all of these classical covering spaces has the following property: Each point y of Y has some neighborhood N_y such that $N_x = \pi(N_y)$ is a neighborhood of $x = \pi(y)$ and $\pi\colon N_y \to N_x$ is a homeomorphism. Actually N_y can be chosen such that both N_y and N_x are open sets. For instance if $X = \mathbb{R}_1$ and $Y = \mathbb{R}_m$ or $Y = \mathbb{R}$ it is sufficient to choose an open interval $(y - \varepsilon, y + \varepsilon)$ of length $2\varepsilon < 1$.

Definition 4. *The map $\pi: Y \to X$ of the topological space Y onto the topological space X is called a local homeomorphism if every point y of Y has some neighborhood N_y which is homeomorphically mapped by π onto a neighborhood of $\pi(y)$.*

It follows that every local homeomorphism is a continuous map. This shows that one can choose for N_y an open neighborhood O_y such that its homeomorphic image under π is an open neighborhood of $\pi(y)$: For if $\pi(y)=x$ and if $\pi: N_y \to N_x = \pi(N_y)$ is a homeomorphism then $O_y = \pi^{-1}(N_x^i)$ is an open set in Y and lies in N_y. Thus π maps O_y homeomorphically onto the open set $\pi(O_y)=O_x=N_x^i$. We also see that the topology induced on the fiber $\pi^{-1}(x)$ by the topology of Y is discrete: For if $y \in \pi^{-1}(x)$ and $\pi: O_y \to O_x$ is a homeomorphism then $\pi^{-1}(x) \cap O_y$ consists of the single point y and O_y, being open in Y, it is an open set of $\pi^{-1}(x)$. We can also show that a local homeomorphism is an open map: Let O be an open set in Y and let $y \in O$. We can choose O_y such that $\pi: O_y \to O_x$ is a homeomorphism. Since $O \cap O_y$ is an open set in O_y its image $\pi(O \cap O_y)$ is an open set in O_x and so it is open also in X. Now we have

$$\pi(y) = x \in \pi(O \cap O_y) = \pi(O) \cap O_x \subseteq \pi(O)$$

where $\pi(O \cap O_y)$ is open and y is an arbitrary point in O. Thus $\pi(O)$ is an open set.

We wish to specially distinguish those generalized covering spaces for which $\pi: Y \to X$ is a local homeomorphism. These are of great importance because we shall see that there is a natural one-to-one correspondence between these special covering spaces (Y, π, X) and sheaves of sets over the base space X. In view of this correspondence we shall call these covering spaces *Sheaves* with capital S.

Definition 5. *A Sheaf of sets over a base space X is a generalized covering space (Y, π, X) such that $\pi: Y \to X$ is a local homeomorphism.*

A covering space in the classical sense or a *smooth covering space* is a Sheaf (Y, π, X) having the property that for each point x in X there is an open set O_x and there are corresponding disjoint neighborhoods O_y around each $y \in \pi^{-1}(x)$ such that π maps each O_y homeomorphically onto O_x. The local homeomorphism property implies the existence of a pair O_x, O_y for each fixed y in $\pi^{-1}(x)$ but it might be necessary to vary O_x with y. Thus a Sheaf is not necessarily a covering space in the classical sense. Moreover one often adds a further restriction on smooth covering spaces, namely that the fibers $\pi^{-1}(x)$ be denumerable.

Let (Y,π,X) be a generalized covering space, let $\pi\colon Y\to X$ be continuous and suppose that X admits a subset B having no accumulation point in X with the following property: If $x\notin B$ and $y\in\pi^{-1}(x)$ then there exist open neighborhoods O_x and O_y such that π maps O_y homeomorphically onto O_x. Then (Y,π,X) is called a *Sheaf with branch points* and B is the set of its branch points. If in addition a common O_x can be chosen for every y in $\pi^{-1}(x)$ then we speak about a classical covering space with branch points. A simple example is furnished by any one of the maps $\pi_n\colon z\mapsto z^n$ $(n=1,2,\ldots)$ mapping the complex plane $\mathbb{C}$ onto itself. For $n=1$ there is no branch point and for $n\geqslant 2$ there is only one branch point, the origin. For any other point $x=a\exp i\alpha$ we may choose for instance $O_x=\{r\exp i\varphi\colon r>0 \text{ and } |\varphi-\alpha|<\pi\}$ and $O_{y_k}=\{r\exp i\varphi\colon r>0 \text{ and } |\varphi-(\alpha+2\pi k)/n|<\pi/n\}$ where we used the notations $y_k=a^{1/n}\exp(\alpha+2\pi k)/n$ and $k=0,1,\ldots,n-1$.

It is possible that every fiber $\pi^{-1}(x)$ of a covering space is endowed with an algebraic structure of the same kind. For instance in the foregoing example $\pi^{-1}(0)=\{0\}$ and also for $x\neq 0$ the elements $y_0,\ldots,y_{n-1}$ form a cyclic group. If this is the case then we speak about a covering space of algebraic systems. In particular if (Y,π,X) is a Sheaf then we are dealing with a Sheaf of algebraic systems e. g. a Sheaf of groups, rings or algebras. Trivial examples of such Sheaves can be constructed for instance by using projections in product spaces: We let $Y=X\times A$ where X is a topological space, A is a topological algebraic system and π maps (x,a) into x. A slightly less trivial example of a Sheaf of groups consists of the helix

$$Y=\{(\cos\varphi,\sin\varphi,\varphi)\colon -\infty<\varphi<\infty\},$$

the unit circle

$$X=\{(\cos\varphi,\sin\varphi)\colon 0\leqslant\varphi<2\pi\}$$

and the map $(\cos\varphi,\sin\varphi,\varphi)\mapsto(\cos\varphi,\sin\varphi)$. For every $x=(\cos\varphi,\sin\varphi)$ the fiber $\pi^{-1}(x)$ admits the structure of an infinite cyclic group.

Now we begin the study of the equivalence of sheaves and Sheaves. Let X and Y be topological spaces and let $\pi\colon Y\to X$ be surjective but not necessarily continuous. We also suppose that the fibers $\pi^{-1}(x)$ $(x\in X)$ are endowed with an algebraic structure all of which are of the same type e. g. a commutative group or a ring with identity. However the algebraic structure might be empty in which case $\pi^{-1}(x)$ is merely a non-void subset of Y. Starting from the triple (Y,π,X) we are going to construct a sheaf, called the *sheaf associated with the map* $\pi\colon Y\to X$ or the *sheaf of sections* (Y,π,X). The process is the same which gave the sheaf of continuous functions with values in a topological space A.

At present Y will play the role of A and only certain special continuous functions $s: O \to Y$, called sections will be included in $S(O)$.

Definition 6. *A section of* (Y, π, X) *means a continuous function* $s: O \to Y$ *mapping an open subset* O *of* X *into* Y *such that* $(\pi s)(x) = x$ *for every* $x \in O$.

Note. We call s a *section* over O; if $O = X$ then we speak about a *cross-section.* The section s could be defined also when the subset O is not open but such sections are less important.

By definition the only section over the empty set $\emptyset$ is the symbol $\emptyset$. If we associate with every open set O of X the subset $S(O)$ of sections over O then $S(O)$ carries an algebraic structure of the same type as the fibers $\pi^{-1}(x)$ $(x \in X)$. For every pair of open sets $O \supseteq Q$ we let $\rho(O,Q)s$ denote the restriction of $s: O \to Y$ to Q so that $\rho(O,Q): S(O) \to S(Q)$ is a homomorphism. Since these restrictions satisfy the transitivity law $\rho(O,R) = \rho(Q,R)\rho(O,Q)$ for every triple $O \supseteq Q \supseteq R$ we see that (X, S, ρ) is a presheaf.

As every s is a genuine map and every $\rho(O,Q)$ means restriction in the usual sense the validity of axiom (S.1) is trivial: If $x \in O$ then $x \in O_i$ for some $i \in \mathscr{I}$ and so $s'(x) = \rho(O,O_i)s'(x) = \rho(O,O_i)s''(x) = s''(x)$. Thus $s'(x) = s''(x)$ for every x in O and $s' = s''$. The proof of (S.2) is the same as in the case of the sheaf of continuous functions: If we define s by the rule $s(x) = s_i(x)$ for every $x \in O_i$ and let i vary over $\mathscr{I}$ then the compatibility condition expressed in the hypothesis of (S.2) implies that $s: O \to Y$ is single valued. Moreover $s_i: O_i \to Y$ is continuous and fulfills the additional requirement $\pi s_i(x) = x$ for all $x \in O_i$. Hence $s: O \to Y$ is continuous and $\pi s: O \to O$ is the identity map. We proved that (X, S, ρ) is a sheaf.

Definition 7. *A presheaf* $\mathscr{A} = (X, A, \rho)$ *is called a subpresheaf of a presheaf* $\mathscr{B} = (X, B, \sigma)$ *if the following conditions are satisfied:*

1. *Each* $A(O)$ *carries the same type of algebraic structure as* $B(O)$ *and* $A(O) \subseteq B(O)$ *for every open subset* O *of* X.

2. *For every pair* $O \supseteq Q$ *the image of* $A(O)$ *under* $\sigma(O,Q)$ *lies in* $A(Q)$ *i.e.* $\sigma(O,Q)A(O) \subseteq A(Q)$.

3. *The restriction of* $\sigma(O,Q)$ *to* $A(O)$ *is* $\rho(O,Q)$ *for every pair* $O \supseteq Q$.

If $\mathscr{A}$ is a subpresheaf of $\mathscr{B}$ then the transitivity relation $\rho(O,R) = \rho(Q,R)\rho(O,Q)$ is a consequence of the corresponding relation for σ. Similarly if $\mathscr{B}$ satisfies axiom (S.1) then (S.1) holds also for $\mathscr{A}$. However $\mathscr{A}$ can be a subpresheaf of a sheaf $\mathscr{B}$ without being a sheaf. For example let X be the space of reals, let $A(O)$ be the ring of bounded, real valued, continuous functions on the open set $O \subseteq X$ and let $\rho(O,Q)\colon A(O) \to A(Q)$ be restriction in the usual sense. Then $\mathscr{A}$ is a subpresheaf of the sheaf of continuous real valued functions $f\colon O \to \mathbb{R}$ but $\mathscr{A}$ is not a sheaf because axiom (S.2) fails.

If the subpresheaf $\mathscr{A}$ of $\mathscr{B}$ is a sheaf then we speak about a *subsheaf* $\mathscr{A}$ of $\mathscr{B}$. Here $\mathscr{B}$ might be only a presheaf e. g. $\mathscr{A}$ can be the sheaf of functions mapping sets of $X = \mathbb{R}$ into zero while $\mathscr{B}$ is the presheaf of bounded, real valued, continuous functions.

Theorem 8. *For every (X,π,Y) with surjective π the sheaf of sections $s\colon O \to Y$ is a subsheaf of the sheaf of all continuous functions with values in Y.*

Proof. This is an immediate consequence of the definition of a subsheaf and the sheaf of continuous functions with values in Y.

Let X, Y, and Z be topological spaces, let $\pi\colon Y \to X$ and $\tau\colon Z \to X$. A *homomorphism* of (Y,π,X) into (Z,τ,X) is a continuous map $f\colon Y \to Z$ satisfying $\pi = \tau f$. The condition $\pi = \tau f$ implies that f maps fibers into fibers, more precisely $f\colon \pi^{-1}(x) \to \tau^{-1}(x)$ for every $x \in X$. Now suppose that Y and Z are not merely topological spaces but a fixed type of algebraic structure is given on each of their fibers. Then a homomorphism of $\mathscr{A} = (Y,\pi,X)$ into $\mathscr{B} = (Z,\tau,X)$ is defined to be a continuous map $f\colon Y \to Z$ such that $\pi = \tau f$ and $f\colon \pi^{-1}(x) \to \tau^{-1}(x)$ is an algebraic homomorphism for every $x \in X$.

Let (Y,π,X) and (Z,τ,X) be Sheaves and let $f\colon Y \to Z$ be such that $\pi = \tau f$. We prove that f is a local homeomorphism of Y into Z. In fact given y in Y by the local homeomorphism property of π we can find open sets O'_y and O'_x containing y and $x = \pi(y)$ respectively such that $\pi | O'_y$ is a homeomorphism onto O'_x. Similarly there are open sets O''_z and O''_x containing $z = f(y)$ and x respectively such that $\tau | O''_z$ is a homeomorphism onto O''_x. We let $O_x = O'_x \cap O''_x$ and define $O_y = \pi^{-1}(O_x)$ and $O_z = \tau^{-1}(O_x)$. Then $\pi\colon O_y \to O_x$ and $\tau^{-1}\colon O_x \to O_z$ are homeomorphisms and therefore so is $f = \tau^{-1}\pi\colon O_y \to O_z$. We proved that $f\colon Y \to Z$ is a local homeomorphism and as such is continuous. Thus in the definition of *Sheaf homomorphisms* it is sufficient to require that $\pi = \tau f$. The continuity of $f\colon Y \to Z$ will follow. By the above reasoning we also have:

Theorem 9. *If $f\colon Y \to Z$ is a homomorphism of the Sheaf (Y,π,X) into the Sheaf (Z,τ,X) then $f\colon Y \to Z$ is a local homeomorphism.*

By our earlier results we have the following consequence of this theorem:

Corollary. *The Sheaf homomorphism* $f: Y \to Z$ *is a continuous and open map. Thus if* f *is one-to-one then it is a homeomorphism.*

Two triples (Y, π, X) and (Z, τ, X) with surjective π and τ are called *isomorphic* if there is a homeomorphism $f: Y \to Z$ of Y onto Z which is a homomorphism. A map f satisfying these requirements is called an *isomorphism* of (Y, π, X) onto (Z, τ, X). In particular two Sheaves are isomorphic if there is a homomorphism of one onto the other which is a one-to-one map of Y onto Z.

Definition 10. *A homomorphism* h *of a presheaf* $\mathscr{A} = (X, A, \rho)$ *into another* $\mathscr{B} = (X, B, \sigma)$ *is a family of homomorphisms* $h(O): A(O) \to B(O)$, *one for each open subset of* X, *such that for every pair of open sets* $O \supseteq Q$ *one has* $\sigma(O, Q)\, h(O) = h(Q)\, \rho(O, Q)$.

If each $h(O)$ *is an isomorphism of* $A(O)$ *onto* $B(O)$ *then* h *is called an isomorphism of* $\mathscr{A}$ *onto* $\mathscr{B}$ *and* $\mathscr{A}$ *and* $\mathscr{B}$ *are said to be isomorphic presheaves.*

Two special cases are worth mentioning separately: First if every $h(O)$ is a one-to-one map of $A(O)$ into $B(O)$ then $h: \mathscr{A} \to \mathscr{B}$ is called an *injective homomorphism.* Similarly if every $h(O)$ maps $A(O)$ onto $B(O)$ then $h: \mathscr{A} \to \mathscr{B}$ is said to be a *surjective homomorphism.* If $h: \mathscr{A} \to \mathscr{B}$ is both injective and surjective that is to say if each $h(O): A(O) \to B(O)$ is a bijection then h is an *isomorphism* of $\mathscr{A}$ onto $\mathscr{B}$. The significance of the commutativity relation $\rho(O, Q)\, h(O) = h(Q)\, \rho(O, Q)$ will become obvious later.

Theorem 11. *A homomorphism* $f: Y \to Z$ *of* (Y, π, X) *into* (Z, τ, X) *induces a homomorphism* $h: \mathscr{A} \to \mathscr{B}$ *of the associated sheaves of sections. The image of the section* $a: O \to Y$ *under* $h(O)$ *is the section* $fa: O \to Z$.

Proof. This follows right away from the definitions of the concepts involved in the theorem.

If $f: Y \to Z$ is an injection then so is $h: \mathscr{A} \to \mathscr{B}$ and similarly if f is surjective then so is h. In particular if (Y, π, X) and (Z, τ, X) are isomorphic then so are the associated sheaves of sections.

Next starting from a presheaf $\mathscr{S} = (X, S, \rho)$ we are going to construct a Sheaf (Y, π, X) called the Sheaf associated with $\mathscr{S}$. We shall construct (Y, π, X) such that if $\mathscr{S}$ is a sheaf then the sheaf associated with (Y, π, X) is isomorphic to $\mathscr{S}$. We shall see that if one starts from isomorphic presheaves then the associated Sheaves will be isomorphic. Given the presheaf $\mathscr{S} = (X, S, \rho)$ we want to determine a set Y, a projection map

$\pi: Y \to X$ and introduce a topology on Y such that π becomes a local homeomorphism. First for every x in X we are going to construct the set $\pi^{-1}(x)$. This will define not only Y but also the projection map $\pi: Y \to X$. We notice that the family of open neighborhoods O_x of x in X is a directed set under the ordering by inverse inclusion i. e. $O_x \leqslant Q_x$ meaning $O_x \supseteq Q_x$: In fact any two elements, say O'_x, O''_x have a common successor, namely $O_x = O'_x \cap O''_x$. Since $\mathscr{S} = (X, S, \rho)$ is a presheaf an algebraic system $S(O_x)$ is associated with each element of this directed set and a homomorphism $\rho(O_x, Q_x): S(O_x) \to S(Q_x)$ is defined for every pair $O_x \leqslant Q_x$. Hence we are given a directed system of a certain type of algebraic systems and so we may define $\pi^{-1}(x)$ as the direct limit of this directed system: $\pi^{-1}(x) = \varinjlim S(O_x)$. Thus if $\mathcal{O}_x$ denotes the family of all open neighborhoods of x in X then $\pi^{-1}(x)$ consists of the equivalence classes of the set $S_x = \bigcup \{S(O_x): O_x \in \mathcal{O}_x\}$ two elements $s' \in S(O'_x)$ and $s'' \in S(O''_x)$ being equivalent if they have a common successor i. e. if $\rho(O'_x, O_x)s' = \rho(O''_x, O_x)s''$ for some O_x satisfying $O'_x, O''_x \leqslant O_x$. Moreover $\pi^{-1}(x)$ carries an algebraic structure of the same kind as the sets $S(O_x)$. We let $Y = \bigcup \{\pi^{-1}(x): x \in X\}$ and define $\pi(y) = x$ for all y in $\pi^{-1}(x)$. The definition is satisfactory if X is a T_0-space because then $x \neq y$ implies that $\pi^{-1}(x) \cap \pi^{-1}(y) = \emptyset$. If X does not satisfy the T_0 separation axiom then in order to make the fibers $\pi^{-1}(x)$ $(x \in X)$ disjoint we have to make a slight modification in the definition: We let $\pi^{-1}(x)$ consist of all ordered pairs such that the first element is x and the second is an equivalence class of S_x.

If $s \in S(O)$ then O is an O_x for every x in O and so we have $s \in S_x$. Let $s(x)$ be the equivalence class containing s so that $s(x)$ is an element of $\pi^{-1}(x)$. Then $\{s(x): x \in O\}$ is a subset of Y which will be denoted by the symbol $O(s)$. It will be called the *open set generated in* Y *by* s or the image of the section s. Sometimes one writes s instead of $O(s)$ but we will avoid this simplified notation. The function $f_s: O \to Y$ with values $f_s(x) = s(x)$ has the property $\pi f_s(x) = x$ for all $x \in O$ which is characteristic of sections. Hence it is natural to replace f_s by s and use $O(s)$ to denote the image set $\{s(x): x \in O\}$.

Now comes the introduction of a suitable topology on the set Y. If (Y, π, X) is a Sheaf then the sections $s: O \to Y$ are open maps so we have to choose the topology of our set Y such that the sets $O(s)$ are all open. On the other hand sections are continuous maps and the inverse image of any open set under a continuous map is open. Using this fact we can show that if we have a topology on Y such that (Y, π, X) is a Sheaf with sections f_s where $s \in S(O)$ for some open set O in X then every open set of Y is a union of sets $O(s)$: For if A is an open set in Y and $s: O \to Y$ then $Q = s^{-1}(O(s) \cap A)$ is an open set in X. If $y \in O(s) \cap A$ then

y is in the image set of $s|Q$ which is $\rho(O,Q)s(Q)$. Conversely if $y\in\rho(O,Q)s(Q)$ then $s^{-1}(y)\in Q$ and so $s^{-1}(y)=s^{-1}(z)$ for some $z\in O(s)\cap A$. But $y=z$ by $s^{-1}(y)=s^{-1}(z)$ and so $y\in O(s)\cap A$. Hence we have $O(s)\cap A=\rho(O,Q)s(Q)$. This shows that the intersection of the open set A with an $O(s)$ is itself the image of a section. By the local homeomorphism property of π we see that A is the union of sets $O(s)\cap A$ and so A is a union of images of sections. Thus if a proper topology exists on Y then it is generated by the base consisting of the sets $O(s)$ where $s\in S(O)$ for some O and O varies over the open subsets of X.

There remains to show that the axioms of a base for open sets are satisfied by the family described above: First if $y\in Y$ then $y\in\pi^{-1}(x)$ for some $x\in X$ and so y is an equivalence class of S_x. If $s\in S(O)$ is an element of this equivalence class then $s(x)=y$ and so $y\in O(s)$. This shows that each point of Y belongs to some $O(s)$. Next suppose that $y\in O(s')\cap O(s'')$ where $s'\in S(O')$ and $s''\in S(O'')$. Then $\pi(y)=x\in O'\cap O''=O$ and $y\in O(s)\subseteq O(s')\cap O(s'')$ where $s=\rho(O',O)s'=\rho(O'',O)s''$. The existence and uniqueness of a suitable topology on Y has been proved.

Theorem 12. *Let $\mathscr{A}=(X,A,\rho)$ and $\mathscr{B}=(X,B,\sigma)$ be presheaves of the same algebraic type and let (Y,π,X) and (Z,τ,X) be the associated Sheaves. Then every homomorphism $h\colon\mathscr{A}\to\mathscr{B}$ induces a Sheaf homomorphism $f\colon Y\to Z$. If $a\in A(O)$ belongs to the equivalence class $y\in\pi^{-1}(x)$ then $f(y)\in\tau^{-1}(x)$ is the equivalence class containing $h(O)(a)\in B(O)$.*

Proof. Since (Y,π,X) and (Z,τ,X) are Sheaves and $\pi=\tau f$ the map $f\colon Y\to Z$ is an algebraic homomorphism of $\pi^{-1}(x)$ into $\tau^{-1}(x)$ for every x in X. This is the point where the commutativity relation $\sigma(O,Q)h(O)=h(Q)\rho(O,Q)$ is needed.

Suppose that $h\colon\mathscr{A}\to\mathscr{B}$ is injective i.e. every $h(O)\colon A(O)\to B(O)$ is a one-to-one map. Let $a^i\in A(O^i)$ $(i=1,2)$ be such that $h(O^1)(a^1)$ and $h(O^2)(a^2)$ are equivalent elements of $\bigcup\{B(O_x)\colon O_x\in\mathscr{O}_x\}$. This means the existence of an open set O containing x such that $\sigma(O^1,O)h(O^1)(a^1)$ $=\sigma(O^2,O)h(O^2)(a^2)$. Since $\sigma h=h\rho$ we obtain $h(O)\rho(O^1,O)a^1$ $=h(O)\rho(O^2,O)a^2$ and h being injective this gives $\rho(O^1,O)a^1=\rho(O^2,O)a^2$. Thus a^1 and a^2 are equivalent elements of $\bigcup\{A(O_x)\colon O_x\in\mathscr{O}_x\}$ and so we proved that $f\colon Y\to Z$ is an injective map. Hence if the presheaf homomorphism $h\colon\mathscr{A}\to\mathscr{B}$ is injective then so is the associated Sheaf homomorphism $f\colon Y\to Z$.

A similar situation prevails for surjective homomorphisms $h\colon\mathscr{A}\to\mathscr{B}$. Given $z\in\tau^{-1}(x)$ there is a $b\in B(O)$ such that $x\in O$ and b belongs to the equivalence class z. Since $h(O): A(O) \to B(O)$ is surjective we can choose a in $A(O)$ such that $h(O)(a) = b$. Thus if y denotes the equivalence class containing a then $f(y) = z$ and so $f: Y \to Z$ is surjective. We proved that if $h: \mathscr{A} \to \mathscr{B}$ is surjective then so is the associated Sheaf homomorphism $f: Y \to Z$.

An immediate corollary of these remarks is:

Theorem 13. *If $\mathscr{A} = (X, A, \rho)$ and $\mathscr{B} = (X, B, \sigma)$ are isomorphic presheaves then their associated Sheaves are isomorphic.*

Let us now start from a sheaf $\mathscr{A} = (X, A, \rho)$, construct the associated Sheaf (Y, π, X) and then pass to the sheaf of sections $\mathscr{S} = (X, S, \sigma)$ of (Y, π, X). It is natural to ask what kind of a relation exists between $\mathscr{A}$ and $\mathscr{S}$. Using the same kind of reasoning as in the introduction of the topology of Y we can prove that $\mathscr{A}$ and $\mathscr{S}$ are isomorphic sheaves:

First of all every $a \in A(O)$ determines a section $s_a: O \to Y$ of (Y, π, X), namely if $x \in O$ then we let $a(x) \in \pi^{-1}(x)$ denote that equivalence class in S_x which contains a and define $s_a(x) = a(x)$ for every $x \in O$. The map s_a constructed in this fashion is continuous and satisfies $\pi s_a(x) = x$ for every $x \in O$. So far we did not use the fact that $\mathscr{A}$ is a sheaf.

Next we prove that every section $s: O \to Y$ of (Y, π, X) is a section of the type s_a for a suitable a in $A(O)$: For $s(O)$ is an open set in Y and so by the definition of the topology of Y it is a union of images $O(a_i)$ where $a_i \in A(O_i)$ and $i \in \mathscr{I}$ for some index set $\mathscr{I}$. Let two of these elements, say a_i and a_j, be given. If $x \in O_i \cap O_j$ then $s(x)$ is the equivalence class containing both a_i and a_j. Hence a_i and a_j have a common successor, that is $\rho(O_i, O_x) a_i = \rho(O_j, O_x) a_j$ for a suitable open set O_x satisfying $x \in O_x \subseteq O_i \cap O_j$. Using the transitivity property of the ρ's we obtain

$$\rho(O_i \cap O_j, O_x)\, \rho(O_i, O_i \cap O_j) a_i = \rho(O_i \cap O_j, O_x)\, \rho(O_j, O_i \cap O_j) a_j .$$

Since $O_i \cap O_j = \bigcup \{O_x : x \in O_i \cap O_j\}$ we can apply axiom (S.1) and conclude that

$$\rho(O_i, O_i \cap O_j) a_i = \rho(O_j, O_i \cap O_j) a_j .$$

Now we can use axiom (S.2) to derive the existence of an element a in $A(O)$ such that $\rho(O, O_i) a = a_i$ for every O_i $(i \in \mathscr{I})$. This shows that $s(O) = \bigcup O(a_i) = O(a)$ and $s = s_a$.

Our last task is to prove that if $s_{a'} = s_{a''}$ then necessarily $a' = a''$. However $s_{a'}(x) = s_{a''}(x)$ means that a' and a'' belong to the same equivalence class of S_x. In other words we have $\rho(O, O_x) a' = \rho(O, O_x) a''$ for some open set O_x satisfying $x \in O_x \subseteq O$. The union of these sets O_x $(x \in O)$

is O and so by axiom (S.1) we have $a' = a''$. We proved the following fundamental result:

Theorem 14. *There is a natural one-to-one correspondence between the isomorphism classes of sheaves $\mathscr{A} = (X, A, \rho)$ and those of Sheaves (Y, π, X).*

Note. In order to see the actual one-to-one correspondence it is necessary to read the foregoing proof.

The isomorphic sheaf $\mathscr{S}$ constructed from the sheaf $\mathscr{A}$ via a Sheaf (Y, π, X) can be very different from $\mathscr{A}$ but nonetheless it inherits all the characteristic properties of $\mathscr{A}$. For instance let X be a non-trivial topological space e.g. let X be the plane and let $\mathscr{A}$ be its sheaf of real valued continuous functions $a: O \to \mathbb{R}$. Then we can find functions a', a'' in $A(O)$ such that $a'(x) = a''(x)$ for some x in O but $\rho(O, Q_x)a' \neq \rho(O, Q_x)a''$ for every open neighborhood Q_x of x lying in O. On the other hand if $s', s'' \in S(O)$ are sections in the sheaf of germs of continuous functions and $s'(x) = s''(x)$ for some x in O then there is a suitable open subset Q_x of O containing x and such that $\rho(O, O_x)s' = \rho(O, O_x)s''$. In fact we have seen that $s' = s_{a'}$ and $s'' = s_{a''}$ for suitable $a', a'' \in A(O)$. Moreover $s'(x) = s''(x)$ means that $a'(y) = a''(y)$ for every $y \in O_x$ for some open neighborhood O_x of x lying entirely in O. Thus we have $s'(y) = s''(y)$ for every $y \in O_x$.

It is very instructive to study the sheaf obtained from the *presheaf of a simple sheaf*: Let (X, A, ρ) be the presheaf consisting of the sets $A(O) = A$ where A is a fixed non-void set and O is a non-void open set in X and the identity maps $\rho(O, Q): A \to A$ where $\emptyset \neq Q \subseteq O$. First we construct the Sheaf (Y, π, X) associated with (X, A, ρ): Since $A(O_x) = A$ for every open neighborhood O_x of $x \in X$ and $\rho(O, Q)$ is the identity map the direct limit $\pi^{-1}(x) = \varinjlim A(O_x)$ can be identified with A itself. Thus to each $x \in X$ there corresponds one copy A_x of A and so Y can be considered as $Y = X \times A$. In order to determine the topology of Y we first determine the elements $O(a)$ of the base for the open sets of Y. Any $a \in A(O)$ is an element of A and the equivalence class $a(x)$ has been identified first with a and then more precisely with the pair (x, a). Thus $O(a) = \{(x, a): x \in O\}$ which is an open set of the product space $Y = X \times A$ when A is topologized by the discrete topology. From this we see that the Sheaf topology of Y is the product topology. The sections of (Y, π, X) are continuous functions $s: O \to Y$ subject to the condition $\pi s(x) = x$ $(x \in O)$ and so they can be identified with those functions $s: O \to A$ which are continuous with respect to the discrete topology of A. A function $s: O \to A$ has this property if and only if it is constant on the components of O.

4. Differentiable Manifolds

A topological space M is called *locally Euclidean* if every point m of M has some open neighborhood O which is homeomorphic to a finite dimensional Euclidean space $\mathbb{R}^{n(m)}$. The dimension $n(m)$ can vary from point to point but the set of points m at which $n(m)$ is a fixed value n is an open set O^n and $\{O^n\}$ $(n=0, 1, 2, \ldots)$ is a decomposition of M into disjoint open sets. Hence the value of $n(m)$ is constant on each component of M and if M is connected then $n(m)$ assumes only one value.

Contrary to the first impression a locally Euclidean space need not be a Hausdorff space. For instance let the space S consist of the open intervals $(-1, 1)$ and $(2, 3)$ and the point 2. A subset O of S not containing 2 will be considered open if it is open in the usual sense. If 2 belongs to O then it will be called open provided

$$\{O \cap (-1, 1)\} \cup \{x: x+2 \in O \cap [2, 3)\}$$

is open in the ordinary sense. This space is locally Euclidean and arcwise connected but the points 0 and 2 can not be separated by a pair of disjoint open neighborhoods.

Definition 1. *By a manifold we mean a locally Euclidean Hausdorff space M.*

We have seen that the dimension $n(m)$ is the same for all points of a component of M. This common value is called the *dimension of the component* and the least upper bound of these dimensions is the *dimension of the manifold.* The dimensions of the components are all finite but the manifold itself might be infinite dimensional. In some of the earlier literature manifolds were always assumed to be connected spaces and consequently were finite dimensional. If the manifold is compact then it is necessarily finite dimensional. If every component has the same dimension then we speak about a *homogeneous manifold.* If its dimension is one it is called a *curve* and if it is two dimensional then it is a *surface.* Some people speak about a surface only if the family of open sets has a denumerable base. Since a manifold is locally metrizable this holds if and only if it is a separable space.

By definition a manifold is a special kind of topological space without any additional structure. Differentiable manifolds which we are going to introduce now are of a more restricted type. Actually we distinguish among various grades of restriction: We speak about *k-manifolds* where $k=0,1,2,\ldots$ or ∞ and about *real analytic* and *complex analytic manifolds.* A 0-manifold means a manifold without any further restriction and the degree of constraint increases with k. A complex analytic manifold for instance is already so specialized that the dimension of each of its components is even.

Definition 2. *A differentiable structure is said to be defined on a topological space M if there is given a system of pairs (O, φ) consisting of an open subset O of M and a homeomorphism φ of O onto an open subset of a finite dimensional Euclidean space $\mathbb{R}^n$ subject to the condition that the sets O of the system cover the underlying space M. The dimension n can depend on the open set O.*

In terms of this definition a manifold is a Hausdorff space admitting a differentiable structure. A distinction between various differentiable manifolds will be made by introducing a grading of the systems which define the differentiable structure. Let O be an open set of $\mathbb{R}^n$ and let f map O into $\mathbb{R}^n$. We denote by $f(x)_\nu$ the ν-th coordinate of the image point $f(x)$ so that f_ν is a real valued function of n real variables $x_1, \ldots, x_n$ and is defined on the open set O of points $x=(x_1, \ldots, x_n)$. If every f_ν has continuous mixed derivatives on O of all orders up to and including k where $k=0, 1, 2, \ldots$ then $f: O \to \mathbb{R}^n$ is called a *k-times differentiable map or transformation* of O into $\mathbb{R}^n$. Similarly we can define *infinitely differentiable maps* and *real analytic maps*. The latter means a map f such that every point x of O has a suitable open neighborhood in which every f_ν admits a convergent power series representation

$$f(y)_\nu = \sum_{r \geqslant 0} a_r (y-x)^r$$

where $r=(r_1, \ldots, r_n)$ is an ordered n-tuple of non-negative integers and $(y-x)^r=(y_1-x_1)^{r_1} \ldots (y_n-x_n)^{r_n}$. The existence of these power series developments implies that f is infinitely differentiable. If n is even, say $n=2m$ then it is meaningful to speak about complex analytic transformations: Introduce the notations $z_\nu = x_\nu + i x_{m+\nu}$ and $w_\nu = f_\nu + i f_{m+\nu}$ for $\nu=1, \ldots, m$. Then w_ν is a complex valued function of the complex variables $z_1, \ldots, z_m$. If is possible that every point $(z_1, \ldots, z_m)$ has an open neighborhood in which each w_ν can be represented by a convergent power series of the type $w(\zeta)_\nu = \sum_{r \geqslant 0} a_r (\zeta - z)^r$ where the coefficients a_r are complex numbers. If this is so then $f: O \to \mathbb{R}^{2m}$ is called a *complex analytic transformation*. It is clear that every complex analytic transformation is necessarily real analytic.

Definition 3. *A system $\{(O, \varphi)\}$ defining a differentiable structure is called a system of class k where $k=0,1,2,\ldots$ or ∞ if for any two pairs (O, φ) and (Q, ψ) of the system the map $\psi \varphi^{-1}: \varphi(O \cap Q) \to \psi(O \cap Q)$ is* *k-times differentiable. Similarly if all the maps $\psi\varphi^{-1}$ are real (complex) analytic then $\{(O, \varphi)\}$ is said to be a real (complex) analytic structure.*

A topological space M admitting a differentiable structure is necessarily locally Euclidean. It is called of class k if it admits a system of class k. If M is a Hausdorff space then we speak about a *differentiable manifold of class k*, a *real analytic* or a *complex analytic manifold.* The system $\{(O, \varphi)\}$ is said to define a differentiable structure $\mathscr{C}^k$, $\mathscr{C}^\infty$, $\mathscr{C}^r$ or $\mathscr{C}^c$ on M. We agree that k in general denotes a non-negative integer or one of the symbols ∞, r or c. A space M with a given differentiable structure is more than a topological space. For it rarely happens that only one differentiable structure is compatible with a given topology.

We notice a fine point in the definition: A system $\{(O, \varphi)\}$ is said to define a differentiable structure but is not considered to be identical with the structure which it defines. Thus the definition of a structure $\mathscr{C}^k$ is a further task. One can proceed in two different directions: The first, more primitive method is to decide when two systems define the same differentiable structure. This introduces an equivalence relation on the set of admissible systems and each equivalence class can be considered as a differentiable structure $\mathscr{C}^k$. The second, more refined method consists of determining a characteristic invariant of these equivalence classes and identifying $\mathscr{C}^k$ with this invariant.

Definition 4. *The systems $\{(O, \varphi)\}$ and $\{(\tilde{O}, \tilde{\varphi})\}$ of a locally Euclidean space M are called k-equivalent if for any two pairs (O, φ) and $(\tilde{O}, \tilde{\varphi})$ the maps $\tilde{\varphi}\varphi^{-1}: \varphi(O \cap \tilde{O}) \to \tilde{\varphi}(O \cap \tilde{O})$ and $\varphi\tilde{\varphi}^{-1}: \tilde{\varphi}(O \cap \tilde{O}) \to \varphi(O \cap \tilde{O})$ are k-times differentiable. Every k-equivalence class $\mathscr{C}^k$ of the set of all k-times differentiable systems of M is defined to be a k-times differentiable structure on M.*

Let M be a locally Euclidean space and let $\mathscr{C}^k$ be a differentiable structure which is compatible with the topology of M. We fix a system of pairs $\{(O, \varphi)\}$ and consider the set of all pairs (Q, ψ) of those homeomorphisms ψ mapping open sets Q of M into $\mathbb{R}^n$ which satisfy the following requirements: The maps $\varphi\psi^{-1}: \psi(O \cap Q) \to \varphi(O \cap Q)$ and $\psi\varphi^{-1}: \varphi(O \cap Q) \to \psi(O \cap Q)$ are k-times differentiable transformations for every choice of the pair (O, φ) from the given system $\{(O, \varphi)\}$. It is clear that the set of all these pairs (Q, ψ) includes the system $\{(O, \varphi)\}$ and also every other system which is k-equivalent to it. Moreover using the chain rule of partial differentiation we can show that the set of pairs (Q, ψ) is itself a system of class k which is clearly k-equivalent to $\{(O, \varphi)\}$: Given (Q, ψ) and $(\tilde{Q}, \tilde{\psi})$ such that $Q \cap \tilde{Q} \neq \emptyset$ for every point m in $Q \cap \tilde{Q}$ one can find a pair (O, φ) such that $m \in O$. In terms of this homeomorphism φ we have

$$\tilde{\psi}\psi^{-1} = (\tilde{\psi}\varphi^{-1})(\varphi\psi^{-1}) \colon \psi(O \cap Q \cap \tilde{Q}) \to \tilde{\psi}(O \cap Q \cap \tilde{Q})$$

where $\varphi\psi^{-1}$ and $\tilde{\psi}\varphi^{-1}$ are k-times differentiable transformations. Therefore $\tilde{\psi}\psi^{-1}$ is also k-times differentiable. This proves that every class $\mathscr{C}^k$ of M contains a unique maximal system $\{(Q, \psi)\}$ and so one can identify $\mathscr{C}^k$ with this maximal representative system. The elements (Q, ψ) of the maximal representative system are called *coordinate neighborhoods, local coordinate systems* or *local charts.* The *local coordinates* of a point $m \in Q$ relative to the coordinate system (Q, ψ) are the coordinates $\psi(m)_1, \ldots, \psi(m)_n$ of the image point $\psi(m)$.

In practice one is usually interested in the other extreme and one wishes to determine a representative system for $\mathscr{C}^k$ which consists of as few coordinate neighborhoods as possible. In most instances there is no minimal system and even if such a system exists it is not uniquely determined by its class. If $\mathscr{C}^k$ contains a system consisting of a single coordinate neighborhood (M, φ) then we say that $\mathscr{C}^k$ admits global coordinates and (M, φ) is a global coordinate system. Any open set M of $\mathbb{R}^n$ can be endowed with a real analytic structure $\mathscr{C}^r$ which admits a global coordinate system.

We are going to give another definition of differentiable structures which is more elegant and considerably more natural than the foregoing definition. Namely we shall associate with the structure $\mathscr{C}^k$ a sheaf of functions $\mathscr{F}^k$ called the *structure sheaf of* $\mathscr{C}^k$, determine which sheaves of functions arise in this manner and prove that there is a natural one-to-one correspondence between the classes $\mathscr{C}^k$ and these special sheaves $\mathscr{F}^k$. For $k=0, 1, \ldots, \infty$ or r the functions of $\mathscr{C}^k$ are continuous real valued functions defined on open sets of the underlying topological space M and if $k=c$ then $\mathscr{F}^k$ consist of continuous complex valued functions defined on open sets.

Let M be a locally Euclidean space and let $\{(O, \varphi)\}$ be a system of coordinate neighborhoods of a differentiable structure $\mathscr{C}^k$ of class $k=0, 1, \ldots, \infty$ or r. Let j be an integer subject to the restriction $0 \leqslant j \leqslant k$, let S be an open subset of M and let $f \colon S \to \mathbb{R}$ be such that $f\varphi^{-1}$ is j-times differentiable for every coordinate neighborhood (O, φ) of the given system. We recall that this means not only the existence of mixed derivatives of all orders up to j but also their continuity. A function f satisfying this requirement is called a *j-times differentiable* real valued function on the open set S. The definition involves the system $\{(O, \varphi)\}$ but the concept itself depends only on the class $\mathscr{C}^k$. For if $\{(Q, \psi)\}$ is equivalent to $\{(O, \varphi)\}$ then $\varphi\psi^{-1}$ and $\psi\varphi^{-1}$ are k-times differentiable maps and so applying the chain rule of partial differentiation we see that $f\varphi^{-1}$ is j-times differentiable for every (O, φ) if and only if $f\psi^{-1}$ is j-times differentiable for every choice of (Q, ψ).

Let O be an open subset of the k-manifold M. The family of k-times differentiable real valued functions $f: O \to \mathbb{R}$ is an algebra over the reals which we denote by $F^k(O)$. If O is void we let $F^k(\emptyset) = \{\emptyset\}$. Let $\rho(O, Q)f$ denote the restriction of $f: O \to \mathbb{R}$ to the open subset $Q \subseteq O$. Since differentiability and continuity are local properties $\rho(O, Q)$ is a homomorphism of $F^k(O)$ into $F^k(Q)$. Therefore the algebras $F^k(O)$ and the restriction maps $\rho(O, Q)$ define a presheaf $\mathscr{F}^k = (M, F^k, \rho)$. Every presheaf of functions satisfies axiom (S.1) and so (S.1) holds for $\mathscr{F}^k$. Let O_i and f_i $(i \in \mathscr{I})$ denote the open sets and functions given in the hypothesis of axiom (S.2). Then we know the existence of a function f on $O = \bigcup O_i$ such that $\rho(O, O_i)f = f_i$ for every O_i but in general f need not belong to $F(O)$. At present however the elements of $F(O)$ are selected by requiring a local property, namely k-times differentiability. Since every f_i is k-times differentiable so is $f: O \to \mathbb{R}$. Thus $f \in F(O)$ and axiom (S.2) holds. Therefore $\mathscr{F}^k = (M, F^k, \rho)$ is a sheaf which is called the structure sheaf of the differentiable structure $\mathscr{C}^k$ given on the manifold M.

In the case of a complex analytic structure $\mathscr{C}^a$ it is more natural to consider complex valued functions defined on open subsets of M. The function $f: S \to \mathbb{C}$ is called complex analytic if $f\varphi^{-1}$ is an analytic function of the complex local coordinates $z_\nu = x_\nu + i x_{m+\nu}$ $(\nu = 1, \ldots, m)$ for every choice of the local chart (φ, O). For each open subset O of the complex analytic manifold M we let $F^c(O)$ denote the algebra of all complex analytic functions $f: O \to \mathbb{C}$. As earlier we let $\rho(O, Q)$ be the homomorphism of $F^c(O)$ into $F^c(Q)$ which maps $f: Q \to \mathbb{C}$ into $f|Q$. The triple $\mathscr{F}^k = (M, F^c, \rho)$ is the structure sheaf of the complex analytic structure $\mathscr{C}^a$ given on M.

Not every sheaf of continuous functions appears as a structure sheaf of some differentiable structure $\mathscr{C}^k$. In fact if (O, φ) is a local chart of a $\mathscr{C}^k$ then the coordinate functions $\varphi_1, \ldots, \varphi_n: O \to \mathbb{R}$ of the transformation

$$\xi \mapsto \varphi(\xi) = (\varphi_1(\xi), \ldots, \varphi_n(\xi))$$

belong to $F^k(O)$ that is to say $\varphi_1, \ldots, \varphi_n$ are k-times differentiable. Thus for instance $\mathscr{F}^k$ can not be the simple sheaf of locally constant real valued continuous functions on open sets of M.

Structure sheaves have a further distinguishing property: Namely if (O, φ) is a coordinate neighborhood of a structure $\mathscr{C}^k$ then the elements of $F(O)$ can be characterized in terms of (O, φ) as follows: $f: O \to \mathbb{R}$

belongs to $F(O)$ if and only if $f\varphi^{-1}$ is k-times differentiable in the usual sense. Indeed if this condition is satisfied and if (Q, ψ) is an arbitrary neighborhood of $\mathscr{C}^k$ then by the chain rule of partial differentiation $f\psi^{-1} = (f\varphi^{-1})(\varphi\psi^{-1})$ is k-times differentiable. This property is characteristic of structure sheaves. For we have the following:

Theorem 5. *Let M be a topological space and let $\mathscr{F} = (M, F, \rho)$ be a sheaf of functions f mapping open sets of M into the reals. Then $\mathscr{F}$ is the structure sheaf of a differentiable structure $\mathscr{C}^k$ of class $k = 0, 1, \ldots, \infty$ or r if and only if there is a family of open sets O of M such that the following conditions are fulfilled:*

1. *The union of these open sets covers M.*

2. *For each O there exist finitely many functions $\varphi_1, \ldots, \varphi_n$ in $F(O)$ such that the map $\varphi: O \to \mathbb{R}^n$ given by $\xi \mapsto (\varphi_1(\xi), \ldots, \varphi_n(\xi))$ is a homeomorphism of O onto an open set $\varphi(O)$ of $\mathbb{R}^n$.*

3. *For every open set O of this family and every open set S of M the set $F(S \cap O)$ is identical with the set of those functions $f: S \cap O \to \mathbb{R}$ for which $f\varphi^{-1}|S \cap O$ is k-times differentiable.*

Note. In general to every O there correspond several maps $\varphi: O \to \mathbb{R}^n$. If $\mathscr{F}$ is a structure sheaf then 3. holds for every choice of φ. As far as the sufficiency of these conditions is concerned it is enough to require 3. for only one φ.

Proof. We already know that a structure sheaf fulfills these requirements. Thus it is sufficient to show that any sheaf of functions $\mathscr{F} = (M, F, \rho)$ satisfying 1., 2. and 3. is the structure sheaf of a differentiable structure $\mathscr{C}^k$ on M of class k. We define $\mathscr{C}^k$ by specifying a system of coordinate neighborhoods. Namely we consider those open sets O whose existence is stipulated in the theorem and for each O we choose a homeomorphism $\varphi: O \to \mathbb{R}^n$ as is described in condition 2. At least one such homeomorphism exists for every O. The pairs (O, φ) form a system of coordinate neighborhoods for a differentiable structure of class k: If (O, φ) and (Q, ψ) are such pairs then the restrictions of $\psi_1, \ldots, \psi_n$ to $O \cap Q$ all belong to $F(O \cap Q)$ and so by 3. the functions $\psi_\nu \varphi^{-1}: \varphi(O \cup Q) \to \mathbb{R}$ $(\nu = 1, \ldots, n)$ are k-times differentiable.

There remains to prove that $\mathscr{F}$ is the structure sheaf associated with the differentiable structure defined by the system $\{(O, \varphi)\}$. First if S is an open set in M and if $f \in F(S)$ then for every O of the given system $\{(O, \varphi)\}$ we have $\rho(S, S \cap O) f \in F(S \cap O)$. By 3.

$$(\rho(S, S \cap O) f)\varphi^{-1}: \varphi(S \cap O) \to \mathbb{R}$$

is k-times differentiable and so $\rho(S, S \cap O) f$ belongs to the structure sheaf of $\mathscr{C}^k$. The union of these open sets O covers S and so by the (S.2) axiom f belongs to the structure sheaf of $\mathscr{C}^k$. Conversely if $f: S \to \mathbb{R}$ is in the structure sheaf of $\mathscr{C}^k$ then $(\rho(S, S \cap O) f) \varphi^{-1}$ is k-times differentiable for every choice of (O, φ). Hence $\rho(S, S \cap O) f$ is in $F(S \cap O)$ for every O whence by axiom (S.2) we have $f \in F(S)$.

The case of a complex analytic structure is similar to the others: The underlying space M admits a cover by open sets O such that for every O there exist complex valued functions $\varphi_1, \ldots, \varphi_n \in F(O)$ such that the map $\varphi: O \to \mathbb{C}^n$ defined by

$$\xi \mapsto (\varphi_1(\xi), \ldots, \varphi_n(\xi))$$

is a homeomorphism of O onto an open subset $\varphi(O)$ of $\mathbb{C}^n$. If O belongs to this cover then so does every open set $Q \subseteq O$. Moreover $F(O)$ is identical with the set of complex analytic functions $f: O \to \mathbb{C}$ with respect to the complex analytic structure defined on M by the coordinate neighborhoods (O, φ).

Let M and $\tilde{M}$ be locally Euclidean spaces with differentiable structures of class k and $\tilde{k}$, respectively. A map $\Phi: M \to \tilde{M}$ will be called *j-times differentiable* if for every choice of the coordinate neighborhoods (O, φ) and $(\tilde{O}, \tilde{\varphi})$ the transformation $\tilde{\varphi} \Phi \varphi^{-1}$ is j-times differentiable in the usual sense. We are primarily interested in the case when $j \leqslant \min\{k, \tilde{k}\}$. If $j=1$ then Φ will be called a *differentiable map.* On some occasions one speaks about differentiable maps also in the case when $j=k=\tilde{k}=\infty$. The fact that Φ is meant to be differentiable as often as we please or only once should be clear from the context.

We consider now the situation when $k=\tilde{k}$ and Φ is an invertible map of M onto $\tilde{M}$ such that both Φ and Φ^{-1} are k-times differentiable. In this case we say that M and $\tilde{M}$ have isomorphic differentiable structures and $\Phi: M \to \tilde{M}$ is a diffeomorphism of M onto $\tilde{M}$. Since Φ and Φ^{-1} are continuous the underlying topological spaces M and $\tilde{M}$ are homeomorphic. Let $\mathscr{F}^k$ and $\tilde{\mathscr{F}}^k$ denote the structure sheaves of M and $\tilde{M}$ and let $\Phi: M \to \tilde{M}$ be a diffeomorphism. By the definition of differentiability $\tilde{f} \in \tilde{F}(\tilde{O})$ if and only if $\tilde{f} \Phi \in F(O)$ where $O = \Phi^{-1}(\tilde{O})$. The map $\tilde{f} \mapsto \tilde{f} \Phi$ is an isomorphism of $\tilde{F}(\tilde{O})$ onto $F(O)$ and the totality of these isomorphisms is an isomorphism of $\tilde{\mathscr{F}}^k$ onto $\mathscr{F}^k$. Hence we see that *isomorphic differentiable structures have isomorphic structure sheaves* and actually every diffeomorphism Φ induces a natural isomorphism of the corresponding structure sheaves.

In the elements of classical differential geometry one deals with lines, surfaces and hypersurfaces which are smoothly embedded in some finite dimensional Euclidean space. The smoothness of the embedding means that tangent lines, planes or hyperplanes exist at each point of the geometric object under study. These linear subspaces of the envelopping Euclidean space are used to approximate the surface in the neighborhood of their respective points of contact. In the more advanced chapters of differential geometry one studies more general geometric objects which need not be subsets of Euclidean spaces but are such that the concept of an approximating linear space is meaningful at each of their points. Here we shall prove that any locally Euclidean space can be a candidate for such studies because if it admits a differentiable structure $\mathscr{C}^k$ of class $k>0$ then the approximating linear spaces can be constructed in a natural way.

If the hypersurface is smoothly embedded in a Euclidean space then the approximating hyperplane can be interpreted as the linear space of the tangent vectors emitting from the point of contact. This is a vector space over the reals whose dimension is the same as that of the hypersurface in the neighborhood of the point of contact. Let $\boldsymbol{t}$ be one of these vectors and let $t_1, \ldots, t_d$ be its coordinates with respect to a basis $\boldsymbol{e}_1, \ldots, \boldsymbol{e}_d$ in the envelopping Euclidean space $\mathbb{R}^d$. If we change coordinates in $\mathbb{R}^d$ by introducing another basis $\boldsymbol{e}'_1, \ldots, \boldsymbol{e}'_d$ then the old and new coordinates of a point are related by $x'_i = \sum_{j=1}^{d} a_{ij} x_j \, (i=1, \ldots, d)$ where $\boldsymbol{e}_j = \sum_{i=1}^{d} a_{ij} \boldsymbol{e}'_i$ for $j=1, \ldots, d$. In the new coordinate system $\boldsymbol{t}$ has coordinates $t'_i = \sum_{j=1}^{d} a_{ij} t_j$ where $i=1, \ldots, d$. Since the same matrix appears in the expression of t'_i in terms of $t_1, \ldots, t_n$ as in the expression of x'_i in terms of $x_1, \ldots, x_n$ we say that $\boldsymbol{t}$ is *contravariant*.

Starting from a differentiable structure $\mathscr{C}^k$ of class $k>0$ on the locally Euclidean space M we shall construct for each point m of M a vector space M_m over $\mathbb{R}$ such that M_m inherits the above mentioned properties: M_m is finite dimensional over $\mathbb{R}$ and its dimension is the same as the local topological dimension of M in the neighborhood of m. Every coordinate neighborhood (O, φ) containing m determines a natural basis. Contravariance with respect to the local coordinates can be defined and it actually holds for the coordinates of t relative to the basis naturally associated with (O, φ).

One way to introduce the space of tangent vectors is by considering the Sheaf associated with the structure sheaf $\mathscr{F}^k$ of the locally Euclidean space M. We recall that two k-times differentiable functions $f_i\colon O \to \mathbb{R}$ $(i=1, 2)$ are called equivalent relative to a point $m \in O_1 \cap O_2$ if there is a sufficiently small neighborhood of m in which f_1 and f_2 coincide i.e.

if $\rho(O_1, O) f_1 = \rho(O_2, O) f_2$ for a suitable open set O such that $m \in O \subseteq O_1 \cap O_2$. The equivalence classes relative to m are called *function germs* or more precisely *germs of k-times differentiable functions* and the germ containing the function f will be denoted by $\tilde{f}(m)$. These germs make up the fiber $\pi^{-1}(m) = \mathscr{F}_m$ of the Sheaf associated with $\mathscr{F}^k$. Every fiber $\mathscr{F}_m$ is an algebra over $\mathbb{R}$: Multiplication by scalars is defined by $\lambda \tilde{f}(m) = (\lambda f)^{\sim}(m)$ and addition and multiplication by the rules

$$\tilde{f}_1(m) + \tilde{f}_2(m) = (\rho(O_1, O) f_1 + \rho(O_2, O) f_2)^{\sim}(m)$$

and

$$\tilde{f}_1(m) \tilde{f}_2(m) = (\rho(O_1, O) f_1 \, \rho(O_2, O) f_2)^{\sim}(m).$$

Let a k-times differentiable function f be defined in an open neighborhood of m and let (O, φ) be a coordinate neighborhood containing m. We shall denote by $(f\varphi^{-1})_i(\varphi(m))$ the partial derivative of $f\varphi^{-1}$ with respect to the i-th local coordinate at the point $\varphi(m)$. Suppose f is such that $(f\varphi^{-1})_i(\varphi(m)) = 0$ for $i = 1, \ldots, n$ where n denotes the local dimension of M at m. If (Q, ψ) is a second coordinate neighborhood covering m then $f\psi^{-1} = (f\varphi^{-1})(\varphi\psi^{-1})$ in a small neighborhood of $\psi(m)$ and so by the chain rule $(f\psi^{-1})_i(\psi(m)) = 0$ for $i = 1, \ldots, n$. Thus the equations

$$(f\varphi^{-1})_i(\varphi(m)) = 0 \qquad (i = 1, \ldots, n)$$

either hold for all or for none of the coordinate neighborhoods containing m. If they hold then we say that f is *stationary* at m or has *zero differential* at m and write $df(m) = 0$.

It is clear that if $df(m) = 0$ then $(d\lambda f)(m) = 0$ for every real λ and if $(df_i)(m) = 0$ for $i = 1, 2$ then $d(f_1 + f_2)(m) = 0$ and $d(f_1 f_2)(m) = 0$. Thus the germs of those functions f which satisfy $df(m) = 0$ form a subalgebra of $\mathscr{F}_m$ which we denote by $\mathscr{F}_m^o$. If we ignore the multiplicative structures and regard $\mathscr{F}_m$ and $\mathscr{F}_m^o$ as vector spaces over the reals then we can consider the quotient space $M_m^* = \mathscr{F}_m / \mathscr{F}_m^o$. Therefore M_m^* is a vector space over $\mathbb{R}$ whose elements are equivalence classes of function germs $\tilde{f}(m)$, two germs $\tilde{f}_1(m)$ and $\tilde{f}_2(m)$ being equivalent if $f_1 - f_2$ has zero differential at m. We call M_m^* the space of *tangent covectors* at m. Every k-times differentiable function f defined in a neighborhood of m determines a unique element of M_m^*, namely the class modulo $\mathscr{F}_m^o$ containing $\tilde{f}(m)$. This covector will be denoted by $df(m)$ and called the *differential* of f at m or the *value of the differential* of f at m.

At this point it is necessary to introduce some terminology and clarify notations. We shall use the symbols $\not{f}$ and $\not{f}(m)$ to denote an arbitrary element of $\mathscr{F}_m$. The second, more descriptive notation should be carefully distinguished from $\tilde{f}(m)$ which is the germ associated at m with the function $f: O \to \mathbb{R}$ where $m \in O$. The section associated with the function f is denoted by $\tilde{f}$ and its value at m is exactly the germ $\tilde{f}(m)$ lying in $\mathscr{F}_m$. Equivalent functions, i. e., functions having the same germ at m coincide in a neighborhood of m and so it is meaningful to speak about the *value of a function germ* $\not{f}$ at m: It is defined as the common value at the point m of all functions f belonging to the class $\not{f}$. A convenient notation for this germ value is $\not{f}(m)$. If $\tilde{f}(m)$ is the germ of a function $f: O \to \mathbb{R}$ then the value of $\tilde{f}(m)$ at m is $f(m)$.

Equivalent functions have equal differentials at m and so we can also speak about $d\not{f}(m)$ for any germ $\not{f}$ in $\mathscr{F}_m$. Let $f, f_1, f_2: O \to \mathbb{R}$ be k-times continuously differentiable functions and let λ be a real number. Then we have

$$d\lambda f(m) = \lambda\, d f(m)$$

$$d(f_1 + f_2)(m) = d f_1(m) + d f_2(m)$$

$$d(f_1 f_2)(m) = f_1(m) d f_2(m) + f_2(m) d f_1(m).$$

For instance in order to prove the last formula it is sufficient to check that the function $f_1 f_2 - f_1(m) f_2 - f_2(m) f_1$ has zero differential at m: This can be done by choosing a local chart (O, φ) and computing the partial derivatives of

$$\begin{aligned}(f_1 f_2 - f_1(m) f_2 - f_2(m) f_1)\varphi^{-1} \\ = (f_1 \varphi^{-1})(f_2 \varphi^{-1}) - f_1(m) f_2 \varphi^{-1} - f_2(m) f_1 \varphi^{-1}\end{aligned}$$

at $\varphi(m)$. These formulas can be interpreted also in terms of the sections $\tilde{f}, \tilde{f}_1, \tilde{f}_2$ associated with the functions f, f_1, f_2. For instance the second becomes

$$d(\tilde{f}_1 + \tilde{f}_2)(m) = d\tilde{f}_1(m) + d\tilde{f}_2(m)$$

where $\tilde{f}_1 + \tilde{f}_2$ is the sum of the sections $\tilde{f}_1, \tilde{f}_2 \in F(O)$.

Let (O, φ) be a coordinate neighborhood covering the point m and let n be the local dimension of M at m so that φ is the map

$$\xi \mapsto (\varphi(\xi)_1, \ldots, \varphi(\xi)_n).$$

The functions $\varphi_1, \ldots, \varphi_n: O \to \mathbb{R}$ are k-times differentiable because if we test differentiability by using the neighborhood (O, φ) then $\varphi_i \varphi^{-1}: \varphi(O) \to \mathbb{R}$ is identical with the i-th projection map

$(\varphi_1, \ldots, \varphi_n) \mapsto \varphi_i$. In particular we see that $(\varphi_i \varphi^{-1})_j(\varphi(m)) = \delta_{ij}$, the Kronecker delta. Thus for every choice of the reals $\lambda_1, \ldots, \lambda_n$ and every $j = 1, \ldots, n$ we have

$$\left(\sum_{i=1}^{n} \lambda_i \varphi_i \varphi^{-1}\right)_j (\varphi(m)) = \sum_{i=1}^{n} \lambda_i (\varphi_i \varphi^{-1})_j (\varphi(m)) = \lambda_j .$$

This shows that $d(\sum \lambda_i \varphi_i)(m) = 0$ if and only if $\lambda_1 = \cdots = \lambda_n = 0$. In other words $\sum \lambda_i d\varphi_i(m) = 0$ if and only if $\lambda_1 = \cdots = \lambda_n = 0$. We proved the following:

Lemma 6. *The covectors $d\varphi_1(m), \ldots, d\varphi_n(m)$ associated with the local chart (O, φ) are linearly independent elements of M_m^*.*

If g is a k-times differentiable real valued function defined in an open neighborhood of m then

$$g = f - \sum_{i=1}^{n} (f \varphi^{-1})_i (\varphi(m)) \varphi_i$$

is defined in a small neighborhood of m and is k-times differentiable there. Moreover

$$(g \varphi^{-1})_j (\varphi(m)) = (f \varphi^{-1})_j (\varphi(m)) - \sum_{i=1}^{n} (f \varphi^{-1})_i (\varphi(m)) (\varphi_i \varphi^{-1})_j (\varphi(m)) = 0$$

for every $j = 1, \ldots, n$ and so $dg(m) = 0$. Consequently

$$df(m) = \sum_{i=1}^{n} (f \varphi^{-1})_i (\varphi(m)) d\varphi_i(m)$$

where $(f \varphi^{-1})_1(\varphi(m)), \ldots, (f \varphi^{-1})_n(\varphi(m))$ are real numbers. Since every element of M_m^* can be represented in the form $df(m)$ we see that $d\varphi_1(m), \ldots, d\varphi_n(m)$ is a basis of the vector space M_m^*. Our conclusions can be summarized as follows:

Proposition 7. *If M is a k-manifold, $m \in M$ and (O, φ) is a coordinate neighborhood covering m then $d\varphi_1(m), \ldots, d\varphi_n(m)$ is a basis of the space of covectors M_m^*.*

Corollary. *The real vector space M_m^* is finite dimensional and its dimension is equal to the local topological dimension of the space underlying M.*

If V is a finite dimensional vector space over a field F then the set of linear functions mapping V into F is endowed with a natural vector space structure over F, addition and multiplication by elements of F being defined as pointwise operations. This vector space is the dual of V and it is usually denoted by V^*. Its elements are called the dual vectors and they will be denoted by $\boldsymbol{v}^*$. The dimensions of V and V^* are the same: Actually if $\boldsymbol{v}_1,\dots,\boldsymbol{v}_n$ is a basis in V then the equations $\boldsymbol{v}_i^*(\boldsymbol{v}_j)=\delta_{ij}$ $(i,j=1,\dots,n)$ define a set of linear maps $\boldsymbol{v}_1^*,\dots,\boldsymbol{v}_n^*: V\to F$ which is a basis of V^*, called the dual basis of $\boldsymbol{v}_1,\dots,\boldsymbol{v}_n$. Thus V and V^* are isomorphic vector spaces.

Lemma 8. *If $\boldsymbol{v}_1,\dots,\boldsymbol{v}_n$ and $\boldsymbol{v}'_1,\dots,\boldsymbol{v}'_n$ are bases of the vector space V and*

$$\boldsymbol{v}'_i = a_{i1}\boldsymbol{v}_1 + \cdots + a_{in}\boldsymbol{v}_n \qquad (i=1,\dots,n)$$

then the corresponding dual bases are connected by the equations

$$\boldsymbol{v}'^*_j = b_{1j}\boldsymbol{v}_1^* + \cdots + b_{nj}\boldsymbol{v}_n^* \qquad (j=1,\dots,n)$$

where (b_{ij}) denotes the inverse of the $n\times n$ regular matrix (a_{ij}).

Proof. For any i,j satisfying $1\leqslant i,j\leqslant n$ we have

$$\sum_{r=1}^{n} b_{rj}\boldsymbol{v}_r^*(\boldsymbol{v}'_i) = \sum_{r=1}^{n} b_{rj}\boldsymbol{v}_r^*\left(\sum_{s=1}^{n} a_{is}\boldsymbol{v}_s\right) = \sum_{r,s=1}^{n} a_{is}b_{rj}\boldsymbol{v}_r^*(\boldsymbol{v}_s).$$

Hence by applying $\boldsymbol{v}_r^*(\boldsymbol{v}_s)=\delta_{rs}$ we obtain

$$\sum_{r=1}^{n} b_{rj}\boldsymbol{v}_r^*(\boldsymbol{v}'_i) = \sum_{r=1}^{n} a_{ir}b_{rj} = \delta_{ij} = \boldsymbol{v}'^*_j(\boldsymbol{v}'_i).$$

Since this holds for every basic vector $\boldsymbol{v}'_i$ we see that the formula given above holds.

By definition the *tangent space* M_m of the k-manifold M at the point $m\in M$ is the dual of M_m^*. Hence the elements of M_m are real valued linear functionals on M_m^*. A typical element of M_m will be denoted by the symbol $\boldsymbol{t}$ and called a *tangent* or *tangent vector* to M at m. By the corollary of Proposition 7 the dimension of M_m is equal to the local topological dimension of M in the neighborhood of m. If $\mathfrak{f}\in\mathscr{F}_m$ then we shall use $\boldsymbol{t}(\mathfrak{f})$ to abbreviate $\boldsymbol{t}(\mathfrak{f}+\mathscr{F}_m^o)$. Similarly if the real valued differentiable function f is defined in a neighborhood of m then we define $\boldsymbol{t}(f)$ as $\boldsymbol{t}(\tilde{f}(m)+\mathscr{F}_m^o)$ where $\tilde{f}(m)$ is the germ in $\mathscr{F}_m$ containing f.

Let (O,φ) be a coordinate neighborhood containing m and let $x_i=\varphi(p)_i$ $(i=1,\dots,n)$ be the local coordinates of an arbitrary point $p\in O$. If $\mathfrak{f}\in\mathscr{F}_m$ and $f\in$ then $f\varphi^{-1}$ is defined in a small neighborhood of $\varphi(m)$. By the definition of $\mathscr{F}_m$ and $\mathscr{F}_m^o$ the value of $\partial f\varphi^{-1}/\partial x_i$ at

$\varphi(m)$ depends only on the coset $\not{f}+\mathscr{F}_m^o$. If we denote this value by $\partial/\partial x_i\,(\not{f}+\mathscr{F}_m^o)$ then $\partial/\partial x_i$ is a linear functional on $\mathscr{F}_m/\mathscr{F}_m^o=M_m^*$ and so $\partial/\partial x_i$ is an element of M_m.

Proposition 9. *Let (O,φ) be a coordinate neighborhood containing $m\in M$ and let $x_i=\varphi(p)_i$ $(i=1,\ldots,n$ and $p\in O)$ be the local coordinates. Then $\partial/\partial x_1,\ldots,\partial/\partial x_n$ is a basis of the tangent space M_m which is the dual of the basis $d\varphi_1(m),\ldots,d\varphi_n(m)$ in M_m^*.*

Note. The basis $\partial/\partial x_1,\ldots,\partial/\partial x_n$ is called the *basis associated with* (O,φ) or the *basis associated with the local coordinates* $x_1,\ldots,x_n$.

Proof. By the definition of the dual basis $(d\varphi_i(m))^*(d\varphi_j(m))=\delta_{ij}$ $(i,j=1,\ldots,n)$. Hence it is sufficient to show that $\partial/\partial x_i(d\varphi_j(m))=\delta_{ij}$ for all $i,j=1,\ldots,n$. By definition $d\varphi_j(m)$ is the coset $\tilde{\varphi}_j(m)+\mathscr{F}_m^o$ and so

$$\frac{\partial}{\partial x_i}(d\varphi_j(m))=\frac{\partial \varphi_j\varphi^{-1}}{\partial x_i}=\frac{\partial x_j}{\partial x_i}=\delta_{ij}.$$

We shall now further clarify the concepts of *contravariance* and *covariance*, the first of which was briefly mentioned in the beginning. Let (O,φ) and (Q,ψ) be coordinate neighborhoods containing the point m and let $x_1,\ldots,x_n$ and $y_1,\ldots,y_n$ denote the corresponding coordinates of a variable point $p\in O\cap Q$. If f is a differentiable function defined in the neighborhood of m then

$$f\psi^{-1}(y_1,\ldots,y_n)=f(p)=f\varphi^{-1}(x_1,\ldots,x_n)$$

and so

$$\frac{\partial}{\partial y_i}(f)=\frac{\partial f\psi^{-1}}{\partial y_i}\bigg|\varphi(m)=\sum_{j=1}^{n}\frac{\partial x_j}{\partial y_i}\bigg|_{\psi(m)}\frac{\partial}{\partial x_j}(f).$$

This shows that the bases associated with the coordinate neighborhoods (O,φ) and (Q,ψ) are related by

$$\frac{\partial}{\partial y_i}=\sum_{j=1}^{n}\left(\frac{\partial x_j}{\partial y_i}\right)\frac{\partial}{\partial x_j} \tag{1}$$

where $i=1,\ldots,n$ and the partial derivatives are evaluated at $\psi(m)$.

Let $t_{1\varphi},\ldots,t_{n\varphi}$ and $t_{1\psi},\ldots,t_{n\psi}$ denote the coordinates of a vector $t\in M_m$ with respect to the bases $\partial/\partial x_1,\ldots,\partial/\partial x_n$ and $\partial/\partial y_1,\ldots,\partial/\partial y_n$, respectively. Then by (1) we have

$$t_{i\psi}=\sum_{j=1}^{n}\left(\frac{\partial y_i}{\partial x_j}\right)t_{j\varphi} \tag{2}$$

where $i=1,\ldots,n$. Conversely by interchanging the roles of (O,φ) and (Q,ψ) we see that (2) implies (1). Hence the vector space attached to the point m of M has the following properties:

1. Every local coordinate system defined in the neighborhood of m determines a basis M_m denoted by $\partial/\partial x_1,\ldots,\partial/\partial x_n$.

2. If we introduce a new local coordinate system $y_1,\ldots,y_n$ in the neighborhood of m then the bases associated with these coordinate systems are related by (1) or equivalently the coordinates of an arbitrary vector $\boldsymbol{t}\in M_m$ with respect to the old and new bases are connected by (2). These facts are summarized in classical differential geometry by saying that *the elements of M_m are contravariant vectors.* The only inessential deviation from the classical terminology is that now we speak about coordinate neighborhoods, local charts or local coordinates instead of *curvilinear coordinates.*

Using Lemma 8 and (1) we see that the bases $d\varphi_1(m),\ldots,d\varphi_n(m)$ and $d\psi_1(m),\ldots,d\psi_n(m)$ of M_m^* are connected by the formulas

$$d\psi_i(m) = \sum_{j=1}^{n} \left(\frac{\partial y_i}{\partial x_j}\right) d\varphi_j(m) \tag{3}$$

where $i=1,\ldots,n$. From (3) we obtain the following transformation formula for the coordinates of a vector $\boldsymbol{v}\in M_m^*$:

$$v_{i\psi} = \sum_{j=1}^{n} \left(\frac{\partial x_j}{\partial y_i}\right) v_{j\varphi} \tag{4}$$

where $i=1,\ldots,n$. Conversely from (4) we can derive (3). Since the bases of M_m^* which are naturally associated with local coordinate systems are such that (3) and (4) hold we say that *the elements of M_m^* are covariant vectors.* This explains the expression *covector* introduced earlier.

The foregoing definitions of the vector spaces M_m^* and M_m are meaningful for every k-manifold M such that $1\leqslant k\leqslant\infty$. If M is a real or complex analytic manifold then of course M is a ∞-manifold and so we can define M_m^* and M_m by viewing M as an ∞-manifold and use the above definitions with $k=\infty$. More generally if a differentiable manifold M of class $k>1$ is given and if $1\leqslant i\leqslant k$ then we can consider M as an i-manifold and define the spaces of covectors and tangents accordingly. Let $\mathscr{F}_m^i$ denote the vector space $\mathscr{F}_m$ when M is viewed as a i-manifold and let $\mathscr{F}_m^{io}$ be defined similarly. We let M_m^{i*} and M_m^i be the vector spaces of corresponding covectors and tangents. Then for every i and j satisfying $1\leqslant i<j\leqslant k$ there is a natural isomorphism of M_m^{j*} onto M_m^{i*}. For there is a natural injection $\mathscr{F}_m^j\to\mathscr{F}_m^i$ which maps

the germ of a j-times differentiable function f into the germ of f when we look at f as an i-times differentiable function. For every f in $\mathscr{F}_m^j$ we let the same symbol f denote its image in $\mathscr{F}_m^i$ under this injection. It is obvious that $\mathscr{F}_m^{jo}$ is mapped into $\mathscr{F}_m^{io}$. Then

$$f+\mathscr{F}_m^{jo}\mapsto f+\mathscr{F}_m^{io} \tag{5}$$

is the natural isomorphism of M_m^{j*} onto M_m^{i*}. The map is a well defined homomorphism of M_m^{j*} into M_m^{i*}. If $f\in\mathscr{F}_m^j$ is such that its image under $\mathscr{F}_m^j\to\mathscr{F}_m^i$ belongs to $\mathscr{F}_m^{io}$ then $f\in\mathscr{F}_m^{jo}$. This shows that the kernel of (5) is trivial. Siqce M_m^{j*} and M_m^{i*} are both n dimensional (5) is surjective.

By the existence of these natural isomorphisms we can define the space of covectors M_m^* and the space of tangents M_m by viewing our k-manifold as an i-manifold for any i satisfying $1\leqslant i\leqslant k$. Very often it is convenient to choose $i=1$.

Suppose that M and N are differentiable manifolds of the same class. Let (O,φ) and (Q,ψ) denote typical coordinate neighborhoods of M and N, respectively. We consider the sets $O\times Q$ which lie in $M\times N$ and the maps $\varphi\times\psi: O\times Q\to\mathbb{R}^{m+n}$ where m and n denote the dimensions of M and N in O and Q, respectively. Then the sets $O\times Q$ cover $M\times N$ and the maps $\varphi\times\psi$ are homeomorphisms of these sets onto open subsets of $\mathbb{R}^{m+n}$. Moreover if (O_i,φ_i) and (Q_i,ψ_i) $(i=1,2)$ are such pairs then $(\varphi_1\times\psi_1)(\varphi_2\times\psi_2)^{-1}$ is a differentiable map of

$$(\varphi_2\times\psi_2)(O_1\times Q_1\cap O_2\times Q_2)$$

into $\mathbb{R}^{m+n}$. Therefore the pairs $(O\times Q,\varphi\times\psi)$ define a differentiable structure on $M\times N$ of the same class as M and N. The topology underlying this structure is the product topology of $M\times N$. It is called the *product* of the manifolds M and N. The definition of the *product manifold* is meaningful also in the case when the differentiable structures of M and N belong to different classes. Then the differentiable structure of $M\times N$ belongs to the weaker one of these two classes. Later in Section 6 we shall define what is meant by a submanifold of a given manifold. Once the concept is understood it is easy to prove that the set $\{(m,n): m\in M\}$ together with the differentiable structure inherited from M is a submanifold of $M\times N$ which is diffeomorphic to M. Similarly $\{(m,n): n\in N\}$ gives an injection of the manifold N into $M\times N$ when the differentiable structure is defined by using the sets $\{(m,n): n\in Q\}$ and the maps $(m,n)\mapsto\psi(n)$. The torus $T^2=T\times T$ gives a simple example of a product manifold. If more generally we define the C^∞ structure of the toroids T^n by considering the products $T\times T^{n-1}$ then one can easily verify that $T^m\times T^n$ is diffeomorphic to T^{m+n} for all choices of $m,n=1,2,\ldots$

5. Lie Groups and their Lie Algebras

We are going to define Lie groups and prove that a real Lie algebra $\mathfrak{g}$ can be associated with every such group G. The definiton of $\mathfrak{g}$ involves a few fundamental concepts from the theory of C^∞ manifolds such as C^∞ vector fields X and their bracket $[X,Y]$. All these concepts will be introduced in the sequel. Near the end of the section we shall give some practical methods for the determination of $\mathfrak{g}$ and we shall work out some examples.

Definition 1. *A Lie group is a set G which is a group and a C^∞ manifold such that the group operations $(x,y)\mapsto xy$ and $x\mapsto x^{-1}$ are C^∞ functions.*

Note. It would be no real restriction to suppose that the two group operations are real analytic. Also it would be sufficient to suppose that the differentiable structure of G is C^k for some large k. Moreover due to the positive answer to Hilbert's fifth problem one could require this only with $k=0$.

It is obvious that instead of the C^∞ nature of the group operations one can require that $(x,y)\mapsto xy^{-1}$ is a C^∞ map of $G\times G$ onto G. We see also that $x\mapsto x^{-1}$ is a C^∞ isomorphism of G onto itself and similarly if a is a fixed element in G then $x\mapsto ax$ and $x\mapsto xa$ are C^∞ diffeomorphisms of G onto itself. Since a Lie group is a topological group which is a Hausdorff space all the results proved for such groups apply to Lie groups. The component G_o containing the identity element e of G is an open, invariant, connected subgroup of G and the components of G are the cosets of G with respect to G_o. The subgroup G_o is called the identity component of G.

Lemma 2. *If N_e is a neighborhood of the identity e of the connected group G then every element x of G can be represented in the form $x=x_1\ldots x_n$ where $x_1,\ldots,x_n\in N_e$.*

Note. One says that the connected group G is generated by N_e.

Proof. We may suppose that N_e is symmetric i.e. $N_e=N_e^{-1}$. Let S be the set of all products $x_1\ldots x_n$ where $x_1,\ldots,x_n\in N_e$. It is open because if $x\in S$ then $N_x=xN_e$ lies in S. Next we suppose that x is in the closure of S. Then we can choose a point y in $S\cap xN_e$. Since N_e is symmetric we have $y^{-1}x\in N_e\subseteq S$ and so by $y\in S$ we obtain $x\in S$. This shows that S is a closed set. Since S is both open and closed and G is connected it follows that $S=G$.

Lemma 3. *Every Lie group is paracompact.*

Proof. By the foregoing lemma the identity component G_o is generated by any neighborhood N_e lying in G_o. More precisely given N_e and $x \in G_o$ there is an exponent $n = n(x)$ such that N_e^n contains x. The neighborhood filter $\mathcal{N}(e)$ has a basis $\mathcal{B}(e)$ which is denumerable and consists of open subsets B_m $(m = 1, 2, \ldots)$ of G_o. The set of all powers B_m^n $(m, n = 1, 2, \ldots)$ is a base for the open sets of G_o. Hence G_o and G are paracompact.

We are already familiar with many Lie groups because more than half of the examples discussed in Sections V.1 and V.2 are Lie groups. The simplest are the additive group of the reals $\mathbb{R}^+$ and the one dimensional torus $T = S^1$ which is analytically isomorphic to $\mathrm{Sl}(2, \mathbb{R})$. From these we can construct $(\mathbb{R}^+)^m \times T^n$ $(m, n = 1, 2, \ldots)$ which is actually the most general commutative Lie group up to C^∞ isomorphism. Among the non-commutative ones we find $\mathrm{Gl}(n, \mathbb{R})$, $\mathrm{Gl}(n, \mathbb{C})$ and their subgroups $\mathrm{Sl}(n, \mathbb{R})$, $\mathrm{Sl}(n, \mathbb{C})$, $\mathrm{U}(n)$, $\mathrm{SU}(n)$, $\mathrm{O}(n, \mathbb{R})$ and $\mathrm{SO}(n, \mathbb{R})$. The group of proper motions of $\mathbb{R}^n$ which we encountered in the form $\mathbb{R}^n @ \mathrm{SO}(n, \mathbb{R})$ is also a Lie group.

The fact that $\mathrm{Gl}(n, \mathbb{R})$ and $\mathrm{Gl}(n, \mathbb{C})$ are Lie groups is obvious because these groups have global coordinate systems which define the C^∞ structure. For example in the real case we let $O = \mathrm{Gl}(n, \mathbb{R})$ and we define $\varphi(p) = p - I$ for every $p \in O$ where I is the $n \times n$ identity matrix. Then φ is a one-to-one map of $\mathrm{Gl}(n, \mathbb{R})$ onto an open subset of $\mathbb{R}^{n^2}$. This choice of φ has certain advantages due to the fact that $\varphi(e) = 0$. All but one of the other groups listed above are closed subgroups of $\mathrm{Gl}(n)$. Later on in Section 7 we shall see that every closed subgroup of a Lie group is also a Lie group. In the case of $\mathrm{Sl}(n, \mathbb{R})$ there is a natural local coordinate system which will be discussed near the end of this section where the Lie algebra of $\mathrm{Sl}(n, \mathbb{R})$ is determined. The C^∞ structure of the underlying manifold can be introduced in a natural way also in several special cases which include $\mathrm{SO}(2, \mathbb{R})$, $\mathrm{O}(2, \mathbb{R})$ and $\mathrm{SU}(2)$.

We are going to associate a Lie algebra $\mathfrak{g}$ with every Lie group G. We recall that $\mathfrak{g}$ is a vector space with a multiplication called the bracket which is not associative except in very special cases. The vector space of $\mathfrak{g}$ can be identified with the real vector space of tangent vectors at the neutral element e of G. Thus the question is reduced to the definition of the product operations. Although one can describe the bracket as an operation on tangent vectors we obtain more insight by proceeding quite differently. The elements of $\mathfrak{g}$ will not be tangent vectors but other objects called *invariant vector fields* on G and the operations will be defined on this set $\mathfrak{g}$. We shall see that there is a natural isomorphism between this vector space and the vector space G_e of tangent vectors at e.

A *vector field* on a differentiable manifold M is a function X defined on M with values $X(m) \in M_m$ $(m \in M)$ where M_m is the tangent space of M at the point m. One could consider more general vector fields e.g. we could suppose that X is defined only on an open subset of M but for our purposes it is sufficient to restrict our attention to those vector fields which are everywhere defined on M. It is obvious that the set of all vector fields $X: M \to \bigcup M_m$ is a vector space under the operations $(X, Y) \mapsto X+Y$ and $(\lambda, X) \mapsto \lambda X$ where $(X+Y)(m) = X(m) + Y(m)$ and $(\lambda X)(m) = \lambda X(m)$.

Let M and N be differentiable manifolds and let $\Phi: M \to N$ be a map which is differentiable at $m \in M$. Then we define the *differential of Φ at m* as a certain map $d\Phi_m: M_m \to N_{\Phi(m)}$. Namely let $\mathscr{F}_{\Phi(m)}$ be the vector space of C^1 function germs at $\Phi(m)$ and let $\mathscr{F}^o_{\Phi(m)}$ be the subspace of stationary function germs. If f_1 and f_2 are equivalent C^1 functions in the neighborhood of $\Phi(m)$ then $f_1 \Phi$ and $f_2 \Phi$ are also equivalent C^1 functions defined in some neighborhood of m. Hence if $\mathscr{f} \in \mathscr{F}_{\Phi(m)}$ then $\mathscr{f}\Phi$ is a well defined element of $\mathscr{F}_m$. We can now define $d\Phi_m$ for every $\boldsymbol{t} \in M_m$ by

$$(1) \qquad d\Phi_m(\boldsymbol{t})(\mathscr{f} + \mathscr{F}^o_{\Phi(m)}) = \boldsymbol{t}(\mathscr{f}\Phi + \mathscr{F}^o_m).$$

If Φ is everywhere differentiable then $d\Phi_m$ is defined for every m in M and one writes $d\Phi$ for the family of these maps $d\Phi_m$ $(m \in M)$. It is the *differential* of the differentiable map $\Phi: M \to N$.

We suppose now that $M = N = G$ is a Lie group and $a \in G$. Then $x \mapsto ax$ $(x \in G)$ is a differentiable map $L_a: G \to G$ and so we can consider dL_a. By (1) it is given by

$$(2) \qquad (dL_a)_x(\boldsymbol{t})(\mathscr{f} + \mathscr{F}^o_{ax}) = \boldsymbol{t}({}_a\mathscr{f} + \mathscr{F}_x^o)$$

where $x \in G$ and $\boldsymbol{t}$ is in the tangent space G_x. Since the left translates of equivalent differentiable functions are equivalent the translated germ ${}_a\mathscr{f}$ is well defined for every $\mathscr{f} \in \mathscr{F}_{ax}$.

A vector field X defined on the Lie group G is called a *left invariant vector field* if $(dL_a)_x X(x) = X(ax)$ for every a and x in G. By (2) this gives

$$(dL_a)_x X(x)(\mathscr{f} + \mathscr{F}^o_{ax}) = X(x)({}_a\mathscr{f} + \mathscr{F}^o_x) = X(ax)(\mathscr{f} + \mathscr{F}^o_{ax})$$

where $\mathscr{f} \in \mathscr{F}_{ax}$. If g is a stationary function in the neighborhood of ax then ${}_ag$ is a stationary function defined in the neighborhood of x. Hence if $\boldsymbol{f} \in M^*_{ax} = \mathscr{F}_{ax}/\mathscr{F}^o_{ax}$ then ${}_a\boldsymbol{f}$ is a well defined element of $M^*_x = \mathscr{F}_x/\mathscr{F}^o_x$. Therefore, if X is a left invariant vector field then $X(x)({}_a\boldsymbol{f}) = X(ax)(\boldsymbol{f})$ for every $\boldsymbol{f}$ in M^*_{ax}. We can specialize this equation by letting $x = a^{-1}$

and obtain for every a in G the relation $X(a)({}^a f)=X(e)(f)$ where f is in G_e^*. Since $f\mapsto {}^a f$ is an isomorphism of G_e^* onto G_a^* we obtain:

Proposition 4. *Every left invariant vector field is uniquely determined by its value at e, namely*

$$X(a)({}^a f)=X(e)f$$

for every $a\in G$ and $f\in G_e^*$. Moreover, given any vector $\boldsymbol{t}$ *in G_e this equation defines a vector field $X_{\boldsymbol{t}}$ on G such that $X_{\boldsymbol{t}}(e)=\boldsymbol{t}$.*

The left invariant vector fields form a real vector space and $\boldsymbol{t}\mapsto X_{\boldsymbol{t}}$ is an isomorphism of G_e onto this vector space. This clarifies our introductory remark according to which we can choose either G_e or the vector space of left invariant vector fields as the vector space of the Lie algebra of G.

We shall now turn to the definition of the operation $(X,Y)\mapsto[X,Y]$. The bracket $[X,Y]$ is meaningful not only in the case of Lie groups but for certain vector fields X, Y defined on any C^∞ manifold M. First we define what is meant by Xf when X is a vector field on M and $f:M\to\mathbb{R}$ is a C^1 function on M: We let $Xf:M\to\mathbb{R}$ be such that $Xf(m)=X(m)(f+\mathscr{F}_m^o)$ for every $m\in M$. The vector field X is called differentiable if Xf is a C^1 function for every C^1 function f. Thus if X is a differentiable vector field and Y is an arbitrary vector field then $YXf=Y(Xf)$ is defined for every C^1 function f. If X and Y are both differentiable then $(XY-YX)f$ is a C^1 function on M which will be denoted by $[X,Y]f$.

We call X a C^∞ *vector field* if Xf is a C^∞ function for every C^∞ function $f:M\to\mathbb{R}$. Then $f\mapsto Xf$ is a linear operator $D_X:C^\infty(M)\to C^\infty(M)$ where $C^\infty(M)$ is the real vector space of C^∞ functions $f:M\to\mathbb{R}$. By Proposition 4.9 we have

$$(Xfg)(m)=X(m)(fg+\mathscr{F}_m^o)=(Xf)(m)g(m)+f(m)(Xg)(m)$$

for every m in M and so $X(fg)=(Xf)g+f(Xg)$ for all f,g in $C^\infty(M)$. Hence D_X is a derivation of $C^\infty(M)$.

Proposition 5. *The map $X\mapsto D_X$ is an isomorphism of the vector space of C^∞ vector fields X of the C^∞ manifold M onto the vector space of derivations of the real algebra $C^\infty(M)$. The derivation D of $C^\infty(M)$ is the image of that C^∞ vector field X_D for which $X_D(m)f=Df(m)$ for every $m\in M$ and $f\in C^\infty(M)$.*

Proof. It is clear that the map $X\mapsto D_X$ is linear. Let X be such that $Xf=0$ for every f in $C^\infty(M)$. Then given any m in M we have $Xf(m)=X(m)(f+\mathscr{F}_m^o)=0$ for every $f\in C^\infty(M)$. We note that if $f\in M_m^*$ then by Lemma 8 given below there is an f in $C^\infty(M)$ such that $\tilde{f}(m)+\mathscr{F}_m^o \in \boldsymbol{f}$. It follows that $X(m)$ is the zero functional on M_m^*. This shows that the kernel of $X \mapsto D_X$ is zero and so the map is an isomorphism. The proof will be completed by showing that given a derivation D of $C^\infty(M)$ there is a C^∞ vector field X such that $D_X = D$. We know that for every $\boldsymbol{f}$ in M_m^* there is a function f in $C^\infty(M)$ such that $\tilde{f}(m)+\mathscr{F}_m^o \in \boldsymbol{f}$. Hence we can define

$$X(m)\boldsymbol{f} = X(m)(\tilde{f}(m)+\mathscr{F}_m^o) = D f(m)$$

provided the right hand side is independent of the choice of f. If this is really so then it is easy to verify that $X(m)$ is a linear functional on M_m^* i.e. $X(m) \in M_m$ and so $m \mapsto X(m)$ defines a vector field X. If $f \in C^\infty(M)$ then

$$D_X f(m) = X f(m) = X(m)(f+\mathscr{F}_m^o) = D f(m)$$

and so $D_X = D$. We also see that $Xf = Df$ where $Df \in C^\infty(M)$ by $D\colon C^\infty(M) \to C^\infty(M)$. This shows that X is a C^∞ vector field. Therefore in order to complete the proof it will be sufficient to prove the following:

Proposition 6. *If $D\colon C^\infty(M) \to C^\infty(M)$ is a derivation and f in $C^\infty(M)$ is stationary at $m \in M$ then $Df(m) = 0$.*

The proposition is the direct consequence of a few basic results on the derivations of $C^\infty(M)$ and $C^\infty(O)$ where O is an open subset of M. We start with the following.

Lemma 7. *Given $b > a > 0$ there is a C^∞ function $f\colon \mathbb{R}^n \to \mathbb{R}$ such that $0 \leqslant f(x) \leqslant 1$ for every $x \in \mathbb{R}^n$ and*

$$f(x) = \begin{cases} 1 & \text{if } \|x\| \leqslant a \\ 0 & \text{if } \|x\| \geqslant b. \end{cases}$$

Note. Here $\|x\|$ denotes the norm derived from the Euclidean inner product of $\mathbb{R}^n$.

Proof. If we define $h(t) = \exp((1/t-b)-(1/t-a))$ for $a < t < b$ and $h(t) = 0$ for $t \notin (a,b)$ then $h\colon \mathbb{R} \to [0,1]$ is a C^∞ function. Thus

$$g(u) = \left(\int_a^b h(t)\,dt\right)^{-1} \int_u^b h(t)\,dt$$

defines a C^∞ function on $\mathbb{R}$ such that $0 \leqslant g(u) \leqslant 1$ everywhere, $g(u) = 1$ for $u \leqslant a$ and $g(x) = 0$ for $u \geqslant b$. If we let

$$f(x) = f(x_1, \ldots, x_n) = g(\|x\|)$$

then f satisfies the requirements.

Lemma 8. *Let C be a compact subset and O an open subset of the C^∞ manifold M such that $C \subset O$. Then there is a C^∞ function $f: M \to [0,1]$ such that $f(c)=1$ for every $c \in C$ and f identically vanishes on the complement of O.*

Proof. For every point $c \in C$ we can find a coordinate neighborhood (Q, φ) such that $c \in Q \subseteq O$. Given m and (Q, φ) by Lemma 7 we can select a function $g: M \to [0,1]$ such that g is 0 on the complement of Q and it is 1 on some compact neighborhood S of c where $S \subset Q$. Since C is compact there are finitely many points $c_1, \ldots, c_m$ such that if $Q_1, \ldots, Q_m$ and $S_1, \ldots, S_m$ denote the corresponding neighborhoods then $C \subseteq S_1 \cup \cdots \cup S_m$ and $Q_1 \cup \cdots \cup Q_m \subseteq O$. Let $g_1, \ldots, g_m$ be the corresponding C^∞ functions. Then $f = 1-(1-g_1)\ldots(1-g_m)$ is a C^∞ function on M which satisfies all the requirements.

Proposition 9. *Let M be a C^∞ manifold and let D be a derivation of $C^\infty(M)$. If f is in $C^\infty(M)$ and it identically vanishes in an open subset O of M then Df also vanishes in O.*

Proof. Given a point p in O by Lemma 8 there is a C^∞ function $g: M \to [0,1]$ such that $g(p)=0$ and $g(m)=1$ for all $m \notin O$. Thus by the properties of f we have $fg(m) = f(m)$ for every m in M. Hence

$$Df(p) = Dfg(p) = Df(p)g(p) + f(p)Dg(p) = 0$$

where p is an arbitrary point in O.

Corollary. *If $f_1, f_2 \in C^\infty(M)$ and $f_1 = f_2$ on an open subset O of M then $Df_1 = Df_2$ in O.*

We can now turn to derivations of $C^\infty(O)$ where O is an open subset of the C^∞ manifold M:

Proposition 10. *If O is an open subset of the C^∞ manifold M and D is a derivation of $C^\infty(M)$ then D defines a derivation $D(O)$ of $C^\infty(O)$. If $f \in C^\infty(M)$ then $Df(m) = D(O)(f|O)(m)$ for every $m \in O$.*

Proof. If $f \in C^\infty(O)$ is given and $m \in O$ we choose h in $C^\infty(M)$ such that f and h coincide in a neighborhood of m and we define $D(O)f(m) = Dh(m)$. The existence of such a function h follows from Lemma 8. By the corollary of Proposition 9 $D(O)f(m)$ is independent of the choice of h. Since the same h can be used for a whole neighborhood of m it follows that $D(O)f \in C^\infty(O)$. It is easy to verify that $D(O): C^\infty(O) \to C^\infty(O)$ is a derivation. If $f \in C^\infty(M)$ then one can use f itself as a function h for $f|O$ and so $D(O)(f|O)(m) = Df(m)$ for every $m \in O$.

Note. We see that $D \to D(O)$ is a transitive process: If O, Q are open sets in M and $O \supseteq Q$ then $D(O)(Q) = D(Q)$. We call $D(O)$ the *derivation induced on $C^\infty(O)$ by D.*

Lemma 11. *Let B be a ball in $\mathbb{R}^n$ with center c and let f belong to $C^\infty(B)$. Then there exist $g_1, \ldots, g_n$ in $C^\infty(B)$ such that $g_i(c) = f_i(c)$ where f_i $(i = 1, \ldots, n)$ is the i-th partial derivative of f and*

$$f(x) = f(c) + \sum_{i=1}^{n} (x_i - c_i)\, g_i(x)$$

for every $x = (x_1, \ldots, x_n)$ in B.

Proof. For every x in B we have

$$f(x) = f(x_1, \ldots, x_n) = f(c) + \int_0^1 \frac{d}{dt} f(c_1 + t(x_1 - c_1), \ldots, c_n + t(x_n - c_n))\, dt\,.$$

Therefore expressing the derivative in terms of the partial derivatives of f we obtain

$$f(x) = f(c) + \sum_{i=1}^{n} x_i \int_0^1 f_i(c_1 + t(x_1 - c_1), \ldots, c_n + t(x_n - c_n))\, dt\,.$$

Now we let

$$g_i(x) = g_i(x_1, \ldots, x_n) = \int_0^1 f_i(c_1 + t(x_1 - c_1), \ldots, c_n + t(x_n - c_n))\, dt$$

for every x in B so that $g_i \in C^\infty(B)$. By the first mean value theorem of the integral calculus $g_i(x) = f_i(c + \tau_i(x - c))$ for some τ_i $(0 \leqslant \tau_i \leqslant 1)$. Hence by letting $x \to c$ we obtain $g_i(c) = f_i(c)$.

Proposition 12. *Let M be a C^∞ manifold and let (O, φ) be a local chart of M with local coordinates $x_1, \ldots, x_n \in \mathbb{R}$. Then for every derivation D of $C^\infty(M)$, every f in $C^\infty(M)$ and every m in O we have*

$$D f(m) = \sum_{i=1}^{n} D(O)\, x_i(m) (f \varphi^{-1})_i(\varphi(m))\,.$$

Note. The following notations are used: $D(O)$ is the derivation induced on $C^\infty(O)$ by D. The function $m \mapsto x_i(m)$ $(m \in O)$ is the composition of φ with the map $x \mapsto x_i$ of $\mathbb{R}^n$ onto $\mathbb{R}$ and $(f \varphi^{-1})_i$ is the i-th partial derivative of $f \varphi^{-1}$.

Proof. Given m in O we can choose a ball B with center $c=\varphi(m)$ in $\mathbb{R}^n$ such that $Q=\varphi^{-1}(B)\subset O$. Now we use Lemma 11 with B and $f\varphi^{-1}|B$ and write down the result in terms of f and Q: There exist functions $g_1,\ldots,g_n$ in $C^\infty(Q)$ such that $g_i(m)=(f\varphi^{-1})_i(\varphi(m))$ for every $i=1,\ldots,n$ and

$$f(q)=f(m)+\sum_{i=1}^{n}(x_i(q)-c_i)g_i(q)$$

for every q in Q. If we apply $D(Q)$ to the equation

$$f|Q=f(m)+\sum_{i=1}^{n}(x_i-c_i)g_i$$

then we obtain

$$D(Q)(f|Q)(m)=\sum_{i=1}^{n}D(Q)(x_i|Q)(m)g_i(m).$$

By Lemma 10 and the note following its proof we have $D(Q)(f|Q)(m)=Df(m)$ and $D(Q)(x_i|Q)(m)=D(O)x_i(m)$. Therefore

$$Df(m)=\sum_{i=1}^{n}D(O)x_i(m)g_i(m)$$

for every m in O. The result follows by substituting the value of $g_i(m)$ in this equation.

Now let $f\in C^\infty(M)$ be stationary at $m\in M$. Then by definition $(f\varphi^{-1})_i(\varphi(m))=0$ for every $i=1,\ldots,n$ and so by Proposition 12 $Df(m)=0$. Therefore Propositions 5 and 6 are proved.

Lemma 13. *For every f in $C^\infty(G)$ and $\mathbf{t}$ in G_e the correspondence $p\mapsto\mathbf{t}({}_pf)$ defines a C^∞ function on G.*

Proof. Let (O,φ) and (Q,ψ) be local charts such that $e\in O$ and $p\in Q$. Let x and y be points in $\varphi(O)$ and $\psi(Q)$ with components $x_1,\ldots,x_n$ and $y_1,\ldots,y_n$, respectively. We can find sufficiently small open neighborhoods $O_{\varphi(e)}$ and $Q_{\psi(p)}$ such that $O_{\varphi(e)}\subseteq O$, $Q_{\psi(p)}\subseteq Q$ and $\psi^{-1}(y)\varphi^{-1}(x)$ lies in Q for every $x\in O_{\varphi(e)}$ and $y\in Q_{\psi(p)}$. Hence we can define $\mu\colon O_{\varphi(e)}\times Q_{\psi(p)}\to\psi(Q)$ by $\mu(x,y)=\psi(\psi^{-1}(y)\varphi^{-1}(x))$. Since G is a Lie group multiplication it is a C^∞ operation and so μ is a C^∞ function. By $\mathbf{t}\in G_e$ we have

$$\mathbf{t}=c_1\frac{\partial}{\partial x_1}+\cdots+c_n\frac{\partial}{\partial x_n}$$

for suitable $c_1, \ldots, c_n$ in $\mathbb{R}$. If $x \in O_{\varphi(e)}$ and $y = \psi(q) \in Q_{\psi(p)}$ then

$$_qf(\varphi^{-1}(x)) = f(q\varphi^{-1}(x)) = f(\psi^{-1}(y)\varphi^{-1}(x)) = f\psi^{-1}(\mu(x, y)).$$

Therefore the function value of the composition of ψ with $q \mapsto \boldsymbol{t}(_qf)$ at y is

$$\boldsymbol{t}(_qf) = \boldsymbol{t}(_{\psi^{-1}(y)}f) = \sum_{i=1}^{n} c_i \frac{\partial}{\partial x_i} f\psi^{-1}\mu(x, y)\bigg|_{x=\varphi(e)}.$$

We know that $f\psi^{-1}$ and μ are C^∞ functions and so $f\psi^{-1}\mu$ is a C^∞ function on $O_{\varphi(e)} \times Q_{\psi(p)}$. Hence the terms of the foregoing sum are all C^∞ functions of the variable $y \in Q_{\psi(p)}$. This shows that $q \mapsto \boldsymbol{t}(_qf)$ is a C^∞ function in the neighborhood of p where $p \in G$ is arbitrary.

Corollary. *Every left invariant vector field is a C^∞ vector field.*

In fact if X is a left invariant vector field and $f \in C^\infty(G)$ then by Proposition 4 we have $Xf(p) = X(e)(_pf)$ for every p in G. Hence applying the foregoing lemma with $X(e) = \boldsymbol{t}$ we see that Xf is a C^∞ function on G.

A derivation D of $C^\infty(G)$ is called a *left invariant derivation* on the Lie group G provided $D_af = {}_aDf$ for every $a \in G$ and $f \in C^\infty(G)$. It means that D commutes with the translation operator $f \mapsto {}_af$ induced on $C^\infty(G)$ by the diffeomorphism $L_a: G \to G$ where $L_a x = ax$ for all $x \in G$. If X is a left invariant vector field then by the foregoing corollary X is a C^∞ vector field and so we can consider the derivation D_X associated with X according to Proposition 5. For every $a, p \in G$ we have $D_X(_af)(p) = X(p)(_af) = X(e)(_p(_af)) = X(e)_{ap}f = X(ap)f = D_Xf(ap) = {}_aD_Xf(p)$. Therefore the left invariance of X implies that D_X is left invariant. Conversely let us suppose that D is a left invariant derivation on G. By Proposition 5 $D = D_X$ where the C^∞ vector field $X = X_D$ is defined by $X(p)f = Df(p)$ for every $p \in G$ and $f \in C^\infty(M)$. Thus we have

$$X(p)f = Df(p) = {}_pDf(e) = D(_pf)(e) = X(e)_pf$$

for every $p \in G$ and so X is a left invariant vector field. Hence Proposition 5 can be supplemented with the following:

Proposition 14. *If G is a Lie group then the map $X \mapsto D_X$ is an isomorphism of the vector space of left invariant vector fields onto the vector space of left invariant derivations on G.*

We can now define the bracket operation $[X, Y]$ for any two left invariant vector fields X, Y as follows: By Proposition 5 X and Y determine two derivations D_X and D_Y of the algebra $C^\infty(G)$. By Proposition 1.11 $[D_X, D_Y] = D_X D_Y - D_Y D_X$ is a derivation of $C^\infty(G)$ and by Proposition 5

$[D_X, D_Y]$ is the image under the map $Z \mapsto D_Z$ of a C^∞ vector field $[X, Y]$. We prove that $[X, Y]$ is left invariant.

The bracket of two invariant derivations D_1 and D_2 is left invariant because L_a commutes with D_1 and D_2 and so clearly it commutes with $[D_1, D_2]$.

Therefore $[D_X, D_Y]$ is a left invariant derivation and so by Proposition 14 $[X, Y]$ is a left invariant vector field. The bracket operation is now defined. Our results can be summarized as follows:

Theorem 15. *Let G be a Lie group. Those derivations* $D\colon C^\infty(G) \to C^\infty(G)$ *which commute with the translation operators* $f \mapsto {}_a f$ $(a \in G$ *and* $f \in C^\infty(G))$ *form a subalgebra of the real Lie algebra of all derivations of* $C^\infty(G)$. *This is the Lie algebra of left invariant derivations on G. It is naturally isomorphic to the Lie algebra* $\mathfrak{g}$ *of left invariant vector fields X on G i.e. those which satisfy* $X(p) f = X(e)_p f$ *for all* $p \in G$ *and* $f \in C^\infty(G)$. *The image of D is the vector field* X_D *defined by* $X_D(p) f = D f(p)$ *where* $p \in G$ *and* $f \in C^\infty(G)$. *The product* $[X, Y]$ *is defined by this isomorphism* $X \mapsto D_X$. *The Lie algebra of left invariant vector fields is by definition the Lie algebra of G. The map* $X \mapsto X(e)$ *is a vector space isomorphism of* $\mathfrak{g}$ *onto* G_e. *If we define the product* $[\boldsymbol{t}_1, \boldsymbol{t}_2]$ *of two tangent vectors* $\boldsymbol{t}_1$ *and* $\boldsymbol{t}_2$ *by this isomorphism we obtain a Lie algebra structure on* G_e.

Note. The Lie algebra G_e of tangent vectors or the Lie algebra of left invariant derivations on G can also be considered as the Lie algebra of G. There are situations when it is more advantageous to use one of these alternate definitions.

We shall now discuss a method by which the Lie algebra of a given Lie group G can be determined. Let e be the identity of G and let (Q, φ) be a local chart containing e. It is no restriction to suppose that $\varphi(e)$ is the origin 0 of $\mathbb{R}^n$ where n is the dimension of G. Since multiplication is continuous there is an open set O containing e such that $pq \in Q$ for every $p, q \in O$. Clearly $e \in O \subseteq Q$ and so $O_o = \varphi(O)$ is an open neighborhood of the origin and $\varphi(O) \subseteq \varphi(Q)$. Let $\varphi(p) = x = (x_1, \ldots, x_n)$ and $\varphi(q) = y = (y_1, \ldots, y_n)$. We define $\mu\colon O_o \times O_o \to \varphi(Q)$ by

$$\mu(x, y) = \varphi(p\, q) = \varphi(\varphi^{-1}(x)\varphi^{-1}(y)).$$

Since $eq = q$ and $pe = p$ we have $\mu(0, y) = y$ and $\mu(x, 0) = x$ for all $x, y \in O_o$. Similarly by the associativity of the product we have $\mu(\mu(x, y), z) = \mu(x, \mu(y, z))$ for all $x, y, z \in O_o$. If $\mu(x, y)_i$ $(i = 1, \ldots, n)$ denote the coordinates of $\mu(x, y)$ then $(x, y) \mapsto \mu(x, y)_i$ is a C^∞ function for every $i = 1, \ldots, n$. The system consisting of (Q, φ), O, O_o and μ will be called a *germ of the Lie group G*.

We note that the expression germ has another, related meaning. For let O_o be an open subset of $\mathbb{R}^n$ containing the origin 0 and let $\mu: O_o \times O_o \to \mathbb{R}^n$ be a function satisfying the following conditions:

1. The functions $(x, y) \mapsto \mu(x, y)_i$ $(i=1, \ldots, n)$ all belong to $C^2(O_o \times O_o)$.
2. For all $x, y \in O_o$ we have $\mu(x, 0) = x$ and $\mu(0, y) = y$.
3. For all $x, y, z \in O_o$ we have $\mu(\mu(x, y), z) = \mu(x, \mu(y, z))$.

Then the system consisting of O_o and μ is called a *Lie group germ*. Sometimes it is supposed that the function described in 1. belongs to $C^\infty(O_o)$.

Let a germ $(Q, \varphi), O, O_o, \mu$ of G be given. Let X_1, X_2 be left invariant vector fields on G. By Proposition 4 and Theorem 15 in order to find $[X_1, X_2]$ it is sufficient to determine $[X_1, X_2](e)$. Since G is a C^∞ manifold we also know that for every function germ $\not{f} \in \mathscr{F}_e$ there is a function $f \in C^\infty(G)$ such that $f \in \not{f}$. If $p, q \in O$ and $x = \varphi(p)$, $y = \varphi(q)$ then

$$ {}_pf(\varphi^{-1}(y)) = f(p\varphi^{-1}(y)) = f\varphi^{-1}(\varphi(\varphi^{-1}(x)\varphi^{-1}(y))) = f\varphi^{-1}\mu(x, y). $$

Therefore we have

$$ {}_pf\varphi^{-1}(y) = f\varphi^{-1}\mu(x, y) \tag{3} $$

for all $x = \varphi(p) \in O_o$, $y \in O_o$. Let $\partial/\partial X_i$ and also $\partial/\partial Y_i$ denote those directional derivations in $\mathbb{R}^n$ which correspond to the tangent vector $X_i(e)$ $(i=1, 2)$ in the local chart (Q, φ). Then from (3) we obtain

$$ X_i(e)({}_pf) = \frac{\partial}{\partial Y_i} f\varphi^{-1}\mu(x, \cdot)\Big|_{y=0} $$

and so considering p as a variable we have

$$ X_j(e)(X_i(e)_p f) = \frac{\partial^2}{\partial X_j \partial Y_i} f\varphi^{-1}\mu\Big|_{x=0, y=0}. $$

Using this with $(i, j) = (2, 1)$ and $(1, 2)$ we obtain

$$ [X_1, X_2](e)(f) = \left(\frac{\partial^2}{\partial X_1 \partial Y_2} - \frac{\partial^2}{\partial X_2 \partial Y_1}\right) f\varphi^{-1}\mu(0, 0). $$

Since the tangent vectors $\boldsymbol{t}_i$ $(i=1, \ldots, n)$ corresponding to the partial derivations $\partial/\partial x_i$ $(i=1, \ldots, n)$ form a basis for G_e the multiplication table of $\mathfrak{g}$ can be determined from the equations

$$ [\boldsymbol{t}_i, \boldsymbol{t}_j](f) = \left(\frac{\partial^2}{\partial x_i \partial y_j} - \frac{\partial^2}{\partial x_j \partial y_i}\right) f\varphi^{-1}\mu(0, 0) \tag{4} $$

where $i,j=1,\ldots,n$. For example if G is commutative then $\varphi^{-1}(x)\varphi^{-1}(y)=\varphi^{-1}(y)\varphi^{-1}(x)$ and so $f\varphi^{-1}\mu(x,y)=f\varphi^{-1}\mu(y,x)$ for all $x,y\in O_o$. Therefore from (4) we obtain $[\boldsymbol{t}_i,\boldsymbol{t}_j]=0$ for every $i,j=1,\ldots,n$. Hence the Lie algebra $\mathfrak{g}$ is abelian and so we proved the following:

Proposition 16. *The Lie algebra of a commutative Lie group G is abelian and in particular the Lie algebra of $\mathbb{R}^m\times T^n$ is $\mathbb{R}^{m+n}$.*

Since the Lie algebra of $\mathbb{R}^m\times T^n$ depends only on $m+n$ we see the existence of non-isomorphic, connected Lie groups with isomorphic Lie algebras.

We shall now use (4) to compute the Lie algebra of $\mathrm{Gl}(n,\mathbb{R})$. Let the $n\times n$ real matrices $x=(x_{ij})$ and $y=(y_{ij})$ be interpreted as elements of $\mathbb{R}^{n^2}$ and let I be the $n\times n$ real identity matrix. We let $p=x+I$ and $q=y+I$ so that p and q are elements of $\mathrm{Gl}(n,\mathbb{R})$ if $\det(x+I)\neq 0$ and $\det(y+I)\neq 0$. Since $\mathrm{Gl}(n,\mathbb{R})$ has a global coordinate system we can choose $O=\mathrm{Gl}(n,\mathbb{R})$ and let $\varphi(p)=p-I$ for every $p\in O$. Then we have

$$\mu(x,y)=x+y+xy\,.$$

Let f be a C^∞ function on $\mathrm{Gl}(n,\mathbb{R})$ and let $\tilde{f}=f\varphi^{-1}$ so that $\tilde{f}$ is a C^∞ function of n^2 real variables. We have

$$(5)\qquad \frac{\partial}{\partial y_{kl}} f\varphi^{-1}\mu=\sum_{r,s=1}^{n}\tilde{f}_{rs}\frac{\partial}{\partial y_{kl}}(x+y+xy)_{rs}$$

where $\tilde{f}_{rs}$ denotes a partial derivative of $\tilde{f}$ and

$$(x+y+xy)_{rs}=x_{rs}+y_{rs}+\sum_{t=1}^{n}x_{rt}y_{ts}$$

is an entry of the matrix $x+y+xy$. Hence the right hand side of (5) is

$$\sum_{\rho=1}^{n}\tilde{f}_{\rho l}(\delta_{k\rho}+x_{\rho k})\,.$$

Differentiating once more we obtain

$$\frac{\partial^2}{\partial x_{ij}\partial y_{kl}} f\varphi^{-1}\mu=\sum_{\rho=1}^{n}\sum_{r,s=1}^{n}\tilde{f}_{\rho l,rs}(\delta_{ir}\delta_{js}+\delta_{ir}y_{js})(\delta_{k\rho}+x_{\rho k})+\delta_{jk}\tilde{f}_{il}\,.$$

Hence we have

$$(6)\qquad \frac{\partial^2}{\partial x_{ij}\partial y_{kl}} f\varphi^{-1}\mu(0,0)=\tilde{f}_{kl,ij}+\delta_{jk}\tilde{f}_{il}$$

for all $i, j, k, l=1, \ldots, n$. We let $\boldsymbol{t}_{ij}$ denote that tangent vector to G at e which corresponds to the partial derivation $\partial/\partial x_{ij}$ in the local chart (O, φ). These vectors $\boldsymbol{t}_{ij}$ $(i, j=1, \ldots, n)$ form a basis for the Lie algebra $\mathfrak{g}$ of $\mathrm{Gl}(n, \mathbb{R})$. The relations given in (6) yield

$$[\boldsymbol{t}_{ij}, \boldsymbol{t}_{kl}](f)=\delta_{kj}\boldsymbol{t}_{il}(f)-\delta_{il}\boldsymbol{t}_{kj}(f)$$

where f is an arbitrary function in $C^\infty(G)$. Therefore multiplication in $\mathfrak{g}$ is uniquely determined by the equations

$$[\boldsymbol{t}_{ij}, \boldsymbol{t}_{kl}]=\delta_{kj}\boldsymbol{t}_{il}-\delta_{il}\boldsymbol{t}_{kj} \tag{7}$$

where $i, j, k, l=1, \ldots, n$.

Let E_{ij} denote the $n\times n$ real matrix which has 1 at the $i-j$ place and zero everywhere else i.e. let $E_{ij}=(\delta_{ir}\delta_{js})$. These matrices form a basis for the vector space of all $n\times n$ real matrices and we have $E_{ij}E_{kl}=\delta_{jk}E_{il}$ for all choices of $i, j, k, l=1, \ldots, n$. Hence using (7) we see that $\boldsymbol{t}_{ij}\mapsto E_{ij}$ $(i, j=1, \ldots, n)$ defines an isomorphism of $\mathfrak{g}$ onto $\mathfrak{gl}(n, \mathbb{R})$. We proved the following:

Theorem 17. *The Lie algebra of* $\mathrm{Gl}(n, \mathbb{R})$ *is isomorphic to* $\mathfrak{gl}(n, \mathbb{R})$.

The computation of the bracket can be considerably simplified by using a device due to Lie. It associates a Lie algebra $\mathfrak{g}$ with every Lie group germ $\mu: O_o\times O_o\to\mathbb{R}^n$ such that if μ is the germ of a Lie group G then $\mathfrak{g}$ is isomorphic to the Lie algebra of G. The elements of $\mathfrak{g}$ are the vectors $\boldsymbol{t}\in\mathbb{R}^n$ considered as bound vectors emitting from the origin. We shall use curves $\alpha, \beta, \ldots$ in $\mathbb{R}^n$ determined by a real parameter t $(t\geqslant 0)$ such that at $t=0$ the curve is at the origin i.e. $\alpha(0)=0$ and it has a tangent $\boldsymbol{t}_\alpha$ there. Hence $\boldsymbol{t}_\alpha=\alpha'(0)$. The interval in which t varies can be different for different curves but we suppose that the curve lies entirely within a neighborhood Q_o of 0 whose existence and properties are given in the following:

Lemma 18. *If* $\mu: O_o\times O_o\to\mathbb{R}^n$ *is a Lie group germ then there is an open neighborhood* Q_o *of* 0 *such that for every* x *in* Q_o *there is a unique* y *in* O_o *satisfying* $\mu(x, y)=\mu(y, x)=0$.

Proof. If the Lie group germ comes from a Lie group then the existence of Q_o follows from the continuity of the inverse operation. If μ is a general Lie group germ then one uses the implicit function theorem: Since $\mu(x,0)=x$ the Jacobian matrix of the map $x\mapsto\mu(x,0)$ is the identity at $x=0$. Hence in some neighborhood of the origin for each x there is a unique $b=b(x)$ such that $\mu(x,b)=0$. Similarly using $\mu(0,y)=y$ one sees the existence of a unique $a=a(y)$ such that $\mu(a,y)=0$ in an open neighborhood of 0. We let Q_o be the intersection of those two neighborhoods. Then we have

$$a=\mu(a,0)=\mu(a,\mu(x,b))=\mu(\mu(a,x),b)=\mu(0,b)=b.$$

Hence $y=a(x)=b(x)$ is the unique vector satisfying the requirements.

Given two curves α and β we can define several new curves by using the vector operations in $\mathbb{R}^n$ and the function μ:

1. We define $-\alpha$ by $(-\alpha)(t)=-\alpha(t)$. Then $-\alpha$ is a curve and $\boldsymbol{t}_{-\alpha}=-\boldsymbol{t}_\alpha$.
2. We define $\alpha+\beta$ by $(\alpha+\beta)(t)=\alpha(t)+\beta(t)$. Then $\alpha+\beta$ is a curve and $\boldsymbol{t}_{\alpha+\beta}=\boldsymbol{t}_\alpha+\boldsymbol{t}_\beta$.
3. We define $\alpha\beta$ by $\alpha\beta(t)=\mu(\alpha(t),\beta(t))$. Then $\alpha\beta$ is a curve and $\boldsymbol{t}_{\alpha\beta}=\boldsymbol{t}_\alpha+\boldsymbol{t}_\beta$.
4. We define α^{-1} by choosing $\alpha^{-1}(t)$ such that $\mu(\alpha^{-1}(t),\alpha(t))=\mu(\alpha(t),\alpha^{-1}(t))=0$. Then α^{-1} is a curve and $\boldsymbol{t}_{\alpha^{-1}}=-\boldsymbol{t}_\alpha$.

In order to see 3. we note that in the neighborhood of (0,0) the k-th coordinate of μ satisfies

$$\mu(x,y)_k=x_k+y_k+O(\|x\|^2+\|y\|^2). \tag{8}$$

In fact $\mu(0,0)=0$ and by $\mu(x,0)=x$ and $\mu(0,y)=y$ we have

$$\frac{\partial\mu_k}{\partial x_i}(0,0)=\frac{\partial x_k}{\partial x_i}=\delta_{ik}\quad\text{and}\quad\frac{\partial\mu_k}{\partial y_i}(0,0)=\frac{\partial y_k}{\partial y_i}=\delta_{ik}.$$

The fact that α^{-1} is a curve with $\boldsymbol{t}_{\alpha^{-1}}=-\boldsymbol{t}_\alpha$ also follows from (8).

Proposition 19. *If α and β are curves then $t\mapsto(\alpha\beta-\beta\alpha)(t^{\frac{1}{2}})$ defines a curve $[\alpha,\beta]$ and the k-th coordinate of $\boldsymbol{t}_{[\alpha,\beta]}$ is*

$$\sum_{i,j=1}^{n}\left((\boldsymbol{t}_\alpha)_i(\boldsymbol{t}_\beta)_j-(\boldsymbol{t}_\alpha)_j(\boldsymbol{t}_\beta)_i\right)\alpha_{k,ij}$$

where $a_{k,ij}$ is the value of $\partial^2\mu_k/\partial x_i\partial y_j$ at (0, 0).

Proof. By the definition of $a_{k,ij}$ in the neighborhood of (0,0) we have

$$\mu(x,y)_k=x_k+y_k+\sum_{i,j=1}^{n}a_{k,ij}x_iy_j+o(\|x\|^2+\|y\|^2). \tag{9}$$

Since α and β are curves we know that $\alpha(t)=(\boldsymbol{t}_\alpha)t+o(t)$ and $\beta(t)=(\boldsymbol{t}_\beta)t+o(t)$ as $t\to 0+$. Therefore

$$\mu(\alpha(t),\beta(t))_k=\alpha(t)_k+\beta(t)_k+\sum_{i,j=1}^{n}a_{k,ij}(\boldsymbol{t}_\alpha)_i(\boldsymbol{t}_\beta)_jt^2+o(t^2)$$

and a similar expression holds for $\mu(\beta(t),\alpha(t))_k$. By subtracting we obtain

$$\begin{aligned}[\alpha,\beta](t^2) &= \mu(\alpha(t),\beta(t))_k - \mu(\beta(t),\alpha(t))_k \\ &= \sum_{i,j=1}^{n} (\boldsymbol{t}_\alpha)_i(\boldsymbol{t}_\beta)_j(a_{k,ij}-a_{k,ji})\,t^2 + o(t^2).\end{aligned}$$

Note. We remark that $t \mapsto \alpha\beta\alpha^{-1}\beta^{-1}(t^{\frac{1}{2}})$ also defines a curve γ and $\boldsymbol{t}_\gamma = \boldsymbol{t}_{[\alpha,\beta]}$.

We can now return to the vector space $\mathfrak{g}$ of bound vectors $\boldsymbol{t}$. For every $\boldsymbol{t}$ there is a curve α such that $\boldsymbol{t}=\boldsymbol{t}_\alpha$ because for instance we can choose $\alpha(t)=t\cdot\boldsymbol{t}$. From Proposition 19 we see that $\boldsymbol{t}_{[\alpha,\beta]}$ depends only on the tangent vectors $\boldsymbol{t}_\alpha, \boldsymbol{t}_\beta$ and not on the curves themselves. Hence given $\boldsymbol{t}_1$ and $\boldsymbol{t}_2$ in $\mathfrak{g}$ we can choose curves α_i with tangent vectors $\boldsymbol{t}_i$ $(i=1,2)$ and define $[\boldsymbol{t}_1,\boldsymbol{t}_2]=\boldsymbol{t}_{[\alpha_1,\alpha_2]}$. The vector $[\boldsymbol{t}_1,\boldsymbol{t}_2]$ is in $\mathfrak{g}$ and it depends only on $\boldsymbol{t}_1$ and $\boldsymbol{t}_2$.

Theorem 20. *If $\mu\colon O_o\times O_o\to\mathbb{R}^n$ is a Lie group germ then the set $\mathfrak{g}$ of bound vectors $\boldsymbol{t}$ emitting from the origin of $\mathbb{R}^n$ is a Lie algebra with respect to the vector space operations of $\mathfrak{g}$ and the bracket $[\boldsymbol{t}_1,\boldsymbol{t}_2]$.*

Proof. If $\boldsymbol{t}_i$ $(i=1,2,3)$ and $\boldsymbol{t}$ are elements of $\mathfrak{g}$ then there are curves α_i $(i=1,2,3)$ and α such that $\boldsymbol{t}_i=\boldsymbol{t}_{\alpha_i}$ and $\boldsymbol{t}=\boldsymbol{t}_\alpha$. Now each Lie algebra axiom is an immediate consequence of a corresponding property of the associated curves which can be verified by considering the situation at an arbitrary $t\geqslant 0$. For instance $[\boldsymbol{t}_1+\boldsymbol{t}_2,\boldsymbol{t}_3]=[\boldsymbol{t}_1,\boldsymbol{t}_3]+[\boldsymbol{t}_2,\boldsymbol{t}_3]$ because $[\alpha_1+\alpha_2,\alpha_3]=[\alpha_1,\alpha_3]+[\alpha_2,\alpha_3]$ which can be seen from

$$\begin{aligned}&\mu(\alpha_1(t)+\alpha_2(t),\alpha_3(t))_k - \mu(\alpha_3(t),\alpha_1(t)+\alpha_2(t))_k \\ &\quad= \sum_{i,j=1}^{n} a_{k,ij}(\alpha_1(t)+\alpha_2(t))_i\alpha_3(t)_j - \sum_{i,j=1}^{n} a_{k,ij}\alpha_3(t)_i(\alpha_1(t)+\alpha_2(t))_j + o(t^2).\end{aligned}$$

Note. The Lie algebra $\mathfrak{g}$ will be called the *Lie algebra of the Lie group germ* $\mu\colon O_o\times O_o\to\mathbb{R}^n$.

Theorem 21. *The Lie algebra of the Lie group G is isomorphic to the Lie algebra $\mathfrak{g}$ of any germ of G.*

Proof. We consider the tangent vectors $\boldsymbol{t}_i$ $(i=1,\dots,n)$ in G_e which correspond to the partial derivations $\partial/\partial x_i$ $(i=1,\dots,n)$ in the local chart (Q,φ) determining the germ $(Q,\varphi), O, O_o, \mu$. The brackets $[\boldsymbol{t}_i,\boldsymbol{t}_j]$ are given in equation (4). Let $f\in C^\infty(G)$ so that

$$f\varphi^{-1}(x) = f(e) + \sum_{k=1}^{n} c_k x_k + \sum_{k,l=1}^{n} c_{kl}x_kx_l + o(\|x\|^2)$$

in the neighborhood of $0 \in \mathbb{R}^n$. By combining this approximation with (9) we obtain

$$\tilde{f}(\mu(x,y)) = \sum_{k=1}^{n} c_k(x_k+y_k) + \sum_{k=1}^{n} \sum_{i,j=1}^{n} c_k a_{k,ij} x_i y_j$$

$$+ \sum_{k,l=1}^{n} c_{kl}(x_k+y_k)(x_l+y_l) + o(\|x\|^2+\|y\|^2).$$

Therefore

$$\frac{\partial^2 f \varphi^{-1} \mu}{\partial x_i \partial y_j}(0,0) = \sum_{k=1}^{n} c_k a_{k,ij} + c_{ij} + c_{ji}.$$

Using (4) we obtain

$$[\boldsymbol{t}_i, \boldsymbol{t}_j](f) = \sum_{k=1}^{n} c_k(a_{k,ij} - a_{k,ji}).$$

Since $c_k = \boldsymbol{t}_k(f)$ we see that

(10) $$[\boldsymbol{t}_i, \boldsymbol{t}_j] = \sum_{k=1}^{n} (a_{k,ij} - a_{k,ji}) \boldsymbol{t}_k$$

for every $i,j=1,\dots,n$. We associate with each $\boldsymbol{t}_i$ the bound unit vector emitting from the origin and lying on the positive half of the coordinate axis x_i in $\mathbb{R}^n$. This determines an isomorphism of G_e onto $\mathfrak{g}$. If the bound unit vector associated with $\boldsymbol{t}_i$ is denoted by the same symbol $\boldsymbol{t}_i$ then by Proposition 19 the k-th coordinate of $[\boldsymbol{t}_i, \boldsymbol{t}_j]$ is $a_{k,ij} - a_{k,ji}$ and so $[\boldsymbol{t}_i, \boldsymbol{t}_j]$ satisfies (10). Hence our map of G_e onto is a Lie algebra isomorphism.

As an illustrative example we use Theorem 21 to prove the following result:

Theorem 22. *The Lie algebra of* $\mathrm{Sl}(n,\mathbb{R})$ *is isomorphic to* $\mathfrak{sl}(n,\mathbb{R})$, *the subalgebra consisting of those elements of* $\mathfrak{gl}(n,\mathbb{R})$ *which have zero trace.*

Proof. Consider that n^2 dimensional Euclidean vector space whose elements are the $n \times n$ real matrices $x=(x_{ij})$ and the inner product is $(x,y) = \operatorname{tr} x y^t = \sum_{i,j=1}^{n} x_{ij} y_{ij}$. The orthogonal complement of the identity matrix I in this space is a hyperplane which shall be denoted by $\mathfrak{sl}(n,\mathbb{R})$. It is the n^2-1 dimensional vector space of all $n \times n$ real matrices x with trace $\operatorname{tr} x = 0$. The image of $\mathrm{Sl}(n,\mathbb{R})$ in our vector space under the map $p \mapsto p - I$ is the hypersurface $\det(I+x)=1$ which contains the origin. The normal to this hypersurface H at 0 is I and so

$\mathfrak{sl}(n,\mathbb{R})$ is the tangent space at 0. The orthogonal projection of H onto the tangent plane is the map

$$(11) \qquad x \mapsto x - \frac{\operatorname{tr} x}{n} I.$$

We can find a sufficiently small open neighborhood N_0 of 0 in the tangent plane and a corresponding open neighborhood of 0 in H such that (11) is a one-to-one map of the second onto the first. The composition of the inverse of this map with $x \mapsto x+I$ determines an open neighborhood Q of e in $\mathrm{Sl}(n, \mathbb{R})$. The inverse of this composition is the map φ such that

$$\varphi(p) = p - \frac{\operatorname{tr} p}{n} I$$

and φ is a homeomorphism of Q onto N_0. Hence (Q,φ) is a local chart such that $e \in O$ and so it can be used to define the C^∞ structure of $\mathrm{Sl}(n,\mathbb{R})$. We consider a group germ $(Q,\varphi), O, O_o, \mu$. Hence O_o lies in $\mathfrak{sl}(n,\mathbb{R})$ and the function values $\mu(x,y)$ are $n \times n$ real matrices with zero trace.

If $x \in O_o$ then $\varphi^{-1}(x) = x - \lambda_x I$ where λ_x is the unique scalar such that $\det(x - \lambda_x I) = 1$. Hence if $x, y \in O_o$ then $\varphi^{-1}(x)\varphi^{-1}(y) = (x-\lambda_x I)(y - \lambda_y I)$ and so we see that

$$(12) \qquad \mu(x,y) = -\frac{\operatorname{tr} xy}{\mathrm{n}} I - (\lambda_y) x - (\lambda_x) y + xy.$$

Now suppose that $\boldsymbol{t}_1$ and $\boldsymbol{t}_2$ are unit vectors in $\mathfrak{sl}(n,\mathbb{R})$. Then $\alpha_i(t) = t \cdot \boldsymbol{t}_i$ defines a curve such that $\boldsymbol{t}_i = \boldsymbol{t}_{\alpha_i}$ $(i=1,2)$. Be definition we have

$$[\alpha_1, \alpha_2](t^2) = \mu(\alpha_1(t), \alpha_2(t)) - \mu(\alpha_2(t), \alpha_1(t)).$$

Hence using (12) and Proposition II.5.13 we obtain $[\alpha_1, \alpha_2](t) = (\boldsymbol{t}_1 \boldsymbol{t}_2 - \boldsymbol{t}_2 \boldsymbol{t}_1) t$ and so $[\boldsymbol{t}_1, \boldsymbol{t}_2] = \boldsymbol{t}_1 \boldsymbol{t}_2 - \boldsymbol{t}_2 \boldsymbol{t}_1$. Thus by Theorem 21 the Lie algebra of $\mathrm{Sl}(n,\mathbb{R})$ is isomorphic to the subalgebra $\mathfrak{sl}(n,\mathbb{R})$ of $\mathfrak{gl}(n,\mathbb{R})$.

Theorem 23. *The Lie algebra of* $\mathrm{O}(n,\mathbb{R})$ *is isomorphic to* $\mathfrak{o}(n,\mathbb{R})$, *the subalgebra consisting of the skew-symmetric elements of* $\mathfrak{gl}(n,\mathbb{R})$.

Proof. The method is the same which was used to determine the Lie algebra of $\mathrm{Sl}(n,\mathbb{R})$. The image H of $\mathrm{O}(n,\mathbb{R})$ under the map $p \mapsto p - I$ is the intersection of the hypersurfaces H_{ij} defined by

$$\sum_{k=1}^{n} (\delta_{ik} + x_{ik})(\delta_{jk} + x_{jk}) = 0$$

where $i=1,\ldots,j$ and $j=1,\ldots,n$. The normal to H_{ij} is the matrix $N_{ij}=\frac{1}{2}(E_{ij}+E_{ji})$ where E_{ij} has 1 in the $i-j$ place and zeros elsewhere. The elements of the tangent space of H are those matrices x which are orthogonal to all these vectors N_{ij}. This means that the tangent space, which we call $\mathfrak{o}(n,\mathbb{R})$, is the set of all skew symmetric $n\times n$ real matrices. The map

$$\varphi(p)=\tfrac{1}{2}(p-p^t) \qquad (p\in \mathrm{O}(n,\mathbb{R}))$$

is the composition of $p\mapsto p-I=x$ and all the maps $x\mapsto x-(x_{ij}+x_{ji})N_{ij}$ taken in any prescribed order. The first is one-to-one and it maps $\mathrm{O}(n,\mathbb{R})$ onto H. The remaining ones being orthogonal projections onto a succession of tangent spaces terminating in $\mathfrak{o}(n,\mathbb{R})$ are one-to-one maps if we restrict p to a sufficiently small open neighborhood Q of e in $\mathrm{O}(n,\mathbb{R})$.

Now let $\boldsymbol{t}_1,\boldsymbol{t}_2$ be unit vectors in $\mathfrak{o}(n,\mathbb{R})$, let $u_i=t\boldsymbol{t}_i$ for small $t\geqslant 0$ and let $\varphi^{-1}(u_i)=I+x_i$ for $i=1,2$. Then by the tangential property

$$x_i=u_i+o(t) \quad \text{as } t\to 0+. \tag{13}$$

Moreover $\mu(u_1,u_2)=\varphi((I+x_1)(I+x_2))=u_1+u_2+\varphi(x_1x_2)$ and so

$$\begin{aligned}[\alpha_1,\alpha_2](t^2)&=\mu(u_1,u_2)-\mu(u_2,u_1)=\varphi(x_1x_2-x_2x_1)\\&=\varphi(u_1u_2-u_2u_1)+o(t^2).\end{aligned}$$

It is easy to see that if A, B are skew symmetric $n\times n$ matrices then $AB-BA$ is also skew symmetric. Therefore $\varphi(u_1u_2-u_2u_1)=u_1u_2-u_2u_1$ and so $[\alpha_1,\alpha_2](t)=t(\boldsymbol{t}_1\boldsymbol{t}_2-\boldsymbol{t}_2\boldsymbol{t}_1)+o(t)$. This implies that $[\boldsymbol{t}_1,\boldsymbol{t}_2]=\boldsymbol{t}_1\boldsymbol{t}_2-\boldsymbol{t}_2\boldsymbol{t}_1$ i.e. multiplication in the Lie algebra $\mathfrak{o}(n,\mathbb{R})$ of $\mathrm{O}(n,\mathbb{R})$ is the same as that in $\mathfrak{gl}(n,\mathbb{R})$.

It is possible to reformulate Proposition 19 and Theorems 20 and 21 in terms of curves lying in the Lie group G. By a *curve* α in G usually we mean a differentiable curve and such curves play an important role which is discussed in the next section. At present however it is sufficient to consider maps α of a half-open interval $[0,\varepsilon)$ into G such that $\alpha(0)=e$ and α has a tangent at e; this means that if (O,φ) is a local chart containing e then $(\varphi\alpha)'(0)$ exists. If this holds for one chart (O,φ) then by the differentiability of $\psi\varphi^{-1}$ it is true for all local charts (Q,ψ) containing e. Every curve α defines a unique tangent vector $\boldsymbol{t}_\alpha$ in the tangent space G_e: For if $f\in\mathscr{f}$ where $\mathscr{f}\in\mathscr{F}_e$ then $f\mapsto(f\alpha)'(0)$ defines a linear functional $\boldsymbol{t}_\alpha$ on G_e^*. Moreover every $\boldsymbol{t}\in G_e$ can be represented in the form $\boldsymbol{t}=\boldsymbol{t}_\alpha$. In order to find a suitable α we choose a local chart (O,φ) such that $e\in O$, express $\boldsymbol{t}$ in terms of partial derivations, say

$\boldsymbol{t}=\sum_{k=1}^{n} a_k(\partial/\partial x_k)$ and let α be the image of the curve $t\mapsto(a_1 t,\ldots,a_n t)$ under φ^{-1}.

Now let α and β be curves on G and let (Q,φ) be a local chart containing e such that $\varphi(e)=0\in\mathbb{R}^n$. As earlier we consider a group germ $(Q,\varphi), O, O_o, \mu$. Then $\tilde{\alpha}=\varphi\alpha$ and $\tilde{\beta}=\varphi\beta$ are curves in $\mathbb{R}^n$ with $\tilde{\alpha}(0)=\tilde{\beta}(0)=0$. Using $\tilde{\alpha}$ and $\tilde{\beta}$ we see that $t\mapsto\alpha(t)^{-1}$ and $t\mapsto\alpha(t)\beta(t)$ define curves on G. We are going to prove that the formula

$$[\alpha,\beta](t)=\alpha\beta\alpha^{-1}\beta^{-1}(t^{\frac{1}{2}})=\alpha(t^{\frac{1}{2}})\beta(t^{\frac{1}{2}})\alpha(t^{\frac{1}{2}})^{-1}\beta(t^{\frac{1}{2}})^{-1}$$

defines a curve on G which will be called the *commutator* of the curves α, β and it will be denoted by $[\alpha,\beta]$. We remark right away that $[\alpha,\beta]$ is in general different from the curve $\varphi^{-1}[\tilde{\alpha},\tilde{\beta}]$ but both curves define the same tangent vector $\boldsymbol{t}_{[\alpha,\beta]}$ at e.

By (9) and $\mu(\tilde{\alpha}(t),\tilde{\alpha}(t)^{-1})=0$ we have

$$\tilde{\alpha}(t)_k+\tilde{\alpha}^{-1}(t)_k+\sum_{i,j=1}^{n} a_{k,ij}\tilde{\alpha}(t)_i\tilde{\alpha}^{-1}(t)_j+o(t^2)=0$$

as $t\to 0+$. A similar relation holds for $\tilde{\beta}$. Using

$$\text{(14)}\qquad \tilde{\alpha}(t)_k=a_k t+o(t) \quad\text{and}\quad \tilde{\beta}(t)_k=b_k t+o(t)$$

we obtain

$$\tilde{\alpha}(t)_k+\tilde{\beta}(t)_k+\tilde{\alpha}^{-1}(t)_k+\tilde{\beta}^{-1}(t)_k=\sum_{i,j=1}^{n} a_{k,ij}(a_i a_j+b_i b_j)t^2+o(t^2).$$

Hence applying (9) three times we see that

$$\begin{aligned}\varphi(\alpha(t)\beta(t)\alpha^{-1}(t)\beta^{-1}(t))_k &= \mu(\mu(\tilde{\alpha}(t),\tilde{\beta}(t)),\mu(\tilde{\alpha}^{-1}(t),\tilde{\beta}^{-1}(t))_k)\\ &= \sum_{i,j=1}^{n} a_{k,ij}(a_i a_j+b_i b_j)t^2+\sum_{i,j=1}^{n} a_{k,ij}\tilde{\alpha}(t)_i\tilde{\beta}(t)_j\\ &\quad+\sum_{i,j=1}^{n} a_{k,ij}\tilde{\alpha}^{-1}(t)_i\tilde{\beta}^{-1}(t)_j\\ &\quad+\sum_{i,j=1}^{n} a_{k,ij}(\tilde{\alpha}(t)_i+\tilde{\beta}(t)_i)(\tilde{\alpha}^{-1}(t)_j+\tilde{\beta}^{-1}(t)_j)+o(t^2).\end{aligned}$$

If we substitute from (14) and replace t^2 by t then we obtain

$$\varphi[\alpha,\beta]_k(t)=\sum_{i,j=1}^{n} a_{k,ij}((a_i a_j+b_i b_j)+2a_i b_j-(a_i+b_i)(a_j+b_j))\,t+o(t).$$

Therefore $\varphi[\alpha,\beta]$ has a tangent at 0 and the k-th coordinate of its tangent vector is the same as that of $[\tilde{\alpha},\tilde{\beta}]$. We proved the following:

Theorem 24. *If α and β are curves in the Lie group G with tangents $\boldsymbol{t}_\alpha$ and $\boldsymbol{t}_\beta$ at e then $[\boldsymbol{t}_\alpha,\boldsymbol{t}_\beta]=\boldsymbol{t}_{[\alpha,\beta]}$ where $[\alpha,\beta]$ is the commutator curve of α and β.*

For example if $G=\mathrm{Sl}(2,\mathbb{R})$ then the curves

$$\alpha_1(t)=\begin{pmatrix}1 & t\\ 0 & 1\end{pmatrix},\quad \alpha_2(t)=\begin{pmatrix}1 & 0\\ t & 1\end{pmatrix}\quad \text{and}\quad \alpha_3(t)=\begin{pmatrix}1+t & 0\\ 0 & (1+t)^{-1}\end{pmatrix}$$

define three tangent vectors $\boldsymbol{t}_1$, $\boldsymbol{t}_2$ and $\boldsymbol{t}_3$ at e which form a basis for the tangent space at e. We find that

$$[\alpha_1,\alpha_2](t^2)=\begin{pmatrix}1+t^2+t^4 & -t^3\\ t^3 & 1-t^2\end{pmatrix}$$

and so $[\boldsymbol{t}_1,\boldsymbol{t}_2]=\boldsymbol{t}_3$. Similarly $[\boldsymbol{t}_1,\boldsymbol{t}_3]=-2\boldsymbol{t}_1$ and $[\boldsymbol{t}_2,\boldsymbol{t}_3]=2\boldsymbol{t}_2$. If we let

$$A_1=\begin{pmatrix}0 & 1\\ 0 & 0\end{pmatrix},\quad A_2=\begin{pmatrix}0 & 0\\ 1 & 0\end{pmatrix}\quad \text{and}\quad A_3=\begin{pmatrix}1 & 0\\ 0 & -1\end{pmatrix}$$

then $\boldsymbol{t}_i\mapsto A_i$ $(i=1,2,3)$ gives an isomorphism of the Lie algebra of $\mathrm{Sl}(2,\mathbb{R})$ onto $\mathfrak{sl}(2,\mathbb{R})$. We note that the curves $\alpha_1,\alpha_2,\alpha_3$ define one parameter subgroups of $\mathrm{Sl}(2,\mathbb{R})$. The use of such special curves in the determination of the Lie algebra of a Lie group is of definite advantage. The existence of sufficiently many such subgroups to determine the Lie algebra of G will be proved in the next section.

Two Lie groups G and H are called *isomorphic* if there is a group isomorphism Φ of G onto H such that Φ and Φ^{-1} are both diffeomorphisms. More generally a C^∞ group homomorphism $\Phi\colon G\to H$ is called a *Lie group homomorphism* or simply a *homomorphism*. These maps Φ are the morphisms of the set $\mathrm{Hom}(G,H)$ of the category of Lie groups.

Let G be a Lie group and let $\mathfrak{g}$ be its Lie algebra. We call G *semisimple* if $\mathfrak{g}$ is a semisimple Lie algebra. Similarly G is called *nilpotent* or *solvable* according as $\mathfrak{g}$ is a nilpotent or solvable Lie algebra. The homomorphic image of a nilpotent Lie group is nilpotent and similarly the image of a solvable group is solvable. These results follow from the corresponding Lie algebra results and from the fact that if $\Phi\colon G\to H$ is a surjective homomorphism then the differential $d\Phi_e$ is a homomorphism of $\mathfrak{g}$ onto $\mathfrak{h}$. Later we shall define what is meant by a Lie subgroup H of a

Lie group G. (See Definition 6.10.) Then using Proposition 7.2 one can immediately show that the Lie subgroups of a nilpotent Lie group are nilpotent. A similar result holds for solvable Lie groups too.

Let S be a Lie subgroup of $\mathrm{Gl}(n, \mathbb{R})$ and suppose that the elements of S are all upper triangular matrices. Then one can prove that S is solvable. If in addition the diagonal entries of these matrices are all equal to 1 then S is nilpotent.

6. The Exponential Map and Canonical Coordinates

The first main object of the following discussion is to define a mapping of the Lie algebra $\mathfrak{g}$ associated with the Lie group G into the group G. This map is called the *exponential map* and the image of $X \in \mathfrak{g}$ under it is denoted by $\exp X$. The real, finite dimensional Lie algebra is a C^∞ manifold. We shall see that $\exp 0 = e$, the identity of G and $X \mapsto \exp X$ is a C^∞ diffeomorphism of a sufficiently small open neighborhood of 0 in $\mathfrak{g}$ onto an open neighborhood O of e in G. The inverse of this diffeomorphism is the *logarithmic map* and $(O, \log)$ is called the *canonical* or *normal coordinate neighborhood* of e in G. Often one speaks also about the *logarithmic local chart* and about *logarithmic coordinates*. The importance of exp will be seen when we shall investigate the relation between subalgebras of $\mathfrak{g}$ and Lie subgroups of G in the light of the exponential map.

We start with a few elementary facts about differentiable curves on C^∞ manifolds. First of all by a curve we shall understand a parametrized curve and not only a set of points lying in the manifold M. Hence a *curve* γ is a map $t \mapsto \gamma(t)$ of an open interval I of $\mathbb{R}$ into M. By a *segment* of γ we mean the restriction of $\gamma: I \to M$ to a closed subinterval of I. Thus $\gamma: [a,b] \to M$ is a curve segment only if it has an extension to an open interval containing $[a,b]$. The curve γ is called *infinitely differentiable* or C^∞ if γ is a C^∞ map: This means that for every local chart (O, φ) of M the composition $\varphi\gamma$ is a C^∞ map of I into $\mathbb{R}^n$. It is a special case of the concept of a C^∞ map $\Phi: M \to N$ of one C^∞ manifold M into another N. The latter means that $\psi\Phi\varphi^{-1}$ is a C^∞ map of $\mathbb{R}^m$ into $\mathbb{R}^n$ for each pair of coordinate neighborhoods (O, φ) and (Q, ψ) in M and N, respectively. We shall always tacitly suppose that all curves under consideration are infinitely differentiable.

Since our curves γ are differentiable for every t in I there is a unique tangent to γ at $m = \gamma(t)$ which one denotes by $\gamma'(t)$, $\gamma_*(t)$, $\dot{\gamma}(t)$ or

$(d\gamma/dt)|_t$: Namely for every $\mathscr{f}\in\mathscr{F}_m$ one lets

$$\gamma'(t)(\mathscr{f})=(f\gamma)'(t)=\left.\frac{df\gamma}{dt}\right|_t$$

where f is an arbitrary function in the germ $\mathscr{f}$. This defines a linear functional $\gamma'(t)\colon \mathscr{F}_m/\mathscr{F}_m^o\to\mathbb{R}$ and so $\gamma'(t)$ is an element of the tangent space $M_m=(\mathscr{F}_m/\mathscr{F}_m^o)^*$. We can regard $\gamma'(t)(f)$ as the directional derivative of f in the direction of γ at t which we identify with the direction of the tangent vector $\gamma'(t)$. It is obvious that

$$\gamma'(t)(fg)=\gamma'(t)(f)g(m)+f(m)\gamma'(t)(g)$$

for every pair of differentiable functions f and g defined in the neighborhood of $m=\gamma(t)$. If $\gamma'(t)\neq 0$ then we say that the curve γ has the tangent vector $\gamma'(t)$ at $\gamma(t)$. Briefly we say that γ has a tangent at $\gamma(t)$. One also speaks about the existence of a tangent at the parameter value t.

Lemma 1. *If $\gamma\colon I\to M$ is an infinitely differentiable curve and $g\colon J\to I$ is an infinitely differentiable map of the open interval J into I then $\gamma g\colon J\to M$ is an infinitely differentiable curve and $(\gamma g)_*(\tau)=g'(\tau)\gamma_*(t)$ with $t=g(\tau)$ for every τ in J.*

Proof. If $\mathscr{f}\in\mathscr{F}_{\gamma(t)}$ and $f\in\mathscr{f}$ then by definition $(\gamma g)_*(\tau)(f)=(f\gamma g)'(\tau)$ and so by the chain rule

$$(\gamma g)_*(\tau)(f)=g'(\tau)(f\gamma)'(t)=g'(\tau)\gamma_*(t)(f).$$

Corollary. *If $g'(\tau)\neq 0$ and γ has a tangent at $t=g(\tau)$ then γg has a tangent at τ.*

We note in passing that $t\mapsto(\gamma(t),\gamma'(t))$ maps the open interval I into the set

$$T(M)=\{(m,\boldsymbol{t})\colon m\in M \text{ and } \boldsymbol{t}\in M_m\}$$

which is called the *tangent manifold* of M. For $T(M)$ can be given a C^∞ structure as follows: Let $\pi\colon T(M)\to M$ be the map such that $\pi((m,\boldsymbol{t}))=m$ for all $m\in M$ and $\boldsymbol{t}\in M_m$. Every coordinate neighborhood (O,φ) of M determines a coordinate neighborhood (Q,ψ) of $T(M)$; namely $Q=\pi^{-1}(O)$ and

$$\psi(m,t)=(x_1,\ldots,x_n,y_1,\ldots,y_n)$$

where $x_1,\ldots,x_n$ are the local coordinates of m in the system (O,φ) and $y_1,\ldots,y_n$ are the coordinates of $\boldsymbol{t}$ in M_m with respect to the basis

$\partial/\partial x_1, \ldots, \partial/\partial x_n$ of M_m. It is not hard to verify that $t \mapsto (\gamma(t), \gamma'(t))$ is a C^∞ curve in the tangent manifold $T(M)$. This fact however is not essential for our purposes.

We shall now define what is meant by a geodesic or a geodesic curve. This concept is meaningful for those differentiable manifolds for which one can speak about two tangent vectors being parallel. This always holds for Lie groups and we shall consider geodesics only in this case. The definition involves only the function $m \mapsto \gamma'(m)$ $(m \in \gamma(I))$ which can be regarded as a vector field defined on the subset $\gamma(I)$ of M. If $a \in G$ and $t \in G_e$ then at is defined by $at(f) = t({}_a f)$ where f varies over the elements of $\mathcal{F}_a$. Therefore at is a tangent vector belonging to G_a and $t \mapsto at$ is an isomorphism of G_e onto G_a. We could more generally consider the map $t \mapsto at$ where $t \in G_p$ for some $p \in G$ and obtain an isomorphism of G_p onto G_{ap}. Then $(p, t) \mapsto (ap, at)$ defines an action of the group G on the tangent manifold $T(G)$.

Definition 2. *A C^∞ curve $\gamma: I \to G$ on the Lie group G is called a geodesic if*

$$\gamma(t_1)^{-1} \gamma'(t_1) = \gamma(t_2)^{-1} \gamma'(t_2)$$

for every $t_1, t_2 \in I$.

It is obvious that if we have

$$\gamma'(t) = \gamma(t)\gamma(t_0)^{-1}\gamma'(t_0) \tag{1}$$

for some fixed $t_0 \in I$ and every $t \in I$ then γ is a geodesic. In particular if $0 \in I$ and $\gamma(0) = e$ then

$$\gamma'(t) = \gamma(t)\gamma'(0) \qquad (t \in I) \tag{2}$$

is a necessary and sufficient condition that γ be a geodesic curve. We see that a geodesic has a tangent vector at each of its points and these tangents are interrelated in a simple manner.

We shall now prove that the geodesic curve $\gamma: I \to G$ is essentially uniquely determined by the set of points $\gamma(t)$ $(t \in I)$. More precisely we prove the following:

Lemma 3. *The parametrization of a geodesic curve is unique within a linear change of variable, say $t = a\tau + b$ where a and b are reals and $a \neq 0$.*

Proof. Let $\gamma: I \to G$ be a geodesic curve and let $g: J \to I$ be an infinitely differentiable function such that $\gamma g: J \to G$ is also a geodesic. Then g' does not vanish and g is invertible. Let $\tau_0, \tau \in J$ and let $t_0 = g(\tau_0)$ and

$t=g(\tau)$. By Lemma 1 and (1) we obtain

$$(\gamma g)_*(\tau)=g'(\tau)\gamma_*(t)=g'(\tau)\gamma(t)\gamma(t_0)^{-1}\gamma_*(t_0).$$

On the other hand γg being also a geodesic by (1) we have

$$(\gamma g)_*(\tau)=\gamma g(\tau)\gamma g(\tau_0)^{-1}(\gamma g)_*(\tau_0)=\gamma(t)\gamma(t_0)^{-1}(\gamma g)_*(\tau_0).$$

Using Lemma 1 we obtain

$$(\gamma g)_*(\tau)=\gamma(t)\gamma(t_0)^{-1}g'(\tau_0)\gamma_*(t_0).$$

Therefore combining these two expressions of $(\gamma g)_*(\tau)$ we obtain

$$(g'(t)-g'(t_0))\gamma(t)\gamma(t_0)^{-1}\gamma_*(t_0)=0.$$

Since $\gamma_*(t_0)\neq 0$ it follows that $g'(t)=g'(t_0)$.

A geodesic $\gamma: I\to G$ is called *maximal* if it is not part of a larger geodesic $\delta: J\to G$, that is to say $\gamma(I)\subseteq\delta(J)$ implies that $\gamma(I)=\delta(J)$. The last result shows that if $I=(-\infty,\infty)$ then the geodesic $\gamma: I\to G$ is maximal. It is also meaningful to speak about a *finite geodesic* $\gamma: I\to G$; this means that I is a finite interval. We see that every parametric representation of γ will remain finite.

Theorem 4. *If G is a Lie group with Lie algebra $\mathfrak{g}$ and $\boldsymbol{t}\in\mathfrak{g}$ then there is a unique maximal geodesic $\gamma:(-\infty,\infty)\to G$ such that $\gamma(0)=e$ and $\gamma'(0)=\boldsymbol{t}$. The map $(\boldsymbol{t},t)\mapsto\gamma(t)$ is a C^∞ map on $\mathfrak{g}\times\mathbb{R}$ with values in G.*

Note. The elements of $\mathfrak{g}$ are interpreted at present as tangent vectors $\boldsymbol{t}$ at e. The dependence of γ on $\boldsymbol{t}$ can be indicated by using the notation $\gamma_{\boldsymbol{t}}$ instead of γ.

By Proposition 5.4 the tangent vector $\boldsymbol{t}$ determines a left invariant vector field $X(\cdot)$ on G. By (2) the geodesic γ whose existence is stated above is a C^∞ curve such that $\gamma'(t)=\gamma(t)X(e)$ for every $t\in(-\infty,\infty)$. Since $X(\cdot)$ is a left invariant $\gamma(t)X(e)=X(\gamma(t))$ and so γ is a geodesic satisfying $\gamma(0)=e$ and $\gamma'(0)=\boldsymbol{t}$ if and only if $\gamma(0)=e$ and $\gamma'(t)=X(\gamma(t))$ for every t in $(-\infty,\infty)$. Therefore an alternate formulation of Theorem 2 states the existence of a unique integral curve γ for the left invariant vector field $X(\cdot)$ determined by the tangent vector $\boldsymbol{t}$ such that $\gamma(0)=e$. In order to prove this modified result first we look at the problem locally by choosing a coordinate neighborhood (Q,φ) containing e.

Let $(Q,\varphi), O, O_o, \mu$ be a germ of the Lie group G and let

$$\boldsymbol{t}=c_1\frac{\partial}{\partial x_1}+\cdots+c_n\frac{\partial}{\partial x_n}. \tag{3}$$

We let the vector emitting from the origin correspond to the tangent vector $\boldsymbol{t}$. Next we determine the vector which corresponds to $X(p)$ where $p \in O$. Let $f: G \to \mathbb{R}$ be a C^∞ function and let $\tilde{f} = f\varphi^{-1}$. Then for any p and q in O we have

$$_p f(q) = {_p}(\tilde{f}\varphi)(q) = \tilde{f}\varphi(pq) = \tilde{f}\varphi(\varphi^{-1}(x), \varphi^{-1}(y))$$

where $x = \varphi(p)$ and $y = \varphi(q)$. Hence $_p f(q) = \tilde{f}\mu(x, y)$ for all $x, y \in O$ and so

$$X(p)f = X(e)(_p f) = \sum_{k=1}^{n} c_k \frac{\partial \tilde{f}\mu(x, \cdot)}{\partial y_k}\bigg|_{y=0}. \tag{4}$$

By the chain rule we have

$$\frac{\partial \tilde{f}\mu}{\partial y_k}\bigg|_0 = \sum_{l=1}^{n} \tilde{f}_l(0) \frac{\partial \mu(x, \cdot)_l}{\partial y_k}\bigg|_{y=0} \tag{5}$$

where $\tilde{f}_l(0)$ is the l-th partial derivative of $\tilde{f}$ computed at $0 \in \mathbb{R}^n$ and the second factor is the partial of the l-th component of μ with respect to the k-th variable in the second group of variables computed at $y = 0$. By (4) and (5) we have

$$X(p)f = \sum_{l=1}^{n} \left(\sum_{k=1}^{n} c_k \frac{\partial(x, \cdot)_l}{\partial y_k}\bigg|_0 \right) \tilde{f}_l(0).$$

Therefore the vector corresponding to $X(p)$ in the germ $(Q, \varphi), \ldots, \mu$ is

$$\left(\ldots, \sum_{k=1}^{n} c_k \frac{\partial(x, \cdot)_l}{\partial y_k}\bigg|_0, \ldots \right) \tag{6}$$

with starting point at x. Since the germ belongs to the Lie group G the function is C^∞ and so the vector field defined by (6) in the open set O lying in $\mathbb{R}^n$ is also C^∞. We can now solve the problem of finding γ locally in the group germ $(Q, \varphi), \ldots, \mu$ by using the following result from the theory of ordinary differential equations:

Theorem 5. *Let O be an open set in $\mathbb{R}^n$ containing the origin 0 and let $x \mapsto X(x)$ be a C^∞ vector field in O. Then there is an interval $(-\varepsilon, \varepsilon)$ and a C^∞ curve $\gamma: (-\varepsilon, \varepsilon) \to O$ such that $\gamma(0) = 0$ and $d\gamma/dt = X(\gamma(t))$ for every $t \in (-\varepsilon, \varepsilon)$. The curve γ is uniquely determined in the following sense: If γ_1 and γ_2 are both solutions then $\gamma_1(t) = \gamma_2(t)$ for every t in the intersection of their domains of definition.*

Proof of Theorem 4. Let $(Q, \varphi), O, O_0, \mu$ be a germ of G, let $x \mapsto \tilde{X}(x)$ be the vector field defined on O by (6) and let $\tilde{\gamma}: (-\varepsilon, \varepsilon) \to O$ be the C^∞

curve satisfying $\tilde{\gamma}(0)=0$ and $d\tilde{\gamma}/dt=\tilde{X}(\tilde{\gamma}(t))$; the existence of $\tilde{\gamma}$ is assured by Theorem 5. We let $\alpha=\varphi^{-1}\tilde{\gamma}$ so that $\alpha:(-\varepsilon,\varepsilon)\to G$ is a C^∞ curve on G with $\alpha(0)=e$ and such that $\alpha'(0)=\boldsymbol{t}$. Using α we shall construct γ as the limit of a sequence of geodesics γ_n $(n=0,1,2,\ldots)$ each γ_n being an extension of the ones preceeding it. We let $\gamma_0=\alpha$ and choose δ such that $\varepsilon/2<\delta<\varepsilon$.

Now suppose that $n\geqslant 0$ and $\gamma_n(t)$ is already defined for $-(n-1)\delta-\varepsilon<t<(n-1)\delta+\varepsilon$ such that $\gamma_n'(t)=\gamma_n(t)\boldsymbol{t}$ in its domain of definition. We define

$$\gamma_{n+1}(t)=\begin{cases}\gamma_n(-n\delta)\alpha(t+n\delta) & \text{for } -n\delta-\varepsilon<t<-n\delta\\ \gamma_n(t) & \text{for } -n\delta\leqslant t\leqslant n\delta\\ \gamma_n(n\delta)\alpha(t-n\delta) & \text{for } n\delta<t<n\delta+\varepsilon\,.\end{cases}$$

Then we have for $n\delta<t<n\delta+\varepsilon$

$$\begin{aligned}\gamma_{n+1}'(t)=\gamma_n(n\delta)\alpha'(t-n\delta)&=\gamma_n(n\delta)(\alpha(t-n\delta)X(e))\\ &=(\gamma_n(n\delta)\alpha(t-n\delta))\boldsymbol{t}=\gamma_{n+1}(t)\boldsymbol{t}\,.\end{aligned}$$

Similarly we see that $\gamma_{n+1}'(t)=\gamma_{n+1}(t)\boldsymbol{t}$ holds also for every t satisfying $-n\delta-\varepsilon<t<-n\delta$. The same can be proved at the values $t=-n\delta$ and $t=n\delta$ by considering the left and right derivatives of γ_{n+1} there. The curve γ_{n+1} is an extension of γ_n by the uniqueness of $\tilde{\gamma}$ and α as described in Theorem 5. Hence $\gamma:(-\infty,\infty)\to G$ can be defined by $\gamma(t)=\gamma_n(t)$ where n is any natural number satisfying $|t|<n\delta$.

The uniqueness of γ is proved as follows: Let $\gamma_i:(-\infty,\infty)\to G$ $(i=1,2)$ be geodesic curves such that $\gamma_i(0)=e$ and $\gamma_i'(0)=\boldsymbol{t}$ for $i=1,2$. We let S be the set of those t in $(-\infty,\infty)$ for which $\gamma_1(t)=\gamma_2(t)$. By hypothesis we have $0\in S$ and S is obviously a closed set. Since the solution of the local problem treated in Theorem 5 is unique $t\in S$ implies that the values of γ_1 and γ_2 coincide in a small neighborhood of t. This shows that S is an open set. Since S is a non-void open and closed subset of the real line it follows that $S=(-\infty,\infty)$. Thus $\gamma_1(t)=\gamma_2(t)$ for all $t\in(-\infty,\infty)$.

Let G be any Lie group and let $\mathfrak{g}$ be its Lie algebra. For any left invariant vector field $X\in\mathfrak{g}$ we define $\exp X=\gamma(1)$ where $\gamma:(-\infty,\infty)\to G$ is the unique maximal geodesic such that $\gamma(0)=e$ and $\gamma'(0)=X(e)$. The existence and uniqueness of γ is assured by Theorem 4. Hence $X\to\exp X$ is a well defined map of the Lie algebra $\mathfrak{g}$ into the Lie group G which is called the *exponential map*.

Lemma 6. *Let M be a differentiable manifold and let $\gamma: I\to M$ be a curve which has a tangent $\gamma'(t_0)$ at $\gamma(t_0)$ for some $t_0\in I$. Let $\tau\mapsto g(\tau)$ be a real valued function of the real variable τ such that $g(\tau_0)=t_0$ and $g'(\tau_0)$*

exists. Then $\alpha(\tau)=\gamma g(\tau)$ *defines a curve in the neighborhood of* τ_0 *which has tangent* $\alpha'(\tau_0)=g'(\tau_0)\gamma'(t_0)$ *at* $\alpha(t_0)$.

Proof. Let the real valued C^1 function f be defined in the neighborhood of $\gamma(t_0)$. Then we have

$$g'(\tau_0)(f\gamma)'(t_0)=(f\gamma g)'(\tau_0)=(f\alpha)'(\tau_0)=\alpha'(\tau_0)(f).$$

Lemma 7. *If* $X\in\mathfrak{g}$ *and* $\gamma:(-\infty,\infty)\to G$ *is the geodesic curve such that* $\gamma(0)=e$ *and* $\gamma'(0)=X(e)$ *then* $\gamma(t)=\exp tX$ *for every* t *in* $(-\infty,\infty)$.

Proof. Let s be a fixed real number and let $\alpha(t)=\gamma(st)$ for all t in $(-\infty,\infty)$. Then by Lemma 6 we have

$$\alpha'(t)=s\gamma'(st)=s\gamma(st)X(e)=\alpha(t)(sX)(e).$$

Choosing $t=0$ we obtain $\alpha'(0)=\alpha(0)(sX)(e)=sX(e)$. Hence α is the geodesic satisfying $\alpha(0)=e$ and $\alpha'(0)=sX(e)$. Therefore we have $\exp sX=\alpha(1)=\gamma(s)$ where $s\in(-\infty,\infty)$ is arbitrary.

Lemma 8. *Let* G *be a Lie group and let* $\gamma: I\to G$ *be a curve which has tangent* $\gamma'(t_0)$ *at* $\gamma(t_0)$ *for some* $t_0\in I$. *Then for every* $a\in G$ *the curve* $t\mapsto a\gamma(t)$ *has tangent at* $a\gamma(t_0)$ *and it is* $a\gamma'(t_0)$.

Proof. If f is any real valued C^1 function defined in the neighborhood of $a\gamma(t_0)$ then ${}_af$ is defined in the neighborhood of $\gamma(t_0)$ and we have

$$\gamma_*(t_0)({}_af)=({}_af\gamma)'(t_0)=(fa\gamma)'(t_0)=(a\gamma)_*(t_0)(f).$$

The result follows by noticing that the left hand side is $a\gamma_*(t_0)(f)$.

The importance of the exponential map becomes obvious from the following result:

Theorem 9. *For every Lie group* G *and every element* X *of its Lie algebra* $\mathfrak{g}$ *the map* $t\mapsto\exp(tX)$ *is a homomorphism of the additive group* $\mathbb{R}^+$ *into* G.

Note. The image group is the geodesic γ satisfying $\gamma(0)=e$ and $\gamma'(0)=X(e)$. It is called the *one parameter subgroup* generated by the infinitesimal generator X.

Proof. If we fix the real s and let $\alpha(t)=(\exp sX)^{-1}\exp((s+t)X)$ for all $t\in(-\infty,\infty)$ then by Lemmas 6, 7 and 8

$$\alpha'(t)=(\exp sX)^{-1}\frac{d}{dt}\exp(s+t)X=(\exp sX)^{-1}(\exp(s+t)X)X(e)$$

and so $\alpha'(t)=\alpha(t)X(e)$ for all t in $(-\infty, \infty)$. Hence α is the unique geodesic satisfying $\alpha(0)=e$ and $\alpha'(0)=X(e)$. Therefore by Lemma 7

$$(\exp s X)^{-1}\exp(s+t)X=\exp t X$$

where s and t are arbitrary real numbers.

Since $t\mapsto\exp t X$ is a geodesic it is a differentiable curve and so it is a locally bicontinuous map of $(-\infty, \infty)$ into G which can be used to introduce a C^∞ structure on the image group $\exp t X(-\infty<t<\infty)$. Therefore the one parameter subgroup generated by X is equivalent to $\mathbb{R}^+$ or $T=S^1$ according as the kernel of $t\mapsto\exp t X$ is trivial or not. In any case $\exp t X$ is a commutative, one dimensional, connected Lie group. It is important to realize that the topology induced on the group $\exp t X$ by the topology of G need not be the one which comes from the manifold structure of the Lie group $\exp t X$. We shall give an example: For instance let G be the torus obtained from the plane $\mathbb{R}^2$ by identification modulo integers i.e. let $G=\mathbb{R}^2/\mathbb{Z}^2$. We realize G as a point set lying in $\mathbb{R}^2$ by identifying it with the square $0\leqslant x, y<1$. We consider $t\mapsto(t+\mathbb{Z}, \alpha t+\mathbb{Z})$ where $t\in(-\infty, \infty)$ and α is irrational. The image of this one parameter subgroup in the unit square $0\leqslant x, y<1$ is a dense family of parallel straight line segments with slope α. Every open planar neighborhood of a point on one of these segments intersects an infinity of the other segments and so the induced topology is not the topology of the one parameter subgroup in question. An alternate way to see the difference is by noticing that the subgroup is homeomorphic to the real line and the group G is compact. If the slope α is rational then the topology of the one parameter subgroup is the one induced by G.

Theorem 9 has a number of trivial consequences such as the formula

$$(\exp t_1 X)(\exp t_2 X)=(\exp t_2 X)(\exp t_1 X) \tag{7}$$

where $t_1, t_2\in\mathbb{R}$ and $(\exp X)^n=\exp n X$ where $n=0, \pm 1, \pm 2, \ldots$ We note that for $a\in G$ and $\boldsymbol{t}\in G_p$ we can define $\boldsymbol{t}a$ by $\boldsymbol{t}a(\not{f})=\boldsymbol{t}(\not{f}_a)$ where $\not{f}$ varies over $\mathscr{F}_{pa}$. This is in complete analogy with the definition of the left action of G on $T(G)$. By applying Lemma 8 and its right sided analogue to the identity $(\exp X)(\exp t X)=(\exp t X)(\exp X)$ by Lemma 7 we obtain $(\exp X)\gamma(t)X(e)(\exp X)$. By the invariance of the vector field X we have $\gamma(t)X(e)=X(\gamma(t))$ and so we see that

$$(\exp X)X(\gamma(t))=X(\gamma(t))(\exp X) \tag{8}$$

for every real t and every X in the Lie algebra of G. Parallelism and geodesics can also be considered from the right and one can study the right handed analogue of Theorem 4. The result is very simple because (8) shows that the unique curve $\gamma: (-\infty, \infty) \to G$ satisfying $\gamma(0) = e$ and $\gamma'(t) = X(e)\gamma(t)$ is $\gamma(t) = \exp tX$.

Our next main object is to prove that there is a natural one-to-one correspondence between the connected Lie subgroups H of a Lie group G and the Lie subalgebras $\mathfrak{h}$ of its Lie algebra $\mathfrak{g}$. This requires some preparation consisting of various definitions and preliminary results including the existence of canonical neighborhoods which were already mentioned earlier.

Let M and N be differentiable manifolds of the same class and let Φ be a differentiable map of M into N. In the beginning of Section 5 we defined the differential of Φ at a point $p \in M$ and we denoted it by $d\Phi_p$. The map $\Phi: M \to N$ is called *regular* at p provided Φ is differentiable at p and $d\Phi_p$ is an injection of M_p into $N_{\Phi(p)}$. If this is not the case we say that Φ is *singular* at p. We recall that the differentiability of Φ at p means that $\Phi \varphi^{-1} \in C^k$ for some and hence all coordinate neighborhoods containing p where k is the class of M and N. We are primarily interested in the case $k = \infty$.

Using the concept of regularity we can now define submanifolds. We start with ordinary submanifolds: Let M and N be differentiable manifolds of the same class and let M be a subset of N. We call M a *submanifold* of N provided the inclusion map $I: M \to N$ is regular at each point of M. More generally we can consider the situation when the manifold M is not necessarily a subset of the manifold N but instead a regular bijection $\Phi: M \to N$ is given. The differentiable structure of M can be transferred onto $\Phi(M)$ by using Φ, namely the local charts $\Phi(M)$ are $(\Phi(O), \varphi \Phi^{-1})$ where (O, φ) is an arbitrary local chart of M. We say that $\Phi(M)$ is a *submanifold of N with respect to the injection* $\Phi: M \to N$ or briefly we say that $\Phi: M \to N$ *is a submanifold of* N. Thus an ordinary submanifold M is a submanifold with respect to the inclusion map I. Since Φ is regular it is continuous and so the topology defined on $\Phi(M)$ by its manifold structure is at least as strong as the one induced on it by the topology of N.

If G is a Lie group and $\gamma: (-\infty, \infty) \to G$ is a one parameter subgroup of G then the image group $\gamma(-\infty, \infty)$ is a submanifold of G with respect to the map γ. We have seen that the topology of the image group can be strictly stronger than the one induced on it by G. Whenever we shall speak about a *one parameter subgroup* we shall always mean the map γ defining the image group. Thus if we say that H is a one parameter subgroup of G then there is given, perhaps tacitly, a geodesic $\gamma: I \to G$ with image set $\gamma(I) = H$.

Definition 10. *A subgroup H of the Lie group G is called a Lie subgroup of G provided H is a submanifold of G and H is a Lie group with respect to its manifold structure and the group structure inherited from G.*

One expects that if $\Phi: M \to N$ is a submanifold of N then there is some simple relation between the local charts of M and those of N. In fact we have the following:

Lemma 11. *Let $\Phi: M \to N$ be a submanifold of N and let $p \in M$. Then there is a local chart (Q, ψ) containing $\Phi(p)$ in N such that $\psi\Phi(p) = 0 \in \mathbb{R}^n$ and the open set*

$$O = \{q : q \in Q \text{ and } \psi(q)_{m+1} = \cdots = \psi(q)_n = 0\}$$

together with $\varphi = \psi | O$ is a local chart of $\Phi(M)$.

Proof. We choose a coordinate neighborhood (O_p, y) in M such that p belongs to O_p and $y(p) = 0 \in \mathbb{R}^m$. The local coordinates will be denoted by $y_1, \ldots, y_m$ so that $y_1(p) = \cdots = y_m(p) = 0$. Similarly we choose a coordinate neighborhood $(O_{\Phi(p)}, z)$ in N such that $\Phi(p)$ lies in $O_{\Phi(p)}$ and $z(\Phi(p)) = 0 \in \mathbb{R}^n$. The local coordinates in this case will be $z_1, \ldots, z_n$ so that $z_1(\Phi(p)) = \cdots = z_n(\Phi(p)) = 0$. We let $f = z\Phi y^{-1}$ so that

$$z_j = f(y_1, \ldots, y_m)_j \qquad (1 \leqslant j \leqslant n) \tag{9}$$

are differentiable functions defined on the open set $y(O_p)$ lying in $\mathbb{R}^m$. The meaning of differentiability depends on the common class of M and N, but since we are primarily interested in the case of C^∞ manifolds we can suppose that f is a C^∞ map. Since Φ is regular at p the rank of the Jacobian matrix $(\partial z_j / \partial y_i)$ $(1 \leqslant i \leqslant m$ and $1 \leqslant j \leqslant n)$ is m in the neighborhood of $0 \in \mathbb{R}^m$. By changing the ordering of the coordinates y_i and z_j, if this is necessary, we can achieve that the determinant of the minor $(\partial z_j / \partial y_i)$ $(1 \leqslant i, j \leqslant m)$ is different from zero at the origin $0 \in \mathbb{R}^m$. By the implicit function theorem there exists an inverse map. Thus there are m differentiable functions

$$y_i = g_i(z_1, \ldots, z_m) \qquad (1 \leqslant i \leqslant m) \tag{10}$$

defined in a small neighborhood of $0 \in \mathbb{R}^m$ such that $(y_1, \ldots, y_m) \mapsto (z_1, \ldots, z_m)$ is the inverse of $(z_1, \ldots, z_m) \mapsto (y_1, \ldots, y_m)$. In a sufficiently small open neighborhood Q of $\Phi(p)$ in N we can now define the map ψ with values in $\mathbb{R}^n$ by letting $\psi(q)_i = z(q)_i$ for $1 \leqslant i \leqslant m$ and

$$\psi(q)_i = z(q)_i - f(\ldots, g_i(z(q)_1, \ldots, z(q)_m), \ldots)_i$$

for $m < j \leqslant n$. If $q = \Phi(o)$ for some $o \in O_p$ and q is such that ψ is defined then by (10) we see that $g_i(z(q)_1, \ldots, z(q)_m)$ is y_i, the i-th coordinate of o. Hence by (9) $f(\ldots, g_i(z(q)_1, \ldots, z(q)_m), \ldots)$ is $z(q)_i$ and so we have $\psi(q)_i = 0$ for this point q and all indices i satisfying $m < i \leqslant n$. For the sake of simplicity let $x_i = \psi(q)_i$ for $1 \leqslant i \leqslant n$. Then using $x_i = z_i$ $(1 \leqslant i \leqslant m)$ and $x_i = z_i - f(\ldots, g_i(z_1, \ldots, z_m), \ldots)$ we see that $\partial x_i / \partial y_i = 1$ for $1 \leqslant i \leqslant n$ and $\partial x_i / \partial y_j \neq 0$ can hold only if $m < i \leqslant n$ and $1 \leqslant j \leqslant m$. Therefore the determinant of the Jacobian matrix $(\partial x_i / \partial y_j)$ $(1 \leqslant i, j \leqslant n)$ is 1 and so Q can be chosen so small that $\psi\colon Q \to \mathbb{R}^n$ is a one-to-one map. Therefore (Q, ψ) is a coordinate neighborhood in N such that $\Phi(p) \in Q$ and $\psi(\Phi(p)) = 0 \in \mathbb{R}^n$.

Since $g_1, \ldots, g_m$ are continuous at $0 \in \mathbb{R}^m$ making Q smaller we can achieve that $q \in Q$ implies that

$$(g_1(z_1, \ldots, z_m), \ldots, g_m(z_1, \ldots, z_m)) \in y(O_p)$$

where $z_i = z(q)_i$ for $1 \leqslant i \leqslant n$. Hence there is a unique point o in O_p such that $y_i = y(o)_i = g_i(z_1, \ldots, z_m)$ for $1 \leqslant i \leqslant m$. Now if $q \in Q$ is such that $x_i = \psi(q)_i = 0$ for $m < i \leqslant n$ then we have

$$z_i = f(\ldots, g_i(z_1, \ldots, z_m), \ldots)_i = f(\ldots, y_i, \ldots)_i$$

for $m < i \leqslant n$. Hence by the definition of f we obtain $q = \Phi(o)$. We already proved that if $q \in Q \cap \Phi(M)$ then $x_i = \psi(q)_i = 0$ for all $m < i \leqslant n$. Hence (Q, ψ) is the coordinate neighborhood having the required properties.

Lemma 12. *Let $\Phi\colon M \to N$ be a submanifold of N and let $p \in M$. If f is a real valued differentiable function defined in a neighborhood of $\Phi(p)$ in $\Phi(M)$ then f coincides in a small neighborhood of $\Phi(p)$ with a differentiable function defined in a neighborhood of $\Phi(p)$ in N.*

Note. The conclusion holds for every k-times differentiable function f where k does not exceed the class of M and N.

Proof. We choose a coordinate neighborhood (Q, ψ) satisfying the conditions of Lemma 11. Then $\tilde{f} = f\varphi^{-1}\colon \varphi(O) \to \mathbb{R}$ is a k-times differentiable function which can be extended to a k-times differentiable function $\tilde{g}\colon \psi(Q) \to \mathbb{R}$ by defining $\tilde{g}(x_1, \ldots, x_n) = \tilde{f}(x_1, \ldots, x_m)$ for every $(x_1, \ldots, x_n) \in \psi(Q)$. Then $g = \tilde{g}\psi$ is the desired extension of f.

Lemma 13. *If* $\Phi: M \to N$ *is a submanifold of N then its differential* $d\Phi_p$ *is an isomorphism of* M_p *into* $N_{\Phi(p)}$ *for every* $p \in M$.

Note. Equivalently we can say that the differential of the inclusion map $I: \Phi(M) \to N$ is an isomorphism of $\Phi(M)_p$ into N_p for every $p \in \Phi(M)$.

Proof. In general, if the map $\Phi: M \to N$ is differentiable at $p \in M$ then its differential defined by formula (1) of Section 5 is a homomorphism of M_p into $N_{\Phi(p)}$. Now let $\Phi: M \to N$ be a submanifold of N and let $\boldsymbol{t}_1, \boldsymbol{t}_2 \in M_p$ be such that $d\Phi_p(\boldsymbol{t}_1) = d\Phi_p(\boldsymbol{t}_2)$. We are going to prove that $\boldsymbol{t}_1 = \boldsymbol{t}_2$. For let $\not{f} \in \mathscr{F}_p$ and let $f \in \not{f}$. Then by Lemma 12 there is a differentiable function g which is an extension of $f\Phi^{-1}$ to a neighborhood of $\Phi(p)$ in N. Then we have

$$\boldsymbol{t}_1(\not{f}) = \boldsymbol{t}_1(f + \mathscr{F}_p^o) = d\Phi_p(\boldsymbol{t}_1)(g + \mathscr{F}_{\Phi(p)}^o) = d\Phi_p(\boldsymbol{t}_2)(g + \mathscr{F}_{\Phi(p)}^o)$$
$$= \boldsymbol{t}_2(f + \mathscr{F}_p^o) = \boldsymbol{t}_2(\not{f}).$$

Since $\not{f}$ is an arbitrary element of $\mathscr{F}_p$ we obtain $\boldsymbol{t}_1 = \boldsymbol{t}_2$.

Lemma 14. *If* $\Phi: M \to N$ *is a differentiable map and* $d\Phi_p$ *is an isomorphism of* M_p *onto* $N_{\Phi(p)}$ *then there is an open set O in M which contains p and is such that* $\Phi(O)$ *is an open set in N and* $\Phi|O$ *is a diffeomorphism of O onto* $\Phi(O)$.

Note. The result holds for all classes k but we need it only in the special case when M, N and Φ are of class C^∞.

Proof. The local dimension n of M near p is the same as that of N near $\Phi(p)$. We choose coordinate neighborhoods (O_p, x) and $(Q_{\Phi(p)}, y)$ such that $p \in O_p$ and $\Phi(p) \in Q_{\Phi(p)}$ and let $x_1, \ldots, x_n$ and $y_1, \ldots, y_n$ denote the corresponding local coordinates. Then by hypothesis the Jacobian matrix $(\partial y_i / \partial x_j)$ $(1 \leq i, j \leq n)$ has rank n at $x(p) \in \mathbb{R}^n$. Hence by the implicit function theorem the restriction of $y\Phi x^{-1}$ to a sufficiently small open neighborhood of $x(p)$ is a diffeomorphism. The inverse image of this open set under x^{-1} is the desired open set O in M.

We can now return to the study of the exponential map. Let G be an n dimensional Lie group with Lie algebra $\mathfrak{g}$ the elements of $\mathfrak{g}$ being interpreted as tangent vectors $\boldsymbol{t} \in G_e$. Since $\mathfrak{g}$ is isomorphic to $\mathbb{R}^n$ as a vector space $\mathfrak{g}$ inherits the C^∞ structure of $\mathbb{R}^n$. Therefore we can consider $\mathfrak{g}$ as a commutative Lie group the group operation being vector addition. By Theorem 4 the exponential map is a C^∞ map of the manifold $\mathfrak{g}$ into G. The tangent space of $\mathfrak{g}$ at 0 is naturally isomorphic to the vector space $\mathfrak{g}$ and so $d(\exp)_0$ is an isomorphism of $\mathfrak{g}$ onto itself. Hence by applying Lemma 14 we obtain the following:

Proposition 15. *If G is a Lie group with Lie algebra $\mathfrak{g}$ then there is an open set O in $\mathfrak{g}$ containing 0 such that $\exp O$ is an open set in G which contains e and the exponential map is a C^∞ diffeomorphism of O onto $\exp O$.*

The restriction of exp to the above given open set has an inverse which we denote by log. If we let $Q = \exp O$ then log is a C^∞ diffeomorphism of Q onto an open set lying in $\mathbb{R}^n$ and containing 0. Hence $(Q, \log)$ is a coordinate neighborhood of G containing the identity element e. It is called a canonical or normal coordinate neighborhood and the corresponding local coordinates are the *canonical, normal* or *logarithmic coordinates* in G. They depend on the choice of the basis $X_1, \ldots, X_n$ of $\mathfrak{g}$ which is used to map $\mathfrak{g}$ isomorphically onto $\mathbb{R}^n$. Although O and Q are not uniquely determined by G and $\mathfrak{g}$ the values assumed by exp and log are independent of the choice of O and Q, respectively.

Let $m \in \mathbb{Z}$ and let g in the Lie group G be so close to e that both g and g^m belong to a canonical neighborhood $(Q, \log)$. Then by $g \in Q$ there is an X in the Lie algebra $\mathfrak{g}$ of G such that $g = \exp X$. Let $X_1, \ldots, X_n$ be a basis of $\mathfrak{g}$ and let $X = x_1 X_1 + \cdots + x_n X_n$ so that $x_1, \ldots, x_n$ are the canonical coordinates of g. We have

$$g^m = (\exp X)^m = \exp mX = \exp(m x_1 X_1 + \cdots + m x_n X_n)$$

where $g^m \in Q$. This shows that the canonical coordinates of g^m are $m x_1, \ldots, m x_n$.

Lemma 16. *Let M and N be differentiable manifolds of the same class, let $\Phi: M \to N$ be a differentiable map and let $\gamma: I \to M$ be a differentiable curve. Then for every $t \in I$ we have*

$$d\Phi_{\gamma(t)}(\gamma_*(t)) = (\Phi\gamma)_*(t).$$

Proof. For every f in $\mathscr{F}_{\gamma(t)}$ we have

$$d\Phi_{\gamma(t)}(\gamma_*(t))(f) = \gamma_*(t)(f\Phi) = (f\Phi\gamma)'(t) = (\Phi\gamma)_*(t)(f).$$

Lemma 17. *Let H and K be Lie groups and let $\Phi: H \to K$ be a C^∞ homomorphism. Then for all $a, b \in H$ and $\mathbf{t} \in H_b$ we have*

$$d\Phi_{ab}(a\mathbf{t}) = \Phi(a)\, d\Phi_b(\mathbf{t}) \quad \textit{and} \quad d\Phi_{ba}(\mathbf{t}a) = d\Phi_b(\mathbf{t})\,\Phi(a).$$

Proof. If $f \in \mathscr{F}_{\Phi(ab)}$ then on the one hand

$$d\Phi_{ab}(a\mathbf{t})(f) = a\mathbf{t}(f\Phi) = \mathbf{t}({}_a(f\Phi)) = \mathbf{t}(f({}_a\Phi)).$$

On the other hand we have

$$(\Phi(a)\,d\Phi_b(\boldsymbol{t}))\,(\not f)=d\,\Phi_b(\boldsymbol{t})({}_{\Phi(a)}\not f)=\boldsymbol{t}(({}_{\Phi(a)}\not f)\,\Phi)\,.$$

The first identity follows by noticing that $\not f({}_a\Phi)=({}_{\Phi(a)}\not f)\,\Phi$. The second identity can be proved similarly.

Lemma 18. *Let H and K be Lie groups, let $\Phi\colon H\to K$ be a C^∞ homomorphism, let $\boldsymbol{t}\in H_e$ and $\boldsymbol{u}=d\Phi_e\boldsymbol{t}$. Denote by X and Y the left invariant vector fields generated by $\boldsymbol{t}$ and $\boldsymbol{u}$, respectively and let D_X and D_Y be the corresponding derivations. Then we have*

$$D_X\,f\Phi = X(f\Phi) = (Yf)\,\Phi = (D_Y\,f)\,\Phi$$

for every f in $C^\infty(K)$.

Proof. For every $h\in H$ and $\not f\in\mathscr{F}_{\Phi(h)}$ we have

$$d\Phi_h\,X(h)\,(\not f)= X(h)\,(\not f\Phi) = X(e)\,({}_h(\not f\Phi)) = \boldsymbol{t}({}_h(\not f\Phi))$$

and also

$$Y(\Phi(h))(\not f)= Y(e)({}_{\Phi(h)}\not f)=\boldsymbol{u}({}_{\Phi(h)}\not f)=d\,\Phi_e\,\boldsymbol{t}({}_{\Phi(h)}\not f)=\boldsymbol{t}(({}_{\Phi(h)}\not f)\,\Phi)\,.$$

For any x in H we have

$${}_h(f\Phi)\,(x) = f\Phi(hx) = f(\Phi(h)\,\Phi(x)) = ({}_{\Phi(h)}f)\,\Phi(x)\,.$$

Hence from the above relations we obtain

$$d\Phi_h\,X(h) = Y(\Phi(h))$$

for every h in H. On the one hand we have

$$d\Phi_h\,X(h)\,(\not f)= X(h)\,(\not f\Phi) = X(\not f\Phi)\,(h)$$

and on the other hand

$$Y(\Phi(h))(\not f)= Y\not f(\Phi(h))=(Y\not f)\,\Phi(h)\,.$$

Thus we obtain $X(f\Phi)=(Yf)\Phi$ for all f in $C^\infty(K)$. Next be the definition of D_Y we have

$$D_Y\,f(\Phi(h)) = Y(\Phi(h))\,f = Yf(\Phi(h))$$

for every $h\in H$ and $f\in C^\infty(K)$. Hence $(D_Y\,f)\,\Phi=(Yf)\,\Phi$ for all f in $C^\infty(K)$. Similarly

$$D_X\,f\Phi(h) = X(h)\,f\Phi = X(f\Phi)\,(h)$$

and so $D_X\,f\Phi=X(f\Phi)$ for all f in $C^\infty(H)$.

Proposition 19. *Let H and K be Lie groups, let $\mathfrak{h}$ and $\mathfrak{k}$ denote their Lie algebras and let $\Phi: H \to K$ be a C^∞ homomorphism. Then $d\Phi_e$ is a homomorphism of $\mathfrak{h}$ into $\mathfrak{k}$ and $\Phi(\exp X) = \exp d\Phi_e X$ for every X in $\mathfrak{g}$.*

Proof. In order to obtain the first conclusion it is sufficient to prove that

$$d\Phi_e[\boldsymbol{t}_1, \boldsymbol{t}_2] = [d\Phi_e \boldsymbol{t}_1, d\Phi_e \boldsymbol{t}_2]$$

for all $\boldsymbol{t}_1$, $\boldsymbol{t}_2$ in H_e. Let $X_{\boldsymbol{t}_i}$ denote the left invariant vector field determined by $\boldsymbol{t}_i$ and let $D_{\boldsymbol{t}_i} = D_{X_{\boldsymbol{t}_i}}$ be the corresponding derivation ($i=1,2$). For the sake of simplicity let $\boldsymbol{u}_i = d\Phi_e \boldsymbol{t}_i$ and let $D_{\boldsymbol{u}_i}$ be the associated derivation ($i=1,2$). Then by Lemma 18 for any f in $C^\infty(K)$ we have

$$(D_{\boldsymbol{u}_1}(D_{\boldsymbol{u}_2} f))\Phi = D_{\boldsymbol{t}_1}((D_{\boldsymbol{u}_2} f)\Phi) = D_{\boldsymbol{t}_1}(D_{\boldsymbol{t}_2}(f\Phi)) = D_{\boldsymbol{t}_1} D_{\boldsymbol{t}_2} f\Phi .$$

Therefore

$$D_{\boldsymbol{t}_1} D_{\boldsymbol{t}_2} f\Phi(e) = D_{\boldsymbol{u}_1} D_{\boldsymbol{u}_2} f(\Phi(e))$$

and a similar expression holds for $D_{\boldsymbol{t}_2} D_{\boldsymbol{t}_1} f\Phi(e)$. Now using $\not{f}$ to denote $f + \mathscr{F}_e^o$ by the definition of $d\Phi$ we have

$$\begin{aligned} d\Phi_e[\boldsymbol{t}_1, \boldsymbol{t}_2](\not{f}) &= [\boldsymbol{t}_1, \boldsymbol{t}_2](\not{f}\Phi) = X_{[\boldsymbol{t}_1, \boldsymbol{t}_2]}(e)(\not{f}\Phi) = D_{[\boldsymbol{t}_1, \boldsymbol{t}_2]}\not{f}\Phi(e) \\ &= (D_{\boldsymbol{t}_1} D_{\boldsymbol{t}_2} - D_{\boldsymbol{t}_2} D_{\boldsymbol{t}_1}) f\Phi(e) = (D_{\boldsymbol{u}_1} D_{\boldsymbol{u}_2} f - D_{\boldsymbol{u}_2} D_{\boldsymbol{u}_1} f)(\Phi(e)) \\ &= D_{[\boldsymbol{u}_1, \boldsymbol{u}_2]} f(\Phi(e)) = X_{[\boldsymbol{u}_1, \boldsymbol{u}_2]}(e)(\not{f}) = [\boldsymbol{u}_1, \boldsymbol{u}_2](\not{f}) . \end{aligned}$$

Hence $d\Phi_e[\boldsymbol{t}_1, \boldsymbol{t}_2] = [\boldsymbol{u}_1, \boldsymbol{u}_2]$.

In order to prove the relation $\Phi(\exp X) = \exp d\Phi_e X$ we consider the curves $\alpha(t) = \exp t\, d\Phi_e X(e)$ and $\beta = \Phi\gamma$. The first is a maximal geodesic satisfying $\alpha(0) = e$ and $\alpha_*(0) = d\Phi_e X(e)$. We also have $\beta(0) = e$ and by Lemmas 16 and 17

$$\beta_*(t) = (\Phi\gamma)_*(t) = d\Phi_{\gamma(t)}\gamma_*(t) = d\Phi_{\gamma(t)}(\gamma(t) X(e)) = \Phi(\gamma(t))\, d\Phi_e X(e) .$$

Therefore $\beta_*(0) = d\Phi_e X(e) = \alpha_*(0)$ and $\beta_*(t) = \beta(t)\beta_*(0)$ proving that β is also a maximal geodesic satisfying the same initial conditions as α. By the uniqueness property expressed in Theorem 4 it follows that $\alpha = \beta$. The desired relation is the same as $\alpha(1) = \beta(1)$.

We shall now state and prove a result which can be obtained also as a corollary of a deeper theorem due to Campbell and Hausdorff. If $\mathfrak{g}$ is a real Lie algebra and $X_1, \ldots, X_n$ is a basis of $\mathfrak{g}$ then we let $\|X\| = (x_1^2 + \cdots + x_n^2)^{\frac{1}{2}}$ where $x_1, \ldots, x_n$ are the coordinates of X in this basis. Let μ map a neighborhood of $(0,0)$ in $\mathfrak{g} \times \mathfrak{g}$ into $\mathfrak{g}$. We say that $\mu(X, Y) = O(\|X\|^3 + \|Y\|^3)$ in the neighborhood of $(0,0)$ if there is a constant $c > 0$ such that $\|\mu(X, Y)\| \leqslant C(\|X\|^3 + \|Y\|^3)$ for every (X, Y) in a neighborhood of $(0,0)$ in $\mathfrak{g}\times\mathfrak{g}$. We wish to point out that the property of μ expressed by this order relation is independent of the choice of the basis $X_1,\ldots,X_n$. This follows from the fact that the norms of $\mathbb{R}^n$ are all equivalent. We also know that the above relation means the same as $\mu(X,Y)=O(\|(X,Y)\|^3)$ where $\|(X,Y)\|=(\|X\|^2+\|Y\|^2)^{\frac{1}{2}}$.

Proposition 20. *Let G be a Lie group with Lie algebra $\mathfrak{g}$. Then*

$$\log(\exp X \exp Y)=X+Y+\tfrac{1}{2}[X,Y]+O(\|(X,Y)\|^3)$$

in the neighborhood of $(0,0)$ in $\mathfrak{g}\times\mathfrak{g}$.

First we are going to establish a differentiation formula which can be used to prove the result in a special case. We recall that the elements X of $\mathfrak{g}$ are left invariant vector fields and $X(p)\mathscr{f}=X(e)({}_p\mathscr{f})$ for every $p\in G$ and $\mathscr{f}\in\mathscr{F}_e$. Hence if f is a real valued C^∞ function on G then $Xf(p)=X(p)f$ defines a function $Xf\colon G\to\mathbb{R}$ which is C^∞ by the corollary of Lemma 5.13. Consequently $X^m f$ can be defined for every C^∞ function $f\colon G\to\mathbb{R}$ and $m=0,1,2,\ldots$ as a real valued C^∞ function on G.

Lemma 21. *If $p\in G$, $X\in\mathfrak{g}$ and $t\in(-\infty,\infty)$ then*

$$X^m f(p\exp tX)=\frac{d^m}{du^m}f(p\exp uX)\bigg|_{u=t} \tag{11}$$

for every $m=0,1,2,\ldots$

Proof. If $m=0$ then (11) is trivial. Let γ be the geodesic curve $\gamma(t)=\exp tX$ where $-\infty<t<\infty$. In the beginning of this section we have seen that $\gamma'(t)(f)=(f\gamma)'(t)$ for every real valued f in $C^\infty(G)$ and $t\in(-\infty,\infty)$. Therefore

$$Xf(p)=X(e)({}_pf)=({}_pf\gamma)'(t)=\frac{d}{du}f(p\exp uX)\bigg|_0$$

and so (11) holds with $m=1$ and $t=0$. Hence more generally

$$Xf(p\exp tX)=\frac{d}{du}f(p\exp tX\exp uX)\bigg|_{u=0}=\frac{d}{du}f(p\exp(t+u)X)\bigg|_{u=0}$$

which shows that (11) holds for $m=1$ and every real t. Next we suppose that $m>1$ and (11) holds with $m-1$ instead of m. Then we have

$$X^m f(p)=X(X^{m-1}f)(p)=X(p)(X^{m-1}f)=\frac{d}{dt}(X^{m-1}f)(p\exp tX)\bigg|_0$$
$$=\frac{d}{dt}\left\{\frac{d^{m-1}}{du^{m-1}}f(p\exp uX)\bigg|_t\right\}\bigg|_0=\frac{d^m}{du^m}f(p\exp uX)\bigg|_0.$$

Therefore (11) follows in the special case when $t=0$ and the order of differentiation is m. Similarly as in the case $m=1$ we can now obtain (11) in general.

Lemma 22. *If* $X, Y \in \mathfrak{g}$ *and* $f \in C^\infty(G)$ *then*

$$X^m Y^n f(e) = \frac{\partial^{m+n}}{\partial u^m \partial v^n} f(\exp u X \exp v Y)\Big|_{(0,0)} \tag{12}$$

for every $m, n = 0, 1, 2, \ldots$

Proof. First we apply (11) with $p=e$ and $t=0$: We obtain

$$X^m Y^n f(e) = \frac{d^m}{du^m} (Y^n f)(\exp u X)\Big|_0 .$$

Thus using (11) with $p = \exp u X$ and $t=0$ we get

$$X^m Y^n f(e) = \frac{d^m}{du^m} \left\{ \frac{d^n}{dv^n} f(\exp u X \exp v Y)\Big|_0 \right\}\Big|_0 .$$

Proposition 23. *If* $p \in G$ *and* f *is real analytic in the neighborhood of* p *then there is a neighborhood* N_0 *of* 0 *in* $\mathfrak{g}$ *such that*

$$f(p \exp X) = \sum_{m=0}^{\infty} \frac{1}{m!} (X^m f)(p)$$

for every X *in* N_0.

Proof. We choose a basis $X_1, \ldots, X_n$ in $\mathfrak{g}$ and let $(Q, \log)$ denote the canonical coordinate neighborhood corresponding to this choice of basic elements. Then $(pQ, {}^p\log)$ is a coordinate neighborhood in G containing p. Since the inverse map of ${}^p\log$ is $p\exp$ and f is analytic at p the map

$$(x_1, \ldots, x_n) \mapsto f(p(\exp(x_1 X_1 + \cdots + x_n X_n)))$$

can be developed into a power series which is convergent for every $(x_1, \ldots, x_n)$ such that $x_1^2 + \cdots + x_n^2 < \varepsilon^2$. Hence given X with coordinates $x_1, \ldots, x_n$ satisfying this inequality we have $f(p \exp t X) = \sum_{m=0}^{\infty} a_m t^m$ for every real t subject to the condition $|t| < 1$. Since $m! a_m$ is the m-th derivative of $t \mapsto f(p \exp t X)$ at $t=0$ the desired result follows from (11).

Lemma 24. *If* $X, Y \in \mathfrak{g}$ *then*

$$\log(\exp tX \exp tY) = t(X+Y) + \tfrac{1}{2}t^2[X, Y] + O(t^3) \tag{13}$$

in the neighborhood of $0 \in (-\infty, \infty)$.

Proof. Let $f: G \to \mathbb{R}$ be a C^∞ function. Then $(u, v) \mapsto f(\exp uX \exp vY)$ is a C^∞ function on $\mathbb{R}^2$ and so we can compute its value in the neighborhood of $(0, 0)$ by using the finite Taylor-Maclaurin formula with Lagrange's remainder term. Since the coefficients can be determined from (12) we obtain

$$f(\exp tX \exp tY) = f(e) + t(X+Y)f(e) + \tfrac{1}{2}t^2(X^2 + 2XY + Y^2)f(e) + O(t^3) \tag{14}$$

in the neighborhood of 0.

If t is sufficiently close to 0 then we can define

$$Z(t) = \log(\exp tX \exp tY)$$

so that $Z(t)$ is an element of $\mathfrak{g}$. By Proposition 15 we have

$$Z(t) = tA + t^2 B + O(t^3) \tag{15}$$

where $A, B \in \mathfrak{g}$. Now let $X_1, \ldots, X_n$ be a basis of $\mathfrak{g}$ and let $f: G \to \mathbb{R}$ be defined by $f(\exp(x_1 X_1 + \cdots + x_n X_n)) = x_k$ where k $(1 \leqslant k \leqslant n)$ is fixed. Then by (15) we have

$$f(\exp Z(t)) = f(tA + t^2 B) + O(t^3).$$

Applying Proposition 23 we see that

$$f(\exp Z(t)) = \sum_{m=0}^{\infty} \frac{1}{m!}((tA + t^2 B)^m f)(e) + O(t^3).$$

Comparing this relation with (14) we obtain

$$(X+Y)f(e) = Af(e) \quad \text{and} \quad (X^2 + 2XY + Y^2)f(e) = (A^2 + 2B)f(e).$$

If $X(e) = \sum x_i(\partial/\partial X_i)$ then $X(e)f = \sum x_i(\partial \tilde{f}/\partial X_i)$ where $\tilde{f}$ is the composition of f with exp. Thus $X(e)f = x_k$, the k-th coordinate of $X(e)$. Therefore we have $X+Y = A$ and $X^2 + 2XY + Y^2 = A^2 + 2B$ and so it follows that $A = X+Y$ and $B = \frac{1}{2}[X, Y]$. Hence (15) is the same as (13) and the lemma is proved.

Proof of Proposition 20. Let (Q, φ) be any coordinate neighborhood containing the identity e of G and let $\mu: O_o \times O_o \to \mathbb{R}^n$ be the associated

Lie group germ. Thus $O_o = \varphi(O)$ where O is an open subset of Q such that $e \in O$ and $pq \in Q$ for every $p, q \in O$. We let $x_1, \ldots, x_n$ denote the local coordinates. We shall use formulas (4), (5) and (10) of Section 5 and we shall identify the tangent space G_e with $\mathfrak{g}$. Given X and Y in $\mathfrak{g}$ let $X = \sum x_i \boldsymbol{t}_i$ and $Y = \sum y_j \boldsymbol{t}_j$ where $\boldsymbol{t}_i$ denotes the tangent vector corresponding to $\partial/\partial x_i$ $(i = 1, \ldots, n)$. By (5.10) the k-th coordinate of $[X, Y]$ with respect to the ordered basis $\boldsymbol{t}_1, \ldots, \boldsymbol{t}_n$ is

(16) $$[X, Y]_k = \sum_{i,j=1}^{n} (a_{k,ij} - a_{k,ji}) x_i y_j$$

where $k = 1, \ldots, n$.

Now we suppose that (Q, φ) is a canonical coordinate neighborhood with respect to the ordered basis $X_1 = \boldsymbol{t}_1, \ldots, X_n = \boldsymbol{t}_n$ of $\mathfrak{g}$. If we choose X and Y sufficiently close to 0 then the k-th coordinate of $\mu(tX, tY) = \log(\exp tX \exp tY)$ is on the one hand given by (5.9) and on the other hand it can be determined also from (13). By comparing the corresponding terms and using (16) we obtain

$$\frac{1}{2} \sum_{i,j=1}^{n} (a_{k,ij} - a_{k,ji}) x_i y_j = \sum_{i,j=1}^{n} a_{k,ij} x_i y_j .$$

Hence if (Q, φ) is a canonical coordinate neighborhood then the coefficients $a_{k,ij}$ occuring in (5.9) satisfy the identity $a_{k,ij} + a_{k,ji} = 0$ where $i, j, k = 1, \ldots, n$. Therefore (16) and (5.9) imply the result stated in Proposition 20.

In Section III.4 $\exp x$ has been defined for every element x of a complex Banach algebra A with identity e as an element of A and several properties of the exponentiation were discussed there. In particular $(\exp x)(\exp -x) = e$ and so $\exp x$ belongs to the group of units $A^\times$ of A. We also saw that $\lambda \mapsto \exp \lambda x$ is differentiable in the complex sense at $\lambda = 0$ and the derivative is x there. It follows that $\exp \lambda x$ is everywhere differentiable and $(\exp \lambda x)' = x \exp \lambda x$. The same holds also for real Banach algebras provided differentiation is interpreted in the real variable sense.

The full matrix algebra $M_n(\mathbb{C})$ becomes a complex B^* algebra if we introduce the inner product $(A, B) = \operatorname{tr} B^* A = \operatorname{tr} A B^*$. Indeed $M_n(\mathbb{C})$ is star isomorphic to $\mathscr{L}(\mathbb{C}^n)$ under the map $A \mapsto L_A$ where the linear transformation $L_A: \mathbb{C}^n \to \mathbb{C}^n$ is defined by $L_A V = AV$ for every column vector $V \in \mathbb{C}^n$. Then it is easy to verify that (A, B) is equal to the HS inner product (L_A, L_B). It follows that $\exp X$ is defined for every complex square matrix X and $\exp X$ belongs to $\mathrm{Gl}(n, \mathbb{C})$.

On the other hand we can consider the $2n$ dimensional Lie group $\mathrm{Gl}(n, \mathbb{C})$, the real Lie algebra $\mathfrak{gl}(n, \mathbb{C})$ and define $\exp X$ by using the exponentiation introduced in this section. We prove that the two definitions give the same value for $\exp X$. Indeed given the matrix X consider the curve γ defined by $\gamma(t)=\exp tX$ where $-\infty<t<\infty$ and exp denotes exponentiation in the Banach algebra sense. Then γ is a curve in $\mathrm{Gl}(n, \mathbb{C})$ and $\gamma(0)=I$. The curve γ is differentiable, $\gamma'(0)=X$ and $\gamma'(t)=\gamma(0)X$ everywhere. Hence γ is a maximal geodesic in $\mathrm{Gl}(n, \mathbb{C})$. Since $\exp X$ in the Lie group sense is $\gamma(1)$ we see that the two definitions of the exponential coincide. Of course the same holds for the Lie group $\mathrm{Gl}(n, \mathbb{R})$ and its Lie algebra $\mathfrak{gl}(n, \mathbb{R})$.

Lemma 25. *If D is a complex diagonal matrix with diagonal entries $d_{11}, \ldots, d_{nn}$ then $\exp D$ is also diagonal and its diagonal entries are $\exp d_{11}, \ldots, \exp d_{nn}$.*

Proof. The diagonal entries of D^n are d_{ii}^n $(i=1, \ldots, n)$ and so the entries of the power series of $\exp D$ can be determined explicitely.

Lemma 26. *If $X \in \mathfrak{gl}(n, \mathbb{C})$ and $A \in \mathrm{Gl}(n, \mathbb{C})$ then $\exp AXA^{-1} = A(\exp X)A^{-1}$.*

Proof. We have $(AXA^{-1})^n = AX^nA^{-1} = AX^nA^{-1}$ and so the identity follows from the power series representation of the exponential map.

We note that more generally by using a lemma to be presented in the next section we can prove:

Lemma 27. *If G is a Lie group with Lie algebra $\mathfrak{g}$ then $\exp aXa^{-1} = a(\exp X)a^{-1}$ for every $X \in \mathfrak{g}$ and $a \in G$.*

Proof. By Lemma 7.7 $t \mapsto a(\exp tX)a^{-1}$ is the maximal geodesic $t \mapsto \exp taXa^{-1}$. Hence the result is obtained by letting $t=1$.

We are going to give two related topological applications. Let $\mathscr{S}$ denote the set of all symmetric matrices in $\mathfrak{gl}(n, \mathbb{R})$ and let $\mathscr{P}$ be the set of all symmetric, positive definite elements. We topologize $\mathscr{P}$ and $\mathscr{S}$ by the subspace topology inherited from the topology of $\mathfrak{gl}(n, \mathbb{R})$. Thus $\mathscr{P} \subset \mathscr{S}$ and $\mathscr{S}$ is homeomorphic to $\mathbb{R}^{\frac{1}{2}n(n+1)}$. The elements of $\mathscr{P}$ are distinguished by the fact that all their characteristic values are positive. From this it follows that $\mathscr{P}$ is an open subset of $\mathscr{S}$. Therefore $\mathscr{P}$ and $\mathscr{S}$ are both differentiable manifolds.

Proposition 28. *If M and N are manifolds and $\Phi: M \to N$ is a continuous bijection then Φ is a homeomorphism.*

This is an easy consequence of a classical topological result due to Brouwer and known as the *theorem on the invariance of domain.* For Brouwer's theorem implies the following: If O and Q are homeomorphic subsets of a manifold and O is open then Q is also an open set. For our purposes it would be sufficient to know only the following: If M and N are C^∞ manifolds and $\Phi: M \to N$ is a differentiable bijection then Φ is a homeomorphism. Since the proof amounts to showing the continuity of Φ^{-1} this result follows from the implicit or inverse function theorem.

Proposition 29. *The exponential map is a homeomorphism of $\mathscr{S}$ onto $\mathscr{P}$.*

Proof. First we prove that $\exp: \mathscr{S} \to \mathscr{P}$ is an injection. For let X_1 and X_2 in $\mathscr{S}$ be such that $\exp X_1 = \exp X_2$. Then $X_i = A_i D_i A_i^{-1}$ where A_i is in $\mathrm{Gl}(n, \mathbb{R})$ and D_i is a diagonal matrix. By Lemma 26 we have

$$\exp D_1 = A(\exp D_2) A^{-1} \tag{17}$$

where $A = A_1^{-1} A_2$. By Lemma 25 $\exp D_1$ and $\exp D_2$ are diagonal matrices. Since similar matrices have the same characteristic values we obtain $D_1 = D_2 = D$. Let $A = (a_{ij})$ and let $d_{11}, \ldots, d_{nn}$ be the diagonal entries of D. Then by (17) we have $a_{ij} \exp d_{jj} = a_{ij} \exp d_{ii}$ for every $i, j = 1, \ldots, n$. If $d_{ii} \neq d_{jj}$ then we obtain $a_{ij} = 0$ and so $a_{ij} d_{jj} = d_{ii} a_{ij}$. Of course the same holds if $d_{ii} = d_{jj}$. Hence $AD = DA$ that is to say $X_1 = X_2$.

The map $\exp: \mathscr{S} \to \mathscr{P}$ is surjective. For given $P \in \mathscr{P}$ we have $P = ADA^{-1}$ where D is a diagonal matrix with diagonal entries $d_{11}, \ldots, d_{nn}$. We let $\log D$ be the diagonal matrix with diagonal entries $\log d_{ii}$ $(i = 1, \ldots, n)$ and define $X = A(\log D) A^{-1}$. Then by Lemmas 25 and 26 we have $\exp X = P$. Since $\mathscr{S}$ and $\mathscr{P}$ are manifolds and the exponential map is continuous by Proposition 28 it follows that $\exp: \mathscr{S} \to \mathscr{P}$ is a homeomorphism.

Proposition 30. *The map $(O, P) \mapsto OP$ is a homeomorphism of $O(n, \mathbb{R}) \times \mathscr{P}$ onto $\mathrm{Gl}(n, \mathbb{R})$.*

Proof. First we prove that the above map is injective. Indeed if $OP = A$ then by $O^t O = I$ we have $A^t A = P^2$ and so P and O are uniquely determined by the image A: Namely $P = +\sqrt{A^t A}$ and so $O = AP^{-1}$. The map is surjective: If A is in $\mathrm{Gl}(n, \mathbb{R})$ then we let $P = +\sqrt{A^t A}$ and $O = AP^{-1}$. Then P is positive definite and O is orthogonal because

$$OO^t = AP^{-1}(P^{-1})^t A^t = AP^{-2} A^t = A(A^t A)^{-1} A^t = I.$$

Since $(O, P) \mapsto OP$ is a continuous bijection by Proposition 28 it is a homeomorphism.

Similarly one can prove that $\mathrm{Gl}(n, \mathbb{C})$ is homeomorphic to the product space $U(n) \times \mathscr{P}$ where $U(n)$ is the unitary group and $\mathscr{P}$ has the same meaning as before.

Proposition 31. *If the characteristic roots of the* $n \times n$ *complex matrix* A *are* $\lambda_1, \ldots, \lambda_n$ *then* $\exp\lambda_1, \ldots, \exp\lambda_n$ *are the characteristic roots of* $\exp A$.

Proof. One can prove this by induction on $n = 1, 2, \ldots$ The case $n = 1$ is trivial. Suppose that the proposition holds for $n-1$. Let $e = (1, 0, \ldots, 0)^t$ and let x be a complex column vector such that $Sx = e$. We obtain $SAS^{-1}e = \lambda_1 e$ i.e.

$$SAS^{-1} = \begin{pmatrix} \lambda_1 & * \\ 0 & B \end{pmatrix} \tag{18}$$

where B is in $\mathfrak{gl}(n-1, \mathbb{C})$. Hence $SA^mS^{-1} = \begin{pmatrix} \lambda_1^m & * \\ 0 & B^m \end{pmatrix}$ for every $m = 0, 1, 2, \ldots$ and so

$$S(\exp A)S^{-1} = \begin{pmatrix} \exp\lambda_1 & * \\ 0 & \exp B \end{pmatrix}. \tag{19}$$

If $\lambda_2, \ldots, \lambda_n$ denote the characteristic roots of B then those of SAS^{-1} are $\lambda_1, \ldots, \lambda_n$. Hence $\lambda_1, \ldots, \lambda_n$ are the characteristic roots of A. By hypothesis $\exp B$ has roots $\exp\lambda_2, \ldots, \exp\lambda_n$ and so by (19) $\exp\lambda_1, \ldots, \exp\lambda_n$ are the characteristic roots $S(\exp A)S^{-1}$ and $\exp A$.

Proposition 32. *For every* X *in* $\mathfrak{gl}(n, \mathbb{C})$ *we have* $\det\exp X = \exp\operatorname{tr} X$.

Proof. The determinant of $\exp X$ is the product of its characteristic values and $\operatorname{tr} X$ is the sum of the characteristic values of X. Hence the result follows from Proposition 31.

The ordinary exponential function $e^t\,(-\infty < t < \infty)$ maps $\mathbb{R}$ onto $\mathbb{R}^\times$ and similarly e^z maps $\mathbb{C}$ onto $\mathbb{C}^\times$. The proper generalization of these results is that $\exp\mathfrak{g}$ generates the identity component of the Lie group G. This follows immediately from Lemma 5.2 and Proposition 15. In order to see an example where exp is properly contained in G consider $\mathrm{Sl}(2, \mathbb{R})$. Every group element of the form $\exp X$ has a square root, namely $\exp\frac{1}{2}X$. On the other hand it is easy to verify that $\begin{pmatrix} x & 0 \\ 0 & 1/x \end{pmatrix}$ has no square root if $x < 0$ and $x \neq -1$.

7. Lie Subgroups and Subalgebras

Let G and H be Lie groups. A map $\Phi\colon G \to H$ is called a *homomorphism*, or more precisely a *Lie group homomorphism* provided Φ is a C^∞ map and a group homomorphism in the algebraic sense. If Φ is a one-to-one map then it is called a *Lie group isomorphism* or simply an *isomorphism.* The groups G and H are called *isomorphic* if there is an isomorphism Φ mapping G onto H. A *local homomorphism* of G into H is a map $\Phi: Q \to H$ such that

1) Q is an open set in G containing the identity element and similarly $\Phi(Q)$ is open in H and contains the identity of H;

2) there is an open set O in G such that $e \in O$, for every $p, q \in O$ the product $pq \in Q$ and $\Phi(pq) = \Phi(p)\Phi(q)$.

A *local isomorphism* is a local homomorphism $\Phi: Q \to H$ such that for some open set O containing e the restricted map $\Phi|O$ is injective. For instance if G is the unit circle and H is the reals then $z \mapsto (\frac{1}{2}\pi i)\log z$ with $z = \exp 2\pi i t$ and $-\infty < t < \infty$ is a local isomorphism of G into H. If such a Φ exists then the Lie groups G and H are called locally isomorphic.

Proposition 1. *Locally isomorphic Lie groups have isomorphic Lie algebras.*

Proof. This follows from the fact that the Lie algebra operations are uniquely determined by any germ of the Lie group in question.

It follows that there exist non-isomorphic Lie groups with isomorphic Lie algebras. For example the unit circle and the real line are such analytic groups. Therefore the Lie algebra $\mathfrak{h}$ of the Lie group H does not determine H within isomorphism. The situation however is completely different if we restrict our attention to analytic subgroups H of a fixed Lie group G. Indeed we have the following two important results:

Proposition 2. *If H is a Lie subgroup of the Lie group G then the Lie algebra $\mathfrak{h}$ of H can be identified with a subalgebra of $\mathfrak{g}$, the Lie algebra of G. Namely if $I: H \to G$ is the identity mapping then dI_e is an isomorphism of H_e into G_e. An element X of $\mathfrak{g}$ belongs to $dI_e\mathfrak{h}$ if and only if the curve $t \mapsto \exp tX$ lies entirely within the subgroup H.*

Proof. Since H is a submanifold of G the map I is regular and so by Lemma 6.13 dI_e is a vector space isomorphism of H_e into G_e. By Proposition 6.19 dI_e is a Lie algebra homomorphism. More generally we see that if $\Phi: H \to G$ is a Lie subgroup of G, i.e. Φ is a submanifold and a group homomorphism, then $d\Phi_e$ identifies $\mathfrak{h}$ with a Lie subalgebra of $\mathfrak{g}$. In the special case $I: H \to G$ the maps $\exp: \mathfrak{h} \to H$ and $\exp: \mathfrak{g} \to G$ coincide on $\mathfrak{h}$ by the uniqueness property of maximal geodesics described in Theorem 6.4. Using the first one of these maps we see that if $X \in \mathfrak{h}$ then the geodesic $t \mapsto \exp tX$ lies in H. Conversely if $X \in \mathfrak{g}$ and the geodesic $t \mapsto \exp tX$ lies in H then the image under dI_e of the tangent to this curve in H is the tangent to it in G.

Theorem 3. *Let G be a Lie group with Lie algebra $\mathfrak{g}$ and let $\mathfrak{h}$ be a subalgebra of $\mathfrak{g}$. Then there is a unique analytic subgroup H of G such that $\mathfrak{h}$ is the subalgebra corresponding to the Lie algebra of H under the isomorphism $dI_e: H_e \to G_e$ where $I: H \to G$ is the identity mapping, $G_e = \mathfrak{g}$ and H_e is the Lie algebra of H. It is the smallest subgroup containing* $\exp \mathfrak{h}$.

Note. We recall that an analytic group means a connected Lie group.

Proof. Let H denote the smallest subgroup of G containing $\exp \mathfrak{h}$. We shall introduce a manifold structure on H such that H becomes a connected Lie subgroup satisfying the requirements. We start by introducing a topology on H which will turn it into a topological group. The construction will be such that in addition to the topology a manifold structure will be simultaneously introduced on H. It will be obvious that H is then a submanifold of G and we will prove that H is a Lie group with respect to this manifold structure. Let $X_1, \ldots, X_m$ be a basis of $\mathfrak{h}$ and let $X_1, \ldots, X_m, X_{m+1}, \ldots, X_n$ be a basis of $\mathfrak{g}$. If $X \in \mathfrak{g}$ and $X = x_1 X_1 + \cdots + x_n X_n$ then we let $\|X\| = (x_1^2 + \cdots + x_n^2)^{\frac{1}{2}}$. Let $\eta > 0$ be so small that the restriction of $\exp: \mathfrak{g} \to G$ to the set $B = \{X : X \in \mathfrak{g} \text{ and } \|X\| < \eta\}$ is a diffeomorphism of B onto an open subset O_e of G where $e \in O_e$. The existence of such η follows from Proposition 6.15. Let $V = \exp(\mathfrak{h} \cap B)$ so that V lies in H and $e \in V$. Given any x in V there is a unique X in $\mathfrak{h}$ such that $x = \exp X$; we let $\varphi(x) = (x_1, \ldots, x_m)$ where $X = x_1 X_1 + \cdots + x_m X_m$. Then $\varphi(V)$ is the open ball of radius η centered around the origin of $\mathbb{R}^m$ and the map φ is a bijection of V onto this ball $\mathscr{B}$.

We associate with each $h \in H$ the set Vh and the map φ^h where $\varphi^h(x) = \varphi(xh^{-1})$ for every $x \in Vh$. Then $h \in Vh$ by $e \in V$ and $\varphi^h: Vh \to \mathscr{B}$ is a bijection. Let $h_1, h_2 \in H$ be such that $Vh_1 \cap Vh_2 \neq \emptyset$ and consider the map $\varphi^{h_2}(\varphi^{h_1})^{-1}$: If $h \in Vh_1 \cap Vh_2$ then

$$\begin{aligned}(x_1, \ldots, x_m) &= \varphi^{h_1}(h) \xmapsto{\varphi^{-1}} \exp(x_1 X_1 + \cdots + x_m X_m) = \varphi^{-1}(\varphi^{h_1}(h)) \\ &= \varphi^{-1}(\varphi(hh_1^{-1})) = hh_1^{-1} \xmapsto{g \mapsto gh_1} h \xmapsto{g \mapsto gh_2^{-1}} hh_2^{-1} \\ &= \varphi^{-1}(\varphi(hh_2^{-1})) = \varphi^{-1}(\varphi^{h_2}(h)) \xmapsto{\varphi} \varphi^{h_2}(h) = (y_1, \ldots, y_m).\end{aligned}$$

The maps φ^{-1}, $g \mapsto gh_1$, $g \mapsto gh_2^{-1}$ and φ are C^∞ on $\mathscr{B}$, G, G and V, respectively. Thus $(x_1, \ldots, x_m) \mapsto (y_1, \ldots, y_m)$ is a C^∞ map of $\varphi^{h_1}(Vh_1 \cap Vh_2)$ into $\mathscr{B}$.

We topologize Vh by requiring that φ^h be a homeomorphism of Vh onto $\mathscr{B}$. We prove that the intersection of an open subset O_1 of Vh_1 with an open subset O_2 of Vh_2 is an open subset of both Vh_1 and Vh_2. This will show that the open subsets of Vh $(h \in H)$ form a basis for a topology on H. Since $\varphi^{h_2}(\varphi^{h_1})^{-1}$ is a C^∞ map it is continuous. Thus given h in $O_1 \cap O_2$ there is an open subset O_x of $\mathscr{B}$ such that

$$\varphi^{h_1}(h) = (x_1, \ldots, x_m) \in O_x \subseteq \varphi^{h_1}(O_1) \tag{1}$$

and $\varphi^{h_2}(\varphi^{h_1})^{-1}(O_x) \subseteq \varphi^{h_2}(O_2)$. Clearly $h \in (\varphi^{h_1})^{-1}(O_x)$ by $\varphi^{h_1}(h) \in O_x$. By (1) we have $(\varphi^{h_1})^{-1}(O_x) \subseteq O_1$. Furthermore $(\varphi^{h_1})^{-1}(O_x)$ is an open set in Vh_1 because O_x is open in $\mathscr{B}$. We proved that $O_1 \cap O_2$ is a union of open sets of Vh_1 and so it is an open subset of Vh_1. Consequently we can topologize H as follows: A subset O of H is open if it is a union of open subsets of sets Vh $(h \in H)$. This topological space is a differentiable manifold because the pairs (Vh, φ^h) $(h \in H)$ are the coordinate neighborhoods of a C^∞ structure on H.

We prove that the manifold H is connected. It is sufficient to prove that given h in H there is a connected set containing both h and e. Since H is generated by $\exp \mathfrak{h}$ we have $h = \exp Y_1 \ldots \exp Y_r$ for suitable $Y_1, \ldots, Y_r$ in $\mathfrak{h}$. Therefore by using the relation $\exp a Y = (\exp Y)^a$ $(a = 1, 2, \ldots)$ we see that h can be written in the form $h = h_1 \ldots h_s$ where $h_1, \ldots, h_s \in V$. Now

$$V, Vh_s, Vh_{s-1}h_s, \ldots, Vh_i \ldots h_s, \ldots, Vh_1 \ldots h_s = Vh$$

are all homeomorphic to the ball $\mathscr{B}$ and so they are connected sets. By $e \in V$ and $h_i \in V$ we have

$$h_{i+1} \ldots h_s, h_i \ldots h_s \in V(h_{i+1} \ldots h_s) \qquad (1 \leqslant i < s)$$

and similarly $e, h_s \in V$. Hence the union of the sets $V, \ldots, V(h_1 \ldots h_s) = Vh$ is a conntected set and it contains e and h.

Since $\mathfrak{h} \cap B$ is a submanifold of B and B is a submanifold of O_e under exponentiation it follows that V is a submanifold of O_e. Thus O_e being a submanifold of G we see that V is a submanifold of G. More generally Vh is a submanifold of G for every h in H and so H is a submanifold of G under the inclusion map $I: H \to G$. Therefore by Lemma 6.13 $dI_e: H_e \to G_e$ is a vector space isomorphism. Let us now suppose that H is known to be a Lie subgroup of G. Then by Proposition 6.19 it follows that dI_e is a Lie algebra isomorphism. If $X \in \mathfrak{h}$ then the curve $t \mapsto \exp tX$ lies entirely within H and so by Proposition 2 X belongs to the range of dI_e. Therefore $dI_e(H_e) \supseteq \mathfrak{h}$. On the other hand $\dim dI_e(H_e) = \dim H_e = \dim H = \dim \mathfrak{h}$ and so $dI_e(H_e) = \mathfrak{h}$. Hence the Lie algebra of H is isomorphic to $\mathfrak{h}$ under dI_e. Thus in order to complete the existence part of Theorem 3 it is sufficient to prove that H is a Lie subgroup of G. This will easily follow from the following general result:

Lemma 4. *Let M and N be differentiable manifolds of the same class and let $\Phi: M \to N$ be a differentiable map such that $\Phi(M) \subseteq S$ where S is a submanifold of N and $\Phi: M \to S$ is continuous. Then $\Phi: M \to S$ is a differentiable map.*

Note. We need the result only in the special case when all the differentiable structures and maps are C^∞.

Proof. Let $p \in M$ and let s and n denote the dimensions of S and N in the neighborhood of $\Phi(p)$. By Lemma 6.11 we can choose the coordinate neighborhood (Q, ψ) in N such that $\psi\Phi(p) = 0 \in \mathbb{R}^n$ and

$$O = \{q: q \in Q \text{ and } \psi(q)_{s+1} = \cdots = \psi(q)_n = 0\}$$

together with $\varphi = \psi | O$ is a coordinate neighborhood of S. By the continuity of $\Phi: M \to S$ there is a coordinate neighborhood (P, ω) in M such that $p \in P$ and $\Phi(P) \subseteq M$. The coordinates of $\Phi(q)$ $(q \in P)$ in the system (Q, ψ) depend differentiably on the coordinates of q in the system (P, ω) because $\Phi: M \to S$ is a differentiable map. Hence the coordinates $\varphi_i(\Phi(q)) = \psi_i(\Phi(q))$ $(1 \leqslant i < s)$ of $\Phi(q)$ $(q \in P)$ depend differentiably on the coordinates of q in (P, ω). This shows that $\Phi: M \to S$ is differentiable in the neighborhood of p.

Lemma 5. *Let G be a Lie group, let H be a subgroup of G and suppose that a manifold structure is given on H which makes it a topological group and a submanifold of G. Then H is a Lie subgroup of G.*

Proof. Let Φ denote the restriction of $(x, y) \mapsto xy^{-1}$ $(x, y \in G)$ to $H \times H$. Then $\Phi: H \times H \to G$ is a C^∞ map because G is a Lie group. By hypothesis $\Phi: H \times H \to H$ is continuous. Therefore we can apply the last lemma with $M = H \times H$, $N = G$ and $S = H$. It follows that $\Phi: H \times H \to H$ is a C^∞ map and so H is a Lie group.

Lemma 6. *Let H be a group and a topological space such that the left and right translates of open sets are also open and $(x, y) \mapsto xy^{-1}$ is continuous at (e, e). Then H is a topological group.*

Proof. We are going to show that $(x, y) \mapsto xy^{-1}$ is continuous at (h, k) where h, k are arbitrary elements of H. Given an open set O containing hk^{-1} the object is to find an open set O_h containing h and another O_k containing k such that $\xi \in O_h$, $\eta \in O_k$ imply $\xi\eta^{-1} \in O$. By hypothesis the set $h^{-1}Ok$ is open and contains e so there is an open set Q containing e such that $x, y \in Q$ implies $xy^{-1} \in h^{-1}Ok$. We let $O_h = hQ$ and $O_k = kQ$. If $\xi \in O_h$ and $\eta \in O_k$ then $h^{-1}\xi \in Q$ and $k^{-1}\eta \in Q$ and so $(h^{-1}\xi)(k^{-1}\eta)^{-1} = h^{-1}(\xi\eta^{-1})k \in h^{-1}Ok$ i.e. $\xi\eta^{-1} \in O$.

Lemma 7. *If* $t \mapsto \gamma(t)$ *is a geodesic satisfying* $\gamma(0) = e$ *then so is the curve* α *where* $\alpha(t) = a\gamma(t)a^{-1}$ *and* $a \in G$. *Moreover* $\alpha(0) = e$ *and* $\alpha'(0) = a\gamma'(0)a^{-1}$.

Proof. We have $\alpha'(t) = a\gamma'(t)a^{-1} = a\gamma(t)\gamma'(0)a^{-1}$ and so $\alpha'(0) = a\gamma'(0)a^{-1}$ by $\gamma(0) = e$. Hence $\alpha'(t) = (a\gamma(t)a^{-1})(a\gamma'(0)a^{-1}) = \alpha(t)\alpha'(0)$. Thus α is a geodesic satisfying $\alpha(0) = a\gamma(0)a^{-1} = e$.

We can now prove that the subgroup H constructed in the proof of Theorem 3 is indeed a Lie subgroup of G. In view of Lemma 5 it is sufficient to prove that H is a topological group. We shall do this by showing that H satisfies the hypotheses of Lemma 6.

First we prove that the map $(x, y) \mapsto xy^{-1}$ $(x, y \in H)$ is continuous at (e, e). Given an open set Q in H such that $e \in Q$ we are going to find another open set O in H such that $x, y \in O$ implies $xy^{-1} \in Q$. We may suppose that

$$Q = \{x : x \in H \text{ and } \|\log x\| < \varepsilon\}$$

where $\varepsilon < \eta$. The set

$$\tilde{Q} = \{x : x \in G \text{ and } \|\log x\| < \varepsilon\}$$

is open in G because exp is a C^∞ homeomorphism of B onto O_e and $\varepsilon < \eta$. Since $e \in \tilde{Q}$ there is an open set $\tilde{O}$ in G such that $x, y \in \tilde{O}$ implies $xy^{-1} \in \tilde{Q}$. We can suppose that $\tilde{O}$ is of the form

$$\tilde{O} = \{x : x \in G \text{ and } \|\log x\| < \delta\}$$

for some δ $(0 < \delta < \eta)$. Now

$$O = \{x : x \in H \text{ and } \|\log x\| < \delta\}$$

is an open set in H because O is the inverse image under φ of $\{X : X \in \mathfrak{h} \text{ and } \|X\| < \delta\}$ which is an open set in $\mathfrak{h}$ in the topology induced by the map $X \mapsto (x_1, \ldots, x_m)$ where $x \in \mathfrak{h}$. If $x, y \in O$ then $x, y \in \tilde{O}$ and so $xy^{-1} \in \tilde{Q}$ and $xy^{-1} \in Q$ by $x, y \in H$.

It is clear from the definition of the topology of H that right translations of open sets are also open. If we prove that hOh^{-1} is open for every open set O of H and every h in H then it will follow that left translations of open sets are also open. It will be sufficient to show that given h and an open set O containing e there is another open set Q in V such that $e \in Q$ and $hQh^{-1} \subseteq O$.

Let h in H be given and let $\Phi: G \to G$ be such that $\Phi(x) = hxh^{-1}$ for every $x \in G$. Applying Proposition 6.19 with $H = K = G$ and Φ we see that $d\Phi_e: \mathfrak{g} \to \mathfrak{g}$ is a homomorphism. Its kernel is trivial because $d\Phi_e X(f) = X(f\Phi_e)$ for every f in $\mathscr{F}_e$ and the range of the map $f \mapsto f\Phi_e$ is $\mathscr{F}_e$. Thus $\mathfrak{g}$ being finite dimensional $d\Phi_e$ is an automorphism of $\mathfrak{g}$. If we let $X' = d\Phi_e X$ then by Proposition 6.19 we have

$$(2) \qquad h(\exp tX)h^{-1} = \exp tX'.$$

We are going to prove that $X' \in \mathfrak{h}$ for every $X \in \mathfrak{h}$. By definition we have

$$X'(f) = d\Phi_e X(f) = X(f\Phi) = X({}_h f^h)$$

for every f in $\mathscr{F}_e$. Moreover $h = (\exp Y_1) \ldots (\exp Y_r)$ where $Y_1, \ldots, Y_r \in \mathfrak{h}$. Thus in order to prove that $X' \in \mathfrak{h}$ for every $X \in \mathfrak{h}$ it is sufficient to prove the following: Given $X, Y \in \mathfrak{h}$ and $k = \exp Y$ the formula $Z(f) = X({}_k f^k)$ $(f \in \mathscr{F}_e)$ defines an element Z of $\mathfrak{h}$. In other words we have to prove that

$$(3) \qquad (\exp Y)X(\exp -Y) \in \mathfrak{h} \quad \text{for every } X, Y \in \mathfrak{h}.$$

Let the curve α be defined by

$$\alpha(t) = (\exp Y)(\exp tX)(\exp Y)^{-1} \qquad (-\infty < t < \infty).$$

By Lemma 7 $\alpha'(0) = (\exp Y)X(\exp Y)^{-1}$. Since $X, Y \in \mathfrak{h}$ by the definition of H the group elements $\exp Y$ and $\exp tX$ belong to H and so α lies entirely in H. Hence dI_e identifies $\alpha'(0)$ with an element of $\mathfrak{h}$ and so (3) holds.

For $\varepsilon > 0$ let $B_\varepsilon = \{X : X \in \mathfrak{g} \text{ and } \|X\| < \varepsilon\}$. Since G is a topological group $h^{-1}(\exp B_\varepsilon)h$ is an open set in G for every $\varepsilon < \eta$. Since O is open in H and $e \in O$ we can choose δ such that $\exp(\mathfrak{h} \cap B_\varepsilon) \subseteq O$. Now $\exp: \mathfrak{g} \to \mathfrak{g}$ being a continuous map there is a δ such that $\delta < \eta$ and $\exp B_\delta \subseteq h^{-1}(\exp B_\varepsilon)h$. Thus by (3) we have

$$h \exp(\mathfrak{h} \cap B_\delta)h^{-1} \subseteq \exp(\mathfrak{h} \cap B_\varepsilon) \subseteq O.$$

If we let $Q = \exp(\mathfrak{h} \cap B_\delta)$ then Q is an open set in H, it contains e and $hQh^{-1} \subseteq O$. We proved that left translates of open sets of H are open. Hence Lemma 6 can be applied and so H is a Lie subgroup of G.

We complete the proof by showing that H is uniquely determined by $\mathfrak{h}$. Let us suppose that H is an analytic subgroup of G such that its Lie algebra is identified with the Lie subalgebra $\mathfrak{h}$ of $\mathfrak{g}$. Then we know that exponentiation in H is the restriction of $\exp: \mathfrak{g} \to G$ to $\mathfrak{h}$ and so $\exp \mathfrak{h}$ is uniquely determined. By Proposition 6.15 $\exp \mathfrak{h}$ is a neighborhood of e in H so by Lemma 5.2 $\exp\mathfrak{h}$ generates H. This shows that H as an abstract subgroup of G is uniquely determined by $\mathfrak{h}$. Now suppose that two C^∞ structures are defined on H which turn it into Lie subgroup H_i $(i=1,2)$ of G. We prove that $H_1=H_2$ as Lie groups. For by the second half of Proposition 2 their Lie algebras $\mathfrak{h}_i$ $(i=1,2)$ are both identified by $\mathfrak{h}$ and by Proposition 6.15 exp is a diffeomorphism of a neighborhood of 0 into a neighborhood of e in H_i $(i=1,2)$. This completes the proof of Theorem 3.

Corollary. *Let H_1 and H_2 be Lie subgroups of the Lie group G such that $H_1=H_2$ as abstract and topological groups. Then H_1 and H_2 are the same also as Lie groups.*

In fact by the second half of Proposition 2 H_1 and H_2 have identical Lie algebras. Thus by Theorem 3 their identity components H_{io} $(i=1,2)$ coincide as Lie groups.

Lemma 8. *Let G be a Lie group with Lie algebra $\mathfrak{g}$ and suppose that $\mathfrak{g}=\mathfrak{a}\oplus\mathfrak{b}$ where $\mathfrak{a}$ and $\mathfrak{b}$ are vector subspace of $\mathfrak{g}$. Then the map $X+Y \mapsto \exp X \exp Y$ where $X\in\mathfrak{a}$ and $Y\in\mathfrak{b}$ is a local isomorphism of $\mathfrak{g}$ into G.*

Proof. Let f denote the given map of the additive Lie group $\mathfrak{g}$ into G. Since the exponential map is infinitely differentiable it is clear that f is a C^∞ map. Hence it will be sufficient to show that the Jacobian determinant of f does not vanish in a neighborhood of $0\in\mathfrak{g}$. We choose a basis $X_1,\dots,X_a$ of $\mathfrak{a}$ and another $X_{a+1},\dots,X_n$ of $\mathfrak{b}$. Then $X_1,\dots,X_n$ is a basis of $\mathfrak{g}$. Let $x_1,\dots,x_n$ denote the canonical coordinates in G with respect to this basis. Then we have

$$f\left(\sum_{i=1}^{n} x_i X_i\right)=\left(\exp \sum_{i=1}^{a} x_i X_i\right)\left(\exp \sum_{j=a+1}^{n} x_j X_j\right).$$

Let f_k $(1\leqslant k\leqslant n)$ denote the k-th coordinate of f in the given canonical coordinate system. In order to determine $\partial f_k/\partial x_i$ at 0 we can choose $x_j=\delta_{ij}t$ where δ_{ij} is the Kronecker delta and t is real. Then $f(x_1,\dots,x_n)$ $=\exp tX_i$ and so $f(x_1,\dots,x_n)_k=\delta_{ij}t$ and $\partial f_k/\partial x_i=\delta_{ij}$. Therefore the Jacobian is not zero at 0 and so by its continuity it does not vanish in the neighborhood of 0.

Theorem 9. *Every closed subgroup H of a Lie group G is a Lie subgroup. More precisely if the subgroup H is a closed subset of G then there is a unique C^∞ structure on H which turns H into a Lie subgroup of G and induces the subspace topology on it.*

Proof. We know that H is a topological group under the topology induced on it by the topology of G. Let H_o denote the identity component of H. Then by hypothesis H_o is a closed subgroup of G. We let $\mathfrak{g}$ denote the Lie algebra of G and consider

$$\mathfrak{h}=\{X: X\in\mathfrak{g} \text{ and } \exp tX\in H_0 \text{ for every } t\in(-\infty,\infty)\}.$$

We are going to prove that $\mathfrak{h}$ is a Lie subgroup of $\mathfrak{g}$: If $X\in\mathfrak{h}$ and $\lambda\in\mathbb{R}^{\times}$ then $\lambda X\in\mathfrak{h}$ by $t(\lambda X)=(\lambda t)X$. Next let X and Y in $\mathfrak{h}$ and a real number t be given. If $n=1,2,3,\ldots$ is sufficiently large then by Proposition 6.20 we have

$$\left(\exp\frac{tX}{n}\exp\frac{tY}{n}\right)^n=\left\{\exp\left(t\frac{X+Y}{2}+O(n^{-2})\right)\right\}^n=\exp(t(X+Y)+O(n^{-1})).$$

The left hand side shows that these elements all belong to H_0 and the right hand side indicates that their sequence approaches $\exp t(X+Y)$ as $n\to\infty$. Hence H_o being closed this shows that $\exp t(X+Y)\in H_o$ for every $t\in(-\infty,\infty)$ and so $X+Y\in\mathfrak{h}$. Similarly we see that

$$\left(\exp\frac{tX}{n}\exp\frac{tY}{n}\exp\frac{-tX}{n}\exp\frac{-tY}{n}\right)^{n^2}=\left\{\exp\left(\left(\frac{t}{n}\right)^2[X,Y]+O(n^{-3})\right)\right\}^{n^2}$$

$$=\exp(t[X,Y]+O(n^{-1})).$$

These elements belong to H_o and their sequence converges to $\exp t[X,Y]$. Therefore $\exp t[X,Y]$ is in H_o for every t and so $[X,Y]\in\mathfrak{h}$.

Let K be the connected Lie subgroup of G corresponding to the Lie subalgebra $\mathfrak{h}$ of $\mathfrak{g}$ in accordance with Theorem 3. Then $K\subseteq H_o$ because K is the smallest subgroup of G containing $\exp\mathfrak{h}$ and by the definition of $\mathfrak{h}$ we have $\exp X\in H_o$ for every X in $\mathfrak{h}$. Suppose that we can find a neighborhood of e in H_o which lies in K. Then by Lemma 5.2 $H_o\subseteq K$ and so $H_o=K$. Since K is known to be a Lie subgroup of G it will follow that H_o and H are also Lie subgroups of G. Hence it is sufficient to find a suitable neighborhood of e in H_o.

Let $X_1,\ldots,X_m,X_{m+1},\ldots,X_n$ be a basis of $\mathfrak{g}$ such that $X_{m+1},\ldots,X_n$ is a basis of $\mathfrak{h}$. Then the subspace $\mathfrak{a}$ generated by $X_1,\ldots,X_m$ is such that $\mathfrak{g}=\mathfrak{a}\oplus\mathfrak{h}$. Let $\|X\|$ denote the norm in $\mathfrak{g}$ with respect to $X_1,\ldots,X_n$ i.e. let $\|X\|^2=x_1^2+\cdots+x_n^2$ if $X=x_1X_1+\cdots+x_nX_n$. If $\varepsilon>0$ then by Proposition 6.15

$$\{\exp X: \exp X\in\mathfrak{g} \text{ and } \varepsilon>0\}$$

is a neighborhood of e in G. Since H_o is topologized by the subspace topology inherited from G it follows that

$$\{\exp X : X \in \mathfrak{g} \text{ and } \|X\| < \varepsilon \text{ and } \exp X \in H_o\}$$

is a neighborhood of e in H_o. Thus it is enough to prove that for some $\varepsilon > 0$ this set lies entirely within K. We are going to prove this by contradiction.

We suppose that for every $n = 1, 2, \ldots$ there is an X_n in $\mathfrak{g}$ such that $\|X_n\| < 1/n$ and $\exp X_n$ is in H_0 but not in K. If n is sufficiently large then we can apply Lemma 8 with $\mathfrak{g} = \mathfrak{a} \oplus \mathfrak{h}$ and obtain $\exp X_n = \exp A_n \exp H_n$ where $A_n \in \mathfrak{a}$ and $H_n \in \mathfrak{h}$. Moreover by the lemma $A_n \to 0$ and $H_n \to 0$ in the norm topology of $\mathfrak{a}$ and $\mathfrak{h}$, respectively. By the definition of K we have $\exp H_n \in K$ and so using $K \subseteq H_o$ we see that $\exp A_n$ is in H_o but not in K. Then $A_n \neq 0$ and so there exist an integer $a_n \geqslant 0$ and a fraction b_n $(0 \leqslant b_n < 1)$ such that $\|A_n\|^{-1} = a_n + b_n$. Therefore multiplying this equation by A_n we obtain

$$1 - \|A_n\| \leqslant \|a_n A_n\| \leqslant 1 + \|A_n\| . \tag{4}$$

Since $\|A_n\| \to 0$ the sequence of $a_n A_n$ vectors is bounded and so $\mathfrak{a}$ being finite dimensional we can select a convergent subsequence. For the sake of simplicity we change subscripts and let $a_n A_n$ $(n = 1, 2, \ldots)$ denote this convergent subsequence. Let X be its limit so that $X \in \mathfrak{a}$ and $X \neq 0$ by (4).

We prove that $X \in \mathfrak{h}$ which gives a desired contradiction because $\mathfrak{a} \cap \mathfrak{h} = 0$. Indeed given $b = 1, 2, \ldots$ let $a_n = b q_n + r_n$ where q_n and r_n are non-negative integers and $0 \leqslant r_n < b$. Then $(r_n/b) A_n \to 0$ and $b q_n A_n \to X$. Thus

$$\begin{aligned} \exp b^{-1} X &= \lim \exp a_n b^{-1} A_n = \lim \exp(r_n b^{-1} A_n + q_n A_n) \\ &= (\lim \exp r_n b^{-1} A_n)(\lim \exp q_n A_n) = \lim (\exp A_n)^{q_n} . \end{aligned}$$

Since $\exp A_n$ is in H_o and H_o is closed it follows that $\exp b^{-1} X \in H_o$ where $b = 1, 2, \ldots$ is arbitrary. Thus $\exp a b^{-1} X \in H_o$ for every $a = 0, \pm 1, \pm 2, \ldots$ and $b = 1, 2, \ldots$ and so by continuity $\exp t X \in H_o$ for all real t. By the definition of $\mathfrak{h}$ this shows that $X \in \mathfrak{h}$.

The uniqueness of the manifold structure of the closed Lie subgroup H follows from the corollary of Theorem 3 is stated before Lemma 8.

Élie Cartan's Theorem 9 is very important because it can be used to show that many groups are Lie groups without explicitely introducing a differentiable structure on them. For instance $\mathrm{Sl}(2, \mathbb{R})$ is a closed subgroup of $\mathrm{Gl}(n, \mathbb{R})$ and so it is a Lie group. Other important Lie groups are obtained by using the following simple result:

Lemma 10. *If R is a commutative ring with identity and M is an $n \times n$ matrix whose entries belong to R then*

$$\{A: A \in \mathrm{Gl}(n, R) \text{ and } A^t M A = M\}$$

is a subgroup of $\mathrm{Gl}(n, R)$.

Note. We recall that $\mathrm{Gl}(n, R)$ denotes the group of units of $M_n(R)$.

Proof. Using the identity $(AB)^t = B^t A^t$ one can easily show that the above set is closed under matrix multiplication. Similarly using $(A^t)^{-1} = (A^{-1})^t$ one sees that if A belongs to the set then so does A^{-1}.

We let $R = \mathbb{R}$ and consider various matrices M in $M_n(\mathbb{R})$. Every choice of M gives a closed subgroup of $\mathrm{Gl}(n, \mathbb{R})$ and so the subgroups obtained in this manner are all Lie groups. First by choosing $M = I$ we obtain the *orthogonal group* $\mathrm{O}(n, \mathbb{R})$. We note that $\mathrm{O}(n, \mathbb{R})$ consists of those matrices A which leave the bilinear form $x_1 y_1 + \cdots + x_n y_n$ invariant under the transformations $x \to Ax$ and $y \to Ay$. Since this form is $x^t M y$ its invariance follows from the relation $A^t M A = M$ and conversely the invariance of this form implies this relation. Since $\mathrm{Sl}(n, \mathbb{R})$ is also a closed subgroup of $\mathrm{Gl}(n, \mathbb{R})$ by Theorem 9 it follows that the *special orthogonal group* $\mathrm{SO}(n, \mathbb{R}) = \mathrm{Sl}(n, \mathbb{R}) \cap \mathrm{O}(n, \mathbb{R})$ is also a Lie group. The same conclusion could be obtained also by considering $\mathrm{SO}(n, \mathbb{R})$ as a closed subgroup of $\mathrm{O}(n, \mathbb{R})$. We also know that the subgroup $\mathrm{SO}(n, \mathbb{R})$ is the identity component of $\mathrm{O}(n, \mathbb{R})$ and so these two groups have the same Lie algebra which is described in Theorem 5.23.

Next let p and q be non-negative integers such that $p + q = n$ and consider the matrix $M = \begin{pmatrix} I_p & 0 \\ 0 & -I_q \end{pmatrix}$ where I_p and I_q denote the identity matrices in $M_p(\mathbb{R})$ and $M_q(\mathbb{R})$, respectively. The groups obtained in this manner are called the *pseudo orthogonal groups* and one denotes them by $\mathrm{O}(p, q, \mathbb{R})$. The elements can be characterized as those matrices A in $M_n(\mathbb{R})$ for which the bilinear form

$$x_1 y_1 + \cdots + x_p y_p - x_{p+1} y_{p+1} - \cdots - x_n y_n$$

is left invariant under the transformations $x \to Ax$ and $y \to Ay$. By requiring also that $\det A = 1$ we obtain the *special pseudo orthogonal group* $\mathrm{SO}(p, q, \mathbb{R})$. The group $\mathrm{O}(4, 1, \mathbb{R}) = \mathrm{O}(1, 4, \mathbb{R})$ is the *de Sitter group* and sometimes $\mathrm{O}(p, q, \mathbb{R})$ is called a *generalized de Sitter group.* The Lie algebra of this group is denoted by $\mathfrak{o}(p, q, \mathbb{R})$ or $\mathfrak{so}(p, q, \mathbb{R})$ since $\mathrm{SO}(p, q, \mathbb{R})$ has the same Lie algebra as $\mathrm{O}(p, q, \mathbb{R})$. It consists of all $n \times n$ real matrices of the form $X = \begin{pmatrix} X_{11} & X_{12} \\ X_{21} & X_{22} \end{pmatrix}$ where X_{11} and X_{22}

are skew symmetric matrices of orders p and q, respectively, and $X_{21}=X_{12}^t$. It follows that the dimension of all these groups and Lie algebras is the same, namely $\frac{1}{2}n(n-1)$.

Now we suppose that n is even and actually we replace n by $2n$ where the new $n=1,2,\ldots$ This time we choose $M=\begin{pmatrix} 0 & I \\ -I & 0 \end{pmatrix}$ where I is the $n\times n$ identity matrix. The Lie group obtained in this manner is the *real symplectic group* $\mathrm{Sp}(n,\mathbb{R})$. The elements are those A for which the linear form

$$(x_1 y_{n+1}-y_1 x_{n+1})+(x_2 y_{n+2}-x_{n+2} y_2)+\cdots+(x_{n-1} y_{2n-1}-x_{2n-1} y_{n-1}) \\ +(x_n y_{2n}-x_{2n} y_n)$$

stays invariant under the substitutions $x\to Ax$ and $y\to Ay$. One often defines the group $\mathrm{Sp}(n,\mathbb{R})$ by considering the bilinear form

$$(x_1 y_2-x_2 y_1)+\cdots+(x_{2n-1} y_{2n}-x_n y_{2n-1})$$

and the corresponding $2n\times 2n$ diagonal block matrix M having $\begin{pmatrix} 0 & 1 \\ -1 & 0 \end{pmatrix}$ along its main diagonal. The Lie algebra $\mathfrak{sp}(n,\mathbb{R})$ consists of those $2n\times 2n$ real matrices $X=\begin{pmatrix} X_{11} & X_{12} \\ X_{21} & X_{22} \end{pmatrix}$ in which X_{12} and X_{21} are $n\times n$ symmetric real matrices and $X_{22}=-X_{11}^t$. Therefore the dimension of this Lie group and its Lie algebra is n^2+n+1.

The determination of these Lie algebras is a trivial task because we have the following general result:

Proposition 11. *If M is an $n\times n$ real matrix then the Lie algebra of the linear Lie group $\{A\colon A\in \mathrm{Gl}(n,\mathbb{R})$ and $A^t M A=M\}$ consists of those $n\times n$ real matrices X which satisfy the identity $X^t M+MX=0$.*

Proof. If X is in the Lie algebra then for all values of the real variable u we have $(\exp uX)^t M \exp uX=M$. Hence

$$(\exp uX^t)M=M\exp -uX\,.$$

Since $\exp uY=I+uY+o(u)$ for every matrix Y as $u\to 0$ the above identity implies that $X^t M+MX=0$. Conversely let X be any real matrix satisfying this relation. Then by induction on $k=1,2,\ldots$ we can easily show that $(X^t)^k M=(-1)^k M X^k$. Hence for every real u we have

$$(\exp uX^t)M=M(\exp -uX) \quad \text{i.e.}\ (\exp uX)^t M(\exp uX)=M\,.$$

This shows that the curve $u \mapsto \exp uX$ lies entirely within our Lie group. Hence by Proposition 2 X belongs to its Lie algebra.

Of course all these groups are meaningful also when R is an arbitrary commutative ring with identity but the groups obtained in this manner are generally not Lie groups. Nevertheless $\mathrm{O}(n,\cdot)$, $\mathrm{SO}(n,\cdot)$, $\mathrm{S}(p,q,\cdot)$ and $\mathrm{Sp}(n,\cdot)$ are well defined entities. We obtain Lie groups if we let $R=\mathbb{C}$ in Lemma 10 because $\mathrm{Gl}(n,\mathbb{C})$ is a Lie group. There are two standard ways in which it can be given a differentiable structure: First, $\mathrm{Gl}(n,\mathbb{C})$ is an open subset of $M_n(\mathbb{C})$ and so it has a global coordinate neighborhood. The manifold structure obtained in this manner is complex analytic and for this reason $\mathrm{Gl}(n,\mathbb{C})$ is called a *complex Lie group*. Another way of obtaining essentially the same differentiable structure is by injecting $\mathrm{Gl}(n,\mathbb{C})$ into $\mathrm{Gl}(2n,\mathbb{R})$ via the isomorphism $z=u+iv \mapsto \begin{pmatrix} u & v \\ -v & u \end{pmatrix}$ and using Theorem 9.

Other important Lie groups can be defined by considering a variant of Lemma 10. Namely we take complex $n\times n$ matrices A. As usual we let $A^* = \bar{A}^t$, the Hermitian adjoint of A. Then we have:

Lemma 12. *If M is an $n\times n$ complex matrix then the set*

$$\{A : A\in \mathrm{Gl}(n,\mathbb{C}) \text{ and } A^* M A = M\}$$

is a subgroup of $\mathrm{Gl}(n,\mathbb{C})$.

Note. We could also use the defining equations $AMA^* = M$ and $A^{-1}MA = M$.

Proof. One uses the identities $(AB)^* = B^* A^*$ and $(A^{-1})^* = (A^*)^{-1}$ to prove the closure property under matrix multiplication and matrix inversion.

It is easy to show that a matrix A belongs to the subgroup if and only if the transformations $x\to Ax$ and $y\to Ay$ leave the Hermitian form $H(x,y)=y^* M x$ invariant: Indeed if A is in the subgroup then $(Ay)^* M(Ax) = y^*(A^* M A)x = y^* M x$ for any pair of complex column vectors x, y. Conversely if $y^*(A^* M A)x = y^* M x$ holds for all these vectors then $A^* M A = M$.

By considering the case $M=I$ we see that the *unitary group* $U(n)$ is a closed subgroup of $\mathrm{Gl}(n,\mathbb{C})$ and so by Theorem 9 $U(n)$ is a real Lie group. The *pseudo unitary group* $U(p,q)$ is obtained by choosing $M = \begin{pmatrix} I_p & 0 \\ 0 & -I_q \end{pmatrix}$. Consequently the elements of $U(p,q)$ are those matrices

A for which the Hermitian form

$$x_1\overline{y_1}+\cdots+x_p\overline{y_p}-x_{p+1}\overline{y_{p+1}}-\cdots-x_n\overline{y_n}$$

is left invariant under the transformations $x\to Ax$ and $y\to Ay$. It is clear that $U(n,0)=U(n,0)=U(n)$. The Lie algebra $\mathfrak{u}(p,q)$ can be identified with the Lie algebra of those $n\times n$ complex matrices $Z=\begin{pmatrix} Z_{11} & Z_{12} \\ Z_{21} & Z_{22}\end{pmatrix}$ in which Z_{11} and Z_{22} are skew Hermitian matrices of orders p and q, respectively, and $Z_{21}=Z_{12}^*$. Therefore it follows that the dimension of all these groups and algebras is n^2. The elements of the algebra $\mathfrak{su}(p,q)$ of $\mathrm{SU}(p,q)=U(p,q)\cap \mathrm{Sl}(n,\mathbb{C})$ are obtained by requiring also that $\operatorname{tr} Z_{11}+\operatorname{tr} Z_{22}=0$. Thus $\mathfrak{su}(p,q)$ and $\mathrm{SU}(p,q)$ are n^2-1 dimensional. Due to their importance we are going to prove these results in the special case $p=n$, $q=0$.

Proposition 13. *The Lie algebra* $\mathfrak{u}(n)$ *of* $U(n)$ *consists of all skew Hermitian* $n\times n$ *complex matrices and such a matrix belongs to* $\mathfrak{su}(n)$, *the Lie algebra of* $\mathrm{SU}(n)$ *if and only if its trace is zero.*

Proof. The groups $U(n)$ and $\mathrm{SU}(n)$ are connected but the result can be obtained without knowing this fact. Let Z denote an $n\times n$ complex matrix or if we wish the $2n\times 2n$ real matrix obtained from such a complex matrix via the correspondence $u+iv\mapsto \begin{matrix} u & v \\ -v & u\end{matrix}$. By Lemma 6.24 for small values of t we have

$$\log(\exp tZ \exp tZ^*)=t(Z+Z^*)+o(t)\,. \tag{5}$$

If Z is in the Lie algebra of $U(n)$ then for all real t we have $(\exp tZ)(\exp tZ)^*=I$ where $(\exp tZ)^*=\exp tZ^*$. Since $\log I=0$ using (5) we obtain $Z+Z^*=0$. If the group is $\mathrm{SU}(n)$ then in addition $\det\exp tZ=1$ and so by Lemma 6.32 we have $\operatorname{tr} Z=0$. Conversely if $Z+Z^*=0$ then by $tZ=-tZ^*$ we have $\exp tZ(\exp -tZ)^*=I$ and so $\exp tZ\in U(n)_o$ for every $t\in(-\infty,\infty)$. Hence by Proposition 2 the matrix Z belongs to the Lie algebra of $U(n)_o$ which is the same as that of $U(n)$. If in addition $\operatorname{tr} Z=0$ then by Lemma 6.32 $\exp tZ=1$ and so $\exp tZ\in\mathrm{SU}(n)_o$ for all t. Consequently Z is in the Lie algebra of $\mathrm{SU}(n)_o$.

There is a general result corresponding to Proposition 11 also in the complex case. Namely by applying the same method as the one used to obtain Proposition 11 one proves the following:

The Lie algebra of the Lie group defined in Lemma 12 consists of those $n\times n$ *complex matrices* Z *which satisfy the identity* $Z^*M+MZ=0$.

By Proposition 11 the Lie algebra $\mathfrak{sp}(n, \mathbb{C})$ is isomorphic with the Lie algebra of those $n \times n$ complex matrices $Z = \begin{pmatrix} Z_{11} & Z_{12} \\ Z_{21} & Z_{22} \end{pmatrix}$ in which Z_{12} and Z_{21} are symmetric $n \times n$ matrices and $Z_{22} = -Z_{11}^t$. By definition $\mathrm{Sp}(n) = \mathrm{Sp}(n, \mathbb{C}) \cap U(2n)$. The Lie algebra $\mathfrak{sp}(n)$ consists of those $2n \times 2n$ complex matrices $Z = \begin{pmatrix} Z_1 & Z_2 \\ -\overline{Z_2} & \overline{Z_1} \end{pmatrix}$ in which Z_1 is skew Hermitian and Z_2 is symmetric. It follows that the dimension of $\mathrm{Sp}(n, \mathbb{C})$ is $4n^2 + 2n$ and $\mathrm{Sp}(n)$ is $2n^2 + n$ dimensional.

The groups $U(2)$ and $\mathrm{SU}(2)$ deserve special attention. If $A = \begin{pmatrix} \alpha & \beta \\ \gamma & \delta \end{pmatrix}$ is in $U(2)$ then

$$\begin{pmatrix} \overline{\alpha} & \overline{\gamma} \\ \overline{\beta} & \overline{\delta} \end{pmatrix} = \begin{pmatrix} \alpha & \beta \\ \gamma & \delta \end{pmatrix}^* = \begin{pmatrix} \alpha & \beta \\ \gamma & \delta \end{pmatrix}^{-1} = \mathrm{sgn}(\alpha\delta - \beta\gamma) \begin{pmatrix} \delta & -\beta \\ -\gamma & \alpha \end{pmatrix}.$$

Hence $\overline{\alpha} = \pm \delta$ and $\overline{\gamma} = \mp \beta$. Consequently

$$U(2) = \left\{ \pm \begin{pmatrix} \alpha & \beta \\ -\overline{\beta} & \overline{\alpha} \end{pmatrix} : \alpha\overline{\alpha} + \beta\overline{\beta} = 1 \right\}$$

and by choosing only the $+$ sign we obtain the elements of $\mathrm{SU}(2)$.

The stereographic projections associated with the poles $(0, 0, 0, 1)$ and $(0, 0, 0, -1)$ define a C^∞ structure on the sphere S^3. The group $\mathrm{SU}(2)$ can be identified with the sphere S^3. For we let

$$\Phi(a_1, a_2, b_1, b_2) = \begin{pmatrix} \alpha & \beta \\ -\overline{\beta} & \overline{\alpha} \end{pmatrix}$$

where $a_1^2 + a_2^2 + b_1^2 + b_2^2 = 1$ and $\alpha = a_1 + i a_2$, $\beta = b_1 + i b_2$. Then on the one hand a group structure is defined on S^3, namely

$$(a_1, a_2, b_1, b_2)(c_1, c_2, d_1, d_2) = (\alpha, \beta)(\gamma, \delta) = (\alpha\gamma - \beta\overline{\delta}, \alpha\delta + \beta\overline{\gamma}).$$

On the other hand $\mathrm{SU}(2)$ inherites the C^∞ structure of S^3. By Theorem 9 $\mathrm{SU}(2)$ admits only one C^∞ structure which makes it a Lie subgroup of $\mathrm{Gl}(2, \mathbb{C})$. Hence if we prove that the injection $\Phi: S^3 \to \mathrm{Gl}(2, \mathbb{C})$ is everywhere regular then it follows that the Lie group structure of $\mathrm{SU}(2)$ is the one inherited from S^3. Let p in S^3 and $\mathfrak{v}$ in $(S^3)_p$ be such that $d\Phi_p \mathfrak{v} = 0$. Then $\mathfrak{v}(f\Phi) = 0$ for every C^∞ function $f: \mathrm{Gl}(2, \mathbb{C}) \to \mathbb{R}$. If $f: S^3 \to \mathbb{C}$ then f can be extended to a map $f: \mathrm{Gl}(2, \mathbb{C}) \to \mathbb{R}$ by defining $f(A) = f(U)$ where U is unitary, $A = UP$ and P is positive definite. Hence we see that $\mathfrak{v}(f) = 0$ for every C^∞ function $f: S^3 \to \mathbb{R}$. Therefore $\mathfrak{v} = 0$ and the map Φ is regular at p where p is an arbitrary point in S^3. Hence $\Phi: S^3 \to \mathrm{Gl}(2, \mathbb{C})$ is a submanifold. We proved the following:

Proposition 14. *The differentiable structure of the Lie group* SU(2) *is the one induced by the* C^∞ *structure of the sphere* S^3 *under the map* $(\alpha, \beta) \mapsto \begin{pmatrix} \alpha & \beta \\ -\bar{\beta} & \bar{\alpha} \end{pmatrix}$ *where* $\alpha = a_1 + i a_2$, $\beta = b_1 + i b_2$ *and* $a_1^2 + a_2^2 + b_1^2 + b_2^2 = 1$.

8. Invariant Lie Subgroups and Quotients of Lie Groups. The Projective Groups and the Lorentz Group

Let $\mathfrak{g}$ be a real, finite dimensional Lie algebra and let $X_1, \ldots, X_n$ be an ordered basis in $\mathfrak{g}$. We associate with every linear transformation $T: \mathfrak{g} \to \mathfrak{g}$ its matrix with respect to this ordered basis. This defines an isomorphism Φ of $\mathscr{L}(\mathfrak{g})$ onto $M_n(\mathbb{R})$ and so the same map Φ is an isomorphism of $\mathfrak{gl}(\mathfrak{g})$ onto $\mathfrak{gl}(n, \mathbb{R})$. Since there is a natural bijection of $\mathfrak{gl}(n, \mathbb{R})$ onto $\mathbb{R}^{n^2}$ it is a real analytic manifold and $\mathrm{Gl}(n, \mathbb{R})$ is a submanifold which is a Lie group. By using Φ the manifold structure of $\mathfrak{gl}(n, \mathbb{R})$ can be transferred onto $\mathfrak{gl}(\mathfrak{g})$ and $\mathrm{Gl}(\mathfrak{g})$ where $\mathrm{Gl}(\mathfrak{g})$ is the group of invertible linear transformations $A: \mathfrak{g} \to \mathfrak{g}$. Then $\mathrm{Gl}(\mathfrak{g})$ is a submanifold of $\mathfrak{gl}(n, \mathbb{R})$ and it is a Lie group. The real analytic structures introduced in this manner on $\mathfrak{gl}(n, \mathbb{R})$ and $\mathfrak{gl}(\mathfrak{g})$ are independent of the choice of $X_1, \ldots, X_n$. The restricted map

$$\Phi: \mathrm{Gl}(\mathfrak{g}) \to \mathrm{Gl}(n, \mathbb{R})$$

is an analytic diffeomorphism and so denoting by $\mathfrak{gl}$ the Lie algebra of $\mathrm{Gl}(\mathfrak{g})$ the map

$$d\Phi_e: \mathfrak{gl} \to \mathfrak{gl}(n, \mathbb{R})$$

is a surjective Lie algebra isomorphism by Theorem 5.17 and Lemma 6.13. Let $\mathbb{R}^n$ be the vector space of column vectors $X = (x_1, \ldots, x_n)^t$. We can identify $\mathfrak{gl}(n, \mathbb{R})$ with $\mathfrak{gl}(\mathbb{R}^n)$ by associating with the matrix T in $\mathfrak{gl}(n, \mathbb{R})$ the linear transformation $X \mapsto TX$ where TX is a matrix product. The same symbol T will be used to denote the $n \times n$ matrix in $\mathfrak{gl}(n, \mathbb{R})$ and the corresponding transformation in $\mathfrak{gl}(\mathbb{R}^n)$ and the map itself is

$$\Psi: \mathfrak{gl}(n, \mathbb{R}) \to \mathfrak{gl}(\mathbb{R}^n).$$

We can now construct the following commutative diagram:

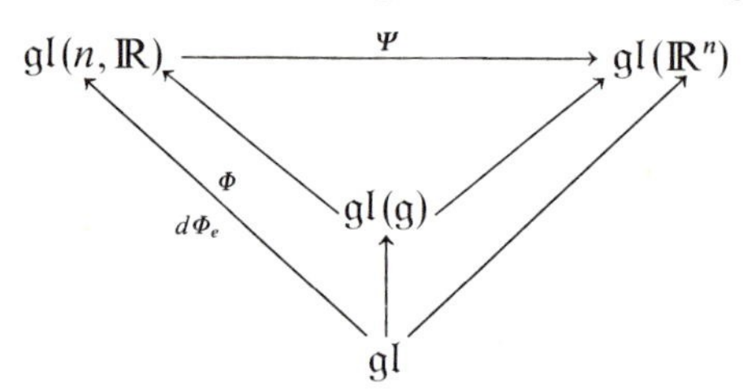

Here $\mathfrak{gl}(\mathfrak{g}) \to \mathfrak{gl}(\mathbb{R}^n)$ is the composition $\Phi \circ \Psi$ and $\mathfrak{gl} \to \mathfrak{gl}(\mathfrak{g})$ is defined as $d\Phi_e \circ \Phi^{-1}$. This uniquely determines $\mathfrak{gl} \to \mathfrak{gl}(\mathbb{R}^n)$ as the composition of $\mathfrak{gl} \to \mathfrak{gl}(\mathfrak{g})$ with $\mathfrak{gl}(\mathfrak{g}) \to \mathfrak{gl}(\mathbb{R}^n)$. The map $\mathfrak{gl} \to \mathfrak{gl}(\mathfrak{g})$ identifies the Lie algebra of $\mathrm{Gl}(\mathfrak{g})$ with $\mathfrak{gl}(\mathfrak{g})$.

Let $\mathrm{Aut}(\mathfrak{g})$ be the group of automorphisms of $\mathfrak{g}$. Thus an element A of $\mathrm{Gl}(\mathfrak{g})$ belongs to $\mathrm{Aut}(G)$ if and only if $A[X,Y]=[AX,AY]$ for every X and Y in $\mathfrak{g}$. It is clear that $\mathrm{Aut}(\mathfrak{g})$ is a closed subgroup of $\mathrm{Gl}(\mathfrak{g})$ and so by Theorem 7.9 $\mathrm{Aut}(\mathfrak{g})$ is a Lie subgroup of $\mathrm{Gl}(\mathfrak{g})$. By Proposition 7.2 the Lie algebra $\mathfrak{aut}(\mathfrak{g})$ of $\mathrm{Aut}(\mathfrak{g})$ is isomorphic to a subalgebra of $\mathfrak{gl}$ and so $\mathfrak{aut}(\mathfrak{g})$ can be identified with a Lie subalgebra of $\mathfrak{gl}(\mathfrak{g})$.

Proposition 1. *The Lie algebra $\mathfrak{d}(\mathfrak{a})$ of all derivations of the real Lie algebra $\mathfrak{g}$ is the Lie algebra $\mathfrak{aut}(\mathfrak{g})$ of the group of automorphisms* $\mathrm{Aut}(\mathfrak{g})$.

Proof. Let $\mathbb{R}^n$ be the space of column vectors $X=(x_1,\ldots,x_n)^t$. The ordered basis $X_1,\ldots,X_n$ chosen in the beginning determines an isomorphism of $\mathfrak{g}$ onto $\mathbb{R}^n$ and so we can define $[X,Y]$ for any two column vectors in $\mathbb{R}^n$ by using the corresponding operation in $\mathfrak{g}$. This turns $\mathbb{R}^n$ into a Lie algebra which is isomorphic to $\mathfrak{g}$. We shall use the same symbol to denote an element of $\mathfrak{g}$ and the corresponding column vector in $\mathbb{R}^n$. The map Φ can be used to identify the elements of $\mathfrak{gl}(\mathfrak{g})$ with those of $\mathfrak{gl}(n,\mathbb{R})$ and for the sake of simplicity we shall use the same symbol T for the matrix $\Phi(T)$ and the linear transformation $T\colon \mathfrak{g} \to \mathfrak{g}$. Given such a T and a real number t the matrix e^{tT} belongs to $\mathrm{Gl}(n,\mathbb{R})$. Thus it corresponds to a unique element in $\mathrm{Gl}(\mathfrak{g})$ which is $\exp tT$ by Proposition 6.19. By using the power series representation of e^{tT} we obtain

(1) $$e^{tT} = I + tT + o(t) \quad \text{as } t \to 0.$$

Given X and Y in $\mathfrak{g}$ let $f(t)=e^{tT}[X,Y]$ for all t in $(-\infty,\infty)$. Using (1) we see that

$$f'(0) = \lim_{t\to 0} \frac{e^{tT}[X,Y]-[X,Y]}{t}$$

exists and $f'(0)=T[X,Y]$. If T is in $\mathfrak{aut}(\mathfrak{g})$ then by Proposition 7.2 $\exp tT$ lies in $\mathrm{Aut}(\mathfrak{g})$ for every t in $(-\infty,\infty)$. Hence by (1) and the identity

$$e^{tT}[X,Y]-[X,Y] = [e^{tT}X - X, e^{tT}Y] + [X, e^{tT}Y - Y]$$

we obtain $f'(0)=[TX,Y]+[X,TY]$. Thus every $T \in \mathfrak{aut}(\mathfrak{g})$ is a derivation of $\mathfrak{g}$.

Conversely let us suppose that D is a derivation of $\mathfrak{g}$. Then the corresponding matrix D in $\mathfrak{gl}(n,\mathbb{R})$ is such that

$$D[X,Y]=[DX,Y]+[X,DY] \tag{2}$$

for any two column vectors X and Y in $\mathbb{R}^n$. From (2) we obtain

$$D^k[X,Y]=\sum_{i=0}^{k}\binom{k}{i}[D^iX,D^{k-i}Y]$$

for every $k=0,1,2,\ldots$ Therefore

$$e^{tD}[X,Y]=\sum_{k=0}^{\infty}\frac{1}{k!}D^k[X,Y]=\sum_{k=0}^{\infty}\sum_{i+j=k}\frac{1}{i!j!}[D^iX,D^jY]=[e^{tD}X,e^{tD}Y].$$

This proves that $\exp tD$ is in $\operatorname{Aut}(\mathfrak{g})$ for every t in $(-\infty,\infty)$ and so D belongs to $\mathfrak{aut}(\mathfrak{g})$ by Proposition 7.2. This completes the proof.

Now we consider the Lie derivatives $\theta(A)\colon\mathfrak{g}\to\mathfrak{g}$ where $\theta(A)X=[A,X]$ for all $X\in\mathfrak{g}$. By Proposition 1.12 the set of these inner derivations of $\mathfrak{g}$ is an ideal $\operatorname{ad}\mathfrak{g}$ of the derivation algebra $\mathfrak{d}(\mathfrak{g})$. Since $\mathfrak{d}(\mathfrak{g})$ is the Lie algebra of $\operatorname{Aut}(\mathfrak{g})$ by Theorem 7.3 there is a unique analytic subgroup which corresponds to the ideal $\operatorname{ad}\mathfrak{g}$ of $\mathfrak{d}(\mathfrak{g})$. This invariant Lie subgroup of $\operatorname{Aut}(\mathfrak{g})$ is called the *adjoint group* and it is denoted by $\operatorname{Ad}(\mathfrak{g})$.

Let G be a Lie group and let $\mathfrak{g}$ be its Lie algebra. If $a\in G$ then $x\to axa^{-1}$ $(x\in G)$ is a diffeomorphic automorphism of G which we denote by $\Phi(a)$. Then $\operatorname{Ad}(a)=d\Phi(a)_e$ is an automorphism of $\mathfrak{g}$. Since

$$\begin{aligned}(d\Phi(ab)_eX)(f)&=X(f\Phi(ab))=X(f\Phi(a)\Phi(b))=(d\Phi(b)_eX)(f\Phi(a))\\&=\big(d\Phi(a)_e(d\Phi(b)_eX)\big)(f)=(d\Phi(a)_e d\Phi(b)_eX)(f)\end{aligned}$$

we see that $a\mapsto\operatorname{Ad}(a)$ is a homomorphism of G into $\operatorname{Aut}(\mathfrak{g})$. It is called the *adjoint representation* of G and it can be proved that it is a C^∞ map. However we shall not use this last property.

Lemma 2. *For every* $a\in G$ *and* $X\in\mathfrak{g}$ *we have* $\exp(\operatorname{Ad}(a)X)=a(\exp X)a^{-1}$.

Proof. This follows from the definition of $\operatorname{Ad}(a)$ by Proposition 6.19.

Lemma 3. *For every* X *in* $\mathfrak{g}$ *we have* $\operatorname{Ad}(\exp X)=\exp(\operatorname{ad}X)$.

Proof. Using Proposition 6.20 we see that

$$\begin{aligned}&\log(\exp X\exp Y\exp Z)\\&=(X+Y+Z)+\tfrac{1}{2}([X,Y]+[X,Z]+[Y,Z])+O(\|(X,Y,Z)\|^3)\end{aligned}$$

in the neighborhood of $(0,0,0)$ in $\mathfrak{g}\times\mathfrak{g}\times\mathfrak{g}$. Therefore by Lemma 2

$$\begin{aligned}\exp(\operatorname{Ad}(\exp tX)tY) &= \exp tX \exp tY \exp -tX \\ &= \exp(tY + t^2[X,Y] + O(t^3)).\end{aligned}$$

This shows that $\operatorname{Ad}(\exp tX)Y = Y + t[X,Y] + O(t^2)$. Using this with $Y = Y_1, \ldots, Y_n$ where $Y_1, \ldots, Y_n$ is a basis of $\mathfrak{g}$ we obtain:

$$\operatorname{Ad}(\exp tX) = I + t \operatorname{ad} X + O(t^2). \tag{3}$$

where I is the identity of $\mathrm{Gl}(\mathfrak{g})$ and t varies in the neighborhood of 0.

Now we consider the curve $\gamma(t) = \operatorname{Ad}(\exp tX)$ where $-\infty < t < \infty$. We have $\gamma(0) = \operatorname{Ad}(e) = I$ and by (3) $\gamma'(0) = \operatorname{ad} X$. Moreover for any real t we obtain

$$\gamma'(t) = \operatorname{Ad}(\exp tX) \lim_{\tau\to t} \frac{\operatorname{Ad}(\exp(\tau - t)X) - I}{\tau - t} = \gamma(t)\gamma'(0).$$

Hence γ is a maximal geodesic with $\gamma'(0) = \operatorname{ad} X$. Since $\exp(\operatorname{ad} X) = \gamma(1)$ the lemma is proved.

The following result should be considered as an addition to the results expressed in Proposition 7.2 and Theorem 7.3:

Proposition 4. *Let G be a Lie group with Lie algebra $\mathfrak{g}$ and let H be an analytic subgroup with Lie algebra $\mathfrak{h}$ so that $\mathfrak{h} \subseteq \mathfrak{g}$. Then H is an invariant subgroup of the identity component G_o if and only if $\mathfrak{h}$ is an ideal of $\mathfrak{g}$.*

Proof. First we suppose that H is an invariant subgroup of G_o. Then for every $a \in G$ we have $\Phi(a)H = aHa^{-1} \subseteq H$ and so by Proposition 7.2 the Lie algebra of $\Phi(a)H$ lies in $\mathfrak{h}$, which is the Lie algebra of G_o and G. Since $\Phi(a)\colon H \to \Phi(a)H = H$ is a diffeomorphic isomorphism $d\Phi(a)_e = \operatorname{Ad}(a)$ maps $\mathfrak{h}$ onto the Lie algebra of $\Phi(a)H$ and so $\operatorname{Ad}(a)\mathfrak{h} = \mathfrak{h}$. Now given $X \in \mathfrak{g}$ and $Y \in \mathfrak{h}$ we have $\operatorname{Ad}(\exp tX)Y \in H$ for every t and so

$$\frac{d}{dt}(\operatorname{Ad}(\exp tX)Y) = (\operatorname{ad} X)Y = [X,Y] \in \mathfrak{h}.$$

This proves that $\mathfrak{h}$ is an ideal in $\mathfrak{g}$.

Conversely let us suppose that $[X,Y] \in \mathfrak{h}$ for every $X \in \mathfrak{g}$ and $Y \in \mathfrak{h}$ or in other words $(\operatorname{ad} X)\mathfrak{h} \subseteq \mathfrak{h}$. This implies that $(\operatorname{ad} X)^n \mathfrak{h} \subseteq \mathfrak{h}$ for every X and so using the power series representation of $\exp(\operatorname{ad} X)$ we obtain $(\exp \operatorname{ad} X)Y \in \mathfrak{h}$ for every $X \in \mathfrak{g}$ and $Y \in \mathfrak{h}$. Thus by applying Lemma 3 we obtain $\mathrm{Ad}(\exp X) Y \in \mathfrak{h}$ and so by Theorem 7.3 we see that $\exp(\mathrm{Ad}(\exp X) Y)$ is in H for all $X \in \mathfrak{g}$ and $Y \in \mathfrak{h}$. By applying Lemma 2 it follows that

$$(\exp X)(\exp Y)(\exp X)^{-1} \in H \qquad (X \in \mathfrak{g} \text{ and } Y \in \mathfrak{h}).$$

Now we keep Y fixed in $\mathfrak{h}$ and let X vary in $\mathfrak{g}$. By Proposition 6.15 and Lemma 5.2 we obtain from this $a(\exp Y)a^{-1} \in H$ for the arbitrary, fixed $Y \in \mathfrak{h}$ and for every $a \in G_o$. Hence by the same proposition and lemma we arrive at the relation $aha^{-1} \in H$ for every $h \in H$ and $a \in G_o$. This proves that H is an invariant subgroup of G_o.

Theorem 5. *Let G be a Lie group and H a closed subgroup of G. Then a C^∞ structure can be introduced on G/H such that the underlying topology is the quotient topology of G/H and if H is a normal subgroup then G/H is a Lie group with Lie algebra $\mathfrak{g}/\mathfrak{h}$ where $\mathfrak{g}$ and $\mathfrak{h}$ denote the Lie algebras of G and H, respectively.*

Proof. We choose a vector subspace $\mathfrak{a}$ of $\mathfrak{g}$ such that $\mathfrak{g} = \mathfrak{a} \oplus \mathfrak{h}$. By Lemma 7.8 there is an open neighborhood O_0 of 0 in $\mathfrak{a} \oplus \mathfrak{h}$ such that $(A, X) \mapsto \exp A \exp X$ is an isomorphism of O_0 onto an open neighborhood O_e of e in G. Let denote this isomorphism. Then there are open sets $O_{0\mathfrak{a}}$ and $O_{0\mathfrak{h}}$ in $\mathfrak{a}$ and $\mathfrak{h}$, respectively, such that $O_{0\mathfrak{a}} \times O_{0\mathfrak{h}} \subseteq O_0$ and $Q_e = \Phi(O_{0\mathfrak{a}} \times O_{0\mathfrak{h}})$ is an open set containing e and $Q_e^{-1} Q_e \subseteq O_e$. We prove that $S = \exp O_{0\mathfrak{a}}$ intersects xH in exactly one point for every $x \in Q_e$. Indeed if $x \in Q_e$ then $x = \exp A \exp X$ where $A \in O_{0\mathfrak{a}}$ and $X \in O_{0\mathfrak{h}}$. By $A \in O_{0\mathfrak{a}}$ we have $\exp A \in S$ and by $\exp X \in H$ we have $\exp A \in xH$. Thus S intersects every fiber xH $(x \in Q_e)$. Let us suppose that the vectors A_i $(i = 1, 2)$ are such that $A_i \in O_{0\mathfrak{a}}$ and $\exp A_i \in xH$ for some $x \in Q_e$. Then $(\exp A_1)^{-1} \exp A_2$ is in $H \cap O_e$ and so $(\exp A_1)^{-1} \exp A_2 = \exp X$ for a suitable X in $\mathfrak{h} \in O_0$. Then $\exp A_2 \exp O = \exp A_1 \exp X$ and so using the isomorphism property of the map Φ we see that $X = 0$ and $A_1 = A_2$.

We define $\varphi: Q_e \to \mathfrak{a}$ by $\varphi(xH) = \log(S \cap xH)$ where $x \in Q_e$. This gives a one-to-one correspondence between the set $O = \{xH : x \in Q_e\}$ and the open set $O_{0\mathfrak{a}}$ in $\mathfrak{a}$. Thus for every p in G the map $\varphi^p : pO \to \mathfrak{a}$ is a bijection of pO onto $O_{0\mathfrak{a}}$. We can easily check that the coordinate neighborhoods (pO, φ^p) $(p \in G)$ define a C^∞ structure on G/H. For if p and q are such that $pO \cap qO \neq \emptyset$ then $\varphi^q(\varphi^p)^{-1}$ is a C^∞ map because by Proposition 6.15 $(\varphi^p)^{-1} = p\exp$ and $\varphi^q = {}_y\log$ are infinitely differentiable.

If H is a closed subgroup of G then G/H is a Lie group with respect to the C^∞ structure constructed above. The natural map $\Psi: G \to G/H$

is a C^∞ group homomorphism and so $d\Psi_e:\mathfrak{g}\to\mathfrak{k}$ where $\mathfrak{k}$ denotes the Lie algebra of G/H. The Lie algebra homomorphism $d\Psi_e$ is surjective and one can show that its kernel is $\mathfrak{h}$. This will then prove that $\mathfrak{k}$ is isomorphic to $\mathfrak{g}/\mathfrak{h}$. We omit the details.

We can use Theorem 5 to prove for instance that the group of projective transformations $\mathrm{Pl}(n,\mathbb{R})$ is a Lie group. We review briefly the concepts involved: Let F be an arbitrary field and let $P^n(F)$ be the n dimensional projective space over F. By definition it is the set of all one dimensional subspaces of F^{n+1}. In other words we consider those $(n+1)$-tuples $x=(x_0,x_1,\ldots,x_n)^t$ $(x_0,x_1,\ldots,x_n\in F)$ which are not $(0,0,\ldots,0)^t$ and call any two of them equivalent if they differ by a factor $\lambda\in F^\times$. It will be convenient to interpret these $(n+1)$-tuples as column vectors. We let ξ denote the equivalence class containing x. Then $P^n(F)$ is the set of these equivalence classes. If $A=(a_{ij})\in\mathrm{Gl}(n+1,F)$ then Ax is in F^{n+1} and $A\lambda x=\lambda Ax$ which shows that Ax and λAx belong to the same equivalence class. Hence $A\xi$ is defined for every ξ in $P^n(F)$ as an element of $P^n(F)$. The map $\xi\mapsto A\xi$ is called the *projective transformation, projective linear transformation* or *projectivity* induced on $P^n(F)$ by A. Let P_A denote this transformation. Then it is clear that $P_A P_B=P_{AB}$ and P_I is the identity map on $P^n(F)$. Therefore the set of projective transformations is a group under composition. This group is called the *group of projective transformations* and it will be denoted by $\mathrm{Pl}(n,F)$.

By a *hyperplane* in $P^n(F)$ we mean the set of those elements $\xi=(x)$ which satisfy a linear equation

$$a_0x_0+a_1x_1+\cdots+a_nx_n=0.$$

It is clear that a projective transformation P maps hyperplanes into hyperplanes. This is the meaning of linearity for projective transformations and for this reason they used to be called *collineations* in the case when $F=\mathbb{R}$. If F is arbitrary then there might exist collineations i.e. transformations $T\colon P^n(F)\to P^n(F)$ mapping hyperplanes into hyperplanes which are not projective transformations. This occurs already in the case $F=\mathbb{C}$ due to its automorphism $z\mapsto\bar{z}$.

Since $P_AP_B=P_{AB}$ it is clear that $A\mapsto P_A$ is a homomorphism of $\mathrm{Gl}(n+1,F)$ onto $\mathrm{Pl}(n,F)$. If the projectivity P_A is the identity then there is a $\lambda\in F^\times$ such that

$$a_{i0}x_0+a_{i1}x_1+\cdots+a_{in}x_n=\lambda x_i \qquad (i=0,1,\ldots,n)$$

for every x in $P^n(F)$ and so it follows that $A=\lambda I$. Conversely it is clear that $P_{\lambda I}$ is the identity map. Hence the kernel of $A\mapsto P_A$ is $F^\times\cdot I=\{\lambda I:\lambda\in F^\times\}$ and so $\mathrm{Pl}(n,F)$ is isomorphic to $\mathrm{Gl}(n+1,F)/F^\times\cdot I$.

We note that $F^{\times} \cdot I$ is the center of $\mathrm{Gl}(n+1,F)$. Similarly $\mathrm{Pl}(n,F)$ is isomorphic to $\mathrm{Sl}(n+1,F)/K$ where K is the group of those $\lambda \in F^{\times}$ for which $\lambda^n = 1$. In particular $\mathrm{Pl}(2,\mathbb{R})$ and $\mathrm{Sl}(3,\mathbb{R}) \Big/ \begin{pmatrix} \pm 1 & 0 \\ 0 & \pm 1 \end{pmatrix}$ are isomorphic groups. Since $\mathrm{Sl}(3,\mathbb{R})$ is a Lie group by Theorem 5 it follows that $\mathrm{Pl}(2,\mathbb{R})$ is a Lie group and its Lie algebra is $\mathfrak{sl}(3,\mathbb{R})$. More generally we see that $\mathrm{Pl}(n,\mathbb{R})$ is a Lie group for every $n = 1,2,\ldots$ with Lie algebra $\mathfrak{sl}(n+1,\mathbb{R})$. Similarly $\mathrm{Pl}(n,\mathbb{C})$ is a Lie group and its Lie algebra is $\mathfrak{sl}(n+1,\mathbb{C})$ because the group of complex n-th roots of unity is finite and so its Lie algebra is 0.

It is considerably easier to prove that the underlying sets $P^n(\mathbb{R})$ and $P^n(\mathbb{C})$ are differentiable manifolds. For example in case of the reals we consider the sets O_i $(i=0,1,\ldots,n)$ where $(x) \in O_i$ if and only if $(x) \in P^n(\mathbb{R})$ and $x_i \neq 0$. Then we define

$$\varphi_i(\xi)_j = \varphi_i((x))_j = \frac{x_j}{x_i} \quad \text{where } j=1,\ldots,i-1,\, i+1,\ldots,n.$$

This defines a map φ of O_i onto $\mathbb{R}^n \setminus \{0\}$. The pairs (O_i,φ_i) $(i=0,1,\ldots,n)$ define a differentiable structure on $P^n(\mathbb{R})$: Indeed if $i \neq j$ and $x_j \neq 0$ then

$$\begin{aligned} \varphi_j \varphi_i^{-1}(x_0,\ldots,x_{i-1},x_{i+1},\ldots,x_n) &= \varphi_j(x_0,\ldots,x_{i-1},1,x_{i+1},\ldots,x_n) \\ &= \left(\frac{x_0}{x_j},\ldots,\frac{x_{i-1}}{x_j},\frac{1}{x_j},\frac{x_{i+1}}{x_j},\ldots,\frac{x_{j-1}}{x_j},\frac{x_{j+1}}{x_j},\ldots,\frac{x_n}{x_j}\right). \end{aligned}$$

Since the functions x_k/x_j $(k \neq i,j)$ and $1/x_j$ are real analytic it follows that $P^n(\mathbb{R})$ is an n dimensional real analytic manifold. Similarly one can see that $P^n(\mathbb{C})$ is a complex analytic manifold of real dimension $2n$. It is easy to show that $P^1(\mathbb{R})$ and the circle S^1 are diffeomorphic. Similarly one proves that $P^1(\mathbb{C})$ and the Riemann sphere are diffeomorphic.

The group $\mathrm{Pl}(n,\mathbb{R})$ acts on $P^n(\mathbb{R})$ and $(P,\xi) \mapsto P\xi$ is a C^∞ mapping of the product manifold $\mathrm{Pl}(n,\mathbb{R}) \times P^n(\mathbb{R})$ onto $P^n(\mathbb{R})$. We say that $\mathrm{Pl}(n,\mathbb{R})$ is a Lie transformation group acting on the manifold $P^n(\mathbb{R})$. In general we call a Lie group G a Lie transformation group if G acts on a differentiable manifold M such that $(g,p) \mapsto gp$ is a differentiable mapping of $G \times M$ onto M. Thus for every fixed group element g the map $p \mapsto gp$ is a diffeomorphism of M onto itself. Similarly $\mathrm{Pl}(n,\mathbb{C})$ is a Lie transformation group acting on $P^n(\mathbb{C})$. Simpler examples are obtained by considering various Lie subgroups of $\mathrm{Gl}(n,\mathbb{R})$ as groups acting on $\mathbb{R}^n$.

We shall now explain what is meant by *projective Banach spaces* and *projective representations.* The latter were already mentioned in Section IV.1. By definition a complex projective Banach space $\tilde{\mathscr{X}}$ is the set of all one dimensional subspaces of a complex Banach space $\mathscr{X}$. If ξ is a non-zero vector in $\mathscr{X}$ then we let $\tilde{\xi}=\mathbb{C}\xi$ so that $\tilde{\xi}$ is the one dimensional subspace containing ξ. We topologize $\tilde{\mathscr{X}}$ by the strongest topology which makes the map $\xi\mapsto\tilde{\xi}$ continuous when $\mathscr{X}$ is topologized by the topology derived from its norm. Thus a set $\tilde{\mathscr{O}}$ of $\tilde{\mathscr{X}}$ is open if and only if its inverse image $\mathscr{O}=\{\xi:\tilde{\xi}\in\tilde{\mathscr{O}}\}$ is an open set in $\mathscr{X}$. Similarly one can define also real projective Banach spaces.

A subset $\tilde{\mathscr{Y}}$ of $\tilde{\mathscr{X}}$ is called a projective subspace if it is a closed subset of $\tilde{\mathscr{X}}$ and $\tilde{\xi},\tilde{\eta}\in\tilde{\mathscr{Y}}$ implies that $\widetilde{\xi+\eta}\in\tilde{\mathscr{Y}}$. Hence if $\tilde{\mathscr{Y}}$ is a subspace then $\tilde{\xi},\tilde{\eta}\in\tilde{\mathscr{Y}}$ implies that $\widetilde{\lambda\xi+\mu\eta}\in\tilde{\mathscr{Y}}$ for every $(\lambda,\mu)\neq(0,0)$ where $\lambda,\mu\in\mathbb{C}$. It follows that $\mathscr{Y}=\{\eta:\tilde{\eta}\in\tilde{\mathscr{Y}}\}$ is a linear manifold in $\mathscr{X}$. Since $\mathscr{Y}$ is the inverse image of a closed set it is a subspace of $\mathscr{X}$. Conversely if $\mathscr{Y}$ is a subspace of $\mathscr{X}$ then $\tilde{\mathscr{Y}}=\{\tilde{\eta}:\eta\in\mathscr{Y}\}$ is a projective subspace of $\tilde{\mathscr{X}}$: The algebraic requirement is clearly satisfied by $\tilde{\mathscr{Y}}$. Let $\tilde{\eta}_i$ $(i\in\mathscr{I})$ be a convergent net in $\tilde{\mathscr{Y}}$ with limit $\tilde{\eta}\in\tilde{\mathscr{X}}$. We may suppose that $\|\eta_i\|=1$ for every index $i\in\mathscr{I}$ and $\|\eta\|=1$. Since $\{\tilde{\xi}:\xi\in\mathscr{X}$ and $\|\xi\|=1$ and $\|\xi-\eta\|<\varepsilon\}$ is an open set in $\tilde{\mathscr{H}}$ by $\tilde{\eta}_i\to\tilde{\eta}$ there is an index $i(\varepsilon)$ such that $\|\eta_i-\eta\|<\varepsilon$ for every i satisfying $i(\varepsilon)\leqslant i$. Since $\mathscr{Y}$ is closed this shows that $\eta\in\mathscr{Y}$ and so $\tilde{\eta}\in\tilde{\mathscr{Y}}$ proving that $\tilde{\mathscr{Y}}$ is a closed set in $\tilde{\mathscr{X}}$.

If $A:\mathscr{X}\to\mathscr{X}$ is an invertible continuous linear operator then we can define an operator $\tilde{A}:\tilde{\mathscr{X}}\to\tilde{\mathscr{X}}$ by $\tilde{A}\tilde{\xi}=\widetilde{A\xi}$. We let $\mathscr{P}\ell(\tilde{\mathscr{X}})=\{\tilde{A}: A\in \mathrm{Gl}(\mathscr{X})\}$ so that $\mathscr{P}\ell(\tilde{\mathscr{X}})$ is a group with identity $\tilde{I}$. The elements of $\mathscr{P}\ell(\tilde{\mathscr{X}})$ are all homeomorphisms of $\tilde{\mathscr{X}}$: For every $\tilde{A}:\tilde{\mathscr{X}}\to\tilde{\mathscr{X}}$ is a continuous bijection: In fact if $\tilde{\mathscr{O}}$ is an open set in $\tilde{\mathscr{X}}$ and $\mathscr{O}$ is its inverse image in $\mathscr{X}$ then $\mathscr{O}$ is open and so by the continuity of $A:\mathscr{X}\to\mathscr{X}$ we see that $A^{-1}\mathscr{O}$ is an open set in $\mathscr{X}$. However

$$A^{-1}\mathscr{O}=\{\xi: A\xi\in\mathscr{O}\}=\{\xi:\widetilde{A\xi}\in\tilde{\mathscr{O}}\}=\{\xi:\tilde{A}\tilde{\xi}\in\tilde{\mathscr{O}}\}$$

and so $A^{-1}\mathscr{O}$ is the inverse image of $\tilde{A}^{-1}\tilde{\mathscr{O}}$ under the map $\xi\mapsto\tilde{\xi}$. Therefore by the definition of the topology of $\tilde{\mathscr{X}}$ the set $\tilde{A}^{-1}\tilde{\mathscr{O}}$ is open in $\tilde{\mathscr{X}}$. Thus $\tilde{A}$ is a continuous map. Since several topologies are available on $\mathrm{Gl}(\mathscr{H})$ we have several choices when we define the topology of $\mathscr{P}\ell(\tilde{\mathscr{H}})$.

Let us now suppose that $\mathscr{H}$ is a complex Hilbert space. Then the inner product of $\mathscr{H}$ defines a complex valued function on $\tilde{\mathscr{H}}\times\tilde{\mathscr{H}}$, namely we let $(\tilde{\alpha},\tilde{\beta})=(\alpha,\beta)/\|\alpha\|\,\|\beta\|$ so that $|(\tilde{\alpha},\tilde{\beta})|\leqslant 1$ for all $\tilde{\alpha},\tilde{\beta}\in\mathscr{H}$. We let $\mathscr{U}(\tilde{\mathscr{H}})$ denote the image of $\mathscr{U}(\mathscr{H})$ under the map $A\mapsto\tilde{A}$. Thus $\mathscr{U}(\tilde{\mathscr{H}})$ is a closed subgroup of $\mathscr{P}\ell(\tilde{\mathscr{H}})$.

Lemma 6. *A necessary and sufficient condition that* $\tilde{U}\in\mathscr{P}\ell(\tilde{\mathscr{H}})$ *be an element of* $\mathscr{U}(\tilde{\mathscr{H}})$ *is that* $(\tilde{U}\tilde{\alpha},\tilde{U}\tilde{\beta})=(\tilde{\alpha},\tilde{\beta})$ *for every* $\tilde{\alpha},\tilde{\beta}\in\mathscr{H}$.

Proof. Suppose that U is the image of $U \in \mathrm{Gl}(\mathscr{H})$ under the map $A \mapsto \tilde{A}$. Then

$$(\tilde{U}\tilde{\alpha}, \tilde{U}\tilde{\beta}) = \frac{(U\alpha, U\beta)}{\|U\alpha\| \, \|U\beta\|}$$

for all $\alpha, \beta \in \mathscr{H}$. If U is unitary then the right hand side is $(\alpha, \beta)/\|\alpha\| \, \|\beta\| = (\tilde{\alpha}, \tilde{\beta})$. Conversely if $\tilde{U}$ is such that the identity $(\tilde{U}\tilde{\alpha}, \tilde{U}\tilde{\beta}) = (\tilde{\alpha}, \tilde{\beta})$ holds then we obtain

$$(4) \qquad (\|U\alpha\|^{-1} U\alpha, \|U\beta\|^{-1} U\beta) = (\|\alpha\|^{-1}\alpha, \|\beta\|^{-1}\beta)$$

for any non-zero α and β in $\mathscr{H}$. Hence by Euclidean plane geometry $k^2 = \|U\alpha\|/\|\alpha\|$ is independent of the choice of $\alpha \neq 0$ in $\mathscr{H}$. Therefore by (4) $U = kV$ where V is a unitary operator. This proves that $\tilde{U} = \tilde{V} \in \mathscr{U}(\tilde{\mathscr{H}})$.

A unitary projective representation of a group G is a group homomorphism $\tilde{\rho}: G \to \mathscr{U}(\tilde{\mathscr{H}})$ where $\tilde{\mathscr{H}}$ is a projective Hilbert space. If G is a topological group then $\tilde{\rho}$ is called strongly continuous provided $x \mapsto \tilde{\rho}(x)\tilde{\alpha}$ is a continuous map for each fixed $\tilde{\alpha}$ in $\tilde{\mathscr{H}}$. It is weakly continuous if $x \mapsto (\tilde{\rho}(x)\tilde{\alpha}, \tilde{\beta})$ is continuous for every choice of $\tilde{\alpha}$ and $\tilde{\beta}$ in $\tilde{\mathscr{H}}$. Every unitary representation $\rho: G \to \mathscr{U}(\mathscr{H})$ defines a unitary projective representation $\tilde{\rho}$, namely $\tilde{\rho}(x) = \widetilde{\rho(x)}$ for all $x \in G$. If ρ is strongly continuous then by the continuity of the map $\xi \mapsto \tilde{\xi}$ the associated projective representation $\tilde{\rho}$ is also strongly continuous. Similarly the weak continuity of ρ implies that of $\tilde{\rho}$. If given $\tilde{\rho}: G \to \mathscr{U}(\tilde{\mathscr{H}})$ there is a $\tilde{\rho}: G \to \mathscr{U}(\tilde{\mathscr{H}})$ such that $\tilde{\rho}$ is associated with ρ then we say that $\tilde{\rho}$ can be *lifted* or admits a *lifting* ρ. We speak about a *lifting in the strict sense* only if ρ has the same continuity properties as $\tilde{\rho}$. Let $\tilde{\rho}: G \to \mathscr{U}(\tilde{\mathscr{H}})$ be given and let a unitary operator $\rho(x)$ be chosen for every x in G such that $\widetilde{\rho(x)} = \tilde{\rho}(x)$. This defines a map $\rho: G \to \mathscr{U}(\mathscr{H})$. Since $\tilde{\rho}(xy) = \tilde{\rho}(x)\tilde{\rho}(y)$ we see that $\rho(xy) = \sigma(x,y)\rho(x)\rho(y)$ for some $\sigma(x,y) \in \mathbb{C}^{\times}$. Therefore $\rho: G \to \mathscr{U}(\mathscr{H})$ is a unitary projective group representation in the sense given on pp. 147—148.

We discuss another important example of a Lie transformation group, the *proper Lorentz group*. Algebraically this group can be defined as $\mathrm{Sl}(2, \mathbb{C}) \Big/ \begin{pmatrix} \pm 1 & 0 \\ 0 & \pm 1 \end{pmatrix}$ and so it is a Lie group by Theorem 5. Since its Lie algebra is $\mathfrak{sl}(2, \mathbb{C})$ we see that it is six dimensional. The manifold on which the group act is $\mathbb{R}^4$ with elements (t, x_1, x_2, x_3) or $(t, \boldsymbol{x})$ where $\boldsymbol{x} = (x_1, x_2, x_3)$. Of course t stands for time and x_1, x_2 and x_3 are the space coordinates. It will be convenient to identify $\mathbb{R}^4$ with the real vector space of 2×2 self adjoint, complex matrices. Namely we let $(t, \boldsymbol{x})$ correspond to the matrix

$$S = \begin{pmatrix} t + x_1 & x_2 + i x_3 \\ x_2 - i x_3 & t - x_1 \end{pmatrix}. \tag{5}$$

First we define an action of $\mathrm{Sl}(2, \mathbb{C})$ on $\mathbb{R}^4$ by letting $A \cdot S$ for any A in $\mathrm{Sl}(2, \mathbb{C})$ be the self adjoint matrix $A S A^*$. It is plain that $A \cdot (B \cdot S) = A B \cdot S$ for any A, B in $\mathrm{Sl}(2, \mathbb{C})$ and since A is invertible it is also obvious that $S \mapsto A \cdot S$ is a bijection. Moreover A and $-A$ define the same transformation. Hence it is in fact the Lorentz group which is acting on $\mathbb{R}^4$. The elements of the transformation group constructed in this manner are called *Lorentz transformations.* It is simple to prove that the group of these transformations is isomorphic to the Lorentz group and so it can be identified with it.

It is obvious that $\det A \cdot S = \det S$ where

$$\det S = t^2 - (x_1^2 + x_2^2 + x_3^2)\,.$$

Therefore every Lorentz transformation preserves the Lorentz inner product

$$((t, \boldsymbol{x}), (t', \boldsymbol{x}')) = t t' - x_1 x_1' - x_2 x_2' - x_3 x_3'\,.$$

We recall that $\mathrm{O}(1, 3, \mathbb{R})$ consists exactly of those 4×4 real, invertible matrices T for which the above inner product is invariant under the linear transformations $(t, \boldsymbol{x}) \mapsto T(t, \boldsymbol{x})^t$ and $(t', \boldsymbol{x}') \mapsto T(t', \boldsymbol{x}')^t$. Since $\det T = \det T I = \det A I A^* = 1$ we see that every Lorentz transformation determines an element T of $\mathrm{SO}(1, 3, \mathbb{R})$. Clearly this correspondence is an isomorphism of the Lorentz group into $\mathrm{SO}(1, 3, \mathbb{R})$. We are going to prove that the image group is $\mathrm{O}(1, 3, \mathbb{R})_o$, the identity component of the de Sitter group $\mathrm{O}(1, 3, \mathbb{R})$.

In order to find a 4×4 matrix T corresponding to a matrix $A = \begin{pmatrix} \alpha & \beta \\ \gamma & \delta \end{pmatrix}$ in $\mathrm{Sl}(2, \mathbb{C})$ it is best to compute the matrix of the linear transformation $S \mapsto A S A^*$ with respect to the ordered basis $\begin{pmatrix} 1 & 0 \\ 0 & 1 \end{pmatrix}, \begin{pmatrix} 1 & 0 \\ 0 & -1 \end{pmatrix}, \begin{pmatrix} 0 & 1 \\ 1 & 0 \end{pmatrix}, \begin{pmatrix} 0 & i \\ -i & 0 \end{pmatrix}$. After some routine computation one finds that the desired matrix is

$$\begin{pmatrix} \frac{1}{2}(\alpha\bar\alpha + \beta\bar\beta + \gamma\bar\gamma + \delta\bar\delta) & \frac{1}{2}(\alpha\bar\alpha - \beta\bar\beta + \gamma\bar\gamma - \delta\bar\delta) & \mathcal{R}e\,(\bar\alpha\beta + \gamma\bar\delta) & -\mathcal{I}m(\alpha\bar\beta + \gamma\bar\delta) \\ \frac{1}{2}(\alpha\bar\alpha + \beta\bar\beta - \gamma\bar\gamma - \delta\bar\delta) & \frac{1}{2}(\alpha\bar\alpha - \beta\bar\beta - \gamma\bar\gamma + \delta\bar\delta) & \mathcal{R}e\,(\alpha\bar\beta - \gamma\bar\delta) & -\mathcal{I}m\,(\alpha\bar\beta - \gamma\bar\delta) \\ \mathcal{R}e\,(\alpha\bar\gamma + \beta\bar\delta) & \mathcal{R}e\,(\alpha\bar\gamma - \beta\bar\delta) & \mathcal{R}e\,(\alpha\bar\delta + \beta\bar\gamma) & -\mathcal{I}m(\alpha\bar\delta - \beta\bar\gamma) \\ \mathcal{I}m\,(\alpha\bar\gamma + \beta\bar\delta) & \mathcal{I}m(\alpha\bar\gamma - \beta\bar\delta) & \mathcal{I}m(\alpha\bar\delta + \beta\bar\gamma) & \mathcal{R}e(\alpha\bar\delta - \beta\bar\gamma) \end{pmatrix}$$

For the sake of brevity we call this matrix and the Lorentz transformation which it represents L_A.

We already know that L_A is in $O(1, 3, \mathbb{R})$. Hence we have $(L_A)^t M L_A = M$ where M is the 4×4 diagonal matrix with diagonal entries $1, -1, -1, -1$. It follows from this that $\det L_A$ is ± 1. The group $Sl(2, \mathbb{C})$ is connected because for instance if $A = \begin{pmatrix} \alpha & \beta \\ \gamma & \delta \end{pmatrix}$ is in $Sl(2, \mathbb{C})$ then

$$(1-(\alpha+\delta)\tan t + \tan^2 t)^{-\frac{1}{2}} \begin{pmatrix} \alpha + \tan t & \beta \\ \gamma & \delta + \tan t \end{pmatrix} \quad \left(0 \leqslant t \leqslant \frac{\pi}{2}\right)$$

is an arc connecting A with I. The map $A \mapsto L_A$ is clearly continuous as a function of the complex variables $\alpha, \beta, \gamma, \delta$ and so the set $\{\det L_A : A \in Sl(2, \mathbb{C})\}$ is also connected. Therefore we see that $\det L_A = 1$ for every A in $Sl(2, \mathbb{C})$ and so the image group lies in $SO(1, 3, \mathbb{R})$. We know that $\alpha\bar{\alpha} + \delta\bar{\delta} \geqslant 2\alpha\delta$ and $\beta\bar{\beta} + \gamma\bar{\gamma} \geqslant -2\beta\gamma$. Hence $\alpha\bar{\alpha} + \beta\bar{\beta} + \gamma\bar{\gamma} + \delta\bar{\delta} \geqslant 2(\alpha\delta - \beta\gamma) = 2$ and so the upper left entry of L_A is at least 1. Since this entry is positive the linear transformation L_A *preserves the positive time direction*. Similarly by $\det L_A > 0$ we know that L_A *preserves the orientation* of $\mathbb{R}^4$.

The time direction and orientation preserving properties of L_A can be expressed by saying that L_A is a *proper Lorentz transformation*. Then any linear transformation preserving the Lorentz inner product is called a generalized Lorentz transformation. The group of these is isomorphic to the full group $O(1, 3, \mathbb{R})$.

Next we are going to determine the identity component of $O(1, 3, \mathbb{R})$. Since no simplification is gained by restricting ourselves to this special case we are going to study $O(1, n, \mathbb{R})$ where $n = 1, 2, \ldots$ is arbitrary. If $A = (a_{ij})$ $(i, j = 0, 1, \ldots, n)$ is in $O(1, n, \mathbb{R})$ then by $A^t M A = M$ we have $\det A = \pm 1$. Since the identity component is connected from this we derive that $\det A = 1$ for every A in $O(1, n, \mathbb{R})_o$. By computing the upper left hand entries of $A^t M A$ and M we see that

$$a_{00}^2 = 1 + a_{10}^2 + \cdots + a_{n0}^2 \tag{7}$$

for every A in $O(1, n, \mathbb{R})$. Hence $a_{00} \geqslant 1$ or $a_{00} \leqslant -1$ and so by the connectedness property we have $a_{00} \geqslant 1$ for every A in $O(1, n, \mathbb{R})_o$.

It is easy to prove that the matrices A satisfying $a_{00} \geqslant 1$ form a subgroup of $O(1, n, \mathbb{R})$. Indeed if $B^t M B = M$ then by $M^2 = I$ we have $(MB)^t(BM) = I$ and so $(BM)(MB)^t = I$ and $BMB^t = M$. By comparing the upper left hand entries of BMB^t and M we obtain

$$b_{00}^2 = 1 + b_{01}^2 + \cdots + b_{0n}^2 . \tag{8}$$

Now if $a_{00} \geqslant 1$ and $b_{00} \geqslant 1$ then from (7) and (8) it follows that

$$\begin{aligned} a_{00}b_{00}+a_{01}b_{10}+\cdots+a_{0n}b_{n0} &\geqslant a_{00}b_{00}-(a_{01}^2+\cdots+a_{0n}^2)^{\frac{1}{2}}(b_{01}^2+\cdots+b_{0n}^2) \\ &= a_{00}b_{00}-(a_{00}^2-1)^{\frac{1}{2}}(b_{00}^2-1)^{\frac{1}{2}} \geqslant 1 \end{aligned} \tag{9}$$

because $(ab-1)^2 \geqslant (a^2-1)(b^2-1)$ by $(a-b)^2 \geqslant 0$. Hence if $a_{00} \geqslant 1$ and $b_{00} \geqslant 1$ then $c_{00} \geqslant 1$ holds for $C=AB$. Since $A^t M A = M$ implies $A^{-1} = M A^t M$ it is clear that $a_{00} \geqslant 1$ implies the validity of the corresponding inequality for A^{-1}. Hence the condition $a_{00} \geqslant 1$ characterizes a subgroup of $\mathrm{O}(1, n, \mathbb{R})$.

The foregoing reasoning can be extended to prove considerably more: For every A in $\mathrm{O}(1, n, \mathbb{R})$ let $\Phi(A) = \operatorname{sgn} a_{00}$. Since $|a_{00}| \geqslant 1$ we have $\Phi(A) = 1$ or -1 according as $a_{00} \geqslant 1$ or $\leqslant -1$. If $\Phi(A) = \Phi(B) = -1$ then (9) shows that $\Phi(AB) = 1$. If we suppose that $\Phi(A) \neq \Phi(B)$ then $a_{00} b_{00} < 0$ and so

$$\begin{aligned} a_{00}b_{00}+a_{01}b_{10}+\cdots+a_{0n}b_{n0} &\leqslant a_{00}b_{00}+(a_{01}^2+\cdots+a_{0n}^2)^{\frac{1}{2}}(b_{10}^2+\cdots+b_{n0}^2)^{\frac{1}{2}} \\ &= (|a_{00}|^2-1)^{\frac{1}{2}}(|b_{00}|^2-1)^{\frac{1}{2}} - |a_{00}|\,|b_{00}| \leqslant -1 \end{aligned}$$

because $(a^2-1)(b^2-1) \leqslant (ab-1)^2$ by $(a-b)^2 \geqslant 0$ for any $a, b \geqslant 1$. (Actually for any two real numbers a, b.) Hence if $\Phi(A) \neq \Phi(B)$ then $\Phi(AB) = -1 = \Phi(A)\Phi(B)$. We proved that Φ is a homomorphism of $\mathrm{O}(1,n,\mathbb{R})$ into the multiplicative group $\{-1,1\}$. Its kernel is the set $\{A : A \in \mathrm{O}(1,n,\mathbb{R}) \text{ and } a_{00} \geqslant 1\}$ and so this is an invariant subgroup of $\mathrm{O}(1, n, \mathbb{R})$. The group of the corresponding linear transformations on $\mathbb{R}^4$ is called the *ortochronous Lorentz group.*

Proposition 7. *The identity component of* $\mathrm{O}(1, n, \mathbb{R})$ *consists of those matrices satisfying* $A^t M A = M$ *for which* $\det A = 1$ *and* $a_{00} \geqslant 1$.

Note. The first condition is equivalent to $\det A > 0$ and the second to $a_{00} > 0$.

Proof. We already proved that $\mathrm{O}(1, n, \mathbb{R})_o$ lies in the intersection of the two subgroups characterized by the conditions $\det A = 1$ and $a_{00} \geqslant 1$, respectively. We shall now construct two connected subgroups of $\mathrm{O}(1, n, \mathbb{R})_o$. The first is

$$\{A_\alpha : -\infty < \alpha < \infty\} \tag{10}$$

where

$$A_\alpha = \begin{pmatrix} \operatorname{ch}\alpha & \operatorname{sh}\alpha & 0 \\ \operatorname{sh}\alpha & \operatorname{ch}\alpha & 0 \\ 0 & 0 & I \end{pmatrix}.$$

Here I denotes the $(n-1)\times(n-1)$ identity matrix and 0 is used to denote zero rows and columns of length $n-1$. It is simple to verify that $(A_\alpha)^t M A_\alpha = M$ and $A_\alpha A_\beta = A_{\alpha+\beta}$. Since the group given in (10) is connected it lies in $\mathrm{O}(1, n, \mathbb{R})_o$.

Next given a matrix O in $\mathrm{O}(n, \mathbb{R})$ we let $A_O = \begin{pmatrix} 1 & 0 \\ 0 & O \end{pmatrix}$ where 0 denotes a zero row and column of length n. It is clear that $(A_O)^t M A_O = M$ and $A_O A_Q = A_{OQ}$. Hence by the connectedness of $\mathrm{SO}(n, \mathbb{R})$ we see that

$$\{A_O : O \in \mathrm{SO}(n, \mathbb{R})\} \tag{11}$$

is a connected subgroup of $\mathrm{O}(1, n, \mathbb{R})$ and so it lies in $\mathrm{O}(1, n, \mathbb{R})_o$.

Let A in $\mathrm{O}(1, n, \mathbb{R})$ be such that $(1, 0, \ldots, 0)^t$ is a fixed point of the linear transformation $\boldsymbol{x} \to A\boldsymbol{x}$ where $\boldsymbol{x} = (x_0, x_1, \ldots, x_n)^t$ and $x_0, x_1, \ldots, x_n \in \mathbb{R}$. Then $a_{00} = 1$ and $a_{1i} = 0$ for $1 \leqslant i \leqslant n$. Since $(A^t M)(A M) = I$ we also see that $a_{0j} = 0$ for $1 \leqslant j \leqslant n$. It follows that the bilinear form $x_1 y_1 + \cdots + x_n y_n$ is left invariant by the transformations $\boldsymbol{x} \to A\boldsymbol{x}$ and $\boldsymbol{y} \to A\boldsymbol{y}$. Hence $A = A_O$ where $O \in \mathrm{O}(n, \mathbb{R})$. It is obvious that every matrix of the form A_O is such that $(1, 0, \ldots, 0)^t$ is left invariant by the transformation $\boldsymbol{x} \to A\boldsymbol{x}$. Hence the connected group described in (11) is naturally isomorphic to the stability subgroup of the point $(1, 0, \ldots, 0)^t$ in the group of transformations $\boldsymbol{x} \to A\boldsymbol{x}$ where $A \in \mathrm{O}(1, n, \mathbb{R})$.

Let $A = (a_{ij})$ belong to $\mathrm{O}(1, n, \mathbb{R})$ and suppose that $a_{00} \geqslant 1$. We are going to prove the existence of a real α and matrices O and Q in $\mathrm{O}(n, \mathbb{R})$ such that

$$A = A_O A_\alpha A_Q \,. \tag{12}$$

If $\det A = 1$ then we may also suppose that $\det O = \det Q = 1$. We let $a_0 = (a_{10}, \ldots, a_{n0})^t$ and choose $n-1$ pairwise orthogonal vectors $o_i = (o_{1i}, \ldots, o_{ni})^t$ $(i = 2, \ldots, n)$ in $\mathbb{R}^n$ in the hyperplane which is orthogonal to $(a_0)^t$. If we let o_1 be one of the unit vectors in the direction of $(a_0)^t$ then $O = (o_1 \ldots o_n)$ is in $\mathrm{O}(n, \mathbb{R})$. Moreover by our construction

$$A_{O^t} A (1, 0, \ldots, 0)^t = (a_{00}, s, 0, \ldots, 0)^t \tag{13}$$

where $a_{00}^2 - s^2 = 1$. Now if $a_{00} \geqslant 1$ then there is a real α such that $a_{00} = \operatorname{ch} \alpha$ and $s = \operatorname{sh} \alpha$. Then by (13) we have

$$A_{-\alpha} A_{O^t} A (1, 0, \ldots, 0)^t = (1, 0, \ldots, 0)^t .$$

Therefore $A_{-\alpha} A_{O^t} A = A_Q$ for some Q in $\mathrm{O}(n, \mathbb{R})$. This proves the existence of the decomposition given in (12).

Now it is easy to finish the proof: The subgroup generated by the connected subgroups given in (6) and (11) is contained in $O(1, n, \mathbb{R})_o$. By the existence of the decomposition described in (12) it follows that $O(1, n, \mathbb{R})_o$ contains every $A=(a_{ij})$ from $O(1, n, \mathbb{R})$ for which $a_{00} \geqslant 1$ and $\det A = 1$. Since the opposite inclusion is already known the proof of Proposition 7 is completed.

We can now return to the study of the group of proper Lorentz transformations acting on the space of vectors $(t, \boldsymbol{x})$ where $t \in \mathbb{R}$ and $\boldsymbol{x}=(x_1, x_2, x_3) \in \mathbb{R}^3$. We identified these vectors with 2×2 self adjoint matrices S by the formula given in (5). Let $\mathscr{S}$ denote the vector space of these matrices S. We defined the Lorentz transformation L_A associated with A by the rule $S \mapsto A S A^*$ where $S \in \mathscr{S}$ and we associated with every Lorentz transformation a 4×4 real matrix. Since this association is an isomorphism we used the same symbol L_A to denote the corresponding 4×4 matrix. We proved that L_A belongs to $O(1, 3, \mathbb{R})$, its determinant is 1 and its upper left hand entry is at least 1. Hence by Proposition 7 L_A belongs to $O(1, 3, \mathbb{R})_o$. The following result implies that actually

$$\{L_A : A \in \mathrm{Sl}(2, \mathbb{C})\} = O(1, 3, \mathbb{R})_o. \tag{14}$$

Proposition 8. *The map* $A \mapsto L_A$ *is a homomorphism of* $\mathrm{Sl}(2, \mathbb{C})$ *onto* $O(1, 3, \mathbb{R})_o$ *and its kernel is* $\{I, -I\}$. *The image of* $\mathrm{SU}(2)$ *under this map is isomorphic to* $\mathrm{SO}(3, \mathbb{R})$, *namely it is the group of all matrices* $A_O = \begin{pmatrix} 1 & 0 \\ 0 & O \end{pmatrix}$ *where* $O \in \mathrm{SO}(3, \mathbb{R})$.

Proof. We already know that $A \mapsto L_A$ is a homomorphism of $\mathrm{Sl}(2, \mathbb{C})$ into $O(1, 3, \mathbb{R})_o$. Let us suppose that A belongs to the kernel i.e. L_A is the identity map on $\mathbb{R}^4$. Then choosing $S=I$ we obtain $A A^* = I$ and so $A \in \mathrm{SU}(2)$, say $A = \begin{pmatrix} \alpha & \beta \\ -\bar{\beta} & \bar{\alpha} \end{pmatrix}$ for suitable α, β in $\mathbb{C}$. By choosing $S = \begin{pmatrix} 0 & 1 \\ 1 & 0 \end{pmatrix}$ we obtain $\alpha^2 - \beta^2 = 1$ and if we let $S = \begin{pmatrix} 0 & i \\ -i & 0 \end{pmatrix}$ then it follows that $\alpha^2 + \beta^2 = 1$. Hence $\alpha = \pm 1$ and $\beta = 0$ proving that $A = I$ or $-I$.

In order to prove that the map $A \mapsto L_A$ is surjective by (12) it is sufficient to prove that the subgroups described in (10) and (11) lie in the image. If $A = \begin{pmatrix} \exp \alpha & 0 \\ 0 & \exp \alpha \end{pmatrix}$ then L_A is A_α and so we see that the first group is indeed contained in the image. Let O denote the matrix of a rotation around the coordinate axis x_1 by the angle φ. Then one sees that A_O is the image of $A = \begin{pmatrix} \exp \frac{1}{2} i \varphi & 0 \\ 0 & \exp -\frac{1}{2} i \varphi \end{pmatrix}$ under the map $A \mapsto L_A$. Similarly if O is a rotation about the x_2 axis by the angle φ then $A_O = L_A$ with $A = \begin{pmatrix} \cos\frac{1}{2}\varphi & -i\sin\frac{1}{2}\varphi \\ -i\sin\frac{1}{2}\varphi & \cos\frac{1}{2}\varphi \end{pmatrix}$. Finally if the axis of rotation is the x_3 axis and its angle is φ then $A_O = L_A$ where $A = \begin{pmatrix} \cos\frac{1}{2}\varphi & -\sin\frac{1}{2}\varphi \\ \sin\frac{1}{2}\varphi & \cos\frac{1}{2}\varphi \end{pmatrix}$. Thus in order to see that the group given in (11) lies in the image of the map $A \mapsto L_A$ it is sufficient to show that the set of rotations about the three coordinate axes generate the group SO(3, $\mathbb{R}$). Let $\boldsymbol{e}$ denote a unit vector in the axis of the arbitrary rotation O and let φ denote the angle of rotation. By applying first a suitable rotation O_1 about the first coordinate axis and another O_3 about the third we can transform $\boldsymbol{e}$ into $\boldsymbol{e}_1 = (1, 0, 0)^t$. Hence

$$O = (O_1)^t (O_3)^t Q O_3 O_1$$

where Q denotes the rotation about the x_1 axis with angle φ. If φ_1 and φ_3 denote the angles of the rotations O_1 and O_3, respectively, then the triple $(\varphi_1, \varphi, \varphi_3)$ determines a local coordinate neighborhood on SO(3, $\mathbb{R}$) in the neighborhood of 0. We mention that the Euler angles of O could also be used to express it as a product of three rotations about coordinate axes.

We complete the proof of Proposition 8 by showing that the image of SU(2) under $A \mapsto L_A$ is $\{A_O : O \in \mathrm{SO}(3, \mathbb{R})\}$. If $A \in \mathrm{SU}(2)$ then $A I A^* = A A^* = I$ and so $L_A(1, 0, 0, 0)^t = (1, 0, 0, 0)^t$. Thus $L_A = A_O$ for some O in SO(3, $\mathbb{R}$). Conversely if $L_A = A_O$ for some $O \in \mathrm{SO}(3, \mathbb{R})$ then from (6) it follows that A is in SU(2).

The use of the 0-th coordinate as the distinguished coordinate had definite advantage during some of the proofs. However in physics it is customary to use the first three coordinates for the space and the fourth one for time. The space of these vectors $(t, \boldsymbol{x})$ is called the *space-time world* and the Lorentz group acts effectively on this space. In terms of this modified notation our results can be summarized as follows:

Theorem 9. *The proper Lorentz group* $\mathscr{L}$ *is the group of those linear transformations of the space-time world* $\mathbb{R}^3 \times \mathbb{R}$ *which preserve the orientation and the positive time direction. A linear transformation* $L: \mathbb{R}^3 \times \mathbb{R} \to \mathbb{R}^3 \times \mathbb{R}$ *belongs to* $\mathscr{L}$ *if and only if its matrix with respect to the ordered basis* $(1, 0, 0, 0)^t, \ldots, (0, 0, 0, 1)^t$ *belongs to the identity component of* O(3, 1, $\mathbb{R}$). *Since* O(3, 1, $\mathbb{R})_o$ *is a closed subgroup* $\mathscr{L}$ *is a Lie transformation group. It is diffeomorphic to the Lie group* $\mathrm{Sl}(2, \mathbb{C})/\{I, -I\}$.

The group $\mathscr{L}$ is sometimes called the *restricted Lorentz group* or the *restricted homogeneous Lorentz group.* The *proper inhomogeneous*

Lorentz group is obtained by taking into account also the translations of the space $\mathbb{R}^3$. It is the semidirect product of $\mathscr{L}$ with the group of translations of $\mathbb{R}^3$.

We also add that the following connection exists between the groups SU(2) and Sl(2, $\mathbb{C}$) which, as we have seen, play a significant role in the study of $\mathscr{L}$. The Lie algebra $\mathfrak{su}(2)$ of SU(2) has a *complexification*, namely the Lie algebra of all complex 2×2 matrices which have trace 0. This Lie algebra $\mathfrak{sl}(2, \mathbb{C})$ is a subalgebra of $\mathfrak{gl}(2, \mathbb{C})$ and so by Theorem 7.3 there is a connected real Lie group corresponding to it. This is of course the group Sl(2, $\mathbb{C}$). Hence in this sense Sl(2, $\mathbb{C}$) is the *complexification* of SU(2).

Remarks

Lie algebras were first studied by Sophus Lie (1, 2) in conjunction with local Lie groups i.e. Lie group germs during the last third of the past Century. He associated a Lie algebra with every Lie group germ and proved that every real Lie algebra belongs to a suitable Lie group germ. The corresponding result for global Lie groups was established by Élie Cartan (1) in 1936. This important result is the topic of Proposition 7.2 and Theorem 7.3. Some simple examples of Lie algebras, such as the Lie algebra defined on $\mathbb{R}^3$ by the vector product $[x, y]$ were considered already earlier. Jacobi's identity first appeared in his works on partial differential operators.

The work of Lie was continued by a number of eminent mathematicians. The most important results among the early contributors to the subject were obtained by Killing and Élie Cartan. In most of these studies Lie groups and Lie algebras occur concurrently and only real and complex Lie algebras are considered. An independent study of Lie algebras over arbitrary fields is of more recent origin and the best summary of this field is available in the monographs of Bourbaki (1, 2) and Jacobson (5). Much valuable material on real and complex Lie algebras can be found also in a few works dealing primarily with Lie groups. The book of Freudenthal and de Vries (1) and the tables of Tits (1) belong to this category. I mention also the very recent book on abstract Lie algebras by Wintner (1) although I am not familiar with it. Similarly I do not know the manuscript of the forthcoming book of Sagle and Walde (1) but is should be a book worth consulting when it becomes available. There are a number of partially finished works published in the form of notes. The publishers whose series contain most of these notes are W. A. Benjamin and Springer-Verlag.

The material presented between Definition 1.16 and Lemma 1.25 is based on Jacobson's elegant treatment of this topic in his book on Lie algebras (5). The same book contains an easy and elegant proof of the extension of Theorem 1.30 to arbitrary infinite fields. The proof of Lemma 1.32 follows the steps of a similar proof which can be found in the book of Helgason (1). The proof of Theorem 2.1 is essentially the same as the corresponding proof in the printed notes of Hausner and J. T. Schwartz on Lie groups and Lie algebras (1). Similarly in the case of Theorem 2.11 instead of going back to the original source I could rely on the exposition which can be found in Hochschield's book (1). A detailed study of the classification of simple Lie algebras is available in the books of Freudenthal and de Vries (1), Jacobson (5) and Hausner and Schwartz (1).

The first exposition of the theory of sheaves appeared in the seminar notes of Henri Cartan (1). Leray's two fundamental papers (1, 2) appeared approximately at the same time. These were followed by the notes on the lectures given by Dowker in the Tata Institute. These notes are available in some mathematics libraries. There are now several books on the subject; the most significant ones are by Godement (8) and Hirzebruch (1). A small and inexpensive volume was written on this subject by Swan (1). The concept of a differentiable manifold, perhaps in a primitive form, was already known to Riemann.

As far as Section 5 is concerned first we mention that a generation ago Lie groups were called *continuous groups*. The main items of the classical literature on this subject are the books of Lie (1, 2) and Élie Cartan (2), a volume by G. Kowalewski (1) and a book by P. Zervos (1) on the problem of Monge. The various volumes and editions of a book by Chevalley (1, 2, 3) are more recent. In addition to the modern literature already mentioned in connection with Lie algebras we mention a small book by P. M. Cohn (1) and the Lie group seminar notes from Paris. There are also several books covering more practical aspects of Lie group theory such as its applications to special function theory. A big book by Vilenkin (1) and another volume by W. Miller (1) are the best known among these. A more recent book by Miller (2) and a small book by R. Hermann (1) deal with applications to physics.

The relevant literature on Hilbert's fifth problem can be traced back from Gleason's paper (1) and the book of Montgomery and Zippin (1). It is not necessary to worry much about the manifold structures of the various special Lie groups which were introduced after Lemma 5.3. For the existence of the appropriate differentiable structures follows in each case from Theorem 7.3. It was pointed out that the differentiability conditions in the definition of a Lie group germ can be relaxed considerably. However if all such conditions are dropped then there exist group germs which do not correspond to global Lie groups. An appropriate example was constructed by Malcev (1).

The Lie algebra $\mathfrak{g}$ of a Lie group G is the Lie algebra G_o of its identity component which is an analytic group. Thus Example 2 of Section 1 and Theorem 5.23 show that the Lie algebra of the rotation group $SO(3, \mathbb{R})$ is isomorphic to the Lie algebra on $\mathbb{R}^3$ with vector product as bracket. We add to the classification given at the end of Section 5 the following: Let G be a Lie group and H a closed subgroup of G so that G/H is a C^∞ manifold. If G is solvable then G/H is called a *solvmanifold* and similarly if G is nilpotent then G/H is a *nilmanifold*. One often uses this terminology only under more restrictive hypotheses e.g. when G is supposed to be connected. A closed subgroup H of a Lie group G is now frequently called a *uniform subgroup* provided G/H is compact. The same terminology can be used also for closed subgroups of topological groups. The definition of *solvability* introduced for topological groups in the beginning of Section VII.5 was quite different from the one given in Section 5. One can prove that the case of Lie groups these two definitions are equivalent. The definition given in Section VII.5 is obviously meaningful also for any abstract group G. There is also a corresponding criterion for *nilpotency*: Let G be an abstract group, let H_1 and H_2 be subgroups of G and let $[H_1, H_2]$ denote the subgroup generated by the elements $h_1^{-1} h_2^{-1} h_1 h_2$ where $h_i \in H_i$ $(i=1,2)$. Let $G^1 = G$ and $G^k = [G^1, G^{k-1}]$ where $k = 1, 2, \ldots$ We call G nilpotent if $G^k = e$ for a sufficiently large value of k. A Lie group G can be proved to be nilpotent if and only if it is nilpotent as an abstract group.

The theorem on the invariance of domain is fully discussed and proved in Hurewicz and Wallman's Dimension theory (1). The example mentioned at the very end of Section 6 was given by Élie Cartan. More recently Malcev (1) proved that if G is a nilpotent, simply connected analytic group then the exponentiation is a homeomorphism of $\mathfrak{g}$ onto G. It is also known that in Malcev's result the word "nilpotent" can not be replaced by "solvable".

The main results given in Section 7 are those of Élie Cartan. However the result expressed in Proposition 7.2 and Theorem 7.3 can be formulated also for Lie group germs and this formulation of the result was first stated and proved by Lie. The Lorentz group is interesting also from pure mathematical point of view. For example it is known that it has no finite dimensional unitary representations except the trivial ones. In addition to Naimark's (2) well known monograph there is another book on the representations of the Lorentz group by Gelfand, Minlos and Shapiro (1).

Since at present I have no chance to add more material to the text I shall finish by mentioning a few names whose work is related to some of the material discussed in Chapter VI but their results concern Lie groups. I am first of all thinking of Harish-Chandra and Langlands whose contributions are very important. I do not wish to say anything about arithmetic questions. Then I like to mention one specific research area which was started by my collegue L. Green (1) and was continued by C. C. Moore (2) and more recently by J. Brezin (1) and L. F. Richardson (1). They use the results of Takenouchi (1) and Kirillov (1) on the unitary representations of solvable and nilpotent Lie groups and their work concerns the decomposition of $L^2(G/H, \mu)$ when G is a simply connected, nilpotent Lie group, H is a discrete, uniform subgroup of G and μ is the invariant measure of the compact homogeneous space G/H.

Bibliography

Aarnes, J. F.: (1) Continuity of group representations, with applications to C^* algebras. J. Funct. Anal. **5**, 14—36 (1970).

Ambrose, W.: (1) Structure theorems for a special class of Banach algebras. Trans. Amer. Math. Soc. **57**, 364—386 (1945).

Arens, R. F.: (1) On a theorem of Gelfand and Neumark. Proc. Nat. Acad. Sci. U.S.A. **32**, 237—239 (1946).
(2) Representations of Banach *-algebras. Duke Math. J. **14**, 269—282 (1947).

Artin, E.: (1) Zur Theorie der hyperkomplexen Zahlen. Abh. Math. Sem. Univ. Hamburg **5**, 251—260 (1927).

Baer, R.: (1) Zur Topologie der Gruppen. J. Reine Angew. Math. **160**, 208—226 (1929).

Banach, S.: (1) Théorie des Opérations Linéaires. New York: Hafner Publishing Co. 1932 and later.
(2) Sur les fonctionelles linéaires I. Studia Mathematica **1**, 211—216 (1929).
(3) Sur les fonctionelles linéaires II. Studia Mathematica **1**, 233—239 (1929).

Bargmann, V.: (1) Irreducible unitary representations of the Lorentz group. Ann. Math. **48**, 568—640 (1947).

Birkhoff, Garrett: (1) Lattice theory. Amer. Math. Soc. Coll. Publ. 25, Providence R.I.: Amer. Math. Soc. 1948.
(2) A note on topological groups. Compositio Mathematica **3**, 427—430 (1936).

Blattner, R. J.: (1) On induced representations. Amer. J. Math. **83**, 79—98 (1961).

Bohnenblust, H. F., Sobczyk, A.: (1) Extensions of functionals on complex linear spaces. Bull. Amer. Math. Soc. **44**, 91—93 (1938).

Bourbaki, N.: (1) Éléments de mathématique. Paris: Hermann 1940.
(2) Elements of mathematics. Paris: Hermann, Éditeurs and Addison-Wesley Publ. Co. 1966.

Brezin, J.: (1) Harmonic analysis on nilmanifolds. Trans. Amer. Math. Soc. **150**, 611—618 (1970).

Bruhat, F.: (1) Sur les représentations induites des groupes de Lie. Bull. Soc. Math. France **84**, 97—205 (1956).
(2) Lectures on Lie groups and representations of locally compact groups. Bombay: Tata Institute of Fundamental Research 1958.

Burrow, M.: (1) Representations theory of finite groups. New York and London: Academic Press 1965.

Calkin, J. W.: (1) Two-sided ideals and congruences in the ring of bounded operators in Hilbert space. Ann. Math. **47**, 839—873 (1941).

Cartan, Élie: (1) Oeuvres completes. Paris: Gauthier-Villars 1952.
(2) La théorie des groupes finis et continus et la géometrie différentielle, traités par la méthode du repère mobile. Leçons professées à la Sorbonne par Élie Cartan rédigées par Jean Leray. Paris: Gauthiers-Villars 1937.

Cartan, Henri: (1) Séminaire de topologie algébrique. École Normal Superieur. Paris 1950—1951.

Chevalley, C.: (1) Theory of Lie groups I. Princeton, N. J.: Princeton Univ. Press 1946.

(2) Théorie des groupes de Lie II. Actualités Sci. et Ind. 1152, Paris: Hermann & Cie. 1951.

(3) Théorie des groupes de Lie. Groupes algébriques, théorèmes généraux sur les algébres de Lie. Paris: Hermann & Cie. 1968.

Cohn, P. M.: (1) Lie groups. Cambridge Tracts No. 46, Cambridge at the University Press: 1957.

Curtis, Ch. W., Reiner, I.: (1) Representation theory of finite groups and associative algebras. New York, London and Sydney: Interscience Publishers 1962.

Daniell, P. J.: (1) A general form of integral. Ann. Math. **19**, 279—294 (1917—1918).

Dantzig, D. van: (1) Zur topologischen Algebra I. Math. Ann. **107**, 587—626 (1932).

Day, M. M.: (1) Normed linear spaces. Berlin, Göttingen, Heidelberg: Springer 1958.

Dickson, L. E.: (1) History of the theory of numbers. Washington: Carnegie Institution of Washington 1919—1923.

Dixmier, J.: (1) Les C^*-algèbres et leur représentations. Paris: Gauthier-Villars Éditeur 1969.

(2) Les algèbres d'opérateurs dans l'espace Hilbertian (Algèbres de von Neumann). Paris: Gauthier-Villars Éditeur 1969.

(3) Sur la réduction des anneaux d'opérateurs. Ann. Sci. École Norm. Sup. **68**, 185—202 (1951).

Dunford, N., Schwartz, J. T.: (1) Linear operators I, II, III. New York, London and Sydney: Interscience Publishers 1958, 1963 and 1971.

Dwork, B. M.: (1) Analytic theory of the zeta function of algebraic varieties. In: Arithmetical algebraic geometry, pp. 18—32. New York: Harper & Row, Publishers 1965.

Edwards, R. E.: (1) Functional analysis. New York, Chicago, San Francisco, Toronto, London: Holt, Rinehart and Winston 1965.

Fakler, R. A.: (1) On induced representations of locally compact groups. Ph. D. Thesis: University of Minnesota 1972.

Federer, H., Morse, A. P.: (1) Some properties of measurable functions. Bull. Amer. Math. Soc. **49**, 270—277 (1943).

Freudenthal, H.: (1) Einige Sätze über topologische Gruppen. Ann. of Math. **37**, 57—77 (1936).

(2) Topologische Gruppen mit genügend vielen fast periodischen Funktionen, Ann. Math. **37**, 57—77 (1936).

Freudenthal, Hans, de Vries, H.: (1) Linear Lie groups. New York and London: Academic Press 1969.

Frobenius, G.: (1) Über Relationen zwischen den Characteren einer Gruppe und ihrer Untergruppen. Sitzungsb. Kön. Preusz. Akad. Wiss. zu Berlin, 501—515 (1898).

(2) Über die Composition der Charactere einer Gruppe. Sitzungsber. Kön. Preusz. Akad. Wiss. zu Berlin, 330—339 (1899).

Fuglede, B.: (1) A commutativity theorem for normal operators. Proc. Nat. Acad. Sci. U.S.A. **36**, 35—40 (1950).

Fukamiya, M.: (1) On a theorem of Gelfand and Neumark and the B^* algebra. Kumamoto J. Sci. Ser. A **1**, 17—22 (1952).

Gelfand, I. M.: (1) On normed rings. Dokl. Akad. Nauk SSSR **23**, 430—432 (1939).

(2) Normierte Ringe. Mat. Sbornik **9**, 3—24 (1941).
(3) Spherical functions in symmetric Riemann spaces. Doklady Akad. Nauk SSSR **70**, 5—8 (1956).
(4) Automorphic functions and the theory of representations. Proceedings of the International Congress of Mathematicians, Stockholm 1962. Djursholm: Institute Mittag-Leffler 1963.

Gelfand, I. M., Graev, M. I., Pyatetskii-Shapiro, I. I.: (1) Representation theory and automorphic functions. Philadelphia, London and Toronto: W. B. Saunders Company 1969. (The original Russian edition was published in Moscow by Nauka Press in 1966.)

Gelfand, I., Minlos, R., Shapiro, Z.: Representations of the rotation and Lorentz groups. London: Pergamon Press 1963.

Gelfand, I. M., Naimark, M. A.: (1) On the imbedding of normed rings into the ring of operators in Hilbert space. Mat. Sbornik, N.S. **12**, 197—213 (1943).

Gelfand, I., Raikov, D.: (1) Irreducible unitary representations of locally compact groups. Mat. Sbornik **13**, 301—316 (1943).

Gelfand, I. M., Silov, G. E.: (1) Über verschiedene Methoden der Einführung der Topologie in die Menge der maximalen Ideale eines normierten Ringes. Mat. Sbornik **9**, 25—39 (1941).

Gleason, A.: (1) Groups without small subgroups. Annals of Math. **56**, 193—212 (1952).

Godement, R.: (1) Les fonctions de type positif et la théorie des groupes. Trans. Amer. Math. Soc. **63**, 1—84 (1948).
(2) La formule des traces de Selberg considerée comme source de problèmes mathématiques. Séminaire Bourbaki. 15e année No. 244, 1962—1963.
(3) Théorie des caractères. I. Algèbres unitaires. Ann. Math. **59**, 47—62 (1954).
(4) Théorie des caractères. II. Définition et propriétés générales des caractères. Ann. Math. **59**, 63—85 (1954).
(5) Sur les rélations d'orthogonalité de V. Bargmann. I. Résultats préliminaires. C. R. Acad. Sci. Paris **225**, 521—523 (1947).
(6) Sur les rélations d'orthogonalité de V. Bargmann. II. Démonstration générale. C. R. Acad. Sci. Paris **225**, 657—659 (1947).
(7) A theory of spherical functions. I. Trans. Amer. Math. Soc. **73**, 496—556 (1952).
(8) Théorie des faisceaux. Actualités Sci. et Ind. No. 1252. Paris: Hermann 1958.

Gramsch, B.: (1) Eine Idealstruktur Banachscher Operatoralgebren. J. Reine Angew. Math. **225**, 97—115 (1967).

Green, L., Auslander, L., Hahn, F.: (1) Flows on homogeneous spaces. Ann. of Math. Studies, No. 53, Princeton Univ. Press 1963.

Greenleaf, F. P.: (1) Amenable actions of locally compact groups. J. Funct. Anal. **4**, 295—315 (1959).

Grosser, S., Moskowitz, M.: (1) On central topological groups. Trans. Amer. Math. Soc. **72**, 317—340 (1967).
(2) Compactness conditions in topological groups. I. J. Reine Angew. Math. **246**, 1—40 (1971).

Gunning, R. C., Rossi, H.: (1) Analytic functions of several complex variables. Englewood Cliffs, N. J.: Prentice-Hall 1965.

Haar, A.: (1) Der Maßbegriff in der Theorie der kontinuierlichen Gruppen. Ann. of Math. **34**, 147—169 (1933).

Hahn, H.: (1) Über lineare Gleichungssysteme in linearen Räumen. J. Reine Angew. Math. **157**, 214—229 (1927).

Halmos, P. R.: (1) Introduction to Hilbert space and the theory of spectral multiplicity. New York: Chelsea 1951.

Harish-Chandra: (1) Representations of semisimple Lie groups. II. Proc. Nat. Acad. Sci. U.S.A. **37**, 362—365 (1951).

(2) Representations of a semisimple Lie group on a Banach space. I. Trans. Amer. Math. Soc. **74**, 185—243 (1953).

(3) Representations of semisimple Lie groups II. Trans. Amer. Math. Soc. **76**, 26—65 (1954).

(4) Representations of semisimple Lie groups III. Trans. Amer. Math. Soc. **76**, 234—253 (1954).

(5) Representations of semisimple Lie groups VI. Amer. J. Math. **78**, 564—628 (1956).

(6) Spherical functions on semisimple Lie group I. Amer. J. Math. **80**, 241—310 (1958).

(7) Spherical functions on semisimple Lie group. II. Amer. J. Math. **80**, 553—613 (1958).

Hausner, M., Schwartz, J. T.: (1) Lie groups. Lie algebras. New York, London and Paris: Gordon and Breach 1968.

Helgason, S.: (1) Differential geometry and symmetric spaces. New York and London: Academic Press 1962.

Hermann, R.: (1) Lie groups for Physicists. New York and Amsterdam: W. A. Benjamin, Inc. 1966.

Hermite, Ch.: (1) Oeuvres publiées sous les auspices de l'Academie des sciences par Émile Picard. Paris: Gauthier-Villars, Imprimeur-Libraire 1905—1917.

Hewitt, E., Ross, K. A.: (1) Abstract harmonic analysis I, II. Berlin, Göttingen, Heidelberg: Springer 1963 and 1970.

Hilbert, D.: (1) Grundzüge einer allgemeinen Theorie der linearen Integralgleichungen IV. Nachr. Akad. Wiss. Göttingen. Math.-Phys. Kl. 157—227 (1906).

Hille, E.: (1) Functional analysis and semigroups. Amer. Math. Soc. Coll. Publ. 31 New York: Amer. Math. Society 1948.

Hille, E., Phillips, R. S.: (1) Functional analysis and semi-groups. Amer. Math. Soc. Coll. Publ. 31 Providence, Rhode Island: Amer. Math. Society 1957.

Hirzebruch, F.: (1) Neue topologische Methoden in der algebraischen Geometrie. Berlin, Göttingen, Heidelberg: Springer 1956.

Hochschield, G.: (1) The structure of Lie groups. San Francisco, London and Amsterdam: Holden-Day, Inc. 1965.

Hopkins, C.: Rings with minimal conditions for left ideals. Ann. of Math. **40**, 712—730 (1939).

Hopkins, C.: (1) Rings with minimal conditions for left ideals. Ann. Math. **40**, 712—730 (1939).

Horváth, J.: (1) Topological vector spaces and distributions. Reading Mass.: Addison-Wesley Publ. Co. 1966.

Hörmander, L.: (1) An introduction to complex analysis in several variables. Princeton, N. J., Toronto and London: D. van Nostrand Company 1966.

Hurewicz, W., Wallman, H.: (1) Dimension theory. Princeton N. J.: Princeton University Press 1948.

Iwasawa, K.: (1) On some types of topological groups. Ann. Math. **50**, 507—558 (1949).

Jacobson, N.: (1) The radical and semisimplicity for arbitrary rings. Amer. J. Math. **67**, 300—320 (1945).

(2) A topology for the set of primitive ideals in an arbitrary ring. Proc. Nat. Acad. Sci. U.S.A. **31**, 333—338 (1945).
(3) Structure of rings. Amer. Math. Soc. Coll. Publ. 37. Providence, R.I.: Amer. Math. Society 1956.
(4) Structure theory of simple rings without finiteness assumptions. Trans. Amer. Math. Soc. **57**, 228—245 (1945).
(5) Lie algebras. New York: Interscience Publ. 1962.

Kakutani, S.: (1) Iteration of linear operations in complex Banach spaces. Proc. Imp. Acad. Tokyo **14**, 295—300 (1938).
(2) On the uniqueness of Haar's measure. Proc. Imp. Acad. Tokyo **14**, 27—31 (1938).
(3) Über die Metrization der topologischen Gruppen. Proc. Imp. Acad. Jap. **12** 82—84 (1936).

Kaplansky, I.: (1) Topological rings. Amer. J. Math. **69**, 153—183 (1947).
(2) Projections in Banach algebras. Ann. Math. **53**, 235—249 (1951).
(3) Dual rings. Ann. Math. **49**, 689—701 (1948).
(4) The structure of certain operator algebras. Trans. Amer. Math. Soc. **70**, 219—255 (1951).
(5) Group algebras in the large. Tohoku Math. J. **3**, 249—256 (1951).
(6) Groups with representations of bounded degree. Canadian J. Math. **1**, 105—112 (1949).

Kelley, J. L.: (1) Note on a theorem of Krein and Milman. J. Osaka Institute Sci. Tech. Part I. **3**, 1—2 (1951).

Kelley, J. L., Vaught, R. L.: (1) The positive cone in Banach algebras. Trans. Amer. Math. Soc. **74**, 44—45 (1953).

Kirillov, A. A.: (1) Unitary representations of nilpotent Lie groups. Uspekhi Mat. Nauk **17**, 57—110 (1962).

Kowalewski, G.: (1) Einführung in die Theorie der Kontinuierlichen Gruppen. Leipzig: Teubner Verlag 1931.

Köthe, G.: (1) Topological vector spaces. Berlin, Heidelberg, New York: Springer 1971.

Kraljevic, H.: (1) Induced representations of locally compact groups on Banach spaces. Glasnik Mat., Ser. III **4**, 183—196 (1969).

Krein, M., Milman, D.: On extreme points of regularly convex sets. Studia Math. **9**, 133—138 (1940).

Kunze, R. A. (1) A note on square integrable representations. Journ. Funt. Anal. **6**, 454—459 (1970).

Langworthy, H.: (1) Imprimitivity in Lie groups. Ph. D. Thesis: University of Minnesota 1970.

Leja, F.: (1) Sur la notion du group abstrait topologique. Fund. Math. **9**, 37—44 (1927).

Leray, J.: (1) L'homologie d'un espace fibré dont la fibre est connexe. J. Math. pur appl. **29**, 169—213 (1950).
(2) L'anneau spectral et l'anneau filtré d'homologie d'un espace localment compact et d'une application continue. J. Math. pur. appl. **29**, 1—139 (1950).

Lie, Sophus: (1) Vorlesungen über continuierliche Gruppen mit geometrischen und anderen Anwendungen. Bearbeitet und herausgegeben von Dr. Georg Scheffers. Leipzig: B. G. Teubner 1893.
(2) Theorie der Transformationsgruppen. Unter Mitwirkung von F. Engel, Leipzig: Teubner Verlag 1888, 1890 und 1893.

Loomis, L. H.: (1) An introduction to abstract harmonic analysis. Princeton, Toronto, London and Melbourne: D. van Nostrand Company 1953.

(2) Positive definite functions and induced representations. Duke Math. J. **27**, 569—580 (1960).

Mackey, G. W.: (1) Commutative Banach algebras. Rio de Janeiro: Livraria Castelo 1959.

(2) The theory of group representations. Chicago: Department of Mathematics, The University of Chicago 1955.

(3) Infinite dimensional group representations. Bull. Amer. Math. Soc. **69**, 628—686 (1963).

(4) Induced representations of locally compact groups. I. Ann. Math. **55**, 101—140 (1952).

(5) Imprimitivity for representations of locally compact groups, Proc. Nat. Acad. Sci. U.S.A. **35**, 537—545 (1949).

(6) Induced representations of locally compact groups, II. Ann. Math. **56**, 193—221 (1953).

(7) On induced representations of groups. Amer. J. Math. **73**, 576—592 (1951).

(8) Induced representations of groups and quantum mechanics. New York and Amsterdam. Benjamin, Inc. 1968.

Malcev, A.: (1) On a class of homogeneous spaces. Izv. Akad. Nauk SSSR Ser. Mat. **13**, 9—32 (1949) Amer. Math. Soc. Translations **39** (1949).

Masterson, J. J.: (1) Structure spaces of a vector lattice and its Dedekind completion. Proc. Kon. Nederl. Akad. Wet. **71**, 468—478 (1968).

Mazur, S.: (1) Sur les anneaux linéaires, C. R. Acad. Sci. Paris **207**, 1025—1027 (1938).

Michal, A. D., Martin, R. S.: (1) Some expansions in vector space. J. Math. pures appl. (9) **13**, 69—91 (1934).

Miller, Willard Jr.: (1) Lie theory and special functions. New York and London: Academic Press 1968.

(2) Symmetry groups and their applications. New York and London: Academic Press 1972.

Montgomery, D., Zippin, L.: (1) Topological transformation groups. New York, N.Y.: Interscience Publishers 1955.

Moore, C. C.: (1) On the Frobenius reciprocity theorem for locally compact groups. Pacific J. Math. **12**, 359—365 (1962).

(2) Decomposition of unitary representations defined by discrete subgroups of nilpotent groups. Ann. Math. **82**, 146—182 (1965).

Moore, R. T.: (1) Measurable, continuous and smooth vectors for semigroups and group representations. Mem. Amer. Math. Soc. No. **78**. Providence R.I.: Amer. Math. Soc. 1968.

Murray, F. J., v. Neumann, J.: (1) On rings of operators I. Ann. Math. **37**, 116—229 (1936).

(2) On rings of operators II. Trans. Amer. Math. Soc. **41**, 208—248 (1937).

(3) On rings of operators IV. Ann. Math. **44**, 716—808 (1943).

Nagumo, M.: (1) Einige analytische Untersuchungen in linearen metrischen Ringen. Jap. J. Math. **13**, 61—80 (1936).

Naimark, M. A.: (1) Normed rings. Groningen, the Nederlands: P. Noordhoff N.V. 1964. (The first ,Russian edition dates from 1955.)

(2) Linear representations of the Lorentz group. Moscow: Fizmatgiz 1958.

v. Neumann, J.: (1) On complete topological spaces. Trans. Amer. Math. Soc. **37**, 1—20 (1935).

(2) Zur Algebra der Funktionaloperationen und Theorie der normalen Operatoren. Math. Ann. **102**, 370—427 (1930).

(3) Über adjungierte Funktionaloperatoren. Ann. Math. **33**, 294—310 (1932).
(4) On rings of operators III. Ann. Math. **41**, 94—161 (1940).
(5) On rings of operators. Reduction theory. Ann. Math. **50**, 401—485 (1949).
(6) The uniqueness of Haar's measure. Mat. Sbornik, N.S. **1**, 106—114 (1934).
(7) Zum Haarschen Masz in topologischen Gruppen. Comp. Math. **1**, 106—114 (1934).

Péter, F., Weyl, F.: (1) Die Vollständigkeit der primitiven Darstellungen einer geschlossenen kontinuierlichen Gruppe. Math. Ann. **97**, 737—755 (1927).

Pontryagin, L. S.: (1) Topological groups. New York, London and Paris: Gordon and Breach, Science Publishers 1966.
(2) Topologische Gruppen, I—II. Leipzig: B. G. Teubner 1957, 1958.

Porta, H.: (1) Ideals and universal representations of certain C^*-algebras. Revista Union Mat. Argentina **25**, 27—36 (1971).

Putnam, C. R.: (1) On normal operators in Hilbert space. Amer. J. Math. **73**, 357—362 (1951).

Radon, J.: (1) Theorie und Anwendungen der absolut additiven Mengenfunktionen. Sitzungsber. math.-naturwiss. Kl. Akad. Wiss. Wien 122 Abt. IIa, 1295—1438 (1913).

Richardson, L. F.: (1) Decomposition of the L^2 space of a general compact nilmanifold. Amer. J. Math. **93**, 173—190 (1971).

Rickart, Ch. E.: (1) General theory of Banach algebras. Princeton, N. J. Toronto and London: D. van Nostrand Company 1960.
(2) The singular elements of a Banach algebra. Duke Math. J. **14**, 1063—1077 (1947).

Riesz, F.: (1) Sur certain systèmes singuliers d'équations integrales. Ann. École Norm. Sup., (3) **28**, 33—62 (1911).
(2) Les systèmes equations linéaires à une infinité d'inconnues. Paris: Gauthier-Villars, Imprimeur-Libraire 1913.
(3) Über lineare Funktionalgleichungen. Acta Math. **41**, 71—98 (1918).

Riesz, F., Sz. Nagy, B.: (1) Leçons d'analyse fonctionelle. Budapest: Akadémiai Kiadó 1952.
(2) Functional analysis. New York: Frederick Ungar Publishing Co. 1955.

Rosenbloom, M.: (1) On a theorem of Fuglede and Putnam. J. London Math. Soc. **33**, 376—377 (1958).

Rudin, W.: (1) Fourier analysis on groups. New York and London: Interscience Publishers 1962.

Sagle, A. A., Walde, R. E.: (1) Introduction to Lie groups and Lie algebras. New York and London: Academic Press 1973.

Sakai, S.: (1) C^*-algebras and W^*-algebras. New York, Heidelberg, Berlin: Springer 1971.
(2) An uncountable number of II_1 and II_∞ factors. J. Funct. Anal. **5**, 236—246 (1970).

Schaefer, H. H.: (1) Topological vector spaces. New York: Springer 1971.

Schatten, R.: (1) Norm ideals of completely continuous operators. Berlin, Göttingen, Heidelberg: Springer 1960.

Schauder, J.: (1) Über lineare, vollstetige Funktionaloperationen. Studia Math. **2**, 183—196 (1930).

Schochetman, I.: (1) Dimensionality and the duals of certain locally compact groups. Proc. Amer. Math. Soc. **26**, 514—520 (1970).

Schreiber, M.: (1) Compactness of the structure space of a ring. Proc. Amer. Math. Soc. **8**, 684—685 (1957).

Schreier, O.: (1) Abstrakte kontinuierliche Gruppen. Abh. Math. Sem. Univ. Hamburg **4**, 15—32 (1925).

Schwartz, J. T.: (1) W^*-algebras. New York, London and Paris: Gordon and Breach 1967.

Segal, I. E.: (1) The group algebra of a locally compact group, Trans. Amer. Math. Soc. **61**, 69—105 (1947).

(2) Irreducible representations of operator algebras. Bull. Amer. Math. Soc. **53**, 73—88 (1947).

(3) Two-sided regular representations of a unimodular locally compact group. Ann. Math. **51**, 293—298 (1950).

(4) The group algebra of a locally compact group. Trans. Amer. Math. Soc. **61**, 69—105 (1947).

(5) Decomposition of operator algebras I, II. Memoirs Amer. Math. Soc. No. 9 New York: Amer. Math. Soc. 1951.

Selberg, A.: (1) Harmonic analysis and discontinuous groups in weakly symmetric Riemannian spaces with applications to Dirichlet series. J. Indian Math. Soc. **20**, 47—87 (1956).

Serre, J.-P.: (1) Endomorphismes completement continus des espaces de Banach p-adiques. Inst. Hautes Études Sci. Publ. Math. No. **12**, 69—85 (1962).

Šilov, G.: (1) Sur la théorie des idéaux dans les anneaux normés de fonctions. Doklady Akad. Nauk SSSR, N.S. **27**, 900—903 (1940).

Simmons, G. F.: (1) Introduction to topology and modern analysis. New York, San Francisco, Toronto and London: McGraw-Hill Company 1963.

Soukhomlinov, G. A.: (1) Über Fortsetzung von linearen Funktionalen in linearen komplexen Räumen und linearen Quaternionräumen. Mat. Sbornik, N.S. **3**, 353—358 (1938).

Stone, M. H.: (1) Application of the theory of Boolean rings to general topology. Trans. Amer. Math. Soc. **41**, 375—481 (1937).

v. Sz. Nagy, B.: (1) Spektraldarstellung linearer Transformationen des Hilbertschen Raumes. Berlin and New York: Springer 1942 and 1967.

Takenouchi, O.: (1) Sur la facteur-représentation d'un groupe de Lie résoluble de type (E). Math. Journ. Okayoma Univ. **7**, 151—161 (1957).

Tamagawa, T.: (1) On Selberg's trace formula. J. Fac. Sci. Univ. Tokyo Sct. I. **8**, 363—386 (1960).

Tits, J.: (1) Tabellen zu den einfachen Lie-Gruppen und ihren Darstellungen. Lecture Notes in Mathematics, 40. Berlin, Heidelberg, New York: Springer 1967.

Trèves, F.: (1) Topological vector spaces, distributions and kernels. New York: Academic Press 1967.

Vilenkin, N. Ja.: (1) Special functions and the theory of group representations. Providence, R.I.: Amer. Math. Society 1968.

Wedderburn, J. H. M.: (1) On hypercomplex numbers. Proc. London Math. Soc., (2) **6**, 77—117 (1908).

Weil, A.: (1) L'intégration dans les groupes topologiques et ses applications. Actualités Sci. et Ind. 869, Paris: Hermann & Cie. 1941.

(2) Sur les espaces a structure uniforme et sur la topologie générale. Actualités Sci. et Ind. 551 Paris: Hermann & Cie. 1941.

(3) Sur les groupes topologiques et les groupes mesurés. C. R. Acad. Sci. Paris **202**, 1147—1149 (1936).

(4) Sur les "formules explicites" de la théorie des nombres premiers. Comm. Séminaire Math. Université de Lund (dédie à M. Riesz), Medd. Lunds Univ. Mat. Sem. Tome Supplémentaire, 252—265 (1952).

Weyl, H.: (1) Theorie der Darstellung kontinuierlicher halb-einfacher Gruppen durch lineare Transformationen. I. Math. Z. **23**, 271—309 (1925).
Wiener, N.: (1) Note on a paper of M. Banach. Fund. Math. **4**, 136—143 (1923).
Wigner, E.: (1) On unitary representations of the inhomogeneous Lorentz group. Ann. Math. **40**, 149—204 (1939).
Wintner, David J.: (1) Abstract Lie algebras. Cambridge Mass.: M.I.T. Press 1972.
Yosida, K.: (1) Mean ergodic theorem in Banach spaces. Proc. Imp. Acad. Tokyo **14**, 292—294 (1938).
Zervos, Panajiotis: (1) Le problème de Monge. Paris: Gauthier-Villars 1932.

Subject Index

Index of Notations and Special Symbols

Numbers and Latin and German alphabets

Greek alphabet

Special Symbols

Die Grundlehren der mathematischen Wissenschaften in Einzeldarstellungen mit besonderer Berücksichtigung der Anwendungsgebiete

Eine Auswahl

---

163. Stoer/Witzgall: Convexity and Optimization in Finite Dimensions I. DM 54,—; US $ 20.00
164. Sario/Nakai: Classification Theory of Riemann Surfaces. DM 98,—; US $ 36.30
165. Mitrinović/Vasić: Analytic Inequalities. DM 88,—; US $ 32.60
166. Grothendieck/Dieudonné: Eléments de Géométrie Algébrique I. DM 84,—; US $ 31.10
167. Chandrasekharan: Arithmetical Functions. DM 58,—; US $ 21.50
168. Palamodov: Linear Differential Operators with Constant Coefficients. DM 98,—; US $ 36.30
170. Lions: Optimal Control Systems Governed by Partial Differential Equations. DM 78,—; US $ 28.90
171. Singer: Best Approximation in Normed Linear Spaces by Elements of Linear Subspaces. DM 60,—; US $ 22.20
172. Bühlmann: Mathematical Methods in Risk Theory. DM 52,—; US $ 19.30
173. F. Maeda/S. Maeda: Theory of Symmetric Lattices. DM 48,—; US $ 17.80
174. Stiefel/Scheifele: Linear and Regular Celestial Mechanics. Perturbed Two-body Motion—Numerical Methods—Canonical Theory. DM 68,—; US $ 25.20
175. Larsen: An Introduction of the Theory of Multipliers. DM 84,—; US $ 31.10
176. Grauert/Remmert: Analytische Stellenalgebren. DM 64,—; US $ 23.70
177. Flügge: Practical Quantum Mechanics I. DM 70,—; US $ 25.90
178. Flügge: Practical Quantum Mechanics II. DM 60,—; US $ 22.20
179. Giraud: Cohomologie non abélienne. DM 109,—; US $ 40.40
180. Landkoff: Foundations of Modern Potential Theory. DM 88,—; US $ 32.60
181. Lions/Magenes: Non-Homogeneous Boundary Value Problems and Applications I. DM 78,—; US $ 28.90
182. Lions/Magenes: Non-Homogeneous Boundary Value Problems and Applications II. DM 58,—; US $ 21.50
183. Lions/Magenes: Non-Homogeneous Boundary Value Problems and Applications III. DM 78,—; US $ 28.90
184. Rosenblatt: Markov Processes. Structure and Asymptotic Behavior. DM 68,—; US $ 25.20
185. Rubinowicz: Sommerfeldsche Polynommethode. DM 56,—; US $ 20.80
186. Wilkinson/Reinsch: Handbook for Automatic Computation II, Linear Algebra. DM 72,—; US $ 26.70
187. Siegel/Moser: Lectures on Celestial Mechanics. DM 78,—; US $ 28.90
188. Warner: Harmonic Analysis on Semi-Simple Lie Groups I. DM 98,—; US $ 36.30
189. Warner: Harmonic Analysis on Semi-Simple Lie Groups II. DM 98,—; US $ 36.30
190. Faith: Algebra: Rings, Modules, and Categories I. DM 54,—; US $ 19.95
191. Faith: Algebra: Rings, Modules, and Categories II. In preparation.
192. Maltsev: Algebraic Systems. DM 88,—; US $ 32.60
193. Pólya/Szegö: Problems and Theorems in Analysis. Vol. 1. DM 98,—; US $ 36.30
194. Igusa: Theta Functions. DM 64,—; US $ 23.70
195. Berberian: Baer ∗-Rings. DM 76,—; US $ 28.20
196. Athreya/Ney: Branching Processes. DM 76,—; US $ 28.20
197. Benz: Vorlesungen über Geometrie der Algebren. DM 88,—; US $ 32.60
198. Gaal: Linear Analysis and Representation Theory. DM 124,—; US $ 45.90
200. Dold: Lectures on Algebraic Topology. DM 68,—; US $ 25.20

Prices are subject to change without notice